Manual of
RHEUMATOLOGY

EDITOR-IN-CHIEF

U R K Rao MD (Med) AIIMS
Consultant Rheumatologist
Director, Sri Deepti Rheumatology Centre, Hyderabad

EDITORS

K M Mahendranath
MD FRCP (London)
Consultant Rheumatologist
Samarpan Arthritis and Rheumatology Clinic, Bengaluru

Ramnath Misra
MD FRCP (London)
Dean
Professor and Head Clinical Immunology
Sanjay Gandhi Post Graduate Institute of Medical Sciences, Lucknow

S J Gupta
FRCP
Consultant Rheumatologist
Apollo Hospital and Sant Parmanand Hospital, New Delhi

Rohini Handa
MD DNB FRCP (Glasgow)
Senior Consultant Rheumatologist, Apollo Indraprastha Hospitals, New Delhi
Formerly, Professor of Medicine, AIIMS

Annil Mahajan
MD Fellow Rheumatology FICP MAMS
Professor, HOD Medicine and In-charge Rheumatology Clinic
Post Graduate Department of General Medicine, Govt. Medical College, Jammu (J & K)

Ved Chaturvedi
MD DM
Major General
Consultant Rheumatologist, Armed Forces Medical Services, New Delhi

V Krishnamurthy
MD DM
Consultant Rheumatologist
Chennai Meenakshi Multispeciality Hospital and Apollo Specialty Hospital, Chennai

Binoy Paul
MD PhD DNB
Professor of Medicine & Rheumatology
KMCT Medical College, Kozhikode, Senior Consultant Rheumatologist, Baby Memorial Hospital, Kozhikode

S Shankar
MD DNB MNAMS FRCP (Glasg)
Professor (Internal Medicine)
Rheumatologist & Clinical Immunologist, Department of Internal Medicine, Armed Forces Medical College, Pune

Firdaus Fatima
MD PGDGM
Consultant Rheumatologist
Sri Deepti Rheumatology Centre, A B Rheumatology Centre, Hyderabad

IRA

Indian Rheumatology Association

FOURTH EDITION

First Edition : 1999
Second Edition : 2002
Third Edition : 2009
Fourth Edition : 2014

Published by : **Indian Rheumatology Association**

Printing and Setting by : **Neel Graphics,** Mumbai

Exclusive Distributor :

Raju Shah

THE NATIONAL BOOK DEPOT

Opp. Wadia Children's Hospital, Parel, Mumbai - 400 012.
Tel : 2416 5274 / 2413 1362 / 2413 2411 Fax : 2413 0877
E-mail : nationalbook55@gmail.com

Notice

Medicine is an ever changing science. New research and clinical experience is accumulating rapidly and may result in additions or changes to the existing treatment and drug therapy. The publisher and authors of the Manual of Rheumatology have taken every effort to provide information that is complete and in accord with the standards of Medical practice acceptable, at the time of publication. Readers are however advised to check the information contained herein with other appropriate sources (such as the product information sheet included with each drug they plan to administer) to apprise themselves of any changes/amendments in the treatment regime of drug dosages.

Printed in India

ISBN No: 978-93-80206-33-2

Price : Rs. 1,800/-

Disclaimer

The editors have checked the information provided in the Manual and to the best of their knowledge, it is as per the standards acceptable at the time of publication. However, in view of the possibility of human errors and changes in medical science, neither the authors nor the publisher nor any other party who has been involved in the preparation of this work warrants that the information contained herein is in every respect accurate or complete and therefore disclaim all responsibility for any errors or omissions or for the results that might be obtained from use of the information contained in this work.

Preface to the Fourth Edition

Rheumatology has made rapid advances in the management of various rheumatological disorders in the past decade. "Cure sometimes, relieve often and comfort always" of Hippocratic aphorism holds true to rheumatology. However, we are coming nearer to find "cure" for some of the chronic rheumatic disorders.

The current edition is completely updated and revised with multicolor two column text with large print. Many chapters have been added by new authors. There are about 90 contributors for 65 chapters. Further reading at the end of each chapter contains carefully selected references to original work through November 2013. The salient feature of this book is the importance given to Indian references.

The highlights of the reset edition include newer modalities of therapy like biologics in both adult and pediatric rheumatological disorders. There is a full section on the outcome measures we perform in practice. Alternative systems of Medicine are increasingly recognized. There is a chapter on Rheumatology and Indian Systems of Medicine. In view of the prevalence of infections in our country, the association of infections such as tuberculosis, brucellosis, leprosy, HIV with rheumatological disorders is emphasized. The epidemiological data of rheumatic disorders in India is scanty; a whole section is dedicated for this topic. With increasing geriatric population, osteoporosis and osteoarthritis have gained much importance. They are highlighted in separate sections. Gout and other crystal arthritides are updated. New chapters on cardiovascular and pulmonary disorders associated with rheumatological diseases are added.

Recent advances made in the classification and management of rheumatoid arthritis, spondyloarthropathies and psoriatic arthritis are added. The section on connective tissue diseases is updated with addition of new classification criteria and latest advances in the management.

We have tried our best to minimize the errors and overlaps in this multi author large volume book. We trust that this manual will continue to change and evolve in future editions. It gives me great pleasure to present to you the much awaited 4th edition of Manual of Rheumatology.

U R K Rao
Editor-in-chief
Manual of Rheumatology
Fourth Edition

Preface to the First Edition

Perhaps nowhere in the world as in India that the arthritic patient is vulnerable to influences and counter-influences of modern medicine, homoeopathy, ayurveda and unani system of medicine, not to speak of acupuncture, copper bangles, magnetic therapy and what have you. In an ancient land such as India, a rheumatologist is competing with these modalities of alternate medicine all the time. Alternate medicine, if practiced in a scientific way may be justifiable, but under this garb outright quackery is practiced as well. Steroids are mixed on the sly with ayurvedic and homoeopathic preparations. Branding of joints with hot iron rods in the hope of curing arthritis still goes on rampantly. Some arthritic patients have lost teeth under the false belief that removal of "septic focus" would cure arthritis. Not surprisingly, even practitioners of modern medicine at times prescribe drugs of alternate medicine without necessarily understanding their role. The stark reality of the Indian situation is that there is such little understanding of rheumatic diseases in the minds of average practitioners. This is attributed to the fact that rheumatology is not taught in medical schools. Perhaps a few orthopaedic surgeons may refer to rheumatic diseases during their interaction with students. But is this enough?

There is thus a paramount need for a "Manual of Rheumatology" intended for practitioners, postgraduates in medicine and orthopaedics and orthopaedic surgeons who mainly treat rheumatic diseases day in and day out. Rheumatologists are so few in India that much of the arthritis care is delivered by these practitioners. Hence, this manual is mainly addressed to this target audience.

No doubt, the Indian Rheumatism Association (IRA) conducts CME programmes for practitioners throughout the country from time to time, but its reach is restricted. The Manual of Rheumatology aims at bringing recent diagnostic and therapeutic skills available in modern medicine to the practitioner. For example, there are few practitioners who realize that there are a hundred different types of arthritis. The manual aims at building their confidence in at least making a tentative diagnosis on clinical grounds. It sets out guidelines on laboratory investigations for arthritis, which to order and how to interpret that results. Radiology of arthritis is well illustrated. The reader should be able to discern between rheumatoid arthritis and seronegative spondyloarthropathies. The practitioner should be able to treat successfully a case of infective arthritis with cure as the outcome. The book emphasizes the wide prevalence of tuberculous arthritis in our country. It brings into focus the new diseases like HIV in arthritis. Complexities of vasculitis and antiphospholipid syndrome are explained in simple terms. Therapeutic modalities are described in a way that the practitioner should be able to choose a rational NSAID and not be afraid to prescribe DMARDs. Finally, the manual raises the challenges of patient education in the management of chronic diseases such as arthritis.

This Manual is a dream come true. It is a labour of love of us rheumatologists in India. It was indeed a privilege to write for this Manual and to edit the various chapters and finally publish it within the stipulated time, for the new millennium. Initially, the print run is a modest five thousand. We hope that the sponsors M/s. Pfizer Ltd. will take up the challenge of reaching the practicing community at large and multiply the print run several times. We pray that the Manual will be received and read by the practitioners with enthusiasm in the hope of better care and treatment for the arthritic patient to whom the book is dedicated.

Prakash K Pispati
Editor-in-chief

Nirupa E Borges
Associate editor

SS Uppal
Associate editor

LIST OF CONTRIBUTORS

Prakash Pispati
MD FRSM (Lon)
Director of Rheumatology: Jaslok Hospital,
Senior Consultant Rheumatologist, Saifee Hospital, Mumbai
Past President, Indian Rheumatology Association
Master, Honorary Member and Past President,
APLAR International Fellow Member, ACR

S J Gupta
FRCP
Consultant Rheumatologist.
Apollo Hospital and Sant Parmanand Hospital, New Delhi

Anand N Malaviya
MD ACR Master, FRCP (Lond)
(Ex-Head, Dept of Medicine & Rheumatology Services, AIIMS, Delhi)
Consultant Rheumatologist "A&R Clinic for Arthritis & Rheumatism" and ISIC Super specialty Hospital, Vasant Kunj, Delhi.

B G Dharmanand
MD DM
Consultant Rheumatologist,
SAKRA world Hospital, Bengaluru

Sapan Pandya
MD DM
Consultant Rheumatologist,
Vedanta Institute of Medical Sciences
Rheumatologist & Immunologist,
VS Municipal Hospital, Ahmedabad.

Gp Capt S Shankar
MD. DNB. MNAMS. FICP. FRCP (Glasg)
Professor (Internal Medicine)
Rheumatologist & Clinical Immunologist
Department of Internal Medicine
Armed Forces Medical College, Pune

Maj Shaman Gill
MD
Assistant Professor (Internal Medicine)
Department of Internal Medicine
Armed Forces Medical College, Pune

Darshan S Bhakuni
MD DM
Consultant & Head, Dept of Rheumatology and Clinical Immunology, Army Hospital (R & R), Delhi

Shanmuganandan K
Rheumatologist, Dept of Rheumatology
AHRR, New Delhi

Sivasami Kartik
Senior Resident
AHRR,New Delhi

Aman Sharma
MD MAMS FRCP (London)
Assistant Professor, Rheumatology wing,
Internal Medicine, PGIMER, Chandigarh

Pradeep Bambery
MD FRCP FRACP
Former Professor of Medicine, PGIMER, Chandigarh
Associate Professor, University of Queensland Rural Medical School
Director of Medicine Bundaberg Hospital, Queensland Australia.

Arvind Chopra
MD DNB
International Fellow American College of Rheumatology
Director and Chief Rheumatologist
Co-ordinator, WHO COPCORD.
Center for Rheumatic Diseases, Pune

Liza Rajasekhar
MD PGIMER Chandigarh
Fellowship in Clinical Immunology Chang Gung Memorial Hospital Taiwan
Professor & Head, Department of Rheumatology
Nizam's Institute of Medical Sciences, Hyderabad

Rajendra Vara Prasad
MD DM
Assistant Professor, Dept. of Rheumatology,
Nizams Institute of Medical Sciences,
Hyderabad.

Sita Naik
MD
Advisor, Apollo Hospitals Educational and Research Foundation,
Former Professor and Head, Department of Immunology
Sanjay Gandhi Postgraduate Institute of Medical Sciences, Lucknow

Radha Madhavan
MD Ph D
Professor and HOD
Department of Microbiology
SRM Medical College Hospital & Research Centre, TN

U R K Rao
MD AIIMS
Consultant Rheumatologist
Director, Sri Deepti Rheumatology Centre, Hyderabad

Kakarla Subbarao
MD
Former Director, NIMS
Chairman, KREST, Hyderabad
President, Musculoskeletal Society, India
President Krishna Foundation for Research, KIMS Hospital, Secunderabad

J Shivanand
MD
Consultant
Sri Deepti Rheumatology Centre, Hyderabad

K M Mahendranath
MD FRCP (London)
Consultant Rheumatologist
Samarpan Arthritis and Rheumatology Clinic, Bengaluru

Ajay Wanchu
MD DM FICP MAMS
Arthritis and Rheumatic Diseases, Oregon Health & Science University, Portland, USA

Neha Garg
MD
Division of Arthritis and Rheumatic Disease,
Oregon Health & Science University,
Portland, USA

VS Negi
MD DM
Professor and Head
Department of Clinical Immunology
Jawaharlal Institute of Postgraduate Medical Education and Research, Puducherry

Paul T Antony
MD DM
Assistant professor
Department of Clinical Immunology
Jawaharlal Institute of Postgraduate Medical Education and Research, Puducherry

Gulati R
MD DM
Assistant Professor
Department of Pediatrics
Jawaharlal Institute of Postgraduate Medical Education and Research, Puducherry

Dhanita Khanna
MD DM
Consultant and Head,
Department of Clinical Immunology and Rheumatology,
Sahara Hospital, Lucknow

Arun Shrivastava
MD DM
Consultant Rheumatologist and Immunologist,
Lucknow

Able Lawrence
MD DM
Additional Professor, Clinical Immunology
Sanjay Gandhi Postgraduate Institute of Medical Sciences
Lucknow

Rajkiran Dudam
MD FIRH
Consultant Rheumatologist
Care Hospital, Hyderabad

Chivukula Lakshmikanth
MD
Consultant Rheumatologist
Sai Rheumatology Centre, Hyderabad

S Pal
MD
Consultant Rheumatologist
Advance Rheumatology Centre, Hyderabad

Bidyut Kumar Das
MD PhD
Associate Professor, Dept of Medicine and
Head of Division of Clinical Immunology and Rheumatology, SCB Medical College, Cuttack

CM Adya
MD
Brigadier
Professor and Head, Medicine and Rheumatology
Gian Sagar Medical College, Patiala

C Shivashankar
MS Orthopedics PG trainee
MNR Medical College
Sangareddy, Medak

Firdaus Fatima
MD PGDGM
Consultant Rheumatologist, Sri Deepti Rheumatology Centre and A B Rheumatology Center, Hyderabad

C Satyavati
MD
Consultant Biochemist
Head, Laboratory Services
Sri Deepti Rheumatology Centre, Hyderabad

UmaKumar
MD
Professor of Medicine
Head Clinical Immunology & Rheumatology division
All India Institute of Medical Sciences
New Delhi

V Shantaram
MD FICP FACCP
Consultant Physician and Cardiologist
Former Dean, Professor & HOD Medicine
NIMS, Hyderabad

Milind Nadkar
MD FICP
Professor of Medicine &
Chief of Rheumatology,
Seth GS Medical College and KEM Hospital, Mumbai

Vikram Londhey
MD FICP
Assoc. Professor of Medicine & Rheumatology Clinic,
TN Medical College and BYL Nair Ch Hospital, Mumbai

Raghava Dutt Mulukutla
MS Ortho FRCS MCh Ortho
Consultant Orthopedic Surgeon and Chief of Spine Surgery
Udai Clinic. Centre for Advanced Orthopedics and
Apollo Health City, Hyderabad

Aditya Kapoor
MS Ortho MRCS (Edin)
Consultant Orthopedic Surgeon
Joint Replacement and Adult Reconstruction Division
Udai Clinic. Centre for Advanced Orthopedics, Hyderabad

Wagh Shrikant
MD MA Sc
Consultant Rheumatologist
Pune Hospital and Jahangir Hospital
Lupus Clinic, Pune

Datta Kumar
BAMS PhD
Consultant
Sri Deepti Rheumatology Centre, Hyderabad

Debashish Danda
MD DM FRCP
Professor & Head
Clinical Immunology and Rheumatology Dept
Christian Medical College and Hospital , Vellore

R. Porkodi
MD DM
Retd.Prof. and HOD, Department of Rheumatology
Madras Medical College,Chennai.

Sanjiv Kapoor
MD DM
Consultant Rheumatologist
Indian Spinal Injuries Centre
Max Super specialty Hospital, New Delhi

Jyostna Oak
MD
Consultant Physician and Rheumatologist,
Kokilaben Dhirubhai Ambani Hospital & Research Centre, Mumbai;
Formerly Professor of Medicine and In-Charge Rheumatology,
LTMG Hospital and LTM Medical College, Mumbai.

Keshav Kale
MD
Lecturer, LTMG Hospital and Medical College
Mumbai

Shabbir Chikani
DNB
Consultant Rheumatologist
Rajkot

Nimish Nanavati
MD
Consultant Rheumatologist
Dr Balabhai Nanavathi Hospital, Mumbai

Ramnath Misra
MD FRCP (London)
Dean
Professor and Head Clinical Immunology
Sanjay Gandhi Post Graduate Institute of Medical Sciences
Lucknow

M R Sivakumar
MD DM PhD FRCP
Program Director- Stroke and Neurocritical Care
GLB Hospitals and Acute Stroke Center,
Chennai

Binoy Paul
MD PhD DNB
Professor of Medicine & Rheumatology
KMCT Medical College, Kozhikode
Senior Consultant Rheumatologist
Baby Memorial Hospital, Kozhikode

Renu Saigal
MD FICP MAMS
Professor Medicine, In-Charge Rheumatology clinic,
Mahatma Gandhi Medical College and Hospital, Sitapura, Jaipur
Former Professor and Head
Dept of Medicine, In-Charge Rheumatology clinic
SMS Medical college and Hospital, Jaipur

Siddharth Kumar Das
MD
Professor & Head, Dept of Rheumatology
K G Medical University, Lucknow

Rohini Handa
MD DNB FRCP (Glasgow)
Senior Consultant Rheumatologist
Apollo Indraprastha Hospitals, New Delhi
Formerly, Professor of Medicine, AIIMS

V. Rajamani
MRCPI FRCPI
Consultant Rheumatologist
Kovai Medical Center and Hospital
Coimbatore

Amita Aggarwal
MD DM
Professor, Department of Clinical Immunology
Sanjay Gandhi Postgraduate Institute of Medical Sciences
Lucknow

C P Rajendran
MD DM
Professor and Head, Dept of Rheumatology
SRM Medical College and Research Centre, Potheri, TN

Sujata Sawhney
MD MRCP CCST
Senior Consultant in Pediatric Rheumatology,
Institute of Child Health
Sir Gangaram Hospital, New Delhi

Surjit Singh
MD DCH (Lon.) FRCP (Lon.)
Professor of Pediatrics, Advanced Pediatrics Centre,
Post Graduate Institute of Medical Education and Research,
Chandigarh

Nutan Kamath
MD
Professor Pediatrics
Kasturba Medical College
Manipal University
Mangalore

Kaushik Bhojani
MD
Consultant Rheumatologist,
Kennisha Rheumatology Care & diagnostics
Bhandup Mumbai and Fortis hospitals, Mumbai

Shubhada Kalke
MD
Consultant Rheumatologist,
Kennisha Rheumatology Care & diagnostics
Bhandup Mumbai and
Lilavati Hospital, Mumbai

Yojana Gokhale
MD
Professor of Medicine & In-Charge of Rheumatology services
Lokmanya Tilak Municipal Medical College, Mumbai
In-Charge of Arthritis Center, Mumbai

Vikas Agarwal
MD DM
Additional Professor
Dept of Clinical Immunology, Sanjay Gandhi Postgraduate
Institute of Medical Sciences, Lucknow

Anupam Wakhlu
MD DM
Associate Professor
Rheumatology
King George's Medical University, Lucknow

Ajay Wanchu
MD DM FICP MAMS
Arthritis and Rheumatic Diseases, Oregon Health & Science University
Portland, USA

VS Negi
MD DM
Professor and Head
Department of Clinical Immunology
Jawaharlal Institute of Postgraduate Medical Education and Research, Puducherry

Paul T Antony
MD DM
Assistant professor
Department of Clinical Immunology
Jawaharlal Institute of Postgraduate Medical Education and Research, Puducherry

Gulati R
MD DM
Assistant Professor
Department of Pediatrics
Jawaharlal Institute of Postgraduate Medical Education and Research, Puducherry

Dhanita Khanna
MD DM
Consultant and Head,
Department of Clinical Immunology and Rheumatology,
Sahara Hospital, Lucknow

Arun Shrivastava
MD DM
Consultant Rheumatologist and Immunologist,
Lucknow

Able Lawrence
MD DM
Additional Professor, Clinical Immunology
Sanjay Gandhi Postgraduate Institute of Medical Sciences
Lucknow

Rajkiran Dudam
MD FIRH
Consultant Rheumatologist
Care Hospital, Hyderabad

Chivukula Lakshmikanth
MD
Consultant Rheumatologist
Sai Rheumatology Centre, Hyderabad

S Pal
MD
Consultant Rheumatologist
Advance Rheumatology Centre, Hyderabad

Bidyut Kumar Das
MD PhD
Associate Professor, Dept of Medicine and
Head of Division of Clinical Immunology and
Rheumatology, SCB Medical College, Cuttack

CM Adya
MD
Brigadier
Professor and Head, Medicine and Rheumatology
Gian Sagar Medical College, Patiala

C Shivashankar
MS Orthopedics PG trainee
MNR Medical College
Sangareddy, Medak

Firdaus Fatima
MD PGDGM
Consultant Rheumatologist, Sri Deepti Rheumatology Centre and A B Rheumatology Center, Hyderabad

C Satyavati
MD
Consultant Biochemist
Head, Laboratory Services
Sri Deepti Rheumatology Centre, Hyderabad

UmaKumar
MD
Professor of Medicine
Head Clinical Immunology & Rheumatology division
All India Institute of Medical Sciences
New Delhi

V Shantaram
MD FICP FACCP
Consultant Physician and Cardiologist
Former Dean, Professor & HOD Medicine
NIMS, Hyderabad

Milind Nadkar
MD FICP
Professor of Medicine &
Chief of Rheumatology,
Seth GS Medical College and KEM Hospital, Mumbai

Vikram Londhey
MD FICP
Assoc. Professor of Medicine & Rheumatology Clinic,
TN Medical College and BYL Nair Ch Hospital, Mumbai

Raghava Dutt Mulukutla
MS Ortho FRCS MCh Ortho
Consultant Orthopedic Surgeon and Chief of Spine Surgery
Udai Clinic. Centre for Advanced Orthopedics and
Apollo Health City, Hyderabad

Aditya Kapoor
MS Ortho MRCS (Edin)
Consultant Orthopedic Surgeon
Joint Replacement and Adult Reconstruction Division
Udai Clinic. Centre for Advanced Orthopedics, Hyderabad

Wagh Shrikant
MD MA Sc
Consultant Rheumatologist
Pune Hospital and Jahangir Hospital
Lupus Clinic, Pune

Datta Kumar
BAMS PhD
Consultant
Sri Deepti Rheumatology Centre, Hyderabad

CONTENTS

SECTION - 1

APPROACH TO RHEUMATIC DISEASES

SECTION - 2

OUTCOME MEASURES IN RHEUMATOLOGICAL DISORDERS

SECTION - 3

EPIDEMIOLOGY

SECTION - 4

EVALUATION OF RHEUMATIC DISEASES

SECTION - 13

INFECTION AND ARTHRITIS

SECTION - 14

MISCELLANEOUS

SECTION - 9
GOUT AND OTHER CRYSTAL ARTHROPATHIES

SECTION - 10
OSTEOARTHRITIS

SECTION - 11
OSTEOPOROSIS

SECTION - 12
RHEUMATIC DISEASES OF CHILDHOOD

SECTION - 13

INFECTION AND ARTHRITIS

SECTION - 14

MISCELLANEOUS

Acknowledgements

We would like to acknowledge the enormous contribution of the authors to the Manual, without whom, this work would not have been possible.

We thank Indian Rheumatology Association for giving us the opportunity to perform the great task of editing this book.

We acknowledge Mr. Raju Shah of 'The National Book Depot' and 'Neel Graphics', Mumbai in bringing out this beautiful book in time. Last, but certainly not the least, we wish to acknowledge our patients who continue to provide challenges to us in our endeavor to manage rheumatic disorders.

Editors
Manual of Rheumatology
Fourth Edition

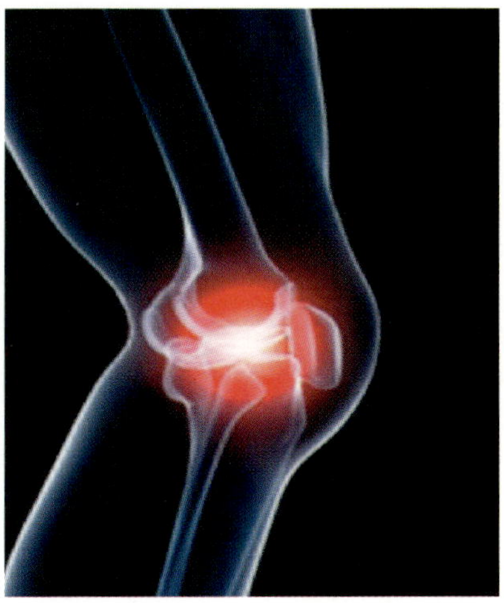

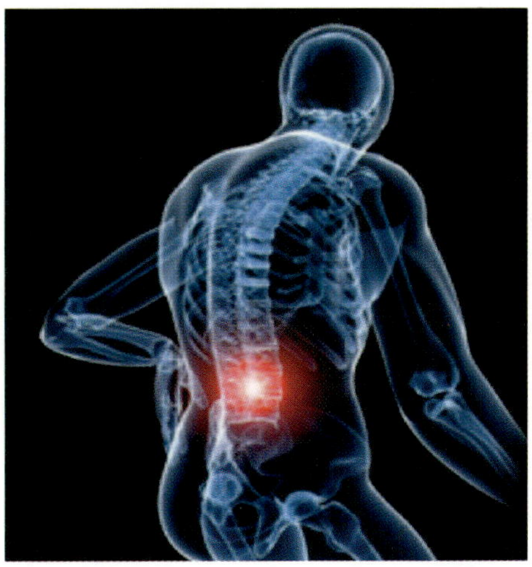

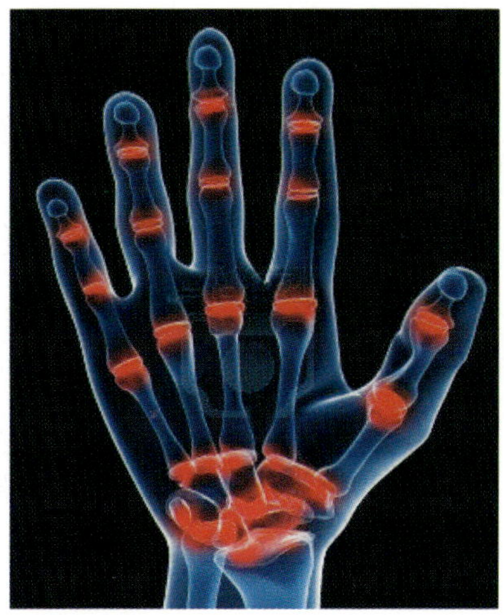

SECTION 1

Approach to Rheumatic Diseases

RHEUMATOLOGY NON-STOP

A short story of History of Rheumatology and History in the making

1

Prakash Pispati

Why History? When at school the subject I dreaded most was mathematics (probably drove me to medicine) and the subject I enjoyed the least was perhaps history. And now just as I enjoy visiting ancient temples, churches, castles, monuments... so do I enlighten myself dwelling into History of Medicine. Man's quest to conquer diseases is as ancient as mankind itself. Can you imagine a human being in his lifetime without complaining of aches and pains casual mostly; crippling at times? No wonder, medical scriptures are replete with accounts of what physicians in earlier times thought about 'rheuma', 'rheumatism' and now rheumatology.... Are such words meaningful today? A young investigator may mock, yet did not Winston Churchill write? "The longer you look back, the further you look forward".

Ancient times and thoughts

Greek physician Hippocrates (460-377BC) developed a medical theory called humoralism, which held that four humors (liquids) coursing through the human body determined one's temperament and state of health. He used the term rheuma, which literally means "flowing", to describe an excess of the watery humor thought to flow down from the brain. The words rheuma and catarrhos ("flowing down") were used interchangeably by ancient Greeks to describe a variety of illnesses including joint problems.

In 13th-Century Europe any joint ailment was called gutta (Latin, meaning "a drop") for a noxious humor falling drop by drop into the joint. Gout and gouty diathesis were terms used broadly to connote arthritis as we understand today. Thomas Sydenham (1624-1689), a London physician who himself suffered from gout, distinguished the episodic disorder from acute arthritis that attacked children (probably rheumatic fever) and from chronic crippling arthritis [probably rheumatoid arthritis (RA)] that came to be called 'rheumatic gout'.

Not until 1642 was the term 'rheumatism' used by the French physician Ballonius, who sought to distinguish noxious humours.

In the Indian literature, Charak Samhita (written in around 300-200 BC) also described a condition that describes pain, joint swelling and loss of joint mobility and function. In our own ancient Ayurveda science, the concept of vata, pitta, kapha (wind, bilious and phlegm) may seem parallel. Interestingly, while Charaka Samhita's texts suggest a joint disease rheumatoid like, it must have been inspired to prepare Suvarna Bhasma perhaps akin to 'gold oxide'. Remember Injection sodium aurothiomalate (Myocrisine)? Regretfully we stopped to use it even though its side effects were manageable, its cost - a fraction of today's biologicals.

Gout understandably is the most vividly described rheumatic disease, why not? Truly, its acute dramatic manifestations must have made "headline news" even in those days as monarchs and ministers be they in office or in churches, authors or poets of romantic lifestyle. Hippocrates, himself a victim of gout described :

> "Eunchs do not take gout, nor become bald
> A woman does not take gout unless her menses are stopped. A young man does not take gout unless he indulges in coitus. In gouty affection inflammation subsides in 40 days".
> - *Hippocrates (5th Century BC, Greece).*

Era and age of reason, experiments and science

Such were powerful descriptions that any joint pain was thought to be 'gouty'. Yet multiple, painful chronic joint deformities were surely noticed by some physicians who correctly described these as "arthritis deformans". In 1858, Sir Alfred Baring Garrod insisted that the term be changed to RA. It didn't make much of an impression. Even Sir William Osler had this to write:

"Arthritis Deformans:
Once established, the disease is rarely curable. Too often it is a slow, but progressive, crippling of the joints, with a disability that makes the disease one of the most terrible of human afflictions". *(from Principles and Practice of Medicine, 1909).*

arthritis, urethritis, dermatitis confirmed Reiter's disease, a name that is now changed to Reactive arthritis. Eponyms galore! 15,000 or more on record. A few are short listed[1]. The last significant eponym seems the "Hughes Syndrome" synonymous with anti phospholipid syndrome Table 1.

Table 1 Common Eponyms in Rheumatology

- Behcet's syndrome
- Caplan syndrome
- Charcot's joints
- Churg-Strauss Syndrome
- Cogan's syndrome
- Crohn's disease
- deQuervain's tenosynoviitis
- Dupuytren's contracture
- Farber's disease
- Felty's syndrome
- Francoise syndrome
- Gaucher's disease
- Handigodu syndrome
- Heberden's nodes
- Henoch-Schönlein purpura
- Hoffa's disease
- Hughes Syndrome (Antiphospholipid syndrome)
- Jaccoud's syndrome
- Kashin-Beck disease
- Kawasaki disease
- Leri's pleonosteosis
- Libman Sachs endocarditis
- Lyme arthritis
- Lofgren syndrome
- Morton's Metatasalgia
- Osgood-Schlatter's disease
- Osler's nodes
- Paget's disease
- Pott's disease
- Reiter's disease
- Sever's disease
- Sjogren's syndrome
- Still's disease
- Sudeck's atrophy
- Sweet's syndrome
- Takayasu's disease
- Thiemann's disease
- Tietze syndrome
- Wegener's granulomatosis (granulomatosis with polyangiitis)
- Weil's disease
- Werner's syndrome
- Wilson's disease
- Whipple's disease

In 1923 a meeting of concerned physicians took place at the Royal Society of Medicine where Sir Archibald Edward Garrod (son of Sir Alfred Baring Garrod) a strong plea carried weight and the name was formally changed from "Arthritis Deformans" to "Rheumatoid Arthritis". In 1941, the American Rheumatism Association followed suit and since then these two words seem here to stay permanently[1]. Keen readers should ask themselves, what seems more apt, rheumatoid arthritis or rather Arthritis deformans?

Galen introduced the term rheumatismus in second century AD. Bannatyne described the radiological features of RA. It was in 1940 that Camroe coined the term Rheumatologist; and the term Rheumatology was coined by Hollander in 1949.

The wars had minor salutary side effects. Reiter in the 19th century Russian war noticed that a triad of

SORDID FOUNDATIONS OF MICROBIOLOGY, IMMUNOLOGY AND RHEUMATOLOGY

Contemporary rheumatology is surely immunology based driven, inspired, since 18th century. A general practitioner, Edward Jenner in Yorkshire made a profound observation that those children who suffered from 'cowpox' never developed 'smallpox' (Fig. 1).

Fig.1

Mere annotations were not enough. He experimented, injecting his son with extracts from cowpox. He was condemned and removed from the medical fraternity. Strangely, despite England and France going to war periodically, it was Napoleon Bonaparte who invited Jenner to vaccinate his troops! Later the U.S. accepted this practice and then Britain had little choice to honour Edward Jenner knighted Fellowship of the Royal Society and a reward of £ 30,000. Thus was laid the foundation of infection and immunity upon which the very edifice of immunology is carefully built on scientific experiments over centuries that followed.

A Dutch lensman who wanted to decipher the cause of coughs and common cold to answer everyone's question 'there must be something in the air', curiously led him to find out 'what's in the air' till he tumbled into moving creatures in a drop of water which came under his lens. Excitedly came the cry 'animalcules... eureka!' ...Archimedes repeated! (Fig. 2).

Fig. 2

Microbiology and immunology were harmonised by Louis Pasteur and Robert Koch .... Who knows it not? The Germ theory of disease was then thought sacrosanct. Question thus arose as to how come humans live that long when surrounded by millions of microbes throughout, outside and inside? What protects us? In his figment of imagination, Paul Erhlich drew a pencil sketch, the hypothesis of antigens and antibodies (Fig. 3a & 3b). Experiments proved this to be very true. Every antibody was thought then to be protective till a couple of decades ago. We were then reminded that some unwanted antibodies without identified antigens were in fact damaging auto antibodies that led to the discovery of auto immune diseases. So now we have autoimmune pathophysiology to explain, collagen vascular diseases, the main thrust of contemporary rheumatologists.

Fig. 3a : Paul Ehrlich

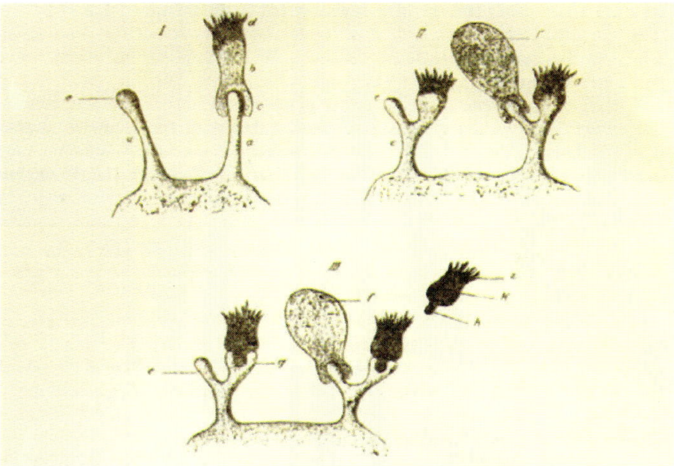

Fig. 3b : Ehrlich's pencil sketch of side chain theory

Foundations of contemporary science

Every simple observation in medicine has a deep-rooted complex message waiting to be decoded (Fig. 4). Eli Metchnikoff's discovery that macrophages devour the invading bacteria as the defending soldiers of the body. His astute description of phagocytosis had a profound impact on every disease. Thus were laid the foundations of both cellular and humoral immunity challenging the sole germ theory of every disease. Later developments of immunology time line are depicted in Table 2. A mischief monger like a pathogen would fool, evade, enter and corrupt some of the soldiers. Thus, we have antigen presenting cells and T cells asking B cells to produce unwanted and damaging auto antibodies, through signals produced by cell to cell talk. These are called cytokines - some exciting, some inhibitory.

Some of these discoveries are enlisted in time-line (Table 2).

Table 2 Immunology - Rheumatology time-line

1798	Smallpox vaccination	Edward Jenner
1876	Validation of germ theory of disease discovering. Bacterial basis of anthrax	Robert Koch
1879	Chicken cholera vaccine development	Louis Pasteur
1890?	Discovery of Diphteria "anti-toxin" in blood	Emil Von Behring Shibasaburo Kitasato
1882	Isolation of tubercule bacillus	Robert Koch
1883	Delayed type of hypersensitivity	Robert Koch
1884	Phagocytosis: Cellular theory of immunity	Elie Metchnikoff
1891	Proposal that antibodies are responsible for immunity	Paul Erhlich
1891	Passive immunity	Emil Roux
1944	Immunological hypothesis of allograft rejection	Peter Medawar
1948	Demonstration of antibody production in plasma	Astrid Fagraeus
1949	Distinguishing self vs. non-self and its role in maintaining Immunological unresponsiveness (tolerance) to self	Macfarlane Burnet & Frank Fenner
1953	Immunological tolerance hypothesis	Rupert Billingham, Leslie Brent, Peter Medawar & Milan Hasek
1954	Rheumatoid Factor	Charles Plotz
1958	Identification of first autoantibody and first recognition of Autoimmune diseases	Henry Kunkel
1968	Distinction of bone marrow and thymus derived lymphocyte populations; Discovery of T and B cell collaboration.	Jacques Miller & Graham Mitchell
1974	HLA-B27 predisposes to an autoimmune disease	Derek Brewerton
1983	Antiphospholipid (Hughes) Syndrome	Graham Hughes
1990.....	Discovery of TNF and Monoclonal Antibodies	Feldmann and Maini

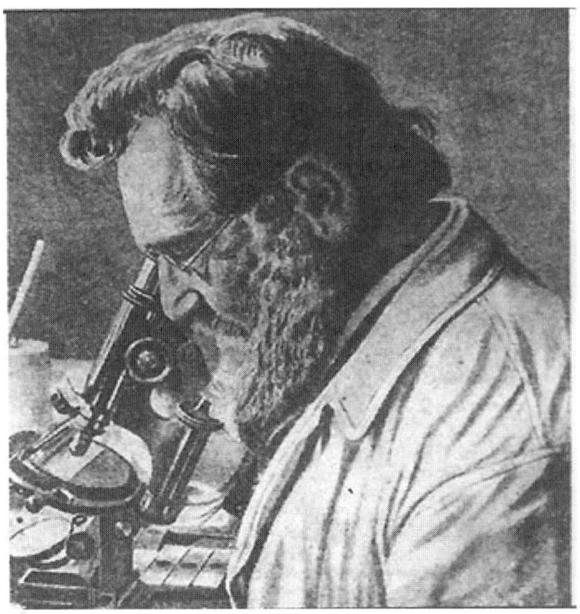

Fig. 4. Eli Metchnikoff

Investigations : Laboratory Diagnostic Approach:

Diagnostic laboratory investigations seem to have evolved much later. Significantly perhaps, Rose and Waaler sheep's red blood cell agglutinate test changed the mindset to look for disease markers. Both were New York based, but never met (Waaler was Norwegian). 1948 recorded the first ever simple but significant lupus phenomenon (LE Cell test)[1,2]. It got replaced in 1960 when Henry Kunkle's antinuclear antibody was detected by brilliant immunoflorescence. Since then, subsets of ANA-ENAS, Anti dsDNA antibody, ANCA and APLA are in vogue. Albeit, progress in diagnostic laboraoty seems at turtle pace, isn't it?

1953 onwards Charles Plotz a famous rheumatologist from New York proved that antibodies could be antigenic and produce anti-antibodies (later called auto-antibodies) and that the latex fixation test could not only determine the presence of rheumatoid factor but that the technique could be used, with differing coatings, to determine the presence of auto-antibodies.

His first exposure to latex particles was to note their use as measures in electron microscopy. They were manufactured in very precise diameters by the Dow Chemical Company in Michigan. He tested various sizes and found that those of 0.9 micra in diameter worked best in the latex fixation test. Dow primarily sold these particles to companies (like DuPont) who dyed them and suspended them in water to make latex paints. While his fame for the invention of the latex fixation test was world-wide, his reasoning that antibodies could now be proven to be antigenic by means of latex fixation (a tube test, not a slide test), first discussed at South East Asia Pacific Area League Against Rheumatism (SEAPAL) Congress has proven to be an important milestone in the development of immunology as a science.

By the way, the first SEAPAL Congress was at the Taj Mahal Hotel, Mumbai organized by the first IRA President, the late Dr. M. M. Desai, under the aegis of Indian Rheumatology Association (IRA) in 1968. In that congress, papers a plenty were presented on rheumatic fever. In those days, ASO titre test was thought to be a marker of rheumatic fever, often over diagnosed and children administered multiple penicillin injections irrationally.

Still's disease was described at the Institute of Child Health at Great Ormond Street, London. Then came in such terms as juvenile rheumatoid arthritis (JRA), juvenile chronic arthritis (JCA), and now juvenile idiopathic arthritis (JIA), a rather apologetic term. Paediatric Rheumatology has come of age.

From willow bark to biotechnology :

- *Rheumatism becomes rheumatology.* While many physicians never wait to work out exact pathophysiology, they choose to alleviate pain and swelling, the motto of every doctor. Since the time of Hippocrates the willow bark was tried. Centuries later the active ingredient was the identified molecule salicylic acid. Its huge chunk of crystals required good volume of water. So bottles of salicylate mixture had to be drunk for relief of pain at the expense of gastric mucosa. A 27 year old brilliant chemist Von Felix Hoffman found that to be impractical. His father had RA. So was synthesized acetyl salicyclic acid. ASA tablets could be swallowed with ease. The drug was named Aspirin. As a caring son, he tried his new invention on a fish in a bowl in his house. Satisfied that the fish could still swim, he administered aspirin to his father. Medical historians noted that to be one of the first drug trial on laboratory animal, though it was only on a fish and not the traditional guinea pig!
- Heyday of Antirheumatics: Thus Germany Bayer's Aspirin became perhaps the first and the best known pharmaceutical brand as surely the best analgesic, anti-inflammatory, antipyretic and hence the first effective antirheumatic (Fig. 5). Its G.I. toxicity prompted synthetic chemists to look for options; phenylbutazone and congenres. The Rhine Valley from Basle to Frankfurt was the buzzword of synthetic, organic pharmaceutical chemistry evolving many potent anti-rheumatics, albeit with side effects.

Fig. 5. Von Felix Hoffman: Willow Bark to Aspirin

No wonder the famous German pharmacologist Adrian Albert said, "*Pharmacology is the science of selective toxicity*". Post World War II came indomethacin, naproxen from the U.S., and British industries came out with blockbuster ibuprofen, ketoprofen, naproxen which led to the heyday of NSAIDs. Relieving pain and reducing swelling became the therapeutic mantra. Sir John Vane's precise work unravelled mechanism of action of aspirin and NSAIDS and paved the way of Cox-1 and Cox-2 inhibitors. Pain relief was surely accomplished with gastric protection, albeit partially.

- Glucocorticoids: Boon or bane? A wonder story of clinical observations: 1925, New Jersey, Philip Hench, an American rheumatologist made an observation that young women with rheumatoid arthritis went into remission should they incidentally develop jaundice or become pregnant. He wanted to know if there was a substance released in the body to explain it (compound X). Others, later termed it compound E. Painstaking biochemical studies in such patients were moving steadily and slowly, till World War II broke out. CIA reported German Luftwaffe pilots were flying higher even in non-pressured fighter planes to evade French and British radars. Rumours galore that they were swallowing a 'mysterious powder' (cortisone) secreted by adrenal cortex[2]. Tons of bovine adrenal glands were imported by Germany from Argentina to extract the powder. That development followed Hench and Kendall's observations that compound E was actually a hormone compound, cortisone. The White House funded generously to accelerate research and manufacture of cortisone and funded even a private drug company, (MSD) with Philip Hench in the driver's seat to push the exciting research, together with Kendall and Reichstein in Basle. They received the Nobel Prize (1950).

Dramatic remissions of arthritis in glucocorticoid treated patients led to its overuse, misuse and abuse. Adverse reactions included even fatalities. Embarrassed by mounting criticism, Philip Hench went into depression to breathe his last in 1965. However, glucocorticoids prescribed sanely are in the therapeutic bag of every rheumatologist.

Fig. 6. Source of Chloroquine: Cinchona

Disease Modifying Anti Rheumatic Drugs (DMARDs) - a new phase in therapeutics

JP Payne used chloroquine first in the treatment of psoriatic arthritis and later in rheumatoid arthritis finding salutary results[3,4] (Fig. 6). (Chloroquine is generally not used in psoriasis or psoriatic arthritis now). Since then, antimalarials are prescribed as DMARDs, as a routine by rheumatologists, at times perplexing patients. An Indian brand of chloroquine on its pack carries a picture of a mosquito. The dispensing chemist has to clarify it is for rheumatoid arthritis!

In early 20th century, sulphanomides were popular antibacterials. Swedish rheumatologists wondered if rheumatoid arthritis was of infective origin. Thus they synthesized sulfasalazine .... widely prescribed even today, isn't it?

In late 1970s, cyclophosphamide anticancer cytotoxic was reported to benefit rheumatoid arthritis patients. In 1985, a landmark paper by Michael Wienblatt has made methotrexate the gold standard for rheumatoid arthritis treatment actually replacing gold salts[5].

And now the Brave New World of Biotechnology

Like Paul Erhlich's pencil hypothetical sketch, now comes the therapeutic breakthrough. Marc Feldmann an Australian immunologist working on thyroiditis, was looking for a good size "sample" to work upon. Feldmann and Ravinder Maini met over lunch in London. Like Paul Erhlich's sketch (the famous side chain theory sketch), they drew their penciled hypothesis: is TNF the culprit cytokine producing synovitis in RA? Would a specific blockade be a mechanism of action result in remission? And thus came in, infliximab and etanercept, and now we even know how to deplete B cells and practice targeted therapy with newer biologicals.... Rheumatism becomes Rheumatology non-stop.

History in the making

Today's rheumatology evolved over a few centuries with inventions and discoveries have made this jigsaw puzzle into a fine canvas. Future direction will enable us to predict and ensure that the utopian dream of deformity free arthritic patient is about to come true. Microbiology and immunology now seem the new propellers to energise new inventions. Microbiota comprising hundred trillion microorganisms residing in each of us are about to be labeled as 'unwanted' and 'wanted'. Nobel laureate (1903) Joshua Lederberg's concept of microbiota will facilitate *in vivo* microbiological homoeostasis to mutate and prevent many rheumatic diseases. Closely linked intracellular signalling codes unravelled will yield tiny new therapeutic small molecules to replace huge proteonic macromolecules and conventional synthetic pharmaceuticals[6].

Finesse in stem cell biotechnology will actually set up functioning tissue factories discarding deranged and degenerate ageing cells and organs. Nano technology will offer bedside mini diagnostic devices replacing spread out laboratory floors housing space occupying machines. Electrostatic bio-waves will be captured on mobile phones displaying biomedical details, challenging and embarrassing our traditional medical personalities. Rheumatologists of tomorrow will be techno-savvy and I predict drastic changes in medical college curricula. Yet, traditional 'feel the pulse' benign family doctor may well outweigh 'caught in between' super specialist unable to parade and march with time onslaught of biomedical technology.

In the evolution of new galloping targeted therapy, we often forget and gloss over strange clinical symptoms and signs in changing environment whereby undiscovered diseases and syndromes wait to surface[7]. A fine example is the vivid description of Antiphospholipid (Hughes) Syndrome[8]. Rheumatology + Immunology = Rheumanology is the modern mantra. We are its disciples and apostles of this inevitable change from history to history in the making. Alvin Toffler's bestseller *Future Shock* is already upon us[9]. Are we ready?

Further Reading :

1. Pispati P. Confusion or Clarity: Nomenclature, Semantics, Jargon, lingo, eponyms, Etymology and terminology in rheumatology. The Rheumatologist. 2012; 6(10): 10, 13.
2. Pispati P. Seven Wonders in the world of rheumatology: a short story of romance, reminiscences and renaissance. International Journal of Rheumatic Diseases 2008; 11: 91-96.
3. Milestones in Rheumatology: Year Book 2001. New York, N.Y.: The Parthenon Publishing Group Inc 1999.
4. Nanavati N. Rheumatoid Arthritis. In: Pispati P, Editor-in-Chief, Manual of Rheumatology, 2nd ed. Mumbai: Indian Rheumatology Association 2002; 72-82.
5. Wienblatt ME, et al. (1985) Efficacy of low-dose methotrexate in rheumatoid arthritis. N. Engl Med 112, 818-22.
6. Sher JU, Abramson SB. The Microbiome. A voyage to (our inner) Lilliput. The Rheumatologist. 2011; 5(11): 1, 32-35.
7. V. R. Joshi. Rhuematology, past, present and future. JAPI 2012; 60: 21-24.
8. Matteson EL, Woywodt A. Eponymophilia in rheumatology. Rheumatology 2006; 45: 1328-30.
9. Toffler Alvin. Future Shock. Random House in 1971.

INTRODUCTION TO CLINICAL APPROACH TO ARTHRITIS 2

Prakash Pispati

It is an all-time truism that the British Medical Journal described Rheumatology as the last remaining bastion of clinical medicine. This primary lesson still holds water. Rheumatic diseases were described with substantial accuracy long before the advent of laboratory medicine and imaging technology. Of course we need these, in fact with newer innovations towards accurate and early diagnosis, with enhanced ratio of specificity and sensitivity. Therefore, mastering clinical approach to the arthritic patient with which any consultation begins is sacrosanct.

When a patient approaches us doctors with a musculoskeletal complaint, we must decide at once, is this soft tissue rheumatism or arthritis? Should it be arthritis, then let us ask ourselves these key questions :

- Is this of acute, subacute or of chronic onset ?
- Is it monoarticular, pauciarticular or polyarticular ?
- What are the extra-articular features, if any ?
- Is it inflammatory, degenerative or metabolic ?
- Which are other systems involved, if any ?
- Is it primary or secondary to any other systemic disease ?

Such key questions should help us proceed on subsequent algorithms, enabling the clinician to make a diagnosis with *simplicity* and *clarity*; add on minimum investigations, and we will most probably hit the bull's eye, to initiate appropriate treatment : therapeutic success in sight !

As we drive our car to a traffic signal we will see green, amber or red lights. So also, in the clinical approach to the rheumatic patient :

Green	Amber	Red
• Insidious onset • No joint involvement (eg. soft tissue rheumatism) • One or few joints (< 5) • Dull ache and pain, mainly localized • Normal ESR, negative CRP • No significant disability	• Chronic onset • Slow progression • Variable joint involvement • Regional pain, discomfort, moderate debility • Moderately elevated ESR, CRP	• Pyrexia • Loss of weight • Significant rash • Lymphadenopathy • Vital organ involvement • Unbearable pain and debility • Weakness and malaise • Vasculitic manifestations • Multisystem involvement • High ESR, CRP

Now let us check out the following seven masquerades, the devils in disguise:

- Depression
- Diabetes mellitus
- Drugs
- Anaemia
- Thyroid disease
- UTI
- Spinal dysfunction

All this is for the doctors. The patient is certain to pose us with multiple questions, relevant or irrelevant, so should we :

- What made you call up the doctor ?
- Where exactly is the pain ? Is this static or does it shift ?
- What brought it on ? Any specific event ?
- Any injury, obvious or silent ?
- Is this pain disturbing your sleep ?
- How do you feel in the morning ?
- As you move or exercise does pain increase ? Are your movements difficult ?
- Any fever ? Any cough ? Any skin rash ?

- Do you feel tired, weak, fatigued, lethargic ?
- Any sore throat ?
- Any burning urine ? What is its colour ?
- Any loose motions ?
- What is your marital / personal life ?
- Any family history of arthritis ?
- What is your occupation, are you overworked ? Is this aggravating your pain ?
- Is your work disturbed by pain ?
- Where do you live ? Do you travel ? How do you travel, is it in overcrowded situations ?

To an experienced rheumatologist, posing such simplistic questions may seem naive, but there are two huge advantages. Firstly, it facilitates to zero on to the diagnosis, and not unimportant, enhance patient confidence towards subsequent investigations and treatment that we may advise, eliciting patient co-operation and therapeutic compliance as well. The next two sections of Clincal Approach to Monoarthritis and Polyarthritis are certain to enhance your routine work for every patient.

Yes, the clinical approach to a rheumatic patient is by far the first golden step in arthritis management, making us better clinicans and patients comfortable.

Further Reading:

John Davis, Kevin Moder, Gene Hunder. History and physical examination of the musculoskeletal system. In Kelley's textbook of Rheumatology (8th Ed.). Gary Firestern et al. (Eds). Saunders Elsevier, Philadelphia 2009; 515-31.

APPROACH TO MONOARTHRITIS

3

S J Gupta

Introduction

Joint inflammation that starts and remains at a single joint is termed as Monoarthritis (MA); inflammation at more than one but less than 5 joints is termed as Oligoarthritis and when inflammation occurs at 5 or more joints, it is termed as Polyarthritis. MA can occur in a previously normal joint or might occur in a joint that is abnormal due to previous joint disease. MA may be: a) Acute MA- which is of recent onset and less than 6 weeks duration, or b) Chronic MA - which has been present for at least 6 weeks. The evaluation of a patient presenting with MA involves a systematic clinical approach, keeping in mind specific clinical features in the individual patient. This chapter will aim to provide an overview of the suggested clinical approach to a patient with MA.

Differential Diagnosis

There can be many possible causes of MA and in order to determine the cause in an individual patient, it is obviously important to be aware of these. Specific conditions may present with and exhibit differing clinical features. Therefore, one should bear in mind the diagnostic possibilities depending upon the presentation of MA; that is, is the MA acute or chronic and is there any history of previous joint disease[1,2] (Tables 1 & 2). Though many causes are not immediately serious or potentially dangerous, the evaluation of a patient with MA needs to be completed fairly soon, as infection is one of the possible causes and if not recognised and treated adequately, can result in joint damage and considerable morbidity. It is also important to remember that occasionally, a polyarticular or systemic disease may present with an initial MA involvement.

Table 1 Possible causes of acute/chronic monoarthritis

ACUTE MA	CHRONIC MA
Infection	***Infection***
Bacterial, Mycobacterial, Fungi, Viral, Lyme disease	Plant thorn (foreign body) synovitis
Crystals	***Sarcoidosis***
MSU (Gout); CPPD (Pseudogout); BCP	***Enteropathic arthritis***
(Destructive arthropathy); Calcium oxalate	(Crohn's disease predominantly)
Trauma	***Pigmented villo-nodular synovitis***
Fracture, Internal derangement, Haemarthrosis	***Synovial chondromatosis***
Osteoarthritis	***Synovial sarcoma***
Presenting as MA- 1st CMC joint, MTP joint, Hip or Knee	***Monoarticular presentation of a***
Foreign body synovitis	***polyarticular disease***
Ischaemic necrosis:	
Tumour	
Metastases, Pigmented villo-nodular synovitis,	
Osteoid osteoma	
Systemic disease presenting as MA	

MSU= mono sodium urate; CPPD= calcium pyro-phosphate dihydrate;
BCP= basic calcium phosphate hydroxy-apatite

Table 2 Causes of Monoarthritis related to presence or absence of previous joint disease

Previously normal joint (no previous joint disease)	**Abnormal joint** (previous joint disease or damage)
Septic arthritis	***Pseudo gout*** in association with osteoarthritis
Juvenile idiopathic arthritis	***Avascular necrosis*** or subchondral collapse or fracture
Crystal synovitis (gout or pseudogout)	***Cartilage tear***
Monoarticular presentation of oligo/polyarthritis (initial) Seronegative Spondarthritis (reactive / psoriatic); Erythema nodosum; Rheumatoid arthritis; Juvenile idiopathic arthritis	***Haemarthrosis***
Trauma (haemarthrosis)	***Septic arthritis***
Foreign body reaction	

A. Acute MA

As the term suggests, acute MA implies the development of inflammation at a single joint fairly rapidly or suddenly. The patient will complain of the development of pain, swelling and stiffness at the affected joint over a matter of hours or at most, a few days, with the duration being less than 6 weeks. (Fig. 1).

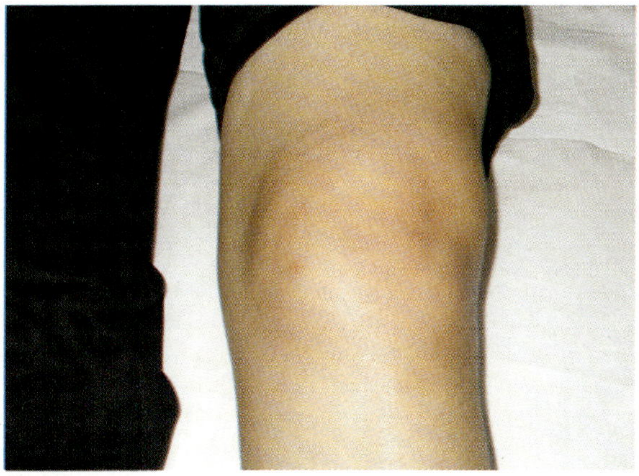

Fig. 1. Acute MA of the left knee: Note the obvious effusion at the joint.

Acute MA constitutes a medical emergency, considering the rapid development of symptoms that can often incapacitate a patient in a short span of time. Prompt and rapid recognition of the cause in any given patient is important, to institute correct treatment and alleviate the clinical symptoms. Infection (septic arthritis) is the most important cause of acute MA, as rapid treatment is essential in order to prevent joint damage and morbidity. Other possible common causes of acute MA that need to be kept in mind and excluded are: Crystal induced synovitis; Osteoarthritis (OA); Trauma; and Mono-articular presentation of a polyarticular or oligoarticular disease (see Table 1).

1. Septic Arthritis

Septic arthritis indicates joint infection or sepsis, caused by pathogenic bacteria, either by the haematogenous route or by direct inoculation of the joint (such as following trauma); this distinguishes septic arthritis from reactive arthritis (ReA) where the joint inflammation is due to an immunological reaction to the organism, which may be elsewhere in the body. A patient with joint infection will typically present with a painful, hot and swollen joint, with restricted movements. The symptoms develop rapidly over a period of days and within about 2 weeks[3]. Delayed treatment of septic arthritis can lead to irreversible joint damage and in addition, there is a significant mortality, with an estimated case fatality rate of 11%[4,5]. It is therefore crucially important that the diagnosis of septic arthritis should be made as rapidly as possible. However, the diagnosis can be difficult and may be delayed, even in the hands of an experienced clinicians. Some clinical situations when associated with the presentation of acute MA, should raise the index of suspicion for septic arthritis as the cause and these include[6]: a) The presence of existing joint disease such as rheumatoid arthritis (RA) or osteoarthritis (OA); b) Acute inflammation in an artificial or prosthetic joint; c) Low socio-economic status; d) Intra-venous drug abuse; e) Chronic alcoholism; f) Other associated medical conditions such as diabetes mellitus; g) Previous intra-articular injections; and h) Ulcers or infection of the skin overlying the affected joint. Usually, the presentation is at a large joint (knee or hip) with acute, pain, swelling, overlying warmth and restricted movement at the affected joint. Occasionally, the presentation may be with more than one joint being affected, so an oligo or polyarticular presentation does not necessarily mean that septic arthritis can be safely excluded. Staphylococcus and Streptococcus are the most common organisms that cause septic arthritis and account for about 90% of cases. Gram-negative organisms are more common in immuno-compromised or older patients and anaerobic bacteria may be responsible for septic arthritis following penetrating injuries[5].

Once septic arthritis is suspected clinically, *prompt aspiration and analysis of synovial fluid from the affected joint is extremely important.* A Gram's stain and routine bacterial culture of the fluid would enable confirmation of septic arthritis and examination for crystals under polarising light microscopy would enable the diagnosis of crystal-induced arthritis[7]. Estimation of the white blood cell (WBC) count in the synovial fluid can also be helpful in identifying septic arthritis. WBC counts of >100,000/mm^3 are more likely to be associated with joint infection and counts of < 50,000/mm^3 are less likely to be due to infection and more likely to be inflammatory in nature[8]. The synovial fluid sample should be transported as rapidly as possible to the laboratory and it would be prudent to warn the microbiologist that a sample from a suspected case of septic arthritis is being sent for analysis, beforehand. In order to improve the diagnostic yield from culture, it has been suggested that the synovial fluid should be inoculated into blood culture bottles in addition to direct culture on agar plates[9]. Blood cultures may help to isolate the causative organism in some cases and the serum WBC count, ESR and CRP levels may help to monitor response to treatment. There are no typical diagnostic imaging techniques to identify septic arthritis; however, isotope scanning may help to distinguish OA from sepsis (but will not necessarily distinguish sepsis from other inflammatory causes) and magnetic resonance imaging (MRI) may help to identify appearances consistent with infection[3]. The diagnosis of septic arthritis therefore rests upon appreciating the clinical features, prompt joint aspiration with appropriate synovial fluid analysis, together with other relevant investigations.

2. Crystal Induced Synovitis

Crystal induced synovitis is an important cause of acute MA. A number of crystals can cause acute joint inflammation, the most common of these being mono sodium mono urate (MSU) crystals that cause Gout. Recognition of gout as the cause of acute MA depends upon recognising the typical clinical features: Acute gout is eight times more common in men than women and the first attack commonly occurs between the third to sixth decades. *It is extremely uncommon in women before the menopause.* The commonest joint to be affected is the first metatarso-phalangeal (MTP) joint, also referred to as podagra. A typical attack is sudden in onset, develops rapidly with peak symptoms developing within 24 hrs and even within 6 to 8 hrs and may often awaken the patient from sleep. The joint becomes red, hot, and swollen, with shiny overlying skin, which may desquamate later. The joint is exquisitely painful and tender and the patient is usually unable to weight bear on it. The attack is not uncommonly misdiagnosed as local infection. During an acute attack there may be associated constitutional symptoms such as anorexia, nausea and fever. Leucocytosis and an elevated ESR may also be present. An elevated serum uric acid is usually (but not invariably) present[10]. These clinical features will often allow a diagnosis of acute gout to be made with fair confidence. Occasionally, the initial attack may be polyarticular and the 1st MTP joint may be spared. To establish the diagnosis of gout beyond doubt, synovial fluid should be aspirated from an affected joint whenever possible and examined under polarising light microscopy (PLM) to look for typical MSU crystals. These crystals will appear as needle shaped, negatively birefringent crystals (see Fig. 2).

Fig. 2 : Needle shaped crystals of MSU seen under polarising light microscopy

Recurrent acute attacks may result in chronic tophaceous gout, with tophii being found typically in the periarticular tissues, cartilaginous helix of the ear, bursae and tendon sheaths. The tophii appear as periarticular swellings, filled with a soft, 'cheesy' or 'chalky' material (see Fig. 3). The presence of tophi in a patient presenting with acute MA and a compatible history is a sure clinical sign confirming the diagnosis of gout and the tophaceous material will reveal numerous MSU crystals under PLM.

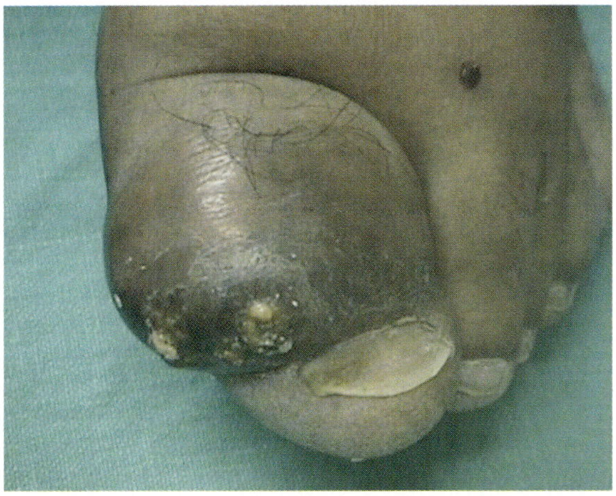

Fig. 3 : Acute synovitis of 1st MTP joint with a large overlying tophus and typical 'cheesy' material extruding from the tophus.

Some calcium containing crystals may also cause acute episodes of synovitis. Calcium pyrophosphate dihydrate (CPPD) and basic calcium phosphate (BCP) are the common calcium crystals. CPPD crystals can cause an acute attack of synovitis, which may be indistinguishable from an attack of gout. This is termed as pseudo-gout. The deposits can often be seen as a fine rim of calcification overlying the articular cartilage of the affected joint (chondrocalcinosis) - see Fig. 4; and the crystals of CPPD can be identified on PLM, confirming the diagnosis[11].

The BCP crystals include - hydroxy-apatite, octacalcium phosphate and tricalcium phosphate. BCP crystals most often cause a periarthritis or calcific tendonitis, but may also cause a synovitis of a large joint such as the knee or shoulder (Milwaukee shoulder). Though this tends to be a more chronic synovitis, it may occasionally present as an acute MA and ought to be kept in mind in the evaluation of a patient with acute MA[11].

3. Osteoarthritis

Osteoarthritis (OA) is the most common joint disease and therefore, should always be borne in mind as a possible cause of joint symptoms including acute MA. Though usually a chronic, slowly progressive disease, OA can occasionally present with acute worsening of symptoms at a joint, with associated swelling[12]. This is often seen at the knee joint, where mild trauma, overuse of the joint or chondrocalcinosis may cause an acute increase in symptoms with clinical signs of inflammation [1]. However, these acute symptoms would be part of the overall clinical spectrum of OA, which commonly affects multiple joints (hands, feet, spine, and hips) and when associated with bony swelling at the DIP joints of the hands (Heberden's nodes) is referred to a Primary Generalised Osteoarthritis (PGOA) or 'Nodal' OA. Occasionally, OA may affect single large joints (knees or hips)[13]. The diagnosis of OA is not too difficult, keeping in mind the clinical picture and can be easily confirmed by radiology- see Fig. 4.

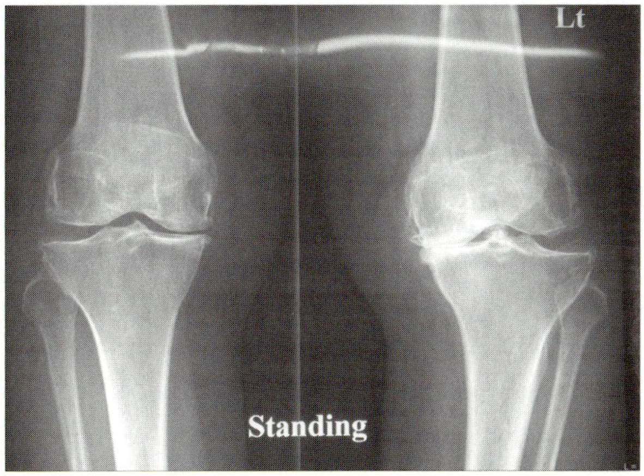

Fig. 4 : Chondrocalcinosis, seen as a thin line of calcification over the articular cartilage of the right knee lateral compartment. Also demonstrates OA changes at both knees, (left > right) with reduced joint space, osteophyte formation and varus deformities.

4. Trauma, haemarthrosis and foreign body reaction

Trauma to a joint may cause acute pain due to internal derangement (particularly at the knee), haemarthrosis or fracture. This may be mistaken for acute synovitis. Milder episodes of trauma may cause a transient (reactive) synovitis at the joint and should be borne in mind in the evaluation of a patient with MA[14]. Patients with coagulation or bleeding disorders (haemophilia or anticoagulant therapy) may present with bleeding into a joint, which clinically present quite like an episode of acute MA. Joint aspiration will however confirm the diagnosis. Penetrating injuries from thorns (plant thorn synovitis), wood fragments and other foreign particles (foreign body synovitis) may cause an acute synovitis, due to a foreign body reaction, which may appear shortly after the injury or even a few weeks or months later[15]. A careful history, eliciting possible trauma or even occupation of the patient (such as gardening) is therefore important, in reaching an accurate diagnosis.

5. MA presentation of a polyarticular disease

Many systemic or polyarticular diseases may initially present with an episode of acute MA. These conditions include: RA, Systemic lupus erythematosus (SLE), the arthritis of inflammatory bowel disease, Psoriatic arthritis (PsA), Behcet's disease, Reiter's disease and other forms of reactive arthritis. Most of these conditions will however not persist as MA and in addition, the associated clinical features of the underlying disorder will enable an accurate diagnosis to be made.

B. Chronic MA

Inflammatory MA persisting beyond 6 weeks may have many causes (see Table 1). As the duration of synovitis increases, infection (particularly chronic infection such as TB) becomes more likely and a synovial biopsy should be undertaken to confirm or exclude infection and also to exclude primary synovial conditions (such as villo nodular synovitis)[2]. However, should there be associated radiological features suggesting the possibility of infection, such as periosteal reaction, osteopenia or bony erosion, the synovial biopsy should not be delayed. Some cases of chronic MA may eventually evolve to a polyarticular disease (OA or RA) but about 1/3rd may remain undiagnosed, particularly when the knee is involved [2].

Approach to the evaluation of a patient with MA

The process of evaluation of a patient presenting with joint symptoms, really commences even before the formal consultation, by careful observation of the patient as he/she enters the consulting room. The gait of the patient, how ill the patient appears, any visible extra-articular features (such as a rash) may all suggest possible diagnosis to an astute clinician[16] (Table 3).

Table 3 Common Eponyms in Rheumatology

	Clinical observation	Possible diagnosis
1.	Middle aged man with an acute, swollen, painful foot (MTP joint) OR Elderly lady, single, swollen and painful joint, on diuretics	Crystal synovitis/Gout
2.	Young male. Knee or ankle pain and enters with stiff neck and back	Spondarthritis/Ankylosing spondylitis
3.	Elderly, ill looking individual with an acute, hot, swollen, painful joint	Septic arthritis

History

The next and certainly the most important step in evaluation, is to take a detailed and full history; this is of course true for the evaluation of all medical problems, but particularly so for the evaluation of musculo-skeletal complaints and acute MA. Onset of symptoms is important in order to determine acute from chronic MA and often, patients may have some difficulty in recalling exactly when their symptoms began. Patients may not associate their present acute symptoms with pain at other joint(s) or even previous episodes of joint pain; however this would be important to elicit, even by specific questioning, as a history of previous similar episodes would suggest a non-infectious cause and the possibility of crystal arthritis. A history of fever; insect or tick bites; sexual contact; pre-existing systemic disease; intravenous drug use; or travel, might all raise the possibility of infection. The presence of associated features, such as skin rash, ocular complaints, urethritis or genital ulcers and diarrhoea, would suggest the possibility of Reiter's disease. History of trauma might suggest the possibility of traumatic synovitis or fracture. A history of concurrent medical problems might be relevant such as the presence of psoriasis would make one consider psoriatic arthritis. History of drug therapy would also be important; recent introduction of diuretics might suggest gout as a possibility. A family history of gout, OA or other inflammatory joint diseases should be sought and might often lead to a diagnosis.

Examination

The most crucial aspect of the physical examination would be to distinguish true synovitis (joint inflammation) from bursitis, cellulitis or tendonitis. The presence of a joint effusion and synovial thickening, tenderness over the joint line and restricted active and passive movements at the joint, indicate joint inflammation. By and large, this is not difficult and one can recognise joint inflammation clinically, particularly at certain joints such as the knee joint. However, some clinical situations pose specific problems: olecranon bursitis may sometimes be difficult to differentiate from synovitis at the elbow; cellulitis around the ankle or knee may simulate synovitis; and deep seated joints such as the hip and sacro-iliac joints can be difficult to evaluate.

Once the primary question is settled that one is dealing with a MA, one should attempt to look for specific clues that might suggest an underlying diagnosis on further examination (Table 4).

Table 4 Diagnostic clues on examination

Findings on Examination	Possible Diagnosis/think of
Significant lymphadenopathy	Infection (tubercular)
Oral ulcers	Behcet's disease; Reiter's syndrome; or SLE
Cutaneous tophii	Gout
Nail pitting or onycholysis	PsA
Specific skin rash:	
- Erythematous facial rash	- SLE
- Psoriatic patches (natal cleft, behind ears or at umbilicus)	- PsA
- Erythema nodosum	- Sarcoidosis, inflammatory bowel disease, SLE
- Keratoderma blennorrhagicum	- Reiter's disease
- Pyoderma gangrenosum	- Inflammatory bowel disease
- Ulcers	- Infection
Sacroiliitis	**Spondarthritis**

Table 5 Possible systemic diseases presenting as acute MA

Clinical presentation	Possible systemic disease(s)
Acute MA	Psoriasis; Reiter's disease; Inflammatory bowel disease
Septic arthritis	Hypogammaglobulinaemia; Bacterial endocarditis; Bacterial pneumonia; HIV/AIDS
Pseudogout	Haemochromatosis; Acromegaly; Hypoparathyroidism;

Investigations

The most important clinical investigation in the evaluation of a patient with MA is synovial fluid examination (particularly if infection is suspected) and virtually every patient presenting with a MA should have the joint aspirated with this objective[1]. Joint aspiration is a simple, safe and easy bedside procedure with virtually no complications, providing a sterile no-touch technique with appropriate disposable equipment is used[17]. The synovial fluid should be examined for the following: 1) Gram's stain; 2) Routine culture; 3) WBC count (total and differential); and 4) Crystals under PLM. Virtually any joint can be aspirated easily and only a few drops of fluid are needed for these tests. Some joints such as the hip and sacro-iliac joints are deep seated and are best needled under image guidance. Synovial biopsy obtained by needle aspiration or arthroscopy, might be indicated for the evaluation of chronic MA rather than in the acute situation. Often, the biopsy reveals "non-specific inflammatory" changes on histology and this is particularly so for the non-infective, inflammatory conditions. Synovial biopsy would be of specific help in identifying conditions such as amyloidosis, sarcoidosis, pigmented villo-nodular synovitis or other synovial tumours. Some recent advances in the processing of synovial biopsy specimens, by polymerase chain reaction (PCR) and immunoelectron microscopy, might help to identify tiny fragments of the aetiological agent and have been found to be useful for the diagnosis of Lyme disease, Gonococci and Chlamydiae[1]. These new techniques may not be easily available but certainly rapid and appropriate synovial fluid analysis should be regarded as a sine-qua-non in the evaluation of MA.

Other investigations might be helpful depending upon the suspected clinical diagnosis. Leucocytosis might suggest infection and cultures of the blood and urine might also help in identifying the causative organism in some situations. Radiology of the affected joint will seldom identify the cause of MA, but might help to identify fracture (suggesting trauma), chondrocalcinosis, sacroiliitis or changes of OA. MRI is helpful in identifying possible infective changes at deep-seated joints (such as the sacro-iliac) and can also help to detect bone and soft-tissue involvement.

Finally, one should keep in mind that MA may occasionally be the first clinical sign of a systemic disease and therefore one may need to consider specific investigations depending upon the clinical features- Table 5 enumerates the systemic conditions that might present as an acute MA.

Key Points

1. Detailed history and examination to: a) confirm joint inflammation; b) confirm single joint inflammation; evaluate for possible underlying causes.
2. Investigate based upon clinical features.
3. Do not delay joint aspiration, if crystal arthritis or joint infection is suspected.
4. Treat as appropriate, based upon above aspects.

Suggested algorithm for the evaluation of MA

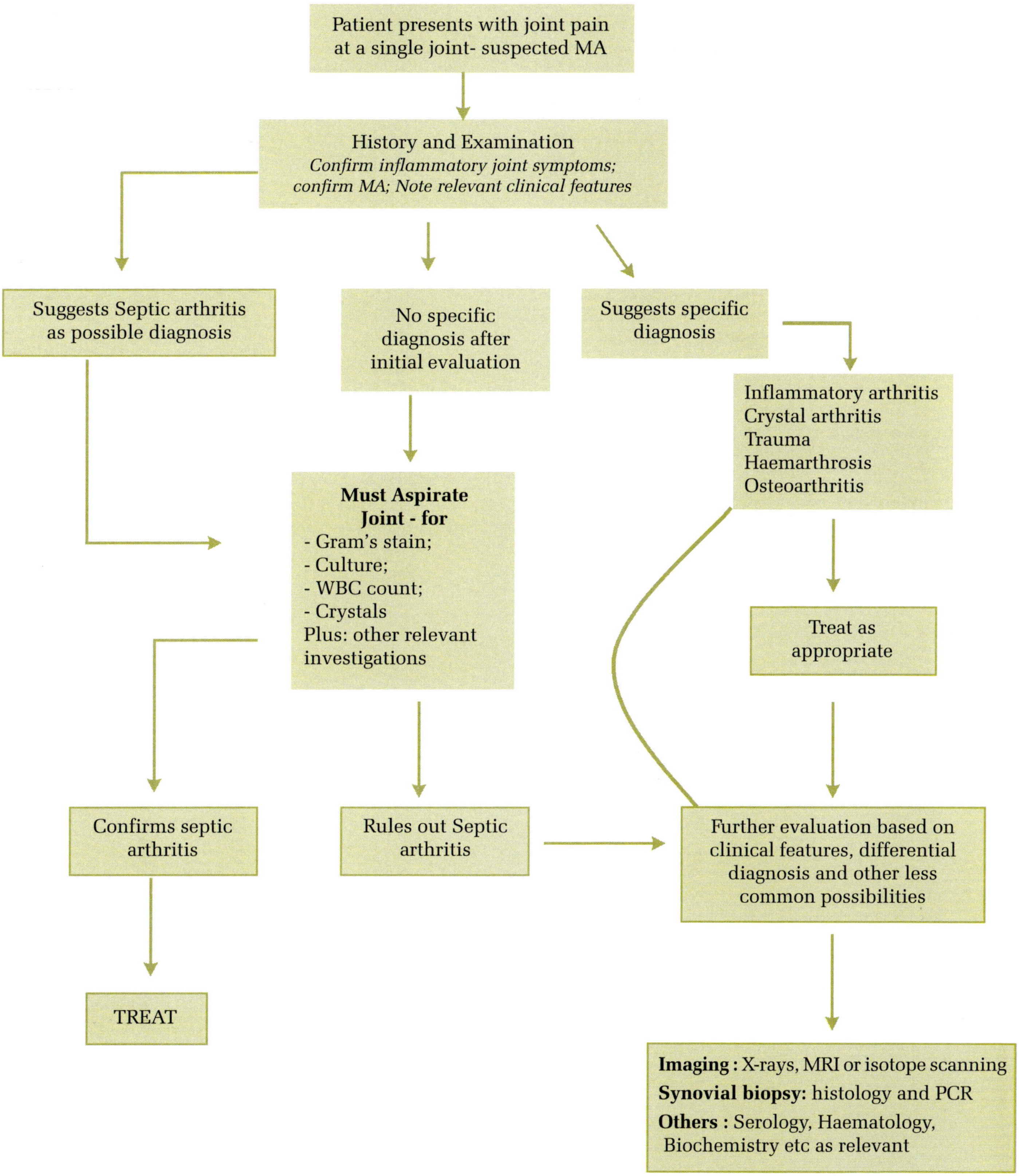

Further Reading :

1. Baker DG, Schumacher HR. Acute Monoarthritis. *N Engl J Med* 1993; 329: 1013-20.
2. Doherty M, Lanyon P, Ralston SH. Presenting Problems in Musculoskeletal Disease. In: *Davidson's Principles and Practice of Medicine*: 2006: Eds. Boon NA, Colledge NR, Walker BR, Hunter JAA. Churchill Livingstone Elsevier. 20th Edition; 1076-77.
3. Mathews CJ, Kingsley G, Field M et al. Management of septic arthritis: a systematic review. *Ann Rheum Dis*: 2007; 66: 440-45.
4. Weston VC, Jones AC, Bradbury N, Fawthrop F, Doherty M. Clinical features and outcome of septic arthritis in a single UK health district 1982-1991. *Ann Rheum Dis* 1999; 58: 214-99.
5. Gupta MN, Sturrock RD, Field M. A prospective 2-year study of 75 patients with adult-onset septic arthritis. *Rheumatology (Oxford)* 2001; 40: 24-30.
6. Sharp JT, Lidsky MD, Duffy J, Duncan MW. Infectious arthritis. Arch Intern Med: 1979; 139: 1125-30.
7. Swan A, Amer H, Dieppe P. The value of synovial fluid assays in the diagnosis of joint disease: a literature survey. *Ann Rheum Dis* 2002; 61: 493-98.
8. Coutlakis PJ. In: Roberts WN, Wise CM, Eds. Another look at synovial fluid leucocytosis and infection. *J Clin Rheumatol* 2002; 8: 67-71.
9. von Essen R, Holtta A. Improved method of isolating bacteria from joint fluids by the use of blood culture bottles. *Ann Rheum Dis* 1986; 45: 454-57.
10. Gupta SJ. Crystal induced arthritis: an Overview. *J Indian Rheumatol Assoc* 2002: 10(1): 5-13.
11. Halverson Paul B, Derfus Beth A. Calcium crystal-induced inflammation. *Curr Opin Rheumatol* 2001. 13: 221-24.
12. Goldenberg DL, Egan MS, Cohen AS. Inflammatory synovitis in degenerative joint disease. *J Rheumatol* 1982; 9: 204-09.
13. Sarzi-Puttini P, Cimmino MA, Scarpa R, et al. Osteoarthritis: An Overview of the Disease and Its Treatment Strategies. *Semin Arthritis Rheum* 2005; 35(suppl 1): 1-10.
14. Rooks YL, Corwell B. Common urgent musculoskeletal injuries in primary care. *Prim Care* 2006; 33(3): 751-77, viii.
15. O'Connor CR, Reginato AJ, DeLong WG Jr. Foreign body reactions simulating acute septic arthritis. *J Rheumatol* 1988; 15: 1568-71.
16. Russell AS. Clinical Presentations in Different Age Groups: The Adult Patient. In: *Oxford Textbook of Rheumatology* 2004; Eds. Isenberg DA, Maddison PJ, Woo P, Glass D and Breedveld F. Oxford University Press. 3rd Edition: 3-8.
17. Gupta, SJ. Intra-articular steroid therapy. *Journal Int Med Sci Aca* 1998; 11: 215-17.

APPROACH TO A PATIENT WITH POLYARTHRITIS 4

Anand Malaviya

Introduction

Clinicians use three 'tools' for making a diagnosis namely (i) clinical history; (ii) physical examination; (iii) laboratory and imaging investigations. Studies have shown that approximately 60-80% of the information towards making a diagnosis comes from clinical history; approximately 7-12% of additional information is gathered from physical examination and laboratory investigations including imaging add only approximately 5-11%[1-3]. Applying the same principle to patient presenting with musculoskeletal symptoms, especially polyarthritis, a thorough clinical history remains the most important diagnostic tool. There are excellent write-ups describing clinical approach to polyarthritis; reader is provided with a short list of the same for further reading[4-11].

Diagnosing polyarthritis is the commonest clinical condition in daily rheumatology practise. Making a diagnosis of polyarthritis is not only intellectually satisfying but most rewarding as well. Using the 'tools' of history and physical examination, a skilled and experienced clinician can formulate a short list of differential diagnosis. Base-line relevant laboratory investigations and specific radiographs are used primarily as adjunct to confirm the clinical diagnosis[10].

Is it Polyarthritis?

Within the category of polyarthritis i.e. involvement of more than one joint, many clinicians would further classify patients with only 2, 3 or 4 joint involvement as oligo- or pauciarthritis. It further narrows down the diagnostic possibilities to only a few conditions (see below). Thus polyarthritis is involvement of more than 4 joints. It is important to point out, however, that quite often the entire attention of the patient is focused towards a single joint simply because that is the most symptomatic joint. A beginner in the field of rheumatology may be misled to believe that the patient has monoarthritis. However, further probing of the history and a thorough physical examination would easily reveal a polyarticular involvement where a single joint is more symptomatic than the others. There are conditions involving musculoskeletal systems that mimic polyarthritis. These must be carefully excluded before the condition is labelled polyarthritis. Table 1 gives the differential diagnosis of musculoskeletal pains that can mimic polyarthritis.

Table 1 Differential diagnosis of Musculoskeletal pains that may mimic polyarthritis.

- Multifocal osteonecrosis, periostitis, other primary bone diseases
- Muscle disorders
- Polymyalgia rheumatica
- Neuropathies
- Tendonitis, fibrositis, multifocal bursitis
- Functional, fibromyalgia and malingering

Is it oligo-pauci-arthritis?

As mentioned above, involvement of 2 to 4 joints is called oligo-/pauci arthritis. Its separation from polyarthritis serves the purpose of narrowing down the diagnostic possibilities. Typically, inflammatory peripheral arthritis seen in the various forms of spondyloarthritis (SpA) is commonly oligoarticular in nature with involvement of joints of the lower segment of the body with prominent asymmetry and associated enthesopathy. Table 2 gives a short list of conditions that commonly present with oligoarthritis. It is to be noted that almost all forms of inflammatory arthritides may evolve through oligoarticular stage but 'true' oligoarthritis would persist as oligoarticular over time.

Table 2 Conditions most commonly present as *oligo-/pauci-arthritis*:

a. Peripheral *oligoarthritis*, often asymmetrical
   i. PsA
   ii. Peripheral involvement in SpA including Reiter's disease and other forms of 'classical' reactive arthritis.
   iii. Enteropathic arthritis
   iv. Behcet's disease
   v. Polyarticular gout and pseudogout
   vi. Infective endocarditis
   vii. Chronic gouty arthritis and CPPD deposition arthropathy

viii. Relapsing polychondritis
ix. Sarcoidosis
x. Bacterial endocarditis
xi. Amyloid arthropathy

b. Peripheral oligo-/pauci-arthritis associated with axial involvement
i. All the diseases in 1b category (above) can have axial disease *except* polyarticular gout and infective endocarditis

(**Note:** 1. PsA as well as sarcoid joint disease have several well defined patterns. Therefore, they could be the cause of inflammatory polyarthritis of different categories and their names may appear in different lists of polyarthritides. 2. For labelling the condition as oligoarthritis (2-4 joints) midfoot and wrist is taken as a single joint. 3. Several conditions mentioned above as 'oligoarthritis' may have a subset that presents mainly as oligoarthritis, in others it may only be a phase in the evolution of a polyarthritis).

Polyarthritis : Clinical analysis for reaching a diagnosis

The following points need to be ascertained:

1. Is it *acute* or *chronic*?
2. Is it *inflammatory* or *noninflammatory*?
3. Past, personal and family history.
4. Review of systems and *extraarticular* manifestations
5. *Pattern* and *topography* of joint involvement.

Is polyarthritis acute or chronic?

By definition involvement of more than 4 joints of less than 6-week duration is classified as *acute polyarthritis*. Patients presenting with acute polyarthritis pose a major diagnostic challenge as this category includes almost every disease of importance in rheumatology some of whom could be potentially life-threatening and associated with high morbidity. Yet, in others it could simply be a benign self-limiting condition e.g. a postviral arthritis that resolves completely within 6 weeks. In the majority of acute polyarthritis cases definitive diagnosis is usually not possible at the outset. The reason is that by-and-large *clinical rheumatology is a science of recognising pattern of joint involvement and its evolution* and that still remains central to making a rheumatological diagnosis. Therefore, in acute polyarthritis the symptoms may not have yet evolved into a recognisable pattern for a definitive diagnosis. In some of them, a more detailed history including past history, family history, review of systems, history of similar disease in the community, physical examination and diagnostic evaluation may give a clue to the diagnosis. Table 3 gives common causes of *acute* polyarthritis.

Table 3 Causes of acute polyarthritis

- RA, SLE, PsA, other CTDs, systemic vasculitides
- Septic (mainly gonococcal),
- Reiter's and other 'classic' or 'non-classic reactive arthritides' (e.g. acute rheumatic fever, Poncet's, leprosy-related)
- Viral arthritis : Chikungunya, rubella, hepatitis-associated arthritis, parvovirus-B19, others
- IBD-related arthritis
- Sarcoidosis (Lofgren's syndrome-type presentation)
- Erythema nodosum and other panniculitis-related arthritis
- Serum sickness
- Haematological causes: Sickle-cell anaemia, leukaemic arthritis, coagulation disorders
- Palindromic rheumatism

(Abbreviations: SLE = systemic lupus erythematosus; PsA = psoriatic arthritis; CTDs = connective tissue diseases; IBD = inflammatory bowel disease)

For example, acute dactylitis of a toe in a young person is strongly suggestive but not diagnostic of psoriatic arthritis. Further evolution of the disease over time and/or history of psoriasis or psoriatic arthritis in a blood relative would confirm the diagnosis. However, in other patients no such clue may be forthcoming in the beginning and the diagnosis is possible only over time. Therefore, most experienced rheumatologists realise that despite their best efforts only further *evolution of pattern of symptoms over time* may give a clue to diagnosis. It may be difficult for a young in-training rheumatologist to accept this 'defeatist' attitude but in clinical rheumatology it is no shame to wait-and-watch till either the symptoms resolve by themselves (benign self-limiting conditions) or evolve over time to become recognisable for making a definitive diagnosis. It may, however, be noted that future may be much brighter. Rapid advances in diagnostic techniques based on advanced technologies may be able to provide a diagnostic clue not only in the early stages of polyarthritis but even in preclinical stages. The best example is the presence of cyclic citrullinated peptides antibody (ACPA) in patients with rheumatoid arthritis[12]. Presence of high titres of these antibodies in a patient with acute polyarthritis is now considered a strong clue that it would evolve into rheumatoid arthritis. The American College of Rheumatology (ACR) took cognisance of this new research finding (and several additional new advances) to propose re-revised classification criteria for rheumatoid arthritis in 2010 that is specifically suited for making an early diagnosis[13]. Using these classification criteria now, most rheumatologists would not wait for 6 weeks before embarking upon specific treatment for RA in such a patient. Thus, it is obvious that clinical rheumatology is an exciting and rapidly evolving field. The prevailing diagnostic paradigm may not hold true for too long. However, for the time being rheumatologists must follow the time-honoured methods of making a diagnosis.

Is the polyarthritis inflammatory in nature?

Possibly the most crucial point related to making a diagnosis in polyarthritis (or for that matter in all forms of arthritis) is to evaluate whether it is *inflammatory* or *non-inflammatory* in nature[8]. This clinical point is pivotal as it determines the management and prognosis of the patient. A beginner or a non-rheumatologist clinician may not realise the importance of this distinction or mistakenly believe that presence or absence of time honoured easily recognisable features of acute inflammation (namely red, hot, tender, swollen and non-functional) would be easily recognisable in joints for classifying polyarthritis as inflammatory or non-inflammatory. Unfortunately, the joints may be tender, swollen and non-functional even in noninflammatory conditions e.g. osteoarthritis (OA) or traumatic arthritis. On the other hand, redness and local heat may not always be present or difficult to perceive even in inflammatory joint disease. For this reason, rheumatologists have devised an entirely different set of clinical features that distinguishes *inflammatory* from *non-inflammatory* joint diseases with high degree of sensitivity and specificity. Table 4 lists the contrasting clinical features that distinguish inflammatory from non-inflammatory joint diseases. It is strongly recommended that clinicians get used to properly quizzing the patients about these symptoms by putting open-ended questions. For example a leading question 'Are you stiff in the morning'? would be a wrong way of putting the question as against an open-ended question 'How do you feel when you get up in the morning'?

Table 4 Clinical features to distinguish inflamm-atory from noninflammatory joint disease.

1. Early morning stiffness that takes at least 30 minutes for maximum improvement, more often it takes about 1 hour for maximum improvement.
2. Activity (gentle movement) improves the symptoms but inactivity does not (inactivity may actually increase the stiffness).
3. Spontaneous 'flares' with fluctuation in the severity of symptoms over time without any obvious extraneous cause.
4. Constitutional symptoms: fatigue and tiredness, feverish feeling or actual fever, night sweats, loss of appetite and weight.

Note: *Presence of one, more or all of these manifestations would strongly suggest the inflammatory nature of the joint problem with increasing probability.*

If the clinical history suggests inflammatory polyarthritis, physical examination may reveal a warm joint. Erythema or redness on the inflamed joints is usually not perceptible especially in those with dark skin. However, in those with intense inflammatory arthritis as seen in periankle-ankle involvement of acute sarcoidosis (Lofgren syndrome) and acute gouty arthritis involving joints in the feet, the overlying skin would show indurated inflammatory oedema with peeling off of the outer layer of skin as the inflammation subsides. However, in most of the common forms of inflammatory polyarthritides (RA, PsA) features of local inflammation may only be minimal or imperceptible. Table 5 gives a comprehensive list of the majority of arthritic conditions classified under 2 major heading namely *inflammatory* or *non-inflammatory* in nature.

Table 5 Common causes of polyarthritis classified as inflammatory vs. non-inflammatory further categorised on the topographic pattern of joint involvement (Modified from Ref. 10)

1. Inflammatory polyarthritis:
   a. Peripheral polyarthritis, mostly symmetrical in distribution
      i. RA
      ii. SLE and other connective tissue diseases
      iii. Wegener's granulomatosis and other systemic vasculitides
      iv. PsA
   b. Peripheral oligoarthritis, often asymmetrical
      i. PsA
      ii. Peripheral involvement in SpA including Reiter's disease and other forms of 'classical' reactive arthritis.
      iv. Enteropathic arthritis
      v. Behcet's disease
      vi. Polyarticular gout and pseudogout
      vii. Infective endocarditis
      viii. Chronic gouty arthritis and CPPD deposition arthropathy
      ix. Relapsing polychondritis
      x. Sarcoidosis
      xi. Bacterial endocarditis
      xii. Amyloid arthropathy
   c. Peripheral arthritis associated with axial involvement

i. All the diseases in 1b category (above) can have axial disease except polyarticular gout and infective endocarditis

(**Note:** 1. As mentioned in table 2, PsA as well as sarcoid joint disease have several well defined patterns. Therefore, they could be the cause of inflammatory polyarthritis of different categories and their names may appear in different lists of polyarthritides. 2. As mentioned in table 2, several conditions mentioned above as 'oligoarthritis' may have a subset that presents mainly as oligoarthritis, in others it may only be a phase in the evolution of a polyarthritis).

2. Noninflammatory polyarthritis:
   a. Hereditary factors
      i. Primary nodular OA of the hands (distal and proximal interphalangeal joint involvement)
      ii. Generalised primary OA
   b. Environmental factors and age-related: life-style, occupational, trauma
      i. Knee OA - usually an obese elderly person
      ii. OA following local injury (including surgery) in a joint
      iii. OA in strenuous exercise-related cartilage damage (chondromalacia)
      iv. OA in elderly
   c. Metabolic conditions, some uncommon diseases
      i. Hypothyroidism
      ii. Acromegaly
      iii. Haemochromatosis
      iv. Ochronosis

Past, personal and family history

Past history may have the clue for diagnosing polyarthritis. A transient episode of heel pain (Achilles tendon enthesitis, plantar fasciitis) at juvenile age or an episode of uveitis in the past may be the clue to the present problem of inflammatory low back pain due to ankylosing spondylitis-spondyloarthritis (AS-SpA) group. History in the recent past, of a febrile episode with or without exanthema, sexual contact or frank venereal disease, episode of diarrhoea, or acute conjunctivitis, may give a clue to the diagnosis of reactive arthritis, gonococcal arthritis or AS related peripheral arthritis. Quite often, past history may also unmask a diagnostic mistake made in the distant past e.g. history of 'fever-of-unknown-cause' in the past that was mistakenly either treated as typhoid or tuberculosis may actually have been an early manifestation of systemic lupus erythematosus that became overt in due course of time. Three most common diagnostic mistakes seen in day-to-day rheumatology practice involve tuberculosis, gout and rheumatic fever. It is quite common that a child or juvenile was given treatment for tuberculosis or rheumatic fever (long-acting penicillin) for months in the past for a mono- or oligoarthritis that evolved into rather typical enthesitis-related arthritis (ERA) over time. The other common mistake is to diagnose and treat the patient as 'gout' (only based upon trivial increase in serum uric acid) over years while patient has been evolving into any of the common polyarthritides. The importance of *personal history* has sharply come under focus in the recent years. Thus, it has now been proven that cigarette smoking may be a trigger for most inflammatory arthritides (RA, SpA, PsA, SLE) as well as accelerate cartilage damage[14-18]. There may be several other personal factors yet to be discovered. *Family history* is equally important in making a diagnosis of polyarthritis. Similar disease (e.g. RA), or closely related autoimmune disease (thyroid, connective tissue disease, inflammatory bowel {IBD}disease, other autoimmune conditions) in the family members could be very useful in reaching a diagnosis. A so-called 'seronegative inflammatory polyarthritis' may actually be PsA (even in the absence of plaque psoriasis in the patient) if a blood relative is known to have plaque psoriasis. The significance of family history is more obvious in frank hereditary conditions e.g. AS and gout where invariably a positive history is elicited.

Review of systems and extraarticular manifestations

Inflammatory polyarthritis has bidirectional association with other organ systems of the body. On the one hand, it may affect different body parts and organs secondarily. On the other hand, by itself it may be part of a multisystem disease. Therefore, a detailed *review of systems* along with a careful general and systemic physical examination is mandatory for making an accurate diagnosis.

History of *skin, mucosal and subcutaneous lesions* is highly relevant for making a diagnosis in a large number of inflammatory polyarthritides so much so that a rheumatologist has to be a good dermatologist. Thus, diagnosis of an otherwise undiagnosed 'seronegative inflammatory polyarthritis' becomes 'PsA' on discovering a psoriatic patch anywhere in the body, how-so-ever small it may be. Even a family history of psoriasis may help similarly. RA may have lesions of pyoderma gangrenosum (a form of neutrophilic dermatosis, also seen in IBD that itself may also be associated inflammatory polyarthritis, see below), subcutaneous nodules, and vasculitic lesions. Facial 'butterfly' rash of systemic lupus erythematosus (SLE) along with mucosal ulcers, alopecia and vasculitic skin lesions is diagnostic of the disease. Oral ulcers are seen in several other polyarthritides including Behcet's disease (very painful commonly associated with genital ulcers), IBD related arthritis, and SpA. Gottron's papules, heliotrope-facial rash, 'mechanic's hands', 'shawl sign' along with periungual vasculitic infarcts are typical of dermatomyositis. Raynaud's phenomenon associated with scleroderma skin changes including telangiectasia are typical of the disease and must be distinguished from scleredema of Bushke (no Raynaud's phenomenon). Sarcoid skin lesions including lupus pernio may come handy in diagnosing the cause of otherwise undiagnosed 'seronegative inflammatory polyarthritis'. Urticarial vasculitis can be suspected on typical skin lesions. Erythema nodosum and several other forms of inflammatory lesions of the subcutaneous tissue (panniculus) is seen in association with a number of inflammatory arthritides the best known being Lofgren syndrome (acute sarcoidosis), already mentioned above, Behcet's disease (the lesion often ulcerates) and IBD. Subcutaneous nodules in association with polyarthritis is seen with several diseases including RA, SLE, OA (Heberden's and Bouchard's nodules), gout (gouty tophi), sarcoidosis, rheumatic fever, polyarteritis nodosa, dislipidaemias, and others. Mucosal ulcers in SLE (moderately painful), SpA (little pain), and Behcet's disease (severely painful) are helpful clinical features for diagnosis. Palpable as well as non-palpable purpuric lesions are seen in several polyarthritides including SLE, Sjögren's syndrome (including hyper-gammaglobulinaemic purpura of Waldenström) and others. Leg ulcers (vasculitic or part of neutrophilic dermatosis e.g. pyoderma gangrenosum) are seen in RA, SLE (subcutaneous inflammation that may ulcerate e.g. lupus profundus), polyarteritis nodosa (including cutaneous polyarteritis), Behcet's disease and others. Keratoderma blenorrhagicum seen with Reiter's syndrome is a typical lesion that helps in making a definitive diagnosis. Recognition of gonococcal skin lesions could come in handy for diagnosing an otherwise idiopathic migratory inflammatory polyarthritis. Gross clubbing in a smoker presenting with painful swelling of distal joints in the extremities (wrists, small joints in the hands, ankles, joints in the feet) should immediately bring in the possibility of carcinoma lung-related hypertrophic pulmonary osteoarthropathy. The list is rather big to be covered here but suffice it would be to note that skin-subcutaneous tissue-mucosa (including nails, scalp) should be carefully evaluated in any patient with inflammatory polyarthritis.

The association of *eye* and *ear-nose-throat-mouth* with polyarthritis is possibly equally important. Sicca symptoms are well known features of Sjögren's syndrome, but also occur secondary to RA, SLE, scleroderma and others. Episcleritis-scleritis is common in RA. Iridocyclitis-uveitis is typically seen in association with SpA group including Reiter's syndrome. Retinal lesions are seen in SLE, systemic vasculitides and others, retrobulbar lesions (causing proptosis) may be seen in granulomatosus with polyangiitis (Wegener's granulomatosus), involvement of rectus muscles may occur as part of inflammatory myositis. Inflammation of cartilage in the pinna and bridge of the nose (with similar cartilage lesions at other sites) is typically seen in relapsing polychondritis causing flail or deformed pinna giving the appearance of a 'boxer's ear' and 'saddle-nose' deformity. The latter is also seen in granulomatosis with polyangiitis (Wegener's granulomatosis) that usually destroys the cartilage at bridge of the nose. Involvement of middle and inner ear is typically seen in granulomatosis with polyangiitis (Wegener's granulomatosis).

Inflammatory polyarthritis also has close association with *genitourinary* and *gastrointestinal systems*. Urethritic form of 'reactive arthritis', the most dramatic subset being Reiter's syndrome, is a well known cause of seronegative inflammatory polyarthritis. Gonococcal infection with urethritis, cervicitis and migratory inflammatory polyarthritis is a well recognised association. IBD also has duel association with inflammatory polyarthritis. Thus, several patterns of inflammatory polyarthritis have been described in association with IBD including peripheral polyarthritis and erythema nodosum (mentioned above). Another form is that of spondyloarthritis some of whom may evolve into frank ankylosing spondylitis. Subclinical/ asymptomatic small segments of lesions of IBD have been demonstrated by endoscopic examination in a significant proportion of patients with spondyloarthritides[19]. Behcet's disease is the other condition where IBD-type lesions may be seen. In fact, there are patients diagnosed as IBD who on detailed clinical evaluation may show features of milder forms of Behcet's disease including orogenital ulcers, erythema nodosum lesions, eye inflammation.

Renal disease and inflammatory polyarthritis combination is among the most serious, often life threatening forms of arthritic disorders. Urinary sediment abnormality indicative of glomerular inflammation (microscopic haematuria, variety of casts and WBC), sub-nephrotic or nephrotic range proteinuria, frank nephrotic syndrome, renal arterial involvement with renal hypertension, rapidly progressive glomerulonephritis, and end-stage renal disease are the manifestations of several diseases that may have associated inflammatory polyarthritis. This group mainly includes SLE and its overlap with other connective tissue diseases, antiphospholipid syndrome, and anti-neutrophil cytoplasmic antibody (ANCA)-associated systemic small and medium vessel vasculitides [granulomatosis with polyangiitis, microscopic polyangiitis, eosinophilic granulomatosis with polyangiitis (Churg-Strauss syndrome) and classic polyarteritis nodosa].

Peripheral as well as central nervous system involvement and inflammatory polyarthritis is another potentially serious, often life threatening group of arthritides. These include SLE, antiphospholipid syndrome, systemic vasculitides of several types (mononeuritis multiplex is a common manifestation in some of them), and Behcet's disease.

Lung involvement is commonly seen with RA, connective tissue diseases, systemic vasculitides, sarcoidosis, relapsing polychondritis, and AS. There is an important clinical subset of *pulmonary-renal syndrome* where kidney and lung are affected in addition to an inflammatory polyarthritis. SLE, scleroderma and other connective tissue diseases, ANCA-associated vasculitides, Goodpasture's syndrome, pauciimmune necrotising and cresentic glomerulonephritis (CGN), Henoch-Schönlein's purpura (HSP), and essential mixed cryoglobulinemia (EMC) are some of its causes[20].

Heart involvement is typically seen in antiphospholipid syndrome, SLE, scleroderma, dermatomyositis, and in a form of systemic vasculitis seen in paediatric age group (Kawasaki disease). Rheumatic fever is well known to be reminded of but, it must be noted that persistent inflammatory polyarthritis is not a feature of this disease. Moreover, it is a disease of paediatric age group. Polyserositis with inflammatory polyarthritis may be seen in SLE, overlap connective tissue diseases and a relatively uncommon condition in our country namely Familial Mediterranean Fever (FMF). Pleuropericarditis is typically seen in SLE but may be present in RA and other systemic conditions.

Obstetric and gynaecological problems may occur in several diseases with inflammatory polyarthritis, mainly due to joint deformities causing biomechanical-physical issues. But, antiphospholipid syndrome is the most notorious for causing foetal wastage and related problems. Increased risk during pregnancy and purpureum is characteristic of SLE.

Haematological abnormalities are universal in inflammatory polyarthritis and include high erythrocyte sedimentation rate (ESR), high platelets, anaemia of chronic disease and occasionally, leukocytosis that could be extreme on occasions. However, some haematological abnormalities may specifically point towards a certain diagnosis. Thus, low platelets may point towards the diagnostic possibility of SLE and antiphospholipid syndrome. Rare cases of leukaemic arthropathies in adults may have not only low platelets but also other abnormalities in the blood that may be a clue to diagnosis. Hypercoagulable state with widespread thrombo-embolic manifestations is rather typical of antiphospholipid syndrome that is often associated with SLE.

What is the pattern of joint involvement?

The next point to note in polyarthritis is the *pattern of evolution* of joint involvement. Basically there are 3 patterns namely *additive*, *intermittent*, and *migratory*.

In *additive* pattern (increasing number) joints are progressively recruited with many joints getting affected over time. Rheumatoid arthritis is a typical example of such a pattern of involvement.

In *intermittent pattern* the joint involvement appears and then disappears till the next such episode with completely asymptomatic period in between these episodes. In its inflammatory form it is typically seen in palindromic rheumatism, acute gouty arthritis, relapsing polychondritis and relapsing seronegative symmetrical synovitis with pitting oedema (RS3PE). The last mentioned condition is associated with marked joint stiffness and symmetrical synovitis in the joints of the hands and feet. A condition called intermittent hydrarthrosis is an idiopathic noninflammatory intermittent arthritis.

In *migratory pattern* the joints become symptomatic then subside while different joints get involved. The tempo of migration may differ for example it could be only hours between one to the next joint involvement in rheumatic fever. On the other hand the tempo of 'migration' may be much slower (days) in gonococcal arthritis. These patterns are not mutually exclusive but the dominant pattern may indicate a specific diagnosis.

The *topography or distribution of joint involvement* (which joints are affected in what distribution) usually helps in narrowing down the diagnostic possibilities[21]. For describing the topography of joint involvement the following anatomical regions are taken in account: *peripheral* joints (in the extremities), *axial* joints (traditionally sacroiliac joints are considered part of axial joints), *root* joints (that overlap between peripheral and axial joints namely shoulder and hip joints), and *upper* segment and *lower* segment of the body (above and below waist). In peripheral joints *symmetry* or *asymmetry* is important to note. Table 3 lists some of the common causes of polyarthritis classified as inflammatory vs. non-inflammatory further categorised on the basis of the *topographic pattern* of joint involvement. Fig. 1 to 5 depict some of the typical topographic patterns of joint involvement in different forms of inflammatory and noninflammatory arthritides. Caines A, et al[21] have recently described pain distribution patterns diagrammatically depicting musculoskeletal involvement in different diseases. Thus inflammatory synovitis of small as well as large joints in the extremities in a symmetrical distribution on both sides of the body, equally distributed in upper and lower segment with predominant involvement of joints of the hands, wrists and feet, with sparing of the distal interphalangeal (DIP) joints, is rather typical of RA. Then, there are certain joints that, if involved in isolation or predominantly, may be a clue to diagnosis. Thus, involvement of DIP joints without synovitis or inflammation is typically seen in primary nodular OA but with synovitis, it is often seen in some subsets of PsA, enteropathic arthritis and sarcoidosis. Non-inflammatory involvement of the first carpometacarpal joint (floor of the 'snuff- box') is

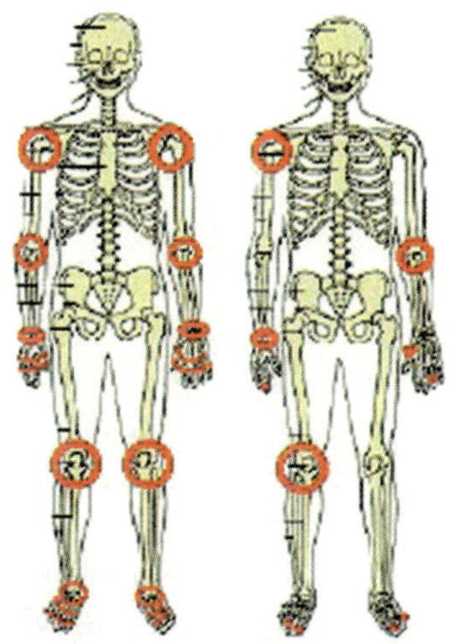

Fig. 1 : Topographic pattern of the joint involvement in RA versus PsA. Note the polyarticular involvement that is symmetrical in rheumatoid arthritis but asymmetrical in psoriatic arthritis.

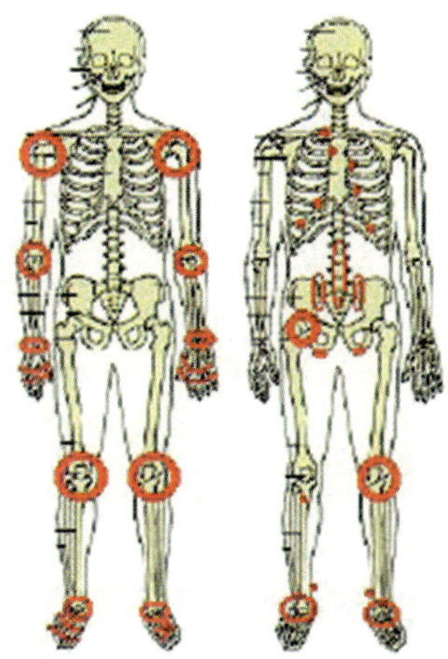

Fig. 2 : Topographic pattern of the joint involvement in RA versus SpA. Note the polyarticular involvement that is symmetrical in RA but asymmetrical in SpA predominantly affecting the joints in the lower segment of the body with the involvement of sacroiliac joints. Involvement of costochondral joints in the thorax and enthesitis at the various sites are depicted with red dots.

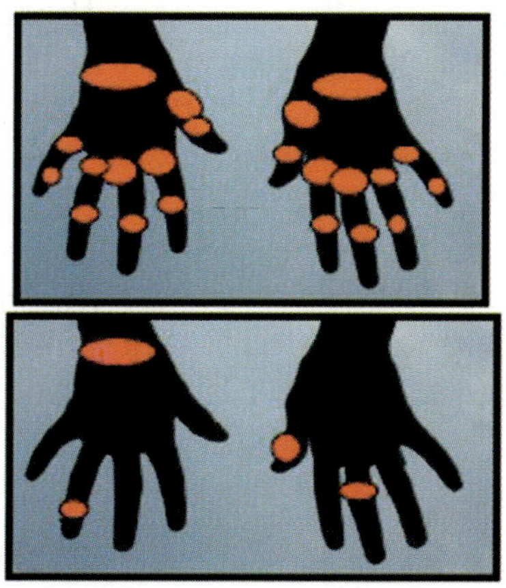

Fig. 3 : Topographic pattern of involvement of the hand joints in RA (marked as red circles or ellipse) versus OA (marked as blue circles or ellipse). Note that despite symmetry, target joints are different except metacarpophalangeal joints that are involved in both the diseases.

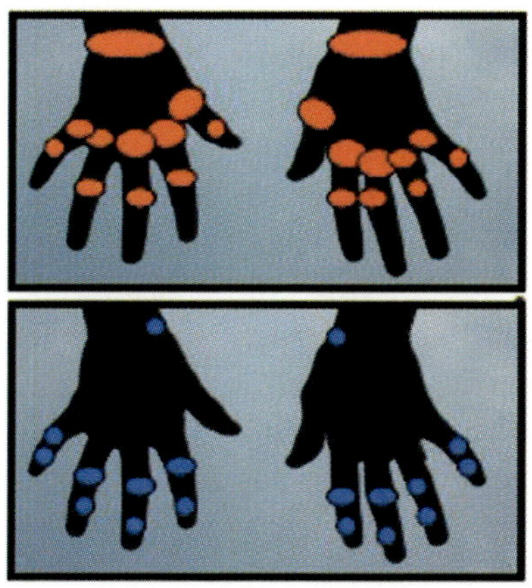

Fig. 4: Topographic pattern of the hand joint involvement in RA versus PsA. Note the asymmetrical involvement as well the involvement of DIP joints.

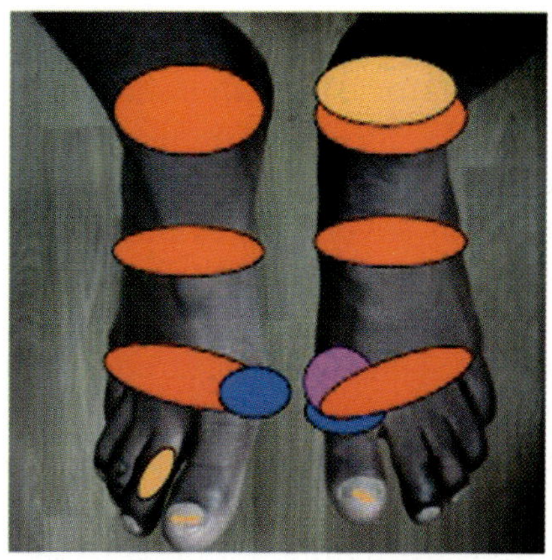

Fig. 5 : Topographic pattern of the feet joint involvement in the various common forms of polyarthritides. The pattern of involvement differs in different diseases but certain joints are involved in more than one disease e.g. 1st metacarpophalangeal joint that is most commonly affected in the commonest form of arthritis namely OA (noninflammatory) but is commonly affected in a relatively uncommon disease namely gout (inflammatory, called podagra). Colour coding is as follows:

RED: RA; BLUE: OA; PURPLE: acute gouty arthritis (podagra); BROWN: PsA

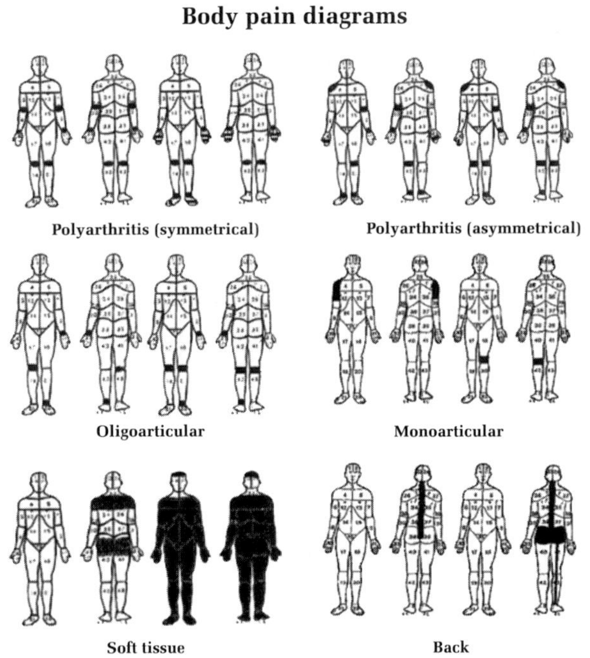

Fig 6 : ***Top row of panels: left*** *- symmetrical polyarthritis: prototype rheumatoid arthritis;* ***right*** *- asymmetrical polyarthritis: prototype spondyloarthritis - peripheral (including psoriatic arthritis).* ***Middle row of panels: left*** *- oligoarticular involvement: prototype spondyloarthritis;* ***right*** *- monoarticular arthritis - infection, crystals, early stage of any of the polyarticular disease.* ***Lower row of panels: left*** *- shoulder and hip girdle involvement (polymyalgia rheumatica), generalised aches and pains (fibromyalgia, hypothyroidism, malignancy - associated symptoms, vitamin D deficiency etc.);* ***right*** *- spine and buttock involvement (ankylosing spondylitis).*

Reproduced with permission from Caines A, Samadi N, Ouimet G, Thompson A, Pope JE. The sensitivity and specificity of pain diagrams in rheumatic disease referrals. Rheumatology (Oxford) 2012; 51: 1093-1098.

typically seen in primary OA. Similarly, noninflammatory first metatarsophalangeal joint involvement is most commonly seen in OA. In acute gouty attack the same joint is often the first one to be affected with severe acute inflammatory response (podagra). Periankle-ankle inflammatory arthritis, (as mentioned earlier) usually associated with erythema nodosum is seen in acute sarcoid arthritis (Lofgren syndrome). Acute dactylitis has already been mentioned to be strongly suspicious of psoriatic arthritis. Enthesitis (inflammation at the sites of insertion of ligaments, tendons to the bone such as Achilles tendonitis, plantar fasciitis and others) associated with inflammatory axial joint involvement, sacroiliitis (classically presents with alternating buttock pain) and lower extremity asymmetrical inflammatory arthritis, often associated with root joint involvement is rather typical of spondyloarthritis (SpA) group. Prominent involvement of metacarpophalangeal joints without synovitis is typically seen in haemochromatosis. It needs to be emphasised, however, that pattern recognition as mentioned above is only a rough guide to diagnosis since there is considerable overlap in the pattern of joint involvement among different types of polyarthritis.

Oligoarthritis

Involvement of 2 to 4 joints is called oligo- /pauci arthritis. Its separation from polyarthritis serves the purpose of narrowing down the diagnostic possibilities. Typically, inflammatory peripheral arthritis seen in the various forms of SpA is oligoarticular in nature with involvement of joints of the lower segment of the body with prominent asymmetry and associated enthesopathy. Table 3 gives a short list of conditions that commonly present with oligoarthritis. It is to be noted that almost all forms of inflammatory arthritides may evolve through oligoarticular stage but 'true' oligoarthritis would remain oligoarticular over time.

Joint count

It has been documented that formal quantitative joint count is not included in most visits of most patients with rheumatoid arthritis by most rheumatologists[22]. However, rheumatologists need to be reminded that with rapid advancements in drug therapy of patients with a variety of musculoskeletal diseases and especially that of inflammatory arthritides, some sort of objective assessment of the disease activity, the simplest method being joint count, must be carried out in every patient at the first encounter[23]. Irrespective of what methodology is adopted, it should not be ignored. One may try to adopt the most suited method for the type of clinical set-up one works in. Original Disease Activity Score (DAS), its modification (DAS28), more recently devised and validated Clinical Disease Activity Index (CDAI) and Simplified Disease Activity Index (SDAI) are widely recommended for use in RA and other inflammatory arthritides[24]. Bath Ankylosing Spondylitis Disease Activity Index (BASDAI), Bath Ankylosing Spondylitis Functional Index (BASFI), Bath Ankylosing Spondylitis Metrology Index (BASMI), are available for objective assessment of different aspects of inflammatory spinal arthritis. Methodology is also available for objective assessment of noninflammatory diseases like osteoarthritis Western Ontario and McMaster (WOMAC) index for osteoarthritis. The importance of objective assessment of disease activity cannot be overemphasised. Over the years it has become clear that tight control of disease activity in inflammatory arthropathies prevents joint damage and disability and reduces morbidity and mortality. But, tight-control of disease activity with 'treat-to-target' approach[25], is possible to achieve only if it is measured routinely. Treatment of diabetes mellitus or hypertension is good example to understand this argument. If blood glucose or blood pressure readings are not documented regularly they cannot be controlled satisfactorily. Unfortunately, most clinicians would treat inflammatory polyarthritis but, without knowing the degree of disease activity to guide them. Such a patient is likely to develop increasing damage and disability over time with increased morbidity and mortality.

Investigations

As mentioned above, the most crucial point towards making a diagnosis in polyarthritis is to differentiate between *inflammatory* and *non-inflammatory* conditions. History and physical examination provide good idea whether the condition is inflammatory in nature. In most situations the inflammatory nature of polyarthritis can be confirmed by routine laboratory investigations for non-specific markers of inflammation namely, raised levels of *acute phase proteins*. Table 4 gives the list of commonly available tests for confirming inflammation. Some exceptions, however, must be pointed out. In certain diseases ESR may not truly reflect the inflammatory state. This is often true of SLE, scleroderma and AS where ESR may be disproportionately high or low. Similarly platelet counts may be low in SLE as part of its clinical manifestation, as also in a relatively uncommon condition in adults namely, leukaemic arthropathies. Examination of *synovial fluid* in polyarthritis only to confirm inflammatory nature of the joint disease is generally not recommended as this can be reasonably differentiated from noninflammatory condition on history and physical examination and confirmed by tests for acute phase reactants. However, on rare occasions there could be some confusion because of poor history, no definite clue in physical examination and inconclusive or borderline results of acute phase reactants. In such cases joint fluid examination (if joints have effusion) for leukocyte count can help. A leukocyte count of > 2000/cmm is a definite sign of inflammatory synovitis. Additional tests including gram staining, bacterial and mycobacterial culture, and examination under polarised light microscopy for crystals would also help in making a definitive diagnosis. Radiological examination of the joints and other routine imaging investigations do not distinguish between inflammatory and non-inflammatory conditions. They may, however, detect typical pattern of radiological abnormalities associated with certain diseases (e.g. typical erosions of RA or PsA in feet and hand joints, typical 'scooped-out' lesion of chronic gouty joint, sacroiliitis of SpA, aggressive destructive lesions of joint tuberculosis and other infections, chondrocalcinosis in calcium pyrophosphate dihydrate {CPPD}crystal deposition disease, etc.). This may indirectly give a clue to the inflammatory nature of the joint disease. Magnetic resonance imaging (MRI) on the other hand can demonstrate bone marrow oedema that has been demonstrated to be the earliest sign of inflammatory nature of the joint disease in recent studies[26]. Being an expensive investigation it cannot be recommended for routine use for confirming inflammatory joint disease. Histopathology is another investigation that may give a definitive diagnosis in difficult cases of polyarthritis. However, only in rare circumstances synovial biopsy may be necessary for making a diagnosis. Thus, it may be required in suspected polyarticular infection (mycobacterial infection, fungal), villonodular synovitis etc. On the other hand skin or subcutaneous lesions may often yield useful diagnostic information (e.g. sarcoidosis, erythema nodosum, vasculitis). Specific tissue biopsy may be indicated in specific diseases e.g. sural nerve biopsy in systemic vasculitis, muscle biopsy in inflammatory myositis. Once a provisional diagnosis has been made, in the second-step of diagnostic work-up, more specific investigations are usually required. Thus, in RA, one may like to have some idea of disease prognosis. Quantitative measurement of rheumatoid factor (RF) and ACPA are useful in this respect. In suspected SLE and related connective tissue disorders screening for immunofluorescent antinuclear antibody (F-ANA) is of great importance. F-ANA has a very high sensitivity for SLE and other connective tissue diseases but it has rather low specificity. This makes F-ANA a very useful test for excluding the diagnosis of these diseases; if F-ANA test is negative it would be most unlikely that the patient would have SLE or related connective tissue disease. Immunofluorescent anti-neutrophil cytoplasmic antibody (ANCA) test is similarly a very useful screening test for patients suspected of systemic vasculitides; a negative immunofluorescent ANCA test would go strongly against the diagnosis of ANCA-associated vasculitides. ANCA titres have also been shown to have some degree of correlation with disease activity. For routine monitoring of drug toxicities standard complete blood counts, liver enzymes, renal parameters, and routine urine examination would suffice in most cases. Baseline electrocardiogram, chest radiograph and stool examination (including occult blood) should be done routinely in patients with polyarthritis. HLA B27 screening is very useful but only in a patient with typical history of inflammatory back pain, its presence would strongly favour the diagnosis of ankylosing spondylitis.

Non-specific arthralgias, myalgias, pain amplification *versus* polyarthritis

There are patients who present with vague pains in the joints, muscles and other parts of the musculoskeletal system. They may also complain of developing numbness in different parts of the extremities on inactivity, neck-nape region pains, pains running down from neck along the arms, low back pain that often runs down along the legs. Usually the severity of the pain fluctuates and is often described as 'unbearable'. They may show features of anxiety, fear of 'getting crippled over time', or worry about some serious unrecognised disease. They often have sleep abnormalities ('I am as tired on getting up in the morning as I am when I go to bed') often associated with some features of fibromyalgia including fibromyalgic tender spots. In most cases the duration of symptom is vaguely 'since long' and they carry bags-full of investigation and imaging reports. They usually have a readymade hand written list of complaints ('I have written it down lest I forget some of them when I visit the doctor') and another list of specialists they have already consulted ('I have already consulted so-and-so', usually a list of famous doctors from 'Who-is-who' list), yet they have not been satisfied as no definite diagnosis has been made and their symptoms are getting worse. They continue to remain worried and frustrated. This is a most puzzling group who should not be over diagnosed or over treated. Occasionally these patients may be mistaken as having true polyarthritis and started on some form of 'arthritis treatment'. This approach of management must be strongly discouraged as it leads to psychological dependence on drugs. Once a label of 'arthritis' is put on such patients it becomes almost impossible to convince them later that they do not have 'arthritis'. However, a complete evaluation is called for in older age persons, in those with disease of more than 6 weeks duration, and in those with constitutional symptoms or symptoms referable to other organ systems. A limited evaluation should include careful history, review of systems, physical examination and minimal investigations including non-specific markers of inflammation (ESR, high sensitivity-C-reactive protein {hs-CRP}), test for thyroid status (minimum thyroid stimulating hormone {TSH}levels TSH), ACPA, 25-hydroxy-vitamin D levels, (low vitamin D levels have been found to cause vague musculoskeletal pains[27]) and routine cancer screening (chest x-ray, prostate specific antigen, mammography, stool occult blood), if these had not already been done in the recent past. In most such cases the problem usually turns out to be functional and related to pain amplification. It is best to address their concerns directly in a forthright manner. Some of them may require help of a counsellor or a psychiatrist. Table 1 gives some of the mimics of polyarthritis.

Summary

As a general rule, in musculoskeletal diseases 80% of the information towards making a diagnosis comes from accurate clinical history, 15% from physical examination and only 5% from laboratory investigations. The same rule holds true for patients with polyarthritis. Polyarthritis and its diagnosis encompass almost the whole field of rheumatology. *Making a diagnosis of polyarthritis is really a test of the knowledge of general internal medicine*; deeper the knowledge of general internal medicine easier it is for the clinical rheumatologist to reach a definitive diagnosis. Polyarthritis is among the commonest problems that a rheumatologist faces in his/her daily practice, it is necessary to follow a definite diagnostic algorithm for convenience and accuracy. Is it *polyarthritis*? → What is the *duration* (acute or chronic)? → What are the *circumstances of onset*? → Is it *inflammatory* or non-inflammatory?

What is the *pattern of evolution*? → What is the *topography* of involvement? → Is there a clue to diagnosis in the past history, family history and review of systems? → Does *physical examination* give a diagnostic clue (e.g. a small psoriatic patch)? This approach narrows down the diagnostic possibilities and helps in requisitioning a limited number of focused relevant investigations to confirm the diagnosis. In the majority, specific drug treatment along with patient education, counselling, occupational and physiotherapy may be initiated even before all the investigation reports become available. However, in doubtful cases it would be prudent to initiate only symptomatic treatment till all the relevant reports become available to make a definitive diagnosis. Acute inflammatory polyarthritis could be a self limiting *benign* condition, the beginning of a *serious chronic* illness with significant morbidity that could even be life threatening or, a rheumatological *emergency* requiring urgent diagnosis and management. Distinction between benign, serious chronic illness and a rheumatological emergency requires deep knowledge of rheumatology as well as general internal medicine; details of past, family and personal history and in-depth review of systems, usually help reaching a working diagnosis. Addition of a few relevant investigations usually confirms the diagnosis that helps in initiating appropriate treatment.

Key Points

Approach to the patient with polyarthritis

- **True Polyarthritis vs. Polyarthralgia or non-specific MSK pains**
  - Pattern of evolution:
    - Intermittent vs. migratory vs. additive
  - Disease duration
    - Acute vs. Chronic
- **Pathological nature of the disease**
  - Inflammatory vs. Non-inflammatory
- **Pattern (topography) of joint involvement**
  - Peripheral vs. axial disease
  - Overlap peripheral and axial involvement
- **Peripheral arthritis**
  - Symmetrical vs. asymmetrical
  - Upper vs. lower segment predominantly involved vs. equally involved
  - Without vs. with axial involvement
- **Unique patterns**
  - Purely DIP (*Inflammatory*: psoriatic arthritis, sarcoidosis, IBD-related; *Non-inflammatory*: Nodular primary OA)
  - Peri-ankle/ankle inflammatory arthritis with erythema nodosum and hilar lymphadenopathy (Lofgren syndrome)

Further Reading :

1. Peterson MC, Holbrook JH, Hales DV, Smith NL, Staker LV. Contributions of the History, Physical Examination, and Laboratory Investigation in Making Medical Diagnoses. The Western Journal of Medicine 1992; 156(2): 163-65.
2. Hampton JR, Harrison MJG, Mitchell JRA, *et al:* Relative contributions of history-taking, physical examination, and laboratory investigation to diagnosis and management of medical outpatients. Br Med J 1975; 2: 486-89.
3. Markert RJ, Rich EC, Sakowski HA, Maio AC, Haist SA, Hillson SD. Medscape General Medicine. 2004; 6(2): 64 (on line).
4. Liang MH, Sturrock RD. Evaluation of musculoskeletal symptoms. In: Klippel JH, Dieppe PA (eds), Practical Rheumatology, Mosby, London, 1995: pp3-20.
5. Pinals RS. Musculoskeletal signs and symptoms B. Polyarticular joint disease. In: Klippel JH (ed), Primer on the rheumatic diseases., Arthritis Foundation, Atlanta, Georgia, 2001: pp160-65.
6. Malaviya AN, Kumar A. Postgraduate Clinic: Polyarthritis. J Assoc Physicians India 1997; 45: 55-58.
7. Malaviya AN. A patient with musculoskeletal (MSK) pains - clinical approach. Bull Kuwait Inst Med Spec 2004; 3: 72-81.
8. Malaviya AN. Clinical approach to patients with joint disease: importance of distinguishing inflammatory from non-inflammatory conditions. APLAR Journal of Rheumatology 2006; 9: 11-17.
9. Lipsky PE. Algorithms for the diagnosis and management of musculoskeletal complaints. Introduction. Am J Med 1997; 103: 1S-2S.
10. Sergent JS. Howard Fuchs. Polyarticular arthritis. In. Gary Firestein, et al. (eds), Kelly's Textbook of Rheumatology (8th ed.), WB Saunders, Elseiver Philadelphia, 2009, pp 545-53.
11. Ellrodt AG, Cho M, Cush JJ, Kavanaugh AF, Lipsky PE. An evidence-based medicine approach to the diagnosis and management of musculoskeletal complaints. Am J Med 1997; 103: 3S-6S.
12. Jansen ALMA, van der Horst-Bruinsma I, van schaardenburg D, *et al.* Rheumatoid factor and antibodies to cyclic citrullinated peptide differentiate rheumatoid arthritis from undifferentiated polyarthritis in patients with early arthritis. J Rheumatol 2002; 29: 2074-76.
13. Aletaha D, Neogi T, Silman AJ, Funovits J, Felson DT, Bingham CO, III et al. 2010 Rheumatoid arthritis classification criteria: an American College of Rheumatology/European League Against Rheumatism collaborative initiative. Arthritis Rheum 2010; 62: 2569-81.
14. Amin S, Niu J, Guermazi A, Grigoryan M, Hunter DJ, Clancy M *et al.* Cigarette smoking and the risk for cartilage loss and knee pain in men with knee osteoarthritis. Ann Rheum Dis 2007; 66: 18-22.
15. Davies-Tuck ML, Wluka AE, Forbes A, Wang Y, English DR, Giles GG *et al.* Smoking is associated with increased cartilage loss and persistence of bone marrow lesions over 2 years in community-based individuals. Rheumatology (Oxford). 2009; 48: 1227-31.
16. Kiyohara C, Washio M, Horiuchi T, Tada Y, Asami T, Ide S, *et al*; Kyushu Sapporo SLE (KYSS) Study Group. Cigarette Smoking, STAT 4 and TNFRSF1B Polymorphisms, and Systemic Lupus Erythematosus in a Japanese Population. Lupus. 2009; 18: 630-38.
17. Fisher MC, Hochberg MC, El Taha M, Kremer JM, Peng C, Greenberg JD. Smoking, Smoking Cessation, and Disease Activity in a Large Cohort of Patients with Rheumatoid Arthritis. J Rheumatol. 2012 Mar 15. (Epub ahead of print) doi: 10.3899/jrheum. 110852.
18. Chung HY, Machado P, van der Heijde D, D'Agostino MA, Dougados M. Smokers in early axial spondyloarthritis have earlier disease onset, more disease activity, inflammation and damage, and poorer function and health-related quality of life: results from the DESIR cohort. Ann Rheum Dis Ann Rheum Dis 2011; 70: 1930-36.
19. Mielants H, Veys EM, Goemaere S, *et al.* Gut inflammation in the spondyloarthrorpathies: clinical, radiologic, biologic and genetic features in relation to the type of histology. A prospective study. J Rheumatol 1991; 18: 1542-51.
20. Budhiraja S, Malaviya AN, Al-Assousi A, Kaushik P, Francis IM. ANCA-positive Pulmonary-renal Syndrome: Wegener's Granulomatosis? J Ind Rheumat Assoc 2000; 8: 45-48.
21. Caines A, Samadi N, Ouimet G, Thompson A, Pope JE. The sensitivity and specificity of pain diagrams in rheumatic disease referrals. Rheumatology 2012; 51: 1093-98.
22. Pincus T, Segurado OG. Most visits of most patients with rheumatoid arthritis to most rheumatologists do not include a formal quantitative joint count. Ann Rheum Dis 2006; 65: 820-22.
23. Scott DL, Antoni C, Choy EH, van Riel PCLM. Joint counts in routine practice. Rheumatology 2003; 42: 919-23.
24. Anderson J, Caplan L, Yazdany J, Robbins Ml, Neogi T, Michaud K, *et al.* Rheumatoid Arthritis Disease Activity Measures: American College of Rheumatology Recommendations for Use in Clinical Practice. Arthritis Care & Research 2012; 64, 640-47.
25. Smolen JS, Aletaha D, Bijlsma JW, Breedveld FC, Boumpas D, Burmester G *et al.* Treating rheumatoid arthritis to target: recommendations of an international task force. Ann Rheum Dis 2010; 69: 631-37.
26. Jimenez-Boj E, Nobauer-Huhmann I, Hanslik-Schnabel B, Dorotka R, Wanivenhaus AH, Kainberger F *et al.* Bone erosions and bone marrow edema as defined by magnetic resonance imaging reflect true bone marrow inflammation in rheumatoid arthritis. Arthritis Rheum 2007; 56: 1118-24.
27. de Torrente de la Jara, Pecoud A, Favrat B. Female asylum seekers with musculoskeletal pain: the importance of diagnosis and treatment of hypovitaminosis D. BMC Family Practice 2006; 7: 4-11.

APPROACH TO REGIONAL PAIN

5

B G Dharmanand

Regional pain syndromes are defined as pain and dysfunction affecting a single musculoskeletal area (e.g.: shoulder, elbow or knee). Four major groups of problems can present as regional pain syndrome[1].

1. Periarticular pain.
2. Joint pain/Arthritis.
3. Neurogenic pain.
4. Referred pain.

Certain type of pathology occurs more frequently in certain joint regions, for example, periarticular problems dominate regional pain in shoulder region. It is prudent to record as pain in a joint region rather than pain in the joint during the initial evaluation. This would stimulate one to consider non- articular lesions in the differential diagnosis.

Periarticular pain

Periarticular pain originates in anatomical structures that are involved in joint motion but are located outside the joint capsule. Interestingly the conditions affecting the capsule have similar clinical findings as articular diseases.

Inflammation of tendons, ligaments and enthesis can occur in isolation or could occur as a feature of inflammatory arthritis like seronegative spondylarthritis and rheumatoid arthritis. Trauma and repetitive use could also trigger inflammation in these structures.

A periarticular origin is suggested by pain that is elicited or worsened by specific movements, which compress or stretch the inflamed structure. Other movements may not be painful or restricted (selectivity of painful movements). The rhythm of pain is usually mechanical (made worse by movements and relieved with rest), though the pathology may be inflammatory. On clinical examination, the pain will be more on active movement than passive movement. Conversely, articular lesions show pain and restriction of motion both during active and passive movements. Resisted movements cause pain of periarticular origin but would be pain free in purely intra articular disease. (Fig. 1) For example, resisted abduction would be painful in supraspinatus tendonitis and relatively pain free in shoulder arthritis and frozen shoulder. Direct palpation of inflamed tissue would be painful, whereas joint line tenderness is typical of articular pathology. Except in acute inflammation in superficial bursae, swelling and warmth are not usually seen in periarticular disease.

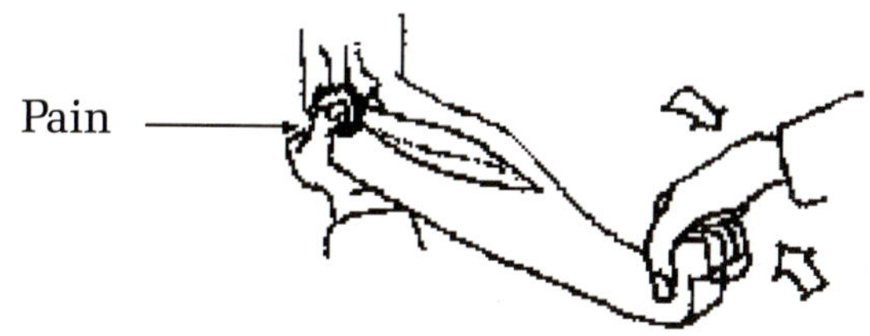

Fig. 1 : Resisted extension of wrist reproducing the pain in Lateral epicondylitis.

Monoarthropathy

A lesion that affects a single joint would result in a regional pain syndrome. Pain related to use of the joint and painful restriction of movements in all the directions of movement are characteristic of articular diseases. In mild form of arthritis, the pain will be elicited at full or extreme range of movement in all the allowed directions of movement (on stressing the joint). This is called stress pain and is typical of arthritis. Active and passive movements would be similarly restricted. Resisted movements would not be painful as the joint remains still during this maneuver. Crepitus, swelling and warmth, when present would make the diagnosis obvious. Clicking, locking, buckling and instability on testing would point towards an intra-articular pathology like loose body or ligament tear.

Neurogenic pain

Neurogenic pain is caused by the compression or irritation of nerve roots or peripheral nerves. Carpal tunnel syndrome and tarsal tunnel syndrome can present as regional pain in hands and feet respectively. L2, L3 root compression can present as regional pain in the knee. Neurogenic pain has dysaesthetic quality (burning, tingling, shock like and numbness). The pain and discomfort can be very severe and do not respond to usual measures that relieve nociceptive pain like heat and NSAID. Positive symptoms (pain, shock like sensation, paresthesia) usually precede negative symptoms and signs (numbness, loss of sensation and motor defect). There could be neurogenic edema and erythema, which could mimic cellulitis and acute inflammatory arthritis. Eliciting historical clues to the neuropathic quality of the pain and examining the nervous system are the keys to the diagnosis. This is particularly true when the regional musculoskeletal examination fails to reveal any clues to the diagnosis. A sensation of wrist bracelet is typical of median nerve lesion. The pain that radiates below elbow in cases of shoulder pain is likely to be of neurological etiology[2].

Complex regional pain syndrome can also present as regional pain. The pain here too has neuropathic quality. The swelling is more neurogenic edema than inflammatory one.

The differences and distinctive features between these causes of regional pains are given above (Table 1).

clinical approach

When confronted with a patient with pain in a single musculoskeletal region, it is essential to take a thorough history including the site, nature and rhythm of pain, relieving and aggravating factors. History of trauma and recent surgery are common in complex regional pain syndromes. Trauma, repetitive injuries and unaccustomed exertion can precipitate regional pain in the involved region. Occupational history is also important. The disability and the functional difficulties also need to be documented. The impact of the regional pain on the day to day activities, professional and personal life has to be understood. Clinical examination should include local musculoskeletal and neurological evaluation. Observation of skin colour changes, sensitivity (hyperesthesia, allodynia), and temperature would give clues regarding inflammation and nerve involvement.

Table 1 Characteristcs of various regional pain syndromes[1]

	Articular pain	Periarticular pain	Neuropathic pain	Referred pain
Enquiry	All joint movements are painful	Selective painful movements	Dysaesthetic quality	Unrelated to movement Visceral timing
Pain on motion	Active and passive. In all directions	Active > passive. Selected directions	Normal	Normal
Range of motion	Global reduction. Both active and passive	Active-limited. Passive- full	Normal	Normal
Resisted motion	No effect	Pain on specific maneuvers	No effect	No effect
Local palpation	Joint line tenderness Swelling, effusion Crepitus	Pain upon affected structure	Normal Nerve tenderness	Normal
Neurological exam	Normal	Normal	May be abnormal	Normal

Referred pain

Referred pain infers symptoms felt at a distance from their anatomical origin. This can present as regional pain syndrome. The pain can be referred from the internal organs or from other joints (proximal joint pain can be felt at the distal joint). Local examination is likely to be normal and the rhythm of pain would relate to the physiology of the internal organ. Cardiac and sub-diaphragmatic pathology presenting as shoulder tip pain is the best appreciated of referred pain syndromes. Hip arthritis presents as knee and thigh pain not infrequently.

Passive and active joint motions along with resisted movements have to be performed.

A nociceptive pain, arising from tendons, ligaments and capsular structures is fairly localized[3]. A pain that is worse in the early morning and late night and also the rest pain indicate inflammatory origin.

Muscles around the region of pain have to examined for trigger points. Myofascial syndrome can mimic neuropathic and arthritic conditions. Pain can be felt at areas remote to the trigger point. So, knowledge about trigger points and the radiation patterns would be useful[4] (Fig. 2 & 3).

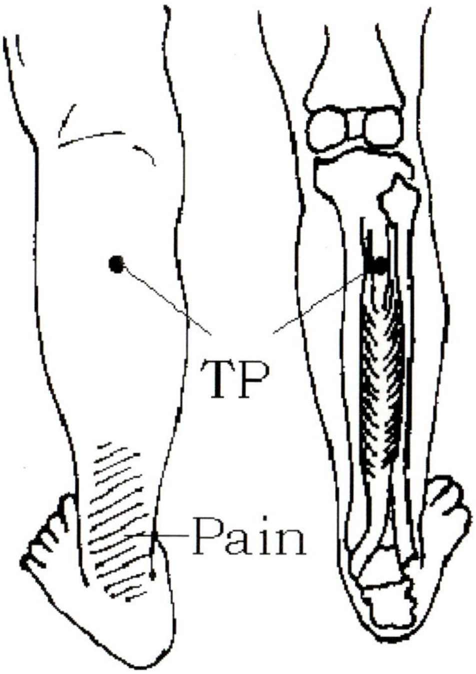

TP - Trigger Point

Fig. 2 : MFPS in the leg.

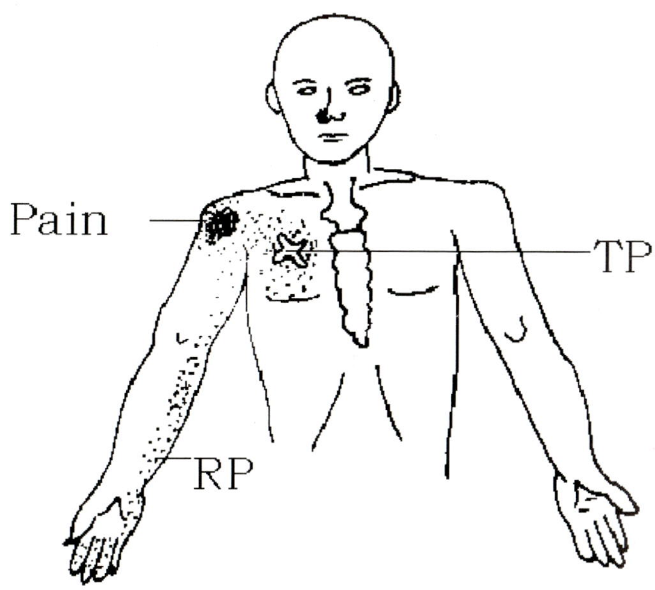

TP - Trigger Point
RP - Reffered Pain

Fig. 3 : MFPS -Upper limb.

Regional pains may be a part of a systemic disorder or widespread pain syndrome. So, a quick review of past medical history, systemic and organ specific symptoms is important. Pain in other joint regions and wide spread tender points need to be looked for. Sometimes serious life threatening illness can present as regional pain syndrome. It is essential to look for red flags so that hospitalization and appropriate referrals could be made[2].

Red flags

Fever
Night sweats
Weight loss
Major trauma
Past or current cancer
Infection
Radiotherapy
Metabolic diseases
Wide spread neuropathy
Immunosuppression
Unrelenting pains/nocturnal pains

Regional pain syndromes are common in the community. It is not clear why some seek medical advice and some do not!. Some painful conditions may be a window on the predicament of life. Regional pains may be the surrogates for distress[5]. Occupational history should also screen for work satisfaction, since this seem to determine the symptoms and dysfunction more than the intensity and nature of the work or job[6]. Threshold by which palpation induces pain is important to notice. Low pain threshold can occur with distress. The clinician has to be vigilant for abnormal illness behaviors like, constantly holding the painful part still or massaging and stroking the region of pain, usage of disability aids like collor, wheel chair etc and dramatic description of pain. These along with healthcare seeking behaviour, job absenteeism form yellow flag symptoms. When identified, appropriate interventions like referral to mental health professionals would be facilitated.

COMMON CAUSES OF REGIONAL PAIN

Some of the common regional pains encountered in various regions are listed in the following table (Table 2)[1].

Table 2 Common causes of regional pain

Region	Arthropathy	Periarticular	Neuropathic	Reffered pain
Chest wall	Sternoclavicular syndrome Manubro-sternal arthritis Teitz syndrome	Costochondritis	Pre-Herpetic neuralgia Post-Herpetic neuralgia	Pleural pain Myofascial pain Referred from fracture ribs
Shoulder	Frozen shoulder Other arthritis	Rotator cuff tendonitis Myofascial pain	Cervical radiculitis	Heart, Sub-diaphragmatic lesions Cervical spine Myofascial pain (MFPS)
Elbow	Arthritis	Tennis elbow Medial epicondylitis	Posterior interosseous nerve entrapment	Shoulder MFPS
Wrist & Hands	Arthritis	Tenosynovitis De Quervain's Ganglion	Carpal tunnel syndrome	Cervical spine MFPS
Hip	Osteoarthritis AVN	Psoas abscess Trochanteric bursitis	L1 radiculopathy	Sacroiliac Renal Intra-abdominal
Buttock	Sacroiliitis	Gluteal bursitis Metastasis in bones	Radiculopathy	Hip Facet joints MFPS
Knee	Arthritis Intra-articular pathology	Bursitis Enthesitis	L2, L3 radiculopathy	Hip MFPS
Ankle & Feet	Gout Other arthritis	Plantar fascitis Achilles tendonitis	Tarsal tunnel syndrome Morton's neuroma	MFPS L5, S1 radiculopathy

MFPS : Myofascial pain syndrome
AVN : Avascular necrosis

Further Reading:

1. Jose da Silva, Karen L Faarvang, Catia Duarte in EULAR online course on rheumatic disorders, module n" 1 April 2007; 1-6.
2. Bernard Fourquet, clinical examination as a tool identifying the origin of regional musculoskeletal pain, Best practice & research- Clinical rheumatology 2003; 17: 1-15.
3. Rice JR & Pisetsky DS. Pain in the rheumatic diseases. Practical aspects of diagnosis and management. Rheumatic disease clinics of North America 1999; 25: 15-30.
4. Mike Cummings. Myofascial syndromes. Soft Tissue Rheumatology Ed. Hazleman B, Riley G, Speed C (Oxford) 2004. 6.12: 509-22.
5. Nortin M. halder. Regional pain syndromes, Non-articular musculoskeletal disorders and fibromyalgia. Year book of Rheumatology, arthritis and musculoskeletal disease 2006 (Mosby); 247-48.
6. Lipscomb L, Trikoff A, Brady B, et al Health care system changes and reported musculoskeletal disorders among registered nurses. American journal of public health 2004, 94: 1431-35.

SECTION 2

Outcome Measures in Rheumatological Disorders

OUTCOME ASSESSMENT IN RHEUMATIC DISEASES 1

Sapan Pandya

Introduction

Outcome assessment is the measurement process by which consequences of diseases and health care management decisions are evaluated. There are two types of measures : process measures which include the analysis of histological, biochemical or imaging procedures (e.g. MRI) or organ function tests (e.g. pulmonary function test) and outcome measures which measure clinical consequence of a disease. The Fries's paradigm [5 Ds: Death, Disability, Discomfort, Drug (or iatrogenic reactions) and Dollar (economic) costs] forms the basis of the various domains used for assessing outcome in different diseases[1]. Outcome measures can also be broadly divided into two categories : observer dependent (rated by the assessor e.g. interview scored questionnaires, physical findings or performance based tests like grip strength, walk time etc) and observer independent (self reported eg. self reported questionnaires).

Criteria that outcome measurement techniques must fulfil

Ethicality

The outcome measurement procedure must be ethical. The inherent risk involved in using the technique/procedure must be weighed against any new information that would be acquired by using it. This should be explained to all patients/ participants and accepted by them.

Validity

Is the extent to which an instrument specifically measures what it is supposed to measure.

Reliability

Is the extent to which a measurement technique yields the same result on repeated determination.

Responsiveness

The outcome measurement technique must be aimed to detect even minimal change in a parameter using scaling methods. More recent publications have used the Standardized Response Mean (SRM) or Effect Size (ES) to capture these differences[2].

The OMERACT and the OMERACT filter

The outcome measures in rheumatology clinical trials (OMERACT) is an informal group of health professionals from across the world that organizes meetings on outcome assessment across randomized controlled trials and longitudinal observational studies. It works on a stepwise approach which is a data driven and iterative process. The OMERACT has devised a "filter" for the applicable measures. The OMERACT filter comprises of three attributes : Truth (is the measure truthful - measure what it is intended to measure) Discrimination (discriminate between two situations of interest - sensitivity to change) and Feasibility (the measure can be applied easily under the constraints of time, money and interpretability).

Outcome assessment techniques

These can be broadly divided into clinical measurements and laboratory or imaging based measurements.

Pain

Can be rated both subjectively by the patient and pain behaviour can be rated by the assessor. It must be mentioned what aspect of pain is being measured eg. at rest or on movement. The time frame during which it has been measured (last week, 48hrs or so) also must be mentioned. Different scales that have been used are Likert scale (very bad – bad – average – good – very good), VAS (Visual Analog Scale – usually numbered 1 to 100mm on a horizontal line), ladder scale, chromatic continuous scale, pain face scale etc. Most commonly used are Likert and VAS scales.

Impairment

Any loss or abnormality of pyschologic, physiologic or anatomic structure (organ level) e.g. swollen joint count, tender joint count. Joint counts can be binary (present or absent), graded, weighted indices or total vs reduced joint counts (e.g. EULAR 28 joint count).

Disability

Lack of ability to perform an activity in a manner considered normal (personal level). Measures for assessing disability can be both performance based (e.g. hand grip) or capacity based (most of the validated scales eg. HAQ for RA,WOMAC Index for OA(2), ASFI for ankylosing spondylitis)[3]. Disability can be physical,social or emotional. Many scales (e.g. SF 36) do address the latter two parameters also and newer scales for the same (social, emotional disability) are being devised.

Handicap

Due to the disability or impairment an individual cannot perform appropriate role (considered normal) in the society (social level). The Disease Repercussion Profile is a valid and reliable measure of patient perceived handicap[4].

Global assessments

Patient global assessment rated by the patient – It is essential to mention which aspect e.g. symptom severity or overall health status, is being evaluated and also the time frame (last 1 week).

Physicans global assessment : Adds little to the patient global but the physician may have more insight into the disease severity (biochemical abnormality, radiologic change etc. Here also both the aspects being assessed and time frame (usually mentioned "today" as the physician has to evaluate on the basis of "that" day).

Psychosocial issues

Clinical depression has been found to range between 8 to 22% in RA in different studies. Depression is assessed mainly by self administered questionnaires (Beck depression Inventory, Zung depression scale, Middlesex Hospital questionnaire, the hospital anxiety and depression scale)[5]. Standardized psychiatric interviews are the other form of assessment of depression. Psychosocial assessment is included in most studies now as an outcome measure. Pain can also be influenced by depression. Cognitive and emotional processes contribute to individual differences in the perception of pain. Pain, catastrophising and depression can lead to lack of response and even early mortality in these diseases and can be important targets to assess outcome.

Sleep

Fatigue is an important symptom of inflammatory rheumatic diseases. While it may be a marker of inflammation, there are other causes of fatigue too like depression, drugs, sleep disturbance etc. Sleep can be a good measure of fatigue. There can be co-morbid conditions that affect sleep eg restless leg syndrome, obstructive sleep apnoea etc. Also, sleep deprivation can lead to a worsening of the functional ability in these patients. The assessment of sleep by the conventional polysomnography and sleep laboratory techniques can be difficult in rheumatologic patients because of issues of mobility restriction and pain apart from high expense. Patient reported outcome of sleep would be a good tool to guage sleep problems. Sleep scales like the Epworth Sleepiness Scale (daytime sleepiness), Functional outcome of Sleep Questionnaire (sleep related quality of life), Insomnia Severity Index (subjective symptoms and difficulty maintaining sleep) and the Pittsburg Sleep Quality Index (perceived sleep quality) were not developed for assessing rheumatologic conditions but have been used varyingly and can be a starting point to derive a more specific one.

Multidimensional indices

Several have been developed that have greater validity, reliability and responsiveness. They can be segregated eg HAQ where there are different subscales for different aspects of the condition or aggregated e.g. the DAS score (disease activity score[6]) in rheumatoid arthritis where different dimensions have been weighted and condensed down to a single score.

Adverse reactions

Two systems for categorizing adverse events in clinical trials are : Coding symbols for thesaurus of adverse reaction terms (COSTART) and WHO adverse reaction dictionary (WHOARD)[7]. The attribution is often graded as none, possible, probable and definite and severity rated as mild, moderate (affecting normal life) or severe (incapacitating, hospital admission). Outcome of adverse event also needs to be categorized (resolved, worsened, hospitalized etc.). The OMERACT has formed a special Drug Safety Working Group which in 1999 developed the rheumatology common toxiciy criteria (RCTC) based on the Oncology Criteria[8]. The OMERACT has been working in conjunction with the ILAR, EULAR and ACR to validate the RCTC so that both investigators and pharmaceutical companies can use them in future trials.

Instruments which are both patient self reported and investigator reported have been designed and are to be used alongside current methods of adverse event reporting in clinical trials.

Concomitant therapy as outcome measure

Therapies other than the study drug are called concomitant therapies. These could be in the form of rescue therapy (given as a backup in case of necessity eg. paracetamol for pain in an NSAID trial), concomitant therapy that predates the study drug and is continued throughout the study period but the dosage may be modulated (e.g. corticosteroids in most DMARD/biologic studies in RA) or alternative therapy which when instituted, results in primary endpoint considered as a failure of the study drug eg. total joint arthroplasty for osteoarthritis. Collection of information regarding its consumption (e.g. pill count, keeping a diary) is very important in any study.

Self management interventions

In recent years it is being increasingly realized that the main responsibility for managing a chronic illness on a daily basis lies with the patients themselves and so self management interventions that would allow them to manage their illness more effectively should be designed and evaluated. They are also called "psychoeducational" interventions. A Special Interest Group (SIG) for these measures is working on this and outcome and mediating variables for the same have been identified.

The patient's perspective

Most of the outcome measures that are assessed today are more important from the investigator point of view. Outcomes important from the patients perspective are now being looked at. In a study on patients perspective of outcomes in RA, patients found important not only physical outcomes such as pain and disability but also fatigue and a general sense of wellness[9]. The relative importance of these depended on the stage of the disease (e.g. in early stage even small differences in therapeutic efficacy are satisfying from the patients perspective but in late stages, large differences are needed). Patient satisfaction depended on communication with the health care professional, access to treatment and efficacy. Calibration is needed in existing measures to take into account different outcomes at different stages and different magnitudes of change needed for the same outcome at different stages.

The 10th OMERACT conference had conducted 4 workshops on incorporating patient's perspective in outcome measures. Most of the patients reported outcome measures include visual analog scales which are difficult to interpret at an individual level. At the individual level, four concepts must be taken into consideration : improvement, status of well-being, onset of action and sustainability. Use of IRT (Item response theory) and CAT (Computerised adaptive testing) can make instruments to assess patients perspective more sensitive and widely applicable. Separate domains need to be defined for non-pharmacologic interventions which not infrequently are used by patients.

Economic analysis

It becomes very relevant to include economic analysis especially with the advent of costly therapies like the biologics. Cost effectiveness and cost utility analysis both should be included and quality of life scales should be used in such assessment. Direct costs borne mainly by the health care sector must be assessed e.g. lab tests, radiology, out patient visits, aids and devices etc. Indirect costs such as loss of work on the patients part also will have to be included in such measures.

Outcome assessment in clinical practice

There are some differences in the setting of clinical practice when compared to clinical trial setting, patients have different disorders, their numbers are not fixed, they may be taking co therapies, there might be co morbidities. Patients in clinical trials are likely to be from a higher socioeconomic class and, in practice, are followed up for longer periods of time. Patient self reported questionnaires are the best tool in such a setting since they can fill up while sitting in the waiting room. It has been proven that such data (patient filled) are highly reliable and correlate well with joint counts, radiology, laboratory tests, functional status, long term outcome and mortality in RA. The questionnaires have to be short, cost effective and clinically useful. They could be general purpose eg. HAQ, CLINHAQ (Clinical Health Assessment Questionnaire) or disease specific e.g. WOMAC, BASFI. In routine clinical practice, the goal may be simply to track the record of an individual patient. The best combination is concomitant use of disease specific questionnaires, patient and physician global assessments, generic health related quality of life measures and, in the case of inflammatory arthritis, swollen and tender joint counts. Flow charts to document clinical status and drug tolerability are particularly useful.

Outcome measures in individual rheumatic diseases

Rheumatoid arthritis

Discussed in the following chapter.

Juvenile idiopathic arthritis (JIA)

Measures of outcome in children especially with juvenile idiopathic arthritides have also been validated. Physician's global assessment, parents global assessment, childhood HAQ (CHAQ), Child Behaviour Check List (CBCL), overall well being and more generic health status instruments like child health questionnaire (CHQ) which is divided into two parts the Physical Score (PhS) and the Psychosocial Score (PsS) are some of the instruments used to assess outcome in juvenile idiopathic arthritides[16,29]. There are issues related to mental,emotional and social aspects in children and "hidden morbidity" which are all included in instruments to assess outcome in children. Newer scales/ instruments are being developed to measure outcome in different pediatric rheumatic diseases like juvenile dermatomyositis, juvenile lupus, vasculitides, juvenile spondyloarthritides esp. Juvenile ankylosing spondylitis, psoriasis etc[17].

Psoriatic arthritis (PsA)

Assessment of PsA would include peripheral joints, axial disease, enthesitis, dactylitis as well as skin manifestations (including nail). Patient reported features like fatigue, quality of life, pain, disease activity and assessment of damage and tissue histology are also important outcome measures. The group for research and assessment of psoriasis and PsA (GRAPPA) along with the classification for PsA group (CASPER) have identified key domains for assessment of psoriatic arthritis, 6 core set domains are peripheral joint activity, skin activity, pain, patient global assessment, physical function and health related quality of life. Other domains considered important but not mandatory are spinal disease, dactylitis, enthesitis, fatigue, nail disease, radiography, physician global assessment and acute phase reactants.

Peripheral arthritis

The psoriatic arthritis response criteria (PsARC) measures tender joint scores, swollen joint scores, patient global (0 to 5, Likert scale) assessment of disease activity and physicians global (0 to 5, Likert scale) the joint counts are a total of 78 joints instead of 68 to include the DIP joints which are not infrequently involved in psoriatic arthritis[18]. Improvement in joint score is determined by 30% or greater benefit; overall response is indicated by an improvement of 2 out of the 4 (improvement in global assessements is at least by 1 on Likert scale) items, one of them MUST be joint scores and no worsening in any of the 4 items. The DAS scores used in RA has also been used in psoriatic arthritis.

Dactylitis

In previous trials dactylitis was just mentioned as present or absent but in recent trials, the leeds dactylitis instrument (LDI) is designed to provide a less subjective measure of dactylitis. It is based on diameter of the digits on both sides (> 10% difference defining dactylitis). There is also a tenderness score.

Enthesitis

Instruments quantifying enthesitis have been used earlier in ankylosing spondylitis by the ASAS working group (the Asessment in AS). The Newcastle enthesitis index (NEI) has 66 sites and was found too tedious to follow. The Maastricht AS enthesitis score (MASES) used a modified index with 13 sites. The spondyloarthritis research consortium of Canada(SPARCC) used a simpler index with 8 sites with similar reliability. A simple enthesitis measure evaluating only plantar fascia and Achilles tendon was used in the IMPACT 1 and 2 trials. The Leeds Enthesitis Index (6 sites : 2 Achilles tendon, 2 medial epicondyles and 2 lateral epicondyles) also showed good responsiveness.

Axial skeleton

The bath ankylosing spondylitis metrology index (BASMI), the BASRI (radiological index), the stoke AS spine score (SASSS) and a modified SASSS have all been used in psoriatic spondyloarthritis to measure axial disease[19].

Skin scores

Are based on patient and physician global assessment of skin activity and objectively on scores derived on body surface area involved and degree of induration (thickness of lesion). Specific tools include psoriasis area severity index (PASI), lattice system psoriasis global assessment (LSPGA), national psoriasis foundation psoriasis score (NPFPS). The PASI requires an improvement by 75 to 90% to define significant benefit of the therapy tried. The international psoriasis council (IPC) is currently developing a new instrument to assess skin in psoriasis.

Nail

The nail psoriasis severity index (NAPSI) has been developed to standardize assessment of nail disease in psoriasis. It has not yet been validated. Patient global assessment of disease activity, health related quality of life, fatigue and physical function are important from the patient's perspective.

Ankylosing Spondylitis

Discussed in the following chapter.

Systemic lupus erythematosus (SLE)

Discussed in the following chapter.

Scleroderma

Skin score (Rodnan's), measures of Raynaud's, pulmonary function tests, blood pressure, pain, HAQ and Medical Outcomes Survey Short Form SF – 36 are validated outcome measures in scleroderma. Newer methods like HRCT for ILD, 6 minute walk test for PAH and ILD, patient reported measures in SSc – Physical function and health related quality of life are being reviewed as outcome measures. A combined index from 13 domains - skin, pulmonary, cardiac, combined cardiopulmonary, gastro-intestinal, renal, musculoskeletal, Raynaud's and digital ulcers, quality of life and function, global health, biomarkers and others or organizing them into 3 physiologic groupings – fibrotic, immune and vasculopathic, is being worked out for assessing scleroderma activity/damage in a patient[22].

Vasculitis

Discussed in the following chapter.

Fibromyalgia

The core domains for evaluation of fibromyalgia treatment are : pain, fatigue, patient global, sleep quality, HRQOL, physical function and side effects. Other domains thought important are tender point intensity, dyscognition, anxiety and clinician global. The fibromyalgia impact questionnaire (FIQ) developed by Burckhardt and coworkers consists of questions and visual analog scales addressing these domains[25]. The Fibromyalgia Intensity Score is calculated by taking the average of scores on a scale of 0 to 10 of all 18 tender points. Objective measures or biomarkers being evaluated for fibromyalgia activity are evoked potentials (visual, auditory, somatosensory), neuroimaging (MRI, PET, SPECT), autoantibodies (anti serotonin, antiganglioside), plasma cortisol levels, substance P in CSF. Psychophysical testing is the best supported objective measure today. Pressure pain thresholds, heat pain threshold, tender point intensity are used to differentiate patients from controls. Diminished diffuse noxious inhibitory controls (DNIC) is noted in FMS patients compared to controls. Tilt table testing and heart rate variability and sleep studies have been used as measures of activity in fibromyalgia patients. Another important domain listed by patients was stiffness. A responder index is being developed from these domains. Response criteria include a 30% reduction in pain intensity score as clinically important and 50% as highest degree of improvement. Improved measures of function, QOL and participation are being developed by the coordination of WHO's international classification of functioning (ICF) disability and health and US National institute of health's Patient reported outcome measures information system network (PROMIS).

Osteoarthritis (hip, knee and hand OA)

A responder index for outcome evaluation in osteoarthritis has recently been proposed by the osteoarthritis research society (OARS) and the OMERACT. Efficacy in clinical trials in osteoarthritis is evaluated by the WOMAC (Western Ontario McMaster University OA Index[26]) and global evaluation. The OMERACT-OARSI responder index needs the following to judge response in OA : High improvement in pain or in function > (or equal to) 50% and absolute change > (or equal to) 20 or improvement in at least 2 of 3 of the following :

Pain > (or equal to) 20% and absolute change > (or equal to) 10%, Function > (or equal to) 20% and absolute change > (or equal to) 10%, Patients global assessment > (or equal to) 20% and absolute change > (or equal to) 10%.

In studies > 1 year duration, evaluating efficacy of structure modifying OA drug, imaging was also included as a core set measure.

Time to need total joint replacement (TJR) of the hip/knee is also taken as an outcome measure. Pain, function (hip : HOOS : Hip dysfunction and osteoarthritis outcome score and for the knee, KOOS: Knee injury and Osteoarthritis Outcome Score and also using the WOMAC questionnaire) and structure (Kellgren Lawrence, OARSI stages and joint space width) are the three domains from which the OMERACT/OARSI working group is proposing a composite index to define OA severity and a surrogate measure of "need for replacement surgery".

Gout

For the diagnosis of acute gout, the Special Interest Group of the OMERACT recommends using the 7 ARA criteria for gout and for diagnosis of chronic gout, MSU crystal identification is needed. 9 domains have been identified for outcome evaluation of chronic gout : serum urate, gout flares, tophus size, joint imaging, health related quality of life (HRQOL), musculoskeletal functioning, patient global, participation and safety/tolerability. As a part of validation of HRQOL measurement in gout, 4 scales were listed in the gout specific PRO (patient related outcome) instrument : GI (Gout impact), Gout pain and severity between flares (GPSB), Well being impact (WBI) and Productivity (P)[27,30]. Defining gout flares, instruments to guage tophus size (ultrasound/ MRI), using the S-VdH (Sharp-van der Heijde) erosion and narrowing scores for gout, defining patient oriented domains are issues to be resolved. 5 core domains for acute gouty arthritis outcome assessment have been identified : pain (Likert scale, at 2, 4 ,8, 12, 36, 48 hours), inflammation (0-3 scale, erythema and heat marked as present or absent), function at the joint (3 = total disability, 2 = movement possible, 1 = weight bearing possible, 0 = painless full function), patient global and safety. The above listed projects will undergo validation through the OMERACT filter before being used in trials as outcome measures of gout.

Miscellaneous

Low backache and osteoporosis are other problems where outcome measures have been defined. Special issues related to worker productivity outcome measures, patient related outcome (PRO) measures, single joint assessment and process measures radiological like plain x-ray or ultrasound or MRI and histopathological assessment including the new biomarkers (e.g. MMP3, synovial macrophages) are some other areas where new instruments have been devised (not discussed here as they are outside the scope of this article).

CONCLUSION

Two or three things emerge clearly out of the different exercises done for evaluating outcome in patients with different rheumatic diseases. One, patient's perception is becoming increasingly important in defining outcome and valid scales to measure the same will have to be included in all instruments. Second, information techonology and the use of computers especially in radiology will make outcome assessment much more objective and accurate subjective reading error will now be lesser and lesser. All these are aimed at trying to shorten the trial period or decreasing the number of patients needed in clinical trials to make them economically more viable apart from making outcome assessment more objective.

The Indian version for the Health Assessment Questionnaire for Rheumatoid arthritis has been validated. Few studies on the economic and social impact of these diseases on our patients have been published. The future will require clinical trials wherein these instruments will be used to assess outcome and we will need to acquaint ourselves and train ourselves to use them. We also need to have our own instruments and scales in some diseases and especially to assess socio-economic aspects in our setting and to validate them for use in trials and clinical practice.

Further Reading:

1. Fries JF, Bird HA. The assessment of disability : from first to future principles. Br J Rheum 1983; 22(s): 48-58.
2. Bellamy N, Buchanan WW, Goldsmith CH et al. Validation study of WOMAC : A health status instrument for measuring clinically important relevant outcomes to antirheumatic drug therapy in patients with osteoarthritis of hip or knee. J Rheum 1988; 15: 1833-40.
3. Calin A. The individual with ankylosing spondylititis: defining disease status and the impact of illness. Br J Rheum 1995; 34: 663-72.
4. Carr A. A patient centered approach to evaluation and treatment in rheumatoid arthritis: the development of a clinical tool to measure patient – perceived handicap. Br J Rheum 1996; 35: 921-32.
5. DeVillis BM : Depression in rheumatological diseases. Baillieres Clin Rheum1993; 7: 241-58.
6. Van der Heijde DMFM, van't Hof MA, van Riel PLCM et al. Validity of single variables and composite indices for measuring disease activity in rheumatoid arthritis. Ann Rheum Dis 1992; 51: 177-81.
7. WHO. The WHO adverse reaction dictionary. Uppsala: WHO Collaborating Center for International Drug Monitoring; 1990.
8. Woodworth TG, Furst DE, Strand V et al. Standardising assessment of adverse effects in rheumatology clinical trials. Status of OMERACT Toxicity Working Group March 2000: towards a common understanding of comparative toxicity/safety profiles for antirheumatic therapies. J Rheum 2001; 28: 1163-69.
9. Minnock P, Bresnihan B. Pain outcomes and fatigue levels reported by women with established rheumatoid arthritis Arth Rheum 2004; 50(S): 471.
10. OMERACT Committee : Conference on outcome measures in Rheumatoid arthritis clinical trials, Maastricht, The Netherlands, April 29-May 2, 1992 J Rheum 1993; 20: 525-91.
11. Wressle E, Lindstrand J, Neher M, Marcusson J, Henriksson C. The Canadian Occupational Performance Measure as an outcome measure and team tool in day treatment program. Disabil Rehabil 2003; 25: 497-506.
12. Felson DT, Anderson JJ, Boers M, et al. American College of Rheumatology preliminary definition of improvement in rheumatoid arthritis. Arth Rheum 1995; 38: 727-35.
13. Van Gestel AM, Prevoo MLL, van't Hof MA, et al. Development and validation of the European League Against Rheumatism response criteria for rheumatoid arthritis. Arth Rheum 1996; 39: 34-40.
14. Gabriel SE, Tugwell P, Drummond M. Progress towards an OMERACT-ILAR guideline for economic evaluation in rheumatology. Ann Rheum Dis 2002; 61: 370-73.
15. Felson TD, Zang B, Siegel JN : Trials in rheumatoid arthritis : choosing the right outcome measure when minimal disease is achievable. Ann Rheum Dis 2008; 67: 580-83.
16. Duffy CM. Measurement of health status, functional status and quality of life in children with juvenile idiopathic arthritis : clinical science for a pediatrician. Pediatr Clin N Amer 2005; 52: 359-72.
17. Tucker LB. Outcome measures in childhood rheumatic diseases. Rheum Rep 2000; 2: 349-54.
18. Taylor WJ. Preliminary identification of core domains for outcome studies in psoriatic arthritis using Delphi methods. Ann Rheum Dis 2005; 64: 110-12.
19. Dawes PT. Stoke Ankylosing Spondylitis Spine Score. J Rheum 1999; 26: 993-96.
20. Ruof J, Stucki G. Comparison of the Dougodas Functional Index (DFI) and Bath Ankylosing Spondylitis Functional Index (BASFI) - a literature review. J Rheum 1999; 26: 955-60.
21. Gladman DD, Ginzler E, Goldsmith C, et al. The development and initial validation of the SLICC/ACR Damage index for SLE. Arth Rheum 1996; 39: 363-69.
22. Pope JE, Bellamy N. Outcome measurement in scleroderma clinical trials. Semin Arth Rheum 1993; 23: 22-33.
23. Stone JH, Hoffman GS, Merkel PA et al. A disease specific activity index for Wegener's granulomatosis : modification of the Birmingham vasculitis activity score. International network for the study of systemic vasculitides (INSSYS). Arth Rheum 2002; 44: 912-20.
24. Exley AR, Bacon PA, Luqmani RA, et al. Development and initial validation of the vasculitis damage index for the standardised clinical assessment of damage in systemic vasculitides. Arth Rheum 1997; 40: 371-80.
25. Burckhardt CS, Clark SR, Benett RM. The Fibromyalgia Impact Questionnaire : development and validation J Rheum 1991; 18: 728-33.
26. Bellamy N, Buchanan WW, Goldsmith CH, Campbell J, Stitt LW. Validation study of WOMAC : a health status instrument for measuring clinically important patient relevant outcomes to antirheumatic drug therapy in patients with osteoarthritis of the hip or knee J Rheuml 1988; 15: 1833-40.
27. Osterhaus JT, Patel P, Palo WA, Bakst A, Joseph-Ridge N. Patient reported outcomes associated with gout : baseline results from two clinical trials. Arth Rheum 2005; 52(S): 401.
28. A Kumar, A N Malaviya, A pandhi, R singh. Validation of an Indian version of the Health Assessment Questionnaire in patients with rheumatoid arthritis. Rheumatology 2002; 41(12): 1457-59.
29. Flora McErlane, Michael Beresford, Eileen Baildam, et al. Recent developments in disease activity indices and outcome measures for juvenile idiopathic arthritis Rheumatology (2013) 52 (Nov): 1941-51 doi:10.1093/rheumatology/ket150.
30. Priyanka Chandratre, Edward Roddy, Lorna Clarson, et al.Health-related quality of life in gout: a systematic review Rheumatology (2013) 52 (Nov): 2031-40 doi:10.1093/rheumatology/ket265.

OUTCOME MEASURES IN RHEUMATOID ARTHRITIS 2

Gp Capt S Shankar, Maj Shaman Gill

Rheumatoid Arthritis (RA) is a chronic inflammatory disease affecting multiple joints and has a multifactorial pathogenesis. It is different from other chronic diseases like hypertension, diabetes and coronary artery disease in that a gold standard measure is not available for diagnosis, prognosis, management and monitoring. Hence, extensive arrays of disease-specific quantitative measures and indices have been developed to assess the patient status for clinical trials, clinical research and clinical care.

In RA, disease outcomes can be a manifestation of the underlying disease process or activity (synovitis, acute phase response, pain), measures of discomfort (pain, fatigue, stress), measures of disability, measures reflecting joint damage (clinical or radiological), and eventually death. Outcome measures have applications in clinical trials and observational studies, as well as in clinical practice. Recently, there have been lot of advances in outcome measures in rheumatoid arthritis which has led to better treatment options and better disease outcomes. The objective of this chapter is to describe recent developments in outcome measurement for rheumatoid arthritis for trials and clinical practice.

Disease Activity

In RA, active disease is associated with worsening of joint damage and disability. Hence it is important to develop disease activity measures which can be easily performed in the outpatient department. Also, no single measure can correctly assess disease activity in RA, hence several composite indices of disease activity have been developed in recent years which are characterized by the integration of various single measures into one summary number and incorporate both objective (i.e. acute phase reactants like ESR, CRP and radiographic imaging results) and subjective (i.e. joint count and patient assessment of disease activity) parameters.

Measures to assess disease activity in RA fall into two distinct groups. 'Status measures' assess disease activity at a specific point of time and 'response measures' assess how disease activity changes over time (for example, response to medication). Composite indices such as the Disease Activity Score (DAS), DAS28, DAS28-CRP (DAS28 using C-reactive protein instead of ESR), Simplified Disease Activity Index (SDAI) and Clinical Disease Activity Index (CDAI) are examples of status measures (Table 1).

Table 1 : Composite indices of disease activity

Index	How calculated	Cut off points			Remarks
		Remission	Low disease activity	High disease activity	
DAS	DAS44 = 0.53938* √(Ritchie) + 0.06465* (SJC44) + 0.330* ln (ESR) + 0.00722* (General Health)	< 1.6	≤ 2.4	> 3.6	
DAS28-ESR	DAS28-4 = 0.56* √(TJC28) + 0.28* (SJC28) + 0.70* ln (ESR) + 0.014* (General Health)	≤ 2.4	≤ 3.2	> 5.1	Most widely used Needs calculator
DAS28-CRP	0.56* √(TJC28) + 0.28* √ (SJC28) + 0.014* GH + 0.36* ln (CRP+1) + 0.96	< 2.6	≤ 3.2	> 5.1	
CDAI	SJC + TJC + PGA + PhGA	≤ 10	≤ 10	> 22	Easiest to use. Robust for defining remission
SDAI	SJC + TJC + PGA + PhGA + CRP	≤ 11	≤ 11	> 26	

Ritchie - Ritchie articular index, ESR: erythrocyte sedimentation Rate, CRP - C Reactive Protein, SJC - swollen joint count, TJC - tender joint count, PGA - Patients Global Assessment, PhGA - Physicians Global Assessment

These indices are important in evaluating treatment strategies in individual RA patients. Response measures evaluate change in clinical status over time in clinical trials to determine efficacy but can also be implemented in longitudinal observational studies to evaluate clinical change over time.

The response measures developed by the American College of Rheumatology are depicted in (Table 2)[1,2].

Table 2 : Response measures, ACR improvement criteria

≥ 20% improvement in swollen joint count

≥ 20% improvement in tender joint count

≥ 20% improvement in at least three of the following five measures:

a. Patients global assessment of disease activity
b. Physicians global assessment of disease activity
c. Patients assessment of pain
d. Acute phase reactant (ESR or CRP)
e. Patients self assessed disability

Due to the limitations of the ACR response criteria and the need for a continuous measure of disease. Disease activity score (DAS) was developed in 1993[3]. The DAS incorporates weighted values of the Richie Articular Index (a joint count that grades the tenderness of each joint on a scale of 0-3), a swollen joint count (based on a 44 joint count), erythrocyte sedimentation rate (ESR), and patient assessment of general health, to provide a value between 0-10. Later, DAS 28 was developed in which the Richie Articular Index and the 44-joint count was replaced with swollen and tender joint counts based on a 28-joint count[4,5]. In DAS28-CRP, ESR has been replaced by C-reactive protein (CRP) measurements and constant values have been substituted in place of the patient assessment of general health.

The DAS based European League Against Rheumatism (EULAR) response criteria were developed to measure individual response in clinical trials. The EULAR response criteria classify individual patients as non-moderate or good responders, dependent on the extent of change and the level of disease activity reached (Table 3).

DAS, DAS 28 scores have been widely used as an outcome measure in clinical trials incorporating ESR or CRP. However they may be time consuming to perform in routine clinical practice.

For the ease of calculation, Smolen and colleagues devised the Simplified Disease Activity Index (SDAI) in 2003[6]. SDAI is the sum of the tender joint count, swollen joint count, patient global assessment, physician global assessment, and CRP. Later CDAI (clinical disease activity index) was developed which does not include an acute-phase reactant and is based on a swollen/tender joint count of 28 joints and patient and physician global assessments of disease activity and perform equally well as the SDAI and DAS/DAS28[7].

The GAS (Global arthritis score) is a measure that is achieved by adding patient pain (measured 0-10), the raw scored from the Modified Health Assessment Questionnaire (MHAQ) (questions with total possible score of 8-24) and the tender joint count (0-28). When added together these scores range from 0-62 and have been validated and perform comparably to the SDAI and CDAI.

The RA Disease Activity Index (RADAI) is a self-report questionnaire of symptoms, including a self-assessed joint count[8]. The Routine Assessment of Patient Index Data (RAPID) and the Patient Activity Scale (PAS) are other patient self report questionnaires consisting of the three patient-reported outcomes of the core set of endpoints for RA clinical trials: pain, patient global assessment of disease activity, and the Health Assessment Questionnaire (HAQ)[9,10]. RADAI takes only average 3 minutes to be calculated and patients can themselves score while waiting in the OPD.

Damage

Damage in rheumatoid arthritis can be assessed by clinical evidence of damage and by radiological evaluation. Radiographic progression is an important outcome measure in rheumatoid arthritis. It correlates well with disease activity and functional outcome and is widely used in clinical trials and observational studies. Its widespread use is due to easy availability of the radiographs of hands and feet

Table 3 : The EULAR response criteria using the DAS and DAS28

DAS at endpoint	DAS28 at endpoint	Improvement in DAS or DAS28 from baseline		
		≤ 1.2	> 0.6 and ≤ 1.2	≤ 0.6
2.4 and ≤ 5	≤ 3.2	Good	Moderate	None
> 2.4 and ≤ 3.7	> 3.2	Moderate	Moderate	None
> 3.7	> 5.1	Moderate	None	None

and due to the fact that the disease activity leads to radiographic progression which correlates with functional capability of the patient. A longitudinal relationship has been demonstrated between disease activity and radiological progression[11]. Various scoring methods are used to assess the radiological damage amongst which the commonly used are the vander Heijde modification of the Sharp Scoring system (SvDH) and the Larsen scoring system (Table 4 and Fig. 1).

sharp scores done at two different time points. Since Sharp scoring system is time consuming, Simplified Erosion and Narrowing Score (called "SENS") is developed which is based on SvDH method and presence or absence of erosions and/or joint space narrowing per joint is scored , thus arriving at a sum score of 86 instead of 448[12]. The SENS was shown to be reliable with respect to intra-and interreader reliability, and is sensitive to change.

Table 4 : Radiographic scoring methods

Method	Erosion	JSN*	Osteoporosis	Malignment	Soft tissue swelling	(Sub)luxation	Ankylosis	Cyst
Sharp (1971)	?	?					?	?
Sharp (1985)	?	?				Excluded †	?	
Genant (1983)	?	?			?			
Genant (1998)	?	?				?	?	
Kaye/Sharp (1986)	?	?		?		?	?	
van der Heijde/Sharp (1989)	?	?				?	?	
SENS (1999)	?	?				?	?	
Larsen (1977)	?	?	?		?	Excluded†	Excluded †	
Larsen (1995)	?	?	Excluded†		Excluded †			
Scott/Larsen (1995)	?	?	?		?	?		?
Rau/Larsen (1995)	?	?	?		?			
Ratingen (1998)	?	?	Excluded †		Excluded †			
SES (2000)	?	?	Excluded †		Excluded †			

*JSN = joint space narrowing , ? Included in the scoring system; †not included in the scoring system

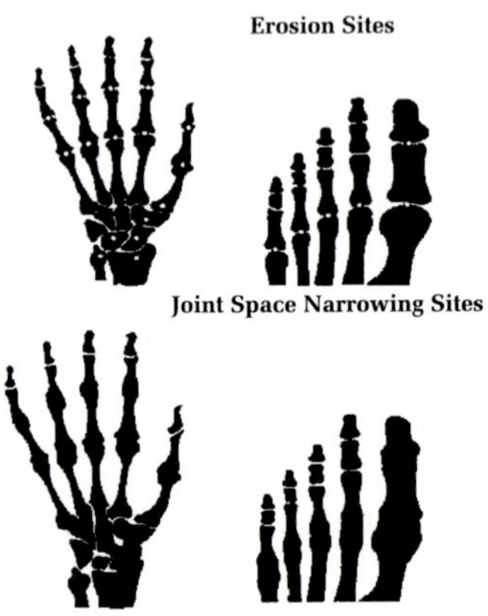

Fig. 1 : Scoring sites of the Vander Heijde modification of the sharp score with regard to erosions and joint space narrowing. Erosions in the hands are from zero to 5, erosions in the feet are from zero to 10, joint space narrowing in the hands and feet scored from zero to 4.

In SvDH method, scoring is done for the presence of erosions in 16 joints of hands and wrists which are graded from 0 to 5, and in 6 joints of the feet which are graded from 0 to 10, and the presence of joint space narrowing in 15 joints of the hands and wrists (graded from 0 to 4) and in 6 joints of the feet (graded from 0 to 4). The maximal range is 280 units for erosion and 168 units for joint space narrowing, summing up to 448 units for the total Sharp score (TSS). Progression is calculated by substracting the

Functional outcome measures

RA can lead to significant discomfort and functional disability due to disease activity. Patient can have persistent pain, fatigue and depression. Disability is a central concept in rheumatic disorders and there are many disorder-specific questionnaires that measure the level of patient-perceived disability. Patient related outcomes are helpful as they provide important information about patient's health, functional status and quality of life. Health assessment questionnaire (HAQ) is the commonly used disability measure however, similar information is also provided by generic health measures like Short Form 36 and Nottingham Health Profile. HAQ, introduced in 1980, is based on five parameters: disability, pain, medication effects, costs of care, and mortality[13]. It assesses multiple dimensions based on patient-centered values and thus the emphasis is on the outcomes that are relevant to the patients rather than laboratory values. HAQ has two versions - Full HAQ, which assesses all five dimensions, and the short or 2-page HAQ, which contains only the HAQ disability index (HAQ-DI) and the HAQ's patient global and pain visual analogue scales (VAS). HAQ-DI is composed of 20 items in 8 categories: dressing and grooming, hygiene, arising, reaching, eating, gripping, walking, and common daily activities and provide important information about the patient's functional status. The Multidimensional HAQ (MDHAQ) is a modification of HAQ which also scores for fatigue, psychological distress, morning stiffness and change in status on one side of one page, and review of systems and medications on the reverse side of the page[14]. The MDHAQ consists of 10 items from the original 20-item HAQ items, and 21-count visual analogue scales for both pain and function that are separated by a patient-reported joint count (16 of 18 items which are scored). It is more useful in clinical care as it fits on a single side of paper, is easily filled and easily scored and provides vital information to the health care provider[21].

Minimal Disease activity (MDA) in RA

The Outcome Measures in Rheumatology Clinical Trials (OMERACT) group has proposed definitions of minimal disease activity that is most generally defined as the disease state that lies between active disease and remission[15]. The preliminary definition for MDA was derived by consensus and includes: patients with a DAS $\leq$ 2.85 or patients fulfilling 5 of the following 7 disease activity criteria: pain score $\leq$ 2 on a 0-10 visual analogue scale; swollen joint count $\leq$ 1 based on a 28-joint count; tender joint count $\leq$ 1 based on a 28-joint count; HAQ score of $\leq$ 0.5; physician global assessment of disease activity $\leq$ 1.5 on a 0-10 visual analogue scale; patient global assessment of disease activity $\leq$ 2 on a 0-10 visual analogue scale; ESR $\leq$ 20[16].

Outcome measures

Fatigue in RA : The Bristol RA Fatigue analog scale and Bristol RA Multidimensional Questionnaire formed the basis of developing a new scale in the 10th OMERACT meet. There are four dimensions of measuring fatigue in RA : physical fatigue, emotional fatigue, cognitive fatigue and living with fatigue (impact). A consensus technique that would merge qualitative and quantitative data by including patients themselves will be the future direction in this outcome measure.

Sleep in RA : The Medical Outcome Study Sleep Module has been found to detect clinically significant changes in sleep improvements in a recent study and a new sleep scale may not be needed.

Remission, patients perspective : Most of the patients were interested in suggesting measures that they thought were important in defining remission, there being a lack of data on some of them. Patient global assessment and RADAI were amongst those items that needed more clear definitions of what was included in it.

Flare in RA : More validated measure of defining flare in RA is needed. There are four components to this : Patient self report using a 'Patient Global question', Patient self report using a validated questionnaire for flare, Physician assessment of disease activity and physician's decision to change treatment. The ultimate goal is to develop a standardized method to measure flare in RA.

The Synoviomics Project has been started to study synovial biomarkers - both diagnostic and prognostic.

Both ultrasound score to create a GLOSS (Global Scoring System for Synovitis) and an MRI score to validate Joint Space Narrowing in wrist and MCP joints have been worked upon. Subclinical inflammation (osteitis and synovitis by MRI) in those in clinical remission explaining the continuing radiographic progression in these cases. Excluding some of wrist bones in the Rheumatoid Arthritis MRI Scoring System (RAMRIS), it improved the discriminative power.

Death

Most of the studies investigating mortality among patients with RA, have demonstrated reduced life expectancy ranging from 5 to 15 years compared with the general population[17,18]. In RA, it appears that mortality has not improved over time[19]. Observing improving trends of survival over calendar time in rheumatoid arthritis is an important outcome of the quality of rheumatology care for these patients. Death (survival time) is hence a relevant outcome for clinical trials that include patients who are treated for life-threatening complications of their rheumatic disorder[20].

Further Reading:

1. Felson DT, Anderson JJ, Boers M, et al. The American college of rheumatology preliminary core set of disease activity measures for rheumatoid arthritis clinical trials. The committee on outcome measures in rheumatoid arthritis clinical trials. Arthritis Rheum 1993; 36: 729-40.
2. Felson DT, Anderson JJ, Boers M, et al. American college of rheumatology. Preliminary definition of improvement in rheumatoid arthritis. Arthritis Rheum 1995; 38: 727-35.
3. Van der Heijde DM, van't Hof M, van Riel PL, van de Putte LB. Development of a disease activity score based on judgment in clinical practice by rheumatologists. J Rheumatol 1993; 20: 579-81.
4. Prevoo ML, van't Hof MA, Kuper HH, et al. Modified disease activity scores that include twenty-eight-joint counts. Development and validation in a prospective longitudinal study of patients with rheumatoid arthritis. Arthritis Rheum 1995; 38: 44-48.
5. Smolen JS, Breedveld FC, Eberl G, et al. Validity and reliability of the twenty-eight-joint count for the assessment of rheumatoid arthritis activity. Arthritis Rheum 1995; 38: 38-43.
6. Smolen JS, Breedveld FC, Schiff MH, et al. A simplified disease activity index for rheumatoid arthritis for use in clinical practice. Rheumatology (Oxford) 2003; 42: 244-57.
7. Aletaha D, Nell VP, Stamm T, et al. Acute phase reactants add little to composite disease activity indices for rheumatoid arthritis: validation of a clinical activity score. Arthritis Res Ther 2005; 7: 796-806.
8. Stucki G, Liang MH, Stucki S, Brühlmann P, Michel BA: A self administered rheumatoid arthritis disease activity index (RADAI) for epidemiologic research. Psychometric properties and correlation with parameters of disease activity. Arthritis Rheum 1995; 38: 795-98.
9. Pincus T, Yazici Y, Bergman M, Swearingen C, Harrington T: A self-report questionnaire Routine Assessment of Patient Index Data (RAPID) score as a guide to a 'continuous quality improvement'. Clin Exp Rheumatol 2006, 24 (6 Suppl 43): S-60-65; quiz S-66-73.
10. Wolfe F, Michaud K, Pincus T: A composite disease activity scale for clinical practice, observational studies, and clinical trials: the Patient Activity Scale (PAS/PAS-II). J Rheumatol 2005, 32: 2410-15.
11. Welsing PM, Landewe RB, Van Rielpl etal.: The relationship between disease activity and radiologic progression in patients with rheumatoid arthritis: a longitudinal analysis. Arthritis Rheum 2004; 50: 2082-93.
12. Van Der Heijde D, Dankert T, Nieman F, Rau R, Boers M: Reliability and sensitivity to change of a simplification of the Sharp/van der Heijde radiological assessment in rheumatoid arthritis. Rheumatology (Oxford) 1999; 38: 941-47.
13. Bruce B, Fries JF. The Stanford Health Assessment Questionnaire: dimensions and practical applications. Health Qual Life Outcomes. 2003; 1: 20.
14. Pincus T, Swearingen C, Wolfe F. Toward a Multi-Dimensional Health Assessment Questionnaire (MDHAQ): assessment of advanced activities of daily living and psychological status in the patient friendly health assessment questionnaire format. Arthritis Rheum 1999; 42: 2220-30.
15. Wells GA, Boers M, Shea B, et al. Minimal disease activity for rheumatoid arthritis: a preliminary definition. J Rheumatol 2005; 32: 2016-24.
16. Khanna D, Oh M, Furst DE, et al. Evaluation of the preliminary definitions of minimal disease activity and remission in an early seropositive rheumatoid arthritis cohort. Arthritis Rheum 2007; 57: 440-47.
17. Cobb S, Anderson F, Bauer W: Length of life and cause of death in rheumatoid arthritis. N Engl J Med 1953, 249: 553-56.
18. Sokka T, Pincus T: Mortality in the rheumatic diseases. Introduction. Clin Exp Rheumatol 2008, 26 (Suppl 51): S1-4.
19. Boers M, Dijkmans B, Gabriel S, Maradit-Kremers H, O'Dell J, Pincus T: Making an impact on mortality in rheumatoid arthritis. Targeting cardiovascular comorbidity. Arthritis Rheum 2004; 50: 1734-39.
20. Wong WK, Furst DE, Clements PJ, Streisand JB: Assessing disease progression using a composite endpoint. Stat Meth Med Res 2007; 16: 31-49.
21. A Kumar, A N Malaviya, A pandhi, R singh. Validation of an Indian version of the Health Assessment Questionnaire in patients with rheumatoid arthritis. Rheumatology 2002; 41(12): 1457-59.

OUTCOME MEASURES IN SPONDYLOARTHRITIS

3

Darshan Bhakuni, Shanmuganandan Krishnan, Kartik Sivasami

Introduction

Spondyloarthritis is characterized by the presence of inflammatory back pain, asymmetrical peripheral arthritis, enthesitis, dactylitis and uveitis. What was initially an heterogeneous group of diseases labeled as seronegative spondyloarthritides have now been grouped into Axial and Peripheral Spondyloarthritis based on clinical features[1]. Outcome measures are important in measuring disease activity, to decide the appropriate therapy, documenting the response to therapy and to know the extent of damage or functional limitation caused by the disease. Multiple outcome measures have been developed by the Assessment in Spondyloarthritis international Society (ASAS) (formerly Assessment in Ankylosing Spondylitis) working group along with Outcome Measures in Rheumatology (OMERACT).

Classification Criteria for Spondyloarthritis

One of the first steps in disease assessment is to classify the disease as per the existing criteria. These classification criteria are devised for the the purposes of recruiting a patient into a clinical study as patient population are required to be homogeneous for the studies. The present Assessment in Spondyloarthritis international Society (ASAS) for Axial Spondyloarthritis and the proposed criteria for Peripheral Spondyloarthritis are presented in Table 1 and 2 respectively.

Table 1 ASAS Classification criteria for Axial Spondyloarthrtitis

Imaging arm: Presence of sacroiliitis by imaging or by Magnetic Resonance Imaging (MRI) plus at least one Spondyloarthritis (SpA) feature.

Clinical Arm: Presence of HLA-B27 and at least 2 SpA features.

SpA Features	Definition
1. Inflammatory back pain	Age at onset < 40 yr Insidious onset Improvement with exercise No improvement with rest Pain at night (with improvement on getting up) 4 of the above 5 should be present
2. Arthritis	Past or present active synovitis diagnosed by a physician
3. Enthesitis (heel)	Past or present spontaneous pain or tenderness at examination of the site of the insertion of the Achilles tendon or plantar fascia at the calcaneus
4. Uveitis	Past or present uveitis anterior, confirmed by an ophthalmologist
5. Dactylitis	Past or present dactylitis, diagnosed by a physician
6. Psoriasis	Past or present psoriasis, diagnosed by a physician
7. Inflammatory bowel disease	Past or present Crohn's disease or ulcerative colitis diagnosed by a physician
8. Good response to NSAIDs	24-48 hours after a full dose of a non-steroidal anti-inflammatory drug (NSAID): the back pain is no longer present or much better.

9. Family history for SpA	Presence in first-degree (mother, father, sisters, brothers, children) or second-degree (maternal and paternal grandparents, aunts, uncles, nieces, and nephews) relatives of any of the following: (a) ankylosing spondylitis, (b) psoriasis, (c) acute uveitis, (d) reactive arthritis, or (e) inflammatory bowel disease
10. Elevated C-reactive protein	CRP concentration above upper normal limit in the presence of back pain; after exclusion of other causes for elevated CRP concentration
11. HLA-B27-positive	Testing according to standard laboratory techniques
12. Sacroiliitis by radiographs	Bilateral grade 2-4 or unilateral grade 3-4 sacroiliitis on plain radiographs, according to the modified New York criteria
13. Sacroiliitis by MRI	Active inflammatory lesions of sacroiliac joints with definite bone marrow edema/osteitis, suggestive of sacroiliitis associated with spondyloarthritis

Table 2 Proposed ASAS Classification criteria for Peripheral Spondyloarthritis

Arthritis, Dactylitis or Enthesitis plus either of the 2 columns	
Plus ≥ 1 of	Plus ≥ 2 of the remaining
Psoriasis	Arthritis
Inflammatory bowel disease	Enthesitis
Preceding infection	Dactylitis
HLA-B27	IBS in the past
Uveitis	Positive familyhistory for SpA
Sacroiliitis on imaging (radiographs or MR)	

Measurements of disease activity in ankylosing spondylitis

The **Bath Ankylosing Spondylitis Disease Activity Index (BASDAI)** is a questionnaire filled in by the patient which includes fatigue, back pain, morning stiffness and peripheral features of arthritis and enthesial pain and is used for the assessment of disease activity (Table 3). The disadvantages of BASDAI are that it only includes patient reported measures, there is no weightage of individual clinical manifestations (the variables are simply summed without taking the relative importance and dependency into account) and lacks any objective measures of inflammation. To circumvent this, a new **Ankylosing Spondylitis Disease Activity Score (ASDAS)** has been proposed by ASAS (Table 4) which includes, besides clinical parameters, C-reactive protein (CRP) and Erythrocyte Sedimentation Rate (ESR). Online calculator is available at the ASAS website for ASDAS-CRP and ASDAS-ESR scores: www.asas-group.org/research/asdas_calculator/asdas.html. Based on these scores the various disease activity states are defined and cutoff scores are given to define clinically significant improvement (Table 5). Apart from these, Patient Global Assessments and Pain scale based on Numerical Rating Scale (NRS) or Visual Analogue Scale (VAS) can also be used to assess disease activity (Table 6).

Table 3 The components of Bath Ankylosing Spondylitis Disease Activity Index (BASDAI)

1. How would you describe the overall level of fatigue/tiredness you have experienced ?
2. How would you describe the overall level of AS neck, back or hip pain you have had ?
3. How would you describe the overall level of pain swelling in joints other than neck, back or hips you have had ?
4. How would you describe the overall level of discomfort you have had from any areas tender to touch or pressure ?
5. How would you describe the overall level of morning stiffness you have had from the time you wake up ?
6. How long does your morning stiffness last from the time you wake up ?

The BASDAI is evaluated by computing the mean of Question 5 & 6, and adding them to the sum of Questions 1-4 and dividing the result by 5.

Table 4 ASDAS -CRP AND ASDAS-ESR formulae.

ASDAS-CRP: 0.121 x total back pain + 0.110 x patient global + 0.0736 peripheral pain/swelling + 0.0586 duration of morning stiffness + 0.5796Ln (CRP+1).

ASDAS-ESR: 0.113 x patient global + 0.293 x √ESR + 0.086 x peripheral pain/swelling + 0.069 x duration of morning stiffness + 0.0796 x total back pain.

ASDAS-CRP is preferred, but the ASDAS-ESR can be used in case CRP data are not available. CRP in mg/ litre; all patient assessments on a 10 cm scale.

Table 5 Interpretation of ASDAS measures

Disease activity states	
Inactive disease	< 1.3
Moderate disease activity	1.3-2.1
High disease activity	2.1-3.5
Very high disease activity	> 3.5
Cut off scores for improvement	
Clinically important improvement	≥ 1.1
Major improvement	≥ 2.0

Table 6 Numerical Rating Scale for patient global disease activity

Please tick a box to indicate your overall assessment of your disease activity **during the last week**.

0	1	2	3	4	5	6	7	8	9	10	
None	1	2	3	4	5	6	7	8	9	10	severe

Table 7 The components of Bath Ankylosing Spondyliitis Functional Index (BASFI)

1. Putting on your socks or tights without help or aids (e.g. sock aid).
2. Bending forward from the waist to pick up a pen from the floor without an aid
3. Reaching up to a high shelf without help or aids (e.g. helping hand)
4. Getting up out of an armless dining room chair without using your hands or any other help
5. Getting up of the floor without help from lying on your back
6. Standing unsupported for 10 minutes without discomfort
7. Climbing 12-15 steps without using a handrail or walking aid one foot on each step
8. Looking over your shoulder without turning your body
9. Doing physically demanding activities (e.g. physiotherapy exercise, gardening or sports)
10. Doing a full days activities whether it be at home or at work

Functional and quality of life assessment in spondyloarthritis

The Bath Ankylosing Spondylitis Functional Index (BASFI) (Table 7) is a validated tool for assessment of functional impairment in Ankylosing Spondylitis. There is also a Health Assessment Questionnaire-Spondylitis (HAQ-S) developed specifically for spondyloarthritis. The Ankylosing Spondylitis Quality of Life Index (AS-QoL) and Medical Outcome Study Short Form-36 (MOS-SF-36) are the best studied disease specific measures available to assess Health related quality of life in spondyloarthritis cohorts[2].

Metrometry

Bath Ankylosing Spondylitis Metrometry Index (BASMI)

BASMI is a composite index of the measures of Cervical rotation, Tragus to wall distance (for involvement at cervical spine), modified Schober's and Lateral Flexion (measuring the movements at the lumbar spine) and intermalleolar distance (for hip involvement). Each of these are scored on a scale of 0-2 (3 points) or 0-10 (11 points) based on conversion tables (BASMI 3 is shown in Table 8).

Individual tests are covered in detail in the succeeding paragraphs and depicted in corresponding figures.

Table 8 The Bath Ankylosing Spondylitis Metrology Index (BASMI) 3 point answer scale

	0	1	2
	Mild	**Moderate**	**Severe**
Lateral lumbar flexion (cm)	> 10cm	5-10 cm	> 5 cm
Tragus to wall distance (cm)	< 15 cm	15-30 cm	> 30 cm
Lumbar flexion (modified Schober) (cm)	> 4 cm	2-4 cm	< 2 cm
Maximal intermalleolar distance (cm)	> 100 cm	70-100cm	< 70 cm
Cervical rotation (°)	> 70	20-70	< 20

Modified Schober's test (Fig. 1)

Schober's test measures the forward flexion at the lumbar spine. The original Schober's test measured the increase on maximal forward flexion in a 10 cm segment with lumbosacral junction as the inferior point. This was initially modified to include 5 cm below, modified again to use the anchor as the upper limit of the dimples of Venus for the lumbosacral junction. In the final modification, the lower 5cm has been dropped and this is used in BASMI measurement. A horizontal line is drawn on the spine at the level of posterior superior iliac spine with the patient standing straight. Another mark is placed 10 cm above this line. Patient bends forward maximally without bending the knees and the increase in the 10 cm segment is recorded to the nearest 0.1 cm. Normal increase is a minimum of 4 cm.

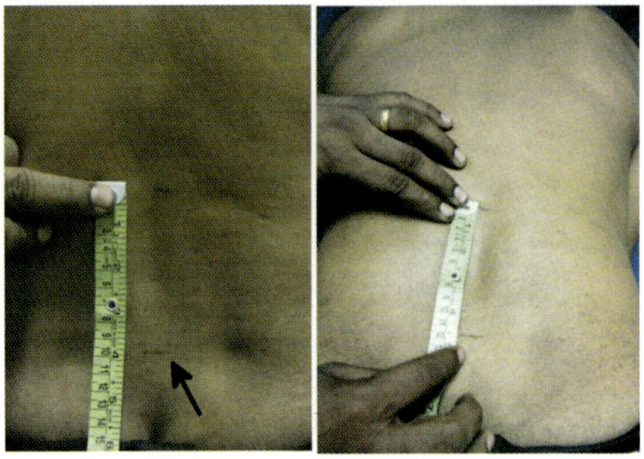

Figs. 1a & b. Modified Schober's test: 2 points are marked 10cms apart with the lower one (arrow in Fig. 1a) at the level of posterior superior iliac spines 'dimple of venus'. The distance between the points are again measured with the patient bending forward with knees straight (Fig. 1b). The difference between the 2 readings is noted (2 cms in this case).

Occiput to wall or Tragus to Wall distance (Fig. 2)

This measures the limitation in cervical extension. The patient stands straight with the back to the wall and the heels touching the wall. Effort is made to touch the head against the wall with the chin at its usual position. The distance between the wall to the occiput (normal 0 cm) or the wall to the tragus is measured. The better of two attempts is recorded to the nearest 0.1 cm.

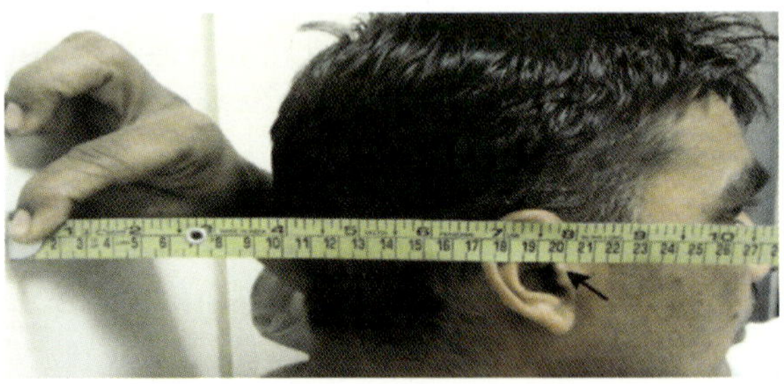

Fig 2. Tragus to wall distance. Measured between the wall and the tragus (arrow) with the patient in normal stance standing opposed to a wall with chin in neutral position.

Lateral bending (Fig. 3)

This measures the lateral flexion at the lumbar spine. Here the patient stands with the heels and back against the wall. There is no flexion of the knees, nor bending forward. A mark is placed at the level of patient's middle fingertip on the thigh and another after the patient bends sideways without bending the knees or lifting the heels. The difference between the two is recorded. The best of two tries is recorded for left and right. The mean of left and right gives the final result (in centimeters to the nearest 0.1 cm).

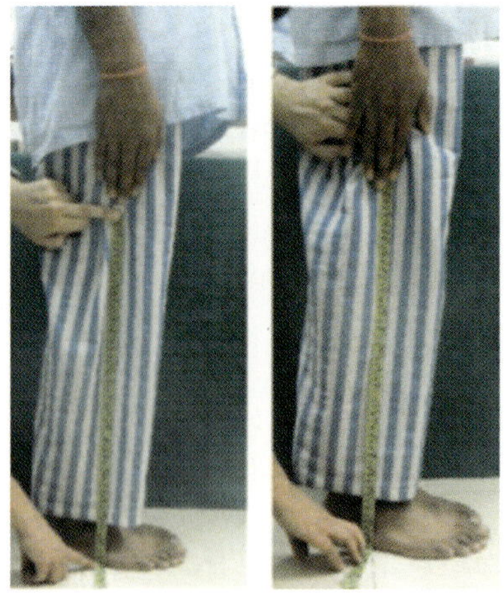

Fig 3a & b. The distance between the tip of the middle finger and the floor is measured with the patient standing straight (Fig 3a). The patient slides his fingers across while leaning to the ipsilateral side without bending the knees or leaning forward and the difference is noted. (9.5cm in this case) (Fig 3b)

Cervical rotation (Fig. 4)

Cervical rotation is measured by goniometer and the mean of the readings of both sides is taken.

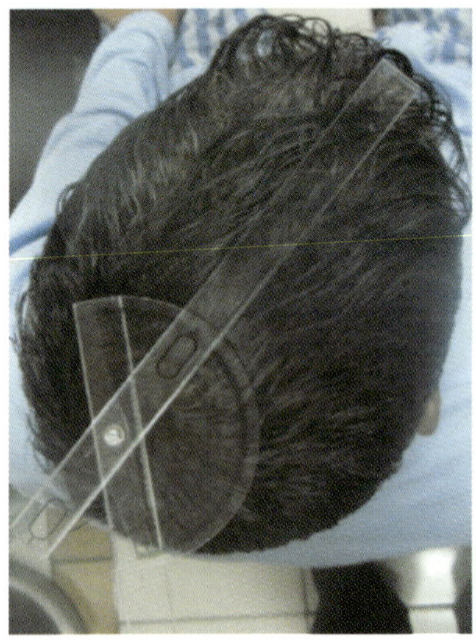

Fig 4. Measuring cervical rotation with the help of a goniometer.

Intermalleolar distance (Fig. 5)

This measures the hip abduction and can indicate hip involvement in spondyloarthrtitis.

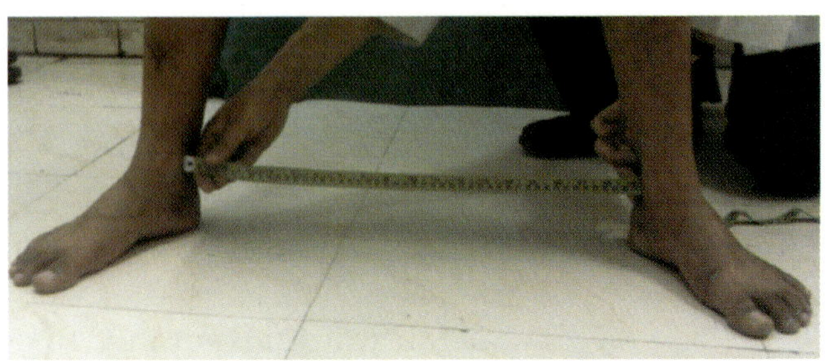

Fig 5. Measuring the intermalleolar distance.

Chest Expansion

Chest wall expansion with inspiration is measured with a tape placed circumferentially around the chest wall at the level of fourth intercostal space.

Other measurements like finger to floor and chin to chest distance are infrequently used in routine clinical practice.

Measurement of peripheral disease

Peripheral arthritis and enthesitis are the commonest extraaxial manifestations of spondyloarthritis. A 44 joint count to enumerate the number of swollen joints (without grading) is recommended in ASAS handbook. **Maastricht Ankylosing Spondylitis Enthesitis Score (MASES)**[3,4] is a 13 point scoring system based on the commonest enthesial sites involved in Spondyloarthritis. This includes bilateral 1st and 7th costochondral sites, anterior superior iliac spine, iliac crests, posterior superior iliac spines, the proximal insertions of Achilles tendon and spinous process of L5 vertebra.

Imaging measurements

X Ray

Sacroiliitis is graded as per the New York criteria into 4 grades (Table 9)[5]. These show the chronic changes and will not aid in early diagnosis of spondyloarthritis. Lateral X Rays of Cervical and Lumbar spine are obtained to score chronic changes in the spine using the modified Stoke Ankylosing Spondylitis Spinal Score (mSASSS).

Table 9 Grading of radiographic sacroiliitis

Grade 0	Normal
Grade 1	Suspicious changes
Grade 2	Minimal abnormality - small localised areas with erosion or sclerosis, without alteration in the joint width
Grade 3	Unequivocal abnormality - moderate or advance sacroiliitis with one or more of : erosions, evidence of sclerosis, widening, narrowing, or partial ankylosis
Grade 4	Severe abnormality - total ankylosis

MRI

Several scoring methods for assessing inflammatory activity in the spine and sacroiliac joints using MRI have been used in the past and have also recently been compared with each other however none has been validated so far. However attempts should be made to differentiate between active inflammatory lesions like bone marrow edema, capsulitis, synovitis and enthesitis v/s chronic lesions like sclerosis, erosions, fat deposition and bony bridges and ankylosis.

Composite measures

ASAS has developed core sets for clinical record keeping (Table 10), symptom modifying antirheumatic drugs (SM-ARD) and physical therapy and disease-controlling antirheumatic treatments (DC-ART). There are also criteria for response to treatment: ASAS 20, ASAS40, ASAS 5/6 improvement criteria and ASAS partial remission criteria.

Table 10 Assessment of Spondyloarthritis International Society (ASAS) core set for clinical record keeping

Domain	Instrument
Function	BASFI
Pain	NRS/VAS (last week/spine/at night due to AS) NRS/VAS (last week/spine/due to AS)
Spinal mobility	Chest expansion Modified Schober's test Occiput to wall Cervical rotation Lateral spinal flexion or BASMI
Patient global	NRS/VAS (global disease activity last week)
Peripheral joints and entheses	Number of swollen joints (44-joint count) Validated enthesitis score, such as MASES, San Francisco and Berlin
Stiffness	NRS/VAS (duration of morning stiffness/spine/last week)
Acute phase reactants	C-reactive protein (CRP) or erythrocyte sedimentation rate (ESR)
Fatigue	Fatigue question BASDAI

Conclusion

Outcome measures are an integral part of objective assessment in spondyloarthropathies to guide treatment and assess response to therapy. Incorporating such outcome measures as enumerated above in clinical practice will translate into better care for these patients.

Biomarkers like IL-6, MMP3, VEGF have been evaluated and ASDAS was found to correlate better than BASDAI reflecting inflammatory burden in a more appropriate way.

MRI findings confirming inflammatory lesions (spondylitis) and structural changes (fat deposition) are being validated to diagnose the disease early and also as an outcome measure for clinical studies.

The DISQ (Dudley Inflammatory Bowel Symptom Questionnaire) has been validated to measure bowel symptoms (should be included as a domain) in Spondyloarthritides.

Summery

1. Spondyloarthritis is characterized by inflammatory back pain, asymmetrical peripheral arthritis, enthesitis, dactylitis and uveitis.
2. Multiple outcome measures have been developed by the Assessment in Spondylo Arthritis international Society (ASAS) (formerly Assessment in Ankylosing Spondylitis) working group along with Outcome Measures in Rheumatology (OMERACT).
3. The Bath Ankylosing Spondylitis Disease Activity Index (BASDAI) is a questionnaire which includes fatigue, back pain, morning stiffness and peripheral features of arthritis and enthesial pain and is used for the assessment of disease activity.
4. The Bath Ankylosing Spondylitis Functional Index (BASFI) is a well validated tool for assessment of functional impairement in Ankylosing Spondylitis.
5. Bath Ankylosing Spondylitis Metrometry Index is a composite index of the measures of cervical rotation and Tragus to wall distance (for involvement at cervical spine), modified Schober's and Lateral Flexion (measuring the movements at the lumbar spine) and intermalleolar distance (for hip involvement).
6. Schober's test measures the forward flexion at the lumbar spine.
7. Other measures are Occiput to wall or Tragus to Wall distance, Lateral bending, Cervical rotation, Intermalleolar distance and Chest Expansion.
8. Maastricht Ankylosing Spondylitis Enthesitis Score (MASES) is a 13 point scoring system based on the commonest enthesial sites involved in Spondyloarthritis.
9. Imaging modalities like MRI helps in distinguishing active inflammatory lesions like bone marrow edema, capsulitis, synovitis and enthesitis v/s chronic lesions like sclerosis, erosions, fat depostion and bony bridges and ankylosis,
10. Outcome measures are an integral part of objective assessment in Spondyloarthropathies to guide treatment and assess response to therapy. Incorporating such outcome measures as enumerated above in clinical practice will translate into better care for these patients.

Further Reading:

1. Sieper J, Rudwaleit M, Baraliakos X, Brandt J, J Braun, R Burgos-Vargas, et al. The Assessment of SpondyloArthritis international Society (ASAS) handbook: A guide to assess spondyloarthritis. Ann Rheum Dis. 2009; 68: ii1-ii44.
2. Doward LC, Spoorenberg A, Cook SA, Whalley D, Helliwell PS, Kay LJ, et al. Development of the ASQol: a quality of life instrument specific to ankylosing spondylitis. Ann Rheum Dis 2003; 62: 20-26.
3. Ware JE Jr, Sherbourne CD. The MOS 36-item short-form health survey (SF 36). Conceptual framework and item selection. Med Care 1992; 30: 473-83.
4. Heuft-Dorenbosch L, Spoorenberg A, van Tubergen A, Landewe R, van ver Tempel H, Mielants H, et al. Assessment of enthesitis in ankylosing spondylitis. Ann Rheum Dis 2003; 62: 127-32.
5. van der Linden S, Valkenburg HA, Cats A. Evaluation of diagnostic criteria for ankylosing spondylitis. A proposal for modification of the New York criteria. Arthritis Rheum 1984; 27: 361-68.

OUTCOME MEASURES IN SYSTEMIC LUPUS ERYTHEMATOSUS

4

Shanmuganandan Krishnan, Kartik Sivasami

Introduction

Systemic Lupus Erythematosus (SLE) is a multisystem disease of autoimmune etiology with heterogeneous manifestations. Outcome measures are important in measuring disease activity, to decide the appropriate therapy, documenting the response to therapy and to know the extent of damage or functional limitation caused by the disease. Multiple outcome measures have been developed by various organisations which include more than 40 disease activity instruments reflecting the complexity and the challenges in measuring this multifaceted disease.

The need for outcome measures in lupus

SLE is a disease affecting predominantly females of child bearing age. Outcomes in SLE can be described by three domains, namely disease activity, accumulated damage and health status of patients. In the past, outcomes in SLE were measured in terms of survival. However, with the advent of targeted treatment and better outcomes, survival is more than 80% in 10 years. A good outcome measure should be able to distinguish between disease activity for example active 'renal lupus' and disease damage due to chronic disease e.g. chronic kidney disease as a sequelae of renal lupus and the resultant functional impairment. Patient may be assessed for various purposes including: routine care- quantifying disease and planning a therapeutic strategy, assess response to the therapy, to verify that the disease is in remission and thus the patient may plan her pregnancy, to include the patient in a clinical trial and to compare the responses of various subpopulations in a clinical trial scenario.

Classification of SLE

SLE can manifest in various ways ranging from subtle laboratory abnormalities eg. immune thrombocytopenia, to overt clinical manifestations like classic malar rash or as a life threatening emergencies eg. diffuse alveolar haemorrhage. Classification criteria of SLE are being constantly revised to incorporate these diverse presentations. A new Classification Criteria for Systemic Lupus Erythematosus has been validated by Systemic Lupus International Collaborating Clinics (SLICC). They identified seventeen criteria. The requirement for fulfillment of SLICC criteria includes the presence of either : 1) Fulfillment of at least four criteria, with at least one clinical criterion AND one immunologic criterion OR 2) Lupus nephritis as the sole clinical criterion in the presence of ANA or anti-dsDNA antibodies.

Assessment of disease activity

Assessment of Disease Activity in SLE includes a thorough head to toe clinical examination as practically every system in the body can be potentially involved, appropriate investigations and relevant serological studies. The questions to be asked are: how long does a particular finding/ abnormality has been there in existence, whether it can be attributed directly to SLE, is it a new finding, or has the condition improved or worsened or remained the same since initial review. It is important to determine indicators of active lupus, adverse drug effects, complications of disease such as infection or cardiovascular disease, or other comorbidities such as fibromyalgia, depression, cancer or thyroid disease.

Use of laboratory tests to assess disease activity

The so called 'routine' blood counts and urine routine and microscopic examinations are the most important laboratory tests in a patient with lupus and can give a clue to the disease activity. Lymphopenia is the commonest hematological manifestation. Other features include thrombocytopenia, anemia and leucopenia. Most of the cytotoxic immunosuppressive regimens will also affect the blood counts. However presence of these abnormalities indicating an active disease may necessitate a more aggressive management.

Active renal disease is assessed by the presence of active urinary sediments. Infection, stone or menstrual blood loss need to be excluded.

Acute Phase reactants can give a clue to the disease activity in lupus. ESR is usually raised in a setting of active disease while C Reactive protein is not. Modest elevations of CRP can be present when associated with serositis or arthritis and coexisting infections can cause a marked increase (usually > 50 mg/l) and such patients should be screened for appropriate infections before embarking upon an aggressive immunosuppressive regimen.

Serological tests may be useful in certain subsets of patients with Lupus. However these test have to be correlated with the presence of clinical findings. These corelate with disease activity especially renal lupus. Anti double stranded(ds) DNA titres can be a useful marker in individuals who manifest these antibodies (about 60% of SLE patients). These correlate with the disease activity, however the levels can be falsely low if there is deposition of the antibodies in tissues. Low Complement levels C3 or C4 are seen especially in patients with renal, hematological and vasculitis disease.

Disease activity measures

There are various standardized measures to assess disease activity in Lupus. The commonly used activity indices are Systemic Lupus Erythematosus Disease Activity Index (SLEDAI) (Table 1), the British Isles Lupus Assessment Group (BILAG), Systemic Lupus Activity Measure (SLAM) and the European Consensus Lupus Activity Measure (ECLAM). Each of these indices have been validated in different population subsets and can be used in adults as well as children. The various clinical and laboratory measures to be assessed vary amongst the indices. The Lupus Activity Index is a concise measure comprised of a 0-3 visual analog scale for 4 symptoms (fatigue, rash, joint involvement, serositis) and 4 signs (neurologic, renal, pulmonary, and hematologic involvement). While its list of SLE features is not comprehensive and its measure of disease activity is not weighted, it provides a simple, broad, and accurate assessment of activity in much the same fashion as the Physicians'Global Assessment. The SLEDAI and its modifications: SLEDAI 2000 and SELENA-SLEDAI are the simplest of indices to use in a clinical setting. Twenty-four features that are attributed to lupus are listed, with a weighted score given to any one that is present. The more serious manifestations (such as renal, neurologic, and vasculitis) are weighted more than others (such as cutaneous manifestations). The maximum possible score is 105. The BILAG is an organ-system score index compared to other 'global' indices and it records clinical disease activity in 8 different organ systems for a total of 86 items. It is a more comprehensive index and calculation of the scores are aided by a software. It may be cumbersome to use in a routine clinical setting and is more apt for clinical trials and clinical research. Later modification (BILAG - 2004 index) includes in addition ophthalmologic and gastrointestinal features which were already covered in some of the other indices.

Assesment of damage

With better management and resultant increasing survival in Lupus patients, more number of patients have long term disease and treatment related complications and resultant organ damage and death. A bimodal pattern of deaths have been reported with patients dying of disease activity in early stages and cardiovascular disease later. The Systemic Lupus International Collaborating Clinics - American College of Rheumatology (SLICC-ACR) (Table 2) Damage Index (DI) includes 41 items covering 12 systems. Damage tends to accumulate over time and it is an important outcome measure supplementing disease activity measures. A multi-national study of more than 1000 patients, who were followed-up longitudinally, showed that higher SLICC-ACR DI scores were at higher risk of death. Therefore it is important to identify complications early and prevent their development.

Functional and quality of life assessment in SLE

The indices above measure the physicians assessment of the disease. However it is also important to know the patients perception of her health and the overall quality of life. No specific measure has been developed for assessment of Quality of Life (QoL) and functional status of patients with SLE. The Short Form-36 (SF36) is most widely used. It includes a single multi-item scale that assesses 8 domains : limitations in physical activities because of health problems; limitations in social activities because of physical or emotional problems; limitations in usual role activities because of physical health problems; bodily pain; general mental health; limitations in usual role activities because of emotional problems; vitality (energy and fatigue); and general health perception. These measures have been shown to be impaired in patients with Lupus. They have been found to correlate more with psychosocial factors and less with disease activity/damage. Hence there may be a set of patients who may require non-pharmacological approaches to therapy including providing education, social support and counselling.

Conclusion

SLE can present itself in various forms, with periods of clinical inactivity, remission and intermittent flares. Long standing disease activity can cause organ damage leading to morbidity and mortality. The use of standardized and validated instruments for monitoring disease activity, organ damage and functional status will aid in the assessment of this multifaceted disease in to and lead to better disease outcomes to the satisfaction of patient as well as the physician.

The SLE Responder Index (SRI) is a composite index derived from the SELENA-SLEDAI, BILAG (2004) and the physicians global assessment and has been found useful to study responses to biologics.

The CLASI (Cutaneous Lupus Disease Area and Severity Index) activity and damage scores correlates with physician assessed cutaneous damage and is a validated measure to evaluate cutaneous lupus.

LupusPRO (patient reported outcome) has been validated in all ethnic groups and also men with lupus.

Key Points

- SLE is a multisystem disease with varied manifestations and periods of disease activity/ flares and remission.
- Clinical evaluation of SLE includes a thorough head to toe examination and appropriate laboratory assessment.
- Clinical features should be probed to ascertain whether they are manifestation of active SLE, accrued chronic damage, a comorbidity or a treatment related complication.
- Use of disease activity index, SLICC-ACR damage index and QoL assessment should be part of routine monitoring of SLE patients.

TABLE 1 SYSTEMIC LUPUS DISEASE ACTIVITY INDEX (SLEDAI)

(adapted from 3)

Descriptor	Definition	Weightage
Seizure	Recent onset. Exclude metabolic, infectious, or drug-related causes.	8
Psychosis	Altered ability to function in normal activity due to severe disturbance in the perception of reality. Includes hallucinations; incoherence; marked loose associations; impoverished thought content; marked illogical thinking; bizarre, disorganized or catatonic behavior. Exclude the presence of uremia and offending drugs.	8
Organic brain syndrome	Altered mental function with impaired orientation or impaired memory or other intellectual function, with rapid onset and fluctuating clinical features. Includes a clouding of consciousness with a reduced capacity to focus and an inability to sustain attention on environment and at least two of the following: perceptual disturbance, incoherent speech, insomnia or daytime drowsiness, increased or decreased psychomotor activity. Exclude metabolic, infectious and drug-related causes.	8
Visual	Retinal changes from systemic lupus erythematosus: cytoid bodies, retinal hemorrhages, serous exudates or hemorrhages in the choroid, optic neuritis (not due to hypertension, drugs, or infection).	8
Cranial nerve	New onset of a sensory or motor neuropathy involving a cranial nerve.	8
Lupus headache	Severe, persistent headache; may be migranous; unresponsive to narcotics.	8
Cerebrovascular accident	New syndrome. Exclude arteriosclerosis.	
Vasculitis	Ulceration, gangrene, tender finger nodules, periungual infarction, splinter hemorrhages. Vasculitis confirmed by biopsy or angiogram.	8
Arthritis	More than 2 joints with pain and signs of inflammation.	4
Myositis	Proximal muscle aching or weakness associated with elevated creatine phosphokinase/aldolase levels, electromyographic changes, or a biopsy showing myositis.	4
Casts in urine	Heme, granular, or erythrocyte.	4
Hematuria	More than 5 erythrocytes per high power field. Exclude other causes (stone, infection).	4

Proteinuria	More than 0.5 grams of urinary protein excreted per 24h. New onset or recent increase of > 0.5 g/24h.	4
Pyuria	More than 5 leukocytes per high-power field. Exclude infection.	4
New malar rash	New onset or recurrence of an inflammatory type of rash.	2
Alopecia	New or recurrent. A patch of abnormal, diffuse hair loss.	2
Mucous membranes	New onset or recurrence of oral or nasal ulcerations.	2
Pleurisy	Pleuritic chest pain with pleural rub or effusion, or pleural thickening.	2
Pericarditis	Pericardial pain with at least one of rub or effusion. Confirmation by electro- or echocardiography.	2
Low complement	A decrease in CH50, C3, or C4 level (to less than the lower limit of the laboratory-determined normal range).	2
Increased DNA binding	More than 25% binding by Farr assay (to > the upper limit of the laboratory-determined normal range, e.g. 25%).	2
Fever	More than 38°C after the exclusion of infection.	1
Thrombocytopenia	Fewer than 100,000 platelets	1
Leukopenia	Leukocyte count of < 3000/mm3 (not due to drugs)	1

TABLE 2 Systemic Lupus International Collaborating Clinics/American College of Rheumatology (SLICC/ACR) Damage Index.

(adapted from 5)

System	Item	Score
Ocular (either eye, by clinical assessment)	Any cataract ever	1
	Retinal change or optic atrophy	1
Neuropsychiatric	Cognitive impairment (e.g. memory deficit, difficulty with calculation, poor concentration, difficulty in spoken or written language, impaired performance level) OR major psychosis	1
	Seizures requiring therapy for 6 months	1
	Cerebrovascular accident ever (score 2 if > 1)	1,2
	Cranial or peripheral neuropathy (excluding optic)	1
	Transverse myelitis	1
Renal	Estimated or measured glomerular filtration rate < 50%	1
	Proteinuria > 3.5g/24h	1
	OR End-stage renal disease (regardless of dialysis or transplantation)	3
Pulmonary	Pulmonary hypertension (right ventricular prominence, or loud P2)	1
	Pulmonary fibrosis (physical and radiograph)	1
	Shrinking lung (radiograph)	1
	Pleural fibrosis (radiograph)	1
	Pulmonary infarction (radiograph)	1
Cardiovascular	Angina OR coronary artery bypass	1
	Myocardial infarction ever (score 2 if > 1)	1,2
	Cardiomyopathy (ventricular dysfunction)	1
	Valvular disease (diastolic murmur or systolic murmur > 3/6)	1
	Pericarditis for 6 months, OR pericardectomy	1
Peripheral vascular	Claudication for 6 months	1
	Minor tissue loss (pulp space)	1
	Significant tissue loss ever (e.g. loss of digit or limb)(score 2 if > 1 site)	1,2
	Venous thrombosis with swelling, ulceration, OR venous stasis	1
Gastrointestinal	Infarction or resection of bowel below duodenum, spleen, liver or gallbladder ever, for any cause (score 2 if > 1 site)	1,2
	Mesenteric insufficiency	1
	Chronic peritonitis	1
	Stricture OR upper gastrointestinal tract surgery ever	1
	Chronic pancreatitis	1

Musculoskeletal	Muscle atrophy or weakness	1
	Deforming or erosive arthritis (including reducible deformities, excluding avascular necrosis)	1
	Osteoporosis with fracture or vertebral collapse (excluding avascular	1
	necrosis)	1
	Avascular necrosis (score 2 if > 1)	1
	Osteomyelitis	1
	Tendon rupture	
Skin	Scarring chronic alopecia	1
	Extensive scarring of panniculum other than scalp and pulp space	1
	Skin ulceration (excluding thrombosis for > 6 months)	1
Premature gonadal failure		
Diabetes	(regardless of treatment)	1
Malignancy	(exclude dysplasia) (score 2 if >1 site)	1,2

Further Reading:

1. Petri M, Ana-Maria Orbai, Alarcón GS, Gordon C, Merrill JT, Paul R. F.et al Derivation and Validation of Systemic Lupus International Collaborating Clinics Classification Criteria for Systemic Lupus Erythematosus. Arthritis & Rheumatism. DOI 10.1002/art.34473.
2. Liang MH, Socher SA, Larson MG, Schur PH. Reliability and validity of six systems for the clinical assessment of disease activity in systemic lupus erythematosus. Arthritis Rheum 1989; 32: 1107-18.
3. Bombardier C, Gladman DD, Urowitz MB, Caron D, Chang CH. Derivation of the SLEDAI. A disease activity index for lupus patients. The Committee on Prognosis Studies in SLE. Arthritis Rheum 1992; 35: 630-40.
4. Symmons DP, Coppock JS, Bacon PA, et al. Development and assessment of a computerized index of clinical disease activity in systemic lupus erythematosus. Members of the British Isles Lupus Assessment Group (BILAG). Q J Med 1988; 69: 927-37.
5. Gladman DD, Goldsmith CH, Urowitz MB, Bacon P, Fortin P, Ginzler E, et al. "The Systemic Lupus International Collaborating Clinics/American College of Rheumatology (SLICC/ACR) Damage Index for Systemic Lupus Erythematosus International Comparison", J. Rheumatol, 2000; 27(2): 373-76.
6. Isenberg Dand, Ramsey-Goldman RAssessing patients with lupus: towards a drug responder index. Rheumatology 1999; 38: 1045-49.
7. Mosca M, S Bombardieri S, Disease-specific quality indicators, guidelines, and outcome measures in systemic lupus erythematosus (SLE). Clin Exp Rheumatol 2007; 25 (Suppl. 47): S107-S113.

DISEASE ASSESSMENT IN SYSTEMIC VASCULITIS

5

Aman Sharma, Pradeep Bambery

Introduction

The systemic vasculitides are a clinically heterogeneous group of uncommon autoimmune diseases characterized by the inflammation of the vessel wall[1,2,3,4]. This inflammatory process can either result in partial or total occlusion of the involved vessel causing tissue ischemia/infarction; or can result in hemorrhage due to rupture of the weakened, sometimes aneurysmally dilated vessel wall. Over the last fifty years, there has been significant advancement in the available treatment options and these previously invariably fatal diseases have been converted to chronically manageable disorders. However, these therapeutic options are not without safety concerns for their continued use.

The clinical spectrum of these disorders may vary from mild limited/localized disease to a life threatening multi organ involvement. Thus, for the critical evaluation of presentation, response to treatment and outcome, it is important to have objective tools, responsive to change for the assessment of disease activity. As these disorders can result in permanent, irreversible damage due to healed disease and drug toxicity, it is important to differentiate this damage from disease activity. These measures not only help in clinical decision-making and prognostication in an individual patient, but have been invaluable tools to judge outcomes in various multicenter randomized controlled therapeutic trials. As the therapeutic options increase due to global multicentre initiatives, these tools of assessment of disease activity and damage can help in moving towards tailored treatment regimes for different clinical presentations rather than a single 'one shoe fits all strategy'.

Assessment of vasculitis

The Importance of assessment

The purpose of assessment of the severity of a potentially chronic disease process like the vasculitides are manifold; while disease activity assessment is of immediate therapeutic importance for initial presentation in deciding the treatment protocol, differentiating damage (irreversible scars) accumulated over a period of time from disease activity is important during the course as it may necessitate a different therapeutic approach.

The difficulties in assessment are due to the multi system nature of these disorders and also to variability in disease activity in different organ systems. These can be further confounded by the closely mimicking presentations of drug toxicity and infection. For example, hematuria may be caused by relapse of the basic disease or hemorrhagic cystitis as a side effect of cyclophosphamide or due to urinary tract infection. Hence the comprehensive assessment of the current activity depends upon thorough clinical evaluation supplemented by investigations. The absence of a valid biomarker of the disease activity makes this even more important for rational therapeutic decision-making. The structured clinical tools provide a qualitative and quantitative assessment of systemic vasculitis acting on surrogate markers of disease activity.

Tools for assessment

Although histologic demonstration and grading of vasculitic injury would appear to be a gold standard, the patchy nature of involvement and the impracticality of repetitive biopsy, make this unviable. Even the yield of biopsy from different sites varies considerably, varying from 60% sensitivity of nasal biopsy in Granulomatosis with Polyangiitis (GPA, previously known as Wegener's Granulomatosis) to almost 100% sensitivity of renal biopsy in a patient with renal involvement[5,6].

Radiological assessment

Radiological investigations have a vital role in the diagnosis and assessment of the vasculitides[7]. Chest X-ray and X-ray of the paranasal sinuses are helpful in making the initial diagnosis and monitoring of Anti-neutrophilic cytoplasmic antibodies (ANCA) associated vasculitis. Conventional arteriography can demonstrate aneurysms, occlusion and stenosis.

However, this method has a limitation of being invasive and carries a substantial radiation exposure besides posing technical difficulties in patients with lengthy stenotic lesions. It is also unable to image the vessel wall.

Computerised Tomography (CT)

CT is not only helpful in assessing structural lesions but also monitoring the response to therapy. Most patients with GPA have paranasal sinus or lung involvement. CT findings of paranasal sinus involvement may be in the form of nodular thickening of the nasal mucosa, punctate bony destruction mainly in the midline, periantral soft tissue infiltration associated with bone demineralization, orbital extension and sclerosing osteitis of the mastoid aircells and maxillary sinuses. CT scan can delineate the type, extent and position of lung involvement in GPA. The various findings can be lung nodules or masses, seen in about 90% of patients. These are usually multiple, bilateral and subpleural, but can occasionally be peribroncho-vascular. Bronchial wall thickening, large airways abnormality, patchy consolidation and ground-glass shadowing can be the other findings. Pleural effusion is uncommon[8]. Nodular lesions are common in GPA, whereas infiltrative lesions are more common in Microscopic polyangiitis and Churg-Strauss syndrome. Severely abnormal chest CT scans with involvement of more than 80% of the lung parenchyma is an independent predictor of mortality[9].

CT scanning of the chest is of value in the monitoring of disease in GPA (Wegener's Granulomatosis). While the nodular infiltrates respond very well to treatment and often resolve completely. Atelectasis, fibrotic bands and areas of bronchiectasis represent the damage/scars.

CT angiography is superior to conventional angio-graphy because both mural as well as luminal changes can be studied. Thus it has a role in diagnosis and monitoring of Takayasu arteritis. CT angiography can demonstrate stenosis, occlusions, aneurysms, concentric arterial wall thickening and heavy transmural calcification (differentiating from atherosclerosis) in chronic disease. CT angio-graphy can also demonstrate microaneurysms or end organ changes in Polyarteritis nodosa.

Magnetic Resonance Imaging and Angiography (MRI & MRA)

It provides information regarding the state of the vessel wall, e.g. vessel wall oedema and progressive vessel wall thickening. The major limitations of MRA are an over estimation of the degree of stenosis; and poor correlation between gadolinium enhancement of the vessel wall and disease activity[10]. High resolution MRI can demonstrate the distribution of vessel involvement in Giant cell arteritis even beyond the temporal arteries.

Ultrasonography

Ultrasonography is useful in the diagnosis of temporal arteritis by demonstrating a 'halo' around the temporal artery. The effect of glucocorticoids can also be demonstrated by resolution of this 'halo'. Flow abnormalities on color doppler can provide information regarding luminal narrowing. Ultrasonography has a good sensitivity and specificity for the diagnosis of giant cell arteritis. It is 95% and 93% respectively when compared to histology, and 88% and 97% when compared to clinical diagnosis[11,12]. It may play an important role in management, as the histological involvement of temporal arteries is patchy and can result in low sensitivity of temporal artery biopsy.

Positron Emission Tomography (PET)

Positron Emission Tomography (PET) can assess metabolic activity of an organ with radio labelled 18-fluorodeoxyglucose. Vascular inflammation also leads to increased metabolic activity hence, PET can be used to assess vessels with a caliber of greater than 4mm. Although it is not suitable for imaging temporal arteries due to their superficial location and intense fluorodeoxyglucose signal from the brain, yet it has demonstrated widespread, often clinically silent involvement of other areas in temporal arteritis like the involvement of subclavian arteries and the aorta[13].

Diagnosis of large vessel vasculitis, particularly in the abdominal aorta should not be solely based on PET as inflammatory atherosclerotic plaques may exhibit isotope uptake. The reported sensitivity and specificity of PET for diagnosing large vessel vasculitis (Giant cell arteritis and Takayasu arteritis) is 60% to 89% and 90.9% to 99.8% respectively[14,15].

One has to remember that though all these imaging techniques discussed are of value for the monitoring and/or diagnosis of the systemic vasculitides, but these are expensive and operator dependent.

Serological markers in disease assessment

Anti-neutrophil cytoplasmic antibodies (ANCA) are directed against primary granules of neutrophils and monocytes[16]. Discovery of ANCA has been a major advancement in the diagnosis and understanding of the immunopathogenesis of small vessel vasculitis[17,18]. Two patterns are identified using an indirect immunofluorescence technique namely, C ANCA (diffuse cytoplasmic uptake), and P ANCA (perinuclear uptake). These antibodies are specifically targeted against antigens in the primary granules of neutrophils.

Proteinase 3 (PR3) and myeloperoxidase (MPO) are the two most relevant antigens and the ANCA to those antigens can be measured using an ELISA technique. ANCA alone should not be the basis of diagnosis of systemic vasculitis. The international consensus statement on reporting of ANCA recommends that both IIF and ELISA testing should be done when seeking to identify ANCA but has also stressed the importance of histological confirmation as gold standard[19]. ANCA is important as a diagnostic tool when used in an appropriate clinical setting but its role in monitoring disease activity is controversial. Although it has been shown that PR3 ANCA levels rise in patients with reactivation of GPA, upto one third of patients may have a rise in these titers without any clinical consequence[20].

ESR and CRP are non-specific markers of inflammation and cannot differentiate between disease activity and infection. While normal ESR and CRP cannot exclude systemic vasculitis, their levels can be used for following up the therapeutic response in established systemic vasculitis.

von Willebrand factor is released by platelets and endothelial cells in response to injury or activation. The levels, however, do not correlate well with disease activity measures. Role of other surrogate markers like soluble adhesion molecules is uncertain.

Clinical indices of disease assessment

The failure of serological and other tests to provide reliable disease activity and damage assessment has necessitated the development of clinical assessment tools. The three main aspects of clinical assessment of vasculitis are (1) disease activity (2) disease damage and (3) functional and social consequences of having vasculitis and its treatment. These aspects are common to other inflammatory diseases such as rheumatoid arthritis and systemic lupus erythematosus. The tools help in bringing uniformity to clinical assessment and objectivity to the assessment of outcomes of various therapeutic interventions.

Disease activity

Clinical evaluation remains a cornerstone of disease activity assessment. The tools help in recording disease activity in a reproducible and meaningful way, by (a) providing a template for clinical assessment with regards to disease activity, (b) giving a numeric score, which can be used to follow up disease activity, and (c) by standardizing data collection for research studies. These clinical assessment tools use clinical and laboratory data.

Instruments for assessing disease activity

The various disease activity assessment tools are given in table 1.

Table 1 Disease activity assessment tools in vasculitis

Sr. No.	Instrument
1.	Five Factor score (FFS)
2.	Disease extent index (DEI)
3.	Groningen Index
4.	Vasculitis activity index (VAI)
5.	Birmingham vasculitis activity index (BVAS)
6.	Indian Takayasu activity score (ITAS)

Five Factor Score

Loic Guillevin et al proposed a simple yet effective tool, factor five score (FFS), in 1988[21]. The five factors taken into account were proteinuria of > 1 g/day, renal insufficiency, gastrointestinal, cardiac and CNS involvement, with each component getting a score of one. This tool is a simple measure of disease activity that identified the involvement most likely to adversely affect outcome and thus, helps in prognostication. When FFS = 0, mortality at five years is 12%. When FFS = 1, mortality was 26% and when FFS was > 2, mortality was 46%. Although it has not been an established basis for treatment decisions, it seems logical to consider it during risk stratification. This score has been used for measuring disease activity in Polyarteritis nodosa and Churg Strauss syndrome.

Groningen Index

It takes into account the clinical and histological features for determining disease activity. This is an disease activity assessment tool specific for GPA but its requirement of biopsy limits its value in serial assessment of disease activity[22].

Disease Extent Index (DEI)

The disease extent index (DEI) is again a GPA specific tool[23]. This is an extension of old ELK (ear/nose/throat, lung, kidney) system for GPA which was first hinted at by Godman et al in 1954, but was established in 1976 by the Mayo group led by De Remee. DEI is based on a score given to active disease in various organ systems namely ENT and upper airways, inflammatory eye lesions, heart, lung and lower airways, kidney, gastrointestinal tract, peripheral nervous system, central nervous system, skin, arthralgia/arthritis and also constitutional symptoms. Each organ system involvement is given a score of 2 irrespective of one or more than one manifestation in that system. The constitutional symptoms are given a score of 1. So the maximum score is 21. It is important for the observer to note that all symptoms in a given organ system are due to active vasculitis and other causes like infection/malignancy and damage have been excluded. DEI was developed for long term outcome studies to include full extent of GPA (WG).

Vasculitis Acitivity Index (VAI)

It contains nine rating scales for the amount of disease activity in each organ system[24]. The laboratory abnormalities are also taken into account. This score was introduced by a group in Baltimore. There is a possibility of observer bias but it can be a useful tool in the hands of experienced observers.

Birmingham Vasculitis Activity Score (BVAS)

This is the most widely used tool to quantify disease activity in systemic vasculitis at this time. The original version (version 1) was developed by consensus expert opinion in 1994[25]. It consisted of 59 items grouped into nine organ systems. The BVAS was subsequently modified (version 2) for use in the European Vasculitis Study Group (EUVAS) trials and has been modified again, to its current version (version 3; BVAS v. 3)[26,27]. The main difference between BVAS v. 3 and v. 2 is that the persistent boxes for each variable in version 2 have been replaced by a single box for the whole form, which is only ticked if all the items are due to persistent disease. Though there has been a reduction in the number of items from 64 to 56, the overall maximum score has been maintained. The weighting of items in the original version has remained relatively unchanged between the three versions. The maximum score is 63. The BVAS v. 3, which was initially validated in a cohort of 313 patients with mixed primary and secondary vasculitis from the UK, has also been revalidated in other cohorts[28]. BVAS is an attempt to produce a tool with a strong emphasis on items determined clinically. A simplified disease specific version of BVAS was also developed for GPA. BVAS calculator can be accessed online at http://www.epsnetwork.co.uk/BVAS/bvas_flow.html.

Indian Takayasu Activity Score (ITAS)

As the standard tools BVAS and Vasculitis Damage Index (VDI) did not perform well for the different disease patterns in Takayasu arteritis (TA). MRI and PET are expensive and lack established correlation with active inflammation. The Indian Rheumatology Association's Vasculitis (IRAVAS) group devised and validated a clinical index to capture the activity and damage of TA[29,30,39]. Six domains are scored in the ITAS where new/worsening of disease activity is present in the last three months. These domains are systemic symptoms, abdominal, genitourinary, renal, neurological and cardiovascular.

Damage

These chronic multisystem diseases leave permanent residual damage/scar on clinical resolution. This damage may not necessarily be because of vasculitis *per se* but may also be due to side effects of drugs, surgical procedures or other concurrent events. It is important to understand that the scars/damage accumulated over a period of time are not going to heal so a damage index can only either remain static, or worsen, over a period of time.

Vasculitis Damage Index (VDI)

VDI is the most comprehensive damage index available and is the most widely used instrument for assessing damage[31]. It represents a list of 64 damage items grouped into 11 organ systems. The items are not weighed; damage has to be present for at least 3 months before it is scored to avoid scoring disease activity. Some damage items represent single event like myocardial infarction and stroke whereas others may represent chronic ongoing damage like persistent skin ulcers. VDI has been widely used in trials of vasculitis and has prognostic value. A 6 month VDI score ≥ 4 has been shown to be associated with increased mortality. The VDI compares favourably with other damage indices developed for systemic lupus erythematosus or reported in a comparison of patients with Polyarteritis nodosa and Churg Strauss syndrome. The regular use of VDI has shown that damage occurs early in the course of vasculitis[32]. Hence it is important to understand that in contrast to disease activity, which fluctuates especially in response to therapy, the damage never gets better.

Disease Extent Index-Takayasu (DEI.Tak)

For assessment of damage in TA, DEI.tak has been developed and validated by the IRAVAS group[30]. It provides a comprehensive damage index which confirmed the different patterns observed in TA in a large cross sectional study. It did not correlate with ESR and physicians global assessment. This has also been used for assessing disease activity in other populations[33].

Quality of life assessment

Current therapies have improved survival in patients with vasculitis and transformed these conditions into chronic disease states. However, there is still significant morbidity due to disease relapses and immunosuppressive treatment; and there is continuing suffering in spite of these patients being considered to be in remission due to a significant impairment in the quality of life. Improved survival without improvement in functional status is not a satisfactory outcome. There is no disease specific tool to assess quality of life in vasculitis. The clinical instrument used is the Medical Outcomes Study 36 item Short-form General Health Survey (SF-36)[34]. It can be summarized as mental health functioning and physical functioning into two specific numeric scores. It consists of 36 individual questions. Answers to 35 questions are used to score 8 subscales from 0-100, higher values are indicative of better health related quality of life. The eight subscales are physical function, role-physical, bodily pain, general health, vitality, social functioning, role-emotional, and mental health. The 36th item is a self reported health transition over the previous 1 year. It has been translated and validated into several other languages. Since it is of a generic nature, it might fail to capture items more specific for patients with systemic vasculitis. There is no significant correlation between the VDI and any of the SF 36 domains. A highly abnormal SF 36 with a normal or low BVAS would argue against use of aggressive immunosuppressive therapy.

Patient reported outcomes

Outcome assessment in vasculitis is based on items which the physician consider relevant, whereas the patients subjective experiences may represent key domains of illness that differ from clinicians views. In a study of 264 patients from three countries, fatigue (75%), followed by pain (31%), musculoskeletal symptoms (24%), difficult breathing (19%), financial aspects (13%), nasal discharge/crusting (14%) were the most common items reported from the free text sections. Severe manifestations like dialysis, seizures and oxygen dependency were ranked lower[35]. This study suggested that there are various manifestations, which are important to the patients but are not measured in the present outcome measures used in various clinical trials. This highlights the need for development and validation of patient reported outcome instrument in vasculitis.

Summary

The primary systemic vasculitides are a group of rare disorders, which have now become more chronic, manageable diseases due to advancement in treatment options[36]. But these treatment options are not without side effects/risks. Hence the proper assessment of vasculitides has become an integral part of management as differentiating damage from activity can save the patient from unnecessary immunosuppressive therapy. As there are no validated biomarkers to assist treatment decisions, clinical decisions have to be taken on the basis of detailed history and clinical examination and supported by laboratory and imaging investigations.

BVAS (version 3) is the clinical tool of choice for assessment of disease activity. ITAS is a validated tool for assessing disease activity in TA. VDI is the most commonly used tool for assessment of damage in vasculitis and also has a role in assessing prognosis. There is no disease specific tool for assessment of quality of life in patients with vasculitis. The studies on patient reported outcomes have highlighted the need for development and validation of patient reported outcome instrument in vasculitis.

A Japanese VAS has also been developed and in a study was found to be simpler and more reliable than BVAS in predicting death in MPO-ANCA vasculitis[37].

A pediatric vasculitis activity score has also been developed for assessing activity in children with vasculitides[38].

Further Reading:

1. Bambery P, Sakhuja V, Gupta A, et al. Wegener's granulomatosis in north India. An analysis of eleven patients. Rheumatol Int 1987; 7: 243-47.
2. Bambery P, Sakhuja V, Bhusnurmath SR, et al. Wegener's Granulomatosis: Clinical experience with eighteen patients. J Assosc Physicians India 1992; 40: 597-600.
3. Malaviya AN, Kumar A, Singh YN, et al. Wegener's granulomatosis in India: not so rare. Br J Rheumatol. 1990; 29: 499-500.
4. Kumar A, Pandhi A, Menon A, et al. Wegener's granulomatosis in India: clinical features, treatment and outcome of twenty-five patients. Ind J Chest Dis Allied Sci 2001; 43, 197-204.
5. Aasarod K, Bostad L, Hammerstrom J, et al. Renal histopathology and clinical course in 94 patients with Wegener's granulomatosis. Nephrol Dial Transplant 2001; 16: 953-60.
6. Devaney KO, Travis WD, Hoffman G, et al. Interpretation of head and neck biopsies in Wegener's granulomatosis. A pathologic study of 126 biopsies in 70 patients. Am J Surg Pathol 1990; 14: 555-64.
7. Bambery P, Kataria S, Sakhuja V, et al. Wegener's granulomatosis in North India. Radiological manifestations in eleven patients. Acta Radiol 1988; 29: 11-13.
8. Bambery P, Sakhuja V, Behera D, et al. Pleural effusions in Wegener's granulomatosis: report of five patients and a brief review of the literature. Scand J Rheumatol 1991; 20: 445-47.
9. Lohrmann C, Uhl M, Kotter E, et al. Pulmonary manifestations of Wegener granulomatosis: CT findings in 57 patients and a review of the literature. Eur J Radiol 2005; 53: 471-77.
10. Andrews J, Al-Nahhas A, Pennell D, et al. Non-invasive imaging in the diagnosis and management of Takayasu's arteritis. Ann Rheum Dis 2004; 63: 995-1000.
11. Schmidt WA, Gromnica E. Incidence of temporal arteritis in patients with polymyalgia rheumatica: a prospective study using colour doppler ultrasonography of the temporal arteries. Rheumatology (Oxford) 2002; 41: 46-52.
12. Schmidt WA, Blockmans D. Use of ultrasonography and positron emission tomography in the diagnosis and assessment of large-vessel vasculitis. Curr Opin Rheumatol 2005; 17: 9-15.
13. Blockmans D, Stroobants S, Maes A, et al. Positron emission tomography in giant cell arteritis and polymyalgia rheumatica: evidence for inflammation of the aortic arch. Am J Med 2000; 108, 246-49.
14. Kobayashi Y, Ishii K, Oda K, et al. Aortic wall inflammation due to Takayasu arteritis imaged with 18F-FDG PET coregistered with enhanced CT. J Nucl Med 2005; 46: 917-22.
15. Walter M A, Melzar RA, Schindler C, et al. The value of [18F] FDG-PET in the diagnosis of large-vessel vasculitis and the assessment of activity and extent of disease. Eur J Nucl Med Mol Imaging 2005; 32: 674-81.
16. Bambery P, Chugh KS. Anti-neutrophil cytoplasmic antibodies (ANCA) in Wegener's Granulomatosis. Int J Artif Organs 1989; 12: 68.
17. van der Woude F J, Rasmussen N, Lobatto S, et al. Autoantibodies against neutrophils and monocytes: tool for diagnosis and marker of disease activity in Wegener's granulomatosis. Lancet 1985; 1: 425-29.
18. Davies DJ, Moran JE, Niall J, et al. Segmental necrotising glomerulo nephritis with antineutrophil antibody: possible arbovirus etiology. Br Med J (Clin Res Ed) 1982; 285: 606.
19. Savige J, Paspaliaris B, Silvestrini R, et al. A review of immunofluorescent pattern associated with anti neutrophil cytoplasmic antibodies (ANCA) and their differentiation from other autoantibodies. J Clin Pathol 1998; 51: 568-75.
20. Boosma MM, Stegeman CA, van der Leij MJ, et al. Prediction of relapse in Wegener's Granulomatosis by measurement of anti neutrophil cytoplasmic antibody levels: a prospective study. Arthritis Rheum 2000; 43: 2025-33.
21. Guillevin L, Lhote F, Gayraud M, et al. Prognostic factors in polyarteritis nodosa and Churg-Strauss syndrome. A prospective study in 342 patients, Medicine (Baltimore) 1996; 75: 17-28.
22. Kallenberg CG, Tervaert J W, Stageman CA, et al. Criteria for disease activity in Wegener's granulomatosis: a requirement for longitudinal studies. APMIS Suppl 1990; 19: 37-39.
23. de Groot K, Gross WL, Herlyn K, Reinfold-Keller E, et al. Development and validation of disease extent index for Wegener's granulomatosis. Clin Nephrol 2001; 55: 31-38.
24. Whiting-O'Keefe QE, Stone JH, Hellman DB, et al Validity of vasculitis activity index for systemic necrotizing vasculitis. Arthritis Rheum 1999; 42: 2365-71.
25. Luqmani RA, Bacon PA Moots RJ, et al. Birmingham vasculitis activity score (BVAS) in systemic necrotizing vasculitis. Quart J Med 1994; 87: 671-78.
26. Luqmani RA, Exley AR, Kitas GD, et al. Disease assessment and management of the vasculitides. Baillieres Clin Rheumatol 1997; 11: 423-46.
27. Mukhtyar C, Lee R, Brown D et al. Modification and validation of the Birmingham vasculitis activity score (version 3). Ann Rheum Dis 2009; 68: 1827-32.
28. Suppiah R, Mukhtyar C, Flossman O, et al. A cross-sectional study of the Birmingham Vasculitis Activity Score version 3 in systemic vasculitis. Rheumatology (Oxford) 2011; 50: 899-905.
29. Sivakumar MR, Misra RN, Bacon PA , et al. The Indian perspective of Takayasu arteritis and development of a disease extent index (DEI.TAK) to assess Takayasu arteritis. Rheumatology 2005; 44 (Suppl. 3), iii6-iii7.
30. Misra R, Dhanda D, Jayaseelan L, et al. ITAS and DEI.TAK-scores for clinical disease activity and damage extent in Takayasu aorto-arteritis. Rheumatology 2008; 47: (Suppl. 2), ii101.
31. Exley AR , Bacon PA, Luqmani R, et al. Development and initial validation of the vasculitis damage index (VDI) for standardized clinical assessment of damage in systemic vasculitides. Arthritis Rheum 1997; 40: 371-80.
32. Exley A, Carruthers DM, Luqmani RA, el al. Damage occurs early in systemic vascultis and is an index of outcome. Quart J Med 1997; 90: 391-99.

33. Aydin SZ, Yilmaz N, Akar S, et al. Assessment of disease activity and progression in Takayasu's arteritis with Disease Extent Index-Takayasu. Rheumatology (Oxford) 2010; 49: 1889-93.

34. Ware JE, Scherbourne CD. The MOS 36-item short-form health survey (SF-36). 1. Conceptual framework and item selection. Med Care 1992; 30: 473-83.

35. Herlyn K, Hellmich B, Seo P, et al. Patient reported outcome assessment in vasculitis may provide important data and a unique perspective. Arthritis Care Res 2010; 62: 1639-45.

36. Sharma A, Kumar S, Wanchu A, et al. Successful treatment of hypertrophic pachymeningitis in refractory Wegener's granulomatosis with rituximab. Clin Rheumatol. 2010; 29: 107-10.

37. Koike K. Fukami K. Yonemoto K. et al. A New Vasculitis Activity Score for Predicting Death in Myeloperoxidase-Antineutrophil Cytoplasmic Antibody-Associated Vasculitis Patients. Am J Nephrol 2012; 35(1): 1-6.

38. Demirkaya E, Luqmani R, Ayaz NA, et al. FMF Arthritis Vasculitis and Orphan Disease Research in Paediatric Rheumatology (FAVOR). Pediatr Rheumatol Online J. 2011 Sep 26; 9(1): 29.

39. Ramnath Misra, Debashish Danda, Sivakumar Rajappa, et al on behalf of the Indian Rheumatology Vasculitis (IRAVAS) group. Development and initial validation of the Indian Takayasu Clinical Activity Score (ITAS2010) Rheumatology 2013; 52 (Oct): 1795-1801 .doi:10.1093/rheumatology/ket128.

SECTION 3

Epidemiology of Rheumatic Diseases in India

EPIDEMIOLOGY OF RHEUMATIC DISEASES IN INDIA 1

Arvind Chopra

Population and community based epidemiological surveys are best suited to provide estimate of disease burden. Logistically and socio-economically, the surveys are rather challenging. The sample size is statistically derived so as to capture the true extent of burden in the population. Though cumbersome, randomized methods are preferred for better representation. However, a well planned executed convenience sample survey may suffice. The success for a good survey depends upon the devotion and expertise of the investigators and honest and complete filing of data by the grass root paramedic field workers. Meticulous pre-survey planning is required to ensure survey completion in a systematic and timely fashion. While maintaining high scientific standards, the surveys must remain feasible and goal oriented.

Good epidemiologic data is hard to find in India for several reasons ranging from poor incentives to difficult logistics and socioeconomics. In our setting, surveys are often limited by shoe-string budgets. Also, a single survey is not enough to account for the intense diversity in India. Several surveys in time and place and using uniform standard protocols would be required to provide credible data for estimating national disease burden. However, the situation is considerably improved for musculoskeletal (MSK) and rheumatic disorders in India. In the last two decades or so, several surveys were carried out in different regions[1,2,3,4,5,6,7,8].

The maiden WHO ILAR (International League of Associations for Rheumatology) COPCORD (community oriented program for control of rheumatic diseases) India was launched in village Bhigwan (district Pune) in 1996[4,5,7,9]. COPCORD was conjointly launched by WHO and ILAR in 1981 to fill the void created by lack of MSK data from the developing countries[10]. Over 21 countries have completed COPCORD and in several instances with multiple surveys. Table 1 shows the frequency/ prevalence of selected rheumatic musculoskeletal rheumatic disorders in some well designed Asian-Pacific COPCORD surveys[2,11,12,13,14,15,16]. Detail descriptions of COPCORD are available on a recently launched website[2]. The Bhigwan survey achieved a unique distinction in evolving a fast tract COPCORD model[2,7,17] that was adopted by several countries and sites in India (Fig. 1).

Two non-randomized surveys, the Delhi study[3] and the Pune-Bhigwan COPCORD[4,7] have been frequently resourced to describe the Indian scenario. The results of these surveys were not comparable because of significant methodology differences. The screening in the Delhi study targeted 'painful swollen joints' while in COPCORD the question was essentially 'pain in joints and/or MSK soft tissues (current or past)'. The Delhi survey, conducted by Malaviya and colleagues from AIIMS, was primarily meant to collect data on SLE/lupus and Rheumatoid Arthritis (RA). The actual survey covered a near 40,000 semi-urban/rural population from the fringe villages of Delhi. For lupus study[19], the Delhi workers in addition included a larger number of blood samples from an ongoing malaria program in Delhi. Though point prevalence for several rheumatic disorders was published from the Delhi survey3, important limitations were; lack of appropriate screening methods and an unduly prolonged survey period (of three years).

In 1996, Chopra and colleagues completed the WHO-ILAR Bhigwan COPCORD rural survey (district Pune) in 5 weeks time. About 7000 villagers (children included) were screened. Overall, the Bhigwan COPCORD study[5,7,9] unequivocally demonstrated for the first time that (1) MSK rheumatic disorders were the predominant community health problem (2) the community suffered (Fig. 2) mostly from soft tissue rheumatism, ill-defined MSK symptoms, and degenerative arthritis (3) and that about 10% cases suffered from some form of inflammatory arthritis (including RA and seronegative spondyloarthritis). The Bhigwan survey was continued as a prospectively designed longitudinal observational follow up study for over

Fig. 1 : COPCORD POPULATION SURVEY SITES (Community Oriented Program for Control of Rheumatic Diseases) 1996 - 2011: Population sample size shown in parenthesis on the map

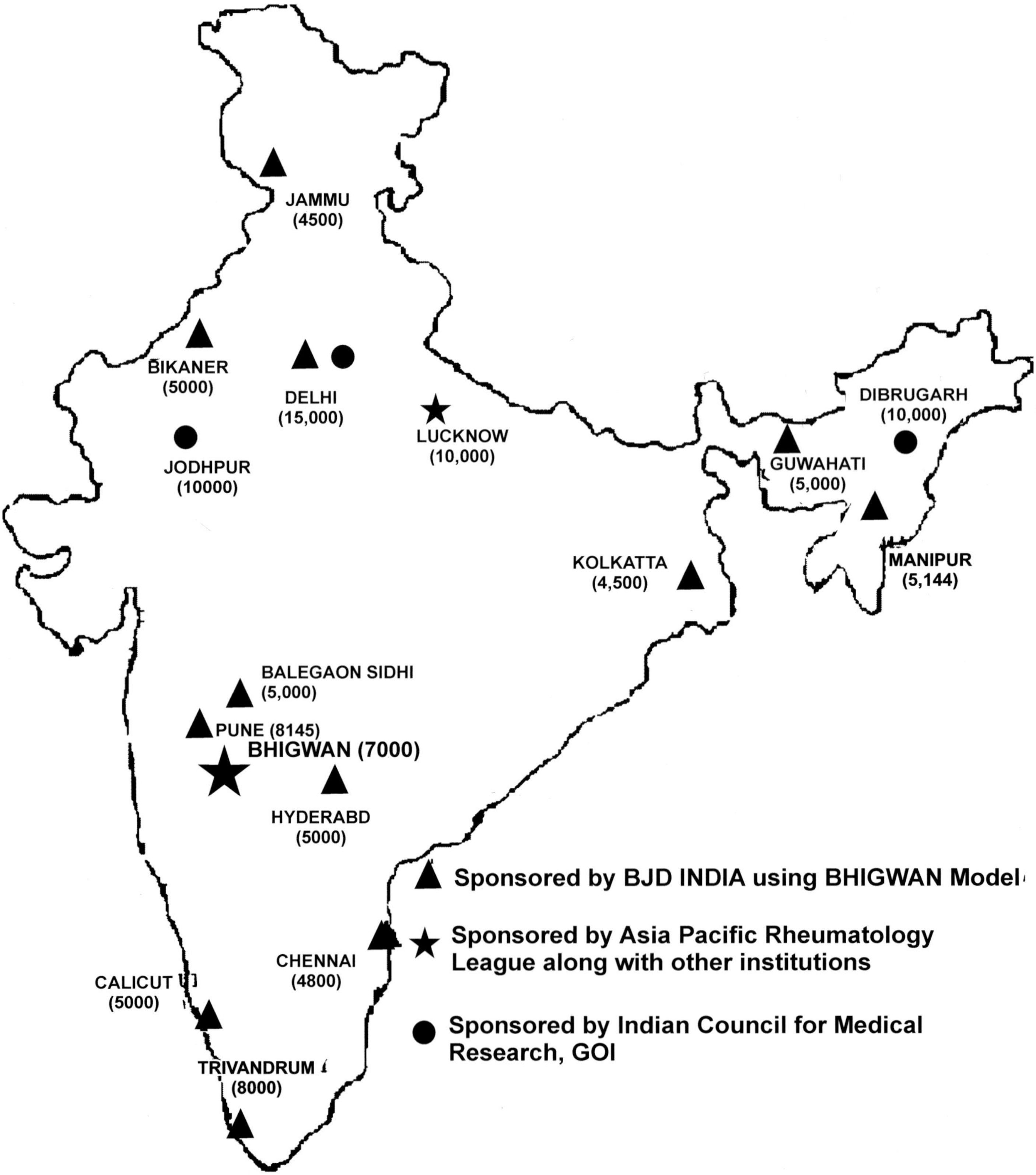

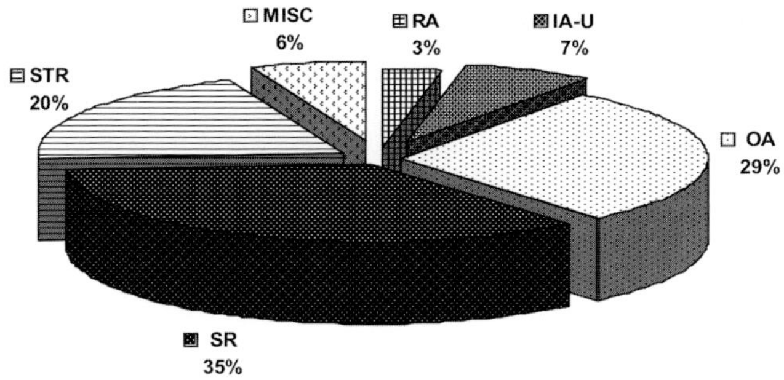

Fig. 2 : RA: Rheumatoid Arthritis; OA: Osteoarthritis; IA-U: Undifferentiated Inflammatory Arthritis; STR: Soft Tissue Rheumatism; SR: Ill Defined Symptoms; Misc: Miscellaneous

17 years till date. During the latter period, the investigators provided free rheumatologic services and health education to the villagers and further carried out evaluation (including case-control nested cohort) to determine disease profile, therapy response, incidence cases and risk factors[12,18,20,21,22]. The Bhigwan data was also used by the WHO to project the burden of illness in the South East Asia region[23]. The Bhigwan program is the longest running COPCORD in the World[18].

Following the Bhigwan survey, Das and colleagues completed a nearly 10,000 population COPCORD survey (divided into urban and rural components) in the Lucknow region and detail report is not yet published[6,24].

Using COPCORD Bhigwan model, the Bone and Joint Decade India (BJD) and Indian Council of Medical Research (ICMR) launched population surveys[8,25] at several sites (Fig. 1) during the period 2006-2011 to study about 90,000 population, both urban and rural. Preliminary results were presented in the annual meeting of the American College of Rheumatology 2012; almost one fifth of the Indian population suffered from some MSK pain and disorder and millions suffered from serious rheumatic diseases; about 5 million patients of RA was estimated.

COPCORD was not designed to identify and estimate uncommon disorders like lupus which have been essentially described in the India scenario by case series (see later). Traditionally medical camps are popular in our setting and have contributed to MSK and rheumatic disease data[27,28] but are grossly limited from an epidemiological view point.

Soft Tissue Rheumatism (STR)

STR encompasses a broad range of diverse conditions with different causal anatomical (tendon, fascia, bursa) sites and patho-physiological mechanisms (? Inflammatory ? mechanical and traumatic &/or ? neurovascular). It is an extremely common cause of pain in the community and can sometimes cause disability and absence from work as demonstrated in Bhigwan survey[7,9]. In Bhigwan survey, 14% non specific arthralgias (NSA) and 11% STR cases also suffered from a moderately severe disability as assessed by a modified HAQ (health assessment questionnaire) validated for Indian use; correspondingly there were 37% rheumatoid arthritis (RA) and 23% osteoarthritis (OA) cases[29].

Fibromyalgia is probably a variant with more widespread and diffuse musculoskeletal pain, well defined tender points, and associated with undue fatigue and sleep disturbance. The latter is generally considered to be a psycho functional form of STR. However, several rural patients from the Bhigwan COPCORD with soft tissue and non-specific articular pains, in the absence of any obvious functional overlay, fulfilled the ACR criteria of fibromyalgia[30]. The latter were considered to be due to occupation, posture and work environment (the so-called occupational overuse syndrome or cumulative tissue trauma) rather than a psyche related cause. In the Lucknow Survey 3.8% and 4.5% of the urban and rural population respectively were found to suffer from fibromyalgia (ACR criteria)[24].

Regional forms of STR (fascitis, bursitis, enthesitis, tendonitis, epicondylitis) are commonly encountered in clinical practice. Foot and heel pains, due to local causes, were common in rural surveys[9]. Painful tender swollen tendoachilles may be a forerunner of SSA and needs to be carefully evaluated. Shoulder, knee, neck and back are common site for regional STR and NSA in COPCORD surveys and if not carefully evaluated may be misdiagnosed as inflammatory rheumatism and/or arthritis. Several cases of NSA and STR in COPCORD surveys were considered to be due to occupational overuse and trauma in a socioeconomically challenged community[7,9].

Osteoarthritis (OA)

Osteoarthritis is the second most common rheumatological problem in the community (Fig. 2) and a leading cause of disability and poor quality of life beyond the fifth decade of life. OA commonly affects the knees and spine. Nodal and erosive forms of generalized OA are also seen. It has been often observed that OA hip is uncommon in the Indian population. Almost 15-20% of population, equally distributed between sexes, in the age range 60-80 years, were affected by OA knees in the Bhigwan Survey (Table 2) 23 and the pattern was similar in urban and rural communities[7].

Rheumatoid arthritis (RA)

The worldwide prevalence of clinical RA is believed to be about 1% but this is supported by few studies Worldwide[12].

The prevalence varied considerably amongst the several population surveys in the recently completed BJD India study but a pooled age sex adjusted (India census 2001) point prevalence for RA was reported 0.3% (95% confidence interval 0.08, 0.79) and is likely to be a robust estimate for India[26].

Interestingly, the Delhi semi urban/rural study (1994)[3] reported a much higher prevalence of 0.75 % and the patients had allegedly fulfilled the American College of Rheumatology (ACR) 1987 criteria. An urban population study from West Bengal (1997)[31], using an index case linked cluster survey model, reported a crude unadjusted clinical prevalence of 0.3%. The COPCORD Bhigwan (1996)[5,7] reported a clinical prevalence of 0.7% but on applying the ACR 1987 criteria and standardization (age sex adjustment), the rate was reduced to 0.55%. The neighboring urban COPCORD Pune (2004)[7] reported an adjusted prevalence (ACR) of 0.3%. The prevalence of RA in women aged 30-44 years in COPCORD Bhigwan was 1639/100,000 (Table 3), which was several times higher than that reported from other countries (300 in Norway and USA, 237 in China, 275 in Indonesia, and 1042 in West Africa; all rates are per 100,000) in a recent WHO Technical Research Series[23],

After a ten year follow up in village Bhigwan (assuming geometric growth pattern; adult population about 4545), the period prevalence and incidence of RA was 1.17% and 0.044% (44/100,000) respectively[12].

The natural history of RA in Indian patients[32,33,34] was found to be generally similar to that reported from Caucasian populations. However, some differences were reported. In Indian patients, the disease was (i) an earlier onset in younger women[23] (ii) with fewer extra-articular features, including subcutaneous nodules (less than 10%) (iii) with a lower seropositivity for rheumatoid factor (RF) at 55-80%.

It is the author's contention[35], that Indian patients suffer disproportionately from early severe crippling affection of the knee joints which is often made worse by the persistence to follow the Indian habit of squatting and sitting cross legged on the floor. The rheumatoid knee (and RA) is further worsened by the lack of basic toilet facilities in several Indian homes. The suffering is intense when a young village woman has to relieve herself somewhere in the open fields between sunset and dawn. There seems to be a difference in the MSK pain and disorders between the urban and rural settings[7,35]. Overall, RA is likely to be more severe and functionally disabling in a rural setting.

82% patients of RA in the Delhi study 3, were seropositive rheumatoid factor (RF) which was surprisingly similar to that reported from hospital-based Indian series[32,34]. Seropositive RF is generally considered to be much less in community RA. Amongst clinical RA patients, 45% patients in rural COPCORD Bhigwan, 62% patients in urban COPCORD Pune and 57% patients in urban community referral practice (CRD, Pune) were seropositive for RF[7,35]. 100% RA patients in COPCORD Pune and 60% RA patients in COPCORD Bhigwan were seropositive for anti-CCP (Cyclic Citrullinated Peptide) antibodies, second generation ELISA)[7,35]. 54% of the clinical RA patients in COPCORD Bhigwan demonstrated radiological erosions[35].

HLA DRB1* alleles are considered to be an important factor in the aetiopathogenesis of RA and may determine both the susceptibility and severity of disease[36]. The association of HLA DR B1* alleles (in particular *0401, *0404) or the shared epitope with RA, though largely documented in Caucasian population studies, is well established in reports from several parts of the World[36,37] and this has also been supported by hospital based studies in India[32,38]. However, the rural COPCORD Bhigwan[20] reported a modest association of HLA DRB1*1001 (odds ratio 1.85) with RA; none was seen with *0404 and *0401. Subsequently, in a better designed larger sample study[39,40] of immunogenetics in patients (both survey and incident cases) suffering from RA and undifferentiated Inflammatory Arthritis (IA) identified in the COPCORD Pune Bhigwan region (rural and urban), a significant association of RA was shown for the shared epitope (including only HLA DRB1*1001); other alleles significantly associated were HLA DQA1*0103, DQB1*0303, DPB1*0201 and DP B1*0401. None of the haplotypes analysed showed significant association with RA in the latter study.

Tobacco use and smoking in particular is significantly associated with RA[41] and other IA and adds to the premature risk of atherosclerosis and coronary artery disease[42]. COPCORD Bhigwan rural survey[9] showed a rampant use of oral tobacco (called 'misri' in regional Marathi language) and showed a significant association (odds ratio 1.71

univariate analysis, 95% CI 1.44, 2.04) with rheumatic MSK pain and this was further endorsed by urban COPCORD Pune[7] (odds ratio 3.16 univariate analysis, 95% CI 2.88, 4.05) and rural COPCORD Ralegaon[43]; village Ralegaon is about 70 kms east of Pune. However, a significant association between tobacco use and RA could not be demonstrated in the Pune region study[43] because of a relatively smaller number of RA patients. Being an extremely important public health preventive measure, tobacco use in all its forms, must be thoroughly condemned and stopped in patients suffering from RA[35].

The use of steroids (and analgesics-NSIAD) is rampant and often unsupervised and surreptitious. Often these drugs are sold mixed with herbal formulations under the false premise of Ayurvedic drugs by unscrupulous elements and quacks. Patients often do not reveal their medications. All this makes the recognition of early RA and inflammatory arthritis difficult as the signs of inflammatory arthritis get masked, and this was the experience of the author in several COPCORD community surveys[12,18].

RA is now a recognized cause of premature atherosclerosis and coronary artery disease and related morbidity and mortality[42]. The author has recently reported mortality in RA[35] in a cross sectional analysis of a case series of 60 fatalities (cases selected in no particular order). Infections (23%) and cardiac disorders (40%) were the leading causes of death; other causes classified were 10% respiratory, 9% hepatic, 8% stroke and 10% miscellaneous. Al least 30% deaths could be due to events associated with ischemic heart disease/ myocardial infarction. Multi-organ failure and hemodynamic shock was the terminal event in patients dying of septicemia. Microbial culture reports of staphylococci and gram negative bacilli were available in 11 cases of septicemia (4 pneumonia, 1 total knee replacement, 1 dental surgery root canal, 1 pancreatitis, 1 fracture neck of femur with bed sores, 3 peripheral limb gangrene). Methotrexate associated severe hepatic dysfunction and pancytopenia was recorded in 3 cases (two males) of septicemia; none were diabetic or immunosuppressed due to any other cause. Disseminated intravascular coagulation complicated the terminal event in one patient of septicemia. Other clinically important contributory factors recorded in this cohort were : 5 diabetes, 5 hypertension, 1 cerebrovascular disease, 1 coronary artery bypass surgery, 2 gastrointestinal bleed, 2 peptic ulcer (probably non steroidal anti-inflammatory drugs related), and 1 chronic renal failure. Though details of medication in the 3 month period prior to death could not be confirmed in the large majority, at least half of the cases were on long term methotrexate (generally 15 mg weekly and oral steroids (5-10 mg daily).

Spondyloarthritis (SpA/SSA)

The commonest form of SSA[43,44,45] reported in Indian patients is the undifferentiated type characterized by a dominant asymmetric lower limb oligoarthritis and often accompanied with enthesitis. The latter are probably forme-fruste of the classical Reiter's syndrome and are believed to be triggered by a wide array of gut and genitourinary related infections[45].

The prevalence of AS (ankylosing spondylitis) in the Caucasian population is 0.1%. A large hospital based population study from South India[46] reported a prevalence of 0.6% AS with HLA B27 positive in 87% patients. In a similar hospital based study of AS from North India[47], HLA B27 was positive in 94% patients. The profile of AS and other SSA was similar to that seen in the Caucasian populations and association with HLA B27 varied in different SSA subsets at different investigation sites (Table 4)[48].

Psoriatic arthritis is not too uncommon in the hospital setting. The Pune-Bhigwan and the Delhi survey did not report a single case (see above). During the ten year follow-up of about 5000 adult population, the author has diagnosed two cases (both males) of psoriatic arthritis in Bhigwan (unpublished). However, a South Indian retrospective study[49] reported a hospital prevalence of 0.83% for psoriatic arthritis amongst the 16,293 rheumatology referral cases seen during the period 1981-94. The latter reported asymmetrical oligoarthritis (63%), symmetrical polyarthritis (32%), dominant distalinterphalangeal (DIP) joint arthritis (25%) and predominant spondyloarthropathy (15%). Amongst these patients, the skin lesions preceded arthritis in over 60% patients, the skin and joints were simultaneously affected in 15% patients, and arthritis preceded the skin lesion in 19% patients, respectively; almost 50% of the patients had some form of nail changes that were often associated with DIP joint diseases.

Systemic connective tissue disorders (CTD) and vasculitis

The extent of CTD in the Indian population is still unclear though many believe it to be uncommon. Twenty one cases of CTD were diagnosed amongst

68273 inpatients (civilians & services) during 1982-1986 in a large Armed Forces referral-teaching hospital in Pune which gave an incidence of 0.03% in hospitalized patients[50]. Numerous hospital based series (see below) of systemic lupus erythematosus (SLE)[40,41], scleroderma, mixed connective tissue disease, vasculitis are published and the descriptions are generally similar to the Western profiles.

The Delhi study[19] reported a point prevalence of 3 per 100,000 for SLE. This was much lower than that reported from the west (varying from 12.5 per 100,000 adults in England to 124 per 100,000 in USA) or even that observed in China (Table 1). The Pune-Bhigwan COPCORD follow-up study found a crude incidence rate for SLE of 1 per 25,000 person years i.e. 4 per 100,000 populations per year. A large amount of referral non-standardized rheumatology/ hospital based data on SLE collected from different regions has been recently published in the IRA (Indian Rheumatology Association) guidelines on SLE[51,52]. 63% and 32% patients of SLE from North[51] and South India[53] respectively had neuropsychiatric SLE. In another hospital based inpatient rheumatology emergency case series, 65% of the SLE patients, often in the age group 20-30 years, were admitted with a neurological complication[54].

All forms of Vasculitides[55,56,57,58,59,60] have been described but the most frequent form is systemic necrotizing vasculitis that appears to be an overlap of collagen vascular diseases or an unclassifiable undifferentiated multisystemic variety. 100 patients (1.3%) were found to suffer from vasculitis in a retrospective study[57] of 7700 patients attending a rheumatology referral outpatient during the period 1979-1994 in a large referral hospital in Mumbai. In the latter, the majority was small vessel affection (Henoch Schonlein purpura being the commonest) followed by medium vessel disease (polyarteritis nodosa followed by SLE) and large vessel disease (5 cases; aorto-arteritis in 4 and temporal arteritis in 1) but interestingly not a single case of Wagener's Granulomatosus (WG) was reported. This study also reported vasculitis in 0.9% of RA cases which was almost similar to that reported from hospital studies in Delhi[32] and Chennai[56].

The diagnosis of WG was been facilitated by the advent of ANCA auto-antibody and per bronchoscopic biopsy of pulmonary tissue. And several cases treated for pulmonary TB were actually found to be that of WG[59,60]. Uncommonly, polyarteritis nodosa and Churg-Straus syndrome (probably more frequent in chest disease outpatient and the diagnosis is missed) have been reported. Recently, vasculitis leading to avascular necrosis in the joints was reported[61].

Takayasu's arteritis or aorto-arteritis has been uncommonly reported[62,63] from all regions and unlike elsewhere in the World it is an important cause of renovascular hypertension in India. The tubercular etiology continues to be debated. 5 patients were diagnosed as Takayasu's arteritis out of a retrospective case series of 8600 rheumatology patients seen during the period 1992-96 in a large referral hospital in South India[62].

A case series[64] of 50 patients attending rheumatology outpatients in AIIMS, Delhi were diagnosed as Behcet's disease during the period 1978-2002. The disease, at least in its complete form of oculomucocutaneous ulcers and neurovascular complications, is believed to be uncommon in India though HLA B5, a marker of Behcet's disease, is reported in almost 30% Indians.

Sjogren's syndrome

In a retrospective analysis of 24,500 cases seen over 20 years, only 3 cases were found to have primary Sjogren's syndrome[65]; 9 patients (RA in 8, SLE in 1) met the criteria for secondary Sjogren's syndrome. 26 patients (21 women) of primary Sjogren's syndrome were reported over a 10 year period from a large tertiary rheumatology referral center in Lucknow[66] and the authors concluded that it was an uncommon disease; presenting symptoms included dry eyes, dry mouth, and arthritis/ arthralgias and extra-glandular manifestations were glomerulonephritis, vasculitis, renal tubular acidosis and peripheral neuropathy. The important laboratory abnormalities in the Lucknow cohort were hypergammaglobulinaemia, antinuclear antibodies, anti-La and anti-Ro and in 60% cases a definitive diagnosis was provided by a minor salivary gland biopsy. Danda and coworkers presented a large case series from the rheumatology department of Christian Medical College, Vellore, in the annual 2012 meeting of the Indian Rheumatology Association (unpublished) and reported protean manifestations and further concluded that the prevalence was much higher and that the diagnosis was probably missed in the Indian setting. In the author's opinion, secondary sicca in rheumatology outpatients is not uncommon but needs to be searched and is probably missed in population studies because of lack of awareness and poor facilities in field conditions for detail eye and oral examination and definitive investigation tests.

Hyperuricemia and Gout

It is generally stated that gout is probably less common in the Indians (Table 1 & 2) as compared to the Caucasians. In the ongoing Pune-Bhigwan COPCORD rural population (about 5000 adults) study (see above), the author observed classical acute gout episodes in only 4 patients during the 10 year follow-up period (unpublished). In a retrospective study, 101 cases (95 males) of acute gout were diagnosed out of 20,000 patients attending rheumatology outpatients in a Government hospital in Chennai during the period 1989-2001[67]. Similar series were reported from Delhi[68] and Kolkata[69]. Overall the pattern of Indian gout was similar to that seen in the Caucasians with perhaps a lesser prevalence of classical first metatarsophalangeal affection (35-75%; maximum reported from Chennai and least reported from Kolkata versus about 90% in Caucasians) and tophi (5-27%; maximum reported from Delhi and least reported from Kolkata versus about 35% in Caucasians). Almost 20% patients in the Chennai study had polyarticular onset. 20% patients in the Delhi study had renal calculi as compared to about 10% found amongst the Caucasians suffering from gout. All forms of gout, ranging from the classical monoarticular podagra to the polyarticular tophaceous gout were described. Interestingly, a large number of patients in the Delhi study had auricular tophi, which is uncommon in the experience of several Indian workers including the current author (unpublished). The majority of Indian patients, as reported by the above-mentioned studies, were under-excretors of uric acid. In the author's observation, asymptomatic hyperuricemia is over diagnosed and unnecessarily treated. We do see patients of gout who are strict vegetarians and have never consumed alcohol, and further, hail from a family lineage of vegetarians.

Infections

Chandrasekaran reported[70] an extensive experience regarding infections in patients attending a rheumatology clinic in a Government hospital in Chennai during the period 1976-92 (i) Septic arthritis, usually due to staphylococcus aureus, was diagnosed in 9 out of 24,500 patients (ii) In almost 1.7% patients with post diarrhea asymmetrical polyarthritis and lumbago, salmonella typhi (not reported in the Western literature) and salmonella typhimurium were ascertained to be the causative microbes; HLA-B27 was positive in 72% of these patients (iii) Shigella shigae was responsible for at least 14% of the cases with acute polyarthritis following diarrhea; HLA-B27 was positive in 86% patients (iv) almost 1.9% patients with polyarthritis had significant Brucella titer; fever, hepato-splenomegaly and lymphadenopathy was reported in all (v) almost 5% patients had a significant VDRL titer and features to suggest secondary syphilis (vi) almost a quarter of patients with monoarticular affection of the knee had tubercular etiology (vii) Poncet's disease, a form of acute painful arthritis and considered to be a hypersensitivity phenomenon to tubercular infection elsewhere, was diagnosed in 0.23% patients.

Several reports on infection and arthritis were recently published in a special supplement issue of the Indian Journal of Rheumatology[71].

In a retrospective analysis[72] of 1,900 in patients over 2 years in a medical college in Chennai, 0.2% had septic arthritis, 0.68% had tuberculosis, 0.95% had reactive arthritis and 0.57% had post viral arthritis. In a similar inpatient study[73] from Hyderabad, almost one fifth of all admissions (1,236 cases) in a rheumatology ward had some discernible infection; 84 had SLE (E. coli was the commonest isolate) and 72 suffered from RA and about one third of patients had lower respiratory tract infection. Almost a quarter of all rheumatologic cases in the Hyderabad study were complicated by a TB infection. In a report from Lucknow[74], common infections in rheumatology practice were septic arthritis, tropical pyomyositis, tuberculosis and leprosy associated arthritis; S. aureus caused majority of the pyogenic rheumatic infections. MSK infections pose challenging diagnosis and community acquired methicillin resistant staphylococcus aureus (with activated panton valentine leucocidine gene) seems to be an important emerging pathogen[75].

Usually less than one-tenth of all patients of TB in developing countries develop osteoarticular TB[76]. Case series of TB arthritis and rheumatism were reported from Pune[77] and Kolkata[78] which included several atypical forms of articular and spinal TB that were identified early due to excellent imaging modalities (MRI) and newer mycobacterium detection methods (especially PCR) in recent times. In a retrospective cross sectional analysis of 23,651 patient records in central database in Pune from 2005-2009, TB associated arthritis was found in 0.35% cases; 0.4% cases of RA were complicated by concurrent TB. In this report[77], 12 of the 16 cases with tubercular MSK infection were considered to have Poncet's disease, a hypersensitivity form of TB; 2 cases showed multifocal lesions which also included non contiguous spine involvement.

Leprous arthritis, usually a component of the lepra reaction, and mimicking seronegative RA or asymmetrical oligoarthritis, is sometimes encountered in rheumatology practice and needs a high index of clinical suspicion[79]. Several Indian workers have reported leprosy associated rheumatic complications[80,81]. Uncommonly, leprosy can present with a fairly acute onset and progressive rheumatic phenotype that can be mistaken for RA or SSA[82]; RF and other autoantibodies may be false positive. Uncommonly, leprosy may complicate patients with RA on long term immunosuppressive therapy or be a manifestation of an immune reconstitution syndrome.

HIV is reported to present a varied spectrum of rheumatic syndromes - from reactive arthritis to polymyositis like CTD profile. A hospital based case series from Mumbai[83] reported that while 4% of HIV cases suffered from some form of rheumatism (commonly bone pains and arthralgias), almost 5% of rheumatology patients tested positive for HIV. 46% of the 102 consecutive patients of HIV seen in an Armed Forces referral hospital had some form of rheumatic complaints, commonest being non-specific myalgia and arthralgia[84].

The re-emergence of Chikungunya (CHIK) epidemic[85] in 2006 in central and south regions was unprecedented. Though several reports[86,87] (mostly retrospective) emerged from hospital and/or community based studies, some rheumatology driven studies[88,89,90,91,92,93] evaluated the long term MSK and rheumatism sequel. Following the acute illness several thousand patients suffered from chronic rheumatic MSK pain and rheumatism. Though the overall clinical phenotype was not different from classical text book descriptions, there were several differences especially regarding complications and the magnitude of long term sequel. Predominantly, this acute febrile arboviral illness, characterized by severe and sometimes excruciating musculoskeletal articular aches and pains and arthralgias, was overall benign and self limiting in a rural population study (Bavi, Sholapur) and almost two third cases resolved completely within three weeks[92]. Chikungunya ought to be differentiated from Dengue (also an arbovirus), which generally confines to a severe back ache, and may cause life threatening thrombocytopenia. The author described[88] a wide spectrum of rheumatology phenotype in naïve patients (for arthritis) following an acute chikungunya examined in a busy rheumatology outpatient clinic; from a typical RA like illness to a psoriatic illness, AS and undifferentiated SSA and speculated that no other virus had shown such a propensity for rheumatic disorders. Intriguingly, the frequency of RF and anti CCP in patients with otherwise typical RA profile was low and the patients showed good therapeutic response to short term steroids and supervised disease modifying antirheumatic drugs[88]. The Bavi[92] and Modnimb[89] (Sholapur) community studies also demonstrated a prolonged elevation of IL-6, anti CHIKV IgM and IgG antibodies despite somewhat benign rheumatic sequel; 4% cases had persistent MSK pains and arthralgias at 2 years and none suffered from significant persistent arthritis. An intriguingly questionable report[94] of 'erosive arthritis' following acute CHIKV ought to be validated by rheumatologists. What is CHIKV arthritis? Is it a distinct entity or just a known phenotypic rheumatic disease precipitated by the CHIKV. Several Indian rheumatologists and the author believe that some of the chronic arthritis may be unique in that it is RA like with absent RF and anti-CCP and a dominant periarticular soft tissue inflammation (prayer hand sign). Does the virus persist in the MSK tissues[92] and the behavior of the virus may be akin to that of its close cousin, the Ross River virus which is well known to cause chronic MSK sequel. However, the virus seems to cause an autoimmune arthritis like clinical picture but the author did not observe an increase frequency of RF or other autoantibodies in community studies[88,92]. Will the Indian epidemic in 2006 lead to an increased incidence of rheumatic MSK disorders in years to come? Six years after the re-emergence, unlike the previous epidemics, the CHIKV continues to transmit and infect in several parts of India. There are several other epidemiological questions about the chronic CHIKV MSK disorders that need to be addressed. However, post CHIKV arthritis and rheumatism continues to overload rheumatology outpatients and provide an excellent opportunity to probe the role of arboviruses in chronic inflammatory arthritis and other connective tissue syndromes.

Several other infections[71] (bacterial, viral, fungal, parasitic) are also known to cause arthritis and sometimes are transient (e.g. respiratory viruses, chicken pox, measles, rubella, hepatitis). Hepatitis C is known to cause several rheumatological syndromes but its extent in India is not yet reported but likely to be uncommon. Brucellosis, affecting spine and large joints, is reported[95].

Childhood onset & juvenile idiopathic arthritis (JIA):

There are few population based studies worldwide but JIA is rather uncommon as compared to it's adult counterpart. The COPCORD Bhigwan (survey included children)[4] reported a point prevalence of 0.26% (an unusually high figure) and 0.09% for JCA and rheumatic fever respectively.

Pediatric rheumatology has slowly but surely emerged as an important clinical service and a recent review of Indian publications and reports was published[96]. Several hospital/clinic outpatient based case series (Table 5) from Lucknow[97], Delhi[98] and Pune[99] were published; polyarticular forms and enthesitis related arthritis were predominant. A recent update review with a large clinical experience from Delhi[100] and Mumbai[101] also endorsed the latter observation. In the Mumbai experience[101] (2003-11) based on 1365 children examined in a pediatric rheumatology outpatient in a popular corporate hospital, 32% cases were JIA, 11% were CTD, 10% were vasculitis, 5% were pain syndromes and 1% was auto inflammatory; several children also suffered from mechanical problems. The recent ILAR classification system (Edmonton 2001)[102] for JIA was found to be valid and relevant in the Indian setting also[99]. Long term sequel of JIA was described[103].

Rheumatic fever and its cardiac sequel continue to be an important communicable and preventable disorder[104]. In recent times, a post-streptococcal arthritis variant which is truly a reactive polyarthritis and can mimic juvenile rheumatoid arthritis and its adult form was described[104]. A case series of Kawasaki's disease was recently described from North India[105]. A case series of macrophage activation syndrome in patients with JIA (systemic onset) was recently reported[106].

Miscellaneous

An unusual familial endemic form of crippling arthritis, usually affecting the lower spine and hips, called Handi-Godu disease, in young adults, was reported from Malnad area of Karnataka, South India[107] and the overall picture suggested a primary spondylo-epiphyseal dysplasia presenting as a degenerative disorder. Endemic fluorosis[108] in central parts of India continues to produce disabling musculoskeletal axial and dental complications and can in early stages present as non-specific rheumatic syndrome. Thorn prick arthritis (non-infective inflammatory and usually in hand fingers) has been observed by the author during the rural COPCORD Bhigwan.

Amyloidosis can complicate RA, AS, JIA and other chronic inflammatory disorders. A recent report[109] showed a relatively high frequency of amyloid deposits present in abdominal fat aspirate in patients with RA.

Kumar and coworkers (2012) demonstrated low serum vitamin D in a hospital based study of patients with non specific MSK pains and arthralgias and who improved symptomatically with Vitamin D supplements[110]. Vitamin D is being increasingly reported to be low in the Indian community and its MSK profile needs to be evaluated validated in controlled community based studies.

Conclusion

The Indian population suffers from a large spectrum and burden of rheumatic musculoskeletal disorders and this has been clearly demonstrated in several COPCORD population surveys[113]. We still do not have enough data on their socioeconomic impact which by all logic and standards is bound to be immense. These disorders are bound to adversely impact national productivity and health. This paves way for the national health policy makers to recognize the disease burden[8] and rheumatology as an important super specialty. The time is ripe to include rheumatic disorders in public health programs[111]. Control of rheumatic diseases are likely to raise the standard of living[112].

Table 1 : Some Important Features and Point Prevalence Rates (Percent) In Selected Published WHO ILAR COPCORD Surveys- Global Comparison

COPCORD: community oriented program for control of rheumatic diseases; RA: rheumatoid arthritis; OA: osteoarthritis; STR: soft tissue rheumatism; AS: ankylosing spondylitis; SSA:

Country	Year	Sample Size	Type	Design	MSK Pain	RA	OA	OA Knee	STR	AS	Gout	SSA	Low Back Pain	Knee Pain	Remark
India Bhigwan	2001	4092	R	C, S	19.5	0.54	6.25	4.42	3.77	0.1	0.13	0.3	12.6	13.7	adjusted)age-sex),India Census population 2001;except RA, rest clinical diagnosis; Bhigwan IA-Un 0.76 %; Pune IA-Un 0.32 %
India Pune	2009	8145	U	C, S	11.5	0.19	4.01	3.41	1.2	0.07	0.06	0.27	4.6	6	
China Beijing	1987	4192	R	C, S	40.3	0.34 *	NR	NR	NR	0.26	NR	0.26	35	29	adjusted *1958 ARA criteria for RA, else clinical ; SLE 0.07 %
China Shanghai	2003	6584	U	R, S	13.3	0.28	NR	4.1	3.4	0.11	0.22	0.11	5.6	7	adjusted to China national population
Lebanon	2012	3530	R	R,S	32.9	1	4	3	5.8	0.1	0.01	0.3	13.6	14.2	adjusted to national population; ACR criteria when applicable; FM 1%
Egypt	2004	5120	R	C	16.2	0.29	8.5	NR	6.6	0.09	NR	0.15	4.9	9.1	Except RA, rest clinical diagnosis; FM 1.3 %; CTD 0.15%
Indonesia	1992	4683	U+R	C	24	0.2	5.1	NR	15	NR	1.7	NR	15.1	12.2	not standardized
Iran	2008	10291	U	R, S	41.9	0.33	16.6	15.3	4.6	0.12	0.13	0.23	21.7	25.2	adjusted to national population; all diagnosis clinical including RA
Bangladesh urban	2005	1259	U	C	27.9	0.2	NR	10.6	3.3	NR	NR	NR	18.4	15.8	not standardized; clinical diagnosis; FM 4.4% rural, 3% urban
Bangladesh rural	2005	2635	R	C	26.9	0.7	NR	7.5	2.6	NR	NR	NR	20.1	14	
Mexico	2011	19213	U	R, S	25.5	1.49	10.24	NP	3.8#	0.15	0.35	NP	5.8	NP	adjusted to national population; ACR criteria when applicable; # Rheumatic regional pain syndrome; FM 0.68%; SLE 0.06%; Scl 0.02%

seronegative spondyloarthritis; FM: fibromyalgia; SLE: systemic lupus erythematosus ; Un: undifferentiated; IA: inflammatory arthritis; CTD: connective tissue disease; Scl: scleroderma; U: urban; R: rural; R: randomized sample; C: convenience sample; S: standardized/adjusted age-sex; NR: not reported/studied; RA classified as per American College Rheumatology (ACR) 1987 criteria unless shown otherwise (see remarks); See text for references (also available at www.copcord.org)

Table 2 : Prevalence of Osteoarthritis of the knees per 100,000 population in India (South East Asia Region).

Age bands	15–24	25–34	35–44	45–59	60–69	70–79	80+
Males	-	-	-	4644	15385	20000	6250
Females	-	-	2247	6587	14371	19608	14286

Case, based on clinical ACR criteria.
Source: A Chopra, Bhigwan COPCORD study (India)
(Reproduced with permission from WHO - Publications; text reference 23)

Table 3 : Prevalence of rheumatoid arthritis per 100,000 population in India (South East Asia Region).

Age bands (years)	15–29	30–44	45–59	60–74	75+
Males	-	479	-	1136	-
Females	113	1639	1775	1914	3846

Case, based on clinical ACR criteria.
Source: A Chopra, Bhigwan COPCORD study (India)
(Reproduced with permission from WHO- Publications; text reference 23)

Table 4 : Frequency and relative risk of B27 locus antigen in SSA in Indian patients

Subset	Present study		Delhi		Chennai		Caucasians	
	f	R	f	R	f	R	f	R
AS	73	26	92	193	83	1091	90-100	90-118
RS	67	20	80	66	57	307	65–100	17-40
SSA-U	41	7	84	83	21.4	70	60-70	18
Healthy	9		6		0-2.5		9+	

Note: f: frequency percent ; R: relative risk; AS: Ankylosing Spondylitis;
RS:Reiter's Syndrome; SSA-U: Seronegative Spondarthritis
Undifferentiated. Note: see text references (48)

Table 5 : Frequency Distribution of Inflammatory disorders of Childhood Arthritis : Outpatients/clinic based series.

	Lucknow* (n=89)	Delhi* (n=361)	Pune ** (n=272)		
			Subset	RF (%)	ANA(%)
SA	14	24	9	0	27
OLA	34	30	26	0	19
P	-	-	20	0	19
E	-	-	6	0	0
PA	52	46	31	29	26
SN	-	-	22	0	16
SP	-	-	9	100	10
PsA	-	-	1	0	0
EnA	-	-	27	0	7
UnA	-	-	6	88	41

*JRA Subsets; **ILAR- JIA Subsets; SA: Systemic arthritis; PsA: Psoriatic arthritis; OLA: Oligoarthritis; P: persistent; -E: extended; PA: polyarthritis; SP: seropositive; SN: seronegative; EnA: Enthesitis; UnA: Undifferentiated arthritis;
Note: See Text Reference For Lucknow (97), Delhi (98), Pune (99)

Further Readings

1. http://www.rheumatologyindia.org
2. http://www. copcord.org
3. Malaviya AN, Singh RR, Kapoor SK, Sharma A, Kumar A, Singh YN. Prevalence of Rheumatic diseases in India: Results of a population survey. J Indian Rheumatism Assoc. 1994; 2: 13-17.
4. Chopra A, Patil, Billampelly V, Relwani J, Tandle HS. The Bhigwan (India) COPCORD: methodology and first information report. APLAR J Rheumatology 1997; 1: 145-54.
5. Chopra A, Patil J, Billampelly V, Relwani J, Tandale HS. Prevalence of Rheumatic diseases in a rural population in Western India: A WHO-ILAR COPCORD Study. J Assoc Physicians India 2001(2); 49: 240-46.
6. Das SK, Kumar P, Srivastava R, et al. Distribution of various rheumatological diseases in rural and urban population of Lucknow, North India. J Indian Rheumatol and urban population of Lucknow, North India. J Indian Rheumatol Assoc. 2005; V13 (Suppl): O17 (abstract).
7. Joshi VL, Chopra Arvind. Is there an urban rural divide? Population surveys of rheumatic musculoskeletal disorders in the Pune region of India using the COPCORD Bhigwan model. J Rheum 2009; 36: 3; 614-22.
8. Chopra A, Ghorpade R, Sarmukkadam S, Joshi VL, Mathews A, Gaur L et al. A Staggering Burden of Pain and Rheumatic Disorders in India: A National Bone & Joint Decade India Community Oriented Program for Control of Rheumatic Disease Survey 2006-2011. Arthritis & Rheumatism 2012; 64(10) (suppl): S 23.
9. Chopra A, Saluja M, Patil J, Tandle H. Pain and Disability, Perceptions and Beliefs of a Rural Indian Population: A WHO-ILAR COPCORD Study. J Rheumatol 2002; 29(3): 614-21.
10. Muirden KD. The origins, evolution and future of COPCORD. APLAR J Rheumatol 1997; V1: 44-8.
11. Wigley RD. Rheumatic problems in the Asia pacific region. In: Wigley RD, editor. The primary prevention of rheumatic diseases. New York: The Parthenon Publishing Group Inc 1994; 21-25.
12. Chopra A, Abdel-Nasser A. Epidemiology of rheumatic musculoskeletal disorders in the developing World. Best Practice & Res Clinical Rheumatol 2008; 22: 583-604.
13. Haq SA, Rasker JJ, Darmawan J, Chopra A (2008) WHO ILAR-COPCORD in the Asia-Pacific: the past, present and future. Int J Rheum Dis 11, 4-10.
14. Davatchi F, Jamshidi A, Banihashemi AT et al. WHO-ILAR COPCORD Study (Stage 1, Urban Study) in Iran. J Rheumatol 2008; 35: 1384-90.
15. Chopra Arvind. The WHO ILAR COPCORD Latin America. Consistent With the World and Setting a New Perspective. J Clin Rheumatol 2012; 18 (4): 167-69.
16. Chopra A. The COPCORD World of Musculoskeletal Pain and Arthritis. Deeply Rooted in the Community. Rheumatology (Oxford) 2013 (accepted for publication).
17. Chopra A. The WHO ILAR COPCORD Bhigwan (India) model- Foundation for a future COPCORD design and data repository. Clin Rheumatol 2006; 25: 443-47.
18. Chopra Arvind. Community rheumatology in India. A COPCORD driven perspective. Indian J Rheumatology 2009; 4: 119-26.
19. Malaviya AN, Singh RR, Singh Y, Kapoor SK, Kumar A. Prevalence of systemic lupus erythematosus in India. Lupus 1993; 2: 115-18.
20. Chopra A, Poulton K, Silman A, Thomson W. HLA DRB1 associations in a community based study of inflammatory arthritis in India (abstract). Arthritis Rheum 2000; 43: 9: S71.
21. Saluja M, Chopra A. Bare Foot Applications of HAQ, and Identification of Some Risk Factors in WHO-ILAR COPCORD Bhigwan (India) Stages II and III: an Ongoing Longitudinal Population Based Study 1996-2004. In: Nilganuwong Surasak, editor. Proceedings of the 10th Asia Pacific League of Associations for Rheumatology Congress. Bangkok: Supjaroon Printing Co Ltd; 2002: 335-40.
22. Chopra A, Chaturvedi S, Saluja M, et al. Oral tobacco use is significantly associated with rheumatic musculoskeletal (RMSK) pain: result of a WHO ILAR COPCORD (Community Oriented Program for Control of Rheumatic Diseases) rural population study in India. Arthritis Rheum. 2010; 62: 1551.
23. Report of a WHO scientific group. The burden of musculoskeletal conditions at the start of the new millennium. Geneva (Switzerland): World Health Organization 2003. WHO Technical Report Series 919.
24. Das SK, Kumar P, Srivastava R, et al. High prevalence of fibromyalgia in both rural and urban areas of Lucknow. Indian J Rheumatol Assoc 2006; V1(3): P57(abstract).
25. www.bjdindia.org
26. Chopra A, Ghorpade R, Sarmukkadam S, Joshi VL, Mathews A, Gaur L et al. 5 Million Patients and Not 0.34% Is Worrisome: Burden of Rheumatoid Arthritis in India Based On a Bone and Joint Decade India Community Oriented Program for Control of Rheumatic Disease. Arthritis & Rheumatism 2012; 64 (10) (suppl): S 23.
27. Chopra A. Rheumatology: Made in India (Camps, COPCORD, HLA, Ayurveda, HAQ, WOMAC and Drug Trials). J Indian Rheumatol Assoc 2004; 12: 43-53.
28. Chopra A, Pispati P, Sancheti KH, et al. Medical camps in 'arthritis and rheumatism': a community oriented prospective study, with emphasis on diagnosis and education. Bull Indian Rheumatism Assoc 1991: 5: 31-37.
29. Chopra Arvind, Saluja Manjit. Validation and usefulness of Indian version (CRD Pune) Health Assessment Questionnaire: Drug Trials, Community practice and COPCORD Bhigwan population study (1994-2012). Indian J Rheumatol 2012 : doi: 10.1016/j.injr.2012.04.009.
30. Wolfe F, Smythe HA, Yunus MB, et al: The American college of Rheumatology 1990 criteria for the classification of fibromyalgia. Report of the multicenter criteria committee. Arthritis Rheum 1990, 33: 160-72.
31. Ganguly KS, Roy M. An epidemiological study of rheumatoid arthritis in an urban population in a district of West Bengal. J Indian Rheumatism Assoc 1997; 5: 91-100.
32. Malaviya AN, Mehra NK, Dasgupta B, et al. Clinical and immunogenetic studies in rheumatoid arthritis in India. Rheumatol Int 1983; 3: 105-08.
33. Chopra A, Raghunath D, Singh A, Subramanian AR. Pattern of rheumatoid arthritis in the Indian Population: A Prospective study. British J. Rheumatol.1988; 27(6): 454-56.

34. Chandrasekaran AN, Radhakrishna B, Porkodi R, Madhavan R, Parthiban M. Spectrum of clinical and immunological features of systemic rheumatic disorders in a referral hospital in South India: Rheumatoid Arthritis. J Indian Rheumatism Assoc 1994; V2: 1: 18-26.

35. Chopra A, Epidemiology of rheumatoid arthritis, Monograph on Rheumatoid Arthritis. Ed Mukherjee S, Ghosh A. Indian College of physicians. Academic Wing of API India 2012. Kolkata: Marksman Media Service. 1-9.

36. Gregersen PK, Silver J, Winchester RJ. The shared epitope hypothesis: an approach to understanding the molecular genetics of susceptibility to rheumatoid arthritis. Arthritis Rheum 1987; 30: 1205-13.

37. Ghodke Y, Joshi K, Chopra A, Patwardhan B. HLA and disease. Eur J Epidemiol. 2005; 20(6): 475-88.

38. Taneja V, Mehra NK, Chandrasekaran AN, Ahuja RK, Singh YN, Malaviya AN.HLA-DR4-DQw7 haplotypes occur in Indian patients with rheumatoid arthritis. Rheumatol Int 1992; 11: 251-55.

39. Anuradha V, Bharadwaj R, Chopra A. HLA Class II associations Differ in Indian (Asian) patients suffering from RA: Regional population (urban and rural) surveys using COPCORD Bhigwan model. Arth Rheum 2010; Vol 62 Number 10 (Suppl); S684.

40. Anuradha V. Defining the Immunogenetic Markers in Patients with Inflammatory Arthritis in a Rural Population of Western Maharashtra. Ph.D dissertaion, University of Pune, 2010.

41. Harel-Meir M, Sherer Y, Shoenfeld Y. Tobacco smoking and autoimmune rheumatic diseases. Nat Clin Pract Rheumatol. 2007 Dec; 3(12): 707-15.

42. Gabriel SE, Michaud K. Epidemiological studies in incidence, prevalence, mortality, and comorbidity of the rheumatic diseases. Arthritis Research & Therapy 2009, 11: 229.

43. Malaviya AN, Mehra NK, Adher GC, et al. The clinical spectrum of HLA B27 related rheumatic diseases in India. J Assoc Physicians India 1979; 27: 487-92.

44. Mahadevan R, Chandrasekaran AN, Parthiban M, Achutan, Porkodi R, Rajendran CP. HLA profile of seronegative spondarthropathies in a referral hospital in South India. J Indian Rheumatism Assoc 1996; V4; 3: 91-95.

45. Aggarwal A, Misra R, Chandrasekara S, Prasad KN, Dayal R, Ayyagiri A. Is undifferentiated seronegative spondyloarthropathy a forme fruste of reactive arthritis? Br J Rheumatol 1997; 36: 1001-04.

46. Chandrasekaran AN, Porkodi R, Achutan K, Madhavan R, Parthiban M. Spectrum of clinical and immunological features of systemic rheumatic disorders in a referral hospital in South India: Primary Ankylosing Spondylitis. J Indian Rheumatism Assoc 1994; V2: 4: 149-52.

47. Prakash S, Mehra NK, Malaviya AN et al. Ankylosing Spondylitis in North India: A clinical and immunogenetic study. Ann Rheum Dis 1984; 43: 381-85.

48. Chopra A, Raghunath D, Singh A. A spectrum of seronegative spondarthritis (SSA) with special reference to HLA profiles. J Assoc Physicians India 1990; 38: 351-55.

49. Chandrasekaran AN, Ramakrishnan S, Radhakrishnan B, Parthiban M. Spectrum of clinical and immunological features of systemic rheumatic disorders in a referral hospital in South India: Psoriatic Arthropathy. J Indian Rheumatism Assoc 1995; V3: 1: 7-13.

50. Jagadish TK, Kasthuri AS, Chopra A et al. Clinical profile of connective tissue diseases in a referral service hospital. J Assoc physicians India 1988; 36: 602-05.

51. Kumar A. Indian Guidelines on The Management of SLE. J Indian Rheumatol Assoc 2002; 10: 80-96.

52. Patwardhan M, Pradhan V, Rajadhyaksha A, Umare V, Rajendran V, Surve P et al. Clinical and serological features of male Systemic Lupus Erythematosus patients from Western India. Indian J Rheumatol 2012; 7: 204-08.

53. Chandrasekaran AN, Porkodi R, Radhakrishanan B, Ramakrishnan S, Radha Mahadevan, Thenmozhi Valli PR. Neuropsychiatric manifestations of systemic lupus erythematosus. J Indian Rheumatism Assoc 1994; V2: 2: 90-93.

54. Bichile LS, Vishwanath, Rajdhyaksha. Rheumatologic emergencies in Indian patients: Experience from a rheumatology unit. APLAR J Rheumatol 2000; 4: 114-16.

55. Kumar A, Malaviya AN, Bhat A, Misra R, Banerjee S, Sindhwani R et al. Clinicopathological profile of Vasculitides in India. J Assoc Physicians India 1985; 33: 694-98.

56. Chandrasekaran AN, Venugopal B, Porkodi R et al. Spectrum of clinical and immunological features of systemic rheumatic disorders in a referral hospital in South India: Vasculitis. J Ind Rheumatism Ass; 1995; 3: 84-92.

57. Samant R, Vaidya S, Nadker M, Borges NE. Spectrum of clinical features of Vasculitides in a referral hospital from Western India. J Indian Rheumatism Assoc 1997; 5: 3: 69-74.

58. Gupta R, Kumar A, Malviya AN. Outcome of polyarteritis nodosa in northern India. Rheumatol Int 1997; 17: 101-03.

59. Bamberry P, Sakhuja V, Bhusnurmath SR, Jindal SK, Deodhar SD, Chugh KS. Wegener's granulomatosis: clinical experience with eighteen patients. J Assoc Physicians India 1992; 40: 597-600.

60. Kumar A, Pandhi A, Menon A, Sharma SK, Pande JN, Malaviya AN. Wegener's granulomatosis in India: Clinical features, treatment and outcome of 25 patients. Indian J Chest Dis Allied Sci 2001; 43: 197-204.

61. Mittal G, Mangat G, Samant R, Pathan E, Sule A, Rajdhyaksha S, Gaitonde S, Joshi VR. Avascular Necrosis in Systemic Rheumatic Diseases: An analysis of 29 cases. J Indian Rheumatism Assoc. 2000; 8: 91-96.

62. Chandrasekaran AN, Rajasekhar G, Porkodi R, Rajendran CP, Madhavan R, Parthiban M. Spectrum of clinical and immunological features of systemic rheumatic disorders in a referral hospital in South India: Takayasu's Arteritis. J Indian Rheumatism Assoc 1996; V4: 3: 77-82.

63. Uppal SS, Kakker R, Malviya AN, Sharma S. Clinical aspects of Takayasu arteritis: Current concepts. J Indian Rheumatism Assoc. 2000; 8: 34-41.

64. Kumar A, Malaviya AN, Grover R, Marwaha V, Talwar D, Kanga U, Mehra NK. Clinical and Immunogenetic features of Behcet's disease in India. In Nilganuwong Surasak, editor. Proceedings of the 10th Asia Pacific League of Associations for Rheumatology Congress. Bangkok: Supjaroon Printing Co Ltd. 2002: 207-11.

65. Chandrasekaran AN, Rajendran CP, Porkodi R, Ramakrishana S, Venugopal B, Thenmozhi Valli PR, et al. Spectrum of clinical and immunological features of systemic rheumatic disorders in a referral hospital in South India: Sjogren's Syndrome. J Indian Rheumatism Assoc 1994; V2: 2: 71-75.

66. Misra R, Hissaria P, Tandon V, Aggarwal A, Krishnani N, Dabadghao S. Primary Sjogren's syndrome: rarity in India. J Assoc Physicians India. 2003 Sep; 51: 859-62.
67. Porkodi R, Parthiban M, Rukmangatharajan S, Kanakarani P, Rajendran CP. Clinical spectrum of Gout in South India. J Indian Rheumatism Assoc 2002; 10: 61-63.
68. Kumar A, Malviya AN, Singh YN et al. Clinical profile, therapeutic approach and outcome of gouty arthritis in Northern India. J Assoc Physicians India. 1990; 38: 400-02.
69. Gupta SK. Primary gout in an orthopedic practice in Calcutta. J Indian Rheumatism Assoc 1994; 2: 153-56.
70. Chandrasekaran AN. Infections and arthritis:our experience - a retrospective and prospective study. J Indian Rheumatism Assoc 1993, 1, 1, 7-20,
71. Chopra Arvind. Bugs and we: who is chasing whom? Indian J Rheumatology 2011; 6(1): 1-4.
72. Porkodi R, Rathnasamy R. Infections in a rheumatology setting-Chennai experience. Indian J Rheumatology 2011; 6(1): 31-35.
73. Irlapati RVP, Nagaprabu VN, Suresh K, Agrawal S, Gumdal N. Infections in rheumatology practice: an experience from NIMS, Hyderabad. Indian J Rheumatology 2011; 6(1): 25-30.
74. Agarwal V, Aggarwal A. Acute and chronic bacterial infections in rheumatology practice. Indian J Rheumatology 2011; 6(1): 69-74.
75. Gupta N, Shetty A, Soman R,Raodrigues C. Musculoskeletal infections: selected cases from a referral hospital in Mumbai. Indian J Rheumatology 2011; 6(1): 55-58.
76. Shanmugasundaram TK, in Current Concepts in bone and joint tuberculosis, 1985.
77. Chopra A, Joshi VL, Joshi V, Rane R, Anuradha V. Old enemy on a difficult terrain: tuberculosis in rheumatology practice with focus on uncommon extrapulmonary affection. Indian J Rheumatology 2011; 6: 44-54.
78. Haldar S, Ghosh P, Ghosh A, Tuberculous arthritis-the challenges and opportunities: observations from a tertiary center. Indian J Rheumatology 2011; 6: 62-68.
79. Samant R, Nadkar MY, Vaidya SS, Chhugani SJ, Balaji R, Borges NE. Leprosy: a close mimic in a rheumatology clinic. J Assoc Physicians India. 1994; 47: 576-79.
80. Haroon N, Agarwal V, Agarwal P, Lawrence A, Agarwal A, Misra R. Rheumatic complaints as the presenting manifestation of leprosy. J Indian Rheumatol Assoc. 2005; V13 (suppl): P52 (abstract).
81. Sarkar RN, Phaujdar S, Banerjee S, Siddhanta S, Bhattacharyya K, De D et al. Musculoskeletal involvement in leprosy. Indian J Rheumatology 2011; 6(1): 20-24.
82. Sheetal S, Arvind C. Lest we forget Hansen's disease (leprosy): an unusual presentation with an acute onset of inflammatory polyarthritis and the rheumatology experience. Int J Rheum Dis. 2009; 12(1): 64-69.
83. Vaidya S, Samant RS, Nadkar MY, Koppikar GV, Kulkarni MG, Wadhva SL, Padte SV, Borges NE. HIV infections and Rheumatological disorders. J Indian Rheumatism Assoc 1996, 4, 3: 83-87.
84. Achuthan K, Uppal SS. Rheumatological manifestations in 102 cases of HIV infection. J Indian Rheumatism Assoc. 1996; 4: 43-47.
85. Lahariya C, Pradhan SK. Emergence of chikungunya virus in Indian subcontinent after 32 years: A review. J Vector Borne Dis. 2006;43(4): 151-60.
86. Ray P, Ratagiri VH, Kabra SK, Lodha R, Sharma S, Sharma BS, Chikungunya infection in India: results of a prospective hospital based multi-centric study. PLoS One. 2012; 7(2): e30025.
87. Nagpal BN, Saxena R, Srivastava A, Singh N, Ghosh SK, Sharma SK et al. Retrospective study of chikungunya outbreak in urban areas of India. Indian J Med Res. 2012 135: 351-58.
88. Chopra A, Anuradha V, Lagoo-Joshi V, Kunjir V, Salvi S, Saluja M. Chikungunya virus aches and pains: an emerging challenge. Arthritis Rheum. 2008 ; 58(9): 2921-22.
89. Chopra A, Venugopalan A. Persistent rheumatic musculoskeletal pain and disorders at one year post-chikungunya epidemic in south Maharashtra-a rural community based observational study with special focus on naïve persistent rheumatic musculoskeletal cases and selected cytokine expression. Indian J Rheumatology 2011; 6(1): 5-11.
90. Paul BJ, Pannarkady G, Moni SP, Thachil EJ. Clinical profile and long-term sequel of Chikungunya fever. Indian J Rheumatology 2011; 6(1) 12-19.
91. Ganu MA, Ganu AS. Post-chikungunya chronic arthritis-our experience with DMARDs over two year follow up. J Assoc Physicians India. 2011; 59: 83-86.
92. Chopra A, Anuradha V, Ghorpade R, Saluja M. Acute Chikungunya and persistent musculoskeletal pain following the 2006 Indian epidemic: a 2-year prospective rural community study. Epidemiol Infect. 2012; 140(5): 842-50.
93. Mathew AJ, Goyal V, George E, Thekkemuriyil DV, Jayakumar B, Chopra A; Trivandrum COPCORD Study Group. Rheumatic-musculoskeletal pain and disorders in a naïve group of individuals 15 months following a Chikungunya viral epidemic in south India: a population based observational study. Int J Clin Pract. 2011; 65(12): 1306-12.
94. Manimunda SP, Vijayachari P, Uppoor R, Sugunan AP, Singh SS, Rai SK et al. Clinical progression of chikungunya fever during acute and chronic arthritic stages and the changes in joint morphology as revealed by imaging. Trans R Soc Trop Med Hyg. 2010; 104(6): 392-99.
95. Pandit D. Brucella arthritis-an update. Indian J Rheumatology 2011; 6(1): 75-79.
96. Dhir V. Important publication of Indian Pediatric Rheumatology (original articles only) for past 5 years (2006-2012). Indian J Rheumatology 2012; 7(1): 119-22.
97. Agarwal A, Mishra R, Juvenile chronic arthritis in India. Is it different from that seen in Western countries. Rheumatol Int 1994; 15: 53-56.
98. Seth V, Kabra Sk, Semwal OP, Jain Y, Clinico-immunological profile in junvenile rheumatoid arthritis - an Indian experience. Indian J Pediatr 1996 May- June; 63(3): 293-300.
99. Kunjir V, Venugopalan A, Chopra A. Profile of Indian patients with juvenile onset chronic inflammatory joint disease using the ILAR Classification Criteria for JIA: A Community-Based Cohort Study. J Rheumatol 2010; 37: 1756-62.

100. Sawhney S. Juvenile idiopathic arthritis: Classification, clinical features, and management. Indian J Rheumatology 2012; 7(1): 11-21.
101. Khubchandani RP, Hasija RP. Spectrum of paediatric rheumatologic disease: The Mumbai experience. Indian J Rheumatology 2012; 7 (1): 7-10.
102. Petty RE et al. ILAR Classification of Juvenile Idiopathic Arthritis (second revision), Edmonton, 2001. J Rheumatol 2003; 31: 390-92.
103. Aggarwal A, Shenoy S, Gupta R. Long-term outcome in juvenile idiopathic arthritis. Indian J Rheumatology 2012; 7(1): 22-26.
104. Viswanathan V. Acute rheumatic fever. Indian J Rheumatology 2012; 7(1): 36-43.
105. Gupta a, Singh S. Kawasaki disease-A preventable cause of cardiac morbidity. Indian J Rheumatology 2012; 7(1): 87-91.
106. Singh S, Chandrashekaran S, Ahluwalia J, Suri D, Rawat A, Ahmed N et al.Macrophage activation syndrome in children with systemic onset juvenile idiopathic arthritis: clinical experience from northwest India. Rheumatol Int. 2012 Apr; 32(4): 881-86.
107. Krishnamachari KA, Bhat RV. Endemic familial arthritis of Malnad an outbreak in Southern India. Trop Geogr Med. 1978 Mar; 30(1): 33-37.
108. Rawlani S, Rawlani S, Rawlani S. Assessment of Skeletal and Non-skeletal Fluorosis in Endemic Fluoridated Areas of Vidharbha Region, India: A Survey. Indian J Community Med. 2010; 35(2): 298-301.
109. Wakhlu A, Krisnani N, Hissaria P, Aggarwal A, Misra R. Prevalence of secondary amyloidosis in Asian North Indian patients with rheumatoid arthritis. J Rheumatol. 2003; 30(5): 948-51.
110. Kumar A, Gopal H, Khamkar K, Prajapati P, Mendiratta N, Misra A et al. Vitamin D deficiency as the primary cause of musculoskeletal complaints in patients referred to rheumatology clinic: A clinical study.Indian J Rheumatology 2012; 7(4): 199-203.
111. Mathew AJ, Rahim A, Bina T, Paul BJ, Chopra A. A call to integrate musculoskeletal disorders into the national programme for non-communicable diseases (NCD). Indian J Med Res 2012; 136: 499-501.
112. Wigley R, Chopra A, Wigley S, Wigley A. Does control of rheumatic disease raise the standard of living in developing countries? Clin Rheumatol 2009; 28: 621-22.
113. Arvind Chopra. The COPCORD world of musculoskeletal pain and arthritis. Rheumatology 2013; 52 (Nov): 1925-28.

SECTION 4

Evaluation of Rheumatic Diseases

AUTOIMMUNITY IN RHEUMATOLOGY 1

Liza Rajasekhar, I. Rajendra Varaprasad

Introduction

Autoimmune diseases (AD) are diseases resulting from the breakdown of immune tolerance due to complex interplay of genetic predisposition of the individual and environmental triggers over time. Immunopathological mechanisms of these diseases are complex and often overlap.

Classification of autoimmune diseases

Traditionally AD are classified as organ specific and systemic diseases.

Organ specific diseases involve a specific tissue like endocrinologic (diabetes mellitus type 1, Hashimoto's thyroiditis, Addison's disease etc.), dermatologic (pemphigus vulgaris), neurologic (multiple sclerosis), or hematologic (autoimmune haemolytic anemia). It is rare to find evidence of systemic inflammation or production of autoantibodies to antigens other than those found in the target organ.

Systemic AD with chronic activation of adaptive and innate immune mechanisms target multiple organs; include diseases like Systemic lupus erythematosus (SLE), Sjogren's syndrome, Scleroderma, Rheumatoid arthritis (RA), and Dermatomyositis. These conditions tend to be associated with autoantibodies to antigens which are not tissue specific. The disease manifestations vary depending on the target cell and the affected organ. Signs and symptoms such as arthralgia, arthritis, alopecia, fatigue, photosensitivity, Raynaud's phenomenon, non-specific autoantibodies (i.e. antinuclear antibodies, rheumatoid factor, anti-Ro antibodies) and high levels of cytokines (i.e. TNF, IL-1, IL-6, IL-10, IL-17, etc.) are shared among the diseases.

AD can also be classified into primary T or B lymphocyte mediated diseases but it seems a simplistic view considering that all cellular elements of the immune system i.e. the T lymphocyte, the B lymphocyte, the dendritic cells (DC) and macrophages are intimately involved in the activation and survival of each other.

Mechanisms of autoimmunity

There may be two ways by which autoimmunity develops. One is the activation of pre-existing cells which are primed for self antigens and have escaped tolerance education and the other is by breakdown of tolerance.

To understand why autoimmunity develops and how it leads to autoimmune rheumatic disease it is imperative to understand how the immune system trains to recognize and tolerate self antigens while simultaneously retaining an ability to recognize foreign antigens.

Human immune system

The two arms of the human immune system are the innate and adaptive immunity. In reality they are highly integrated and interdependent. Innate immunity represents the primary, prompt but less specific barrier against damage and infection, while adaptive immunity is a more efficient but slower defense mechanism[1].

Innate immunity

Innate immunity is played out by the macrophages, dendritic cells and B cells. The prompt response against microbes is possible due to the presence of germ line coded receptors called pattern recognition receptors (PRR) on these cells. These receptors choose to focus on certain highly conserved sequences called pathogen associated molecular patterns (PAMPs) and damage associated molecular patterns, (DAMPs) which occur exclusively only on microbial pathogens and damaged cells respectively. By recognizing these molecules macrophages, dendritic cells and B cells act as antigen presenting cells (APC). Three classes of PRR have been identified: the Toll-like receptors (TLRs), the retinoic acid-inducible gene-I (RIG-I)-like receptors (RLRs) and the nucleotide-binding oligomerization domain (NOD)-like receptors (NLRs). TLRs help to take up the microbial peptides and process them before presentation to the adaptive arm of the immune system.

TLR Receptors

The TLR family is a typical signaling PRR. TLRs are expressed on the plasma membrane or in the endosome and play major roles in activating APCs. Toll like receptors recognize a wide variety of microbial pathogens or certain endogenous molecules released during inflammation by PAMPs. Their activation initiates protective Th-1 immune response, prevent dendritic cell maturation and hence control autoimmunity in humans[2]. TLR5 is highly expressed on "natural T regulatory cells (nTReg)" cells, and its ligand flagellin, augments their suppressive capacity, possibly due to enhanced FoxP3 expression. The ability of TLRs to recognize host-derived danger signals, which are produced on cell damage and necrosis, is also recognized. Studies have found that the immune complexes containing self-RNA and/or self-DNA can act as endogenous triggers for the activation of TLRs. Release of endogenous TLR ligands during inflammation and the consequent activation of TLR signaling pathways have been indicated and polymorphism of the TLR4 and TLR9 gene have been reported to be associated with many autoimmune diseases, such as SLE, atherosclerosis, asthma, type 1 diabetes, multiple sclerosis, and RA[3].

Adaptive immunity

Adaptive immunity needs 3 to 5 days to mature. It involves B cells, T cells, cytotoxic T cells and antibody production. It is characterized by highly specific antigen recognition through specific antigen receptors, among which the most important are B and T cell receptors (BCR and TCR). TCR is engaged by a complex on APC consisting of antigenic peptide and a MHC molecule and CD28 by B7 molecule. Similarly the B cell receptor (BCR) which is an immunoglobulin molecule on the surface of B cell is engaged prior to either activation or apoptosis of the B cell. In addition to antibody production activation of the lymphocyte leads to production of cytokines and chemokines to contain the infective/inflammatory stimuli.

Tolerance and its breakdown

Recognition and removal of microbial pathogens is also essential, yet tolerance to self is also crucial. This fine balance between immunity and self-tolerance is due to the presence of multiple check points in the immune system. Crucial to understanding autoimmunity is the concept of tolerance and its failure. Tolerance is the process by which self reactivity is abolished or kept for becoming pathogenic in the immune system. Cells of the immune system undergo this training (central training) in the central lymphoid organs (thymus and bone marrow) and peripheral lymphoid organs (spleen, lymph nodes etc). The process continues in the peripheral germinal centers of spleen and lymph nodes to eliminate antigen activated self reactive B and T lymphocytes. This is called peripheral tolerance.

Central tolerance

Central T cell tolerance is achieved by low affinity binding of TCR with endogenous self peptide Major Histocompatibility Complex (MHC). This leads to failure to deliver antiapoptotic signals to the T lymphocyte resulting in death by neglect. When a developing lymphocyte receptor interacts with intermediate affinity with a self antigen it receives a survival signal also referred to as positive selection. These self derived antigens are presented in the context of self MHC molecules and so positive selection ensures the survival of only those lymphocytes which will be able to recognize a peptide only when complexed with self MHC. These lymphocytes then migrate to the periphery. If the antigen is presented to the T lymphocyte by the APC in the absence of inflammatory signals or cytokines when no co-stimulatory molecules are present then an inactivating signal is generated in the lymphocyte. However this is also true that self reactive T lymphocytes are found in the circulation. This happens because of escape from negative selection or by positive selection by certain auto antigens. Tolerance maybe achieved by either low levels of antigen, physical separation of antigen from T lymphocyte or absence of co-stimulatory signals. B cell tolerance is predominantly due to removal or negative selection of autoreactive B cells by one of the following three processes[4-6]:

1. Receptor Editing : B cells during maturation require close interaction with the bone marrow stromal cells which generates an in-frame sequence that leads to the generation of pre B cell receptor (BCR). BCR cross-linking by high avidity self-antigens drives B cells to receptor editing - the rearrangement of light-chain Ig genes allows the substitution of self-reactive receptors with non-self-reactive receptors. In the primary lymphoid tissue when a naïve B cell encounters an antigen which binds strongly to

the BCR there is light chain rearrangement leading to a new BCR immunoglobulin molecule. If the resulting new BCR is autoreactive then the B cell is removed from circulation by deletion or anergy.

2. Anergy : When an immature B cell undergoes a moderate degree of engagement of the BCR it is anergised. This means early apoptosis, exclusion from entering lymphoid follicles and death if anergy is not reversed within a narrow time frame.
3. Deletion : When the BCR of an immature cell undergoes extensive cross linking by antigen it undergoes programmed cell death and it is also deleted.

Peripheral tolerance

Lymphocytes which exit the thymus circulate continuously in the blood after going through the peripheral lymphoid tissues in each circuit. Lymphocytes reactive to self antigen are eliminated during the generation of peripheral tolerance. TCR triggering by self antigens can eliminate the T lymphocyte after a brief period of activation and cell death, a process called activation induced cell death (AICD), mainly through the CD95 pathway. In T cells the process of AICD mediated by TNF receptors and Fas ligand (FasL) (CD95) plays an important role in the control of peripheral tolerance[7].

Another mechanism of peripheral tolerance of T cells is anergy. In the peripheral lymphoid tissues if the lymphocytes encounter their receptor ligands or antigen presenting cell they undergo differentiation. In the absence of either they recirculate till death as long lived naïve or immature lymphocytes.

Differentiation of a T lymphocyte entails the secretion of cytokines, chemokines, immunoglobulin and certain inducible costimulatory molecules. The latter may act as regulatory check points to cut off the process of differentiation and activation. Like in the central lymphoid tissues in the periphery, encounter of the mature B lymphocyte with a strongly cross linking antigen in the periphery leads to direct apoptosis of the B lymphocyte. Encounter of the mature B lymphocyte with abundant soluble antigen also leads to anergy. In the presence of inflammatory cytokines, surface co-stimulatory molecules lymphocytes are activated. In the absence of inflammatory cytokines an antigen (likely to be a self-antigen) presented to a lymphocyte will lead to AICD or anergy. In the periphery despite the presence of BCR with antigen specificity there may be a state of inertia due to sequestration from self antigens or low concentration of the antigen. Conditions which reverse this situation like over expression of sequestered antigens or activation of B lymphocyte in an inflammatory milieu can lead to autoimmunity. Negative selection i.e. elimination of lymphocytes with weak affinity to self antigens, may continue from the primary lymphoid tissue in the spleen too. The stringency of the process is influenced by antigen concentration, hormones, co-stimulatory pathways and cytokines.

The failure of both central and peripheral tolerance maturations lead to an increased number of circulating self-reactive B cells, favoring the development of autoimmunity if interacted with helper T cells. Hence the potential for B cell growth in response to self-antigen is limited by dependence on two cooperating cells, a B cell and a T cell, both of which must bypass multiple checkpoint mechanisms.

Apoptosis

Apoptosis is an integral part of housekeeping in an organism. It is programmed cell death and removes unwanted cell. In the process certain self antigens undergo either relocation or biochemical alteration but are eliminated as the apoptotic cells are cleared. During the process of change in the intracellular and often nuclear antigen, the antigen and hence apoptotic cell may become immunogenic. The apoptotic cell undergoes one of three processes.

1. Clearance by macrophages : If the macrophage recognizes phosphatidylserine deriving from the loss of plasma lipid asymmetry and modifications of membrane carbohydrate groups on the surface of an apoptotic cell it secretes anti-inflammatory cytokine TGF-α. This prevents an immune response consequently the apoptotic cell is ignored.
2. Clearance by immature dendritic cells : The dendritic cell which takes up an apoptotic cell travels through the bloodstream and presents self antigens acquired from these apoptotic cells to T lymphocyte in the absence of co-stimulatory molecules. This results in tolerance to those antigens. This presentation to T cell in the absence of co-stimulatory molecules happens only if the DC acquired the apoptotic cell in the absence of inflammatory stimuli like TNF-α, IFN-α or bacterial LPS. It is also proposed that presence of regulatory complement proteins in the apoptotic cell helps to keep the DC tolerised.

3. If incomplete signals are delivered to macrophages and DC this results in recognition as non-self. It has been seen that absence of some complement components like C1q, iC3b and C4 on apoptotic cells causes the macrophages and DC to sense them as foreign and the former to secrete inflammatory cytokines TNF-α. This causes DC maturation making them very efficient in presenting the antigens to T cells and generating an autoimmune response. The defect may also be in the receptors present on the surface of macrophages, leading to essentially the same process again.

Regulation of B lymphocyte activation

B cell activation can occur by T cell dependent and independent pathways. Auto-antibodies are secreted by autoreactive B cells which escape negative selection. This may be due to intrinsic B cell defects or by excessive co-stimulation by T cells or dendritic cells. Increasing co-stimulatory engagement increases the degree of auto-reactivity in the repertoire[8].

Antigen specific T lymphocyte expressing antigen can activate B cells. Engagement of the T lymphocyte receptor with the MHC associated processed antigen and the complex interactions of a wide variety of surface co-stimulatory molecules on the T and B lymphocyte leads to T cell activation[9] (Fig. 1).

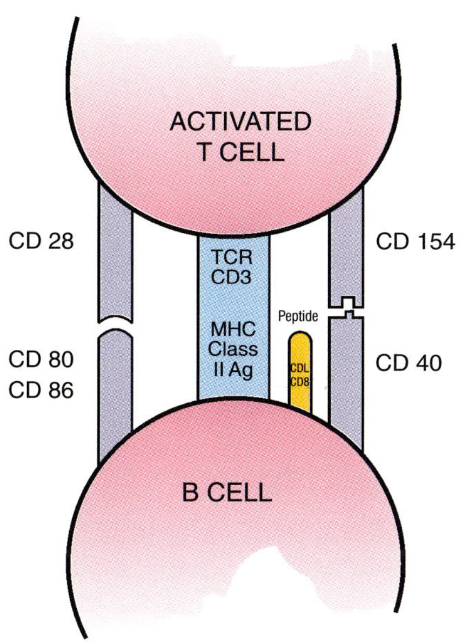

Fig. 1 : T cell activation

The CD28:CD80/86 families of molecules and the CD40:CD40L pair of molecules are considered as critical co-stimulatory elements. This cascade of activation leads to expression of more epitopes of the processed antigen, a process called epitope spreading which is very important in autoimmunity. Interaction of CD40L (CD154) and CD40 results in inhibition of B cell apoptosis, increased expression of co-stimulatory molecules, and promotion of class switching with formation of germinal centers and memory B cells. Down regulation of this interaction shuts off B cell interaction. In SLE over expression of CD80/86 on B cells and CD40L on T cells has been noted[10]. T lymphocyte independent B lymphocyte regulation is affected by B1 and marginal zone B cells. Follicular B cells are responsible for most autoimmunity. The B lymphocyte stimulator (BLyS also termed BAFF) family of ligands and receptors plays a central role in B lymphocyte development, selection, and homeostasis[11]. Since the receptors (BR3, BCMA and TACI) are differentially expressed among naïve and experienced B cells and the ligands have differential affinities for the receptors there may be many functionally distinct B cell subsets. BLyS signaling via BR3 is the dominant homeostatic regulator of primary B cell pools, whereas APRIL interactions with BCMA likely govern memory B cell populations[12]. BAFF has been found on fibroblast like synoviocytes in inflamed joints of RA, in salivary glands of Sjogren's syndrome and in CNS lesions of multiple sclerosis.

Over expression of BLyS is accompanied by the appearance of pathogenic antibodies and lupus like autoimmune disease[13]. Therapies targeting this family of ligands and receptors thus may affect the different B cell populations nonhomogenously. CD22, FcgII, PD1 and CD5 (a unique surface marker of B1B cells) are negative regulators of B cell function. FcgRIIb is an inhibitory receptor on B cell and dendritic cells. Reduced expression of this receptor can lead to expansion of self reactive plasma cells[14]. PD-1 is a member of the Ig superfamily expressed on activated T and B cells. It binds to PDl-L which is constitutively expressed in the kidney, heart and lung. This ligation inhibits proliferation of both B cells. If autoreactive B lymphocytes enter the circulation and get activated by self antigens, then expression of PD-1 and its ligand in the various organs suppress the activation of the autoreactive B lymphocytes in the target organ. With cytokine exposure there is upregulation of MHC class I molecules in the tissues which is a prerequisite for autoimmunity. So the state of the target organ may also be a determinant of whether the tissue will suppress or sustain the autoimmune response.

Regulation of T lymphocyte activation

Self-reactive T cells undoubtedly exist and are constituents of the normal T-cell repertoire. The question is how they are regulated such that they do not elicit unwanted pathologies. The activation or otherwise of T cell depends on the type of signal it receives. Presentation of antigenic peptide in the absence of CD28 leads the T lymphocyte to become anergic and incapable of differentiating to helper or suppressor T lymphocyte. After T cell activation T-lymphocyte associated antigen 4 (CTLA-4) is induced on T cell surface. It also binds to CD80/86 but shuts off the activation of T lymphocyte and promotes T cell tolerance. The expression of CD80/86 on APCs is enhanced by the presence of microbes and by cytokines that are produced in response to microbes.

As a part of normal immune physiology there exists a subgroup of CD4 + T helper cells called the normally occurring T regulatory cells or the nTreg cells. They are characterized by the expression of two unique cell surface markers: CD25 and FoxP3, a transcription factor[15]. The latter molecule directly affects the ability of the T lymphocyte to proliferate, produce cytokines and express cell surface markers. Naturally occurring CD4 + CD25 + Treg cells develop in the normal thymus assuming a T cells repertoire which is broad enough to recognize a vast repertoire of self and non-self antigens. The development of CD4 + CD25 + Treg cells is genetically determined, influenced by antigen specific and nonspecific signals, co-stimulation and cytokines that control their activation, expansion and suppressive activity. These cells are notorious as hyporesponsive cells. They have a profound inability to produce IL-2 a critical T cell growth factor. CD28 present on T lymphocyte promotes the survival of nTreg cells via induction of antiapoptotic molecules and or cytokines predominantly IL-2.These cells help to maintain self non-reactivity in the periphery by patrolling antigens. In the presence of these cells the APC function is dampened by decreased production of surface co-stimulatory molecules. This assures peripheral T cell tolerance. Functional abrogation of these cells in vivo or genetic defects that affect their development of function, like mutations in FoxP3 gene has been associated with severe autoimmunity. When local APC secretes IL-12 in the presence of IFN-γ, CD4 T lymphocyte differentiate into Th-1 cells. These cells produce more IFN-g and promote intracellular pathogen killing. If IL-4 is present in the antigen presenting milieu CD4 T lymphocyte become Th-2 cells producing IL-4, IL-5 and IL-13 facilitating a humoral response to extracellular pathogens. When the cytokine milieu includes TGF-β, IL-6 and IL-23 CD4 T lymphocyte differentiate into Th-17 cells which secrete IL-17 and IL-17F and TNF, GMCSF etc which may help in the clearance of certain pathogens[16]. If this T cell activation is not regulated well, persistence of T cell activation leads to autoimmune diseases. There is increasing evidence that IL-23 driven differentiation of T lymphocyte into Th-17 cells which produce IL-17 is important in many ADs as shown in the Table 1. Th17 cells are crucial in mediating host defense and in leading to the onset of ADs in genetically-susceptible individuals. Novel factors have been proposed to influence Th17 cells including hormones such as aldosterone which enhances CD8 + T cell activation as well as Th17 cell polarization, toll-like receptors (TLR) 9 and 4 to modulate pristane-induced lupus nephritis, and the cytotoxic T-lymphocyte-associated protein 4 (CTLA-4) interaction with the major histocompatibility complex B27 in autoimmune myocarditis[17].

Table 1 : Cytokines of relevance in autoimmune rheumatic diseases[1]

Increased serum levels	Autoimmune diseases
TNF-α	RA, SLE
IL-1	RA, PM-DM
Il-2	SLE
IL-4	SSc
IL-6	RA, SLE
IL-8	SLE
IL-10	SLE, SSc
Il-12	SLE, SSp
Il-18	Cutaneous LE, PM-DM

How tolerance checkpoints are subverted leading to autoimmunity

Heat shock proteins (HSPs)

Heat shock proteins (HSPs) have highly conserved protein sequences through eukaryotic and prokaryotic development. They have housekeeping functions in the cell. They help in proper folding and translocation of other proteins.

HSPs are strong immune modulators and are able to influence the impact and direction of immune responses. HSPs can be pro-inflammatory, for instance, by eliciting an interferon (IFN)α response by T cells, or tumour necrosis factor (TNF)α production by macrophages. However, HSPs can also induce an anti-inflammatory response, for example, production of IL-10 also promotes interactions of these cells with other immune cells, eventually leading to a tolerogenic response. HSP60 peptides influence Tregs via the TCR, or via TLRs, which are present on both T cells and antigen-presenting cells such as DCs[18].

TLR Receptors : have already been discussed in the section on Innate immunity.

T regulatory cells

It is now clearly established that a subset of CD4+ T cells called natural suppressor/regulatory T (Treg) cells, expressing the Forkhead Box Protein 3 (FoxP3), is essential for the maintenance of self tolerance and of immune homeostasis. FoxP3 expressing Treg cells are capable of suppressing the activation, the proliferation and the effector functions like the production of cytokines of a wide variety of immune effector cells in vivo and in vitro including CD4+ and CD8+ T cells, NK and NK T cells, B cells, and antigen presenting cells i.e. dendritic cells or macrophages. These regulatory T cells encompass IL-10 secreting TR1 cells, TGF-α producing TH-3 cells, TCR expressing T cells and CD8 + CD28 - T cells.

FoxP3 + Treg cells are naturally occurring regulatory T cells that are developmentally programmed under the control of the transcription factor FoxP3[17].

CD95 pathway

The importance of CD95 in controlling autoreactive T cells is illustrated by the autoimmune lymphoproliferative syndrome (ALPS). In ALPS, various components of the CD95 signaling pathway are mutated such that T cells become refractory to AICD which results in a lymphoproliferative disease. Breakdown of T lymphocyte tolerance can occur by any of the following mechanisms:

1. Genetic programming of the extent of deletion of autoreactive T lymphocyte in the thymus.
2. Activation of self reactive T lymphocyte in the periphery by infectious agents. Infectious agents may cause sequestered autoantigens to be exposed to autoreactive T lymphocyte or may be mistakenly recognized as self (molecular mimicry) or induce inflammatory cytokines and co-stimulatory molecules.

Key players implicated in breakdown of tolerance

Sex

The prevalence of ADs is about 5%. Eighty percent of these patients include women and is one of the common cause for disability. This bias is due to various genetic and hormonal factors.

Genetic defects

Most recent genomewide association studies (GWAS) report significant associations between human diseases and specific genes, though are found of limited clinical utility at present. More than 200 genetic loci have been shown to be associated with one or more autoimmune disorders. The concept of quantitative threshold has emerged, which is culmination of effects of multiple genetic factors predisposing to a state of autoimmunity[19,20].

For most of these disorders, genes within the major histocompatibility complex (MHC) have by far the strongest single genetic effect. Unlike the rare single gene defects as a cause of autoimmunity the genetic aberration usually involve a particular functional pathways in immunity including the intracellular signaling (PTPN22, genes involved in ubiquitation), that drives the activation of T and B cells, signaling by cytokines and cytokine receptors (IL2RA), and pathways that mediate innate immunity and microbial responses.

Environmental factors

Infections, smoking, drugs, silica dust, cooking oil, sun exposure are factors often implicated as environmental triggers.

Latitude

Several environmental factors are common risk factors for autoimmune diseases. Geographic clustering not explained by genetic variation strongly suggests an environmental effect. The more recent research field of geoepidemiology illustrates a latitudinal gradient in the prevalence and incidence of AD thus supporting that environmental factor may be crucial to disease onset[15].

Infection

Infectious agents are important in the pathogenesis of AD since they are a major part of the environmental autoimmunity trigger. Epstein-Barr virus and cytomegalovirus infections are consistently associated with multiple autoimmune diseases. There is now enough evidence to say periodontitis due to porphyromonas gingivalis is an important factor in the pathomechanism of RA.

Smoking

Nowhere is environment as a trigger for AD is more evident than in the role of smoking as a risk factor for seropositive RA. Smoking induces the appearance of citrullinated autoantigens in the lung, thus providing the beginnings of a causal pathway linking an environmental exposure (smoking, and possibly air pollution) to the production of pathogenic antibodies (anticitrullinated autoantibodies) in genetically predisposed persons.

Vitamin d deficiency

Recent evidence shows a crucial role of Vitamin D in immunity. Vitamin D deficiency impairs the immune system, especially the peripheral adaptive immune system, i.e. T and B lymphocytes. Vitamin D deficiency status has been shown to effect diseases including multiple sclerosis, diabetes mellitus type I, SLE, vasculitis, RA and other autoimmune rheumatological disorders.

Stress

The hypothalamic- pituitary-adrenal axis modulates immune functions. Experimental stress polarizes immune activity toward a Th2-type humoral response. Stressors trigger pro-inflammatory cytokines predisposing to autoimmunity.

How autoantibodies produce damage in autoimmune diseases

Mechanisms of autoantibody mediated tissue damaged in the most common autoimmune conditions involve

1. Cell surface binding and lysis.
2. Interaction with cell surface receptors (blockage/ stimulation) : Autoantibodies show reactivity to membrane receptors and can both active or block selective pathways. e.g. myasthenia gravis, autoimmune thyroiditis.
3. Immune complex mediated damage.
4. Intracellular antigens : Two mechanisms have been suggested to explain the paradox: cross reaction with membrane antigens and exposure of the antigens during apoptosis. some evidence suggests that autoantibodies can penetrate living cells, i.e. anti-RNP or IgA-AMA can transcytose through biliary cells.
5. Binding to extracellular molecules : extracellular protein interference resulting in a systemic clinical syndrome. e.g. anti-phospholipid antibodies.
6. Autoantibody transfer across the placenta : Neonatal lupus syndromes, Graves disease, myasthenia gravis, pemphigus vulgaris and thrombocytopenic purpura.

Different phases in the development of ADs have been proposed. In the first phase, normal immunity is present in the context of susceptibility genetic background. In the second phase, abnormal immune response occurs, but without clinical manifestations (preclinical autoimmunity). In this phase, autoantibodies or cellular abnormalities may be evidenced. These anomalies may eventually increase in the following months or years, leading to initial clinical symptoms and then to diagnosis.

Multisystem disorders as an example of the complexity of mechanisms involved in generation of autoimmunity

PATHOGENESIS OF SLE (Fig. 2).

SLE is a systemic autoimmune disease of unknown etiology with variable clinical manifestations and immunological dysfunction well illustrated by complement abnormalities, changes in the function of B cells, T cells both CD4+ and CD8+, and the innate immune compartment. Plasmacytoid dendritic cells (pDCs) in response to viral infection, and DNA or RNA containing particles including demethylated CpG rich dsDNA produce Type I IFN. This occurs via activation of Toll receptors present intracellularly or intraendosomally IFN in turn induces the TLR9 receptors present on predendritic cells. TLR9 induce surface expression of co-stimulatory molecules on the DC it also helps to load the antigenic peptide onto the MHC molecule for presentation to the T lymphocyte[21]. Normally these TLR9 receptors are negatively regulated. TLRs have been demonstrated to be crucial for SLE pathogenesis with TLR4 involved in the development of serum autoantibodies while TLR7 and TLR9 in the disease onset and severity. Interferon-over expression, APC-T lymphocyte interaction and B cell hyper reactivity resulting in autoantibody production are consistently observed in SLE patients. DNA and RNA complexes generated during apoptosis are helped by auto-antibodies to be taken up by BCR or FcgRII. Naïve B cell when activated by ligation of their TLR9 receptors undergo class switching to produce anti-nucleosomal IgG2a and IgG2b. If repeatedly acted on by their ligands circulating in the

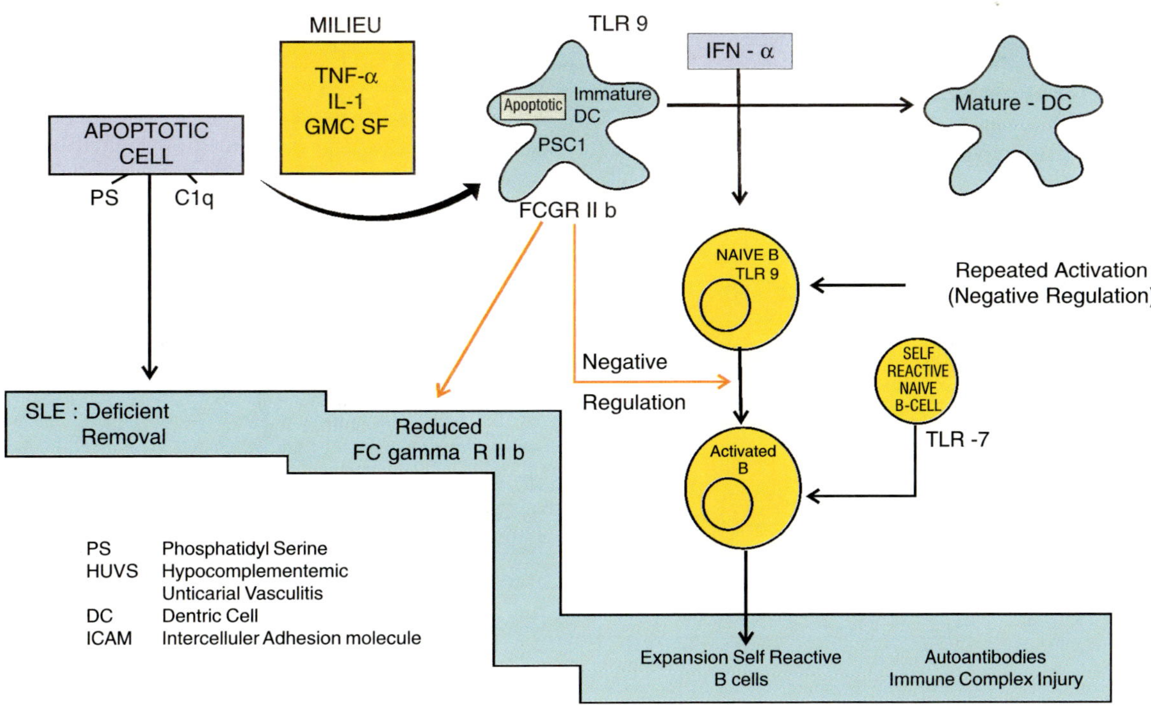

Fig. 2 : Pathogenesis of SLE

bloodstream may induce tolerance. In the face of loss of peripheral tolerance mechanisms there is expansion of self reactive IgG2a and IgG2b plasma cells. This situation occurs when there is reduced expression of the FcgRIIb on B cells and dendritic cells, where they normally function as inhibitory receptor. Genetic mutations in FcgRIIb are the prerequisite for the activation and differentiation of self-reactive B cells and development of autoimmunity in mice and humans. This may have a role in SLE (Fig. 2).

TLR7 present on self reactive naïve B cell promotes self reactive B cell expansion. Some molecules have been recently developed that block all downstream signaling events in TLR9-responsive cells. Some of these also target TLR7 signaling pathways[22]. Prolonged expression of CD40L which is normally expressed only transiently on T lymphocyte leads to state of persistent activation of B cells and the resultant prolificity of antibodies in SLE.

PATHOGENESIS OF RA (Fig. 3 & 4)

In genetic studies, RA is strongly linked to the MHC class II antigens HLA DRB1*0404 and DRB1*0401[23]. These antigens present an unknown arthritogenic peptide to the CD4+T lymphocyte. The latter, causes activation of macrophages, synoviocytes, B cells by secretion of interferon-γ, IL-6 and IL-17. B lymphocytes then produce auto antibodies including IgM RF, while the other cells secretes IL-1, 6, TNF-α and matrix metallo-proteinases which mediate the joint destruction in RA[24]. Recent studies have convincingly demonstrated that citrunillated amino acids are a specific antigenic determinant frequently recognized by auto antibodies in RA patients[25]. In most rheumatic autoimmune disease, spontaneous onset of autoimmunity is noted suggesting that the first reason is more important i.e. specific activation of self reactive lymphocytes by self antigens.

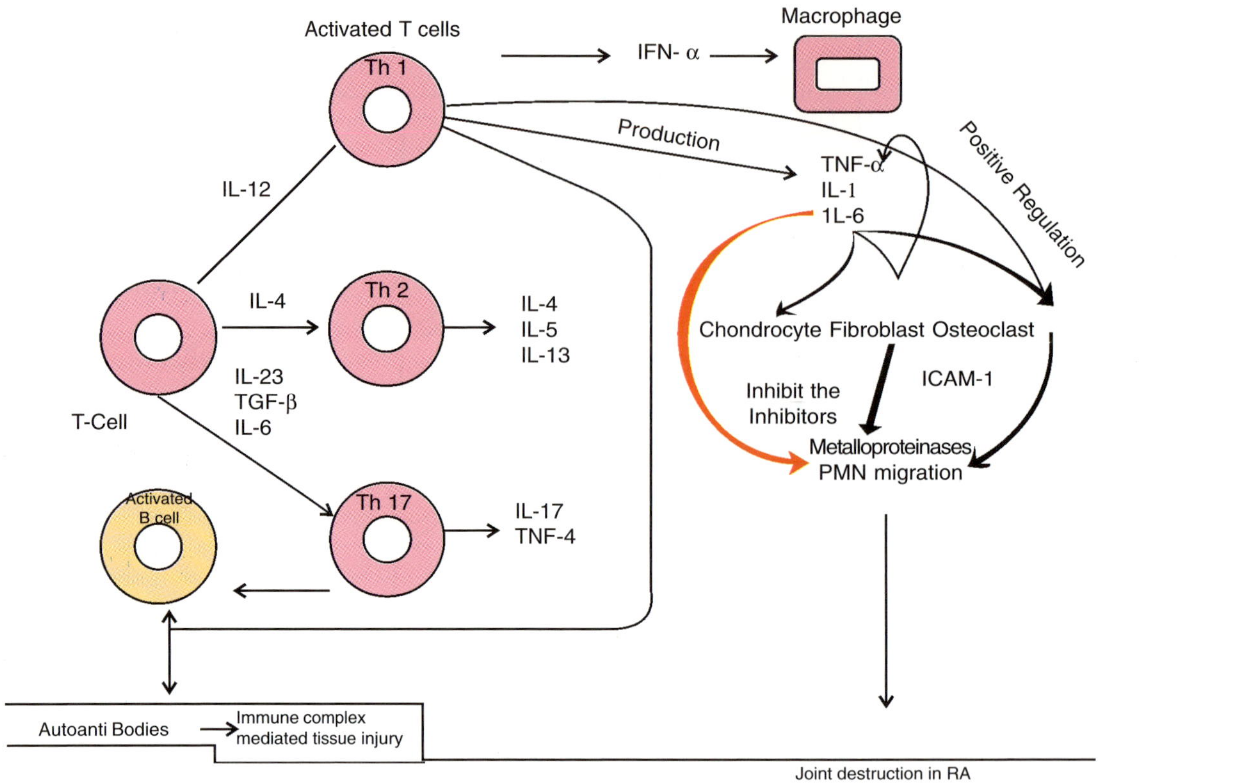

Fig. 3 : Pathogenesis of RA

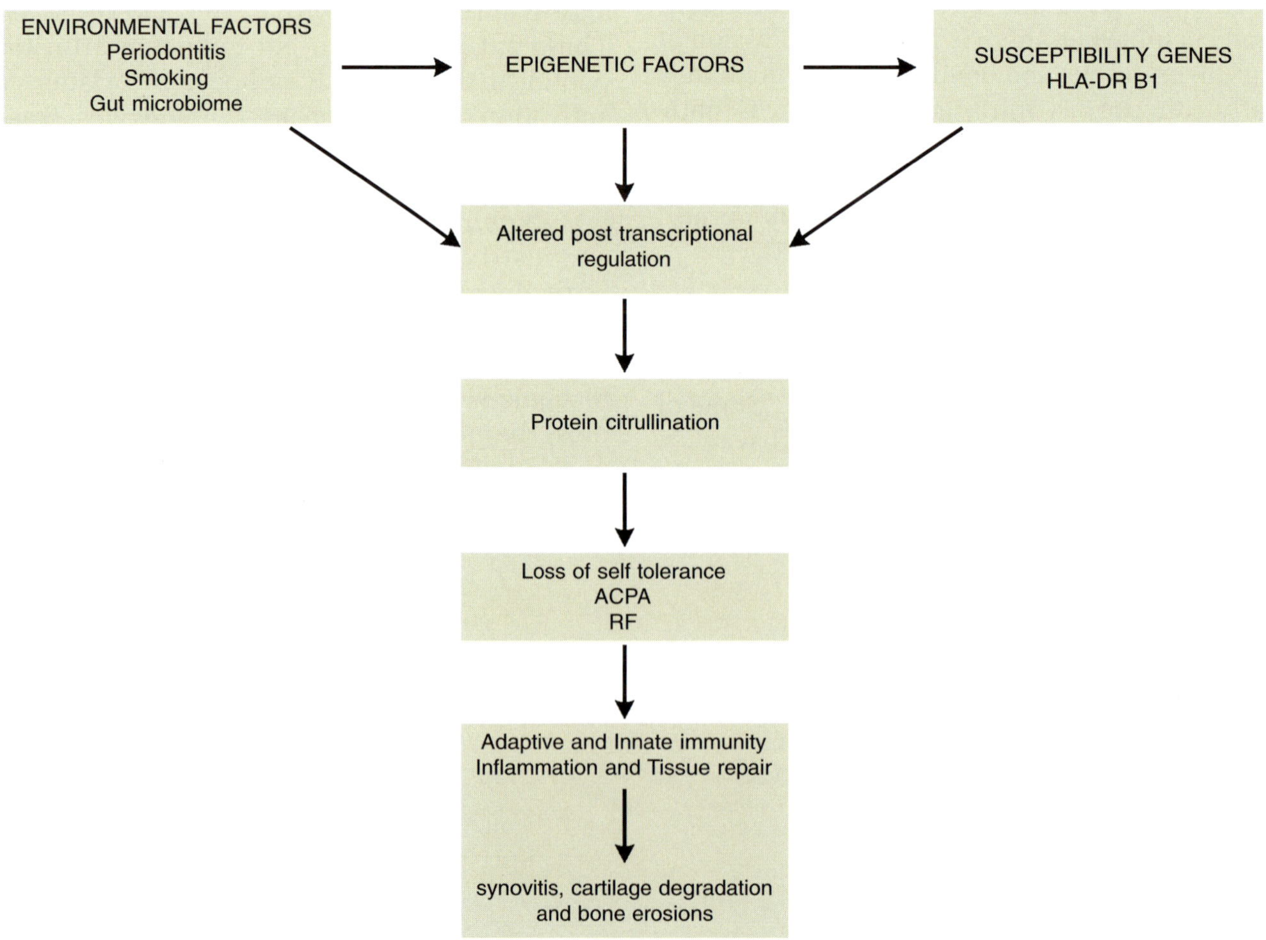

Fig. 4 : Schematic representation of pathogenesis of RA (adapted from 26).

Further Reading:

1. Doria A, Zen M, Bettio S, et al. Autoinflammation and autoimmunity: Bridging the divide. *Autoimmun Rev.* 2012;12(1): 22-30.
2. Ehlers M, Ravetch JV. Opposing effects of Toll-like receptor stimulation induce autoimmunity or tolerance. *Trends Immunol.* Feb 2007; 28(2): 74-79.
3. Marc C. Hochberg M, MPH, Alan J. Silman, MD, Josef S. Smolen, MD, Michael E. Weinblatt, MD and Michael H. Weisman, MD ed. *The Scientific Basis of Rheumatic Disease: Textbook of Rheumatology.* 5th ed; 2011; No. 1.
4. Tiegs SL, Russell DM, Nemazee D. Receptor editing in self-reactive bone marrow B cells. *J Exp Med.* Apr 1 1993; 177(4): 1009-1020.
5. Goodnow CC, Cyster JG, Hartley SB, et al. Self-tolerance checkpoints in B lymphocyte development. *Adv Immunol.* 1995; 59: 279-368.
6. Mackay F, Woodcock SA, Lawton P, et al. Mice transgenic for BAFF develop lymphocytic disorders along with autoimmune manifestations. *J Exp Med.* Dec 6 1999; 190(11): 1697-1710.
7. Zhang J, Xu X, Liu Y. Activation-induced cell death in T cells and autoimmunity. *Cell Mol Immunol.* Jun 2004; 1(3): 186-92.
8. Diamond B, Bluestone J, Wofsy D. The immune tolerance network and rheumatic disease: immune tolerance comes to the clinic. *Arthritis Rheum.* Aug 2001; 44(8): 1730-35.
9. Reif K, Ekland EH, Ohl L, et al. Balanced responsiveness to chemoattractants from adjacent zones determines B-cell position. *Nature.* Mar 7 2002; 416(6876): 94-99.
10. Nagafuchi H, Shimoyama Y, Kashiwakura J, Takeno M, Sakane T, Suzuki N. Preferential expression of B7. 2 (CD86), but not B7.1 (CD80), on B cells induced by CD40/CD40L interaction is essential for anti-DNA autoantibody production in patients with systemic lupus erythematosus. *Clin Exp Rheumatol.* Jan-Feb 2003; 21(1): 71-77.
11. Moore PA, Belvedere O, Orr A, et al. BLyS: member of the tumor necrosis factor family and B lymphocyte stimulator. *Science.* Jul 9 1999; 285(5425): 260-63.
12. Treml LS, Crowley JE, Cancro MP. BLyS receptor signatures resolve homeostatically independent compartments among naive and antigen-experienced B cells. *Semin Immunol.* Oct 2006; 18(5): 297-304.
13. Cancro MP. The BLyS/BAFF family of ligands and receptors: key targets in the therapy and understanding of autoimmunity. *Ann Rheum Dis.* Nov 2006; 65 Suppl 3: iii34-36.
14. Ehlers M, Fukuyama H, McGaha TL, Aderem A, Ravetch JV. TLR9/MyD88 signaling is required for class switching to pathogenic IgG2a and 2b autoantibodies in SLE. *J Exp Med.* Mar 20 2006; 203(3): 553-61.
15. Sakaguchi S. Naturally arising Foxp3-expressing CD25 + CD4 + regulatory T cells in immunological tolerance to self and non-self. *Nat Immunol.* Apr 2005; 6(4): 345-52.
16. Kikly K, Liu L, Na S, Sedgwick JD. The IL-23/Th(17) axis: therapeutic targets for autoimmune inflammation. *Curr Opin Immunol.* Dec 2006; 18(6): 670-75.
17. Selmi C. Autoimmunity in 2010. *Autoimmun Rev.* Oct; 10(12): 725-32.
18. Vercoulen Y, van Teijlingen NH, de Kleer IM, Kamphuis S, Albani S, Prakken BJ. Heat shock protein 60 reactive T cells in juvenile idiopathic arthritis: what is new? *Arthritis Res Ther.* 2009; 11(3): 231.
19. Anaya JM. Common mechanisms of autoimmune diseases (the autoimmune tautology). *Autoimmun Rev.* 2010; 11(11): 781-84.
20. Cho JH, Gregersen PK. Genomics and the multifactorial nature of human autoimmune disease. *N Engl J Med.* 2011; 365(17): 1612-23.
21. Sospedra M. and R. Martin, When T cells recognize a pattern, they might cause trouble. *Curr Opin Immunol.* 2006; 18: 697-703.
22. Lenert P.S. Targeting Toll-like receptor signaling in plasmacytoid dendritic cells and autoreactive B cells as a therapy for lupus. *Arth Res Ther.* 2006; 8: 203.
23. Lanchbury, J.S., The HLA association with rheumatoid arthritis. *Clin Exp Rheumatol.* 1992; 10: 301-04.
24. Choy E.H. and G.S. Panayi, Cytokine pathways and joint inflammation in rheumatoid arthritis. *N Engl J Med.* 2001; 344: 907-16.
25. Schellekens G.A., et al., Citrulline is an essential constituent of antigenic determinants recognized by rheumatoid arthritis-specific autoantibodies. *J Clin Invest.* 1998; 101: 273-81.
26. McInnes IB, Schett G. The pathogenesis of rheumatoid arthritis. *N Engl J Med.* 2011; 365(23): 2205-19.

2 MHC AND NON-MHC GENES AND RHEUMATIC DISEASES

Sita Naik

Major histcompatibilty complex

The products of major histocompatibilty complex (MHC) in man is termed human leukocyte antigens (HLA). HLA antigens are controlled by a cluster of highly polymorphic genes located on the short arm (p21.3) of chromosome 6[1,2]. The region is arbitrarily subdivided into three gene clusters. HLA-A, HLA-B and HLA-C loci form the Class I cluster while HLA-DR, HLA-DQ and HLA-DP loci form the Class II gene cluster. The class I and class II gene products differ in structure, function and tissue distribution. Genes which encode molecules that function as peptide transporters are also mapped within class II region. These proteins transport peptides from cytoplasm to endoplasmic reticulum, enabling the peptide to interact with newly synthesized class I molecules. A third set of genes, called class III genes, encode for a group of proteins that include complement components C2, C4 and factor B, TNF, 21-Hydroxlase and heat Shock Protein-70. These are structurally unrelated to class I and II gene products and are grouped with them only because they are located in the same chromosomal region (Fig. 1).

MHC class II molecules are expressed constitutively only on the surface of interstitial dendritic cells, macrophages, and B cells. MHC class II expression is strongly upregulated on epithelial cell and vascular endothelial cell after exposure to the pro-inflammatory cytokine IFN. MHC class II molecules bind peptides derived from exogenous proteins.

Structure of HLA antigens

HLA molecules are cell surface-bound glycoproteins that contain four immunoglobulin- like domains. (Fig. 2) MHC class I molecules consist of MHC encoded heavy chain and β2 micro- globulin, the gene for which is located on chromosome 15 in humans. The heavy chain has 3 extra cellular domains, α1, α2 and α3, with α3 domains closest to the plasma membrane. β2 microglobulin is associated with the extra cellular portion of the β chain. The class II molecules consist of two covalently bound chains, termed α and β chains, each consisting of 2 extra cellular domains. The peptide binding grooves of Class I and Class II molecules are structurally analogous and consist of

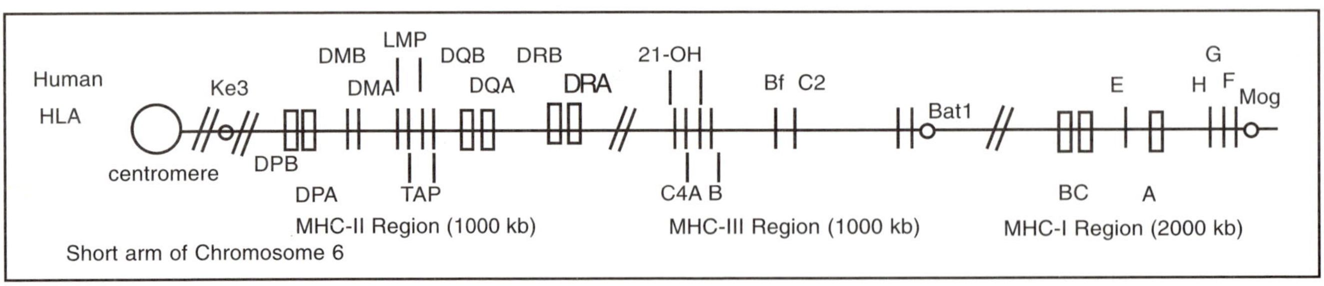

Fig. 1. The arrangement of HLA loci on chromosome 6

The products of the MHC are expressed on the surfaces of a variety of cells. The major function of HLA molecules is presentation of foreign or self antigens to the cells of the immune system. Antigen specific T cells do not recognize antigen in free or soluble form or as intact protein but recognize the processed peptides bound to MHC molecules.

MHC class I molecules are present on all nucleated cells and present cytoplasm derived peptides originating from the normal components of the same cell, or of intracellular parasites, principally viruses.

two parallel α helical amino acid chains, resting on an eight β pleated polypeptide sheets. It is made of the 1 and 2 domains of the heavy chain of class I molecules and the α1 and β1 domains of class II molecules. The class I and II genes are highly polymorphic with, most of the polymorphism occurring in HLA-A, B and C and HLA-DRB genes.

Nomenclature

All class II genes are prefixed by a letter for the sub region (N, O, P, Q, R) and then the letters A or B for

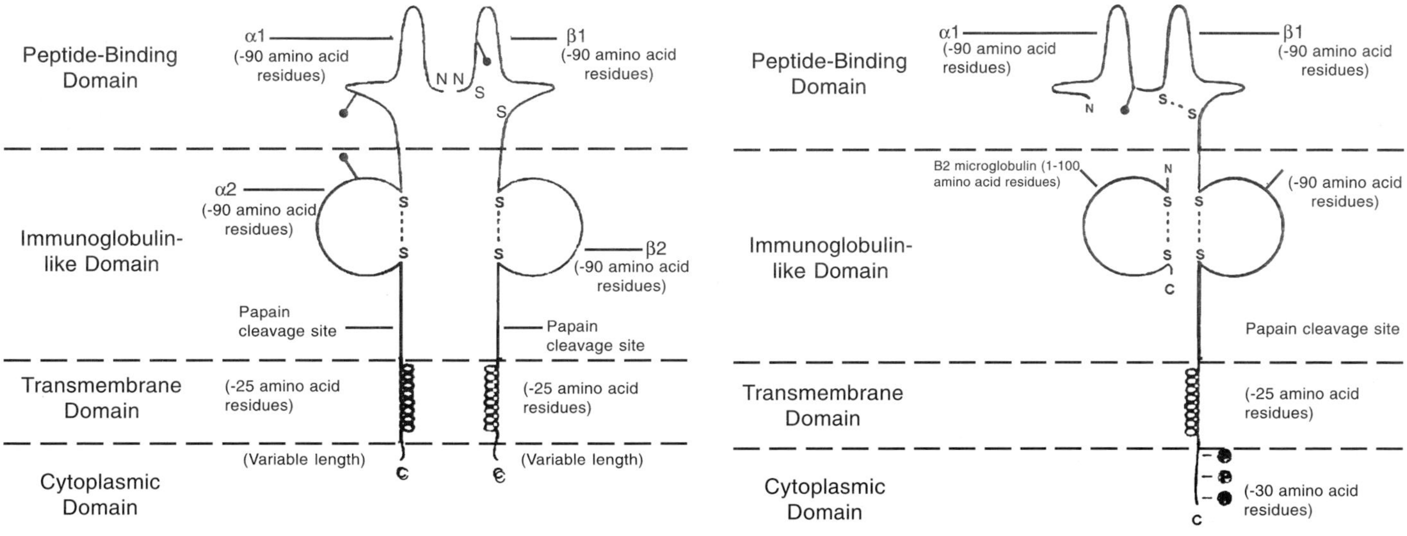

Fig. 2. Schematic diagram of HLA Class II (Left) and Class I (Right) molecules

α or β chains an Arabic number for the different genes follows this. For example, DQB1 designates the gene that encodes the expressed chain of the DQ molecule and DQB1 codes for chain of DQ region. Alleles for each gene are numbered using four digits. The first two digits describe the most closely associated serologic specificity and the next two the subtypes within a serological specificity. An asterisk is used to separate genes and alleles. For example, serologically defined specificity DR7 will be written as DRB1*07, since this is the allele of the β1 locus of DR gene. The subtypes of DRB1*07 will be designated as DRB1*0701, or DRB1*0704. DR4 subtypes would be written as DRB1*0401, DRB*0403 and DRB1*0405, while DQB1*0302 is a DQB subtype of the DQ3 specificity

Method of HLA typing

HLA typing for clinical purposes is now performed only by molecular techniques. Specified stretches of DNA, which reflect the allelic differences between the HLA antigens, are amplified by PCR and hence the method is named Sequence Specific primer (SSP) based typing. This procedure consists of four steps, the extraction of chromosomal DNA from blood, PCR amplification using the extracted DNA and the supplied, primers from recognized companies, separation of the PCR fragments by agarose gel electrophoresis followed by documentation and interpretation of the result. The whole procedure takes about two hours and 30 minutes. Many reliable SSP kits are available which have multiple primer sets in separate reactions to type for many alleles at a locus or loci (Fig. 3).

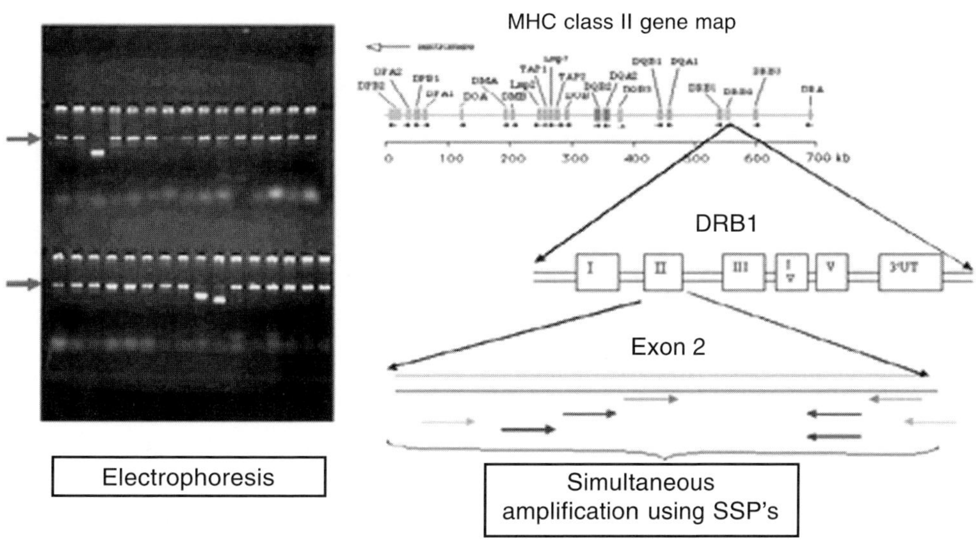

Fig. 3. Low-resolution PCR with sequence-specific primers (PCR-SSP)

HLA typing in rheumatic diseases

The strength of genetic association is reflected in terms of relative risk (RR), which is a statistical odds ratio representing the risk of disease in an individual carrying a particular genetic marker compared with the risk in individuals in that population without the marker. Although an increased RR has been found with a number of rheumatic diseases, such as rheumatoid arthritis (RA), ankylosing spondylitis (AS), systemic lupus erythematous (SLE) etc. these give an indication of the contribution of genetic factors towards the etiology[3,4]. The highest relative risk is seen for HLA-B27 in AS and other spondyloarthropathies[5]. However, the presence of HLA-B27 has no value for diagnosis of ankylosing spondylitis (AS). A positive HLA-B27 test has limited utility in young patients with inflammatory back pain in whom the positive B27 test increases the likelihood of having AS, particularly if imaging of SI joints does not provide conclusive results.

HLA DR4 has a very high relative risk both for development of RA and disease severity. However, HLA is not included in the diagnostic criteria for RA and typing in a suspected case does not in any way assist in establishing diagnosis.

Non-MHC Genes

The rheumatic diseases have an immune basis for their pathogenesis. Hence it would not be surprising that genetic variations in genes for many immunologically important molecules like cytokines, cytokine receptors, adhesion molecules, integrins, signal transduction molecules etc. should be associated with disease susceptibility. The current practice is to look at the total genomic differences between patients and normal controls, an approach that is termed genome wide association scan (GWAS). Single nucleotide polymorphisms in PTPN22 IRF5 (interferon regulatory factor 5), TNFα-induced protein-3 (TNFAIP3; also known as A20), the region containing TRAF1 (TNF receptor associated factor 1) and C5 (complement component 5), Integrin-α-M (ITGAM) have been reported to show association with RA and/or SLE[7,8]. Whereas, no single association has consistently shown the degree of association which would lead to its application in clinical practice, the field is very dynamic and new findings are being reported regularly. Besides, genetic variants associated with disease susceptibility, there is also the possibility that markers for disease severity, progression, treatment response etc will also be discovered. Hence, this is an exciting time and there is every hope that testing for genetic markers or combination of markers will become part of management of patients in the near future. Hence, it is important to understand these developments in order to be able to use them when they become available.

Further Reading :

1. Kindt TJ, Goldsby RA, Osborne BA. The major histocompatibility complex and antigen presentation. In Kuby Immunology. W H Freeman and Co; New York. 2007; 189-217.
2. Trowsdale J. HLA genomics in the third millennium Curr Opin Immunol 2005; 17: 498-504.
3. Olsson T, Jagodic M, Piehl F, Wallstrom E.Genetics of autoimmune neuroinflammation. Curr Opin Immunol. 2006; 18: 643-49.
4. Pearce SH, Merriman TR. Genetic progress towards the molecular basis of autoimmunity. Trends Mol Med 2006; 12: 90-98.
5. Gaston H. Mechanisms of Disease: the immuno-pathogenesis of spondyloarthropathies. Nat Clin Pract Rheumatol. 2006; 2: 383-92.
6. Cowell JK, Hawthorn L. The application of microarray technology to the analysis of the cancer genome. Curr Mol Med. 2007; 7: 103-20
7. Serrano NC, Millan P, Paez MC. Non-HLA associations with autoimmune diseases. Autoimmun Rev 2006; 5: 209-14.
8. Clarke A, Vyse TJ. Genetics of rheumatic disease. Arthritis Res Ther. 2009; 11: 248.

SEROLOGIC EVALUATION OF RHEUMATIC DISEASES 3

Radha Madhavan

Introduction

Most of the systemic rheumatic diseases are characterized by

a) Chronic inflammation
b) Acute exacerbation and spontaneous remission of disease activity
c) Auto immune response
d) Intercurrent infection
e) Damage to organs like liver and kidney as a part of the disease activity or as side effects of prolonged therapy required for these type of rheumatic diseases

Serum serves as an excellent source to assess these events, which take place in various tissues and organs of the body.

Serological markers of inflammation

Acute phase response is a major patho-physiological phenomenon that accompanies not only acute inflammation but also chronic inflammatory events. Acute phase proteins (APP) are those proteins which are associated with acute phase response. Plasma concentration of APPs change by at least 25% in the first 7 days of inflammation. There are approximately 30 APPs. The concentration of some proteins increase (positive reactants) while some decrease (negative reactants) in response to inflammation. APPs are produced by liver cells induced by several cytokines which are derived from activated macrophages at the site of injury and many other cell types such as fibroblasts and endothelial cells. The properties of major APPs are shown in Table 1[1,2].

APPs most commonly measured for evaluation of inflammatory status are C Reactive Protein (direct measurement) and fibrinogen [indirectly as Erythro sedimentation rate (ESR) and Plasma viscosity].

The relative merits of ESR, CRP and plasma viscosity are shown in Table 2.

Table 1 : Properties of major APPs

Type of Reactants	Group	Acute Phase Proteins	Response Time (Hrs)	Adult Reference Range[1,2]
Positive reactants	**Group I**	Ceruloplasmin	48-72	0.20-0.60 g/l
	Mild elevation	Complement C3	48-72	0.75-1.65 g/l
	< 1 fold increase	Complement C4	48-72	0.20-0.65 g/l
	Group II	α_1 acid glycoprotein	24	0.6-1.28 g/l
	Moderate elevation	α_1 antitrypsin	10	1.1-2.18 g/l
	2-4 fold increase	Haptoglobin	24	0.5-2.08 g/l
		Fibrinogen	24	2-4 g/l
	Group III	CRP	6-10	0.06-8.0 mg/l
	Marked elevation	Serum Amyloid A	6-10	–

Albumin and Transferrin are negative reactants

Table 2 : Relative merits of CRP, ESR and plasma viscosity[3]

	ESR	Plasma Viscosity	CRP
Method	Simple, inexpensive, difficult to standardise	More rapid test, easily standardised	Semi quantitative and quantitative methods available
Variables	Age,anaemia, large asymmetric plasma proteins (fibrinogen, α2 macroglobulin, immunoglobulins)	Plasma proteins similar to ESR; Independent of age and anaemia	Determined solely by rate of synthesis
Kinetics	Rises after 24-48 hrs of tissue damage; subsides with half life of 4-6 days	Same as ESR	Rises 6 hrs after tissue damage; subsides with half life of 48 hrs
Value of the test	Integrates the effect of anaemia, acute phase response and immune response; for detecting and monitoring chronic inflammation	Reflects changes in plasma proteins; for chronic inflammation	Reflects ongoing inflammation; unsuitable for detecting mild inflammation

CRP is commonly measured by latex agglutination test, using latex particles coated with antibodies to CRP serum containing CRP of 6mg/l or more, when mixed with latex particles will agglutinate latex particles. The highest dilution of serum (titre) causing agglutination can be converted to CRP level in mg/l by multiplying with a factor provided in the commercial kit. Thus it is a semi quantitative measure. CRP measurement by turbidimetry or naphelometry give the exact measure of CRP in sera.

Measure of these APPs are of clinical use

- As diagnostic tools
- to assess the disease activity, the course of disease and response to therapy
- to detect inter current infection

CRP and/or ESR: as diagnostic tools

- CRP along with ESR-screening test to differentiate inflammatory arthritis. Osteoarthritis is not associated with acute phase response
- Markedly elevated ESR (> 100 mms) - very specific for organic pathology and bacterial infection
- 95 to 100% polymyalgia rheumatica (PMR)/ giant cell arteritis (GCA) patients present with markedly elevated ESR

An ESR > 50 is included as one of 5 diagnostic criteria for classification of GCA helps to differentiate PMR from other non inflammatory causes of myalgia.

- Patients presenting with back pain and markedly elevated ESR: Consider myeloma, infective discitis, osteomyelitis and malignancy
- Elevated ESR and normal CRP: Anaemia, hyper lipidaemia, hypergammaglobulinaemia to be considered (such situation is characteristic of SLE and Sjogren's)

CRP/ESR : Other clinical uses

Rheumatoid arthritis (RA), is generally associated with marked acute phase response. CRP levels are generally in the range of 30-40mg/l with moderate disease activity. In severe disease, it may reach in excess of 100mg/dl. CRP of more than 150 mg may be an indication of inter current infection. Elevated CRP correlates with generalized osteoporosis hence treatment of RA is aimed at bringing down and maintaining the CRP level to the lowest possible. The American College of Rheumatology has recommended that ESR or CRP should be one of the core set of disease activity measures in clinical trials. Rather than the non steroidal anti inflammatory drugs, disease modifying drugs have more effect on bringing down the level of acute phase proteins.

In clinically active SLE, ESR may be elevated but comes down with remission but CRP is rarely elevated in active disease. This point is useful to differentiate the cause of fever in SLE due to disease activity from that due to bacterial infection when CRP will be markedly elevated.

In Ankylosing Spondylitis, CRP or plasma viscosity are more sensitive markers of active disease than ESR.

In rheumatic fever raise in CRP and ASO titre signifies reactivation of disease.

In giant cell arteritis, measurement of either CRP or ESR may be used, together with clinical assessment to monitor the initial effectiveness of steroid therapy, the normal level being possible within 7-14 days after start of steroid.

Serological markers of auto immune response - auto antibodies

Auto antibodies are characteristic feature of auto immune rheumatic diseases. Though auto immune response is considered as an abnormal immune response, in auto immune rheumatic diseases, auto antibodies are directed against highly specific targets.

It should be made clear that mere presence of auto antibodies (auto immunity) is not synonym with auto immune diseases. Auto antibodies can be observed in otherwise healthy individuals. But when the presence of auto antibodies is associated with characteristic clinical manifestations, then, that condition is synonym with auto immune disease. Hence selection of specific auto antibody tests and interpretation of test results need a clear understanding of validity and reliability of the test requested.

Validity of A test can be assessed by many parameters:

- Sensitivity (test with 100% sensitivity does not miss detection of antibody even in a single patient)
- Specificity (test with 80% specificity detects antibody in 20% of normal individuals without disease)
- Positive predictive value (PPV) (chances of having a disease when a test is positive)
- Negative predictive value (NPV) (chances of not having a disease when the test is negative). It is desirable to have high values for all these parameters

The positive and negative predictive values are affected by what are known as pre test disease probability and the prevalence of a disease in a population.

To explain with an example, SLE is not a common condition (prevalence < 0.001% i.e. 1 in 1000 individuals) whereas the Anti Nuclear Antibody (ANA), an auto antibody, specific for SLE can be present in 5 to 8% of normal population. Therefore if the ANA test is performed randomly without any clinical indication, and if the result is positive, the PPV will work out to be < 1% and NPV will be > 99%. But if the test is performed in a young female patient who presents with clinical manifestations characteristic of SLE (fever, rash, arthralgia, oral ulcers, serositis, leucopenia etc.) then the PPV reaches > 90%[4].

Table 3 : Auto antibodies and associated auto immune diseases

Auto antibody	Target antigen	Autoimmune disease
Rheumatoid Factor	Fc fraction of IgG molecule	Rheumatoid Arthritis
Anti Nuclear Antibody	Nucleic acids, nucleoproteins and cytoplasmic components	ANA Diseases
Anti Neutrophil Cytoplasmic Antibodies	Cytoplasmic components of neutrophils	One group of vasculitides
Anti Phospholipid Anti body	Phospholipids and β2 microglobulin	APL syndrome

Many autoimmune diseases are associated with a specific autoantibody at a significantly high titre compared to the level present in individuals without that particular disease. Thus, interpretation of a test report requires an understanding of the cut off value for a particular antibody. The upper limit of antibody level which is present in not more than 5% normal population is fixed as the cut off level of an auto antibody for the specific disease.

Important autoantibodies of clinical importance, the autoantigens which are the targets of autoantibodies and the associated auto immune diseases are listed in Table 3.

Rheumatoid factor (RF)

Rheumatoid factors are auto antibodies directed against Fc portion of IgG. Though RF may belong to any isotype, IgM RF is the one which is currently measured in clinical practice. RF can be detected by latex agglutination method, ELISA or nephelometry. Latex agglutination is routinely used in clinical lab for the diagnosis of Rheumatoid arthritis. The reciprocal of highest dilution of the serum which is able to agglutinate IgG coated latex particles is expressed as titre. The titre can be converted to International units (IU) by multiplying the titre with the conversion factor provided in the kit. For eg if the highest dilution is 1/20, the titre is 20 and if the conversion factor is 8, then RF level is calculated as 20 x 8 = 160 IU.

Testing for RF is primarily used for the diagnosis of rheumatoid arthritis. 70-80% of RA patients are positive for IgM RF at a titre more than 20 IU.

Table 4 : Association of RF with conditions other than RA

Category of disease	Disease conditions
Rheumatic diseases other than RA	SLE, SSc, SS, MCTD
Active viral infections	Mono nucleosis Hepatits Influenza After vaccination
Parasitic infestations	Trypanasomiasis Malaria Schistosomiasis
Chronic infective inflammatory disease	Tuberculosis Leprosy Syphilis Sub acute bacterial endocarditis
Neoplasms	After irradiation or chemo therapy
Other hyperglobulinaemic diseases	Chronic liver disease Sarcoidosis Cryglobulinaemia

However, RF can be positive in conditions other than RA (Table 4).

Hence interpretation of results should be made only in the back ground of clinical presentation:

- If RF is positive along with characteristic clinical presentation, the patients are categorized as sero positive RA and require aggressive therapy, because presence of RF early in the course of disease is associated with erosion and extra articular features.
- If RF is negative consistently in a patient presenting with typical clinical features of RA, the patient is categorized as sero negative RA. It is for this group that other serological diagnostic markers are to be tested for early intervention.
- If RF is positive in patients without typical features of RA, other causes for RF positivity are to be ruled out.

Anti Cyclic Citrullinated Peptide Antibodies (Anti CCP) has emerged as another serological marker for Rheumatoid Arthritis (RA). The generation of Anti CCP is an early event during the disease course of RA[5]. The diagnostic accuracy of anti CCP is as good in early disease as in established RA. Moreover, Anti CCP positive patients have a more erosive disease than Anti CCP negative patients[5]. Anti CCP has improved the ability to predict the risk of progression of early inflammatory arthritis to RA[6]. Aggressive therapy with disease modifying anti rheumatic drugs during this early phase of arthritis can change the natural course of diseases for better outcome. Anti CCP-2 ELISA assay has been shown to be more predictive than Anti ccp-1 assay[7].

Anti CCP has been included as one of the serological markers in 2010 ACR classification criteria for RA[8,9]. It is suggested that each lab should set up their own Upper Limit of the Normal (ULN) value expressed as International Units (IU). Level of Anti CCP less than ULN is considered to be negative (score 0); value greater than ULN but less than 3 times ULN is considered to be moderate positive (score 2) and value greater than 3 times ULN is considered to be highly positive (score 3).

Antinuclear Antibodies (ANA)

ANA are associated with many systemic connective tissue disorders (ANA diseases)[10] including systemic lupus erythematosus (SLE), systemic sclerosis, mixed connective tissue disorders (MCTD), Sjogren's syndrome (SS), polymyositis/dermatomyositis (PM/DM).

Methods of detection of ANA

1. *Fluorescent ANA (FANA)* - Mouse liver cells/ Hep2 cells fixed on glass slides are used a sources of antigens; sera containing ANAs will bind to specific nuclear antigens the ag-ab binding is traced with anti human immuno globulin tagged with fluorescent dye (Fluorescent Iso Thio Cyanate). When viewed under fluorescent microscope, the targeted antigen will appear apple green in colour. The specific pattern of fluorescence is diagnostic of different ANA diseases (Fig. 1).
2. *ELISA* - Various nuclear and cytoplasmic antigens are separated and purified and coated onto wells of ELISA plates. Sera containing specific antibody will bind to the target antigens and the binding is traced with enzyme tagged anti human immuno globulin antibodies. The intensity of colour change in a positive reaction is proportional to antibody titre and is measured as optical density (OD) values which can be expressed as units of activity.
3. *Immunoblot* - The nuclear and cytoplasmic antigens are separated in a gel according to their molecular weights and then transferred to membranes. Principle similar to ELISA is used to view the distinct bands of reactivity. The advantage of this method is that the presence of many auto antibodies in a patient can be simultaneously detected as set of bands (auto antibody profile).
4. *Immuno precipitation* - Radio labeled cell extracts are immuno precipitated by reaction with test sera and then resolved by electrophoresis and analysed. This technique is restricted to research laboratories.

Table 5 provides information about various auto antibodies commonly detected in laboratory. Some are useful as screening test (to rule out the possibility of auto immune rheumatic diseases), and others are very useful for diagnosis, for assessing prognosis and disease activity of specific rheumatic diseases.

Antinuclear antibodies of dagnostic significance

Fluorescent Anti Nuclear Antibodies (FANA) : A screening test

It is a valuable screening test to diagnose ANA diseases. Rodent liver cells or Hep2 cells (malignant cell line) are used as antigen source. Use of Hep2 cells increases the sensitivity of ANA detection, because of the larger size of nucleus and a good representation of most of the target antigens. Hep2

Fig. 1 : Fluorescent antinuclear antibodies - patterns

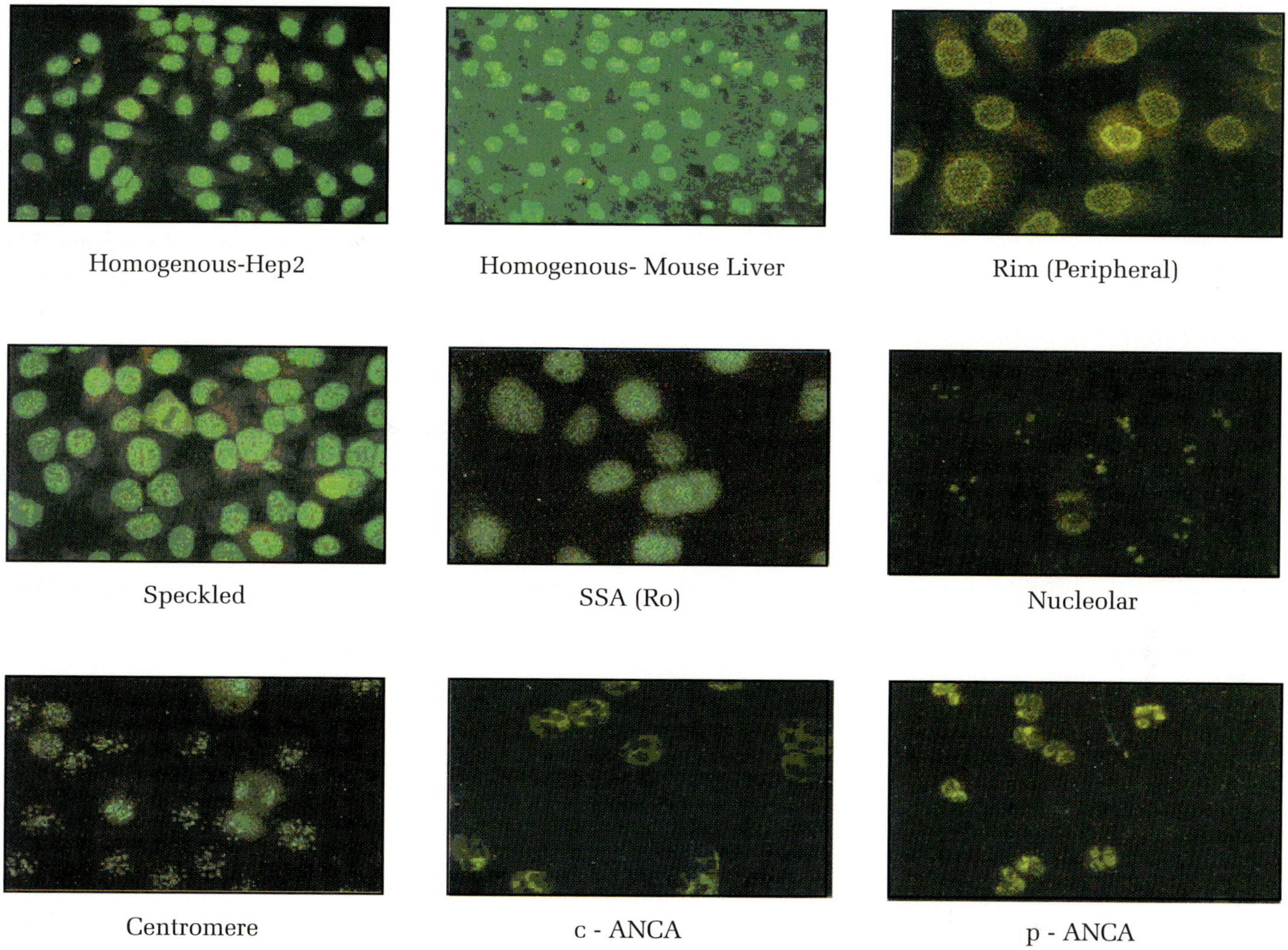

cells coated on to slides are available commercially. Rodent cells as antigenic source can be prepared in the laboratory at a considerably lower cost, provided animal house facilities are available[11].

The pattern of fluorescence gives a clue to differentiate among ANA diseases.

Almost 95% untreated SLE are positive for ANA at higher titre of serum. A significant cut off titre for the diagnosis of SLE should be established for each laboratory by testing on SLE, non SLE connective tissue diseases and other non rheumatic diseases. Moreover interpretation of test should be made only in the context of presenting clinical presentation.

The sensitivity of ANA for a particular ANA disease can vary widely Table 6.

When FANA is positive, depending on the clinical presentation, additional tests are to be ordered for confirmation of diagnosis (Table 7). These confirmatory tests are mostly performed by ELISA using commercial kits. If FANA is negative, these tests have no value as confirmatory tests. However, upto 5% of SLE and certain percentage of patients with PM/DM and Anti Phospholipid Antibody syndrome may be ANA negative. When there is a strong clinical suspicion, Anti ds DNA, Anti tRNA synthetase antibodies and APA may be estimated for these patients respectively.

Anti ds DNA assay by ELISA is valuable for confirming the diagnosis of SLE since in other ANA diseases anti ds DNA is present only in $< 5\%$. Anti Sm is detected only in SLE sera adding their value in confirmation of SLE.

Though SLE patients may show a set of ANAs, Anti u1 RNP, Anti SSA, anti SSB, MSA and Anti Scl70 and anti centromere antibodies are present in higher titre and in higher percentage of MCTD, SS, PM/DM and Systemic sclerosis patients respectively, thus are valuable as confirmatory tests for corresponding diseases[12,13].

Table 5 : Value of ANA tests

Antibody specificity	Method of detection	Clinical relevance				Disease association
		Screening test	Diagnosis	Disease pattern/ Prognosis	Disease activity	
ANA	IIF	Yes				ANA diseases
Ds DNA	IIF, ELISA		Yes	Yes	Yes	SLE
U1 RNP	ELISA, IB			Yes		SLE, MCTD
Sm	ELISA, IB		Yes	Yes		SLE
Ro (SSA)	ELISA, IB			Yes		SLE, SS, Neonatal lupus
La (SSB)	ELISA, IB			Yes		SLE, SS, Neonatal lupus
Topoisomerase 1 (scl 70)	ELISA, IB		Yes	Yes		Systemic sclerosis
Histidyl-tRNA synthetase (Jo1)	ELISA, IB, IP		Yes	Yes		Polymyositis
Anti centromere	IIF, ELISA		Yes	Yes		(CREST Syndrome) Systemic sclerosis
Anti neutrophils Cytoplasmic antibodies	IIF, ELISA		Yes	Yes	Yes	Wegener's Granulomatosis Microscopic poly angiitis, Churg Straus
Myositis specific antibodies	IF, immuno precipitation		yes	Yes		Antisynthetase syndrome*, polymyositis, Dermatomyositis

IIF: Indirect immuno fluorescence; IB; Immuno blot; IP: immunoprecipitation
* Antisynthetase syndrome : interstitial lung disease, fever, arthritis, Raynaud's phenomenon, and mechanic hands.

Table 6 : FANA- Diagnostic tool

Category	Disease	FANA Percentage	Value of ANA Testing
ANA diseases	SLE	95-100	Helpful for diagnosis
	Systemic sclerosis	60-80	Helpful for diagnosis
	Mixed Connective Tissue Disease	100	Required for diagnosis
	Sjogren's syndrome	40-70	Helpful for diagnosis
	Polymyositis/Dermatomyositis	40-60	Helpful for diagnosis
	Rheumatoid Arhtritis	52	Not useful for diagnosis
	Poly arteritis nodosa	18	Not useful for diagnosis
	Drug induced lupus #	100	Required for diagnosis
Other rheumatic diseases	Juvenile rheumatoid arthritis	20-50	May be useful for prognosis
	Anti phopholipid antibody syndrome	40-50	May be useful for prognosis
Non Rheumatic diseases $	Auto immune hepatitis	100	Required for diagnosis
	Hashimoto's thyroiditis	46	Not useful for diagnosis
	Graves' disease	50	Not useful for diagnosis
Normal population		8	

$ ANA can also be associated with malignanacy, chronic infectious diseases, sub acute bacterial endocarditis, some lymphoproliferative disorders
# Drugs like procainamide and hydralazine

Anti neutrophil cytoplasmic antibodies (ANCAs)

ANCAs are closely associated with Wegener's granulomatosis, microscopic poly arteritis and Churg Straus syndrome. They are detected by 2 methods. One method is by immuno fluorescence using ethanol fixed and formalin fixed neutrophils as substrate. Two types of fluorescent patterns can be observed.

Cytoplasmic ANCA (cANCA) stain the cytoplasm diffusely in ethanol fixed neutrophils and are targeted against a proteinase called PR3. This antibody is mainly seen in patients with diffuse Wagener's granulomatosis and renal involvement (90%) and microscopic poly arteritis (50%)[14].

Perinuclear ANCAs (pANCA) stain the rim of nucleus in ethanol fixed neutrophils and diffuse granular cytoplasmic staining with formalin fixed neutrophils. These antibodies are directed against myelo peroxidase (MPO) and many other cytoplasmic antigens of neutrophils and are usually associated with many rheumatic and non rheumatic diseases. The second method of measuring Anti PR3 and anti MPO is by ELISA.

The predictive value of ANCA by any or both methods depends on clinical presentation of patients. An elevated ANCA titre in a patient presenting with acute or rapidly progressive glomerulo nephritis predicts the presence of Wegener's granulomatosis, microscopic poly arteritis or idiopathic necrotizing glomerulo nephritis with 98% accuracy. But the accuracy is low for patients who are tested for the possible diagnosis of vasculitis. cANCA is more specific than pANCA[15]. However diagnosis of vasculitis is confirmed only by biopsy.

Table 7 : Guidelines for additional ANA tests for evaluation

Clinical indication for FANA test	FANA pattern	Antibody specificity	Sensitivity %	Clinically associated subgroups
SLE	Homogenous	Ds DNA	70-80	Disease activity, ?Lupus nephritis
	Speckled nuclear	Sm	30-50	Active disease
	Coarse speckled nuclear	U1 RNP	40	Myositis, Raynaud's phenomenon, fibrosing alveolitis
	Fine speckled/ Negative	SSA(Ro)60/ SSB (La)	SSA(Ro) 60kd/52kd - 30% SSB	Sjogren's syndrome, Skin manifestations Complement deficiencies
	Fine speckled/ negative	SSA(Ro)52/ SSB(La)?	(La) - 10%	Congenital heart block
Systemic Sclerosis	Grainy nuclear, nucleolar	TopoisomeraseI[12]	22	Diffuse disease and fibrosing alveolitis
	Diffuse Nucleolar	Th, RNA polymerase		
	Coarse speckled nuclear	U1 RNP	10	Diffuse disease, myositis, pulmonary hypertension
	Centromere	CENP A,B,C	40	Limited disease, pulmonary hypertension
PM/DM	negative	tRNA synthetase (Jo1)[13]	20	Fibrosing alveolitis, anti-synthetase syndrome
	Diffuse cytoplasmic	SRP		Severe disease, distal myositis
	Fine speckled nuclear	Mi2		Classical dermatomyositis
SS	Fine speckled/ negative	SSA(Ro), SSB (La)	60-75	Primary & secondary SS High titre associated with extra glandular features (vasculitis)
MCTD	Coarse speckled nuclear	U1 RNP	100	RA, SLE, SScl, PM overlap

Anti phospholipid antibodies (aPL antibodies)

Antiphospholipid antibodies are a heterogeneous group of antibodies that bind serum proteins such as 2 GP1 or prothrombin,cardiolipin and negatively charged phospholipids. Patients with these antibodies may have a variety of clinical manifestations including venous and arterial thrombosis, recurrent fetal losses or thrombocytopenia referred to as anti phospholipid Syndrome (APS). APS may be primary when it occurs alone or secondary when it is associated with SLE or other auto immune disorders.

aPL : Measured by 2 types of tests

One is a functional assay known as lupus anti coagulant (LAC) test which measures the ability of a subset of aPL antibodies to prolong clotting tests such as partial thromboplastin time (PTT) or Russel Viper Venum test.

Anti cardiolipin antibodies (aCL), the other subgroup of aPL can be measured by ELISA. The antibodies of different isotypes such as IgG, IgM or IgA are measured and reported as GPL, MPL or APL units respectively. 10-20 GPL or MPL represent low positive 20-80 moderately positive and > 80 GPL or MPL represent high positive values. A moderate to high GPL, present on more than one occasion at least 6 weeks apart is most specific for a diagnosis of APS.

aPL is found with increased frequency in SLE patients and associated with hypercoagulable state, thrombo cytopenia and late trimester abortion. Approximately 31% have LAC and 40-47% have aCL. 50% of patients with LAC have SLE[16].

Both LAC and aCL have also been found in patients with a variety of auto immune and rheumatic diseases, idiopathic thrombocytopenic purpura, certain infections and after administration of some drugs.

Serological markers of infection

Anti Streptolysin O antibody (ASO)

It is an useful serological evidence of preceding streptococcal infection and useful in the evaluation of acute rheumatic fever. Recovery of the causative organism, beta haemolytic streptococci from the primary site of infection (throat) is usually negative by the time rheumatic fever appears. After streptococcal pharyngitis, the antibody peaks at about 4-5 weeks, which corresponds to second to third week of rheumatic fever. Thus in a paired serum sample (one at the time when rheumatic fever is suspected and the second 1 to 2 weeks later), a four fold rise in titre is diagnostic of rheumatic fever. In healthy school going children the upper limit of the normal is 400 IU/ml and detected by latex agglutination test.

Since acute rheumatic fever is an inflammatory condition, a rise in CRP and ESR are associated with acute rheumatic process, if not suppressed by anti rheumatic drugs.

Other markers of infection

Whenever reactive arthritis or inter current infection is suspected, appropriate serological tests for specific antibodies may give a clue to diagnosis.

Biochemical markers

Apart from a few tests which are essential for confirming clinical diagnosis, biochemical tests are very important for

- Detecting damage to organs like kidney and liver at the time of diagnosis or during the follow up of patients suffering from rheumatic diseases like SLE, PAN, SS, gout etc.
- Detecting toxic effect of drugs used for therapy. Several drugs used in the treatment of rheumatic diseases, particularly disease modifying drugs may cause nephro toxicity. Drugs like methotrexate which are excreted mainly by kidney may cause toxicity. Several other drugs such as sulphasalazine, azathioprine, cyclophosphamide, and methotrexate may be hepatotoxic. Hence it is essential that hepatic and renal functions are assessed before commencing therapy and at regular intervals therafter (Table 8).

Key points

- Serum is a rich source of markers of inflammation, auto immune response, inter current infection and damage to organs as part of disease or due to drugs.
- Important screening tests are ESR and CRP to differentiate inflammatory disorders from non inflammatory conditions.
- Screening for rheumatoid factor can help to diagnose rheumatoid arthritis the result to be interpreted only in the context of clinical presentation considering the possible presence of rheumatoid factor in conditions other than rheumatoid arthritis.
- FANA is an important screening test when patients present with symptoms suggestive of ANA diseases.

- If FANA is positive, the pattern of fluorescence and the clinical features can be the guide to request for additional specific ANA tests by ELISA. The pre test disease probability should be considered before requesting for any specific auto antibodies.
- Interpretation of ANAs should always be in the context of clinical features considering the fact that they may be present in a small percentage of normal population.
- Once the diagnosis is established and once the treatment is started continuous monitoring to assess the disease activity, course of the disease and toxic effects of medications are required.

Table 8 : Relevance of biochemical tests in rheumatology practice

Clinical Condition	Biochemical Assay	Relevance of the test
Gout	Uric acid level Regular Renal function tests	Diagnostic confirmation To assess the renal damage
Adult Still's disease	Serum ferritin level[17,18]	Diagnostic confirmation Dramatic elevation to > 4000mg/l may even go to > 10,000 mg /l Normal level being 300mg/l
Polymyositis/dermatomyositis	Creatine kinase	Plasma level may correlate with disease activity
Rheumatoid arthritis	Renal and hepatic profile	Renal damage as part of disease or during therapy
SLE	Renal and hepatic profile	High risk for renal involvement. Regular monitoring required
Systemic juvenile rheumatic disease	Liver function tests	May be impaired at the time of presentation or after salicylate therapy
Primary & secondary vasculitis	Renal function tests	Glomerular dysfunction during course of disease

Further Reading:

1. A. Milford Ward PG Riches, R. Fifield, Smith AM. Protein reference unit. Reference ranges in PRU handbook of clinical immuno chemistry 4th ed. PRU publications Sheffield 1993; 77-89.
2. Claus DR, Osmand AP, Gewurz H. Radio immuno assay of C-reactive protein and levels in normal sera. Jo Clini Labo Medi 1976; 87: 120-28.
3. Zlonis M. The mystique of the erythrocyte sedimentation rate. A reappraisal of one of the oldest laboratory tests still in use. Clini Labo Medi 1993; 13: 787-800.
4. VR Joshi, Rational Interpretation of auto antibodies . The Association of Physicians of India's Medicine Update 2005; 781-82.
5. Walther J. van Venrooij, Joyce J. B. C. van Beers & Ger J. M. Pruijn. (2011) Anti-CCP antibodies: the past, the present and the future. Nature Reviews Rheumatology 7, 391-98.
6. Finckh, Axel 2009. Early inflammatory arthritis versus rheumatoid arthritis. Current Opinion in Rheumatology: Volume 21 - Issue 2 - 118-23.
7. K P Liao, K L Batra, L Chibnik 2008 Anti-cyclic citrullinated peptide revised criteria for the classification of rheumatoid arthritis Ann Rheum Dis; 67: 1557-61.
8. Daniel Aletaha, Tuhina Neogi, Alan J. Silman et al. 2010 Rheumatoid arthritis classification criteria: An American College of Rheumatology/European League Against Rheumatism collaborative initiative. Arthritis & Rheumatism 2010; Vol 62: 2569-81.
9. Steiner G and Smolen J. Autoantibodies in rheumatoid arthritis and their clinical significance. Arth Res 2002; 4: S1-S5.
10. Stanford Peng and Joseph Craft. Anti nuclear antibodies In: Kelley's Text Book of Rheumatology 6th ed. Ed. Gary Firestein. Saunders, Elsevier, Philadelphia 2009; 741-54.
11. Madhavan R, Chandrasekaran A.N., Thenmozhivalli P.R., Parthiban M. Spectrum of auto antibodies in systemic rheumatic diseases in a referral hospital in South India I. Fluorescent anti Nuclear Antibodies and Anti ds DNA antibodies. J Ind Rheum Asso 1994; 2: 127-32.
12. Rothfield NF., Auto antibodies in scleroderma. Rheum Dis Clin N Amer 1992; 18: 483.
13. Targoff I.N. Immunologic aspects of myositis. Curr Opin in Rheuma 1989; 1: 432-42.
14. Rao J.K., Weinberger M., Oddone EZ et al . The role of anti neutrophils Cytoplasmic antibody (cANCA) testing in the diagnosis of Wegener's Granulomatosis. A literature review and meta analysis. Ann Intern Med 1995; 123: 925-32.
15. Jennette J.C., Wilkman A.S., Falk R.J. Diagnostic predictive value of ANCA serology. Kidney Int 1998; 53: 796-98.
16. H Schur, Robert H Shmerling. Laboratory tests in rheumatic disorders In : Rheumatology 3rd Ed. Eds. Marc Hochberg, Alan J Silman, Josef S somlen etal Mosby Edinburg 2003; 208.
17. Gonzalez- Hernandez T., Martin- Mola E., Fernandez-Zamoeano A., balso- Criado. Serum Ferritin can be useful for diagnosis in Adult onset still's disease. J Rheumo 1989; 16: 412-13.
18. Ohta A, Higashi S, Suzuki H, et al. Increased serum ferritin levels in Adult Still's Disaese. Lancet 1987; I: 562-63.
19. Sita Naik. Laboratory test and pediatric rheumatologist: what does he need to know? Ind J Rheumat 2012; 7 (Suppl Ped Rheuma): 106-11.

4 CONVENTIONAL RADIOLOGY IN RHEUMATOLOGICAL DISORDERS

U R K Rao, Kakarla Subbarao, J Shivanand

Conventional radiographs are the corner stone in the evaluation of patients with various rheumatological disorders. Plain radiographs are inexpensive, widely available and can be easily obtained. Hence, imaging of rheumatological disorders should start with plain radiographs. Well performed good quality Radio graphs depict the extent and progression of the disease and are helpful in patient follow up and assessment[1]. They include bone erosions, joint space narrowing, subluxation, misalignment and ankylosis of the joints. The disadvantages of conventional radiography include insensitivity in detecting early bone disease or soft tissue manifestations[2]. Today digital radiography has become popular as the resolution is much better than old conventional radiography. The following mnemonic is easy to remember and helpful in the evaluation of a joint on a plain radiograph.

A Alignment
B Bone density
C Cartilage
D Distribution of lesions
E Erosions/enthesopathy/eburnation
S Soft tissue evaluation

Appropriately selected radiographic projections would serve the purpose of delineating the lesions in various joints (Table 1). In general the symptomatic joint should be radiographed in two different positions. The second radiograph taken at an angle of 90 degrees to the first view can show some abnormalities. On occasion, special radiographs in different positions are taken.

Table 1: Commonly requested radiographic projections

Joint	Projection
Hand and wrist	Posteroanterior (PA) Anteroposterior (AP). Oblique
Elbow	AP, lateral
Shoulder	AP (with internal and/or external rotation of humerus)
Hip	Pelvis AP [internal rotation and/ or external rotation of leg (frog leg)]
Knee	AP (standing), lateral (AP in semi flexion-tunnel view and sunrise view)
Ankle	AP, lateral/oblique
Foot	AP (lateral/oblique)
Sacroiliac joints	Pelvis AP (SI oblique)
Lumbar spine	AP, lateral (oblique, coned views)
Thoracic spine	AP, lateral
Cervical spine	AP, lateral (flexion, extension and open mouth view)
Chest	PA, lateral

(The projections mentioned in brackets are for special circumstances)

Chest radiograph (CxR) should be included among the investigations requested in a patient with rheumatological disorders. Despite various advances in the field of radiology, radiograph of chest continues to be an important screening modality, especially in Indian patients.

CxR is taken in erect position in postero anterior view (PA).

The first step in interpreting a CxR or any radiograph is to check if it has been labelled properly - name, medical record number and date. Identify the right side of the body before placing it on the view box. Presence of previous radiographs will help in comparisons. The technique and positioning of the patient are quite important.

Start the analysis by comparing the right and left side of the chest in a sequence beginning with the chest wall, soft tissues, mediastinum and ending with the lungs, pleura, heart.

Evaluation of a chest radiograph in rheumatology may help in the following :

I. Pleural effusions (Fig. 1).
II. Interstitial lung disease (reticulo nodular).
III. Pulmonary infiltrates/nodules/mediastinal nodes (Fig. 2).
IV. Pulmonary hemorrhage (ground glass).
V. Drug induced complications eg. methotrexate pneumonitis shows infiltrates mostly at bases and out of proportion to chest signs[3].
VI. Calcification.

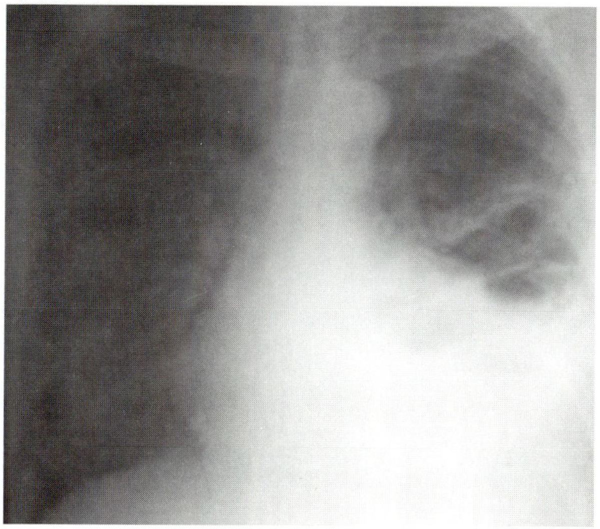

Fig.1 : Chest (PA): Left pleural effusion - Pulmonary Kochs

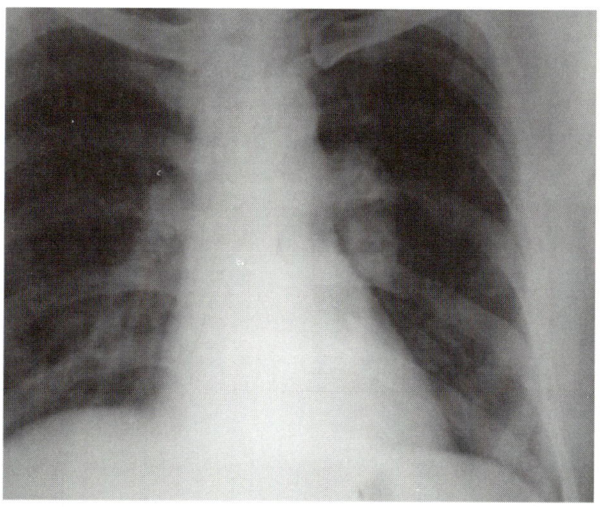

Fig. 2 : Chest (PA): Bilateral hilar lymphnodes: Sarcoidosis

Rheumatoid arthritis

Rheumatoid arthritis (RA) generally presents as a symmetric arthritis involving peripheral joints with sparing of axial skeleton except for the upper cervical spine. Radiographs of the hands are helpful in the diagnosis and assessment of the disease progression. Radiological changes are seen in the most commonly involved joints - metacarpophalangeal (MCP) and proximal interphalangeal (PIP) joints of hands, metatarsophalangeal (MTP) and interphalangeal joints of feet, wrists, elbows, knees and hips. Soft tissue swelling around the joints and paraarticular osteopenia are early manifestations of RA (Fig. 3, 4) (Table 2).

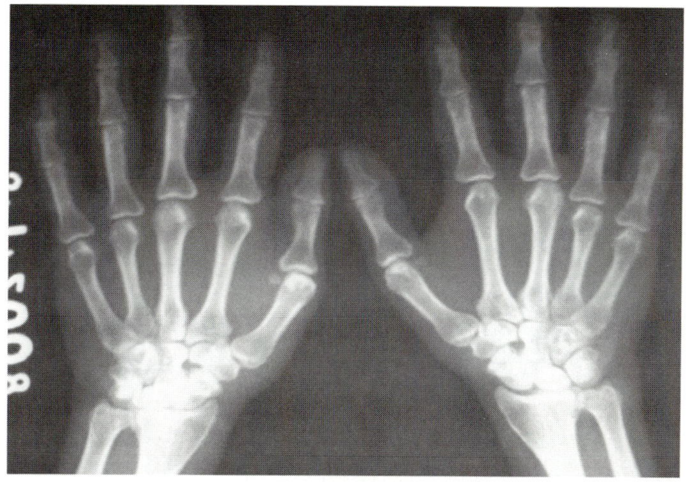

Fig. 3 : Hands (AP): Early RA with bilateral symmetrical soft tissue swelling at PIP joints

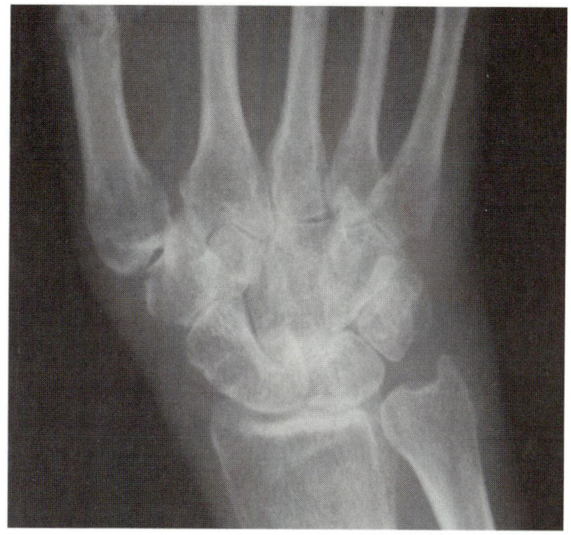

Fig. 4 : Wrist (AP): Early RA with subarticular osteopenia and joint space narrowing at carpo - metacarpal, intercarpal and radio-carpal joints

Table 2 : Features of rheumatoid arthritis

Early
Soft tissue prominence around the joint
Joint space narrowing
Paraarticular osteopenia
Delayed
Paraarticular erosions
Cystic spaces
Subluxations and deformities
Ankylosis

The swelling of the PIP joint is usually fusiform. Bilateral and symmetrical swelling of MCP joints, especially 2nd and 3rd are very common. The distal interphalangeal (DIP) joints are spared in RA. As the disease progresses, the erosions appear at synovial attachment sites (Fig. 5). Subarticular cystic lesions, bony ankylosis or subluxation of MCP joints and ulnar deviation of fingers are noted in the more advanced disease (Fig. 6). The ulnar styloid is commonly affected with erosive lesions. In the knee, the joint space is narrowed in both (medial and lateral) compartments of knee with juxtra articular osteoporosis (Fig. 7).

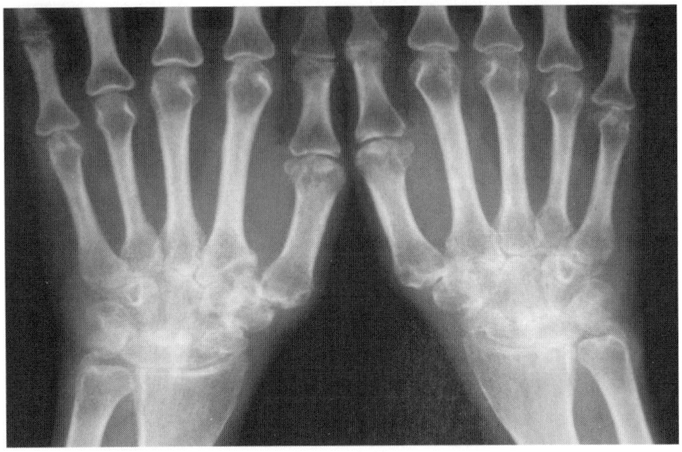

Fig. 5 : Hands (AP): RA with erosions and narrow joint space in bilateral 2nd, 3rd MCP joints and wrist joints

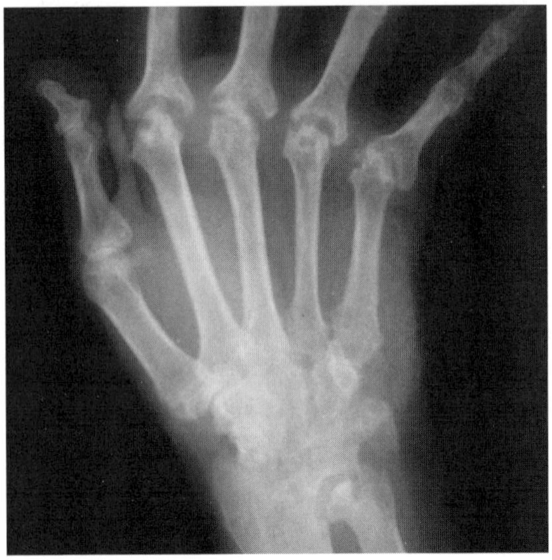

Fig. 6 : Hand (AP): Advanced RA with erosions, subluxation, and ulnar deviation at MCP joints

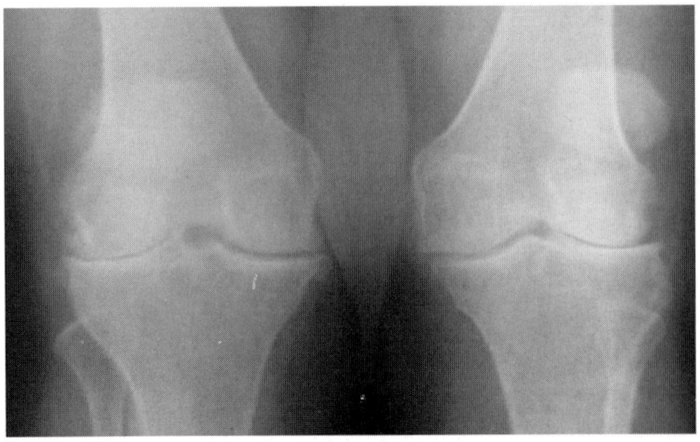

Fig. 7 : Knees (AP): RA with both compartments narrow, lateral more than medial

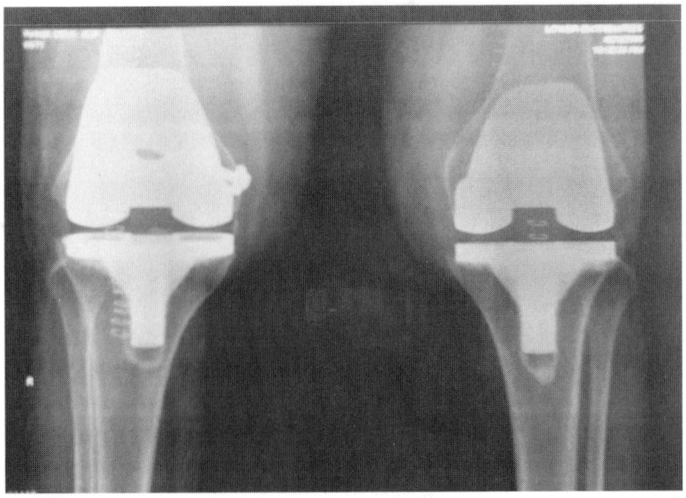

Fig. 8 : Knees (AP): RA-Total knee replacement

It may need surgical intervention (Fig. 8) if it is advanced and disabling. In the feet, early involvement shows juxta-articular osteopenia and erosions of the bare areas on the heads of the metatarsals. As the disease progresses erosions increase and subluxation of proximal phalanges occur. Dorsiflexion deformities of joints with hallux valgus/varus deformities of big toe are also seen (Fig. 9). Feet involvement in RA generally accompanies hand involvement or may be involved later. The earliest erosive change in radiograph of feet is seen on the lateral aspect of 5th metatarsal head. The other metatarsals are eroded medially and laterally. In progressive disease the cartilage loss is uniform.

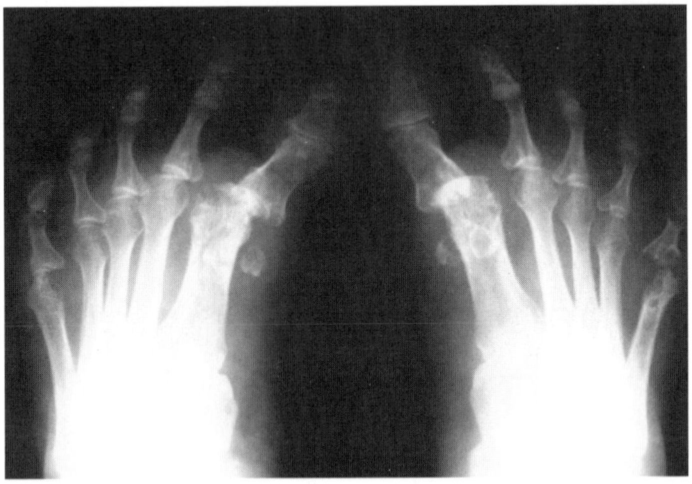

Fig. 9 : Feet (AP): RA - Dorsi flexion deformities of IP joints and hallux varus deformities of big toe with erosions noted at 1st and 5th MTP joints.

Hip joint involvement is less common than the knee, seen in less than 50% of patients. Rheumatoid involvement of the pelvis is seen as bilateral symmetrical involvement of hip joints. The cartilage loss is uniform and the femoral head moves axially within the acetabulum. There are varying degrees of joint space narrowing (Fig. 10) with erosions and synovial cyst formation seen in head of femur. It may need surgical intervention if it is advanced and disabling (Fig. 11).

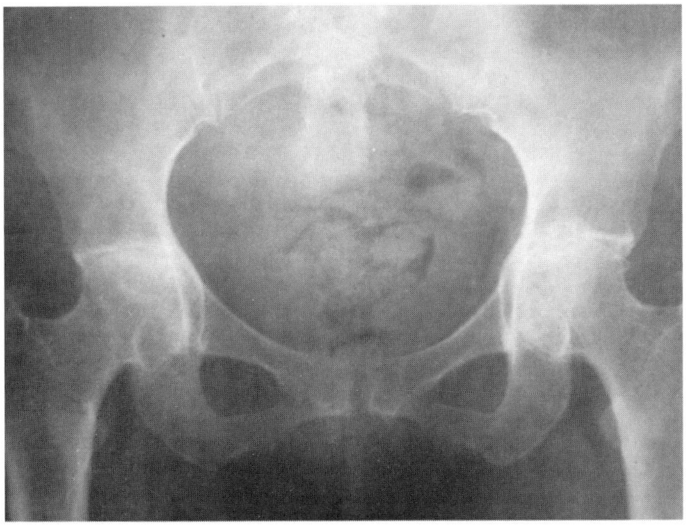

Fig.10 : Pelvis (AP): RA-bilateral uniform hip space narrowing with erosions.

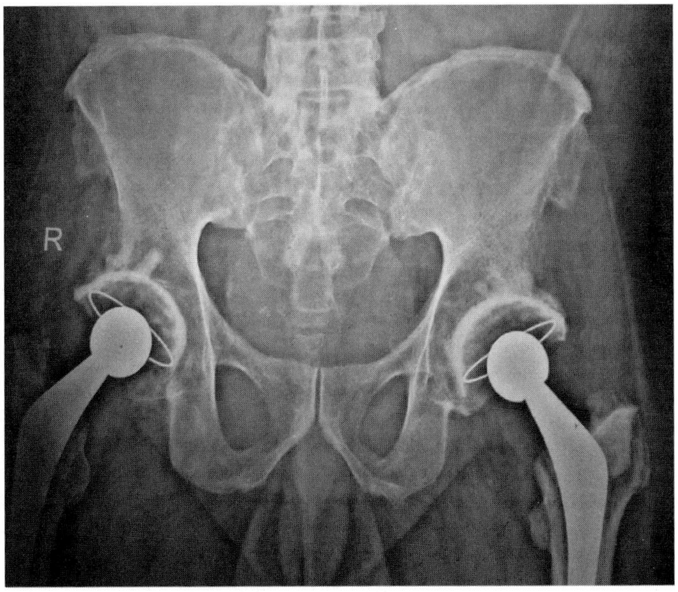

Fig.11 : Pelvis (AP): RA -bilateral hip replacement

Shoulder involvement is seen in 60% of patients with RA. Radiograph shows uniform narrowing of all compartments of shoulder joints (glenohumeral, acromiohumeral and acromioclavicular joint). Except for upper cervical spine, axial skeleton is seldom involved in RA. Subluxation at atlanto-axial (C1-C2) joint due to progressive erosive RA affecting the odontoid process is seen in some of the patients with chronic RA (Table 3) (Fig. 12).

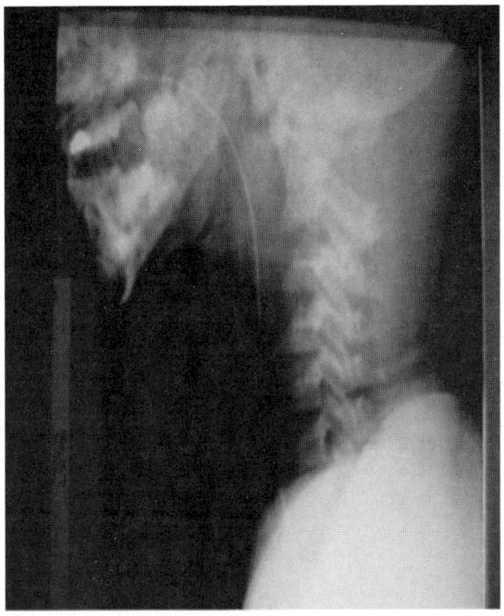

Fig.12 : Cervical spine (Lateral): Subluxation at atlanto-axial joint

Juvenile Idiopathic Arthritis (JIA) is most common connective tissue disorder of the childhood. Radiologically, soft tissue swelling around an affected joint is typical with paraarticular osteoporosis. Erosions are late finding in JIA (Fig. 13), since erosions in the cartilaginous epiphysis will not be visible until ossification is complete. In the advanced disease, complete cartilage loss with bony ankylosis occurs. In the cervical region, there may be fusion of segments of 2-4 vertebrae. Atlanto-axial subluxation is common in seropositive, polyarticular RA. Abnormalities in growth and maturation may lead to advanced skeletal maturation, enlargement of secondary centers of ossification, increased longitudinal growth, shortening of small tubular bones, undergrowth of mandibular condyles and facet joint ankylosis in cervical spine. In the knees, enlargement of epiphyses and the metaphyseal ends of the shafts are characteristic findings.

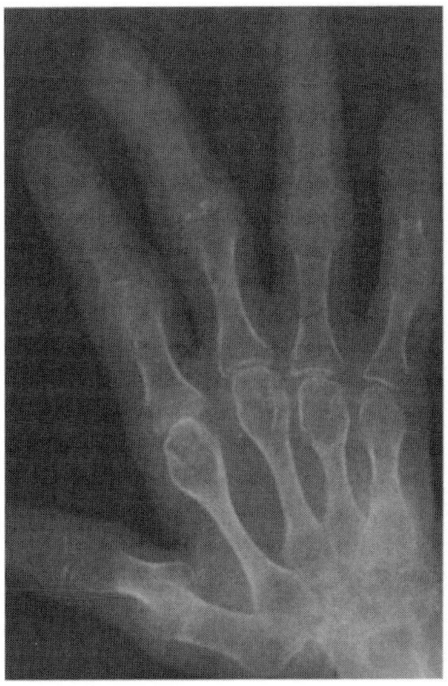

Fig.13 : Hand (AP): JIA-diffuse osteoporosis. Erosions noted 2nd and 3rd PIP joints

Table 3 : Deformities in rheumatoid arthritis

Site	Abnormality	Name
Hand		
PIP and DIP joints	Flexion PIP, extension DIP	Boutonniere deformity
	Extension PIP, flexion DIP	Swan-neck deformity
MCP joints	Ulnar deviation, Volar subluxation	
Wrist		
Radio carpal joint	Radial deviation	
Hip	Acetabular wall medial to Ilioischial line	Protrusio acetabuli
Knee	Outward deviation of knee	Genu varus
	Inward deviation of knee	Genu valgum
Feet		
First MTP joint	Lateral deviation	Hallux valgus
	Deviation away from 2nd toe	Hallux varus
PIP and DIP	Hyperextension and dorsal Subluxation of MTP	Cock-up toe
	Hyper-flexion PIP or DIP	Hammer toe
Cervical spine		
Atlanto-axial joint	Atlanto-odontoid space > 3 mm	Atlanto-axial subluxation
All levels	Subluxation of apophyseal joints	Stair-step deformity

Seronegative spondyloarthropathy

The seronegative spondyloarthropathies (SpA) are a group of multi-system inflammatory joint disorders that share a variety of genetic, clinical and radiological features. They include ankylosing spondylitis (AS), psoriatic arthritis (PsA), Reiter's disease (RD), arthritis associated with inflammatory bowel disease and unclassified SpA. In these disorders, spinal predilection is common resulting in sacroiliitis and spondylitis (Fig. 14). Inflammation at the bony insertions of ligaments and tendons (enthesopathy) is radiologically characterized by bony erosions followed by bony proliferation and eventually ankylosis of the adjacent bones.

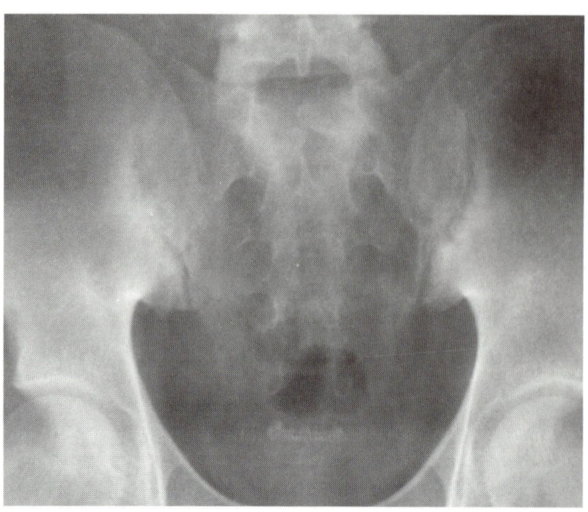

Fig. 14 : Pelvis (AP): Sacroiliitis-Early Sclerosis and erosions of both SI joints

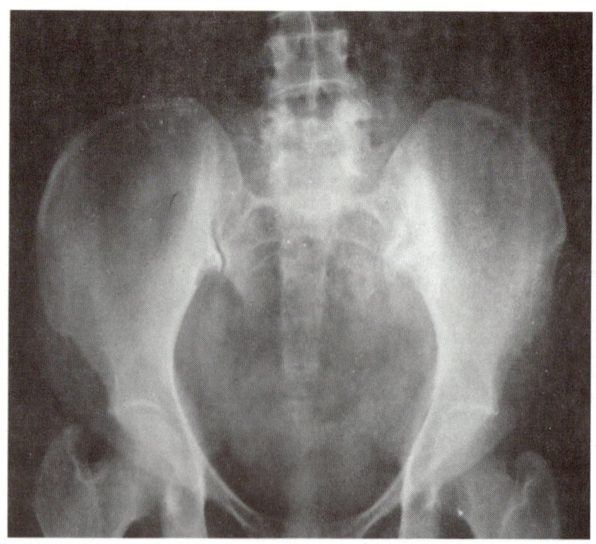

Fig. 15 : Pelvis (AP): Osteitis condensans ilii-triangular homogeneous sclerosis involving the iliac side of left SI joint

Ankylosing spondylitis

Ankylosing spondylitis (AS) is the prototype of SpA, primarily affecting the axial skeleton. The disease starts in the sacroiliac (SI) joints with sub-chondral bony erosions on the iliac side of the joint with "saw-teeth" appearance. The erosions are followed by bony proliferation and sclerosis. The changes start appearing in the lower third of the SI joint which is synovial. Initially, the joint may look widened. The changes gradually affect whole of the SI joint and the joint eventually fuses. Once the fusion is complete, sclerosis disappears. The sacroiliitis in AS is usually symmetrical and can easily be differentiated from osteitis condensans ilii (Fig. 15). Spine is initially involved at lumbosacral and thoracolumbar segments. In Psoriasis and Reiter's disease, no such orderly spread is seen and paravertebral ossifications tend to appear in a random fashion.

The enthesopathy in the spine manifests as small erosions surrounded by bony sclerosis at the vertebral corners seen as "shiny corners". This is followed by ossification of the outer annular fibers to form marginal and symmetric syndesmophytes (Fig. 16). They are different from osteophytes seen with degenerative disc disease in that syndesmophytes are parallel to the spinal axis whereas osteophytes are perpendicular (grow outwards). Squaring of the vertebral bodies is characteristic of early AS caused by the corner erosions and anterior apposition of periosteal new bone formation. Squaring is easier to appreciate in the lumbar spine. The complete fusion of the syndesmophytes is the so called "bamboo-spine" which is a late feature (Fig. 17, 18). Erosions are seen in the apophyseal and costovertebral joints followed by ankylosis of these joints. Following the spinal fusion, the spine becomes osteoporotic and is vulnerable to stress. Fractures in AS occur most commonly at the thoracolumbar and cervicothoracic junctions. It may proceed to non-union with pseudoarthrosis, an appearance resembling disc space infection (Andersson lesion) (Fig. 19). Enthesopathy of ischial tuberosities is described as "WHISKERING".

The changes of AS are less severe and delayed in appearance in women with AS, and in children with AS changes in appendicular joints are noted more frequently. Peripheral arthritis occurs in about 30 percent of the patients with AS. They include hips, shoulders and knees. Hip disease in AS is characterized by uniform joint space narrowing and axial migration of the femoral head.

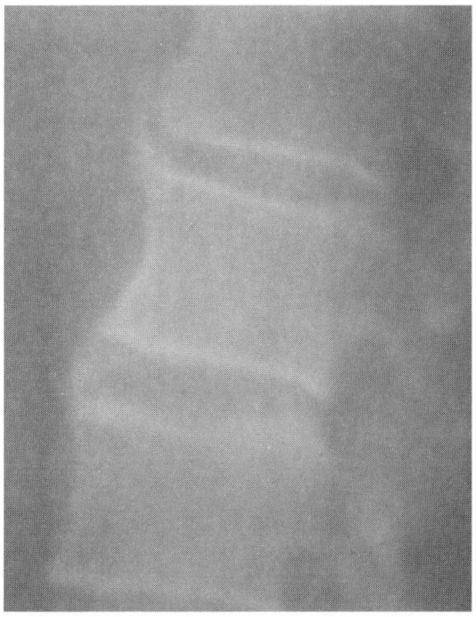

Fig. 16 : Lumbar spine (Lateral): AS-early syndesmophytes noted L2-L3, L3-L4

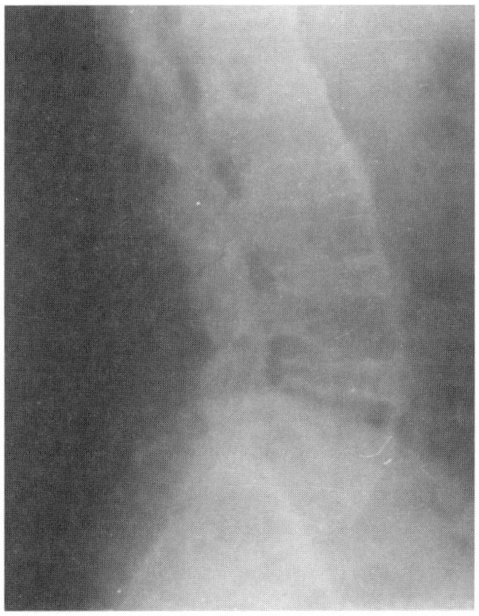

Fig.17 : Lumbar spine (Lateral): AS-syndesmophytes joining anterior vertebral margins

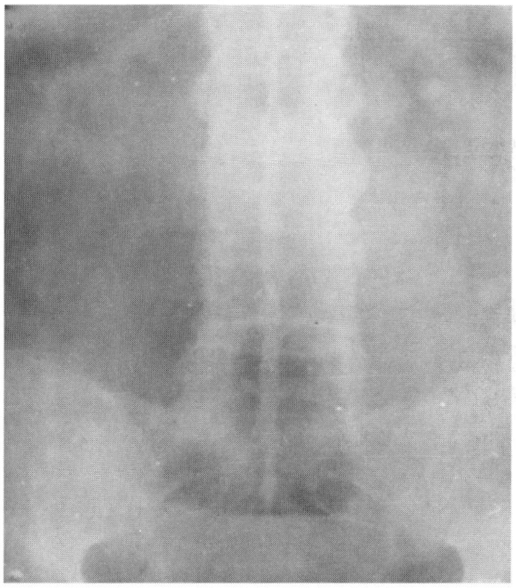

Fig.18 : Lumbar spine (AP): AS-with bamboo spine

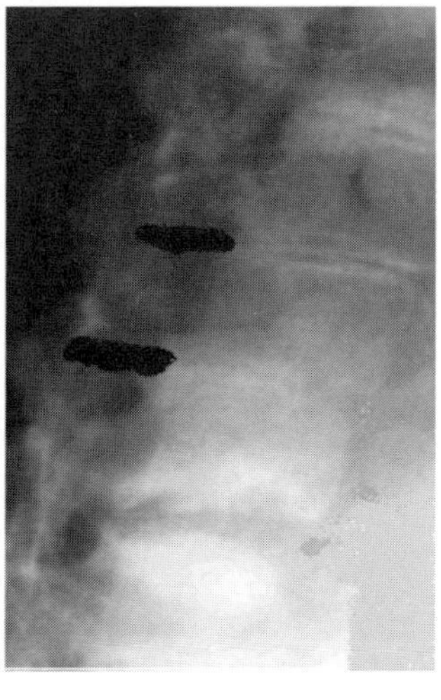

Fig. 19 : Dorsolumbar spine (lat): AS with pseudoarthrosis D11-D12 (Andersson lesion)

Psoriatic arthritis (PsA) and Reiter's disease (RD)

SI joint and spine involvement occur in 20 to 40 percent of patients with PsA. The sacroiliitis in PsA and RD may or may not be symmetrical and spondylitis occurs in a more random fashion. It is associated with presence of coarse, symmetric paravertebral ossification (non-marginal syndesmophytes). The joint involvement in psoriasis can be variable (Table 4).

Table 4 : Modes of presentation in psoriatic arthritis

Distal interphalangeal joint involvement
Proximal interphalangeal and metacarpophalangeal joint involvement that resemble seronegative RA
Asymmetric oligoarthritis affecting large joints
Destructive form of arthropathy (arthritis mutilans)
Seronegative spondyloarthropathy

Erosions in the phalanges are seen in the bare areas with associated fluffy bone formation (mouse ears) and sometimes followed by fusion. The distal erosions (Fig. 20) can evolve into the so called 'pencil and cup' appearance (Fig. 21). In the feet, erosions appear in metatarsophalangeal (MTP) and interphalangeal joints, especially in great toe. Tuftal proliferation of the digit results in a shaggy dense distal phalanx (ivory phalanx). Periosteal reaction can be seen early in the disease along with shafts of the phalanges and metatarsals and is associated in the feet with sausage digits. Reiter's disease (RD) typically affects the lower extremities. Inflammatory spurs are commonly present in Reiter's diseases which are distinguished by their fluffy appearance [Fig. 22 (a)] from degenerative calcaneal spurs [Fig. 22 (b)].

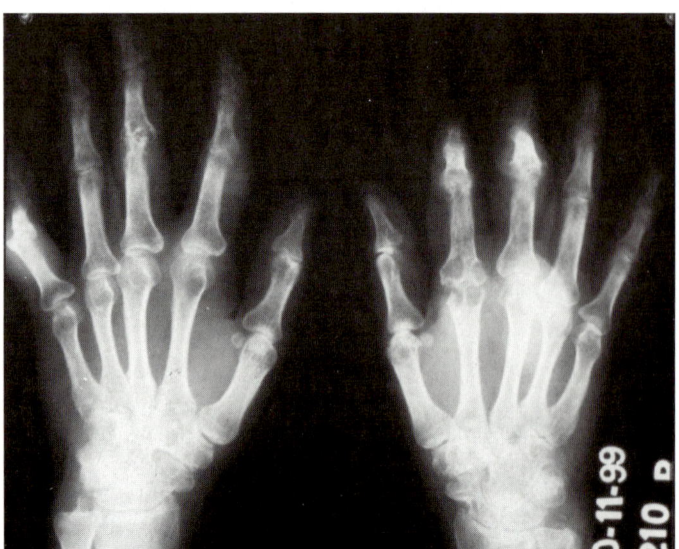

Fig. 20 : Hands (AP): Left 2nd, 3rd and right 5th DIP joint involvement in PsA

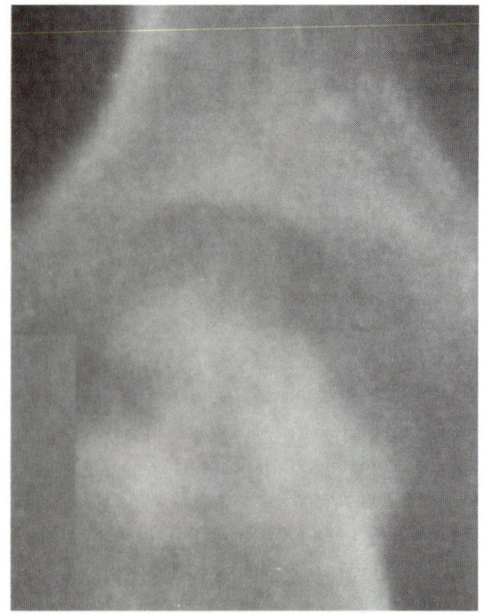

Fig. 21 : DIP (AP): Pencil in cup shape in PsA

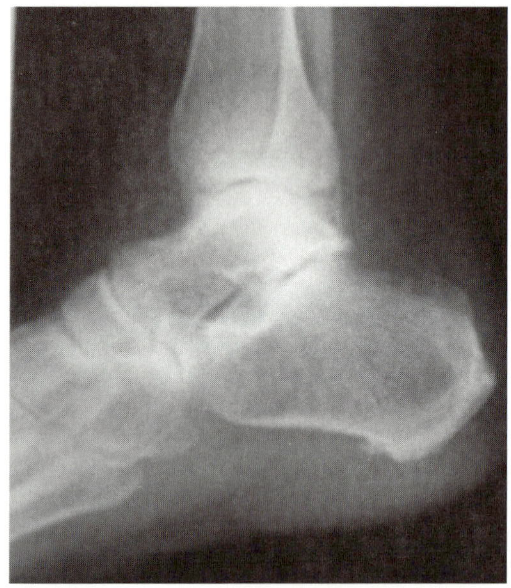

Fig. 22 (a) : Ankle (Lateral): Reiter's disease : Fluffy appearance of calcaneal spur

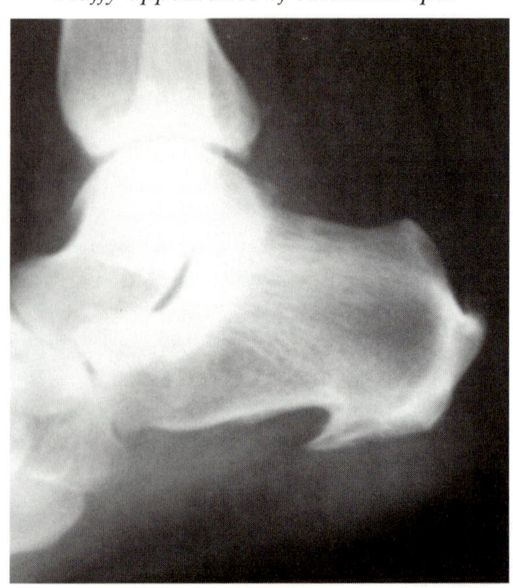

Fig. 22 (b) : Ankle (Lateral): Degenerative calcaneal spur

Osteoarthritis (Degenerative joint disease)

The common joints involved in osteoarthritis (OA) are knees, hips, DIP, first carpometacarpal (CMC) joints, first MTP joint and spine (cervical, lumbar and lumbosacral). The radiographic appearance of OA in different joints varies, being dependent on anatomic relationships and on the stress to which each joint is subjected. The most typical pattern is the presence of reparative changes in both the "stressed and non-stressed" segments of the joints. In the stressed segment, there is narrowing of joint space due to thinning of cartilage with the development of erosions and eburnations. Sub-chondral bone sclerosis with formation of cystic lesions follows. In non-stressed segment, reparative changes lead to osteophytosis. Osteophytes are classified as marginal, central, periosteal and capsular. Degeneration of surrounding tissues such as ligaments and capsule may lead to malalignment and subluxation of the joint. In advanced OA, fragmentation of the cartilaginous or osseous surface can lead to the formation of intraarticular "loose bodies" (Table 5).

Table 5. Features of OA

- Narrowing of joint space
- Osteophytosis
- Subarticular Sclerosis
- Deformities

Hip

Asymmetric joint space narrowing is the most reliable sign of OA in the hip joint. The femoral head displacement can be categorized into three different migration patterns - superior migration, medial migration and axial migration. The superior migration pattern in turn is of 2 types - superolateral and superomedial

Knee

Weight bearing AP view of the knee is better for accurate assessment. This view shows the extent of joint space loss and the degree of angulation (varus or valgus) and subluxation. The knee joint is divided into three compartments - the medial femorotibial, lateral femorotibial and patellofemoral. Most frequently, the medial compartment is involved in primary OA of knees (Fig. 23). Both or only lateral compartment are/is involved in secondary OA knees due to inflammatory arthropathies such as RA. The patellofemoral compartment (Fig. 24) is affected usually in association with medial compartment disease. Best of taking AP-view of the knees is in standing position.

The cardinal features of OA are asymmetric joint space narrowing accompanied by subchondral cysts and sclerosis. Marginal osteophytes are frequent finding at the femoral and tibial margins of the joints in advanced cases at the intercondylar tubercles. Features of patellofemoral OA include joint space loss with deformity at the patellar articular surface and in severe cases formation of osteophytes. Additional signs of OA are angulation and subluxation with varus (primary OA) or valgus (secondary OA) deformities, small joint effusions and presence of intraarticular bodies (loose bodies), called secondary osteocondromatosis.

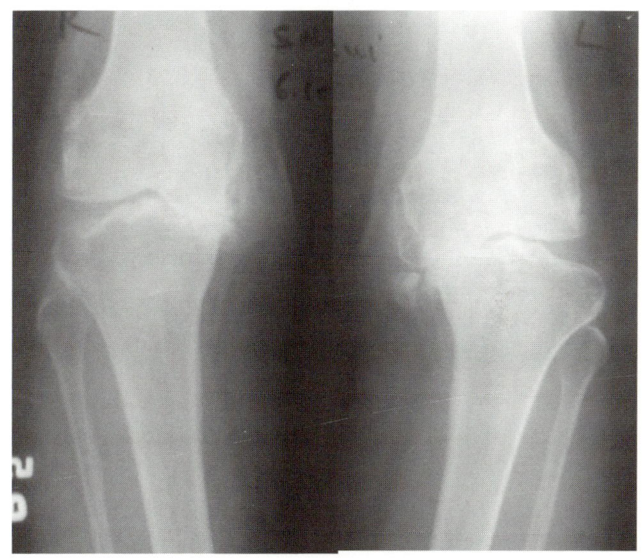

Fig. 23 : Knees (AP): Gross osteoarthritis knees with obliteration of medial compartments (OA).

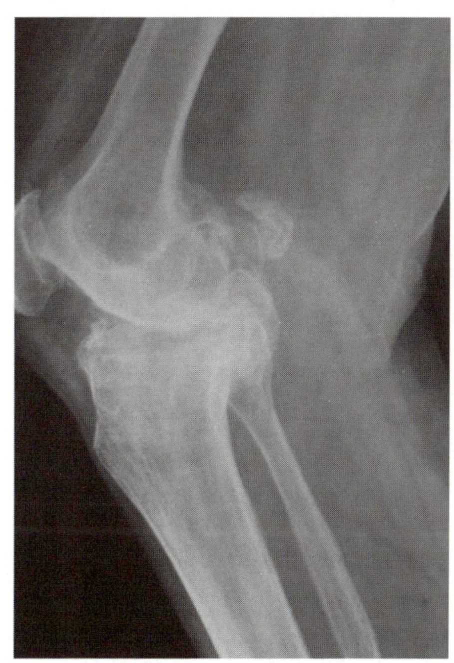

Fig. 24 : Knee (Lateral): Patellofemoral joint space narrowing with loose bodies (OA).

Spine

In the axial skeleton, degenerative joint disease is mainly observed at the sites of maximum weight bearing and movement. In the spine, cervical (C5-6, C6-7) (Fig. 25) and lumbosacral (L4-5, L5-S1) (Fig. 26) areas are most commonly affected.

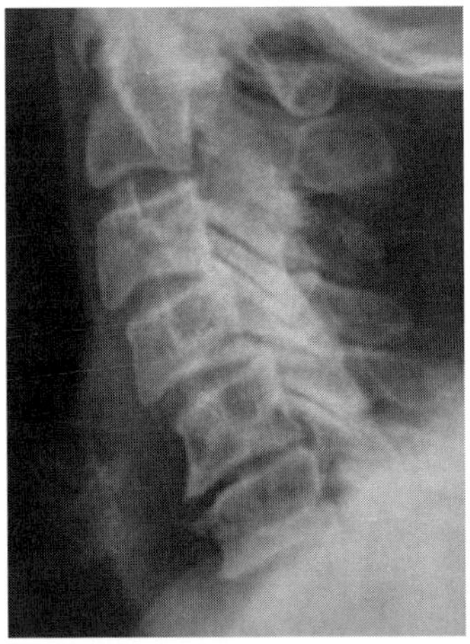

Fig. 25 : Cervical spine (Lateral) : cervical spondylosis with narrow C5-6 intervertebral space and osteophytosis.

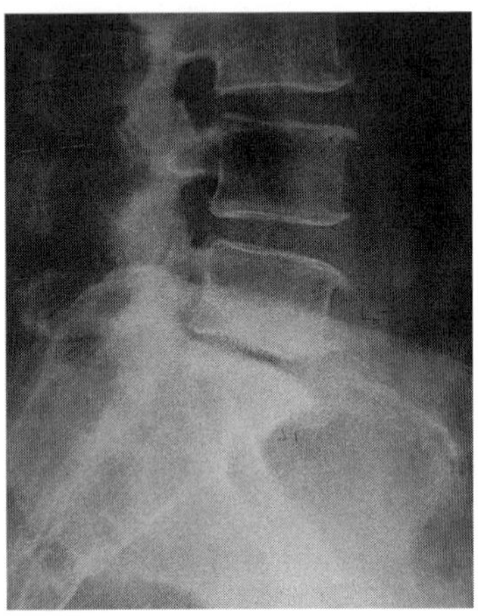

Fig. 26 : Lumbosacral spine (Lateral): Lumbar spondylosis with narrow L5-S1 and anterior osteophytes noted at L5 with vacuum sign L5-S1

In intervertebral (disc) spaces, marginal osteophytes are observed early. As the disease advances, thinning of the disc space with sclerosis of subchondral bone takes place with subluxation of various degrees leading to instability and pain. The hyaline cartilage of the facet joints may also be involved by the degenerative process.

Hands and feet

The degenerative joint disease can occur primarily in non weight bearing joints such as DIP joints (Fig. 27), first CMC joint (Fig. 28) and first metatarsal joints (Fig. 29). The erosive and nodal osteoarthritis may mimic RA or PsA. However, subarticular osteoporosis is not encountered in OA.

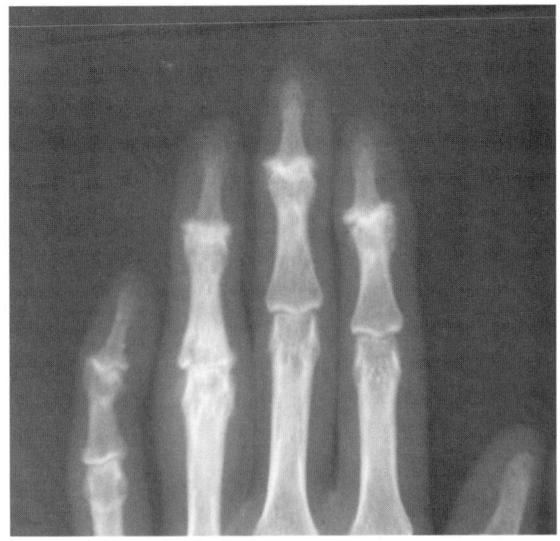

Fig. 27 : Hand (AP) : OA DIP joints with narrow joint space, subarticular sclerosis and erosions

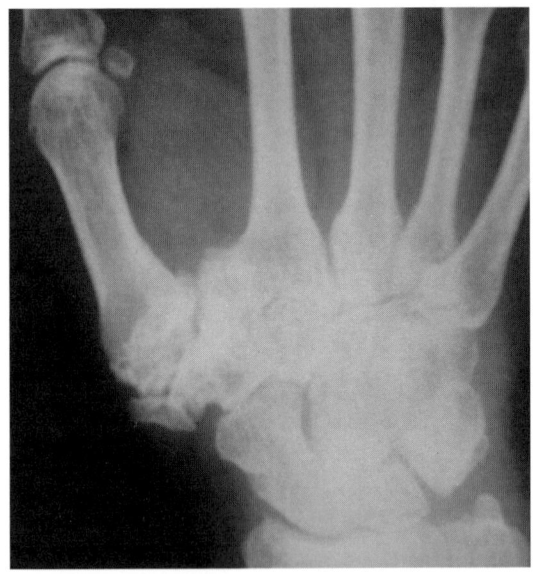

Fig. 28 : wrist (AP): OA first CMC joint with narrow joint space, subarticular sclerosis and osteophytes

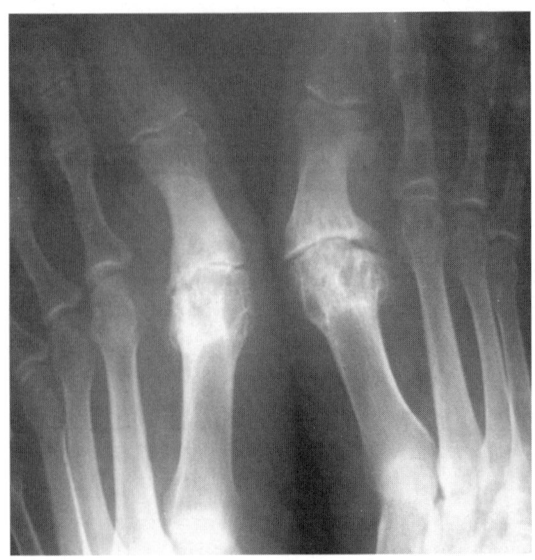

Fig. 29 : Feet (AP): First MTP joints with narrow joint space, subarticular sclerosis and osteophytes (OA)

Crystal arthropathies

The crystal induced arthropathies are a heterogeneous group of biochemical disorders that share the common features of crystal deposition in and around joints. The common crystals are monosodium urate (MSU), calcium pyrophosphate dihydrate (CPPD) and calcium hydroxyapatite (CHA). Radiography often demonstrates characteristic changes in these disorders and they delineate the extent of articular and periarticular disease.

Gout : The radiographic findings of acute gout are non-specific. Soft tissue swelling, osteopenia or a joint effusion may be seen. In chronic tophaceous gout, the radiological features include loss of joint space, preservation of bone density leading to osteopenia in late stages and osseous erosions (Table 6).

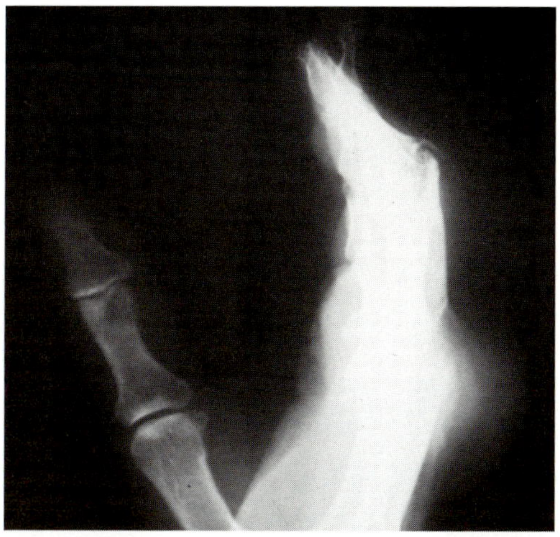

Fig. 30 : Hand (Oblique): Gouty tophus at 2nd MCP joint with mouse ears (overhanging margins) at 2nd PIP

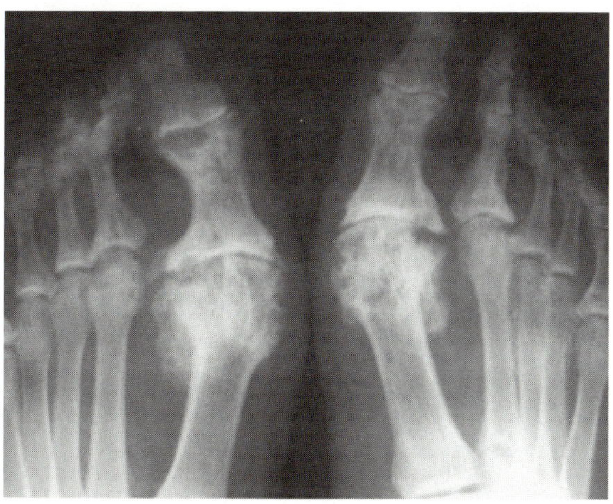

Fig. 31 : Feet (AP): Interosseous osteolysis at both 1st MTP joints with overhanging margins. Punched out erosions are also noted near IP joints

Though lesion is more common in first metatarsophalangeal joint, similar lesions are also seen in ankles, elbows, wrists and hands (Fig. 30). Intraarticular erosions tend to involve the joint margins before extending centrally. An elevated margin of the bone may extend outside the erosions as "overhanging edge" which is very suggestive of gout (Fig. 31). Intraosseous MSU crystals deposition and tophus formation may give rise to osteolytic lesions that simulate neoplastic disease.

Table 6 : Characteristic radiographic features of gouty arthritis.

Asymmetrical involvement	Relative preservation of joint space
Eccentric soft tissue prominence	Relative preservation of bone density
Punched out erosions	Intra-osseous calcification/ lytic lesions
Overhanging margins	

CPPD deposition disease : Articular calcification related to CPPD is most frequent in the knee followed by symphysis pubis, wrist, elbow, shoulder and hip. Calcification in the hyaline cartilage and/ or fibrocartilage (chondrocalcinosis) is present (Fig. 32a, b). Fibrocartilage calcification is coarse and irregular, commonly occurring within the meniscus of the wrist and the pubic symphysis. Hyaline cartilage calcification is fine and linear following the contour of underlying subchondral bone. Apart from the cartilage, CPPD deposition may also occur in the capsule, synovium, ligaments, and soft tissue mostly appearing as curvy linear calcifications.

Calcium hydroxyapatite : CHA crystal depositon commonly occurs in shoulder and is called peritendinitis calcarea. But it may also be seen in hip, elbow and wrist joints. Radiographically amorphous calcific deposits are noted (Fig. 32c).

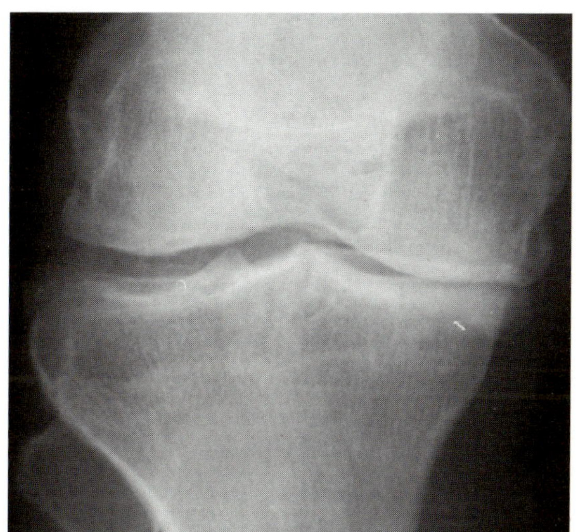

Fig. 32(a) : Knee (AP): Linear calcification seen in hyaline cartilage at lateral compartment (CPPD).

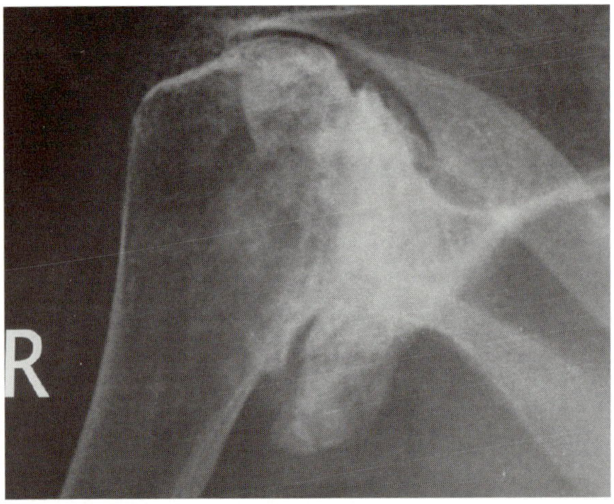

Fig. 32(b) : Shoulder (AP): Shoulder with CPPD arthropathy. There is loss of the gleno-humeral joint space with osteophyte formation, subchondral sclerosis and cyst formation.

Other Collagen Vascular Diseases

Systemic lupus erythematosus (SLE): The radiological features of SLE may mimic early RA, but erosions are uncommon in SLE. However, deforming arthropathy can occur due to soft tissue involvement. Osteonecrosis may occur in patients with SLE especially on glucocorticosteroids (GCs) (Fig. 33).

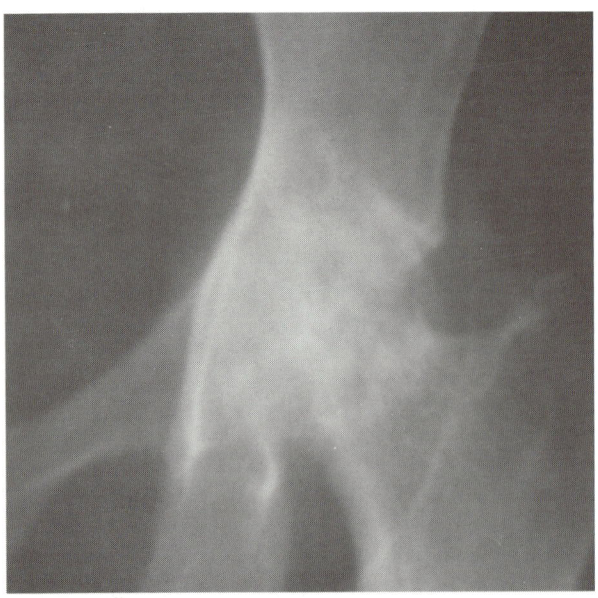

Fig.33 : Hip (AP): SLE on GCs - osteonecrosis with secondary OA.

Systemic sclerosis (SS) (Scleroderma): Articular or non-articular features are commonly seen in scleroderma especially with Raynaud's phenomena (Fig. 34) (Table 7). CREST syndrome in localized scleroderma includes calcinosis, Raynaud's phenomenon, esophageal hypomotility, sclerodactyly and telangiectasia.

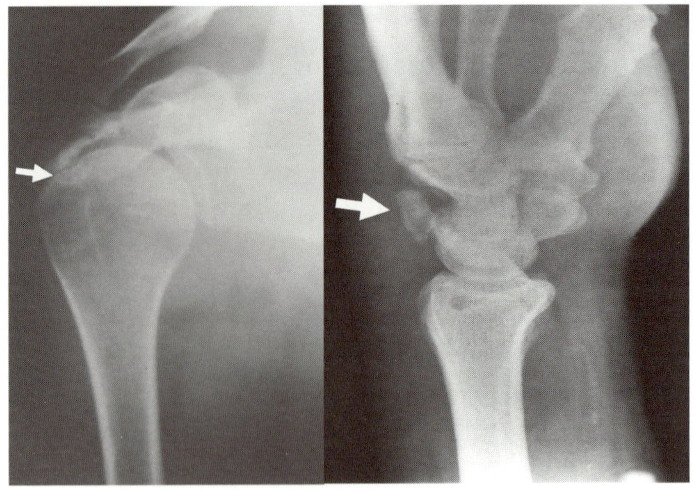

Fig. 32(c): Shoulder (AP) and wrist (oblique) : CHA crystal depositions.

Table 7 : Radiographic features in systemic sclerosis

Soft tissue resorption, most commonly at finger tips.

Soft tissue calcification - subcutaneous tissue, joint capsule, tendons or ligaments

Osteolysis - resorption of terminal phalanges (acroosteolysis)

Erosive articular disease

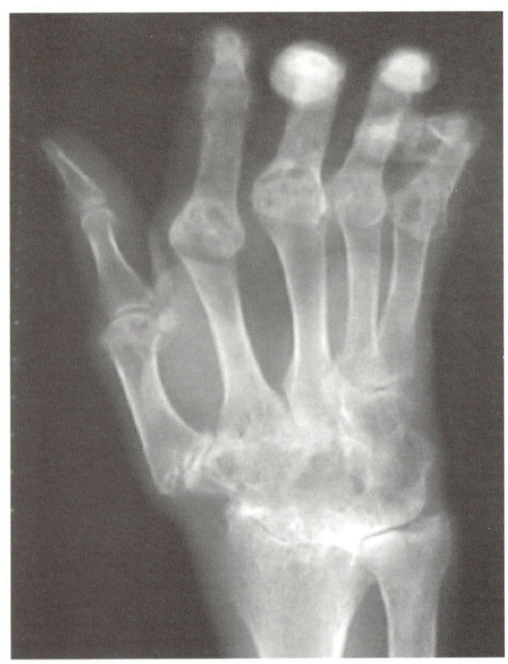

Fig. 34: Hand (AP): Scleroderma - acroosteolysis, subcutaneous calcification finger tufts with erosive articular disease of wrist.

Inflammatory Myositis : The most characteristic soft tissue abnormality in dermato/polymyositis is calcification in subcutaneous tissue and intermuscular fascia.

Miscellaneous disorders

Osteonecrosis : The most common sites of osteonecrosis (avasular necrosis) are the femoral heads, femoral condyles, humeral head (Fig. 35), talus, scaphoid and lunate. In children, the idiopathic form of osteonecrosis of the femoral head is called Legg-Calve-Perthes disease. Secondary osteonecrosis is commonly seen in adults. Radiographic changes are not evident in early epiphyseal osteonecrosis. As the disease progresses, arc like radiolucent subchondral band (crescent sign) and the formation of patchy subchondral lucent and sclerotic foci are seen. Fragmentation and collapse of the articular surface follows. The joint space is usually preserved until late in the course of the disease when secondary osteoarthritis may develop (Fig. 33).

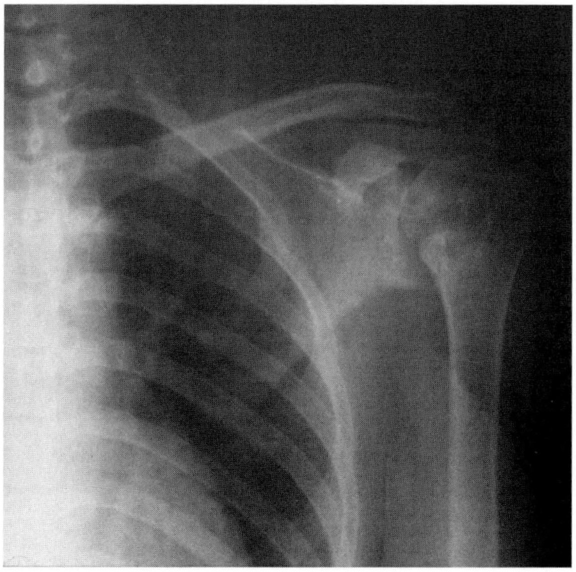

Fig.35 : Shoulder (AP): Osteonecrosis of humeral head in a patient with SLE on GCs.

Fluorosis: Excessive ingestion of fluoride through water or food causes toxicity or a debilitating disease called 'fluorosis'. Excessive intake during pre-eruptive stage of teeth leads to dental fluorosis and further continued ingestion over years and decades causes bony or skeletal fluorosis. Patient complains of pains in the small joints of the limbs and back, which are often mistaken for RA or AS. Radiological findings show calcification and ossification of various bones (Fig. 36), joints, spine (Fig. 37), related ligaments and tendinous attachments to the bones and interosseous membranes (Fig. 38).

Diffuse Idiopathic Skeletal Hyperostosis (DISH) : DISH radiologically resembles fluorosis. Since it is seen in older people, density of the bones is not that of fluorosis, flowing hyperostosis is added feature (Fig. 38a).

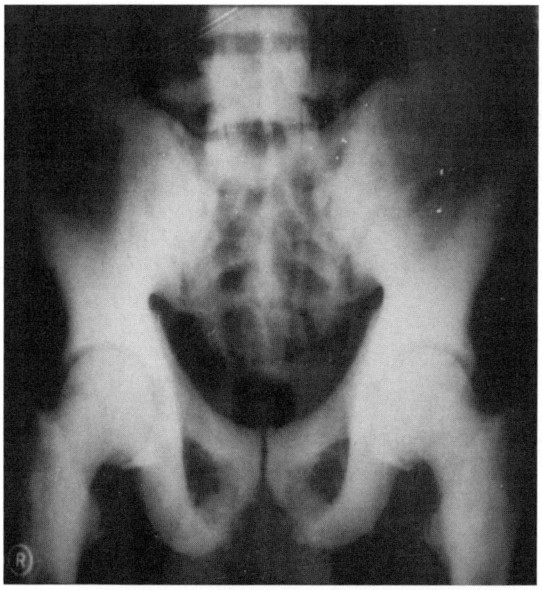

Fig. 36 : Pelvis (AP): Dense bones in fluorosis.

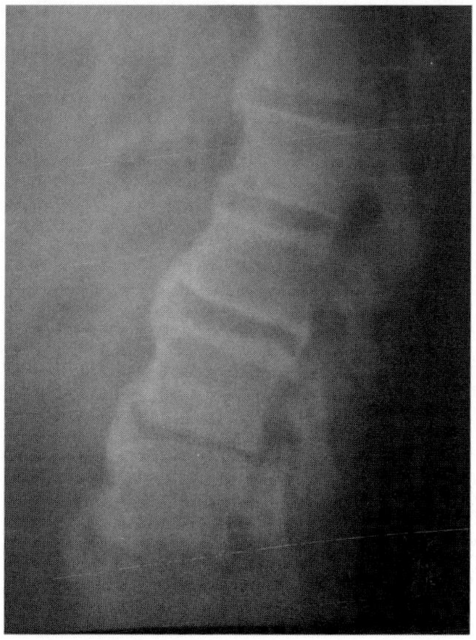

Fig. 37 : Lumbar Spine (Lateral): Dense vertebrae with bridging osteophytes (Flurosis).

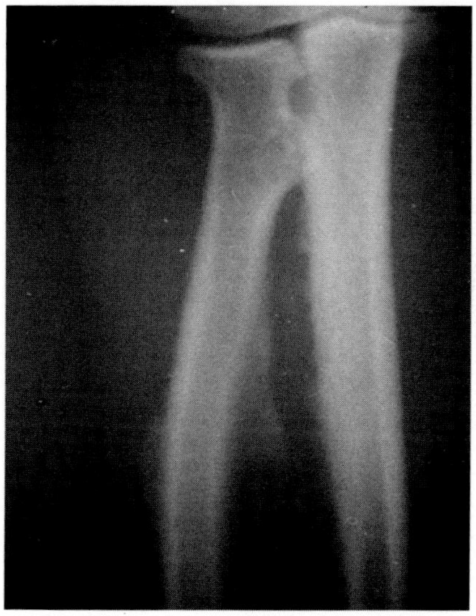

Fig. 38 : Forearm (AP): Calcification of interosseous membrane (Flurosis).

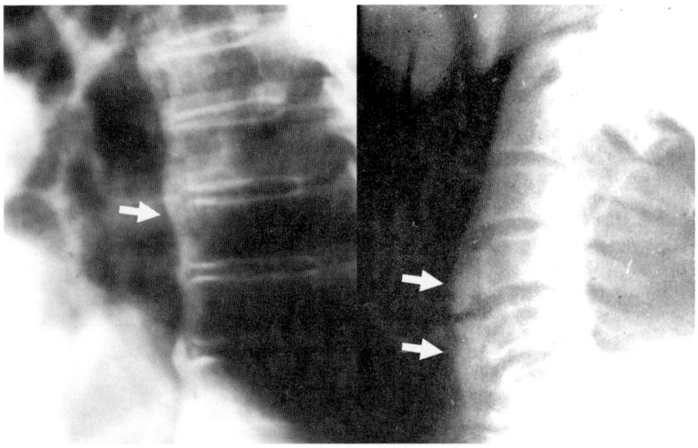

Fig. 38(a) : Lumbar spine (lateral) and cervical spine (lateral): DISH with flowing hyperostosis.

Infective arthritis: Infective arthritis is a form of joint inflammation caused by a live or inactive, degraded organism. The organism can be a bacterium, a virus or a fungus. Infection of the joints usually occurs after a previous infection elsewhere in the body. Usually only one joint is involved, though sometimes more joints can become infected. Mostly, infectious arthritis affects the large joints (shoulders, hips, knees), but smaller joints (fingers, ankles) can also be involved (Fig. 39a & b).

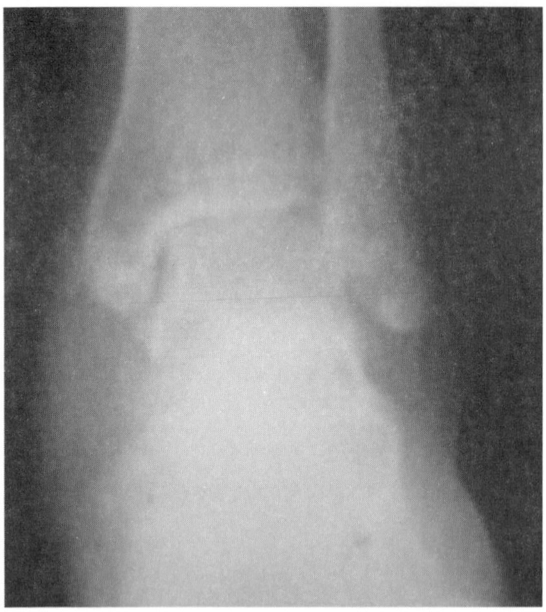

Fig. 39(a): Ankle (AP): infective

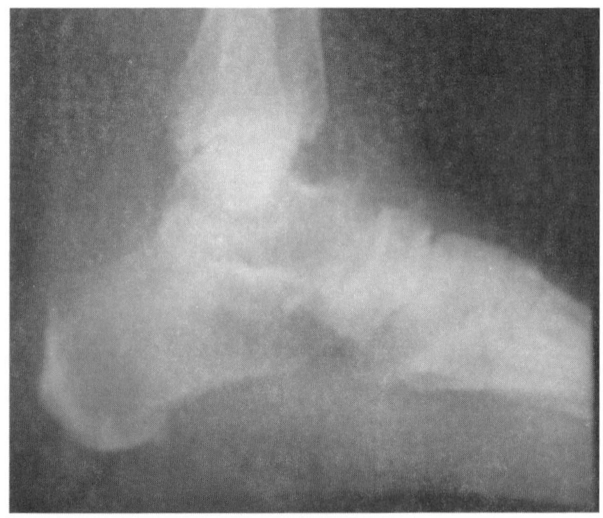

Fig. 39(b) : Ankle (Lateral): infective

Soft tissue swelling with narrow joint space, talotibial and subtalar arthritis with erosions, biopsy showed granulomatous synovitis with caseous necrosis.

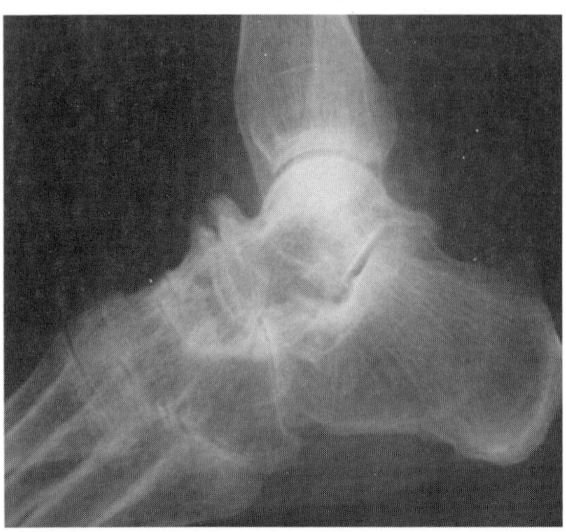

Fig. 40 : Ankle (Lateral) Osteoporosis with vertical striations at calcaneal bone.

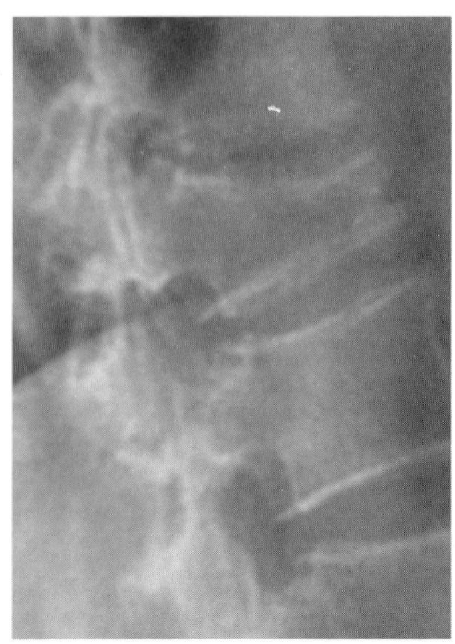

Fig. 41 : Dorsal spine (Lateral): Wedge compression at D10 due to osteoporosis.

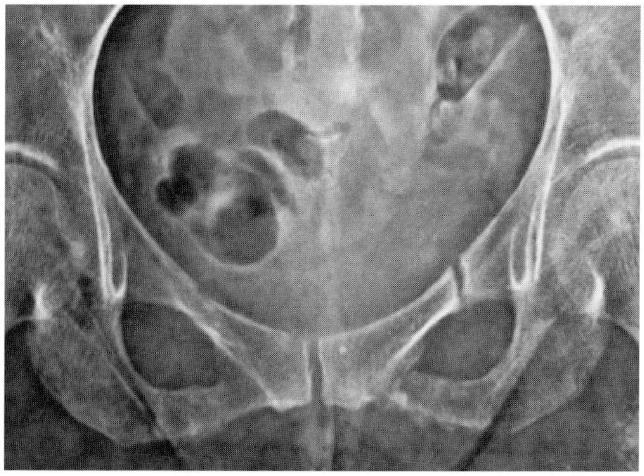

Fig. 42 : Pelvis (AP): Osteomalacia with pseudofractures at superior ramus of pubic bone on both sides

Osteoporosis and Osteomalacia : Osteoporosis is a systemic skeletal disease with low bone mass, increase in bone fragility and susceptibility to fracture. It is seen most commonly in post menopausal women. The risk factors are age, genetic, personal habits (smoking. alcohol etc), drugs (GCs, thyroxine etc.) and medical problems (RA, hyperparathyroidism etc). Radiologically reduced bone density is observed in vertebrae, wrist, hips, humerus and tibial bones. The bones have prominent vertical striations resulting from a relatively greater loss of the horizontally oriented trabeculae (Fig. 40). In severe osteoporosis vertebral fractures (wedge compression) can be seen (Fig. 41).

Defective mineralization of the organic bone matrix in adults leads to osteomalacia. Most common clinical features are bone pain and tenderness in spine, shoulders, ribs and pelvis. The characteristic feature of osteomalacia is ground glass appearance without differentiation between cortex and spongiosa the detection of Looser's zone or pseudofractures commonly seen in upper femur, pelvis and scapula (Fig. 42).

Key points

1. Radiographs are requested after obtaining the patient's complete history and physical examination.
2. Plain radiography remains the core imaging technique. It is simple, convenient and economical.
3. Clinical indications will guide which joints are to be investigated.
4. Radiographs are helpful in the diagnosis, prognosis and management of rheumatological disorders.

Further Reading

1. Subba Rao K. Arthritides (disorder of joints) in diagnostic radiology and imaging. Subba Rao K, Banerjee S, Aggarwal S, Bhargava S (ed), Jaypee Bro New Delhi, 1997; pp269-71.
2. Ostergaard M, Ejbjerg B, Szkudlarek M. Imaging in early rheumatoid arthritis; roles of magnetic resonance imaging, ultrasonography, conventional radiography and computed tomography. Best Pract Res Clin Rheum 2005; 19: 91-116.
3. Barrera P Laan R, van Riel et al. Methotrexate related pulmonary complications in rheumatoid arthritis, Ann of Rheum Dis 1994; 53: 434-39.

ROLE OF OTHER IMAGING MODALITIES IN RHEUMATOLOGY

5

K M Mahendranath

Since the inception (evolution) of Rheumatology as a super speciality, plain X-Rays (Radiographs) remained as the sole imaging modality in diagnosing and assessing the damage and disease progression or remission till the Computerized Tomography (CT) arrived two decades ago to be followed by Magnetic Resonance Imaging (MRI), Nuclear Bone Scan and later Ultrasonography in the past decade.

The modern imaging modalities have helped the patients and Rheumatologists in diagnosing and assessing the disease and monitoring the progress.

1. Nuclear Medicine
2. Computer Tomography (CT)
3. Magnetic Resonance Imaging (MRI)
4. Bone densitometry
5. Single Photon Absorptiometry
   Dual Photon Absorptiometry
6. Dual energy X-ray absorptiometry
7. Quantitative bone scan
8. Digital Radiography
9. Ultra Sonography (Musculoskeletal)

Nuclear medicine[1]

The primary advantage of nuclear medicine techniques is that they provide morpho functional images giving valuable information on the metabolic activity of the structure in question. Although they are less specific than conventional radiologic methods, they are also more sensitive, in many cases revealing early stage lesions that cannot be observed using conventional methods. In a patient with polyathralgia a normal bone scan excludes inflammatory joing disease where as an abnormal bone scan suggests that inflammatory disease is present which may affect treatment. Isotope bone scanning is at least 70% more sensitive than conventional radiography and also a good prognosticator for subsequent erosion. Inactive erosion may be negative on phosphate scanning.

Bone scintigraphy (or radionuclide imaging) is generally performed with phosphonated compounds that have been labeled with technetium 99m [Tc 99m). It is an ideal scanning agent because of its short 1/2 life and gamma energy of 141 kev which is ideal for gamma camera.

Conventional or static plane, scinitigraphy

The image, which represents uptake, is recorded 2-3 hours (Delayed phase) after intravenous injection of the radioisotope. In this delayed type of scinitigraphy, bone tissue uptake of the radionuclide is primarily a reflection of reactive osteogenesis (osteoblastic reaction). The study can be segmental or "total body".

Advantages: Conventional scinitigraphy provides useful information on the presence of biological activity of bone disease.

Dynamic scinitigraphy (angioscinitigraphy, sequential scinitigraphy, triphasic scinitigraphy)

Following the injection of a radionuclide, a series of closely sequenced images are acquired with a modern gamma camera. A computer is used to obtain uptake curves, which can then be broken down to analyze the vascularization of the region being studied, the bone extraction index of the radionuclide and the degree of osteoblastic remodeling.

Advantages: This method offers all of the advantages of static radionuclide imaging, but it can also reveal local vascular changes (eg. hyperemia, inflammation, ischemia) which help in determining the chronicity of abnormalities, because increased in vascularity returns to normal sooner than increased in uptake evident on delayed images.

Quantitative scintigraphy

It provides a quantitative numerical index that is expressed as the ratio between two surfaces [eg. sacrum and sacroiliac joints (SI)].

Advantages: This method is used to monitor the evolution of bone disease.

Computed tomography

Computed tomography (CT) is a technically complex method of imaging that allows the study of axial (or transverse) sections of the human body. A close derivative of conventional tomography, CT differs from the latter primarily in the manner in which the radiogenic tube moves about the patient and the method by which the images are obtained.

Advantages: Computed tomography offers a number of advantages over plain flim radiography:

Thanks to its axial and multiplanar reconstructed images, which allow investigation of axial skeleton (craniovertebral junction, atlanto axial joint, vertebral column) or sacroiliac joints, that are diffcult to visualize on standard X-rays because of their particular anatomical configuration.

CT can also detect minimal differences between the densities of different structures (intervertebral disks, nerve radicals, menisci, articular cartilage, muscles, tendons, ligaments, etc.) so that the result image is extremely clear (axial scans, as well as sagittal and coronal reconstructions).

Contrast media can be administered intravenously (CT-angiograhpy) or injected into the subarachnoid space (CT-myelography) or the joint itself (CT-arthrography). CT-myelography is helpful for evaluating the vertebral canal, particularly when there is extradural pathology (herniated disc, ostephytes, tumors, etc.)

Cervical spine abnormalities in rheumatoid arthritis

Occipitoatlantoaxial articulation

Atlantoaxial subluxation
Odontoid erosion and fracture
Apophyseal joint erosion, sclerosis, and fusion

Subaxial Articulation

Subaxial subluxation
Apophyseal joint erosion, sclerosis, and fusion
Intervertebral disc space narrowing
Erosion ad sclerosis of vertebral body margins
Spinous process erosion
Osteoporosis

CT scanning[2,3]

CT scanning peri articular (capsular) and para articular (ligamental).

Calcifications can contribute in the accurate detection and differential diagnosis of rheumatoid arthritis, chondrocalcinosis, seronegative spondylo-arthritis, scleroderma and septic arthritis (where it can show contour irregularities).

HRCT is used to detect and evaluate interstitial lung disease in systemic rheumatic diseases.

The SI joints show symmetrical erosions in ankylosing spondylitis. In other seronegative spondyloarthropathies like Psoriatic arthritis, ulcerative colitis, reiters syndrome, asymmetrical erosions are seen.

CT is also useful in diagnosing calcium pyrophosphate deposition disease (CPPD), diffuse idiopathic skeletal hyperostosis (DISH) and Paget's disease.

CT is useful in therapeutic/interventional procedures, particularly involving facet joints and SI joints.

Magnetic resonance imaging[4,5,6]

Magnetic resonance imaging (MRI) does not utilize X-rays, but rather magnetic fields and radio frequency waves. The patient is placed within a magnetic field, which interacts with the most abundant element in the human body hydrogen ion.

MRI is ideal for imaging diarthordial joints. It provides direct multiplanar tomography with high spatial resolution and clear details of soft tissue structures.

Using MRI synovium, synovial fluid, cartilage, ligaments, tendon and bone can all be identified.

MRI is used to, assess the activity/severity of rheumatoid arthritis, response to variety of therupatic regiments, analysis of the extent of articular, para articular changes and complication.

On routine spin echo studies without injecting contrast, low intensity signals are derived from both fluid and pannus on T1 weighted images, whereas high intensity signals characterise both fluid and pannus on T2 weighted images. Following intra venous injection of gadolinium containing agent, the effusion remains of low signals intensity on T1 weighted spin echo images, and the synovium demonstrates enhancement with increased signal intensity. Synovial enhancement is greater in regions of hypervascularity.

Proliferative pannus is less in areas of fibrotic synovial tissue. On delayed images due to seepage of contrast agent, there is increased signal intensity of joint fluid on T1 weighted spin echo images.

The following musculoskeletal complications of rheumatoid arthritis can well be evaluated with MRI.

1. Ischemic Necrosis of bone.
2. Bursitis.
3. Synovial cyst formation.
4. Rheumatoid nodules.
5. Carpal tunnel syndrome and other entrapment neuropathies.
6. Tendon injury and disruption.
7. Insufficiency fractures.
8. Spinal cord disruption/myelopathy.

MRI is the current gold standard for imaging neuro psychiatric systemic lupus erythematosus (NPSLE). It is very sensitive for lobar infarcts and intracranial haemorrhage. Small white matter hyperintensities in fronto parietal regions are most common. Cortical atrophy and ventricular dilatations are also seen.

Bone densitometry

For quantitative assessment of osteoporosis, several methods are used to determine the mineral contents of the cortical and cancellous bone compartments. These techniques are of proven value for the quantitative diagnosis and follow-up of osteoporosis and in estimating the risk of stress fractures (particularly common among post menopausal women, patients with chronic renal failure, and those on bone depleting drugs). They vary to some extent in their precision, accuracy and sensitivity. Precision refers to method's ability to provide similar values with repeated measurements (reproducibility) in actual patients or in "reference dummies". This aspect is particularly important during the evaluation of response to treatment. Accuracy refers to the method's ability to furnish data that reflect the actual mineral content of the volume or surface unit of bone being examined and sensitivity refers to the capacity to differentiate between a normal and a pathological subject and to recognize alterations that occur over time in a given individual or within a study population.

Single-photon absorptiometry

This technique is mainly used in the follow-up of patients treated with cortisone, which has osteopenic effects.

Advantages: Single-photon absorptiometry combines high sensitivity with low cost and limited radiation dosage.

Disadvantages: Absorption of radiation by extra skeletal tissues (soft tissues, subcutaneous fat, calcified blood vessels, etc.) can reduce the accuracy of measurements of bone density.

Dual-photon absorptiometry

This method is based on the measurement of transmission of two photon beams of different energies through bone and soft tissues.

Advantages: The use of two photon beams corrects the soft-tissues interference seen in single-photon absorptiometry. The method is both sensitive and accurate.

Dual energy X-ray absorptiometry

The principles are the same as those of dual-photon absorptiometry except that a single beam of X-rays with two different energies is used. It is currently the most commonly used method for quantitative assessment of osteoporosis.

Advantages: This method is characterized by high accuracy and precision and extremely reliable measurements can be obtained at both the hip and spine.

Digital radiography

Digital radiography is the direct descendent of standard radiography and it is based on the same physical principles. The main difference is that the image is not recorded on radiographic film, but rather on a special plate (Photostimulable phosphor plate) that is sensitive to radiation.

These plates are more sensitive to X-rays than conventional radiographic film, and the radiation dose can thus be considerably reduced, which is particularly useful for examination of children or pregnant women, the images can be stored on a computer for later consultation (while films themselves are often misplaced), superior contrast resolution, provides greater latitude of exposure.

Musculoskeletal ultrasound

Musculoskeletal ultrasound has been described as the "Stethoscope of the joint". Musculoskeletal ultra sonography is one of the most exciting important comprehensive clinical tool developed in the recent years in the practice of clinical rheumatology. It is safe, simple, reliable, cost effective, and bedside tool in the diagnosis, follow-up, prognosis and prediction of many rheumatological conditions. In addition, the real time nature of this technology enables the performance of provocative maneuvers, for visualisation during interventional procedures involving soft tissues.

Its advantage is, it can be used as an office practice equipment like stethoscope or echocardiography. It does not require too much of time, can be repeated frequently and does not need preparation.

The main advantages of sonography are the absence of ionizing radiation and its ability to distinguish solid structures from cystic structures. It can be used to examine joint effusions, synovial and capsular structures, ligaments, muscles and tendons (both at rest and during contraction). It is particularly useful for soft tissue studies to determine the nature (solid, cystic, mixed) of opacities that appear on the standard X-rays and to better define extension and relations with adjacent structures.

Ultrasound technique is constantly being refined and perfected and this method is now being used even in the study of articular cartilage of the hip, the elbow and the knee. It is not considered reliable for evaluation of the menisci. Ultrasound can be used to show:

1. The capsule, synovium, effusion and loose bodies.
2. Surrounding ligments and tendons.
3. Leaks of synovium into the muscles eg. ruptured baker's cyst.
4. Tears and tumors in muscles.
5. US guided therapy (joint injections).

The principle underlying the function of an ultrasonographic apparatus is fairly complex. A special piezo electric is used to limit a beam of ultrasound waves at selected frequency. As a result of their interaction with various anatomic structures, these waves are variably absorbed or reflected. The reflected waves, or echoes are then picked up by the same crystal that emitted them and electronically elaborated into images.

How much of the beam is absorbed by the body tissues will depend on a number of factors, one of the most important is the Emission frequency. The ultrasound beams penetrating power decreases as the frequency increases and vice versa.

Examination Technique

Use linear-array transducer.

Use highest frequency available (5 to 12 MHz or 7 to 15MHz).

Use a stand-off pad only for very superficial structures and for combined palpation and sonography.

Identify and correct anisotropy-related artifacts (false hypoechogenicity) due to improper angle of insonation of the tendon.

Always combine longitudinal and transverse scans.

Check contralateral tendon for reference.

Perform dynamic examination during flexion and extension maneuvers.

Use power Doppler imaging.

US in rheumatoid arthritis:

Abnormal changes in synovium, tendon sheath, cartilage can be detected at early stages to diagnose and initiate treatment particularly in small joints of hand and feet. Due to its multiplanar capability it can detect bony erosions very early.

US is superior to CT in detecting small erosions in large joints like shoulder, knees etc.

US has the capacity to directly visualize both synovitis and bone damage.

Ultrasound can be used to assess two aspects of Synovitis: its Morphology and Quantity, using Grey Scale (GS) and Synovial Vascularity by Colour or Power Doppler (PD). It is latter component that has attracted particular attention as this has been shown to better correlate with inflammatory activity than GS alone.

TURA 16 Trial has been designed to assess whether, the addition of Ultrasound guided therapy to standard Clinical Management of RA using T2T principles, can further improve outcomes in the patients with stable, clinical low disease activity/ remission. The primary objective is to determine whether Therapy Modification guided by Ultra Sound findings and DAS28 scores can reduce the level of inflammation assessed by PD Ultra Sound, compared to Clinical Assessment of DAS28 alone. Other key objectives are to establish whether addition of Ultra Sound guided therapy can reduce radiological programming and improve functional outcomes, and identify predictive factors for sustained remission, absence of structural damage and improvement in functional outcomes.

It is hoped that results from the TURA trial will contribute to the development of new and improved disease management principles for RA.

US in osteoarthritis:

US can demonstrate loss of joint space, cartilage thinning, osteophytes, synovitis and cysts. It can also show osteochondral loose bodies.

In major weight bearing joints like the knee and the hip only a small portion of articular surface can be visualised.

US in Spondyloarthritis

Enthesis is the point of attachment between a tendon, ligament, capsule and bone.

Enthesitis is inflammation of this site is a significant feature in spondyloarthritis.

It also occurs in osteoarthritis secondary to mechanical stress after trauma and in gout. Ultrasound can help in diagnosing bursal diseases like retrocalcaneal and infrapatellar bursitis.

In early disease ultrasound shows smooth fusiform swelling of the tendon or ligament and loss of normal fibrillar pattern representing oedema.

In late disease ultrasound can show increased echogenicity, irregular thickening, bone spurs and periosteal new bone growth.

Ultrasound in crystal arthropathies :

Gouty tophi can be seen as hyperechoic nodules. Calcium pyrophosphate deposit in the cartilage is visualized as a dense white line. Gouty crystals can be seen only if they are calcified.

Ultrasound in Lupus :

Ultrasound is widely used in the assessment of Lupus nephritis. In view of its low cost and safety it can be used frequently to assess the disease activity and response.

Ultrasound in muscle disease:

Polymyositis shows atrophy and increased echointensity predominantly of lower extremity muscles. In dermatomyositis echointensities were high in forearm muscles. Severe muscle atrophy is more prominent in inclusion body myositis.

Ultrasonography is sensitive tool for monitoring localized scleroderma.

It can detect abnormalities in active lesion including increased blood flow, increased echogenicity and loss of subcutaneous fat and reduction in lesion size during treatment.

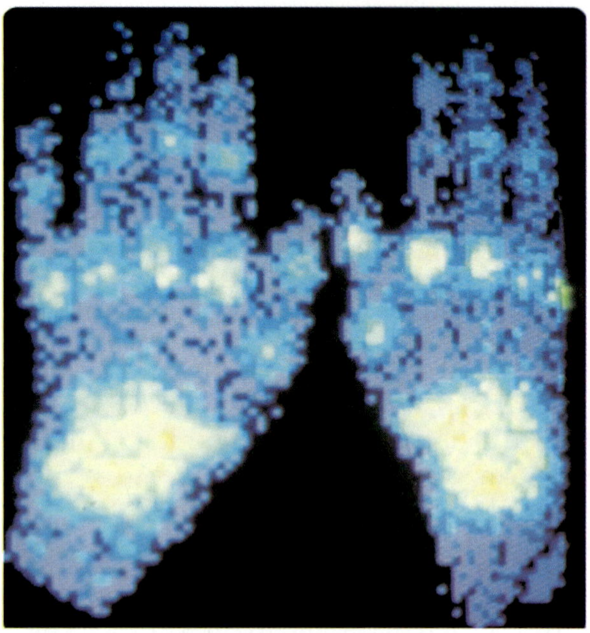

Fig. 1 : Bone scan showing typical bilateral and symmetric involvement of the metacarpophalangeal joints and those of the wrists in rheumatoid arthritis.

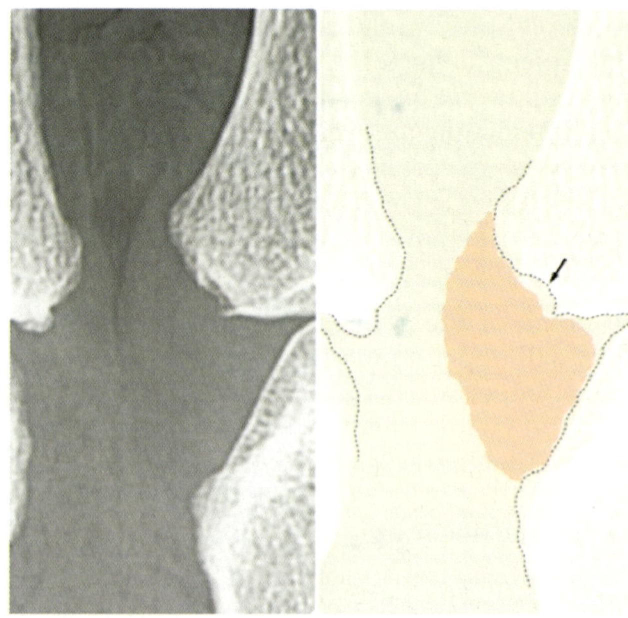

Fig. 2 : Digital radiographic image showing early rheumatoid arthritis with fusiform swelling of the finger and very subtle signs of erosion (arrow).

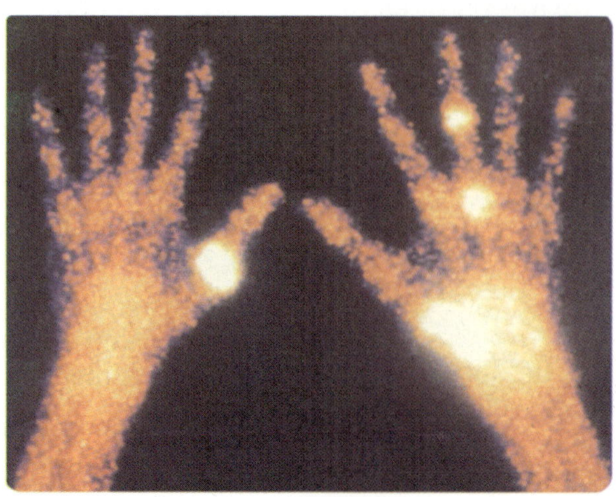

Fig. 3 : Bone scan of the hands in psoriatic arthritis showing casual asymmetric joint involvement.

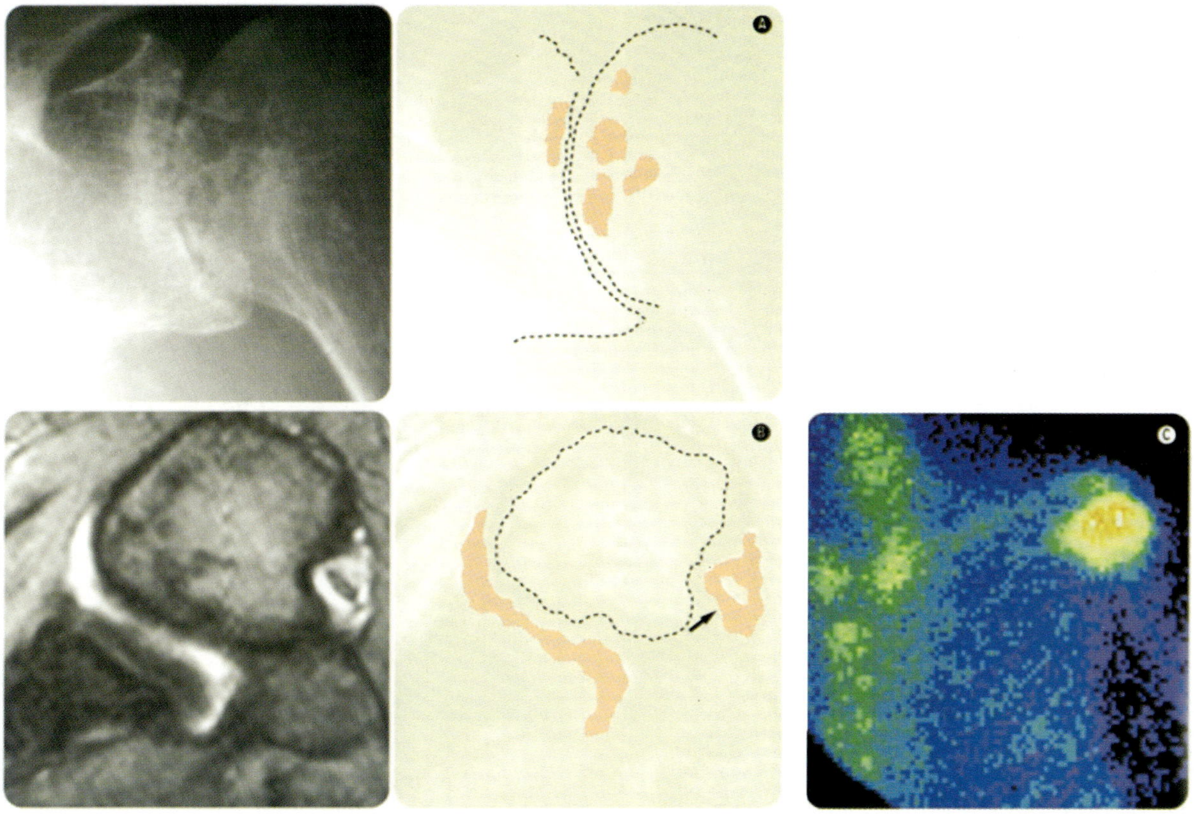

Fig. 4 : Plain-film (A), magnetic resonance (B) and scintigraphic (C) appearances of shoulder arthritis characterized by bone erosions (A), inflammatory involvement of the synovial membrane extending to the long head of the biceps (arrow) (B), and increased uptake in the articular heads (C).

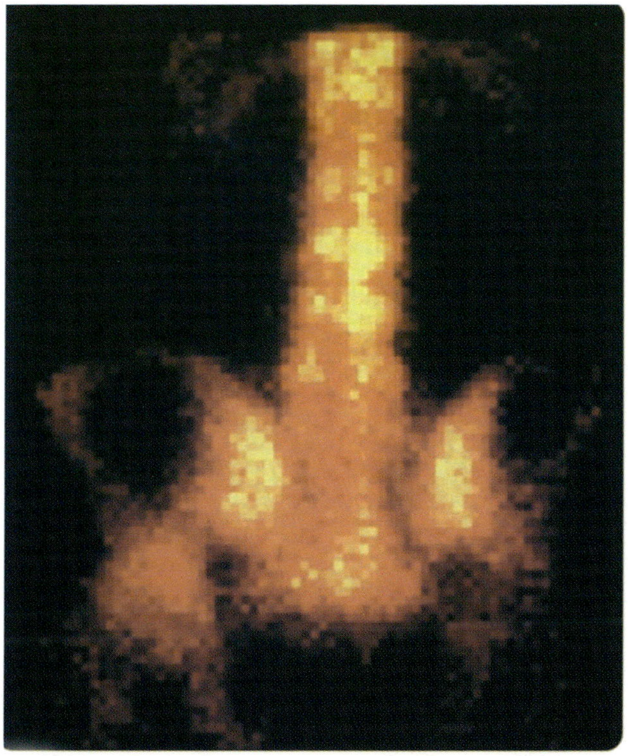

Fig. 5 : Bone scan showing ankylosing spondyloarthropathy with spondylitis and bilateral symmetric sacroiliitis.

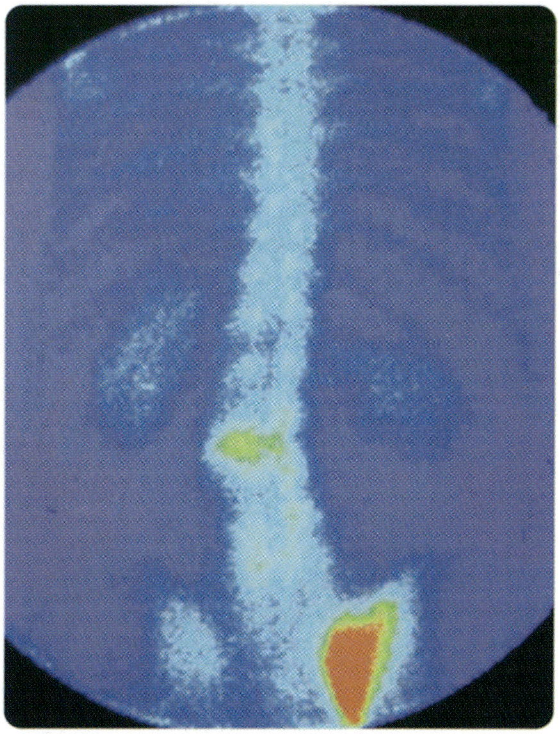

Fig. 6 : Bone scan showing sacroiliitis and spondylitis in a patient with Reiter's syndrome.

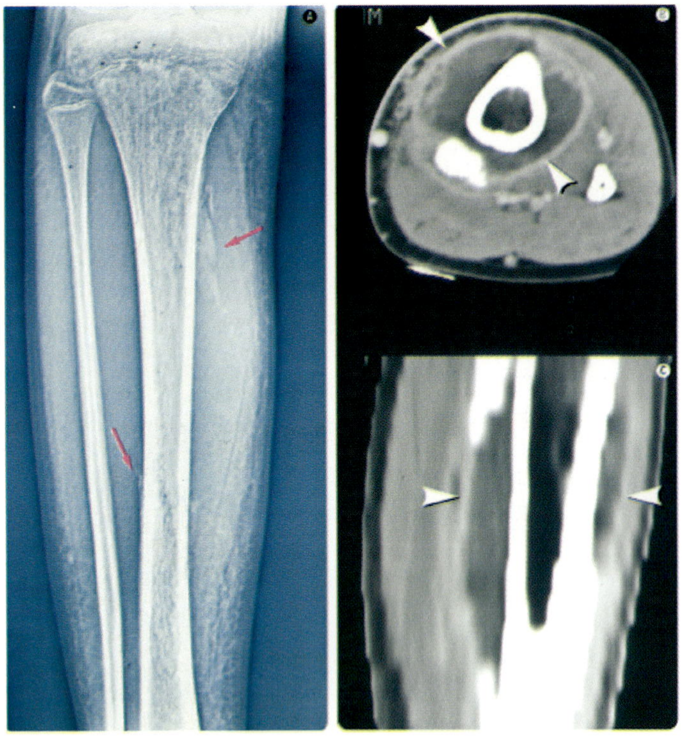

Fig. 7 : Xeroradiographic (A) and CT images (axial -B, sagittal reconstruction - C) of acute tibial osteomyelitis with signs of lamellar periosteal apposition (red arrows)

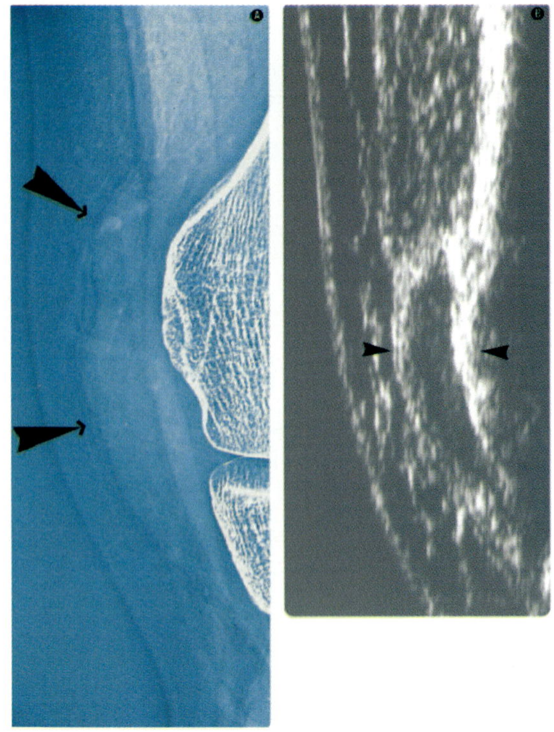

Fig. 8 : Xeroradiographic (arrow, A) and sonographic (arrowhead, B) images of bursitis.

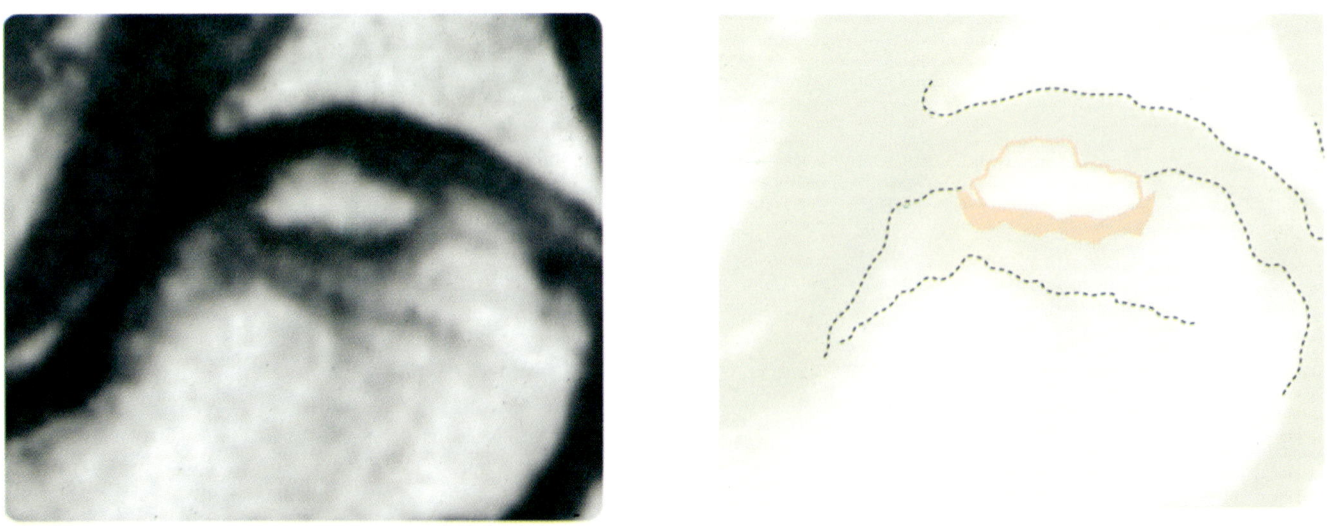

Fig. 9 : MRI image of aseptic osteonecrosis of the femoral head. The hypointense band surrounding the necrotic zone is probably caused by revascularization phenomena. (See previous figure.)

The author expresses his gratitude to Dr. Mahboob Pasha, Consultant Radiologist, Bangalore Hospital and Dr. Ramesh Jois, Consultant Rheumatologist, Fortis Hospital, for their material and intellectual inputs.

Further Reading :

1. Francesco Prioli Bone and Joint imaging in Rheumatic and orthopaedic disease. Published by Rohapharm 2006 (circulated at EULAR symposium Amsterdam).
2. A.K. Brown, T.E. Roberts, PJ Connor, R.J. Wakefield. Z. Karim and P. Emery. The development of an evidence based educational framework to facilitate the training of competent Rheumatologists Ultrasonographers. Published in Rheumatology 1007 March 2007; 40: 391-97.
3. Cardinal E, Chhem, R.K., Beauregard C.G. et al (1996), Planter FaciitisACHTIS : Sonographic Evaluation Radiology 201. 257-59.
4. Gibbon VV.W. and Long G (1999) Ultrasounds of the plantar aponeurosis (fascia) Skeletal Radiol 28. 21-26.
5. LI.S.C. Liebling M.S. and Haines K.A. Ultrasonography is a Scientific tool for monitoring localized scleroderma. Rheumatology 1007; 106: 1316-19.
6. Dihlmann W. (1983) Criticism of Sacroiilitis staging in Ank. Spond (German) Z. Rheum 41: 49-57.
7. EL Maghroui A. Lecoules S. Lechevalier D.Magnin J and Eulry F (1999). Acute Sacroiliitis as a manifestation of calcium pyrophosphate dihydract crystal deposition disease. Clinic. Exp Rheum 17: 477-78.
8. Jorgenson C, Cyeral C. Ananya J. Baron, Lamanque J. and Sany J (1993) Sensitivity of Magnetic Resonance imaging of the wrist in very early Rheumatoid disease Br J Rheum 32: 3-8.
9. Emery P and Luqmani R (1993) - The validating of surrogate markers in Rheumatic diseases Br J Rheum 32: 3-8.
10. Pamela L.P. Axford john S. Isemberg D.A. Imaging in CNS Lupus, Clinical Rheumatology - Best practice and research. Editor in Chief Woolf A.D. Guest Editor Gordon C. Ramsey Goldman. R. Oct 2005; 19.
11. Diagnostic Ultrasound By Carol M. Rumack, Third Edition, Published by Elsevier Mosby.
12. Bone and Joint Imaging By Donald Resnick, Third Edition, Published by Elsevier Saunders.
13. Text book of Radiology and Imaging by David Sutton, Sixth Edition, Published by Churchill Livingstone.
14. Richard J Wakefield, Maria Antonietta D'Agostino, Esperanza Naredo et al. After treat-to-target: can a targeted ultrasound initiative improve RA outcomes? Ann Rheum Dis 2012; 71: 799-803.
15. Koski JM, Saarakkala S, Helle M, et al. Power Doppler ultrasonography and synovitis : correlating ultrasound imaging with histopathological findings and evaluating the performance of ultrasound equipments. Ann Rheum Dis 2010; 69: 1789-95.
16. Paul Emery. Abstract Book; EULAR 2012, Sat Symposium, 8 June 2012, Berlin, Germany.

SYNOVIAL FLUID ANALYSIS 6

Sanjeeb Kakati

Background

Synovial joints consist of discrete elements capsule, cartilaginous articular surface, synovial fluid and synovium. Synovial fluid is regarded as body fluid lying within the tissue bounded by synovium. It is often discussed in similar terms to those used for lining and contents of other body cavities[1,2].

The synovial fluid is composed of glycoprotein and hyaluronate in addition to small filtrate molecules from plasma. Passage of larger molecules like fibrinogen results from increased vascularity i.e. injury or inflammation. Lymphatics are the drainage path of these molecules back to plasma and plays a least role in disease condition.

The main function of synovial fluid is to provide lubrication, joint stability and supply of nutrition to the cartilage and it also has antibacterial properties[3,21].

Synovial fluid analysis consists of a group of tests that detect changes in synovial fluid that may indicate the presence of diseases that affect joint structure and function.

Indications of joint fluid analysis[4,5]

Diagnostic	Septic Arthritis Crystal deposition disease When coexistence of two pathological condition suspected i.e. Rheumatoid arthritis (RA) and Osteoarthritis (OA)
Therapeutic	Large joint effusion with pain and swelling.

Timing of the arthrocentesis is of utmost importance. Institution of antibiotic therapy may alter the finding of synovial fluid and hence delay or never establish the diagnosis in septic arthritis. On the other hand simple rest may decrease the effusion sufficiently and thereby an opportunity of diagnosis may be lost. In crystal induced arthritis specially the level of Monosodium urate (MSU) or Hydroxyapatite crystals goes low rapidly and hence the synovial fluid analysis may become inconclusive. Therefore once the decision of synovial fluid analysis is taken it should be carried out promptly[6].

Contraindication of Arthrocentesis

There is no absolute contraindication for synovial fluid aspiration if it is strongly indicated in a given case. However relative contraindications are : (1) Bleeding diathesis eg. hemophilia, anticoagulant therapy, thrombocytopenia etc. (2) Cellulitis overlying the joint[2].

Selection of site

Knee, elbow, ankle, shoulder, wrist including olecranon or prepatellar bursae are preferred site for synovial fluid aspiration. Arthrocentesis of the hip needs fluoroscopic guidance[1,2,7].

Procedure

Under all aseptic precautions arthrocentesis should be carried out. A topical antiseptic preferably Povidone iodine is to be applied over the area. Use of sterile gloves is mandatory. 25 gauze needle is preferred for administration of local anesthetic (e.g. 1% lidocaine). With the help of a 20-30cc syringe and 18 gauze needle aspiration should be carried out[7].

Complications of arthtrocentesis

Infection (<1 in 10,000), pain, haemarthrosis, cartilage injury can occur during the procedure of joint fluid aspiration.

Causes of dry tap

Very thick fluid, obstruction of the needle lumen with debris, deposition of thick layer of fat in chronically inflamed synovium, obese patient with thick layer of medial fat in the knee joint and faulty technique are the common causes of dry tap[7].

Collection and transportation

Inflamed synovial fluid tends to clot. Therefore it is to be collected in Sodium heparin container to prevent clotting. For culture and sensitivity the synovial fluid should be sent in the syringe itself. While transporting the syringe it should be wrapped with a sterile plastic cap. The fluid should be analysed at the earliest otherwise there may be reduction in the cell count, decrease in the number of crystals or appearance of artefactual crystals[1,7].

Examination of synovial fluid

According to the type of examination that is performed, the tests can be grouped into:

- Physical characteristics-an evaluation of the appearance of the fluid
- Microscopic examination- cells and crystals that may be present are counted and identified by type under a microscope.
- Chemical tests-detect changes in the chemical constituents of the fluid
- Infectious disease tests-detect and identify microorganisms, if present.

Normal synovial fluid is a clear, yellowish fluid and transparent enough to read newsprint through.

Microscopic analysis of synovial fluid

Microscopic study of synovial fluid is aimed at cytoanalysis and detection of crystals. Specimen requirement for cell count and differential is 1 ml of synovial fluid transferred to an EDTA tube. Leucocytes are counted in the same way as it is done in peripheral blood but normal saline is used as diluent instead of Acetic acid. White cell counts (WBC) are normally less than 200/mm^3 with fewer than 25% neutrophils. A WBC count of 2,000/mm^3 and a neutrophil count of 75% serve as useful cutoff points to distinguish inflammatory from non-inflammatory disease. However, there is much overlap in synovial fluid WBCs between the inflammatory, crystal-induced and sepsis categories.

Table 1 : Joint fluid characteristics[7,21,22]

Test	Normal	Non-inflammatory	Inflammatory	Sepsis	Crystal	Hemo
Clarity	Clear	Slightly turbid	Turbid	Turbid	Turbid	Bloody
Color	Clear	Straw	Straw to opalescent	Opaque	Yellow-milky	Red-Brown
Viscosity	Very High	High	Low	Variable	Low	Reduced
Mucin clot	Firm	Firm to friable	Friable	Friable	Friable	Friable
Clotted	No	Occasional	Occasional	Often	Occasional	Yes
WBC/mm^3	0-200	200-2000	2000-50,000	2000-> 50,000	200-> 50,000	50-10,000
% Polys	10	< 20	20-70	> 70	< 90	< 50
Glucose difference	0-10	0-10	0-40	20-100	0-80	0-20
Crystals	Absent	Absent	Absent	Absent	Present	Absent
Culture	Sterile	Sterile	Sterile	Positive	Sterile	Sterile
Diseases and Differential Diagnosis		Osteoarthritis, Osteochondritis, Pigmented villonodular synovitis, Sickle cell disease, Neuropathic	Rheumatoid arthritis, SLE, Reiters syndrome, Ankylosing spondylitis, Ulcerative colitis, Psoriasis Rheumatic fever Gouty arthritis Pseudogout	Bacteria, Tubercular arthritis Gonococcal arthritis Fungi	Gout, pseudogout	Trauma, Hemophilia, Hemangioma, Pigmented villonodular synovitis, Anticoagulant therapy, Tumors

Physical Characteristics

Volume, clarity, color and viscosity of synovial fluid may give an indication of the underlying cause. Normal fluid is either colorless or straw color and transparent. In septic arthritis fluid is either translucent or, opaque. Physical characteristic of three groups of fluid viz. non-inflammatory, inflammatory and septic arthritis are shown in Table 1.

WBC is greater than 50,000/ mm^3 in 70% of patients with septic arthritis, 15% with gout, 10% with pseudogout and 4% with rheumatoid arthritis. Under light microscopy abnormal cells like Reiter's cells i.e. macrophages with phagocytosed neutrophils or ragocytes i.e. neutrophils with large intracytoplasmic inclusion bodies can be detected[1]. These cells with large inclusion bodies consist of complement and immunoglobulin are found in rheumatoid arthritis which are poor responder to therapy. The joint fluid leucocyte count and its interpretation is shown in Table 1. Joint fluid eosinophilia is observed in -

Hypereosinophilic syndrome, Lyme disease, parasitic arthritides, psoriatic arthritis, RA etc.

Presence of apoptotic polymorphs, erythrophagocytic mononuclear cells and mast cells distinguish seronegative arthropathy from rheumatoid arthritis as the latter does not have cytophagocytic cells in the synovial fluid[2]. Joint fluid may be haemorrhagic either due to trauma during arthrocentesis or, other disease conditions viz. bleeding diathesis, tumor, joint prosthesis, haemangioma, A-V fistula etc. In case of trauma during arthrocentesis the fluid will be nonhomogenously stained and clump rapidly. Demonstration of crystals in the joint fluid is the mainstay of diagnosis of crystal induced arthritis.

Table 3 : Causes of joint effusion[1,3,7]

Non inflammatory	Inflammatory
Osteoarthritis	Rheumatoid arthritis
Trauma	Crystal induced arthritis
Internal derangement	Psoriatic arthritis
Osteochondritis dessicans	Reactive arthritis
Charcot's arthropathy	Whipples disease
Villonodular synovitis	Polymyalgia rheumatica
Myxoedema	Polychondritis
Acromegaly	Sarcoidosis
Haemochomatosis	Behcet's syndrome
Ochronosis	Ankylosing spondylitis
Sickle cell disease	Lyme disease
Milwaukee shoulder	Leukaemia

Table 2 : The difference of gout and pseudogout crystals

	Gout	Pseudogout
Composition	Monosodium Urate (MSU)	Calciumpyrophosphate dihydrate (CPPD)
Shape	Needle shaped	Rhomboid or rectangular
Birefringence	Negative	Positive
Colour	Yellow	Blue

Crystals like MSU can be identified under light microscope, but if they are too small they may go undetected. Compensated polarized microscope help in the identification of variety of crystals. Simple presence of crystals in the joint fluid does not suggest the presence of disease rather. Intracellular phagocytosed crystals like Monosodium urate or, calcium pyrophosphate dihydrate crystals are pathognomonic of gout and pseudogout respectively[1]. Synovial fluid analysis of previously inflamed, asymptomatic knees or first metatarsophalangeal joints is a simple bedside procedure for the diagnosis of intercritical gout. Such an approach may facilitate the diagnosis of gout and help to avoid unnecessary diagnostic procedures[22].

Other crystals like basic calcium phosphate (BCP) which is the pathological agent in "Milwaukee shoulder disease" is seen only on electron microscopy. They form clump and appear like shiny coin. However it is the responsibility on the part of the concerned clinician to determine the pathognomonic significance of the presence of crystal in joint fluid reported in a disease condition[9].

Chemical Tests

Tests that may be performed on synovial fluid samples may include:

- Glucose-typically a bit lower than blood glucose levels; may be significantly lower with joint inflammation and infection.
- Protein-increased with bacterial infection
- Lactate dehydrogenase (LDH) - increased level may be seen in RA, infectious arthritis or gout.
- Uric acid-increased with gout

Table 4 : Chemical Tests

Test	Normal Result
Protein	Less than 3 g/dL (approx. 20gm/L)
Glucose	Less than 10 mg/dL of blood level
Uric acid	Parallels serum level
RF	Parallels serum level
ANA	Parallels serum level
C3	Parallels serum level

Synovial fluid protein concentration is usually 25% of serum (1-3 g/dL). Synovial fluid glucose is normally 10mg/dl lower than the corresponding serum glucose. Neither synovial fluid glucose nor protein provides much useful diagnostic information. Glucose levels below 20mg/100ml suggest infection and also occur in RA. Rheumatoid factor (RF) may be positive in synovial fluid even if it is negative in serum but it is not disease specific. Synovial fluid biochemical and immunological assays have been used widely as research tools but not routinely advocated for clinical use[10,18]. Currently there is interest in new type of Biomarkers eg. Cartilage oligometric matrix protein considered to be raised in OA[11,19]. The total amount of protein in synovial fluid is approximately 20gm/l. Electrophoretic pattern is similar to the serum. Investigations of synovial proteins are largely concentrated in inflammatory joint diseases. Increased level of immunoglobulin with lowered level of complement is found in RA. However their level do not correspond to the disease activity. Complement level remains normal in most non rheumatoid situations[12,13].

Test for Infectious Diseases

In addition to chemistry tests, other tests may be performed to look for microorganisms if infection is suspected. Gram stain along with both aerobic and anaerobic cultures of synovial fluid are indicated whenever infection is suspected in a joint. Gram stain allows for the direct observation of bacteria or fungi under a microscope. There should be no microorganisms present in synovial fluid. In presence of strong suspicion of infection culture must be done even if the fluid appears clear. Bacterial cultures require 2-3 ml of fluid in a tube. If bacteria are present, susceptibility testing against certain antibiotics can be performed to guide antimicrobial therapy. If there are no microorganisms present, it does not rule out an infection; they may be present in small numbers or their growth may be inhibited because of prior antibiotic therapy. In suspected case of tuberculous arthritis ZN stain supported by culture for mycobacteria should be carried out. At times PCR test for mycobacteria is helpful for early diagnosis of joint tuberculosis[17]. Coexistence of septic and crystal arthritis is possible, therefore one must look for crystals even after diagnosis of septic arthritis[14,15].

Summary

- Synovial fluid analysis is an important tool for diagnosis in joint diseases specially in case of monoarthritis.
- Methodical collection followed by physical analysis including volume, viscosity, color should be noted.
- Prompt laboratory investigations should be carried out subsequently. These include:
  - cell count,
  - detection of crystals,
  - gram stain and
  - culture and sensitivity, etc.
- Presence of intracellular crystals in the synovial fluid confirms the diagnosis of gout or pseudogout.

Further Reading :

1. Houk J. Lawrence. "Joint Fluid" Oxford Text Book of Rheumatology Vol-1 ed. Maddison P.J, Isenberg David A, Woo Patricia, Glass David N. Newyork Oxford University Press: 1993; 677-85.
2. Freemont A. Role of cytological analysis of synovial fluid in diagnosis and research : Annals of the Rheumatic diseases 1991; 50: 120-23.
3. Simkin P.A. "Physiology and Pathophysiology" Practical Hand Book of Joint Fluid Analysis (2nd edition) editors- Gattar R.A, Schuwacher H.R. Lee and Febiger, Philadelphia 1991; 8-23.
4. Dieppe P, Campion G, Doharty M. "Mixed Crystal deposition" Rheumatic disease clinics of N Amer 1988; 14: 415-26.
5. Cohen A.S., Brendt K.D, Krey P.R "Synovial Fluid" Laboratory Diagnstic Procedure in Rheumatic diseases. (2nd edition) edit. Cohen A.S. 1-62 Little Brown and co. 1975.
6. Kerolus G. Clayburne G. Schumacher H.R. "It is mandatory to examine synovial fluid promptly after arthtrocentesis". Arthritis and Rheumatism 1998; 32: 271-78.
7. Gattar R.A, Schumacher H.R "Joint aspiration : Indications and Technique " Practical Hand Book of Joint Fluid Analysis" (2nd edition) edit Gattar RA. 1991; 14-23.
8. Hollander J.L, Mc Carty. DS. J., Rawson A.J. R.A. cell "Ragocyte" or, "Inclusion body cell" Bulletin of the Rheumatic Diseases 1965; 16: 382-83.
9. McCarty D.J., Kahn N.N, Faires J.S. The significance of Calcium phosphate crystals in the synovial fluid of arthritis patients, the "Pseudo gout syndrome". Annals of Int. Medicine 1962; 56: 711-37.
10. Schumacher H.R. Reproducibility of Synovial fluid analysis. Arthritis and Rheumatism 1986; 29: 770-74.
11. Swan A, Amer H, Dieppe P. The value of synovial fluid assay in the diagnosis of joint disease: A literature survey; Ann Rheum Dis 2002; 61: 493-98.
12. Bunch T.W., Hunder, G.G., Offord K. and McDuffie. F.C. Synvial fluid complement : usefulness in diagnosis and classification of Rheumatoid arthritis. Ann. Intern Med 1974; 81: 32-35.
13. Crecchiolo A. and Barnett E. The role of immunological tests in routine synovial fluid analysis. J Bone. Joint surg 1972; 54: 828-40.
14. Baer, P.A, Tenenbaum J, Little H. Coexistent septic and crystal arthritis. Report of four cases and Literature review. Jour Rheum 1986; 13: 604-07.
15. Pispati P.K. Rao U.R.K. (1999) Laboratory investigations for Arthritis, J Asso Phy Ind 47: 298-304.
16. Draeger, H.T., Twining J.M, Johnson C.R. et al. "A randomized controlled trial of the reciprocating syringe in arthrocentesis" Ann Rheum South Asian edition 2006; v4: 270-73.
17. Negi S.S, Gupta S, Lal S. "Comparison of various microbiological tests including PCR for diagnosis of osteoarticular Tuberculosis" Indian Jour of Med Microbiology 2005; V23: 245-48.
18. Caspi D, Anonk M, Golan I, Parah D, et al. "Synovial fluid levels of anti cyclic citrullinated peptide antibodies and IgA Rheumatoid factor in Rheumatoid arthritis, Psoriatic arthritis and Osteoarthritis" Arthritis Rheum. 2006; 15: 55: 681-82.
19. Chaturvedi V, Handa R, Rao D.N., WaliJ.P. "Estimation and significance of serum and synovial fluid malondialdehyde levels in Rheumatoid arthritis" Ind J Med Res 1999; 109: 170-74.
20. Choudhury V, Agarwal A, Naik S. Laboratory Investigations that aid in the diagnosis of Rheumatic diseases. Bombay Hosp Jour 1997; 39: 234-39.
21. Gruber BF, Miller BS, Onnen J, Welling RD, Wojtys EM Antibacterial properties of synovial fluid in the knee. Synovial Fluid - Wheeless' Textbook of Orthopaedics, J Knee Surg. 2008; 21(3): 180-5.
22. Eliseo Pascual, Enrique Batlle-Gualda, Agust´n Mart´nez, Jose´ Rosas, Paloma Vela, Synovial Fluid Analysis for Diagnosis of Intercritical Gout : *Annals of Internal Medicine*, 1999; 131: 756-59.

SECTION 5

Therapy

NON-PHARMACOLOGICAL THERAPY IN RHEUMATIC DISEASES

1

S Chandrashekara

Non-pharmacological therapy in rheumatic disease

Rheumatoid arthritis (RA) and other rheumatic diseases require both pharmacological and non pharmacological interventions for their effective management. Being chronic diseases, they have significant impact on quality of life; in addition they require continued therapy. The interventions to improve the quality of life, alleviate symptoms and to achieve a better compliance to therapy contribute significantly in overall disease control. Thus non-pharmacological interventions play crucial role in achieving a better outcome. They are discussed in four important areas:

1. Education or counseling of the patients and relatives.
2. Psychosocial management.
3. Physiotherapy and Orthotics.
4. Rehabilitation and occupational therapy.

Education and counseling

The patient compliance to therapy has been demonstrated to be directly proportional to his or her concerns and information about the disease. The knowledge base about the disease can be enhanced by educating and counseling the patients. A few minutes spent by physician or his team member in explaining about the disease and simultaneously advising the patient about the disease adds a lot in patient's management. The entire process is considered as educating and counseling the patients.

The counseling and educating patients though individualized, a structured approach will help in achieving the purpose. The process of counseling should be clear on the objective, target and the purpose. Majority of systemic analysis as well as the reviews agree upon the fact that contents of the counseling vary based on the disease, the patient and the health care settings. However the purpose of the counseling as currently agreed upon is to educate, empower and elucidate the patient about the disease and treatment. Thus the steps commonly include, providing patients and their relatives factual information about the disease; subsequently to explain the issues of concern to both patients and their relatives like uncertainty of results of treatment, adverse reactions etc. The most critical step is to involve both patients and their relatives in decision making with reference to treatment and other related issues.

In diseases like RA and other rheumatic diseases it is important to address the prevailing confusion among the patients and their relatives regarding diseases. The issues like seeking cure where it does not exist and misunderstanding such as no cure means no treatment, need to be addressed effectively. Patients need to be explained about the available option of therapy as well as the risk and benefits of therapy. Participation of patients and their relatives in decision making regarding the treatment will help to improve compliance. While discussing about the treatment, physician should volunteer to disclose the possible therapeutic toxicity, benefit, cost and other compliance problems.

Counseling is not just passing on the information verbatim and should have the characters listed in Table 1. A few points should be considered in a counseling session.

a. Only possible result should be conveyed either a falsely optimistic or pessimistic statement will not yield good result.
b. The contents should be scientific and clear; should use the language what patient can understand.
c. Adjectives, exaggeration of the facts and condemning any opinions of the patients are not a good act in counseling.
d. A good counselor is also a good listener.

If these simple rules are followed an effective communication can bridge the gap between the patient and physician. This improves the confidence of patients in the treatment, enhances his or her self-care attitude and improves the compliance.

Table 1. Points in effective communication

1. General clarity,
2. elicitation of and responsiveness to patient problems, concerns and expectations,
3. explanation of condition, progress, and prognosis,
4. explanations of processes of care,
5. explanations of self-care,
6. empowerment, and
7. decision-making.

Psychosocial Interventions

Rheumatic diseases often impact on both psychological and social aspects of the patients. The chronicity of the diseases, complicated by fear of unpredictable course, pain and disfiguring potentials, often produce distressing depressive emotion to the patients. Persisting pain over the joints often disturb patients. The response may shift from anxiety, guilt and later to depression. The depressions as estimated from various studies depend upon variety of factors including socio-economical factors, social support, and educational status and intrinsic factors of patients. The work disabilities often have been attributed to factors other than disease activity. de Croon EM et al in their systemic review found inconsistent evidence for the direct effect of biomedical variables like pain on work disability, and confirm the notion "that work disability is biopsychosocially determined by the deficit between work environment demands and individual capability". This suggests the fact that the effective intervention in psychosocial spheres of the patients suffering from rheumatic disease improves significantly his or her quality of life. Psychosocial approaches enhance medical regimes of care. There is evidence that psychosocial interventions improve coping and self-efficacy, reduce psychological distress, and reduce pain, at least in the short term.

The psychosocial intervention should consider the patient as an individual with his or her pre-defined role. The distress and the required intervention depend on the role and requirement of an individual. There are no single well organized studies which suggest - which is the best method to intervene the psychosocial aspects of rheumatic disease? Psychosocial approaches in managing arthritis pain include educational programs, coping skills training (CST), and cognitive behavioral therapy (CBT). The education and the content of the same are discussed in previous sections. CST to face the pain, disability and other forced limitation because of disease, will help in reducing the distress faced by the patients. Adequate concept about the disease, the potential events and how to plan and face situations or consequences arising out of disease will help the patients. The adoptability of the patients to situation depends on several factors like, role of the patients at his or her work, family and society, his or her family and social backup, available financial and other resources. The coping can be defined as process of responding and managing or contending with life stresses and difficulties arising out of stressful situations which are beyond the control of individual patient. Coping strategies are 'active coping' (problem-based coping including learning and enacting behaviors designed to face the problems), 'passive coping' (avoiding situations or giving up to situation) and 'catastrophizing' (exaggerating the perceived threat and focusing on the worst that can happen). Both physician and his or her team including the counselor should assist in managing these issues actively. These are achieved by CBT, based on training to alter the thought processes which influence feelings and behaviors. It involves, learning how to recognize destructive thoughts or ineffective responses, reflection and to set goals, and practicing new strategies.

Studies have revealed a significant improvement in quality of life of patients suffering from rheumatoid arthritis with a good psychosocial therapy. This can also be achieved, in addition to individualized therapy, using family support, group therapy, group discussions, support group and association. This will help the patients to learn about the disease, to share and cope about the disease there by improve the quality of life.

Rehabilitation

"Someone to look after the whole patient, not just the part that hurts" was how one consultant summarized the medical needs of patients with rheumatic disease (Gloag D, BMJ 1985). The process of rehabilitation is expected to encompass all therapeutic domains other than drugs. Rehabilitation in rheumatic diseases almost begins from the day of diagnosis. The rehabilitation programme incorporated depends upon diagnosis, patients' conditions and psychosocial variables. There are many steps and effective ways by which rehabilitation is managed. Rehabilitation may involve a simple step like guiding how to use a stick or may be as extreme as replacing a damaged joint. Several factors influence the choice of procedures and steps adopted in rehabilitation (Fig.1). The key factors which influence the decision making are age of the patients, extent of disability, reversibility of the disability, expectation of the patients as well as the cost to benefit ratio to the patients. For example, a young executive, with a hip disease, where even if the deformities are minimal, if he or she is limited by more than 50%, it is more practical to go for replacement of the joint rather than deferring it for little more period. On the contrary in an elderly, with a comorbid condition, better choice would be to improve his or her mobility with a splint and support. The study evolving the utilization of rehabilitation services by rheumatologists was significantly low. The outcome especially the quality of life measures were significantly better in the patients who utilized rehabilitation strategies.

In some countries the rehabilitation in rheumatic disease has been considered as a separate specialty. In India it is less frequently utilized and is often limited to physiotherapy and usage of splints and replacement. The vocational training, joint restoration and other active interventions in maintaining functionality of the patients are expected to reduce the loss of job and improve overall quality of life. Rehabilitation is both an art and science. Though no well-organized studies appreciating the utility of the methods are available, systemic studies have demonstrated its usefulness.

The simple principle of rehabilitation is to make life little better. The plan for rehabilitation should be individualized and a step wise approach is required.

Physiotherapy:

Physiotherapy aims essentially to relieve the stiffness and improve the flexibility of the joints and surrounding muscles. Joint inflammation leads to muscle wasting and contractures. Physiotherapy plays a critical role in the management of musculoskeletal manifestation of rheumatic disease. In principle, physiotherapy relays on physical agents, manipulation and exercise for effectively improving both the symptoms and functions. In diseases like ankylosing spondylitis, late rheumatoid arthritis and osteoarthritis physiotherapy has a significant role. The symptoms such as pain relief, maintenance of posture, reduction of disability and overall improvement in quality of life are the main contributions of physiotherapy.

Fig.1. Steps in Rehabilitation

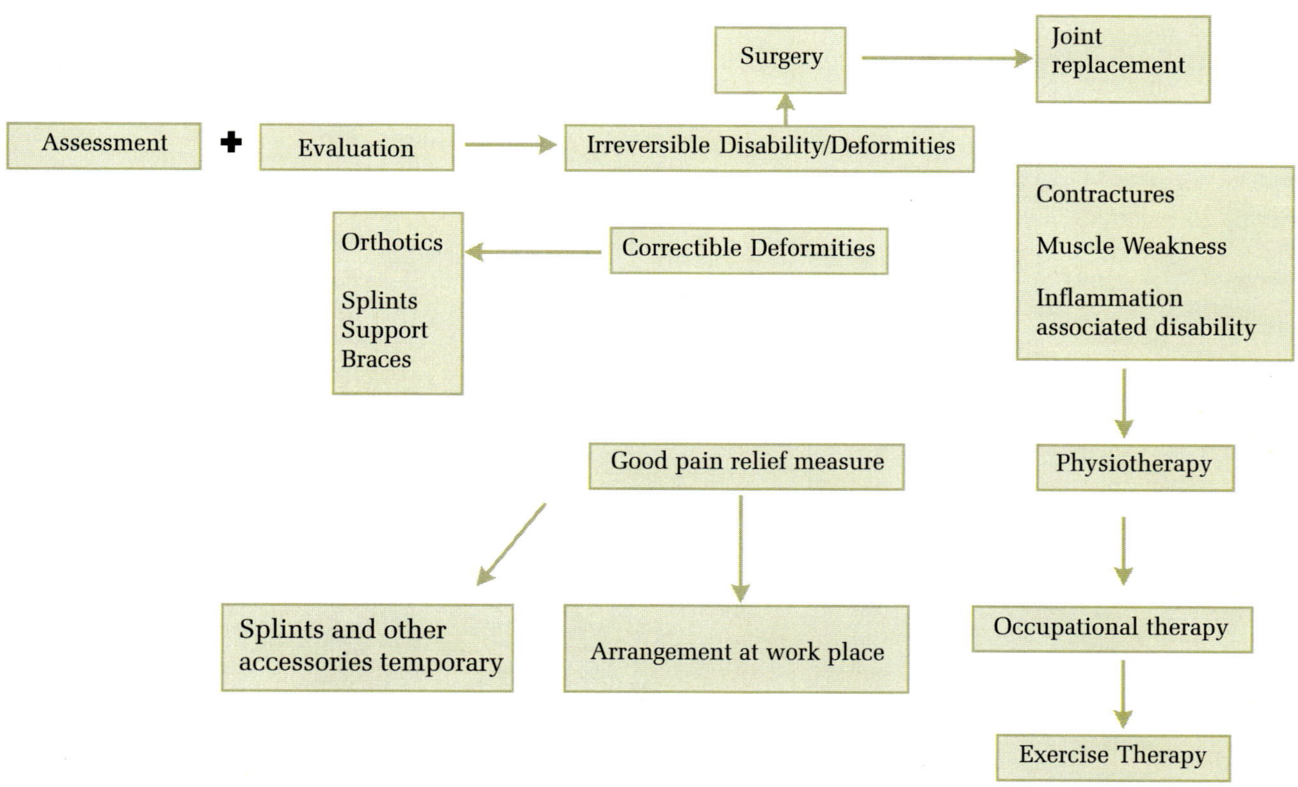

Physical therapy:

Most commonly includes short wave diathermy, interface therapy, transcutaneous electrical nerve stimulations (TENS), ultrasound therapy, wax therapy, muscle stimulation therapy, heat and cold application and few other modalities. Heat pad is the commonest house hold therapy which is used in alleviation of pain. The skilful combination of these modalities helps in reducing pain and swelling. Wax therapy and heat and cold application, with gradated stretching can be useful in releasing the contractures and improving the range of motion. The physiotherapy is useful in both early as well as in later part of rheumatic disease. They are used both during active inflammation as well as when it is quiet. Different modalities and their usefulness are listed in Table 3.

Exercise therapy:

In rheumatic diseases rest should be used less frequently and discreetly. Prolonged rest often increases stiffness and also atrophy of the muscles. Even prolonged splinting especially in odd positions may result in contractures adding to disability of the disease. Regular exercise will help the patients to keep the muscle strength, reduce contracture and keep the range of motion in the joints better. Stamina will be better if the patient continue to maintain exercise. Patients with excess of fatigue and pain an hour of rest may help to improve their performance. The aims of exercise therapy are in Table 2. While guiding the patients on exercise therapy, one has to consider the patient's physical status especially of joints and muscles. Assessment of posture and gait should also be considered while planning the exercise therapy. In addition one has to focus on associated comorbid condition especially cardiovascular status. Exercise though built on a common principle, may not essentially be same to all patients. They have to be tailored to each patient and may need to be constantly reviewed from both therapist and the patient end. For example, in a patient suffering from ankylosing spondylitis with predominant axial diseases, the focus of exercise is on the spine rather than on the hands.

Table 2. Aims and objectives of exercise therapy

1. Decrease pain.
2. Increase or maintain range of motion.
3. Increase or maintain muscle strength.
4. Increase endurance and reduce biomechanical strain in all joints.
5. Promote independence in activity of daily living (ADL).
6. Improving efficient and safe gait pattern.
7. Educating the patients and families on the importance of exercise.

Table 3. Physiotherapy plan of treatment in rheumatic diseases

Active inflammatory stage of musculoskeletal disorder (RA)	
To relieve pain and muscle guarding and promote relaxation	Electrotherapy Gentle massage Immobilization in splint Relaxation techniques
Minimize joint stiffness and maintain available motion	Active assistive exercise ROM within limits of pain Gradual progression as tolerated Mobilization
Minimize muscle atrophy Prevent deformity and protect joint structures	Isometrics exercises to muscles Use of supportive and assistive equipment Good bed positing Avoidance of activities that stress the joints
Inactive stage (RA)	
Decrease effects of stiffness	Patient education Active range of movement exercise Mobilization
Decrease pain from mechanical stress	Supportive / assistive equipment to minimize stress Increase strength in supportive muscles
Increase range of movement (ROM) Decrease pain at rest if present Prevent deformities	Strengthening of muscle Electrotherapy Splinting and patient education Non impact or low impact aerobic

Orthotics and splints:

Many well designed braces and splint supports have improved the care of patients with musculoskeletal disability. These devices help patients by reducing the load on the joint concerned as well as permitting the movement of the joints. They are often used for providing rest to the joint. Splints and braces permits the movement in a restricted and a guided manner there by they can effectively reduce the pain. Splints are very useful to reduce and correct the deformities.

Other complementary therapies:

Few of the traditional system if appropriately used are useful adjuvant as a non-pharmacological intervention in rheumatic disease. A good practice of acupuncture and acupressure has been demonstrated to help patients to improve their pain as well as the quality of life. Traditional yoga and meditation practice have been demonstrated to improve the quality of life and few of the measures of disease activity such as functional score as well as the pain. In addition a brief knowledge about the possible consumption of alternative drugs is required. The patients may discuss the issue or sometime may not, divulge the full information.

In conclusion several non-pharmacological interventions help in the management of the patients suffering from rheumatological disease. Skillful utilization will complement patient care.

Further Reading:

1. Hammond A, Freeman K. One-year outcomes of a randomized controlled trial of an educational-behavioural joint protection programme for people with rheumatoid arthritis. Rheumatology (Oxford) 2001 Sep; 40: 1044-1051.
2. Lineker SC, Bell MJ, Wilkins AL, Badley EM. Improvements following short term home based physical therapy are maintained at one year in people with moderate to severe rheumatoid arthritis. J Rheumatol 2001; 28: 165-168.
3. Keefe FJ, Rumble ME, Scipio CD, Giordano LA, Perri LM. Psychological aspects of persistent pain: current state of the science. J Pain 2004 May; 5: 195-211.
4. Peter Oesch, Stefan Bachmann. Introduction to physical medicine and rehabilitation. In Kelley's Text book of Rheumatology. Eds Gary Firestein et al. 8th Ed Saunders Elsevier. 2009; 1023-33.
5. Backman CL. Arthritis and pain. Psychosocial aspects in the management of arthritis pain. Arthritis Res Ther 2006; 8: 221.
6. de Croon EM, Sluiter JK, Nijssen TF, Dijkmans BA, Lankhorst GJ, Frings-Dresen MH: Predictive factors of work disability in rheumatoid arthritis: a systematic literature review. Ann Rheum Dis 2004 Nov; 63: 1362-1367.

PHYSIOTHERAPY 2

Shashikala A, Archana Rani S, Mariam Younis

Physiotherapy uses physical approaches to promote, maintain and restore physical, psychological and social well-being, taking account of variations in health status. Physiotherapy and rehabilitation applications significantly augment medical and surgical therapy by improving the management of arthritis and reducing the disability in daily living. Once the patient with arthritis is referred to a physical therapist, functional assessment is done to plan the treatment program. Patients are evaluated for their level of fitness or disability, their ability to participate in routine activities of daily living.

Aims of Physiotherapy

- Relief of pain and inflammation
- Prevention / correction of deformity
- Restoration and maintenance of joint motion
- Improvement of muscle strength and endurance
- Prevent disability and increase functional capacity
- Patient education

Depending upon the patients functional assessment within the home and work environment, the physiotherapy planning consist of

a) Physiotherapy Modalities
b) Therapeutic exercise
c) Rehabilitation

a) Physiotherapy Modalities

i) Hot/Cold Applications
ii) Electrotherapy
iii) Hydrotherapy

i) Hot and Cold applications:

Hot or cold applications for10-20 minutes once / twice a day are most commonly used physical agents in arthritis treatment. Cold application is mostly used in acute stages and replaced joints whereas hot is used in chronic stages of arthritis. Ice or heat is never applied directly over the skin.

By using heat, analgesia is accomplished, muscle spasm is relieved and elasticity of periarticular structures is obtained. Heat can be used before exercise for maximum benefit. Types of thermotherapy include superficial hot pack, infrared radiation, paraffin wax, fluidotherapy or hydrotherapy.

Cold application is preferred in active joints where intra articular heat increase is undesired. It numbs pain and helps relieve swelling. Cold pack, ice, nitrogen spray and cryotherapy are different methods of applying cold therapy.

ii) Electrotherapy:

Electrotherapy is used to stimulate nerves and muscles to alleviate pain. It is a non-invasive pain relief therapy. Electrotherapy is used for relaxation of muscle spasms, prevention and retardation of disuse atrophy, muscle rehabilitation, to maintain and increase range of motion and management of chronic pain e.g. Rheumatoid arthritis (RA).

Transcutaneous electrical nerve stimulation, interferential therapy and ultrasound therapy are most commonly used modalities in the treatment of arthritis. The therapy is tailored according to the patient needs.

iii) Hydrotherapy:

Hydrotherapy is an effective way of treating arthritis as it provides heat whilst encouraging exercise. The buoyancy property of water provides the patient with relative weightlessness and joint unloading, allowing active motion with increased ease. Hydrotherapy helps to increase range of motion, to strengthen muscles, to relieve painful muscle spasms and to improve the patient's well-being. In addition, it also has a sedating effect.

b) Therapeutic exercise

Therapeutic exercise[1] is a systemic and planned performance of bodily movements, postures, or physical activities intended to provide a patient with the means to

- Remediate or prevent impairments
- Improve, restore, or enhance physical function
- Prevent or reduce health related risk factors
- Optimize overall health status, fitness, or sense of well-being

Exercise is a part of the evidence based recommendations for the treatment of arthritis. Muscle weakness in patients with arthritis may occur because of immobilization or reduction in activities of daily living. Maintenance of normal muscle strength is important not only for physical function but also for stabilization of the joints and prevention of deformities. Exercise should be adapted to the individual. In the same way that a medication is always prescribed at a named and appropriate dosage, exercises should be clear and have a definite aim and repetitions. 'Go and exercise!' is not an acceptable approach. People with arthritis who exercise regularly have less pain, more energy, improved sleep and better functions. Even mood will improve and depression will decrease with regular exercises (of any chronic illness)[2].

More than 2 of 5 adults with arthritis were physically inactive[3]. Inactivity was associated with lack of strong motivation for physical exercises and a disbelief in the benefits of activity. Inactivity and lack of mobility aggravates debility and stiffness of joints. Sufficient evidence has accumulated to show significant benefit of exercise over no exercise in patients with osteoarthritis (OA).An approach combining exercises to increase strength, flexibility, and aerobic capacity is likely to be most effective in the management of the same[4].

The Goals of Exercises

- Joint Protection
- Range of Motion
- Strength
- Cardiac fitness
- Weight reduction

Components of exercise therapy are:

1. **Free exercises[5]:** Free exercises are those which are performed by the patient's own muscular efforts without the assistance or resistance of any external force, other than the gravity[5]. With free exercises relaxation is induced; muscle tone is maintained and power is increased according to the speed, leverage and duration of exercise; confidence in the ability to perform and control movement is established. While doing free exercises the starting position (to ensure postural efficiency), speed and duration of the exercise are to be considered. Instruction has to be given to understand both the pattern and purpose of the exercise. Range of motion exercises (ROM) and aerobic exercises are some of the free exercises as shown in Fig. 1.

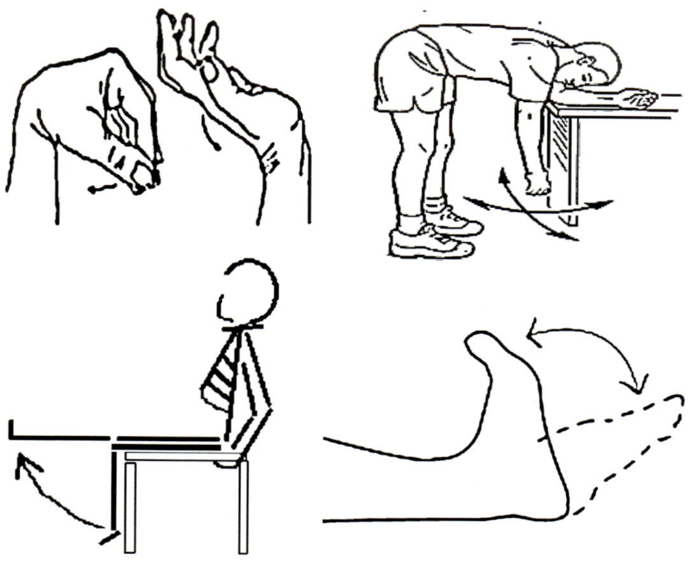

Fig. 1 : Free exercises

2. **Assisted exercises[5]** : Assisted exercises are those which are performed when the muscle strength or coordination is inadequate to perform a movement. The external force will be applied to compensate the deficiency in the direction of the muscle action. Few of the assisted exercises are shown in Fig. 2. The range of effective joint movement may be increased by assisted exercises. Confidence in the ability to move will be established. While performing assisted exercises; the starting position, the pattern of movement, adequate fixation of the bone of origin of the prime movers and support of the joint to be moved are to be considered. Every effort has to be made to reduce tension in the muscles which are antagonistic to the movement. The force used to augment the action of the muscle should be in the direction of the movement. Passive manipulation and mobilization are some of the assisted exercises.

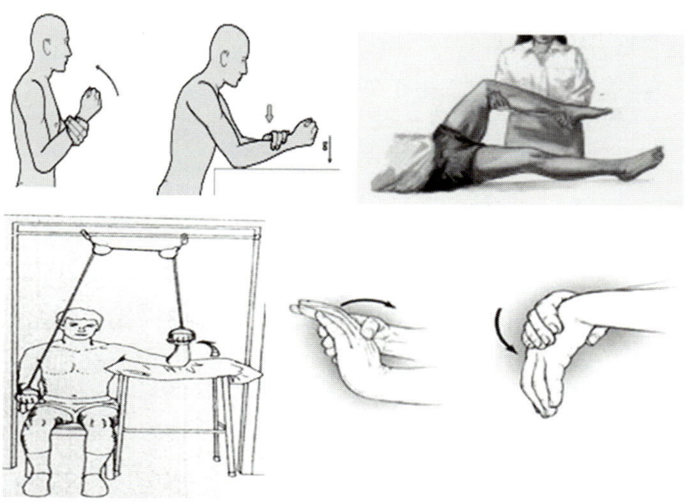

Fig 2 : Assisted exercises

Assisted - Resisted exercises are those exercises that constitute both the assistance and resistance during a single movement.

3. **Resisted exercises**[5] : Resisted exercises are those exercises where an external force is applied to the body to oppose the force of muscular contraction. Resisted exercises are used to strengthen weak muscles and to improve the endurance. Endurance develops as a result of repetitive contraction. Low repetition, high resistance training enhances strength development, whereas high repetition, low intensity training optimizes muscle endurance. Strengthening exercises are the examples of resisted exercise. Some of the hand grip exercises are shown in Fig. 3. In arthritis minimal and mild resistance is given; minimal resistance at large joints for example, to the knee joint while doing static cycling, mild resistance at the small joints example, hand joints to improve the atrophy of the lumbricals. Resisted exercises using heavy weights are not suggestible in acute stage of arthritis.

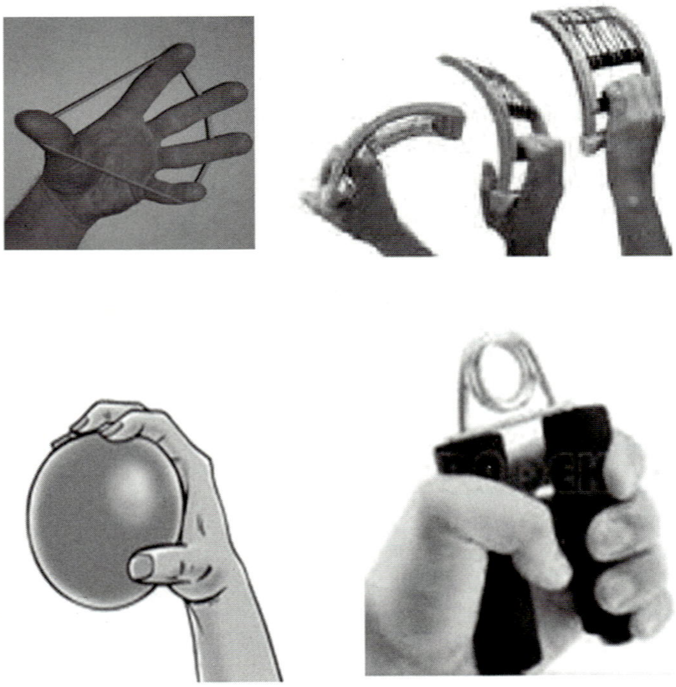

Fig 3 : Hand grip exercises

4. **Isometric Exercises**[5] : An isometric exercise constitutes an increase in intramuscular tension without change in the length of the muscle. Isometric exercises are used to minimize the muscle atrophy. These exercises are suggested during acute stage of the disease. Isometric exercises of neck and knee are shown in Fig. 4.

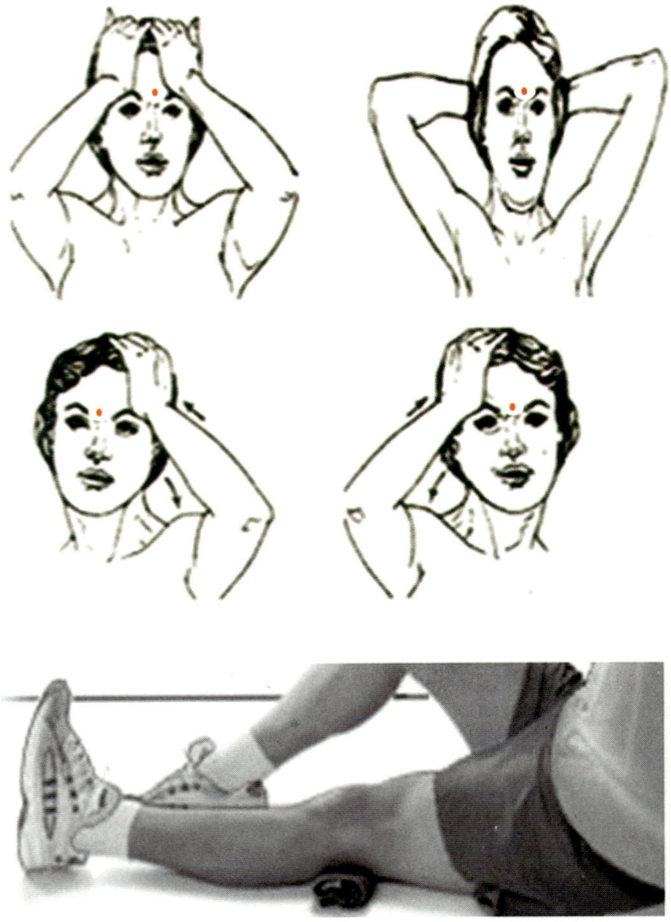

Fig 4 : Isometric exercises

5. Isotonic Exercises[5] : Isotonic Exercises constitute an increase in intra muscular tension accompanied by a change in the length of the muscle. The change in length may either shorten or lengthen the muscle. Isotonic exercises are performed by using very low weights. These are suggested during inactive stage of arthritis. Some of the isotonic exercises are shown in Fig 5.

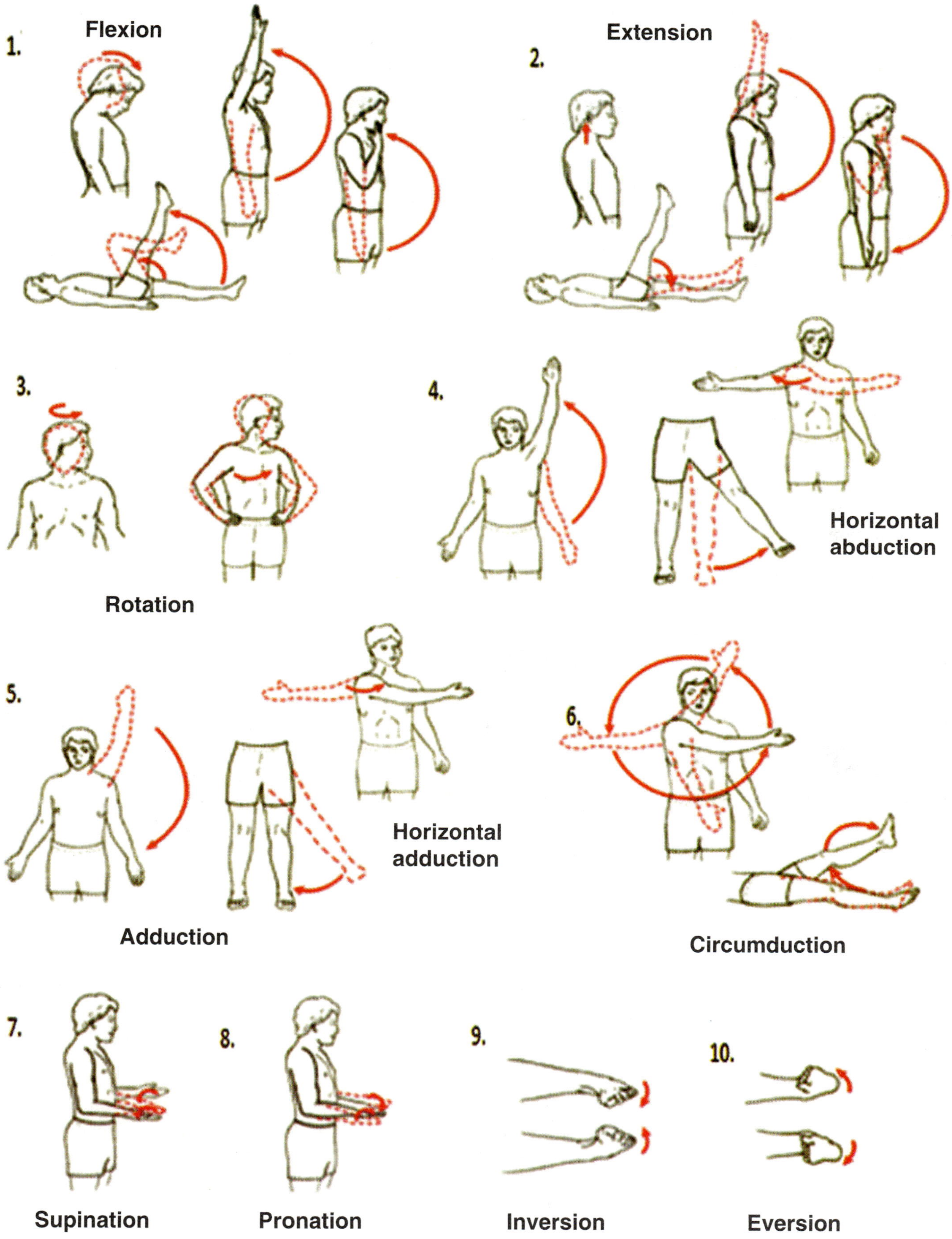

Fig 5: Isotonic exercises

6. Aerobic Exercises : Aerobic exercises aim at improving the condition of heart and lungs, needed for an optimal oxygen uptake of the muscles. Aerobic exercises increase the nutrition of muscles and thereby decrease the muscle soreness. In arthritis patients the aerobic exercises in combination with muscle strengthening exercises, reduce pain and activity limitation. Examples of aerobic are walking, cycling and swimming.

Prior to establishing an exercise program the following characteristics should be considered:

- Whether the involvement of the joints is local (e.g. trauma, OA) or systemic (e.g. RA)
- stage of the disease
- age of the patient
- compliance of the patient
- Duration and severity of the exercise are adjusted according to the patient needs

c) Rehabilitation

The aim of rehabilitation is to maximize function and minimize the limitation of activity. Rehabilitation starts with the initial visit to doctor and extends throughout the course of the disease. The rehabilitation program consists not only of specialized therapies delivered by physical therapists or other health professionals but also includes education, home exercises and environmental interventions.

Self management activities such as participation in education programs and physical activity are central to the success of physiotherapy and rehabilitation. It could be improved by the systematic provision of education and supportive measures by health care staff to increase patient ability in managing their health problems. The five A's model described by Glasgow and others for this strategy are[6] :

Assess

The therapist should assess the patients' confidence and their ability and intentions to participate in physical activity programs.

Advice

Patients are advised to be active for the need to lose weight and the risk of not making the recommended changes in the life style.

Agree

Mutually agreeable action plans should be implemented for the change in lifestyle for betterment eg weight loss to improve physical activity and increase physical activity to lose weight.

Assist

The therapist should assist the patients to develop the skills and confidence to achieve their action plan by providing educational materials.

Arrange Follow-up

Follow up should be arranged for the ongoing support and assistance to provide reinforcement and reassess the patient's status.

Components of Rehabilitation are:

- **Rest and splinting :** The joints should be put into rest during the acute stage of the disease. Bed rest relieves the pain in cases of extensive joint involvement.

  A splint is a device used for holding a part of the body stable and motionless to decrease pain and prevent further deformity. These splints include resting hand splints, wrist supports, finger splints, and special shoes and shoe inserts.

- **Assistive devices and adaptive equipment :** Assistive devices are used in order to reduce functional deficits, decrease pain, and keep independence and self-efficiency. Few of the assistive devices are
  - Walking stick which reduces loading forces on the joint by 30% and is associated with a decrease in pain and improvement of function.
  - Micro Cellular Polymer (MCP) footwear for plantar calcaneal spur, flat feet, limb length discrepancy etc
  - Knee caps while walking, using stair case and for prolonged standing

Certain adaptations like elevated toilet seats, widened gripping handles, arrangements related with bathrooms, work table, computer monitor, keypad etc. might all facilitate the daily life. Examples of adaptive equipments are in Fig 6.

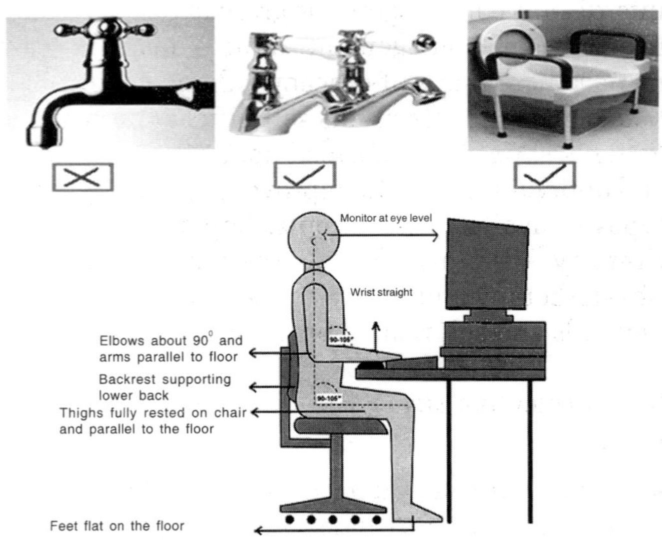

Fig. 6 : Adaptive Equipments

- **Exercises :** During active inflammatory stage, passive mobilization of joints within pain limit, isometric exercises and active assisted exercises to reduce joint stiffness, maintain joint movement and to prevent contractures are recommended. During inactive stage of arthritis, active range of motion exercises to reduce joint stiffness, isotonic exercises, strengthening exercises to improve muscle strength, endurance, joint range of motion and conditioning/aerobic exercises with adequate resting periods are recommended.

  Massage therapy : Gentle massage can be done to relieve pain and promote relaxation. The Modes of Therapeutic Massage commonly used in arthritis patients are Stroking, Effleurage and Kneading (thumb and finger).

Patient education : Education of the patient will reduce apprehension, improve compliance and help to avoid disease or drug related complications. All clinics that deal with the treatment of rheumatic diseases should provide education and information to their patients about their condition, various physical therapy and rehabilitative options that are available to improve their quality of life. Many patients have misconceptions about the nature of the arthritis and its cause. Expecting cure where it does not exist and misunderstanding that no cure means no treatment and frequent changing of various systems of medicine in the fond hope of cure should be dealt with. The education programme must be multidisciplinary in its presentation and have literature to support the presentations to remind patients in the future. The discussion should be scientific and clear in the language patient understands. The PRISM study (Program for Rheumatic Independent Self-Management) was an interdisciplinary programme given to patient groups to teach self-management, adult - learning case management and self-efficacy enhancement. They had beneficial effects on self-efficacy and pain[7]. Patients should be informed about their condition and options available to improve their quality of life, types of available treatments, importance of drug therapy - benefits/risks, therapeutic role of physiotherapy and orthotics, psychological coping methods and self relaxation techniques.

Joint protection strategies:

DO's

- Every joint should be moved in the ROM (range of movement) as many times as possible in a day in order to prevent contractures. Exercises can be started in a graded manner; a few minutes each day, adding more time as days go by. Aim is to exercise at a moderate pace, 30 to 60 minutes, most days each week.
- Balance periods of rest and activity during the day. Pace the activities, work in short bursts.
- In the case of acutely inflamed joints, isometric exercises provide adequate muscle tone without exacerbation of clinical disease activity.
- Isotonic Exercises if the disease activity is low, by using very low weights
- In chronic stage with inactive arthritis, conditioning exercises such as swimming, walking and cycling with adequate resting periods are recommended.
- Good posture should be maintained when standing, walking, lifting or sitting as shown in Fig 8. Poor posture causes uneven weight distribution and may strain the ligaments and muscles.
- The load over several joints is spread over evenly. Use the strongest joint rather than smaller joints while working.
- Splints are used whenever recommended - but patients are advised to have periods without them to prevent muscular weakness and wasting.
- A firm mattress with as few pillows as possible is used while lying down.

Fig. 8 : Posture

Don'ts

- There should be no straining exercises during acute arthritis.
- Keeping the joints in the same position for a prolonged period of time is avoided.
- To avoid the positions which push the joints towards deformity and which strain the muscles badly.
- Do not grip things too tightly. Tight grips produce a lot of strain on knuckles and thumb joints.
- When knees are affected, activities such as squatting, climbing stairs or heavy weight lifting are better minimized. The patello-femoral Joint-reaction force (JRF) increases with increasing knee flexion. It is observed that the JRF during walking (10-15° of flexion) is approximately half of the bodyweight, while climbing upstairs (60° of flexion), the JRF is 3.3 times the bodyweight and during squats (130° of flexion), the JRF is 7.8 times bodyweight[8] (Fig 9).
- Avoid excessive stress over the tendons during the stretching exercises.

Fig. 9 : Patellofemoral Joint-reaction Force.

Key points

Stage of the disease	Osteoarthritis	Rheumatoid Arthritis	Spondyloarthropathy
Acute Stage	• Rest • Use of cane, knee caps, MCP footwear etc. • Passive ROM exercises to maintain joint function • Isometric exercises • Active assisted exercises • Pain management	• Rest • Use of assistive/ adaptive equipments • Deep breathing exercises • Passive ROM exercises to maintain joint function • Isometric exercises • Active assisted exercises • Pain management	• Hydrotherapy to relieve pain and muscle spasm • Postural awareness and ergonomics • Breathing exercises • ROM exercises
Chronic Stage	In addition to above, add • Exercises to improve joint ROM and strength of the muscles • Aerobics like Walking • Education & Joint protection strategies. • Maintenance of balance and coordination	In addition to above, add • Exercises to improve joint ROM, strength and endurance of the muscles • Aerobics like Walking, bicycling etc. • Stretching of the joints • Education & Joint protection strategies.	In addition to above, add • Exercises to improve joint ROM, flexibility strength and endurance of the spinal and pelvic muscles. Few of the spinal/pelvic exercises are given in Fig 10. • Aerobics like Walking, swimming etc. • Stretching of the joints • Education & Joint protection strategies.

Fig. 10 : pelvic/spinal exercises

Diet

Diet seems to have little effect on most types of arthritis but maintaining ideal body weight and good nutrition are important. However, there is a conflicting advice regarding diet for arthritis. In most cases, patients find an improvement from **including** or **excluding** certain foods in their diets. Some diets are beneficial which are able to reduce inflammation, ease the pain and allow the joints to move more freely. In general, balanced nutritious diet is all that required.

For a patient with arthritis, in fact for any disease, weight should be optimal. Diet restrictions are essential to avoid putting extra kilos. Being overweight burdens the joint, increases pain and inflammation. By reducing the weight by 5 kilos in an obese individual the burden on the knees can be reduced by 50%.

In a study, among overweight and obese adults with knee OA, after 18 months, participants in the diet + exercise and diet groups had more weight loss and greater reductions in IL-6 levels than those in the exercise group; those in the diet group had greater reductions in knee compressive force also[9].

Foods that may help

- Oily fish - such as cod liver oil supplements, tuna, salmon and sardines.
- Fruits, vegetables, grains and legumes are high in phytonutrients. These are chemicals in plants that have disease - fighting properties and immune-boosting antioxidants such as vitamin C, vitamin E, selenium, and the carotenoids.
- Turmeric and Ginger (anti inflammatory).
- Omega-3, in doses of 3 grams or more per day, has been found effective for those with rheumatoid arthritis.
- Some report that vegetarian or vegan diet helps.

Foods that may aggravate

- Vegetables from Solanum family may cause problems in some individuals e.g. potatoes, capsicum, tomato, egg plant etc.
- Foods high in saturated fats - such as full fat dairy, fatty meat, baked foods etc.
- Red meat particularly in gouty arthritis.
- As Methotrexate and leflunomide are the commonly prescribed drugs in rheumatoid arthritis, consumption of alcohol damages the liver which would be a serious side effect. Alcohol consumption also increases the uric acid deposition in patients with gout.

Further Reading :

1. American Physical Therapy Association: Guide to Physical Therapist Practice, ed 2. Phys Ther 2001; 81: 9-744.
2. Matthew PH - effect of exercise training on depressive symptoms among patients with chronic illness-Meta analysis of RCT. Arch Intern Med. 2012; 172: 101-11.
3. Lee Jet al. The public health impact of risk factors for physical inactivity in adults with RA. Arthritis care and research 2012. DOI: 10.
4. Olalekan Uthman, Danielle vander Windt, Joanne Jordan et al. Exercise for lower limb osteoarthritis: systematic review incorporating trial sequential analysis and network meta-analysis. BMJ 2013; 347: f5555 doi: 10.1136/ bmj.f5555.
5. M Dena Gardiner: The Principles of Exercise Therapy 2000; 4.
6. Glasgow RE, Seth Emont, Miller DC. Assessing delivery of the 5 "A"s for patient centered counseling. Health Promot Int 2006; 21:245-55.
7. Alderson M, Starr L, Gow S, Moreland J.The program for rheumatic independent self management: a pilot evaluation. Clin Rheumatol 1999;18: 283-92.
8. Grelsamer RP and Weinstein CH. Applied biomechanics of the patella Clin Orthop Relat Res 2001; 389: 9-14.
9. Stephen P. Messier, PhD et al. Effects of Intensive Diet and Exercise on Knee Joint Loads, Inflammation and Clinical Outcomes Among Overweight and Obese Adults With Knee Osteoarthritis.The IDEA Randomized Clinical Trial, JAMA. 2013; 310(12): 1263-73. doi:10.1001/jama.2013.277669.

NON-STEROIDAL ANTI INFLAMMATORY DRUGS

3

Annil Mahajan, Vishal Tandon

Introduction

Nonsteroidal anti-inflammatory drugs (NSAIDs) are among the most widely prescribed and used medications in the world because of their demonstrated efficacy in reducing pain and inflammation in patients with various rheumatological disorders. More than 70 million prescriptions for NSAIDs are written each year in United States and more than 30 billion over the counter tablets are sold annually[1]. NSAIDs, are a necessary choice in pain management and symptomatic relief of inflammatory conditions[2]. This chapter will provide an update on NSAIDs and their role in rheumatology (Table 1).

Table 1. Recent classification of NSAIDs on the basis of COX selectivity[3,4]

Non selective NSAIDs	Average adult dose/ frequency
Aspirin	600-900mg 4-6 hrly
Diclofenac	50 mg 8hrly
Aceclofenac	100mg 12 hrly
Indomethacin	50 mg 8 hrly
Ketorolac	10 mg 6 hrly
Sulindac	150-200 mg twice daily
Ibuprofen	600-800 mg 8 hrly
Flurbiprofen	100 mg 8 hrly
Naproxen	500 mg 12 hrly
Mefenamic acid	500 mg 8 hrly
Piroxicam	20 mg once daily
Tenoxicam	20 mg once daily
Preferential COX-2 (Show preferential COX-2 inhibition over COX-1)	
Nimesulide	100 mg 12 hrly
Nabumetone	200-400mg once daily
Etodolac	7.5-15mg twice daily
Meloxicam	7.5 mg once daily
Selective COX-2 Inhibitors	
Celecoxib (FDA alert)	100-200mg 12 hrly
Etoricoxib	60-120 mg once daily
Lumiracoxib	400 mg once daily
Parecoxib	10mg/kg daily dose (Inject able)
Zaltoprofen	80 mg thrice daily

MECHANISM OF ACTION[3,4,5] (Table 2): NSAIDs work by interfering with the cyclooxygenase pathway, which involves the conversion of arachidonic acid, by the enzyme cyclooxygenase (COX) to prostaglandins (PGs) (Fig. 1). There are two types of COX i.e. COX-1 and COX-2. The COX-1 enzyme is ***constitutive***, meaning that its concentration in the body remains stable. It is present in most tissues and it generates 'good' PGs for 'housekeeping' functions such as gastric mucosal integrity, platelet homeostasis and regulation of renal blood flow. In contrast, the COX-2 enzyme is ***induced*** dramatically by the action of macrophages, the scavenger cells of the immune system. COX-2 is involved in producing 'bad' PGs involved in inflammatory reactions (i.e. arthritis, pain, hyperalgesia). This hypothesis suggest that the anti-inflammatory actions of NSAIDs are due to inhibition of COX-2 enzyme, whereas the unwanted side effects of conventional NSAIDs such as gastrotoxicity, are due to the inhibition of the constitutive COX-1 isoenzyme (Fig. 2, Table 3). Whereas, selective inhibition of COX-2 causes CVS toxicity[3-8] and inhibition of both COX-1 and COX-2 causes renal toxicity[4].

COX-3: A new isoform of the COX family has been discovered very recently, which is expressed abundantly in cerebral cortex and heart and it is thought to contribute for central mechanism by which paracetamol decrease fever. The relevance of COX-3 in humans is still debatable[3].

Fig. 1. COX Dependent mechanism of action[4,5]

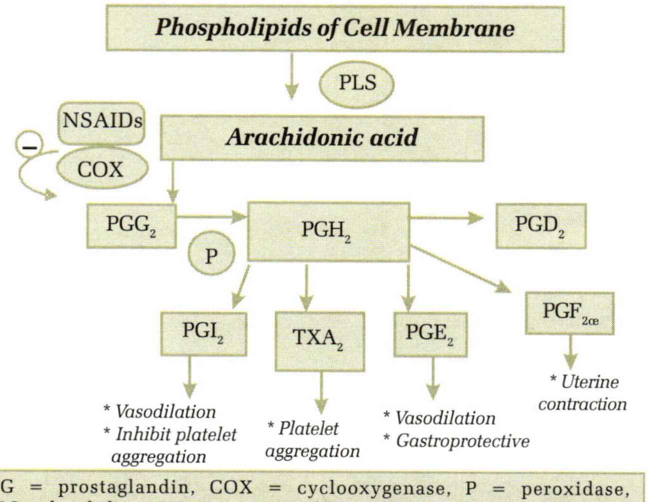

PG = prostaglandin, COX = cyclooxygenase, P = peroxidase, PLS = phospholipase, NSAID = nonsteroidal anti-inflammatory drug, - = inhibition.

Fig. 2 : Cox-2 : Cox-1 selectivity of NSAIDs

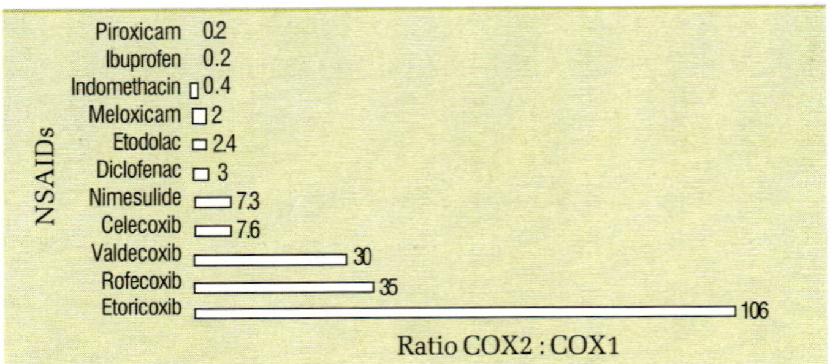

Reference[3,4,5]

Table 2. COX independent mechanisms of NSAIDs[4,5,9]

- Inhibitor effects on lipooxgenase products (sulindac, diclofenac)
- Inhibition of superoxide formation (indomethacin, piroxicam)
- Inhibition of neutrophil aggregation, adhesion and enzyme release (salicylates, indomethacin, aceclofenac)
- Depression of lymphocyte transformation (salicylates)
- Inhibition of rheumatoid factor production (piroxicam)
- Inhibition of cytokine production by inhibiting factor kappa B (salicylates)
- Suppression of proteoglycan production in cartilage (salicylates, piroxicam, ibuprofen, fenoprofen)
- Inhibit mononuclear cell phospholipase C activity and inhibit inducible nitric oxide synthase (Aspirin)
- Stimulation of adenosine receptors (salicylates)
- Peroxisome proliferators activated receptor (PPAR) Activation
- COX independent apoptosis and angiogenesis (COX-2 inhibitors)
- Inhibition of IL-beta and TNF in inflammatory cells and stimulation of synthesis of the extra-cellular matrix of human articular cartilage (Aceclofenac)

Table 3. Beneficial pharmacological actions of NSAIDs[4,5]

- Anti-inflammatory effects (Utilizes various mechanism but main is PGs inhibition)
- Analgesic effects (most effective against inflammation associated pain)
- Antipyretic effect (by inhibiting synthesis of PGE2 in hypothalmus)
- Anti-platelet effect (Low dose aspirin selectively inhibits TXA2)
- Closure of ductus arteriosus (Indomethacin by inhibiting PGE2 and I2)
- Cancer chemoprevention (Familial adenomatous polyposis, colorectal cancer/adenoma)
- Alzheimer's disease (Epidemiological data supports-unknown mechanism)

Toxicity

Gastrointestinal Toxicity[2] : 35 million people consume NSAIDs on a daily basis, and about 30% of these users may develop GI toxicity. Approximately 1,07,000 patients are hospitalized annually for NSAID-related GI complications, and at least 16,500 NSAID-related deaths occur each year among arthritis patients alone[10].

Topically it is related to the acidic nature of most NSAIDs, which promotes the accumulation of ionised molecules (ion trapping) within the mucosal cells. *Systemically*, GI toxicity is mediated through (COX-1) inhibition and a subsequent decrease in gastroprotective prostaglandins.

NSAID induced GI injury include superficial damage such as mucosal haemorrhages and erosions, endoscopically documented non-symptomatic ('silent') ulcers and symptomatic ulcers causing complications such as GI haemorrhage.

Renal Toxicity : There is no indication of difference between COX-2 specific and non specific NSAIDs with regard to renal toxicity[11]. Thereby suggesting, COX-1 and 2 are required for normal renal function. Locally synthesized PGI_2, PGE_2 and PGD_2 cause vasodilatation, decrease vascular resistance and enhance perfusion with redistribution of blood flow from the renal cortex to nephrons in the juxta-medullary region. Moreover, PGE_2 and $PGF_{2\alpha}$ cause diuresis by inhibiting the transport of Na and Cl in the thick ascending limb of loop of Henle and collecting ducts, thus their inhibition by NSAIDs can cause renal toxicity. Nephrotoxicity with NSAIDs includes acute tubular necrosis, acute tubulo-interstitial nephritis, glomerulonephritis, renal papillary necrosis, salt and water retention, chronic renal failure, hypertension, hyperkalemia[12].

Cardiovascular Toxicity : The CLASS trial[13], VIGOR trial[14], case-control study[15], Adenomatous Polyp Prevention on Vioxx trial (APPROVe) trial[16], clinical trial (APC) for prevention of colorectal adenoma continued exposure to celecoxib[17] and another recent study[18], demonstrated that rofecoxib, celecoxib, valdecoxib and parecoxib increase the incidence of thrombotic events (myocardial infarction, unstable angina, cardiac thrombus, resuscitated cardiac arrest, sudden or unexplained death, ischemic stroke, and transient ischemic attacks Table 4).

Traditional NSAIDs have varying associations with risk for myocardial infarction (MI). In nested case-control study diclofenac was shown to increase the risk of MI[19]. In another population based study, a significant increase (24-55%) in risk of MI was observed in patients receiving rofecoxib, diclofenac and ibuprofen[20]. In a recent metanalysis, among the seven NSAIDs (naproxen, ibuprofen, diclofenac, celecoxib, etoricoxib, rofecoxib, lumiracoxib) analyzed, naproxen seemed least harmful for cardiovascular safety[21]. **CVS toxicity of COX-2 inhibitors** can be prevented to some extent by combining essential fatty acids eicosapentaenoic acid (EFAs)[22].

Table 4. Mechanisms of CVS toxicity[3-8]

- COX-2 inhibitors decrease vascular prostacyclin (PGI 2) production and may affect the balance between prothrombotic and antithrombotic eicosanoids (thromboxane A2)
- COX-2 may also play a role in destabilizing plaques
- Inhibition of COX-2 cause an increase in the flux of arachidonate through the LO (lipooxygenase) pathways, which is important in the setting of inflammation in the atheromatous plaque. Furthermore, generated 12-LO and 15-LO contribute to LDL oxidation
- COX-2 has been recognized cardioprotective in ischemia-reperfusion injury and has antiproliferative effects toward vascular smooth muscle cells in conjunction with NO
- COX-2 inhibitors have also been shown to increase blood pressure (BP), due to alterations in the renin angiotensin pathway, change in sodium and water retention by the kidney, inhibition of vasodilating PG's and production of various vasoconstricting factors, including endothelin-1

Hepatic : NSAIDs can cause transient rise in liver enzymes. Reye's syndrome has been reported with aspirin, whereas, cholestasis is seen with sulindac and diclofenac therapy and fulminant hepatitis with the use of nimesulide in children has been reported[4,5].

Bone : NSAIDs also interfere with and suppress bone repair and remodeling[4,5].

Sleep Pattern : The drugs can decrease sleep[4,5].

Endocrinological effects : Aspirin and ibuprofen augment insulin secretion[4,5].

Psychiatric : NSAIDs have also been reported to induce adverse psychiatric reactions like depressive disorders and suicidal tendencies[4,5].

CNS : Headache, aseptic meningitis, drowsiness, altered mood, tinnitus etc. are commonly reported with NSAIDs therapy[4,5].

Hematologic, dermatologic and respiratory: Phenylbutazone or oxyphenbutazone can cause aplastic anemia and thrombocytopenia. However, COX-2 specific inhibitors do not affect platelet functions. Macular eruptions, urticaria, pseudoporphyria, erythema multiforme have been reported with their use: Asthma can be precipitated with NSAIDs especially aspirin[4,5].

Factors to be considered while prescribing NSAIDs

DRUG FACTORS

Efficacy: Although all NSAIDs are nearly equal in efficacy. Nonspecific NSAIDs or COX-2 specific NSAIDs show comparable efficacy in different rheumatological conditions[1,4].

Safety: Greater the potency more is the chance to cause side effects. Long term use of nonspecific NSAIDs known risk factor for GI toxicity unlike the COX-2 inhibitors. Nimesulide remains safe therapeutic choice in asthamatics; however should not be used in children less than 12 years[24]. Sulindac has lower incidence of renal side effects. Increased risk of acute myocardial infarction has been associated with selective COX-2 inhibitors and treatment with diclofenac and ibuprofen, but not with naproxen[23].

Pharmacokinetics: The shorter the half life, more the frequency of administration. The compliance is better with drugs like piroxicam (once daily) or naproxen (twice daily) than drugs like ibuprofen having shorter half life. It is advantageous to start with a high dose of a short-life drug (eg., ibuprofen) and then adjust the dose downward when analgesic efficacy has been achieved. For management of chronic pain, administration of NSAIDs with long half-lives (naproxen & COX-2 inhibitors) has clear advantages in allowing once-or twice-a-day dosing[2,4].

Dosage/formulation: Diclofenac sodium softgel has also been shown to provide a very rapid onset of analgesic activity and prolonged analgesic duration compared with conventional diclofenac potassium and better tolerated[2,4].

Concomitant treatment[4,5,25]: The likelihood of NSAID-induced GI toxicity is increased with concomitant use of corticosteroids, NSAIDs can decrease the clearance of methotrexate and lithium. Warfarin competes with NSAIDs for protein binding site and can increase the risk of GI bleed. NSAIDs can also decrease antihypertensive efficacy of beta-blockers, diuretics, angiotensin converting enzyme inhibitors, calcium channel blockers etc.

PATIENT'S FACTORS

Disease: Indomethacin is the traditional choice in gout. Anklyosing spondylitis (AS) is better treated with indomethacin or naproxen. RA is better treated with drug having long half life such as piroxicam & naproxen[4,5,25].

Target symptom: Morning stiffness may be better controlled with a long half life drugs such as piroxicam or a night time dose of indomethacin[4,5,25].

Concomitant disease: GI dysfunction, renal disease, cardiovascular dysfunction, diabetes mellitus and hypertension influence the choice in between COX selective vs non selective[4,5,25].

Old Age: NSAIDs may be able to protect neurons directly by reducing cellular response to glutamate and have potential to reduce the risk of Alzheimer's disease in old patients. However, the chances of drug interaction and serious adverse effects are more[4,5,25,26].

Pregnancy[27]: *Aspirin high dose:* should be avoided during all stages of pregnancy, particularly in last trimester, as it can inhibit uterine contractility and prolong labour. It can cause premature closure of ductus arteriosus, kernicterus, haemorrhage and renovascular complications. *Non selective COX-Inhibitors:* can be used in the first and second trimester of pregnancy, if necessary to control maternal disease. Paracetamol may be considered as safe analgesic. COX-2 *Selective Inhibitors:* theoretically the effect of COX-2 selective inhibitors on duration of labour and clotting should be less, whereas effect on fetal vasculature and renal function should be equal. However, there is lack of data on these drugs, hence should be avoided as far as possible.

Lactation: Most NSAIDs are safe during breast feeding as amount of drug excreted in breast milk is not significant, although there is potential risk of jaundice and kernicterus. There is insufficient data on COX-2 inhibitors and lactation. Drugs like indomethacin and aceclofenac should be avoided in lactation[28].

NSAIDs in rheumatology

Role in RA: NSAIDs are the most often used and perhaps the most effective adjunctive therapy in RA, providing both analgesic and anti-inflammatory benefits. However, they do not cure or alter the course of the disease and articular cartilage damage may continue despite symptomatic relief[29]. In a very recent study only naproxen showed significant decrease in CRP whereas lumiracoxib increased CRP[30].

Gout: NSAIDs are preferred in patients with uncomplicated gout. Indomethacin is the traditional choice. An initial dose of 50 to 75mg followed by 50mg every 6 to 8 hours with maximum of 200mg in the first 24 hours and then tapering has been recommended. Oral naproxen, ibuprofen, sulindac, piroxicam, ketoprofen as well as intramuscular ketorolac are also effective[4,5].

Osteoarthritis[31,32]: NSAIDs are commonly prescribed in OA and are considered only for patients who do not obtain adequate pain relief with paracetamol and have associated swelling and inflammation however, they have no structural or disease modifying effect. Aceclofenac by inhibition of IL-beta and TNF in inflammatory cells and stimulation of synthesis of the extra-cellular matrix of human articular cartilage can be another useful option in OA patients. Opioids (codeine and Tramadol) and paracetamol in combination provide better analgesia than paracetamol alone.

Low back Pain: NSAIDs prescribed at regular intervals provide effective pain relief for simple acute low back pain, but do not affect return to work[4,5].

Juvenile Rheumatoid Arthritis: Naproxen (15-20mg/kg/day twice daily) and ibuprofen (35mg/kg/day four times daily) are the approved drugs. An unusual toxic result of naproxen and other NSAID has been occurrence of pseudoporphyria. The response of NSAIDs varies among children with JRA but satisfactory treatment of arthritis is achieved in about 50%[4].

Anklyosing spondylitis: Many NSAIDs are effective in patients with AS. No NSAIDs has documented clear superiority in terms of efficacy. AS is better treated with indomethacin or naproxen. It is not well known whether continuous treatment with NSAIDs has long term advantage over intermittent therapy in term of functional ability and prevention of structural damage[4,5,25].

Reiter's syndrome, Undifferentiated Spondylo-arthritis and Reactive Arthritis: With the advent of anti-TNF alpha therapy effective treatment of inflammation of both peripheral joints and the axial skeleton is now possible. Nevertheless, first line treatment for all these patients is still NSAIDs. NSAIDs provide maximum suppression of pain and inflammation within 2 weeks[4,5,25].

Psoriatic arthritis: The treatment of psoriatic arthritis is directed at controlling the inflammatory process. Initial treatment is NSAIDs for joint disease[4,5,25].

SLE: Patients with mild symptoms, administration of analgesics, NSAIDs or salicylates may provide adequate relief. Ibuprofen, sulindac can cause aseptic meningitis. Through their effects on renal prostaglandin all NSAIDs can reduce GFR and renal blood flow especially in patients who have associated clinical or sub-clinical nephritis. Salt retention secondary to NSAIDs use in SLE patients may elevate blood pressure and cause pedal edema. Finally patients with SLE have a higher incidence of hepatotoxicity. Thus, SLE patients treated with NSAIDs should be monitored regularly for renal, GI and hepatic side effects[4,5,25].

Others: The anti-inflammatory effects of NASIDs have also been demonstrated in rheumatic fever. Although not as rigorously proven their efficacy is also widely accepted in the treatment of acute and chronioc bursitis, tendinitis, radiation synovitis and sport injuries[4,5,25].

A stepped care approach has been recommended in chronic pain disorders with simple analgesics (acetaminophen or nonsteroidal anti-inflammatory drugs) followed by tricyclic antidepressants (if neuropathic, back or fibromyalgia pain) or gabapentin, duloxetine or pregabalin if neuropathic pain and cyclobenzaprine, pregabalin, duloxetine, or milnacipran for fibromyalgias[33].

FDA Expert Advisory Committee and Multidisciplinary European expert panel recommendations[34,35]

- COX-2 inhibitors and NSAIDs should be prescribed with the lowest effective dose and for the shortest duration.
- They should not be prescribed for high risk patients. Two or more NSAIDs in combination should not be give.
- Cyclooxygenase-2 selective inhibitors (C2SI) alone and non-selective NSAID plus PPI should be preferred for patients with elevated gastrointestinal risk and low cardiovascular risk.
- Naproxen plus PPI is favoured in patients with high cardiovascular risk.
- For the combination of high gastrointestinal/ high cardiovascular risk the use of any NSAID was discouraged; if needed, naproxen plus PPI or a C2SI plus PPI could be considered

NO-NSAID[2,3,6,8,36]: Nitric oxides releasing non-steroidal anti-inflammatory drugs are a novel group of drugs synthesized by the ester linkage of an NO-releasing moiety to conventional (NSAID) such as NO-aspirin, NO-naproxen and NO-ibuprofen. NO-NSAID cause markedly diminished gastrointestinal and hepatic toxicity. NO is a potent inhibitor of platelet activation and adhesion to the vessel wall. NO-NSAID'S theoretically, are expected to exhibit a greater degree of inhibition of platelet function. In addition, the vasodilator activity of NO may assist in counteracting the vasoconstriction caused by platelet-derived mediators such as TXA2. Thus, it may prove cardio protective with selective COX-2 inhibitors. However, the existing data is from preclinical studies clinical clinical trials yet to prove these roles.

NSAIDs and Cancer protection

Some recent studies suggested that, NSAID use is associated with a reduced colorectal cancer risk[37], prostate cancer risk[38], pancreatic cancer risk[39] and breast cancer risk on long term use[40]. In rheumatology practice, patients requiring NSAIDs for longer duration may have some blessing effects in the form of chemo protection (Table 3).

Recent causations and advances

Frequent use of NSAIDs but not aspirin seemed to be associated with increased absolute incidence of Crohn Disease and Ulcerative Colitis[41]. Recently, thiazolo [3, 2-b]-1, 2, 4-triazole-5 (6H) - derivatives of ibuprofen has been shown to improve analgesic and anti-inflammatory activities and lower ulcerogenic risk[42]. Stable gastric pentadecapeptide BPC 157 is an anti-ulcer peptidergic agent, has been proposed to be used as an antidote against NSAIDs induced toxicity[43].

key points and Conclusion

- NSAIDs are among the most widely prescribed for pain and inflammation in various rheumatological disorders.
- Logical therapeutic approach is to choose low dose of NSAID for shorter duration.
- During this initial stage an analgesic such as paracetamol may be added to supplement analgesic activity.
- However, the choice of NSAID for a particular patient is based upon a number of factors, including relative efficacy, toxicity, concomitant drug use, concurrent disease states, the patient's age, renal function, cost and to a certain extent, on the prescriber's preference.
- It is important to give each NSAID an appropriate therapeutic trial before an alternative is tried.

Furder Reading :

1. Genovese MC, Harris ED. Treatment of Rheumatoid arthritis. In : (Editors) Harris ED, Budd RC, Genovese MC, Firestein GS, Sargent JS, Sledge CB, Ruddy S. Kelley's Textbook of Rheumatology. Vol-I, 7th edition, Elsevier Saunders 2005. pp. 1079-1100.
2. Ong CKS, Lirk P, Tan CH, Seymour RA. An Evidence-Based Update on Nonsteroidal Anti-Inflammatory Drugs. Clin Med Res 2007; 5(1): 19-34.
3. Kulkarni SK, Jain NK. Coxibs: The new super aspirins or unsafe pain killers? Ind J Pharmacol 2005; 37(2): 86-89.
4. Robert LJ, Morrow D. Analgesic-antipyretic and anti-inflammatory agents and drugs employed in the treatment of gout. In: (Editors) Hardman JG, Limbird LE, Gilman AG. Goodman & Gilman's The Pharmacological basis of Therapeutics, 10 ed. McGraw-Hill New York 2005; 687-731.
5. Crofford LJ. Nonsteroidal anti-inflammatory drugs. In : (Editors) Harris ED, Budd RC, Genovese MC, Firestein GS, Sargent JS, Sledge CB, Ruddy S. Kelley's Textbook of Rheumatology. Vol-I, 7th edition, Elsevier Saunders 2005. 839-59.
6. Mahajan A, Sharma R. COX-2 Selective Nonsteroidal Anti-inflammatory Drugs: Current Status. JAPI 2005; 53: 200-4.
7. Mahajan A, Sharma R. Cox-2 Inhibitors: Cardiovascular Safety. JK Science 2005; 7(2): 61-62.
8. Mahajan R, Sharma R, Singh JB, Hussain J. Non-steroidal Anti-inflammatory Drugs: Controversies JIACM 2007 8(1): 65-72.
9. Collier DH. Nonsteroidal anti-inflammatory Drugs (NSAIDs). In: (Ediotors) West SG. Rheumatology Secrets, 2nd edition. Hanley & Belfus, Elseviers 2003. pp. 561-71.
10. Graumlich JF. Preventing gastrointestinal complications of NSAIDs: risk factors, recent advances, and latest strategies. Postgrad Med 2001; 109(5): 117-28.
11. Schneider V, Levesque LE,Zhang B, Hutchinson T, Brophy JM. Association of Selective and Conventional Nonsteroidal Anti inflammatory Drugs with Acute Renal Failure: A Population-based, Nested Case-Control Analysis. Am J Epidemiol 2006 Sep 27; [Epub ahead of print].
12. Ejaz P, Bhojani K, Joshi VR. NSAIDs and Kidney. JAPI 2004; 52: 632-39.
13. White WB, Faich G, Whelton A, et al. Comparison of thromboembolic events in patients treated with celecoxib, a cyclooxygenase-2 inhibitor, versus ibuprofen or diclofenac. Am J Cardiol 2002; 89: 425-30.
14. Bombardier C, Laine L, Reicin A, et al. Comparison of upper gastrointestinal toxicity of rofecoxib and naproxen in patients with rheumatoid arthritis: VIGOR Study Group. N Engl J Med 2000; 343: 1520-28.
15. Graham DJ, Campen D, Hui R, et al. Risk of acute myocardial infarction & sudden cardiac death in patients treated with COX-2 selective & non- selective non-steroidal anti-inflammatory drugs: nested case-control study. Lancet 2005; 365: 475-81.
16. Antman EM, DeMets D, Loscalzo J. Cyclooxygenase Inhibition and Cardiovascular Risk Circulation 2005; 112: 759-70.
17. Solomon SD, McMurray JJV, Pfeffer MA, et al. Cardiovascular risk associated with celecoxib in a clinical trial for colorectal adenoma prevention. N Engl J Med 2005; 352: 1071-80.
18. Nussmeier NA, Whelton AA, Brown MT, et al. Complications of the COX-2 inhibitorss Parecoxib and valdecoxib after cardiac surgery. N Engl J Med 2005; 352: 1081-91.
19. Hippisley-Cox J, Coupland C. Risk of myocardial infarction in patients taking cyclo-oxygenase-2 inhibitors or conventional non-steroidal anti-inflammatory drugs: Population based nested case control study. Br Med J 2005; 330: 1366-69.
20. Johnsen SP, Larsson H, Tarone RE. Risk of hospitalization for myocardial infarction among users of rofecoxib, celecoxib, and other NSAIDs: a population-based case-control study. Arch Intern Med 2005; 165: 978-84.
21. Trelle S, Reichenbach S, Wandel S. Cardiovascular safety of non-steroidal anti-inflammatory drugs: network meta-analysis. BMJ 2011; 342: c7086.
22. UN Das. Can COX-2 Inhibitor-induced Increase in Cardiovascular Disease Risk Be Modified by Essential Fatty Acids? JAPI 2005; 53: 623-27.
23. Singh G. Risk of acute myocardial infarction with nonselective non-steroidal anti-inflammatory drugs: a meta-analysis. Arthritis Res Ther 2006; 8: R153.
24. Ransfer KD, Members of Consensus Report Group on Nimesulide. Nimesulide-Amultifactorial approach to inflammation and pain: Scientific and Clinical Consensus. Curr Med Res Opin 2006; 22(6): 1161-70.
25. Woodfork KA, Dyke KV. Anti-inflammatory and antirheumatic drugs. In Craig CR, Stitzel RE, (eds). Modern Pharmacology with clinical applications, 6th edition, Lippincott Williams and Wilkins-Philadelphia 2004: 423-39.
26. Kokki H. Nonsteroidal anti-inflammatory drugs for postoperative pain: a focus on children. Paediatr Drugs 2003; 5: 103-23.
27. Tandon VR, Sharma S, Mahajan A, Khajuria V, Kumar A. Pregnancy and Rheumatoid Arthritis. Indian Journal of Medical Sciences 2006; 60(8): 334-44.
28. Temprano KK, Bandlamudi R, Moore TL. Antirheumatic drugs in pregnancy and lactation. Semin Arthritis Rheum 2005; 35(2): 112-21.
29. Suresh E. Management of early Rheumatoid arthritis. JAPI 2007; 55: 355-62.
30. Tarp S, Bartels EM, Bliddal H et al. Effect of non steroidal antiinflamatory drugs on C-reactive protein in rheumatoid arthritis: A meta analysis of randomized controlled trials. Arthritis Rheum 2012 Jul 25 Ahead of Pub.
31. Mahajan A, Verma S, Tandon VR. Osteoarthritis. JAPI 2005; 53: 634-41.
32. Marc C. Hochberg MC, Altman RD, Toupin D et al. American College Of Rheumatology 2012 Recommendations For The Use Of Nonpharmacologic And Pharmacologic Therapies In Osteoarthritis Of The Hand, Hip, And Knee. Arthritis Care & Research 2012; 64(4): 465-74.
33. Kroenke K, Krebs EE, Bair MJ et al. Pharmacotherapy of chronic pain: a synthesis of recommendations from systematic reviews General Hospital Psychiatry 2009; 31: 206-19.
34. Young D. FDA labors over NSAID decisions: panel suggests COX-2 inhibitors stay available. Am J Health Syst Pharm 2005; 62: 668-72.
35. Gerd Burmester, Angel Lanas, Luigi Biasucci, et al. The appropriate use of non-steroidal antiinfl ammatory drugs in rheumatic disease: opinions of a multidisciplinary European expert panel Ann Rheum Dis 2011; 70: 818-822.
36. Keeble JE., Moore PK. Pharmacology and potential therapeutic applications of nitric oxide-releasing non-steroidal anti-inflammatory and related nitric oxide-donating drugs Br J Pharmacol 2002;137:295-310.

37. Elizabeth H. Ruder 1, 2, Adeyinka O. Laiyemo 3, Barry I. Graubard 1. Non-steroidal anti-inflammatory drugs and colorectal cancer risk in a large, prospective cohort. Am J Gastroenterol. 2011 July; 106(7): 1340-50.

38. Salaheddin M. Mahmud 1, 2, 3*, Eduardo L. Franco 1, 4, Donna Turner 3.. Mahmud SM, Franco EL, Turner D, Platt RW, Beck P, et al. (2011) Use of Non-Steroidal Anti-Inflammatory Drugs and Prostate Cancer Risk: A Population-Based Nested Case-Control Study. PLoS ONE 6(1): e16412. doi: 10.1371/journal.pone.0016412.

39. MC Bradley 1, 2, CM Hughes 2, MM Cantwell 1, G Napolitano 3 and LJ Murray*, 1 Non-steroidal anti-inflammatory drugs and pancreatic cancer risk: a nested case-control study. British Journal of Cancer (2010) 102, 1415-21.

40. Theodore M. Brasky 1, Matthew R. Bonner 1, Kirsten B. Moysich 2, Non-steroidal anti-inflammatory drug (NSAID) use and breast cancer risk in the Western New York Exposures and Breast Cancer (WEB) Study. Cancer Causes Control. 2010 September; 21(9): 1503-12.

41. Ashwin N. Ananthakrishnan, MD, MPH, Leslie M. Higuchi, Aspirin, Nonsteroidal Anti-inflammatory Drug Use, and Risk for Crohn Disease Ulcerative Colitis: Ann Intern Med. 2012 March 6; 156(5): 350-59.

42. Uzgören-Baran A, Tel BC, Sarigöl D, Oztürk EI. Thiazolo [3, 2-b]-1, 2, 4-triazole-5(6H)-one substituted with ibuprofen: Novel non-steroidal anti-inflammatory agents with favorable gastrointestinal tolerance. Eur J Med Chem. 2012 Jul 14. [Epub ahead of print].

43. Sikiric P, Seiwerth S, Rucman R, Turkovic B, et al. Toxicity by NSAIDs. Counteraction by stable gastric pentadecapeptide BPC 157. Curr Pharm Des. 2012 Aug 23. [Epub ahead of print].

GLUCOCORTICOID THERAPY IN RHEUMATIC DISEASES

4

Vinod Ravindran

Introduction

Glucocorticoids (GCs); also called corticosteroids or more loosely, "steroids" are one of the most widely used groups of medicines. They are used in the management of inflammatory disorders such as allergies, asthma, sepsis and several autoimmune diseases including of skin, bowel and joints. They have a special place in the management of autoimmune diseases such as rheumatoid arthritis (RA) and vasculitides. After the extremely successful initial clinical trial published in 1948, where cortisone was used in a patient with RA to dramatic improvement, GCs rapidly consolidated their position as a viable therapeutic option in the management of RA[1]. Since then uses of GCs in many more autoimmune rheumatic diseases have been established. However, therapeutic advances in the management of these conditions and concerns regarding the safety of GCs have led to continuous re-appraisal of their relevance. At the same time effective strategies to deal with adverse effects of GCs have been developed. Newer GCs molecules hold promise for optimum use of GCs in future. In this chapter, up to date summary of some of these issues is presented.

Mechanisms of action

The main endogenous GC in man is cortisol (cortisone in the inactive form), the majority of which is bound in the circulation to corticosteroid-binding globulin (CBG). It is secreted from the zona fasciculata of the adrenal gland and regulated by the adrenocorticotropic hormone (ACTH). Endogenous control and regulation comes from the interactions of the hypothalamic-pituitary-adrenal (HPA) axis. There are recognised associations between the immune system, inflammation and HPA control such that there is effectively a second negative feedback loop.

The efficacy of GCs in alleviating inflammatory disorders results from the pleiotropic effects of the glucocorticoid receptor (GCR) on multiple signalling pathways. Evidence indicates that the GCs inhibit inflammation through all three following mechanisms: direct and indirect genomic effects and non-genomic mechanisms.

Genomic pathways

At physiological or low GC doses after easily passing through the cell membranes into the cytoplasm, cortisol binds with high affinity to a cytoplasmic GCR-alpha (cGCRα). On binding, molecular chaperones including heat shock proteins (HSP) are released. Within the cell, cortisol acts in two ways. First, the cortisol-GCR complex moves to the nucleus and activates genomic signalling by interacting with GC response elements (GREs). The resulting complex recruits either coactivator (for transcriptional activation or *transactivation*) or corepressor (for transcriptional repression or *transrepression*) proteins that modify the structure of chromatin, there by facilitating or inhibiting assembly of the basal transcription machinery and the initiation of transcription by RNA polymerase II[2]. GCs induce synthesis of some entities through transcriptional activation, and transcriptional repression switches off synthesis of other entities. These entities include cytokines [interlukin (IL) IL-1, IL-2, IL-6, IL-8, tumour necrosis factor -alpha (TNF-α) and interferon gamma (IFN-γ)], cytokine receptors, chemotactic proteins, major histocompatibility complexes (MHC) Class I and II, adhesion molecules, collagenases and matrix metalloproteinases.

Broadly speaking, transrepression leads to anti-inflammatory effects of GCs and transactivation leads to many of the recognised side-effects of GCs. This distinction is not clear-cut, however, and exceptions include osteocalcin and osteoprotegerin, proteins that are critical to new bone formation and are repressed.

Second, regulation of other GC-responsive genes involves interactions between the cGCR complex and other transcription factors, such as nuclear factor-*k*B (NF-*k*B)[3]. In the nucleus, NF-*k*B binds DNA sequences called NF-*k*B elements and stimulates the transcription of cytokines, chemokines, cell adhesion molecules, complement factors, and receptors for these molecules. NF-*k*B also induces the transcription of cyclooxygenase 2, an enzyme essential for prostaglandin production. Direct interactions between the GCR and NF-*k*B probably account for most of the inhibitory effects of glucocorticoids on NF-*k*B signaling.

Non genomic pathways

Recent work suggests that GCs can have rapid effects on inflammation that are not mediated by changes in gene expression. At higher concentrations GCs are thought to interact with membrane-bound receptors or directly with the membrane to induce non-genomic effects. The exact mechanisms of non-genomic signalling remain under debate but are probably responsible for very rapid effects of high-dose GC such as inhibition of arachidonic acid release. Non-genomic effects themselves are currently thought to come in two varieties: specific non-genomic effects related to the activation of membrane GC receptors (not normally engaged in physiological circumstances), and non-specific non-genomic effects engendered when very high concentrations of GC swamp all the available receptors and begin to dissolve directly in the cell membrane (and perhaps the membranes of mitochondria), initiating other intracellular metabolic changes[4]. These effects are all engaged when high-dose therapy is used, such as pulsed intravenous methylprednisolone.

Therefore, GCs have a mixture of therapeutic effects depending on the dose used (Table 1). This has important implications for consideration of the potential for adverse effects, as a simple extrapolation from effects observed in high-dose regimens to the expectation of similar but lesser effects at lower doses may not be appropriate if the adverse effect is related to mechanisms which are not employed at lower doses.

Table 1. Key differences between genomic and non-genomic actions of glucocorticoids

Characteristics	Mechanism	
	Genomic	Non-genomic
Glucocorticoid dose	Low	High
Onset of action	Hours to days	Minutes
Efficacy	Moderate	High
Adverse effects	Low	High

Definitions of "doses"

The recognition of different mechanisms of action also provides an underlying rationale to the clinically recognisable differences in effect for different doses, and for the definition of 'low', 'medium' and 'high' doses of GC. In the clinical setting, GC doses can be clustered using the following scheme: low doses are less than 7.5mg of prednisone-equivalent doses; medium doses are between 7.5 and 30mg; high doses > 30mg and up to 100mg prednisone-equivalent per day; very high doses > 100mg prednisone-equivalent per day; finally, when doses are > 250mg of prednisone-equivalent per day for few days, usually no more than 5 days, it is termed as pulse therapy[5]. It is recommended that a treatment with ≤ 7.5 mg prednisolone equivalent a day should be considered as low dose GC treatment because this dose occupies less than 50% of the GC receptors, this dose is often used for the maintenance therapy for many rheumatic diseases requiring GCs and relatively few adverse effects are expected. The recognition that a relative hypocortisolism is present in chronic inflammatory conditions such as RA lead observers to regard low dose GC treatment as a means of replacement therapy for reduced adrenal production. Therefore most authors now regard ≤ 7.5mg prednisolone /day to be near the physiologic range.

Types of glucocorticoids and their potency

It has been suggested that equivalent GC doses should be calculated using prednisone as the comparator of the group regarding the different potencies. The traditional ways to measure GC and mineralocorticoid (MC) potencies of steroids refer to anti-inflammatory (rat ear oedema test, McKenzie vaso-constriction test) or metabolic (liver glycogen assay) effects and sodium retention. Depending on the testing conditions, absolute potency values vary considerably. It is therefore extremely difficult, if not impossible, to arrange the individual results from different research groups to compile a comprehensive list (Table 2). Some of the recent studies using more standardised ways of comparing MC and GC properties of different steroids, eg. in the same in vitro test system [luciferase trans-activation assay in CV-1 cells transfected with either mineralocorticoid receptor (MCR) or GCR expression vectors] complemented by a system to test the steroid binding affinities at the human MCR (protein expression in T7-coupled rabbit reticulocyte lysate) have found clinically relevant differences in the potencies such as for deflazacort[6]. Deflazacort, an oxazoline derivative of prednisolone, was initially thought to be as effective as prednisone while inducing fewer adverse events (less bone wasting, less significant alteration in the pattern of growth hormone (GH) secretion, no negative on the affect the overall amount of GH secreted, and on the carbohydrate metabolism). Whilst evidence for some of these beneficial effects comes mainly from the animal models[7], apart from the potency issues, availability of studies of relatively poor quality using very small number of patients has limited the precise appraisal of the role of deflazacort in clinical rheumatology in particular in the management of RA[8,9].

Table 2. Potency comparision of selected therapeutic glucocorticoids

Agent	GC Potency	MC Potency
Hydrocortisone	1	0.002
Deflazacort	4	No data
Prednisolone	4	0.002
Methylprednisolone	5	0.0013
Betamethasone	25	0
Dexamethasone	25	0

GC; glucocorticoid, MC; mineralocorticoid

Glucortioid use in rheumatological conditons

In rheumatology GCs are utilized in several different ways and by various routes (Table 3). Prior to discussing some of the more specific indications it would be useful to consider the following EULAR evidence based recommendations[10] on the management of systemic GC therapy in rheumatic diseases;

1. The adverse effects of GC therapy should be considered and discussed with the patient before GC therapy is started. This advice should be reinforced by giving information regarding GC management. If GC are to be used for a more prolonged period of time, a "glucocorticoid card" is to be issued to every patient, with the date of commencement of treatment, the initial dosage and the subsequent reductions and maintenance regimens.
2. Initial dose, dose reduction and long-term dosing depend on the underlying rheumatic disease, disease activity, risk factors and individual responsiveness of the patient. Timing may be important, with respect to the circadian rhythm of both the disease and the natural secretion of GCs.
3. When it is decided to start GC treatment, comorbidities and risk factors for adverse effects should be evaluated and treated where indicated; these include hypertension, diabetes, peptic ulcer, recent fractures, presence of cataract or glaucoma, presence of (chronic) infections, dyslipidaemia and comedication with non-steroidal anti-inflammatory drugs.
4. For prolonged treatment, the GC dosage should be kept to a minimum, and a GC taper should be attempted in case of remission or low disease activity; the reasons to continue GC therapy should be regularly checked.
5. During treatment, patients should be monitored for body weight, blood pressure, peripheral oedema, cardiac insufficiency, serum lipids, blood and/or urine glucose and ocular pressure depending on individual patient's risk, GC dose and duration.
6. If a patient is started on prednisone > 7.5 mg daily and continues on prednisone for more than 3 months, calcium and vitamin D supplementation should be prescribed.
7. Antiresorptive therapy with bisphosphonates to reduce the risk of GC-induced osteoporosis (GIOP) should be based on risk factors, including bone-mineral density measurement.
8. Patients treated with GCs and concomitant non-steroidal anti-inflammatory drugs (NSAIDs) should be given appropriate gastro-protective medication, such as proton pump inhibitors or misoprostol, or alternatively could switch to a cyclo-oxygenase-2 selective inhibitor.
9. All patients on GC therapy for longer than 1 month, who will undergo surgery, need perioperative management with adequate GC replacement to overcome potential adrenal insufficiency.
10. GCs in low to moderate doses during pregnancy have no additional risk for mother and child.
11. Children receiving GCs should be checked regularly for linear growth and considered for growth-hormone replacement in case of growth impairment.

Table 3. Glucocortiocoid utilization in rheumatic diseases

Routes	Strategies
Oral	Short term, high dose Medium term, low to moderate dose Long term, low dose
Intramuscular	At initiation and for flares
Intravenous	Pulse therapy
Intra-articular	Single or more affected joints, refractory joint
Intra-lesional	Epicondylitis, bursitis, tendinosis etc.

Rheumatoid arthritis

Rheumatoid arthritis (RA) is a chronic inflammatory disease that leads to disability, impaired functioning and premature death. Though many patients now receive disease-modifying antirheumatic drugs (DMARDs) and biologics, since 1950s GCs are commonly used in the treatment of RA. Recently a large registry based study in the UK has shown that remarkably, the long-term use of GCs for RA between 1991 and 2008 did not decrease, although several new (synthetic and biological) DMARDs have entered the market since 1991[11]. German national database on arthritis has also shown that the percentage of GC treated patients with RA is fairly constant over the years at about 56%, but with an increase in the use of GC dosages of ≤ 7.5 mg/prednisone equivalent per day[12]. Several meta-analyses have confirmed the therapeutic advantages of short term, medium term and long term GC therapy in RA in combinations with DMARDs[13,14,15]. In early RA (disease duration of less than 2 years), the use of low-dose GCs is based not solely on disease symptoms, but also on joint sparing effects on the long-term, as GCs can be categorised as DMARDs; based on category IA level of evidence presently available[16]. The current recommendations are therefore is to "use GCs added at low to moderately high doses to synthetic DMARD monotherapy (or combinations of synthetic DMARDs) to provide benefit as initial short-term treatment, but should be tapered as rapidly as clinically feasible"[17]. Scenarios of GC use in RA are presented in the Box 1.

Box 1. Indications for the possible use of glucocorticoids in RA

- Bridge therapy for patients who have experienced a severe functional decline with limitations interfering with necessary daily living or vocational activities.
- Patients with NSAID contraindication who have acute inflammation unlikely to respond rapidly enough to second -line agents.
- Rheumatoid vasculitis or other serious life threatening or organ damaging manifestation.
- Men or women not at reduced risk of bone loss, for whom GCs may be reasonable in combination therapy regimens with a single daily dose of less than 7.5mg/day.
- Pregnant or lactating women with active RA.

Small and medium vessel vasculitis

Vasculitis is defined by inflammation of the blood vessel wall and forms the pathological foundation of a diverse group of individual disease entities. The primary systemic vasculitides are heterogeneous, multi-system disorders characterized by inflammation and necrosis of blood vessels of various sizes, some of them are associated with anti-neutrophil cytoplasmic antibodies (ANCA). Examples are granulomatosis with polyangiitis (Wegener's), Churg Strauss syndrome, microscopic ployangiitis, polyarteritis nodosa, cutaneous leucocytoclastic vasculitis etc. Several of these have the potential of multi organ involvement with significant morbidity.

The use of high-dose GCs is an important part of remission induction therapy[18]. There are no clinical trials examining the role of GC therapy but every clinical trial or cohort study conducted has used GC therapy in combination with immunosuppressive therapy. It is common practice to commence prednisolone or prednisone at 1 mg/kg/day as in recent clinical trials. Current recommendations state that the initial high dose should be maintained for 1 month, and should not be reduced to less than 15 mg/day for the first 3 months[18]. The GC dose should then be tapered to a maintenance dose of 10 mg/day or less during remission. When a rapid effect is needed, intravenous pulsed methylprednisolone may be used in addition to the oral prednisolone as part of remission induction therapy. Furthermore, GC in low doses has been recommended in the remission-maintenance therapy in combination with immunosuppressive therapy such as azathioprine, leflunomide or methotrexate[18].

Temporal arteritis and polymyalgia rheumatica

The signs and symptoms of polymyalgia rheumatica (PMR) and temporal arteritis (TA) overlap and can often be confused, but both are treated with GCs. Early intensive therapy with high-dose GC induces remission in patients with large vessel vasculitis[19]. Visual loss in one eye is prevalent in 18% of patients at diagnosis. It is usually irreversible and pulsed intravenous methylprednisolone may be of benefit to some patients who present early following the onset of visual symptoms. The initial dose of prednisolone is 1 mg/kg/day (maximum 60 mg/day) and the initial high-dose should be maintained for a month and tapered gradually. The taper should not be in the form of alternate day therapy, as this is more likely to lead to a relapse of vasculitis. At 3 months, the GC dose in clinical trials has been between 10-15 mg/day. The duration of GC therapy for patients with TA is variable and can extend to several years, but some patients may not be able to tolerate discontinuation of GC therapy due to recurrent disease or secondary adrenal insufficiency. The high doses of GC required initially and the duration of treatment can be associated with

GC side effects, and epidemiological studies have shown a high incidence of fractures. These risks can be substantially reduced with concurrent osteoporosis prophylaxis and this should now be routine practice.

Systemic lupus erythematosus (SLE)

In the treatment of SLE without major organ manifestations, GCs are indicated[20]. Concomitant use of anti-malarials and immunosuppressive drugs may help keep daily doses of prednisone < 7.5 mg/day, or even to withdraw it. In patients with severe lupus activity higher dose of GC (1 mg/kg/day) is being used, though recent evidence suggest that significantly lower doses (0.5 mg/Kg/day) might be equally effective and have less adverse effects. In patients with proliferative lupus nephritis, GCs in combination with immunosuppressive agents are effective against progression to end-stage renal disease[20]. With regard to pulse methylprednisolone, a recent prospective study reported similar efficacy of 500 mg/day during three consecutive days vs. 1000 mg/day for three to five consecutive days in treating severe lupus flares[21]. The lower-dose group suffered a significant lower number of serious infections. It is likely that lower dose of GCs in "pulse therapy" would become a standard practice in near future.

Gout

GCs [and in the past Adreno corticotropic hormone (ACTH)] have been used in gout mainly in treating an acute attack. In acute gout, GC can either be given systemically, intra-venous, intra-muscular or intra-articular routes if one or two joints are affected. GCs are a good alternative where NSAID and colchicine cannot be used or in refractory cases. A recent randomized control trial (RCT) involving 90 patients compared the analgesic efficacy of oral prednisolone (30mg daily for 5 days) plus paracetamol vs. oral indomethacin (50 mg three times daily for 2 days and 25 mg three times daily for 3 days) plus paracetamol[22]. Pain reduction was similar but the steroid group experienced fewer adverse events (AEs). Similar results were found in a randomized double-blind trial demonstrating that 35 mg of prednisolone gives equivalent pain relief to naproxen 500 mg twice a day[23]. Glucocorticoids may have fewer AEs than other acute treatments when used for short term, particularly in the elderly.

Glucocortioid resistance

It is generally accepted that a significant proportion of patients will be resistant to the therapeutic effects of GC. In fact, GC resistance may occur in up to 25-30% of those taking the drug[24]. This is not restricted to patients with RA and has been observed in other conditions such as asthma and inflammatory bowel disease.

Some patients will exhibit a genetic predisposition to GC resistance as a consequence of familial or sporadic mutations in the GCR gene. GC signal transduction becomes impaired and tissues become less sensitive to GC. As a result, there is up regulation of the HPA axis to compensate, with subsequent unwanted clinical manifestations of hyper-androgenism, fatigue and mineralocorticoid excess[25]. This is rare and accounts for only a very small proportion of patients with GC resistance.

The alternative GC receptor, GCR-α, has been implicated in GC resistance. GCR-α does not bind GC but does compete with GCR-α for binding sites, thereby attenuating GCR-α action. It can be thought of as a dominant negative inhibitor of GCR-α. Over-expression of GCR-α and its association with GC resistance has been demonstrated in a number of autoimmune and inflammatory conditions. It is also clear that during periods of inflammation, pro-inflammatory cytokines such as TNF-α and IL-1 can lead to increases in the expression of GCR-α, which may in part explain why GC resistance may become apparent with worsening inflammatory disease[26].

Mechanisms of acquired GC resistance are a current 'hot topic' with a variety of theoretical models postulated. One emerging theme appears to be the role of T-helper cells and the balance between IL-10 and IL-17. IL-17 producing T-helper cells (T_h-17 cells) seem to play a significant role, with in-vitro studies revealing the link between expanded T_h-17 cell subsets and GC resistance[27]. Increased levels of IL-17 ultimately lead to over-expression of GCR-α. At the same time, increased differentiation of induced regulatory T-cells can shift the balance in favour of IL-10 production and subsequent expression of GCR-α. This reciprocal relationship is important as the underlying balance may determine the clinical phenotype of GC resistance in a particular patient. The role of T_h-17 cells will become clearer over the next few years and this could have implications in deciding whether patients will benefit from GC or not.

Adverse effects of GC use

GCs may cause a variety of adverse effects. Recently the evidence has emerged regarding two patterns of GC associated AEs namely "threshold" and "linear"[28]. For cushingoid phenotype, ecchymosis, leg oedema, mycosis, parchment-like skin, shortness of breath and sleep disturbance, a "linear" increase in the frequency with increasing dose has been noted. The term "threshold" pattern describes an elevation in the frequency of health problems beyond a certain threshold value. The threshold for the increase in glaucoma, as well as depression/listlessness and an increase in blood pressure was a dosage of more than 7.5 mg/day[28]. Dosages of 5 mg/day or more was identified to be relevant for epistaxis and weight gain. A very low threshold was seen for eye cataract (less than 5 mg/day)[28].

Whereas the clinical benefits of low dose GC and their ability to inhibit the progression of radiological damage is generally accepted in RA, the balance of benefits and risks associated with prolonged use is less certain because of concerns over toxicity. A comprehensive review of four prospective trials in RA and the COBRA study concluded that definitive associations of low-dose GC with many adverse effects remain elusive[29]. In general, the precise data on the frequency and severity of AEs of GCs and methods of assessment are not well-established. By using patient withdrawal as a marker of adverse events, a recent meta-analysis of six randomised controlled trials utilizing low-dose GC lasting $\geq$ 2 years found no significant difference in toxicity compared to placebo[15]. Nevertheless, the long-term use of GC in RA continues to generate debate.

Several areas of GC therapy such as patient selection, duration, dose, timing and measures to prevent side effects are a part of the currently agreed research agenda at international level. Initiatives to improve monitoring and documentation of adverse events of GCs have recently been developed and the use of these approaches in future clinical trials may help clarify the issue[30]. Standardization of assessment of AEs may enable aetiological and casual inferences[30]. Also, evidence-based recommendations for the safer use of systemic GC therapy in rheumatic diseases have been published recently[10]. These include patient education, monitoring for AEs and use of concomitant therapy to reduce unwanted side effects and special safety advice. Adherence to these guidelines might reduce the AEs in patients on GC therapy.

Table 4 lists some of the important adverse effects of GCs. A special mention has to be made of GC induced osteoporosis (GIOP). There is no doubt that long-term GC use even in moderate doses is associated with reduced bone mineral density, particularly loss of trabecular bone, and increased fracture risk at the hip and spine. The mechanisms of GIOP are multiple but include disturbed calcium homeostasis, osteoblast dysfunction and osteocyte apoptosis. Although fracture risk rises soon after GCs are commenced it also rapidly declines when treatment is stopped and this is independent of effects on bone mineral density. Chronic inflammatory conditions such as RA are themselves associated with an increased fracture risk and bone density loss, and therefore distinguishing disease effects from GC effects at low doses is complicated. Several studies including from India have confirmed much higher prevalence of reduced bone mass in RA[31].

Table 4. Toxicity of glucocorticoids

Organ system		Dose	
	Probably with high dose only	High or low dose	
		Probably	Definitely
General	Infections		Weight gain
			Redistribution of body fat
		Facial odema	
			Withdrawal symptoms
Musculoskeletal	Osteonecrosis	Osteoporosis	
Gastrointestinal	Pancreatitis	Peptic ulcer	
	Bowel rupture		
Cardiovascular	Hypertension		
	Dyslipidemia		
	Atherosclerosis		
Reproductive	Foetal wastage		
Neurological	Psychosis		
Cutaneous	Delayed wound healing		Purpura
			Skin atrophy
			Acne
			Easy bruising
Eye			Glaucoma
			Cataract
Endocrine		HPA dysfunction	Diabetes

Guidelines in several countries and by several different organisations have now been produced for managing GIOP. The Indian Rheumatology Association (IRA) has produced guidelines for the management of GIOP which have also been endorsed by the Endocrine Society of India and Indian Society for Bone and Mineral Research[32]. The key recommendations[32] are as follows:

1. Use minimum dose of oral GC for shortest period of time.
2. Use GC sparing immunosuppressive drugs like Methotrexate, Azathioprine whenever indicated.
3. Use inhalational route in patients with asthma and enema in ulcerative colitis.
4. Use supportive therapy like balanced nutrition, adequate calcium, regular exercise, avoidance of smoking and alcohol.
5. Assess fracture risk at initiation of therapy by using FRAX tool.
6. Start primary prevention in high risk group (postmenopausal women, men > 65 years and those with previous fragility fracture).
7. In those without high risk factors do a baseline BMD if T-score <-1.5 start primary prevention.
8. If baseline T-score >-1.5, follow-up and assess BMD after 1-2 years if therapy with GC is being continued.
9. Assess fracture risk at initiation of therapy by using FRAX tool (FRAX tool for Singapore Indians is available and can be applied to our Country). (http://www. sheffield.ac.uk/FRAX).

Newer GCS

From the forgoing discussion it is apparent that GCs have several desirable clinically beneficial effects. At the same time ever growing concerns regarding their toxicity combined with issue of GC resistance have driven research in to newer GC molecules. A logical approach to diminish the adverse effects of GCs is the development of novel drugs that induce transrepression but stimulate very little, or virtually no, transactivation. Novel drug delivery systems and combinations are also being experimented.

Selective GC-receptor agonists (SEGRAs)

Selective GC-receptor agonists (A276575, its four enantiomers, and ZK216348) or dissociating GC exert their therapeutic effects by 'dissociating' genomic trans-repression from trans-activation, thus aiming to optimise therapeutic benefit with minimal detriment to adverse effects. ZK216348 has showed clear dissociation between trans-activation and trans-repression with anti-inflammatory effects in a murine model of skin inflammation. These effects were comparable with prednisolone but ZK216348 was associated with reduced increases in blood glucose and reduced skin atrophy[33]. More recently ZK245186 has demonstrated efficacy in a murine model of skin inflammation comparable to conventional GC. The safety profile was much improved, when compared to GC, with reduced growth inhibition, reduced skin atrophy and hyperglycaemia. Human studies of SEGRAs are now required.

Nitrosteroids

Nitrosteroids are compounds which combine GC with nitric oxide (NO) in a single molecule. At the site of action the compound is broken down, releasing GC and NO (or other nitrate species), both of which have anti-inflammatory properties. This combination appears to act synergistically, resulting in potent anti-inflammatory effects in rat models. One such nitrosteroid, NCX-1015 (prednisolone and NO), appears to have protective properties on bone resorption in-vitro and in-vivo[34]. Nitration of the GCR can potentiate the anti-inflammatory effects of this steroid ligand via augmentation of GC binding, heat shock protein-90 dissociation and nuclear translocation. Human studies of nitrosteroids are yet to be done.

Combination of prednisolone and dipyridamole

Combination treatment with dipyridamole has provided promise in its ability to produce a dissociated GC profile[35]. Dipyridamole remains commonplace in the management of cerebrovascular disease as an anti-platelet agent, but in this context it can be thought of as a 'steroid amplifier'. It appears to enhance the ability of GC to influence trans-repression of NF-kB without enhancing factors that can lead to toxicity. This has been supported by more recent work in which a dipyridamole-GC combination was shown to have a synergistic effect on pro-inflammatory cytokine suppression without any increase in certain adverse effects such as osteoporosis or HPA-axis suppression.

Liposomes containing GCs

Liposomes preferentially accumulate at sites of inflammation such as the synovium in active RA. Liposomes containing GC will therefore act as a carrier for the drug, shielding it from the general circulation and allowing it to exert local rather than systemic effects. In a study a single injection of liposomal prednisolone phosphate (10mg/kg) resolved joint inflammation in a rat model of arthritis for one week, compared with un-encapsulated prednisolone which did not reduce inflammation and only provided a slight benefit even when given as repeated injections[36]. Improved results in rat and dog models of arthritis have also been reported.

Non-steroidal GCR ligand

Another novel substance, LGD-5552, a non-steroidal GCR ligand that bind the cGCR efficiently and strongly represses inflammatory gene expression. This compound has potent anti-inflammatory activity *in vivo*, but demonstrates differential gene regulation in comparison with commonly prescribed GCs. It has fewer adverse effects on fat distribution and bone than currently used GCs[37].

Chronotherapy

A clear circadian variation in interleukin-6 (IL-6) levels in RA patients with peak values occurring in the morning has been noted. In RA patients treated with low-dose GC (5mg or 7.5mg prednisolone) at either 2 am or 7.30 am it was found that those given GC in the morning had significant improvements in morning stiffness, pain and serum concentrations of IL-6. Explanation for this is that the IL-6 levels peak in the early morning, roughly 1-2 hours before the peak of cortisol. It was suggested therefore that, by administrating GC at around 2 am, the early morning peak of the pro-inflammatory cytokine IL-6 can be suppressed. This led to the development of a timed release formulation of prednisone (Lodotra, Horizon Pharma) which is taken at 10 pm with the active ingredient being released 4 hours post ingestion (i.e. at 2 am). Pivotal RCT of this molecule published in 2008 (CAPRA-1), looking at the efficacy of simply switching RA patients already taking regular morning prednisone to the timed-release formulation at the same dose[38]. Significant improvements in morning stiffness were achieved with the timed-release preparation. Mechanistic studies showed that the increase in IL-6 was indeed almost completely suppressed by 2 am prednisone. Concerns regarding a potential increase in HPA axis suppression with low-dose GC have also been addressed by this study[38] and in the analyses of findings from the CAPRA-1 study[39]. HPA axis responsiveness was not suppressed and may even be enhanced.

Inhibition of Bax protein

A recent study has suggested that the pro-apoptotic protein Bax has a key role in the chondrocyte dysfunction underlying these issues. Mice lacking Bax are totally resistant to GC-induced bone growth impairment. Furthermore, in dexamethasone-treated wild-type mice, bone growth retardation was associated with decreased proliferation and increased apoptosis of chondrocytes, and reduced matrix apoptosis of chondrocytes, and reduced matrix content and integrity of growth plates; Bax gene knockout mice were protected from these effects[40]. It is therefore conceivable that suppression of Bax using small molecules or peptides could also prevent the undesired effects of GCs on bone. One such peptide is humanin (originally discovered as a neuronal rescue factor in patients with Alzheimer disease), which acts as a negative regulator of Bax, but further research is warranted.

Conclusions

GCs remain an important treatment option in many conditions in rheumatology, and new understanding of disease-specific and dose-specific effects has resulted in renewed effort to make full use of their potential. Box 2 summarises some important points presented in this review.

Box 2. Key learning points

- Glucocorticoids (GCs) are important therapeutic option in several rheumatic diseases.
- Although no dose of GC is totally without adverse effects, low dose and high dose GC have different mechanisms of action, therapeutic effects and adverse effects.
- Better understanding of GC mechanism of actions (genomic and non genomic) helps in understanding issues with GC adverse effects observed at a particular dose.
- A disease modifying effect of GC therapy in RA has become established.
- Several newer GCs molecules have the potential to enhance the current role of GCs by selective augmentation of the beneficial effects.
- Adherence to relevant guidelines such as for GC induced osteoporosis is mandatory in minimizing the GC adverse effects.

Further Reading :

1. Hench PS, Kendall EC, Slocumb CH, Polley HF. The effect of a hormone of the adrenal cortex (17-hydroxy-11-dehydrocorticosterone: compound E) and of pituitary adrenocorticotropic hormone on rheumatoid arthritis. Mayo Clin Proc 1949; 24: 181-97.
2. Hebbar PB, Archer TK. Chromatin remodelling by nuclear receptors. Chromosoma 2003; 111: 495-504.
3. Nissen RM, Yamamoto KR. The glucocorticoid receptor inhibits NF-kappaB by interfering with serine-2 phosphorylation of the RNA polymerase II carboxy-terminal domain. Genes Dev 2000; 14: 2314-29.
4. Rhen T, Cidlowski JA. Antiinflammatory actions of glucocorticoids-new mechanisms for old drugs. N Engl J Med 2005; 353: 1711-23.
5. Buttgereit F, da Silva JA, Boers M, Burmester GR, Cutolo M, Jacobs J et al. Standardised nomenclature for glucocorticoid dosages and glucocorticoid treatment regimens: current questions and tentative answers in rheumatology. Ann Rheum Dis 2002; 61: 718-22.
6. Grossman C, Scholz T, Rochel M, Bumke-Vogt C, Oelkers W, Pfeiffer AFH et al. Transactivation via the human glucocorticoid and mineralocorticoid receptor by therapeutically used steroids in CV-1 cells: a comparison of their glucocorticoid and mineralocorticoid properties. European Journal of Endocrinology 2004; 151: 397-406.
7. Latta K, Krieg Jr RJ, Hisana S, Veldhuis JD, Chan JCM. Effect of the synthetic glucocorticoid, deflazacort, on body growth, pulsatile secretion of GH and thymolysis in the rat. European Journal of Endocrinology. 1999; 140: 441-46.
8. Avioli LV. Potency ratio-a brief synopsis. Br J Rheumatol 1993; 32: 24-26.
9. Kroggsgaard MR, Thamsborg G, Lund B. Changes in bone mass during low-dose corticoid treatment in patients with polymyalgia rheumatica: a double blind, prospective comparison between prednisolone and deflazacort. Ann Rheum Dis 1996; 55: 143-46.
10. Hoes JN, Jacobs JWG, Boers M, Boumpas D, Buttgereit F, Caeyers N et al. EULAR evidence-based recommendations on the management of systemic glucocorticoid therapy in rheumatic diseases. Ann Rheum Dis 2007; 66: 1560-67.
11. Fardet L, Petersen I, Nazareth I. Prevalence of long-term oral glucocorticoids prescriptions in UK over the last 20 years. Rheumatology 2011; 50: 1982-90.
12. Buttgereit F. Do the treatment with glucocorticoids and/or the disease itself drive the impairment in glucose metabolism in patients with rheumatoid arthritis? Ann Rheum Dis 2011; 70: 1881-83.
13. Gotzsche PC, Johansen HK. Short term low dose corticosteroids vs. placebo and non-steroidal anti-inflammatory drugs in rheumatoid arthritis. Cochrane Database Syst Rev 2005; (1): CD000189.
14. Criswell LA, Saag KG, Sems KM, Robinson V, Shea B, Wells G et al. Moderate-term, low-dose corticosteroids for rheumatoid arthritis. Cochrane Database Syst Rev 1998; (3): CD001158.
15. Ravindran V, Rachapalli S, Choy EH. Safety of medium- to long-term glucocorticoid therapy in rheumatoid arthritis: a meta-analysis. Rheumatology (Oxford) 2009; 48: 807-11.
16. Kirwan JR, Bijlsma JWJ, Boers M, Shea BJ. Effects of glucocorticoids on radiological progression in rheumatoid arthritis. Cochrane Database Syst Rev 2007; (1): CD006356.
17. Smolen JS, Landewé R, Breedveld FC, Dougados M, Emery P, Gaujoux-Viala C et al. EULAR recommendations for the management of rheumatoid arthritis with synthetic and biological disease-modifying antirheumatic drugs. Ann Rheum Dis 2010; 69: 964-97.
18. Mukhtyar C, Guillevin L, Cid MC, Disrupt B, de Grout K, Gross W et al. EULAR recommendations for the management of primary small and medium vessel vasculitis. Ann Rheum Dies 2009; 68: 310-17.
19. Mukhtyar C, Guillevin L, Cid MC, Dasgupta B, de Groot K, Gross W et al. EULAR recommendations for the management of large vessel vasculitis. Ann Rheum Dis 2009; 68: 318-23.
20. Bertsias G, Ioannidis JPA, Boletis J, Bombardieri S, Cervera R, Dostal C et al. EULAR recommendations for the management of systemic lupus erythematosus. Report of a Task Force of the EULAR Standing Committee for International Clinical Studies Including Therapeutics. Ann Rheum Dis 2008;67: 195-205.
21. Kong KO, Badsha H, Lian TY, Edwards CJ, Chng HH. Low-dose pulse methylprednisolone is an effective therapy for severe SLE flares. Lupus 2004; 13: 212-13.
22. Man CY, Cheung I, Cameron I, Peter A, Rainer TH. Comparison of oral prednisolone/paracetamol and oral indomethacin/paracetamol combination therapy in the treatment of acute gout like arthritis: a double-blind, randomized, controlled trial. Ann Emerg Med 2007; 49: 670-77.
23. Janssens HJ, Janssen M, Van de Lisdonk EH, van Riel PL, van Weel C. Use of oral prednisolone or naproxen for the treatment of gout arthritis: a double-blind, randomised equivalence trial. Lancet 2008; 371: 1854-60.
24. Sliwinska-Stanczyk P, Pazdur PJ, Ziolkowska M, Jaworski J, Kaminska-Tchorzewska E, Lacki JK. The effect of methylprednisolone on proliferation of PBMCs obtained from steroid-sensitive and steroid-resistant rheumatoid arthritis patients. Scan J Rheumatol 2007; 36: 167-71.
25. Charmandari E, Kino T, Ichijo T, Chrousos GP. Generalised glucocorticoid resistance: clinical aspects, molecular mechanisms and implications of a rare genetic disorder. J Clin Endocrinol Metab 2008; 93: 1563-72.
26. Webster JC, Oakley RH, Jewell CM, Cidlowski JA. Proinflammatory cytokines regulate human glucocorticoid receptor gene expression and lead to the accumulation of the dominant negative isoform: a mechanism for the generation of glucocorticoid resistance. Proc Natl Acad Sci USA 2001; 98: 6865-70.
27. Lee RWJ, Creed TJ, Schewitz LP, Newcomb PV, Nicholson LB, Dick AD et al. CD4 + CD25 int T-cells in inflammatory diseases refractory to treatment with glucocorticoids. J Immunol 2007; 179: 7941-48.
28. Huscher D, Thiele K, Gromnica-Ihle E, Hein G, Demary W, Dreher R et al. Dose-related patterns of glucocorticoid-induced side effects. Ann Rheum Dis 2009; 68: 1119-24.
29. Da Silva JA, Jacobs JW, Kirwan JR, Boers M, Saag KG, Inês LB et al. Safety of low dose glucocorticoid treatment in rheumatoid arthritis: published evidence and prospective trial data. Ann Rheum Dis 2006; 65: 285-93.

30. van der Goes MC, Jacobs JW, Andrews T, Blom-Bakkers MA, Buttgereit F et al. Monitoring adverse events of low-dose glucocorticoids therapy: EULAR recommendations for clinical trials and daily practice. Ann Rheum Dis 2010; 69: 1913-19.

31. Jain S, Ravindran V, Mathur DS. Detection of low bone mass using quantitative ultrasound measurements at calcaneus: comparative study of an Indian rheumatoid arthritis cohort. Int J Rheum Dis. 2008; 11: 414-20.

32. Krishnamurthy V, Sharma A, Aggarwal A, Kumar U, Amin S, Rao URK et al. Indian rheumatology association guidelines for management of glucocorticoid-induced osteoporosis. Ind J Rheumatol 2011; 6: 68-75.

33. Schacke H, Schottelius A, Docke W-D, Strehlke P, Jaroch S, Schmees N et al. Dissociation of transactivation from transrepression by a selective glucocorticoid receptor agonist leads to separation of therapeutic effects from side effects. Proc Natl Acad Sci USA 2004; 101: 227-32.

34. Paul-Clark MJ, Mancini L, Del Soldato P, Flower RJ, Perretti M. Potent antiarthritic properties of a glucocorticoid derivative, NCX-1015, in an experimental model of arthritis. Proc Natl Acad Sci USA 2002; 99: 1677-82.

35. Zimmermann GR, Avery W, Finelli AL, Farwell M, Fraser CC, Borisy AA. Selective amplification of glucocorticoid anti-inflammatory activity through synergistic multi-target action of a combination drug. Arthritis Res Ther 2009; 11: R12.

36. Metselar JM, Wauben MH, Wagenaar-Hilbers JP, Boerman OC, Storm G. Complete remission of experimental arthritis by joint targeting of glucocorticoids with long-circulating liposomes. Arthritis Rheum 2003; 48: 2059-66.

37. Lo´pez FJ, Ardecky RJ, Bebo B, Benbatoul K, De Grandpre L, Liu S et al.LGD-5552, an Antiinflammatory Glucocorticoid Receptor Ligand with Reduced Side Effects, in Vivo. Endocrinology 2008; 149: 2080-89.

38. Buttgereit F, Doering G, Schaefller A, Witte S, Sierakowski S, Gromnica-Ihle E et al. Efficacy of modified release versus standard prednisone to reduce duration of morning stiffness of the joints in rheumatoid arthritis (CAPRA-1): a double-blind, randomised controlled trial. Lancet 2008; 371: 205-14.

39. Kirwan JR. Targeting the time of day for glucocorticoid delivery in rheumatoid arthritis. Int J Clin Rheumatol 2011; 6: 273-79.

40. Zaman F, Chrysis D, Huntjens K, Fadeel B, Sävendahl L. Ablation of the pro-apoptotic protein Bax protects mice from glucocorticoid-induced bone growth impairment. PLoS ONE 2012; 7: e33168.

Following recent excellent review is recommended to those interested in the issues related to GC use in children: Godbole TR, Dabadghao. Glucortiocid use in children: The problems and solutions. Ind J Rheumatol 2012; 7 (Suppl 1): 112-16.

DISEASE MODIFYING ANTI RHEUMATIC DRUGS

5

Ashok Kumar

Introduction

Steady progress in therapeutics of rheumatoid arthritis (RA) in recent years has culminated in a radical change in our approach to the treatment of this disease. Today, the goal of drug therapy of RA is complete remission or very low disease activity[1]. Although small at present, the possibility of achieving 'drug free' remissions has also been realised[2]. Disease-modifying antirheumatic drugs (DMARDs) are the key therapeutic agents for RA (Table 1). They act at some crucial steps in the etiopathogenetic mechanisms thereby retarding the progression of disease to negligible level. To be designated a DMARD, a drug must change the course of RA for at least 1 year as evidenced by sustained improvement in physical function, decreased inflammatory synovitis, and slowing or prevention of structural joint damage. Low doses of systemic steroids have also been shown to have a disease modifying activity in RA[3,4]. Most rheumatologists do include them in the initial treatment regimen because they rapidly reduce the burden of inflammation and facilitate induction of remission with routinely used DMARDs[18]. The latter are divided into conventional and biological DMARDs. The biological DMARDs represent an ever expanding domain of rheumatology and will be discussed in a separate chapter. The present chapter deals with so called conventional, synthetic or traditional DMARDs, i.e. non-biological DMARDs.

Table 1. 'Conventional' disease modifying antirheumatic drugs

1. Methotrexate
2. Sulphasalazine
3. Antimalarials (hydroxychloroquine, chloroquine)
4. Leflunomide
5. Minocycline
6. Cyclosporine
7. Gold (sodium aurothiomalate, auranofin)
8. Miscellaneous group (levamisole, mycophenolate ofetil, tacrolimus)

Relevant Clinical Pharmacology of DMARDs

Methotrexate

Methotrexate (MTX) is currently the sheet anchor of treatment of RA. Therapeutic dose ranges from 7.5mg to 25 mg orally, once a week. Depending on the grade of disease activity, the starting dose could be 7.5 mg/week to 15 mg/week. Dose may be escalated by 2.5 mg increments at intervals of 4 weeks with careful monitoring of disease activity score (DAS28) and drug toxicity. Absorption falls by at least 30% when the weekly dose is 15 mg or more[5]. Therefore, parenteral administration (subcutaneous or intramuscular) is preferred for doses in the range of 15-25 mg/week[6]. Beneficial effect is appreciable in 1-2 months and peaks at 6 months. Most importantly, patients show long-term adherence to methotrexate monotherapy: > 70% continue to use the drug for 5 years. Such high retention rates were not known with older DMARDs such as injectable gold salts[7]. Anorexia and nausea in the 24 hours after the dose is the commonest side effect. This can be reduced by administering folic acid (5-15 mg) per week or ondansetron. Other side-effects include diarrhoea, oral ulcers, stomatitis, elevated liver enzymes, hepatic fibrosis, skin rash, pneumonitis and risk of opportunistic infections. The drug is contraindicated in pregnancy, renal insufficiency and liver disease. Caution is needed in obesity, diabetes and alcohol abuse. In frail, elderly patients lower doses (7.5-10 mg/week) are recommended because cytopenias are frequent in this age group. Concerns have been raised about possible increased risk of lymphoma with long-term use of methotrexate. However, the results of one large retrospective study do not support this possibility[8]. Non-steroidal anti-inflammatory drugs can reduce the renal clearance of methotrexate thereby increasing its toxicity. Hydroxychloroquine when combined with methotrexate, enhances the effect of methotrexate by increasing the 'area under the curve' i.e. plasma concentration x time.

Baseline evaluation before starting methotrexate comprises blood counts, serum creatinine, liver function tests and chest x-ray. Patients at risk should also have hepatitis B and C serology.

Periodic monitoring includes blood counts and liver function tests every 4-8 weeks and serum creatinine every 6 months. If liver enzymes are increased less than twice the upper normal value, it is recommended to repeat these tests in 2-4 weeks. If liver enzymes are increased 2-3 times the upper normal value, it is recommended to decrease the dose and repeat measurements every 4 weeks. If Alanine transaminase (ALT) and Aspartate Aminotransferase (AST) concentrations are greater than 3 times the upper limit of normal on 2 occasions, methotrexate should be stopped[9]. The possibility that enzyme elevation may be due to other medications (e.g. NSAIDs, antituberculous drugs), alcohol, viral hepatitis or autoimmune hepatitis should not be overlooked and carefully ruled out before abandoning methotrexate.

Sulphasalazine

This drug is a poorly absorbed sulphonamide which splits into 5-aminosalicylic acid (5-ASA) and sulphapyridine after reaching the colon. Sulphapyridine is responsible for efficacy in RA while 5-ASA is the component which benefits ulcerative colitis. Sulphasalazine (SSZ) is started at 500 mg/day and then increased by 500 mg weekly to 2-3 grams daily. Side effects include nausea, abdominal pain, skin rashes, CNS disturbances, reversible oligospermia, discoloration of urine and sweat, blood dyscrasias and elevated liver enzymes. It is generally a well tolerated drug with mild side-effects. Its efficacy is lower than that of methotrexate. Blood counts and LFTs should be regularly monitored every 4 weeks for 3 months and then every 3 months. If the results are normal during first year, monitoring may be done 6 monthly during second year. Thereafter, the monitoring may be discontinued[10].

Antimalarials

Antimalarials are used either in mild RA or in combination with other DMARDs. Both chloroquine (CQ) and hydroxychloroquine (HCQ) are equally efficacious for the treatment of RA. HCQ is preferred because ocular toxicity is rare with its use. Both have very long half-lives and get accumulated in tissues (steady state reached only after 3 to 4 months). Recommended dose for CQ is 5 mg/kg/day and that for HCQ is 7.5 mg/kg/ day. Maintenance dose is 50% of initial dose (some patients may require higher dose). Length of time to onset of benefit varies from 1-6 months. Retinal toxicity is rare at doses below 6 mg/kg/day. Fundoscopic examination and visual field testing is recommended annually for HCQ[11]. Other side effects include nausea, indigestion, pruritus, skin rash, hyperpigmentation of exposed parts and rarely, hypoglycemia.

Leflunomide

Leflunomide inhibits de novo pyrimidine synthesis (dihydro-orotate dehydrogenase inhibitor). It is comparable to MTX and SSZ in efficacy and retards the development of erosions. The practice of administering a loading dose of 100 mg/day for 3 days is now considered unnecessary[12]. Leflunomide is started directly at 20mg/day, orally. Beneficial effect starts appearing after 4 to 6 weeks. The drug has a good safety profile. Liver toxicity, nausea, diarrhoea, alopecia, weight loss (sometimes impressive) and occasionally hypertension are known to occur. Many of these side-effects tend to decrease with continued use. The drug tends to accumulate in the body and may not get eliminated till 2 - 3 years. It was found to be highly teratogenic in animal studies and is, therefore, contraindicated in patients who are pregnant or who can become pregnant. Similar restrictions apply to males who may be contemplating fathering a child. Monitoring for this drug is done on the same lines as for methotrexate. In case of severe liver toxicity and bone marrow toxicity, cholestyramine (8 gm TDS for 11 days) is administered for elimination of the drug. The goal should be to achieve plasma concentrations below 0.02 mg/l.

Minocycline

Minocycline has been shown to be beneficial in patients with early and mild RA. The antirheumatic effect of this semi-synthetic derivative of tetracycline is related to its immunomodulatory and anti-inflammatory, rather than to its antibacterial properties. Its dose is 100 mg BD. The common adverse effects include GI side effects, dizziness, rash and headaches. Persistent skin and mucosal hyperpigmentation, lupus-like syndrome and acute hepatic injury are less common. In a clinical trial, it was shown to be superior to hydroxychloroquine in its efficacy[13].

Cyclosporine

Cyclosporine is a potent immunosuppressant approved for use at a dose of 3-5 mg/kg in RA. It selectively blocks the release of IL-1 from monocytes and IL-2 from helper T cells. Its absorption is usually incomplete and reduced by food. 'Go slow and go low' is the key to use of this drug. Complete blood counts, liver enzymes, serum creatinine and urine analysis should be done weekly until stable dose is reached and then monthly.

Rise in serum creatinine and blood pressure are common adverse effects, particularly at doses higher than 3 mg/kg/day. Elevation of serum creatinine > 30% over baseline is an indication to stop the drug. Other adverse effects include hyperkalemia, gum hyperplasia, hypertrichosis and hepatotoxicity. Cyclosporine is efficacious but expensive as well as toxic. It is recommended for patients refractory to methotrexate. Most patients are unable to continue the drug beyond 2-3 years because of renal toxicity.

Gold salts

Gold compounds are of 2 types: water soluble thiolates and the fat soluble phosphine derivatives. Gold is eliminated very slowly from the body. Significant tissue concentrations have been found even 20 years after the last dose. Injectable gold is started at a test dose of 10 mg intramuscularly and then 50 mg weekly until a cumulative dose of 1 gram has been administered. If there is no improvement, the drug is stopped. If there is significant improvement, the frequency of injections is reduced to once in a fortnight for next 3 months, once in 3 weeks for 3 months and then continued at once a month. Oral gold (auranofin) is given at a dose of 3-6 mg/day.

Very few patients are able to continue gold beyond 5 years. Adverse reactions to gold are the major reason for discontinuation. These include skin rashes, oral ulcers, proteinuria, thrombocytopenia and pulmonary toxicity. Aplastic anaemia is rare but potentially fatal. Eosinophilia may predict adverse reactions. Monitoring of blood counts and urine protein should be done before each injection for 4-8 weeks and then before every 3rd or 4th injection. Oral gold is much less toxic but dose-related diarrhoea is a common side effect[14]. It is less efficacious. Gold salts are not freely available at present.

PRACTICAL APPROACH TO THE USE OF DMARDs

The current strategy of treatment of RA is 'treat-to-target'. The treatment is targeted at achieving remission/low disease activity. According to 2012 update of ACR recommendations, DMARD therapy should commence as soon as the diagnosis of RA is made[15]. Thus both early RA (< 6 months) and established RA (> 6 months) qualify for DMARD therapy. If treatment is delayed, beneficial effect will still occur but it may not be possible to avoid some irreversible joint damage[16]. In fact, the new ACR-EULAR classification criteria for RA have been formulated primarily to facilitate early diagnosis of RA and obtain best outcome of treatment by instituting it early[17].

Summary of clinical pharmacology of conventional DMARDs

DRUG	DOSAGE	ADVERSE EFFECTS	MONITORING
Methotrexate	7.5-25 mg per week (orally or subcutaneously)	GI, cytopaenias, ↑AST and ALT, alopecia, stomatitis, infection, pneumonitis	CBC, LFT every 4-8 weeks
Sulphasalazine	2-3 gms daily (build up the dose slowly, over 1 month)	GI, cytopaenia, ↑AST and ALT, photosensitivity	CBC, LFT every 4 weeks for 3 months, then 3 monthly; if stable for 2 years, may stop monitoring
Chloroquine	150 mg (base) daily for 3 months, then alternate day	GI, hyperpigmentation, pruritus, retinopathy	Fundoscopy and visual field testing, 6 monthly
Hydroxychloroquine	200 mg BD for 3 months, then OD	Retinopathy is very rare	Fundoscopy and visual field testing annually
Leflunomide	20 mg OD, orally (loading dose not needed)	GI, alopecia, mild weight loss, ↑AST and ALT	CBC, LFT every 4-8 weeks
Cyclosporine	2.5-5 mg/kg/day, orally	↑BP, ↑hair, hypertrophy of gums, impaired renal function and LFT	CBC, LFT, creatinine, urinalysis weekly until dose is stable, then monthly; monitor BP
Gold thiomalate (Inj.)	10 mg I.M. stat (test dose) then 50 mg per week; frequency reduced after 1 gm total dose	Stomatitis, oral ulcers metallic taste, rashes, photosensitivity, pruritus, cytopaenias, proteinuria	CBC and urine protein before each injection for 1 month and then before every 4th injection
Auranofin (oral gold)	3-9 mg OD	Diarrhoea, stomatitis, rash	CBC and urine protein every month for 3 months and then every 3 months

DMARD combinations

The basic aim of combining DMARDs is to increase the efficacy of treatment. Multiple drugs act synergistically at multiple points in the cascade of inflammation. Since mechanisms of action of most DMARDs remain poorly understood, combinations have been worked out quite empirically. Examples of clinical trials of efficacy of combination DMARD therapy include TICORA, COBRA and FinRaco[19-21].

DMARDs may be combined in 3 different ways:

Step-up : Start with single agent and then keep on adding more DMARDs if the disease does not improve in 3 months. For example, in TICORA trial, patients were started on sulphasalazine in a step-wise escalating treatment protocol. Assessment was done every month to ensure tight control of disease. MTX and HCQ were added if the set goal of keeping DAS in 'low disease activity' range was not achieved. The dose of SSZ and MTX was increased to maximum to achieve the goal. In addition, intra-articular steroid injections were used frequently (up to 3 joints per month). If this did not work, cyclosporine and methotrexate combination was substituted. Next step included substitution with leflunomide and Injection gold.

Step-down : Start with 3 or 4 DMARDs, achieve remission and later withdraw one drug at a time, leaving one DMARD at the end for maintenance of remission. For example, in COBRA trial, patients were given a combination of sulphasalazine (2 g/day), methotrexate (7.5 mg/week) and prednisolone (initially 60 mg/day, tapered in 6 weekly steps to 7.5 mg/day). By 28 weeks, steroid had been withdrawn and at 40 weeks, methorexate was discontinued. Sulphasalazine was then continued as a single DMARD maintenance therapy.

Parallel : Here multiple agents are combined from the start and are continued on a long-term basis. An agent is stopped in the event of adverse effect. In case of lack of efficacy of the regimen in about 3 months, all drugs may be stopped. For example, in Fin-RACo trial, combination therapy (sulphasalazine, methotrexate, hydroxychloroquine, and low-dose prednisolone) was continued for 2 years. The adverse effects of combination therapy were comparable with those of monotherapy with sulphasalazine + prednisolone.

When to stop DMARD therapy?

Attempts to discontinue DMARDs are usually met with relapse of disease. The current policy is to continue DMARDs indefinitely.

DMARDs during pregnancy and lactation

Sulphasalazine and hydroxychloroquine have been found safe during pregnancy as well as lactation[22].

Complete Remission Rates

Drug-sustained remission rates with conventional DMARDs have been steadily rising in recent years and are currently in the range of 60%-65%[23].

Role of vaccinations

According to the recommendations of the American College of Rheumatology[15], all patients commencing DMARD therapy (whether conventional or biological), should be vaccinated against pneumococcus, influenza and herpes zoster. Hepatitis B vaccination should also be administered in patients at increased risk.

Key point

1. Remission is the goal of treatment of RA in modern practice.
2. Available DMARDs, if used properly, can help achieve remission in majority of patients with rheumatoid arthritis.
3. The old term "second-line agents" should not be used anymore because all patients with RA need to be treated with DMARDs.
4. Sooner the DMARDs are started better the chances of remission.
5. Commonly used DMARDs include methotrexate, sulphasalazine, leflunomide and hydroxychloroquine.
6. Methotrexate (7.5-25 mg/week) is the sheet anchor of DMARD therapy (doses greater than 15 mg/week are best administered parenterally.
7. Combinations of DMARDs are most appropriate for RA with poor prognostic factors in early as well as established disease.
8. Drugs such as azathioprine, cyclophosphamide, chlorambucil and d-penicillamine are no longer recommended for treatment of RA because of poor cost benefit ratio.
9. All patients with RA commencing DMARD therapy should receive vaccinations against pneumococcus, influenza, herpes zoster and where indicated, hepatitis B.
10. Sulphasalazine and hydroxychloroquine can be safely used during pregnancy, if required.

Further Reading:

1. Smolen JS, Aletaha D, Bijlsma JW et al. Treating rheumatoid arthritis to target: recommendations of an international task force. Ann Rheum Dis 2010; 69: 631-37.
2. Van der Woude D,Young A, Jayakumar K, Mertens BJ, Toes REM, van der Heijde D, Huizinga TWJ, van der Helm-van Mil AHM. Prevalence of and predictive factors for sustained disease-modifying antirheumatic drug-free remission in rheumatoid arthritis. Results from two large early arthritis cohorts. Arthritis Rheum 2009; 60: 2262-71.
3. Van Everdingen A, Jacobs J, Siewertsz van Reesema D, Bijlsma J. Low-dose prednisolone therapy for patients with early active rheumatoid arthritis: clinical efficacy, disease-modifying properties and side effects. Ann Intern Med 2002; 136: 1-12.
4. Kirwan J. The effect of glucocorticoids on joint destruction in rheumatoid arthritis. The Arthritis and Rheumatism Council Low-Dose Glucocorticoid Study Group. N Engl J Med 1995; 333: 142-46.
5. Hamilton RA, Kremer JM. Why intramuscular methotrexate works better than oral drug in patients with rheumatoid arthritis. Br J Rheumatol 1997; 36: 86-90.
6. Braun J, Kastner, P. Flaxenberb, P. et al. Comparison of the clinical efficiency and safety of subcutaneous versus oral administration of methotrexate in patients with active rheumatoid arthritis. Arthritis Rheum 2008; 58: 73-81.
7. Fleischmann R. Don't forget traditional DMARDs. Old friends are still useful. Rheumatology 2011; 50: 429-30.
8. Moder KG, Tefferi A, Cohen MD, Menke DM, Luthra HS: Hematologic malignancies and the use of methotrexate in rheumatoid arthritis: a retrospective study. Am J Med 1995; 99: 27, 281.
9. Cardiel MH. First Latin American position paper on the pharmacological treatment of rheumatoid arthritis by the Latin American Rheumatology Associations of the Pan-American League of Associations for Rheumatology (PANLAR) and the Grupo Latinoamericano de Estudio de Artritis Reumatoide (GLADAR). Rheumatology 2006; 45: ii7-ii22.
10. Pullar T, Bamji A. National Guidelines For the Monitoring of Second Line Drugs. Produced by the British Society for Rheumatology July 2000. Website address: www.rheumatology.org.uk.
11. Easterbrook M. An ophthalmological view on the efficacy and safety of chloroquine versus hydroxychloroquine. J Rheumatol 1999; 26: 1866-68.
12. Kremer J, Genovese M, Cannon GW, et al. Combination leflunomide and methotrexate (MTX) therapy for patients with active rheumatoid arthritis failing MTX monotherapy: open-label extension of a randomized, double-blind, placebo controlled trial. J Rheumatol 2004; 31: 1521-31.
13. O'Dell JR, Blakely KW, Mallek JA, Eckhoff PJ, Leff RD, Wees SJ et al. Treatment of early seropositive rheumatoid arthritis: a two-year, double-blind comparison of minocycline and hydroxychloroquine. Arthritis Rheum. 2001; 44: 2235-41.
14. Malaviya AN, Rohtagi K, Kabra P, Kumar A. An open trial of auranofin (oral gold preparation) in rheumatoid arthritis. J Assoc Physicians India 1986; 34: 625-27.
15. Singh JA, Furst DE, Bharat A, Curtis JR, Kavanaugh AF, Kremer JM, et al. 2012 update of the 2008 American College of Rheumatology recommendations for the use of disease-modifying antirheumatic drugs and biologic agents in the treatment of rheumatoid arthritis. Arthritis Care Res (Hoboken) 2012; 64: 625-39.
16. Désirée van der Heijde. Long-term effects of early treatment in patients with rheumatoid arthritis. Nature Clinical Practice Rheumatology 2007; 3: 258-59.
17. Aletaha D, Neogi T, Silman AJ, Funovits J, Felson DT, Bingham CO III, et al. 2010 Rheumatoid Arthritis Classification Criteria: an American College of Rheumatology/European League Against Rheumatism collaborative initiative. Arthritis Rheum 2010; 62: 2569-81.
18. S Carette. All patients with rheumatoid arthritis should receive corticosteroids as part of their management. J Rheumatol 2007; 34: 656-60.
19. Grigor C, Capell H, Stirling A, et al. Effect of a treatment strategy of tight control for rheumatoid arthritis (the TICORA study): a single-blind randomised controlled trial. Lancet 2004; 364: 263-69.
20. Boers M, Verhoeven AC, Markusse HM, van de Laar MA, Westhovens R, van Denderen JC, et al. Randomised comparison of combined step-down prednisolone, methotrexate and sulphasalazine with sulphasalazine alone in early RA. Lancet 1997; 350: 309-18.
21. Mottonen T, Hannonen P, Leirisalo-Repo M, Nissila M, Kautiainen H, Korpela M et al. Comparison of combination therapy with single-drug therapy in early rheumatoid arthritis: a randomised trial. FIN-RACo trial group. Lancet 1999; 353: 1568-73.
22. Hazes JMW, Coulie PG, Geenen V, Vermeire S, Carbonnel F, Louis E et al. Rheumatoid arthritis and pregnancy: evolution of disease activity and pathophysiological considerations for drug use. Rheumatology 2011; 50: 1955-68.
23. Sokka T, Makinen H. Remission makes its way to rheumatology. Arthritis Res Ther 2010; 12: 129-30.

BIOLOGICS IN RHEUMATIC AND AUTOIMMUNE DISEASES

6

Ved Chaturvedi

Introduction

With the understanding of the pathogenesis of autoimmune disease and the contribution of the different pathways of activation of inflammation and tissue damage, new targets for therapy were identified. These agents modify the immune response by blocking the effect of pro-inflammatory cytokines or by acting on various immune cells such as the B lymphocyte or the interaction between the T cell and the antigen presenting cell (APC).

In rheumatic diseases, the balance between endogenous anti-inflammatory mediators and pro-inflammatory mediators is disturbed. The production of pro-inflammatory cytokines, TNFα, IL-1 and others in rheumatic disease is overwhelming. New therapeutic strategies for blocking effects of inflammatory cytokines include antibodies directed against TNFα, IL-1 and IL-6 receptor antagonist, B cell depletion therapy with anti-CD20 antibody agent and others. In the last few years new drugs have emerged which directly act and antagonize the action of pro-inflammatory cytokines like TNFα, IL-1 and IL-6, leading to rapid control of disease activity and improvement in quality of life.

The term biologic therapies have emerged to describe these agents with biologic properties like monoclonal antibodies and soluble cytokine receptors. Clinical trials indicate that these biological agents may be more effective than traditional agents because they can alter joint remodeling in addition to attenuating symptoms.

It has been found that nearly 50% of patients with rheumatoid arthritis (RA) are disabled within 10 years of the onset of disease and survival is reduced. With the advent of more potent drugs, the strategy of treating RA has changed considerably. In clinical trials with diseases like RA and psoriatic arthritis (PsA), outcome measures for disease activity are ACR 20, 50 and 70, which indicate 20%, 50% or 70%, improvement in various parameters such as number of tender and swollen joints, acute phase reactants, pain on visual analogue scale and physician and patient global assessment.

Trials in RA use various indices like DAS 28 (Disease Activity Score 28), SDAI (Simplified Disease Activity Index) and CDAI (Clinical Disease Activity Index) for assessment. One of the common outcome measures that determines the extent of joint damage is the Van der Heijde modified sharp score which quantifies erosions and joint space narrowing. Finally, trials also assess the impact on functional disability and quality of life. The common tools used are the health assessment questionnaire (HAQ) and short form 36 (SF-36). The latter has the added dimension of measuring general mental health (psychological distress and well being). There are also various scores and indices relevant to psoriasis like psoriasis area severity index(PASI) and in ankylosing spondylitis like bath ankylosing spondylitis disease activity index (BASDAI), bath ankylosing spondylitis functional activity index (BASFI) and bath ankylosing spondylitis metrological index (BASMI).

Biological agents

1. Anti-TNFα e.g. Infliximab, Etanercept, Adalimumab, Golimumab.
2. Anti-CD20 antibody agent e.g. Rituximab.
3. IL-1 receptor antagonist e.g. Anakinra.
4. Interleukin (IL)-6 antagonist e.g. Tocilizumab.
5. Others: Abatacept, (CTLA-4 Ig), Belimumab, Pegylated Certolizumab, Anti -CD22 and Antilymphostat B.

Anti-TNF Therapy

1. Infliximab

(A) Rheumatoid arthritis

Infliximab, a TNFα inhibitor has emerged as a key treatment for RA and other rheumatic diseases. It is a chimeric anti-TNFα antibody with a predominance of human amino acid sequences. It has been used successfully with remarkable results in RA, AS, giant cell arteritis and polymyalgia rheumatica.

In RA it is used in the dosage of 3 mg/kg/dose at 0, 2, 6, weeks and thereafter every 8 weeks. Premedication with 100 mg of methylprednisolone is preferred by a few to prevent any anaphylaxis. Mantoux test, chest x-ray are prerequisite to infliximab therapy to rule out tuberculosis. Before each infusion, disease activity score by DAS28 and functional activity by Health Assessment Questionnaire (HAQ) are assessed and compared with previous results to determine the efficacy of the drug.

The clinical and radiological efficacy of repeated infusions of infliximab in RA was evaluated in a number of drug trials but the Anti-TNF Trial in RA with concomitant Therapy (ATTRACT)[1,2] needs to be highlighted to show its efficacy. The study enrolled 428 eligible patients with advanced disease, as evidenced by a mean duration of disease ranging from 9-12 years. These patients were randomly allocated to placebo or 3 mg/kg or 10 mg/kg of infliximab at week 0, 2 and 6 followed by infusions every four or eight weeks. The MTX dose was kept constant. Infliximab treatment was efficacious at all doses compared with placebo. The ACR20 response rates were achieved in 50%-60% by the 30 week end point. Significantly more patients in the infliximab treatment groups also achieved ACR50 (26%-31% v 5%) or ACR70 (8%-18% v 0%) responses than those in the MTX placebo group. The tender and swollen joint counts, pain scores and serum concentrations of CRP improved by comparable values. The ACR response rates were generally sustained through one year of treatment.

A recent study compared the occurrence of drug free remission, functional ability and radiological damage after 4 years of response driven treatment according to four different treatment strategies for rheumatoid arthritis (RA). Patient with recent onset, active RA (n=508) were randomly assigned to four different treatment strategies: (1) sequential monotherapy; (2) step-up combination therapy; (3) initial combination therapy with prednisone and (4) initial combination therapy with infliximab. Treatment was adjusted based on 3 monthly disease activity score (DAS) assessments, aiming at a DAS < 2.4. From the third year, patients with a sustained DAS < 1.6 discontinued treatment. In total 43% of patients were in remission (DAS < 1.6) including 18% in the group 4 and 14%, 12% and 8% in group 1-3 respectively at 4 years[3].

Another study included patients with active refractory and erosive rheumatoid arthritis (RA) who were treated with intravenous infusions of infliximab in combination with methotrexate. The objectives of this study were to evaluate the continuation rate of infliximab and its clinical effect over a 7 year period and to document the reasons for discontinuation. After 7 years, 160 of 511 patients (31%) were still on infliximab treatment. The major reasons for infliximab discontinuation included lack of efficacy (104 patients) and adverse events (107 patients). Mean DAS for patients still on treatment with infliximab decreased from 5.7 (standard error [SE] 0.1) at baseline to 3.0 (SE 0.1) at year 4 and remained that low until year 7 (3.0 [SE 0.1]). Low disease activity (defined as DAS28 < 3.2) was present in 60.9% of patients, and 45.5% achieved remission (DAS28 < 2.6)[4].

Since treatment with infliximab incurs a huge economic burden, every patient on treatment often asks the question as to when to stop the therapy? A recent study published has shown that infliximab may be discontinued with disease remission in a few patients[5]. Serum antibodies to infliximab called human antichimeric antibody (HACA) were detected in the range of 17.4% of all infliximab treated patients. However, higher doses of infliximab and concomitant MTX treatment were associated with lower incidences of serum antibodies to infliximab[6].

The British society of rheumatology (BSR) recommends use of infliximab when patient of RA has failed to respond to more than one DMARDs (disease modifying anti-rheumatic drug). However American College of Rheumatology (ACR) recommends use of biologicals like infliximab in early RA.

(B) Ankylosing Spondylitis

In ankylosing spondylitis where there are a very few options like nonsteroidal anti-inflammatory drugs only, infliximab has been used at a dosage of 5 mg/kg/dosage at 0, 2, 6 and then every 8 weeks. The potential clinical efficacy of infliximab therapy for ankylosing spondylitis was examined in an open label study of 21 patients with active spondyloarthritis (ankylosing spondylitis, 10; psoriatic arthritis, 9; undifferentiated spondyloarthritis, The study showed that infliximab 5 mg/kg given at weeks 0, 2 and 6 could produce substantial improvements in several different measures of disease activity indices like BASDAI, BASFI, and BASMI. Moreover, a maintenance regimen of 5 mg/kg every 14 weeks was able to maintain improvement in some cases[7]. One study concluded that Infliximab is an effective therapy for early sacroilitis, providing a reduction in disease activity by week 16. This study was the first to show that infliximab is effective for reducing clinical and imaging evidence of disease activity in patients with MRI-determined early axial spondylarthritis[8]. One study assessed the effect of infliximab on structural changes in AS over 4 yrs. Conventional radiographs of the cervical and the lumbar spine of 33 AS patients at baseline (BL), after 2 (FU1) and after 4 yrs (FU2) of infliximab therapy were scored by the modified Stokes ankylosing spondylitis spinal score (mSASSS). The mean change over 4 yrs was 1.6 ± 2.6 mSASSS units ($P = 0.001$), (0.9 ± 2.3 for BL-FU1 *vs* 0.7 ± 1.6 for FU1-FU2). There was less radiographic progression in comparison with published data from the OASIS cohort (4.4 within 4 yrs)[9].

2. Etanercept

Etanercept is a genetically engineered fusion protein consisting of two identical chains of the recombinant extracellular human TNF-receptor p75 monomer fused with the Fc domain of human IgG1. It effectively binds TNF and lymphotoxin-α inhibiting their activity. It is administered in the dosage 25mg subcutaneous twice a week or 50mg once a week. A pre-filled syringe with diluents and the drug is available for self administration. Pulmonary or extra-pulmonary tuberculosis has to be excluded in the same manner as infliximab.

(A) Rheumatoid arthritis

Etanercept had been found to have significant clinical benefit with minimal toxicity in adult with active RA that did not respond to other disease modifying drugs. Efficacy of etanercept in RA has been successfully documented in the Enbrel ERA (early rheumatoid arthritis)[10] and the TEMPO trial[11] with improvement in ACR response and decrease in disease activity and laboratory parameters.

Enbrel ERA trial[10] evaluated clinical and radiographic outcomes in 632 patients with early, active RA patients who received monotherapy with either etanercept or methotrexate (MTX) for 2 years. These patients were randomized to receive either twice-weekly subcutaneous etanercept (10 mg or 25 mg) or weekly oral MTX (means dosage 19 mg per week) for at least 1 year in a double-blind manner. Following the blinded phase of the trial, 512 patients continued to receive the therapy to which they had been randomized for up to 1 additional year, in an open label manner. At 24 months, 72% patients receiving 25 mg etanercept achieved ACR 20 response compared to 59% receiving methotrexate. The mean changes in radiological erosion score in the etanercept group were significantly lower than those in the MTX group. 55% in etanercept group compared to 37% in methotrexate group had improvement in quality of life. In one study, RA patients with an average moderate disease activity despite previous methotrexate monotherapy, combination treatment with etanercept and methotrexate inhibited radiographic progression and improved radiographic outcomes. The study supports the use of combination therapy in RA patients with moderate disease activity[12]. Another study evaluated safety, efficacy, and radiographic progression in patients with early rheumatoid arthritis (RA) undergoing longterm treatment with etanercept. Rates of serious adverse events and serious infections did not increase with long term exposure to etanercept, and were similar to rates reported for the blinded portion of the efficacy study. Efficacy was sustained in patients who completed 5 years of etanercept treatment a (N = 201), even in those who decreased or discontinued use of MTX or corticosteroids. No radiographic progression (change in TSS < or = 0) was seen in 55% of patients with 5 year radiographs; negative change (TSS < 0) was seen in 11%[13].

(B) Ankylosing spondylitis

Evidence from MRI scans collected as part of a 6 month open-label descriptive study[14] suggested that etanercept may indeed help control entheseal lesions in patients with AS. At the end of the study 86% of the patients treated had resolution or improvement in MRI sites of inflammation[14]. In one study, patients with early axial spondyloarthritis having active inflammatory lesions detected by whole-body MRI improved significantly more with etanercept versus sulfasalazine-treated patients. This effect correlated with a good clinical response in the etanercept group[15]. Another study evaluated long-term safety and efficacy of etanercept treatment in patients with ankylosing spondylitis (AS) : A total of 257 of 277 patients (92%) enrolled in the open lable extension study (OLE). After up to 192 weeks of treatment with etanercept, the most common AEs were injection site reactions, headaches and diarrhoea. Of patients who received etanercept in both the RCT and OLE and were still in the trial, 71% were ASAS20 responders at week 96, and 81% were responders at week 192. ASAS 5/6 response rates were 61% at week 96 and 60% at week 144, and partial remission response rates were 41% at week 96 and 44% at week 192. Placebo patients who switched to etanercept in the OLE showed similar patterns of efficacy maintenance[16].

(C) Psoriatic arthritis and juvenile arthritis

Etanercept has been successfully tried in Psoriatic arthritis[17] and Juvenile Idiopathic Arthritis[18] with significant reduction in disease activity and improvement in quality of life. Few case reports have mentioned development of psoriasis in patients after etanercept therapy.

A study conducted on a total of 672 patients randomized 1:1:1:1 to receive placebo or etanercept 25 mg once weekly, 25 mg twice weekly, or 50 mg twice weekly SC injection for 12 weeks. Etanercept recipients who completed the initial treatment period continued on their assigned dosage for an additional 12 weeks. Among patients randomized to etanercept 25 or 50 mg twice weekly, 32% and 47% of patients, respectively, achieved PASI-75 at week 12, increasing to 41 and 54% at week 24. In another study, a sustained benefit of etanercept treatment, including inhibition of radiographic progression, in patients with PsA was noticed even after 2 years[20]. A study evaluated the clinical efficacy and tolerability of etanercept in the treatment of active and progressive psoriatic arthritis (PsA) in patients who have previously demonstrated an inadequate response to standard treatments, such as disease-modifying anti-rheumatic drugs (DMARDs) and non-steroidal anti-inflammatory drugs. Twenty seven (27/32) patients (84.3%) completed 3 years (144 weeks) of continuous treatment, while 5/32 (15.6%) patients were withdrawn from the study. At week 144, a significant improvement in DAS-28 was registered with a reduction in mean DAS-28 from 5.3 at baseline to 1.8, while 25/27 patients (92.5%) achieved PASI 75 with a mean PASI score of 0.7; the mean pain visual analogue scale (pain-VAS) score decreased from 64.2 at baseline to 2 at week 144, corresponding to an improvement of 94.7%[21].

Currently, Etanercept is the best evaluated biologic for the treatment of JIA. The first trial of etanercept in JIA was published in 2000[22]. During the first open part of this study, improvement was seen in 51 of 69 children with JIA (74%). The responders took part in a placebo controlled trial that confirmed the efficacy of the drug by achieving at least 50% improvement in 70% of children compared with 25% of patients on placebo. This study included children who failed or did not tolerate methotrexate. Following this trial, Etanercept was approved in Europe in 2000 for the treatment of children ages 4-17 years with refractory, polyarticular JIA. Another study was undertaken to evaluate the long term safety and effectiveness of etanercept alone or in combination with methotrexate (MTX) in children with selected categories of juvenile idiopathic arthritis (JIA). Patients aged 2-18 years received MTX alone (> or =10 mg/m(2)/week [approximately 0.3 mg/kg/week], maximum dosage 1 mg/kg/week), etanercept alone (0.8 mg/kg/week, maximum dose 50 mg), or etanercept plus MTX for 3 years in an open-label, nonrandomized study. Exposure-adjusted rates of adverse events were similar among the 3 treatment groups (18.3, 18.7, and 21.6 per 100 patient-years in the MTX, etanercept, and etanercept plus MTX groups, respectively). Respective rates per 100 patient-years of serious adverse events (4.6, 7.1, and 6.0) and medically important infections (1.3, 1.8, and 2.1) were also similar among the 3 treatment groups. Scores for physician's global assessment and total active joints improved from baseline, and improvement was maintained for 3 years in those continuing to receive medication[23, 25].

3. Adalimumab

Other drugs like Adalimumab, a human anti-TNF monoclonal antibody has also been found to inhibit the progression of structural joint damage, reducing the signs and symptoms, and improve physical function in patients with active RA receiving concomitant treatment with methotrexate (MTX)[24]. It is not available in India at present.

4. Certolizumab

Use of pegylated certolizumab has also been found in causing significant benefit in reducing the signs and symptoms as well as reduction in radiological score in RA[26]. It is also used in combination with methotrexate. It is not available in India at present.

5. Golimumab

Golimumab is a human immunoglobulin G1 mAb that is specific for human TNF[27,28]. It was created using genetically engineered mice that were immunized with human TNF, resulting in an antibody with human-derived variable and constant regions[29]. Golimumab binds to both the soluble and transmembrane bioactive forms of human TNF, preventing the binding of TNF to its receptors and thereby inhibiting the biological activity of TNF[19]. Golimumab, in combination with MTX, is approved by the FDA for the treatment of adult patients with moderately to severely active rheumatoid arthritis[30]. It is also approved for the treatment of adult patients with active psoriatic arthritis (alone or in combination with MTX) and for the treatment of adult patients with active AS[30]. The efficacy of Golimumab in RA was proven in three studies, denoted RA-1 (also known as GO-AFTER), RA-2 (also known as GO-FORWARD), and RA-3, involving 1,542 adult patients with moderately to severely active RA [31,32]. It is not available in India at present.

6. Cell Depletion Therapy

B lymphocytes are an essential component of the adaptive (acquired) immune response. They develop initially in the fetal liver and then transferred to the bone marrow at around 12-16 weeks of fetal life where they undergo maturation with the assistance of stromal cells and cytokines. The surface antigen, CD20 acts as a cross membrane calcium ion channel and plays a key role in B cell activation. Therefore targeting CD20 may prove therapeutic in the management of diseases that are largely B-cell mediated. A drug commonly used is Rituximab.

Rituximab is a genetically engineered chimeric anti CD20 monoclonal antibody. It comprised a human IgG1 kappa constant region and a variable region from a murine anti-CD20 antibody IDEC-C2B8, which binds with high affinity to cells expressing the CD20 antigen. Rituximab can selectively deplete B cells by antibody-dependent cell mediated cytotoxicity and direct induction of B cell apoptosis.

(A) Rheumatoid arthritis

Rituximab is generally indicated in patients who had previously shown resistance to anti-TNFα. REFLEX33 and DANCER34 trials have shown improvement in clinical as well as laboratory response in patients of RA managed with Rituximab[33-36]. It is used in the dosage 1,000mg on day 1 and 15 with a premedication of 100 mg of methylprednisolone. Remission in disease activity has been found as long as 18 months with a very few patients requiring additional therapy. Rituximab depleted peripheral CD20+ B cells, without decreasing the mean immunoglobulin levels (IgG, IgM, and IgA). Most adverse events occurred with the first rituximab infusion and were of mild-to-moderate severity.

Studies to compare the effectiveness of RTX with that of an alternative anti-TNF agent in the management of patients with RA who had an inadequate response to anti-TNF therapy have shown better outcomes with RTX[37]. This prospective cohort study was nested within the Swiss clinical quality management RA cohort and included all patients who had an inadequate response to at least 1 anti-TNF agent and subsequently received either 1 cycle of RTX or an alternative anti-TNF agent. 116 patients with RA were included; 50 patients received 1 cycle of RTX, and 66 patients were treated with a second or a third alternative anti-TNF agent. At base line, there were no significant differences between the 2 groups in age, sex, disease duration, and disease activity. Evolution of the DAS28 was more favorable in the group that received RTX compared with the group that received an alternative anti-TNF agent (P=0.01). At 6 months, the mean decrease in the DAS28 was -1.61 among patients receiving RTX and -0.9 among those receiving subsequent anti-TNF therapies. In a systematic review, rituximab is found to be more efficacious in patients with rheumatoid arthritis. Rituximab was found to improve symptoms of patients who had inadequate response to or are intolerant to TNF-alpha inhibitors[38]. One study evaluated the long term safety of rituximab in rheumatid arthritis clinical trials. Safety analyses were based on 5013 patient-years of rituximab exposure. The most frequent AE was infusion-related reactions (25% of patients during the first infusion of Course 1). Less than 1% of infusion-related reactions were considered serious. Rates of AE and serious AE (SAE; 17.85 events/100 patient-yrs, 95% CI 16.72, 19.06) were stable following each course. The overall serious infection rate was 4.31/100 patient-years (95% CI 3.77, 4.92). Infections and serious infections over time remained stable across 5 courses at 4-6 events/100 patient-years[39].

(B) Systemic lupus erythematosus

RTX has been widely used in the treatment of patients with Systemic lupus erythematosus (SLE) who have failed to respond to conventional immuno-suppression[40]. Even though two double-blind controlled trials, Explorer[41] and Lunar[42], failed to meet their end points, increasingly large studies continue to indicate the utility of this approach[43,44]. Although the Lunar and Explorer trials failed to show superiority of RTX-treated SLE patients over those treated more conventionally, many smaller open-labelled studies have shown a good response to RTX[45]. A recent study has shown that RTX is an effective therapy when used early in the disease and offers a useful method of reducing the cumulative steroid burden in patients with predominantly non-renal SLE[46]. In another study, thirty-one sequential patients with relapsing or refractory systemic lupus erythematosus, 11 of whom had active lupus nephritis, received rituximab [either 375 mg/m^2/week × 4 (n = 16) or 1000 mg × 2 (n = 15)]. The median follow-up was 30 months. Thirty of 31 (97%) patients had depleted peripheral B cells. Twenty-seven of 31 (87%) patients achieved remission (17 complete, 10 partial). Renal response occurred in 10/11 patients (4 complete, 6 partial) with active glomerulonephritis. Clinical improvement was reflected by reductions of disease activity, proteinuria and daily prednisolone dose. Eighteen of 27 (67%) patients relapsed after a median of 11 months. Re-treatment with rituximab was effective. Infusion reactions were common (18/31; 58%), and infections occurred in 8/31 (26%) patients[47].

(C) Other clinical conditions

Rituximab has also been found to be effective in the treatment of type II mixed cryoglobulinaemia, dermatomyositis and ANCA associated vasculitis like Wegener's granulomatosis and microscopic polyangiitis.

In Wegener's granulomatosis (now termed as Granulomatosis with Polyangiitis) two randomized controlled trials reported that initial induction therapy with RTX achieved similar remission rates to cyclophosphomide (CYC)[48,49]. In the Rituximab in Vasculitis (RITUXVAS) study (n = 44), complete remission was achieved in 82% of patients who received RTX, and 91% who received CYC (non-significant). In the Rituximab for ANCA-associated vasculitis (RAVE) study (n = 197), 64% of those who received RTX were in remission and steroid free at 6 months, vs 54% of those who received CYC (non-significant). The duration of follow-up was 12 and 6 months, respectively, and the effect of rituximab induction on longer term outcomes is not known. No cost-effectiveness comparison between CYC and rituximab has been performed and no clear safety benefit, apart from CYC sparing was demonstrated. Further data are required before rituximab can be recommended in preference to CYC in the setting of ongoing infection. CYC avoidance is desirable in the presence of ongoing chronic infection, known CYC intolerance or hypersensitivity, or when there is a high risk of infertility or malignancy. Consequently, where CYC avoidance is desirable, rituximab is an effective alternative.

7. Belimumab

Recently published data showed patients with SLE with considerable response in disease activity when treated with belimumab measured by improvement in SLE disease activity index (SLEDAI)[50]. Belimumab is a human monoclonal antibody that inhibits B-cell activating factor(BAFF)[51]. It is approved in the United States, Canada and Europe for treatment of SLE, and is being tested for use in other autoimmune diseases. Belimumab has also undergone phase II clinical trials for rheumatoid arthritis[52]. Preliminary results in November 2005 were encouraging[53]. Belimumab binds primarily to circulating soluble BAFF, therefore not inducing antibody-dependent cellular cytotoxicity that could be expected from this IgG1-type antibody. Belimumab does reduce the number of circulating B cells, but seemingly less deeply and durably than anti-CD20 monoclonal antibodies[54]. The BLISS-52 and BLISS-76 trials showed that belimumab plus standard SLE therapy resulted in significant improvement outcomes than did placebo plus standard therapy at 52 weeks in antinuclear antibody-positive or anti-dsDNA-positive SLE patients[55].

Other drugs

Tocilizumab

Interleukin (IL)-6 receptor antagonist tocilizumab[56] represents a promising new approach to the treatment of rheumatoid arthritis since IL-6 plays a pivotal role in mediating both local and systemic manifestations of RA. It is available in India.

Abatacept

Abatacept is a fusion protein linking the extracellular domain of human cytotoxic T-lymphocyte associated antigen 4 (CTLA-4) to the Fc portion of human IgG1. It works by competing for the binding between CD28 on the T cell and CD80/86 on the antigen-presenting cell. This is an important co-stimulatory signal that is essential for T cell activation. Blocking T cell activation results in a reduction of the autoimmune process and clinical improvement is demonstrated. It is effective in patients who have an inadequate response to either MTX or 1 or more TNF-inhibitors.It is available in India.

Anakinra

The earliest approved biologic response modifier is anakinra, an IL-1 receptor antagonist. It is administered as a subcutaneous injection dosage 100mg (0.6ml) once daily and has efficacy in treating RA when used alone or in combination with MTX. Studies have shown it reduces erosion in RA[57]. It is not commonly used since the TNF-inhibitors show much greater efficacy in comparison and have a more convenient dosing regimen. However, it is a valuable drug in the treatment of some disease such as Adult onset Still's disease and systemic-onset juvenile idiopathic arthritis, where it works much better than the TNF inhibitors. A multicentre, randomised, double-blind, placebo-controlled trial with the IL-1 receptor antagonist anakinra in patients with systemic-onset juvenile idiopathic arthritis (ANAJIS trial) has shown short term efficacy[58].

Other IL- 1 inhibitors

Rilonacept

Rilonacept, also known as IL-1 Trap(marketed by Regeneron Pharmaceuticals under the trade name Arcalyst), is an interleukin 1 inhibitor[59]. Rilonacept is a dimeric fusion protein consisting of the extracellular domain of human interleukin-1 receptor and the Fc domain of human IgG1 that binds and neutralizes IL-1[60]. Rilonacept is used for the treatment of cryopyrin-associated periodic syndromes (CAPS), including familial cold autoinflammatory syndrome, Muckle-Wells syndrome and neonatal onset multisystem inflammatory disease(not approved in the U.S. on this indication). It was given an "Orphan Drug" status by the FDA. It was approved by the FDA for use in CAPS on February 2008[61].

Rilonacept has been also tried in gout based on various murine models and the role of IL-1 in the pathogenesis[62]. The FDA accepted for review the Company's supplemental Biologics License Application (sBLA) for Arcalyst Injection for subcutaneous Use for the prevention of gout flares in patients initiating uric acid-lowering therapy. Under the Prescription Drug User Fee Act (PDUFA), the goal for a standard review of an sBLA is ten months from submission, for a target action date of July 30, 2012. The sBLA submission for Arcalyst is based on positive efficacy data from PRE-SURGE 1, a North American Phase 3 trial, and PRE-SURGE 2, a similarly designed global Phase 3 trial 62. Both trials met the primary endpoint of reduction in the mean number of gout flares per patient during the 16-week treatment period in patients initiating uric acid-lowering therapy. Arcalyst was generally well tolerated with the incidence of serious adverse events generally well-balanced across the placebo and Arcalyst groups. Injection site reactions, usually considered mild, were reported more commonly with Arcalyst than with placebo.

Canakinumab

The results of two phase III randomized trials indicate that canakinumab, a fully human monoclonal antibody directed at IL-1β, could provide fast and effective relief from acute gout flares in patients who cannot take standard anti-inflammatory therapies.

In pooled 12-week results from the two trials, β-RELIEVED and β-RELIEVED II[63], canakinumab was more effective than the corticosteroid triamcinolone acetonide both in reducing pain and in delaying time to a first new flare. Importantly, patients were enrolled in the trials up to 5 days after the onset of an acute flare, and the cohorts included those for whom NSAIDs and/or colchicine were contraindicated, not tolerated or ineffective. For patients with acute gout and limited treatment options, canakinumab might be an important route to relieving pain and inflammation.

Denosumab

Denosumab[64] is a fully human monoclonal antibody for the treatment of osteoporosis, treatment-induced bone loss, bone metastases, rheumatoid arthritis, multiple myeloma, and giant cell tumor of bone[65]. Denosumab is designed to target RANKL (RANK ligand), a protein that acts as the primary signal for bone removal. In many bone loss conditions, RANKL overwhelms the body's natural defenses against bone destruction. In a Phase III clinical trial ('FREEDOM') involving 7,808 women aged 60 to 90, there were significant improvements in the subset of women with more severe disease (two or more prevalent vertebral fractures and/or one or more prevalent vertebral fractures with moderate or severe deformity) at the beginning of the study. Researchers reported a 35% risk reduction with denosumab compared to placebo (17% vs 49%). Within this subset, only 31% of those taking denosumab developed new vertebral fractures, versus 71% of those receiving placebo[66].

Side Effects

Infections are serious potential side effects of any drug that modifies the immune response. For the anti-TNF agents, one particular infection that has to be looked for is tuberculosis (TB). Screening for TB before starting treatment is prudent since there is a high risk for reactivation of latent TB early in the course of anti-TNF therapy. Any latent TB should be treated preferably for a month before initiating anti-TNF therapy. Recent data have documented exacerbations of hepatitis in chronic hepatitis B patients and carriers[67,68]. Prolonged suppression of the immune response may allow for the re-activation of slow viruses such as the JC polyoma virus. Two cases of a fatal brain disease caused by JC virus, called progressive multifocal leukoencephalopathy (PML), have been reported in patients with SLE who were administered RTX[69]. Many of the usual signs of sepsis may be suppressed in patients treated with TNF-inhibitors, since their ability to mount anti-inflammatory response may be seriously compromised. Current available evidence suggesting anti-TNF agents increasing the risk of lymphoma is very small[70,71].

Impact in Current Clinical Practice

Biological response modifiers represent advancement in the treatment of autoimmune disease. In the past, there were patients in whom all existing therapies were unable to control the inflammation and subsequent destruction of their joints. For such patients, these new drugs have given them a new lease on life. For the first time, disease activity can be well controlled to the extent that function improves almost to normal and X-ray changes of destruction stop progressing. However, there are still non-responders and newer biologicals are being tried in these patients[77].

Drawbacks

A major problem in the use of these agents is cost. Insurance companies and state-funded healthcare systems set various criteria for reimbursement for these agents in an attempt to control spiraling healthcare expenditure. Nevertheless, the efficacy of these agents prompts their widespread acceptance by physicians and currently 40% of patients with RA in the United States are receiving a biologic agent[72].

Newer biologics

Secukinumab is a humanized monoclonal antibody, targeting IL-17A; this specific target inhibits the migration of monocytes and neutrophils around the site of inflammation[73]. This particular agent has been investigated as 150 mg daily subcutaneous psoriasis treatment[74]. Based on its mechanism of action, secukinumab may be an option for RA due to a potential role on autoimmune diseases[75].

Ustekinumab is a human monoclonal antibody that targets the p40 subunit of interleukin-12 (IL-12) and interleukin-23 (IL-23), preventing them from binding to their receptors on T-cells and natural-killer cells. The IL-12/23 proteins are believed to be important in regulating the immune system and to play a role in inflammatory diseases such as PsA. Ustekinumab is administered by subcutaneous injection at doses of 45 to 90mg (doses used in trials 1) in weeks 0 and 4, with a maintenance dose every 12 weeks thereafter. A double blind control trial has shown significant reduction in signs and symptoms of psoriatic arthritis and diminished skin lesions compared with placebo, and the drug was well tolerated[76]. Larger and longer term studies are needed to further characterise ustekinumab efficacy and safety for treatment of psoriatic arthritis. Ustekinumab is licensed in EU for the treatment of moderate to severe plaque psoriasis after failure of systemic therapies. It is also in phase II/IIIs trial for Crohn's disease.

Future Trends

In the future, improved versions of existing drugs will become available. Anti-TNF preparations that are given as monthly subcutaneous injections are currently being developed. Other targets for therapy that are under study include anti-CD22, anti-CD40-CD40L and many others. It would be ideal if pathogenesis is found to be driven solely by one of these key molecules. However, this seems increasingly unlikely and there may be better ways in the future of choosing the best therapy for an individual patient based on predictive factors. Current best practice is to optimize the use of DMARDs including the newer agents. This would involve frequent titration of therapy and the use of combination DMARDs if necessary, to ensure good control of inflammation. The issue of cost needs to be addressed and in the future, every patient with inflammatory arthritis will have therapeutic options which will bring their illness into a prolonged, if not, permanent remission.

Table 1. Common biologics in clinical practice, dosage, indication, side effects

Drugs	Indications	Dosage	Side effects
Infliximab	RA, Ankylosing spondylitis, Cutaneous PAN, Crohn's disease	3-5 mg/kg/dose IV 0, 2, 6 followed by every 8 weeks	Infections, Tuberculosis, Flulike symptoms, Malignancy
Etanercept	RA, Ankylosing spondylitis, Psoriasis, JIA, PsA	25 mg sc biweekly or 50mg/wk	Myalgia, Urticaria, Tuberculosis, demyelinating disorders, pancytopenia, congestive heart failure
Rituximab	RA, SLE, ANCA associated Vasculitis, Dermatomyositis	375mg/1.73m^2 1000 mg 2 infusions at day1 and 15	Headache, Infections Malignancy

Further Reading :

1. Maini R, St Clair EW, Breedveld F, Furst D, Kalden J, Wesman M, et al. Infliximab (Chimeric anti-tumor necrosis factor α monoclonal antibody) versus placebo in rheumatoid arthritis patients receiving concomitant methotrexate : a randomized phase III trial. Lancet 1999; 354: 1932-39.
2. Lipsky PE, van der Heijde DMFM, St Clair EW, et al. Infliximab and methotrexate in the treatment of rheumatoid arthritis. N Engl J Med 2000; 343: 1594-602.
3. Drug-free remission, functioning and radiographic damage after 4 years of response-driven treatment in patients with recent-onset rheumatoid arthritis S M Van der Kooij, Y P M Goekoop Ruiterman et al. Ann Rheum Dis 2009; 68: 6 914-21.
4. Seven-year follow-up of infliximab therapy in rheumatoid arthritis patients with severe long-standing refractory disease: attrition rate and evolution of disease activity. Vander Cruyssen et al. Arthritis Research & Therapy 2010, 12: R77.
5. Van der Bijl AE, Goekoop-Ruiterman YP, de Vries-Bouwstra JK et al. Infliximab and methotrexate as induction therapy in patients with early rheumatoid arthritis. Arthritis Rheum 2007 Jun 28 56: 2129-34. 3aDrug-free remission, functioning and radiographic damage after 4 years of response-driven treatment in patients with recent-onset rheumatoid arthritis S M Van der Kooij, Y P M Goekoop Ruiterman et al. Ann Rheum Dis 2009; 68: 6 914-21.
6. Maini RN, Breedweld FC, Kalden JR et al. Therapeutic efficacy of multiple intravenous infusions of anti-tumor necrosis factor α monoclonal antibody combined with low dose weekly methotrexate in rheumatoid arthritis. Arthritis Rheum 1998; 41: 1552-63.
7. Van den Bosch F, Kruithof E, Baeten D, de Kayser F, Mielants H, Veys EM. Effects of a loading dose regimen of three infusions of chimeric monoclonal antibody to tumor necrosis factor α (infliximab) in spondyloarthopathy: an open pilot study. Arth Rheum 2000: 59: 427-33.
8. Clinical and imaging efficacy of infliximab in HLA-B27-Positive patients with magnetic resonance imaging-determined early sacroiliitis. Barkham N, Keen HI, Coates LC, O'Connor P, Hensor E, Fraser AD, Cawkwell LS, Bennett A, McGonagle D, Emery PArthritis Rheum. 2009; 60(4): 946.
9. Radiographic progression in patients with ankylosing spondylitis after4 yrs of treatment with the anti-TNF-a antibody infliximab X. Baraliakos, J. Listing, J. Brandt, H. Haibel, M. Rudwaleit, J. Sieperand J. Braun. Rheumatology 2007; 46: 1450-53.
10. Mark C, Genovese, Joan M Bathon, Richard W Martin et al. Etanercept versus Methotrexate in Patients with early Rheumatoid Arthritis. Two-Year Radiographic and Clinical Outcomes. Arth Rheum 2002; 46: 1443-50.
11. De' sire' e van der Heijde, Lars Klareskog, Vicente Rodriguez-Valverde et al. Comparison of Etanercept and Methotrexate, Alone and Combined in the treatment of Rheumatoid Arthritis Two-Year Clinical and Radiographic Results from the TEMPO study, a Double-Blind, Randomized Trial Arth Rheum 2006; 54 : 1063-74.
12. Inhibition of radiographic progression with combination etanercept and methotrexate in patients with moderately active rheumatoid arthritis previously treated with monotherapy. Ann Rheum Dis. 2009; 68(7): 1113-18.
13. Longterm safety, efficacy, and radiographic outcome with etanercept treatment in patients with early rheumatoid arthritis. Genovese MC, bathon JM, et al: Rheumatol. 2005 Jul; 32(7): 1232-42.
14. Marzo-Ortegah, Mcgonagle D, O'connor P, Emery P: Efficacy of etanercept in the treatment of the entheseal pathology in resistant spondylarthropathy: A clinical and magnetic resonance imaging study. Arthritis Rheum 2001; 44: 2112-17.
15. Effects of etanercept versus sulfasalazine in early axial spondyloarthritis on active inflammatory lesions as detected by whole-body MRI (ESTHER): a 48-week randomised controlled trial.Song IH, Hermann K, Haibel H, Althoff CE, Althoff C, Listing J, Burmester G, Krause A, Bohl-Bühler M, Freundlich B, Rudwaleit M, Sieper J Ann Rheum Dis. 2011; 70(4): 590.
16. Efficacy and safety of up to 192 weeks of etanercept therapy in patients with ankylosing spondylitis J C.
17. Davis Jr, D M van der Heijde, J Braun, M Dougados, D O Clegg, A J Kivitz, R M Fleischmann, D Inman, L Ni, S-L Lin, W H Tsuji. Ann Rheum Dis 2008; 67: 346-52.
18. Mease PJ, Goffe BS, Metz J, Vander Stoep A, Fink B, Burge DJ, Etanercept in the treatment of psoriatic arthritis and psoriasis : a randomized trial. Lancet 2000; 356: 385-90.
19. Daniel J Lovell, Edward et al Etanercept in children with polyarticular juvenile rheumatoid arthritis. For The pediatric Rheumatology Collaborative Study Group N Engl J Med 2000; 342: 763-69.
20. Continued inhibition of radiographic progression in patients with psoriatic arthritis following 2 years of treatment with etanercept. Mease PJ, Kivitz AJ, Burch FX, Siegel EL, Cohen SB, Ory P, Salonen D, Rubenstein J, Sharp JT, Dunn M, Tsuji WJ Rheumatol. 2006; 33(4): 71.
21. Long-term experience with etanercept in psoriatic arthritis patients: a 3-year observational study. Mazoota A, EspositoM, Schipani C, Chimenti S J Dermatalog Treat 2009; 20(6): 354-58.
22. Lovell DJ, Giannini EH, Reiff A, et al. Etanercept in children with polyarticular juvenile rheumatoid arthritis. N Engl J Med 2000; 342: 763-69.
23. Long-term safety and effectiveness of etanercept in children with selected categories of juvenile idiopathic arthritis. Giannini EH, Ilowite NT, Wallace CA, et al Arthritis Rheum. 2009 Sep; 60(9)2794-804.
24. Keystone EC, Kavanaugh AF, Sharp JT et al. Radiographic, clinical and functional outcomes of treatment with adalimumab (a human anti-tumor necrosis factor monoclonal antibody) in patients with active rheumatoid arthritis receiving concomitant methotrexate therapy: a randomized, placebo-controlled, 52 week trial. Arthritis Rheum.
25. Long-term safety and effectiveness of etanercept in children with selected categories of juvenile idiopathic arthritis. Giannini EH, Ilowite NT, Wallace CA, et al Arthritis Rheum. 2009 Sep; 60(9) 2794-804.
26. Keystone E, van der Heijde D, Mason DH, Landewé R, van Vollenhoven R, Combe B, et al. Certolizumab pegol plus methotrexate is significantly more effective than placebo plus methotrexate in active rheumatoid arthritis. Arthritis Rheum. 2008; 58: 3319-29.

27. Shealy, D. et al. Characterization of golimumab (CNTO 148), a novel monoclonal antibody specific for human TNF. Ann. Rheum. Dis. 2007; 66 (Suppl 2): 151.

28. Zhou, H. et al. Pharmacokinetics and safety of golimumab, a fully human anti-TNF- monoclonal antibody, in subjects with rheumatoid arthritis. J. Clin. Pharmacol. 2007: 47: 383-96.

29. Lonberg, N. Fully human antibodies from transgenic mouse and phage display platforms. Curr. Opin. Immunol. 2008; 20: 450-59.

30. Food and Drug Administration. FDA labelling information. FDA web site [online] < http://www.accessdata.fda.gov/drugsatfda_docs/label/2009/125289s000lbl. pdf > (2009).

31. Smolen, J. S. et al. Golimumab in patients with active rheumatoid arthritis after treatment with tumour necrosis factor inhibitors (GO-AFTER study): a multicentre, randomised, double-blind, placebo-controlled, phase III trial. Lancet 2009; 374: 210-21.

32. Keystone, E. C. et al. Golimumab, a human antibody to tumour necrosis factor given by monthly subcutaneous injections, in active rheumatoid arthritis despite methotrexate therapy: the GO-FORWARD study. Ann. Rheum. Dis. 2009; 68: 789-96.

33. Cohen SB, Emery P, Greenwald MW et al REFLEX trial group. Rituximab for rheumatoid arthritis refractory to anti-tumor necrosis factor therapy: Result of a multicenter, randomized, double-blind, placebo-controlled, phase III trial evaluating primary efficacy and safety at twenty-four weeks. Arthritis Rheum 2006; 54: 2793-806.

34. Emery P, Fleischmann R, Filipowicz-Sosnowska A et al. The efficacy and safety of rituximab in patients with active rheumatoid arthritis despite methotrexate treatment : results of a phase IIB randomized, double-blind, placebo-congrolled, dose-ranging trial. DANCER study Group. Arth Rheum 2006; 54: 1390-400.

35. Roll P, Dorner T, H-P Tony Anti-CD 20 Therapy in patients with rheumatoid arthritis. Arths Rheum 2008; 58: 1566-75.

36. Keystone E, Burmester GR, Furie R et at. Improvement in patients-reported outcomes in a rituximab trial in patients with severe rheumatoid arthritis refractory to anti-tumor necrosis factor therapy. Arth Rheum 2008; 59: 785-93.

37. Finckh A, Ciurea A, Bruhart L et al. B cell depletion may be more effective than switching to an alternative anti-tumor necrosis factor agent in rheumatoid arthritis patients with inadequate response to anti-tumor necrosis factor agents. Arth Rheum 2007; 56: 1417-23.

38. Rituximab in patients with rheumatoid arthritis: systematic review Brodszky V; Czirják L; Géher P; Hodinka L; Kárpáti K; Péntek M; Poór G; Szekanecz Z; Gulácsi LBudapesti Corvinus Egyetem, Közszolgálati Tanszék, Egészségügyi Gazdaságtani és Technológiaelemzési Kutatóközpont, Budapest Orv Hetil. 2007; 148(40): 1883-93.

39. Long term safety of patients receiving Rituximabin Rheumatoid arthritis Clinical Trials. Ronald F von Vollenhaven, Paul Emery, lifton O bingham, Ravi Rao. J Rheumatol 2010; 37(3): 558-67.

40. Leandro MJ, Ehrenstein MR, Edwards JC et al. Treatment of refractory lupus nephritis with B lymphocyte depletion Arth Rheum 2003; 48: S378.

41. Merrill JT, Neuwelt CM, Wallace DJ et al. Efficacy and safety of rituximab in moderately-to-severely active systemic lupus erythematosus: the randomized,double-blind, phase II/III systemic lupus erythematosus evaluation of rituximab trial. Arthritis Rheum 2010; 62: 222-33.

42. Furie R, Looney RJ, Rovin B et al. Efficacy and safety of rituximab in subjects with active proliferative lupus nephritis (LN): results from the randomized, double-blind phase III LUNAR study. Arthritis Rheum 2009; 60(Suppl.10): 1149.

43. Lu TY, Ng KP, Cambridge G et al. A retrospective seven-year analysis of the use of B cell depletion therapy in systemic lupus erythematosus at University College London Hospital: the first fifty patients. Arthritis Rheum 2009; 61: 482-87.

44. Terrier B, Amoura Z, Ravaud P et al. Safety and efficacy of rituximab in systemic lupus erythematosus: results from 136 patients from the French AutoImmunity and Rituximab registry. Arthritis Rheum 2010; 62: 2458-66.

45. Favas C, Isenberg DA. B-cell-depletion therapy in SLE-what are the current prospects for its acceptance? Nat Rev Rheumatol 2009; 5: 711-16.

46. Early treatment with rituximab in newly diagnosed systemic lupus erythematosus patients: a steroid-sparing regimen. Amara N. Ezeonyeji and David A. Isenberg Rheumatology (2012) 51 (3): 476-481.doi: 10.1093/rheumatology/ker33.First published online: November 16, 2011.

47. Long-term efficacy and safety of rituximab in refractory and relapsing systemic lupus erythematosusFausta Catapano, Afzal N. Chaudhry, Rachel B. Jone, Kenneth G.C. Smith and David W. Jayne Nephrol Dial Transplant 2010; 25: 3586-92.

48. Jones RB, Cohen Tevaert JW, Hauser T et al. Randomised trial of rituximab versus cyclophosphamide in ANCA associated renal vasculitis 'RITUXVAS'. N Engl J Med 2010; 363: 211-20.

49. Stone JH, Merkel PA, Seo P et al. Rituximab versus cyclophosphamide for induction of remission in ANCA-associated vasculitis: a randomized controlled trial (RAVE). N Engl J Med 2010; 363: 221-32.

50. Paula Moyer. Belimumab in Systemic Lupus Erythematosus. EULAR 2007; Abstract OP0018.

51. Bossen, C; Schneider, P (October 2006). "BAFF, APRIL and their receptors: structure, function and signaling". Semin. Immunol. 18 (5): 263-75.

52. ClinicalTrials. gov NCT00071812 A Safety and Efficacy Study of LymphoStat-B (Monoclonal Anti-BLyS Antibody) in Subjects With Rheumatoid Arthritis.

53. Human Genome Sciences Reports Phase 2 Results For Lymphostat-B (Belimumab) In Patients With Rheumatoid Arthritis (Press release). Human Genome Sciences. 17 November 2005.

54. Halpern WG, Lappin P, Zanardi T, Cai W, Corcoran M, Zhong J, et al. Chronic administration of belimumab, a BLyS antagonist,decreases tissue and peripheral blood B-lymphocyte populations in cynomolgus monkeys: pharmacokinetic, pharmacodynamic, and toxicologic effects. Toxicol Sci 2006; 91: 586-99.

55. Cheema GS, Roschke V, Hilbert DM, Stohl W. Elevated serum B lymphocyte stimulator levels in patients with systemic immune-based rheumatic diseases. Arthritis Rheum 2001; 44: 1313-19.

56. Paula Moyer, Tocilizumab in Rheumatoid Arthritis Phase III trial, EULAR 2007: Abstract OP117.

57. RM Fleischmann, J Tesser, MH Schiff, J Schechtman, G-R Burmester, R Bennett, D Modafferi, L Zhou, D Bell and B Appleton, Safety of extended treatment with anakinra in patients with rheumatoid arthritis Ann Rheum Dis 2006; 65: 1006-12.

58. Pierre Quartier, Florence Allantaz, Rolando Cimaz. A multicentre, randomised, double-blind, placebo-controlled trial with the interleukin-1 receptor antagonist anakinra in patients with systemic-onset juvenile idiopathic arthritis (ANAJIS trial). Ann Rheum Dis 2011; 70: 747-54 doi: 10.1136/ard. 2010.134254.

59. http://ard.bmj.com/cgi/content/short/ard. 2009. 108936v1?rss=1 "The IL-1 inhibitor rilonacept in treatment of chronic gouty arthritis: results of a placebo-controlled, monosequence crossover, nonrandomized, single-blind pilot study " Oct 2009.

60. Molecule of the month. Rilonacept. Drug News Perspect. 2008; 21 (4): 232.

61. "Drugs.com, FDA Approves Arcalyst - New Orphan Drug for Treatment of Cryopyrin-Associated Periodic Syndromes (CAPS)". Retrieved 2008-03-15.

62. "Rilonacept (IL-1 Trap) in the Prevention of Acute Gout Flares During Initiation of Urate-Lowering Therapy: Results of a Phase 2 Clinical Trial." H. R. Schumacher, Jr., John S. Sundy, Robert Terkeltaub, Howard R. Knapp, Scott J. Mellis, Neil Stahl, George D. Yancopoulos, Yuhwen Soo, Shirletta King-Davis, Steven P. Weinstein, Allen R. Radin on behalf of the 0619 Study Group. Arthritis & Rheumatism; Published Online: January 5, 2012 (DOI: 10.1002/art. 33412).

63. Schlesinger, N. et al. Canakinumab for acute gouty arthritis in patients with limited treatment options: results from two randomised, multicentre, active-controlled, double-blind trials and their initial extensions. Ann. Rheum. Dis. doi: 10.1136/annrheumdis-2011-200908.

64. Pageau, StevenC.(2009). "Denosumab". MAbs 1 (3): 210215. DOI: 10.4161/mabs.1.3.8592.

65. McClung, Michael R.; Lewiecki, E. Michael; Cohen, Stanley B.; Bolognese, Michael A.; Woodson, Grattan C.; Moffett, Alfred H.; Peacock, Munro; Miller, Paul D. et al. (2006). "Denosumab in Postmenopausal Women with Low Bone Mineral Density". New England Journal of Medicine 354 (8): 821.

66. Donald A. Bergman (September 16, 2009). "Denosumab: Fracture risk reduced in high-risk subset in FREEDOM". Endocrine Today.

67. Esteve M, Saro C, Gonzalez-Huix F, Suarez F, Forne M, Viver JM, Chronic hepatitis B reactivation following infliximab therapy in Crohn's disease patients : need for primary prophylaxis. Gut 2004; 53: 1363-65.

68. Wendling D, Auge B, Bettinger D, Lohse A, Le Huede G, Bresson-Hadni S, et al. Reactivation of a latent pre core mutant hepatitis B virus related chronic hepatitis during infliximab treatment for severe spondyloarthopathy. Ann Rheum Dis 2005; 64: 788-89.

69. Available at http://www.fda.gov/medwatch/safety/2006/safety06.htm#Rituxan.

70. Gebored P, Baldstrom A, Turesson C, Gulfe A, Petersson IF, Saxne T, et al. Tumor necrosis factor blockers do not increase overall tumor risk in patients with rheumatoid arthritis, but may be associated with an increased risk of lymphoma. Ann Rheum Dis 2005; 64: 699-703.

71. Baecklund E, Ekbom A, Sparen P, Feltelius N, Klareskog L. Disease activity and risk of lymphoma in patients with rheumatoid arthritis : nested case-control study. BMJ 1998; 317: 180-81.

72. Michaud K, Wolfe F, (FRI0090) Trends in medication use by 10,982 community rheumatoid arthritis patients in the United States from 1998 to 2005 : biologic usenow at 40%. Amsterdam : EULAR 2006.

73. Kurzeja M, Rudnicka L, Olszewska M. New interleukin-23 pathway inhibitors in dermatology: ustekinumab, briakinumab, and secukinumab. Am J Clin Dermatol. 2011; 12(2): 113-25.

74. Efficacy, safety, and tolerability of secukinumab in patients with rheumatoid arthritis taking methotrexate. Available at http://clinicaltrials.gov/clinicaltrials.gov/http identifier: NNCT0135994.

75. Efficacy at 24 weeks and safety, tolerability and long term efficacy up to 2 years of secukinumab (AIN457) in patients with active rheumatoid arthritis and an inadequate response to anti-TNFα agents. Available at:http://clinicaltrials.gov/http://clinical trials.gov/identifier: NCT01377012.

76. Gottlieb A, Menter A et al. Ustekinumab, a human interleukin 12/23 monoclonal antibody, for psoriatic arthritis: randomised, double-blind, placebo-controlled, crossover trial. Lancet. 2009 Feb 21; 373(9664): 633-40. Epub 2009 Feb 11.

77. Singh JA, Furst DE, Bharat A, Curtis JR, Kavanaugh AF, Kremer JM, et al. 2012 update of the 2008 American College of Rheumatology recommendations for the use of disease-modifying antirheumatic drugs and biologic agents in the treatment of rheumatoid arthritis. Arthritis Care Res (Hoboken) 2012; 64: 625-39.

SECTION 6

Rheumatoid Arthritis

CLINICAL FEATURES AND ORGAN INVOLVEMENT IN RA

1

Lata Bichile

Clinical features

Rheumatoid arthritis (RA) is a chronic, multisystemic, immuno-inflammatory disease characterized by articular and extra-articular manifestations alongwith systemic manifestations in the form of generalized malaise and fatigue. These systemic features are attributed to circulating cytokines. In Indian patients, the disease prevalence is approximately 0.5%. Female : male ratio is 3:1. The peak incidence of RA occurs in individuals aged 40-60 years. It exhibits a chronic fluctuating course which may result in progressive joint destruction, deformity and disability.

The disease may be diagnosed as early as within 3 months of clinical onset to 2 years when the disease is established. However, this time duration has relevance to the concept of therapeutic window.

- 0-3 months - very early RA (VERA),
- 3 months - 1 year early established RA,
- 1-2 years - late established RA (LERA),
- more than 2 years established stable RA.

To be more specific for the classification of RA, ARA criteria have been laid down which take into consideration the clinical, laboratory and radiological features. These are often useful for epidemiological and research purposes[16,17].

In most patients the clinical presentations are subacute in nature with insidious onset of fatigue, morning stiffness and arthralgias. More acute onset of disease is also described. The clinical features can be divided into articular and extra articular.

Articular

RA is principally a disease with a major impact on the synovial lined joints. Tendon and bursa involvements are frequent and often clinically dominant in early disease. RA usually affects metacarpophalangeal, proximal interphalangeal and metatarsophalangeal joints, as well as the wrists and knees. Articular and periarticular manifestations include joint swelling and tenderness to palpation, with morning stiffness and severe motion impairment in the involved joints.

Morning stiffness in and around the joints, lasting for at least 1 hr before maximal improvement is a typical sign of RA. Joints with effusions may be extremely painful due to high intra-articular pressures that lead to excessive stresses on the periarticular supporting structures. During the physical examination, it is important to assess the following signs and symptoms.

Stiffness: It manifests by limitation of motion. Joint stiffness is due to accumulation of synovial fluid. Stiffness is most effectively relieved with active exercise. It is an important manifestation of inflammatory process and forms an important parameter for assessing therapeutic response. Stiffness lasting for 45 min or more is characteristic of inflammatory arthritis.

Tenderness: Elicited by direct palpation. The enlarged synovial membrane, periarticular ligaments, and supporting structures are the major pain sensitive structures. Pain on motion is often a surrogate for tenderness in joints like shoulder and hip.

Joint swelling: Physical examination will show thickened synovium which is most evident in the small joints of the hands and feet. Joints appear spindle shaped. A boggy feeling is typical.

Deformities: Deformity of the joint results due to destruction of cartilage by inflamed synovium and erosions in the underlying bones. The small joints of hand and foot are most vulnerable. The characteristic deformities are Swan neck, Boutonniere, Z deformity, Hammer toes and Hallux valgus.

Extra articular features and antibodies

Extra-articular features are typical of "sero positive RA" i.e. RA associated with the presence of Rheumatoid factor in the serum. Some of these features are rheumatoid nodules, secondary Sjogren's syndrome, interstitial lung disease (ILD), scleritis & rheumatoid vasculitis. Approximately 70% of patients of RA are rheumatoid factor positive. Equal percentages of patients have antibodies directed against cyclic citrullinated peptide i.e. anti CCP antibodies. High titers of RF are a marker for patients at risk for aggressive, destructive joint disease and extra-articular complications of RA such as ILD and rheumatoid vasculitis. Anti CCP antibodies have prognostic value. They identify patients more likely to suffer erosive disease and treatment response to methotrexate or tumor necrosis factor inhibitor. A study from Netherland demonstrated convincingly that presence of anti CCP antibodies in early undifferentiated arthritis gives definite diagnosis of RA. Only 25% of patients, who are anti CCP negative developed RA by three years[1].

Interestingly some patients have RA which in every way appears to be typical disease, yet do not have either rheumatoid factor or anti CCP antibodies. These patients are said to have "sero negative RA".

Organ involvement in RA

RA is a chronic immuno-inflammatory disorder affecting the synovial joints. The inflammation may extend beyond the joints, tendons and synovial structures and involves various organs. Degree and duration of inflammation and presence of rheumatoid factors are important deciding factors for various organ involvement. Longstanding and untreated RA patients are prone to develop secondary amyloidosis and related vital organ involvement. Organ involvement occurs in 40% of patients during the disease course. Estimates of incidence of specific organ manifestations are highly variable and reported as case series. Subcutaneous nodules, pulmonary fibrosis, pleuritis, pericarditis, vasculitis, amyloidosis, cardiac involvement and Felty's syndrome represents a specific organ involvement. These manifestations are associated with decreased lifespan compared to the cohort without these manifestations. HLA-DR3 and DRH genes are associated with a greater frequency of extraarticular diseases.

Rheumatoid nodules

Rheumatoid nodules are usually not present within first few months of the onset of synovitis. These occur in less than 10% of patients during first year of disease. Their presentation or absence does not reliably exclude the diagnosis of RA. Rheumatoid nodules occur in 20% of RA patients with rheumatoid factors. They are rare in seronegative patients. Nodules are more commonly present with severe disease activity. Nodules develop in pressure areas such as elbows, finger joints, ischeal and sacral prominences, occipital area and forehead. Intra articular nodules are also been reported. They are firm and adherent to the underlying periosteum. Rheumatoid nodules represent necrotizing vasculitis of small vessels with fibrinoid necrosis with fibroblast deposition[2].

Presence of rheumatoid nodules is a bad prognostic marker of RA and warrants judicious use of steroids and DMARDs. Patients on methotrexate do develop multiple nodules while ameliorating the underlying illness.

Hematological abnormalities

Anemia in RA is usually normocytic and normochromic. Moderate anaemia, thrombocytopenia, eosinophilia and raised ESR are common abnormalities in RA patients.

Anemia

Iron deficiency is very common in Indian patients, more so in females. Also it would be most appropriate to evaluate menstrual history in patients with severe anemia. In patients with polymenorrhoea and menorrhaegia uterine fibroids or other treatable causes should be ruled out. Nutritional deficiency, chronic infections, and infestations need consideration while evaluating anemia in these patients.

Anemia associate with RA has multifactorial etiology[3]

1. Inhibition of haemoglobin synthesis.
2. Impaired iron utilization.
3. Direct effect of proinflammatory cytokines such as TNF-α IFN-β and interleukin-6 on precursor cells of bone marrow.
4. Reduced erythropoietin synthesis.
5. Ineffective erythropoiesis leading to RBC destruction, and reduced survival.
6. Increased phagocytosis of red blood cells in the spleen and synovium.
7. Effect of various medications used to treat RA e.g. NSAIDs and corticosteroids causing blood loss due to erosive gastritis, gastric ulcer, duodenal ulcer, methotrexate causing bone marrow depression, etc.

The degree of anemia in RA usually correlates with the degree and duration of disease activity/synovial inflammation. Serum ferritin levels of less than 50mg/ml and more than 100mg/ml is predictive of iron deficiency and anemia of chronic inflammation respectively.

Therapy: As the rheumatoid inflammation improves with DMARDs the anemia of inflammatory origin can improve. Erythropoietin supplementation improves red blood cell counts. All the other factors causing anaemia should receive routine treatment. Thrombocytosis, a marker of inflammatory process is a frequent finding in RA. It correlates with the number of joints involved with active synovitis[4]. Thrombocytopenia is rare in RA except in situations such as drug toxicity and Felty's syndrome. Eosinophilia is commonly associated with rheumatoid nodules and patients receiving gold therapy. Eosinophilia has no bearing on the outcome of RA.

Lymphadenopathy: Lymphadenopathy is frequent in patients with active RA. Lymph nodes are firm, mobile and nontender. Lymphadenopathy resolves with therapy for RA and has no bearing on outcome of RA.

Felty's syndrome: Felty's syndrome consists of RA with splenomegaly and leucopenia. The syndrome occurs in patients with long standing, seropositive nodular, deforming RA. Lower extremity ulcerations, hyperpigmentation and presence of ANA are often observed in these patients. With the fall of WBC count < 100/cu mm bacterial infections occur resulting in death in these patients. Neutropenia in Felty's syndrome is due to inhibition of granulopoiesis by cytokines such as IL-1, TNF-α, and IFN-β. Treatment of rheumatoid disease improves neutropenia. Splenectomy is indicated in patients with recurrent serious infections. Splenectomy rarely helps underlying disease. In selected cases granulocyte colony stimulating factor (G-CSF) administration is recommended[5].

Pulmonary Involvement

Six different pulmonary syndromes are known in patients with RA.

These are:

- Pleurisy with or without effusions
- Diffuse interstitial pleuritis
- Necrobiotic nodule with or without pneumoconiosis (Caplan's syndrome)
- Obstructive airway disease including bronchiolitis obliterans
- Pulmonary hypertension secondary to small pulmonary arteritis
- Apical fibrocavitatory disease[6]

Lung involvement is common in RA with approximately 33% of patients having pulmonary manifestations. Common symptoms are dry cough, breathlessness and chest pain. However in majority the clinical features may be subtle. Pleura, interstitium and pulmonary alveoli are affected in RA. Usually males predominate over females and often they have presence of rheumatoid factor.

Pulmonary involvement in RA increases the morbidity and mortality in RA.

Rheumatoid pleural effusion is often unilateral and rarely bilateral. It precedes the clinical manifestations of synovitis. High degree of clinical suspicion is required for the clinical diagnosis[7]. The typical pleural fluid is exudative with lymphocytic predominance, high protein content of more than 4 gm/dl and typically low in glucose (less than 25 mg/dl) with high LDH. At times it is confused with empyema but rheumatoid fluid does not have predominance of neutrophil. RF may be present in pleural fluid. Pleural effusion usually responds to treatment of RA in combination with steroids.

Pulmonary parenchymal lesions

Pulmonary nodules

Generally asymptomatic, these are observed in patients with active synovitis, with widespread rheumatoid nodules elsewhere in the body. Pulmonary nodules tend to be peripheral in location. They vary in size from less than 1cm to 6-8cm. On cavitation they may cause pleural effusion and bronchopleural fistulae. Single peripheral nodule may need a biopsy to differentiate it from malignancy, secondaries, TB and fungal granuloma. Treatment of RA resolves pulmonary nodules.

Caplan's syndrome

Several pulmonary nodules with cavitation and pneumoconiosis constitutes Caplan's syndrome.

Individuals with extensive exposure to coal dust are prone to develop these lesions. Similar syndrome is reported in individuals exposed to asbestos and silica.

Diffuse interstitial pulmonary fibrosis

The prevalence of ILD in RA varies with diagnostic modalities used. On conventional chest radiography it is evident in 2-6% whereas on HRCT it is seen in 10-47% of RA patients. Joint disease precedes the lung involvement in most cases[8,9]. Clinical presentation and investigations of ILD are similar to idiopathic pulmonary fibrosis. Smoking contributes to the risk of developing ILD in RA. The pathogenesis of rheumatoid pulmonary inflammation is multifactorial and probably related to genetic and

HLA constitution, local production of inflammatory mediators, tobacco chewing, smoking and alpha-1 Antitrypsin phenotype. PFT, HRCT chest, BAL cytology and lung biopsy are useful to assess severity, extent and stage of lung damage.

Cytology or BAL aspirate identifies patients with active lung inflammation. Ground glass appearance on HRCT suggests active lung inflammation. These patients should be treated with aggressive immunosuppressive therapy.

Bronchiolitis obliterans with organizing pneumonia

BOOP is inflammatory in nature. Patients present with fever and constitutional symptoms. They respond dramatically to corticosteroids. High degree of clinical suspicion and judicious use of HRCT, BAL and corticosteroids is rewarding for good clinical outcome. On the other hand constrictive or obliterative bronchiolitis has poor prognosis with poor response to treatment. This is a severe and rapidly progressive entity. Pulmonary biopsy is indicated to demonstrate peribronchial fibrosis and extrinsic narrowing of the bronchiolar lumen and fibrosis.

Genetic susceptibility to airway infections and airway obstructions is also prevalent in RA patients. Upper airway obstruction may occur due to acute inflammation of the cricoarytenoid joint[10]. Patients complain of recurrent upper respiratory infections, foreign body sensation in upper throat, hoarseness of voice and difficulty in inspiration. Laryngoscopy and CT scanning have good diagnostic sensitivity. Therapy with local corticosteroid injections and anti-inflammatory medications are indicated. Surgical intervention is required in fixed upper airway obstruction. Drugs such as methotrexate, D-penicillamine and gold compound do cause pulmonary complications[12]. Secondary pulmonary hypertension occurs in those with pulmonary fibrosis. Shriveling lung syndrome is described in RA.

Cardiovascular features

Coronary vasculitis

Cardiovascular disease is a leading cause of death in RA. In a recent meta analysis risk of cardiovascular death was 50% higher in patients with RA when compared to general populations. Methotrexate and TNF-α inhibitors decrease cardiovascular morbidity and mortality in RA[11]. However there are no prospective studies or randomized controlled clinical trials to document these. In Indian context risk factors such as hypertension, dyslipidemia, body mass index, diabetes mellitus and family history predispose to and enhance cardiovascular morbidity in RA patients. RA patients have an increased risk of subclinical CVS disease and also have a high prevalence of carotid disease, peripheral arterial disease and acute MI.CVD is recognized as leading cause of death in RA. RA must be recognized as a marker of increased risk for MI.

The coronary involvement is either atherosclerotic or secondary to diffuse vasculitis. Coronary vasculitis is reported in 20% of elderly RA patients at postmortem. Differentiation of the two entities is of utmost clinical importance as vasculitis is life-threatening and should be treated promptly with immunosuppressive. Coronary vasculitis is occasionally seen in children, young adults and is attributed to inflammation of coronary vessels seen in early RA disease. Coronary artery inflammation progresses to premature atherosclerotic coronary artery disease as an accompaniment of chronic RA disease progression. Control of inflammation alters the important mechanisms of atherothrombogenesis to a favourable direction which is possible by early institution of DMARDs and biologics.

RA pericarditis

Incidence of pericarditis in RA patients is shown to be 30-50% in postmortem studies. High incidence is also detected on echocardiography. However the clinical incidence is much lower at < 10% in severe RA. It is more frequent in males with rheumatoid nodules and destructive RA. Prognosis in pericarditis is worse in 1st year after diagnosis, advanced age and poor cardiac status of the individual patient.

Constrictive pericarditis and rapidly progressive effusive pericarditis are associated with increased morbidity and high mortality. Evaluation strategy includes screening the patients for ANA, IgM rheumatoid factor and anti citrullinated peptide. NSAIDs, corticosteroids, immunosuppressive drugs are appropriate in the majority of patients and in severe cases pericardiectomy is warranted. Early diagnosis and effective treatment improve the outcome.

RA cardiomyopathy

This is a secondary type of cardiomyopathy. It can be focal, diffuse necrotizing or granulomatous myocarditis. This histological evidence of cardiomyopathy is found in 3-30% of RA patients at postmortem studies. Certain drugs such as corticosteroids and antimalarial are also incriminated as aetiologic factors. 2D-ECHO and cardiovascular MRI help to evaluate cardiac morphology and cine cardiac MR is performed to assess myocardial perfusion thus differentiate ischemic from non-ischemic cardiomyopathy. "Spa tine" pattern on echo-cardiography suggests amyloidosis of heart. Amyloid heart gives rise to high frequency of heart failure which increases the mortality[13].

Arrhythmias

It is an important cause of mortality in RA and may be secondary to ischemia, rheumatoid nodule, amyloidosis or congestive heart failure. Ventricular tachyarrhythmias are due to increased sympathetic activity leading to QT-dispersions.

Valve disease

Regurgitant lesions such as mitral regurgitation and aortic regurgitation are common in RA patients.

Congestive heart failure and IHD

CHF and IHD contribute to mortality among RA patients. RA patients develop CHF twice more frequently than persons without RA. Cardiac mortality is associated with markers of systemic inflammation in RA i.e. increased ESR, rheumatoid vasculitis, rheumatoid lung disease[14]. Most importantly RA patients do not report symptoms of angina and hence are more likely to experience unrecognized myocardial infarction and sudden death. Patients with RF positive status have short disease duration and predispose to myocardial infarction. This is also being reported in patients prior to the diagnosis of RA and such patients have hyperlipidemia independent of inflammatory parameters. In Indian patients smoking, increased body mass index (BMI), diabetes, IHD and hypertension might influence the risk of cardiovascular disease. Also RA severity and atherosclerosis go hand in hand and depicted on carotid plaques and PVD. Besides CHF, LVSD has been reported three times common than the general population such patient can be screened with an abnormal electrocardiogram. Corticosteroids, TNF blockers, methotrexate, NSAIDs and Cox 2 inhibitors are also incriminated in the causation of cardiac dysfunction.

Since these agents have found an important place in the therapeutic armamentarium of RA, it would be most appropriate to have more knowledge on possible deleterious effects on the function of the heart due to these drugs. The prolonged use of over-the-counter available Cox-II inhibitors and corticosteroids needs vigilance in Indian patients.

Ocular involvement

Keratoconjunctivitis sicca is the most common ocular involvement in RA. Common symptoms are dry eyes, burning sensation, foreign body sensation, mucoid discharge, reduced tear secretion, and episcleritis. Episcleritis, an acute painful red eye condition either nodular or diffuse correlates with activity of RA. Scleritis is less common and represents rheumatoid vasculitis; commonly seen in long-standing and active joint inflammation. Untreated scleritis may progress to scleromalacia.

Avascular nature of sclera contributes to resistance to treatment and poor outcome. Other unusual ocular findings in RA are uveitis and ulcerative keratitis known as corneal melt which is usually central. Ulcerative keratitis also occurs at the periphery of cornea and is secondary to scleral inflammation. It is usually bilateral. The scleral and corneal conditions are associated with poor vision outcome and mortality as well. Systemic and local corticosteroids and systemic cytotoxic agents are administered for salvaging the vision.

Neurological involvement

Neurological involvement maybe in the form of entrapment neuropathies, atlanto axial subluxation, basilar invagination, subaxial subluxation, extradural nodule compressing the nerve roots and cauda equine syndrome.

Entrapment neuropathies occur as a result of proliferative synovitis entrapping the nerves at the various joint areas where the nerves are in close proximity. Carpal tunnel syndrome is one of the common manifestations. Atlanto axial subluxation is an important rheumatological emergency in RA. Pannus formation, erosion and ligament instability lead to cord compression. Hyperreflexia, weakness bilateral Babinski's sign alarms this complication. Surgical intervention is required in many patients. Same mechanisms are applicable at the other vertebral sites for cord compression or cauda equina syndrome.

Rheumatoid vasculitis

Rheumatoid vasculitis prevalence varies from 1-5% but postmortem studies have shown higher percentage upto 25%. HLADRB1 showed epitope genotype i.e. strongly associated with extra articular disease manifestations and vasculitis being one of them. The presence of vasculitis in RA patients is associated with poor survival[15]. Small vessel vasculitis is identified as pathogenetic mechanism in many of the manifestations of RA.

There is a strong association with the RF positive status. Rheumatoid nodule develops from vasculitis of small and medium sized arteries. Acute arterial lesions are immunemediated. Small vessel vasculitis of RA is similar to polyarteritis with the exception of renal involvement, small vasculitis affects skin, nail beds and manifests as nail fold infarcts and proximal skin ulceration. Distal sensory neuropathy is another manifestation of small vessel vasculitis.

Systemic vasculitis in RA is uncommon and occurs in RA patients with long standing disease. It causes poor prognosis, p-ANCA is a good marker for the presence of vasculitis. RA synovitis may not be active when systemic vasculitis manifestations are present. Systemic vasculitis may involve coronary vessels.

Leg ulcer is initiated with small vessel vasculitis. Ulcer expansion and its chronicity are influenced by venous insufficiency, arterial insufficiency, dependent edema, cryoglobulins, low C3, trauma, infection and use of steroid. Corticosteroids are blamed in the pathogenesis of vasculitis. Vasculitis blossoms when steroids are tapered. Steroid therapy leads to atherosclerosis and an obliterative vasculopathy. High dose steroids are indicated in the initial stages, cytotoxic agents and anti cytokines can be administered. Aspirin, antiplatelet agents and pentoxiphlyllin should also be considered.

Organ involvement due to amyloidosis in RA

Secondary amyloidosis is well recognized as a severe complication in the late stages of rheumatoid arthritis. The usual manifestations are nephritic syndrome, abdominal pain and ascites. Acute abdominal pain may be the first symptom of amyloid vasculitis presenting as hemorrhagic ascites. The amyloid vasculitis is of necrotizing type.

Amyloid vasculitis occurs in under treated and rheumatoid factor positive patients. Amyloid vasculitis affects skin and nerves. Bowel involvement is associated with a significantly high mortality partly due to its presentation and late recognition. AA amyloid is a rare and long-term concomitant of untreated RA. Anti-inflammatory therapy along with Methotrexate/Cyclophosphamide results in clinical and pathological remission. Secondary amyloidosis is one of the most relentless of the extraarticular features of RA. Serum amyloid levels (S.A.A1.3 Genotype) are a risk factor for secondary amyloidosis. In Japanese RA patients institution of prednisolone and cyclophosphamide ameliorated laboratory markers such as ESR, CRP, and serum albumin, RF factor, serum creatinine and clinical rheumatoid activity. These laboratory markers serve as surrogate markers of therapeutic efficacy.

Reactive AA amyloid develops in long standing RA patients and causes an unfavorable prognosis due to dysfunction of the vital organs. Amyloid fibrils are deposited in subepithelial and subendothelial space of the glomerular capillary wells. Amyloid fibrils filtered from the circulatory precursors and deposited in glomerular vessels result in proteinuria and renal impairment. Cyclophosphamide has a potential to suppress disease activity and degradation of amyloid fibrils in glomeruli.

Early diagnosis of amyloidosis is possible through abdominal fat aspiration biopsy. Sites of biopsy for amyloid deposits are kidney, stomach mucosa and salivary glands.

Modern therapies with biologics and methotrexate right at the initial stages of disease and treat to target approach should lead to improvement in disease control and thereby amyloidosis.

Summary

Rheumatoid Arthritis is a chronic systemic immunoinflammatory disease with a tendency to frequent remissions and relapse. T, B cells and various cytokines are involved in the disease pathogenesis and sustaining disease activity that is detrimental to the joints, lungs, heart, small and medium sized vessels and leads to chronic morbidity and early mortality as compared to the healthy controls. There is a link between chronicity of disease, seropositivity and various organ involvement. But this can be brought down by the targeted treatment approach. Current recommendation is early accurate diagnosis, suppress inflammation, sustain low disease activity, or induce early remission within 3 to 6 months thus preventing erosions. Thus the focus is changed from prevention of erosions to suppression of inflammation.

This is possible with the institution of DMARDs in combinations or methotrexate and biologic agent given together. If inflammation is controlled at its onset and suppressed throughout the course of illness, morbidity due to premature atherosclerosis at coronary, peripheral and CNS will be brought down.

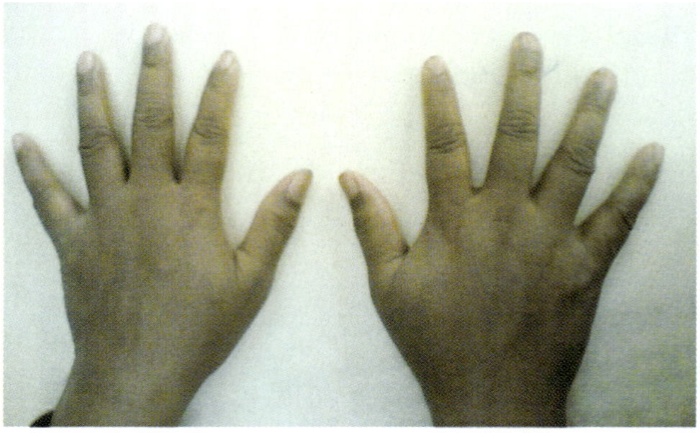

Fig. 1 : Early RA

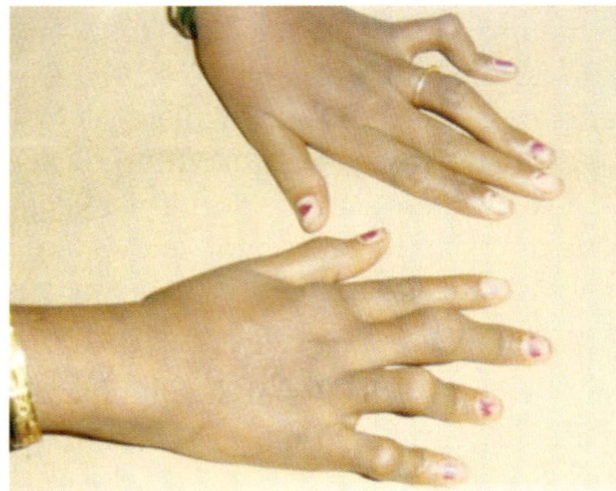

Fig. 2 : Established RA with deformities

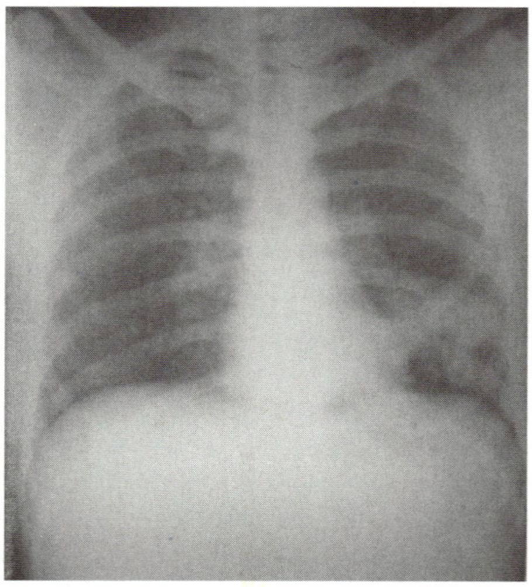

Fig. 3 : RA with lung nodules

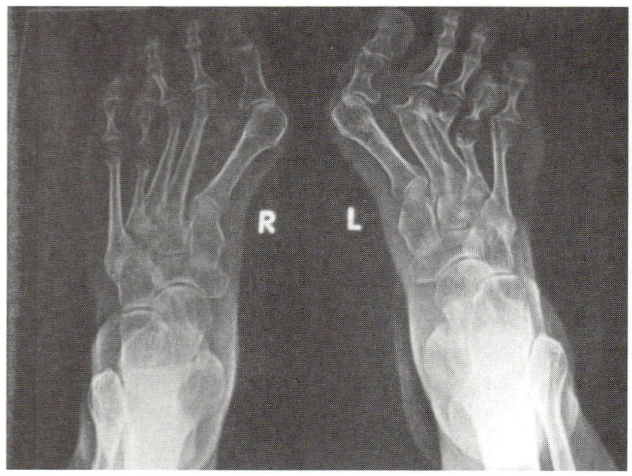

Fig. 4 : RA feet with deformities

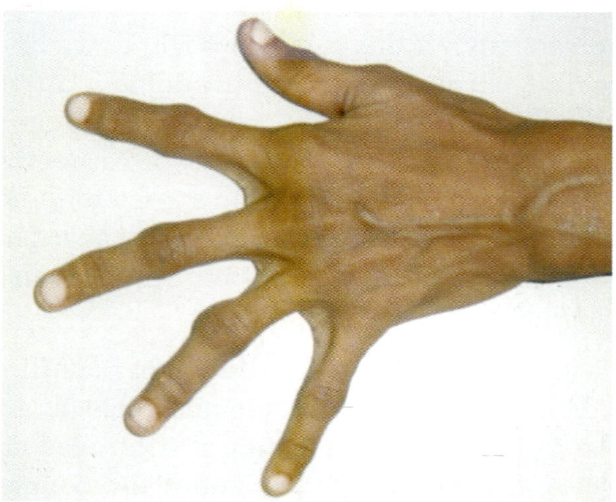

Fig. 5 : RA with rheumatoid nodules

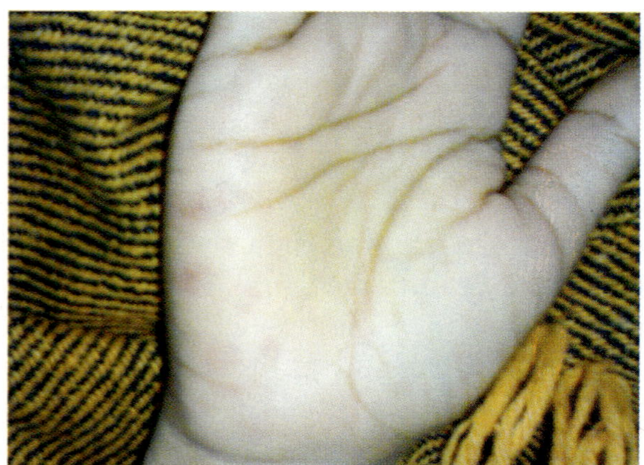

Fig. 6 : RA with vasculitic hands

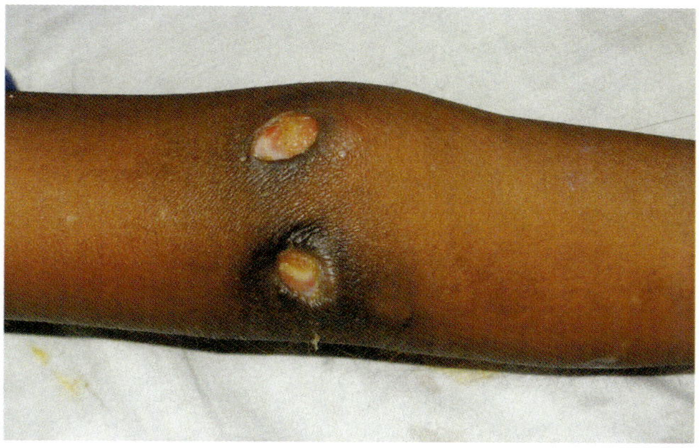

Fig. 7 : RA with vasculitic ulcers over the elbow

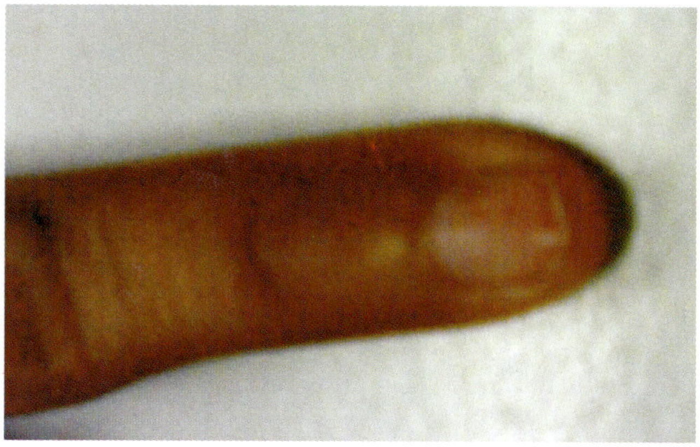

Fig. 8 : RA with splinter haemorrhages

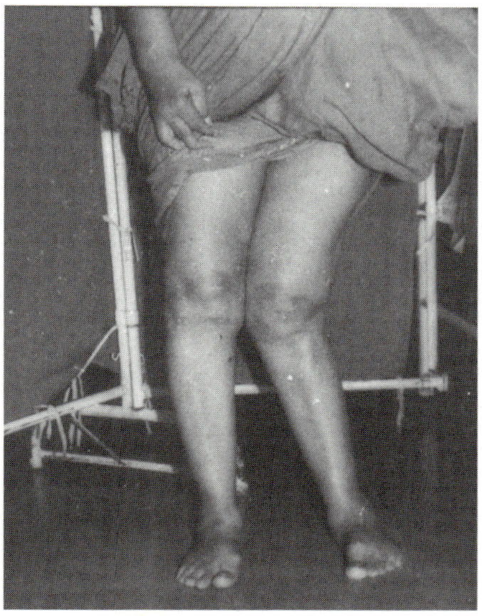

Fig. 9 : RA with knee involvement - genu valgum

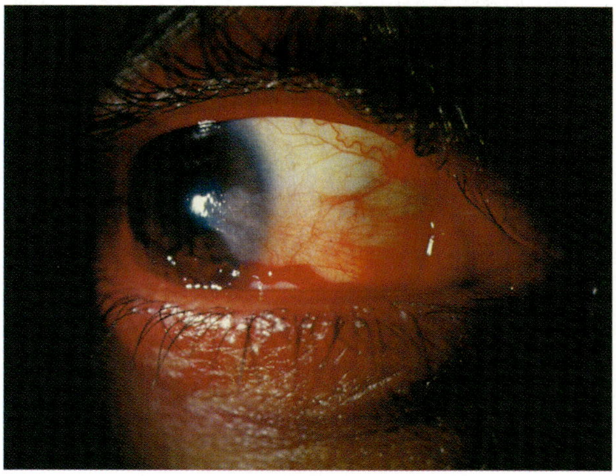

Fig. 10 : RA with eye involvement - dry eye

Further Reading :

1. van Gaalen FA, Linn- Rasha Vintooji WJ. Autoantibodies to cyclic citrullinated peptide predict progression to RA in patients with undifferentiated polyarthritis a prospective cohort study. Arthritis Rhe 2007; 50: 709-15.
2. Voskuyl AE; Zwinderman AH; Westedt ML; Vandenbroucke JP; Breedveld FC; Hazes JM. Factors associated with the development of vasculitis in rheumatoid arthritis: results of a case-control study. Ann Rheum Dis Mar 1996; 55: 190-92.
3. Glossop JR; Dawes PT; Hassell AB; Mattey DL. Anemia in rheumatoid arthritis: association with polymorphism in the tumor necrosis factor receptor I and II genes. J Rheum Sep 2005; 32: 1673-78.
4. Kacena, Melissa A; Horowitz, Mark C. The role of megakaryocytes in skeletal homeostasis and rheumatoid arthritis. Current Opinion in Rheum 2006; 18: 405-10.
5. Balint GP, Balint PV. Felty's syndrome. Best Pract Res Clin Rheum Oct 2004; 18: 631-45.
6. Walkar WC, Wright V, Pulmonary lesions and rheumatoid arthritis. Medicine (Baltimore) 1968; 472-501.
7. Balbir-Gurman A, Yigla M, Nahir AM, Braun-Moscovici Y. Rheumatoid pleural effusion. Semin Arth Rheum Jun 2006; 35: 368-78.
8. Castagnaro A, Chetta A, Marangio E, Zompatori M, Olivieri D. The lung in immune-mediated disorder: rheumatoid arthritis. Curr Drug Targets Inflamm Allergy Dec 2004; 3: 449-54.
9. Horton MR. Rheumatoid arthritis associated interstitial lung disease. Crit Rev Comput Tomogr. 2004; 45: 429-40.
10. Langford, CA, Waes, CV. Upper airway obstruction in the rheumatic diseases: life-threatening complications of autoimmune disease. Rheum Dis Clin N Amer 1997; 23: 346-63.
11. Gon zale A, Kremers HM, Crowson CS. Do cardiovascular risk factors confer the same risk for cardiovascular outcomes in rheumatoid arthritis as in non rheumatoid arthritis patients? Ann Rheum Dis 2006; 67: 64-69.
12. Adamson IY.Drug-induced pulmonary fibrosis. Environ Health Perspect. Apr 1984; 55: 25-36.
13. Solomon DH, Karlson EW, Rimm EB, et al. Cardiovascular morbidity and mortality in women diagnosed with rheumatoid arthritis Circulation 2003; 107: 1303-07.
14. Voskuyl. The heart and cardiovascular manifestations in rheumatoid arthritis Rheum 2006; 45: iv4-iv7.
15. Snowden N, Kay RA. Immunology of systemic rheumatoid disease. Br Med Bull. 1995; 51: 437-48.
16. Arnett FC, Edworthy SM, Bloch DA, et al. The American Rheumatism Association 1987 revised criteria for the classification of rheumatoid arthritis. Arthritis Rheum 1988; 31: 315.
17. Aletaha D, Neogi T, Silman AJ, et al. 2010 Rheumatoid arthritis classification criteria: an American College of Rheumatology/European League Against Rheumatism collaborative initiative. Arthritis Rheum 2010; 62: 2569.

EARLY ARTHRITIS - RATIONALE FOR EARLY DIAGNOSIS AND TREATMENT

2

Banwari Sharma

Background

Rheumatoid arthritis (RA) is the most common and most serious inflammatory arthritis. Untreated it results in joint destruction, functional impairment and increased mortality. The outcome of the disease, however, has improved considerably in recent years with early diagnosis, the availability of effective therapies and the recognition that early intensive treatment strategies result in better outcomes.

In practice, early inflammatory arthritis patients should be seen early and treated at the earliest opportunity. Clinicians must differentiate between the early features of RA (an inflammatory arthritis that has the potential to become progressive and destructive) from those that will remit spontaneously, and from diseases that may present with similar clinical features, e.g. systemic lupus erythematosus (SLE) or the spondarthropathies. Decisions with regard to optimal therapeutic strategies are also required. The goal of treatment is early suppression of inflammation and ideally establishing remission in order to prevent joint damage, disability, and the long-term complications of the disease.

Spontaneous remission in true RA is rare (11%-14%). The joint damage and loss of function in RA occur early in the disease process. Two-thirds of recent onset RA patients develop bony erosions within the first 3 years[1]. Furthermore, within 3 months of disease onset 25% of patients have erosions evident on X-ray[2] and newer imaging techniques such as magnetic resonance imaging (MRI) and ultrasound (US) have confirmed evidence of damage within weeks of onset of symptoms[2-4].

Studies have also shown that a delay of even few months in treatment is associated with poorer outcomes[5]. Looking to the available evidences, it is clear that joint damage occurs early, and delay in starting treatment results in poor outcome.

Why it is important to diagnose inflammatory arthritis early?

Early inflammatory arthritis is a critical period of the disease, during which therapy may have a durable effect and change the natural course of the condition. This period is known as 'window-of-opportunity'.

Most recent studies show that this period may be limited to weeks or months. There is also increasing evidence that very early RA, i.e. within the first 12 weeks, may be an immunopathologically distinct phase compared to later disease[6].

Thus, identifying patients at high risk of developing RA at a very early stage should be an essential objective of the clinical care of patients with undifferentiated inflammatory arthritis. At an early stage, not all patients with inflammatory polyarthritis can be reliably diagnosed. Early polyarthritis may develop into RA or another definite arthropathy, may resolve spontaneously, or may remain undifferentiated, means the outcome of early arthritis patients is highly variable. Approximately only one-third of the patients with a recent onset undifferentiated arthritis (UA) progress to RA. The severity of the progression of joint destruction in RA is highly variable as well, as only a minority will become severely destroyed. Current treatment guidelines for the management of early arthritis recommend that patients at risk of developing persistent and/or erosive arthritis should be started with DMARDs as early as possible, even if they do not fulfill established classification criteria for inflammatory rheumatological diseases.

What is early RA?

The term, "early RA", is not defined precisely in the literature and has become a moving target as more tests for early diagnosis become available. For the practical purposes of this discussion, an early diagnosis of RA is considered to be within three to six months of the onset of joint symptoms.

Inflammatory arthritis may even be identified at an earlier stage. It is now recognised that the steps towards the development of RA occur along a disease continuum from patients who may have a genetic predisposition for the disease, to those presenting with non-specific symptoms to early undifferentiated arthritis and RA. Rheumatoid factor (RF) and antibodies to citrullinated peptide (ACPA) have been detected in patients with RA years before the onset of symptoms - evidence that the disease process begins in the preclinical phase i.e. before the onset of symptoms. Elevated levels of high sensitivity C-reactive protein (CRP) have also been shown before the onset of clinical disease[7]. Complementary to the serologic changes, imaging with ultrasound and MRI, and arthroscopy detect synovitis in clinically normal joints of patients with early RA[8].

New classification criteria for RA

RA patients demonstrate considerable variability in both their presenting symptoms and initial laboratory results. For new-onset RA, the sensitivity and specificity of the ACR criteria (1987) are notably low. The need for better tools for the identification of patients with early RA, from among individuals presenting with any type of inflammatory synovitis, led to the development in 2010 of the American College of Rheumatology (ACR)/European League Against Rheumatism (EULAR) classification criteria for RA, which have greater sensitivity for the diagnosis of RA in early disease than the widely used 1987 ACR revised criteria for the classification of RA[9-11]. The 2010 criteria also incorporate testing for anti-citrullinated peptide antibodies (ACPA) and measurement of acute phase reactants, both of which were not included in the older criteria. Additionally, individual criteria can be counted cumulatively over time, rather than having to coexist at a single time point. Neither the 1987 nor the 2010 sets of classification criteria were designed to serve as diagnostic criteria.

When to suspect RA in a patient of recent onset inflammatory arthritis?

Clues to the presence of early RA can be found in the history, physical examination, laboratory tests, and imaging studies.

History

- Polyarticular involvement at presentation favors development of RA. Established RA tends to involve multiple joints over a period of months or years. However, at the time of first presentation, many patients eventually diagnosed with RA may have only a small number of joints affected. In contrast to the arthritis of rheumatoid patients, the peripheral arthritis of most "seronegative" spondyloarthropathies, such as reactive arthritis, arthritis of inflammatory bowel disease, or ankylosing spondylitis, is usually oligoarticular or monoarticular. Nevertheless, as mentioned above, RA may start with involvement of only few joints, and the 2010 ACR/EULAR classification criteria account for this fact by different weighting for high versus low joint counts.
- Morning stiffness is a typical symptom of inflammatory disease involving the finger joints. Stiffness after awakening that lasts at least 60 minutes is mostly seen in rheumatoid arthritis. By comparison, osteoarthritis patients typically have morning stiffness that does not exceed 30 minutes in duration.
- RA is a chronic disease. In contrast, a variety of conditions can lead to transient joint inflammation that resolves of its own accord within days to weeks. Forms of arthritis induced by viral infections (e.g., parvovirus B19, Alphavirus) are particularly prominent among such conditions. Definite classification as having RA using the 1987 criteria had required that patients have arthritis for at least six weeks, the new criteria allow for classification as RA in some patients with arthritis present less than six weeks, if a number of other criteria are met.

Physical examination

- Synovitis of the metacarpophalangeal (MCP) and metatarsophalangeal (MTP) joints is often recognized by the finding of pain with compression. In a report of 524 consecutive patients with early arthritis, a positive compression test of the bilateral MTP joints in patients with early arthritis was a strong predictor of progression to erosive RA, second only to seropositivity for ACPA[12].
- Symmetrical arthritis is typical of RA. However, in early disease, symmetry is often not apparent. Asymmetric and even monoarticular presentations are often observed. Patients with monoarthritis of a large joint, however, will not meet the 2010 ACR/EULAR classification criteria for RA, even in the presence of other characteristics.
- Involvement of both upper and lower extremities is a characteristic of RA. The finding of both upper and lower extremity joint involvement is more predictive of RA than the involvement of either the upper or lower extremity alone.

Laboratory testing

Assays for autoantibodies and acute phase reactants are helpful in the early diagnosis of RA. The most reliable early predictors of both chronic and erosive disease are the presence of high titer RF and ACPA. The levels of these biomarkers are also important: the higher the level, the greater the likelihood of disease chronicity, bone erosions, and joint destruction.

Rheumatoid factor - The classic rheumatoid factor (RF) is an IgM autoantibody directed against the Fc region of IgG. RF is not pathognomonic for RA and frequently occurs in other disorders, such as connective tissue diseases and chronic infections. High RF titers (e.g. > 50 IU/ml) discriminate well between RA and other types of inflammatory arthritis.

Anti-citrullinated protein antibodies - Anti-citrullinated protein antibodies (ACPA) are more specific but slightly less sensitive for RA than are positive RF assays. Both ACPA and RF are present in approximately one-half of patients with early arthritis who are diagnosed subsequently as having RA.

Acute phase reactants - acute phase reactants such as the erythrocyte sedimentation rate and serum C-reactive protein are typically elevated in early arthritis patients who are eventually determined to have RA. However, there is considerable variability from patient to patient. This variability limits the predictive value of these test results in early diagnosis. On the other hand, clinical disease activity and joint damage correlate well with acute phase reactant levels.

Imaging studies

Plain radiographs - the presence of erosions is strongly indicative of but not specific for RA .They can also occur in gout, psoriatic arthritis, and a number of other inflammatory joint conditions.

In early RA, erosions seldom occur within the first three months of the onset of clinical disease. Thus, the institution of therapy should not be postponed if erosions are absent but RA remains the likely diagnosis, given the importance of early intervention to prevent joint damage.

Other imaging modalities, such as magnetic resonance imaging (MRI) and ultrasonography, have been evaluated for their utility in patients with RA. Both allow detection of synovitis, synovial effusions and tendovaginitis even if clinically not apparent, in addition to joint damage. However, more outcomes data are needed before they can be broadly recommended for routine clinical use.

Early rheumatology referrals

Two sets of European expert panel recommendations regarding early referral of suspected RA have been developed. One suggests clinical suspicion of RA and rapid referral to a rheumatologist, if any of three characteristics are present; these are: the presence of three or more swollen joints, pain with compression of the metatarsal phalangeal joints or metacarpophalangeal joints, and morning stiffness of more than 30 minutes[13]. The other proposal generally suggests referral of patients with one or more swollen joints within six weeks of symptom onset, following careful history, physical, and laboratory examinations to exclude other diseases[14]. Such early referral to a specialist is important to ensure adequate and timely institution of disease modifying antirheumatic drugs, which should be started as soon as the diagnosis of RA is made.

SUMMARY

Early diagnosis of rheumatoid arthritis (RA) is critical to the prevention of long-term joint damage because of the availability of effective therapy and the knowledge that joint destruction occurs within two years of the onset of arthritis in the vast majority of patients and accumulates over time. For practical purposes, an early diagnosis of RA should be considered to be one rendered within three to six months of the onset of joint symptoms.

The American College of Rheumatology (ACR) revised criteria for the classification of RA, issued in 1987, and was developed in studies of patients with established RA. New criteria, published in 2010 by ACR and EULAR, have been developed, in large part, to improve the recognition of patients with early RA for inclusion in clinical studies, but can aid in diagnosis. Classic early RA is associated with a polyarthritis involving the small joints of the hands and feet, at least 60 minutes of morning stiffness, and symptoms that have persisted for at least six weeks.

RA is typically associated with symmetric joint involvement and involvement of both upper and lower extremities. However, exceptions to these rules are common in the early stages of RA.

Assays for rheumatoid factor (RF) and (especially if RF is < 50 IU/mL) for antibodies to citrullinated peptides, and measurement of acute phase reactants should be performed in patients suspected of RA on the basis of historical features and physical examination findings.

Plain radiographs have a low sensitivity for joint erosions in early RA. Magnetic resonance imaging (MRI) studies are superior at detecting erosions early, but the precise role of MRI in the diagnosis and longitudinal management of patients with RA remains to be defined.

Box 1. Features suggestive of inflammatory arthritis

- History of joint swelling, early morning stiffness lasting > 30 minutes, systemic symptoms such as tiredness, malaise, low-grade fever or weight loss, improvement of symptoms with anti-inflammatory medication
- Objective evidence of joint swelling and tenderness on examination
- Raised ESR or CRP, normocytic normochromic anaemia, thrombocytosis, low albumin, raised alkaline phosphatase

Box 2. Conditions that can present as polyarthritis and mimic RA

- Postviral arthritis - e.g. parvovirus, mumps, rubella, hepatitis B and C, alphavirus
- Seronegative spondyloarthritis - e.g. psoriatic arthritis, inflammatory bowel disease
- Connective tissue diseases - e.g. systemic lupus erythematosus, scleroderma, vasculitis, myositis
- Acute Rheumatic fever
- Acute sarcoidosis
- Osteoarthritis
- Crystal arthritis - e.g. polyarticular gout, pseudogout
- Miscellaneous - e.g. Thyroid disease, infective endocarditis, malignant disease

Box 3. Difficulties in diagnosing inflammatory polyarthritis early

- Objective signs may be lacking or have been suppressed by anti-inflammatory medication
- Joint swelling can be difficult to identify in obese patients
- The feeling that joints are swollen may be reported even by some patients with fibromyalgia
- Osteoarthritis as well as RA can cause morning stiffness, though in osteoarthritis it usually lasts less than 30 minutes
- Inflammatory markers such as the ESR or C-reactive protein (CRP) are normal in about 60% of patients with early RA
- In a patient with preceding osteoarthritis, radiographic Changes can be misleading, especially if those suggestive of inflammatory arthritis have not yet developed

Box 4. Investigations that may help in diagnosing common underlying causes of polyarthritis

- Rheumatoid factor-positive in only 70% of patients with RA and present in various other inflammatory diseases and sometimes in health
- Anti Citrulinated peptide antibody - Anti-CCP is present in around 60-70% patients of RA. It is more specific than RF. PPV of both RF and anti-CCP antibody positivity in early undifferentiated arthritis is around 100% for the future development of RA
- Antinuclear antibody-good screening test for SLE but sometimes positive in conditions including RA and in health
- Urinalysis-microscopic haematuria/proteinuria can indicate connective tissue disease
- Viral antibody titers-parvovirus, hepatitis, chikungunya antibodies
- Serum urate/synovial fluid analysis-to exclude gout
- Plain radiographs of hands and feet-can be normal in early RA or show periarticular soft tissue swelling/ osteopenia/marginal erosions; erosions occur earlier in feet, so the feet should be X-rayed even in patients without foot symptoms

Box 5. Practical points to be remembered for RA

- RA is a polyarthritis. Never diagnose RA in a patient with monoarthritis
- Do not diagnose RA unless hands are involved
- Distal interphalangeal joint involvement is exceedingly uncommon in RA. If DIP joints are involved, suspect psoriatic arthropathy, or osteoarthritis
- Lumbar spine is not involved in RA. The presence of inflammatory low back ache with mono or oligoarticular involvement especially in lower limbs should arouse suspicion of seronegative spondyloarthropathy
- Mere presence of rheumatoid factor in blood is not enough to make a diagnosis of RA
- Once RF is positive in a given patient, it need not be repeated, since it correlates poorly with clinical response to treatment
- Titres of RF do not help in monitoring treatment efficacy

Box-6. Predictors of disease persistence in early arthritis

- Female gender
- Duration of symptoms (more than 12 weeks)
- High tender and swollen joint count
- Hand involvement
- Cigarette smoking
- Acute phase response
- Rheumatoid factor
- ACPA
- Erosions on x-ray
- Fulfillment of 1987 ACR diagnostic criteria for RA (sensitivity 88%; specificity 73%)

Box-7. Predictors of disease severity in early arthritis

- Female gender
- High tender and swollen joint count
- High HAQ (health assessment questionnaire) score
- High acute phase reactants
- Base-line rheumatoid factor positivity
- Base-line ACPA positivity
- Shared epitope
- Erosive disease

Further Readings :

1. van der Heijde DM, van Leeuwen MA, van Riel PL, van de Putte LB. Radiographic progression on radiographs of hands and feet during the first 3 years of rheumatoid arthritis measured according to Sharp's method (van der Heijde modification). The Journal of rheumatology. 1995; 22(9): 1792-96.
2. Nell VP, Machold KP, Eberl G, Stamm TA, Uffmann M, Smolen JS. Benefit of very early referral and very early therapy with disease-modifying anti-rheumatic drugs in patients with early rheumatoid arthritis. Rheumatology (Oxford). 2004; 43(7): 906-14.
3. McGonagle D, Conaghan PG, O'Connor P, Gibbon W, Green M, Wakefield R, et al. The relationship between synovitis and bone changes in early untreated rheumatoid arthritis: a controlled magnetic resonance imaging study. Arthritis Rheum. 1999; 42(8): 1706-11.
4. Wakefield RJ, Gibbon WW, Conaghan PG, O'Connor P, McGonagle D, Pease C, et al. The value of sonography in the detection of bone erosions in patients with rheumatoid arthritis: a comparison with conventional radiography. Arthritis Rheum. 2000; 43(12): 2762-70.
5. Finckh A, Liang MH, van Herckenrode CM, de Pablo P. Long-term impact of early treatment on radiographic progression in rheumatoid arthritis: A meta-analysis. Arthritis Rheum. 2006; 55(6): 864-72.
6. Raza K, Falciani F, Curnow SJ, Ross EJ, Lee CY, Akbar AN, et al. Early rheumatoid arthritis is characterized by a distinct and transient synovial fluid cytokine profile of T cell and stromal cell origin. Arthritis research & therapy. 2005; 7(4): R784-95.
7. Nielen MM, van Schaardenburg D, Reesink HW, Twisk JW, van de Stadt RJ, van der Horst-Bruinsma IE, et al. Increased levels of C-reactive protein in serum from blood donors before the onset of rheumatoid arthritis. Arthritis Rheum. 2004; 50(8): 2423-27.
8. Conaghan PG, Ostergaard M, McGonagle D, O'Connor P, Emery P. The validity and predictive value of magnetic resonance imaging erosions in rheumatoid arthritis: comment on the article by Goldbach-Mansky et al. Arthritis Rheum. 2004; 50(3): 1009-11.
9. Arnett FC, Edworthy SM, Bloch DA, et al. The American Rheumatism Association 1987 revised criteria for the classification of rheumatoid arthritis. Arthritis Rheum 1988; 31: 315.
10. Aletaha D, Neogi T, Silman AJ, et al. 2010 Rheumatoid arthritis classification criteria: an American College of Rheumatology/European League Against Rheumatism collaborative initiative. Arthritis Rheum 2010; 62: 2569.
11. Aletaha D, Neogi T, Silman AJ, et al. 2010 rheumatoid arthritis classification criteria: an American College of Rheumatology/European League Against Rheumatism collaborative initiative. Ann Rheum Dis 2010; 69: 1580.
12. Visser H, le Cessie S, Vos K, et al. How to diagnose rheumatoid arthritis early: a prediction model for persistent (erosive) arthritis. Arthritis Rheum 2002; 46: 357.
13. Emery P, Breedveld FC, Dougados M, et al. Early referral recommendation for newly diagnosed rheumatoid arthritis: evidence based development of a clinical guide. Ann Rheum Dis 2002; 61: 290.
14. Combe B, Landewe R, Lukas C, et al. EULAR recommendations for the management of early arthritis: report of a task force of the European Standing Committee for International Clinical Studies Including Therapeutics (ESCISIT). Ann Rheum Dis 2007; 66: 34.

MANAGEMENT OF RHEUMATOID ARTHRITIS

3

C Balakrishnan, V R Joshi

Introduction

Perception of management of rheumatoid arthritis (RA) has undergone a major paradigm shift, from therapeutic nihilism to a positive attitude. The reasons being :

- Realization that, RA is not a benign disease[1,2].
- Joint damage often develops very early in the disease course (within a few weeks of disease onset).
- The loss of function may be irreversible[1,2].
- Early aggressive intervention improves long term outcomes[3,5].
- (Traditional) Disease modifying anti-rheumatic drugs (DMARDs), used either singly or in combination are no more toxic than non-steroidal anti-inflammatory drugs (NSAIDs)[3,4,5].
- Addition of biologic agents to therapeutic armamentarium.
- Steroids have disease modifying effect especially when used in early arthritis (of < 6 months duration).

The aims of treatment are - i) early diagnosis, ii) control of inflammation, iii) retard joint damage (disease progression), and iv) attempt to induce remission induction. The mainstay of treatment is pharmacotherapy.

The therapeutic approach depends upon i) disease duration, ii) severityof disease activity, iii) joint damage, iv) extra-articular manifestations, and iv) associated comorbidities.

i) For clinical and therapeutic purposes RA is considered early (< 6 months duration), established (> 6 months duration), and chronic (> 3 years duration).

ii) Disease activity is assessed using one of the validated scores/indices [Disease activity score 28 (DAS 28), simplified disease activity index (SDAI), clinical disease activity index (CDAI)]. The activity measures are also useful to monitor therapy and 'define' remission[5a].

Common RA disease activity measures and their interpretation

Instruments	Remission	Low activity	Moderate activity	High activity
DAS-28 (range 0 to 9.4)	< 2.6	≥ 2.6 to < 3.2	≥ 3.2 to ≤ 5	> 5.1
SDAI (range 0 to 86.0)	≤ 3.3	> 3.3 to ≤ 11	> 11 to ≤ 26	> 26
CDAI (range 0 to 76)	≤ 2.8	> 2.8 to 10.0	> 10.0 to 22	> 22

iii. Radiological joint damage is assessed(and monitored) using van-der-Hiejde modification of Sharp score. Hand and feet joints are evaluated[5b].

iv. Quality of life (QOL) and Health assessment questionnaire (HAQ) are important tools to assess functional status and outcome. An Indian version of HAQ is simple and easy to administer[5c].

Essentials of RA management

- Early diagnosis 1987 ACR criteria have low sensitivity to diagnose early RA. 2010 ACR/EULAR classification criteria perform better. Early RA provides the best window of opportunity to effectively control the disease and improve long-term outcomes[6,7]. See previous chapter on Early Arthritis.
- Upon diagnosis of RA prompt institution of effective pharmacotherapy is obligatory.
- The aim should be i.e. treat to target. To attain clinical remission or low disease activity with frequent (initially monthly) monitoring of disease activity and advising therapy as per response.
- Methotrexate either alone or in combination is the DMARD of first choice unless contraindicated or not tolerated by the patient.

- Glucocorticosteroids are useful (a) to treat aggressive disease, (b) as bridge therapy, and (c) to tide over acute exacerbations. Intra articular steroid injections are useful to treat isolated joint disease.
- Biologics are presently reserved for patients failing to respond to standard combination DMARD therapy. Cost is a major deterrent.
- Non pharmacotherapy - Patient education, adequate rest, physiotherapy, rehabilitation and nutrition, are important components of management.
- Surgery is required to treat local problems like carpal tunnel syndrome and atlanto-axial subluxation. Joint replacement is indicated for damaged joints.
- DAS 28 (< 2.6), SDAI (< 3.3) and CDAI (< 2.8) define clinical remission. More rigid criteria are applied to clinical trials.

Therapeutics algorithm

- Very early undifferentiated arthritis (< 3 week duration) observe, treat symptomatically.

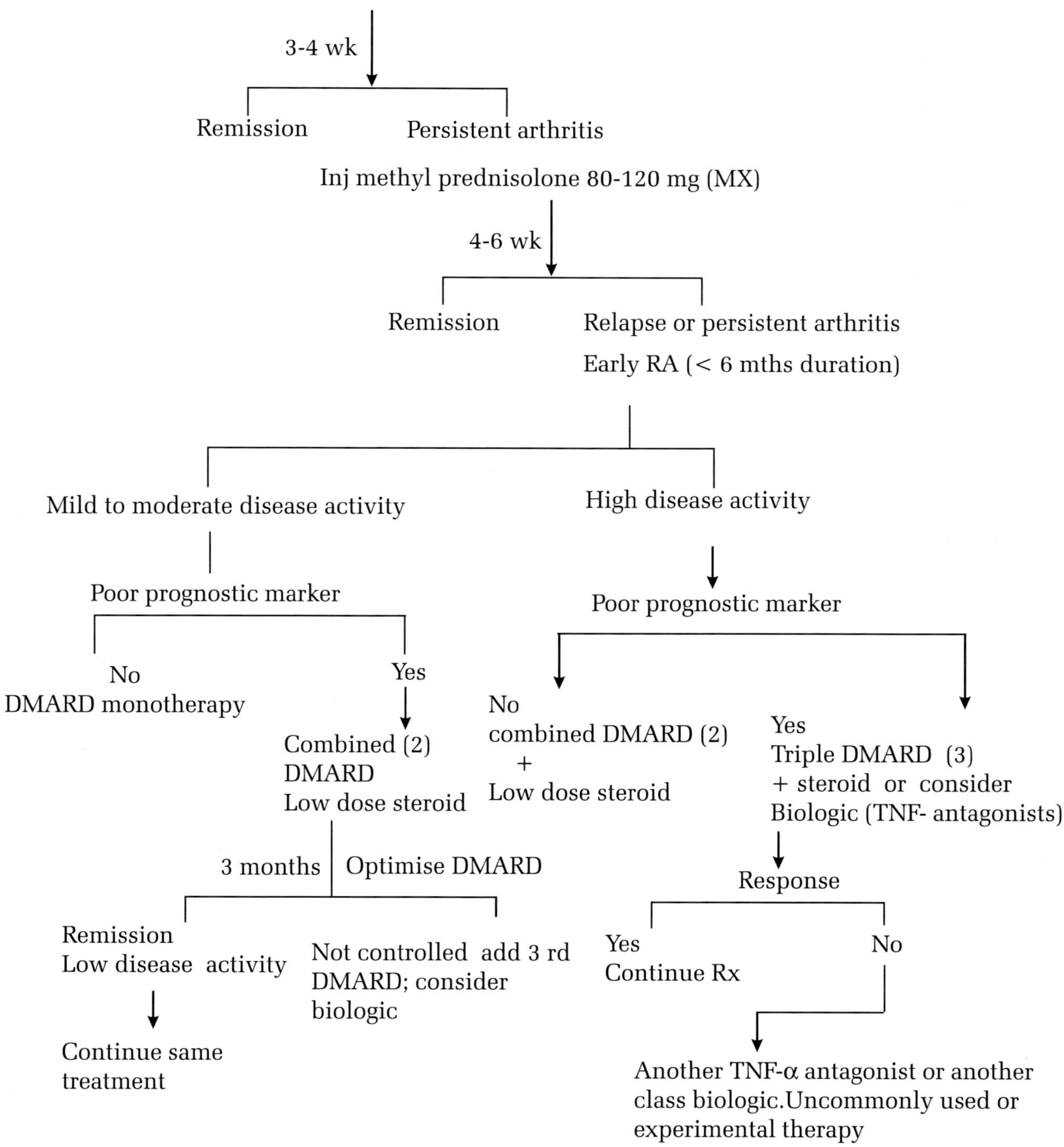

NSAIDs

NSAIDs reduce inflammation and provide symptomatic relief. Though they are almost always a part of initial treatment, they should never be used in isolation; DMARDs must be used simultaneously. The choice of NSAID depends mainly on cost considerations and the gastrointestinal and cardiovascular risk profile of the patient. In a patient with no or little gastrointestinal or cardiovascular risk, any NSAID can be used keeping the cost factor in mind. With significant GI risk, Cox-2 specific NSAIDs are preferred; alternatively a nonselective NSAID can be prescribed along with a PPI; with a past history GI bleed, a Cox-2 specific inhibitor should be used along with a PPI; with significant cardiovascular risk but little GI risk naproxen is preferred[(8)]. All NSAID display renal toxicity and have an effect on BP. These parameters should be monitored regularly[8a]. Initially full doses are prescribed; with disease control the dose of NSAID is reduced and discontinued if possible. A recent consensus statement has made recommendations regarding use of coxibs[9].

Analgesics and spasm relievers can be used along with or as the only medications if NSAIDs are contraindicated.

DMARDs

The choice of DMARD should take into account patient preferences, issues of pregnancy, breast feeding, and co-existing morbidities. Amongst the commonly used DMARDs, (Table-1) weekly methotrexate, in the absence of contraindications, is the drug of choice because of its efficacy, low toxicity profile, better compliance, the fact that patients stay on it for longer period, and the evidence that it reduces mortality[10-11]. It has been shown to retard structural damage.

Methotrexate (MTX)

Methotrexate (Tables 1 and 2) is usually started in a dose of 10 to 12.5 mg once a week, taken orally as a single or divided dose. The dose is increased by 2.5 to 5.0 mg once in two-four weeks till a maximum dose of (i) 25 mg once a week is reached, or (ii) the disease is well controlled or (iii) the drug is not tolerated. At higher doses (15 mg) methotrexate is usually administered parenterally (SC or IM) to overcome the problem of erratic absorption. Its major side effects and monitoring are depicted in table 2 Liver biopsy is indicated only if transaminases are persistently elevated or serum albumin remains persistently low. MTX is teratogenic and should be discontinued at least three months before planned pregnancy, else an MTP is indicated. All patients receiving methotrexate should receive folate supplementation (minimum 5 mg/wk to 5 mg OD daily except on the day of methotrexate). It should be continued during pregnancy if the patient had been on methotrexate prior to conception.

Table 1. Characteristics of commonly used DMARDs*

Drug	Route	Daily Dose	Dosing schedule	Time to benefit
Hydroxychloroquine (HCQ)	Oral	200-400 mg (max 6.5 mg/kg/d)	OD	2-4 months
Chloroquine (CQ)	Oral	150 mg base (5 mg/kg/d)	OD	2-4 months
Methotrexate (MTX)	Oral/parenteral (SC, IM or IV)	7.5-25 mg	Weekly once	2 months
Sulfasalazine (SSZ)	Oral	1.5 to 3.0 g (usual 1.5-2 g) (max 40 mg/kg/d)	Daily in divided doses	2-3 months
Leflunomide	Oral	10-20 mg/d	Single dose	2 months

* Source - IRA guidelines for the management of rheumatoid arthritis (JAPI 2002; 50: 1-2)

Table 2. Toxicity monitoring of common DMARDs ♦

Drug	Toxicity	Base-line	Periodic monitoring	Action***
Chloroquine/ Hydroxychloroquine	Retinal toxicity	Eye check up	Every 3-6 months	Macular change - stop CHQ
Methotrexate	Hepatotoxicity, Myelosuppression, Megaloblastosis, Pulmonary toxicity	CBC with platelet count, S. creatinine, SGPT, S. albumin, S. Alk. phosphatase HBV, HCV, HIV in high-risk patients	Every 4-8 weeks X-ray chest SOS	CBC* Dropping S. albumin** New or increasing cough/ dyspnoea** > 2 N SGPT** Rise in S. creatinine- (reduce dose)
Sulfasalazine	Blood dyscrasias, Haemolytic anaemia, Decrease in sperm count	CBC SGPT G6PD in high-risk patients Baseline sperm count	Monthly for 3 months and then every 3 months sperm count	CBC* SGPT, SGOT, > 2 N**
Leflunomide	Hepatotoxicity, Skin toxicity, Pulmonary toxicity	CBC with platelet count, S. creatinine, SGPT, S. albumin, S. Alkaline phosphatase, HBV, HCV, HIV in high-risk patients	Every 4-8 weeks X-ray chest SOS	CBC* Dropping S. albumin** New or increasing cough/ dyspnoea** > 2 N SGPT** Rise in S. creatinine - (reduce dose)

For more details on toxicity, etc, relevant source material should be referred to.

* CBC with WBC < 4000/mm3, and/or neutrophils < 2000/mm3, and/or platelets < 1 lac/mm^3, MCV > 105 fl - hold therapy, consult a rheumatologist.

** Withhold drug. Consult a rheumatologist.

♦ Source IRA guidelines for the treatment of rheumatoid arthritis.

Sulfasalazine

Sulfasalazine (SSZ) (Tables 1 & 2) may be preferred as the DMARD of choice in milder cases, or in younger patients in whom fertility and pregnancy are a consideration. Sulfasalazine need not be discontinued during pregnancy. It may rarely cause diarrhoea, rash, and haemolytic anaemia especially in G6PD deficient infants/individuals[12]. Adequate folate supplementation to mother is advocated with SSZ. It has been shown to retard joint damage. SSZ is started at a dose 0.5 gm OD and increased over 3-4 weeks to the maximum dose.

Leflunomide

Leflunomide (Tables 1 & 2) is as efficacious as methotrexate[13]. It has been shown to retard structural damage. Though a loading dose of 100 mg OD x 3 d is advocated, many rheumatologists start with the maintenance dose, to avoid GI intolerance associated with the loading dose.

Leflunomide has a long-half life and is teratogenic. It is absolutely contraindicated during pregnancy and lactation. Women must discontinue the drug for at least 2 years before planning conception. Alternatively wash out with cholestyramine 8 gm TDS x 11 days or activated charcoal 50 gm QID x 11 days is advised. Ideally blood levels of leflunomide should be checked after the washout, but the test is not easily available. Leflunomide should be used with caution in patients with mild renal insufficiency and totally avoided in patients with significant renal insufficiency.

Hydroxychloroquine, chloroquine

Hydroxychloroquine, chloroquine (HCQ) (Table 1 & 2) are mild DMARDs, most often used in combination with methotrexate and sulfasalazine. HCQ is safe during pregnancy. HCQ is preferred to chloroquine because of its lesser toxicity. There is no data to suggest that HCQ retards radiological progression[14].

Combination Therapy

Though methotrexate therapy is effective, a fair number of patients fail to achieve maximum disease control[15]. Over the last few years combination DMARD therapy has been shown to be more effective, and associated with improved outcome[16,22]. However, the last word on how to use combination therapy is not said[17]. More importantly, combination therapy is possibly as effective as biologics. Two strategies are advocated. (i) Step up therapy in which patient is started on a single DMARD and additional DMARDs are added if disease cannot be controlled. (ii) In step-down approach combination therapy is started right at the beginning and once remission or adequate control is achieved drugs are withdrawn one by one and patient maintained on a single DMARD. Hydroxychloroquine, sulfasalazine and low dose prednisolone are usually combined with methotrexate either singly or all together. Although FDA has not approved combination therapy of methorexate with leflunomide, there are reports that addition of leflunomide provides significant additional clinical benefit in patients with active rheumatoid arthritis[17]. The combination is generally well tolerated and can be used safely with regular monitoring of liver functions and blood counts.

Steroids

Systemic steroids provide rapid disease control and relief from pain and stiffness[18] Steroids are mainly advocated as part of early therapy until DMARDs effectively control the synovitis. They have been shown to retard radiographic progression[19], but should not be used on a long-term basis because of their side-effects. Steroids can be administered in different ways (i) starting with oral prednisolone dose of 10 mg/d, and tapered off over a few months (ii) as inj methyl-prednisolone 80-120 mg IM, every four weeks for 2-4 months (iii) Intra-articular steroids to treat disease limited to a few joints, or to treat acute flare in a joint. Over enthusiastic and frequent use of IA steroids should be avoided. (not more than 3 joints at a time, not the same joint more than 3 times a year, and a minimum interval of 3 months between successive injections in the same joint).

If steroids are to be used on a long-term basis, prophylactic measures against osteoporosis (daily supplements of 1 gm of elemental calcium and 400 IU Vit D plus oral bisphosphonate e.g. alendronate 35 mg once a week) should be instituted simultaneously. The indications for long-term use of steroids are (i) pregnancy and breast feeding (ii) elderly patients and (iii) patients who do not tolerate or fail to respond to conventional DMARD therapy and cannot afford newer therapies.

Biologics

Biologics are highly effective in the control of disease activity especially when used in combination with a DMARD (usually methotrexate, but could be leflunomide or sulfasalazine)[20,21]. The efficacy of biologics extends to extra articular manifestations. There is retardation of radiological damage. There is no definite proof that these agents induce healing of erosions. The risk of cardiovascular disease is effectively controlled, but susceptibility to infections is increased. Cancer risk does not seem to be increased. The main draw back is cost which makes this mode of therapy beyond the reach of most of the Indian patients.

Biologics are indicated (i) in early disease (< 6 months duration) if disease activity is high. (ii) For established or chronic RA (a) DMARD failure (b) poor prognostic markers present (c) disease activity high.

The experience with biologics in India is limited. A registry has been initiated and it should prove a useful Indian resource (www.irabr.org).

This is an extremely dynamic field with new molecules getting added at a fairly rapid pace. The approved biologics to treat RA are TNF-α antagonists infliximab-(remicade),etanercept-(enbrel), adalimumab-(humira), golimumab-(symponi), certolizumab PE Gol - (cimzia), rituximab (anti CD 20 mab), tocilizumumab-anti-IL-6 receptor mab, abatacept (CTLA 4 IgG 1 fusion protein) and anakinra (IL1 antagonist). Reditux is rituximab biosimilar. Tofacitinib (JAK inhibitor) is a new oral small molecule approved in USA, yet to come to India.

Guidelines for the use of these agents have been suggested[23,33] (Table 3). Infliximab, etanercept, rituximab, tocilizumab and abatacept are currently available in India. Infliximab is given as infusion of 3 mg/kg at 0, 2 and 6 weeks followed by maintenance dose every 8-12 weeks as needed. Higher dosage (5mg/kg) or more frequent dosing (6 weekly) is sometimes needed. Etanercept is given as a subcutaneous injection of 50 mg once or 25 mg twice a week. Infections, aplastic anaemia, demyelinating neurological diseases, ANA positivity, malignancy and rarely drug induced lupus have all been reported with TNF alpha blockers[24]. These agents are used in combination with methotrexate.

Recent studies have demonstrated superiority of combination therapy of biologicals + methotrexate over methotrexate monotherapy alone in early RA[25a,b]. Cost, side-effects remain the main issues. A recent trial has confirmed that combination DMARD therapy is as effective as biologic (etanercept) + methotrexate for treating early RA.

Table 3. Guidelines for the use of biologicals

- Must have failed at least two DMARDs including methotrexate (20-25 mg/wk)
- Must have active disease
- Have no major infections in preceding six months
- Have no malignancies within 5 years
- Non pregnant, non breast feeding women
- To be used in combination with a DMARD (MTX/LEF/SSZ)
- All biologics have similar response rates. TNF-antagonists are the usual or the first choice, mainly because of longer experience and proven safety
- If significant disease control is not achieved with a biologic within 12-24 weeks, the biologic should be changed to another either of the same class or of another class

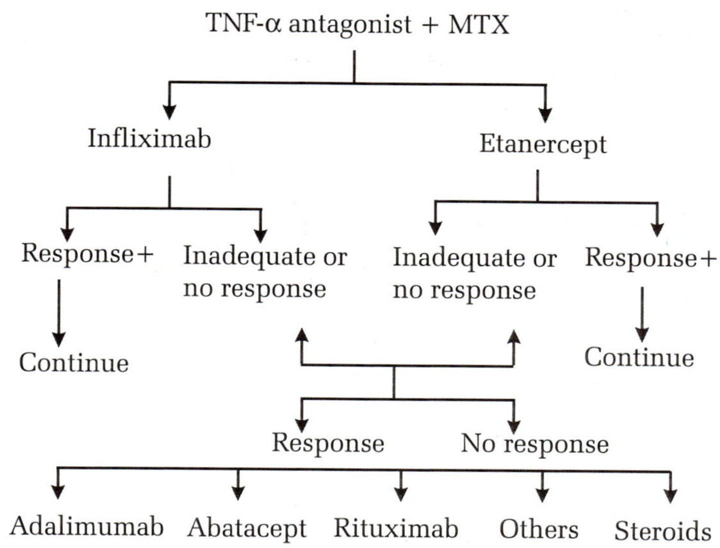

Fig. 1a &1b. - Algorithmic approach to drug treatment of RA.

Indian experience with anti-TNF alpha therapy is limited[26]. Although the experience was with all forms of arthritides, reactivation of latent TB occurred in 10% of patients. Recommendations have been published to tackle the TB problem associated with anti-TNF therapy[27,27b].

It is important to inform patients that biologicals are not a cure. About 40% patients may not respond to a biological, when other preparation/agents can be tried.

Fig 1a and b depict algorithmic approach to drug treatment

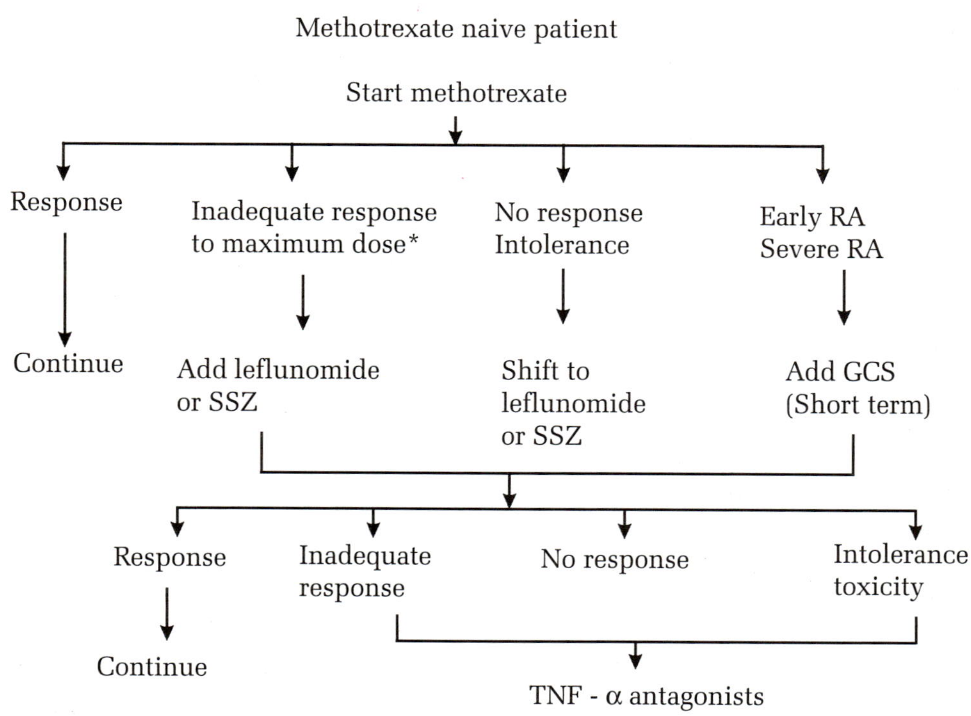

* including parenteral methotrexate

Fig 1 a - Algorithmic approach to drug treatment of RA

Others

Cyclosporine, an immunomodulatory drug, has been shown to be effective in combination with methotrexate. Side effects (renal toxicity, hypertension, hyperkalaemia) limit its use.

Minocycline in clinical trials has shown efficacy. The drug is used infrequently to treat RA patients.

Staphylococcal protein A column therapy : clinical efficacy of this therapy has been documented in patients with severe, refractory RA. Cost, logistics, and adverse events are main limitations to its use.

Gold, d-penicillamine were once the sheet anchors of therapy. Toxicity and availability of safer DMARDs has significantly reduced their use. These may be considered in resistant patients.

Monitoring response

Assessment of disease activity, damage and functional status should be evaluated at base-line and periodically to evaluate the efficacy of treatment. Validated instruments of assessment are available (See Sec. 2) along with certain disease activity tests[34].

Extra-articular manifestations

RA is a systemic disease with many extra-articular manifestations. Management of some is discussed briefly.

Fibrosing alveolitis - It is essential to exclude MTX toxicity. If the condition is progressive, treatment with high dose steroids with or without azathioprine or cyclophosphamide is necessary. Patients with stable disease can be observed[28].

Bronchiolitis obliterans responds poorly to immunosuppressive therapy. Corticosteroids, Azathioprine may be added.

Pleural effusion - large effusions need to be drained. For recurrent effusions tetracycline pleurodesis is indicated. If effusion is part of systemic vasculitis, immunosuppressive therapy is necessary. Pain is managed symptomatically.

Secondary Sjogren's syndrome : Ocular and oral manifestations are treated on the same lines as primary Sjogren's syndrome.

Scleritis, peripheral corneal melting are serious manifestations and need aggressive combination immunosuppressive therapy (Steroids + cytotoxic agents).

Pericarditis - high dose steroid treatment is indicated if features of systemic vasculitis are present or there is evidence of carditis. Chronic constrictive pericarditis needs surgical treatment.

CCF is treated on usual lines along with treatment of RA.

Systemic vasculitis with organ involvement needs to be treated with high dose corticosteroids along with cytotoxic drugs (Cyclophosphamide or azathioprine).

Leg ulcers can be due to vasculitis, Felty's syndrome, or venous stasis. The treatment is as per the cause. Stasis ulcers can recur and are often difficult to treat.

Complications of rheumatoid arthritis

Atlanto-axial subluxation - Presence of neurologic symptoms and signs of cord compression as shown by MRI are definite indications for surgery. Other patients are treated symptomatically (NSAIDs, traction, collar etc). Subaxial subluxation which may follow surgery of atlanto-axial fixation needs surgery. Joint infection - is treated as discussed in chapter on joint infection. If infected prosthesis cannot be removed, life-long antibiotic treatment may be necessary.

Stress fractures may be due to osteoporosis, osteomalacia or a combination of both. These need to be treated in addition to treatment of fracture.

Ruptured synovial cyst - Rupture is most commonly seen with the so called Baker's cyst of knee joint. It can mimic deep vein thrombosis. Sonography is useful to make a correct diagnosis. Treatment consists of intraarticular steroids along with rest.

Carpal tunnel syndrome is treated first with immobilisation (splinting) and local corticosteroid injection. Surgery (decompression) is indicated with failure of conservative treatment or with evidence of motor deficit.

Painful feet may be due to joint inflammation, joint damage, entrapment neuropathy, ischaemia, vasculitis, fracture, or pressure ulcers. Pain due to damage and deformity and due to posterior-tibial nerve entrapment is treated with appropriate footwear.

Tendon rupture commonly involves extensor tendons of fingers. Sonography and MRI help to confirm the diagnosis. Prophylactic dorsal tenosynovectomy with or without excision of head of ulna may prevent progressive extensor tendon damage. Ruptured tendon is treated surgically.

Prevention of long-term complications

Osteoporosis - Vit D deficiency is common in RA patients[29]. All patients should receive calcium and Vit D supplements. Exercises should be encouraged. With anticipated long-term use of steroids bisphosphonates should be started prophylactically.

Cardio-vascular complications[30] (Ischaemic heart disease) - the most important preventive measure is good control of disease activity. NSAIDs and corticosteroids are potential risk factors. Smoking should be discouraged. Healthy diet, exercise, control of comorbidities like hypertension, diabetes and dyslipidemia are important.

Pregnancy Drug Therapy

Paracetamol is safe. NSAIDs and selective Cox-2 inhibitors can be used post implantation; should be discontinued before delivery. Chloroquine, hydroxychloroquine, sulfasalazine, azathioprine and cyclosporine can be continued during pregnancy. Others i.e. MTX, leflunomide, gold, d-penicillamine, are contraindicated. Safety of biologics is not established yet.

During breast feeding corticosteroids are safe.

Physiotherapy

Assistive devices are beneficial. Physiotherapy helps to maintain joint movement, muscle strength and function.

Both aerobic and strengthening exercises are of benefit to patients with RA[31]. Moderate to hard intensity aerobic exercises for up to an hour three times a week are recommended. Adequate periods of rest are necessary. These include swimming, walking, cycling and other aerobics. Moderate to hard intensity strengthening exercises (static or dynamic) including resistance training are also prescribed. Physiotherapy should be limited to gentle range of movement exercises when the disease is active. Education on joint protection to conserve energy is also important.

Occupational therapy

Skilled occupational therapy advice is necessary to those experiencing limitations of function. Resting and working splints can be used to provide pain relief, prevent deformities and improve function. Assistive devices are useful.

Role of surgery[32]

Surgery is an integral part of patient management. Persistent synovitis inspite of adequate treatment is an indication for synovectomy. Tenosynovectomies are done to tackle persistent disease. In the later stages, when there is significant joint damage and disability, joint replacement (arthroplasty) is the treatment of choice. Joint fusion is a viable alternative. Other indications for surgery include carpal tunnel syndrome, atlanto-axial subluxation, tendon rupture etc.

Other measures

All patients are advised normal nutritious diet and adequate rest periods during the working day. Patient education and family education and if available, social support are important.

Future

Even though great strides have been made in the understanding of RA, we are still far from the elusive cure. It is quite likely that in the years to come majority of patients will be offered combination DMARD or frequent use of biologicals to control RA. A recent study has demonstrated superiority of combination therapy with biologicals + methotrexate alone in early RA. Cost, long-term benefit, and side-effects remain the main issues. In another study in patients with recent onset RA, therapy including prednisolone or infliximab resulted in earlier clinical improvement and less joint damage progression than initial monotherapy[25a].

We will also have markers at baseline which will predict which patient will have aggressive disease. It is also likely that pharmacogenetics may give clues as to which drug may be the best for a particular patient.

Cell therapy, gene therapy and vaccination are other futuristic options.

Further Reading :

1. Scott DL, Grindulis KA, Struthers GR, Coulton BL, Popert AJ, Bacon PA. Progression of radiological changes in rheumatoid arthritis. Ann Rheum Dis 1984; 43: 8-17.
2. Pincus T, Callahan LF, Sale WG, Brooks AL, Payne LE, Vaughn WK. Severe functional declines, work disability, and increased mortality in seventy-five rheumatoid arthritis patients studied over nine years. Arthritis Rheum 1984; 27: 864-72.
3. Mottonen T, Hannonen P, Leirisalo-Repo M et al. Comparison of combination therapy with single-drug therapy in early rheumatoid arthritis: a randomized trial. FIN-RAC trial group. Lancet 1999; 353: 1568-73.
4. Boers M, Verhoeven AC, Markusse HM et al. Randomised comparison of combined step-down prednisolone, methotrexate and sulphasalazine with sulphasalazine alone in early rheumatoid arthritis. Lancet 1997; 350: 309-18.
5. Grigor C, Capell H, Stirling A et al. Effect of a treatment strategy of tight control of rheumatoid arthritis (the TICORA study): a single-blind randomized controlled trial. Lancet 2004; 364: 263-69.

5(a) S.K. Das, S. Agarwal, S. K. Mukherjee, Alkendu Ghosh Eds. Assessment In rheumatoid arthritis in Monogram on rheumatoid arthritis. Indian College of Physicians, 2012, pp. 101-04.

5(b) Van der Hejde DM. How to read radiographs according to the sharp/Van der Heijde method. J. Rheumatol. 1999; 26: 743-45.

5(c) Kumar A., Malaviya A N, Pandhi Singh. Validation of an Indian version of the Health Assessment Questionnaire in patients with rheumatoid arthritis Rheumatology 2002; 41: 1457-59.

6. Machold K P, Nell V, Stamm T, Aletaha D, Smolen J S. Early rheumatoid arthrits Curr Opin. Rheumatol. 2006; 18: 282-88.
7. Quinn M A, Green M J, Emery P. Evaluation and management of early rheumatoid polyarthritis. In Hochberg M C, Silman A J, Smolen J S, Weinblatt M E, Weisman M H Eds Rehumatology M Mosby Publ. Edinburgh 2003; 885-92.
8. Schieman J M, Frendick A M. Summing the risk of NSAID therapy Lancet 2007; 369: 1580-81.

8a. Pathan E, Gaitonde S, Rjadhysha S et al. A longitudinal study of patients of rheumatoid arthritis on long term NSAID therapy. J. Assoc. Physicians India. 2003; 51: 1045-49.

9. Tannenbaum H, Bombardier C, Davis P, Russell AS et al. An evidence-based approach to prescribing nonsteroidal anti-inflammatory drugs. Thrid Canadian Consensus Conference. J Rheumatol 2006; 33(1): 140-57.
10. Klaus P Machold, Valerie PK Nell, Tanja A Stamm, and Josef S Smolen. Aspects of early arthritis. Traditional DMARD therapy: is it sufficient? Arthritis Res Ther. 2006; 8(3): 211-16.
11. Pincus T, Sokka T. Should aggressive therapy for rheumatoid arthritis require early use of weekly low-dose methotrexate, as the first disease-modifying anti-rheumatic drug in most patients? Rheumatology 2006; 45: 497-99.
12. Bresnihan B. Treatment of early rheumatoid arthritis in younger patients. J. Rheumatol 2001; 28 (Suppl. 69) : 4-9.
13. Emery P., Breedveld F C, Lemmel E M et al. A comparison of efficacy and safety of leflunomide and methotrexate for the treatment of rheumatoid arthritis. Rheumatol 2000; 39: 635-65.
14. HERA study group : A randomised trial of hydroxychloroquine in early rheumatoid arthritis. The HERA study Am J Med 1995; 98: 156-68.
15. Symmons D, Tricker K, Harrison M et al Patients with longstanding rheumatoid arthritis continue to deteriorate despite intensified treatment with traditional disease modifying antirheumatic drugs. Results of British Rheumatoid Arthritis Outcome study group - randomised controlled trial. Rheumatology 2006; 45: 558-65.
16. O'Dell J R, Haire C E, Erickson d et al. Treatment of rheumatoid arthritis with methotrexate alone, sulfasalazine, hydroxychloroquine or a combination of all three N Engl. J. Med. 1996; 334: 1287-91.
17. Kremmer JM, Genovese MC, Cannon GW, Caldwell JR, Cush JJ, Furst DE et al. Concomittant leflunomide therapy in patients with active rheumatoid arthritis despite stable doses of methotrexate. A randomized, double-blind, placebo-controlled trial. Ann Intern Med. 2002; 137(9): 726-33.
18. Lim SS, Conn DL. The use of low dose prednisolone in the management of rheumatoid arthritis. Bull Rheum Dis. 2001; 50 (12): 1-4.
19. Kirwan J R. The effect of glucocorticoids on joint destruction in rheumatoid arthritis. N Engl. J. Med. 1995; 333: 142-46.
20. Weaver AL. The impact of new biologicals in the treatment of rheumatoid arthritis. Rheumatology (Oxford), 2004: 43(suppl 3): iii17-iii23.
21. Hochberg MC, Tracy JK, Hawkins-Holt M, Flores RH. Comparison of the efficacy of tumour necrosis factor alpha blocking agents adalimumab, etanercept and infliximab when added to methotrexate in patients with active rheumatoid arthritis. Ann Rheum Dis 2003; 62 (suppl 2): ii13-ii16.
22. Arshad A, Sulaiman W. Optimizing the use of traditional DMARD in RA : getting the most of what we can afford ! APLAR J. Rheumatol 2007; 10: 3-8.
23. Furst DE, Breedveld FC, Kalden JR, Smolen JS, Burmester GR, Dougados M et al. Updated consensus statement on biological agents for the treatment of rheumatoid arthritis and other immune mediated inflammatory diseases. Ann Rheum Dis. 2003; 62 (suppl 2): ii2-ii9.
24. Bongartz T, Sutton AJ, Sweeting MJ, Buchan I, Matteson EL, Montori V. Anti-TNF antibody therapy in rheumatoid arthritis and the risk of serious infections and malignancies: systematic review and meta-analysis of rare harmful effects in randomized controlled trials. JAMA. 2006; 295(17): 2275-85.
25. St. Clair E W, Van der Heijde D M, Smden J S et al. Combination of infliximab and methotrexate therapy for early rheumatoid arthritis : a randomised controlled trial. Arthritis Rheum. 2004; 50: 3432-43.

25a. Cornelia F, A, Breedveld F C, Dijkmans BAC, Treatment of recent onset rheumatoid arthritis; Lessons from the BeST study. J. Rheumatol, 2007; 34 Suppl. 80: 25-33.

25b. P. Emery, F. C. Breedveld, S. Hall et al. Comparison of methotrexate monotherapy with a combination of methotrexate and etanercept in acitve, early moderate to severe rheumatoid arthritis (COMET) : a randomised double blind trial. Lancet 2008; 372: 375-82.

26. Kumar A: Experience with anti-tumor necrosis factor-[alpha] therapy in India. APLAR Journal of Rheumatology. 2006; 9(2): 136-41.

27. Peter Ormerod, Heather J Milburn, Stephen Gillespie , Jo Ledingham and David Rampton. BTS recommendations for assessing risk, and for managing M.tuberculosis infection and disease, in patients due to start anti-TNF treatment. Thorax. Published Online First: 29 July 2005. doi:10.1136/thx.2005.046797.

27a. Handa R, Misra R, Chaturvedi V P, Pispati P K, Rao U R K, Joshi V R. Guidelines fortuberculosis prophylaxis during anti-tumour necrosis factor treatment - Indian Rheumatology Association APLAR J. Rheumatol. 2006; 9: 181-83.

28. Rajasekaran Brt, Shovlin D, Lord P, Kelly C A. Interstitital lung disease in patients with rheumatoid arthritis : a comparsion with cryptogenic pulmonary fibrosing alveolitis. Rheumatology 2001; 40: 1022-25.

29. Balakrishnan C, Kalke S., Mangat G., Joshi V R. Concomitant osteomalacia in rheumatic disorders. An important reversible cause of recent onset increase in morbidity J. Ind. Rheumatol. Assoc. 1997; 50: 58-60.

30. Chogle A R, Chakravarty A. Cardiovascular events in systemic lupus erythmatosus and rheumatoid arthritis : emerging concepts, early diagnosis and management. J. Assoc. Physicians India 2007; 55: 32-40.

31. Stenstrom CH, Minor MA. Evidence for the benefit of aerobic and strengthening exercise in rheumatoid arthritis. Arthritis Rheum, 2003; 49(3): 428-34.

32. De Pablo P., Anderson R J, Katz J. N. Medical aspects of surgery In. St clair E W, Pisetsky D S, Haynes B F Eds. Rheumatoid arthritis; Philadelphia Lippincott Williams and Wilkins. Publ. 2004 : 440-48.

33. Singh JA, Furst DE, Bharat A, Curtis JR, Kavanaugh AF, Kremer JM, et al. 2012 update of the 2008 American College of Rheumatology recommendations for the use of disease-modifying antirheumatic drugs and biologic agents in the treatment of rheumatoid arthritis. Arthritis Care Res (Hoboken) 2012; 64: 625-39.

34. Curtis JR et al.Validation of a novel multi-biomarker test to assess rheumatoid arthritis disease activity. Arthritis Care Res 2012;64(12):1794-1803.

Vectra DA (Crescendo bioscience) contains 12 biomarkers.

SECTION 7

Spondyloarthropathies

ANKYLOSING SPONDYLITIS - CLINICAL FEATURES AND MANAGEMENT

1

Alakendu Ghosh, Subhankar Haldar

Ankylosing spondylitis (AS) is a chronic inflammatory rheumatic disease, primarily affecting the axial skeleton (sacroiliac joints and spine) but also affects the peripheral joints, entheses and many extraarticular sites like eyes, heart and lungs and ultimately leading to functional disability with premature death[1]. Inflammatory back pain and radiographic sacroiliitis is the hallmark of this disease.

The age of onset is the second or third decade of life[2] and males are affected two to three times more than females[3]. Female has less severe form of disease than male; cervical spine and peripheral joints are more involved in female over male[4]. If the age of onset younger than sixteen years, this group of patients is classified as juvenile ankylosing spondylitis or enthesitis related arthritis[5]. The prevalence of the disease is varied in different geographic areas. The natural history is that this disease does not follow a single definite course i.e. there is no predictable pattern. The course of disease progression is fluctuating and variable. The first ten years determines subsequent outcome[6].

The disease has got strong association with HLA-B27 haplotype, thereby implying genetic predisposition to the illness. HLA-B27 is the most common genetic component in patients with AS, but only 1-6% of adults with the HLA-B27 positive population suffer from AS. Furthermore, positive family history of AS in the first degree relatives confers 10-30% more risk of developing AS[7]. Thus, there must be other genetic factors contributing. In White population 90% of patients with AS posses HLA-B27, while the prevalence among blacks is only 50%. The evidence suggest that there is 5.6-16 times increased chance of developing the disease among HLA-B27 positive relatives[8].

The modified New York classification criteria is the most commonly used classification/diagnostic criteria for AS. The radiographic sacroiliitis is usually a late feature in AS. The limitation of lumbar spine mobility and limitation of chest expansion may not be the first presenting complaints. These clinical features and the radiographic sacroiliitis may undue delay (5-10 years) the diagnosis of AS in many patients[9]. On the other hand early diagnosis of AS with an early initiation of treatment and thereby decreasing the inflammatory burden, has better overall outcome of such patients. Very recently, MRI of sacroiliac joints is incorporated for this purpose to detect sacroiliitis very early in patients presented with inflammatory back pain.

Pathology and pathogenesis

Pathology of AS - Enthesitis is one of hallmark of Ankylosing Spondylitis. John Ball was the first to describe the pathogenesis of AS and he reported inflammation, bone erosion and new bone formation in AS starts at the entheses. Entheses are the sites of attachment of ligaments, tendons and joint capsules to bone and they are present in vertebrae, sacroiliac joints and as well as in peripheral joints. The Achilles enthesis is the most characterized enthesis in the peripheral skeleton and has been described as an organ, termed as Achilles enthesis organ[10]. This organ consists of ligaments, tendons and capsules attached to the cancellous bone of the calcaneum through the fibrocartilaginous connections and the synovium of the retrocalcaneal bursa is also the part of this organ. The new bone formation at sites of entheses of vertebrae is called syndesmophytes which bridge the vertebrae and fuse them across to form so called 'bamboo spine' in AS[11].

Immunologically there is interaction between class I MHC molecular HLA-B27 and T lymphocytes and tumor necrosis factor (TNF-α) has been identified as a key regulatory cytokine in the inflammatory cascade[2]. Recent study indicates that entheseal fibro cartilage is the major target of the immuno response and primary site of immuno pathology.

The T cells, B cells, bone marrow derived macrophages and cells involved in neoangiogenesis are confined at the sites of enthesitis[12-14]. The bone destructive collagen-degrading matrix metalloproteinase 1 (MMP 1) and collagenolytic proteinase cathepsin-K in the osteoclasts are also present at the sites of enthesitis of the vertebrae[15]. So in AS, bone erosion is caused by osteoclastic activity and which is different from bone erosion in RA, where bone destructive MMP 3 and RANK (receptor activator of nuclear factor kB) ligand play the major role[16]. Probably for this reason, anti-TNF arrests the bone erosion less frequently in AS than RA.

Bone formation in AS - Several clinical and animal studies have shown that inflammation and bone destruction are coupled in RA but uncoupled in AS. As because, the studies have revealed that anti-TNF therapy controls the inflammation and can prevent the bone erosion in RA, but not in AS. In AS, there is marked improvement so far the inflammation is concerned but the entheseal bone erosion and new bone formation (mostly syndesmophytes in vertebrae) have continued even with anti-TNF.

The process of new bone/ syndesmophytes formation is presented in simplified way in Fig 1. The roles of the bone morphogenic proteins (BMPs) Wnt (wingless) pathways and the role of TNF are sometimes conflicting[17]. The investigators reported that there is increased BMPs activation at the Achilles enthesis[18,19], markedly decrease of scleroptin (an endogenous inhibitor of bone formation) in osteocytes even in patients with AS who have already developed sydesmophytes[20] and lower serum concentration of DKK1 (Dickkopf-related protein1) in patients with AS[21] lead to continuous new bone formation despite recent development of therapy like anti-TNF.

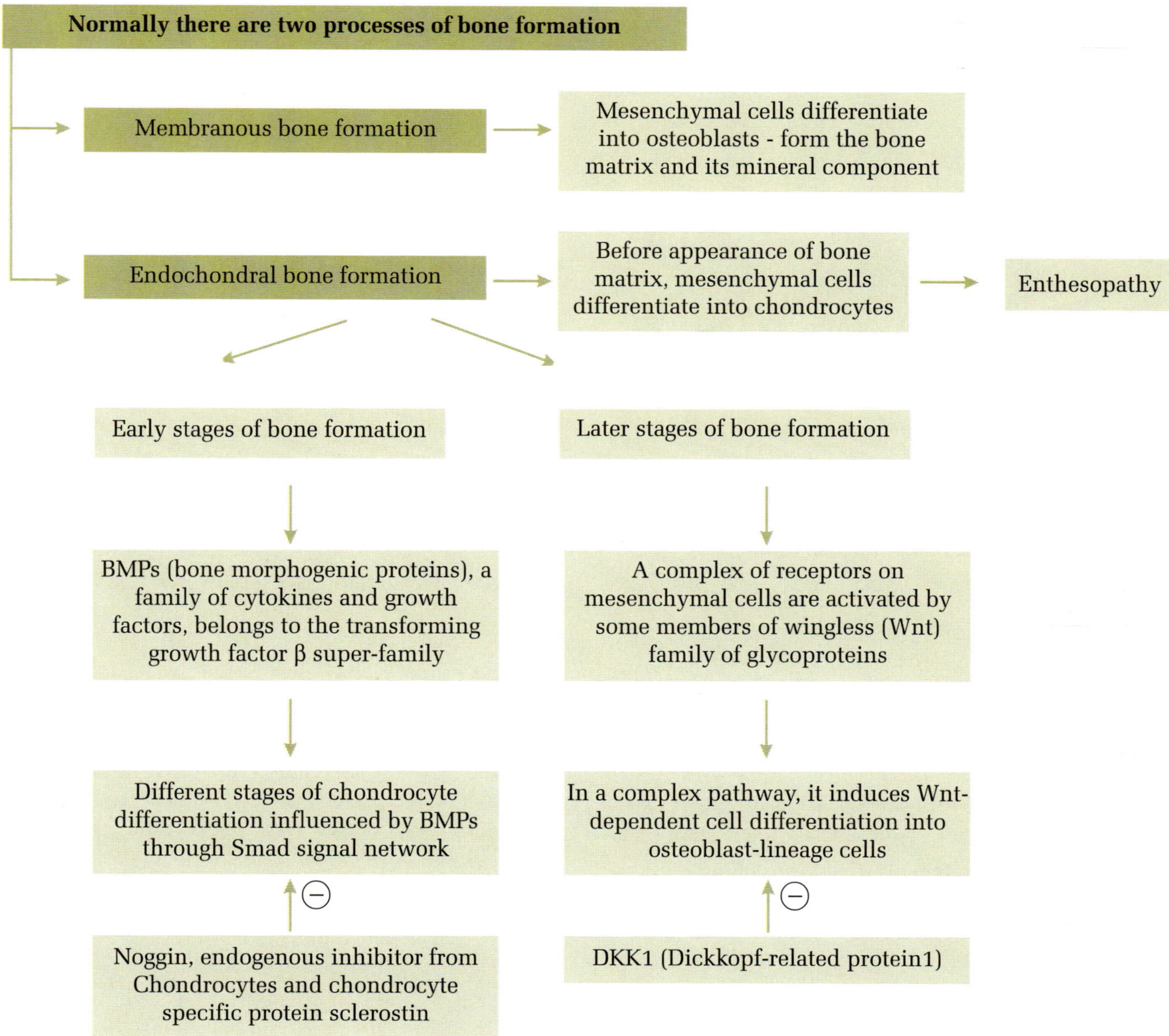

Fig 1 : Process of new bone/syndesmophytes formation.

Synovitis is rare in AS as compared to rheumatoid arthritis but when it is present it is difficult to distinguish. However, immunohistochemistry may differentiate AS from RA. M1 proinflammatory macrophages are present in RA synovia whereas M2 regulatory macrophages consist in high number in synovia of AS[22,23].

The pathogenesis of AS is still poorly understood. The risk factors predisposing to AS are - HLA-B27 seropositivity, positive family history and male sex and frequent GI infections[24]. Among these - HLA-B27 is most important, as there is increased (16 times) chance of developing the disease among HLA-B27 positive relatives[8]. Infection with unknown organism or exposure to an unknown antigen in a genetically susceptible individual (HLA-B27 positive) is hypothesized to result in the clinical expression of AS.

Of the different subtypes of HLA-B27, *2705 is the predominant in most populations, while *2708 is common in Asians[25]. There are a number of hypotheses regarding how the HLA genes cause or predispose to the illness. They are i) arthritogenic peptide theory or molecular mimicry hypothesis where HLA-B27 binds to self or microbial peptides and presents them to CD8- T cells. ii) aberrant folding or processing of the heavy chain of HLA-B27 iii) free heavy chain homodimers hypothesis.

Genome-wide association studies have been identified several genes. In addition to, HLA-B27 these are endoplasmic reticulum aminopeptidase 1 (ERAP 1), interleukin 23 receptor (IL-23R), interlukin - 1 receptor type 2 (IL-1R2) and anthrax toxin receptor 2 (ANTRXR2) or capillary morphogenesis gene 2 protein (CMG2). The role of ERAP 1 in the pathogenesis of AS are described in Fig 2.

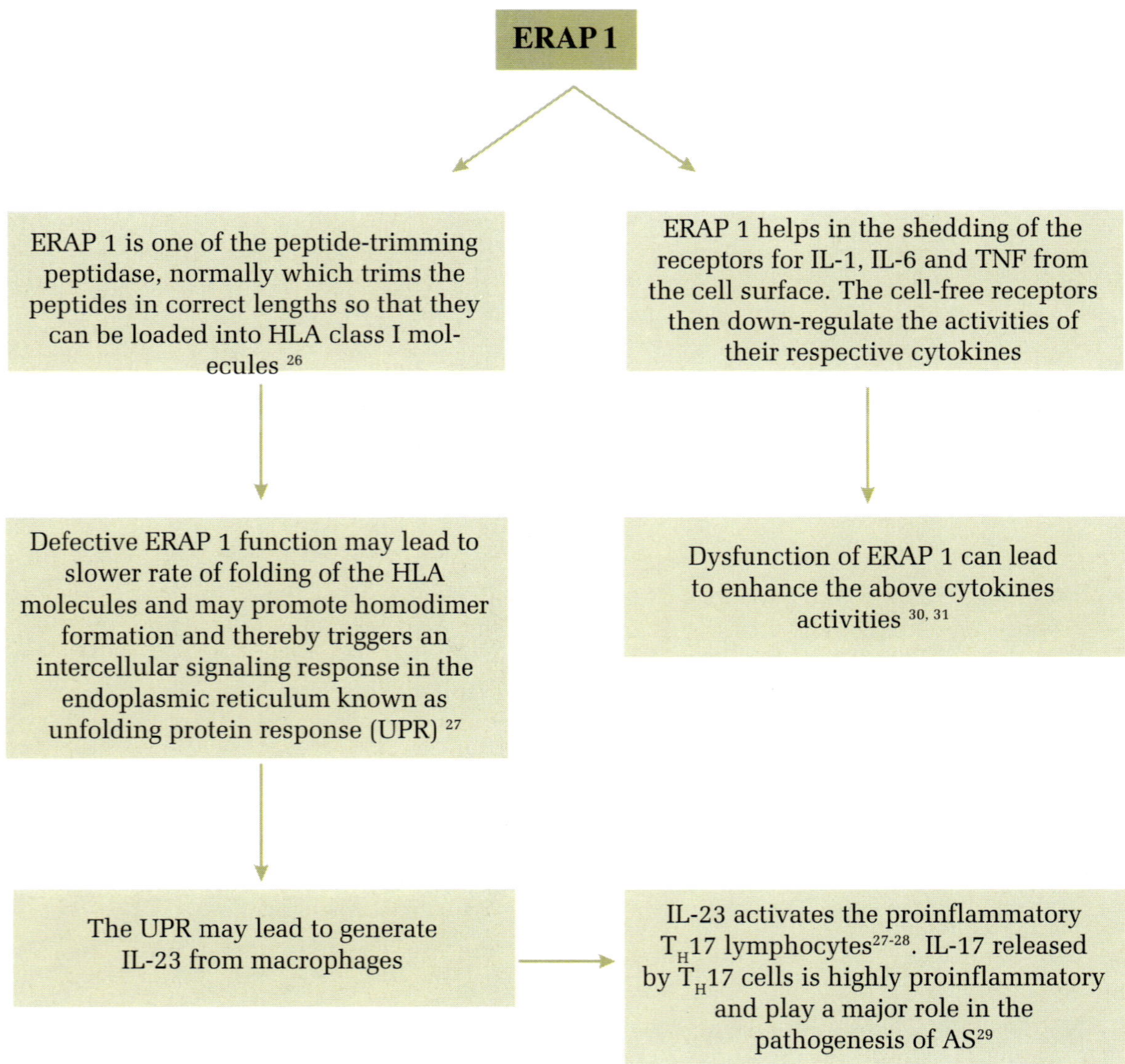

Fig . 2 : The role of ERAP 1 in the pathogenesis of AS.

Cytokines in AS- The TNF is one of the important cytokines which is responsible for morning stiffness, joint pain and swelling, fatigue in patients with AS. The anti-TNF therapy can dramatically improve these symptoms but the clinical remission is often incomplete. Hence, other cytokines might play a role. The role of other cytokines like ERAP 1, IL-23 and IL-17, have been briefly discussed above.

Clinical features

The clinical features of AS are of insidious onset; dull pain felt in lower lumbar region and associated with morning stiffness in the same area lasting for a few hours. There may be asymmetric arthritis of other joints mainly of lower limbs. Neck pain and stiffness are present in advanced cases. Morning back stiffness improves with activity. Buttock pain may be unilateral to start with. But within a few months it becomes bilateral with worsening at night and about 5% of patients having inflammatory back pain at presentation have AS or other spondyloarthropathies[32]. Arthritis in the hips and shoulders occur in some patients. Juvenile AS presents with mainly hip pain. There is an ample evidence to suggest disease heterogenecity of AS. Even though there are a lot of similarities, generally speaking, HLA B27 negative AS is somewhat latter in its onset, significantly less frequently complicated by acute anterior uveitis, and more frequently accompanied by psoriasis, ulcerative colitis and chron's disease and less often shows familial aggregation[33]. The prevalence of vertebral fracture in AS is 10-18% (2% in controls).

The course of disease progression is fluctuating and variable. The first ten years determines subsequent outcome. Onset within 5-10 years, oligoarthritis, hip arthritis, high ESR, poor efficacy of NSAID, lumbar spine limitation, sausage fingers are associated with poor outcome[6]. Patients with AS have 50% increased risk of death. Causes of death are spinal fractures, amyloidosis and cardiovascular lesions.

Physical findings are loss of spinal flexion, extension and diminished chest expansion, loss of lumbar lordosis, exaggerated thoracic kyphosis. Direct pressure over sacroiliac joint will produce increased pain. There is disproportionate limitation of movements with respect to the degree of ankylosis- due to secondary muscle spasm. Disease spectrum may range from mild back stiffness to a totally fused ankylosed spine with variable peripheral arthritis and extra articular involvement. In untreated cases the posture of the patients are exaggerated thoracic kyphosis, neck stooping forward, loss of lumbar lordosis, atrophy of buttocks. Occiput wall test, chest expansion, Schober test, pelvic compression. Gaenslen test and Patrick test are applied for evaluation of sacroiliac joint.

Radiographic changes occurring at SI joint are sacroiliitis characterized by blurring of cortical margin, bone erosion and sclerosis (Fig. 3). As erosion progresses the joint space appears wider and ultimately fibrosis and bony ankylosis of the joint occurs. Radiographic changes are seen primarily in axial skeleton especially in sacroiliac, discovertebral and costotransverse joints. The earliest and most characteristic lesion is sacroiliitis, which is the sine qua non of AS. The sacroiliitis is graded for severity (0=normal, 1=suspicious, 2=minimal abnormality- small localized areas with erosion and sclerosis, without alteration in the joint width, 3=unequivocal abnormality- moderate or advanced sacroiliitis with one or more of the following- erosion, sclerosis, widening, narrowing or partial ankylosis & 4=total ankylosis). In patients with a clinical diagnosis of possible AS, radiographic sacroiliitis may become manifest only after appropriate follow up.

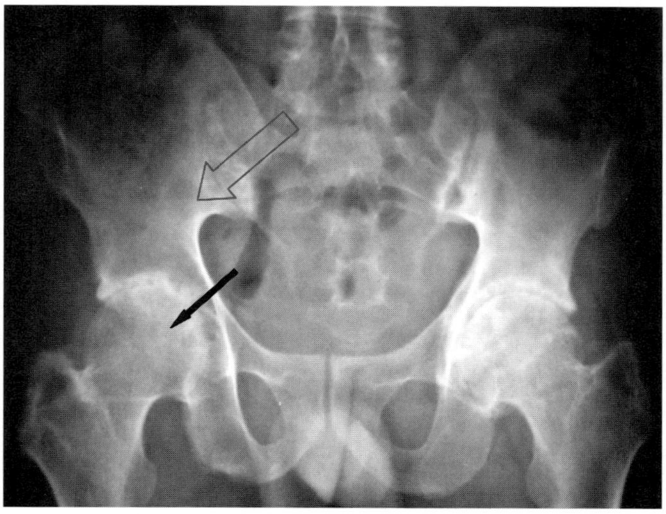

Fig 3: Digital x-ray pelvis shows bilateral radiographic sacroiliitis (wide arrow) and hip joints arthritis (black arrow).

CT and MRI can detect AS lesion earlier with greater consistency than plain radiograph[34]. MRI is most sensitive investigation for detection of early sacroiliitis than radiograph[34]. Bone marrow edema in sacroiliac joint indicates active sacroiliitis, whereas areas of sclerosis in sacroiliac joint indicate chronic sacroiliitis (Fig. 4). CT or MRI is helpful for monitoring progression of SI joint sclerosis, but radiographic findings do not correlate will with disease activity[34].

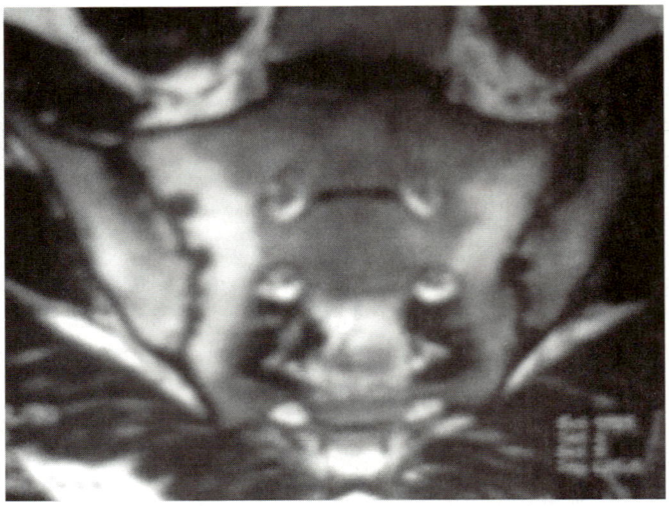

Fig 4 : MRI of sacroiliac joints showing marrow edema and erosion.

MRI features of acute sacroiliitis[35] :

i) Bone marrow oedema (BMO/osteitis)
ii) Synovitis
iii) Enthesitis and
iv) Capsulitis

MRI features of chronic sacroiliitis[35] :

i) Subchondral sclerosis
ii) Erosions
iii) Fatty marrow replacement and
iv) Bony ankylosis

Extra Skeletal Manifestations

Acute anterior uveitis is one of the most common extra skeletal manifestations, particularly in patients with HLA-B27 positive, but if treated promptly permanent visual loss may be avoided[36].

Cardio vascular involvement produces aortitis of ascending aorta leading to fibrosis and aortic insufficiency[36]. Other abnormalities may be arrhythmias including heart block. The incidence of cardiovascular disease in patients with AS is 10 to 40% and it rises with increasing disease duration[37]. Recent studies have shown, there is increased prevalence of cardiovascular diseases and higher cardiovascular mortality in these patients[38]. There may be as high as 4.4 fold increased risk of myocardial infarction in AS[39]. The basic mechanisms of accelerated atherosclerosis and cardiovascular disease is well known in rheumatoid arthritis and it is believed that the same mechanism is involved in AS[40]. Flow-mediated vasodilation (FMD), common carotid intima-medial thickness (ccIMT) and pulse-wave velocity (PWV) are used to assess for the future vascular events. The early detection of risk factors for vascular disorders guide to initiate the preventive measures and to start effective pharmacological therapy.

Patients may have respiratory difficulty due to limitations of chest movement a result of costovertebral and costosternal involvement but more concerning is the development of interstitial lung disease including bronchiolitis obliterans organizing pneumonia (BOOP). HRCT can be helpful in patients with suspected lung involvement.

Secondary renal amyloidosis, IgA nephropathy, mesangioproliferative glomerulonephritis and membranous nephropathy are the relatively rare renal manifestations in AS. Among these, secondary renal amyloidosis is the most common renal involvement.

Rarely patients may have arachnoiditis, which can be severely debilitating and associated with slowly progressive cauda equina syndrome. Of more concern is the spinal fracture specially in patients with advanced AS. Fracture may occur following minimum trauma or may be stress fracture. Acute fracture typically occurs at cervicothoracic and thoracolumbar junction as well as middle to lower cervical spine and typically involved fused intervertebral disc and posterior ligament. This leads to spinal cord compression and compressive myelopathy.

EXTRA SKELETAL MANIFESTATIONS OF ANKYLOSING SPONDYLITIS

Organ Involvement	Manifestations
Eyes	Acute anterior uveitis
Cardiovascular	Aortitis Aortic valve involvement-aortic insufficiency Conduction abnormalities-arrhythmias, heart block Cardiomyopathy Pericarditis
Respiratory system	Limitation of chest movement Upper lobe fibrotic changes
Bronchiolitis obliterans	
Genitourinary	Urethritis
Renal	Secondary amyloidosis IgA nephropathy Mesangioproliferative glomerulonephritis Membranous nephropathy NSAIDs nephropathy
Neurological	Spinal canal stenosis Cauda equina syndrome
Miscellaneous	Amyloidosis Osteoporosis

Diagnosis

The developments of sensitive and specific biological markers or imaging modalities for the diagnosis of early ankylosing spondylitis are one of the major challenges. Most often we face challenges regarding the early diagnosis, disease severity and activity, and refractoriness of the disease. Till date, modified New York Classification criteria are used for the diagnosis of ankylosing spondylitis. But the radiological sacroiliitis on plain radiograph takes long time to develop and for this reason we most often miss the opportunity to treat the disease early. One study had shown that patients with clinically early AS but plain radiograph showed normal sacroiliac joints, 36% had developed sacroiliitis after 5 years and 59% had developed radiographic sacroiliitis after 10 years[41]. Very recently, the ASAS (Assessment of Spodyloarthritis International Society) criteria for axial spondyloarthritis have been developed, where MRI of the sacroiliac joints used instead of conventional radiograph to diagnose the disease early.

The modified New York Classification criteria[42] (1984) for ankylosing spondylitis-

The clinical and radiological criteria as per the modified New York classification criteria are as below.

Clinical criteria	i)	Low back pain and stiffness for more than three months that improves with exercise but is not relieved by rest
	ii)	Limitation of motion of the lumbar spine in both the sagittal and frontal planes
	iii)	Limitation of chest expansion relative to normal values correlated for age and sex
Radiological criteria	i)	Bilateral sacroiliitis ≥ grade 2
	ii)	Unilateral sacroiliitis grade 3 or 4

Definite AS = Patient with one radiological criteria plus at least one clinical criteria
Probable AS = Patients with all three clinical criteria or radiological criteria[42]

There are certain limitations of the modified New York Classification criteria for the early diagnosis of AS. Radiographic sacroiliitis (either bilateral grade 2 or unilateral grade 3 or 4) must be present to diagnose a definite AS. It is fact that radiographic sacroiliitis is most often absent in early stages of the disease. On the other hand, the absence of radiographic sacroiliitis does not necessarily mean that there is absence of ongoing inflammation at the sites of sacroiliac joints and other parts of axial skeleton. Thus radiographic sacroiliitis is a marker of chronicity and or disease severity and it is a consequence of the inflammatory process and does not reflect the ongoing active inflammation.

The diagnosis of early AS, in patients with chronic inflammatory back pain who have absence of radiographic sacroiliitis is really a challenge. Because chronic back pain is a very common symptom in general population but only 5% of all patients with chronic back pain have diagnose to have AS[32]. The typical inflammatory back pain (IBP) is account for 70 to 80% of patients with AS and therefore it does not alone suffice for making the diagnosis.

MRI seems to be an important diagnostic tool for early detection of sacroiliitis in patients who have chronic IBP and absence of radiographic sacroiliitis. But the sensitivity and specificity of MRI findings in patients with early AS are needed to be validated.

The ASAS (Assessment of Spodyloarthritis International Society) criteria for axial spondyloarthritis have been developed where sacroiliitis on MRI and HLA-B27 positivity have been much weight than IBP. The ASAS classification criteria for axial SpA is as below-

In patients with chronic back pain ≥ 3 months and age at onset younger than 45 years	
Sacroiliitis on imaging plus ≥ 1 SpA feature	**HLA-B27 plus ≥ 2 other SpA features**
Sacroiliitis on imaging Active (acute) inflammation on MRI highly suggestive of sacroiliitis associated with SpA OR Definite radiographic sacroiliitis according to modified New York criteria (Adapted from Rudwaleit M, van der Heijde D, Landewe R, et al. The development of Assessment of Spondyloarthritis International Society classification criteria for axial spondyloarthritis (part II): validation and final selection. Ann Rheum Dis.2009; 68: 777-83.)	**SpA features** Inflammatory back pain Arthritis Enthesitis(heel) Uveitis Dactylitis Psoriasis Crohn's/Ulcerative colitis Good response to NSAIDs Family history for SpA HLA-B27 Elevated CRP

Assessment of the disease/outcome measures:

The assessment of disease activity is really difficult in AS, because the laboratory indicators of inflammatory arthritis neither reflect clinical nor radiological progression[43]. For this reason various scoring systems are now used to follow the course and response to treatment in As - i) Disease activity - BASDAI (Bath Ankylosing Spondylitis Disease Activity Index), ASDAS (Ankylosing Spondylitis Disease Activity Score) or ASAS-endorsed disease activity score, VAS of pain, Patient and Physician Global Assessments ii) Function - BASFI (Bath Ankylosing Spondylitis Function Index), DFI (Dougados Functional Index), HAQ-S (Health Assessment Questionnaire-Spondylitis), iii) Global status - BAS-G, iv) chronicity/severity index- BASMI [(Bath Ankylosing Spondylitis Metrology Index) which includes Modified Schober test, tragus to wall distance, Cervical rotation, lumbar lateral flexion, Intermalleolar distance], Occiput-to-wall distance, Chest expansion v) Radiological outcomes - BASRI (Bath Ankylosing Spondylitis Radiographic Index), mSASSS (Modified Stokes Ankylosing Spondylitis Scoring System), Conventional radiograph, MRI (Magnetic Resonance Imaging) vi) Quality of life- ASQoL (Ankylosing Spondylitis Quality of Life Index) which includes Short Form 36 [SF-36] and European Quality of Life Questionnaire [EuroQol].

Modified Stoke Ankylosing Spondylitis Spine Score (mSASSS) : is the most appropriate scoring method for the evaluation of radiographic progression in AS in clinical trials.

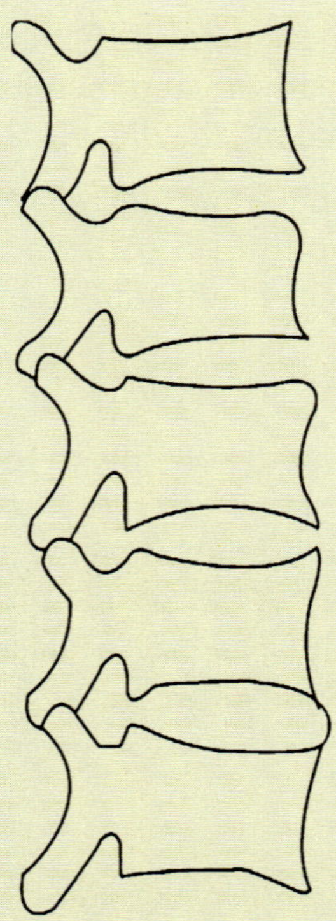

Score	Finding
0	Normal
1	Erosions
1	Sclerosis
1	Squaring
2	Obvious
2	Syndesmophytes
3	Total bony
3	Bridges

It is a 24 sites scored on the lateral cervical and lumbar spine :

Anterior corners of the vertebrae from lower border of C2 to upper border Th1 (including) and from lower border of Th12 to upper border of S1 (including).

Each corner can be scored from 0 to 3, resulting in a range from 0 to 72 for the total mSASSS.

BASRI (Bath Ankylosing Spondylitis Radiographic Index) : is a feasible and user-friendly method for evaluation of radiographic damage in patients with AS and can be used in clinical practice.

New York criteria for sacroiliitis (mean score of both SI joints is used in the BASRI)	0 = Normal 1 = Suspicious (no definite change) 2 = Minimal (minimal sacroiliitis, defined as the loss of definition at the edge of the SI joints, some juxtaarticular sclerosis, minimal erosions, and possible joint space narrowing) 3 = Moderate (moderate sacroiliitis, defined as definite sclerosis on both sides, blurring and indistinct margins, and erosive changes, with loss of joint space) 4 = Severe (complete fusion or ankylosis of the joints)

BASRI-hips (mean score of both hips is used in the BASRI-total)	0 = Normal (no change) 1 = Suspicious (focal joint space narrowing) 2 = Mild (circumferential joint space narrowing > 2 mm) 3 = Moderate (circumferential joint space narrowing 2 mm or bone-on-bone apposition of ≤ 2 cm) 4 = Severe (bone deformity or bone-on-bone apposition ≥ 2 cm) The grade should be increased by 1 if 2 of the following bone changes are present: erosions, osteophytes, and protrusion.
BASRI-spine (for the lumbar spine, AP and lateral views are scored, and the view with the highest score is taken; for the cervical spine, lateral view is scored)	0 = Normal (no change) 1 = Suspicious (no definite change) 2 = Mild (any number of erosions, squaring, or sclerosis, with or without syndesmophytes, on ≤ 2 vertebrae) 3 = Moderate (syndesmophytes on ≥ 3 vertebrae, with or without fusion involving 2 vertebrae) 4 = Severe (fusion involving ≥ 3 vertebrae)
SASSS and modified SASSS (range 0-72) (for the SASSS, the anterior and posterior sites from the lower border of T12 to the upper border of S1 are scored; for the modified SASSS, only the anterior site of the lumbar spine and the anterior site of the cervical spine from the lower border of C2 to the upper border of T1 are scored)	0 = Normal 1 = Erosion, sclerosis, or squaring 2 = Syndesmophyte 3 = Bridging syndesmophyte

SI - Sacroiliac; AP - Anteroposterior; SASSS - Stoke Ankylosing Spondylitis Spine Score.

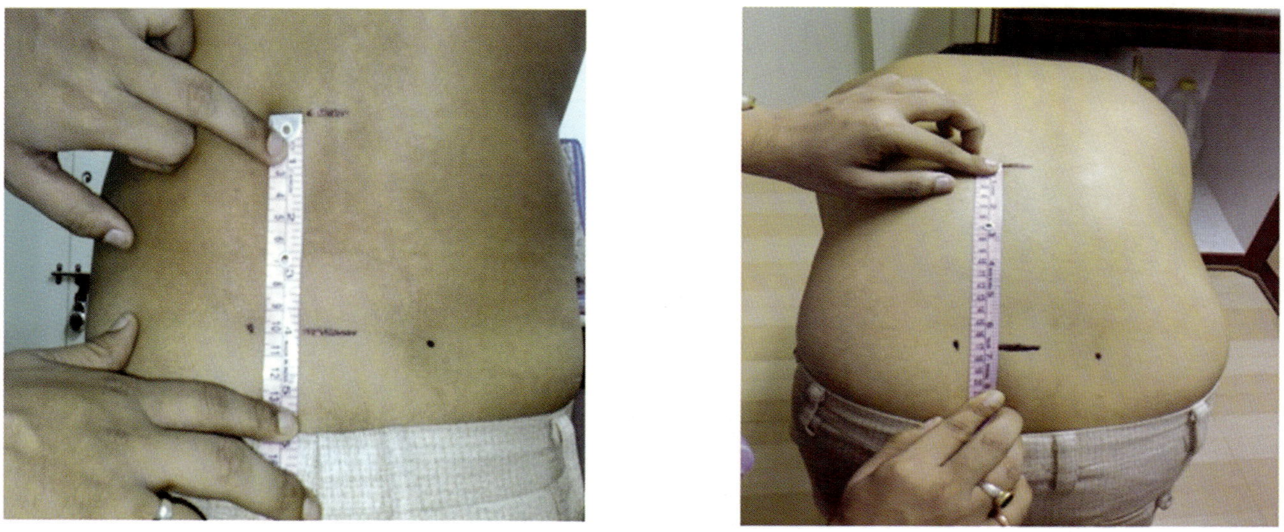

Fig. 5 : Modified Schober test: Patient must be standing erect. Mark an imaginary line connecting both posterior superior iliac spines (close to the dimples of Venus). The next mark is placed 10 cm above. The patient bends forward maximally: measure the difference. Report the increase (in cm to the nearest 0.1 cm). The better of two tries is recorded.

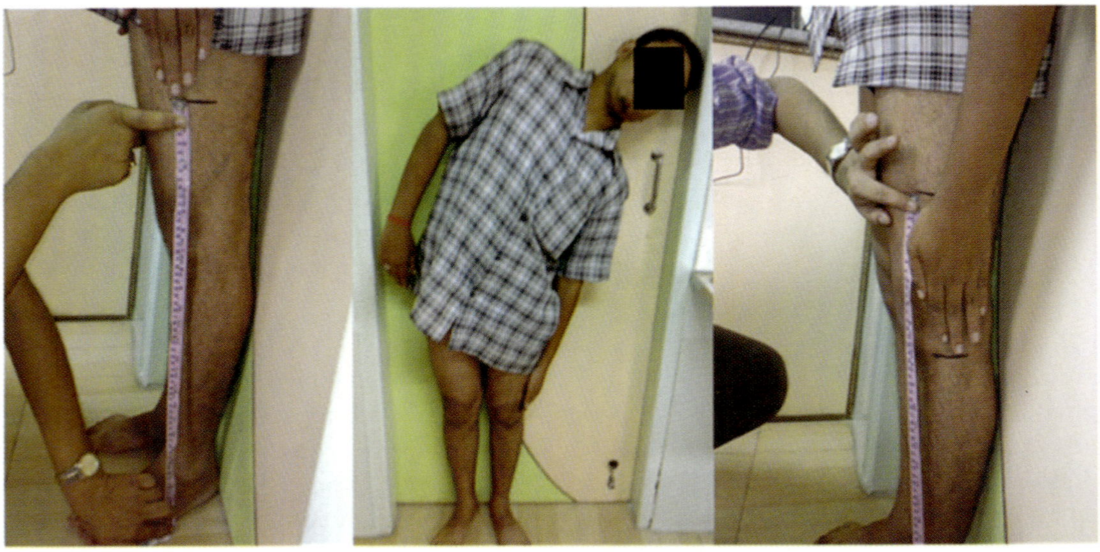

Fig. 6 : Lateral spinal flexion: Patient's heels and back rest against the wall. No flexion in the knees, no bending forward. Place a mark on the thigh (left), bend sideways without bending knees or lifting heels (middle), place a second mark and record the difference (right). The better of two tries is recorded for left and right separately. The mean of left and right is reported for lateral spinal flexion (in cm to the nearest 0.1 cm).

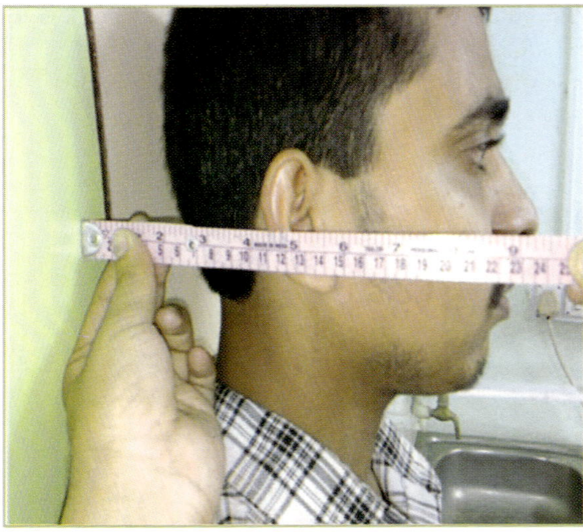

Fig. 7 : Occiput to wall & tragus to wall distance:

Patient's heels and back rest against the wall.
Chin at usual carrying level.
Maximal effort to touch the head against the wall.
Report the better of two tries in cm (e.g., 10.2 cm).
Occiput to wall distance.
Tragus to wall distance.

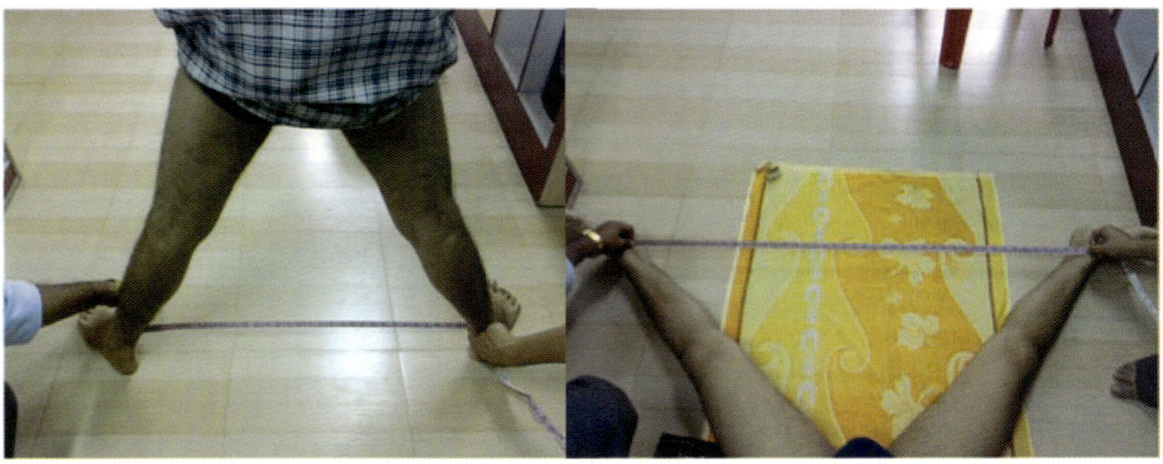

Fig. 8 : Intermalleolar distance: Alternatively, the patient stands (left) and separates the legs as far as possible. The distance between the medial malleoli is measured.

The patient is lying down (right) with the legs separated as far as possible with knees straight and toes pointing upwards (preferred method).

BASDAI is regularly used to assess the disease activity in daily clinical practice. It consists of six questions related to the patient's self assessment, these are fatigue (Q1), spine pain (Q2), joint pain/ swelling (Q3), enthesopathy (Q4) and intensity (Q5) and duration (Q6) of morning stiffness. ASDAS has been developed to improve the objective properties of subjective BASDAI. ASDAS consists of 4 questions from BASDAI and the level of acute phase reactant (CRP, ESR). However, the usefulness of this new index in daily practice needs validation in a large cohort.

In some study showed, arthritis or enthesitis may not present upto 50-70% of patients with AS[44]. As a result the estimated disease activity (by BASDAI) in this group of patients becomes low. In this back ground mini-BASDAI (omitting Q3 and Q4 from original BASDAI) have been proposed by a group of investigators and they concluded that approximately 5% of axial AS patients whose mini-BASDAI is superior to original BASDAI[45]. But till date, the original BASDAI is a valid tool to assess disease activity even in patients of AS without peripheral arthritis.

Disease activity could also be assessed by MRI of sacroiliac joints and the spine. Several scoring systems[46-48] have been developed and they are useful additional outcome measures in clinical trials[49-54] and their added value in clinical practice is unknown.

Treatment

Since no drug is currently available which cures the disease, the objectives of management are: education and counselling, relief of pain with NSAIDs, maintaining good mobility and functional status through physiotherapy and exercise advice, disease modification with drugs (NSAIDs, sulfasalazine, anti-TNF-α etc.), protection from injuries and fractures, corrective surgery (like hip replacement and vertebral osteotomy) and rehabilitation in late and disabling deformities.

NSAIDs

All the nonsteriodal anti-inflammatory drugs are equally efficacious. They are effective in pain relief and relief of stiffness in patients with AS[55] both in short-term and long-term continuous therapy. Recently, a 2 year randomized control trial of 215 patients with AS compares continuous and on demand NSAIDs with celecoxib or another NSAIDs and the result shows continuous treatment resulted in a reduction of radiographic progression without increasing toxicity substancially[56]. These agents along with physiotherapy form 'baseline therapy' and are always included in management protocol for AS. About 75% patients with AS show a good or very good response to a full dose of NSAIDs in 48 hours in contrast with only 15% of patients with mechanical back pain.

DMARDS

Sulfasalazine

Sulfasalazine has been used in AS for last 20 years. Sulfasalazine is superior to placebo in reducing pain and stiffness[57]. Some clinical trials have demonstrated its efficacy in peripheral arthritis and not spinal pain[58,59], while at least one recent study has demonstrated its effectiveness in spinal pain also[60]. Although the studies differ in design and duration, currently it may be summarized that sulfasalazine is effective in AS patients with peripheral arthritis but not in axial disease.

Methotrexate, Leflunomide

Methotrexate has not been found to be of any convincing value in axial AS, but found to have some debatable benefit in peripheral arthritis predominate AS. This has been shown in a recent study as well[61]. Similarly, leflunomide is also having some doubtful role in peripheral arthritis predominate AS[62].

Corticosteroids

It may provide rapid and considerable relief but oral administration is of little value. Intraarticular injection can provide rapid and longstanding relief[63]. As patients with AS already have significant loss of bone density, systemic glucocorticoid has not been recommended[64]. Intra articular corticosteroid injection can produce rapid and long standing pain relief but as the sacroiliac joints are most commonly involved in AS, its complicated anatomy is the only bar but recently CT guided injections of triamcinolone acetonide (40mg) into inflamed joints (SI joints) provide pain relief in about 80% of patients for a period of 8-9 months[65].

These are potent inhibitors of osteoclastic bone resorption and delayed type hypersensitivity characteristic of inflammation. It suppresses generation of proinflammatory cytokines.

Bisphosphonate

Pamidronate is commonly useful in patients with active AS refractory to NSAIDs, but further evaluation is needed to determine appropriate doses and dosing schedules[66]. The use of pamidronate in AS- was prompted by evidences that it suppresses inflammation and cartilage/bone erosion. For this reason pamidronate was commonly used in patients with active AS refractory to NSAIDs. Pulse pamidronate also had a beneficial effect on BASDAT. A randomized, double blind trial comparing 60 mg versus 10 mg IV doses of pamidronate was conducted in 84 patients with NSAID refractory AS (mean disease duration, 15.1 years)[67].

Thalidomide

This drug has immunomodulatory and anti-inflammatory properties. It acts by inhibiting production of IL-12 and TNF-α by monocytes and macrophages by reducing half-life of mRNA, leucocyte chemotaxis into the site of inflammation. It alters density of TNF-α induced adhesion molecules on leucocytes, reduces phagocytosis by PMNL and enhances mononuclear cell production of IL-4 and IL-5.

Different studies showed that the drug in a dose of 100-300 mg/day with a duration of 6 months to 1 year the improvement rate was 60-70%[68].

Anti TNF-α Therapy for AS

Treatment with TNF-α blocker is an important therapeutic advance in ankylosing spondylitis. Patients with shorter disease duration have a better response rate compared with those patients with a longer disease duration.

The drugs that have demonstrated efficacy in the treatment of AS are Chimeric monoclonal antibody Infliximab and the 75 KDa IgG receptor fusion protein etanercept.

In contrast to conventional therapy these biological molecules target specific inflammatory and immunoregulatory molecules. Both the drugs have been used to treat patients with AS in several open pilots and randomized control studies. The rationale for use in AS is significantly higher TNF-α serum levels than in patients with non inflammatory back pain. More recently high amount of TNTR-messenger RNA and protein were detected in sacroiliac joint biopsy specimen[69].

In our Rheumatology Centre at IPGMER, Kolkata 31 patients (Male - 23, Female - 8) with active ankylosing spondylitis were treated with Infliximab at 5mg/kg bodyweight on 0, 2nd and 6th weeks. There were both clinical and serological improvement[70].

MRI has been used to monitor the impact of anti TNF therapy on axial, peripheral and entheseal disease in patients with AS[71].

Braun et al assessed sacroiliitis and spinal MRI changes in 20 patients with AS before and after treatment with infliximab and showed resolution of sacroiliitis and marked improvement in Spondylitis[46].

ASSERT, the largest study of infliximab in AS confirms good clinical response of ASAS 20% (61% vs placebo 19%) and minimal drug toxicity[72].

The adverse effects of anti TNF-α therapy reported are - infections (sepsis, TB), malignancy (lymphoma), hematological disorders (anemia, pancytopenia), demyelinating diseases, auto immune responses, hypersensitivity etc.

Etanercept, is a recombinant 75 Da TNF-α p75 receptor fusion protein that acts competitively to inhibit the cell surface receptor binding of TNF-α Primary outcome measures - ASAS 20% (Study 60% vs Placebo 23%). In our Rheumatology Centre at IPGMER, Kolkata, 10 patients with AS treated with etarnercept (25mg twice weekly for 6 months) out of which 7 responded very well with significant improvement in standard clinical, function and QoL assessment.

Adalimumab (Humira), a humanized monoclonal anti-TNF-α antibody, has also demonstrated effectiveness in AS.

ASAS Improvement Criteria and Partial Remission Criteria

ASAS-20 Improvement Criteria BASFI Morning stiffness Patient global assessment Pain	At least 20% improvement and 10 units improvement in 3 of 4 following domains, without 20% or more worsening and 10 units worsening in the remaining domain
ASAS-40 Improvement Criteria BASFI Morning stiffness Patient global assessment Pain	At least 40% improvement and 20 units improvement in 3 of 4 following domains, without any worsening in the remaining domain
ASAS-5/6 Improvement Criteria BASFI Morning stiffness Patient global assessment Pain Acute-phase reactants Spinal mobility ASAS Partial Remission Criteria	At least 20% improvement in 5 of 6 following domains A value below 20 units in all 4 domains of ASAS-20

Latent tuberculosis infection (LTBI) screening: all patients who are eligible for anti-TNF α therapy (active symptomatic disease, BASDAI $\geq$ 4 and refractory to at least two NSAIDs during one month period) should screen for tuberculosis as because patients with anti-TNF having more chance to flare of tuberculosis. Incidence of TB reactivation is more common in patients treated with infliximab in comparison to etanercept. So our every effort is to exclude LTBI before starting of these drugs. Methods, available for LTBI screening are detailed clinical history (including recent/past contact with patients with active TB, past history of TB infection (TBI), family history of TB infection), routine chest X-ray, Mantoux test and newly introduced Quantiferon-TB Gold in-tube test. Patients with AS having active TBI should be treated promptly with adequate available treatment and anti-TNF is preferred to start after completion of anti-tubercular therapy. There is no current strong recommendation for the treatment of LTBI in our country. We prefer isoniazid plus rifampicin (RH) daily for 4 months for patients with LTBI before initiation of anti-TNFα therapy in our center.

Anakinra, a recombinant human IL-1 receptor antagonist, has shown conflicting results.

Physical therapy including exercise is a necessary adjunct to pharmacotherapy but the lack of controlled data makes it difficult to identify the best application of these interventions[73]. This modality of treatment has immense role in the management of AS in addition to pharmacotherapy. The goals are to improve mobility and strength and to prevent or reduce spinal deformities. But physical exercise is impossible until pain and inflammation are medically controlled.

In AS, the reduction in flexibility renders the spine susceptible to fracture, dislocation, deformity, spinal stenosis and hip disease. Surgery may become necessary in severe cases of AS.

The patient diagnosed with AS must receive education about the nature of disease and the treatment options. Informing about the spinal ankylosis may reinforce compliance with drug treatment, regimes and physiotherapy[74].

Because AS starts usually at an early age, the lifetime socioeconomic impact of the disease can be important for the patient as well as for society.

Conclusions - AS is a complex, unpredictable disease, if not diagnosed and treated at appropriate time and with appropriate drugs it would lead to disability and reduced quality life.

Further Reading :

1. Khan MA. Ankylosing spondylitis: clinical features. Rheumatology. 3rd ed. London, UK 2003: 1161-81.
2. Sieper J, Braun J et al. Ankylosing Spondylitis: an overview. Ann Rheum Dis 2002; 61.
3. Zink A, Braun J et al. Disability and handicap in RA and AS - results from German rheumatological database. J Rheum 2000; 27: 613-22.
4. Lee W, Reveille JD, Weisman MH. Women with ankylosing spondylitis: a review. Arthritis Rheum 2008; 59: 449.

5. Petty RE, Southwood TR, Manners P, Baum J, Glass DN, Goldenberg J, et al. International League of Associations for Rheumatology classification of juvenile idiopathic arthritis: second revision, Edmonton, 2001. J Rheumatol 2004; 31: 390-92.

6. Amor B, Silva-Santos R et al. Predictive factors for the long term outcome of spodyloarthropathies. J. Rheum 1994; 21: 1883.

7. Joel D. Taurog. The Spondyloarthritides. The Harrison's Principles of Internal Medicine. 18th ed. Volume 2; 2774-85.

8. Vander Lindeu SM et al. The risk of developing AS inHLA-B27 +ve individuals. Arthritis Rheum 1999; 17: 614-16.

9. Feldtkeller E, Khan MA, van der Heijde D, van der Linden S, Braun J. Age at disease onset and diagnosis delay in HLA-B27 negative vs. positive patients with ankylosing spondylitis. Rheumatol Int 2003; 23: 61-66.

10. Benjamin, M. & McGonage, D. The enthesis organ concept and its relevance to the spondyloarthropathies. Adv. Exp. Med. Boil. 2009; 649: 57-70.

11. Cawley MI, Chalmers TM, Kellgren JH, Ball J. Destructive lesions of vertebral bodies in ankylosing spondylitis. Ann Rheum Dis 1972; 31: 345.

12. Appel H, Kuhne M, Spiekermann S, et al. Immunohistologic analysis of zygapophyseal joints in patients with ankylosing spondylitis. Arthritis Rheum 2006; 54: 2845-51.

13. Appel H, Kuhne M, Spiekermann S, et al. Immunohistochemical analysis of hip arthritis in ankylosing spondylitis: evaluation of the bone-cartilage interface and subchondral bone marrow. Arthritis Rheum 2006; 54: 1805-13.

14. François RJ, Neure L, Sieper J, Braun J. Immunohistological examination of open sacroiliac biopsies of patients with ankylosing spondylitis: detection of tumour necrosis factor alpha in two patients with early disease and transforming growth factor beta in three more advanced cases. Ann Rheum Dis 2006; 65: 713.

15. Neidhart M, Baraliakos X, Seemayer C, et al. Expression of cathepsin K and matrix metalloproteinase 1 indicate persistent osteodestructive activity in long-standing ankylosing spondylitis. Ann Rheum Dis 2009; 68: 1334.

16. Walsh, N. C. & Gravallese, E. M. Bone remodelling in rheumatic disease: a question of balance. Immunol. Rev. 2010; 233, 301-02.

17. Maksymowych, W. P. Disease modification in Ankylosing spondylitis. Nat. Rev.rheumatol. 2010; 6: 75-86.

18. Lories, R. J., Luyten, F. P. & de Vlam, K. Progress in spondylarthritis. Mechanism of new bone formation in spondyloarthritis. Arthritis Res. Ther. 2009; 11: 221.

19. Lories, R. J., Derese, I. & Luyten, F. P. Modulation of bone morphogenic protein signaling inhibits the onset and progression of ankylosing enthesitis. J. Clin. Invest. 2005; 115: 1571-79.

20. Appel, H. et al. Altered skeletal expression of sclerostin and its link to radiographic progression in ankylosing spondylitis. Arthritis Rheum. 2009; 60: 3257-62.

21. Daoussis, D. Et al. Evidence that Dkk-1 is dysfunctional in ankylosing spondylitis. Arthritis Rheum. 2010; 62: 150-58.

22. Beaten, D. et al. Diagnostic classification of spondyloarthropathy and rheumatoid arthritis by synovial histopathology: a prospective study in 154 consecutive patients. Arthritis Rheum. 2004; 50: 2931-41.

23. Vandooren, B. et al. Absence of classically activated macrophage cytokine signature in peripheral spondyloarthritis, including psoriatic arthritis. Arthritis Rheum. 2009; 60: 966-75.

24. Ckou CT. Factors affecting pathogenesis of AS. Chin Med J (Engl) 2001; 114: 212-13.

25. Ball E, Khan MA. HLA-B27 polymorphism. Bone Joint Spine 2001; 68: 378.

26. Kanaseki T, Blanchard N, Hammer GE, et al. ERAAP synergizes with MHC class I molecules to make the final cut in the antigenic peptide precursors in the endoplasmic reticulum. Immunity 2006; 25: 795-806.

27. Tumer, M. J., Delay, M. L., Bai, S., Klenk, E., & Colbert, R. A. HLA-B27 up-regulation causes accumulation of misfolded heavy chains and correlates with the magnitude of the unfolded protein response in transgenic rats: Implications for the pathogenesis of spondylarthritis-like disease. Arthritis Rheum. 2007; 56: 215-23.

28. Colbert,R. A., DeLay, M. L., Layh-Schmitt, G. & Sowders, D. P. HLA-B27 misfolding and spondyloarthropathies. Prion; 2009: 15-26.

29. Layh-Schmitt, G. & Colbert, R. A. The interlukin-23/ interlukin-17 axis in spondyloarthritis. Curr. Opin. Rheumatol. 2008; 20: 392-97.

30. Cui, X.,Rouhani, F. N., Hawari, F. & Levine, S. J. An aminopeptidase, ARTS-1, is required for interlukin-6 receptor shedding. J. Boil. Chem. 2003; 278: 28677-85.

31. Thomas, G. P. & Brown, M. A. Genetics and genomics of ankylosing spondylitis. Immunol. Rev. 2010; 233: 162-80.

32. Underwood MR, Dawes P. Inflammatory back pain in primary care. Br J Rheumatol 1995; 34: 1074-77.

33. Khan MA, Killnes H. Immunogenetics of Spondyloarthropathy. Rheum Dis Clin North AM 1992; 18: 837-64.

34. Blum V et al. MRI for detection of active sacroiliitis - a positive study comparing conventional radiography. J Rheu 1996; 23: 2107-15.

35. Rudwaleit M, Jurik AG, Hermann KG, Landewé R, van der Heijde D, Baraliakos X et al. Defining active sacroiliitis on magnetic resonance imaging (MRI) for classification of axial spondyloarthritis: a consensual approach by the ASAS/ OMERACT MRI group. Ann Rheum Dis. 2009; 68: 1520-27.

36. Bennett DO, Ohashi K et al. Spondyloarthropathies, ankylosing spondylitis and psoriatic arthritis. Radiologic Clinic of North America 2004; 42: 121-34.

37. Sukenik S, Pras A, Buskila D, Katz A, Snir Y, Horowitz J. Cardiovascular manifestations of ankylosing spondylitis. Clin Rheumatol 1987; 6: 588-92.

38. Zochling J, Braun J. Mortality in ankylosing spondylitis. Clin Exp Rheumatol 2008;5 Suppl 51:S80-4. Heeneman S, Daemen MJ. Cardiovascular risks in spondyloarthritides. Curr Opin Rheumatol 2007; 19: 358-62.

39. Peters MJ, van Eijk IC, Smulders YM, Serne E, Dijkmans BA, van der Horst-Bruinsma IE, et al. Signs of accelerated preclinical atherosclerosis in patients with ankylosing spondylitis. J Rheumatol 2010; 37: 161-66.

40. Peters MJ, Symmons DP, McCarey D, Dijkmans BA, Nicola P, Kvien TK, et al. EULAR evidence-based recommendations for cardiovascular risk management in patients with rheumatoid arthritis and other forms of inflammatory arthritis. Ann Rheum Dis 2010; 69: 325-31.

41. Mau W, Zeidler H, Mau R, Majewski A, Freyschmidt J, Stangel W, et al. Clinical features and prognosis of patients with possible ankylosing spondylitis: results of a 10-year followup. J Rheumatol. 1988; 15: 1109-14.

42. van der Linden S, Valkenburg HA, Cats A. Evaluation of diagnostic criteria for ankylosing spondylitis. A proposal for modification of the New York criteria. Arthritis Rheum. 1984; 27: 361-68.
43. Taylor HG, Wardle T et al. The relationship of clinical and laboratory measurements to radiological changes in AS. Br. J Rheu 1991; 330-35.
44. Heuft-Dorenbosch L, van Tubergen A, Spoorenberg A, et al. The influence of peripheral arthritis on disease activity in ankylosing spondylitis patients as measured with the Bath Ankylosing Spondylitis Disease Activity Index. Arthritis Rheum 2004; 51: 154-59.
45. I H Song, M Rudwaleit, J Listing, J Sieper. Comparison of the Bath Ankylosing Spondylitis Disease Activity Index and modified version of the index in assessing disease activity in patients with in ankylosing spondylitis without peripheral manifestations. Ann Rheum Dis 2009; 68: 1701-07.
46. Braun J, Baraliakos X, Golder W, et al. Magnetic resonance imaging examinations of the spine in patients with Ankylosing spondylitis, before and after successful therapy with infliximab: evaluation of a new scoring system. Arthritis Rheum 2003; 48: 126-36.
47. Maksymowych WP, Imman RD, Salonen D, et al. Spondyloarthritis Research Consortium of Canada magnetic resonance imaging index for assessment of spinal inflammation in ankylosing spondylitis. Arthritis Rheum 2005; 53: 502-09.
48. Madsen KB, Jurik AG. Magnetic resonance imaging grading system for active and chronic spondylarthritis changes in the sacroiliac joint. Arthritis Care Res 2010; 62: 11-18.
49. Weber U, Hodler J, Jurik AG, et al. Assessment of active spinal inflammatory changes in patients with axial spondyloarthritis: validation of whole body MRI against conventional MRI. Ann Rheum Dis. 2010; 69: 648-53.
50. Lambert RG, Salonen D, Rahman P, et al. Adalimumab significantly reduces both spinal and sacroiliac joint inflammation in patients with ankylosing spondylitis: a multicenter, randomized, double-blind, placebo-controlled study. Arthritis Rheum. 2007; 56: 4005-14.
51. Maksymowych WP, Salonen D, Inman RD, et al. Low-dose infliximab (3 mg/kg) significantly reduces spinal inflammation on magnetic resonance imaging in patients with Ankylosing spondylitis: a randomized placebo-controlled study. J Rheumatol. 2010; 37: 1728-34.
52. Pedersen SJ, Sorensen IJ, Hermann KG, et al. Responsiveness of the Ankylosing Spondylitis Disease Activity Score (ASDAS) and clinical and MRI measures of disease activity in a 1-year follow-up study of patients with axial spondyloarthritis treated with tumour necrosis factor alpha inhibitors. Ann Rheum Dis 2010; 69: 1065-71.
53. Barkham N, Keen HI, Coates LC, et al. Clinical and imaging efficacy of infliximab in HLA-B27-positive patients with magnetic resonance imaging-determined early sacroiliitis. Arthritis Rheum 2009; 60: 946-54.
54. Dougados M, Combe B, Braun J, et al. A randomised, multicentre, double-blind, placebo-controlled trial of etanercept in adults with refractory heel enthesitis in spondyloarthritis: the HEEL trial. Ann Rheum Dis 2010; 69: 1430-35.
55. Escalas C, Trijau S, Dougados M. Evaluation of the treatment effect of NSAIDs/TNF blockers according to different domains in ankylosing spondylitis: results of a metaanalysis. Rheumatology (Oxford) 2010; 49: 1317-25.
56. Wanders A, van der Heijde D, Landewe R, Behier JM, Calin A, Olivieri I, Zeidler H, et al. Nonsteroidal antiinflammatory drugs reduce radiographic progression in patients with ankylosing spondylitis: a randomized clinical trial. Arthritis Rheum 2005; 52: 1756-65.
57. Ferraz MB et al. Meta Analysis of sulfasalazine in ankylosing spondylitis. J Rheu 1990; 17: 1481.
58. Clegg DO et al. Comparison of sulfasalazine and placebo in the treatment of axial and peripheral articular manifestations of seronegative spondyloarthropathis. A Department of Veterans Affairs Study.Arth Rheu 1999; 42: 2325-29.
59. Dougdos M et al. Sulfasalazine in the treatment of spondyloarthropathy. Arth Rheu 1995; 38: 618-27.
60. Braun J et al.. A placebo-controlled multicenter study of the efficacy and tolerance of sulfasalazine in early undifferentiated spondyloarthropathy. Ann Rheu Dis 2004; 63(suppl-1): 413.
61. Gonzalez-Lopez L et al. Efficacy of Methotrexate in ankylosing spondylitis:A randomized, double blind, placebo controlled trial. J Rheu 2004; 31: 1568-74.
62. Van Denderen JC et al. Double blind, randomized, placebo controlled study of leflunomide in the treatment of active ankylosing spondylitis. Ann Rheu Dis 2004; 63(suppl- 1): 397.
63. Doughados M, Khan M et al. Conventional treatments for AS. Ann Rheu Dis 2002; 61.
64. Braun J, Sieper J. Therapy of ankylosing spondylitis and other spondyloarthritides: established medical treatment, anti-TNF-alpha therapy and other novel approaches. Arthritis Res 2002; 4: 307.
65. Braun J et al. CT guided cortecosteroid injection of SI joint in patients with spondyloarthropathy with sacroiliitis. J Rheu 1996; 23: 659-64.
66. Maksymowych WP et al. 6 months randomized controlled double blind dose response comparison of IV pamindronate in NSAIDs refractory AS. Arth Rheu 2002; 46: 766-73.
67. Maksymowych WP et al. Clinical and radiological amelioration of refractory peripheral spondyloarthritis by pulse IV pamidronate therapy. J Rheum 2001; 28: 144-55.
68. Breban et al. Thalidomide in AS. Arthritis Rheum 1999.
69. Braun J, Bollow M et al. Use of immunohistologic and in situ hyloridization techniques in the examination of sacroiliac joint biopsy specimen from patients with AS. Arthritis Rheum 1995; 38: 499-505.
70. A. Ghosh et al. Treatment of Ankylosing Spondylitis with special reference to biologics : Single Centre Experience. JIRA 2004; 12: 54-57.
71. Braun J, Brandt et al. Treatment of active AS with infliximab - a double blind, placebo controlled multicenter trial. Lancet 2002; 359: 1187-93.
72. Van Der Heijde D et al. Ankylosing Spondylitis Study for the evaluation of recombinant infliximab therapy study group. Efficacy and safety of infliximab in patients with ankylosing spondylitis, results of a randomized, placebocontrolled trial (ASSERT). Arth Rheu 2005; 52: 582-91.
73. Dagfinrud H, Hagen K. Physiotherapy interventions for ankylosing spondylitis. Cochranc Database Syst Rev 2001; 4.
74. Dougados M, Dijkmaus B et al. Conventional treatments for AS. Ann Rheu Dis 2002; 61.

PSORIATIC ARTHRITIS-CLINICAL FEATURES AND MANAGEMENT

2

V Krishnamurthy, Sripurna Deepti

Introduction

It is an inflammatory arthritis associated with psoriasis and is a member of Spondyloarthritis (SpA) family. The rheumatoid factor and anti cyclic citrullinated peptide are generally negative and HLA-B27 can be positive in those with spinal involvement. It can be associated with dactylitis, peripheral arthritis, enthesitis and iritis.

Among patients with psoriasis 7 to 42% develop psoriatic arthritis (PsA) and in 85% of them skin psoriasis precedes arthritis whereas in 15% psoriasis develops later after the onset of arthritis and rarely never at all (Psoriatic arthritis sine psoriasis). The most common skin lesions are plaque psoriasis and psoriasis vulgaris but other patterns can be seen. Although psoriatic arthritis usually develops in the setting of an established diagnosis of skin psoriasis but some patients might be unaware that they have skin psoriasis. Physicians should carefully look for psoriatic skin lesions in the scalp, intergluteal/ perianal areas and nails since they are at higher risk of developing PsA. The prevalence is 0.04% to 0.1% and the incidence is about 3.6 to 6.6 per 100,000 per annum. The sex distribution is equal and the mean age of onset is 30 to 55 yrs. of age. In a recent study in South India it was found that males were more affected in a ratio of 2:1, the peak incidence was more common in the fourth and fifth decade (69%) and arthritis followed the skin lesion in 50.8% of the patients[1].

First-degree relatives and monozygotic twins have a higher risk for psoriasis and psoriatic arthritis. Psoriasis may be an etiologic and trigger factor for inflammatory arthritis. HIV infection is found to be associated with higher incidence of psoriasis[2].

Aetiology

The definite aetiology of psoriatic arthritis is not clear. Several factors may play a role[2].

Genes

Familial aggregation and paternal transmission are noted with chromosome 16q and indicate a role for genes. HLA class I molecules B-13 and B-17 have been implicated in psoriatic arthritis and Cw6 for psoriasis. HLA B-27 and B-7 are associated with sacroilitis in psoriatic arthritis independent of psoriasis. HLA-B*08, -B*27, -B*38 and C*06 differentiae PsA from psoriasis without arthritis (PsC) The Collaborative Association Study of Psoriasis showed that HLA-C and IL23R SNP's are more strongly associated with psoriasis and IL12B with PsA.

A study of the class I major histocompatibility complex (MHC) chain related gene A (MICA) showed MICA-A9 polymorphism. The corresponding allele MICA-002, was increased in PsA, whereas the Cw*0602 allele was significantly increased in both psoriasis and PsA. Thus MICA-002 may be a possible candidate gene for the development of PsA. There are differences in haplotypes between psoriasis vulgaris and psoriatic arthritis. Certain killer immunoglobulin-like receptor (KIR) alleles have been reported to be associated with PsA.

None of the studies have shown the usefulness of these genetic markers in identifying PsA in psoriasis[3].

Infection

Adequate proof for the role of infection is lacking. Psoriatic arthritis in HIV infection could be due to increase in CD8 T cells rather than infection.

Trauma and stress

Trauma and stress can act by a mechanism of up regulation of proinflammatory cytokines and release of neuropeptides in psoriasis and psoriatic arthritis and do not have a direct action. Psoriatic arthritis can develop following trauma similar to Koebner phenomenon where psoriatic skin lesions develop at sites of skin trauma.

Pathogenesis

T cells

Studies on active skin lesions, synovial fluid and response to biological agents have implicated the role of activated memory T cells expressing CD45RO and HLA-DR antigens in the pathogenesis[2].

Angiogenesis

Thickening and perivascular infiltration of capillaries and small arteries are seen in psoriatic skin lesions and synovium in contrast to straight and branching vessels seen in rheumatoid synovium. Increase in concentration of vascular growth factors and angiogenesis play a role in the pathogenesis.

Cytokines

TNF-α, nuclear factor κB (NFκB), IL-1, IL-12, IL-17 and IL-10 are expressed both in skin and in synovial membrane of patients with PsA in greater concentrations. These cytokines sustain the inflammation and hence the favourable response to TNF blockade. IL-17 and IL-23 potentiate osteoclastogenesis and bone erosion in PsA.

Classification and clinical features

The classification of Moll and Wright is still followed[4] (Table 1). Majority of the oligoarthritis is asymmetrical where as the polyarticular type is predominantly symmetrical in distribution. A new classification has been published called the CASPAR Classification (ClAssification criteria Study for Psoriatic ARthritis)[5]. It has high specificity but less sensitivity and is often used in clinical trials (Table 2). However the pattern of joint involvement is not fixed, it may fluctuate and overlap and can be influenced by treatment.

Table 1 : The moll and wright classification

- DIP arthritis
- Arthritis Mutilans
- Symmetrical Polyarthritis
- Asymmetrical Oligoarthritis
- Spondylitis

Table 2 : CASPAR criteria

Inflammatory Arthritis (Joint or Spine or Enthesitis) with ≥ 3 points from any of the following category

- Psoriasis: current (2), History of (1), Family history of (1)
- Nail Dystrophy (1)
- Negative for Rheumatoid Factor (1)
- Dactylitis: Current (1), History of (1)
- Radiographs of Hands or Feet with juxta articular new bone formation (1)

DIP arthritis

DIP arthritis is characteristic of psoriatic arthritis and occurs early in the disease (Fig. 1, 2). It is often associated with dactylitis and nail dystrophy (Table 3). The prevalence is about 3.9%. Disproportionate swelling of the IP joint of thumb was found to be associated with 28.7% of psoriatic arthritis and may have a positive predictive value of 84 percent[6].

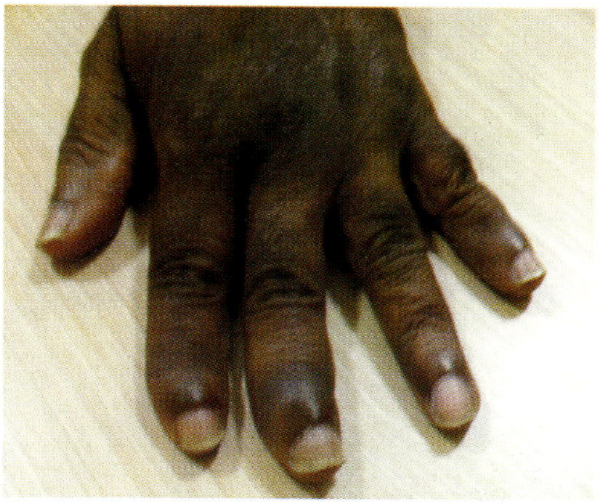

Fig. 1 : DIP arthritis with nail changes.

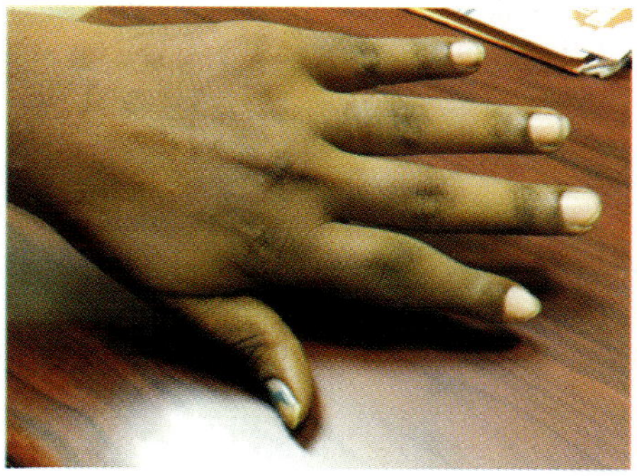

Fig. 2 : Dactylitis or sausaging of digits.

Table 3 : Nail changes in Psoriatic Arthritis

1. Pitting
2. Ridging
3. Onycholysis
4. Yellowish nail margins
5. Dystrophic hyperkeratosis
6. Combination of the above

Mono and oligoarticular arthritis

It is the most common presenting pattern and the prevalence is about 11 to 70%. They progress to involve additional joints and have a male preponderance.

Symmetric polyarthritis

Clinically indistinguishable from rheumatoid arthritis; it involves the small joints of hands and feet with a female preponderance. They often have longer disease duration and are associated with erosive arthritis simulating rheumatoid arthritis.

Arthritis mutilans

This is characterized by erosion and destruction of the joint. Subluxation, flail joints and telescoping of digits results in ('doigt en lorgnette') opera glass finger (Fig. 3). It has female preponderance and the prevalence is 5%. It is often associated with spondyloarthritis[7].

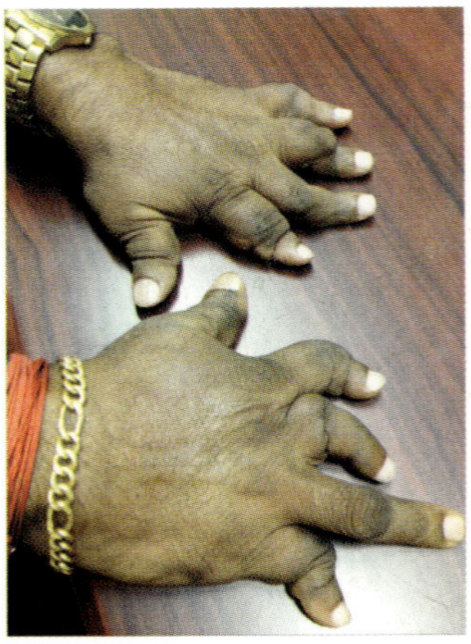

Fig. 3 : Arthritis mutilans.

Spondyloarthritis

This occurs as a predominant manifestation only in about 5% of patients. However spinal involvement can be made out clinically and by radiological methods in almost 40% of patients and its incidence can increase with prolonged disease without treatment. Spinal involvement simulates ankylosing spondylitis in the presence of symmetrical or asymmetrical sacroiliitis and ankylosis of spinal joints but differs by the lower incidence of zygoapophyseal joint involvement, improvement of spinal movement on treatment and atlanto axial subluxation of cervical spine similar to rheumatoid arthritis. Atlanto axial subluxation can be clinically silent and hence special attention should be paid to detect this joint involvement (Table 4). Bilateral sacroiliitis in psoriatic arthritis can be associated with HLA-B27.

Table 4 : Spine in Psoriatic Arthritis

- Unilateral sacroiliitis is common
- Zygoapophyseal fusion is less
- Skip lesions of syndesmophytes
- Atlanto-axial subluxation can occur

Dactylitis

Swelling of whole digit is called dactylitis and it is due to tenosynovitis. It occurs in 30 to 40% of patients and it is often associated with erosive arthritis and DIP involvement. Digits of feet are more commonly involved than the hands (Fig. 4).

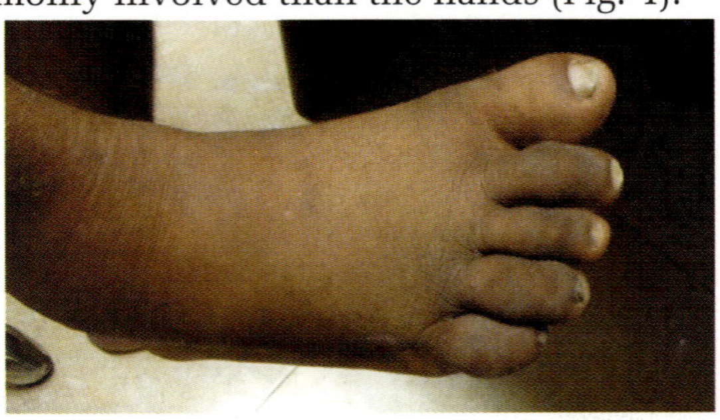

Fig: 4 - Sausaging of toes (Dactylitis).

Enthesitis

Inflammation at the tendon insertion points occur in 20 to 40% of cases and can be a presenting feature in 4% of them. Enthesitis is a characteristic feature of seronegative spondyloarthritis[8]. Tendo Achilles tendonitis, plantar fasciitis and pelvic bone ligament insertion enthesitis, insertions of quadriceps and patellar tendons are the common manifestations seen.

Peripheral oedema

Peripheral oedema of feet can occur due to extensor tenosynovitis and enthesitis and respond to steroids. Peripheral lymphoedema has been reported and can progress with active erosive disease. Its response to treatment with steroids and disease modifying drugs are variable.

Other associated features

The syndrome of synovitis, acne, pustulosis, hyperostosis and osteitis (SAPHO) can occur as a subgroup of psoriatic arthritis in 3% of patients and is associated with psoriasis vulgaris or palmoplantar pustulosis. Anterior chest wall involvement is revealed by scintigraphy in these patients.

Onycho pachydermo periostitis is a rare manifestation and is characterized by severe nail dystrophy, soft tissue swelling, DIP involvement and periostitis of terminal phalanx. Periosteal reaction of the terminal phalanx is radiologically described as Ivory phalanx.

Usually psoriatic arthritis patients have mild to moderate skin lesions. There is no correlation between joint scores and severity of skin lesions whereas there is a close correlation between nail changes and arthritis. 60 to 80% of psoriatic arthritis patients have nail changes. Nail changes include pitting, ridging (Fig. 5), hyperkeratosis and onycholysis. Arthritis occurs within 2 years of onset of nail changes and DIP arthritis is more commonly associated with nail changes. Patients with polyarticular disease and elevated inflammatory markers carry poor prognosis with 47% risk of erosions in first 2 years. There is a significant number of undiagnosed PsA in psoriasis patients and the risk of development of PsA in them does not vary with time.

Extra articular manifestations

Psoriatic arthritis can simulate other seronegative spondyloarthritis. They can manifest with iritis, conjunctivitis, urethritis and aortic valvular diseases and are usually found in patients with spinal involvement. High prevalence of inflammatory bowel disease has been noted in some studies.

Co morbidities

A number of studies had highlighted cardiovascular risk factors and metabolic syndrome in psoriatic arthritis.

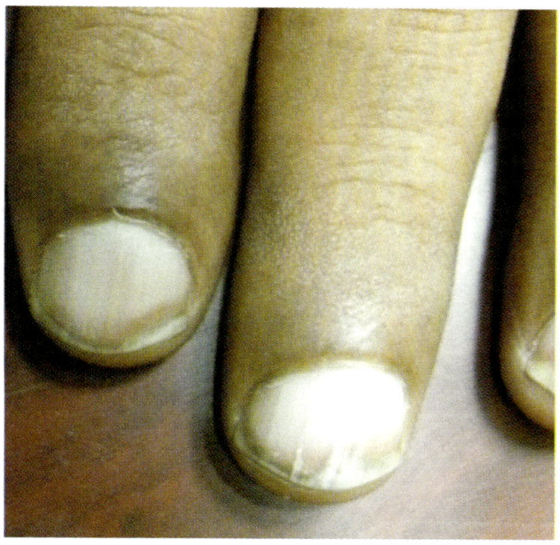

Fig. 5 : Longitudinal ridging of nails with dystrophy.

Juvenile psoriatic arthritis

Psoriatic arthritis occurring before the age of 16 years is labeled as Juvenile psoriatic arthritis. PsA accounts for 8 to 20% of childhood arthritis[9]. The criterion based on Southwood is used for diagnosing patients[9] (Table 5). In this type the arthritis precedes skin lesions in 50% of the patients and eliciting family history of psoriasis is important. There is a female preponderance. Oligoarthritis at presentation may progress to polyarthritis. Looking for nail changes is important in clinching the diagnosis. 5% of adults develop PsA at childhood.

Table 5 : Proposed criteria for the diagnosis of Juvenile Psoriatic arthritis (After southwood et al[10])

DIAGNOSIS	CRITERIA
Definite Psoriatic Arthritis	Arthritis onset before age 16 years and either Typical psoriasis or
	Three minor criteria from: Dactylitis, Nail pitting, Psoriasis like rash, Family history of psoriasis (first or second degree relative)
Probably Psoriatic Arthritis	Arthritis onset before age 16 years and Two minor criteria

Outcome measures

At present there are no valid disease activity scores to measure PsA. ACR 20, 50, 70 response criteria, as used in rheumatoid arthritis are substituted for assessement.

Investigations

Laboratory

There are no specific laboratory markers for psoriatic arthritis. Acute phase reactants are raised and in particular ESR and CRP correlate well with inflammation. The other markers are anaemia of chronic disease, hypoalbuminemia, hypergammaglobulinemia and fibrinogenemia. In 10% of patients low levels of rheumatoid factor and ANA can be positive. A positive test for rheumatoid factor does not rule out psoriatic arthritis. Hyperuricemia can occur and denotes renal impairment and increase in cholesterol. It is not due to increase in cell turnover as was thought earlier.

RADIOGRAPHY

Peripheral Joints

Sites of enthesitis show new bone formation. Erosive changes are more common at DIP joints and there can be associated soft tissue swelling. Subchondral osteoporosis is uncommon. Proliferative changes are due to periostitis and new bone formation along the shafts of metacarpals and metatarsal described as whiskering. Destructive changes or osteolysis result in pencilling or whittling of phalanx causing the pencil in cup deformity.

Spinal joints

Asymmetrical sacroiliitis is more common than symmetrical lesion. Syndesmophytes are thick and not marginal as in ankylosing spondylitis. They occur as skip lesions than as flowing syndesmophytes. In general the spinal lesions are less and not as severe as seen in ankylosing spondylitis but cervical involvement is more common.

MRI and Scintigraphy are useful in picking up subclinical musculoskeletal changes in patients with psoriasis.

Diagnosis

Diagnosis is clinical and a high index of suspicion in a case of inflammatory arthritis is needed to clinch it. Detailed history of psoriasis including family history and looking for skin lesions at the scalp, behind the ears (Fig. 6), umbilicus, gluteal folds and nail changes are essential. Axial signs, enthesitis, dactylitis and radiographic changes help in diagnosing the condition.

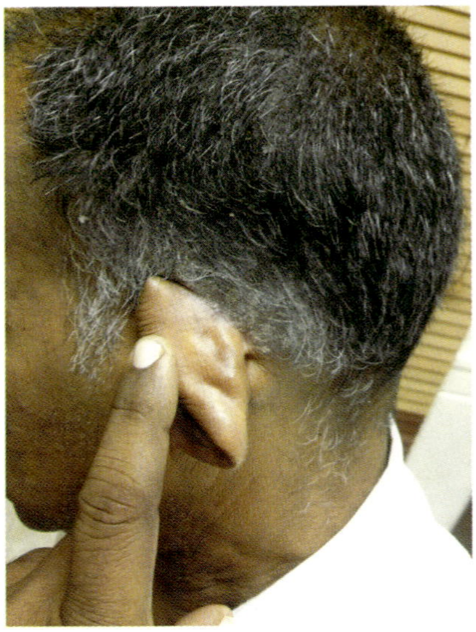

Fig. 6 : Psoriatic lesion behind the ear.

Differential diagnosis

It can be differentiated from rheumatoid arthritis in that it is asymmetrical with DIP involvement and has skin and nail changes. Asymmetrical sacroiliitis with non-marginal syndesmophytes differentiates it from ankylosing spondylitis. Reiter's syndrome can be difficult to differentiate. It involves more of lower limb joints and is associated with eye involvement and occurs in young males. DIP arthritis has to be differentiated from Herberden's nodes of osteoarthritis, gouty tophi in gouty arthritis and multicentric reticulo histiocytosis where periungual nodules are present[11].

Prognosis and mortality

Prolonged disease with polyarthritis is often associated with poor prognosis and increase in mortality. Juvenile onset, extensive skin lesions, onset in young adults, inadequate response to NSAID, association with HIV infection and association with HLA antigens like HLA-B27 (spondylitis), HLA-DR3, DR4 (erosive arthritis) are indicators for poor prognosis in psoriatic arthritis.

Management

Management of psoriatic arthritis is multidisciplinary and consists of non-pharmacological, pharmacological, rehabilitative and surgical therapies along with skin management. Patients with peripheral arthritis respond well to selective and non-selective COX-2 inhibitors. Disease modifying drugs methotrexate and leflunomide alone or in combination is very effective. Biological agents like etanercept and infliximab are very useful in inducing remission and retarding radiological progression. The emotional and social content in the management of psoriatic arthritis is more than that of rheumatoid due to the presence of skin lesions and needs to be addressed with due importance.

NON-PHARMACOLOGICAL MANAGEMENT

This includes patient education, family and employer counselling, rehabilitation and physical therapy.

PHARMACOLOGICAL

NSAIDs

Both selective and non-selective COX-2 inhibitors are useful in relieving pain and swelling of joints.

Steroids

Oral and intralesional corticosteroids have been associated with rebound worsening of psoriasis upon withdrawal of the drug and are therefore to be used with caution.

Disease modifying drugs

Psoriatic arthritis needs aggressive therapy since 20 to 30% of cases can progress to destructive arthritis. Though several drugs are used as DMARDs a few are of proven value.

Methotrexate

Even though convincing data is not available on methotrexate preventing structural damage and disease progression it is still the most commonly prescribed drug for PsA. Methotrexate can be started at 7.5mg to 10mg weekly once with folate supplements and can be escalated upto 25mg per week. It has to be avoided in patients planning for pregnancy, as it is teratogenic. Periodic blood count, renal and liver function tests once in 2 months is indicated to monitor toxicity[12]. Hepatotoxicity has been a concern in methotrexate treated psoriasis patients than in RA patients. It could be attributed to more frequent occurrence of metabolic syndrome and non-alcoholic steato hepatitis in PsA patients than in RA.

Leflunomide

Leflunomide is a prodrug and the active metabolite, A-77-1726, inhibits the enzyme dihydro-orotate dehydrogenase, thereby affecting the de novo production of pyrimidine. It controls T cell proliferation and pyrimidine is needed for this process. It is useful in patients who have not responded to methotrexate as well as a combination with methotrexate (data not convincing in PsA, adopted from RA studies). It is given as 10 to 20mg once daily dosage. It is hepatotoxic and periodic liver enzyme monitoring is necessary. It is to be avoided in women of childbearing age group and in young men who want to father a child since it is highly teratogenic.

Sulfasalazine

It is effective for the peripheral arthritis than the axial disease at a dose of 2gm per day. Gastrointestinal side effects are common with this drug.

Cyclosporine A

It is effective alone or in combination with methotrexate and sulfasalazine in severe resistant psoriatic skin lesions and arthritis. Low dose of 2 to 2.5mg/kg per day is found to be useful. Renal toxicity and high cost are the negative factors with this drug.

Azathioprine

Azathioprine in doses of 2 to 3mg/kg per day is effective. However, bone marrow side effects are high with this drug.

Retinoids

Etritinate is beneficial more in psoriasis than psoriatic arthritis. It is a teratogenic drug and has serious side effects.

Biological agents

Etanercept (soluble receptor for TNF α) 25mg is administered sc twice a week or 50mg sc once a week, Infliximab (chimeric monoclonal anti TNF α antibody) 3-5mg/kg infusion at 8 weeks interval after 3 loading doses at 0, 2, 6 weeks, Adalimumab 40mg sc fornightly, Certolizumab and Golimumab 50mg once a month sc (fully humanised anti-TNF α monoclonal antibody - not available in India) are the TNF α blockers approved for the treatment of psoriasis and psoriatic arthritis. They bring about a marked reduction in the signs of inflammation and delay radiological progression.

Alefacept (a fusion protein),Ustekinumab an (IL12/23 inhibitor), Apremilast(an oral small molecule inhibitor of phosphodiesterase type 4-PDE4) and Brodalumab (a human monoclonal antibody) are the new biologics being studied in psoriasis and psoriatic arthritis. Anti-IL-17 agent, ixekizumab as well the oral JAK inhibitor, tofacitinib, have very limited data.

Disease pattern	Therapy
Skin and nail disease	Topicals, PUVA/UVB, DMARDs (MTX, CsA), Anti TNF Blockers.
Peripheral arthritis	NSAIDS, IA steroids, DMARDs (MTX, LEF, SSZ, but no convincing data on combinations) Anti TNF blockers.
Spinal disease	NSAIDS, Physiotherapy, Anti TNF Blockers.
Dactylitis and Enthesitis	NSAIDS, IA steroids, Anti TNF Blockers.

The European League Against Rheumatism (EULAR) recommendations for the management of PsA are to use NSAIDs for relieving musculoskeletal pain and to use DMARDs at the earliest to induce remission. If there is an inadequate response to these agents TNF blockers are used with or without DMARDs and a switching of TNF blocker is considered only when there is an inadequate response to the first biological agent[13.14].

Summary

Diagnosing and treating psoriatic arthritis can be challenging. Immunosuppressive therapy including biological agents, methotrexate and cyclosporin are contraindicated in psoriatic arthritis secondary to HIV infection. Targeted therapies with biologicals are on the anvil as more information on the disease mechanism is getting unravelled.

Key points

- An inflammatory arthritis
- Elicit history of psoriasis
- Examine for psoriatic skin lesions in hidden areas
- Sex distribution is equal
- HLA-B27 is associated with bilateral sacroiliitis
- No specific laboratory marker
- Characteristic radiological changes are present
- Can simulate several inflammatory arthritic disorders
- NSAIDs and methotrexate play an important role in the management
- Biological agents slow radiological progression

Further Reading :

1. Rajendran CP, Ledge SC, Rani KP, et al. Psoriatic arthritis. J Assoc Phy India 2003; 51: 1065-68.
2. Antoni C, Psoriatic arthritis; aetiology and pathogenesis. In Hocgberg CM, Silman JA, Smolen SJ, Weinblatt EM, Weisman HM (Ed) Rheumatology (3rd Ed) New York: Mosby, 2003; 111: 1253-58.
3. Vinod Chandran. Psoriatic arthritis: How to diagnose it early? J Ind Rheu Assoc 2013; 8: 117-21.
4. Bruce IN, Psoriatic arthritis; Clinical features. In Hocgberg CM, Silman JA, Smolen SJ, Weinblatt EM, Weisman HM (Ed) Rheumatology (4rd Ed) New York: Mosby, 2008; 110: 1165-75.
5. Anthony M. Turkiewicz and Moreland LW. Psoriatic Arthritis- Current Concepts on Pathogenesis- Oriented Therapeutic Options. Arthritis Rheum 2007; 56: 1051-66.
6. Danda D, Cherian AM, Jayaseelan L. Disproportionate swelling of interphalangeal joint of thumb: is it unique to psoriatic arthropathy? J Ind Rheu Assoc 2000; 8: 1-4.
7. Marsal S, Armadans Gil L, Martinez M, et al. Clinical radiographic and HLA associations as markers for different patterns of psoriatic arthritis. Br J Rheu 1999; 38: 332-37.
8. Scarpa R. Peripheral enthesopathies in psoriatic arthritis; J Rheu1998; 25: 2259-90.
9. Ashok Kumar, Rahul Grover. Management of Psoriatic arthritis. J Ind Rheu Assoc 2002; 10: 64-68.
10. Southwood TR, Petty RE, Malleson PN, et al.Psoriatic arthritis in children. Arth Rheu 1989; 32: 1007-13.
11. Taurog DJ. The Spondyloarthritides. In Kasper LD, Fauci SA, Longo LD, Braunwald E, Hauser LS, Larry Jameson J (Ed) Harrisons text book of Medicine(16th Ed). New York: Mc Graw Hill, 2005; 305: 1998-99.
12. Cuellar M, Espinoza KH; Psoriatic arthritis; Management. In Hocgberg CM, Silman JA, Smolen SJ, Weinblatt EM, Weisman HM (Ed) Rheumatology (3rd Ed) New York: Mosby, 2003; 1259-61.
13. Glossec, Smolen JS et al; The European League Against Rheumatism (EULAR) recommendations for the management of Psoriatic Arthritis with pharmacological therapies. Ann Rheum Dis 2012; 71: 4-12.
14. Laura Coates, William Tillett, David Chandler, et al. The 2012 BSR and BHPR guideline for the treatment of psoriatic arthritis with biologics. Rheumatology (Oxford) (doi: 10.1093/rheumatology/ket187). 2013; 52 (Oct):1754-57.

3 ENTEROPATHIC ARTHRITIS

Rajiva Gupta

Enteropathic arthritis

Enteropathic arthritis belongs to the group of seronegative spondyloarthropathies (SSA), which also includes reactive arthritis, psoriatic arthropathy, idiopathic ankylosing spondylitis and unclassified seronegative spondyloarthropathies. Enteropathic arthritis is a form of arthritis associated with chronic inflammatory bowel disease (IBD) like ulcerative colitis (UC) and Crohn's disease (CD)[1]. Nomenclature is more of pragmatic rather than scientific one and is based on clinical observation. Joint involvement is also known to occur in other gastrointestinal diseases like Whipple's disease, intestinal bypass surgery and in celiac disease, although the exact mechanism by which joint disease occurs is not clear. Some rheumatologists also include post dysentery reactive arthritis under this terminology, but in this book it will be classified under reactive arthritis.

Epidemiology

About 3-4 million people worldwide get affected with IBD. Arthropathy is the most common extra-intestinal manifestation seen in IBD, with reported overall prevalence of 17-39%. The prevalence widely varies between studies; these discrepancies may be due to differences in study methods or diagnostic criteria for arthropathy. The prevalence of arthritis may be underestimated due to transient nature of oligoarticular arthritis or the response of arthritis to corticosteroids given for IBD flares. Most of the studies were cross-sectional. As extra-intestinal manifestations develop over time, the true incidence or prevalence cannot be estimated from such studies. Sacroiliitis is the most common manifestation, seen in 2.0-45.7% of patients. AS is seen in 1.0-16% of patients and peripheral arthritis is seen in 2.8-31% of patients. Overall 18.1-45.7% of patients fulfill the ESSG criteria for SpA. Prevalence of enthesitis among IBD patients in the literature is reported to be between 6% and 50%[2-4].

Arthritis may occur more frequently in CD than UC. Risk factors for arthritis may include active disease or family history of IBD[5]. Women may be at greater risk for peripheral arthritis, whereas men tend to have more frequent axial involvement. The presence of other manifestations such as Erythema nodosum or Pyoderma gangrenosum may also be a risk factor[6-8].

Pathogenesis

Although the association between spondyloarthropathy and inflammatory bowel disease (IBD) is largely established, the exact mechanism by which joint disease occurs in bowel inflammation is not known. Interestingly, this association is either ways - approximately 10%-15% of IBD cases are complicated by SSA and ileal inflammation resembling IBD has been reported in up to two thirds of cases of SSA. Familial clustering of IBD and AS, the coexistence of both conditions in a patient, the evidence of an increased risk ratio among first - and second-degree relatives of affected AS or IBD patients and finally, the increased cross-risk ratios between AS and IBD, strongly suggest a shared genetic background. It has been postulated that it is the recirculation of antigen - specific memory T cells from the gut to the joints that is responsible for initiating the joint inflammation. A likely chain of events would involve gastrointestinal infection with a bacterial micro-organism, followed by gut inflammation and inflammation in the joints. This especially applies to the development of peripheral arthritis where arthritis activity runs in parallel to gut activity, but it does not explain the development of axial involvement, which is independent of gut pathology.

Inflammatory bowel disease (IBD)

Arthritis is the most common extraintestinal manifestation of inflammatory bowel disease (IBD) and can have a significant impact on morbidity and quality of life. It occurs in approximately 30% of IBD patients[9,10]. Classically, joint involvement in enteropathic arthritis has two main patterns: 1) an axial involvement resembling idiopathic AS or sacroiliitis, 2) and a peripheral, asymmetric arthritis. These are not mutually exclusive. Other rheumatologic manifestations which are known to occur in IBD are metastatic granulomas of bone and joint, clubbing, periostitis, amyloidosis, psoas abscess leading to septic arthritis of the hip, osteoporosis, osteomalacia, granulomatous vasculitis and complications of steroid therapy[11]. Joint involvement is known to be the most common extra intestinal manifestation of the IBD[12]. A recent study from Brazilian population found that enteropathic arthritis accounted for only 3.5% of all spondyloarthritis[13].

Spondyloarthropathies

The European Spondyloarthropathy Study Group (ESSG) has developed classification criteria for the spondylarthropathies, which enable the inclusion of previously neglected cases of undifferentiated arthropathy[14]. Therefore the clinical spectrum of the spondylarthropathies is wide and includes axial symptoms, sacroiliitis, spondylitis, peripheral synovitis, dactylitis and enthesopathy. Axial involvement in IBD mostly manifests as either isolated radiographic sacroiliitis, which is often asymptomatic or ankylosing spondylitis (AS), which is almost indistinguishable from idiopathic ankylosing spondylitis. Previously, patients with inflammatory back pain (IBP) without radiographic abnormalities were not considered as part of enteropathic arthritis. But more recent studies have also looked into the prevalence of inflammatory back pain in IBD, in addition to sacroiliitis and ankylosing spondylitis[15]. The rate of occurrence of radiographic sacroiliitis (RSI) in IBD varies among series from and ankylosing spondylitis from 3% to 10%[16-20]. In a recent prospective study of 103 patients with IBD, which studied the overall prevalence of spondyloarthropathies in IBD, 30% patients had inflammatory back pain, 18% patients had asymptomatic sacroiliitis and ankylosing spondylitis was found in about 10% patients[21]. These figures contrasted from those seen in a recent population based study of 654 newly diagnosed patients with IBD and followed up for 6 years where the occurrence of ankylosing spondylitis was 3.7%, inflammatory back pain without ankylosing spondylitis was found in 18% patients and RSI was estimated to be present in 2% patients[22]. There was a much lower frequency of RSI in this study compared to hospital based studies probably reflecting the shorter duration of disease of patients and differences in radiological scoring technique and a referral bias in the hospital studies.

Clinical Features

The clinical features of ankylosing spondylitis in IBD are similar to idiopathic ankylosing spondylitis. The low back pain is characteristically insidious in onset and is associated with morning stiffness. Pain is better with exercise, worse with rest and is diffuse, involving the entire low back and buttocks. Although patients with idiopathic ankylosing spondylitis develop symptoms in their 20s, the ankylosing spondylitis in IBD can occur at any age. In idiopathic ankylosing spondylitis, men are more commonly affected while in IBD related ankylosing spondylitis both men and women are affected, although some have shown a male predominance[22]. In contrast to the peripheral arthropathy associated with IBD, the clinical course of ankylosing spondylitis and RSI in IBD runs more or less independently of the bowel disease. In a study designed to identify radiological differences it was observed that the IBD related ankylosing spondylitis has milder changes, squaring of vertebrae was more prevalent and Romanus lesions occurred only in idiopathic ankylosing spondylitis[23]. Peripheral joints may also be involved along with axial disease mainly the hip and shoulders. There is a clear association of HLA-B27 antigen in patients with spondylitis and IBD. While HLA-B27 occurs in 85-90% of patients with idiopathic AS, its frequency is lower in those who have spondylitis related to IBD[24,25]. HLA-B27 in IBD related ankylosing spondylitis is seen in around 50% to 70% patients[22,26,27]. This raises the possibility that there are other genetic or environmental factors, which contribute to disease expression and susceptibility[28]. Isolated sacroiliitis is often asymptomatic and most patients are HLA-B27 negative and often do not progress to ankylosing spondylitis. Magnetic resonance imaging is more sensitive to detect early sacroiliac changes. Symptoms of sacroiliitis when they appear do not correlate with the activity of the underlying bowel disease. Sacroiliitis is often symmetrical in the IBD related disease and in idiopathic ankylosing spondylitis while it is typically asymmetrical or unilateral in reactive arthritis and in psoriatic arthropathy. A recent study from Argentina which looked at differential features of primary ankylosking spondylitis and spondylitis associated with psoriasis and IBD found that patients with primary AS had more severe axial involvement than those with spondylitis associated with psoriasis or IBD. Functional capacity, disease activity, and quality of life were comparable among the groups studied[29].

Peripheral joint involvement

The incidence of peripheral arthritis in IBD has varied from 5 -20%[30-32]. In the past, the incidence of peripheral synovitis in Crohn's disease has been underestimated, probably strict criteria were not employed in distinguishing ulcerative colitis from Crohn's disease. More recently peripheral arthritis occurs slightly more in Crohn's disease than ulcerative colitis. Haslock and Wright in 1973 reported the musculoskeletal findings in 116 Crohn's disease patients. In their study peripheral arthritis was seen in 20%, AS in 6.9%, sacroiliitis alone in 9.5% of Crohn's disease patients[33]. Farmer et al in a systematic study of 615 cases of Crohn's disease found peripheral arthritis in 16% of those with colonic involvement compared with only 4% of those with small intestinal involvement and 3.6% in those with ileocolonic involvement[34].

i) Pattern of peripheral arthritis

The main pattern of joint involvement has been described as a pauciarticular, episodic, non-destructive synovitis involving large more than small and lower limb more than upper limb joints[33]. The classification has never been uniform until Orchard and coworkers from Oxford described the natural history and articular distribution of peripheral arthropathy in 637 patients of ulcerative colitis and 483 patients with Crohn's disease. In this study peripheral joint involvement was divided into two groups according to the number of joints involved (Table 1). The most commonly affected joint in type 1 arthropathy is the knee joint; the small joints of the hands are very rarely involved. In type 2 arthropathy the metacarpophalangeal joints are the most commonly involved. The peripheral arthritis associated with IBD is typically non-erosive and non-deforming in the long term, although erosive disease has also been reported. In a series by McEwen et al, 52 patients were studied, out of which persistent joint damage was seen in 5 patients[35]. In Crohn's disease septic arthritis of the hip has been reported which leads to rapid destruction of the joint and requires aggressive therapy[36].

Table 1. Classification of peripheral arthropathy in enteropathic arthritis[7]

Classification of enteropathic peripheral arthropathy
Type 1 (pauciarticular)
1. Less than 5 joints
2. Acute, self-limiting attacks (<10 weeks)
3. Often coincides with relapses of IBD
4. Strongly associated with extraintestinal manifestations of IBD
5. Association with HLA-B27, B35 and DR 103
Type 2 (polyarticular)
1. Five or more joints
2. Symptoms usually persist for months to years
3. Runs a course independent of IBD
4. Associated with uveitis but not with other extraintestinal manifestations
5. Association with HLA-B44

ii) Relation of peripheral arthritis with intestinal symptoms

Typically joint involvement occurs concurrently or after IBD has been diagnosed, although synovitis may predate the diagnosis of IBD in some percentage of patients. In the study by Orchard et al it was found that 31% of the type 1 arthropathy in ulcerative colitis and 24% in Crohn's disease presented at or 3 years before the diagnosis[5]. In contrast type 2 arthropathy was rarely apparent before diagnosis in both ulcerative colitis and Crohn's diseases. Peripheral arthritis is most commonly seen in extensive disease in ulcerative colitis, but in Crohn's disease the arthritis is more common in patients who have colonic involvement and less in small intestine lesion. In patients with ulcerative colitis, total colectomy may prevent further attacks of peripheral arthritis, whereas such surgery does not typically prevent recurrent arthritis in Crohn's disease[34,35]. It has been postulated that the diseased bowel may provide an immunologic stimulus for the occurrence of arthritis. In case of ulcerative colitis total colectomy provides removal of all diseased bowel therefore eliminating the arthritogenic stimulus. However in Crohn's disease surgical removal of all diseased bowels is unusual.

Other manifestations

Clubbing of fingers, uveitis, and skin manifestations are seen in IBD, with higher frequency in Crohn's disease. (Table 2) Uveitis in IBD is more often bilateral with a tendency for chronicity and poor response to topical steroid therapy. The pathogenesis of these lesions is unclear, but immunologic mechanisms are likely.

Table 2. Extra intestinal manifestations of inflammatory bowel disease (IBD)

	Crohn's Disease	Ulcerative Colitis
Peripheral arthritis	15%	10%
Sacroiliac arthritis	15-20%	10-15%
Uveitis	13%	4%
Nephrolithiasis (oxalate)	< 15%	?
Clubbing	4-3%	1-5%
Erythema nodosum	< 9%	?
Erythema multiformis	Rare	?
Pyoderma gangrenosum	0.5%	0.3-0.4
Aphthous ulcers	Rare	8%
Vasculitis	Takayasu's	< 5%
Liver disease	3-5%	7%
Amyloidosis	Very Rare	?

Gut pathology in spondylo-arthropathy

Spondyloarthropathy and gut inflammation are closely linked. Because the pattern of joint disease seen in patients with IBD and asymptomatic gut inflammation in spondyloarthropathies is similar, it has been postulated that the intestinal inflammation may play a pathogenic role in the arthritis. Mielants and co-workers from Belgium have documented that subclinical gut inflammation is prevalent in more than 50% of patients with spondyloarthropathies (after patients with psoriatic arthropathy and IBD related arthritis were excluded)[37-39]. Other workers have confirmed this finding. They have described two major types of intestinal inflammation in spondyloarthropathies an acute inflammation resembling infectious enterocolitis, with largely intact architecture and neutrophilic infiltration in the lamina propria. The chronic inflammation is more suggestive of early Crohn's disease, with distortion of villi and crypts, apthoid ulceration and mononuclear cell infiltration in the lamina propria[40]. In their follow up study of 217 patients with spondyloarthropathies (excluding psoriatic arthropathy and IBD related arthropathy) who had an initial ileocolonoscopy they found that of the 123 patients regularly followed up, 40 (32%) had normal gut histology, 28 (23%) had acute and 55 (45%) chronic inflammatory gut lesions. Eleven (5%) of these 217 patients went on to develop frank IBD, all these patients had ankylosing spondylitis, and the majority had chronic lesions in the initial ileocolonoscopy and were HLA B27 negative. Therefore, gut inflammation mainly subclinical was demonstrated in 68% of patients with spondyloarthropathies. The risk factors for development of IBD were regular diarrhoea early in the disease, raised inflammatory markers, chronic inflammatory gut lesions, and HLA-B27 negativity in the presence of bilateral sacroiliitis or ankylosing spondylitis[41-43]. Although gut inflammation in SpA has been well documented, it has not been established whether these lesions directly contribute to the pathogenesis of spondyloarthropathies, or whether patients with inflammatory gut lesions have a different disease course than those without. But the presence of gut inflammation does seem to predispose the evolution from non-ankylosing spondylitis/spondyloarthropathies to full blown ankylosing spondylitis[44].

Management

By and large management of peripheral arthritis depends upon controlling the underlying bowel disease, particularly since the natural course of type 1 arthropathy parallels that of the underlying bowel activity. The natural course of axial arthropathy, which is characterized by flares and remissions and the persisting disability of type 2 arthropathy make them more challenging to manage. Therefore in contrast to peripheral type 1 arthropathy that usually responds to treatment of colitis, medical or surgical therapy of the underlying disease does not alter the progressive nature of the axial arthropathy.

Non steroidal anti-inflammatory drugs (NSAIDs)

Most patients with mild spondyloarthropathy will respond to NSAIDS. Although these drugs are known to increase gut permeability and cause inflammation of the small intestine (NSAIDs enteropathy)[45-47], it has been observed that gut inflammation in spondyloarthropathies resolve in some patients who have spondyloarthropathies and who continue to take NASAIDs. NSAIDs enteropathy is mainly asymptomatic, localized to the mid small intestine rather than the ileum and rarely involve colon.

DMARDs

Sulfasalazine (SSZ) - This drug inhibits the function of nuclear factor kappa B (NFkB) that influences the production of pro-inflammatory cytokines. SSZ is of proven value in the induction and maintenance of remission of IBD particularly ulcerative colitis. Its role in remission maintenance in Crohn's disease is less well defined. The beneficial effect of sulfasalazine on the peripheral joint disease in IBD is most probably because of the effect of the 5-aminosalicylate (5-ASA) moiety of the drug on reducing gut inflammation. SSZ reduces peripheral joint disease activity in ankylosing spondylitis, but it has only a modest effect, if any, on the axial disease. In a meta-analysis of five placebo controlled, randomized trials sulfasalazine was effective in reducing morning stiffness and improving well-being in patients who had ankylosing spondylitis[48]. More recently a multicentric randomized controlled study on the role of sulfasalazine in the treatment of undifferentiated spondyloarthropathies (uSpA) and early ankylosing spondylitis showed that sulfasalazine was no better than placebo for the treatment of the signs and symptoms of uSpA however, sulfasalazine was more effective than placebo in the subgroup of patients with IBP and no peripheral arthritis[49].

Methotrexate (MTX) has greater utility in CD than in UC for gut inflammation; further evaluation is ongoing in two randomized multicenter trials (METEOR in Europe and MERIT in U.S.)[50]. MTX, which is in use in rheumatoid arthritis since ages now, may prove beneficial for arthropathy in IBD as well as for IBD itself. Hydroxycholorquine and azathioprine are not effective for arthritis in IBD. Other peripheral musculoskeletal manifestations, such as enthesitis or dactylitis, are also indications for DMARD therapy.

Anti-tumour necrosis factor (anti-TNF) therapy

There are three main biological agents targeting tumour necrosis factor α (TNF-α): the chimeric monoclonal immunoglobulin G1 antibody (with human constant and murine variable regions) infliximab, the recombinant 75-kDa TNF receptor IgG1 fusion protein etanercept and the fully humanized monoclonal antibody adalimumab. TNF inhibitors have been found to be highly effective for IBD patients who are steroid dependent or refractory to conventional treatment. Infliximab is the best studied anti TNF therapy in IBD. Infliximab is highly effective in moderate to severe CD and UC, promoting fistula closure, mucosal healing, and sparing use of steroids. There is also evidence that Infliximab is effective for axial and peripheral joint manifestations of IBD, largely from small case series and open-label studies. Adalimumab also has effectiveness for AS and CD. Results from the CARE study demonstrated a benefit for Adalimumab in treating the EIMs of CD. CARE was a large multi-center phase IIIb open-label clinical trial of 945 CD patients conducted in Europe there was a benefit for arthropathy with incidence of arthritis reduced from 8.7% at baseline (82 patients) to 2.1% (20 patients) at week 20. A reduction in arthralgia and sacroiliitis was also seen. There was no effect seen for ankylosing spondylitis in this study, but the incidence of AS in this population of patients with IBD was very low[51]. The report in 2000 of the successful response of patients with active ankylosing spondylitis to the anti-TNF monoclonal antibody infliximab dramatically expanded the treatment options in the management of ankylosing spondylitis[52,53]. Etanercept has also been studied extensively and has been found to work in patients with ankylosing spondylitis[54,55]. Adalimumab has been proven to be safe and efficacious in RA and psoriatic arthropathy. It has also been shown to reduce spinal symptoms in an open label study of 15 NSAID refractory patients of ankylosing spondylitis with moderate to severe active disease[56]. These biologic agents (infliximab and etanercept) have revolutionized the treatment of Crohn's disease too, especially in fistulous and refractory disease. Infliximab has been found to be useful in controlling not only the bowel inflammation but also the associated articular symptoms[57]. Ustekinumab a fully human monoclonal immunoglobulin (IgG1) against the interleukin (IL)-12/23 shared P40 subunit has been shown in phase 2 studies to have clinical effectiveness in CD gut inflammation, and may be particularly beneficial in patients refractory to Infliximab[58].

Whipple's disease

This is a rare multisystem disease caused by the bacteria Trophyrema whipplei. In its classic form whipple's disease presents with weight loss, diarrhea, fever, joint pain, neurological disease, adenopathy and uveitis[59]. It may also present as a culture negative endocarditis. Joint involvement occurs in up to 90% patients with Whipple's disease[60]. Joint involvement occurs in the form of intermittent migratory polyarthralgia, of the large joints and it is characteristically non-erosive. The articular symptoms are episodic and often followed by periods of remission. Oligoarticular involvement may also occur. The joint involvement is often accompanied by intestinal symptoms and isolated presentation is uncommon. Whipple's disease should be considered in the differential diagnosis of seronegative polyarthralgia in a male patient. Axial involvement with sacroiliitis and lumbar pain may also be seen[61]. Earlier methods of diagnosis relied solely on the detection of diastase resistant, PAS positive, foamy macrophages in biopsy tissue. Diagnosis is now being complemented by the use of PCR to identify the organism in involved tissue like duodenal biopsy[62]. Immunohistological methods to detect T whipplei in involved tissue are currently being developed and will probably have wide applications.

The disease being rare (only 1000 odd cases being reported so far) there are no randomized trials of antibiotic therapy. At presentation patients should be treated with intravenous antibiotics, which cross the blood-brain barrier like streptomycin with either ceftriaxone or penicillin G followed by oral cotrimoxazole. Cotrimoxazole is the oral antibiotic of choice, because it crosses the uninflammed bloodbrain barrier and its administration is observed to be associated with less likelihood of neurological relapse. Patients are usually treated for 12 to 24 months with close clinical, histological and laboratory monitoring by PCR. For patients who do not have neurologic involvement the combination of doxycycline 200mg per day and hydroxychloroquine 200mg three times a day works as bactericidal agents. Interferon gamma can be offered to patients with resistant neurological disease.

Celiac disease

Classically celiac disease is a gluten sensitive enteropathy that occurs in genetically predisposed individuals and responds to the withdrawal of gluten from the diet. The human major histocompatibility molecules DQ2 and DQ8 are essential genetic factors for the development of celiac disease with majority of the patients carrying DQ2 alleles. The triggering antigen is wheat gliadin, which gets deaminidated by tissue transglutaminase. The deaminidated gliadin then binds to the DQ2 molecule of the antigen-presenting cell causing an immune reaction in the lamina propria. The clinical manifestations of celiac disease are protean and it has been called "the great modern day imposter". Adults present with GI symptoms including diarrhoea, weight loss, and abdominal discomfort. The extra-intestinal features results from nutrient malabsorption and include anemia, hypocalcemia, osteopenia, infertility and peripheral neuropathy. Inflammatory arthritis is a well described extraintestinal manifestation seen in up to 26% patients of celiac disease when compared to 7.5% controls[63]. The major pattern of arthritis described is large joint, seronegative, non-erosive, non-deforming oligo or polyarticular type, similar to the other enteroarthritides. In some patients arthritis may be the first presenting complaint and the arthritis may occur in the absence of bowel disease[64,65]. Axial involvement in the form of sacroiliitis has been found in 14 out of 22 adult celiac disease patients studied by bone scintigraphy. Diagnosis is suggested by typical history supported by the presence of IgA anti-endomysial or IgA antitissue transglutaminase antibodies along with small intestinal histology, which shows villous atrophy and intraepithelial lymphocytosis. Gluten free diet is helpful in relieving the arthritis[65].

Collagenous colitis

Collagenous colitis (CC) is a condition characterized histologically by a diffuse thickening (> 10mm) of the subepithelial collagen layer of the colon, along with intraepithelial lymphocytosis and infiltration of the lamina propria with inflammatory cells. It is an inflammatory condition affecting mainly middle aged or elderly females and presents with watery diarrhea, crampy abdominal pain, weight loss and a normal colon by endoscopy[67]. Another related condition with a similar mode of presentation is lymphocytic colitis characterized by diffuse inflammation of the lamina propria and with intraepithelial lymphocytic inflammation. The name microscopic colitis has been given to include both collagenous colitis and lymphocytic colitis[66,68,69]. Although unrelated these conditions are of rheumatological interest because they are associated with many autoimmune conditions. Collagenous colitis has a possible association with autoimmune thyroid disease, idiopathic uveitis, polymyalgia rheumatica, giant cell arteritis, myasthenia gravis, scleroderma, Sjogren's syndrome, idiopathic pulmonary fibrosis, lupus and rheumatoid arthritis. Among the various rheumatological conditions collagenous colitis is most commonly associated with rheumatoid arthritis[70]. Most patients are seropositive and have an erosive disease. Rheumatoid arthritis usually precedes the intestinal symptoms by many years and occurs independent of the exacerbation of the arthritis. Therefore, collagenous colitis should be considered in the differential diagnosis of watery diarrhea in a patient with rheumatoid arthritis. Collagenous colitis has also been reported to occur in patients with SpA who meet the ESSG criteria. Probably the most important hallmark is involvement of the axial skeleton with symptomatic or asymptomatic sacroiliitis[71]. Peripheral joint involvement is either oligoarthritis or polyarthritis of both large and small joints and not necessarily with lower limb predilection. Joint involvement may antedate or coincide with the onset of bowel disease in half of the patients. The symptoms are transient may run an intermittent course and may become chronic in some cases, with the development of ankylosis and erosions on the radiographs. Treatment consists of oral glucocorticoids, sulfasalazine, NSAIDs and anti diarrhoeals.

Arthropathy associated with intestinal bypass surgery

The surgical management of morbid obesity by the induction of an intestinal bypass (to reduce the intestinal mucosal surface) is associated with a large number of metabolic and inflammatory complications. The metabolic complications described are hyperoxaluria, osteomalacia and nephrolithiasis while hepatic degeneration, bypass enteropathy and arthropathy are the major inflammatory complications. These complications occur in both jejuno-colic and jejuno-ileal procedures. Arthropathy occurs in 35 % to 52% of patients post operatively. Arthropathy seems to occur more in jejunocolic bypass than in jejunoileal bypass. Delamere et al found that 5 out of 9 subjects in their study who underwent jejuno-colic bypass had arthropathic symptoms. This could be due to bacterial contamination and overgrowth that occurs in the blind loop[71]. The arthropathy is episodic, migratory nonerosive, involving both large and small joints and reversal of the bypass is associated with complete and permanent remission of the arthritis. The arthritis usually occurs within 3 years after the procedure, although delayed occurrence has also been reported. The pathogenesis is most likely immune related. Increased levels of complement-containing cryoprecipitates and circulating complexes have been demonstrated in patients who develop joint symptoms after bypass procedures[72-74].

Further Reading

1. Holden W, Orchard T, Wordsworth P. Enteropathic arthritis. Rheum Dis Clin N Am 2003; 29: 513-30.
2. Vlam K, Mielants H, Cuvelier C, et al. Spondyloarthropathy is underestimated in inflammatory bowel disease: prevalence and HLA association. J Rheumatol. 2000; 27: 2860-65.
3. Palm O, Moum B, Ongre A, Gran JT. Prevalence of ankylosing spondylitis and other spondyloarthropathies among patients with inflammatory bowel disease: a population study (the IBSEN study) J Rheumatol. 2002; 29: 511-15.
4. Turkcapar N, Toruner M, Soykan I, et al. The prevalence of extraintestinal manifestations and HLA association in patients with inflammatory bowel disease. Rheumatol Int. 2006; 26: 663-68.
5. Gravallese EM, Kantrowitz FG. Arthritic manifestations of inflammatory bowel disease. Am J Gastroenterol 1988; 83: 703-09.
6. Vavricka SR, Brun L, Ballabeni P, et al. Frequency and risk factors for extraintestinal manifestations in the swiss inflammatory bowel disease cohort. Am J Gastroenterol. 2011; 106: 110-19.
7. Orchard TR, Wordsworth BP, Jewell DP. Peripheral arthropathies in inflammatory bowel disease: their articular distribution and natural history. Gut. 1998; 42: 387-91.
8. Veloso FT, Carvalho J, Magro F. Immune-related systemic manifestations of inflammatory bowel disease. A prospective study of 792 patients. J Clin Gastroenterol. 1996; 23: 29-34.
9. Salvarani C, Vlachonikolis IG, Heijde DM, et al. Musculoskeletal manifestations in a population-based cohort of inflammatory bowel disease patients. Scand J Gastroenterol. 2001; 36: 1307-13.
10. Lanna CC, Ferrari Mde L, Rocha SL, et al. A cross-sectional study of 130 Brazilian patients with Crohn's disease and ulcerative colitis: analysis of articular and ophthalmologic manifestations. Clin Rheumatol. 2008; 27: 503-09.
11. Yüksel I, Ataseven H, Basar O, et al. Peripheral arthritis in the course of inflammatory bowel disease. Dig Dis Sci. 2011; 56: 183-07.
12. Lakatos L, Pandur T, David G et al. Association of extraintestinal manifestations of inflammatory bowel disease in a province of western Hungary with disease phenotype: results of a 25-year follow-up study. World J Gastroenterol 2003; 9: 2300-07.
13. Skare TL, Leite N, Bortoluzzo AB, Gonçalves CR, Da Silva JA, et al. Effect of age at disease onset in the clinical profile of spondyloarthritis: a study of 1424 Brazilian patients. Clin Exp Rheumatol. 2012; 30: 351-07.
14. Dougados M, Van der LS, Juhlin R et al. The European Spondyloarthropathy Study Group preliminary criteria for the classification of spondyloarthropathy. Arthritis Rheum 1991; 34: 1218-27.
15. Orchard TR, Wordsworth BP, Jewell DP. Peripheral arthropathies in inflammatory bowel disease: their articular distribution and natural history. Gut 1998; 42: 387-91.
16. Jacoby RK, Jayson MIV. HLA B 27 in Crohn's disease. Ann Rheum Dis 1974; 33: 422-24.
17. Ansell BM, Wrightley RAD. Arthritic manifestations in regional arthritis. Ann Rheum Dis 1964; 23: 64-72.
18. Moll JMH, Haslock I, Macrea IF et al. Associations between ankylosing spondylitis, psoriatic arthropathies, Reiter's disease, the intestinal arthropathies and Behcet's syndrome. Medicine (Baltimore) 1974; 53: 343-64.
19. Acheson ED. An association between ulcerative colitis and regional enteritis and ankylosing spondylitis. Quat J Med 1960; 116: 489-99.
20. Dekker-Saeys BJ, Meuwissen SGM, Van Den Berg-Loonen EM et al. Prevalence of peripheral arthritis, sacroiliitis and ankylosing spondylitis in patients suffering from inflammatory bowel disease. Ann Rheum Dis 1978; 37: 33-35.
21. De Vlam K, Mielants H, Cuvelier C, et al. Spondyloarthropathy is underestimated in inflammatory bowel disease : Prevalence and HLA association. J Rheum 2000; 27: 2861-65.
22. Palm O, Moum B, Ongre A, et al. Prevalence of ankylosing spondylitis and other spondyloarthropathies among patients with inflammatory disease: A population study (The IBSEN Study). J Rheumatol 2002; 29: 511-15.
23. Helliwell PS, Hickling P, Wright V. Do the radiological changes of classical ankylosing spondylitis differ from the changes found in the spondylitis associated with inflammatory bowel disease, psoriasis, and reactive arthritis? Ann Rheum Dis 1998; 135: 135.
24. Schlosstein L, Terasaki PI, Bluestone R, et al. High association of an HL-A antigen, B27, with ankylosing spondylitis. N Engl J Med 1973; 288: 704-06.
25. Morris RI, Metzger AL, Bluestone R et al. HLA-B27 a useful discriminator in the arthropathies of inflammatory bowel disease. N Engl J Med 1974; 290: 1117-19.
26. Jacoby RK, Jayson MI. HLA-27 in Crohn's disease. Ann Rheum Dis 1974; 33: 422-24.
27. Huaux JP, Faisse R, DeBruyere M, et al. HLA-B27 in regional enteritis with and without ankylosing spondylitis or sacroiliitis. J Rheum 1977; 4: 60-63.
28. Pérez Alamino R, Maldonado Cocco JA, Citera G, Arturi P, Vazquez-Mellado J, Sampaio-Barros PD, et al. Differential features between primary ankylosing spondylitis and spondylitis associated with psoriasis and inflammatory bowel disease. J Rheumatol. 201; 38: 1656-60.
29. Enlow RW, Bias WB, Arnett FC. The spondylitis of inflammatory bowel disease: Evidence for a non-HLA linked axial arthropathy. Arth Rheum 1980; 23: 1359-65.
30. Greenstein AJ, Janowitz HD, Sachar DB. The extra-intestinal complications of Crohn's disease and ulcerative colitis: a study of 700 patients. Medicine (Baltimore) 1976; 55: 401-12.
31. Wright V, Watkinson G. The arthritis of ulcerative colitis. BMJ 1965; 2: 670-75.
32. Soren A. joint affections in regional ileiitis. Arch Intern Med 1966; 117: 78-83.
33. Haslock I, Wright V. The musculo-skeletal complications of Crohn's disease. Medicine (Baltimore) 1973; 52: 217-25.
34. Farmer RG, Hawk W, Turnbull RB. Clinical patterns in Crohn's disease: a statistical study of 615 cases. Gastroenterology 1975; 68: 627-35.
35. McEwen C, Lingg C, Kirsner JB. Arthritis accompanying ulcerative colitis. Am J Med 1962; 33: 923-41.
36. Beauvais C, Le Quintrec JL, Prier A, et al. [Destructive arthritis of the hip in Crohn's disease. 10 cases]. Presse Med 1995; 24: 1555-58.

37. De Vos M, Cuvelier C, Mielants H et al. Ileonoscopy in seronegative spondyloarthropathy. Gastroenterology 1989; 34: 339-44.
38. Mielants H, Veys EM, Cuvelier C, et al. HLA-B27 related arthritis and bowel inflammation. Part 2: colonoscopy and bowel histology in patients with HLA-B27 related arthritis. J Rheum 1985; 12: 294-98.
39. Mielants H, Veys EM, Goemaere S, et al. Gut inflammation in the spondyloarthropathies : clinical, radiologic, biologic and genetic features in relation to the type of histology. A prospective study. J Rheum 1991; 18: 1542-51.
40. Cuvelier C, Barbatis C, Mielants H, et al. The histopathology of intestinal inflammation in relation to reactive arthritis. Gut 1987; 28: 394-402.
41. Mielants H, Veys EM, De Vos M, et al. The evolution of Spondyloarthropathies in relation to gut histology. I. Clinical aspects. J of Rheum 1995; 22: 2266-72.
42. Mielants H, Veys EM, De Vos M, et al. The evolution of Spondyloarthropathies in relation to gut histology. II. Histological aspects. J of Rheum 1995; 22: 2273-78.
43. Mielants H, Veys EM, De Vos M, et al. The evolution of Spondyloarthropathies in relation to gut histology. III Relation between gut and joint. J Rheum 1995; 22: 279-84.
44. Sigthorsson G, Tibble J, Hayllar J, et al. Intestinal permeability and inflammation in patients on NSAIDS. Gut 1998; 43: 506-11.
45. Allison MC, Howatson AG, Torrance CJ et al. Gastrointestinal damage associated with the use of nonsteroidal anti-inflammatory drugs. N Engl J Med 1992; 327: 749-54.
46. Bjarnason I, Hayllar J, Macpherson A J, et al. Side effects of nonsteroidal anti-inflammatory drugs on the small and large intestine. Gastroenterology 1993; 104: 1832-47.
47. Ferraz MB, Tugwell P, Goldsmith CH, et al. Meta-analysis of Sulfasalazine in ankylosing spondylitis. J Rheumatol 1990; 17: 1482-86.
48. Carbonnel F. Methotrexate: a drug of the future in ulcerative colitis? Curr Drug Targets. 2011; e-pub, in press.
49. Lofberg R, Louis EV, Reinisch W, et al. Adalimumab produces clinical remission and reduces extraintestinal manifestations in Crohn's disease: Results from CARE. Inflamm Bowel Dis. 2011; epub ahead of print, doi: 10.1002/ibd.21663.
50. Braun J, Zochling J, Barialakos X, et al. Efficacy of sulfasalazine in patients with inflammatory back pain due to undifferentiated spondyloarthritis and early ankylosing spondylitis: a multicentre randomised controlled trial. Ann Rheum Dis 2006; 65: 1147-53.
51. Brandt J, Haibel H, Daniel C, et al, Successful treatment of active ankylosing spondylitis with anti-tumor necrosis factor a monoclonal antibody infliximab. Arth Rheum 2000; 43: 1346-52.
52. Braun J, Baraliakos X, Brandt J, et al. Therapy of ankylosing spondylitis. Part II: Biological therapies in the spondyloarthritides. Scand J Rheum 2005; 34: 178-90.
53. Davis J Jr, Van Der Heijde D, Braun J, et al. Recombinant human tumor necrosis factor receptor (etanercept) for treating ankylosing spondylitis: a randomized, control trial. Arth Rheum 2003; 48: 3230-36.
54. Sandborn WJ, Feagan BG, Fedorak RN, et al. A randomized trial of ustekinumab, a human interleukin-12/23 monoclonal antibody, in patients with moderate-to-severe Crohn's disease. Gastroenterology. 2008; 135: 1130-41.
55. Brandt J, Listing J, Haibel H, et al. Long-term efficacy and safety of etanercept after readminsitration in patients with active ankylosing spondylitis. Rheumatology (Oxford) 2005; 44: 342-48.
56. Haibel H, Rudwaleit M, Brandt HC, et al. Adalimumab reduces spinal symptoms in active ankylosing spondylitis: clinical and magnetic resonance imaging results of a fifty-two -week open label trial. Arth Rheum 2006; 54: 578-81.
57. Van den BF, Kruithof E, De Vos M, et al. Crohn's disease associated spondyloarthropathy: effect of TNF blockade with infliximab on articular symptoms. Lancet 2000; 356: 1821-22.
58. Durand DV, Lecomte C, Cathebras P, et al. Whipple's disease: clinical review of 52 cases. Medicine (Baltimore) 1997; 76: 170-84.
59. Fenollar F, Peuchal X, Raoult D. Whipple's disease. N Engl J Med 2007; 356: 55-66.
60. Khan MA. Axial arthropathy in Whipple's disease. J Rheum 1982; 9: 928-29.
61. Relman DA, Schmidt TM, MacDermott RP. Identification of the uncultured bacillus of Whipple's disease. N Engl J Med 1992; 327: 293-301.
62. Lubrano A, Ciacci C, Ames PR, et al. The arthritis of coeliac disease: prevalence and pattern in 200 adult patients. Br J Rheum 1996; 35: 1314-18.
63. Chakravarty K, Scott DG. Oligoarthritis a presenting feature of occult celiac disease. Br J Rheum 1992; 31: 349-50.
64. Slot O, Locht H. Arthritis as presenting symptom in silent adult celiac disease. Two cases and review of the literature. Scand J Rheum 2000; 29: 260-63.
65. Falcini F, Ferrari R, Simonini G, et al. Recurrent monoarthritis in an 11-year-old boy with occult celiac disease. Successful and stable remission after gluten-free diet. Clin Exp Rheum 1999; 17: 509-11.
66. Chande N, Driman DK, Reynolds RP. Collagenous colitis and lymphocytic colitis: patient characteristics and clinical presentation. Scand J Rheum 2005; 40: 343-47.
67. Schiller LR. Microscopic coloitis syndrome: lymphocytic colitis and collagenous colitis. Semin Gastrointest Dis 1999; 10: 145-55.
68. Nielsen OH, Vainer B, Schaffalitzky de Muckadell OB. Microscopic colitis: a missed diagnosis? Lancet 2004; 364: 2055-57.
69. Bennuci M, Bardazzi G, Magaro L, et al. A case report of a man with rheumatoid factor positive rheumatoid arthritis associated with collagenous colitis [letter]. Clin Exp Rheum 2001; 19: 475.
70. Narvaez J, Montala N, Busquets-Perez N, et al. Collagenous colitis and spondyloarthropathy. Arth Rheum 2006; 55: 507-12.
71. Delamere JP, Baddeley RM, Walton KW. Jejuno-ileal bypass arthropathy: its clinical features and associations. Ann Rheum Dis 1983; 42: 553-57.
72. Clegg DO, Samuelson CO, Williams HJ, et al. Articular complications of jejuno-ileal bypass surgery. J Rheum 1980; 7: 65-70.
73. Fisch C, Schiller P, Harr T, et al. First presentation of intestinal bypass syndrome 18 yr after initial surgery. Rheumatology (Oxford) 2001; 40: 351.
74. Rose E, Espinoza LR, Osterland CK. Intestinal bypass arthritis with circulating immune complexes and HLA-B27. J Rheum 1977; 4: 129-34.

REACTIVE ARTHRITIS 4

Vishesh G, Damodaram P, Narsimulu G

Introduction

Reactive arthritis (ReA) is a sterile joint inflammation that develops after a distant infection[1]. The disease is systemic and not limited to the joints. Triggering infections most commonly originate in the throat, urogenital organs or gastrointestinal tract. The disease also occurs without obvious preceding infection, e.g. in association with inflammatory bowel disease.

ReA is classified with the spondyloarthropathies. There is a common predilection for spondylitis (vertebral inflammation), sacroiliitis, and peripheral arthritis, extra-articular manifestations, and genetic associations with certain HLA-B27 alleles[2]. The spondyloarthropathies include ankylosing spondylitis, psoriatic arthritis, and inflammatory bowel disease associated arthritis. Most patients lack circulating serum rheumatoid factor. The American College of Rheumatology (ACR) and the Third International Workshop on Reactive Arthritis in 1995 established criteria for the diagnosis of ReA[3] (boxes 1 and 2).

The Third International Workshop proposed that the diagnostic criteria should include specific clinical features, such as oligoarthritis or asymmetric arthritis of the lower limb, with or without inflammation of the sacroiliac joint or tendons, and evidence of genitourinary or gastrointestinal infection within the past 4 weeks with known ReA inducing organisms. The ACR criteria require the presence of peripheral arthritis occurring in association with urethritis or cervicitis. Laboratory tests are not necessary for the diagnosis.

Box 1. ACR criteria for reactive arthritis

Arthritis for greater than one month with uveitis or cervicitis.

Arthritis for greater than one month or urethritis or cervicitis or bilateral conjunctivitis.

Episode of arthritis and conjunctivitis.

Episode of arthritis of more than one month, urethritis, and conjunctivitis.

Adapted from ACR slide collection on the rheumatic diseases, 3rd edition.

Atlanta: American college of Rheumatology; 2004.

Box 2. Third international workshop diagnostic criteria

Peripheral arthritis

Predominantly lower limb, asymmetric oligoarthritis

Plus

Evidence of preceding infection

Diarrhoea or urethritis in the prior 4 weeks

No evidence of infection

Exclusion

Other causes of monoarthritis or oligoarthritis excluded

Differential diagnosis

ReA should be a part of the differential diagnosis of undifferentiated oligoarthritis or monoarthritis, particularly because the preceding enteric or genitourinary infection may be mild or even asymptomatic, and likely has resolved by the onset of arthritis. The paramount consideration in such patients is that joint inflammation may represent active infection. Appropriate antibiotics should be given until septic arthritis is excluded based on the results of synovial fluid aspiration, with gram stain and culture.

The differential diagnosis of oligoarthritis or monoarthritis should also include crystalline arthropathy (gout and pseudogout); trauma (acute meniscal or ligamentous tear), other spondylo-arthropathies; HIV arthropathy; Whipple's disease; Lyme arthritis, and immune polyarticular disease[2].

Diagnostic tests must be performed to rule out these diseases before a diagnosis of ReA can be ensured.

Epidemiology and genetics

Because the clinical history of reactive arthritis varies greatly and milder cases apparently go unnoticed, the true incidence of reactive arthritis is hard to assess. The worldwide prevalence of ReA is 1 per 1000, with higher risk tied to specific genetic backgrounds and bacterial exposure[2].

ReA affects all populations. Typically, it's a disorder of young adults, rare in small children and in older people; most patients are aged between 20 and 40 years. Rheumatic fever also can be considered a type of reactive arthritis.

The incidence of ReA in the general population is 4.6 to 13 per 100,000 for genitourinary ReA and 5 to 14 per 100,000 for gastrointestinal infection related ReA. The frequency of cases of ReA induced by a particular organism parallels the rate of primary infection with that organism within the population. A study from USA showed 10-23% patients had arthralgias, 13% arthritis in association with gut infection, most often with yersinia and salmonella[4]. Gastroenteritis in HLA-B27 positive individuals and female sex have increased risk of ReA[5]. Report from western Indian population of 4092 adults no well defined cases of ReA identified despite microbial load[6].

ReA has a male predilection when related to genitourinary infection, with a male to female ratio of 9:1. Enteric related ReA has a male to female ratio of 1:1. This difference may be caused by overlooked urethral symptom in women; misdiagnosis; or a possible deficient immune response to bacterial antigens by males, specifically related to C trachomatis.

Between 65% and 96% of those affected are positive of HLA-B27. Indeed, HLA-B27 is a strong risk factor for developing reactive arthritis, with the increase in risk as high as 50 fold[7]. However, the association is not absolute: HLA-B27 positive individuals do not always develop reactive arthritis, even after a suitable infection, and the arthritis can also occur in HLA-B27 negative individuals. But the disease is more severe and the tendency to chronic development is greater in HLA-B27 positive individuals although HLA-B27 negative individuals can have a prolonged course of the disease. HLA-B27 subtypes associated with spondyloarthropathy include the following[2]

B*2705 : White, African, African-American, Native American (most common subtypes).

B*2702 : Whites.

B*2704 : Eastern Asians. No association with spondyloarthropathy.

B*2703 : West Africa.

B*2705 : Southeast Asians.

There have been claims of association between ReA and HIV infection. It seems that HIV does not cause ReA: the same pathogens that cause ReA are more common in patients with HIV. It is clear, however, that ReA in HIV positive individuals is often severe and explosive, with florid polyarthritis; enthesitis (typically plantar fascitis); and extensive psoriasiform skin lesions. B27 positivity rate of 75% in this population have been reported.

Clinical features

Typically the first symptoms are those of joint discomfort. A careful history may indicate symptoms of enteric and urogenital infection a few days to a couple of weeks previously, and often the patient reports such symptoms in other family members.

Sometimes, abdominal pains in the few days before the appearance of joint symptoms may be suspected to be due to acute appendicitis or salpingo-oophoritis; this is especially the case with young adults. They may even have undergone laprotomy, in which case usually mesenteric lymphadenopathy has been found. A common clinical experience is that the symptoms of the triggering infections have often been mild and, in about 10% of cases, the infection has passed unnoticed[8,9].

Infections most commonly preceding reactive arthritis are those caused by Salmonella, Shigella, Yersinia, Campylobacter, Chlamydia trachomatis or C. pneumoniae, Borrelia, Neisseria and Streptococci. Several viruses have been implicated: rubella, hepatitis, and parvovirus. Salmonella, Borrelia, and Neisseria can cause true septic arthritis in addition to reactive arthritis. Non infectious inflammatory bowel disease may be associated with reactive arthritis[10,11] for example, after intestinal bypass operation for morbid obesity. It is obvious, however, that in these conditions the intestinal microbial flora or the mucosal defense mechanisms are altered. There are reports of occurrence of reactive arthritis following rabies and tetanus vaccination[12].

Musculoskeletal manifestations

Musculoskeletal symptoms usually begin 2 to 4 weeks following a gastrointestinal or genitourinary infection. Arthritis usually the last of the clinical sign to present, usually preceded by the uveitis. By the time articular symptoms appear, the gastrointestinal infection is usually cleared or resolved, so that organism could not be cultured from the stool[2].

The most prominent symptoms are usually those in the joints. They vary from mild arthralgias to severely disabling polyarthritis. Typically, there is asymmetric oligoarthritis or monoarthritis. The weight bearing joints are most often affected: knees, ankles and hips-but quite commonly other large joints, such as the shoulders, elbows, and wrists and small joints of the hands and feet may also be involved.

The inflammation may vary from day to day and shift from one joint to another. The joints involved may be red and warm, and occasionally there is also erythema nodosum. Dactylitis is not uncommon, and sometimes the condition may mimic rheumatoid arthritis, with stiffness on inactivity. Especially in the later stages, patients often complain of pain in the sacroiliac region.

A diagnosis cannot be made on the basis of physical examination alone, because the findings may vary greatly and may mimic any other disease involving joint inflammation, including gout, rheumatoid arthritis or true purulent arthritis[13]. Enthesopathies, tenosynovitis, or both, are present in the high proportion of cases. Thus plantar fasciitis, Achilles tendinitis or bursitis in the ankle region often results in difficulties in walking. Pain on palpation is frequent at the sites where, tendons attach to bones, and this enthesopathy may also limit movement. In severe cases, the muscle wasting may be quite prominent. Usually severe cases may become bedridden. The outcome, however, is not related to the initial severity of the disease.

Mucocutaneous

Cutaneous manifestations in ReA occur in up to 50% of patients. Keratoderma blenorrhagicum is a hyperkeratotic lesion seen in about 25% of affected men, usually on the soles of the feet or palms of the hands.

It begins as a clear vesicle on the erythematous base, developing into nodules. It eventually crusts over, and adjacent lesions coalesce[14]. The cutaneous lesions are difficult to distinguish from pustular psoriasis, both clinically and histologically. Nail changes similar to those of psoriasis are seen in up to 15% of the patients with ReA[2].

Circinate balanitis is a painless shallow erythematous ulcer of the glans penis, seen in up to 25% of men. In circumcised men it is dry, plaque like, and hyperkeratotic, resembling psoriasis or keratoderma. In uncircumcised men it is typically moist, shallow ulcer surrounding the meatus. Superficial, usually painless oral ulcer occurs in approximately 5% to 15% of the patients. They are typically located in the hard palate or tongue but may also be found on the soft palate, gingival and cheeks[15].

Gastrointestinal symptoms

Enteric infections or inflammatory diseases can, like urogenital infections, be the etiologic triggering events. It has become apparent that, as in the urogenital system, the epithelium of the intestine may form part of the reactive disease itself. Patients may complain of occasional abdominal pains and diarrhoea, but at this stage no organism could be isolated from stool cultures. When ileocolonoscopy has been carried out in patients with established ReA, either macroscopically or microscopically detectable lesions resembling ulcerative colitis or Crohn's disease have been seen. A study conducted in India revealed that patients with seronegative SpA with active disease had high incidence of silent gut inflammation.

Ocular lesions

Extra articular manifestations are more common with chlamydial ReA than postdysentery ReA. Conjunctivitis which occurs in 30% to 60% of patients, is usually bilateral, and occurs early in the course of the disease, but typically after urethritis.

Conjunctivitis is more common with sexually acquired and after shigella infection than with other postdysentery ReA syndromes[16]. There is male to female ratio of 1.3:1. Chronic inflammation of the eye can lead to progressive intraocular damage and visual loss[17]. The uveitis is typically characterized by acute unilateral attacks of inflammation of anterior chamber (seen in 5% of patients) presenting as ocular pain, erythema, and photophobia.

The prognosis is favorable, but if untreated up to 11% of those affected may become blind. Concurrence of Sjogren's syndrome in a patient with Chlamydia-induced reactive arthritis has also been described.

Cardiac lesions

Inflammation of the heart and coronary vessels can result in aortic regurgitation. ReA inflammatory aortic disease may be difficult to distinguish pathologically from syphilis, and may require surgical correction[18] Screening echocardiography is warranted only if ReA is chronic[19]. Severe conduction system abnormalities and less commonly mitral regurgitation may be associated with aortic involvement.

Weinberger and coworkers[20] described myocarditis in ReA causing asymptomatic AV node conduction abnormalities. This can be seen in 5% to 14% of patients. First degree AV block is more common, but progression to second degree AV block and even complete heart block can occur[21]. Carditis can occur in carriers of the HLA-B27 allele without being associated with ReA or ankylosing spondylitis.

Pathogenesis

The development of ReA involves infectious and genetic factors. Approximately 4 weeks after the onset of the antecedent gastrointestinal and genitourinary infection, the bacteria have been cleared, but the systemic immune response causes an acute peripheral aseptic synovitis and associated characteristic features. Arthrocentesis and synovial fluid do not reveal live bacteria; a positive culture indicates septic arthritis requiring parenteral antibiotics. In such circumstance, ReA is not the correct diagnosis[2].

Common features of arthritogenic bacteria

These bacteria are all either obligate intracellular or facultative intracellular bacteria. They are aerobic (Campylobacter is microaerophilic) and gram negative, with a Lipopolysaccharide containing outer membrane. They are invasive and infect the gastrointestinal tract, with local colonic inflammation. Perhaps the most critical common feature is that they are all conveyed from the colonic mucosa to the joint by monocytes. Why ReA most commonly affects specific joints, usually of the lower extremity, is yet unclear, but may relate to the fact that circulating mononuclear cells may preferentially exit the circulation at the site of inflammation. There is probably low level inflammation in the repeatedly microtraumatized joints of the weight bearing lower extremity.

There are many other bacteria with similar features that do not result in ReA. Fendler and coworkers[22] investigated 52 patients, all diagnosed with ReA, to find their causative pathogen. Through diagnostic testing, stool cultures, and antibody assays, 56% of the pathogens were identified, being Yersinia, Shigella, and Campylobacter and C trachomatis. Infection with Campylobacter (C.) jejuni and C. coli. can have a long term sequelae of reactive arthritis along with Guillain Barre sndrome.

Box 3. Bacteria associated with reactive arthritis[2]

Genitourinary
- Chlamydia trachomatis
- Chlamydia psittaci
- Possible: Ureaplasma urealyticum

Gastrointestinal
- Shigella flexneri
- Campylobacter jejuni
- Campylobacter fetus
- Salmonella typhimurium
- Salmonella enteritidis
- Shigella sonnei
- Shigella dysenteriae
- Yersinia pseudotuberculosis
- Yersinia enterocolitica 0:3 or 0:9
- Clostridium difficile

Persistence of antigen within the synovial space

In ReA, techniques to identify viable bacteria within the synovium have failed until recently. Studies using PCR have found DNA sequences of chlamydia in the synovium of some ReA patients[23]. In another study, the synovium of 8 patients with Chlamydia associated inflammatory arthritis were inoculated with fluorescein labelled monoclonal antibody to C. trachomatis. Elementary bodies were found in the inflammed synovium[25]. These strands of evidence led to the hypothesis that although viable organisms are no longer present, bacterial antigen persists within the synovium of the affected joints, and that the immune response to the antigens is arthritogenic. Similar evidence has been reported in ReA caused by other bacteria (e.g. Yersinia ReA).

Immune responses

It is believed that ReA is not caused by active infection. Rather, infection with certain gastrointestinal and genitourinary pathogens is thought to precipitate either an autoimmune response to cross reactive host antigens, or an immune response to slowly or incompletely cleared microbial antigens lodged in host tissues[2].

Given the HLA-B27 linkage in ReA, it is tempting to suspect that the immunopathogenesis of ReA is related to CD8+ T cell response. HLA-B27, and other MHC class 1 molecules on antigen presenting cells, typically interacts with CD8+T cells. An attractive theory is that the HLA-B27 molecule is uniquely suited to present pathogen derived peptides. Local immunodysregulation and inflammation can be sustained as long as the supply of such peptides is maintained.

Alternatively, a poor pathogen derived antigen might bind firmly to a molecule found uniquely within the tissues affected in ReA. In such a setting, the presenting foreign antigen could perpetuate long-lasting local inflammation. The absence of a defined antigen, however, has hampered the search for pathogenic T cell clones.

Another theory is that a pathogen derived antigen resembles a host antigen. Such a host antigen might be found in the tissues affected in ReA, leading to a local immune reaction.

Peculiarities of the heavy chain of HLA-B27 have also been implicated in the pathogenesis of ReA. This heavy chain is prone for misfolding, perhaps allowing cryptic self peptides to become available for expression as autoantigens. Further research is required to clarify the precise immunomechanism of ReA[24].

HLA B27

One study demonstrated HLA-B27 positive cells infected with ReA inducing bacteria show an increased presentation of certain self peptides. Intra articular synthesis of IgG by B cells may contribute to the inflammatory response. A study examining the activity of HLA-B27 positive neutrophils involved in Yersinia ReA showed an amplified reaction to chemotactic stimulus, possibly contributing to the more severe inflammatory symptoms in carriers of this allele.

HLA-B27 is not the only allele predisposing to spondyloarthropathies and ReA. In Mexican populations, one study found an association of HLA DR1 and HLA-B15, independent of HLA-B27. HLA B27 was associated with younger age of onset and increased severity of disease, and HLA-DR1 with older age of onset.

The HLA-B27 allele does not predispose the patient to infection or ReA, but increases the risk of severe and prolonged ReA. Patients with this allele have a diminished ability to eliminate the infected bacterium, and may have an increased intracellular survival of pathogens. The expression of B27 does not profoundly alter the gene expression following LPS stimulation. Therefore, additional signals, such as those provided by cytokines or intracellular infection, may be required to reveal any influence of B27 expression on the inflammatory response[26].

Laboratory studies

The ESR and CRP are markedly increased in most cases of acute reactive arthritis. At a later stage, patients with chronic symptoms have normal ESR and CRP values. The blood cell count may reveal moderate leucocytosis and mild anemia.

Urine analysis should be carried out at diagnosis and should be repeated during the follow up, in order to detect possible aseptic pyuria resulting from urethritis. It is advisable to check renal and liver function since this can guide the treatment with NSAIDs. Tests for IgM RF are usually negative.

ECG should be performed and repeated in patients who have a prolonged disease course, to exclude possible conduction disturbances. When endocarditis is suspected, especially in cases of rheumatic fever, it is essential to do echocardiography.

Joint fluid should always be aspirated when possible. Gram stain and bacterial culture should be performed to exclude septic arthritis[1]. In ReA, bacteria are not demonstrable either by staining or even extensive culturing. In early stages of ReA, the cell count in the synovial fluid can be quite high, and neutrophil can dominate the picture. Later, lymphocytosis can develop. Microscopy under polarized light should be performed, to exclude gout or pseudogout. Blood culture is not routinely necessary[1], but should be undertaken if the joint infection is suspected.

Every effort should be made to isolate the causative microorganism from the throat, feces and urogenital tract. For detection of Chlamydia, application of PCR to first voided urine samples is a practical method. The serologic tests to be considered include antibody assays against Salmonella, Yersinia, Campylobacter, and Chlamydia, Neisseria gonorrhoeae, Borrelia burgdorferi and α hemolytic streptococci. Increase in and persistence of IgA class antibodies are especially characteristic.

Imaging in ReA

It should be performed not to confirm the diagnosis of ReA but to rule out erosive diseases. The findings in ReA are usually quite scant, such as juxta- articular osteoporosis and soft tissue swelling. In patients with recurrent and chronic disease, erosions may be observed. More often, periosteal reaction and proliferation, especially at the insertion sites of tendons, are seen. Thus plantar spurs are common sign in chronic cases.

Radionuclide scintigraphy is a sensitive technique for demonstrating inflammation and is superior to radiography, especially in cases of enthesopathy, periosteitis and early sacroiliitis.

Ophthalmologic evaluation

Patients with eye symptoms should be referred to ophthalmologists since uveitis requires prompt diagnosis and appropriate management.

Outcomes

Approximately 75% of patients with ReA have musculoskeletal symptoms persisting for more than 1 year, and up to 30% of those patients have symptoms after 6 years[2].

The duration of joint symptoms is shorter, usually 3 to 5 months, with 15% of patients developing chronic symptoms. In some studies, 20% to 40% of patients have chronic indolent arthritis after 1 year, and 5% of those patients continue to have symptoms at 2 years. Recurrent attacks are more common in patients with Chlamydia ReA and patients with HLA-B27 allele. Patients with full triad of symptoms usually have a longer duration of symptoms and more recurrences. Enthesopathy and skin lesions may persist after joint symptoms resolve. Disease activity and arthritis are linked to increased type 1 collagen degradation, a possible future marker for disease activity of ReA.

Prognostic factors

Hip arthritis and high ESR are associated with a more chronic course. The prognosis is also affected by the inciting bacterium, the presence of HLA-B27, the gender of the patient and the presence of recurrent symptoms. Chronic intestinal inflammation and ReA induced by shigella, salmonella, and chlamydia are associated with greater chronicity. In one study, male gender also influenced the disease course[2]. Men had higher ESR.

Management[2]

Management of ReA includes the following

- Education.
- Physical therapy.
- Occupational therapy.

Acute medication

- NSAIDs to relieve pain and inflammation.
- Antibiotics if infection present.
- Intraarticular steroid injection.

Chronic medication

- Disease modifying antirheumatic drugs (DMARDs).
- Biological agents.

Acute reactive arthritis

The patient may be afraid of having a chronic progressively destructive disease. Correct information is the cornerstone of the treatment, to relieve unnecessary anxiety and to achieve patient's compliance with suggested treatment.

Advice the patient adequate bed rest

Even within a few weeks after recovery, the patient should not use the joints too vigorously as minor traumas may lead to recurrences[1].

Physiotherapy should be advised when several joints are involved to prevent muscle wasting.

Apart from using as analgesics, NSAIDs should be used regularly over some time in order to achieve maximum anti inflammatory effect[2].

During recovery, insufficient analgesia may lead to limited use of the joints and actually prolong the rehabilitation.

There is no definite drug of choice, because patient's responses differ.

In the case of enthesopathies, locally applied NSAID ointments may be useful.

Intraarticular injections are very useful in reducing the pain and can be given after ruling out infection by synovial fluid analysis.

Systemic corticosteroids should be used when NSAIDs have not had a sufficient effect and in whom many joints are affected. Oral corticosteroids should be withdrawn slowly, but with NSAIDs continuing. Corticosteroids should not be used for more than 2-4 months.

Antibiotic treatment

It should be given if a microbe is isolated. This is to tackle the spread of infection than to change the continuing course of the disease. In 2 instances antibiotics have been proven to be useful[1]. In rheumatic fever they are given for primary or secondary prevention. However, they do not have a role in the treatment of rheumatic fever itself. The treatment of borreliosis has been shown to minimize the impact of the Lyme disease.

Clinical and experimental data indicate that, if antibacterial treatment of reactive arthritis can be started very early during the pathogenetic process, the disease can be prevented[1]. All available clinical evidence indicates that, in fully developed reactive arthritis, short term antibacterial treatment has no beneficial effect, and overall the results with long term administration of antibacterial agents are also poor[24-27].

Management of patients with chronic progressive disease

With time, the patients often derive less and less benefit from the various NSAIDs, and side effects become more common with prolonged use. Physiotherapy may give some patients relief, but does not have a long term effect.

In chronic cases corticosteroids are not advisable, because their therapeutic effect in this situation is not very good and the prognosis in any case, is good in almost all patients.

If symptoms last longer than 3 months, DMARD has been used. Sulphasalazine has been shown to be effective in some cases, specifically for the peripheral joint involvement. In controlled trials, it produced improvement in the intensity and duration of symptoms compared with placebo, but only in the short term, and with a significant risk of adverse events, primarily gastrointestinal[26]. Methotrexate and azathioprine have also been used, and in placebo trials show a shortened duration of symptoms[27].

Undifferentiated spondylo arthropathy (USpA)

Definition and criteria[25]

USpA is a syndrome with features consistent with the SpAs, but the patients do not fulfill criteria for any specific SpA (Table 4).

USpA may represent an early phase or incomplete form of ankylosing spondylitis or another SpA. In fact, several studies of USpA included many patients who probably should have been diagnosed with ankylosing spondylitis, RA, ReA, or IBD-associated SpA, which made the clinical description very ambiguous. However, more recent data suggest that these patients may represent a distinct disease entity based on demographic and clinical criteria.

Although no specific criteria are identified, using modified Amor criteria can be helpful in confirming a clinical diagnosis.

Table 4. Diagnostic criteria for USpA using modified Amor criteria

Inclusion criteria		Exclusion criteria
Inflammatory back pain	1 point	Diagnosis of specific SpA
Unilateral buttock pain	1 point	Sacroiliitis on radiograph grade 2
Alternating buttock pain	2 points	Precipitating genitourinary/GI infection
Enthesitis	2 points	Psoriasis
Peripheral arthritis	2 points	Keratoderma blenorrhagicum
Dactylitis (sausage digit)	2 points	IBD (Crohn's disease or ulcerative colitis)
Acute anterior uveitis	2 points	Positive rheumatoid factor
HLA B27 positive or family history of SpA	2 points	Positive antinuclear antibody, titer > 1:80
Good response to nonsteroidal anti-inflammatory drugs	2 points	
Diagnosis of SpA with 6 or more points		

Distinguishing features of USpA[28]

The age of onset has a very wide range, with the peak onset at approximately 50 years. The male-to-female ratio is 1:3. The onset usually is insidious and, even after years of active disease, sacroiliitis and spondylitis is either absent or appears very mild on routine radiographs. Clinical manifestations include inflammatory back pain (90%), buttock pain (80%), enthesitis (85%), peripheral arthritis (35%), dactylitis (17%), and fatigue (55%).

Extra-articular manifestations are uncommon, occurring in fewer than 10% of patients, and include acute anterior uveitis (1-2%), oral ulcers, rash, nonspecific IBD, pleuritis, and pericarditis.

Laboratory studies generally are unremarkable except for the presence of an elevated ESR or C-reactive protein (36%). HLA-B27 antigen is positive only in approximately 20-25% of patients.

These factors, especially the late age of onset, female predominance, and low *HLA-B27* positivity, suggest that USpA is distinct from ankylosing spondylitis and the other classic SpAs.

In addition, when these patients are observed over long periods, they rarely develop clinical manifestations or radiographic changes that result in a change of diagnosis. Occasionally, radiographs show evidence of periosteal new bone formation at sites of enthesitis, especially at the insertion of the Achilles tendon or plantar fascia on the calcaneous, or early syndesmophytes on the lumbar spine without bridging.

Table 5. Clinical and laboratory features of USpA[29]

Clinical feature of USpA	Frequency
Inflammatory back pain	90%
Buttock pain	80%
Enthesitis	75%
Peripheral arthritis	40%
Dactylitis (sausage digits)	20%
Acute anterior uveitis	1-2%
Fatigue	55%
Elevated ESR	32%
HLA-B27 positive	25%

Treatment

Although most patients with USpA have chronic, active disease and require long-term therapy, some patients with USpA have mild and intermittent symptoms that require intermittent symptomatic therapy. These episodes may last from 1-2 weeks to several months, with long asymptomatic periods that do not require therapy.

Most patients (>75%) require long-term therapy for ongoing symptomatic disease[30]. They respond well to nonsteroidal anti-inflammatory drugs (NSAIDs). Most patients maintain good function without progressive disease or clinically significant radiographic changes. A small minority of patients does not respond well to or tolerate NSAIDs. In these patients, treatment progression is similar to patients with ankylosing spondylitis, including the use of sulphasalazine and methotrexate, although no well-designed clinical trials have been conducted on the treatment of USpA.

Key points

- Reactive arthritis (ReA) is a sterile joint inflammation that develops after a distant infection.
- ReA should be a part of the differential diagnosis of undifferentiated oligoarthritis or monoarthritis.
- Musculoskeletal symptoms usually begin 2 to 4 weeks following a gastrointestinal or genitourinary infection.
- Arthritis is usually the last of the clinical sign to present.
- Inflammation of the heart and coronary vessels can result in aortic regurgitation.
- Chronic inflammation of the eye can lead to progressive intraocular damage and visual loss.
- Urine analysis should be carried out at diagnosis and should be repeated during the follow up, in order to detect possible aseptic pyuria resulting from urethritis.
- Joint fluid should always be aspirated when possible.
- Gram stain and bacterial culture should be performed to exclude septic arthritis.
- Hip arthritis and high ESR are associated with a more chronic course.
- Apart from using as analgesics, NSAIDs should be used regularly over some time in order to achieve maximum anti inflammatory effect.
- Sulphasalazine can be given in chronic reactive patients with peripheral joint involvement.
- Methotrexate and azathioprine can also be used.
- USpA may represent an early phase or incomplete form of ankylosing spondylitis or another SpA.
- NSAIDs are the mainstay of treatment in USpA.

Further Reading :

1. Toivanen A, Reactive arthritis: clinical features and treatment, Marc C Hoschberg, Alan J Silman, Josef S Smolen, Michael E Weinblatt, Michel H Weisman, Rheumatology, third edition, Mosby, Spain, 2003; 1233-40.
2. Petersel DL, Leonard H Sigel, Reactive arthritis, Infect Dis Clin N Am 19, 2005; 863-68.
3. Kingsley G, Sieper J. Third international workshop on reactive arthritis, 23-26, September 1995, Berlin, Germany, Ann Rheum Dis 1996; 56: 556-84.
4. Towens JM, DeodharAA, laine ES ,et al. Reactive arthritis following culture - confirmed infections with bacterial enteric pathogens in Minnesota and Oregon:a population based study. Ann Rheum Dis 2008; 67: 1689-96.
5. Schiellerup P, Krogfelt KA, Locht H.A comparision of self-reportd joint symptoms following infection with different enteric pathogens: effects of HLA-B27. J Rheumatol 2008; 35: 480-87.
6. Chopra A, Partil J, Billempelly V, et al. Prevalence of rheumatic diseases in a rural population in Western India: a WHO- ILARCOPCORD study. J ASSoc Physicians India 2001; 49: 240-46.
7. Aho k, Ahvonen P, Lassus et al. Yersinia arthritis and related diseases: clinical and immunogenetic implications. Dumonde DC. Infection and immunology in rheumatic diseases. Blackwell scientific publications, Oxford, 1976: 341-44.
8. Ahvonen P, Human Yersiniosis in Finland, clinical features, Ann Cli Res, 1972; 4: 39-48.
9. Leion R, Toivanen A. Arthritis associated with gastrointestinal disorders. Toivanen A, Toivanen P, Reactive arthritis, Boca Raton, CRC press; 1988: 77-86.
10. Devos M, Cuvelier C, Mielants H et al. Ileocolonoscopy in seronegative spondyloarthropathy, Gastroenterology 1989: 96; 339-44.
11. Narsimulu G, Nausheer Ahmed Khan B, Ajit kumar et al. Silent gut inflammation in seronegative spondyloarthropathy. J Ind Rheumatol Assoc 2004: 12: 0-0.
12. Ashu K, Kaser G, Doganavsargil E. Reactive arthritis following tetanus and rabies vaccinations. Rheumatol Int. 2006 Dec; 24(2): 209-10.
13. Toivanen P, Toivanen A. Bacterial or reactive arthritis? Rheumatol Eur 1995; 24 (suppl 2): 253-55.
14. Hill Gastion JS, Lillicarp MS. Arthritis and Enteric infection. Best Pract Res. Clin Rheumatol 2003; 17: 219-39.
15. Arnette FC, Seronegative Spondyloarthropathies, Bull Rheum Dis. 1987; 37: 1-12.
16. Monnet D, Braban M, Hudry C et al. Ophthalmic findings and frequency of extra articular manifestations in patients with HLA B27 uveitis, a study of 175 cases. Ophthalmology 2004: 111: 802-09.
17. Khan M A, Siepel J, Reactive arthritis. Koopman W J, Moreland L W. Arthritis and allied conditions: a textbook of rheumatology. 15th edition. Philadelphia: Lippincott Williams and Wilkins: 2005. 1336-55.
18. Hoogland Y T, Alexander E P, Patterson R H et al. coronary artery stenosis in Reiter's syndr: a complication of aortitis. J Rheumatol 1994, 21: 757-59.
19. Hannu T, Nieminen M S, Swan H et al. Cardiac findings of reactive arthritis, an observational echocardiographic study. Rheumatol Int 2002; 21: 169-72.
20. Weinberger H W, Ropes M W, Kulka J P et al. Reiter's syndrome: Clinical and pathological observations: a long term study of 16 cases. Medicine (Baltimore) 1962; 42: 35-91.
21. Colmegna I, Cuchacovich R, Espinoza L R. HLA B27 associated reactive arthritis: pathogenetic and clinical considerations. Clin Microbiol Rev 2001; 17: 348-69.
22. Fendler C, Laitkos, Sorensen H et al. Frequency of triggering bacteria in patients with reactive arthritis and undifferentiated oligoarthritis and the relative importance of tests used for diagnosis. Ann Rheum Dis 2001; 60: 337-43.
23. Cuchacovich R, Japa S, Huang WQ, et al. Detection of bacterial DNA in Latin American patients with reactive arthritis by polymerase chain reaction and sequencing analysis. J Rheuamtol 2002; 29: 1426-29.
24. Toivanen A. Managing reactive arthritis. Rheumatol 2000; 39: 117-21.
25. Sieper J, Frendler C, Latiko S et al. No benefit of long term ciprofloxacin treatment in patients with reactive arthritis and undifferentiated oligoarthritis. Arthritis Rheum 1999; 42: 1386-96.
26. Yi- Kerttula T, Luukkainen R, Yi- Kerttula U et al. Effect of a three month course of ciprofloxacin on the outcome of reactive arthritis. Ann Rheum Dis 2000; 59: 565-70.
27. Toivanen A. Bacteria triggered reactive arthritis. Implications for antibacterial treatment. Drugs 2001; 61: 343-51.
28. Hitchon C A, Peschken C A, Shaikh C S, El- Gabalway H S. Early undifferentiated arthritis. Rheum Dis Clin Nor Am. 2005; 31(4): 605-12.
29. Zeidler H, Mau W, Khan MA: Undifferentiated spondyloarthropathies. Rheum Dis Clin North Am 1992; 18(1): 187-202.
30. Rezaian MM, Brent LH: Undifferentiated spondylo-arthropathy: Seven-year follow-up study of 357 patients. Arthritis Rheum 2001; 44: S93.

SECTION 8

Connective Tissue Diseases

SYSTEMIC LUPUS ERYTHEMATOSUS – CLINICAL MANIFESTATIONS AND ORGAN DAMAGE

1

Sukumar Mukherjee, Somnath Bhar

Systemic Lupus Erythematosus (SLE) is a multisystem prototype autoimmune heterogeneous illness characterized by myriad of systemic features primarily due to immune dysregulation at multiple levels of the immune cascade with hyperactivation of B-cell activity and diminished T-cell suppressor activity leading to increased tissue specific and non-tissue specific circulatory antibodies. Having highly variable features like constitutional symptoms, glomerulonephritis, neuropsychiatric disease and cutaneous manifestation SLE is hardly curable, but most of the patients experience remission and their survival has improved over the years[1].

Epidemiology

The annual incidence of SLE in US and Europe has been estimated to be 2-8 per 100,000 and estimates of the total number with this disease generally range between 20 and 60 cases per 100,000. In a study near Delhi the prevalence of SLE was found to be 3.2 per 100,000 population[2]. However, with a large population in India the total burden of SLE is considerable and worldwide more than 85% of SLE patients are female. The female to male ratio as studied by Pande et.al.[3] in 1993 in India is around 9:1. The highest incidence of SLE is often seen in the premenopausal age of 20-50 years. In the US, the incidence and prevalence of SLE is higher among blacks than among whites with a relative risk of 3.0. The same phenomenon has also been observed in Afro-Caribbeans and Indian immigrants in UK. However, low incidence of SLE is observed in West African countries. The genetic factors like HLA B_8 DR_3 DR_2 & DR_4 and complement deficiency are contributory. The common environmental factors like UV light, emotional stress, infections, and drugs including female sex hormones are often incriminated to trigger the event[4].

Pathogenesis

Increased amounts of apoptosis-related endogenous nucleic acids stimulate the production of IFN-α and promote autoimmunity by breaking self-tolerance through activation of antigen-presenting cells (Fig. 1). A seemingly constant feature in SLE patients is this accumulation of unusually large amount of apoptic cells because of defect in the clearance mechanism of phagocytes[5]. Apoptotic cells are non-inflammatory but secondary necrotic changes with disintegration and release of cytosolic compounds make them inflammatory. Subsequently there is activation of autoreactive T & B cells[6]; cellular immunity is also aberrant and is at times characterized by leucopenia, lymphopenia and monocytosis. Ultimately tissue damage is mediated by recruitment of inflammatory cells, reactive oxygen species, production of inflammatory cytokines and modulation of coagulation cascade[7].

Clinical features[4,9]

The onset of SLE may be acute or insidious and any organ system can be involved. Fever, malar rash, arthralgias are common presenting features of SLE. Fatigue is a clinical marker of active disease the mechanism of which is unclear. The major organ involvement like renal and CNS tend to occur early in the natural course of disease and can be life-threatening. At the onset SLE may involve one or several organ systems; but over a period of time cumulative manifestations of disease tend to occur. Severity of SLE varies from mild and intermittent to severe and fulminant. The natural history is characterized by relapse and remission; however permanent complete remission is rare. The comparative data of frequency of occurrence of some common features of SLE is shown in Table 1[10].

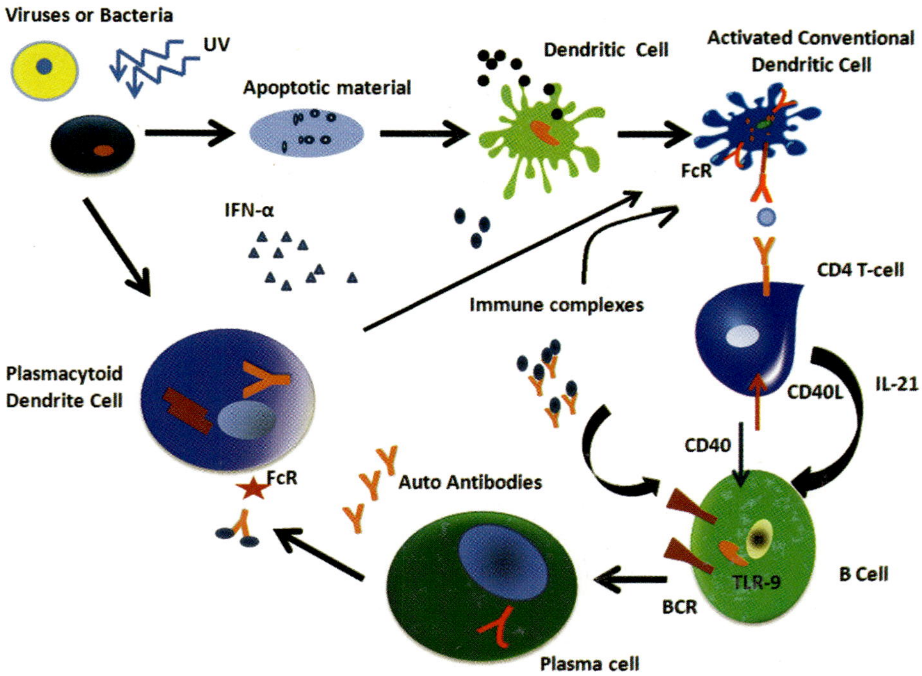

Fig 1.[8] In systemic lupus erythematosus all pathways lead to endogenous nucleic acids-mediated production of interferon α (IFNα). Increased production of autoantigens during apoptosis (UV-related and/or spontaneous), decreased disposal, deregulated handling and presentation are all important for the initiation of the autoimmune response. Nucleosomes containing endogenous danger ligands that can bind to pathogen-associated molecular pattern receptors are incorporated in apoptotic blebs that promote the activation of DCs and B cells and the production of IFN and autoaantibodies, respectively. Cell surface receptors such as the BCR and FcRIIa facilitate the endocytosis of nucleic acid containing material or immune complexes and the binding to endosomal receptors of the innate immunity such as TLRs. At the early stages of disease, when autoantibodies and immune complexes may not have been formed, antimicrobial peptides released by damaged tissues such as LL37 and neutrophil extracellular traps, may bind with nucleic acids inhibiting their degradation and thus facilitating their endocytosis and stimulation of TLR-7/9 in plasmacytoid DCs. Increased amounts of apoptosis related endogenous nucleic acids stimulate the production of IFN and promote autoimmunity by breaking self-tolerance through activation and promotion of maturation of conventional (myeloid) DCs. Immature DCs promote tolerance while activated mature DCs promote autoreactivity. Production of autoantibodies by B cells in lupus is driven by the availability of endogenous antigens and is largely dependent upon T cell help, mediated by cell surface interactions (CD40L/CD40) and cytokines (IL21). Chromatin-containing immune complexes vigorously stimulate B cells due to combined BCR/TLR crosslinking. DC, dendritic cell, BCR, B cell receptor, FcR, Fc receptor, UV, ultraviolet; TLR, toll-like receptor.

Table 1: Frequency of clinical features of SLE[10]

Features (%)	Indian data(%)	Western data (%)
Arthritis	72-92	86-94
Alopecia	52-80	50
Skin rash	74-90	60
Photosensitivity	10-62	33-62
Malar rash	37-76	72-90
Oral ulcers	41-61	30
Fever	74-91	80
Lymphadenopathy	26-47	50
Neuro-psychiatric	19-63	20-45
Renal	35-73	29-73
Cardiac	10-29	20-30
Pleuropulmonary	9-54	36-57

Constitutional symptoms

The spectrum of multitude of symptoms include fever, malaise, headache, lethargy, fatigue, hair loss, weight loss which are often non-specific. Diagnosis may remain a suspect with these set of symptoms.

Mucocutaneous involvement[11,12]

The skin lesions in SLE are classified as specific and nonspecific - the former ones include acute, subacute and chronic skin rashes. The non-specific skin lesions are alopecia, vasculitis, Raynaud's phenomenon, livido reticularis, erythema multiforme and other lesions. The butterfly facial erythematosus rash (Fig. 2) sparing nasolabial furrow is characteristic. Moreover the rash is also seen in the preauricular, auricular and retroauricular areas (Fig. 3). The discoid lesion (Fig. 4), bullous lesions (Fig. 5), urtricaria and vasculitic lesion have also been described in SLE. Discoid lupus are circular with erythematosus margin with surface scaring, follicular plugging. The areas of depigmentation and telangiectasia are also seen. In antiphospholipid antibody syndrome secondary to SLE livedo reticularis is typical. Multiple small painless oral ulcers (Fig. 6) and diffuse alopecia (Fig. 2) are not uncommon. Involvement of the mucous membrane occur in 25-45% of SLE patients. While scalp scarring alopecia (Fig. 7) remains irreversible, otherwise in non-scarring alopecia there is regrowth of hair. 'Lupus hair' is characterized by their hair that easily fractures. Subacute cutaneous lupus erythematosus (Fig. 8) is characterized by the presence of photosensitive, superficial non-indurated non-scaring, annular or papulosquamous psoriasis-like lesions. Photosensitivity occurs in 60-100% of SLE patients. Raynaud's phenomenon is mild and uncommon. In southern India Raynaud's phenomenon is rare where as lymphadenopathy is commonly encountered. Sometimes diffuse hyperpigmentation, urtricaria or diffuse maculopapular rash are also observed. Diffuse hand erythema is not uncommon (Fig. 9). Digital gangrene due to progressive vasculitis is also observed in few patients of active SLE (Fig. 10).

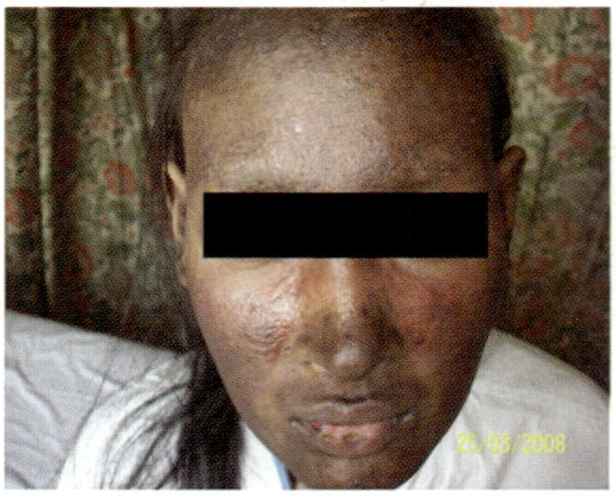

Fig. 2. Malar Rash and alopecia.

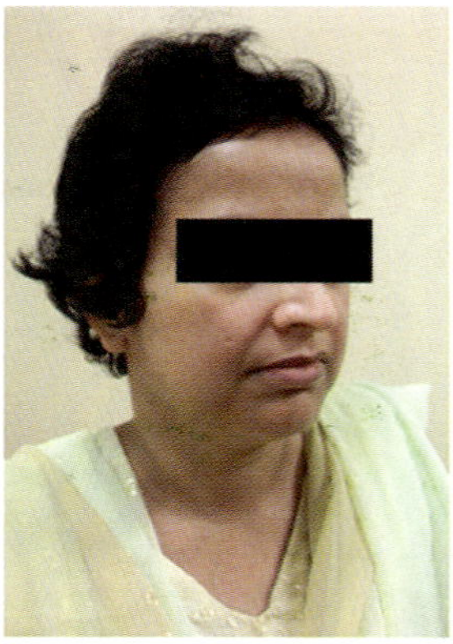

Fig. 3. Wide Spread Facial Rash.

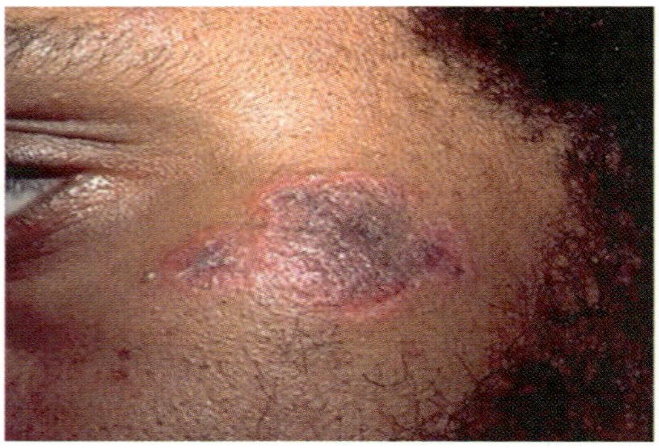

Fig. 4. Discoid Lupus.

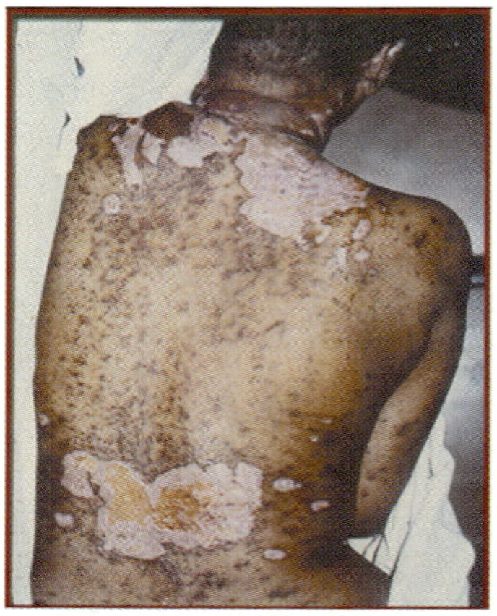

Fig. 5. Bullous eruptions in SLE.

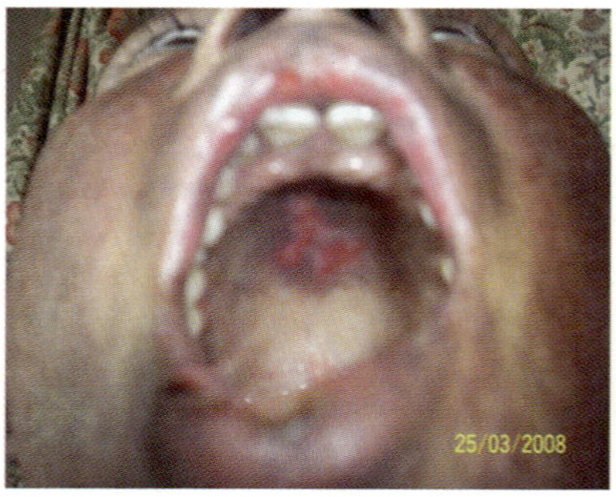

Fig. 6. Oral ulcer.

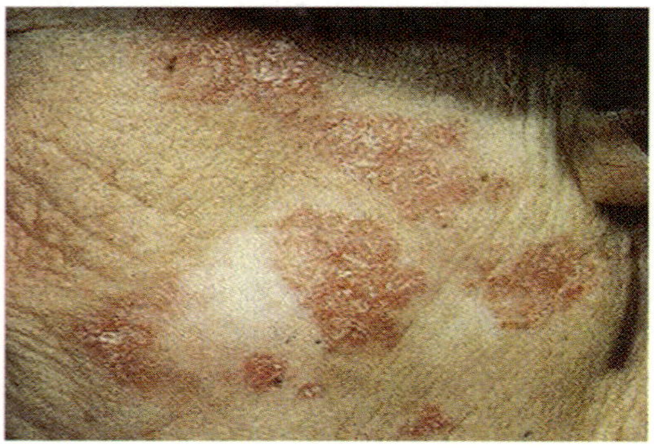

Fig. 7. SLE alopecia.

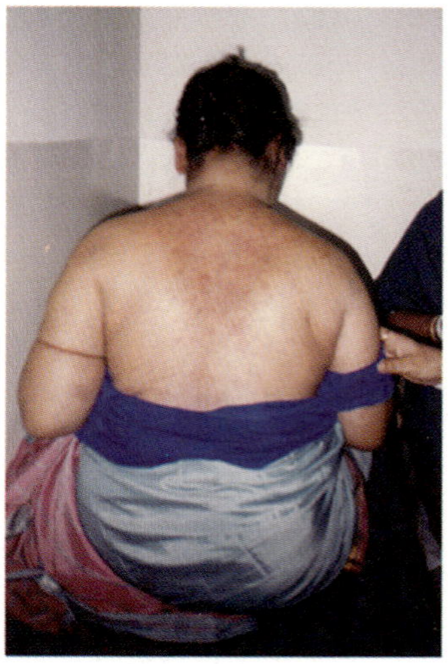

Fig. 8. Sub-acute cutaneous Lupus (Anti Ro positive).

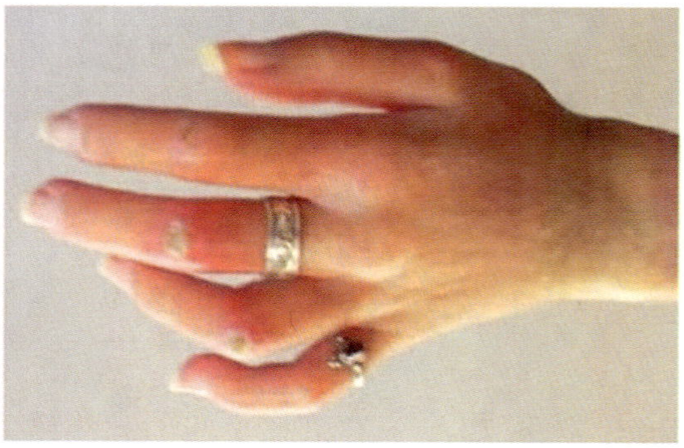

Fig. 9. SLE hand vasculitis.

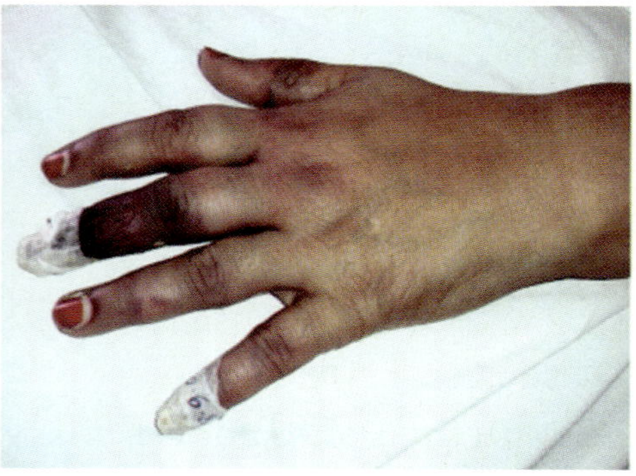

Fig. 10. Digital gangrene.

Musculo skeletal manifestations[13]

Arthralgias are pretty common in 90% patients and may be initial presentation; but non-erosive arthritis is seen in about 10% patients. Arthritis is polyarticular involving hands, wrists, knees and ankles and can be deforming. The synovitis is often mild to moderate with less inflammatory joint effusion. Few patients with impressive synovitis indistinguishable from rheumatoid arthritis are designated as Rhupus. Avascular necrosis of hip joints and uncommonly of shoulder (Fig. 11) and septic arthritis are suspected when there is persistent pain and swelling in one or a few joints. The avascular necrosis is mostly attributed to antiphospholipid syndrome (APS), prolonged steroid therapy or immobilization. Tenosynovitis and tendon rupture can also occur. Myalgia, muscle weakness or tenderness are found in 60% patients whereas inflammatory myositis complicates in 5-10% of patients. Muscle weakness - specifically proximal myopathy - is caused by steroid or chloroquine therapy or rarely due to myasthenia like syndrome. The muscle enzymes are rarely elevated.

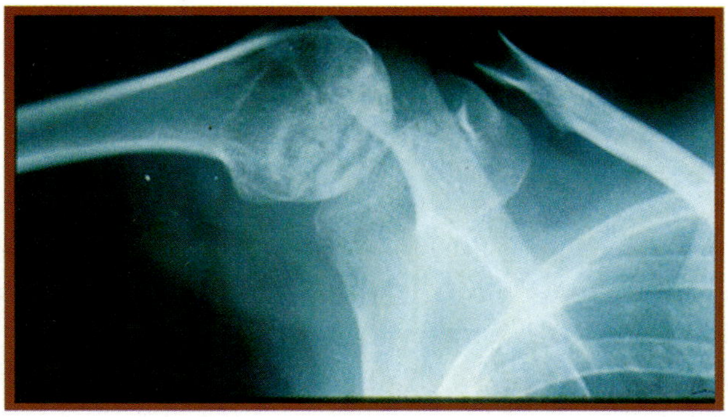

Fig. 11. Avascular necrosis in right shoulder.

Neurological manifestations[14]

The neurological involvement is widespread and tends to occur in nearly 50% patients; however subclinical disease is more than clinical disease. It remains a challenge in terms of pathogenesis, assessment and treatment.

In 1999 a consensus document defined the neurological and psychiatric manifestation in 19 different syndromes as in Table 2[14].

Table 2 : Neuropsychiatric manifestations of SLE[14]

Central nervous system
Aseptic meningitis
Cerebrovascular disease
Demyelinating syndrome
Headache (including migraine and benign intracranial hypertension)
Chorea
Myelopathy
Seizure disorder
Acute confusional state
Anxiety disorder
Cognitive dysfunction
Mood disorder
Psychosis
Peripheral nervous system
Acute inflammatory demyelinating polyradiculopathy
Autonomic disorder
Mononeuropathy, single or multiplex
Cranial neuropathy
Myasthenia gravis
Plexopathy
Polyneuropathy

The most common manifestations of diffuse CNS lupus are cognitive dysfunction (20-30%) headache (20-40%) and seizures (7-10%). There is a clear distinction between CNS manifestations due to active lupus, vasculitis and those due to APS. However opportunistic CNS infections may complicate the clinical picture in an immunocompromised host like SLE. Vascular occlusion may present like a stroke in young with cerebral infarction (Fig. 12).

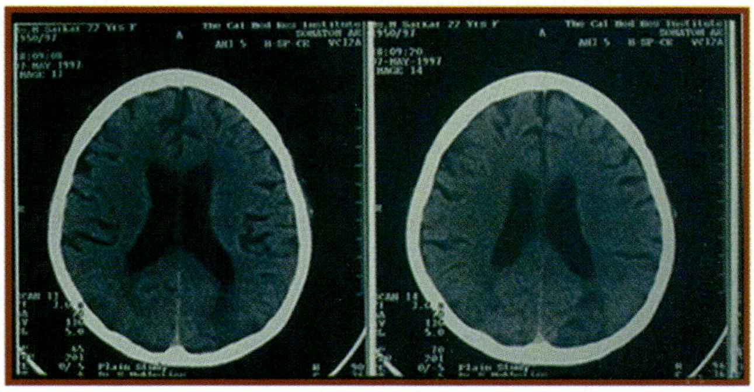

Fig. 12 : Infarct in left occipital lobe in SLE

Overall, CNS disease is associated with significant morbidity and mortality in lupus. Primary neuropsychiatric disease, CNS infection and steroid psychosis are the differential diagnosis to be entertained in patients with cerebral lupus. Multiple peripheral nerves may simultaneously be involved in mononeuritis multiplex (Fig.13).

Renal manifestations[1,4,13]

Nephritis is the most important complication of SLE and occurs eventually in approximately 60% to 75% of patients. The renal disease differs from patient to patient in clinical pattern, severity and prognosis. The spectrum of lupus nephritis may vary from asymptomatic phase to nephritic or nephrotic syndrome. Hence, urinalysis is always recommended in any patient with SLE. The classification of lupus nephritis is essentially histology-based on light and electron microscopy as well as immunofluoroscence staining. Lupus nephritis is thus classified according to World Health Organization (WHO) as in Table 3.

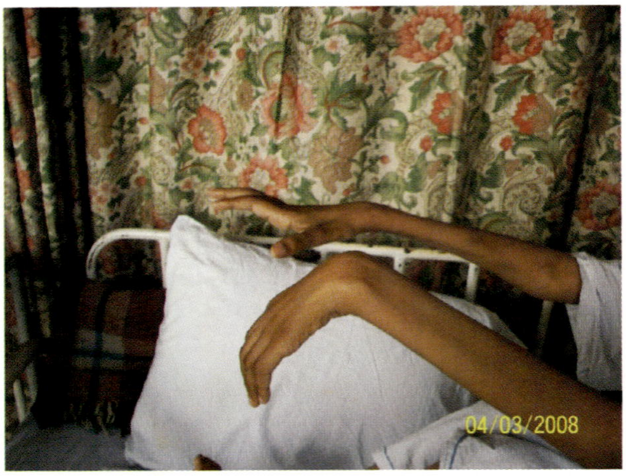

Fig. 13. Left wrist drop and right claw hand (mononeuritis multiplex).

Table 3. The revised 1995 WHO classification of lupus nephritis

1. Normal glomeruli (less than 5% patients)
   (a) Normal by all techniques
   (b) Normal on light microscopy but deposits on immunohistology and/or electron microscopy
2. Pure mesangial alterations (15% patients)
   (a) Mesangial widening and/or mild hypercellularity
   (b) Mesangial cell proliferation
3. Focal segmental glomerulonephritis (associated with mild/moderate mesangial alterations, and/or segmental epimembranous deposits) (20% patients)
   (a) Active necrotizing lesions
   (b) Active and sclerosing lesions
   (c) Sclerosing lesions
4. Diffuse glomerulonephritis (severe mesangial/ mesangiocapillary with extensive subendothelial deposits. Mesangial deposits always present and frequently subepithelial deposits) (50% patients)
   (a) With segmental lesions
   (b) With active necrotizing lesions
   (c) With active and sclerosing lesions
   (d) With sclerosing lesions
5. Diffuse membranous glomerulonephritis (15% patients)
   (a) Pure membranous glomerulonephritis
   (b) Associated with lesions of category II (A or B)
6. Advanced sclerosing glomerulonephritis

This histologic classification is not exclusive and there is likelihood of transformation from one class to other. Therapy is also variable according to class of the disease. The commonest pathology seen on kidney biopsy in India is diffuse proliferative glomerulonephritis (WHO Class IV). Patients with Class IV glomerulonephritis usually have microscopic hematuria and proteinuria (>500 mg per 24 hrs); approximately one-half develop nephritic syndrome and most develop hypertension. If the disease is untreated at this stage, virtually all patients develop ESRD within 2 years of diagnosis. A small group of SLE patients with proteinuria may have membranous glomerulonephritis on renal biopsy. Their outcome is better than those of diffuse proliferative glomerulonephritis. The death due to renal failure tends to occur in 11%, 36%, 6% in WHO Class III, IV, and V respectively. In some cases digital gangrene due to small vessel vasculitis is reported (Fig. 9).

Pulmonary manifestations

The most common pulmonary feature of SLE is pleurisy with or without effusion in 45-60% patients. Pulmonary infiltrates also occur as a manifestation of active SLE and it may be confused with pulmonary infection. The life threatening changes in lungs include interstitial pneumonia leading to fibrosis and intraalveolar haemorrhage; both of these require aggressive and supportive treatment for survival. In this country pulmonary TB, miliary, infiltrative or cavitary forms are common as a result of immunosuppression and need to be distinguished from other lung shadows[15]. Rarely shrinking lung syndrome presenting with dyspnoea on exertion without any ausculatory signs is encountered[16]. Pulmonary hypertension is a rare but potentially life-threatening complication.

Cardiac manifestations[13,17]

Pericarditis is the most frequent form of clinical presentation seen in 25% cases which responds promptly to NSAIDs and on rare occasion it may lead to cardiac tamponade. The other features are myocarditis and Libman-Sacks endocarditis mostly involving mitral or aortic valves. The consequences of these are mostly heart failure, arrhythmias or embolic events. The accelerated atherosclerosis in SLE are related to continuing active disease with coronary vasculitis and prolonged high dose of steroids in combination with other traditional risk factors. The increased risk for vascular events is as much as 50 fold in a women with SLE less than 45 years old when compared to healthy women.

Haematological features[13]

The normocytic normochromic anemia due to active

disease is found in 70% patients. Coombs' positive autoimmune haemolytic anemia and particularly lymphopenia are significantly associated with disease activity and these may be the leading feature. Thrombocytopenia is also recognized and can be due to autoimmune destruction of platelets or because of co-existent antiphospholipid antibody syndrome (APS). This is recognized with purpuric spots. (Fig. 14) In childhood and adolescence idiopathic thrombocytopenic purpura can antedate SLE by years. Bone marrow hypoplasia or aplasia is a rare occurrence. ESR is elevated while CRP mostly remains within normal limits[18]. A rise in CRP denotes either a significant inflammation or associated infection. Lymphadenopathy occurs in approximately 30-40% of patients usually at the onset of disease or during disease flares. Lymph nodes are typically soft, non-tender, discrete usually in neck, axilla and inguinal area; splenomegaly occurs in 10-40% of patients particularly during acute disease and is not necessarily associated with cytopenias.

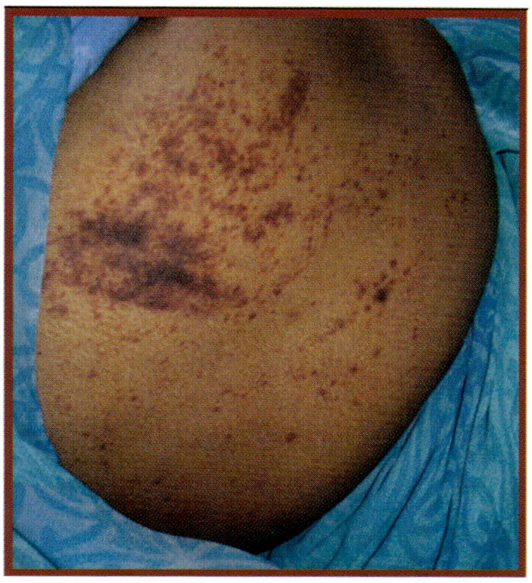

Fig. 14. Purpuric spots.

Gastrointestinal manifestations[19]

The diffuse abdominal pain of autoimmune serositis may be an uncommon presentation. Nausea, vomiting and diarrhea can be due to SLE 'flare'. The subclinical liver dysfunction with raised ALT, AST is sign of active disease. Acute pancreatitis is observed in 2-5% patients presenting with 'acute abdomen'. Vasculitis of intestine in SLE may pose like one of the emergencies with ischaemia, perforation, bleeding and sepsis. Hepatomegaly is reported to occur in 25% and splenomegaly in 10% patients. Parotid enlargement with or without dry mouth is observed upto 8% of patients.

Ocular manifestations[20]

Non-specific conjunctivitis, dry eye (Sicca) syndrome are common in SLE and does not threaten vision. In contrast retinal vasculitis and optic neuritis are vision threatening manifestations of SLE. In patients on long term steroid therapy cataract and glaucoma as complication need to be screened periodically.

Antiphospholipid antibody syndrome (APLA)

A subset of population (around 20-25% antibody positive but 10% having APLA syndrome) with SLE have antiphospholipid antibody syndrome characterized by increased risk of systemic vascular thrombosis (venous and arterial), recurrent abortion, livedo reticularis, thrombocytopenia, ischemic CNS disease and seizures[11]. Premature and accelerated coronary atherosclerosis is often worsened with APLA in SLE patients.

Special problems in SLE

1. ***Pregnancy and SLE*[11] *:*** Fertility is unaffected in SLE except during acute phase of illness. During pregnancy SLE patients may have 'flare'. It is advisable to plan pregnancy while the disease is in remission. Active SLE is associated with increased risk of miscarriage, still birth and low birth weight babies. The SLE patients with antiphospholipid syndrome carry an inordinate risk of foetal wastage. Low dose steroids, azathioprine and hydroxychloroquine are safe in pregnancy.

   New born may have neonatal lupus from SLE mother with Anti Ro-antibodies; the disease is self-limited or the child can have congenital heart block.

   Oestrogen containing hormones and intra-uterine contraceptive devices are better to avoid in lupus patients. Hormone therapy with progesterone alone is permitted. Barrier methods can be safe.

2. ***Paediatric SLE and neonatal lupus syndrome*[19]*:*** The prevalence of paediatric SLE constitutes 20-25% of total SLE population. This special group has higher frequency of hepatosplenomegaly, chorea, renal disease and avascular necrosis. They should be treated as in adult SLE. Most of these patients survive longer with current treatment.

   The neonatal lupus syndrome is characterized by photosensitive skin rash, hepatosplenomegaly, cytopenias and carditis. The serious consequences are endomyocardial fibroelastosis, and congenital complete heart block in infants when born out of anti Ro positive SLE mother. Treatment is supportive when features are transient. Permanent pacing is indicated in symptomatic persistent congenital heart block.

3. ***SLE in elderly***[20,21] : SLE in elderly constitutes around 6-12% patients. The disease is often mild in this age group with insidious onset, longer duration at the time of presentation. There is lower prevalence of renal, CNS and skin involvement.
4. ***Drug induced SLE***[4,7,22] : This is a small subgroup of patients who develop milder SLE on exposure to certain drugs like Procainamide, Hydralazine, INH, Methyldopa, Chlorpromazine and anti-TNF drugs. This is characterized by positive ANF and Antihistone antibodies. Anti dS DNA is absent with the probable exception of disease associated with use of IFNα and anti-TNF therapies. In most patients serological features are accompanied by mild clinical disease including fever, myalgia, rash, arthritis, and serositis. The patient may remain well on withdrawal of culprit agents.
5. ***Infection in SLE***[23,24,25] : The patients with SLE are vulnerable to infections with intracellular organisms like herpes, mycobacteria, candida, salmonella group of organisms[23]. The risk factors for infection includes active disease, long-term disease damage, cytopenias (neutropenia/ lymphopenia), hypocomplementemia, renal involvement, CNS involvement, and immuno-suppressive therapy[24]. At times it is difficult to distinguish between active disease and acute infections. Both can result in anemia and raised ESR but CRP remains negative in active SLE whereas becomes positive in presence of infection[25]. Judicious antibiotics therapy and prophylactic vaccinations are recommended.
6. ***Overlap syndrome*** : This syndrome is said to exist when there is overlap between clinical and serological criteria for two or more connective tissue disorders namely SLE with scleroderma or SLE with RA or dermatomyositis. Mixed connective tissue disease (MCTD) is specifically characterized with Raynaud's phenomenon, swollen dorsum of hands, dermatomyositis and mild polyarthritis and positive Anti U1NP.

Diagnosis

The diagnosis of SLE is based on characteristic clinical features and autoantibodies. The clinical classification criteria has been laid down by Tan et al in 1982[26] and further updated by Hochberg in 1997[27] (Table 4).

Table 4 : Revised ACR classification criteria for SLE: 1997 update

Item	Definition
Malar Rash	Fixed erythema, flat or raised, over the malar eminences, sparing nasolabial folds
Discoid Rash	Erythematous, raised patches with adherent Keratotic scaling and follicular plugging; atrophic scarring may occur in older lesions
Photosensitivity	Skin rash as a result of unusual reaction to sunlight by history or on physical examination
Oral Ulcers	Oral or nasopharyngeal ulceration, usually painless, observed by physician
Non-erosive arthritis	Involving 2 or more peripheral joints, characterized by tenderness, swelling or effusion.
Pleuritis/Pericarditis	a) Pleuritis - convincing h/o Pleuritic chest pain or rub or pleural effusion on physical examination Or b) Pericarditis - documented by ECG, rub or e/o effusion
Renal Disorder	a) Persistent proteinuria > 0.5 gm/ day or > +++ OR b) Cellular casts - may be red cell, Hb, granular, tubular or mixed
Neurological Disorder	a) Seizures - in the absence of offending drugs, or known metabolic derangement, eg. Uraemia, ketoacidosis or electrolyte imbalance, OR b) Psychosis - in the absence of Offending drugs, or known metabolic derangement, eg. Uraemia, ketoacidosis or electrolyte imbalance
Immunological Disorder	a) Anti DNA: antibody to native DNA in abnormal titre, OR b) Anti- Sm: Presence of antibody to Sm nuclear antigen, OR c) Positive finding of aPL antibodies based on: 1. ↑ serum level of IgG or IgM aCL or 2. a positive test result for lupus anticoagulant, using a standard method or 3. a false- positive test for syphilis for at least 6 months and confirmed by TPI or FTA abs test
Haematological Disorder	a) Haemolytic anaemia with reticulocytosis, OR b) Leukopenia < 4000/cu mm on 2 or more occasions, OR c) Lymphocytopenia < 1500 on 2 or more occasions, OR d) Thrombocytopenia < 100,000/cu mm in the absence of offending drugs
Positive ANA	An abnormal titre of ANA by immunofluorescence or an equivalent assay at any point of time in the absence of drug

A person shall be said to have SLE if any 4 or more of the 11 criteria are present, serially or simultaneously during any interval of observation (95% specificity and 85% sensitivity).

The fluorescent antinuclear antibody (FANA) is positive in greater than 95% of patients. The various staining patterns (i.e. homogenous, speckled, rim, nucleolar) can be demonstrated depending on the content of different autoantibodies in the serum. This test has high sensitivity but low specificity. Indeed positive FANA tests occur in most of other connective tissue diseases. To be significant it must have positivity in high titre (1:160 or more) and the patient must have clinical features of SLE. Anti dS DNA are highly specific for SLE but only in about 50 - 60% of patients. It is tested by ELISA, Crithidia, immunofluorescence assay or Farr assay. The anti dS DNA positivity has also got good correlation with progression of lupus nephritis. The reduced value of C3 is closely associated with renal damage. The antibodies to Smith (anti-Sm) has high specificity for SLE but only in about 10% cases. Recently antinucleosome antibody is reported to be associated with progressive lupus nephritis[28,29].

Recent revision of ACR 2012 has been made by the Systemic Lupus Erythematosus Collaborating Clinics (SLICC) after 8 years of study in order to improve clinical relevance and incorporate new knowledge in the field of lupus. However this classification is not meant to be used for the diagnosis of lupus.

SLICC* Classification Criteria for Systemic Lupus Erythematosus[30]

Requirements: ≥ 4 criteria (at least 1 clinical and 1 laboratory criteria) OR biopsy-proven lupus nephritis with positive ANA or Anti-DNA

Clinical criteria	Immunologic criteria
1. Acute Cutaneous Lupus	ANA
2. Chronic Cutaneous Lupus	Anti-DNA
3. Oral or nasal ulcers	Anti-Sm
4. Non-scarring alopecia	Antiphospholipid Ab
5. Arthritis	Low complement (C3, C4, CH50)
6. Serositis	Direct Coombs' test (do not count in the presence of hemolytic anemia)
7. Renal	
8. Neurologic	
9. Hemolytic anemia	
10. Leukopenia	
11. Thrombocytopenia (<100,000/mm^3)	

*SLICC : Systemic Lupus International Collaborating Clinics

The investigations to screen and monitor organ damage in SLE are all advocated depending upon clinical suspicion of organ damage. Besides routine tests the additional investigations recommended are mentioned in Table 5.

Table 5 : Investigation of choice for SLE organ damage

System	Investigations
Renal	24 hours proteinuria Creatinine/albumin ratio Anti DS-DNA C_3, antinucleosome Ab Renal biopsy
Pulmonary	Sputum, CXR, HRCT Chest, PFT with DLCO
Cardiac	ECG, 2D-Echo
Haematological	Coombs' test, bone marrow examination, Anti phospholipids Ab Platelets, Hb, WBC, ESR, CRP
CNS	CSF examination MRI/CT, APL antibody PET/SPECT

Further Reading :

1. Chandrasekharan AN, Ram Krishnan S, Rajendran CP et al. Spectrum of clinical and immunological features of systemic rheumatic diseases in a referral hospital in South India - SLE. J. Ind Rheum Assoc 1994; 2: 35-45.
2. Malaviya AN. Systemic Lupus Erythematosus. In Association of Physician of India Text Book of Medicine Sainani GS (ed) New Delhi (5th ed) 1992; 1125-32.
3. Pande I, Sekharan NG, Kailash S et al. Analysis of clinical and laboratory profile in Indian childhood SLE and its comparison with SLE in adults LUPUS 1993; 2: 83-87.
4. Malaviya AN, Singh RR, Kumar A et al. SLE in Northern India - a review of 329 cases JAPI 1988; 36: 476-80.
5. Mok CC, Lan CS. Pathogenesis of SLE. J. Clin Path. 2003; 56: 481-90.
6. Kalden JR. Defects of clearance mechanisms in the pathogenesis of SLE APLAR 2004, South Korea.
7. Boumpas DT, Carvera R, Bertsias G. Systemic Lupus Erythematosus: Pathogenesis and Clinical Features. 20_Eular_Fpp.indd: 476-505.
8. Bertsias GK, Salmon JE, Boumpas DT. Therapeutic opportunities in systemic lupus erythematosus: state of the art and prospects for the new decade. Ann Rheum Dis 2010; 69: 1603-11.
9. Joshi VR. SLE - a perspective J. Ind Rheum Assoc 1995; 3: 40-42.
10. Joshi VR, Balakrishnan C. Systemic Lupus Erythematosus - Manual of Rheumatology - IRA publication 1999 227-248.
11. Boumpas DT, Austin HA, Fessler BJ et al. SLE - Emerging concepts Part 2 : Dermatologic and joint disease, the antiphospholipid antibody syndrome pregnancy and hormonal therapy, morbidity and mortality and pathogenesis. Ann. Int. Med. 1995; 123: 42-43.
12. A. Kumar. Indian Guidelines on the Management of SLE. J Indian Rheumatol Assoc 2002; 10: 80-96.
13. Boumpas DT, Austin HA, Fessler BJ et al. SLE - Emerging concepts Part I: Renal, neuropsychiatric - cardiovascular, pulmonary and haematologic disease. Ann. Int. Med. 1995; 122: 940-950.
14. ACR Ad HOC Committee on Neuropsychiatric Lupus Nomenclature. ACR nomenclature and case definition for neuropsychiatric lupus syndrome. Arth & Rheum 1999; 42: 599-608.
15. Balakrishna C, Mangat G, Joshi VR. Tuberculosis in patients with SLE. JAPI 1998; 46: 682-83.
16. Balakrishna C, Mangat G, Venkatachalam S et al. Shrinking lungs in SLE. A study of six cases. J. Ind. Rheum. Assoc 1994; 2(Suppl): 85-88.
17. Sturfeit G, Eskilsson J, Nived O et al. Cardiovascular disease in SLE - a study of 75 patients from defined population. Medicine (Baltimore) 1992; 71: 216-23.
18. Joshi AV, Halankar SA, Kulkarni A et al. C-reactive protein in rheumatic diseases. JAPI 1986; 74: 859-60.
19. White PH. Paediatric SLE and neonatal lupus. Rheum Clin North Am 1994; 20: 119-27.
20. Meena HS, Kumar A, Muralidhar R et al. Late onset SLE J. Ind Rheum Assoc. 1995; 3: 14-16.
21. Urowitz MB, Stevens MB, Shulman LE. The influence of age on clinical pattern of SLE Arth & Rheum 1967; 10: 319-22.
22. Kimberley RP. SLE - In Manual of Rheumatology & Outpatients Orthopedic disorders. Paget S, Pellicci P and Beary JF (eds) Boston (3rd ed) 1993; 219-33.
23. Leiopoules AG, Tsokos GC : Immunopathogenesis and spectrum of Infection in SLE. Sem. Arth & Rheum. 1996; 25: 318-36.
24. Gedalia A, Cuchacovich R. Pathophysiology and Clinical Spectrum of Infections in Systemic Lupus Erythematosus. Rheum Dis Clin N Am 2009; 35: 75-93.
25. Evert JB, Horst G, Limburg PC et al. C-RP levels during disease exacerbation and infection in SLE. A prospective longitudinal study J. Rheum 1990; 17: 1642-48.
26. Tan EM, Cohen AS, Fries JF et al. The revised criteria for classification of SLE. Arth & Rheum 1982; 1271-77.
27. Hochberg MC. Updating the ACR revised criteria for the classification of SLE. Arth & Rheum 1997; 40: 1725.
28. Chandrasekharan AN. Clinical and immunological profile of SLE in India. Proceedings of XIX ILAR Congress of Rheumatology, Singapore 1997; 156-64.
29. Pisetsky DS, Gilkeson G, St. Clair EW. SLE - Diagnosis and treatment Med. Clin North Am 1997; 81: 113-28.
30. Petri M, et al. Arthritis and Rheumatism, Aug 2012.

EVALUATION AND MANAGEMENT OF SLE

2

Sukumar Mukherjee, Somanth Bhar

Introduction

Systemic Lupus Erythematosus (SLE) is a complex autoimmune disease with significant heterogeneity and periods of relapse and remission. Recent advances in our understanding of the genetic, molecular and cellular basis of autoimmune pathology have prompted to the application of novel and targeted therapy. Over last few decades it has been found that timely, appropriate and aggressive therapy can delay or prevent organ failure or morbidity and it is cost effective[1]. In view of low prevalence of disease most primary care physicians and general internists will not have sufficient experience in monitoring and management of this life threatening disease. Thus, in most cases, a referral to a rheumatologist may be needed for interpretation of laboratory findings, confirmation of the diagnosis, subsequent monitoring of the disease at its difficult phases of natural history, and thus a comprehensive care of SLE patients. This article reviews the general approach to the therapy of SLE, focusing on current approved therapies and novel approaches that might be used in future. Although there is dramatic improvement in the prognosis for SLE patients, treatment of those with active disease refractory to traditional therapies continues to be a real challenge.

Natural history

SLE is a chronic disease of variable severity with a waxing and waning course. It starts with a preclinical phase characterized by autoantibodies and proceeds with a more disease specific clinically overt autoimmune phase[2]. During its natural history periods of flares intercept periods of remission culminating in disease and therapy related damage. Early damage is mostly related to active disease whereas late damage - namely infections, premature atherosclerosis, osteoporosis and malignancies - is usually related to complications of long standing disease and immunosuppressive therapy.

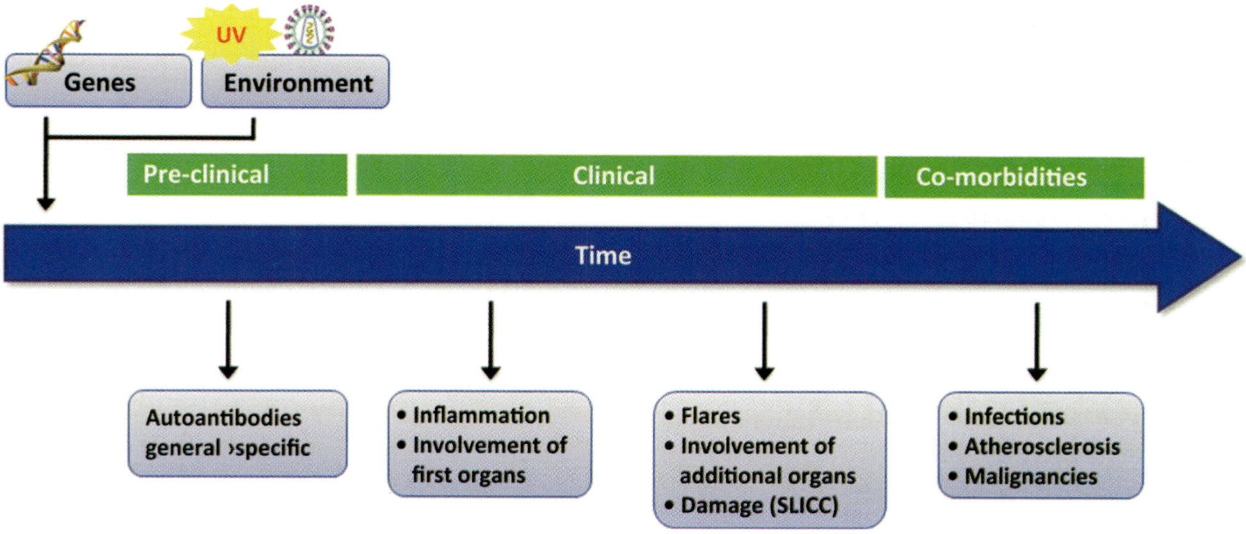

Fig. 1. Natural history of systemic lupus erythematosus

Monitoring

Overall monitoring of SLE is important because of various reasons. The SLE patients are at persistent risk of developing disease flares, as well as permanent organ system damage in more than half the cases[3]. Although prognosis of SLE has improved dramatically, mortality and morbidity still remain a major concern. The mortality rate among patients of SLE is at least three times that of the general population. Survival rates are ~ 80% at 10 years after diagnosis and ~ 65% at 20 years[4-6]. In the late 70s the bimodal mortality pattern was observed in patients with systemic lupus erythematosus. Within first 5 years the main cause of death was active disease or infections. The second peak of mortality was attributable to the consequences of therapy, atherosclerotic vascular disease and malignancy[7].

Early recognition of disease activity or organ damage and life long monitoring are required for most patients, to initiate and modify proactive, aggressive and effective therapy. In milder cases the intensity of therapy may not be so necessary. There is considerable overlap between features of active SLE and symptoms of adverse effects of drugs used in treatment. The drug toxicity needs to be appreciated to salvage the organ at risk. Prediction of 'flare' either in pregnant or non pregnant women is rather important to select and adjust the use of immunosuppressive agents to control inflammation and delay further progression. Concurrent counseling is also rewarding. At times there is possible evolution of SLE in its natural course to other auto immune disorders like antiphospholipid antibody syndrome or Sjögren's syndrome which requires proper screening and recognition for appropriate modification of therapy.

Finally it is important to establish a team relationship among all the healthcare providers, including primary care physicians, specialists, nurses, pharmacists and the patient and family, to allow early detection of manifestations of disease flare and drug toxicity and to identify additional features of the disease, requiring different management approach such as in fibromyalgia.

LEVELS OF MONITORING

Initial monitoring

The monitoring, after the diagnosis, starts with clinical presentation and basic laboratory investigations in regular visits as described in Table 1.

Table 1. Monitoring of patients with SLE[8]

History
Constitutional symptoms such as fever, weight loss of > 5% of body mass, fatigue
New rash over body surface or mouth
Increased hair loss
Pleuritic chest pain
Joint pain and swelling
Physical Examination
Joint swelling
Rash, discoid lesions, alopecia, mucous membrane ulcers
Vasculitic lesions
Fundoscopic examination
Edema
Investigations
Hematology - complete blood count, platelet count, ESR, CRP
Chemistry - serum creatinine
Urinalysis - proteinuria, sediments, 24 hours urine for protein and creatinine clearance
Serology - C3, anti ds-DNA (Crithidia or Farr assay)
Radiology - chest (PA)

The frequency of these evaluations will depend on the activity, the severity and extent of SLE, the response to treatment, the type of treatment, including the need for toxicity monitoring. The ESR, being nonspecific, is not very helpful and CRP is useful sometimes to differentiate infection and the disease activity.

Follow up monitoring

At routine follow up visits every 6 month the following should be obtained depending on clinical features and organ involvement.

- Complete blood count, platelet count, serum creatinine, creatinine clearence, and urinalysis for protein and sediments even if the patient has had normal values in the past and reticulocyte count and Coomb's test in presence of haemolysis.
- Patients with known renal disease should also have serum levels of C3 and anti ds-DNA antibodies every 6-12 months.
- Electrolytes, lipids, calcium, phosphate, and alkaline phosphatase levels should be determined approximately once a month at the onset of nephritis and more often if the condition is unstable.
- Monitoring of laboratory findings, in long standing SLE, may be facilitated by the use of flow charts, or global activity indices.

Mild SLE is characterized by arthritis, arthralgia, myalgia, fatigue, mild mucocutaneous involvement, low-grade fever, mild serositis, 'lupus' headache. Moderate SLE encompasses the features of high-grade fever, toxaemia, severe mucocutaneous manifestations, marked photosensitivity, moderate to severe serositis, lupus pneumonitis, mild to moderate myocarditis, mesangioproliferative or minimal change lupus nephritis, haemolytic anaemia and thrombocytopenia. Severe SLE is considered when there is organ/life-threatening features such as focal/diffuse proliferative glomerulonephritis with or without azotaemia/ hypertension, lupus cerebritis with recurrent seizures, acute confusional state, coma; systemic necrotizing vasculitis such as one causing peripheral gangrene, GI bleeding or mononeuritis multiplex[9].

Patients with mild stable disease may require follow up visits at 3 to 6 months interval. Patients with more severe disease or with complications, patients during pregnancy and post partum period or receiving immunosuppressive medications will need more frequent follow up.

Patients with persistent urinary sediment, such as hematuria and pyuria without adequate explanation, urinary casts, increased serum creatinine level, deteriorating renal function and patients not responding to conventional therapy should have a renal biopsy. The prognostic value of kidney biopsy although controversial, has been demonstrated particularly for patients with normal serum creatinine level[5,8,10]. Patients with proliferative glomerular lesions and chronic changes are at higher risk of end stage renal disease.

Disease activity (reversible changes)

Assessment of disease activity refers to determination of the extent to which immuno-inflammation of SLE contributed to the clinical setting, in regards to morbidity and mortality, at a particular time. The clinical measurement of disease activity in SLE involves an assessment of either presence or absence of specific clinical features (e.g. arthritis, serositis, features of different organ involvement like kidney, CNS, heart), the results of laboratory parameters (e.g. anti ds-DNA antibodies or complement C3,C4,CH50 levels) or by using a global activity index (e.g. SLEDAI, BILAG, SLAM, ECLAM).

Serological monitoring

Anti ds-DNA antibodies are found in over 70% of SLE patients at some point of time, in the disease course and have 95% specificity. They develop early even predating clinical disease onset[11]. Most utilized method for detecting anti ds-DNA antibody is immuno fluorescence against Crithidia luciliae. A number of studies have attested to the fact that anti ds-DNA antibodies parallel the activity of SLE[12,17] or rise in anti ds-DNA antibodies can herald an impending flare[14,16]. The problem with the association of anti ds-DNA antibodies with disease activity is that there may be dissociation between disease activity and levels of anti ds-DNA in natural history of subsets of SLE patients[18]. A negative test for anti ds-DNA was considered as being most helpful in the setting of a low pretest probability of disease activity.

Anti nuclear antibodies are found in up to 99% of SLE patients, however they do not reliably measure disease activity. It does not require repeated ANA estimation once diagnosis is confirmed.

Association of different sub components of antinuclear factors has been well recognized, in different subsets of SLE, as shown in the following Table 2.

Table 2. Antibodies associated with different subsets of disease

Autoantibodies	Disease association
Anti Ro/SSA & anti La/SSB	At risk for sicca complex & photosensitive skin rashes, neonatal lupus
Anti U1-RNP	SLE overlap
Anti ribosomal P	Neuropsychiatric SLE
Anti histone antibody	Drug induced SLE
Anti Scl 70	Pulmonary hypertension

Levels of various complements are reduced during inflammatory process; low complement levels are found in 44% of patients at the onset, and in 77% at any point of time. The C3, C4 and CH50 levels are easily measured and are used to follow the clinical course of the disease; low levels being generally poor prognostic factors. Progressive fall in C3 and C4 may indicate an impending relapse of SLE. Increased levels of inflammatory mediators like C5a and C5b-9 correlated with severe exacerbation or advanced disease[19,20] and C5b-9 levels may be a better predictor of disease activity than C3 and C4. C3b in urine may be helpful in detecting early damage.

Immunological studies are often used as early pre clinical markers of changing lupus activity. There is inverse relationship between ds-DNA, disease activity score and complement level as shown (Fig 2). Whether these immunologic studies predictably identify a window of pre clinical flare in broad population of SLE patients or opportunity of therapeutic intervention to alter the clinical course of the disease, have never been studied in a prospective trial.

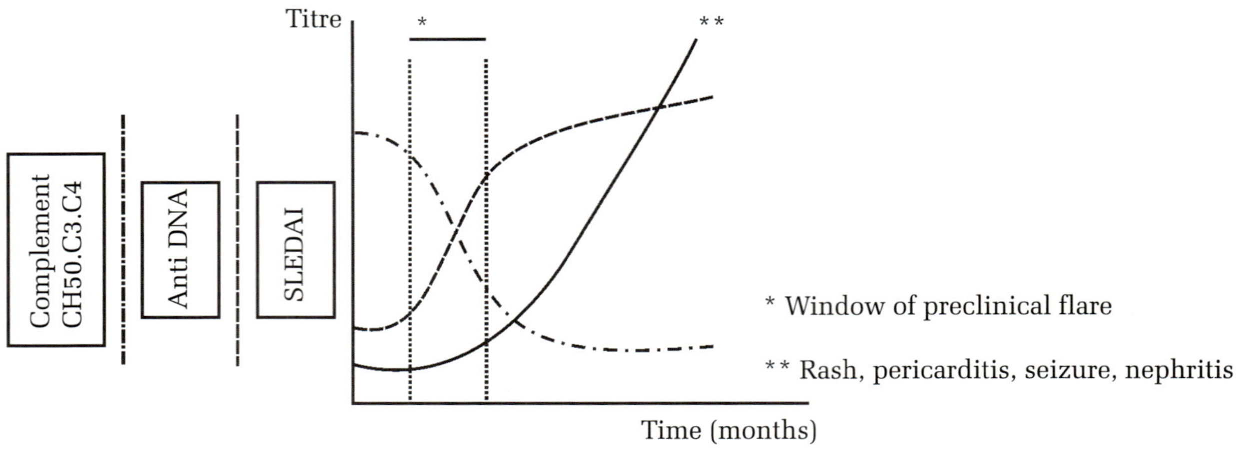

Fig 2. Immune studies and SLE disease activity

A) Disease activity indices

Several instruments of measuring global lupus activity are formally validated including SLAM (Systemic Lupus Activity Measures) from US, SLEDAI (Systemic Lupus Erythematosus Disease Activity Index) from Canada, BILAG (British Isles Lupus Activity Group) from UK, LAI (Lupus Activity Index), ECLAM (European Consensus Lupus Activity Measurement)and SIS (the National Institutes of Health SLE Index Score), of which the first three are most widely used in lifelong follow up of SLE patients and the clinical trials[21].

The SLEDAI Index[22] includes 24 items grouped into 9 organ systems. It uses weights ranging from 1 to 8 with a total possible score of 105. It measures the descriptive disease activity within 10 days period. Fatigue and steroid use were not included and more weightage is given to CNS involvement. It is sensitive to change over time and has been used in prognosis studies[23,24].

Table 3. SLE disease activity index (SLEDAI)[22]

Descriptor	Definition	Score
Seizure	Recent onset, exclude metabolic, infectious or drug causes	8
Psychosis	Altered ability to function in normal activity due to severe disturbance in the perception of reality. Include hallucinations, incoherence, markedly loose associations, impoverished thought content, markedly illogical thinking, bizarre, disorganised or catatonic behaviour. Exclude uraemia and drug causes	8
'Organic brain syndrome'	Altered mental function with impaired orientation, memory or other intellectual function, or Acute confusional state or with rapid onset and fluctuating clinical features, inability to sustain attention to environment, plus at least 2 of the following: perceptual disturbance, incoherent speech, insomnia or daytime drowsiness, or increased or decreased psychomotor activity. Exclude metabolic, drug or infectious causes	8
Visual disturbance	Retinal changes of SLE. Include cytoid bodies, retinal haemorrhages, serous exudates/haemorrhages in choroid or optic neuritis. Exclude hypertension, infection or drug causes	8
Cranial nerve disorder	New onset of sensory or motor neuropathy involving cranial nerves	8
Lupus headache	Severe persistent headache: may be migrainous, but must be non-responsive to narcotic analgesia.	8
CVA	New onset of CVA. Exclude atherosclerosis.	8
Vasculitis	Ulceration, gangrene, tender finger nodules, periungual infarction, splinter haemorrhages, or biopsy or angiogram evidence of vasculitis	8
Arthritis	2 joints with pain and signs of inflammation (tenderness, swelling or effusion)	4

Myositis	Proximal muscle aching/weakness, associated with elevated CPK/aldolase or EMG changes or biopsy evidence of myositis	4
Urinary casts	Haemoglobin, granular or RBC casts	4
Haematuria	> 5 RBC/HPF. Exclude stone, infection or other causes	4
Proteinuria	> 0.5 grams/24 hrs	4
Pyuria	> 5 WBCs/HPF. Exclude infection	4
Rash	Inflammatory type rash	2
Alopecia	Abnormal, patchy or diffuse loss of hair	2
Mucosal ulcers	Oral or nasal ulcerations	2
Pleurisy	Pleuritic chest pain with pleural rub/effusion/pleural thickening	2
Pericarditis	Pericardial pain with at least 1 of the following: rub, effusion or ECG or Echo confirmation	2
Low complement	Decrease in CH50, C3 or C4 below the normal limit of Lab	2
Increased DNA binding	Increased DNA binding using Farr assay	2
Fever > 38 Deg C	Exclude infection	1
Thrombocytopaenia	< 100,000/cu mm, exclude drug causes	1
Leukopaenia	< 3000/cu mm, exclude drug causes	11

The BILAG Index 25 includes a total of 86 items in 8 organs. All features must be attributable to SLE and refer the last four weeks compared with the prior visit's disease activity.

Each feature is scored by: 0 = Not Present, 1 = Improving, 2 = Same, 3 = Worse, 4 = New or Recurrence. BILAG has been shown to function well for each of these organs except CNS disease where it is difficult to decide whether the particular symptom is due to disease activity or a coincidental finding only. It has an overall sensitivity of 87% and specificity of 99%.

Table 4. Parameters to be assessed in BILAG (British Isles Lupus Assessment Group index)[25]

General - Must be SLE related

1. Pyrexia (documented)
2. Weight Loss - unintentional > 5%
3. Lymphadenopathy/Splenomegaly
4. Fatigue/Malaise/Lethargy
5. Anorexia/nausea/vomiting

Mucocutaneous - Must be SLE related

6. Maculopapular eruption - severe, active, (bullous)
7. Maculopapular eruption - mild
8. Active discoid lesions - generalized/extensive
9. Active discoid lesions - localized including lupus profundus
10. Alopecia (severe, active)
11. Alopecia (mild)
12. Panniculitis (severe)
13. Angioedema
14. Extensive mucosal ulceration
15. Small mucosal ulcers
16. Malar erythema
17. Subcutaneous nodules
18. Perniotic Skin Lesions
19. Periungal erythema
20. Swollen fingers
21. Sclerodactyly
22. Calcinosis
23. Telangiectasia

Neurological - Must be SLE related

24. Deteriorating level of consciousness
25. Acute psychosis, delirium, confusion
26. Seizures
27. Stroke or stroke syndrome
28. Aseptic Meningitis
29. Mononeuritis multiplex
30. Ascending or transverse myelitis
31. Peripheral or cranial neuropathy
32. Disc swelling/cytoid bodies
33. Chorea
34. Cerebellar ataxia
35. Headache severe, unremitting
36. Organic depressive illness
37. Organic brain syndrome including pseudotumor cerebri
38. Episodic migranous headaches

Musculoskeletal - Must be SLE related

39. Definite myositis
40. Severe Polyarthritis with loss of function

41. Arthritis
42. Tendonitis
43. Mild chronic myositis
44. Arthralgia
45. Myalgia
46. Tendon contractures and fixed deformity
47. Aseptic necrosis

Cardiovascular & Respiratory - Must be SLE related

48. Pleuropericardial pain
49. Dyspnea
50. Cardiac Failure
51. Friction Rub
52. Effusion (pericardial or pleural)
53. Mild or intermittent chest pain
54. Progressive CXR changes - lung fields
55. Progressive CXR changes - heart size
56. ECG evidence of pericarditis or Myocarditis
57. Cardiac dysrhythmias including tachycardia > 100 in the absence of fever
58. Pulmonary function fall by 20%
59. Cytohistological evidence of inflammatory lung disease

Vascular - Must be SLE related

60. Major cutaneous vasculitis including ulcers
61. Major abdominal crisis due to vasculitis
62. Recurrent thromboembolism excluding strokes
63. Raynaud's phenomenon
64. Livido reticularis
65. Superficial phlebitis
66. Minor cutaneous vasculitis (nailfold vascultitis, digital vasculitis)
67. Thromboembolism (excluding stroke) - first episode

Renal - Must be SLE related

68. Systolic Blood Pressure
69. Diastolic Blood Pressure
70. Accelerated Hypertension
71. Urine dipstick
72. Urinary protein
    a. 24 hr urinary protein
    b. Urine protein-creatinine ratio
73. **Proteinuria**
    a. Newly documented proteinuria of > 1g/ 24 hours
    b. Newly documented protein-creatinine ratio of > 100mg/mmol
74. Nephrotic Syndrome
75. Creatinine (serum)
76. Creatinine clearance/GFR
77. Active urinary sediment
78. Histological evidence of active Nephritis - within 3 months

Hematology - Must be SLE related

79. Hemoglobin (g/dl)
80. Total white cell count
81. Neutrophils
82. Lymphocytes
83. Platelets
84. Evidence of active hemolysis
85. Coombs test positive
86. Evidence of circulating anticoagulant

The SLAM Index[26] includes 32 items divided into 11 organ systems and scores are graded on a scale of 1-3 based on severity, giving a possible total score of 86. It has been later modified as SLAM-R. This index has been found to be reliable and has been used as an outcome measure in therapeutic trials.

The ECLAM Index[28], derived from a study of a large number of real patients of SLE from different centers of Europe, includes 15 items to be recorded over 1 to 3 months period. Fatigue has also been incorporated in it.

These global activity indices have shown sensitivity to change and have been shown to compare with each other. A comparative overview has been given in Table 5.

Table 5. Comparative overview of SLE disease activity indices

Instrument	Strength	Ranking priority	Comments/Limitations
SLEDAI	More specific to lupus	Generally weighs importance of organ	Within organ, degree and change not factored, scored only in presence of symptoms, does not include improvement or worsening, and many serious symptoms
SLAM	Most sensitive	Degree of symptom	Importance of organ not factored, assess disease severity also
BILAG	Most accurate global measure	Importance of organ and degree of symptom	Complexity of instrument, enables assessment of effectiveness of medications for certain organ manifestation, no observer bias, based on 'intention to treat' premise

The effectiveness of all of the SLE disease activity instruments is further limited by the need for intensive investigator training and continuous quality assurance programs to ensure consistency and reproducibility in their use. As the clinical course of lupus is often characterized by periods of stable disease followed by flare or more active disease, relying on fixed time intervals at which to asses disease activity may not provide a true representation of overall disease activity over time and response to therapy.

B) Disease 'flare' and its measurement

Disease activity may be assessed based on the number or time to flare[28,29]. There is no generally accepted single definition of flare as it tends to occur in a variety of ways. Flares may be defined based on changes in a single organ or on a change in therapy. Examples are as follows:

Definitions of a 'flare'

General flare - at least one of the following:

1. Increase in prednisone (> 5mg/day) for at least 14 days since the previous visit.
2. SLE manifestation requiring hospitalization.
3. Addition of new medication or an increase in the dose of an existing medication to specifically treat a manifestation of increase SLE activity.

Renal flare[30,31,32] - nephritic flare include reproducible increase of serum creatinine by > 30% (or, decrease in GFR by 10%) and active urine sediment with increase in glomerular hematuria by 10 red blood cells per high power field, irrespective of change in proteinuria; proteinuric flare include reproducible doubling of UPCR (urine protein: creatinine ratio) to > 100 mg/mmol after complete response or reproducible doubling of UPCR to > 200 mg/mmol after partial response.

Flare based on SELENA trial (the Safety of Estrogen in Lupus Erythematosus National Assessment trial)[28]

Mild / moderate flare

1. Change in SLEDAI >3 points.
2. New/worse skin rash, stomatitis, serositis, arthritis, fever.
3. Increase prednisone < 0.5mg/kg/day.
4. Added NSAIDs.
5. >1.0 increase in a physician global assessment.

Severe flare

1. change in SLEDAI >12.
2. New/worse CNS disease, vasculitis, nephritis, myositis, platelet <60,000, hemolytic anemia with hemoglobin >7gm/dl requiring doubling prednisone.
3. Hospitalisation for SLE.
4. Prednisone >0.5mg/Kg/d.
5. New immunosuppressive use.
6. Increase physician global assessment to >2.5.

Despite all these criteria there is still controversy amongst different laboratory markers like ds-DNA or complement to predict a flare of variable degree. Organ specific measures of activity provide an approach to study therapeutic efficacy in lupus, addressing the particular manifestation of most concern. This is mostly useful in clinical trials but not in practice, as overall assessment is necessary in management of SLE.

C) Disease damage (irreversible changes)

a) SLICC (Systemic Lupus International Collaborating Clinics)/ACR damage index[3] With substantial improved survival figures for patients of lupus, it has become necessary to develop a method of assessing the overall cumulative effect of the disease. The SLICC/ACR index records the number of items of permanent change in an individual patient since the onset of disease regardless of cause (SLE/inter current illness/treatment). The parameter to be recorded must have persisted for at 6 months. It is ascertained by clinical assessment and simple investigations, as available. Damage is distinguished in 12 organ/systems including ocular, neuropsychiatric, renal, pulmonary, cardiovascular, gastrointestinal, musculoskeletal, gonadal, endocrine damage and the occurrence of malignancies. The damage score remains the same or increases with time. The index has been shown to discriminate between changes in disease damage in patients with both active and inactive disease and to have good inter observer reliability. This tool has the capacity to measure the cumulative damage overtime and offers the opportunity to compare treatment arms in clinical trials. The damage score matches well with the weighted average of SLEDAI disease activity score and the number of disease flare, but not with age or steroid dose.

SLICC/ACR damage index[33]

Definition of d0amage : occurring since onset of lupus, ascertained by clinical assessment and present for at least six months unless otherwise stated. Repeat episodes mean at least six months apart to score to the same lesion cannot be scored twice.

Ocular : (Maximum Score 2)

Any cataract ever: score 1

Retinal change OR optic atrophy: score 1

Neuropsychiatric: (Maximum Score 6)

Cognitive impairment OR major psychosis: score 1

Seizures requiring therapy for 6 months or more: score 1

Cerebral vascular accident ever (score 2 if more than 1 x) OR resection not for malignancy: score 1 (2)

Cranial or peripheral neuropathy (excluding optic): score 1

Transverse myelitis: score 1

Renal: (Maximum score 3)

Estimated or measured GFR less than 50%: score 1

Proteinuria 3.5 g/24 hr. or more: score 1

ESR (regardless of dialysis or transplantation): score 1-3

Pulmonary: Maximum score 5

Pulmonary hypertension (right ventricular prominence or loud P2): score 1

Pulmonary fibrosis (clinically and/or by x-ray): score 1

Shrinking lung (by x-ray): score 1

Pleural fibrosis (by x-ray): score 1

Pulmonary infarction (by x-ray) OR resection not for malignancy: score 1

Cardiovascular: Maximum score 6

Angina OR Coronary artery bypass: score 1

Myocardial infarction ever (score 2 if more than 1 x): score 1 (2)

Cardiomyopathy (ventricular dysfunction): score 1

Valvular disease (diastolic murmur, or systolic murmur more than 3/6): score 1

Pericarditis OR pericardiectomy: score 1

Peripheral vascular: Maximum score 5

Claudication: score 1

Minor tissue loss (pulp space): score 1

Significant tissue loss ever (at least loss or resection of a digit) (score 2 if more than 1 x): score 1 (2)

Venous thrombosis with swelling, ulceration OR venous stasis: score 1

Gastrointestinal (Maximum score 6)

Infarction or resection of bowel below duodenum, spleen, liver or gallbladder ever (score 2 for more than 1 site): score 1 (2)

Mesenteric insufficiency: score 1

Chronic peritonitis: score 1

Stricture OR Upper gastrointestinal tract surgery ever: score 1

Pancreatic insufficiency requiring enzyme replacement OR with pseudocyst: score 1

Musculoskeletal: (Maximum score 6)

Muscle atrophy OR weakness: score 1

Deforming or erosive arthritis (including reducible deformities, excluding AVN): score 1

Osteoporosis with fracture or vertebral collapse (excluding AVN): score 1

Avascular necrosis = AVN (score 2 if more than 1x): score 1 (2)

Osteomyelitis: score 1

Skin: (Maximum score 3)

Scarring chronic alopecia: score 1

Extensive scarring of panniculum other than scalp and pulp space: score 1

Skin ulceration (excluding thrombosis) of more than 6 months: score 1

Gonadal: (Maximum score 1)

Premature gonadal failure: score 1

Endocrine: (Maximum score 1)

Diabetes requiring therapy, regardless of treatment: score 1

Malignancy: (Maximum score 2)

Malignancy (exclude dyspepsia) (score 2 if more than 1 site): score 1 (2)

Potential maximum of the total ADI: 46 points

- Adding up to a total renal DS of 3 points

b) Organ specific measures of disease damage include activity and chronicity indices for lupus nephritis based on histology shown in following table 6.

Table 6. Activity and chronicity indices of lupus nephritis

	Activity	Chronicity
Glomerular lesions	1. Proliferation 2. Necrosis/karyorrhexis 3. Hyaline thrombi 4. Cellular crescents 5. Leukocytic exudation	1. Sclerotic glomeruli 2. Fibrous crescents
Tubointerstitial lesions	1. Mononuclear cell infiltration	1. Tubular atrophy 2. Interstitial fibrosis

Individual lesions are graded 0-3. Necrosis and cellular crescents are weighted by a factor of 2.Indices are composite scores of individual components, with activity index score ranging from 0-24, and chronicity index of 0-12. This index has got prognostic and therapeutic implications[34] such as cellularity in biopsy indicates responses to steroid and fibrosis indicates steroid failure.

Health related quality of life measures

Health related quality of life has been assessed in lupus patients using a number of generic instruments including the (Health Assessment Questionnaire) HAQ and its modified form MHAQ, Arthritis impact measurement scale(AIMS), the Medical Outcome Survey Short Form 20 & 36. Short Form SF 36 is the one instrument that has been utilized most frequently to assess health status of SLE patients. It has 8 domains that measure physical as well as mental components and include physical social and emotional function, role limitations, vitality and pain occurring over previous months[35,36]. The score is given as 0-10-0 higher values indicating better health. In SF 36 index different disease activity levels were significantly associated with different quality of life scores and showed an excellent ability to record the continuum from good health to serious illness. In view of the comprehensiveness of the SF 36 and its international validation for a wide range of medical conditions, this is currently the optimal choice of instrument for life impact assessment of patients. SF20 and SF20+, with fatigue included in it, are also used to assess health status.

Economic assessment

This is an important issue often overlooked in health care and clinical trials[37]. Economic impact may be related to cost of therapy, adverse event, resource utilized, lost productivity or all of the above. Economic data can be collected from the patient, reviewing insurance claims or medical records. As per Lupus Foundation of America (LFA) survey done in 2008, the mean annual productivity cost (lost hours of productive work) for participants of employment age (between the ages of 18 and 65) was $8,659 per patient per year. With lupus affecting around 1.5 million Americans, this study estimates that lost hours of work cost these Americans nearly $13 billion annually[38.]

Monitoring of treatment toxicity[39]

All of the medications used in SLE require vigilance to assure their safe use. Monitoring should be started from the day of start of treatment and investigations should be guided by symptoms. It should be kept in mind that many of the symptoms of drug toxicity resemble with the symptoms of SLE (Table 7).

Table 7. Monitoring strategies for drugs commonly used in systemic lupus erythematosus

Drugs	Monitoring of toxicity
NSAIDs	Ulcer dyspepsia, GI bleeding, pedal edema, hypertension, serum creatinine
Glucocorticoids	Ulcer dyspepsia, hypertension, hyperglycemia, lipid profile every 3-6 months, bone mineral density and eye check up yearly.
Hydroxychloroquine	Eye check up for retinal changes and automated perimetry every 6 months
Azathioprine	CBC, platelet count, every 2 weeks initially, ALT, AST every 6-12 weeks
Cyclophosphamide	CBC, platelet count, urinalysis monthly, check for infertility
Methotrexate	CBC, platelet count, ALT,AST, serum albumin every 4-8 weeks
Mycophenolate mofetil	Every 4-8 weeks CBC, platelet count, GI intolerance

In summary overall assessment of disease activity, damage, health perception, drug toxicity measurement, and economic cost should be taken into account while follow up of SLE patients is being done. The new concept of drug responder index has emerged to get comparable outcome in different clinical trials.[40] A responder index would allow a clinical trial to determine directly what proportion of patients had a clinically meaningful improvement from specific drug therapy for SLE.

Monitoring in special situation

Pregnancy[41] - Problems in pregnancy as shown in subsequent table 8, require special consideration.

Table 8. Monitoring of pregnancy issues in SLE

Problems	Comments
Lupus flare, pre-eclampsia	Check BP, complements, urinary sediments, anti ds-DNA titer, disease activity in other organs
Premature rupture of membrane	No association with lupus activity, antiphospholopid antibody
Miscarriage	Anti phospholipid antibody, β2-glycoprotein I, fetal monitoring
Neonatal lupus	Assess anti Ro(SSA), anti La(SSB)

Antiphospholipid antibody (APLA) syndrome

The patients with APLA syndrome secondary to SLE require screening and monitoring of platelet count, anti-cardiolipin antibody(IgM and IgG), lupus anticoagulant and ß2 glycoprotein I. Therapy with Warfarin is monitored with prothrombin time with INR between 2.5 - 3.5. The pregnant mother with SLE and APLA syndrome are associated with adverse consequences like venous thrombosis and foetal loss.

Infection

The common infectious agents are bacteria, fungus and tuberculosis.. Positive CRP with shift to left of WBC predicts infection whereas negative CRP and low complementC3 with fever suggest disease activity when the patients have equivocal symptoms of fever, myalgia and arthralgia.

Drug induced lupus

This is a small subset where low complement or very high titer of anti ds-DNA is unusual. A search for a particular drug is needed such as Hydralazine, Procainamide, Isoniazide, Hydantoins, Chlorpromazine, Methyl dopa, D-penicillamine. Anti histone antibody is found in high titer in this condition.

Accelerated atherosclerosis

SLE patients are always at increased risk of premature atherosclerosis with endothelial dysfunction in the long run, because of the disease process and steroid use in higher doses, besides traditional risk factors. They have significantly increased levels of total cholesterol, triglycerides and small, more atherogenic LDL subfractions. There is also a higher level of lipid peroxidases due to persistent oxidative stress in the SLE patients[42]. Hence they should be screened and monitored regularly for hypertension, lipid profile, blood sugar levels and should be treated accordingly. Active disease also needs to be assessed in progressive atherosclerosis. High sensitivity CRP, an early marker of atherosclerosis and B-mode ultrasonography to study carotid artery plaque should be done for early detection of atherosclerosis.

Management

Systemic Lupus Erythematosus (SLE) have highly variable features like constitutional symptoms, glomerulonephritis, neuropsychiatric disease and cutaneous manifestations. SLE is primarily incurable, but most of the patients experience remission and their survival has improved over the years (Table 9)[43].

Table 9 : Survival rates in SLE[43]

Authors	No. of Patients	5 years (%)	10 years (%)	15 years (%)
Feng PH 1970-80	183	70	60	-
Chen SL 1980-98	50	98	84	76
Chandrasekharan 1985-1994	330	88	82	-
Ward MM 1975-93	408	82	71	63
Abu Shakra 1980-95	665	93	85	79

The causes of death in SLE in first 5-10 years are infections, active disease with renal involvement and vasculitis whereas in later years morbidity and mortality are mostly attributed to osteoporosis and premature coronary atherosclerosis.[44,45]

The principles of management essentially are

- Early diagnosis and assessment
- Patient education
- Preventive measures
- Control of disease activity
- Organ salvage
- Care of comorbid states and emergencies

❖ **Early diagnosis and assessment**

After the confirmation of the diagnosis, the disease needs assessment of its activity and extent of organ damage. The various clinical indices of diseases activity include major organ involvement like CNS, CVS, pulmonary, renal and mucocutaneous lesions; laboratory parameters are proteinuria, urinary sediments, serum anti dS DNA, platelet count and C3 and CH50 assay and finally creatinine clearance. In this context disease activity scores have been adopted by different authors in the name of SLEDAI (SLE Disease Activity Index), BILAG (British Isles Lupus Activity Grade), SLAM (Systemic Lupus Activity Measures).

The disease damage of the organs are often irreversible which are exemplified by advanced glomerulosclerosis (WHO VI), cardiomyopathy, progressive interstitial lung disease, stroke and cognitive dysfunction.

❖ **Patient education**

Patient education is an integral part of comprehensive care. Emotional and psychosocial supports are to be ensured for the success of long term therapy. The counseling about marriage, fertility, contraception need to be addressed.

❖ **Preventive measures :**

1. **Regular evaluation of Lupus activity and damage -** Early identification of these will help to diagnose lupus 'flare' or organ damage. The patients are to be informed in this matter before hand.
2. **Photoprotection -** The patient is to be advised about ill effects of prolonged exposure to sun and use of sunscreens with protection factor (SPF 15 or above) has to be encouraged.
3. **Vaccination -** It is not contraindicated apart from "live" vaccines for patients on more than 10 mg prednisolone a day or other immunosuppressive agents. Regular vaccination against common infections like Pneumonia, Influenza, Hepatitis, Tetanus, Parvo virus infection are recommended.
4. **Osteoporosis prophylaxis -** It is necessary in patients on long term steroids. The supplementation with calcium 1500 mg/day and Vit. D 400-600 IU/day is advisable. Monitoring with DEXA scan of lumbar spine and femoral neck for osteopenia/porosis and timely use of Alendronate 70 mg/wk or Risedronate 35mg once a week is necessary[46,47].
5. **Hormone therapy (HT)**[48] **-** In general, the use of hormones should be tailored to individual risks and benefits with reference to the SLE disease activity profile and symptomatology. The main indication of HT should be severe climacteric symptoms. Hormone therapy should be used when SLE is inactive with no flares for several years, and in women who are antiphospholipid antibody negative and not on high-dose steroids. For such women, low-dose transdermal oestrogen combined with a natural progesterone preparation is recommended because of its less thrombogenic property. Similarly, it has been suggested that micronised progesterone or pregnane derivatives did not result in increased thromboembolic risk, whereas norpregnane derivatives were associated with a 4-fold increase in VTE risk. For those with mildly active disease, a non-oestrogenic agent (eg antidepressant or progestogen) is the preferred first-line treatment. If this is not effective, low-dose transdermal hormone preparation can be used with close monitoring of disease activity. For those with severe active disease or who are antiphospholipid antibody positive, HT should be avoided. The use of HT in those with premature ovarian failure should generally be beneficial and is recommended.

Therapeutic strategy

The broad principle in using various drugs in SLE depends upon extent and severity of organ damage. A special subgroup like antiphospholipid antibody syndrome (APS) has been incorporated in this strategy because of additional therapy with anticoagulants. The approach to therapy is broadly classified (Table 10)

- **Drug Therapy**

The pharmacological treatment of SLE patients consists of four main types of drugs either alone or in combination.[49]

Table 10

SLE confirmed

Minor organ damage e.g. skin, arthralgia, serositis etc.	APS	Major Organ damage e.g. Renal, CNS, CVS Haematological
• NSAIDs • Hydroxychloroquin • Low dose steroid 0.25 mg - 0.5 mg/kg/day	• Anticoagulation • Low dose aspirin ± • Hydroxy-chloroquin	• Highdose prednisolone 1-1.5 mg/kg/day ± Cytotoxics ± "Newer" agents

- Non-steroidal anti-inflammatory drugs (NSAID)
- Hydroxychloroquine
- Corticosteroids
- Cytotoxic drugs
- Newer drugs eg. Biologicals

The broad selection of drugs depending upon organ damage has been shown to be effective in different subgroups. A new study SLICC (Systemic Lupus International Collaborative Clinics) group is currently aiming to define a drug responder index for SLE patient.[50]

- **Non steroidal antiinflammatory drugs -** This is advised to control arthralgia and myalgia. Newer COX II inhibitors are preferred than conventional NSAID for safety of stomach and intestine. However nephrotoxicity of all types of NSAIDs is the limitation in elderly and specially in presence of lupus nephritis.

- **Hydroxychloroquine -** This is used as low potency immunosuppressive agent to control fatigue, arthralgia/arthritis and rash; is shown to be effective to prevent 'flares'[51]. The starting dose is 400mg daily which is tolerated well. The prevalence of retinal toxicity is approximately 0.5%[52,53]. Eye examination every 6 months is advised. Hydroxychloroquine has also been shown to be effective to help lower plasma total triglyceride, VLDL and apolipoprotein CIII levels[54].

- **Corticosteroids -** These drugs are recommended in patients with minor and major organ damage in low and high dosage respectively. It is also used as maintenance therapy. In life threatening and vision threatening situation parenteral methyl prednisolone intravenously is prescribed in a dose 500mg - 1000mg daily for three successive days over a period of 3-4 hours each time. It is followed by oral steroids in a dose of 0.75-1mg/kg body weight daily and tailed off over a period of months.

The major side effects of steroids need to be appreciated like osteoporosis, peptic ulceration, hypertension, hyperlipidemia, risk of infection, alopecia, avascular necrosis of bones and accelerated atherosclerosis. Hence gastric protection with proton pump inhibitors, bone protection with weight bearing exercise, Calcium, Vitamin D and bisphosphonates are essential. It is important to monitor bone mineral density (BMD) every 2 years. The maintenance dose of corticosteroid should not exceed 7.5mg a day on a long term basis.

- **Cytotoxic Drugs -** The mainstay of treating active lupus with major organ damage is broad spectrum immunosuppression with the ultimate aim of preventing irreversible organ damage. The drugs used are Cyclophosphamide, azathioprine,

Mycophenolate Mofetil either alone or in combination with corticosteroids. Methotrexate is also available option in subsets of SLE patients with predominant joint disease.

Cyclophosphamide - This is given orally (50-100 mg/day) daily or intravenously as "pulse" therapy (15-25mg/kg body weight a day) every 3-4 weeks. The adverse effects include nausea, vomiting, alopecia, haemorrhagic cystitis, infertility in child bearing age and bone marrow suppression. It is important to maintain hydration during 'pulse' therapy. Presumably 'pulse' therapy has lesser side effects than maintenance oral therapy.

Azathioprine - It may be used as steroid-sparing agent in the treatment of renal lupus. The dose is 1.5 - 3 mg/kg body weight starting with 50mg per day to maximum 150 mg per day. The adverse effects include myelosuppression, hepatic dysfunction and pancreatitis. Regular blood monitoring is essential.

- **Newer drugs -** Along with the traditional drugs newer drugs have been developed in the recent years. Due to targeted therapeutic approach these drugs are showing good efficacy and minimal immunosuppression related side effects.

Mycophenolate mofetil (MMF)

Mycophenolic acid (MPA) is a potent inhibitor of lymphocyte proliferation via reversible inhibition of the enzyme inosine 5-monophosphate dehydrogenase (IMPDH). This enzyme is critical for the de novo synthesis of guanosine nucleotides. MPA more strongly inhibits the type II IMPDH, an isoform expressed by activated lymphocytes. The selectivity of MPA for lymphocytes is because of the fact that lymphocytes are dependent on the de novo purine pathway for cell proliferation as well as differentiation[56]. Several clinical trials have provided firm evidence for the use of MMF for the induction and maintenance treatment of lupus nephritis. The dosage of mycophenolate mofetil is 1gm twice daily for first 6 months, followed by 500mg twice daily next six months. Aspreva Lupus Management Study[57] (ALMS) of 370 patients was composed of a 6-month induction phase followed by a 36-month maintenance phase. The induction phase compared the efficacy of MMF and monthly pulse IV Cyclophosphamide (IVC) for induction treatment of lupus nephritis. In the maintenance phase, patients who met the response criteria in the induction phase were re-randomized to receive either MMF or Azathioprine for maintenance therapy. The induction phase did not achieve its primary endpoint in that MMF was not superior to IVC. The overall rate of renal response at 6 months was 56% in theMMFgroup and 53% in the IVC group, with only 8% of patients in each arm achieving a complete renal response MMF is generally well tolerated with the most common toxicities being gastrointestinal (nausea, vomiting and diarrhea) and hematological respectively.

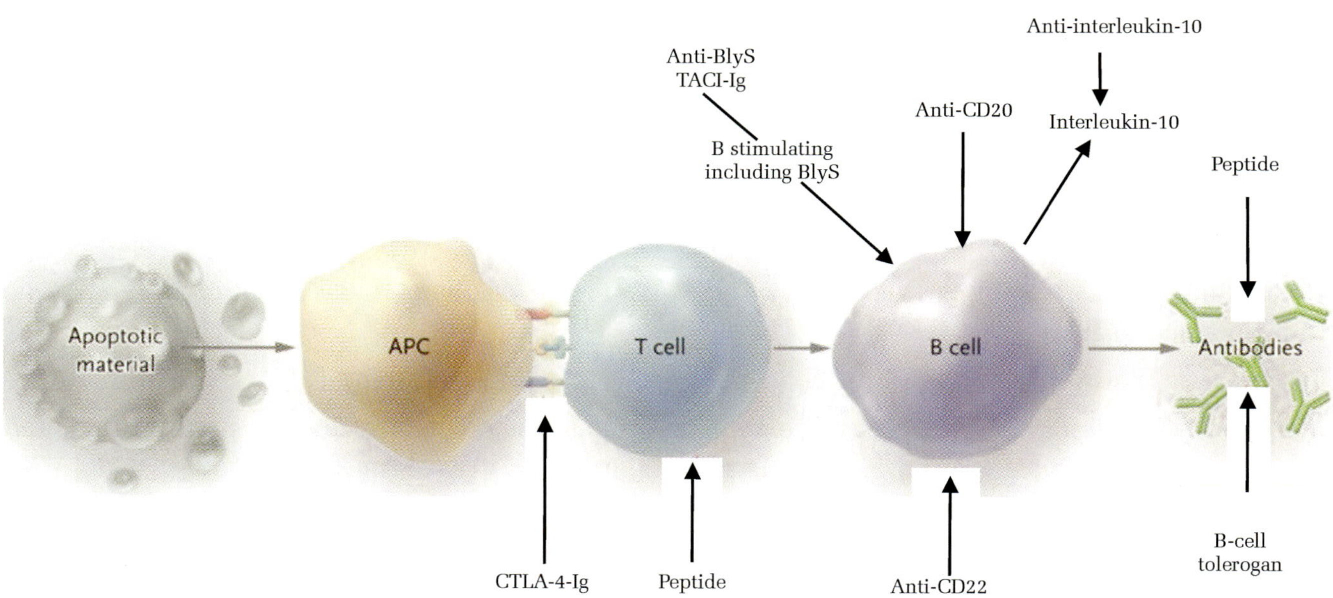

Fig: 3[55] Targeted therapeutic approach in SLE.

Rituximab

The central role of B cells in SLE has been well-identified and its cell-surface antigen CD20 appears early in maturation of B cells. CD20 is expressed on pre-B cells, immature B cells, naïve and memory B cells and germinal centre B cells but not on plasma cells[58]. Hence plasma cells are not depleted during treatment with rituximab. Rituximab - a mouse human IgG IK chimeric anti-CD20 - permits selective elimination of peripheral B cells without preventing the regeneration of B cells from stem cells by ways of antibody dependent, cell-mediated cytotoxicity, complement mediated cell-wall lysis and/or by apoptosis[59]. Looney et al in 200460 published the first results with rituximab in patients with SLE with favourable outcome wherein B cells are depleted. Looney et al[61] furthermore recently completed a phase I/II open dose escalation trial of rituximab added to current therapy for the treatment of 18 patients of SLE. The clinical response showed significant improvement in 10 patients correlating with activity measures of SLAM (Systemic lupus Activity Measure). In an open study of 21 SLE patients treatment with rituximab has been well-tolerated with no serious or opportunistic infection. Five patients had WHO class IV LN and one had class V. The treatment protocol entailed two doses of 100mg of rituximab with 750mg IV CPM and high dose steroid over a two week period. All other immunosuppressives were discontinued with the exception of hydroxychloroquine and prednisolone. The clinical improvement in all patients correlated with improvement in the creatinine, urinary protein/creatinine ratio, complement levels and a drop in the anti dS DNA titres at six months. B cell repopulation occurred between four and eight months (mean seven months)[62]. Similar results have been reported in class III or IV nephritis with rituximab and steroids[63].

The B-cell depletion using this drug offers major benefits in lupus patients with organ involvement including nephritis and systemic improvement lasts for three years. However questions remain about optimal dose and proper use of combination therapy. Long term study is awaited to determine the long-term prognosis with this new drug.

Belimumab[64] :

Belimumab (BenlystaTM), approved by the FDA in 2011 for the reduction of disease activity in adult patients with active, autoantibody positive systemic lupus erythematosus receiving standard therapy, is a human IgG1 antibody that binds to soluble (but not transmembrane) B-lymphocyte stimulator (BLyS; also known as BAFF), a key survival factor for B lymphocytes. Two randomized, double-blind, placebo controlled trials, BLISS-52 and BLISS-76 (with numbers representing the duration of each trial in weeks) showed statistically significant benefit in the primary outcome. Belimumab may be considered in seropositive SLE patients having musculoskeletal, cutaneous, and immunological disease activity despite standard of care. Belimumab has not been studied in severe active lupus nephritis or severe active CNS lupus. Belimumab is approved at a dosage of 10 mg/kg to be given by intravenous infusions at 2-week intervals for the first 3 doses and 4-week intervals thereafter.

During clinical trials Belimumab was not associated with an increase in serious adverse events, infections or malignancy. Severe hypersensitivity reactions, though rare, were reported in belimumab treated patients. However, long term follow up on large numbers of diverse SLE patients is needed for better assessment of this agent's safety.

Management of lupus related special conditions

Renal lupus - The kidney is the major organ at risk and it is common in Indians. Lupus nephritis (LN) may be asymptomatic or may present with nephritis or nephrotic syndrome. It is desirable to have renal biopsy prior to therapy to assess severity, nature and potential reversibility of renal disease. The therapeutic options are matched with WHO histologic staging and classification of renal lupus as in Fig. 4.

Current recommendations[65]:

Renal biopsy: The threshold for performing a renal biopsy should be low because of the potentially aggressive nature of LN. Clinical, serological or laboratory tests fail to predict histological findings accurately. Common indication for renal biopsy includes reproducible proteinuria (>0.5 g/24hr) especially with glomerular hematuria and/or cellular casts. Biopsy may also be considered in

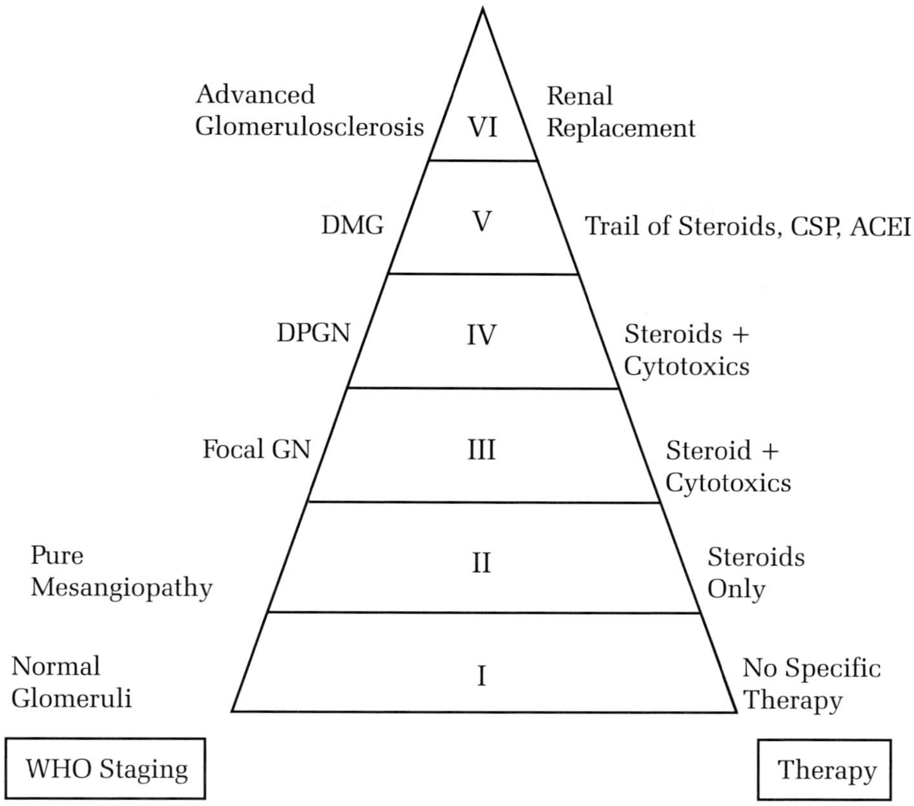

Fig. 4. Traditional management pyramid of lupus nephritis.

persistent isolated glomerular hematuria, isolated leucocyturia (after other causes, such as infection or drugs are excluded) or in unexplained renal insufficiency with normal urinary findings. Biopsy should be performed within the first month of disease onset, preferably before the institution of immunosuppression, unless otherwise contraindicated.

Indications and goals of immunosuppressive treatment in LN: Immunosuppressive treatment is usually not indicated in class I and class VI LN, unless necessitated by extra-renal lupus activity. Treatment should aim for complete renal response which is defined as urinary protein:creatinine ratio UPCR < 50 mg/ml (roughly equivalent to proteinuria <0.5g/24hr) and normal or near normal (within 10% of normal GFR if previously abnormal) GFR. Partial renal response, defined as 50% reduction in proteinuria to subnephrotic levels and normal or near-normal GFR, is to be achieved preferably by 6 months and no later than 12 months following treatment initiation.

Initial treatment of adult LN: Following Aspreva Lupus Management Study (ALMS) trial, the largest trial in LN comparing response rates between MPA (target mycophenolate mofetil (MMF) dose 3g/day) and I.V cyclophosphamide (CY) (monthly pulses 0.5-1 g/m^2), both administered for 6 months, MPA is recommended as initial treatment for most cases of class III-IV LN because of the ease of administration and more favourable gonadal toxicity profile. Low dose I.V cyclophosphamide (total dose 3g over 3 months) in combination with glucocorticoids (0.5 mg/kg/day) is also recommended as initial treatment of class III-IV (±V) in Caucasians as this regime has got better efficacy/toxicity ratio. In pure class V LN many regimes have been tested (glucocorticoids + I.V CY or glucocorticoids + ciclosporin or MPA). Currently MPA is recommended as initial treatment for most cases of class V LN and nephrotic-range proteinuria. For class II LN with proteinuria > 1g/24 hr despite renin-angiotensin-aldosterone system (RAAS) blockade, particularly in presence of glomerular haematuria, low to moderate dose of glucocorticoids (prednisone 0.25-0.5 mg/kg/day) alone or in combination with azathioprine (1-2 mg/kg/day) is recommended. Glucocorticoids alone or in combination with other immunosuppressive agents may also be considered in cases of class I LN with podocytopathy on electron microscopy (minimal change disease) or interstitial nephritis.

Subsequent treatment: For patients improving after initial treatment subsequent immunosuppression is required to consolidate renal response and prevent flares. Azathioprine and MPA are the two popular drugs for maintenance of remission. A large RCT suggested a difference between these two drugs in favour of MPA after induction with either MPA or I.V CY. Moreover, subsequent use of azathioprine after MPA resulted in more treatment failures as compared to MPA followed by MPA. Therefore it is desirable to continue MPA if the drug was successful as initial treatment. Calcinurin inhibitors can be considered in selective cases (lupus membranous nephropathy) with preserved renal function. Intravenous CY, pulsed every 3 months, may be used in selective cases. No clear data is available to guide duration of treatment beyond 3 years. Continuation of treatment for longer time period is to be individualized with an effort first to withdraw glucocorticoids before immunosuppressive agents.

Neuropsychiatric lupus - Intravenous cyclophosphamide is preferred as 'pulse' therapy either alone or in combination with steroids in presence of vasculitis. However it is always important to rule out CNS infection before cytotoxic therapy. Seizures need to be controlled by phenobarbitone, valproate or phenytoin. Cognitive dysfunction needs to be addressed in combination with Psychiatrist. Antiphospholipid syndrome is often associated which requires treatment as below.

Antiphospholipid antibody syndrome - About 10% of SLE patients have secondary antiphospholipid antibody syndrome though 25-30% have antiphospholipid antibodies. In patients with thromboembolic features anticoagulation with heparin to keep activated partial thromboplastin time (APTT) 1.5 to 2 times the baseline is advised[66]. This is usually achieved with heparin dose of approximately 1000 units per hour. Simultaneous oral anticoagulation with warfarin 5-10 mg/d is given for 5-7 days by which time INR reaches the therapeutic levels. Heparin is discontinued once warfarin action is optimized and thereafter it is continued as a maintenance dose to keep INR around 2.5 to 3.67. Concurrent aspirin 75mg/day is also advocated. Immunosuppressives are only needed to control SLE activity.

SLE in pregnancy[68] - Fertility is normal unless the disease is very active. Approximately 25% of patients do not go to full term, well in excess of 10% expected in healthy women. Those with antiphospholipid antibody syndrome are likely to have recurrent miscarriage, still birth and low birth weight babies[69]. Aspirin 75mg a day and/or heparin therapy are advised to salvage foetal wastage. Low dose steroids and azathioprine can be safe during pregnancy but chloroquine, cyclosphosphamide and methotrexate are contraindicated. Medical termination of pregnancy is not advocated. Screening for Anti Ro Ab/Anti La Ab during pregnancy is essential to predict neonatal congenital heart block. Neonatal SLE needs special attention.

Oestrogen containing contraceptives may trigger a 'lupus flare' and should be avoided. Progesterone contraceptives are safe; postmenopausal hormone replacement may be considered under special circumstances.

Infections in SLE - Infection is an important risk in SLE patients on immunosuppressive therapy and it carries considerable morbidity and even mortality. It is difficult to distinguish fever of infection from disease activity. Positive CRP (C-reactive protein) with shift to the left of WBC count predict infection whereas negative CRP and low C3 with fever suggest disease activity. Commonly, tuberculosis[70], candidiasis, pneumocystis carinii are opportunistic infections. Prophylactic antibiotics/immunization are justified in areas of high prevalence. Periodic chest X ray is useful.

Accelerated atherosclerosis[71] - This is considered as one of the delayed complications of SLE. The continuing active vascular inflammation, endothelial dysfunction, antiphospholipid antibody syndrome, abnormal lipids and prolonged steroid therapy are the predisposing factors for such premature, accelerated coronary atherosclerosis with adverse consequences. The management consists of recognition of risk factors and ensuring investigations and treatments. The commonly prescribed options are use of statin 10-20mg/day, aspirin 75-150mg/day, clopidogrel 75mg/day and control of disease with lowest possible dose of Prednisolone 7.5mg to 10mg a day. Statins, in addition to their traditional effect on the suppression of cholesterol synthesis, have direct effect on endothelium, plaque formation and thromboxane synthesis. They also have demonstrated antithrombotic/anti-inflammatory effects on antiphospholipid syndrome[72]. Often hydroxychloroquine 400mg/day helps control of lipids. Traditional risk factors like hypertension and glucose intolerance need to be controlled. Anticoagulation is indicated if there is antiphospholipid antibody syndrome.

Therapy of limited use in SLE

Plasma exchange **-** In late1970 and early 1980 there was great enthusiasm about plasma exchange being used in subsets of renal lupus. However the limitations include "rebound" phenomenon within few weeks, technical difficulties of requiring central venous access with associated complications and patient discomfort with higher cost. Even combination therapy with cyclophosphamide does not offer additional benefit over pulse cyclophosphamide alone[73]. It may have a role in the management of some less common complications of lupus like thrombotic thrombocytopenic purpura, cryoglobulinemia and hyperviscosity[74].

Role of diet **-** There is no specific role of any diet in the management of SLE. But a small study in humans has shown beneficial effect of fish oil supplementation over a six month period[75]. High saturated fat diet is detrimental for premature coronary atherosclerosis seen in later stage of SLE.

IV Immunoglobulins (IVIG)

IVIG have been used successfully in SLE patients with refractory thrombocytopenia, pancytopenia, CNS involvement, secondary APLA syndrome and nephritis. Orbach et al[76] reported that in lupus nephritis IVIG therapy improved proteinuria, nephrotic syndrome and creatinine clearance. Patients with lupus nephritis should receive IVIG under close monitoring and after adequate hydration to correct volume depletion, which together with pre-existing renal disease represents a risk factor for nephrotoxicity. It is observed sucrose-based IVIG therapy is associated with enhanced renal injury.

Cyclosporine

In most patients with SLE membranous glomerulonephritis, cyclosporine - calcineurine - inhibitor - induces remission but relapse is common on drug withdrawal[77]. Clinical and histologic improvement with cyclosporine has been reported in IV-CPM resistant proliferative lupus nephritis. In a randomized controlled trial of proliferative lupus nephritis, maintenance therapy with either cyclosporine or azathioprine are equally effective[78]. Currently it is recommended in membranous lupus nephropathy with preserved renal function.

Tacrolimus

Tacrolimus was isolated in 1984 from the fungus Streptomyces tsukubaensis. It is a macrolide immunomodulator FK506, which acts on T lymphocytes and inhibits interleukin-2 transcription as well as other cytokines[79]. Current recommendation for its use in lupus includes severe drug resistant cutaneous manifestations (topical application) and refractory lupus nephritis[80].

Potential novel therapies

Rapid developments in biotechnology over the past decade have given the opportunity to develop a greater understanding of immune dysregulation characterising lupus and develop targeted therapy at various levels (Table 11)[81].

Table 11 : Summary of treatments and clinical trials in treatment of SLE[81]

Molecular target	Drug(s)	Stage of clinical trial	Result(s)
Immune cell targeted therapies			
Anti-CD20	Rituximab	Phase II/III EXPLORER and LUNAR trial	Failed to meet primary and secondary endpoints
Anti-CD20	Ocrelizumab	Phase III	Terminated due to infectious complications
Anti-CD22	Epratuzumab	Phase IIb	Inadequate information to evaluate study
B cell tolerogen	Abetimus (LJP-394)	Phase III ASPEN trial	Terminated after interim efficacy analysis
BAFF-R, BCMA, TACI	Belimumab	Phase III BLISS-52 and-76	Recommended by FDA Advisory Commitee
TACI	Atacicept	Phase II for Lupus nephritis	Terminated due to infectious complications
	Atacicept	Phase II/III for non-renal lupus	ongoing
Co-stimulatory signaling pathways			
CD40-CD40L	Anti-CD40L Ab (Biogen Hu5c9 antibody)	Terminated	Terminated due to thromboembolic complications
CD28 and CTLA4 receptors and co-ligands	Abatacept	Phase II trial (for Lupus nephritis)	Ongoing
ICOS- B7RP1	Anti-B7RP1 Ab (AMG557)	Phase I	Ongoing
Anti-cytokine therapy			
IL-10	Anti-IL10 mAb (B-N10)	Pilot study showed beneficial effects	No ongoing trials with humanized anti- IL10 mAb
IL-1	Anakinra (IL-1 Ra)	2 small open label trials	No ongoing trials
IL-18	IL-18 blockade	Only mouse studies	No human trials
IL-6	Tocilizumab	Phase I	Larger study is scheduled
Other treatments			
Complement 5	Eculizumab (anti-C5 mAb)	Phase I	No improvement in SLE
NMDA receptor antagonist	Memantine trials	Pilot study failed to improve cognitive function	No ongoing
IFN-α	MEDI-545	Phase I trial	Ongoing
ROI (Reactive Oxygen Intermediates)	N-Acetylcysteine (NAC), Cysteamine (CYST)	Beneficial effect in mice	Ongoing clinical trial in human

It is expected that novel therapies will have better efficacy and/or better side effect profiles for certain manifestation of lupus. In future these may be considered suitable alternative when standard drugs fail to control the disease.

Unmet need :

Despite phenomenal progress in management of SLE unmet needs are still considerable like

a) Better than best and targeted therapies with sustained benefits and fewer side effects are needed.

b) Combination therapy with different biologic agents could be of superior efficacy by synergism.

c) Urgent need for nonimmunosuppressive therapy is warranted.

d) Better tools for the best time of optimal treatment

e) Combining BAFF blockade with B-cell depletion therapy needs more studies in refractory SLE patients.

Conclusion

The management of SLE with appropriate use of steroids and other immunosuppressives has improved the outcome over the period of last 20-30 years. The spectrum of illness varies from mild to aggressive disease and hence therapy has to be tailor-made. However appreciation of disease activity and damage, risk-benefit ratio of drug therapy and meaningful patient counselling are keys to the success of management. Many novel targeted therapy are in the horizon which may give some additional benefit in future. The role of accelerated atherosclerosis is a new challenge to be sorted out.

Further Reading :

1. Esdaile JM, Joseph L, Mackenzie T, Kashgarian M. Hayslett JP. The benefit of early treatment with immunosuppressive agents in lupus nephritis. J Rheumatol 1994; 21: 2046-51.
2. Bertsias GK, Salmon JE, Boumpas DT. Therapeutic opportunities in systemic lupus erythematosus: state of the art and prospects for the new decade. Ann Rheum Dis 2010; 69: 1603-11.
3. Gladman D, Ginzler E, Goldsmith C, Fortin P, Liang M, Urowitz M, et al. The development and initial validation of Systemic Lupus International Collaborating Clinics/ American Collage of Rheumatology Damage Index for systemic lupus erythematosus. Arthritis Rheum 1996; 39: 363-69.
4. Abu-Shakra M, Urowitz MB, Gladman DD, Gough J. Mortality studies in systemic lupus erythematosus: results from a single center. I. Cause of death. J Rheumatol 1995; 22: 1259-64.
5. Gladman DD. Prognosis and treatment of systemic lupus erythematosus. Curr Opin Rheumatol 1996; 8: 430-37.
6. Boumpas DT, Fessler BJ, Austin HA III, Balow JE, Klippel JH, Lockshin MD. Systemic lupus erythematosus: emerging concepts. Part 2. Dermatologic and joint disease, the antiphospholipid antibody syndrome, pregnancy and hormonal therapy, morbidity and mortality, and pathogenesis. Ann Intern Med 1995;123: 42-53.
7. Urowitz MB, Bookman AA, Koelher BE, et al. The bimodal mortality pattern of systemic lupus erythematosus. Am J Med 1976; 60: 221-25.
8. Boumpas DT, Austin HA III, Fessler BJ, Balow JE, Klippel JH, Lockshin MD. Systemic lupus erythematosus: emerging concepts. Part 1.Renal, neuropsychiatric, cardiovascular, pulmonary and hematologic disease. Ann Intern Med 1995; 122: 940-50.
9. A. Kumar. Indian Guidelines on the Management of SLE. J Indian Rheumatol Assoc 2002; 10: 80-96
10. Esdaile JM. Current role of renal biopsy in SLE. Baillieres Clin Rheumatol 1998; 12: 433-38.
11. Arbuckle MR, McClain MT, Rubertone MV, et al. Development of auto antibodies before the clinical onset of systemic lupus erythematosus.N Engl J Med 2003; 349: 1526-33.
12. Ravirajan CT, Rowse L, MacGowan JR, Isenberg DA. An analysis of clinical disease activity and nephritis associated serum autoantibody profiles in patients with systemic lupus erythematosus:a cross sectional study.Rheumatology [Oxford] 2001; 40: 1405-12.
13. Spronk PE, Limberg PC. Kallenberg CG. Serological markers of disease activity in systemic lupus erythematosus.Lupus 1995; 4: 86-94.
14. Ho A, Magder LS, Barr SG, Petri M. Decreases in anti ds DNA levels are associated with concurrent flares in patients with systemic lupus erythematosus. Arthritis Rheum 2001; 44: 2342-49.
15. Mirzayan MJ, Schmidt RE, Witte T. Prognostic parameters for flare in systemic lupus erythematosus. Rheumatology (Oxford) 2000; 39: 1316-19.
16. Horak P, Scudla V, Harmanovo Z et al. Clinical utility of selected disease activity markers in patients of systemic lupus erythematosus. Clin Rheumatol 2001; 20: 337-44.
17. Alarcon GS, Roseman J, Bartolucci AA et al. Systemic lupus erythematosus in three ethnic groups, II: features predictive of disease activity early in its course. Arthritis Rheum 1998; 41: 1173-80.
18. Gladman DD, Hirani N, Inanez D, Urowitz MB. Clinically active serologically quiescent systemic lupus erythematosus. J Rheumatol 2003; 30: 1960-62.
19. Manzi S, Rairie JE, Carpenter AB et al. Sensitivity and specificity of plasma and urine complement split products as indicator of lupus disease activity. Arthritis Rheum 1996; 39: 1178-88.
20. Nagy G, Brozik M, Varga L et al. Usefulness of detection of complement activation products in evaluating SLE activity. Lupus 2000; 9: 19-25.
21. Strand V, Gladman D, Isenberg D, Petri M, Smolen J, Tugwell P,Outcome measures to be usedin clinical trials in systemic lupus erythematosus. J Rheumatol 1999 Feb; 26(2): 490-97.
22. Bombardier Q, Gladman DD, Urowitz MB et al. Derivation of the SLEDAI. A disease activity index for lupus patients. Arthritis Rheum 1992; 35: 630-40.
23. Gladman DD, Goldsmith CH, Urowitz MB et al. Sensitivity to change of 3 systemic lupus erythematosus disease activity indices: international validation. J Rheumatol 1994; 21: 1468-71.
24. McLaughlin JP, Bombardier C, Farewell W et al. Kidney biopsy in systemic lupus erythematosus. III. Survival analysis controlling for clinical and laboratory variables. Arthritis Rheum 1994; 37: 559-67.
25. Symmons DPM, Coppock JS, Bacon PA et al. Development of a computerized index of clinical disease activity in systemic lupus erythematosus. Q J Med 1988; 69: 927-37.
26. Liang M, Socher SA, Roberts WN, Esdaile JM. Measurement of systemic lupus erythematosus activity in clinical research. Arthritis rheum 1988; 31: 817-25.
27. Vitali C, Bencivelli W, Isenberg DA et al. Disease activity in systemic lupus erythematosus: report of the Consensus Study Group of the European Workshop for Rheumatology Research. II. Identification of the variables indicative of disease activity and their use in development of an activity score. Clin Exp Rheumatol 1992; 10: 541-47.
28. Petri M, Buyon J, Kim M. Classification and definition of major flares in SLE clinical trials. Lupus 1999; 8: 685-691.
29. Gordon C, Sutcliffe N, Skan J et al. Definition and treatment of lupus flares measured by BILAG index. Rheumatology 2003; 42: 1-8.
30. Illei GG, Takada K, Parkin D, et al. Renal flares are common in patients with severe proliferative lupus nephritis treated with pulse immunosuppressive therapy : long-term follow up of a cohort of 145 patients participating in randomised controlled studies. Arthritis Rheum 2002; 46: 995-1002.
31. Mok CC, Ying KY, Tang S, et al. Predictors and outcome of renal flares after successful cyclophosphamide treatment for diffuse proliferative lupus glomerulonephritis. Arthritis Rheum 2004; 50: 2559-68.
32. Moroni G, Quaglini S, Maccario M, et al. ' Nephritic Flares' are predictors of bad long-term renal outcome in lupus nephritis. Kidney int 1996; 50: 2047-53.

33. T. Stoll, B. Seifert, D.A Isenberg. SLICC/ACR Damage Index is valid, and renal and pulmonary organ scores are predictors of severe outcome in patients with Systemic Lupus Erythematosus. Br J Rheum 1996; 35: 248-54.
34. Austin HA, Muenz LR, Joyce KM et al. Diffuse proliferative lupus nephritis: identification of specific pathologic features affecting renal outcome. Kidney Int 1984; 25: 689-95.
35. Brazier JE, Harper R, Jones NM et al. Validating the SF 36 health survey questionnaire: new outcome measure for primary care. Br Med J 1992; 305: 160-64.
36. Garratt AM, Ruta DA, Abdalla MI et al. The SF 36 health survey questionnaire: an outcome measure suitable for routine use within the NHS? Br Med J 1993; 306: 1440-44.
37. Gordon C, Clarke A. Quality of life and economic evaluation in SLE clinical trials. Lupus 1999; 8: 645-54.
38. http://www.lupus.org/webmodules/webarticlesnet/templates/new_aboutintroduction.aspx?articleid=247&zoneid=9.
39. Guidelines for referral and management of SLE on adults. Arthritis Rheum. Vol 42, no 9, Sept 1999.
40. Isenberg D, Ramsey-Goldman R. Assessing patients with lupus: towards a drug responder index. Rheumatology 1999; 38: 1045-49.
41. Mok CC, Wong RW. Pregnancy in systemic lupus erythematosus. Postgrad Med J. 2001; 77: 157-65.
42. Gordon C. Long-term complications of systemic lupus erythematosus. Rheumatology 2002; 41: 1095-1100.
43. Chandrasekharan AN : Clinical and immunological profile of SLE in India, Proceedings of XIX ILAR Congress of Rheumatology, Singapore, 1997; 156-64.
44. Malaviya AN, Misra R, Kumar A et al : Systemic connective tissue disease in India, Survival in SLE J. Assoc. Phys. India 1987; 36: 509-11.
45. Ioowwou Y : Sultan S, Isenberg D; Causes of death in SLE : analysis of 266 patients under long term review, Rheumatology 2000; 39 (abstract 265).
46. Saag KG, Emkey R, Schnitzer TJ et al. : Alendronate for the prevention and treatment of glucocorticoid-induced Osteoporosis Intervention Study Group N. Eng J Med 1998; 292-99.
47. Reid DM, Hughes RA, Laan RF et al. Efficacy and safety of daily Risedronate in the treatment of corticosteroid-induced osteoporosis in men and women. J Bone Mineral Res 2000; 15: 1006-13.
48. Raymond HW Li, Ailsa E Gebbie, Raymond WS Wong, Ernest HY Ng, Anna F Glasier, PC Ho. The use of sex hormones in women with rheumatological diseases. Hong Kong Med J 2011; 17: 487-91.
49 Ioannou Y, Isenberg DA : Current concepts for the management of Systemic Lupus Erythematosus in adults: a therapeutic challenge, Postgrad Med J 2002; 78: 599-606.
50. Isenberg DA, Gladman D. The Systemic Lupus International Collaborating Clinics Group - Origins and Outcomes. Lupus 2001; 10: 375-77.
51. The Candian Hydroxychloroquin Study Group. A randomised study of the effect of withdrawing HCQS in SLE, N. Eng J Med 1991; 324 : 150-54.
52. Levy GD, Munz SJ, Paschal J et al. Incidence of Hydroxychloroquin retinopathy in 1,207 patients in a large outpatient practice. Arth. Rheum 1997; 40 : 1482-86.
53. Shipley M, Silman A. Should patients on hydroxychloroquin have their eyes examined regularly? Br. J. Rheumatol 1997; 36: 514-15.
54. Hadis HN, Quismorio FP Jr. Wickham et al. The lipid, tyroprotein and apolipoprotein effects of hydroxychloroquin in patient with SLE; J. Rheumatol 1993; 20: 661-65.
55. Rahman A, Isenberg D. N Engl J Med 2008; 308: 929-39 Maria Dall'Era.
56. Mycophenolate mofetil in the treatment of systemic lupus erythematosus. Curr Opin Rheumatol 23: 454-58. Appel GB, Contreras G, Dooley MA, et al.
57. Mycophenolate mofetil versus cyclophosphamide for induction treatment of lupus nephritis. J Am Soc Nephrol 2009; 20: 1103-12.
58. Sfikakis PP, Boletis JN, Tsokos GC : Rituximab anti B-cell therapy in systemic lupus erythematosus : pointing to the future. Curr. Opin. Rheumatol 2005; 17: 550-57.
59. Analik J, Sauz I, Looney RJ. B cell depletion therapy in systemic lupus erythematosus Curr. Rheumatol Rep 2003; 5: 350-56.
60. Looney RJ, Anolik JH, Campbell D et al. B cell depletion as a novel treatment for systemic lupus erythematosus : a phase I/II dose escalation trial of rituximab. Arth. Rheum 2004; 50: 2580-89.
61. Looney RJ, Anolik J, Sauz I. B lymphocytes in SLE : lessons from therapy targeting B cells. Lupus 2004; 13: 381-90.
62. Leandro MJ, Cambridge G, Edwards JC et al. B cell depletion in the treatment of patients with SLE : a longitudinal analysis of 24 patients. Rheumatology 2005; 44: 1542-45.
63. Khamashta MA, Cuadrado MJ, Mujic F et al. The management of thrombiosis in the antiphospholipid-antibody syndrome. New Eng J Med 1995; 332: 993-97.
64. http://www.rheumatology.org/publications/hotline/2011_03_15_belimumab.pdf
65. Boumpas D T, Bertsias G K, Tektonidou M et al. Joint Europian League Against Rheumatism and Europian Renal Association - Europian Dialysis and Transplant Association (EULAR/ERA-EDTA) recommendations for the management of adult and paediatric lupus nephritis. Ann Rheum Dis 2012; 71: 1771-82.
66. Kamashta MA, Cuadrado MJ, Mujic F, et al. The management of thrombosis in the antiphospholipid antibody syndrome. New Eng J Med 1995; 332: 993-97.
67. Mervill JT. Pathogenesis and treatment of antiphospholipid syndrome. In : Systemic Lupus Erythematosus. Ed R G Lahita 3rd Ed. Academic Pres, Boston. 1999; 887-908.
68. Mok CC, Wong RW: Pregnancy in Systemic Lupus Erythematosus Postgrad Med J 2001; 77: 157-65.
69. Lima F, Khamashta MA, Buchanan NMM et al. A study of sixty pregnancies in lupus with antiphospholipid syndrome Clin Exp Rheumatol 1996; 14: 131-16.
70. Balakrishnan C, Mangat G, Joshi VR. Tuberculosis in patients with SLE. J. Assoc. Physicians India 1998; 46: 682-83.
71. Salman JE, Raman MJ. Accelerated atherosclerosis in SLE: implication for patient management Current Open Rheumatol 2001; 13: 341-44.

72. Lopez-Pedrera C. Global effects of the HMG-CoA reductase inhibitor Fluvastatin on the procoagulant and proinflammatory features of monocytes from APS patients. Annels of Rheumatic Disease. 2009: 45.

73. McClure CE, Isenberg DA. Dows plasma exchange have any part to play in the management of SLE. Controversies in SLE. London, Martin Dunitz 1997; 75-86.

74. Wallace DJ. Apheresis for lupus erythematosus: state of the art. Lupus. 2001; 10: 193-96.

75. Walton AJ, Snaith ML, Locniskar M et al. Dietary fish oil and severity of symptoms in patients with SLE. Ann. Rheum Dis, 1991; 50: 463-66.

76. Orbach H, Tishler M, Shoenfeld Y : Intravenous immunoglobulin and the kidney - a two edged Sword Semin. Arth. Rheum 2004; 34: 393-601.

77. Hu W, Liu Z, Shen S et al. Cyclosporine A in the treatment of membranous lupus nephropathy Chin Med J (Engl) 2003; 116: 1827-30.

78. Ferrario L, Bellone M, Bozzolo E et al. Remission from lupus nephritis resistant to cyclophosphamide after additional treatment with cyclosporine A Rheumatology (Oxford) 2000; 39: 218-20.

79. Homey B, Assmann T, Vohr HW et al. Topical FK506 suppresses cytokine and costimulatory molecule expression in epidermal and local draining lymph node cells during primary skin immune responses. J Immunol 1998; 160: 5331-40.

80. Lee T, Oh KH, Joo KW, et al. Tacrolimus is an alternative therapeutic option for the treatment of refractory lupus nephritis. Lupus 2010; 19: 974-80.

81. J Allergy Clin Immunol. 2011 February; 127(2): 303-14.

82. Ashima Gulati, Arvind Bagga. Management of Lupus nephritis. Ind J Rheumat 2012; 7: 69-79.

UPDATE ON ANTIPHOSPHOLIPID ANTIBODY SYNDROME

3

Vishad Viswanath, Debashish Danda

History

In 1906, Wasserman described a new serological test for syphilis based on complement fixation reaction of sera from syphilitic patients with an antigenic mixture obtained from alcohol extract of bull's heart[1]. Subsequently it was noticed that many individuals, who had no risk factor for syphilis, tested positive by this test. In 1941, Pangborn identified the major component in this mixture of antigens to be a phospholipid known as cardiolipin[2]. In 1955 Moore and Lutz reported that this biological false positivity (BFP) was more common in females and that many of these patients later go onto develop SLE. They also reported that a subset of their patients who had such BFP went on to develop thrombophlebitis[3].

In 1952, a circulating substance leading to prolongation of coagulation time was first described in SLE patients[4]. It was observed that there were increased rates of pregnancy loss in these patients. In 1963 Bowie et al, observed that the circulating anti coagulant in blood caused paradoxical increase in occurrence of thrombosis[5]. It was soon realised that the patients who had the circulating anticoagulant also had biological false positivity for syphilis. The circulating anti coagulant was later termed as Lupus anticoagulant (LAC). Subsequent research showed that Lupus anti coagulant was nothing but an anti phospholipd antibody (aPL ab)[6]. 1985 Hughes formally described the syndrome of arterial and venous thromboses and recurrent pregnancy loss associated with BFP which is now known as the Anti Phospho Lipid Antibody Syndrome (APS)[7]. Sapporo criteria, the first classification criteria for this syndrome was described in 1999[8]. The criterion was modified during the 11th International Congress on Antiphospholipid Antibodies held at Sydney in 2006[9].

The last 3 decades have witnessed remarkable advances in the understanding of manifestations, pathogenesis, diagnosis and treatment of Antiphospholipid syndrome (APS). In this chapter we will endeavour to highlight this information on the pathogenesis, diagnosis and management of APS.

Anti phospholipid antibodies

Definition

Anionic phospholipids were initially described as the target antigens for anti phospholipid antibodies (aPL abs). These phospholipids included cardiolipin (Diphosphatidyl glycerol), phosphatidyl serine, phosphatidyl inositol and phosphatidyl ethanolamine. But subsequent studies showed that the major antigenic targets are in fact certain proteins which complex with these phospholipids. The most important of these proteins was identified as the Beta 2 Glycprotein 1 (β 2 GP1). Later antibodies to about fifty other antigens were identified and these include those reacting with Prothrombin, Thrombin, phosphatidyl serine prothrombin complex, Annexin V, High and Low Molecular Weight Kininogen, Oxidized LDL, Tissue plasminogen activator and coagulation factors XII and VII[10,11]. A subset of these antibodies, however do bind purely to phospholipids. These phospholipid binding antibodies belong to the category of natural antibodies; and probably have role in homeostasis and have been found to be elevated in infections[12,13]. The overall avidity of binding of antibodies to these phospholipid antigens is weaker compared to those antibodies binding with the protein phospholipid complex[13].

Epidemiology

Some studies found upto 10% of healthy blood bank donors positive for anti phsospholid antibodies, however only 2% of them remain persistently positive[14,15]. Studies also found LA in 0.3% and aCL in 2%-9% of normal pregnancies and these figures were comparable to those in healthy non pregnant females[16].

Etiology of antiphospholid syndrome (APS) (Table 1)

APS may occur as a manifestation of another disease (secondary) or idiopathic (Primary). Primary APS constitutes more than 50% of all cases of APS and is by far the most common acquired thrombophilia[17].

The most common cause of secondary APS is SLE[18]. The prevalence of antiphospholipid antibodies in lupus patients range from 12-44% for anti Cardiolipin antibody (aCL), 15%-34% for LAC, and from 10%-19% for antiβ2GPI[19]. The prevalence of antibodies appears to increase with duration of disease. The Hopkins Lupus cohort, a 20 year prospective follow up cohort reported a prevalence of 47%, 26% and 32.5% respectively for aCL, LAC and anti β2GPI[20]. It is also known that aPL abs may be one of the first antibodies to appear in SLE and may predate the diagnosis of SLE by upto 7 years[19]. The presence of these antibodies, especially LAC, is a predictor for thrombosis as upto 50% of patients with LAC develop deep vein thrombosis[20].

Higher than normal prevalence of aPL abs have been described in other auto immune diseases as well, though their relationship to disease manifestations are less well established. Prevalence rates have been between 7 and 37% in RA and along with antibodies to oxidized LDL, antiphospholipid antibodies may have a role in cardiovascular morbidity in RA too[21,22]. However no link with thrombosis have been established till now in RA. Use of Infliximab in RA has also been associated with new occurrence of aPL ab[23]. In Scleroderma, though their prevalence may not be higher than general population, presence of aCL has been reported to be associated with more vascular manifestations as digital pitting scars, Pulmonary Artery Hypertension and capillary changes[21,24]. Presence of LAC in Sjogren's syndrome are associated with increased risk of thrombosis[25]. Presence of the rarer anitphospholipid antibody namely anti phosphatidyl serine has been linked to cutanoeus PAN[26]. There are reports of higher prevalence of Antiphospholipid antibodies in various systemic vasculitis as Wegner's granulomatosis, Giant cell Arteritis, Behcet's syndrome and PAN, but these are not pathogenic and are likely to be an epiphenomenon to systemic inflammation.

Malignancies, both solid and hematological, have been associated with higher prevalence of aPLs[27,28,29]. They may contribute to increased risk of thrombosis in these patients[30]. These antibodies become undetectable following treatment and may even serve as surrogate marker for disease remission. LAC positivity is most common in splenic marginal zone lymphoma and its presence is linked to arterial and venous thrombosis, Catastrophic Anti phospholipid syndrome, immune thrombocytopenia and autoimmune hemolytic anemia[31]. At least two studies have shown that nearly 40% of lymphomas have elevated anti phospholipid antibodies[32,33]. While there was no relation to thromboses, both studies showed that presence of these antibodies is a poor prognostic factor for survival. Up to 20% of unselected APS patients can have malignancies and patients with APS without any CTD appear to have a higher chance of developing malignancy on follow up[27,28]. Hence at least, in elderly patients with high titers of aPLs, it would be prudent to search for underlying malignancies.

Infections and APL positivity:

Infections are associated with aPLantibody positivity, but these are mostly transient, IgM variety, of low affinity and these antibodies seldom cross react to beta 2 GP 1 .

Approximately 20% of HCV positive patients have associated aPL but only a fraction of them have anti beta 2 GP 1 antibody[34,35]. HCV associated aPL ab positivity, however, is not associated with excess of thrombosis. aPL abs are also seen in up to 50% of patients infected with HIV and a fraction of them are even beta 2 GP 1 positive[13]. However no increased rates of thrombosis have been confirmed in these patients. One study had suggested than aPL ab in HIV may be associated with an increased risk of cutaneous necrosis and avascular necrosis[35]. High prevalence of aPLab has been demonstrated in HBV infected individuals, but again this is not associated with thrombosis. Other viral infections as EBV, CMV, Varicalla and Parvo virus have also been associated with aPL ab positivity[34].

aPL ab positivity occurs in many bacterial infections including Leprosy, Tuberculosis, Malaria, Rickettsial infection, syphilis and leptospirosis[19,35]. But no increased risk of thromboses has been demonstrated in these infections. Amongst them, leprosy is reported to have high prevalence of anti beta 2 GP 1 antibodies; but epitopes recognised on Beta 2 GP 1 by these antibodies differ from those in primary APS[36].

Drugs and APS

aPL abs have been known to occur following drug exposure. The list of such offending drugs include Quinidine, Procainamide, Phenthiazines, Phenytoin, Hydralazine, Propranolol, Interferon and Infliximab[38]. These antibodies are usually transient in nature, mostly of IgM isotype and do not cross react with beta 2 GP1. Thromboses, however, has been reported with quinidine, Procainamide and Phenytoin[38]. At least some of the thromboses may be related to the underlying disease or alternate mechanisms as hyperhomocystinemia associated with Phenytoin. Development of anti phospholipid antibodies and rapid development of thromboses has been noted following Interferon alpha treatment[39]. Although formation of aPL abs have been documented with Infliximab, Etanercept and Adalimumab, there is no report of increased risk for thromboses associated with their use[40].

Pathogenesis of anti phospho lipid syndrome

Currently it is understood that the major antigenic target of aPL antibodies is beta 2 Glycoprotein 1 (β-2GPI)[41]. Their role in pathogenesis has been extensively studied. These proteins along with their antibodies that bind to it play a major role in the pathogenesis of APS.

β-2 GPI structure and function

β-2 GPI is a 50kDa protein belonging to the Complement Control Protein (CCP) superfamily[41,42]. The normal concentration in the serum is around 200 microg/dl. It is mainly synthesized in the liver. Although first described in 1961, the exact physiological role of this protein is still not clear. Infact, β-2GP1 deficiency has not been found to cause any phenotypic abnormality in humans[43,44]. In vitro studies have however shown that they have powerful antioxidant effects[41]. It also interferes with many functions of procoagulant proteins and activated protein C[11]. Structurally, β-2GP1 has 5 subunits named from Domain I-V. Domain V is larger than other subunits and is heavily positively charged and have high affinity to anionic phospholipids on cell membrane[41,42]. Domain I is the immune dominant epitope against which most of the functional antibodies are generated[45]. One of the major characteristics of proteins of the CCP family is their ability to exist in different conformations depending on their interaction to the environment. β-2 GP1 thus exists in 2 forms. The oxidized form which has a linear or fish hook configuration and the reduced form which has a circular configuration[46]. 80% of β-2GP1 in plasma is in reduced form[47]. Binding of β-2GP1 to cell surface rich in anionic phospholipids, especially in presence of an oxidative stress induces a conformational change in the protein from the circular form to the linear form (Fig. 1).

β2 GPI as an autoantigen (Fig. 2).

In a study by Ioannou et al, levels of total β-2GP1 were increased in patients with aPL abs in comparison to healthy controls and those with other autoimmune diseases[47]. But the levels of circular or reduced forms are decreased in patients with APS. This indicates that there is an increased conversion of reduced form to oxidized form in patients with APS, possibly reflecting a process to compensate for the oxidative stress in the cells. Indeed in cell culture experiments, oxidative stress increases transcription of β-2GPI through activation of NF-kB and Activated protein 1[48]. In the continuing presence of oxidative stress, the newly synthesized β-2GP1 undergoes further oxidation thus altering the ratio between oxidised and reduced forms. It has been proposed that the linear oxidized forms express more antigenic epitopes in comparison to the circular forms and hence can act as a substrate for generating aPLs. Thus oxidative stress and a cell surface rich in anionic phospholipids together with a predisposition to B cell auto immunity sets up the back ground for generation of anti phospholipid antibodies[46]. Another favored hypothesis has been that apoptotic cells or cells in stress externalize the anionic phospholipids to the cell surface; most noticeably phosphatidyl serine[49,50]. β-2 GP1 then binds with high affinity to these phospholipids potentially leading to expression of cryptic / neo epitopes[51].

Functionally the anti β-2GP1 antibodies can bind to more than one β-2GPI molecule. This binding to anti β-2 GP1 on cell membrane induces clustering of β-2GPI on the cell surface. This binding also enhances the affinity of β-2GP1 for anionic phospholipids on cell surface. These clusters, and not the monomeric forms are in turn capable of activating a myriad of receptors present on the membranes of endothelial cells, monocytes, platelets, trophoblasts and decidual cells[52]. The receptors include ApoER2, Glycoprotein 1 b alpha, Heparin Sulphate, Annexin II, and TLR 2 and 4[52,53,54]. The binding to some of these receptors leads to cell activation through trasnscription pathways as the NF-kB and P38 MAPKs. This in turn triggers procoagulant or inflammatory pathways depending on the type of cell activated.

Presence of benign aPL abs in the sera of a large number of healthy people and transient increase in levels of aPLs following infection has led to the theory that these antibodies may actually have a protective role in immunity. It has, therefore, been suggested that the aPL abs are in reality, natural auto antibodies which become pathogenic in the presence of oxidative stress in individuals predisposed to autoimmunity[55].

Fig. 1 & 2. Two Configurations of beta 2 GP1 and pathogenesis of APS

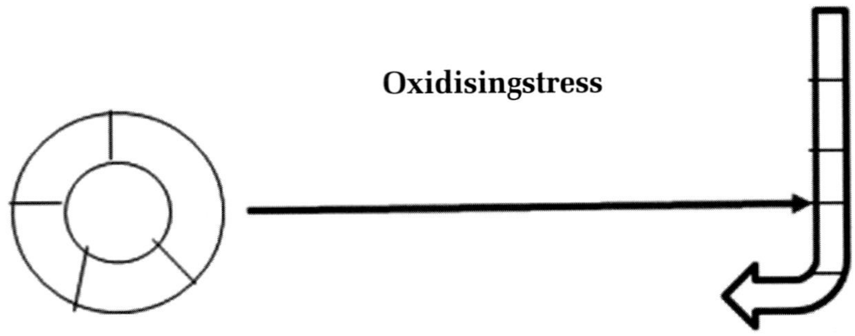

- Reduced form
- Circular structure
- Low levels in APS patients
- Low affinity for phospholipids
- Not recognised by Bea 2 GP 1 antibody

- Oxidised form
- Fish hook or linear structure
- High levels in APS patients
- High affinity for phospholipids
- Recognised by Beta 2 GP1 antibody

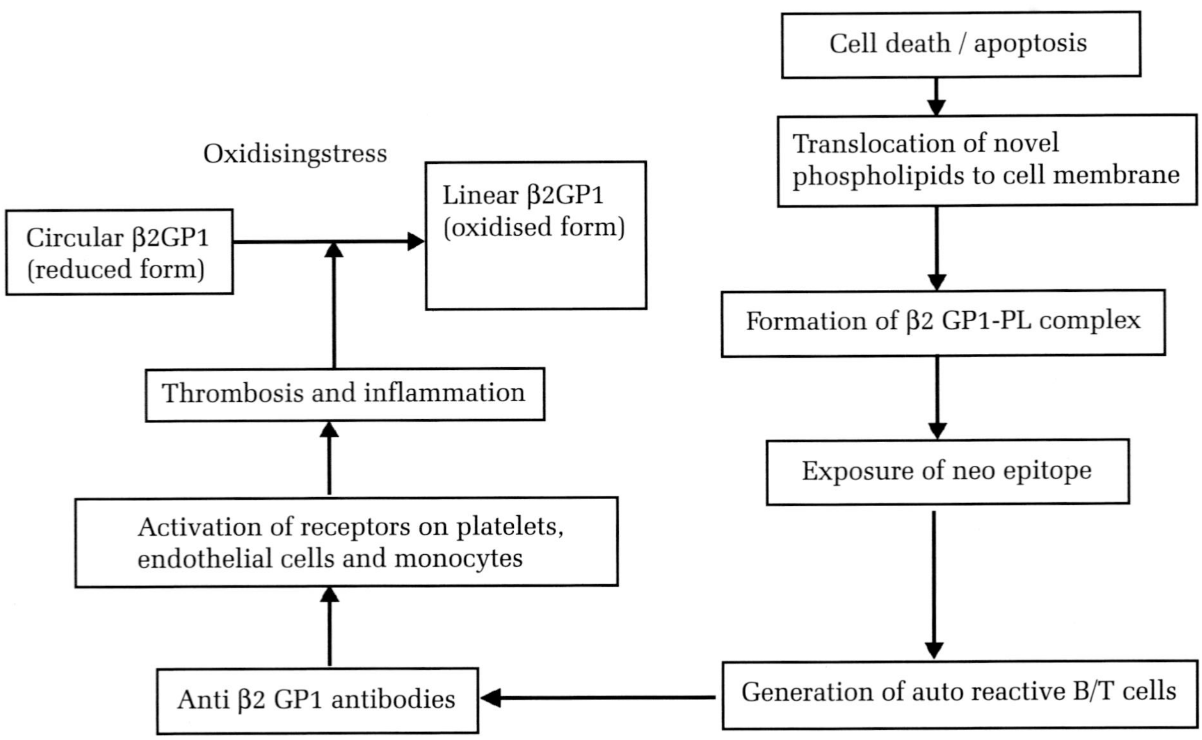

Etiopathogenesis of thromboses and pregnancy loss

Thromboses

Passive infusion of aPLs in mice has been associated with thromboses through increased expression of endothelial adhesion molecules, tissue factor and nitric oxide[56]. The antibody form complexes chiefly with β 2 GPI bound on the cell surface leading to activation of downstream molecules finally leading to endothelial cell activation and thromboses. The role of the recently described anti prothrombin antibodies in thromboses have also been studied and it appears that the binding of the antibody leads to activation of prothrombinase and a 'gain of function of prothrombin' which aids coagulation[58]. The outcome of aPL ab binding to cell will depend on the function of cells. For example anti β-2GP1 binding with β-2GP1 present on platelet surface can activate APO E2 receptor on platelet surface. This can lead to downstream activation of P38/MAPK pathway ultimately leading to increased thromboxane release[57]. It has also been suggested that the Anti phospholipid antibodies are capable of disrupting the Activated Protein C complex, a powerful anti coagulant and this process may contribute to the prothrombotic effect[59]. Other potential causes for thrombosis in APS includes disruption of the anticoagulant Annexin A5 coating on the phospholipid surface by aPL antibodies and interference with fibrinolytic pathways[56].

Pregnancy loss

Four different mechanisms have been suggested for pregnancy loss:

1. Placental infarction

Given the prothrombotic effects of anti phopsholipid antibodies, it is possible that aPL abs lead to formation of thrombus and hence insufficiency in the maternal fetal circulation[60,61,62]. Supporting this theory is the finding that Annxein A5, an anti coagulant protein has high expression on placenta and bind with phospholipids on cell membrane[63]. This coating denies the clotting pathway of the much needed phospholipids for initiating thrombosis. It has been postulated that binding of anti beta 2 GP1 antibodies destabilse the Annexin A5 coating leading to thromboses. However histopathological studies have failed to demonstrate thrombus in placental samples from many of the miscarried fetus from mothers with APS raising questions on this theory for fetal loss.

2. Proinflammatory mileu and complement pathway

Repeated intra peritoneal injections of aPL abs have been found to induce pregnancy loss in mice. It has also been shown that complement C3 deficient mice are resistant to the effects of these intraperitoneal aPL injections[64]. This suggest that a complement mediated inflammatory response may be responsible for the pregnancy loss. First trimester human trophoblastic cells when treated with aPL abs causes production of proinflammatory mediators by the TLR 4/My D 88 pathaway[65]. However histopathological studies again had failed to demonstrate inflammation in the placenta from aborted fetus associated with aPL abs. In addition, apparent inefficacy of corticosteroid in preventing pregnancy loss in APS was considered an argument against a major role of inflammation as a mechanism for pregnancy loss. Recent reports of complement inhibition as major mechanism of action of heparin in preventing foetal loss and efficacy of complement inhibitors in mice in preventing APS related pregnancy loss have brought back the concept of inflammation in obstetric APS to limelight[66,67].

3. Anti trophoblastic effects of aPL abs

High concentration of β-2GP1 has been demonstrated on the surface of trophoblasts. The exact role of this is not known. But binding of aPL abs to this have been shown to cause trophoblastic dysfunction[68]. Binding of anti phosholipid antibodies down regulates IL 6 secretion and reduces the migration of trophoblasts[69]. It has also been demonstrated that trophoblasts internalize anti phospholipid antibodies and causes apoptotic cell death and trophoblast shedding[70]. This leads to reduced invasiveness of trophoblasts and reduced placentation. Further, this causes reduced production of growth factors and hormones like ? HCG from the trophoblasts.

4. Role of beta 2 GPI in fecundity

The role of β-2GPI protein in normal fecundity in mice has been studied. β-2GP1 knock out mice have normal pregnancy rates, but they have higher rates of pregnancy loss and lower birth weight as compared to normal mice. These mice were also detected to have altered placental architecture. To what extent these findings are applicable to human is yet to be explored[71].

Though there is debate on the exact mechanism of pregnancy loss in APS, it is well accepted that the complement pathway plays a major role in its pathogenesis.

The two hit hypothesis

A large number of asymptomatic individuals are seropositive for aPL abs. But only a fraction of them actually develop symptoms. Thus two hit hypotheses has thus been invoked to explain this observation in pathogenesis of APS. Accordingly, an aPL ab positive patient may not develop thromboses unless there is a second insult. Most common insult appears to be infections[72]. In vivo models have identified several augmenting factors including LipoPoly Sacharide, mechanical stimuli and photochemical injury[73].

Clinical features of APS

APS can cause thrombo embolism in any part of the body and the manifestations depend on the organ involved.

Systemic vascular thrombosis

Deep Vein Thromboses (DVT) in the lower extremities is the commonest presentation and up to half of these patients develop pulmonary embolism. In one prospective study, aPL abs were elevated in upto 20% of patients with DVT or pulmonary embolus before the event[74]. Cervera et al reported that venous thromboses are the presenting feature in upto 30% of patients with APS[75]. Superficial vein thromboses can exist concomitantly or as an isolated feature. Spectrum of venous thromboses can range from upper extremity thromboses and cortical venous thrombosis to thromboses involving Superior and Inferior vena cava, hepatic veins (Budd Chiari Syndrome), portal vein, renal veins and retinal veins.

Arterial thromboses may manifest with ischemia or infarction in Central nervous system, Myocardium, kidney, gall bladder, mesentery etc.

Central nervous system (CNS)

Ischemic stroke is the first presentation of APS in up to 30% of patients[75]. They are leading causes of mortality in APS . APS is one of the most common causes of young stroke. In the RATIO study, the odds ratio for ischemic stroke among patients with lupus anti coagulant positivity was 43.1[76]. In one Indian study the prevalence of aPL abs was 29.4% in patients with young stroke[77]. Most often these infarcts involve the Middle Cerebral Arteries. Co-existing risk factors such as age, smoking, OCP use and high CRP levels are associated with an increased risk for stroke.

APS screening hence should be done in all cases of young stroke. Sneddon's syndrome, a triad of livido reticularis, hypertension and stroke is one of the manifestations of APS. Rarely APS can present with dissociated sensory loss and weakness due to anterior Spinal artery involvement. Multi Infarct dementia is another rare but devastating manifestation of APS[78].

Equally important, but less frequently recognized features include the non thrombotic neurological manifestations in APS[79]. The cause of these may be small vessel ischemia, local inflammation or direct neurotoxicity. These manifestations range from cognitive dysfunction, migrainous headache, seizures, chorea, transverse myelitis and optic neuritis. In fact, presence of anti phospholipid antibodies in SLE patients have been found to be the major risk factors for Neuro psychiatric manifestations independent of the presence of thromboses. Asymptomatic LA positivity also has been identified as a predictor of cognitive dysfunction in SLE.

Renal

All renal vasculature can be affected in APS. Both renal artery and renal vein thromboses are known to occur in APS[80,81]. Renal artery thromboses, if unilateral can cause infarction and scarring and if bilateral cause acute renal shut down. Renal artery stenoses without thromboses is an increasingly recognized manifestations of APS. It appears that APS induced endothelial dysfunction and proliferation, rather than frank thromboses is the cause for this. Renal vein thromboses can present with massive hematuria and nephrotic range proteinuria. APS nephropathy is a well recognized manifestation and can closely resemble TTP or lupus nephritis when it occurs in SLE patients. Histology would reveal microvascular thromboses. Alternatively TTP itself can be a manifestation of APS. Proliferative glomerulonephritis has also been associated with APS. In renal transplant recipients, anti phospholipid antibodies are a major risk factor for renal allograft rejection.

Cardiac

Valve thickening especially of the mitral valve can be seen in up to a third of APS patients. This can be associated with regurgitation, stenoses and vegetations[82]. Anti phospholipid antibodies also cause accelerated atherosclerosis and endothelial dysfunction. Myocardial infarction can occur due to premature atherosclerosis or acute thromboses. Recurrent stent re-stenoses is a feature in those patients undergoing coronary stenting. APS can cause intracardiac emboli, dilated cardiomyopathy and both thromboembolic and non thromboembolic pulmonary arterial hypertension.

Pulmonary

Anti phospholipid lung syndrome is a rare and life threatening manifestation of APS and is characterized by pulmonary micro thromboses, thromboembolism of lung arteries, pulmonary hypertension, adult respiratory distress syndrome and diffuse alveolar hemorrhage[83,84].

Dermatological manifestation

There are several non criteria manifestations in skin. This includes livido reticularis, ulcerations, digital gangrene, subungual splinter hemorrhages, superficial vein thromboses, purpura, atrophic blanche, extensive cutaneous necrosis and primary anetoderma[85,86].

Catastrophic Anti Phospholip Syndrome(CAPS) have a higher incidence of skin lesions reflecting generalized vasculopathy.

Obstetrical complications

This includes spontaneous miscarriages, preterm labor, low birth weight, preeclampsia and still birth. Unusually high incidence of recurrent miscarriage at 10 or more weeks of gestation is seen in APS. APS mothers also have increased incidence of thrombocytopenia. The presence of LAC or high titer positivity of β 2 GPI or previous thrombo embolic events are predictors of pregnancy loss. A post partum syndrome has also been described with high spiking fever, pleural effusions and patchy pulmonary infiltrates.

ENT

Autoimmune sensory hearing loss is one of the manifestations in APS. The hearing loss is typically bilateral and rapidly progressive[87].

GIT

APS is the commonest acquired cause of Budd Chiari syndrome. This should be suspected in patients with massive ascitis with or without other features of liver dysfunction. Other GI manifestations include intestinal ischemia and infarction, colonic ulcerations, esophageal necrosis and perforation, hepatic infarction, acalculous cholecystitis with gall bladder necrosis, pancreatitis and mesenteric and portal vein thromboses.

Hematologic complications

APS patients can have leucopenia, immune thrombocytopenia (ITP) and auto immune hemolytic anemia (AIHA) independent of the presence of SLE[88]. ITP in APS is seldom severe enough to cause bleed and if it occurs, possibility of TTP or other causes need to be ruled out. Evans syndrome, a combination of AIHA and Immune thrombocytopenia is a feature of APS and can be notoriously treatment resistant[89]. Occurrence of thrombocytopenia with coexistent venous thromboses is not uncommon, management of which is challenging. A rare but interesting manifestation of APS is acquired hypo prothombinemia and consequent bleed[90]. Levels of prothrombin levels may be less than 10-20% and treatment may require transfusion or corticosteroids. Another rare but devastating complication in APS is gelatinous transformation of bone marrow which presents with multiple areas of bone marrow necrosis and pancytopenia of variable degree.

Endocrine manifestation

Adrenal infarction is another manifestation of APS and patients can present with acute addisonian crisis. When this happens, CAPS should be kept in mind. Acute pituitary infarction and consequent hypopitutarism is another well known manifestation of APS[91]. Auto immune thyroid disease has been associated with APS. In addition, testicular and ovarian dysfunction have also been reported with APS.

Musculoskeletal manifestations.

Primary APS can have non erosive inflammatory arthritis. Another frequent manifestation of APS is avascular necrosis of bone most often involving the hip joints.

Ophthalmic manifestation

Eye manifestations in APS are often under recognized[92]. The following are some of the known eye findings in APS. These include conjunctival telengiectasia, episcleritis, both anterior and posterior scleritis, uveitis, retinal vasculitis, retinal arterial/venous occlusion, arterial and non arterial ischemic optic neuropathy and optic neuritis.

Diagnosis of APS

In order to simplify and standardize the diagnosis of APS, an international consensus statement on preliminary classification criteria for definite APS was drafted following the Eighth International symposium on APA in Sapporo, Japan in 1999[93]. A subsequent meeting in Sydney in 2006 modified the criteria to its present form. The criteria requires satisfaction of one clinical and laboratory criteria separately.

Revised sapporo criteria[94]

Clinical criteria:

There are two clinical criteria namely vascular thrombosis and pregnancy morbidity.

a) Vascular thrombosis refers to one or more clinical episodes of arterial, venous or small vessel thrombosis in any tissue or organ. The thrombosis must be confirmed by imaging or Doppler studies or histopathology. If histopathological confirmation is performed, then there should be thrombosis with no significant inflammation in the vessel wall ruling out vasculitis. Mere occurrence of superficial vein thromboses will not satisfy the criteria.

b) There are three categories of pregnancy morbidity.

Category A: One or more unexplained deaths of a morphologically normal fetus at or beyond the 10th week of gestation. Ultrasound or direct examination of the fetus must document normal fetal morphology.

Category B: One or more premature births of a morphologically normal neonate at or before the 34th week of gestation due to severe pre eclampsia or eclampsia, or severe placental insufficiency.

Category C: Three or more consecutive spontaneous abortions before the 10th week of gestation, following exclusion of both maternal anatomic or hormonal abnormalities and paternal and maternal chromosomal abnormalities.

Laboratory criteria:

One out of following 3 laboratory criteria has to be satisfied:

1. Presence of Lupus Anticoagulant in the plasma detected according to the guidelines of International Society for Hemostasis and Thrombosis.
2. Presence of anti cardiolipin antibody of IgG and/ or IgM isotype in serum or plasma at medium or high titer (> 40 GPL or MPL, or > 99th percentile) on 2 or more occasions measured by a standardized enzyme-linked immunosorbent assay (ELISA).
3. Presence of anti - β2 GP1 antibody of IgG and/or IgM isotype in serum or plasma (titer > 99th percentile) measured by a standardized ELISA according to recommended procedure.

For satisfying the laboratory criteria the tests should be positive on 2 occasions 12 weeks apart.

The major modifications in the revised criteria were the following[95]:

a. The duration of time between 2 tests for antibodies was increased from 6 weeks to 12 weeks. This was in keeping with realization of the fact that this time period may be more optimum to negate the effect of transient positivity of the antibodies which are not pathogenic.

b. The requirement of a β2 GP1 dependent test for the ACL detection was done away with, as currently there is no reliable test to ensure that.

c. Further, detection of anti β2 GP1 was included as a separate laboratory criteria as it was recognized as the major pathogenic antibody in APS.

Laboratory detection of APLs - Methods and fallacies

Lupus anti coagulant (LAC)

Phospho lipid dependent coagulation assays such as APTT, DRVVT, KCT have been shown to be prolonged in APS patients. The reason for this is that the sera of APS patients contain antibodies that coat the phospholipid surfaces and block the access of coagulation factors to the phospholipid surface. Since the coagulation defect is due to an apparent deficiency of phospholipid, addition of plasma which only contains coagulation factors do not correct the deficiency. However extraneous addition of phospholipids corrects the defect. This is the principle for the LAC test. LAC refers to these antibodies which block the coagulation reaction. Thus LAC is a misnomer and inspite of prolonging coagulation reactions in vitro, these antibodies actually predispose individuals to thrombosis. The major targets of the antibodies responsible for the LAC effect are β2 GP 1 and Prothrombin.

The International Society for Thrombosis and Hemostasis (ISTH) has laid down guidelines for detection of LAC[96]. This includes a 3 step process based on the above described principle.

Step 1: Demonstration of prolongation of phospho lipid dependent coagulation assay.

The recommended assays are APTT and DRVVT. Both assays are to be performed and prolongation of any one of the assay will be considered significant. It should be noted that the present guidelines do not recommend the use of other phospholipid dependent assay as Kaolin CT.

Step 2: Addition of platelet poor plasma to confirm that the prolongation is not due to coagulation factor deficiency.

Step 3: Addition of phospholipids to confirm correction of the previously prolonged coagulation time and to demonstrate phospholipid dependence. Platelet poor plasma which has been double centrifuged to remove any residual platelets should be used for the assay as potential source of phospholipids.

The test should ideally not be performed in patients on Vitamin K antagonists or Heparin. In patients on heparin, it is advised to do the test after 12 hours of stopping heparin. In patients on VKAs it is recommended to do the test about 2 weeks after their stoppage. However in exceptional circumstances, the test can be performed by adding double the amount of plasma to supplement the coagulation factors. However in doing so, it is possible to miss low titer LAC positivity due to dilution.

In spite of these guidelines there is still lack of uniformity in reporting. In one study the reproducibility of LAC was as low as 33% between reference laboratories.

Recently a new assay has been developed that measures the anti β2 GP1 dependent LAC which is likely to be more sensitive in detecting pathogenic LAC than the functional assay.

Anti cardiolipin and Anti β2 GP1 antibodies

The detection of these antibodies are based on solid phase ELISA[97]. The sample required for the test are ideally non hemolysed, non lipemic sera. The ELISA plates are coated with the a β 2 GP1, (human or calf origin) with cardiolipin for aCL and with human β2 GPI for anti β2 GPI detection. It is recommended to test for both IgG and IgM isotypes. The reporting is done as MPL or GPL units for aCL. There are no universal standard units for anti β2 GPI. It should be remembered that a positive rheumatoid factor can potentially give a false positive result in IgM ELISA. Like LAC, these assays are also plagued with problems of wide inter assay variability. Differences in the purity of the β2 GPI antigen used, differences in oxidation status of the antigen used, variations in calcium ion concentration in different sera ,lack of standard calibrators for the assay and insufficient definition of positive cut off values are some of the causes of this variability[98].

Laboratory markers as predictors of clinical events

Although the criteria consider positivity of any one of 3 antibodies as essential and sufficient for the diagnosis of APS, the predictive value of each of these antibodies for a clinical event varies. The exact predictive values of the antibodies are not known, but all data suggest that LAC positivity has the highest risk for clinical events[99,100]. However this test is positive in only 20-40% of APS patients. This is understandable as LA is considered to be a functional assay. It measures the ability of APL antibodies to block the phospholipid surface to prevent coagulation. Hence it requires a higher concentration of antibodies to exert this effect. Indeed studies have also shown that LAC positivity correlates with the titer of anti β2 GP1 antibodies.

LA positivity has been associated with increased occurrence of arterial as well as venous thromboses and pregnancy loss[100,101]. Further it has been shown that a test which specifically detects the β-2 GPI dependent Lupus Anticoagulant is more strongly associated with thromboses (OR 10.2 vs 40.3) as compared to standard LA assay[100]. However it should be remembered that some asymptomatic individuals can have LA positivity especially in low titers. The predictive value of other antibodies for clinical events is actually less evident. However it is has been shown that higher concentration of both aCL and anti β2 GPI correlate well with clinical events, albeit to a lesser extent when compared to LAC[102,103].

It is also been increasingly realized that the positivity of more than one of the three standard lab tests has a higher predictive value for events. This is a strong argument in favor of performing all three tests in any given patient[101]. Based on this understanding Miyakis et al have suggested to classify aPL ab positivity into 2[94]:

Class 1 - Two or more laboratory criteria present

Class 2a - isolated LA positive

Class 2b - isolated acL positive

Class 2c - isolated beta 2 GPI positivity

In a recent study on the aPL profile of 175 APS patients by Tincani et al, it was found that 35% , 28% and 37% of patients respectively had all three, two out of 3 and single test positivity[104]. Of the patients with only 1 positive test, 75% had beta 2 GPI antibody, 19% had LA and 6% had aCl positivity.

Newer antibodies and newer assays

Anti phosphatidyl serine prothrombin antibody / anti prothrombin antibody.

Antibodies against prothrombin are detected in patients with APS[105,106]. Two types of antibodies have been detected namely anti prothrombin antibodies (a-PT) and anti Phosphatidyl serine prothrombin antibodies (aPS-PT). These antibodies are different though they may be present in the same patient. Anti-PT antibodies bind to prothrombin as the antigen, where as aPS-PT bind with prothrombin bound to phosphatidyl serine. In vitro studies have shown that these antibodies increase endothelial cell activation. The aPS-PT antibody has been shown to have LAC function as well. But there have been conflicting results regarding the role of these anti bodies in thromboses. In one of the recent studies, Anti PT and anti PS-PT antibodies were present in 52 and 31 % respectively in APS patients with LAC positivity, while their prevalence was 22% and 28% respectively in LAC negative patients. Both aPT and APs-PT antibodies were associated with venous thromboses in this cohort[107]. The link with thromboses appears to be stronger for aPS-PT than for aPT[107,108]. It is possible that these anti bodies may help to identify a subset of patients who are negative by other tests. However at present evidence is insufficient to recommend this test for routine use.

Anti domain 1 antibody

As discussed previously there are 5 domains for the β 2 GP1 protein of which domain 5 is the binding side for the phospholipid surface, where as domain 1 appears to be the epitope against which pathogenic antibodies are raised. Antibodies against the other domains may be seen transiently in conditions as infections and their detection is not predictive of thromboses. Hence an ELISA has been described to detect anti β 2 GPI antibody that specifically targets the Domain I. This assay has been shown to predict thrombotic event better than the conventional assay[109,110].

IgA anti phospholipid antibody

IgA aCL antibody and IgA anti-β2-GP1 are seen in 1-44% and 14-60.9% of SLE patients with APS respectively[111]. There is some evidence to show that IgA aCL and IgA subtype of anti β 2 GPI may correlate with thrombotic manifestations independent of presence of other antibodies. However this association is inconsistent across studies. In addition to APS, IgA aPL abs are also seen in Rheumatoid Arthritis, Sjogren's syndrome and Guillan Barrie syndrome. The international consensus guidelines on aPL ab testing does not recommend this test routinely at the moment, but suggest that they can help in diagnoses of APS in the absence of other antibodies.

Sero negative APS (SNAPS)

A subset of patients present with classical clinical manifestations of APS, are persistently negative for any of the anti phospholipid antibodies tested routinely. These patients can be described to have Seronegative APS (SNAPS). Ever since the first description, the existence of this syndrome has been controversial. It has been argued that the absence of antibodies represent technical issues rather than true absence. Different causes have been suggested. These include:

a) Transient fall in aPL ab levels following thromboses due to consumption due to which if tested immediately following an acute episode of thrombosis, the levels may be falsely low[112].

b) Conventionally only IgM and IgG subtypes of the aCL and anti beta 2 GPI antibodies are measured. However some of the patients with APS have been found to have only IgA subtype and hence their presence is unrecognized[111].

c) Wide intra and inter observer variability has been observed in testing for anti phospholipid antibodies and hence it is possible that the assay errors contribute to a substantial number of SNAPS cases.

d) Antibodies against antigens other than β 2 GP1 have been detected in subsets of patients with APS. Notable among these are antibodies directed against prothrombin and vimentin. The presence of these antibodies have been detected in sera of patients in the absence of the classical antibodies[113].

e) In addition, reduction in antibody titres have been observed in patients with nephrotic syndrome and in those on corticosteroid therapy[114,115].

APS in special scenario

Catastrophic APS (CAPS)

Catastrophic APS is a serious form of this illness presenting with multi organ failure[116]. It is a life threatening medical emergency with a mortality of about 50%. It may be the first presentation of APS in about 50% of this complication and half of the episodes of CAPS are precipitated by infections. It involves at least 3 organs over a period of days or weeks with histological evidence of multiple occlusions of large or small vessels[117,118,119].

The kidney is the organ most commonly involved followed by the lungs, CNS, heart, adrenal and skin. Adult respiratory Distress Syndrome due to non thrombotic microangiopathy is the most common pulmonary manifestation, followed by pulmonary embolism, and alveolar hemorrhage. Cerebral involvement may be in the form of stroke, cortical venous thrombosis or diffuse encephalopathy. The gastrointestinal system could also be involved resulting in gut ischemia and abdominal pain. Patients often have thrombocytopenia and microangiopathic hemolytic anemia. Peripheral smear examination may reveal schistocytes, but this finding is rarer in comparison to that in TTP which is a close differential diagnosis. Disseminated intravascular coagulation (DIC), which is extremely rare in primary or secondary APS, is noted in about 25% of patients with catastrophic antiphospholipid syndrome.

Microvascular manifestations of catastrophic APS include renal thrombotic microangiopathy, adult respiratory distress syndrome, cerebral microthrombi with microinfarctions and myocardial microthrombi.

The pathophysiology of this syndrome is still poorly understood; however, initial thrombosis in a patient with APS may alter the balance of hemostasis and cause a "thrombotic storm", leading to multiple microthrombi in the body.

The precipitating factors include:

1. Infections of upper respiratory tract, intestine, urinary tract and leg ulcers.
2. Surgical procedures including angioplasty, dental extractions, endoscopic retrograde cholangiography, dilatation and curettage, caesarean section, hysterectomy and lung biopsy. But, even trivial things like needle stick injury can be a trigger.
3. **Drugs** : Withdrawal of anticoagulant or introduction of oral contraceptive agents, thiazide diuretics and captopril also can precipitate aCAPS event.

Pseudovasculitis and APS

Due to varied presentations related to thromboembolism, it can mimic several prothrombotic, vasculitic, atherosclerotic or embolic disorders. These include Factor V Leiden mutation, hyperhomocysteinaemia, classic PAN, infective endocarditis, cholesterol embolism and hyperviscosity states including some malignancies.

Thrombocytopenia and autoimmune hemolytic anemia can be confused with microangiopathic process. Catastrophic APS can mimic thrombotic thrombocytopenic purpura (TTP), Sepsis syndrome, disseminated intravascular coagulation etc. Normal prothrombin time and thromboplastin time and presence of schistocytes in peripheral blood support the diagnosis of TTP.

In presence of another connective tissue disease like SLE, vasculitis should be always thought of, before considering the diagnosis of secondary APS.

Infections can cause positive anti cardiolipin antibodies, most of which are non pathogenic. Many of these infections as HIV, HBV, HCV may be associated with vasculitic like presentation mimicking APS and needs to be correctly identified.

Management of Anti phospholipid syndrome (Fig. 3).

The presentation of APS can be varied from calf vein thrombosis to life threatening CAPS and the management strategy will differ based on the severity and spectrum of organ manifestation.

Primary prevention

The link between anti phospholipid antibodies and thromboses has been firmly established. But the risk of thrombosis in otherwise healthy individuals with anti phospholipid antibodies is considered to be small (< 1% / year)[120]. But in aPL ab positive patients with SLE ,without prior thrombotic event, the risk of thromboses is in the range of 0-3.8 % per annum[121]. Of these, 46-76% of the events are associated with a concurrent non aPL thrombotic risk factors. A prospective placebo controlled study (APASS study) addressed the role of aspirin in thromboprophylaxis in asymptomatic APL positive patients[122]. This study included patients who met the serological, but not the clinical criteria for APS with at least 1 year of follow up. The event rate in the aspirin arm was 4.57 per 100 patient years vs 0.86 per 100 patient years in the placebo arm. The difference was however not statistically significant. In this study more than 60% of subjects in either arm had SLE. However this study had a major short coming in that the sample size was too small to derive a meaningful conclusion. Though evidence is scant, based on the available information about the pathophysiology of APS as well as extrapolation of data on usefulness of aspirin in individuals with other cardiovascular risk factors, it is plausible that aspirin has a role in primary prevention of thromboses in asymptomatic aPL ab positivity. A recently concluded meeting of experts on APS has endorsed this view and has suggested that aspirin prophylaxis may be considered especially in high risk patients for thromboses with aPL abs[123]. The high risk factors include SLE, smoking and OCP use.

It appears that premenopausal women with obstetric APS also have an increased risk for subsequent thromboses. One study showed an incidence rate for thromboses of 7.2 per 100 patient years in obstetric APS against an expected back ground rate of 1 per 10,000 patient years in normal premenopausal women[124]. Hence it may be advisable to consider them for long term anti platelet therapy.

Secondary prevention

Patients with one thrombotic event and APS are at an increased rate of recurrent thrombotic events. Schulman et al studied 412 patients with thromboses in whom anticoagulation was stopped after 6 months. aCL was estimated in these patients before discontinuation[125]. Recurrent thromboses were seen in 29% of aCL positive patients in comparison to 14% in aCL negative individuals. In another study stoppage of anti coagulation after 6 months in patients with thrombosis and positive aPL ab (single estimation) was associated with a HR of 7.7 for thrombosis[126]. The risk of thromboses is highest closer to the time of first thromboses and has a predilection to involve the same vessel.

Although currently anti coagulation is the therapeutic option of choice in APS, newer management strategies also have been evolving.

Oral anticoagulation

Oral anticoagulation is the treatment of choice in long term secondary thrombo prophylaxis .However there is much debate on the dose and duration of oral anticoagulant therapy. Although some experts suggest that the duration of anticoagulation should be decided on a case to case basis with a shorter course of anti coagulation for low risk APS (eg:intermediate titre positivity of aCL/Beta 2 GPI or definite correctable precipitating factor for thrombosis), there is good agreement on the fact that it is best to treat all APS patients with past thrombotic event with life long anticoagulation.

Two of the prospective studies comparing the intermediate intensity anti coagulation (target INR 2-3) and high intensity warfarin (INR 3-4) failed to show any advantage of high intensity warfarin in terms of prevention of recurrent thromboses[126,127]. Besides, when the data from the two studies were combined, an increased incidence of minor hemoarrhagic events was seen in the high intensity group although total and major hemorrhagic events were identical in the two groups[128]. Hence in all except refractory venous thrombosis a target INR of 2-3 would be appropriate. However, based on retrospective data, there is some evidence to suggest that high intensity anti coagulation with a target INR between 3 and 4 may be better in arterial thrombosis[129].

Another area of controversy has been the prevention of recurrent stroke in APS. A large population based study, called the WAPS study, failed to demonstrate any advantage for Warfarin over Aspirin (325 mg) in patients with APS in prevention of recurrent stroke[130]. In this study, 1770 (720 APL positive and 1020 negative) patients with stroke were randomized to receive aspirin 325 mg/day vs Warfarin (INR maintained between 1.4-2.8). At the end of 2 years the overall event rate was 22.2% and 21.8% respectively. But this study has been criticized on account that the patients had only a single estimation of aCL, study included those with even low antibody titer positive in the aPL ab group and the achieved INR was suboptimal. Although these patients require some form of thromboprophylaxis (either Aspirin or Warfarin), more data is required before recommending one over the other. However in patients with recurrent thrombosis, multiple antibody positivity and persistent positivity of aPL antibodies, Warfarin with moderate or high intensity anticoagulation may be a more prudent choice. Standard guidelines for anti coagulation need to be followed while initializing and optimizing oral anticoagulation in APS patients. This includes initiating the patient on UFH or LMWH and overlapping with Warfarin for 4-5 days and frequent monitoring of INR.

Hydroxychloroquine (HCQ)

Besides anti inflammatory action, anti platelet action has been demonstrated for HCQ in experimental models. HCQ decreases the surface expression of GP IIb/ IIIa and reduces thrombus size in aPL ab models. HCQ has also been shown to prevent the binding of aPL ab and beta 2 GP 1 complexes with phospholipid surfaces[131]. Further there is preliminary data on the clinical efficacy of HCQ in prevention of thrombosis in SLE patients, in secondary prevention of thrombosis in APS patients as well as in preventing thrombotic events in asymptomatic aCL positive patients[132]. Infact one recent study showed that the use of Hydroxychloroquine in SLE has been associated with lower odds of persisitent positivity of anti phospolipid antibodies[133]. The relatively safe side effect profile; especially absence of bleeding make HCQ an attractive treatment option in thromboprophylaxis. It should be remembered that there is inadequate evidence to recommend HCQ as a single agent in thromboprophylaxis.

Statins

Statins are widely used for the primary and secondary prevention of cardiovascular events. Other than the lipid lowering effect, statins have been known to have pleotropic effects modulating the endothelial activation and coagulation pathways. They cause reduced surface expression of IL 6 and ICAMs (Intercelluar Adhesion Molecules), reduce the tissue factor activity and renders the endothelium less susceptible to the effects of proinflammatory mediators[134]. Recently concluded JUPITER trial has demonstrated that in subjects with high CRP with normolipidemia, addition of Rosuvastatin reduces the incidence of thrombosis but there is no data in patients with APS[135]. Endothelial activation and coagulation are central to the pathogenesis of APS. Hence, statins are an attractive treatment option in these patients, although clinical data is inadequate in APS. Statins should be considered in patients with coexistent cardiovascular risk factors or in management of refractory APS.

Corticosteroids

Contrary to expectations, corticosteroids have not been beneficial in the management of thrombotic manifestation or pregnancy morbidity in APS. However use of corticosteroid is beneficial in some of the non criteria manifestations of APS as Coombs positive hemolytic anemia, Immune thrombocytopenia, demyelinating syndrome and CAPS.

Management of obstetric APS

APS is a major cause for adverse pregnacy outcomes. APLs are present in 15-20% of females with recurrent pregnancy loss (RPL) as against on 5% in non pregnant females[136]. Besides RPL, APS is associated with fetal loss and intra uterine death, IUGR and preeclampsia. Infact only a third of untreated pregnancies with APS and previous recurrent pregnancy loss (RPL) have successful outcome.

Unfractionated Heparin (UFH) is the treatment of choice for prevention of obstetric APS[138,139]. The mechanism of action of UFH in obstetric APS is multifaceted[137]. UFH not only inhibits thrombus formation, but also prevents complement activation, inhibits binding of aPL abs to the trophoblastic surface, promote trophoblast invasiveness and inhibit apoptosis of trophoblasts. UFH is used in doses of 5000-10000U twice daily to reduce pregnancy loss.

The other treatment modalities include Low Dose Aspirin (LDA) and LMWH. Although a few studies had addressed the issue, the effectiveness of LDA alone has not been conclusively proven in APS. Studies comparing LDA alone and LDA with UFH have shown that the combination is better than LDA alone[138,139]. Though evidence is scarce, most experts recommend to use LDA in combination with UFH. A recent Cochrane based systematic review determined that combination of UFH with LDA reduces risk of pregnancy loss in comparison with LDA alone (RR of 0.46, CI 0.29-0.71)[140]. It is recommended to start LDA preconception and Heparin at the first confirmation of pregnancy. Although some centers advocate UFH only for the first 12 weeks if the obstetric morbidity is in the form of first trimester RPL, the practice is not widely accepted. It is recommended to stop Heparin 24 hours prior to the expected delivery and restart 6 hours later to be continued till 6-8 weeks. LDA also should be stopped from the last month of pregnancy till childbirth to prevent patent ductus in the baby.

Recent trials using Enoxaparin and Nadroparin in RPL due to APS have found effects comparable to UFH and may evolve as the drug of choice because of better side effect profile including osteoporosis and heparin induced thrombocytopenia (HITS) and ease of monitoring[141,142]. Since the patients require anticoagulation throughout pregnancy, heparin induced osteoporosis is a major morbidity. In fact heparin induced osteoporosis is seen in 2.2-5% of mothers on this agent. Initial studies show that the effect of LMWH on bone loss may not be that significant. Hence it is a good practice to supplement the mothers with Vitamin D and Calcium.

Management of refractory APS

In all cases of refractory APS a meticulous search for non aPL ab risk factor should be done and addressed accordingly. The compliance with prescribed medications should be ensured.

Refractory thrombotic APS

Recurrent thromboses have been known to occur despite attainment of target INR in APS and management of this is highly challenging and empirical. A high intensity warfarin therapy with target INR 3-4 may be helpful in individuals with no other bleeding risk factors. Addition of Low dose Aspirin, HCQ and or statin can be considered at this stage. In patients not responding or intolerant of high intensity Warfarin, long term anticoagulation with LMWH or heparin has been found to be of use. In one retrospective study, Enoxaparin used over a median duration of 36 months was found to be safe and effective in refractory thromboses.

Oral direct thrombin inhibitors as Dalbigtran and Rivaxoraban have proven their efficacy for thromboprophylaxis in DVT and are used for thromboprophylaxis in patients undergoing joint replacement. Their role in the management of APS is still to be explored. Rituximab is also emerging as a possible treatment option. The result of the first randomized controlled study using Rituximab in APS (RITAPS) is underway, the result of which is much anticipated. Newer targets like Complement pathway inhibitors, inhibitors of NF kB pathway and MAP kinase pathways offer promising therapeutic targets[143].

Management of refractory obstetric APS

In spite of optimum treatment pregnancy loss occurs in about 30% of patients. Previous history of RPL, recurrent DVT and triple antibody positivity are high risks for recurrent adverse pregnancy events.

Initial studies on use of high dose prednisolone showed non favorable obstetric out come as well as higher rate of infections, preterm deliveries, gestational diabetes and hypertension. Bramham et al studied 20 females with history of 89 pregnancies with only 4 of them with favourable outcome in terms of live births. These patients were given prednisolone at 10 mg/day from 14 weeks of gestation. Twenty five pregnancies occurred in them of which 16 (64%) resulted in successful outcome. This dose of Prednisolone used in the study was not associated with any significant side effect[144].

The other potential treatment modality is IVIG. But IVIG is prohibitively expensive and small studies have not shown significant benefits. Hydroxy - chloroquine as an additional agent is also a safe and promising therapeutic option in refractory cases though clear clinical evidence is lacking.

Management of CAPS

CAPS is a life threatening complication of APS. This requires aggressive treatment with high dose pulse Methyl prednisolone, standard anticoagulation with UFH/LMWH and IVIG or plasmapheresis[145]. It should be remembered that 50% of CAPS are precipitated by infections, treatment of which should be instituted at the earliest. Addition of Cyclophosphamide is advisable in SLE related CAPS. In refractory or recurrent CAPS therapeutic options include Rituximab and the anti complement antibody Ecluzimab.

Mortality in APS

The Euro phospholipid cohort is the largest prospective cohort of APS with a total of 1000 APS patients. In the 5 year follow up of this study the mortality was 5.3%. The cause of mortality in the decreasing order of frequency was bacterial infections (20.8%), myocardial infarction (18.9%), stroke (13.2%), hemorrhage (11.3%), malignancy (11.3%), catastrophic APS (9.4%) and pulmonary embolism (9.4%)[75].

The mortality is nevertheless higher in patients presenting with CAPS and can be as high as 50%. Cerebral involvement, followed by cardiac dysfunction and infections are the major causes of mortality in CAPS.

Controversies in APS

Being a relatively common disease in the domain of multiple clinical specialties, there are many controversies in APS. These include characterization of mechanism of thromboses and pregnancy loss, choice of ideal diagnostic assay and their standardization, existence of sero negative APS and issues in management. There is much debate on pathogenesis especially in obstetric APS regarding the differential roles of thrombotic and inflammatory mechanisms. A major limitation in the quality of data in APS is the poor standardization and reproducibility of assays used for detection of aPL abs. Many of the so called Seronegative APS may actually be due to poor standardization of assay. There is conflicting data on the predictive ability of the three major antibody assays in APS. However, it appears that LAC is the most specific for clinical event in overall scenario though the sensitivity is low. Another dilemma to the clinician is in the management of thromboses especially arterial thromboses. The weight of evidence appears slightly in favor of high intensity as against low intensity anticoagulation for arterial thromboses. Similarly the treatment of APS associated stroke also is a controversial issue with the larger studies showing no advantage of oral anticoagulation over Aspirin .However the quality of this data has been severely criticized. Controversies abound in the management of asymptomatic aPL ab carriers too. Recent recommendations, however, have been in favor of treating high risk aPL ab carriers with low dose asprin. With increasing interest in the field , we can expect more rapid advances in diagnosis and management to evolve in the years to come.

Table 1: Causes of Anti phospholipid syndrome

Primary	Idiopathic
Secondary	
Autoimmune diseases	SLE, Sjogren's syndrome, MCTD, Vasculitis
Infections	***Bacterial*** - Tuberculosis, Leprosy, Syphilis, Leptospirosis
	Viral - HCV, HBV, HIV, CMV, EBV,Parvovirus, HTLV
	Parasitic - Malaria, Kala azar
Malignancies	Lymphomas, leukemias, renal cell carcinomas, colonic carcinomas
Transplantations	Bone marrow, Kidney
Drugs	***Anti arrhythmics*** - Procainamide, Quinidine
	Antihypertensives - Hydrallazine, Propranolol, Acebutalol
	Anti epileptics - Phenytoin, Valproic acid
	Antibiotics - Penicillin, Amoxycillin, Streptomycin
	Immunomodulators - Interferon alpha, IL-2, TNF inhibitors
	Antipsychotics - Chlorpromazine, Haloperidol, Fluphenazine, Clozapine
	Others - Oral contraceptives, INH

Table 2. Spectrum of diagnosis in 1000 patients with APS

Underlying condition of patients	Percentage
Primary antiphospholipid syndrome 531 (53.1)	53.1
Systemic lupus erythematosus 362 (36.2)	36.2
Lupus-like syndrome 50 (5.0)	5
Primary Sjogren's syndrome 22 (2.2)	2.2
Rheumatoid arthritis 18 (1.8)	1.8
Systemic sclerosis 7 (0.7)	0.7
Systemic vasculitis 7 (0.7)	0.7
Dermatomyositis 5 (0.5)	0.5

Adapted from Cervera R, Piette JC, Font J, Khamashta MA et al;Euro-Phospholipid Project Group. Arthritis Rheum. 2002 Apr;46(4):1019-27

Fig. 3

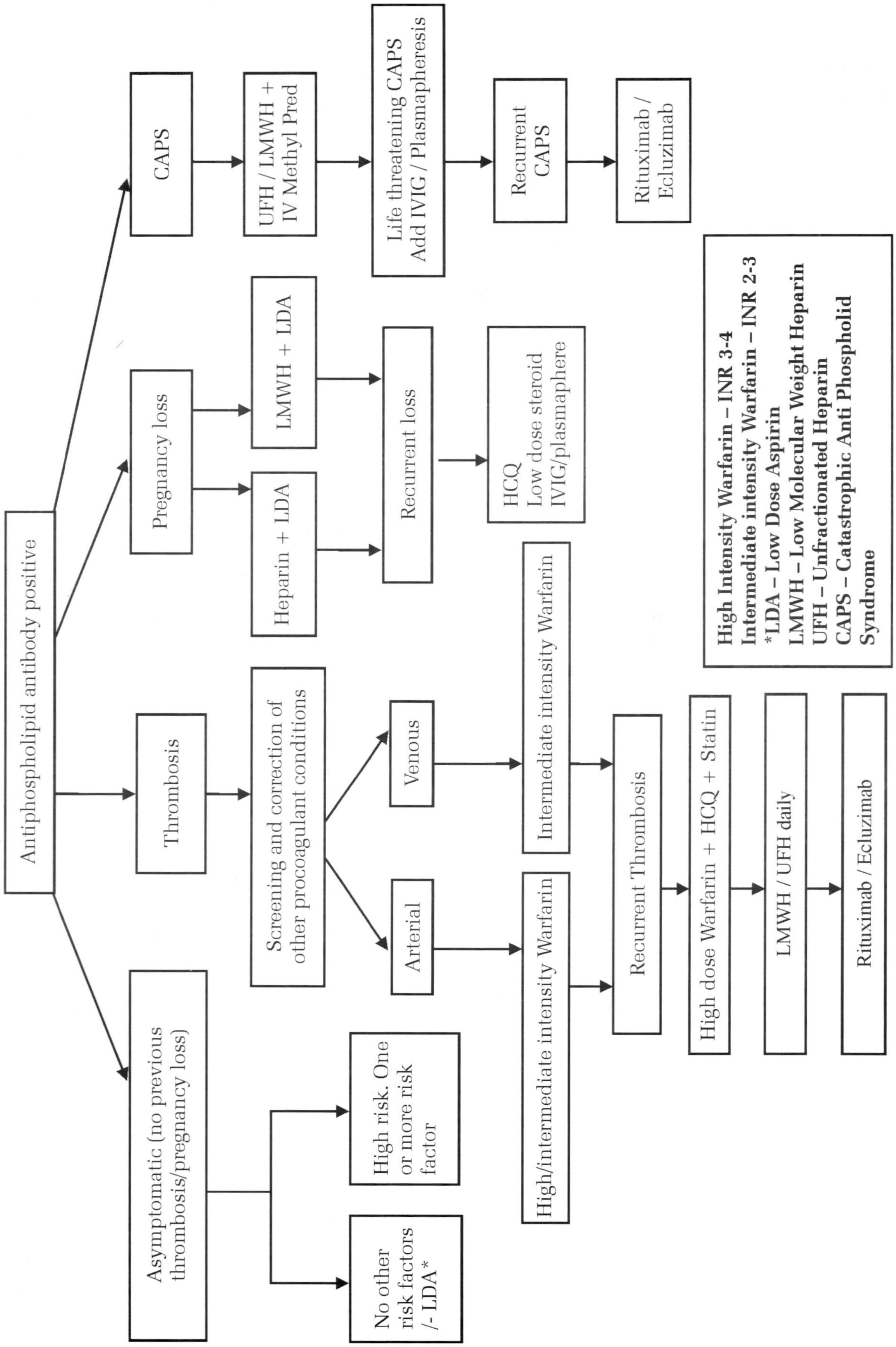
Antiphospholipid antibody positive
Asymptomatic (no previous thrombosis/pregnancy loss)
No other risk factors /- LDA*
High risk. One or more risk factor
High/intermediate intensity Warfarin
Thrombosis
Screening and correction of other procoagulant conditions
Arterial
Venous
Intermediate intensity Warfarin
Recurrent Thrombosis
High dose Warfarin + HCQ + Statin
LMWH / UFH daily
Rituximab / Ecluzimab
Pregnancy loss
Heparin + LDA
LMWH + LDA
Recurent loss
HCQ
Low dose steroid
IVIG/plasmaphere
CAPS
UFH / LMWH + IV Methyl Pred
Life threatening CAPS Add IVIG / Plasmapheresis
Recurrent CAPS
Rituximab / Ecluzimab
High Intensity Warfarin – INR 3-4
Intermediate intensity Warfarin – INR 2-3
*LDA – Low Dose Aspirin
LMWH – Low Molecular Weight Heparin
UFH – Unfractionated Heparin
CAPS – Catastrophic Anti Phospholid Syndrome

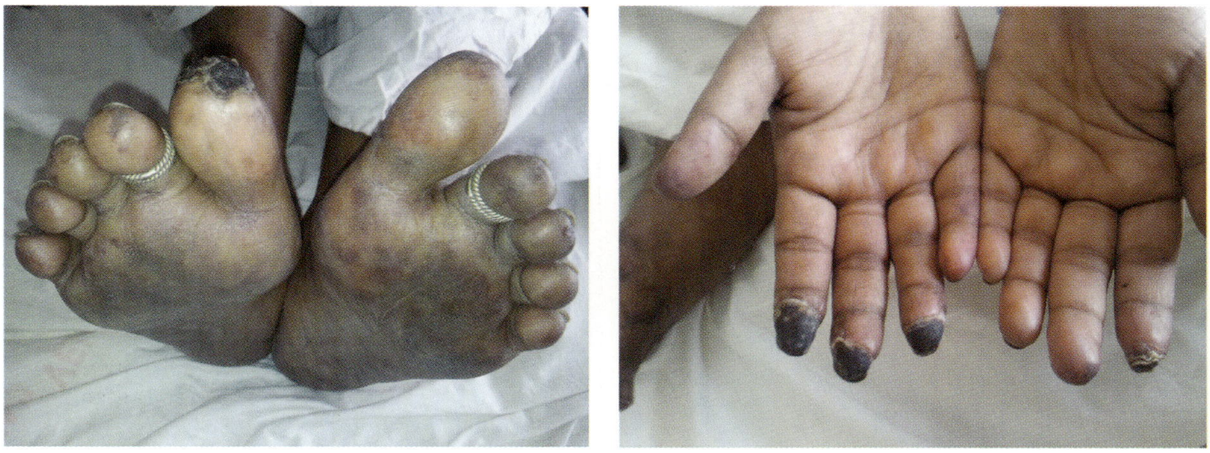

34 year old female with gangrene due to APLA syndrome

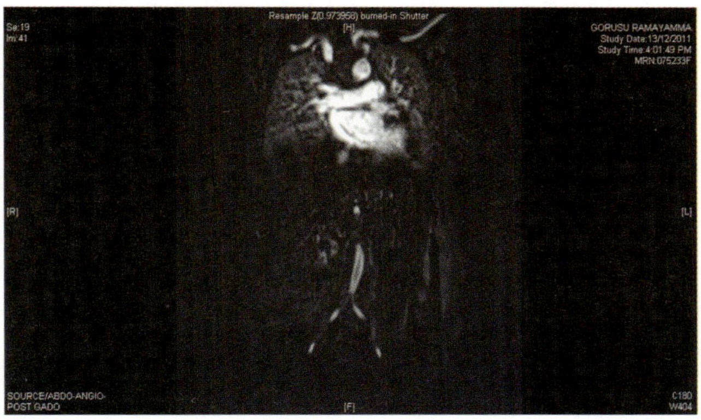

MR angio of a patient with APLA syndrome with mid aortic syndrome

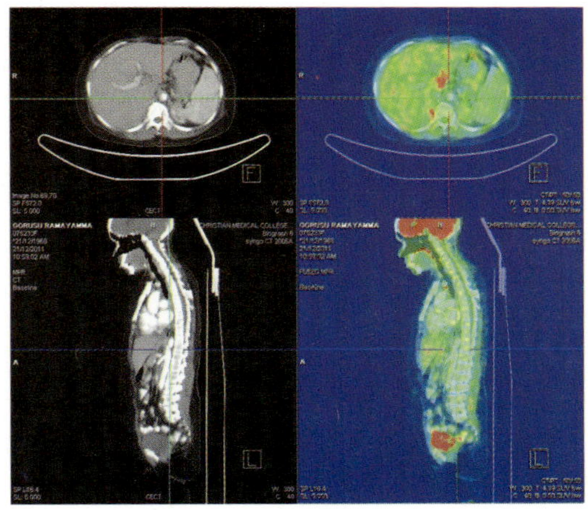

PET CT scan of the same patient showing absent uptake ; distinguishing from Takayasu arteritis syndrome

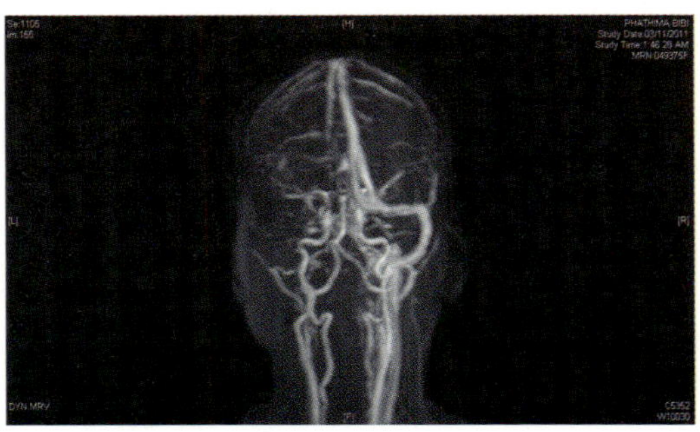

MRV of a patient with anti phospholipid syndrome with left transverse and sigmoid dural venous sinuses. Partial thrombus is also "seen in the right transverse sinus".

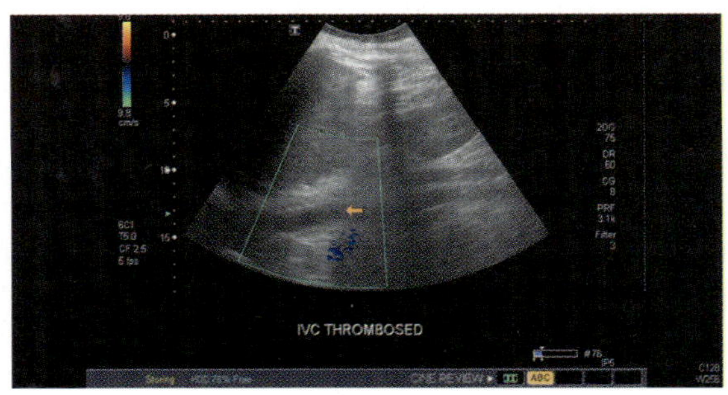

A patient with APLA with thrombosed inferior venacava

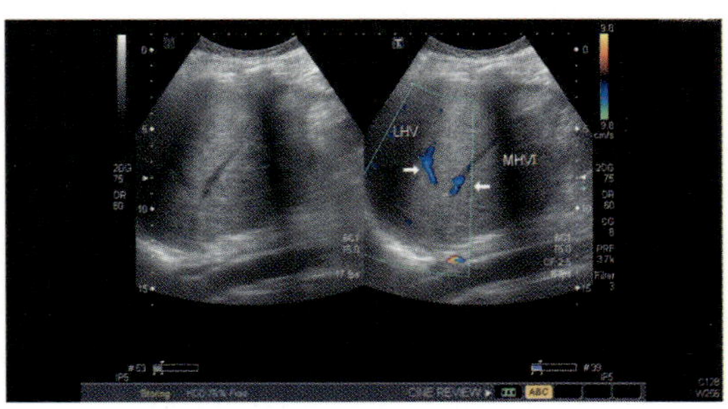

Partial occlusion of left hepatic and main hepatic vein in the same patient

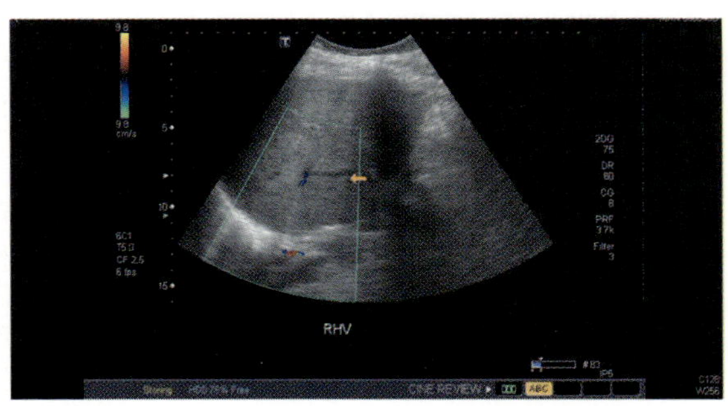

Irregularity of wall of right hepatic vein indicating chronic occlusion

Further Reading :

1. Wasserman A, Neisser A, Bruck C. Eineserodiagnostiche reaction bei syphilis. Dtsch Med Wochenschr 1906; 32: 745-49.
2. Pangborn MC. Cardiolipin and its application in a chemically purified antigen for the serodiagnosis of syphilis. Proc N Y State Assoc Public Health Lab. 1946; 26(1): 26-29.
3. Moore JE, Lutz WB. The natural history of systemic lupus erythematosus: an approach to its study through chronic biologic false positive reactors. J Chronic Dis. 1955; 1(3): 297-316.
4. Conley CL, Hartmann RC. A hemorrhagic disorder caused by circulating anticoagulant in patients with disseminated lupus erythematosus. J Lab Clin Invest 1952; 31: 621-22.
5. Bowie EJ, Thompson JH Jr, Pascuzzi CA, Owen CA Jr. Thrombosis in systemic lupus erythematosus despite circulating anticoagulants. J Lab Clin Med 1963; 62: 416-30.
6. Thiagarajan P, Shapiro SS, De Marco L. Monoclonal immunoglobulin M lambda coagulation inhibitor with phospholipid specificity. Mechanism of lupus anticoagulant. J Clin Invest 1980; 66: 397-405.
7. Hughes GRV. The anticardiolipin syndrome. Clin Exp Rheumatol 1985; 3: 285-86.
8. Wilson WA, Gharavi AE, Koike T, Lockshin MD, Branch DW, Piette JC, BreyR,Derksen R, Harris EN, Hughes GR, Triplett DA, Khamashta MA. International consensus statement on preliminary classification criteria for definite antiphospholipid syndrome: report of an international workshop. Arthritis Rheum. 1999; 42(7): 1309-11.
9. Miyakis S, Lockshin MD, Atsumi T, Branch DW, Brey RL, Cervera R, Derksen RH, DE Groot PG, Koike T, Meroni PL, Reber G, Shoenfeld Y, Tincani A, Vlachoyiannopoulos PG, Krilis SA. International consensus statement on an update of the classification criteria for definite antiphospholipid syndrome (APS). J Thromb Haemost. 2006; 4(2): 295-306.
10. Roubey RA. Antiphospholipid syndrome: antibodies and antigens. Curr Opin Hematol. 2000; 7(5): 316-20.
11. Alessandri C, Conti F, Pendolino M, Mancini R, Valesini G. New autoantigens in the antiphospholipid syndrome. Autoimmun Rev. 2011; 10(10): 609-16. Review.
12. L.L. Horstman, W. Jy, C.J. Bidot, Y.S. Ahn, R.E. Kelley, R. Zivadinov et al. Antiphospholipid antibodies: paradigm in transition. J Neuro inflammation, 6 (2009), p. 3.
13. C. Petrovas, P.G. Vlachoyiannopoulos, T. Kordossis, H.M. Moutsopoulos Anti-phospholipid antibodies in HIV infection and SLE with or without anti-phospholipid syndrome: comparisons of phospholipid specificity, avidity and reactivity with beta2-GPI. J Autoimmun, 13 (1999), 347-55.
14. Shi W, Krilis SA, Chong BH, Gordon S, Chesterman CN. Prevalence of lupus anticoagulant and anticardiolipin antibodies in a healthy population. Aust N Z J Med. 1990; 20(3): 231-36.
15. Vila P, Hernández MC, López-Fernández MF, Batlle J. Prevalence, follow-up and clinical significance of the anticardiolipin antibodies in normal subjects. Thromb Haemost. 1994; 72(2): 209-13.
16. Lockwood CJ, Romero R, Feinberg RF, ClyneLP, Coster B, Hobbins JC. The prevalence and biologic significance of lupus anticoagulant and anticardio-lipin antibodies in a general obstetric population. Am J Obstet Gynecol. 1989; 161: 369-73.
17. Roldan V, Lecumberri R, Muñoz-Torrero JF, Vicente V, Rocha E, Brenner B, Monreal M; RIETE Investigators. Thrombophilia testing in patients with venous thromboembolism. Findings from the RIETE registry. Thromb Res. 2009; 124(2).
18. Cervera R, Piette JC, Font J, Khamashta MA et al; Euro-Phospholipid Project Group. Antiphospholipid syndrome: clinical and immunologic manifestations and patterns of disease expression in a cohort of 1,000 patients. Arthritis Rheum. 2002; 46(4): 1019-27.
19. Biggioggero M, P.L. Meroni P L. Autoimmunity Reviews 9 (2010); Thegeoepidemiology of the antiphospholipid antibody syndrome. A299-A304.
20. Petri M. Update on anti-phospholipid antibodies in SLE: the Hopkins' Lupus Cohort. Lupus. 2010; 19(4): 419-23.
21. Ostrowski RA, Robinson JA. Antiphospholipid antibody syndrome and autoimmune diseases. Hematol Oncol Clin North Am. 2008; 22(1): 53-65, vi. Review.
22. Sherer Y, Gerli R, Vaudo G. Prevalence of antiphospholipid and antioxidized low-density lipoprotein antibodies in rheumatoid arthritis. Ann N Y Acad Sci. 2005; 1051: 299-303.
23. Jonsdottir T,Forslid. J, A. van Vollenhoven et al. Treatment with tumour necrosis factor alpha antagonists in patients with rheumatoid arthritis induces anticardiolipinantibodies. Ann Rheum Dis, 63(9) (2004), 1075-78.
24. Marie I, Jouen F, Hellot MF, Levesque H. Anticardiolipin and anti-beta2 glycoprotein I antibodies and lupus-like anticoagulant: prevalence and significance in systemic sclerosis. Br J Dermatol. 2008; 158(1): 141-14.
25. Pasoto SG, Chakkour HP, Natalino RR, Viana VS, Bueno C, Lianza AC, de Andrade JL, Neto ML, Fuller R, Bonfa E. Lupus anticoagulant: a marker for stroke and venous thrombosis in primary Sjögren's syndrome. Clin Rheumatol. 2012; 31(9): 1331-38.
26. Kawakami T, Yamazaki M, Mizoguchi M, Soma Y. High titer of anti-phosphatidylserine-prothrombin complex antibodies in patients with cutaneous polyarteritisnodosa. Arthritis Rheum. 2007; 57(8): 1507-13.
27. Reinstein E, Shoenfeld Y. Antiphospholipid syndrome and cancer. Clin Rev Allergy Immunol. 2007; 32(2): 184-87 Review.
28. Tincani A, Taraborelli M, Cattaneo R. Antiphospholipid antibodies and malignancies. Autoimmun Rev. 2010; 9(4): 200-02.
29. A. Tincani, M. Taraborelli, R. Cattaneo Antiphospholipid antibodies and malignancies Autoimmunity Reviews, 2010, (9) 200-02.
30. C. Font, L. Vidal, G. et al. Solid cancer, antiphospholipid antibodies, and venous thromboembolism. Autoimmunity Reviews, 2011; 10(4), 222-27.
31. Lechner K, Simonitsch I, Haselböck J, Jäger U, Pabinger I. Acquired immune-mediated thrombophilia in lymphoproliferative disorders. Leuk Lymphoma. 2011; 52(10): 1836-43.
32. Bairey O, Blickstein D, Monselise Y. Antiphospholipid antibodies may be a new prognostic parameter in aggressive non-Hodgkin's lymphoma. Eur J Haematol. 2006; 76(5): 384-91.
33. Zhou YS, Ying ZT, Li R, Zhu J, Li ZG. Clinical and Immunological relevance ofantiphospholipid antibodies in patients with lymphoma. Zhonghua Yi XueZaZhi. 2011; 91(37): 2607-10. (pubmed abstract)

34. Sène D, Jean-Charles Piette J C, Cacoub P.Antiphospholipid antibodies, antiphospholipid syndrome and infections. Autoimmunity Reviews, 2008; 7(4), 272-77.

35. Ramos-Casals M, Cervera R, Lagrutta M, Medina F, Garcia Carrasco M, de la Red G, et al. Clinical features related to antiphospholipid syndrome in patients with chronic viral infections (hepatitis C virus/HIV infection): description of 82 cases. Clin Infect Dis 2004; 38: 1009-16.

36. Martina B, Meroni P L. The geoepidemiology of the antiphospholipid antibody syndrome. Autoimmunity Reviews, Volume 9, Issue 5, March 2010, A299-A304.

37. Arvieux J, Renaudineau Y, Mane I, Perraut R, Krilis SA,Youinou P. Distinguishing features of anti-beta2 glycoprotein I antibodies between patients with leprosy and the antiphospholipidsyndrome. Thromb Haemost 2002; 87: 599-605.

38. Dlott JS, Roubey RA. Drug-induced lupus anticoagulants and antiphospholipid antibodies. Curr Rheumatol Rep. 2012 Feb; 14(1): 71-78. Review.

39. Becker JC, Winkler B, Klingert S, Brocker EB. Antiphospholipid syndrome associated with immunotherapy for patients with melanoma. Cancer. 1994; 73(6): 1621-24.

40. Visvanathan S, Wagner C, Smolen J, St Clair EW, Hegedus R, Baker D, et al. IgG and IgManticardiolipin antibodies following treatment with infliximab plus methotrexate in patients with early rheumatoid arthritis. Arthritis Rheum. 2006; 54(9): 2840-44.

41. Giannakopoulos B, Mirarabshahi P, Krilis SA. New insights into the biology and pathobiology of beta2-glycoprotein I. Curr Rheumatol Rep. 2011; 13(1): 90-95. Review.

42. Passam FH, Giannakopoulos B, Mirarabshahi P, Krilis SA. Molecular pathophysiology of the antiphospholipid syndrome: the role of oxidative post-translational modification of beta 2 glycoprotein I. J Thromb Haemost. 2011: 9 Suppl 1: 275-82. Review.

43. McNeil HP, Simpson RJ, Chesterman CN, Krilis SA. Anti-phospholipid antibodies are directed against a complex antigen that includes a lipid-binding inhibitor of coagulation: beta 2 - glycoprotein I (apolipoprotein H). Proc Natl Acad Sci USA; 87: 4120-24.

44. Takeuchi R, Atsumi T, Ieko M, Takeya H, Yasuda S, Ichikawa K, TsutsumiA, Suzuki K, Koike T. Coagulation and fibrinolytic activities in 2 siblings with beta (2) - glycoprotein I deficiency. Blood. 2000; 96(4): 1594-95.

45. Yasuda S, Tsutsumi A, Chiba H, Yanai H, Miyoshi Y, Takeuchi R, Horita T, Atsumi T, Ichikawa K, Matsuura E, Koike T. beta(2)-glycoprotein I deficiency: prevalence, genetic background and effects on plasma lipoprotein metabolism and hemostasis. Atherosclerosis. 2000; 152(2): 337-46.

46. Ioannou Y. The Michael Mason Prize: Pathogenic antiphospholipid antibodies, stressed out antigens and the deployment of decoys. Rheumatology (Oxford). 2012; 51(1): 32-36. Review.

47. Ioannou Y, Zhang JY, Qi M et al. Novel assays of thrombogenic pathogenicity in the antiphospholipid syndrome based on the detection of molecular oxidative modification of the major autoantigen β2-glycoprotein I. Arthritis Rheum. 2011; 63(9): 2774-82.

48. Chiu WC, Chen CJ, Lee TS, Chen ZJ, Ke PH, Chiang AN. Oxidative stress enhances AP-1 and NF-k B-mediated regulation of β(2)-glycoprotein I gene expression in hepatoma cells. J Cell Biochem. 2010; 111(4): 988-98.

49. Sorice M, Circella A, Cristea IM, Garofalo T, Renzo LD, Alessandri C, et al. Cardiolipin and its metabolites move from mitochondria to other cellular membranes during death receptor-mediated apoptosis. Cell Death Differ 2004; 00: 1-13.

50. Sorice M, Circella A, Misasi R, Pittoni V, Garofalo T, Cirelli A, et al. Cardiolipin on the surface of apoptotic cells as a possible trigger for antiphospholipids antibodies. Clin Exp Immunol 2000; 122: 277-84.

51. Alessandri C, Sorice M, Bombardieri M, Conigliaro P, Longo A, Garofalo T, et al. Antiphospholipid reactivity against cardiolipin metabolites occurring during endothelial cell apoptosis. Arthritis Res Ther 2006; 8: R180.

52. Lutters BC, Derksen RH, Tekelenburg WL, Lenting PJ, Arnout J, de Groot PG.Dimers of beta 2-glycoprotein I increase platelet deposition to collagen via interaction with phospholipids and the apolipoprotein E receptor 2. J Biol Chem. 2003; 278(36): 33831-38.

53. Romay-Penabad Z, Aguilar-Valenzuela R, Urbanus RT etal. Apolipoprotein E receptor 2 is involved in the thrombotic complications in a murine model of the antiphospholipid syndrome. Blood. 2011; 117(4): 1408-14.

54. Alard JE, Gaillard F, Daridon C, Shoenfeld Y, Jamin C, Youinou P. TLR2 is one of the endothelial receptors for beta 2-glycoprotein I. J Immunol. 2010; 185(3): 1550-57.

55. Meroni PL, Borghi MO, Raschi E, Tedesco F. Pathogenesis of antiphospholipid syndrome: understanding the antibodies. Nat Rev Rheumatol. 2011; 7(6): 330-39. Review.

56. Ramesh, S. et al.Antiphospholipidantibodies promote leukocyte-endothelial cell adhesion and thrombosis in mice by antagonizing eNOS via ? 2GPI and apoER2. J. Clin. Invest. 121, 120-31.

57. Vega-Ostertag, M. E. et al.Role of p38 mitogen-activated protein kinase in antiphospholipid antibody-mediated thrombosis and endothelial cell activation. J. Thromb. Haemost. 5, 1828-34.

58. Yang, Y. H. et al. Identification of anti-prothrombin antibodies in the anti-phospholipid syndrome that display the prothrombinase activity. Rheumatology (Oxford) 49, 34-42.

59. Zuily S, AitAissa K, Membre A, Regnault V, Lecompte T, Wahl D. Thrombin generation in antiphospholipid syndrome. Lupus. 2012; 21(7): 758-60.

60. Nayar R. & Lage J. M. Placental changes in a first trimester missed abortion in maternal systemic lupus erythematosus with antiphospholipid syndrome; a case report and review of the literature. Hum. Pathol. 1996; 27, 201-06.

61. Ogishima D, Matsumoto T, Nakamura Y, Yoshida K, Kuwabara Y. Placental pathology in systemic lupus erythematosus with antiphospholipidantibodies. Pathol Int. 2000; 50(3): 224-29.

62. Magid MS, Kaplan C, Sammaritano LR, Peterson M, Druzin ML, Lockshin MD. Placental pathology in systemic lupus erythematosus: a prospective study. Am J Obstet Gynecol. 1998; 179(1): 226-34.

63. Rand J. H., Wu X. X., Quinn A. S. & Taatjes D. J. The annexin A5-mediated pathogenic mechanism in the antiphospholipid syndrome: role in pregnancy losses and thrombosis. Lupus. 2010; 19, 460-69.

64. Holers, V. M. et al.Complement C3 activation is required for antiphospholipid antibody-induced fetal loss. J. Exp. Med. 2002. 195, 211-20.

65. Mulla MJ, Brosens JJ, Chamley LW, Giles I, Pericleous C, Rahman A, Joyce SK,Panda B, Paidas MJ, Abrahams VM. Antiphospholipid antibodies induce a pro-inflammatory response in first trimester trophoblast via the TLR4/MyD88 pathway. Am J Reprod Immunol. 2009 Aug; 62(2): 96-111. 68.
66. Thurman JM. et al. A novel inhibitor of the alternative complement pathway prevents antiphospholipid antibody-induced pregnancy loss in mice. Mol. Immunol. 2010; 42: 87-97.
67. Girardi G., Redecha P. & Salmon J. E. Heparin prevents antiphospholipid antibody-induced fetal loss by inhibiting complement activation. Nat. Med. 2004; 10: 1222-26.
68. Di Simone, N. et al.Pathogenic role of anti- β2-glycoprotein I antibodies in antiphospholipid-associated fetal loss: characterisation of β2-glycoprotein I binding to trophoblast cells and functional effects of anti-β2-glycoprotein I antibodies in vitro. Ann. Rheum. Dis. 2005; 64: 462-67.
69. Chen Q, Viall C, Kang Y, Liu B, Stone P, Chamley L. Antiphospholipid antibodies increase non-apoptotic trophoblast shedding: a contribution to the pathogenesis of pre-eclampsia in affected women? Placenta. 2009; 30(9): 767-73. Placenta. 2009.06.008.
70. Mulla MJ, Myrtolli K, Brosens JJ, Chamley LW, Kwak-Kim JY, Paidas MJ, AbrahamsVM. Antiphospholipid antibodies limit trophoblast migration by reducing IL-6 production and STAT3 activity. Am J ReprodImmunol. 2010; 63(5): 339-48.
71. Robertson SA, Roberts CT, van Beijering E, Pensa K, Sheng Y, Shi T, Krilis SA. Effect of beta 2-glycoprotein I null mutation on reproductive outcome and antiphospholipid antibody-mediated pregnancy pathology in mice. Mol Hum Reprod. 2004; 10(6): 409-16.
72. Shoenfeld Y. et al. Infectious origin of the antiphospholipid syndrome. Ann. Rheum. Dis. 2006; 65: 2-6.
73. Fischetti F. et al.Thrombus formation induced by antibodies to β 2-glycoprotein I is complement dependent and requires a priming factor. Blood. 2005; 106: 2340-46.
74. Ginsburg KS, Liang MH, Newcomer L, Goldhaber SZ, Schur PH, Hennekens CH, Stampfer MJ. Anticardiolipin antibodies and the risk for ischemic stroke and venous thrombosis. Ann Intern Med. 1992; 117(12): 997-1002.
75. Cervera R, Piette JC, Font J, Khamashta MA, et al. Euro-Phospholipid Project Group. Antiphospholipid syndrome: clinical and immunologic manifestations and patterns of disease expression in acohort of 1,000 patients. Arthritis Rheum. 2002; 46(4): 1019-27.
76. Urbanus RT, Siegerink B, Roest M, et al. Antiphospholipid antibodies and risk of myocardial infarction and ischaemic stroke in young women in the RATIO study: a case-control study. Lancet Neurol 2009; 8: 998-1005.
77. Mishra MN, Rohatgi S. Antiphospholipid antibodies in young Indian patients with stroke. J Postgrad Med. 2009; 55(3): 161-64.
78. Sanna G, Bertolaccini ML, Cuadrado MJ, et al. Central nervous system involvement in the antiphospholipid (Hughes) syndrome. Rheumatology (Oxford) 2003; 42: 200-13.
79. Muscal E, Brey RL. Antiphospholipid syndrome and the brain in pediatric and adult patients. Lupus. 2010; 19(4): 406-11. Review.
80. Tektonidou MG. Renal involvement in the antiphospholipid syndrome (APS)-APSnephropathy. Clin Rev Allergy Immunol. 2009 ; 36(2-3): 131-40.
81. Amigo MC. Kidney disease in antiphospholipid syndrome. Rheum Dis Clin North Am. 2006; 32(3): 509-22. Review.
82. Tenedios F, Erkan D, Lockshin MD. Cardiac manifestations in the antiphospholipid syndrome. Rheum Dis Clin North Am. 2006; 32(3): 491-507. Review.
83. Scheiman Elazary A, Cohen MJ, Aamar S, Dranitzki Z, Tayer-Shifman O, Mevorach D, Ben-Chetrit E, Naparstek Y. Pulmonary hemorrhage in antiphospholipid antibody syndrome. J Rheumatol. 2012; 39(8): 1628-31.
84. Kim NH, Lang IM. Risk factors for chronic thromboembolic pulmonary hypertension. EurRespir Rev. 2012.
85. Weinstein S, Piette W. Cutaneous manifestations of antiphospholipid antibody syndrome. Hematol Oncol Clin North Am. 2008; 22(1): 67-77. Review.
86. Francès C, Niang S, Laffitte E, Pelletier Fl, Costedoat N, Piette JC. Dermatologic manifestations of the antiphospholipid syndrome: two hundred consecutive cases. Arthritis Rheum. 2005; 52(6): 1785-93.
87. Wang JG, Xie QB, Yang NP, Yin G. Primary antiphospholipid antibody syndrome: acase with bilateral sudden sensorineural hearing loss. Rheumatol Int. 2009; 29(4): 467-68.
88. Uthman I, Godeau B, Taher A, Khamashta M. The hematologic manifestations of the antiphospholipid syndrome. Blood Rev. 2008; 22(4): 187-94.
89. Font J, Jiménez S, Cervera R, García-Carrasco M, Ramos-Casals M, Campdelacreu J, Ingelmo M. Splenectomy for refractory Evans' syndrome associated with antiphospholipid antibodies: report of two cases. Ann Rheum Dis. 2000; 59(11): 920-23. Review.
90. Mazodier K, Arnaud L, Mathian A, Costedoat-Chalumeau N, Haroche J, Frances C,Harlé JR, Pernod G, Lespessailles E, Gaudin P, Charlanne H, Hachulla E, Niaudet P, Piette JC, Amoura Z. Lupus anticoagulant-hypoprothrombinemia syndrome: report of 8 cases and review of the literature. Medicine (Baltimore).
91. Mehdi AA, Salti I, Uthman I. Antiphospholipid syndrome: endocrinologicmanifestations and organ involvement. Semin Thromb Hemost. 2011; 37(1): 49-57. Review.
92. Yang P, Kruh JN, Foster CS. Antiphospholipid antibody syndrome. Curr Opin Ophthalmol. 2012; 23(6): 528-32.
93. Lockshin MD, Sammaritano LR, Schwartzman S. Validation of the Sapporo criteria for antiphospholipid syndrome. Arthritis Rheum. 2000; 43(2): 440-43.
94. Miyakis S, Lockshin MD, Atsumi T, Branch DW, Brey RL, Cervera R, Derksen RH, DE Groot PG, Koike T, Meroni PL, Reber G, Shoenfeld Y, Tincani A, Vlachoyiannopoulos PG, Krilis SA. International consensus statement on an update of the classification criteria for definite antiphospholipid syndromee (APS). J Thromb Haemost. 2006; 4(2): 295-30.
95. Bobba RS, Johnson SR, Davis AM. A review of the sapporo and revised Sapporo criteria for the classification of antiphospholipid syndrome. Where do there vised sapporo criteria add value? J Rheumatol. 2007; 34(7): 1522-27. Review.
96. Pengo V, Tripodi A, Reber G, Rand JH, Ortel TL, Galli M, De Groot PG; Subcommittee on Lupus Anticoagulant/Antiphospholipid Antibody of the Scientific and Standardisation Committee of the International Society on Thrombosis and Haemostasis. Update of the guidelines for lupus anticoagulant detection. Subcommittee on Lupus Anticoagulant/Antiphospholipid Antibody of the Scientific and Standardisation Committee of the International Society on Thrombosis and Haemostasis. J Thromb Haemost. 2009; 7(10): 1737-40.

97. Lakos G. et al. International consensus guidelines on anticardiolipin and anti β2 glycoprotein I testing: a report from the APL Task Force at the 13th International Congress on Antiphospholipid Antibodies. Arthritis Rheum. http://dx.doi.org/10.1002/art.33349.

98. de Groot PG, Urbanus RT. Screening: Guidelines for antiphospholipid antibody detection. Nat Rev Rheumatol. 2011; 8(3): 125-26.

99. Galli M et al. (2003) Lupus anticoagulants are stronger risk factors for thrombosis than anticardiolipin antibodies in the antiphospholipid syndrome: a systematic review of the literature. Blood; 101: 1827-32.

100. deLaat B, Mertens K, de Groot PG. Mechanisms of disease: antiphospholipid antibodies-from clinical association to pathologic mechanism. Nat Clin Pract Rheumatol. 2008; 4(4): 192-99. Review.

101. Pengo V, Biasiolo A, Pegoraro C, Cucchini U, Noventa F, Iliceto S. Antibody profiles for the diagnosis of antiphospholipid syndrome. Thromb Haemost. 2005; 93(6): 1147-52.

102. Danowski A, de Azevedo MN, de Souza Papi JA, Petri M. Determinants of risk for venous and arterial thrombosis in primary antiphospholipid syndrome and in antiphospholipid syndrome with systemic lupus erythematosus. J Rheumatol. 2009; 36(6): 1195-99.

103. Urbanus RT, de Groot PG. Antiphospholipid antibodies – we are not quite there yet. Blood Rev. 2011; 25(2): 97-106. Review.

104. Tincani A, Andreoli L, Casu C, Cattaneo R, Meroni P. Antiphospholipid antibody profile: implications for the evaluation and management of patients. Lupus. 2010; 19(4): 432-35.

105. Alessandri C, Conti F, Pendolino M, Mancini R, Valesini G. New autoantigens in the antiphospholipid syndrome. Autoimmun Rev. 2011; 10(10): 609-16.

106. Bertolaccini ML. Antibodies to prothrombin. Lupus. 2012; 21(7): 729-31. Review.

107. Bertolaccini ML, Sciascia S, Murru V, Garcia-Fernandez C, Sanna G, Khamashta MA. Prevalence of antibodies to prothrombin in solid phase (aPT) and tophosphatidylserine-prothrombin complex (aPS/PT) in patients with and without lupus anticoagulant. Thromb Haemost. 2012; 109(2).

108. tsumi T, Ieko M, Bertolaccini ML, Ichikawa K, Tsutsumi A, Matsuura E, KoikeT. Association of autoantibodies against the phosphatidylserine-prothrombincomplex with manifestations of the antiphospholipid syndrome and with thepresence of lupus anticoagulant. Arthritis Rheum. 2000; 43(9): 1982-93.

109. De Laat B et al. (2005) IgG antibodies that recognize epitope Gly40-Arg43 in domain I of β2-glycoprotein I cause LAC and their presence correlates strongly with thrombosis. Blood 2005; 105: 1540-45.

110. De Laat HB et al. (2004)β2-glycoprotein I-dependent lupus anticoagulant highly correlates with thrombosis in the antiphospholipid syndrome. Blood; 104: 3598-3602.

111. amarkos M, Asherson RA, Loizou S. The clinical significance of IgA antiphospholipid antibodies. J Rheumatol. 2001; 28(4): 694-97. Review.

112. Drenkard C, Sánchez-Guerrero J, Alarcón-Segovia D. Fall in antiphospholipid anti-body at time of thromboocclusive episodes in systemic lupus erythematosus. JRheumatol 1989; 16: 614-17.

113. Cervera R, Conti F, Doria A, Iaccarino L, Valesini G. Does seronegative antiphospholipid syndrome really exist? Autoimmun Rev. 2012; 11(8): 581-84. Review.

114. Pérez-Vázquez ME, Cabiedes J, Cabral AR, et al. Decrease in serum antiphospholipid antibody levels upon development of nephrotic syndrome in patients with systemic lupus erythematosus: relationship to urinary loss of IgG and other factors. Am J Med 1992; 92: 357-62.

115. Silveira LH, Jara LJ, Espinoza LR. Transient disappearance of serum antiphospholipid antibodies can also be due to prednisone therapy. Clin Exp Rheumatol 1996; 14: 217-26.

116. Cervera R, Espinosa G. Update on the catastrophic antiphospholipid syndrome and the "CAPS Registry". Semin Thromb Hemost. 2012; 38(4): 333-38.

117. Bucciarelli S, Cervera R, Espinosa G, Gómez-Puerta JA, Ramos-Casals M, Font J. Mortality in the catastrophic antiphospholipid syndrome: causes of death and prognostic factors. Autoimmun Rev. 2006; 6(2): 72-75.

118. Cervera R, Asherson RA. Catastrophic antiphospholipid syndrome: therapeutic developments. Expert Rev Clin Immunol. 2007; 3(3): 277-85.

119. Saidi S, Mahjoub T, Almawi WY. Lupus anticoagulants and anti-phospholipid antibodies as risk factors for a first episode of ischemic stroke. J Thromb Haemost. 2009; 7(7): 1075-80.

120. Vila P, Hernandez MC, Lopez-Fernandez MF, Batlle I. Prevalence, follow-up and clinical significance of the anticardiolipin antibodies in normal subjects. Thromb Haemost. 1994; 72: 209-12.

121. Long AA, Ginsberg JS, Brill-Edwards P, et al. The relationship of antiphospholipid antibodies to thromboembolic disease in systemic lupus erythematosus: a cross-sectional study. Thromb Haemost. 1991; 66: 520-24.

122. The APASS Writing Committee. Antiphospho-lipid antibodies and subsequent thrombo-occlusive events in patients with ischemic stroke. JAMA. 2004; 291: 576-84.

123. Ruiz-Irastorza G, Cuadrado MJ, Ruiz-Arruza I, Brey R, Crowther M, Derksen R, Erkan D, Krilis S, Machin S, Pengo V, Pierangeli S, Tektonidou M, Khamashta M. Evidence-based recommendations for the prevention and long-term management of thrombosis in antiphospholipid antibody-positive patients: report of a task force at the 13th International Congress on antiphospholipid antibodies. Lupus. 2011; 20(2): 206-18.

124. ErkanD, Merrill JT, Yazici Y, Sammaritano L, Buyon JP, Lockshin MD. High thrombosis rate after fetal loss in antiphospholipid syndrome: effective prophylaxis.

125. Khamashta MA, Cuadrado MJ, Mujic F, Taub MA, Hunt BJ, Hughes GR. The management of thrombosis in the antiphospholipid antibody syndrome. N Engl J Med. 1995; 332: 993-97.

126. Finazzi G, Brancaccio V, Moia M, et al. Natural history and risk factors for thrombosis in 360 patientswith antiphospholipid antibodies: a four-year prospective study from the Italian Registry. Am J Med. 1996; 100: 530-36.

127. CrowtherMA, Ginsberg JS, Julian J, et al. A comparison of two intensities of warfarin for the prevention of recurrent thrombosis in patients with the antiphospholipid antibody syndrome. N Engl J Med. 2003; 349: 1133-38.

128. Finazzi G, Marchioli R, Brancaccio V, et al. A randomized clinical trial of high-intensity warfarin vs conventional antithrombotic therapy for the prevention of recurrent thrombosis in patients with the antiphos-pholipid syndrome (WAPS). J Thromb Haemost. 2005; 3: 848-53.

129. Lim W, Crowther MA, Eikelboom JW. Management of antiphospholipid antibody syndrome: a systematic review. JAMA. 2006; 295(9): 1050-57. Review.
130. Mohr JP, Thompson JL, Lazar RM, et al. A comparison of warfarin and aspirin for the prevention of recurrent ischemic stroke. N Engl J Med. 2001; 345: 1444-51.
131. Rand JH, Wu XX, Quinn AS, Chen PP, Hathcock JJ, Taatjes DJ. Hydroxychloroquine directly reduces the binding of antiphospholipid antibody-beta2-glycoprotein I complexes to phospholipid bilayers. Blood. 2008; 112(5): 1687-95.
132. Petri M. Use of hydroxychloroquine to prevent thrombosis in systemic lupus erythematosus and in antiphospholipid antibody-positive patients. Curr Rheumatol Rep. 2011; 13(1): 77-80. Review.
133. Broder A, Putterman C. Hydroxychloroquine Use Is Associated with Lower Odds of Persistently Positive Antiphospholipid Antibodies and/or Lupus Anticoagulant in Systemic Lupus Erythematosus. J Rheumatol. 2013; 40(1): 30-33.
134. Ferrara DE, Swerlick R, Casper K, Meroni PL, Vega-Ostertag ME, Harris EN,Pierangeli SS. Fluvastatin inhibits up-regulation of tissue factor expression by antiphospholipid antibodies on endothelial cells. J Thromb Haemost. 2004; 2(9): 1558-63.
135. Glynn RJ, Danielson E, Fonseca FA, Genest J, Gotto AM Jr, Kastelein JJ, Koenig W, Libby P, Lorenzatti AJ, MacFadyen JG, Nordestgaard BG, Shepherd J, Willerson JT, Ridker PM. A randomized trial of rosuvastatin in the prevention of venous thromboembolism. N Engl J Med. 2009 Apr 30; 360(18): 1851-61.
136. Galarza-Maldonado C, Kourilovitch MR, Pérez-Fernández OM, Gaybor M, Cordero C, Cabrera S, Soroka NF. Obstetric antiphospholipid syndrome. Autoimmun Rev. 2012; 11(4): 288-95. Review.
137. Ernest JM, Marshburn PB, Kutteh WH. Obstetric antiphospholipid syndrome: an update on pathophysiology and management. Semin Reprod Med. 2011; 29(6): 522-39. Review.
138. Kutteh WH. Antiphospholipid antibody-associated recurrent pregnancy loss: treatment with heparin and low-dose aspirin is superior to low-dose aspirin alone. Am J Obstet Gynecol. 1996; 174: 1584-89.
139. Rai R, Cohen H, Dave M, Regan L. Randomised controlled trial of aspirin and aspirin plus heparin in pregnant women with recurrent miscarriage associated with phospholipid antibodies (or antiphospho-lipid antibodies). BMJ. 1997; 314: 253-57.
140. Empson M, Lassere M, Craig J, Scott J. Prevention of recurrent miscarriage for women with anti-phospholipid antibody or lupus anticoagulant.Cochrane Database Syst Rev. 2005; (2).
141. Fouda UM, Sayed AM, Ramadan DI, Fouda IM. Efficacy and safety of two doses of low molecular weight heparin (enoxaparin) in pregnant women with a history of recurrent abortion secondary to antiphospholipid syndrome. J Obstet Gynaecol. 2010; 30(8): 842-46.
142. Ruffatti A, Favaro M, Tonello M, De Silvestro G, Pengo V, Fais G, Suma V, Chiarelli S, Todesco S. Efficacy and safety of nadroparin in the treatment of pregnant women with antiphospholipid syndrome: a prospective cohort study. Lupus. 2005; 14(2): 120-28.
143. Scoble T, Wijetilleka S, Khamashta MA. Management of refractory anti-phospholipid syndrome. Autoimmun Rev. 2011 Sep; 10(11): 669-73. Review.
144. Bramham K, Thomas M, Nelson-Piercy C, Khamashta M, Hunt BJ. First-trimester low-dose prednisolone in refractory antiphospholipid antibody-related pregnancy loss. Blood. 2011; 23: 117(2).
145. Erkan D, Espinosa G, Cervera R. Catastrophic anti-phospholipid syndrome: updated diagnostic algorithms. Autoimmun Rev. 2010 Dec; 10(2): 74-79. Review.

SJOGREN'S SYNDROME 4

R. Porkodi

Sjogren's syndrome

Sjogren's Syndrome (SS) is a slowly progressive autoimmune inflammatory disease affecting primarily the exocrine glands and characterized by dry eyes and dry mouth. Secondary Sjogren's syndrome includes the above findings in association with another autoimmune disease like rheumatoid arthritis (RA) or systemic lupus erythematosus (SLE). Prevalence varies from 0.5 to 5 percent. This is a disease predominantly affecting middle aged females, though children and elderly are also affected.

Etiology and pathogenesis

Autoimmunity may be triggered by an environmental factor on a susceptible genetic background. Perpetuation of autoimmune reactivity becomes chronic through immuno regulatory mechanisms and the tissue lesions occur as a consequence of ongoing inflammatory process.

Immunogenetics

SS is associated with increased frequencies of HLA-B8, HLA-DW3 and HLA-DR_3[1]. In Caucasian individuals the HLA DRB1 * 0301 / DQA1 * 0501 / DQB1 *0201 haplotype has the strongest association with the production of SS-A and SS-B antibodies. IRF 5 and STAT 4 gene polymorphisms are involved in the activation of type 1 interferon pathway.

Environmental triggers

Epstein Barrvirus (EBV) infection of the epithelial cells may lead to neo antigen expression and increased rate of apoptosis. The apoptosis is mediated through the Fas-Fas ligand interaction and the perforin / granzyme B pathway. Lymphoid cell activation by EBV results in cleavage of alpha Fodrin to120KD fragments which occurs concomitantly with cellular apoptosis and expression of ZEBRA protein which is a marker for the activation of the life cycle of EBV[2]. Thus EBV induces the formation of auto antigens linked to SS.

Ductal and acinar epithelial cells express HLA class II molecules and act as antigen presenting cells. CD4 cells are attracted and become activated. Proinflammatory cytokines IL-l, IL-6 and TNF-alpha are produced and are responsible for inflammatory perpetuation and stimulation of B lymphocytes.

B cells contain intracytoplasmic immunoglobulins with anti Ro (SS-A) and or anti La (SS-B) reactivity. The most common autoantibodies in SS are directed against these ribonucleoprotein antigens Ro (SS-A) and La (SS-B). Anti La antibodies are more specific for SS while anti Ro antibodies might be present in other diseases like SLE and RA. The Ro antibodies are directed against 52 KD protein in SS and 60 KD protein in SLE[3]. B cell activation is initially polyclonal, later may evolve to oligo clonal or monoclonal activation and may end as malignant monoclonal proliferation. Activation of IFN alpha, B cell activating factor (BAFF) and antibodies to muscarinic acetylcholine receptor are new steps in the road to develop SS.

Clinical manifestations

Ocular manifestations

Symptoms

Patients usually present with red itchy painful eyes, complain of a gritty feeling or have foreign body sensation in the eye. Symptoms include blurring of vision, photophobia, redness and ocular fatigue or inability to tolerate contact lenses.

Physical signs

Dilatation of bulbar conjunctival vessels, pericorneal injections and lacrimal gland enlargement. Thick mucus secretion may cause blurring of vision. Xerophthalmia may lead to corneal abrasions and infections with gram positive organisms.

Tests:

a. Schirmer's test is considered positive when there is less than 5mm of wetting in 5 minutes when a filter paper strip is inserted in the lower eyelid.

b. Slit lamp examination after instillation of dyes over the cornea will show epithelial defects. Fluorescein stains epithelial defects. Rose Bengal binds devitalized cells. Lissamine green is less irritating and equally sensitive.

c. Tear break up time is another clinical test. A drop of fluorescein is applied to the eye and the time between the last blink and the appearance of dark non-fluorescent areas in the tear film is measured. A rapid tear break uptime indicates an abnormality.

Oral manifestations

Symptoms

Patients complain of dry mouth and need a constant supply of water to be comfortable. Eating is difficult without supplemental liquids. Some may have dysguesia. Dry burning throat that leads to dry cough awakening with a need for a drink of water, difficulty in speaking continuously, hoarseness of voice are other symptoms. Dental caries, periodental disease may be the first symptom of dry mouth due to loss of antibacterial action of saliva.

Signs

Enlargement of parotids is frequent and may precede xerostomia by several years. Sublingual and submandibualr glands may also be affected. Absence of pooling of saliva under the tongue is diagnostic of xerostomia. Loss of filiform papillae from the dorsal surface of the tongue, cracked or sore tongue or tongue sticking to the roof, extensive oral erythema, angular chelitis are often found.

Tests

a. Sialometry measures salivary flow rates with or without stimulation for the various salivary glands.

b. Sialography shows sialectasis and is a sensitive test for SS.

c. Scintigraphy measures the 99m Tc pertechnate uptake by the salivary gland.

The uptake is reduced and secretion of labeled saliva is delayed or absent.

In a study from Chennai[4] dry mouth and dry eyes were the presenting symptoms in 55.6% of patients and these were the commonest finding in 89% and 86% respectively, similar to the study from Lucknow (86.4%)[5].

Other xeroses

Dryness of all mucus membrane occur. Dry nose leads to congestion, crusting and epistaxis. Xerotrachea can lead to dry cough and hoarseness. Vaginal dryness leads to pruritis, irritation and dyspareunia. Sweat secretion is reduced leading to pruritis and excoriation. Loss of exocrine function may lead to pancreatic hypofunction and hypochlorhydria.

EXTRA GLANDULAR MANIFESTATIONS

Systemic symptoms

Easy fatiguability, low grade fever, myalgia, arthralgia.

Musculoskeletal symptoms

Arthritis occurs in 54-84% of SS patients. Arthritis may precede overt sicca manifestations and be associated with morning stiffness. Chronic polyarthritis may lead to Jaccoud's arthopathy. Erosions are rare. Arthritis was the commonest extra glandular symptom in 77.8% of patients in the study from Chennai[4].

Skin

Skin manifestations include purpura, annular erythema, pernio like lesions and urticarial lesions. Flat purpura is seen in patients with hypergamma globulinemia. Palpable purpura is a feature of vasculitis. Raynaud's phenomenon is seen in 35% and usually precedes the sicca symptoms by many years.

Pulmonary manifestations

Cough occurs in 40-50% secondary to xerotrachea or to bronchial hyper responsiveness. Bronchiolitis, lymphoid interstitial pneumonia, bronchiectasis and fibrosis can occur.

Gastro intestinal manifestations

Dysphagia is common either due to hyposalivation or due to esophageal dysmotility. Chronic atrophic gastritis leads to epigastric pain and nausea.

Autoimmune hepatitis has been reported in 25% with smooth muscle antibodies being present in 7-33%. Anti mitochondrial antibodies occurs in 7-13% of patients and suggests a close association with primary biliary cirrhosis[6].

Acute or chronic pancreatitis has been reported.

Renal manifestations

Lymphocytic infiltration of the tubulo interstitium results in distal renal tubular acidosis (RTA). The first presenting symptom of SS has been hypokalemic paralysis. In these patients sicca symptoms are mild and previously unrecognized. In a study from Chennai[4], the presenting symptom of hypokalemic paralysis was seen in 5.5% of SS patients. Renal tubular acidosis has also been reported from Mumbai[7].

Glomerular disease is rare. Membranous or membrano proliferative glomerulonephritis has been described. Type 2 mixed cryoglobulinemia and low C_4 level is seen in these patients. Interstitial cystitis may present as nocturia, suprapubic or perineal pain.

Neurological manifestations

Cranial neuropathy involving trigeminal or optic nerve has been reported. Sensorineural loss is noted in one half of SS patients on audiometric testing, Peripheral sensorimotor neuropathy occurs as a consequence of small vessel vasculitis. Various CNS manifestations have been observed in Sjogren's patients including hemiparesis, hemisensory defects, seizures, movement disorders, transverse myelopathy and dementia.

Vasculitis

Vasculitis of small and medium sized vessels is found in 5% of SS patients manifesting as purpura, urticaria, skin ulceration or mononeuritis multiplex. Systemic vasculitis can involve the kidneys, lungs, gastro-intestinal tract, spleen, breast and reproductive tract. Large vessel vasculitis with absent peripheral pulses has been reported from Chennai[8].

Auto immune thyroiditis

Thyroid disease has been seen in more than one third of patients with SS as evidenced by the presence of antithyroid antibodies and elevated levels of thyroid stimulating hormone.

Lymphoproliferative disease

Risk factors for the development of lymphoma in patients with SS are parotid enlargement, splenomegaly lymphadenopathy, low C_4 complement levels and type II mixed cryoglobulinemia. Skin vasculitis, peripheral nerve involvement, fever, anaemia, lymphopenia were observed more frequently in SS patients with lymphoma than in those without lymphoma. Lymphomas are of B cell origin expressing IgMk immunoglobulins and producing rheumatoid factor. Low grade lymphomas remit without therapy, while intermediate and high grade lymphomas have bad prognosis. Rare occurrence of both Non Hodgkins lymphoma and multiple myeloma in a patient with primary SS has been reported[9].

Secondary Sjogren's syndrome

Sjogren's syndrome can be associated with other autoimmune rheumatic diseases like rheumatoid arthritis, systemic lupus erythematosus, systemic sclerosis, mixed connective tissue disease, primary biliary cirrhosis, myositis, vasculitis, thyroiditis, chronic active hepatitis and mixed cryoglobulinemia.

PROGNOSTIC FACTORS

The presence of arthritis, Raynaud's phenomenon, interstitial nephritis, lung and liver involvement early in the disease has a favourable prognosis. On the other hand purpura, glomerulo nephritis, decreased C4 levels and mixed monoclonal cryoglobulinemia predict a bad prognosis.

INVESTIGATIONS

Elevated erythrocyte sedimentation rate is a very common manifestation in 80-90% of patients with SS. Hypergamma globulinemia is found in 80% of patients.

Serological tests

Antibodies to Ro (SS-A) and La (SS-B) occur with high frequency, anti Ro has been reported in 66% and anti La in 62% in a study from Lucknow[5]. Other autoantibodies that may be present include rheumatoid factor, antinuclear antibody, organ specific antibodies like antigastric parietal cell, thyroglobulin, thyroid microsomal, mitochondrial, smooth muscle and salivary duct antibodies.

Minor salivary gland biopsy

Histopathological features include focal aggregates of atleast 50 lymphocytes, plasma cells, macrophages, adjacent to and replacing normal acini and the consistent presence of these foci in all or most of the glands in the specimen.

DIAGNOSIS

Workup for patients with suspected Sjogren's is based on the step wise approach given, by the American European consensus group modification of the European community criteria[10] The six criteria are :

1. Symptoms of dry eye
2. Symptoms of dry mouth or salivary swelling
3. Evidence of dry eye (Schirmer's test or abnormal corneal staining)

4. Evidence of salivary dysfunction (abnormal salivary flow, scintigram or sialogram)
5. Presence of SS-A or SS-B antibody
6. Positive minor salivary gland biopsy

Patients must have 4 of these criteria, atleast 1 of which must be either antibodies or positive biopsy.

Differential diagnosis

This includes diffuse infiltrative lymphocytosis syndrome in HIV patients, HTL V-1, hepatitis-C infections, sarcoidosis, amyloidosis, lymphoma. Dryness due to ageing, previous radiation to head and neck are others. These conditions present with sicca syndrome and salivary gland enlargement, but lack the specific antibodies against Ro or La.

Treatment

Ocular disease

Xerophthalmia can be controlled with artificial tears made of polyvinyl alcohol or methylcellulose. Lubricating ointments and hydroxypropyl cellulose inserts can be tried.

Existing tears may be retained in the eye by blocking their drainage or by inhibiting their evaporation. Punctate occlusion by inserting collagen or silicon plugs or by electro cautery retains the tears in the eyes. Tear evaporation can be prevented by wearing goggles with side chambers. Lateral tarsoraphy is indicated in severe dryness to reduce the ocular surface. Corneal transplantation is advised for corneal perforation. Drugs with anticholinergic side effects like phenothiazine, tricyclic antidepressants, antispasmodics, anti parkinsonian drugs and cigarette smoking should be avoided.

Oral disease

Secretagogues stimulate muscarinic receptors in salivary glands leading to increased salivation, pilocarpine is administered as 5mg (QID), cevemeline is administered as 30mg TID. Methotrexate and cyclosporine A improve the subjective symptoms of dryness.

Other xerosis

Vaginal dryness requires lubricant jellies while moisturizing lotions can be applied for dry skin.

Systemic diseases

Hydroxychloroquine has been used to treat arthralgia, myalgia and fatigue. Systemic cortico steroids and cyclophosphamide are used for vasculitis, peripheral neuropathy, diffuse interstitial pneumonitis, glomerulonephritis. Renal tubular acidosis can be treated with potassium chloride and potassium citrate. Frank lymphomas require chemotherapy and or radiotherapy.

Biologics:

Biological agent infliximab has been found to be useful for treating ocular, oral and systemic manifestations of SS. Treatment with Rituximab an anti CD-20 chimeric humanized monoclonal antibody specific for the B cell surface molecule CD 20 has resulted in significant improvement of subjective symptoms, an increase in salivary gland function and improvement in systemic features. There was rapid depletion of peripheral B cells within a few weeks and a decrease in IgM RF levels[11]. Dose is 375mg/m^2 once weekly for 4 weeks.

Epratuzumab fully humanized monoclonal antibody specific for B cell surface molecule CD 22 given in 4 doses of 360mg/m^2 has shown improvement[12]. SS patients with elevated BAFF levels, hypergammaglobulinemia, elevated levels of auto antibodies might be candidates for anti BAFF therapy like anti BLYS[13].

Conclusion

Sjogren's syndrome is an autoimmune disease affecting the exocrine glands.

Characterized by dry eyes, dry mouth, salivary and lacrimal gland enlargement. Viral infection of the epithelial cells leads to apoptosis, neoantigen expression, and production of auto antibodies by the B cells.

Arthritis, myalgia, purpura, Raynaud' s occur

Pulmonary, GIT and CNS manifestations can occur.

Distal renal tubular acidosis leads to hypokalemic paralysis and renal osteodystrophy.

Elevated ESR, hypergammaglobulinemia, presence of Ro or La antibodies and positive minor salivary gland biopsy showing infiltration with lymphocytes is diagnostic.

Lymphomas of B cell origin can occur.

Dry eyes treated with aritificial tears, dry mouth treated with secretogogues. Systemic symptoms require corticosteroids, hydroxychloroquine, cyclophosphamide and biologics.

Further Reading :

1. Mann D. Immunogenetics of Sjogren's in: Talal N. Moutsopoulas lllM, Kassan SS eds. Sjogren's syndrome: clinical and immunological aspects. Berlin: Springer - Verlag, 1987; 235-43.
2. Inoue H. Tsubota K, Ono M, et al: Possible involvement of EBV - mediated alpha fodrin cleavage for organ - specific autoantigen in Sjogren's syndrome - J Immuno 2001; 166: 5801.
3. Ben-Chetri E, Fox Rl, Tan EM: Dissociation of immune responses to the SS-A (Ro) and 52kd and 60kd polypeptides in systemic lupus erythematosus and Sjogren's syndrome. Arth Rheum 1990; 33: 3449-55.
4. Porkodi R, Rukmanagtharajan S, Kanakarani P, et al: Primary Sjogren's syndrome - clinical and immunological features. JIRA 2003; 11: 63-65.
5. Rissaria P, Aggarwal A, Dabadghao S, et al: Rarity of primary Sjogren's syndrome in India JIRA 2000;8 (suppl) 1): S.36.
6. Lindgren S, Manthorpe R, Eriksson S: Auto Immune liver disease in patients with primary Sjogren's syndrome. J Hepatol1994; 20: 354.
7. EM1, Pathan J.L. Oak, c.G.Jyotish: Extraglandular manifestations in secondary Sjogren's Syndrome - JIRA 1999; 4: 116-17.
8. Porkodi R, Ramakrishnan S, Krishnamurthy V, Rajendran CP, Chandrasekaran A.N., et al: Primary Sjogren's Syndrome with vasculitis JIRA 1993; 1: 42-43.
9. N. Kumar, c.A. Kelly: Primary Sjogren's Syndrome presenting with Non Hodgkins Lymphoma and terminating with myeloma. TIRA 1999; 7 No.4: 114-15
10. Vitalic, Bombardieres S. Johnson R, et al. Classification criteria for Sjogren's Syndrome. A revised version of European criteria proposed by the American-European Consensus Group. Ann Rheum Dis 2002; 61: 554.
11. Pijpe J, Van ImhoffGW, Spij Kervet FKL, et al. Rituximab treatment in patients with primary Sjogren's Syndrome. An open label phase II study Arhritis Rheum 2005; 52 : 2740-50.
12. Steinfield SD, Tant L, Burmeister GR et al. Epratuxamab in primary Sjogren's Syndrome. An open label phase VII study Arthritis Res.Ther. 2006; 8: R 129.
13. Szodoray P, Jonsson R. The BA FF / APRIL system in systemic auto immune disease with a special emphasis on Sjogren's Syndrome. Scand J Immunol2009; 62: 421-28.

MIXED CONNECTIVE TISSUE DISEASE AND OVERLAP SYNDROMES

5

Sanjiv Kapoor

Introduction

A correct and complete diagnosis of a disease heavily depends on the knowledge of its aetiology. Unfortunately, the aetiology of most systemic autoimmune connective tissue diseases remain unknown till date. Therefore, diagnosis is mainly clinical. Reaching a provisional diagnosis depends upon recognition of a recurring set of clinical symptoms and signs. It is then aided by certain specific laboratory investigations to reach a definitive diagnosis. This is the format for diagnosis and classification of most connective tissue diseases for epidemiological and research purposes. Although this approach allows for the selection of reasonably homogeneous patient population for epidemiologic studies, it is of limited value in individual patients and inadequate for the determination of precise treatment protocols and for assessing prognosis.

Systemic rheumatic diseases are characterized by having clinical features that are widely common among them resulting in a clinical continuum that has been given the name of diffuse connective tissue diseases (DCTDs). Within this spectrum there are five well defined diseases with easily recognisable recurring pattern of clinical features to give them a definite diagnostic label. These include rheumatoid arthritis (RA), systemic lupus erythematosus (SLE), systemic sclerosis (SS), dermatomyositis (DM) and polymyositis (PM). Each of these may be associated with Sjögren's syndrome which in its primary form may be the most common connective tissue disease[1].

However, in clinical practice it is not uncommon to come across patients who may have some clinical features of one of these diseases associated with additional clinical features seen in one or more of the other DCTDs. Such patients are labelled as 'overlap connective tissue disease'. For example, features of SS are quite often seen among patients with other DCTDs, making this as the commonest overlap syndromes. In addition to patients with overlap-connective tissue diseases, there are others in whom the clinical features have not fully evolved to be recognisable as one of the defined DCTDs. The commonest example is that of a patient with idiopathic Raynaud's phenomenon where significant titres of antinuclear antibodies are present. Such patients have been traditionally labelled as 'undifferentiated connective tissue disease' (UCTD).

In this chapter clinical and laboratory characteristics of UCTD and different overlap syndromes including mixed connective disease or Sharp's syndrome shall be discussed.

Undifferntiated connective tissue disease

UCTDs are conditions characterized by the presence of clinical and serological manifestations suggestive of systemic autoimmune diseases but not fulfilling the classification criteria for defined CTDs[2]. Most of the patients are females in 3rd-4th decade of life. Clinical features can be arthralgias, Raynaud's arthritis, leukopenias, xerophthalmia, xerostemia, photosensitivity, anemia, serositis, malar rash, oral aphthosis, thrombocytopenias. Most of the patients are females in 3rd-4th decade of life. Many of these patients may evolve over time by developing features which would lead them to fulfill the classification criteria for a specific connective tissue disease. The proportion of patients who evolve into a specific illnesses varies according to the inclusion criteria of the cohort. However generally, only around one third of patients with UCTDs evolve into full blown CTD within the first 2 years[3-5]. Most commonly patients develop systemic lupus erythematosus (SLE), although an evolution to other CTDs such as systemic sclerosis (SSc), Sjogren's syndrome (SS), mixed connective tissue disease (MCTD), systemic vasculitis, polydermatomyositis (PM/DM), rheumatoid arthritis (RA) has been described. The remaining patients who do not evolve into another CTD, i.e. Stable UCTDs, display mild clinical picture and frequently single autoantibody profile. ANA positivity is high ranging from 58% up to 90%; anti-Ro/SSA and anti-RNP antibodies appear to be the most frequent specificities. Classification criteria are still matter of debate. The treatment of patients with UCTD is quite simple with most patients requiring low dose steroids or antimalarials. In fact nearly half the patients in the cohort reported by Mosca et al were not on any treatment at the end of - years of follow up. Thus the patient and the clinician can be reassured that in a confirmed UCTD (i.e. One who has not evolved into a second CTD over at least 3 years of follow up), there is need for minimal therapy and monitoring.

Overlap syndromes

When a patient clearly demonstrates clinical or laboratory features which lead them fulfilling classification criteria for two or more different CTDs the person is said to have overlap syndrome. However it is important to note that features common to the different classification criteria should not be used more than once. (For eg: Arthritis cannot be used for fulfillment of the criteria of RA and SLE.) Major features of more than one rheumatic disease occur in same patient. It occurs in nearly 25% of rheumatic disease patients. True overlap syndromes can be defined in two ways (i) on the basis of clinical involvement and (ii) by the detection of auto-antibodies.

Overlap syndromes based on clinical features

Raynaud's phenomenon, sclerodactyly and alveolitis are common features of a number of autoimmune rheumatic diseases and cannot be used on their own to define a syndrome. Other features, such as thickening of the skin proximal to the fingers in scleroderma or the articular erosion in RA are sufficiently disease specific to suggest that a patient having a combination of both has true overlap. This may be the basis of the early description of RA/ Systemic Sclerosis overlap and RA/SLE overlap or Rhupus.

Overlap syndromes defined by auto antibodies

t RNA synthetase (Jo1, antigen), alanyl and theronyl t RNA synthetases are auto antigens in polymyositis associated overlap syndromes.

Their clinical features are[3] : Raynaud's phenomenon 93%, arthritis 90%, myositis 83%, lung fibrosis 79%, sclerodactyly 72%, sicca syndrome 59%, dermatomyositis rash 38%, facial telangiectasia 31%, dysphagia 31%, soft tissue calcification 24%, tendonitis 17% and clubbing 14%.

The PM/SCL associated overlap syndrome

Antibodies to PM/SCL was found in about 15%[4] of polymyositis patients and appeared to be associated with a polymyositis/scleroderma overlap syndrome with clinical features very similar to those of the anti t RNA associated overlap syndromes. However myositis and fibrosing alveolitis are less frequent, less severe and more likely to respond to treatment with corticosteroids, or immunosuppressive drugs. Every patient has positive HLA DR3.

The SLE/Sjogren's syndrome overlap defined by anti la (SSB)

The La (SSB) antigen is known to shuttle between nucleus and cytoplasm and its antibodies occur in both SLE and Sjogren's syndrome, particularly when there are overlapping features of both diseases[5].

Clinical features

Overlap patients with anti-La antibodies have Sjogren's syndrome with prominent systemic features. Most of the extra glandular features are SLE like, with high frequency of arthritis, rashes, Raynaud's, leucopenia and thrombocytopenia. Characteristic of the anti-La associated overlap syndrome is a purpuric hypergamma globulinemic rash in about 30% of the patients and a low frequency of nephritis as compared to the patients with SLE. However, renal tubular acidosis (often sub clinical) may occur in upto 30% of the patients.

Mixed connective tissue disease

Mixed connective tissue disease (MCTD)/Sharp's syndrome is an apparently distinct rheumatic disease syndrome associated with a specific antibody to extractable nuclear antigen (ENA).

Definition

In 1999, Sharp and Hoffman[6] defined MCTD as patients with autoantibodies against the 70-Kd peptide of the UI snRNP complex or autoantibody against the U1 RNA, have a common gene sequence encoded for MCTD or select genotypes of HLA-DR4 or HLA-DR2 and have common features of Raynaud's phenomenon, swollen hands, sclerodactyly, esophageal hypomotility, polyarthritis, or myositis[7].

Diagnostic criteria for mixed connective tissue disease (MCTD)

Comparisons between the three diagnostic criteria for MCTD are shown in Table 1. The criteria proposed by Alarcon-Segovia and Villanreal are simplest and are likely to be the most acceptable.

Age, sex and genetic prevalence

Age of onset of the disease is in second or third decades. The female to male ratio of the disease is 16:1. The syndrome usually occurs as an isolated finding but familial aggregations have been reported[11]. There is an increased prevalence of HLA-D4, and its specificity for this condition has been confirmed[12].

Table 1. Diagnostic criteria for mixed connective tissue disease (MCTD)

Sharp et al[8]	Kasukawa et al[9]	Alarcon-Segovia and Villanreal[10]
A. Major criteria 1. Myositis, severe 2. Pulmonary involvement a. CO diffusing capacity <70% of normal values b. Pulmonary hypertension c. Proliferative vascular lesion on lung biopsy 3. Raynaud's phenomenon or esophageal hypomotility 4. Swollen hands observed, or sclerodactyly 5. Highest Observed anti-ENA ≥ 1:10,000 and anti-U RNP + and anti-Sm -ve **B. Minor criteria** 1. Alopecia 2. Leukopenia 3. Anemia 4. Pleuritis 5. Pericarditis 6. Arthritis 7. Trigeminal neuropathy 8. Malar rash 9. Thrombocytopenia 10. Myositis, mild 11. Swollen hands	**A. Common symptoms** 1. Raynaud's phenomenon 2. Swollen fingers or hands **B. Anti-nRNP antibody** **C. Mixed findings** 1. SLE like findings a. Polyarthritis b. Lymphadenopathy c. Facial erythema d. Pericarditis or pleuritis e. Leucopenia or thrombocytopenia 2. **PSS like findings** a. Sclerodactyly b. Pulmonary fibrosis, restrictive change of lung, or reduced diffusion capacity c. Hypomotility or dilatation of esophagus 3. **PM like findings** a. Muscle weakness b. Increased serum levels of myogenic enzymes (CPK) Myogenic pattern in EMG	**A. Serologic:** positive anti-RNP at hemagglutination titre of 1:1,600 or higher. **B. Clinical** Oedema of the hands Synovitis Myositis Raynaud's phenomenon Acrosclerosis
Requirements for diagnosis of MCTD		
Four major criteria: anti-U1 RNP ≥ 1:4,000; exclusions; anti-Sm + **Or** Three major criteria; two major criteria from 1, 2 & 3 and two minor criteria with anti-U1 RNP ≥ 1000 **Or** Three major without anti-U1 RNP; two majors, one major and three minors with anti-U1 RNP ≥ 100	A. Positive in either one of two common symptoms B. Positive anti-nRNP antibody C. Positive in one or more findings in two or three disease categories or 1, 2 and 3.	A. Serologic B. At least three clinical findings. C. Association of oedema of the hands, Raynaud's phenomenon, and acrosclerosis requires at least one of the other two criteria

Clinical features

On comparing the incidence of the signs, symptoms and laboratory finding, MCTD occupied a middle ground between systemic sclerosis (SS) and SLE, except for 100% incidence of the presence of a positive ANA (by definition) and greater incidence of myositis and Raynaud's phenomenon.

Early symptoms

The central premise of MCTD is that of an overlap syndrome embracing features of SLE, SS and PM/DM. These overlap features of MCTD seldom occur concurrently; it usually takes several years before enough overlapping features have appeared to be confident that MCTD is the most appropriated diagnosis[13]. The most common clinical associations with U1-RNP antibodies in early phase of the disease are hand edema, arthritis, Raynaud's phenomenon, inflammatory muscle disease and sclerodactyly. Early in the course of the disease most patients complain of easy fatigability, poorly defined myalgia, arthralgia, and Raynaud's phenomenon. At this point of time diagnosis of RA, SLE or UCTD seems appropriate, if such a patient is found to have swollen hands, puffy fingers or both then the serological marker (anti RNP antibodies) should be sought.

Fever

Fever (33%) of unknown origin can be the initial presentation of MCTD. It usually co-exits with myositis, aseptic meningitis, serositis, and lymphadenopathy or inter current infection.

Joints

About 60% of the patients develop arthritis often with deformities commonly seen in Rheumatoid Arthritis such as ulnar deviation, swan neck and boutonnière changes[14]. Radiographs usually show a characteristic absence of severe erosive changes. They often resemble Jaccoud's arthopathy. However, a destructive arthritis, including an arthritis mutilans is also a well-established association.

Skin

Raynaud's phenomenon (85 %) is the most common problem and one of the earliest manifestations of MCTD[15]. It may be accompanied by puffy, swollen digits and sometime total hand edema. Skin changes commonly associated with classic SLE are prominent features.

Muscle

Myalgia is a common symptom. Sometimes an inflammatory myositis (63%) clinically and histologically similar to classic polymyositis may occur.

Heart

Pericarditis is the most common clinical manifestation (10% to 30%).

Lung

Pulmonary disease is evident in 80% of the patients. Most of them are asymptomatic. Interstitial fibrosis, dyspnea, decreased diffusion capability are common. The major cause of death in MCTD is pulmonary hypertension[16]. Pulmonary hypertension in MCTD is usually caused by a bland intimal proliferation and medial hypertrophy of pulmonary arterioles[17].

Kidney

Renal involvement occurs in 25% of patients[18]. However, high titers of anti-U1 RNP antibodies are relatively protective against the development of diffusive proliferative glomerulonephritis. Membranous glomerulonephritis is more common.

Gastrointestinal

Gastrointestinal involvement is a major feature occurring in 60% to 80% of patients, esophageal hypomotility is common but dysphagia is not always evident. Changes that are similar to those of sclerodermal bowel have been reported, along with mesenteric vasculitis.

Nervous system

Neurological lesions that usually are minor are seen in 10% to 15% of the patients. The most common abnormality is trigeminal neuralgia. Changes identical to SLE may occur. Vascular headaches may be pronounced.

Blood

Anemia of chronic inflammation is seen in 75% of the patients. A positive coombs test is seen in 60% of the patients but overt haemolytic anemia is uncommon. Leucopenia affecting mainly the lymphocyte series is seen in about 7% of the patients. Less common associations have been thrombocytopenia, thrombotic thrombocytopenic purpura and red cell aplasia. Hypocomplementemia is not correlated with any particular clinical situation.

Positive RF is seen in 50% of the patients. Presence of RF is associated with severe degree of arthritis.

Anticardiolipin antibodies and lupus anticoagulants have been reported. But β 2-glycoprotein independent ACLA tend to be associated with thrombocytopenia rather than thrombotic events[19].

Serologic features

The basic premise of MCTD is that the presence of high titer anti-U1 RNP antibodies modifies the expression of a DCTD in ways that are relevant to prognosis and treatment. The first laboratory clue to MCTD is usually a positive ANA with high titer, speckled pattern (> 1: 1000). ENA profile (UIRNP, Sm, Ro, and La) of MCTD sera shows predominant U1 RNP reactivity. Antibodies to ds-DNA, Sm, and Ro are occasionally seen as a transient phenomenon. Antibodies to the 70 kDa antigen are most closely associated with the clinical features of MCTD.

Prognosis

MCTD is a dynamic, changing syndrome. It is evident that the course of MCTD is unpredictable. Many patients do follow a relatively benign course but it is major organ involvement that ultimately dictates the morbidity and mortality of the disease. MCTD have a low prevalence of serious renal disease and life threatening neurological problems. However, not all patients with MCTD have a favorable prognosis and death may occur from progressive pulmonary hypertension and its cardiac squealae.

Management

Recommendations for management are based on conventional treatments for SLE, PM/DM, RA and SS. Pulmonary hypertension is the main cause of death in MCTD and patients should be evaluated at regular intervals for development of this complication. Two dimensional echocardiography with doppler flow studies is most useful screening test. For management of Pulmonary hypertension, the reader is referred to excellent reviews published recently as well as the guidelines published by the American Heart Association and the European Society of Cardiology.

Glucocorticoid responsive manifestations of MCTD

Myositis, pleurisy, pericarditis, myocarditis and aseptic meningitis respond to glucocorticoids.

Glucocorticoids resistant manifestations of MCTD

Nephrotic syndrome, Raynaud's phenomenon, deforming arthopathy, acrosclerosis and peripheral neuropathies does not respond to glucocorticoids.

Many of SS like gastrointestinal problems can be managed according to the usual practice in SS. Management of renal crisis is with ACE inhibitors, Raynaud's with calcium channels blockers and gastrointestinal reflux diseases is with proton pump inhibitors as done in patients with systemic sclerosis.

General guidelines for treating specific features of the MCTD spectrum are given in Table 2[20].

Table 2. Management guidelines

Problem	Treatment
Fatigue, arthralgias, myalgias	NSAIDs, antimalarials, low dose prednisone (less than 10 mg/day)
Arthritis	NSAIDs, antimalarials, gold, methotrexate, TNF inhibition
Pleurisy	Indomethacin, short course of prednisolone: about 20 mg/day
Aseptic meningitis	Discontinue NSAIDs, short course of high dose prednisone; about 60mg/day
Myositis	Acute onset/severe; prednisone 30 to 100 mg/day Chronic/low grade; prednisone 10 to 30 mg/day Consider methotrexate and IVIG
Membranous glomerulonephropathy	Mild: no treatment required Progressive proteinuria: trial of angiotensin-converting enzyme inhibitor; trial of low-dose aspirin combined with dipyridamole; Severe: trial of prednisone (15 to 60 mg/day) plus monthly pulse cyclophosphamide or daily chlorambucil.
Nephrotic syndrome	Steroid seldom effective; low dose aspirin combined with dipyridamole to prevent thrombotic complication; trial of prednisone (15-60 mg/day) plus monthly pulse cyclophosphamide or daily chlorambucil; may require dialysis/transplantation.
Raynaud's phenomenon	Keep warm, avoid finger trauma, stop smoking; nifedipine as tolerated; pentoxifyline in severe cases with ischemic eschars. Intra-arterial prostacyclin, local nitroglycerin ointment; consider endothelin receptor antagonist such as bosentan.
Myocarditis	Trial of steroids and cyclophosphamide; avoid digoxin
Incomplete heart block	Avoid chloroquine
Asymptomatic pulmonary hypertension	Trial of steroids and cyclophosphamide, low-dose aspirin and angiotensin-converting enzyme inhibitors; consider endothelin receptor antagonist (oral bosentan)
Symptomatic pulmonary hypertension	Intravenous prostacyclin, angiotensin-converting enzyme inhibitor, anticoagulation; endothelin receptor antagonist (oral bosentan); sildenafil trial; heart - lung transplantation
Vascular headache	Trial of propranolol or alternate-day aspirin (350mg); symptomatic use of sumatriptan
Dysphagia	Mild: no treatment With reflux: proton pump inhibitor; consider Nissen fundoplication Severe: calcium channel antagonist, alone or in combination with an anticholinergic agent
Intestinal dysmotility	Prokinetic agents such as metoclopramide, comperidone, octreotide and erythromycin. Small bowel bacterial over growth: tetracycline, erythromycin

Osteoporosis	Calcium and vitamin D supplements, estrogen replacement or raloxifene; biphosphonates; nasal calcitonin; carboxyl truncated PTH analogues such as hPTH
Heartburn	Raise head of bed, discontinue smoking and avoid caffeine; H2 antagonists, H+ proton blockers; metoclopramide trial pump
Trigeminal neuropathy	None

Summary

- Patients with incomplete feature (e.g. Raynaud's, arthralgia and weak positive ANA) are labeled as **undifferentiated connective tissue disease (UCTD)**.
- Patients having combination of major features of more than one rheumatic disease are labeled as **overlap syndromes**.
- **Clinical overlaps** e.g. SS, RA overlap characterized by skin thickening proximal to metacarpophalangial joints and erosive arthritis; features of SLE with erosive arthritis labeled as rhupus.
- **Auto antibodies defined overlap syndromes**; e.g. tRNA synthetase antibodies and PM/SCL antibodies overlap syndromes, clinically characterized by Raynaud's phenomenon, arthritis, myositis, lung fibrosis and sclerodactyly. Anti-La antibody defined SLE / Sjogren's overlap clinically characterized by purpuric hypergammaglobuliniemic rash.
- **Mixed Connective Tissue Disease (MCTD)**

  Presence of high titer anti-UI RNP antibodies modifies the clinical expression of disease. In early phase of disease it is associated with hand edema, arthritis, Raynaud's phenomenon, inflammatory muscle disease and sclerodactyly. It also seems to be protective against development of diffuse proliferative glomerulonephritis and serious CNS disease expression. However pulmonary hypertension is the most serious manifestation of the disease. Major features of MCTD are in TABLE 3.

Table 3. MCTD at a glance

Parameter	MCTD
Sex incidence	80% female
Age of majority	All ages
Family History	Rarely +
Disease duration	Variable
First changes	Arthritis, Raynaud's phenomenon and swollen puffy hands
Cardiac involvement (Clinical)	Myocarditis of children
Skin and mucous membranes	Rashes of SLE, PSS, and dermatomyosi tis
Ocular	Rare; Sjogren's syndrome
Adenopathy	Minimal
Pleurisy or lung disease	30%
Pericarditis	25%
Generalized abdominal pain and tenderness	Dysphagia common
Hepatomegaly	0
Splenomegaly	0
Joints involved	Erosive arthritis in 30%
Arthritic deformity	20%
Myalgia	50%
Raynaud's phenomenon	80%
CNS involvement	10%
Laboratory	
Urine abnormalities	Lupus nephritis in 10-40%
WBC and differential	Leucopenia in 35%
Anaemia	41%
Uremia	0
Hyperglobulinemia	80%
LE cells	14%
ANA	+ 100% speckled
Muscle biopsy	Often positive
Skin biopsy	+ lupus band test or PSS
Cardinal feature	Swollen hands; anti-RNP antibody

Further Reading :

1. Thomas E, Hay EM, Hajeer A, et al. A community based study of prevalence and impact. Br J Rheum 1998; 37: 1069-76.
2. Williams HJ, Alarcon GS, Joks R, et al. Early undifferentiated connective tissue disease (CTD): VI. An inception cohort after 10 years: disease remissions and changes in diagnoses in well established and undifferentiated CTD. J Rheum 1999; 26: 816-25.
3. Marguerie C, Bunn CC, Beynon HLC, et al. Polymyositis, pulmonary fibrosis and autoantibodies to aminoacyl t RNA synthetases. Q J Med. 1990; 77: 1019-38.
4. Reichlin M, Maddison PJ, Targoff L, et al. Antibodies to a nuclear/nucleolar antigen in patients with plomyositis overlap syndromes. J Clin Immu 1984; 4: 40-44.
5. Price EJ, Venables PJW. The etiopathogenesis of Sjogren's syndrome. Semin Arth Rheum 1995; 25: 117-33.
6. Sharp GC, Hoffman RW, Clinical, immunologic and immunogenetic evidence that mixed connective tissue disease is a distinct entity: comment on article by Smolen and Steiner. Arth Rheum 1999; 42: 190-91.
7. Smolen JS, Steiner G. Mixed connective tissue disease. To be or not to be? Arth Rheum 1998; 41: 768-77.
8. Black C, Isenberg DA. Mixed connective tissue disease: good be to all that. Br. J Rheum 1992; 31: 695-700.
9. Hayashi T, Lee S, Ogasawara H, et al. Exacerbation of systemic lupus erythematosus related to cytomegalovirus infection. Lupus 1998; 7: 561-64.
10. Langgartner J, Andus T, Hemauer A, et al. Parvovirus B19 infection imitating an acute exacerbation of systemic lupus erythematosus. Dtsch Med Waschr 1999; 124: 859-62.
11. Ramos-Niembro F, Alarcon, Segovia D. Familial aspects of mixed connective tissue disease (MCTD). Occurrence of systemic lupus erythematosus in another family member in two families and aggregation of MCTD in another family. J Rheum 1978; 5: 433-40.
12. Gendi NST, Welsh KI, van Vendroou JV, et al. HLA type as a predictor of mixed connective tissue disease differentiation. Ten year clinical and immunogenetic follow up of 46 patients. Arth Rheum 1995; 38: 259-66.
13. Bennett RM, O'Connell DJ : Mixed connective tissue disease: a clinicopathologic study of 20 cases. Semin Arth Rheum 1980; 10: 25-51.
14. Bennett RM, O'Connell DJ : The arthritis of mixed connective tissue disease. Ann Rheum Dis 1978; 37: 397-403.
15. Nedumaran, Rajendran CP, Porkodi R, et al. mixed connective tissue disease : clinical and immunological profile. J Assoc Phys Ind 2001; 49: 412-14.
16. Jones MB, Osterholm RK, Wilson RB, et al. Fatal pulmonary hypertension and resolving immune-complex glomerulonephritis in mixed connective tissue disease. A cases report and review of the literature. Am J Med 1978; 65: 855-63.
17. Singsen BH, Swanson VL, Bernstein BH, et al. A histologic evaluation of mixed connective tissue disease in childhood. Am J Med 1980; 68: 710.
18. Kitridou RC, Akmal M, Turkel SB, et al. Renal involvement in mixed connective tissue disease: a longitudinal clinicopathologic study. Semin Arth Rheum 1986; 16(2): 135-45.
19. Komatireddy GR, wang GS, Sharp GC, et al. Antiphospholipid antibodies among anti-U1-70 kDa autoantibody positive patients with mixed connective tissue disease. J Rheum 1997; 24: 319-22.
20. Robert M. Bennett, et al. Mixed connective tissue disease and other overlap syndrome. In : Edward D. Harris, Jr., Budd RC. Firestein (eds). Kelley's text book of Rheumatology 7th edition: Elsevier, Philadelphia, 2005; 1272.

SCLERODERMA

6

Jyotsna Oak, Keshav Kale

Introduction

The word "Scleroderma" is derived from two Greek words, "sclero" meaning hard and "derma" meaning skin. Scleroderma (Scl) is an autoimmune disorder of unknown etiology characterized by fibrosis and tightening of skin as well as microvascular injury in affected organs. Scl may affect multiple organs simultaneously or at different times and thus has a wide spectrum of clinical manifestations and severity. The involved target organs include skin, muscles, joints, nerves, vasculature, kidney, heart, lung and gastrointestinal tract. The disease can arbitrarily be defined as diffuse cutaneous scleroderma with diffuse cutaneous involvement; skin thickening proximal to the elbows and knees (upper arm, thighs, anterior chest and abdomen) that is documented at any time during the illness. Limited Scl is defined as skin thickening limited to distal extremities and face throughout the illness.

CLASSIFICATION

The classification of scleroderma is given in Table 1. See ref 24 for the latest classification for systemic sclerosis/scleroderma.

Table 1. Scleroderma classification

Presenting feature	Limited	Diffuse
Raynaud's phenomenon	Long duration	Short duration
Skin involvement	Distal limbs and face	Proximal part of limbs and trunk
Tendon friction rub	Absent	Present +/-
Nailfold capillaries	Dilatation	Dilatation and loss
Auto antigen	Centromere	Topoisomerase I

Epidemiology

Scleroderma has a worldwide distribution. The disease is rare in childhood. Women are affected more frequently than men. The highest incidence of scleroderma is in the age group of 30-35 years. The peak age of onset ranges from 25-54 years in Caucasian women and from 34-44 years in African women. Racial factors seem to play a role in disease susceptibility as well as disease expression.

There is no data on the prevalence of scleroderma in India. In U.S.A. the prevalence in the general population has been reported in the range of 50 per 100,000.

Pathogenesis

The characteristic feature of Scl is overproduction and accumulation of collagen and extracellular matrix protein like fibronectin, tenascin, fibrillin-1 and glycosaminoglycans in skin and other organs. The disease process involves immunological mechanisms and vascular endothelial cell activation and/ or injury. It also involves fibroblasts activation resulting in production of excessive collagen (Fig. 1).

An early event in Scl that precedes fibrosis is vascular injury. Small vessel vasculopathy and fibrosis characterize pathology of scleroderma. The vessels involved are not only small arteries, arterioles and capillaries in the skin but also organs like GI tract, kidney, heart and lungs. Microvascular injury leads to proliferation of intima and smooth muscles with deposition of matrix and perivascular fibrosis, leading to obliteration vasculopathy. Endothelial injury plays a major role.

Antiendothelial antibodies (AECA) seen in 25% of the SSc patients have been shown to mediate antibody dependent cytotoxicity against endothelial cells. This also leads to microvascular damage. Pathogenesis involves a complex interplay between endothelial cells, fibroblasts and the immune systems triggered by environmental factors.Injury to endothelial cells and basal lamina lead to proliferation of intima and smooth muscle cells with deposition of matrix and perivascular fibrosis leading to narrowing of lumen and ischemia.

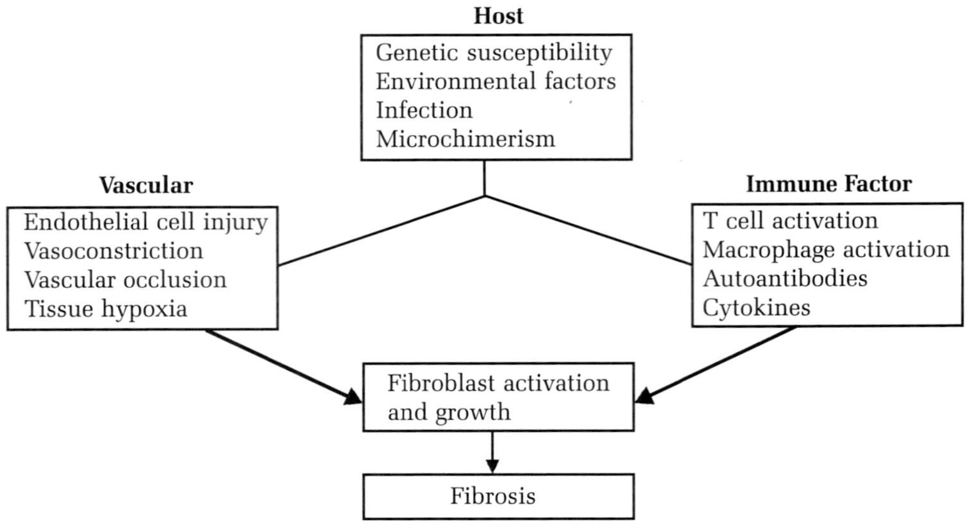

Fig. 1 : Pathogenesis of scleroderma

Clinical features

Raynaud's phenomenon

95% of the patients will experience Raynaud's phenomenon as an initial symptom[1]. It is defined as episodic vasoconstriction of small arteries and arterioles of fingers, toes and sometimes the tip of nose and earlobes.

Patients experience pallor and/or cyanosis followed by rubor on rewarming. Pallor and cyanosis are usually associated with coldness and numbness of fingers and toes.

Episodes are brought by cold exposure, vibration and emotional stress. The pathogenic mechanism of Raynaud's phenomenon is uncertain, but probably represents an imbalance between vaso constrictor and vasodialator mediator in response to physiological level of stimulation by cold or emotions. In diffuse Scl skin changes are seen typically within a year of Raynaud's phenomenon. In case of limited cutaneous Scl it will precede the disease by several months or even years.

The hallmark of scleroderma is taut, hidebound skin. This may take several years to develop. In clinical practice the following subsets are easily identified.

Telangiectasia

These are local dialated loops of small blood vessels in the skin which are prone to hemorrhage on trauma (Fig. 3).

Localized scleroderma

This is usually observed in dermatology practice in the form of morphea and linear scleroderma. Morphea presents as plaques of fibrotic skin and subcutaneous tissue without systemic disease, while the latter is seen as fibrotic bands that occur on the extremities involving skin and deeper tissue.

Limited cutaneous scleroderma

In this type, the scleroderma skin changes affects face, neck, extremities distal to the elbows and knees. Involvement of skin is the hallmark of scleroderma and there are usually three phases, oedematous, indurative and atrophic phase. The oedematous phase is characterized by the complaints of stiff hands with associated puffiness. Subsequently, the skin appears thick and tight, tendon fibrosis results into contracture of joints. The thumb is usually spared. Perioral fibrosis leads to 'pursed lip' and puckered mouth. This along with pinched nose and masked like face constitutes the typical scleroderma facies.

Digital scar/pseudo clubbing; finger contractures, ischemic ulcers calcinosis and telangiectasia are common findings. Calcinosis is uncommon in Indian patients[1,2].

An occasional patient may present with 'CREST' syndrome. (Calcinosis, Raynaud's phenomenon, Esophageal involvement, Sclerodactyly and Telangictasia). Late in the disease course pulmonary hypertension may develop.

Diffuse cutaneous systemic sclerosis

This is characterized by extensive skin involvement extending proximal to the knees and elbows and usually affecting the trunk. The oedematous phase may begin distally in the extremities and advances proximally. The skin becomes firm, thickened and eventually tightly bound to underlying subcutaneous tissue (indutrative phase). In diffuse scleroderma skin changes will become generalized involving extremities followed by trunk and face over a period of time varying from months to years. Rapid progression of these changes over 1-3 years is associated with greater risk of visceral disease particularly heart, lungs and kidneys. Also in diffuse scleroderma, skin changes usually peak for 3-5 years and then slowly improve. On the other hand, the skin of limited cutaneous scleroderma will usually have a more gradual progression.

In the extremities the taut skin over fingers gradually limits full extension and flexion; contractures develop slowly. Ulcers may appear on volar pads of fingertips and over elbows, and malleoli. The volar pads of finger tips develop pitting scars. There is resorption of terminal phalynx. Skin may show diffuse hypopigmentation with sparing of pigment around hair follicles giving a typical 'salt and pepper' appearance. Skin loses hair, oil and sweat glands and appears dry and coarse.

Wide angle microscopy may be helpful in distinguishing patients of limited cutaneous scleroderma who demonstrate enlargement of capillary loops without loss of capillary beds, whereas in diffuse scleroderma there is a loss of capillary loops.

Musculoskeletal features

More than half of the patients with Scl complain of pain, swelling and stiffness of fingers and knees. A symmetrical polyarthritis resembling RA is also observed. In advanced cases leathery crepitations can be palpated over joints. Muscle weakness is observed in patients with severe diffuse disease and a few patients develop myositis which is identical to polymyositis.

Pulmonary involvement

Lung disease is a frequent manifestation and has replaced renal disease as a leading cause of mortality.

Interstitial lung disease is present in almost all patients of diffuse scleroderma and half of these patients may be asymptomatic. Patients present with dry cough and progressive dyspnoea on exertion. Bibasilar velcro rales are typically present on auscultation.

Pulmonary function tests show reduction in vital capacity, low diffusing capacity and low PO_2 with exercise. Chest film shows reticular appearance, honey combing or mottling especially of lower two thirds of lung. HRCT can detect early pulmonary changes showing 'ground glass' appearance due to interstitial alveolitis. Bronchoalveolar lavage (BAL) may detect increased number of alveolar macrophages or neutrophils as evidence of alveolitis.

Patients with diffuse scleroderma who have antitopoisomerse 1 antibody are particularly at risk of developing pulmonary fibrosis. In a subset of patients of limited cutaneous scleroderma after many years of disease, pulmonary hypertension may develop. This is due to narrowing and obliteration of pulmonary arteries and arterioles by intimal fibrosis and medial hypertrophy. It is manifested initially by exertional dyspnoea and eventually by right sided heart failure. The prognosis is poor[3]. With development of pulmonary hypertension, the mean duration of survival is less than 2 years. Other pulmonary complications of scleroderma include aspiration pneumonia, pulmonary hemorrhage, pneumothorax, respiratory failure and increased risk of lung cancer.

The factors signifying poor prognosis of lung disease are male sex, low DLCO, presence of early lung involvement, severe Raynaud's phenomenon, and cigarette smoking.

Cardiac features

Primary cardiac involvement in scleroderma include pericarditis with or without effusion, heart failure, cardiomyopathy (< 10%) to heart blocks or arrythmias. Clinical evidence of cardiac involvement is either subtle or absent. Intermittent vasospasm of coronaries has been demonstrated. Patients may develop left ventricular failure due to systemic hypertension or cor pulmonale due to pulmonary arterial hypertension.

Gastrointestinal involvement

A majority of patients of Scleroderma have symptom due to oesophageal involvement and include epigastric pain or regurgitation of gastric contents. These symptoms are due to reduced pressure of lower oesophageal sphincter. Reflux oesophagitis leads to Barrett's metaplasia. Dysphagia to solid food may occur due to loss of oesophageal motility. Reduced motility of oesophagus can be demonstrated by radionuclide study[5] and oesophageal manometry.

Hypomotility of small intestine produces symptoms of bloating and abdominal pain and can lead to malabsorption with weight loss and diarrhea. Barium study of large intestine may show dilatation, atony and diverticula.

Renal involvement

Renal disease in scleroderma may manifest as proteinuria, azotemia or hypertensive crisis & is seen in 20-30 % of cases with diffuse scleroderma[6,7] and also in those with rapidly progressive and widespread skin thickening in the first 2-3 years of disease. Scleroderma renal crisis (SRC) comes abruptly and is accompanied by severe hypertension, azotemia and microangiopathy. Malignant hypertension may progress to renal failure. New unexplained anemia in patient with diffuse scleroderma is an early warning sign of renal crisis. Kidney histology reveals intimal hyperplasia, fibrinoid necrosis, medial thinning and adventitial fibrosis of interlobular and intralobular arteries. The mechanism of renal crisis is activation of renin angiotensin system. Before the advent of ACE inhibitor drugs, the majority of these patients died within 6 months. A small number of patients may develop renal crisis in the absence of hypertension.

Factors predictive of SRC are diffuse skin involvement, rapid progression of skin disease, disease duration of less than 4 years, new onset of anaemia, cardiac event, and high dose corticosteroid therapy.

Miscellaneous features

These include secondary Sjogren's syndrome, trigeminal neuralgia, depression, and hypothyroidism. Systemic sclerosis in scleroderma is defined by characteristic visceral and serological abnormalities without skin changes. In overlap syndrome one may see manifestastions of scleroderma, SLE, RA, polymyositis or Sjogren's syndrome, whereas the term undifferentiated connective tissue disorder is reserved to classify these patients with Raynaud's phenomenon and clinical or serological features of scleroderma.

Laboratory investigations and diagnosis

Baseline workup includes complete blood count, ESR (erythrocyte sedimentation rate), antinuclear antibody testing. (Other relevant investigations are discussed in management according to specific organ evaluation).

ANA is detected in more than 95% of the patients of scleroderma. Nailfold capillaroscopy is a noninvasive means of detecting vascular changes characteristic of scleroderma. The typical changes include giant loops or a decreased number of capillary loops, whereas in case of primary Raynaud's phenomenon, the capillaroscopy is normal, there is no clinical feature of scleroderma.

Anticentromere antibody is found in 75% of the cases with limited scleroderma while antiScl-70 (antitopoisomerase 1) occurs in 50% of those with diffuse scleroderma.

Diagnosis of scleroderma is fairly simple & clinical. The American College of Rheumatology (ACR) classification criteria 1980[8] are as follows:

- Major
  - Skin thickening proximal to metacarpo-phalangeal joints
- Minor
  - Sclerodactyly (Fig. 4)
  - Digital pitting scars or the loss of subcutaneous tissue of the fingertips
  - Chronic interstitial pulmonary changes on chest radiograph.

Patients exhibiting one major and two minor criteria are considered to have scleroderma. The classification criteria are insensitive in the early, undifferentiated or overlap patients.

Management of scleroderma

Even though scleroderma cannot be cured, treatment of involved organ systems can relieve symptoms and improve function. It naturally endangers great fear in many patients since the disorder can threaten not only their physical and functional abilities but also their professional, social and emotional lives. Thus the overall management of scleroderma requires a high level of education and encouragement of patients in combination with a team approach to care. The team should include the patient, family physicians, physiotherapist, nurse educators, occupational therapists, social workers, counselors and community services. Input from patients support groups can also be helpful.

Initial assessment

The goal of the initial assessment is to determine the distribution of organ involvement and if possible to decide whether organ dysfunction is related to potentially reversible inflammation or vasoconstriction (active disease) or whether the damage is irreversible (eg. Fibrosis). Not all of the following tests are necessary in every patient with scleroderma, but some are necessary to decide the treatment plan.

Skin

- Assess the distribution and severity
- Skin biopsy to differentiate scleroderma from other types of skin thickening when diagnosis is not clear

Kidney

- Baseline Renal Function tests i.e. GFR, 24 hrs urinary protein, Sr. creatinine, urine analysis, BUN, creatinine clearance

GI tract

- Upper GI contrast radiography
- Oesophageal manometry
- Serum muscle enzymes (CPK, aldolase, LDH)
- Assessment of malabsorption (stool examination)

Lungs/heart

- Plain chest radiograph, HRCT, pulmonary function test including DLCO (diffusing capacity for carbon monoxide), arterial blood gas analysis, and echocardiography are recommended tests.
- In selected patients bronchoalveolar lavage, lung biopsy, right heart catheterization, serum CK, troponin, ECG monitoring is necessary
- Assessment of pulmonary hypertension (PAH) by 2-D echo

Musculoskeletal

- Muscle enzymes (CPK; aldolases), Electromyography (EMG), muscle biopsy

The initial assessment, described above is essential for the treatment plan for individual patient.

Organ based treatment

Skin involvement

Lesions of localized scleroderma, including morphea appears to soften with ultraviolet - A (UVA) light therapy, other options include potent topical corticosteroids, calsipotriol and methotrexate.

Many drugs have been used in the treatment of systemic sclerosis without any consistent or prolonged benefit. In uncontrolled studies, penicillamine has been reported to reduce skin thickening and prevent development of significant organ involvement[9]. This drug interferes with inter and intramolecular cross-linking of collagen and is also immunosuppressive which reduces collagen production. Starting dose of D-penicillamine is 125 mg/day and increased to 1.5 gms/day within 1-3 months. There is no major difference between low dose and higher dose of D-penicillamine[10]. It is a toxic drug leading to glomerulonephritis with nephrotic syndrome, aplastic anaemia, leucopenia, thrombocytopenia and myasthenia gravis. Other side effects include fever, rash, anorexia, loss of taste. Patient should have monthly blood count and urine analysis.

Other drugs used in scleroderma are cyclophosphamide, methotrexate, azathioprine and other immunosuppressive drugs like chlorambucil, flurouracil[11]. These are reserved for those patients with rapidly progressive disease. Because of poor prognosis in scleroderma patients who have rapid progressive disease with organ involvement, high dose immunosuppressive therapy with cyclophosphamide followed by autologous stem cell transplantation is under trial.

Currently, no single drug has been proven to successfully control Scl disease process. But immunosuppressive therapy particularly when given at the early inflammatory phase could potentially alter the natural course of the disease. Improvement in modified Rodnan's skin score (mRSS) after treatment with Mycophenolate mofetil (MMF) compared with baseline was seen as early as 3 months and continued through 12 month follow-up[12].

Intense pruritis in early stages of scleroderma is managed with antihistaminics, skin lubricating creams and low dose oral steroids are rarely helpful. Telangiectasia mainly on face can be treated with laser therapy. Calcinosis causes great distress. Pharmacological agents like probenecid, colchicine and warfarin have failed to dissolve or prevent new calcinosis, diltiazem, calcium channel blocker may be useful treatment for calcinosis as it also helps in pulmonary hypertension[13]. Calcinosis can be removed surgically if suitable.

Many other treatment options like high doses of 1, 25 dihydroxy vit D3, retinoid, interferone, cyclosporin, photochemotherapy, plasmapheresis are studied in scleroderma with disappointing results. Other therapies investigated for treatment of skin thickeneing includes the use of humanized anti-transforming growth factor (TGF) beta-1-monoclonal antibody and plant alkaloid halofuginone and thalidomide[14].

Raynaud's phenomenon

The managenent of Raynaud's phenomenon is directed at control of vasospasm. Suggested treatments are as follows:

- Avoid cold and stress, dress warmly and wear mittens and socks
- Avoid smoking and drugs like ergotamine/amphetamine/β-blockers

Drug therapy

- Calcium channel blockers - nifedipine
- Vasodilators - reserpine, prazosin, methyl dopa
- Ketanserin, losartan, bosentan
- Prostacyclin analogues - epoprostenal, iloprost and pentoxyfyllin
- Antiplatelet drugs - aspirin, dipyridamole

Surgical therapy

- Surgical sympathectomy
- Digital sympathectomy (radical microarteriolysis)

Treatment of digital ulcers

- Soak with warm water
- Topical and oral antibiotics
- Debridement and amputation

In above mentioned treatment, calcium channel blockers nifedepine, diltiazem and amlodipine can be effective but side effects like palpitation, light headedness limits their use. Usual dose of nefedipine is 60-90mg.

Nitroglycerine pastes locally and losartan (angiotensin II receptor antagonist) reduces severity and frequency of Raynauds phenomenon. Ketanserine, fluoxetine are oral serotonin antagonist decreases 5-hydroxytryptamine which plays a major role in Raynaud's phenomenon.

Prostacyclin analogs like iloprost given intravenously 0.5-3.0 ng/kg/min for 3 to 5 days or orally (50-150 mcg twice a day) is a potent vasodilator and also inhibits platelet adhesion and aggregation, alters neutrophil function and repairs damaged endothelium. It reduces severity and frequency of Raynaud's phenomenon. Other prostacyclin analogs are epoprostenal and beraprost.

Pentoxifylline and antiplatelets like aspirin and dipyridamole improves the blood flow by reducing platelet adhesion and increasing deformability of red blood cells.

If patient does not respond to medical treatment, surgery may be necessary including surgical sympathectomy (cervical or lumbar), radical microarteriolysis (digital sympthectomy), deb ridement and amputation of digits.

Lung involvement

There are two main types of parenchymal lung disease in scleoderma[15,16]

- Fibrosing alveolitis leading to interstitial lung disease (Fig. 2a, 2b)
- Pulmonary vascular disease (pulmonary hypertension)

Inflammatory fibrosing alveolitis

At initial assessment patient is assessed for having inflammatory alveolitis (active disease) and interstitial fibrosis (chronic state) by HRCT and bronchoalveolar lavage. In active disease drug therapy is helpful. Once extensive lung fibrosis is present, the best response to treatment expected is stabilization of the disease. It should not be regarded as failed treatment.

The major drugs that are currently used are corticosteroids, cyclophosphamide, azathioprine and D-penicillamine. Other drugs with no improvement are colchicine, chlorambucil, methysergide, para aminobenzoic acid.

There are different regimens for active lung disease. One approach is to initiate low dose prednisolone 30-40 mg/day plus cyclophosphamide 600-800 mg/ monthly intravenously or 2mg/kg orally daily. Prednisolone is tapered slowly over 3 months. After 6 months the lung condition is reassessed and dose of cyclophosphamide is individualized. The optimal duration of therapy is unclear. Other approach is to replace cyclophosphamide by azathioprine (1-2 mg/ kg/day) to avoid adverse effects[17,18,19].

In view of RCT, cyclophosphamide (cyc) should be considered for the treatment of scleroderma related ILD. In one study, 158 Scl patients with active alveolitis demonstrated that cyc given orally at a dose of 1-2 mg/kg per day improved lung function test, dyspnoea score and quality of life over 12 months compared with placebo. In another trial, cyc given intravenously at a dose of 600 mg/m^2 was compared with placebo in 45 Scl patients with Scl-ILD. The treatment included 6 infusions of cyc given at 4 weekly interval followed by oral Azathioprine (2.5 mg/kg per day) or placebo for 6 month. Prednisolone was co-administered in active treatment group.

It is a policy to continue treatment for two years before making decision about efficacy. Once disease becomes stable, consideration is given to withdrawing therapy gradually. But once there is lung fibrosis, treatment may not be helpful. Patient should receive polyvalent pneumococcal and influenza vaccination.

Pulmonary hypertension

The therapeutic options for treatment of pulmonary hypertension with scleroderma includes nifedipine, anticoagulation, prostacyclin analogs, endothelin receptor antagonists (Bosentan).

Prognosis of pulmonary hypertension is very poor. If it is refractory to drugs, we can consider heart-lung or single lung transplantation if there is no other major organ involvement.

Nifedipine

It improves pulmonary vascular resistance but there are no definitive studies that shows an improvement in survival dose 60-90 mg/day.

Anticoagulation

Chronic anticoagulation with warfarin improves the survival. Adjust the dose of warfarin to maintain INR - 2.5.

Prostacyclin analogs

Epostenal (2.2-10.0 mg/kg/min) promise when delivered as constant 24 hr infusion via indwelling catheter and pump. It reduces pulmonary resistance and improves cardiac output. The invasiveness of delivery system, adverse effects and high cost limit the use of epoprostenal[20,21].

Iloprost as inhalation therapy is also comparable to epoprostenal. Flushing and jaw pain are more common with iloprost.

Bosantan : It is a nonselective endothelin receptor antagonist. It is safe and effective orally. Dose 125-250 mg bd. The oral route of administration make it an attractive alternative. It produces increase in serum aminotransferase enzyme levels[22]. Bosentan has been proved to improve excercise capacity, functional class and haemodynamic measures in PAH.

Another selective endothelin receptor antagonist called Sitazentan at a dose of 100 mg/day significantly improved exercise capacity and hemodynamics.

Phosphodiesterase inhibitors : Sildenafil (50 mg orally daily) is as efficacious as epoprostenal for reducing pulmonary vascular resistance. It improves ventilation - perfusion matching and oxygenation.

Renal involvement

Renal hypertension crisis in scleroderma is more common in first five years. High blood pressure and proteinuria may be initial presentation. Control of blood pressure is the main stay of treatment of scleroderma renal crisis. ACE inhibitors are first line antihypertensive drugs. Other drugs include calcium channel blockers, beta blockers, clonidine etc. Dialysis may be required in patients with progressive renal failure. Patients are usually not candidates for kidney transplantation because of other system involvement[23]. Steroids are associated with a higher risk of SRC and patients with steroids should be carefully monitored for blood pressure and renal function.

Gastrointestinal involvement

Since 40% of patients with scleroderma related oesophageal reflux are asymptomatic, the emperic use of acid reducing agents is generally recommended. Reflux oesophagitis is treated with frequent meals, antacids, elevation of head of bed, avoiding tea, coffee, alcohol and chocolates which reduces lower oesophageal sphincter tone. H2 blocker (ranitidine, cimetidine), proton pump inhibitor (omeprazole, pantoprazole, etc), metoclopramide and itopride which increases GI motility are useful drugs.

Malabsorption and bacterial overgrowth can be treated with antibiotics like ciprofloxacin, metronidazole and doxycyclin. Parental nutrition may be useful in some patients. Octreotide may sometimes be useful in pseudo-obstruction. Surgical treatment may be necessary in some patients.

Musculoskeletal involvement

Baseline assessment of the patients for joint involvement includes a survey for swollen, tender or deformed joints. Arthritis symptoms require NSAIDs or low dose of steroids to control inflammation and pain. Physiotherapy is important in limiting contractures. It should include both active and passive exercises. Encouragement from the therapist and physician is also critical if a regular programme is to be performed. Surgery has an uncertain role in realignment of contracted fingers, but it does not greatly improve function.

A mild myopathy with little biochemical or histological change is a common feature of scleroderma. Treatment is similar to idiopathic polymyositis. Glucocorticosteroids alone or in combination with methotrexate, azathioprine or other immunosuppressive agents are used. Steroid increases risk of renal crisis.

Cardiac involvement

The direct cardiac manifestations of scleroderma (those not induced by systemic or pulmonary hypertension) includes pericarditis, pericardial effusion, myocarditis, myocardial fibrosis, arrhythmias and coronary artery disease.

Pericarditis without tamponade is treated with NSAIDs and corticosteroids. Cardiac tamponade typically requires pericardiocentesis. If effusion is loculated or there is special need for biopsy or patient had coagulopathy then direct surgical drainage is required.

Myocardial fibrosis is present in upto 90% of autopsies. It produces cardiomyopathies (dilated or restrictive) and arrhythmias. Restrictive cardiomyopathy doesn't respond to steroids. Myocarditis requires many times combination of glucocorticosteroids and cyclophosphamide. Digoxin, diuretics, antiarrhythmatics are required for heart failure.

Patient with dry mouth and dry eyes are treated with pilocarpine hydrochloride tablets and artificial tears respectively.

Pregnancy and Scl

The effect of pregnancy on Scl is greater possibility of renal crisis when scl is of recent onset. There is a risk of prematurity and small full term infant because of Scl induced placental vascular abnormalities.

Prognosis

Systemic sclerosis of either the limited or diffuse cutaneous subset confers substantial increase in the risk of premature death. As the pathophysiology of scleroderma becomes better understood more specific therapies may be developed in future.

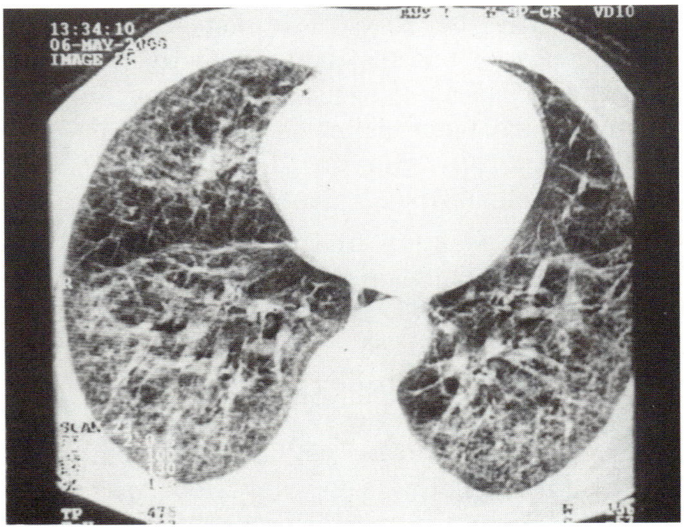

Fig. 2a : HRCT of Scleroderma showing ground glass appearance in interstitial lung disease

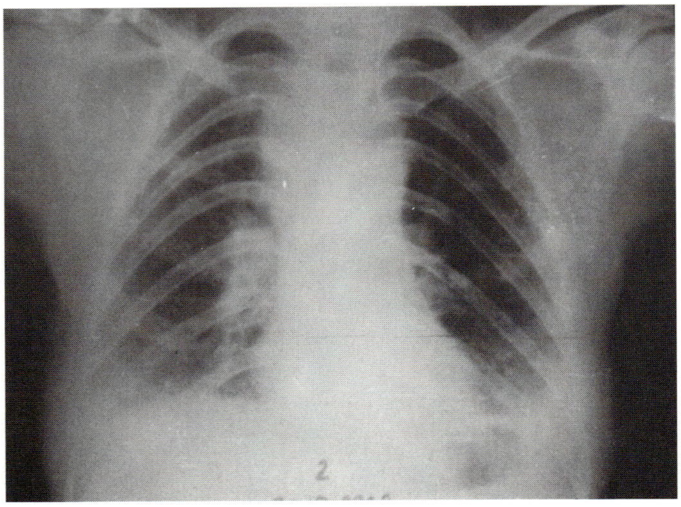

Fig. 2b : HRCT of Scleroderma showing ground glass appearance in interstitial lung disease

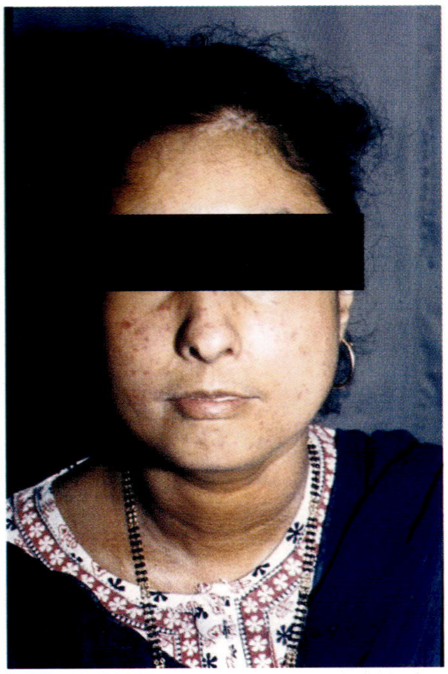

Fig. 3 : Clinical picture of scleroderma showing telangiectasia

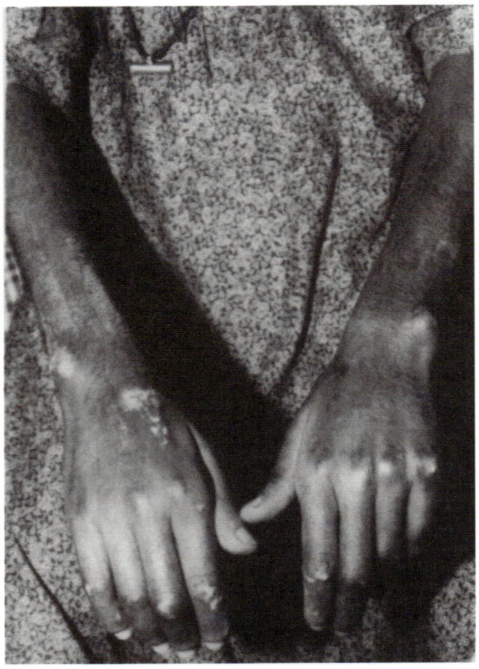

Fig. 4 : Clinical picture of scleroderma showing taut skin over hands and sclerodactyly

Further Reading :

1. Kumar A, Malviya AN, Tiwari SC, et al. Clinical and laboratory profile of systemic sclerosis in northern India, J Assoc. Phy. Ind 1990; 38: 765-68.
2. Krishnamurthy V, Chandrasekharan AN, et al. Progressive systemic sclerosis in south India, J A Assoc. Phy Ind 1991; 39: 254-57.
3. Steen VD, Conte C, Owens GR, Medsger TA, Severe restrictive lung disease in systemic sclerosis. Arthritis & Rheumatism 1994; 37: 1283-89.
4. White B, Moore WC, Wigley FM, Xiao HQ, Wise RD. Cyclophosphamide' is associated with pulmonary function and survival benefit in patients with scleroderma and alveoitis. Ann of Intern Med 2000; 132: 947-54.
5. Garg A, Prakash K, Gopinath PG, Malviya AN. Radionuclide esophageal transit time in progressive systemic sclerosis. Ind J Med Res 1984; 79: 110-13.
6. Steen V D, Renal involvement in systemic sclerosis. Cli Dermatol 1994; 12: 253-58.
7. Desai V, Ghanekar MA, Sequeira RD et al. Renal involvement in scleroderma J A Assoc. Phy Ind 1990; 38: 768-70.
8. Masi AT, Rodnan GP, Medsger TA, et al. Sub-committee for scleroderma criteria of the American Rheumatism Association Diagnostic criteria for the classification of systemic sclerosis (scleroderma) Bull Rheum Dis 1981; 31: 1-6.
9. Steen VD, Medsger TA, Rodnan GP, D-penicillamine therapy in progressive systemic sclerosis (scleroderma), Ann of Intern Med 1984; 97: 652-59.
10. Clements PJ, Furst De: High dose versus low dose D-penicillamine in early diffuse systemic sclerosis: Analysis of two double blind randomized, controlled clinical trial, Arthritis Rheum 1999; 42: 1194.
11. Seibold JR. McCloskey DA, Furst DE: Pilot trial of methotrexate (MTX) in treatment of early diffuse scleroderma (abstract). Arthritis Rheum 1994; 32: R35.
12. Elizabeth N. Le, Fredrich M Wingley; Ann. Rheum. Dis. 2011; 70: 1104-07.
13. Dolan AL, Kassimos D, Gibson T et al. Diltiazam induces remission of calcinosis in scleroderma Br J. Rheumatol 1995; 34: 576-78.
14. Seibold JR et al. Recombinant human relaxin in the treatment of scleroderma. A randomized double blind, placebo controlled trial. Ann intern Med 2000; 132: 871.
15. Black CM: Systemic sclerosis: Management in Rheumatology, 2nd ed, Mosby International 1998, 7: 11.
16. Bolster MB, Silver RM: Assessment and management of scleroderma lung disease Curr Opin Rheumatol 1999; 11: 508.
17. Silver RM, Warrick JH, Kinsella MB et al. Cyclophosphamide and low dose prednisolone therapy in patient with systemic sclerosis with interstitial lung disease, J Rheumatol 1993; 20: 838-44.
18. Steen VD et al. Therapy for severe interstitial lung disease in systemic sclerosis. A retrospective study. Arthritis Rheum 1994; 37: 1290.
19. Donald P, Taskin MD, Robert E Cyclophosphamide versus placebo in scleroderma lung disease, N Engl J Med, 2006; 354: 2655-66.
20. Barst RJ, Rubin LJ, Long WA, et al. A comparison of continuous intravenous epoprostenal (prostacyclin) with conventional therapy for pulmonary hypertension N Engl J Med 1996; 334: 296-301.
21. Kilngs ES et al. Systemic sclerosis associated pulmonary hypertension: Short and long term effects of epoprostenol (prostayclin), Arthritis Rheum 1999; 42: 2638.
22. Renaudineauz et al. Anti endothelial cell antibodies in systemic sclerosis. Clin Diagnosis lab Immunol 1999; 6: 156.
23. Zawada et al. Clinical course of patients with scleroderma renal crisis treated with captopril. Nephron 1981; 27: 74-78.
24. van den Hoogen F, Khanna D, Fransen J et al. 2013 Classification Criteria for Systemic Sclerosis: An American College of Rheumatology/European League Against Rheumatism Collaborative Initiative. Arthritis & Rheumatism. doi: 10.1002/art.38098 also in Ann Rheum Dis 2013;72(Nov):1747-55.

IDIOPATHIC INFLAMMATORY MYOPATHIES

7

Shabbir Chikani, Nimish Nanavati

Idiopathic Inflammatory Myopathies (IIM) are a group of heterogeneous diseases characterised by noninfectious inflammation of skeletal muscles. They are broadly classified into clinical subsets (Table 1), Dermatomyositis (DM) and Polymyositis (PM) are commonest. They will be discussed in detail here. A new disease entity necrotising autoimmune myositis (NAM) will be briefly mentioned. Commonest inflammatory myopathy above the age of 50 in western world, Inclusion body myositis (IBM), is very rarely reported in India. Due to its distinct clinical features, refractoriness to conventional immunosuppressive therapy it is discussed separately. Juvenile Dermatomyositis (JDM) is distinct clinical subset. Subsets of these autoimmune diseases carry good prognosis if diagnosed and treated in early disease course.

Epidemiology

These diseases are relatively rare. Annual incidence is reported to be 2-10 per million. In Canada, prevalence of polymyositis and dermatomyositis was estimated to be 21.5/100,000, in 2003. Prevalence of Inclusion body myositis (IBM) varies from 1-71/ million in western countries. Peak incidence in adults is seen in 35-44 and 55-64 age groups for DM/ PM. Juvenile DM peak is seen at the age of 5-9 years. Necrotising autoimmune myositis (NAM) occurs in adults of middle age and older. Inclusion body myositis (IBM) predominantly occurs above the age of 50. As with other connective tissue disorders there is female predominance (M:F ratio 2-3:1) in DM/PM/ JDM. IBM is more common in males.

Data from India is scarce. One study from south India reported 0.1% of OPD patients/0.6% of indoor patients in neurology department affected by IIMs over 17 years. Age at onset of disease was one to two decade younger as compared to western countries. Sex ratio was similar.

Etiopathogenesis

Etiology of these diseases is unknown. In a genetically susceptible individual with aberrant immune system, exposure to environmental agent leads to autoimmunity against predominantly skeletal muscles. Both innate and adaptive immunity involving humoral and cell mediated immune systems take part in varying degree in different clinical subsets, leading to end result of muscle injury.

HLA genes on chromosome 6 esp. HLA B8, HLA DR3, HLA DRw52 are strongly associated with poly/ dermatomyositis. HLA-DPB1*0101 was associated with IIM overall with odds ratio of 2. HLA DR1, HLA DQ1, HLA DQ2 are associated with inclusion body myositis. Major histocompatibility complex (MHC) class 1 restricted molecules play important role in pathogenesis of PM/JDM/IBM.

Infectious agents are suspected to trigger the onset of these diseases. Clue for these are provided by seasonal variation in the incidence of disease in a subset of patients esp. in NAM. Serological evidence of recent infection of viruses like influenza, coxsackie group, CMV and Toxoplasma was found in a group of patients. NAM is also associated with neoplasia, drugs esp. Statins and autoimmunity. Strong correlation of ultraviolet radiation and Dermatomyositis is recently observed. Diminished proteosomal/ lysosomal protein clearance has important role in IBM leading to accumulation of amyloid like proteins in vacuoles.

Various probable inciting factors listed above (many unknown) leads to activation of innate immunity via toll like receptors (TLR -2,3,4,7,9), interferon system. These in turn leads to recruitment of cell mediated immunity involving T cells, macrophages, B cells. Through effector mechanisms like complement in DM, this leads to myonecrosis. Part of muscle weakness experienced by patient is mediated through nonimmune-mediated mechanisms like an acquired metabolic myopathy and endoplasmic reticulum stress.

Classification

Classifying heterogeneous myositis disorders into mutually exclusive groups by using clinical and immune response features is useful for predicting responses to therapy and prognosis. IIM can be classified in various ways based on clinical features and/ or serology and pathology.

Modified classification of Bohan and Peter's using clinical features (Table 1), originally proposed in 1975 without inclusion body myositis, is still widely accepted. Autoantibody based classification (Table 2) is proposed to better define clinical syndromes and improve prognostication.

Proposed pathologic classification include categories like immune myopathies with perimyseal pathology (IMPP), associated with antisynthetase antibodies; myovasculopathies, including childhood dermatomyositis; immune polymyopathies, active myopathies with little inflammation such as the myopathy with anti SRP antibodies; immune myopathies with endomysial pathology (IM-EP), histiocytic inflammatory myopathies, like sarcoid myopathy; and inflammatory myopathies with vacuoles, aggregates and mitochondrial pathology (IM-VAMP).

Table 1. Clinical classification

1	Primary idiopathic polymyositis
2	Primary idiopathic dermatomyositis
3	PM/DM associated with malignancy
4	PM/DM associated with connective tissue diseases
5	Juvenile dermato (Poly) myositis
6	Inclusion body myositis
7	Rare myositis: Eosinophilic Granulomatous Focal

Table 2. Autoantibody based classification

Autoantibody	Clinical association	Prevalence	Clinical manifestations	Steroid response
Antisynthetase (i.e. anti Jo-1)	Anti synthetase syndrome	20-50%	ILD, arthritis, mechanics hands, Raynaud's phenomenon	Moderate
Anti SRP	Severe resistant PM	< 5%	Cardiac involvement	Variable
Anti Mi-2	Classic DM	5-10%	V-sign, shawl sign, periungual erythema	Good

CININCAL FEATURES

Clinical features can be divided into musculoskeletal, cutaneous and extra musculoskeletal. Clinical hallmark of these diseases is subacute symmetrical proximal weakness with exception of IBM.

Musculoskeletal

- Weakness involving pelvic/shoulder girdle muscle is the earliest symptom. It progresses insidiously over 3-6 months. Rare acute presentation can occur. Patient generally complains of difficulty in getting up from chair, climbing stairs, combing hairs etc.
- Muscle pain is usually minimal or absent.
- Involvement of pharyngeal muscle may result in dysphagia, nasal regurgitation and risk of aspiration pneumonia. If respiratory muscles are involved it can cause dyspnoea.
- Ocular, facial and distal muscle weakness is unusual in poly/dermatomyositis, *if present alternative diagnosis should be sought.*
- Distal weakness involving small muscles of hand and feet can occur in inclusion body myositis.
- Arthralgia/non-erosive arthritis is seen in 20-70% of patients.

Cutaneous (in dermatomyositis)

- Heliotrope rash: (Fig. 1): Purple to erythmatous (violaceous) rash affecting eyelids, malar region, forehead and nasolabial folds. Malar rash of DM crosses nasolabial fold in contrast to SLE rash.

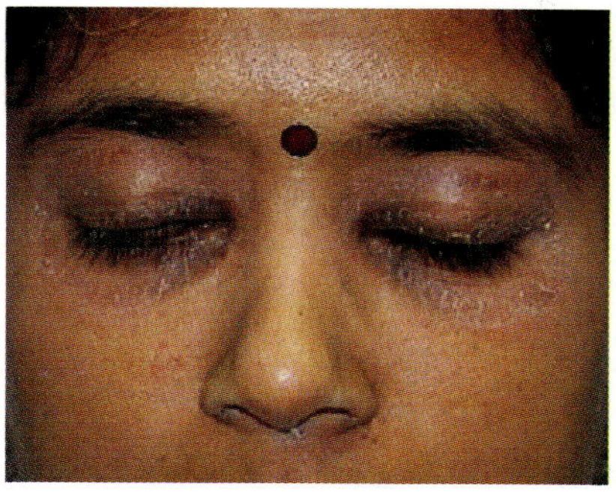

Fig. 1: Heliotrope rash

- Gottron's papules: (Fig. 2): Purple to erythematous, flat or raised lesions over extensor aspects of small joints of hand and occasionally large joints.

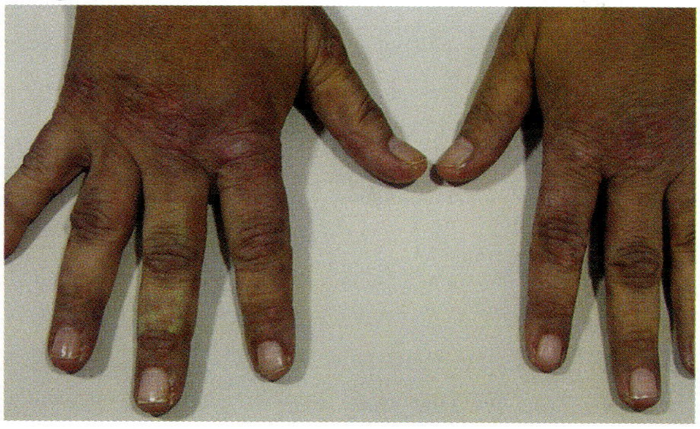

Fig. 2: Gottron's papules

- V-sign: erythematous rash over anterior chest and neck.
- Shawl sign: erythematous rash over shoulder and proximal arms.
- Nail fold abnormalities: periungual erythema, dilated capillary loops
- Subcutaneous calcifications is seen especially in juvenile DM
- Vasculitic ulcers (Fig. 3), livedo reticularis

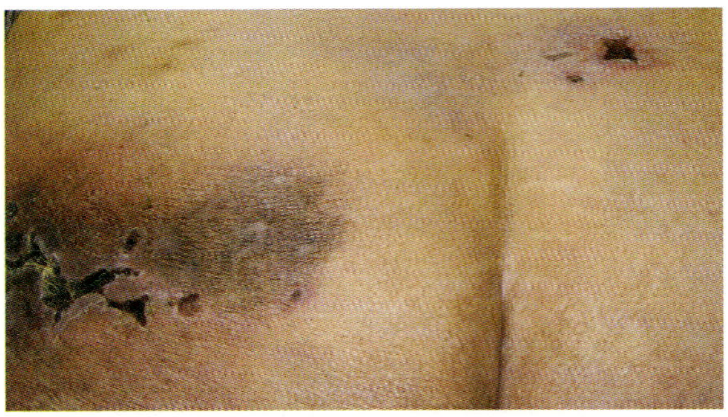

Fig. 3 : Vasculitic ulcers in DM

Extra musculoskeletal

- Dyspnoea on exertion and/or dry cough should alert physician to suspect involvement of pulmonary system. It can occur in the form of ILD (5-10%), aspiration pneumonia, respiratory muscle involvement and rarely pulmonary hypertension.
- Gastrointestinal: dyspepsia due to oesophageal dysmotility (10-30%), intestinal perforation due to vasculitis (rare).
- Cardiac: symptomatic cardiac involvement is rare. Subclinical involvement in the form of ECG abnormalities (dysrhythmias, conduction blocks), myocarditis, congestive heart failure.
- Fatigue, low-grade fever, weight loss may be present.

Association with malignancy

- Increased incidence of malignancy is seen most commonly with DM. Recent studies have shown 6 times increased risk of malignancy with DM and 2 times increased risk with PM.
- Malignancy can be present at diagnosis or can appear up to 5 years after initial diagnosis.
- Breast, ovarian, prostate, stomach, colon, pancreas, lung, nasopharyngeal and non Hodgkin's lymphoma is reported with increased frequency. Nasopharyngeal cancer is more common in Asians.
- Especially patients with age > 50, acute onset of disease, dysphagia, negative autoantibodies, smokers, family history of breast/colon cancers should be carefully searched.

Necrotising autoimmune myositis

- Associated with anti SRP related antibody mediated myositis, statins, cancers, viruses.
- Clinically subacute symmetrical muscle weakness
- Anti 3 HMGCR antibodies detected in subset of statin induced myositis.
- Very high level of CPK > 10 times normal.
- Good response to corticosteroids.

Inclusion body myositis

- Special mention should be made of this unique disease with many atypical features.
- It occurs predominantly in males aged more than 50
- Muscle involvement is predominantly distal. It can be asymmetrical and associated with significant atrophy. Patients give history of frequent falls.
- There may be signs of sensory neuropathy.
- Extraskeletal manifestation is rare.
- Progression of disease occurs over many years.
- Refractory to conventional treatment.

Differential diagnosis

D/D of inflammatory myopathies is large and can be subdivided in to following groups.

Infectious: post viral (commonest i.e. influenza, adenovirus, HIV), bacterial (staphylococcus, streptococcus), parasitic (Toxoplasma, trichinosis)

Drug induced: statins (non immune myopathy), colchicine, antimalarials, AZT

Endocrine: Hypo/hyperthyroid (commonest), Cushing's disease, Addison's disease

Metabolic disorders: hypokalemia, diabetic ketoacidosis

Other rheumatic disorders: polymyalgia rheumatica, fibromyalgia, inflammatory arthritides, systemic vasculitis

Neuromuscular disorders: myasthenia gravis, Eaton-lambert syndrome, muscular dystrophies

Clues pointing to the diagnosis other than myositis are given in Table 3.

Table 3. Clues against diagnosis of Idiopathic Inflammaroty myositis

- Muscle pain out of proportion of weakness.
- Acute onset with preceding history suggestive of infection
- History of drugs causing myopathies
- History of recent trauma, needle stick injury, recent surgery
- Sensory symptoms, fasciculations, abnormal reflexes
- Facial or ocular involvement
- Symptoms related to exercise, fasting
- Family history of similar illness
- Early atrophy or hypertrophy of muscles
- CPK < 2 or > 100 times normal

Diagnosis

High index of suspicion is required to diagnose these disorders, as symptoms are often non-specific and insidious at an early stage. Criteria for diagnosis of poly/dermatomyositis are given in Table 4. Algorithm to diagnose idiopathic inflammatory arthritis is given in the Chart.

Table 4. Criteria for diagnosis of PM/DM

1	Symmetrical proximal weakness developing over weeks to months
2	Elevated serum muscle enzymes i.e. CPK, SGOT, LDH
3	Typical EMG findings: myopathic potentials-low amplitude, polyphasic, short duration, fibrillation
4	Typical muscle biopsy findings
5	Dermatological signs of DM: Gottron's papules, heliotrope rash, V sign, Shawl sign

Once suspected following investigations should be done to confirm the diagnosis.

- Muscle enzymes: CPK, SGOT, LDH, aldolase are elevated in majority of patients. CPK is the most sensitive enzyme. Its level correlates with disease activity in majority of patients. Its elevation alone without associated clinical features is not diagnostic of myositis. Other causes of elevation as mentioned in D/D should be kept in mind.
- EMG/NCV: shows characteristic polyphasic, short duration, low amplitude motor unit action potentials. Helps to rule out alternative diagnoses.
- MRI/USG: Increasing used to confirm/delineate disease process. T2/STIR images show high intensity signals suggestive of inflammation in involved muscles. Helps in identifying biopsy site, as disease process can be patchy. It minimizes the false negative results of muscle biopsy. Recent reports show correlation with therapeutic response.
- Muscle biopsy: Gold standard for diagnosis. Ideally it should be done in all patients. It helps to confirm/classify various diseases in the group of idiopathic inflammatory myopathies. It also guides therapy i.e. high dose steroid may be avoided in patients with inclusion body myositis. Inflammation is hallmark of these diseases. Both cell mediated and humoral immunity is implicated. Biopsy of inflamed muscle shows characteristic changes. In PM, endomysial infiltration of CD8+ T and myeloid derived dendritic cells is seen. In contrast, in DM,

Chart : Algorithm to diagnose idiopathic inflammatory myositis

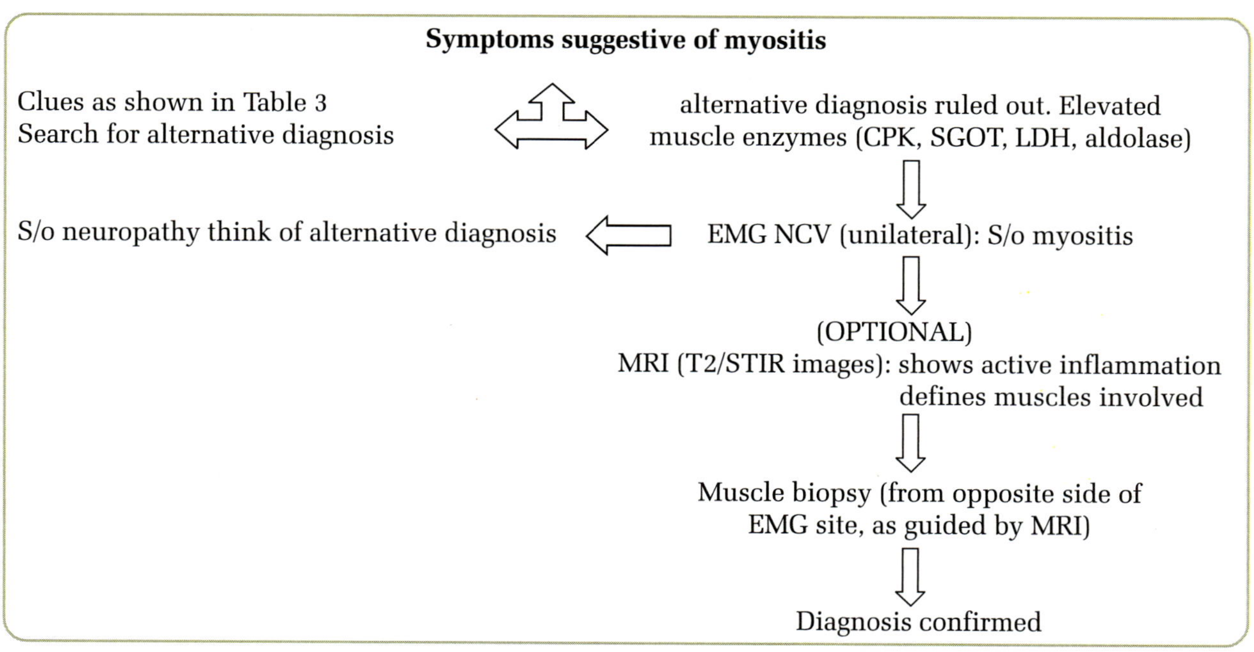

perivascular and perifascicular infiltration of CD4+ T cells and plasmacytoid dendritic cells are seen along with deposition of complement C5b-9 membrane attack complex (MAC) and secondary ischemic changes. Biopsy in NAM shows extensive muscle necrosis with less inflammation mediated by macrophages. Biopsy of IBM shows rimmed vacuoles filled with basophilic inclusions in sarcoplasm, nucleus or both more commonly visible on Gomori trichrome stains present in approximately 1-6% of myofibers. Inflammatory infiltrate is minimal unlike poly/dermatomyositis with predominant T cells. Immunostaining by MHC1 helps to differentiate inflammatory from noninflammatory myopathies reliably.

- Investigations to rule out malignancy: In patients at high risk of malignancy, age/gender specific investigations to rule out malignancy should be done. Apart from routine investigations faecal occult blood should be done in all patients. In women breast examination, Pap smear, mammogram, routine/transvaginal ultrasound and CA-125 should be done. In males above age of 50 digital rectal examination and PSA levels is warranted. High index of suspicion is needed to identify late appearance of malignancy up to 5 years after initial diagnosis.
- Autoantibodies: Though not available widely, these antibodies can help to identify various syndromes as pointed out earlier and can help in monitoring the patient for appearance of internal organ involvement/determine prognosis in early stage of disease. Various myositis specific (i.e. anti t-RNA synthetase, anti SRP antibodies, anti PL 7, anti PL 12) and myositis-associated autoantibodies (anti PM-Scl, Ku, RNP, Ro 52 kDa) are seen in 50% of PM/DM patients. Anti p 155 antibodies are strongly associated with malignancy in DM.
- Routine thorough biochemistry, x-ray chest should be done in all patients. Anaemia of chronic disease is seen in many patients. ESR is elevated in majority during active disease.
- In patients suspected to have ILD-PFT and HRCT thorax should be done.

Management

Being inflammatory diseases (with exception of inclusion body myositis) therapy is directed towards control of inflammation. Corticosteroid is the mainstay of the treatment though objective evidence of its efficacy is lacking. Other immunomodulator agents are used as steroid sparing agents. IVIg, plasmapheresis is used in special circumstances. Choice of drugs depends on severity of disease, presence/absence of extra musculoskeletal features, patient's age, and presence of co morbidities. Supportive therapy to minimise the side effects of steroid is routinely used.

- Corticosteroids are used as first line of drugs. Prednisolone is given at 1-2mg/kg in divided doses initially for 6-8 weeks till disease comes under control. Once disease is under control, steroid is consolidated into single morning dose and then tapered to maintenance dose of 5-10 mg/day over a period of 2-4 months. There are no definite tapering schedules. Therapy is guided by clinical response. Intravenous pulse methyl Prednisolone 1gm/day 3 is used at the time of diagnosis in patients with very severe disease, those with respiratory muscle involvement, dysphagia and ILD. It is also used when there is a flare of disease midway during the therapy.
- Immunomodulators: Methotrexate (up to 25 mg/week), Azathioprine (1-3mg/kg/day), Cyclosporine (3-5 mg/kg/day), Cyclophosphamide (0.5-1 gm/m^2), and Mycophenolate mofetil (1.5-2 gm /day) are used as steroid sparing agents. When to add immunomodulator is not clear. There are two schools of thoughts. Many clinicians add it later when steroid tapering leads to relapse of disease, disease is steroid resistant or there is a high risk of steroid adverse effects. Others commence these drugs from the start for quicker tapering of steroids. Methotrexate and Azathioprine are most commonly used initial drugs. Combination can be used in resistant cases.
- IVIg is given at 0.4mg/kg/day for five days. It is used initially in severe disease or during disease relapse during ongoing therapy. It has a special role in management of inclusion body myositis where it should be considered early. Repeated infusion every 8 week may be required for persistent control.
- Plasmapheresis has doubtful value and should be used as last resort.
- Approximately 25% of patients are nonresponders and continue to experience clinical relapses. In refractory cases anti CD 20 monoclonal antibody Rituximab/Occrelizumab have shown promising results in small case series. Tacrolimus is used in case reports of refractory cases. TNF α blockers should not be used due to lack of efficacy and high risk of infections.

- Biologic drugs targeting T cell mediated pathways like anti CD 52, anti IL 2, antibodies targeting complement are on horizon.
- Sunscreen lotion, topical steroids, antihistaminics are used for dermatomyositis. Hydroxychloroquine (200-400 mg/day) is helpful in rash associated with dermatomyositis.
- Apart from therapy to control inflammation supportive medications to minimize side effects of Corticosteroids must not be forgotten. These include calcium/vit D supplementation (elemental calcium 1 gm/day), bisphosphonates to prevent steroid induced osteoporosis, gastric mucosal protection using PPI and pneumococcal/influenza vaccination. INH prophylaxis is relevant in our setting and is generally used.
- Tight control of Diabetes / Hypertension should be maintained in patients with these co-morbidities.
- Ryle's tube feeding should be given to patients with pharyngeal involvement to minimize the risk of aspiration pneumonia.
- Passive/active assisted ROM exercises should be given in early stages to prevent contractures. Active strengthening programme should be added later once disease comes under control as pharmacotherapy alone does not improve physical function fully.

Duration of therapy

Therapy with maintenance dose of steroids/ immunomodulators should be continued for 1-2 years after the disease is in remission.

Monitoring of disease activity

Objective improvement in muscle strength with reduction in muscle enzyme (CPK) level together is the best indicator of drug response. CPK alone does not correlate with disease activity. There is a lag between reduction in CPK level and improvement in muscle strength. It can also increase transiently with severe exertion, intercurrent viral illness, drugs, IM injections etc. CPK alone should not be used to decide dosage of steroids/immunomodulators. Manual muscle test 8 (MMT 8) is used to record muscle strength objectively. Outcome measures like MITAX, MYOACT, MDI- myositis damage index, DSSI-dermatomyositis skin severity index, CAT-cutaneous assessment tool, IBMFRS - IBM functional rating scale are used for reliable assessment of disease activity and damage.

Prognosis

Patients with adult PM/DM and those associated with connective diseases has good prognosis with 5 year survival rate up to 85%. Patients with malignancy and ILD have poor treatment response and worst prognosis. Amongst patients with autoantibodies, those with anti Mi2 has good prognosis with 5 year survival rate of 90%, with anti SRP has survival rate of only 30%.

Further Reading :

1. Paul Plotz Liza Christopher-Stine. Adult inflammatory myopathies best practice research. Clinical rheumatology 2004; 18(3): 331-44.
2. Okada S, Weatherhead E, Targoff IN, Wesley R & Miller FW. Global surface ultraviolet radiation intensity may modulate the clinical and immunologic expression of autoimmune muscle disease. Arthritis and Rheumatism 2003; 48: 2285-93.
3. Baer AN Differential diagnosis of idiopathic inflammatory myopathies. Curr Rheumatol Rep. 2006 Jun; 8(3): 178-87.
4. George P. Kuo and John A. Carrino Skeletal muscle imaging and inflammatory myopathies Current Opinion in Rheumatology 2007; 19: 530-35.
5. J. Tomasova ´Study´nkova´, F. Charvat, K. Jarosova´ and J. Vencovsky. The role of MRI in the assessment of polymyositis and dermatomyositis. Rheumatology 2007; 46; 1174-79.
6. Marinos C. Dalakas Therapeutic targets in patients with inflammatory myopathies: Present approaches and a look to the future Neuromuscular Disorders 16 (2006); 223-36.
7. Ytterberg SR. Treatment of refractory polymyositis and dermatomyositis. Curr Rheumatol Rep. 2006 Jun; 8(3): 167-73.
8. Merrilee Needham, Frank L Mastaglia Inclusion body myositis: current pathogenetic concepts and diagnostic and therapeutic approaches Lancet Neurol 2007; 6: 620-31.
9. Pierce Gardner, Larry K. Pickering, Walter A. Orenstein, Anne A. Gershon and Kristin L. Nichol Guidelines for Quality Standards for Immunization CID 2002: 35 (1 September)-511.
10. Pestronk A. Acquired immune and inflammatory myopathies: pathologic classification. Curr Opin Rheumatol. 2011 Nov; 23(6): 595-604.
11. Christina Lianga and Merrilee Needham Necrotizing autoimmune myopathy Current Opinion in Rheumatology 2011, 23: 612-19.
12. Madan V, Chinoy H, Griffiths CE, Cooper RG. Defining cancer risk in dermatomyositis. Part I. Clin Exp Dermatol. 2009 Jun; 34(4): 451-55.
13. Madan V, Chinoy H, Griffiths CE, Cooper RG. Defining cancer risk in dermatomyositis. Part II Clin Exp Dermatol. 2009 Jul; 34(5): 561-65.
14. Dalakas MC. Immunotherapy of inflammatory myopathies: practical approach and future prospects. Curr Treat Options Neurol. 2011 Jun; 13(3): 311-23.
15. S Habibi, AV Ramanand. Juvenile Dermatomyositis: Review of Clinical Features and management. Ind J Rheumat 2012; 7: 80-86.

VASCULITIS

8

Ramnath Misra

Introduction

Vasculitides are a group of heterogeneous diseases with diverse etiology characterized by inflammation of blood vessels of any size or any site. The manifestations are varied which depend upon the size, site and number of vessels involved in the process. It may be confined to one organ or vessel type or to multiple organs and vessels and may be clinically insignificant or rapidly fatal unless recognized and prompt treatment instituted. The major manifestations are consequent to ischemia and inflammation of the tissues affected by the vasculitic process. Vasculitis can be primary or secondary to infections, drugs, malignancies or to connective tissue diseases like rheumatoid arthritis, systemic lupus erythematosus etc. The following description shall pertain to the description of primary vasculitides.

Epidemiology

Vasculitides are rare diseases. Epidemiological studies have been done mostly from Western Europe countries and USA involving the white Caucasian population. Epidemiologic studies from England, Spain and Scandinavia show an overall annual incidence of primary systemic vasculitis of approximately 20 per million population. Vasculitis appears more common with advancing age and the prevalence varies from one geographic and ethnic population to the other. Giant cell arteritis is the commonest vasculitis in North America and Western Europe with an incidence of 178 per million described in Minnesota in adults aged 50 or more. Takayasu's arteritis is more common in eastern countries and in females. Classical polyarteritis nodosa is a rare disease with an estimated annual incidence of 2 to 9 per million. But its incidence is more common in Kuwait (45 per million) and in Alaska (77 per million). A study from a predominantly English county showed that the annual incidence of Wegener's granulomatosis is 8.5 per million[1]. Kawasaki disease and Henoch-Schonlein purpura are vasculitic diseases predominantly of the childhood. The annual incidence of Kawasaki's disease was 900 per million children younger than 5 years in Japan in 1991-92 and that of Henoch-Schonlein purpura in UK is 204 per million in children[2].

In India, Henoch-Schonlein Purpura[3,4] is fairly common in pediatric population and Takayasu's arteritis[5] in adults. There are reports of small case series of patients with Wegener's granulomatosis[6-7], Polyarteritis nodosa[8], Kawasaki[9,10] disease from our country. This is attributable to under reporting of vasculitis due to lack of awareness and diagnostic facility. Polymyalgia rheumatica and giant cell arteritis usually occur in the elderly population and since only 5% of our population is above 65 years we tend to see less of these vasculitis.

Classification

The classification is based on vessel size, histology and serology. The American College of Rheumatology (ACR) proposed a classification system in 1990[11] used both vessel size and the type of inflammatory infiltrate. It classified Vasculitides into seven types which included polyarteritis nodosa (PAN), Churg-Strauss syndrome (CSS), Wegener's granulomatosis (WG), hypersensitivity vasculitis, Henoch-Schonlein purpura, giant cell arteritis and Takayasu's arteritis. The sensitivity and specificity varied considerably ranging from 71% to 95.3% and 78.7% to 99.7% respectively. ACR 1990 criteria has 2 drawbacks, it did not recognize microscopic polyangiitis as a distinct entity and not included antibodies to neutrophil cytoplasmic antigen (ANCA). In 1994, the Chapel Hill Consensus Conference (CHCC) proposed definition for ten types of vasculitis[12]. In addition to the 7 ACR subtypes, they have defined microscopic polyangiitis, Kawasaki disease and essential cryoglobulinemic vasculitis. The two classification system does not necessarily classify individual patients into same category. It had also recognized role of ANCA in diagnosis. In 2012, a revised CHCC nomenclature of vasculitides[13] was proposed to include additional categories of vasculitis (Table 1).

Table 1. Nomenclature of systemic vasculitides according to international chapel hill consensus conference (2012)[13]

Type of vasculitis	Diseases
Large vessel Vasculitis (LVV)	Takayasu's arteritis (TAK) Giant cell arteritis (GCA)
Medium Vessel Vasculitis (MVV)	Polyarteritis nodosa (PAN) Kawasaki disease (KD)
Small vessel vasculitis (SVV)	**ANCA associated vasculitis (AAV)** Microscopic polyangiitis (MPA) Granulomatosis with polyangiitis (Wegener's) (GPA) Eosinophilic granulomatosis with polyangiitis (Churg-Strauss) (EGPA) **Immune complex Small vessel vasculitis** Anti-glomerular basement membrane (anti-GBM) disease Cryoglobulinemic vasculitis (CV) IgA vasculitis (Henoch-Schonlein) (IgAV) Hypocomplementemic urticarial vasculitis (HUV)
Variable vessel vasculitis (VVV)	Behcet's disease (BD) Cogan's syndrome (CS)
Single-organ vasculitis (SOV)	Cutaneous leukocytoclastic angitis Cutaneous arteritis Primary central nervous system vasculitis Isolated aortitis Others

Table 2. Common clinical features associated with large, medium and small size vessel involvement

Vasculitis	Blood vessels involved	Clinical features
Large Vessel	Aorta and its major branches	Limb claudication, Asymmetric blood pressures, Absence of pulses, Bruits, Hypertension
Medium Vessel	Main visceral arteries and their initial branches	Mononeuritis multiplex ,Livedo reticularis, Cutaneous nodules, Ulcers, Digital gangrene, Microaneurysms
Small Vessel	Intraparenchymal arteries, arterioles, capillaries, venules	Palpable Purpura, Urticaria, Vesiculobullous lesions, Glomerulonephritis, Alveolar haemorrhage, Splinter haemorrhages, Uveitis/episcleritis/scleritis

Clinical features of vasculitis

The clinical feature of different types of vasculitis depends upon the size of the vessel that is predominantly affected. Table 2 shows the common clinical features associated with large, medium and small size vessel involvement.

Large vessel vasculitis

a. ***Giant cell arteritis (Temporal arteritis)*** is believed to be rare in our country as the demographic pattern of our population is such that only 5 % of our population is above 65 years of age. The disease typically presents abruptly with headaches, Polymyalgia rheumatica and jaw claudication in an elderly individual. Polymyalgia rheumatica refers to marked early morning stiffness of neck and shoulder girdle muscles. The physical signs are that of a thickened tender superficial temporal artery. If not diagnosed and treated early, there is a risk of blindness. The diagnosis is based upon highly raised ESR (close to 100mm/Westergren's method) and histological of superficial temporal artery.

b. ***Takayasu's arteritis:*** Though there are no epidemiological data it is believed to be common in our country. There are a number of reports from our country. Many patients are asymptomatic and are detected by chance detection of absent pulse. Hypertension, stroke or claudication in limbs in a young should be evaluated by conventional or MR angiography of the aorta and its branches for vascular stenosis. It is reported that the pattern of involvement of TA is different in Indian patients as compared to the Japanese patients. In Indians the abdominal aorta, renal artery are initially affected with subsequent involvement of the thoracic aorta whereas in Japanese patients the ascending aorta and its branches are first affected with subsequent affection of its branch, thoracic and abdominal aorta.

Medium size vessel disease

a. ***Polyarteritis nodosa*** (also called as classical PAN to distinguish from microscopic polyangiitis) is less commonly diagnosed now and many cases diagnosed earlier as PAN were indeed cases of microscopic polyangiitis. It is a rare disease. The disease occurs in the middle age and males are more frequently affected. It commonly presents with systemic features such as fever and weight loss associated with ischemic onslaught on one of the target organs like peripheral nerves (mononeuritis multiplex or polyneuropathy), kidneys (hypertension), skin (ulcers, nodules, digital gangrene, livedo reticularis) (Fig. 1), testis (orchitis) or gut (abdominal pain, bleed, perforation). Investigations reveal anemia, leukocytosis and elevated sedimentation rate but are negative for ANCA. The diagnosis is ascertained by angiographic demonstration of aneurysms in the mesenteric or renal vessels or biopsy from skin, muscle, testis or sural nerve. If left untreated it has high mortality.

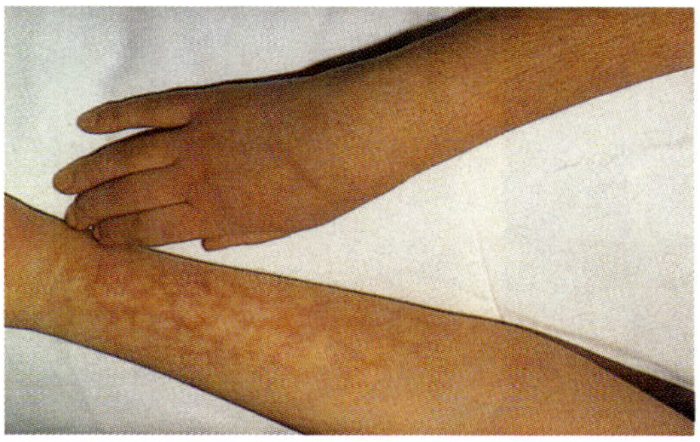

Fig. 1. Livedo reticularis in a patient with polyarteritis nodosa.

b. ***Kawasaki disease:*** Is a disease of children and has been reported from India. There is a concern that many of these cases are underreported. Characteristically the child presents acutely with fever, mucocutaneous ulcerating lesion and conjunctivitis. High mortality due to coronary arteritis which occurs in 25 % of cases.

Small vessel vasculitis

a. ***Microscopic polyangiitis*** – Presents as rapidly progressive glomerulonephritis and pulmonary hemorrhages. Renal disease is seen in all the cases and focal segmental necrotizing glomerulonephritis is typically seen on histology. Shares features with classical PAN and Wegener's granulomatosis.

b. ***Eosinophilic granulomatosis with polyangiitis (Churg Strauss) EGPA*** – EGPA is a systemic vasculitis that develops on a background of late onset asthma. Fever, chest infiltrate, cardiac failure, mononeuritis multiplex along with peripheral and tissue eosinophilia are characteristically present. Histologically there is a granulomatous necrotizing vasculitis.

c. ***Granulomatosis with polyangiitis (Wegener's) GPA*** – It is a rare disease even though small series of cases have been reported from tertiary care centers in our country. Pathologically the disease is characterized by presence of granulomas of upper and lower airways, necrotizing vasculitis and focal glomerulonephritis. In the limited form it involves the airways and the generalized disease presents with systemic features, respiratory tract, skin, gut, kidney and nervous system involvement[14]. Upper airways are commonly involved with features of nasal obstruction, discharge from the ears and nose, sinusitis, nasal septal perforation and saddle nose deformity (Fig. 2), hoarseness or stridor. Uveitis, proptosis and scleromalacia perforans can occur. (Fig. 3) Lung involvement occurs in 87% during the course of the disease. Cough and hemoptysis are the commonest presenting features. The chest X-ray/CT scan shows nodules or cavities mimicking tuberculosis (Fig. 4). Renal involvement is a critical prognostic factor. It can be asymptomatic or present with life threatening rapidly progressive renal failure. Skin is involved in half of the cases with ulcers, palpable purpura, nodules and digital gangrene (Fig. 5). Ocular disease (50%) manifests as conjunctivitis, keratitis, retro-orbital mass causing proptosis, scleritis leading to scleromalacia perforans and retinal vasculitis. In a proportion of patients these patients are serologically positive for HBS antigens.

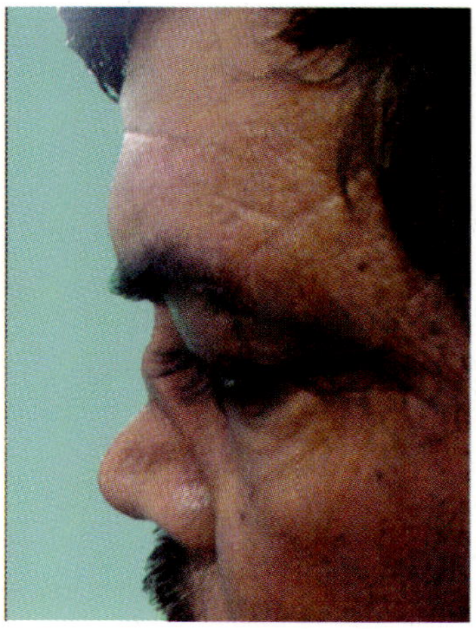

Fig. 2. Saddle-nose deformity in a patient with WG

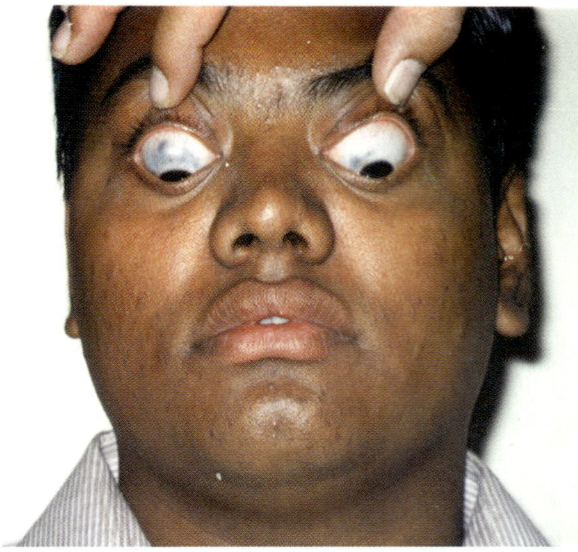

Fig. 3. Proptosis and scleromalacia perforans in a young man with Wegener's granulomatosis

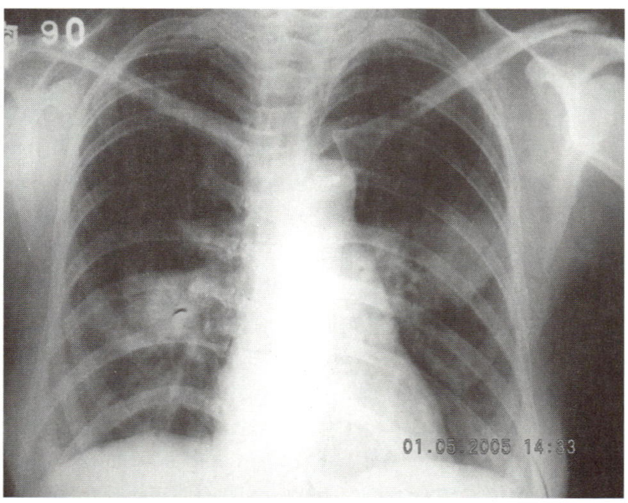

Fig. 4. X-Ray chest showing nodular consolidation in right lung in Wegener's granulomatosis

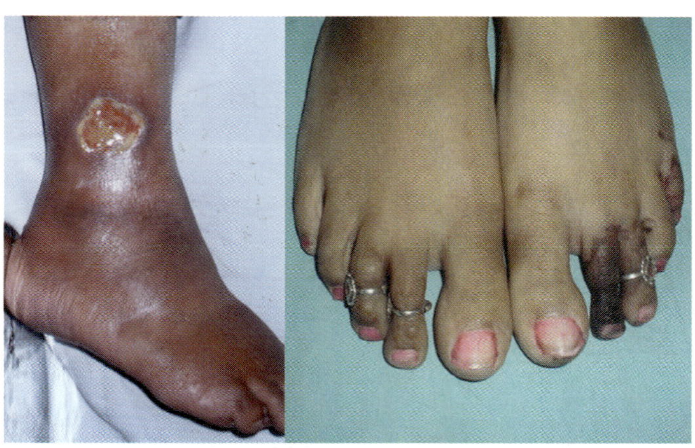

Fig. 5. Non healing ulcer and gangrene of the left second toe in patients with systemic necrotizing vasculitis

d. ***Henoch Schonlein purpura (HSP):*** Is a systemic vasculitis that presents with palpable purpura (Fig. 6) in children and less commonly in adults. Typically the rashes are present on the buttocks and lower limbs. Besides skin rash, there occurs arthralgias/arthritis, microscopic haematuria/proteinuria, haemoptysis and gastrointestinal bleeding. The prognosis is excellent in majority of cases; 10-30% of older children and adults may progress to chronic renal failure. The diagnosis is confirmed by demonstration of IgA immunoglobulin deposits in the skin and kidney on immunofluorescence.

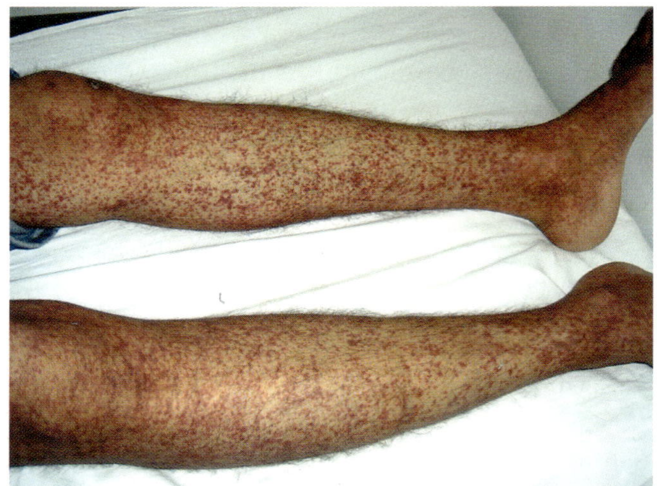

Fig. 6. Palpable purpura in the lower limbs in a patient with Henoch Schonlein purpura

e. ***Hypersensitivity vasculitis:*** The clinical picture is essentially that of palpable purpura in the lower extremities and trunk similar to HSP. The findings are described as leukocytoclastic vasculitis. Immuno-fluorescence studies reveal deposition of all classes of immunoglobulins. Infective endocarditis, hepatitis B, hepatitis C, human immunodeficiency syndrome, drug malignancy, systemic lupus erythematosus and cryoglobulinemia, all cause small vessel vasculitis with similar presentation.

Investigation and approach to diagnosis

It is mandatory to exclude infections and other vaso-occlusive diseases before diagnosing a primary vasculitis, the vasculitis mimickers (Table 3). The investigations are directed to confirm the diagnosis, extent of organ involvement and assessing the disease activity. (Table 4) summarizes the salient features in establishing the diagnosis and assessing both activity and extent of disease. The type and extent of organ involvement helps in determining the type of vasculitis, urgency of initiating treatment

Table 3. Vasculitis mimickers

Infections	Subacute infective endocarditis Neisseria meningitides Viral infections
Malignancies	Metastatic carcinoma
Connective tissue diseases	Systemic lupus erythematosus Rheumatoid arthritis
Occlusive vascular disease	Atrial myxoma Cholesterol embolism Infections (mycotic) Antiphospholipid antibody syndrome Ergotism

Table 4. Assessing extent of disease and salient investigations for the diagnosis of various vasculitides.

Diseases	Routine tests and Serologies	Radiology	Electro-physiology	Biopsy
Temporal arteritis	ESR	-	-	Superficial temporal Art
Takayasu's arteritis	ESR	Angiography Aorta & branches	-	-
PAN	ESR, TLC, Creat	Angiography abdominal vessels	NCV	Sural nerve/ muscle/ testis/ skin
MPA	Urinalysis, Creat pANCA (MPO)	Chest X-ray	-	Renal
WG	TLC, Urinalysis, creatinine, cANCA (PR3)	X-ray chest, PNS CT thorax	-	Nasal mucosa Lung Kidney
CSS	Blood eosinophilia, pANCA (MPO)	X-ray chest	NCV	Sural nerve
HSP	Normal platelet count	-	-	Immunofluorescence (IgA deposits)
Hypersensitivity Vasculitis	-	-	-	Immunofluorescence (IgG, IgM, IgA deposits)

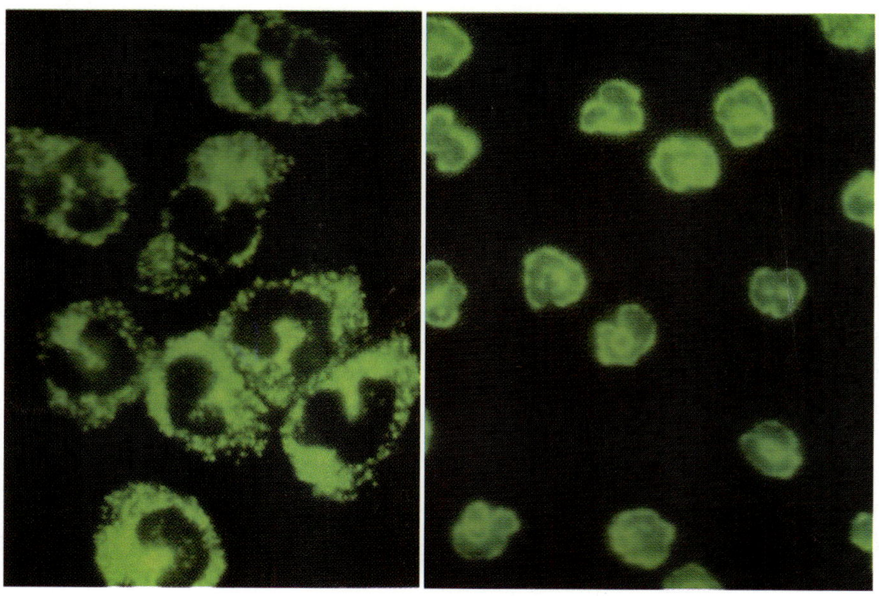

Fig. 7. Cytoplasmic (left) and peripheral (right) ANCA pattern by indirect immunofluorescence.

and planning diagnostic tests. Hepatitis B and C and human immunodeficiency virus serologies should be obtained since these can be associated with vasculitis.

Antineutrophil cytoplasmic antibodies

Antineutrophil cytoplasmic antibodies (ANCA) are useful diagnostic markers for the primary systemic necrotizing vasculitis. These are detected by indirect immunofluorescence which are reported as cytoplasmic or cANCA, perinuclear or pANCA and atypical or xANCA (Fig. 7) cANCA is directed against proteinase-3 and pANCA is directed predominantly against myeloperoxidase. cANCA is present in upto 90% of patients with active, diffuse Wegener's granulomatosis and in upto 70% of patients with limited Wegener's (defined as absence of immediate threat to either the function of a vital organ or patient's life). pANCA is typical of MPA (50-60%) and Churg-Strauss disease (25-30%). Because vasculitides associated with ANCA are rare disorders, the likelihood of false positive result of ANCA is greatly increased if not used in appropriate clinical setting. Enzyme linked immunoassays (EIA) were developed for detection of anti PR3 and anti MPO antibodies. Like cANCA antibodies to PR3 are both specific and sensitive for the diagnosis of WG and similarly antibodies to MPO are specific for microscopic polyanglitis. But EIA was found to have its own problems. Hence, it is advisable to order for both IIF and EIA to confirm the presence and type of ANCA. Indiscriminate ordering of ANCA should be avoided as it would lead to misdiagnosis. It is of value only when there is a high pretest probability of vasculitis. ANCAs of a low titer including borderline PR3 and MPO do not necessarily indicate vasculitis.

Treatment

Treatment of vasculitis depends upon the type of vasculitic syndrome and the seriousness of involvement of organs by the disease. Patients with disease restricted to the skin require only symptomatic treatment. The medical treatment of TA is directed against improving systemic symptoms and controlling hypertension.

Assessment of disease activity

There is no valid assessment of disease activity in TA, the Indian Rheumatology Association Vasculitis Group (IRAVAS) have recently come out with a validated instrument, the Indian Takayasu Activity Score (ITAS) which is a composite disease activity scores[15]. A combination of oral prednisolone starting with 1 mg/kg of body wt along with methotrexate (15-25 mg/week), azathioprine (2mg/kg/day) or mycophenolate mofetil (1.5-3 g/day) generally are employed to remit disease activity. Prednisolone is to be tapered within 2-3 months to a maintenance of 0.15mg/day. Tocilizumab (IL 6 receptor blocker) has also been found to be effective in small case series to lessen disease activity such as ITAS score. Surgery or percutaneous transluminal angioplasty are important considerations for severely occluded vessels. In GCA, oral prednisolone in a dose of 40-60mg/day for 6 weeks is given followed by maintenance for 6-12 months with a dose of 0.15mg/kg/day. Various agents such as azathioprine, cyclophosphamide, dapsone and cyclosporine have been tested as steroid sparing drugs in GCA; however, all were tested in too small or uncontrolled series to permit conclusion.

PAN is considered to be a monophasic illness. Corticosteroids improved the 5 year survival rate of PAN patients from 10-55%. Survival was further prolonged by adding cytotoxic agents, cyclophosphamide or azathioprine, attaining a 5 year survival of 80% for patients given steroids and cyclophosphamide. Patients of PAN who are HBsAg positive are treated with interferon and lamivudine.

In ANCA associated vasculitis (systemic necrotizing vasculitis) disease activity is assessed by clinical scoring instruments. The most widely used is Birmingham vasculitis activity score (BVAS), a simple, practical tool for recording disease activity. It is simple to use and found useful in monitoring treatment. Damage in vasculitis is attributable to disease process and secondary to complication of treatment. Evaluation of damage is important for prognostication and avoidance of overtreatment. Vasculitis damage index (VDI) is most widely used. It is an objective assessment based on eleven organ systems. It records long-term consequences of disease and its therapy. Disease Activity Index, as mentioned above, is also a tool for assessing damage.

The introduction of cyclophosphamide has changed the outlook of treatment with 5 year survival beyond 90%. The disease is treated in 3 phases akin to malignancy. An induction phase lasting for 4-6 months with combination of Cyclophosphamide orally (2-3 mg/kg/day) or intravenous bolus (15 mg/kg) along with oral prednisolone (1-2 mg/kg/day) till remission is achieved. Six intravenous boluses are administered, 3 fortnightly and 3 every 3 weeks. During the consolidation phase (1 year) daily azathioprine is preferred over cyclophosphamide at a dose of 2 mg/ kg/day along with prednisolone (0.15 mg/kg/day) for a period of one year. During the maintenance phase either prednisolone (7.5 mg/day) or azathioprine (1 mg/kg/day) is administered for a period of 1 year. WG with only upper respiratory tract can be managed with cotrimoxazole or methotrexate along with small dose of prednisolone. Almost 50% of WG patients relapse after remission is achieved. The treatment of relapse is similar to that of "remission-induction".

In recent years, rituximab has shown promise for treatment of ANCA associated vasculitis. Two randomized controlled trials (RAVE and RITUXIVAS) have shown that it is not inferior to cyclophosphamide for induction of disease remission and in cases of pulmonary relapses it is rather superior to cyclophosphamide[16]. Presently trials are going to assess the efficacy of B cell depleting therapy in remission maintenance.

Treatment of Henoch Schonlein purpura is largely supportive. NSAIDs help joint pains and do not worsen purpura. Steroids have been shown to be of benefit in managing painful edema. It is also used for GI involvement although its efficacy in this regard is unproven. Treatment of renal involvement is highly controversial. Various treatment modalities have included pulse corticosteroids, cyclophosphamide, plasmapheresis, cyclosporine and azathioprine. Cryoglobulinemic vasculitis has a wide manifestation ranging from mild symptoms to life threatening conditions. The clinical course can be characterized by remissions and exacerbations. Serological parameters do not correlate with disease activity and hence cannot be relied upon for therapeutic decisions. Treatment should therefore not focus upon lowering cryoglobulin levels or normalizing complements but rather on controlling manifestations. Low to moderate dose steroids are sufficient for most minor symptoms but larger doses may be needed for managing renal involvement, neuropathy or serositis. Plasma exchange with or without immunosuppressive may be needed for life threatening conditions. Interferon has been used with good success for inducing clinical and biological remission but the disease relapses in most following cessation.

Conclusion

Primary systemic vasculitides are a group of rare diseases. The etiology and pathogenesis are not well understood. The discovery of ANCA has significantly helped us to further our knowledge with regards to pathogenesis and diagnosis. Immunosuppressive treatment has helped to reduce mortality but morbidity is a major concern in ANCA associated vasculitis.

Further Reading :

1. Watts R, Carruthers D, Scott D. Epidemiology of systemic vasculitis: changing incidence or definition? Semin Arth Rheum 1995; 25: 28-34.
2. Gardner-Medwin JM, Dolezalova P, Cummins C, et al. Ethnic differences in the incidence of Henoch-Schönlein purpura, Kawasaki's disease and rare childhood vasculitides in children of different ethnic origins. Lancet 2002; 360: 1197-1202.
3. Bagga A, Kabra SK, Srivastava RN, et al. Henoch-Schönlein syndrome in northern Indian children. Indian Pediatr. 1991; 28: 1153-57.
4. Kumar L, Singh S. Vasculitis in children. Indian J Pediatr. 1996; 63: 323-34.
5. Jain S, Kumari S, Ganguly NK, et al. Current status of Takayasu arteritis in India. Int J Cardiol. 1996; 54(S): 111-16.
6. Malaviya AN, Kumar A, Singh YN, et al. Wegener's granulomatosis in India: not so rare. Br J Rheum 1990; 29: 499-500.
7. Bambery P, Sakhuja V, Bhusnurmath SR, et al. Wegener's granulomatosis: clinical experience with eighteen patients. J Assoc Phy Ind 1992; 40: 597-600.
8. Handa R, Wali JP, Gupta SD, et al. Classical polyarteritis nodosa and microscopic polyangiitis a clinicopathologic study. J Assoc Phy Ind 2001; 49: 314-19.
9. Pendse RN, Bhandari H, Vats AK, et al, Kawasaki disease Indian perspective. Indian J Pedia 2001; 68: 775-77.
10. Bagyaraj B, Krishnan U, Farzana F. Kawasaki disease in India. Indian J Pedia 2003; 70: 919-22.
11. Fries JF, Hunder GG, Bloch DA, et al. The American college of Rheumatology criteria for the classification of vasculitis: summary. Arth Rheum 1990; 33: 1135-36.
12. Jeanette JC, Falk RJ, Andrassy K, et al. Nomenclature of systemic vasculitides. Proposal of an international consensus conference. Arth Rheum 1994; 37: 187-92.
13. Jennette JC, Falk RJ, Bacon PA et al. 2012 Revised International Chapel Hill Consensus Conference Nomenclature of Vasculitides. Arth Rheum 2013; 65(1): 1-11.
14. Hoffman GS, Kerr GS, Leavitt RY, et al. Wegener's granulomatosis: an analysis of 158 patients. Ann Intern Med 1992; 116: 488-98.
15. Misra R, Danda D, Rajappa SM et al. Development and initial validation of Indian Takayasu Activity Score. Rheumatology, 2013: 10.1093/rheumatology/ket12.
16. Jones RB, Tervaert JW, Hauser T et al. Rituximab versus cyclophosphamide in ANCA-associated renal vasculitis. N Engl J Med 2010; 363: 211-20.

TAKAYASU ARTERITIS

9

M R Sivakumar

History

The first case of Takayasu arteritis was described in 1908 by Dr. Mikito Takayasu at the Annual Meeting of the Japan Ophthalmology Society. Dr. Takayasu described a peculiar "wreathlike" appearance of blood vessels in the back of the eye (retina). Two Japanese colleagues at the same meeting reported similar eye findings in patients whose wrist pulses were absent. The disease was named as Takayasu Arteritis (TA) in 1942[1,2].

Background

TA is a rare form of primary systemic vasculitis that appears to be commoner in Asia than Europe or North America. It is a serious condition that often presents to hospital with acute vascular occlusions or hypertension in young females. TA is a chronic inflammatory disease of the thoracic and abdominal aorta and its major branches and pulmonary arteries that most often affects young women in the second and third decades of life. The initial complaints may be nonspecific constitutional signs and symptoms like fever, malaise, weight loss and lethargy and hence the correct diagnosis may be delayed for months or years. Despite the term pulseless disease, which is a synonym for TA, the predominant findings in TA are asymmetric pulse. Absent peripheral pulses occur late in the course of the disease. While 5 year survival rates exceed 90%, the disease has a high incidence of residual morbidity.

Pathophysiology

TA is characterized by granulomatous inflammation of the aorta and its major branches, leading to stenosis, thrombosis, and aneurysm formation[3]. The lesions of TA are segmental with a patchy distribution. Mononuclear infiltration of the adventitia occurs early in the course of the disease, with cuffing of the vasa vasorum. Granulomatous changes may be observed in the tunica media with Langerhans cells and central necrosis of elastic fibers and smooth muscle cells. A panarteritis with infiltrates of lymphocytes, plasma cells, histocytes, and giant cells is present. Later, fibrosis of the media and acellular thickening of the intima compromise the vessel lumen. Wrinkling of the intima is visible on gross examination. Stenoses are found in many of patients with TA disease. Often patients have poststenotic dilatations and other aneurysmal areas. Stenotic arterial segments result in varied ischemic symptoms. These symptoms may range from abdominal pain after eating secondary to narrowing of the mesenteric arteries, claudication of extremities and Reno vascular hypertension. Endothelial activation leads to a hypercoagulable state predisposing the patient to thrombosis. Congestive heart failure in individuals with TA may occur as a result of hypertension, aortic root dilation, or myocarditis. Transient ischemic attacks, cerebrovascular accidents, mesenteric ischemia, carotidynia, and claudications may occur. Symptoms of vascular compromise may be minimized by the development of collateral circulation with the slow onset of stenoses. Vessel wall dissection or aneurysm may occur in areas weakened by inflammation.

One hypothesis for granulomatous vasculitis development is that antigens deposited in vascular walls activate CD4+ T cells, followed by release of cytokines chemotactic for monocytes. These monocytes are transformed into macrophages that mediate endothelial damage and granuloma formation in the vessel wall. A mouse model supports this hypothesis. Human studies suggesting endothelial cell activation have demonstrated increased expression of intercellular adhesion molecule-1 (ICAM-1) and vascular cell adhesion molecule-1 (VCAM-1) in patients with TA. Humoral immunity also is believed to be involved in this disease anti aorta antibodies and anti endothelial cell antibodies have been found in patients with TA. Immunoglobulin G, Immunoglobulin M, and properdin are found in lesions from pathologic specimens.

Cytokine abnormalities include elevated tumor necrosis factor and interleukin (IL)-6. B cells may also play a role, with plasmablast expansion in patients with active Takayasu arteritis. This increase in B cells is not seen in inactive disease. IL-6 and B-cell activating factor also increase total number of circulating B cells; reduction in disease activity is observed with successful depletion of B cells with anti-CD20.

Incidence

Age and Sex

In adults, females account for 80-90% of patients with Takayasu arteritis. Although sex distribution usually mirrors the 80-90% female preponderance observed in adults, series of studies of Takayasu arteritis in childhood from India and South Africa reported a 2:1 female-to-male ratio.

Takayasu arteritis is the most common large vessel Vasculitis of adolescence. Most cases of Takayasu arteritis present in persons aged 10-30 years. In a series of patients with Takayasu arteritis, 20-35% was younger than 20 years at diagnosis. The youngest patient reported was aged 6 months.

Epidemiology

In Minnesota's Olmstead County, incidence of Takayasu arteritis was estimated at 2.6 cases per million. Hoffman et al[4] described the clinical, laboratory and radiographic manifestations of TA in a cohort of 75 patients and evaluated the response to interventions, remission and relapse rates, and disease progression and compared these observations with those from other cohorts in the US, Japan, India, Italy, and Mexico. 93% of longitudinally followed patients attained disease remission of any duration, but only 28% sustained remission of at least 6 months duration after prednisolone was tapered to < 10 mg daily. Both angioplasty and vascular surgery were initially successful, but recurrent stenosis occurred in 78% of angioplasty and 36% of bypass/reconstruction procedures. More than two thirds of patients had difficulty in performing routine daily activities and approximately one-fourth of all patients were unable to work. Their cohort was similar to the National Institutes of Health, Italian, Japanese, and Mexican cohorts in terms of the predominance of female subjects and disease manifestations, but differed from the Indian cohort in that the latter group had a higher frequency of male subjects, abdominal aorta, renal artery involvement and hypertension. Although improvement of symptoms in TA usually followed glucocorticoid therapy, relapses were common with dosage reduction. Attempts to restore vascular patency were often initially successful, but restenosis was frequent. Chronic morbidity and disability occur in most patients with TA in the US.

Takayasu arteritis is common in developing countries, where the disease is closely associated with tuberculosis. In endemic areas, active tuberculosis may perpetuate Takayasu disease activity through molecular mimicry or chronic antigen stimulation. Takayasu arteritis is more common in Asian populations but has been described in patients of all races. Japanese patients with Takayasu arteritis have a higher incidence of aortic arch involvement. In contrast, series from India report higher incidences of thoracic and abdominal involvement. In US patients with Takayasu arteritis, the most commonly involved vessels are the left subclavian, superior mesenteric and abdominal aorta. In US children with Takayasu arteritis, lesions of the thoracic and abdominal aorta, rather than lesions of the aortic arch, are most common. However, all patterns of vascular involvement have been observed in every country.

Takayasu in India

TA appears to be far common in India, although accurate epidemiology is very difficult to obtain. One group with a major interest reckons to see around 50 new cases annually at a major government hospital in Mumbai[5]. This compares to estimates of 100 new cases per year for the whole of Japan, indicating that the incidence is atleast as high as in that country. The cardiovascular group in Kolkata has published a monograph based on a series of 650 cases[6]. There has been much speculation about the role of Mycobacteria in the etiology of TA, especially in India. Mycobacterium Tuberculosis was seen in a significant proportion of cases in the two large series quoted but it is still a common infection in India. It remains possible that this is simply coincidence of common infection with a rare vasculitis. Another possibility is that some aspects of the pathology share common features.

The Mumbai groups have noted major paraaortic lymphadenopathy as a common feature at Post Mortem[7]. The enlarged nodes may show caseation apparently typical of tuberculosis, yet staining and culture are negative for Mycobacterium Tuberculosis. In India Numano Types IV and V are most frequent.

Clinical presentation

Systemic symptoms include the following: Fever, night sweats, fatigue, weight loss, myalgia, arthralgia/ arthritis.

Common symptoms include headache, dizziness or syncope, congestive heart failure, palpitations, angina and hypertension.

Skin lesions like erythema nodosum, pyoderma gangrenosum, leukocytoclastic vasculitis, and panniculitis have been reported.

Symptoms related to ischemia include the following: Ischemic stroke/transient ischemic attack, visual disturbances (eg, blurred vision, diplopia and amaurosis fugax), carotidynia, abdominal pain, claudications of the jaw, upper or lower limbs.

Neurological manifestations in a cohort of 112 TA patients from India has been reported by Sivakumar[8]. Common symptoms seen are: Confusion, dementia, Seizures (not hypertensive), ischemic Stroke, intracerebral haemorrhage, subarachnoid haemorrhage, Mononeuritis Multiplex, blurred vision and sudden loss of vision.

Neurological manifestations in a cohort of 112 Takayasu arteritis patients

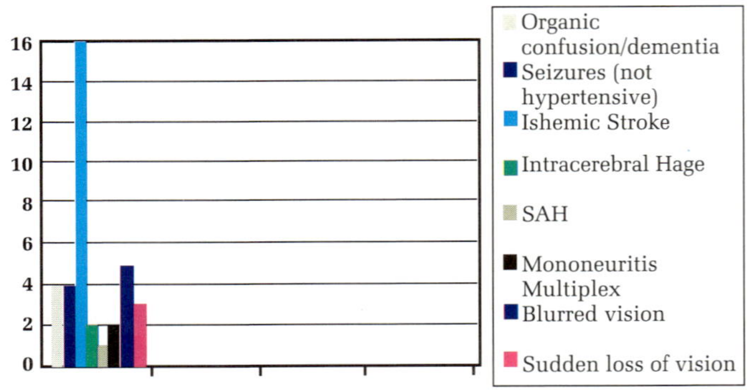

Physical examination

Blood pressure difference greater than 20 mm Hg between arms, asymmetrical pulses, diminished or absent pulses (mid aortic lesions found in children may not affect pulses), Poststenotic dilatations producing what appear to be bounding pulses, hypertension may be paroxysmal since this typically results from renovascular compromise and is a high-renin hypertension, bruits - especially over carotids, subclavian arteries, or the aorta.

Skin lesions include erythema nodosum–like lesions, pyoderma gangrenosum, leukocytoclastic vasculitis, and panniculitis.

Fundus examination may reveal: retinal hemorrhages, cotton-wool exudates, venous dilatation and beading, microaneurysms of peripheral retina, optic atrophy, vitreous hemorrhages and rarely classic wreathlike, peripapillary arteriovenous anastomoses. Ocular findings in Indian subjects are reported by Danda et al[9,10].

Hypertension may be paroxysmal and may only be detected by intra arterial recordings. Carotid thrill, weak carotid pulsations and carotidynia are frequent. Bruits may be present at carotid, supraclavicular, axillary, abdominal and renal angles.

Asymmetrical pulses and absent pulses can be found even in the later stages of the disease.

Diagnostic considerations

Early diagnosis of TA can be difficult due to nonspecific early symptoms. Hence a thorough evaluation to eliminate other diagnoses should be done.

Differential diagnosis

Conditions to consider in the differential diagnosis include: systemic-onset juvenile rheumatoid arthritis, migraine, infection, malignancy, Cogan syndrome, Acute Lymphoblastic Leukaemia, Behcet Syndrome, Fever without a Focus, Hodgkin Disease, Polyarteritis Nodosa and Rheumatic Fever

Etiology : Unknown

TA has been reported in identical twins, leading to hypotheses of a hereditary basis for the disease. In Japan and Korea, TA is associated with human leukocyte antigens HLA A10, B5, Bw52, DR2, and DR4. These associations have not been confirmed in Western studies. TA is associated with HLA B22 in the US. Although early reports associated tuberculosis with TA, these involved populations are endemic for tuberculosis, and population studies have not borne out this association. Others report increased incidence of positive purified protein derivative relative to control patients. Patients with TA do not improve with antituberculous treatment.

Associated diseases

TA is usually primary, but has also been seen in association with other immune related and/or Granulomatous diseases, such as Crohn's disease, ulcerative colitis, granulomatous hypophysitis, Systemic lupus erythematosus, ankylosing spondylitis, rheumatoid arthritis and adult onset Still's disease.

Classification criteria for Takayasu arteritis

The principle limitation of current criteria for the diagnosis of TA is that patients with early phase of non-occlusive disease may fail to fulfill them. The ACR classification remains the most widely applied.

1990 American college of rheumatology criteria for the classification of Takayasu arteritis[11]

1) Age at disease onset <= 40 years. 2) Development of symptoms or findings related to Takayasu arteritis at age <= 40 years. 3) Claudication of extremities Development and worsening of fatigue and discomfort in muscles of 1 or more extremity

while in use, especially the upper extremities. 4) Decreased brachial artery pulse. Decreased pulsations of one or both brachial arteries. BP difference > 10 mm Hg difference of > 10 mm Hg in systolic blood pressure between arms. 5) Bruit over subclavian arteries or aorta bruit audible on auscultation over one or both subclavian arteries or abdominal aorta. 6) Arteriogram abnormality-arteriographic narrowing or occlusion of the entire aorta, its primary branches, or large arteries in the proximal upper or lower extremities, not due to arteriosclerosis, fibro muscular dysplasia, or similar causes changes usually focal or segmental blood pressure (systolic: difference between arms). For purposes of classification, a patient shall be said to have TA if at least 3 of these 6 criteria are present.

The presence of any 3 or more criteria yields a sensitivity of 90.5 % and specificity 97.8 %.

Ishikawa's criteria for the diagnosis of Takayasu arteritis[12]

The proposed criteria consist of one obligatory criterion, two major criteria and nine minor criteria. In addition to the obligatory criterion, the presence of two major criteria, or one major and two or more minor criteria or four more minor criteria suggests a high probability of the presence of TA.

Sharma's modification of Ishikawa's criteria[13]

Few modifications have been proposed in Ishikawa's criteria for the diagnosis of TA. Presence of two major or one major and two minor criteria or four minor criteria suggests a high probability of TA. When applied to 106 Indian patients of angiographically proven TA and 20 control subjects, it had a sensitivity of 92.5% and specificity of 95% that was higher than that of Ishikawa's criteria (sensitivity 60.4%, specificity 95%) and American college of Rheumatology criteria (sensitivity 77.4%, specificity 95%). These criteria had a sensitivity of 96% and a specificity of 96% in 79 Japanese patients of TA and 79 control subjects[12].

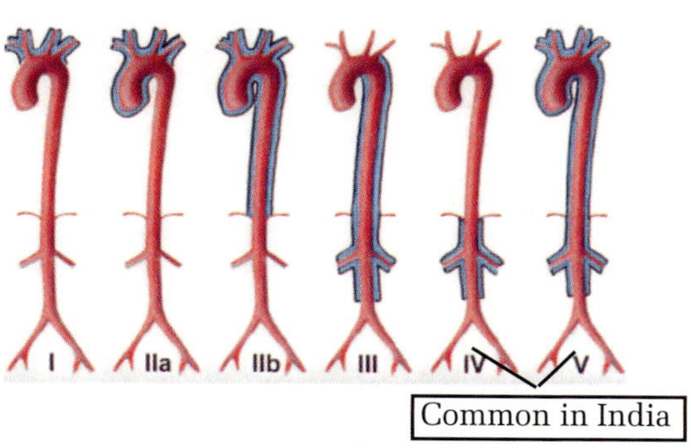

Numano classification by angiographic criteria[14]

Disease monitoring

Monitoring of acute phase reactants as a measure of disease activity has limited value. The acute phase response is neither sensitive nor specific enough to be helpful[15]. Performing regular imaging of affected vasculature and surveillance imaging for new lesions are essential. Treatment may be monitored with MRI and/or MRA or CT scanning. Mural thickening is observed to decrease with corticosteroid treatment.

Patients with apparently inactive disease, who undergo graft placement, have been found to have active aortitis in the resected segment. Periodic imaging may identify an active disease by the appearance of new areas of stenosis, despite normal erythrocyte sedimentation rate and the absence of clinical features. The presence of active disease requires aggressive treatment with corticosteroids and immunosuppressants. However, current markers of disease activity are inadequate to identify all patients with disease flare.

Serial MRI may reveal vessel wall edema, but whether this measures actual inflammation is unclear. Structural changes visible on imaging demonstrate disease progression, but reliable indicators of vessel inflammation prior to structural damage have yet to be identified.

Clinical analysis of TA disease activity, response to treatment and detection of relapse remain suboptimal. Kerr et al[16] defined active TA as the new onset or worsening of two or more of the following features: fever or arthralgia raised ESR (> 20 mm/hr) features of vascular ischemia or inflammation such as claudications, diminished or absent pulse, bruit, vascular pain, asymmetric blood pressure in upper or lower limbs or typical angiographic features however this definition has not been widely validated and is limited in scope.

Assessment of disease activity using clinical instruments like Birmingham Vasculitis Activity Score[17] and Vasculitis Damage index[18] has been extensively used both in trial and clinic setting. There is a need for clinical assessment tools for large vessel disease like Takayasu arteritis, which is seen more frequently in India. An attempt has been made by the Indian Rheumatology Association Vasculitis core group to validate instruments, like Disease Extent Index for Takayasu Arteritis (DEI.Tak), Indian Takayasu Activity Score (ITAS) and Takayasu Arteritis Damage Scores (TADS). Both DEI.Tak and ITAS have received interests from international

DEI.Tak - Disease Extent Index for Takayasu's Arteritis	**Patient Name :**
Tick Box only if abnormality is present (new or worse within 6/12), with duration for each symptom. Tick box only if abnormality is attributed to current vasculities.	**Visit Date :** **Investigator**

		PRESENT	duration
1. SYSTEMIC			
None	☐		
Malaise/Wt. Loss >2Kg		○	
Myalgia/Arthralgia/Arthritis.		○	
Headache		○	
Fever		○	
2. CUTANEOUS			
None	☐		
Gangrene		○	
Other Skin Vasculitis		○	
3. MUCOUS MEMEBRANES			
None	☐		
Present		○	
4. ETES			
None	☐		
Blurred Vision		○	
Sudden Vision Loss		○	
Other		○	
5. ETES			
None	☐		
Present		○	
6. CHEST			
None	☐		
Persistent Cough		○	
Dyspnoea/Wheeze		○	
Haemoptysis/Haemorrhage		○	
Massive Haemoptysis		○	
Respiratory Failure		○	

Chest Radiology	○	
Actove Vasculitis confirmed		○

		PRESENT	duration
8. ABDOMEN			
None	☐		
Severe Abdominal Pain		○	
Bloody Diarrhoea		○	
Gut Perforation/Infarct		○	

Surgical Opinion/tests	○	
Active Vasculitis confirmed		○

		PRESENT	duration
9. RENAL	☐		
None			
Hypertension (Diastole>90)		○	
"" Systolic > 140		○	
Proteinuria (>1+/0.2g/24H)		○	
Haematuria (>1+/10RBC/ml)		○	
Creatinine (125-249 μmol/L)		○	
Creatinine (250-499 μmol/L)		○	
Creatinine (>500 μmol/L)		○	
Rise in creatinine >30% or > 25% fall in creatinine clearance.		○	
10. Nervous System			
None	☐		
Organic Confusion/Dementia		○	
Seizures (not hypertensive)		○	
Stroke		○	
Syncope		○	
Cord Lesion		○	
11. Genitourinary System			
None	☐		
Sexual Impotence		○	
Abortions		○	

5. CARDIOVASCULAR SYSTEM

none	☐	
Bruits (see 7a)		○
Pulse Inequality (See 7B)		○
Pulse Loss (See 7c)		○
Pulse Loss with threatened loss of limb.		○
Claudication (See 7d)		○
Carotidodynia		○
Aortic Incompetence		○
Pericardial Pain/Rub		○
Ischemic Cardiac Pain		○
Congestive Cardiac Failure		○

Cardiology Opinion/Tests	
Active Vasculitis confirmed	○
Pericarditis	○
Myocardial Infarct/Angia	○
Cardiomyopathy	○

7a. Bruits	R		L
Carotid	○		○
Vertebral	○		○
Subclavian	○		○
Renal	○		○
Abdominal		○	
INguinal	○		○
7b. Pulse and BP Inequality			
Present		○	
7c.Pulse Loss			
Carotid	○		○
Subclavian	○		○
Branchial	○		○
Radial	○		○
Femoral	○		○
Popliteal	○		○
Poterior Tibial	○		○
Dorsalis Pedis	○		○
7d. Claudication			
Arm		○	
Leg		○	
Neck		○	

12. Other Vase items:

ESR CRP

13. PGO (Active / Grumbling or persistent / Inactive) :

Revised form by M.R. Sivakumar, R. Mishra, and P.A. Bacon - Dec 2005

ITAS 2010 – Indian Takayasu Activity Score

Tick Box only if abnormality is present and new or worse within the past 3/12. Tick box only if abnormality is ascribed to current, active vasculitis.	Name: Visit Date : Investigator:

	PRESENT
1. SYSTEMIC	
None	☐
Malaise/Wt. Loss>2Kg	○
Myalgia/Arthralgia/Arthritis.	○
Headache	○
2. ABDOMEN	
None	☐
Severe Abdominal Pain	○
3. Genitourinary System	
None	☐
Abortions	○

	PRESENT
4. RENAL	
None	☐
Hypertension (Diastole >90)	[○]
" Systolic >140	○
5. Nervous System	
None	☐
Stroke	[○]
Seizures (not hypertensive)	○
Syncope	○
Vertigo/dizzyness	○

---

6. CARDIOVASCULAR SYSTEM

None	☐
Bruits (see 6a)	[○]
Pulse Inequality (See 6b)	[○]
New Loss of Pulses (See 6c)	[○]
Claudication (See9d)	[○]
Carotidodynia	[○]
Aortic Incompetence	○
Myocardial Infarct/Angina	○
Cardiomyopathy/cardiac failure	○

6a. Bruits	**R**	**L**
Carotid	○	○
Subclavian	○	○
Renal	○	○

6b. Pulse and BP Inequality	
Present	○

6c. Pulse Loss	R	L
Carotid	○	○
Subclavian	○	○
Brachial	○	○
Radial	○	○
Femoral	○	○
Popliteal	○	○
Posterior Tibial	○	○
Dorsalis Pedis	○	○

6d. Claudication	
Arm	○
Leg	○

Other Vasculitis items:

ESR **CRP**

Scoring ITAS ☐ = 0 ○ =1 [○]= 2

PGA (Active / Grumbling or persistent / Inactive):

New Imaging Y / N?
If Yes Any interval change from that of previous visit

2010 Form. Sivakumar, Misra, Danda ,Bacon – Feb'10

TADS – Takayasu's Arteritis Damage Score (Short form)

Record any abnormality that has occurred since the onset of aorto-arteritis currently present or not, as this is a cumulative damage score **Tick Box only if abnormality present for at least 6/12.**	Name or # : Visit Date : Investigator:

	PRESENT
1. EYES	
None ☐	
Visual Loss in one eye	○
Vision Loss in second eye	○
2. CHEST	
None ☐	
Persistent Cough Dyspnoea/Wheeze	○
Respiratory Failure	○
3. RENAL	
None ☐	
Diastolic BP >95 }or requiring	○
Systolic BP >145 }antihypertensives	○
Proteinuria (>1+/0.2g/24H)	○
Creatinine >150	○
End-stage renal failure)	○

	PRESENT
4. NERVOUS SYSTEM	
None ☐	
Organic Confusion/Dementia	○
Seizures (not hypertensive)	○
Stroke	○
2nd Stroke	○
Cord Lesion	○
5. Drug-related and other damage	
None ☐	
Malignancy	○
Infertility	○
Other	○
6. Vascular Interventions	
None ☐	
First dilatation, stent or surgery	○
2nd procedure	○
Blockage/restenosis of above	○
Second reblock	○

7. CARDIOVASCULAR SYSTEM

None ☐	
Bruits	○ ---> 2nd Bruits
Pulse and B.P. Inequality	○
Pulse Loss (See 7a)	○ → 7a
Claudication (See 7b)	○ ---> 7b
Aortic Incompetence	○
Ischemic Cardiac Pain	○
Congestive Cardiac Failure	○
Cardiomyopathy	○

	R		L
2nd Bruits		○	
7a. Pulse Loss			
Carotid	○		○
Brachial	○		○
Radial	○		○
Femoral	○		○
Popliteal	○		○
Posterior Tibial/Dorsalis Pedis		○	
7b. Claudication			
Arm or leg		○	

8.Other Damage Items

TADS short form M.R Sivakumar, R.Misra, & P.A.Bacon – Nov 2010 investigators. Widespread usage of these instruments will pave the way to controlled clinical trials for TA.

TA typically presents with narrowing or occlusion of one or two large vessels, such as subclavian claudication or hypertension in young adults related to renal artery involvement. The Physicians Global Assessment (PGA) is still often the basis for treatment decisions in TA. It is based on clinical features, acute phase reactants, and image abnormalities[19]. The acute phase response is neither sensitive nor specific enough to be helpful as ESR/CRP are not necessarily elevated, even in acute vascular occlusions[20]. Standard vascular imaging shows damage as well as "activity" and may not reverse with medical therapy. Validated outcome assessments specifically relevant to Large Vessel Disease (LVV) are needed and the OMERACT group has noted the particular challenge of developing these for TA[21]. The Vasculitis Study Group of the IRA (IRAVAS) set out to devise and validate assessments for TA, with the intention of producing both an activity score and a damage score that together would document the extent and severity of disease[22].

Disease extent

The first move was agreeing an inclusive organ-based list, modified from the BVAS, of the main clinical features expected in TA. The resultant Disease Extent Index (DEI.Tak) was validated by using it to analyse the actual disease pattern recorded in 145 cases from 2 main Indian centres[22,23]. The cardiovascular system was most frequently scored and 88 cases had overt CVS disease at the time of assessment. 72 had renal involvement, which like the items scored in CVS is again a direct result of major artery disease. 68 had systemic symptoms such as fever, headaches, arthralgias and myalgias – but weight loss was very uncommon, in contrast to its frequency in SSV. Thus the DEI.Tak truly reflects the restricted large vessel involvement of TA. The details of this CVS involvement are important. Thus regular use of DEI.Tak gives a picture of the common clinical pattern of the disease and is a useful teaching aide informing trainees what to look for. The 2-centre Indian data was presented at international meetings and the DEI.Tak has since been used to characterise Turkish cases. This index has real value as a data-base in individual clinics and if widely applied would provide low-level evidence to compare the arbitrary drug regimens adopted at present. It should also have value as an epidemiologic tool to compare disease expression between countries worldwide. However it is not an activity index, as the Turkish group also concluded.

Disease acitivity

The subsequent development of ITAS (the Indian activity index for TA) was data-driven, taking the new disease items (occurring within the past 3 months) that were scored in DEI.Tak in the 2-centre series. It was validated using live patient assessments at a series of group work-shops in major centres. The clinical relevance of the draft format was tested by scoring the large series of patients from a major single clinic in Vellore. Analysis of data recorded in their series of 177 patients confirmed that 5 organ-based systems were rarely or never relevant in TA, so could be omitted from an activity index. Serial assessments allowed further selection of a set of items reflecting active disease, focussing on CVS items, with special emphasis on bruits, pulse loss, and claudication. All individual items scored in < 5% were examined and the majority omitted. The resultant slimmed-down ITAS2010 (Indian Takayasu Activity Score) contains 43 items in 6 systems, is convenient for individual physicians to use in the clinic, and has been extensively validated by IRAVAS. In the group exercise, ITAS again showed a highly significant correlation with BVAS[23,39]. The reproducibility in group usage scoring live patients was excellent, with inter-rater variability equivalent to BVAS. Both were clearly superior in this respect to Physician Global Opinion, (PGO), confirming that careful standardised assessment produces far more consistent agreement than doctors opinion. Weighting has been applied to 7 items related to the major vessel involvement that is clearly most relevant clinically. Importantly, serial assessments in the clinic showed that ITAS both reflects response to therapy and detects flares. It has been used in therapy trials both by the Vellore centre to assess patients on Mycophenolate mofetil and by the Italian group to study the effect of IL-6 receptor blockade[24,25]. In practice, the principle value may be to inform therapy decisions in individual cases.

Damage/disease scars

Damage is conceptually different from disease activity since it reflects scars that will not heal completely. It is clinically important to recognize that it needs to be scored separately as it will not respond to drugs that suppress immune-mediated inflammation. Thus damage scores can only increase or stay static – they do not revert to zero as an activity index should with successful therapy. Accumulating damage adds to the disease burden experienced by patients and may contribute to poor prognosis.

Takayasu Arteritis Damage Score (TADS) has been derived from DEI.Tak to capture this aspect of TA. It contains 42 items in 7 systems, again with a strong focus on CVS items but also including drug-related problems. Disease-related damage resulting from arterial blockage is a common feature of TA that often requires vascular interventions. At onset, arterial blockage results from a combination of inflammation and scarring, so TADS only scores features present for at least six months. When used in another large cohort followed over 2 decades, TADS scores showed that the increase in damage/scars over time was related to disease duration and to features of poor outcome such as pulse loss[26]. One third of cases underwent vascular interventions and over 90% of stents remained patent at 5 years. 18% of TA cases died over 2 decades and the TADS scores in fatal disease were higher than in non-fatal cases. These data show that recording a damage score can delineate features associated with clinically relevant outcomes, including pulse loss, long-term stent patency and mortality. Imaging is the basis of the anatomic classification of TA and that may correlate with the DEI.Tak score but this has not been tested yet. The DEI.Tak has been presented at the ACR meeting at Washington DC in 2006[27].

A glossary has been produced to all the three indices to aid standardization of application. It should also find use as a clinical tool to compare with newer imaging techniques to assess disease involvement such as MR and PET.

Haner et al[28], during follow-up of TA patients in Turkey, observed that most TA patients showed no clinical activity with DEI-Tak. Although the agreement between Kerr's criteria and DEI.Tak seemed very good, using Kerr's criteria instead of DEI.Tak increased the consistency with PGA, which could be explained by the influence of imaging data and acute-phase reactant levels on the physician's decisions.

Laboratory studies :

Takayasu arteritis has no specific markers. Complete blood count (CBC) reveals a normochromic, normocytic anemia in 50% of patients with Takayasu arteritis, acute phase reactants are elevated, with leukocytosis and thrombocytosis. Erythrocyte sedimentation rate is elevated. Comprehensive metabolic profile may indicate elevated transaminase and hypo-albuminemia. The von Willebrand factor–related antigen (factor VIII–related antigen) may be elevated. Antiendothelial antibodies are present. Antinuclear antibody results are usually negative and Rheumatoid factor is elevated in 15% of TA patients. Increased levels of immunoglobulins G, M, and A are present. Hoffman examined multiple serological tests including ESR, CRP, tissue factor, von Willebrand factor, sICAM-1, sVCAM-1 and E-selectin and found that no test reliably distinguished between healthy volunteers and patients with active TA. A recent study, looking for novel disease biomarkers related to pathogenesis, analyzed plasma and serum levels of matrix metalloproteinase (MMP)-2, 3 and 9. MMP-2 was elevated in TA patients versus controls but was not related to disease activity. In contrast, MMP-3 and MMP-9 were only elevated in patients with active disease, as defined by the NIH criteria and fell as disease activity was controlled by corticosteroids.

Imaging studies

Invasive Digital Subtraction Angiography is the gold standard for making the diagnosis of TA. Arteriography may be performed also by CT angiography or by MR angiography (MRA). The use of MRA is increasing rapidly. Drawbacks to arteriography, including morbidity from use of contrast dye in patients with renal disease and cumulative radiation exposure over time, can be avoided by using MRA. Arteriography often demonstrates long, smooth, tapered narrowing or occlusions. Stenoses occur in 90-100% of patients with TA and aneurysm formation in only 27%.

Ideally, non-invasive imaging modalities in TA should do the following: facilitate early diagnosis and accurate assessment of disease extent and lesion anatomy and distinguish lesions associated with TA from atherosclerotic plaque.

Gadolinium-enhanced cardiovascular MRI has also been used to demonstrate myocardial perfusion defects, which may be important in long-term prognosis. Non contrast, T2-weighted, short T1 inversion imaging recovery (STIR) images may be used to monitor edema in the aortic wall, which may be a surrogate for inflammation; edema was found in 94% of patients with clinically active disease. Large vessel edema was found in 56% of patients in clinical remission, similar to the 42-44% of patients in clinical remission who were found to have active vasculitis on pathology from bypass specimens. The prognostic significance of vessel edema is uncertain, as progression of lesions occurs in areas without edema, and progression may be absent from areas with edema on subsequent studies.

Three dimensional magnetic resonance angiography (MRA)[37] imaging of aorta and its branches provide exciting new data that may improve the understanding of the disease. Magnetic resonance imaging (MRI), MRA and computed tomography (CT) scan examinations are useful for serial examinations and diagnosis in the early phase of TA. CT scan and MRI may demonstrate mural thickening of the aorta and luminal narrowing. Use of contrast may reveal inflammatory lesions prior to the development of stenoses, these lesions may be missed by angiography. Aortic lesions including stenosis, dilatation, wall thickening and mural thrombi are well visualized on MRI, which is less adequate in visualizing distal lesions of the subclavian vessels and common carotids.

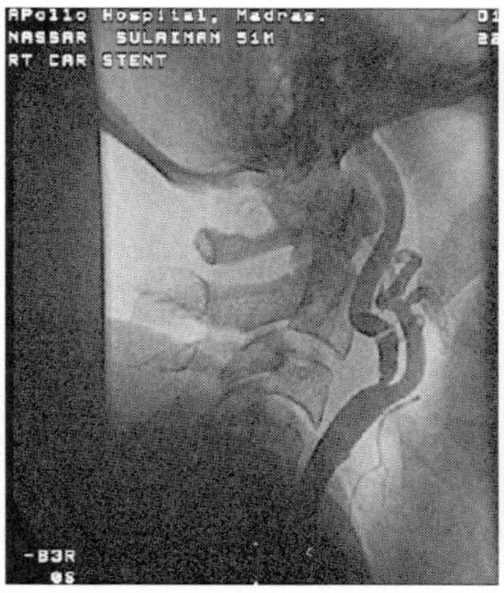

Carotid angiogram showing stenosis

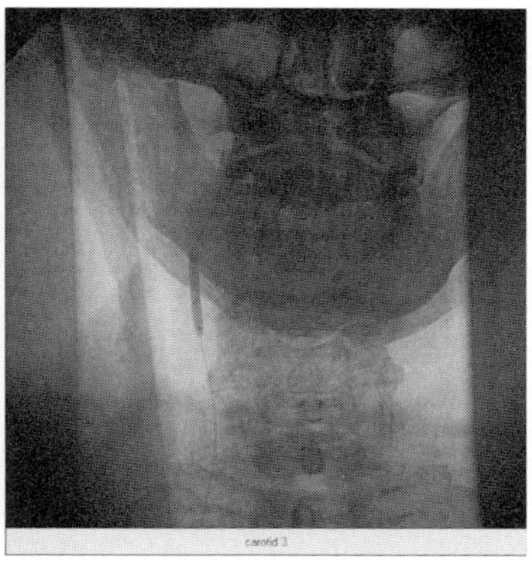

Carotid angioplasty and stenting of the internal carotid at origin

Oedema was found in 56% of the patients in clinical remission, similar to the 42-44% of the patients found to have active vasculitis on pathology from bypass specimens from patients who were in clinical remission.

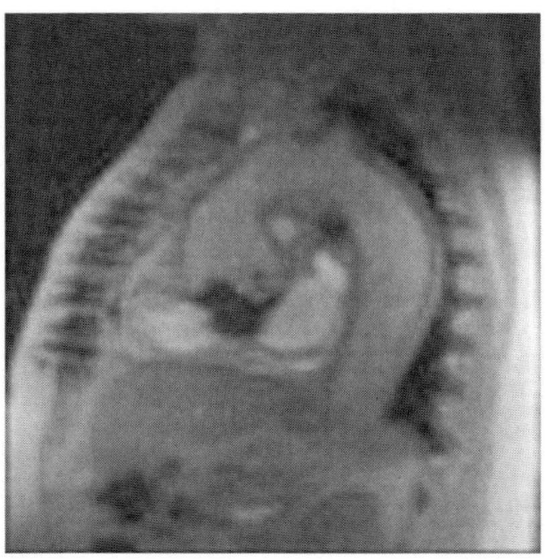

MRI of thoracic aorta showing thickened wall

^{18}F-Fluorodeoxyglucose PET

^{18}F-Fluorodeoxyglucose (^{18}F-FDG)-PET metabolic imaging is increasingly used in the investigation of patients with large vessel vasculitis; however, its precise role in disease management remains to be defined. ^{18}F-FDG-PET makes use of the fact that deoxyglucose is taken up by metabolically active cells and is not highly metabolized. The accumulation of radiolabeled deoxyglucose delineates arterial wall inflammation in giant cell arteritis (GCA) and Takayasu arteritis and reveals the extent of disease. The similarity in the pattern of arterial disease in Takayasu arteritis and GCA is apparent, although involvement of common carotid and pulmonary arteries is seemingly more common in patients with Takayasu arteritis than those with GCA. Combined ^{18}F-FDG-PET-CT scanners are being used, with the co-registered images allowing more precise anatomic location of metabolic activity and hence enhanced sensitivity. At present the principal advantage of ^{18}F-FDG-PET-CT is in the diagnosis of early pre-stenotic disease, an event that can be missed by intra-arterial angiography. However, although 40-60% of patients report extensive systemic symptoms at presentation, the diagnosis is rarely considered at this stage so limiting the utility of PET-CT and demonstrating the need for increased disease awareness. The utility of ^{18}F-FDG-PET-CT in the follow-up of patients remains unclear. Although the intense linear uptake seen in active disease might be reduced by immunosuppressive therapy, it is not unusual for persistent low-grade uptake to be seen in patients considered clinically inactive with normal acute phase reactant levels. Persistent uptake might reflect the residual arterial wall inflammation identified in surgical specimens of many patients with clinically inactive disease. Alternatively, uptake might reflect vascular remodeling, progressive fibrosis, or atheroma. These limitations of ^{18}F-FDG-PET have led to exploration of novel.

PET ligands for imaging vascular inflammation and in particular (11C)-PK11195, a ligand targeting the peripheral benzodiazepine receptor, which is highly expressed on activated monocytes/macrophages. (11C) PK11195 PET-CT allows for sensitive and specific detection of active Takayasu arteritis and GCA, with a high target-to-background ratio[29].

A 21 year old woman with active Takayasu arteritis

(A) Increased 18F-FDG was identified at both vertebral arteries by 18F-FDG PET coregistered with CT (white arrows).

(B) Four sequential 18F FDG PET analyses showed that 18F FDG accumulations in vertebral arteries on both sides disappeared during treatment with prednisolone and cyclosporine (black arrows).

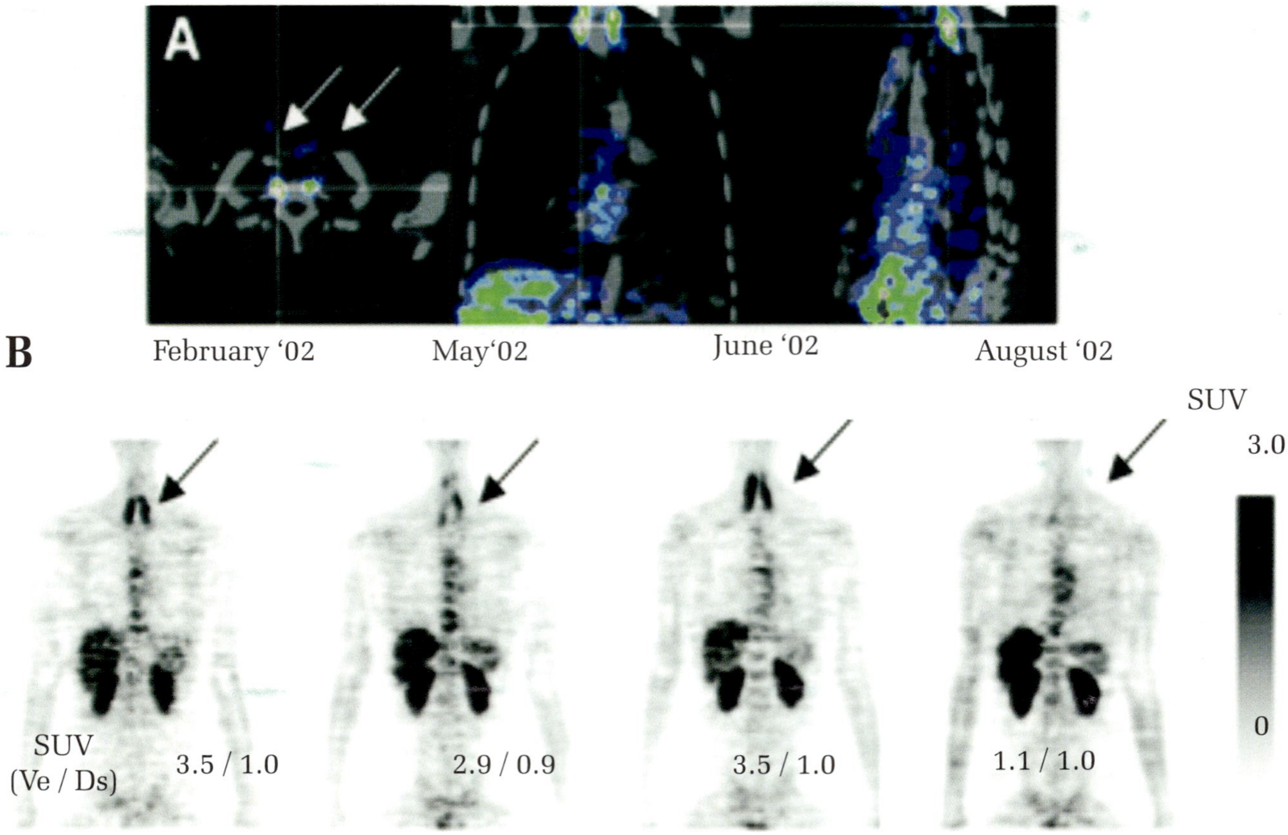

Increased common carotid intima media thickness in all 5 subtypes of Takayasu Arteritis

Gallium-67 radionuclide scan:

This scan may demonstrate increased uptake in the aorta and branches.

Duplex ultrasound

Duplex Doppler may be used to evaluate and monitor disease in the common carotids and subclavian arteries, with carotid evaluation revealing a homogenous, circumferential thickening of the vessel wall that is distinguishable from atherosclerotic thickening; this study is not useful in evaluating the aorta. Duplex Doppler may be used to evaluate and monitor disease in the common carotid and subclavian arteries. The work of Sivakumar et al[30] demonstrated that in TA patients, the mean IMT (0.84mm +/-0.16) was significantly higher than the controls (0.57mm =/-0.15 – p < 0.0001). Increased IMT was seen across all types, as classified by pattern of arteriographic involvement. The risk for a vascular event showed a continuous rise as the IMT increased across the range. The relative risk was 1.52 per increase of IMT by 1 SD (0.16mm). Increased CCA-IMT is a reliable marker of active disease, especially in the absence of angiographically visible carotid disease[31].

Duplex ultrasound showing normal flow in stented Carotid and Subclavian arteries.

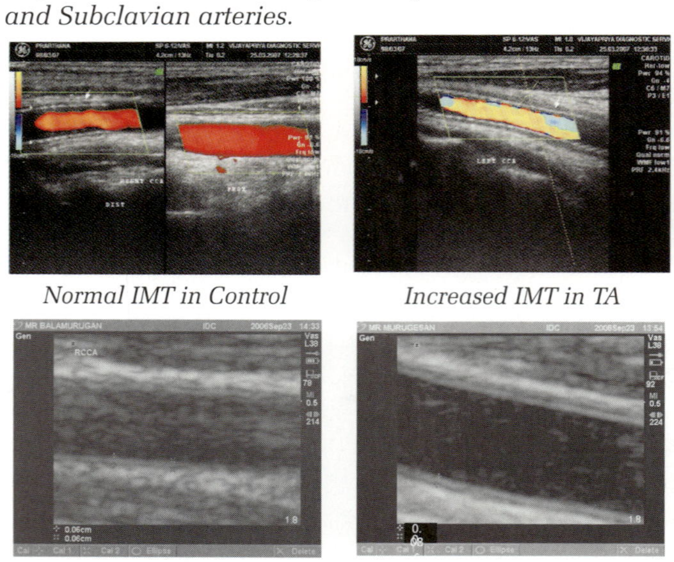

Transcranial Doppler can demonstrate intracranial arterial occlusions and emboli in patients with acute stroke[32].

TCD findings in TA patients

(A) shows 3 arteries with high blood flow velocities in stenotic range associated with an abnormal high flow pulsatility (PI 1.2), suggesting wall stiffness; these findings were observed in patients without extensive involvement of the extracranial cervical arteries.

(B) displays 3 arteries with normal flow velocities and diverse degrees of abnormal waveforms consistent with a dampened or blunted signal associated with low pulsatility flow (PI < 0.65) these patients showed severe obstructive involvement of the extracranial cerebral circulation.

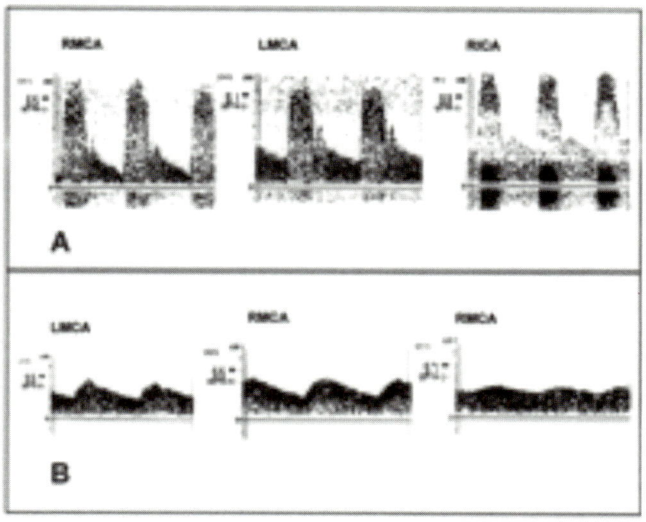

Chest radiography

May reveal widening of the ascending aorta, irregular descending aorta, aortic calcifications, and rib notching (late findings).

Echocardiography

Usually always performed at baseline and follow-up echocardiography is repeated as indicated by the clinical condition.

Histologic findings from arterial biopsies

Mononuclear infiltration of the adventitia with perivascular cuffing of the vasa vasorum occurs early in the disease. Granulomatous changes may be observed in the tunica media with Langerhans cells and central necrosis of elastic fibers and smooth muscle cells. Later, fibrosis of the media and acellular thickening of the intima may compromise the vessel lumen. Grossly, wrinkling of the intima is found. Histologic specimens seldom are available due to the large vessels affected, with the exceptions of specimens obtained during autopsy and bypass surgery.

Prognosis

The relatively poor outcome for patients with this disease reflects a variety of clinical shortcomings. These include a lack of awareness of Takayasu arteritis *per se*. The rarity of the disorder and the heterogeneous nature of its clinical presentation

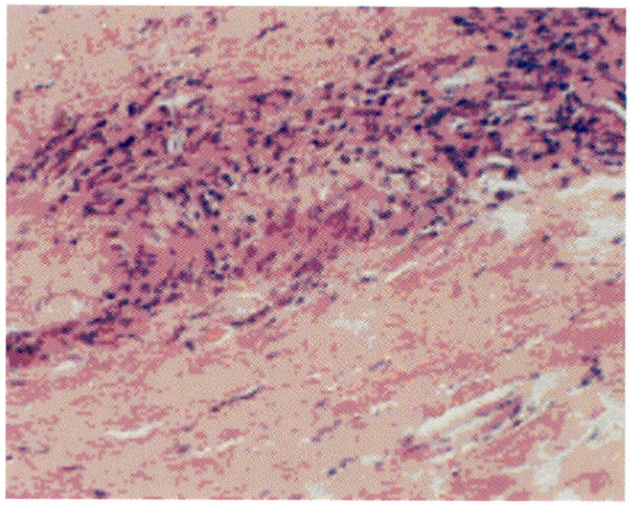

Histopathology from the arterial wall in Takayasu Arteritis

predisposes to late diagnosis and delayed treatment. Since Takayasu arteritis is rare, accurate survival data are uncertain. One study reported a survival rate of 85-95% at 15 years. In a 1994 study, only 2% of deaths were attributed directly to Takayasu arteritis. Japanese studies support 90-95% 5 year survival rates. In contrast, in a series involving 26 Mexican children aged 3-15 years, the 5-year survival rate was only 35%. Deaths resulted from rupture of aorta or aneurysms (2), stroke (2), cardiac failure (2), and peritonitis and ventricular fibrillation.

Morbidities in persons with Takayasu arteritis are related to ischemia and hypertension and include congestive heart failure, transient ischemic attacks, stroke, and visual disturbances.

TREATMENT

Medical

Goals of medical therapy are to control active inflammation and to normalize clinical and laboratory parameters while preventing further vascular damage. Owing to lack of evidence, definitive treatment advice as to dosing and duration of therapy cannot be given. The EULAP 2009 guidelines for large vessel vasculitis recommend an initial prednisolone dose of 1 mg/kg/day (total maximum dose of 60 mg/day). Maintain high dose treatment until all evidence of active disease has resolved, then taper prednisolone dosage gradually. Although 60% of patients respond to this treatment, 40% relapse on steroid taper. Patients not responding to corticosteroids or who relapse during corticosteroid taper require an additional agent. Symptoms of patients who relapse on corticosteroid taper may be controlled with weekly infusions of methylprednisolone (15mg/kg, not to exceed 1 g/wk). However, extensive use of these infusions is associated with significant steroid induced toxicity if continued for any significant period. Immunosuppressive regimens including weekly Methotrexate (7.5-25mg) or daily or monthly intravenous cyclophosphamide in individuals with glucocorticoid resistant TA have been tried. Low dose weekly Methotrexate also has been used as a steroid sparing agent for those TA patients not tolerating corticosteroid taper. Cyclosporine may be an alternative therapy offering lower ovarian toxicity than cyclophosphamide. However, cyclosporine often is associated with decreased renal function and increased blood pressure, which may aggravate the damage to the heart and great vessels. Mycophenolate mofetil is useful to treat individuals with glucocorticoid resistant disease.

Open-label studies from the Cleveland Clinic report encouraging results for the use of the anti-tumor necrosis factor antagonists etanercept and infliximab [33,34]. In a study of 25 patients failing to respond adequately to corticosteroids, and at least one additional immunosuppressive agent, who were given etanercept (50 mg/week) or infliximab 3-5 mg/kg every 6-8 weeks) disease remission and corticosteroid withdrawal was seen in 60%, with a further 28% benefiting through reduction in the prednisone dose to < 10 mg/day.

Cardiovascular risk factors, accelerated atherosclerosis, systemic hypertension and pulmonary hypertension must also be investigated, monitored and treated actively in all patients with Takayasu arteritis, where appropriate. This highlights the need for a multidisciplinary approach to the treatment of patients with this disease. The cardiovascular risk associated with this condition results in most patients receiving low-dose aspirin and a statin, although specific evidence in support of this approach is lacking.

Anecdotal reports of matrix metalloproteinase inhibition using minocycline propose that this may be a useful adjunctive therapy, which may also allow lower doses of corticosteroids and thus reduced toxicity.

New reports demonstrate the use of Rituximab, monoclonal anti-CD20 antibody, for disease control and steroid sparing in patients with Takayasu arteritis. Treatment regimens have been the 4-week lymphoma protocol (375 mg/m^2 weekly for 4 wk) or the 2-week fixed-dose rheumatoid arthritis protocol (1000 mg repeated once 14 d later).

Case reports of Tocilizumab (monoclonal anti–IL-6 receptor antibody) have described remission induction in Takayasu arteritis resistant to other treatments. IL-6 levels are elevated in Takayasu patients and correspond with disease activity, making this an attractive target for therapy.

Myofibroblast proliferation, which predisposes to arterial occlusion and ischemic injury, is an important potential target. The mammalian target of Rapamycin (mTOR) antagonists, Sirolimus and Everolimus, currently used clinically for the prevention of rejection in organ transplantation, are known to inhibit myofibroblast proliferation and their use in Takayasu arteritis should be explored. Likewise, T lymphocyte infiltration is seen in affected arteries and hence Abatacept, a selective modulator of the CD80/CD86:CD28 co-stimulatory signal required for full T-cell activation holds considerable promise for TA.

Improved understanding of the molecular mechanisms involved in disease pathogenesis is required and might also pave the way for novel treatments. Extrapolating from the studies of GCA, dendritic cells and their activation antigens CD83 and CD86, the chemokines CCL19 and CCL21, and the pro-inflammatory cytokine interleukin-18 all have important roles in the development of granulomatous large vessel vasculitis and hence represent potential therapeutic targets. Potential novel therapeutic targets for this granulomatous arteritis include IL-1, IL-6 and IL-18, IFN-γ, activated dendritic cells, Toll-like receptors and myofibroblast proliferation. Drug-eluting stents, such as those containing the antibiotic Rapamycin or the antineoplastic drug Paclitaxel, might also have a role.

Surgery

Following the acute phase, patients with fibrotic changes require surgical treatment of symptomatic stenotic or occlusive disease. This can be achieved by percutaneous angioplasty or stenting or, in severe cases, by resection and placement of a manmade graft. These procedures should always be preceeded and followed by effective immunosuppressive therapy. Percutaneous balloon angioplasty of the aorta[35,36] is reported to normalize systolic and diastolic blood pressures within 24 hours, with improvement of exercise tolerance and restoration of peripheral pulses. Both renovascular hypertension and congestive failure are improved and has been sustained for as long as 5-15 years.

Endovascular stenting is used in patients with severe stenoses, hypertension, or ischemia during the fibrotic phase of the disease. Multiple stents have been used in children to relieve long-segment renal artery stenosis and attendant renovascular hypertension. Children with TA who have received stents have lowered arterial blood pressures and decreased requirement for antihypertensive drugs. Immunosuppressant eluting stents could potentially deliver local treatment at sites of inflammation. Careful pre-procedure analysis of individual lesions is required and trials of stents incorporating antiproliferative drugs such as Sirolimus are indicated.

Follow-up

Measurement of ESR and C-reactive protein may not prove a useful aid in many patients. Regular imaging of affected vasculature as well as surveillance imaging for new lesions is essential. Treatment may be monitored with MRI and/or MRA or CT scan. Mural thickening is observed to decrease with corticosteroid treatment. Recognizing that TA may progress in the absence of clinical findings is important. Periodic imaging may identify an active disease by the appearance of new areas of stenosis, despite normal erythrocyte sedimentation rate and the absence of clinical features.

Key points

1. TA is more commonly seen in women than men in India and Asia, but it also afflicts persons in North America, Europe, Middle East and Africa.
2. The diagnosis of TA is difficult and requires a high index of suspicion.
3. Clinical presentation depends on the location and severity of the aortic branch lesions, ascending aortitis and abdominal aorta and branches are frequently affected in Indian patients.
4. The principal therapeutic intervention for TA is corticosteroids. About 50-60% respond favourably.
5. Therapeutic interventions like Percutaneous Transluminal Angioplasty (PTA) and Stenting, bypass surgeries or surgical reconstruction should be performed when the disease is made inactive by the use of effective immunosuppressive therapy.
6. If undetected early, TA often follows a protracted course with irreversible vascular damage.
7. New non invasive imaging methods have greatly improved the ability to diagnose patients with TA promptly and monitor their level of disease activity more accurately.
8. The Disease Extent Index for Takayasu Arteritis (DEI.Tak), Indian Takayasu Arteritis Score (ITAS) and the Takayasu Arteritis Damage Score (TADS) are valuable tools in documenting the spectrum of TA in epidemiological studies, to assess the severity and presence of active disease and to estimate damage in TA.
9. Strict control of elevated blood pressure which may often be difficult to diagnose in TA is essential to minimize cardiovascular complications.

Further Reading :

1. F Numano. The story of Takayasu Arth Rheum 2002; 41: 103-06.
2. Shinmi Y. A case of Takayasu's arteritis (in Japanese). Sogo Gannka 1942; 36: 1404-10.
3. Hotchi M. Pathological studies on Takayasu's arteritis. Heart Vessels 1992; suppl 7: 11-17.
4. Maksimowicz-McKinnon K, Clark TM, Hoffman GS. Limitations of therapy and a guarded prognosis in an American cohort of Takayasu arteritis patients. Arth Rheum 2007 Mar; 56: 1000-09.
5. Kinare S, Gandhi M, Deshpande J. Non-specific Aorto arteritis (Takayasu disease): pathology and radiology. 1 ed. Quest Publications; 1998.
6. Panja M. Non-specific Aorto-arteritis (Takayasu Arteritis). 1st ed. Kolkata: Marksman Media; 2001.
7. Deshpande J. Pathological features of Takayasu Arteritis (non-specific aorto-arteritis). Rheum 2005; 44: iii8.
8. Sivakumar MR, Sanjeev S. Cerebrovascular manifestations and carotid artery intima medial thickness in Takayasu's Arteritis evaluated by using the Disease Extent Index for TA (DEI.Tak). Clinical and Experimental Rheumatology 2007(S44) Vol 25 no.1 S-120.
9. Peter J, David S, Joseph G, Horo S, Danda D. Hypoperfusive and hypertensive ocular manifestations in Takayasu arteritis. Clin Ophthalmol. 2010 Oct21; 4: 1173-76.
10. Peter J, David S, Danda D, Peter JV, Horo S, Joseph G. Ocular manifestations of Takayasu arteritis: a cross-sectional study. Retina. 2011 Jun; 31(6): 1170-78.
11. Arend WP, Michel BA, Bloch DA, Hunder GG, Calabrese LH, Edworthy SM. The American College of Rheumatology 1990 criteria for the classification of Takayasu arteritis. Arth Rheum 1990; 33: 1129-34.
12. Ishikawa K. Diagnostic approach and proposed criteria for the clinical diagnosis of Takayasu's arteriopathy. J Am CollCardiol. 1988; 12: 964-72.
13. Sharma BK, Jain S, Suri S, Numano F. Diagnostic criteria for Takayasu arteritis. Int J Cardiol. 1996; 54 suppl: S141-47.
14. Moriwaki R, Noda M, Yajima M, Sharma BK, Numano F. Clinical manifestations of Takayasu arteritis in India and Japan new classification of angiographic findings. Angiology 1997; 48: 369-79.
15. Mason JC. Takayasu arteritis – advances in diagnosis and management. Nat Rev Rheumatol. 2010; 6(7): 406–15.
16. Kerr GS, Hallahan CW, Giordano J, Leavitt RY, Fauci AS, Rottem M. Takayasu Arteritis. Ann Intern Med 1994; 120: 919-29.
17. Luqmani RA, Bacon PA, Janssen BA, Moots RJ, Pall A, Adu D et al. Birmingham Vasculitis Activity Score (BVAS). Arth Rheum 1993; 36: S138.
18. Exley A, Banbury P, Carruthers D, Moots R, Luqmani R, Kitas G, et al. The vasculitis damage index (VDI). Arth Rheum 1995; 38: 1124IS.
19. Direskeneli H, Aydin SZ, Merkel PA. Assessment of disease activity and progression in Takayasu's arteritis. Clin Exp Rheumatol. 2011; 29(1 suppl 64): S86–S91.
20. Numano F. Takayasu's arteritis: clinical aspects. In: Hoffman GS, WCM editor. Inflammatory Diseases of Blood Vessels.New York: Marcel Dekker; 2002; p. 455–66.
21. Direskeneli H, Aydin SZ, Kermani TA, et al. Development of outcome measures for large-vessel vasculitis for use in clinical trials: opportunities, challenges, and research agenda. J Rheumatol. 2011; 38(7): 1471–79.
22. Sivakumar MR, Misra R, Bacon PA for the IRAVAS Group Indian prospective of TA and development of disease extent index in Takayasu arteritis. Rheumatology (Oxford). 2005; 44(suppl iii): 6–7.
23. Misra, R. et al. ITAS and DEI.TAK – scores for clinical disease activity and damage extent in Takayasu aorto-arteritis. Rheumatology 47 (Suppl. 2), ii101 (2008).
24. Goel R, Danda D, Mathew J, Edwin N. Mycophenolate mofetil in Takayasu's arteritis. Clin Rheumatol. 2010 Mar; 29(3): 329–32.
25. Salvarani C, Magnani L, Catanoso M, et al. Tocilizumab: a novel therapy for patients with large-vessel vasculitis. Rheumatology (Oxford). 2011.
26. Sivakumar MR. Outcome of vascular interventions in Takayasu arteritis using the Takayasu arteritis damage score. Arthritis Rheum. 2011; 63(10): S588.
27. Sivakumar MR, RamnathMisra, Debashish Danda, Alakendu Ghosh, Amita Aggarwal, Paul A. Bacon Derivation and Validation of a Disease Extent Index for Takayasu Aorto- Arteritis Arth Rheum Volume 54, number 9 (S) September 2006; S758.
28. Sibel Z. Aydin, NeslihanYilmaz, ServetAkar, Haner Direskeneli et al Assessment of disease activity and progression in Takayasu's arteritis with Disease Extent Index-Takayasu. Rheumatology 2010; 49: 1889–93.
29. Hoffman GS, Ahmed AE. Surrogate markers of disease activity in patients with Takayasu arteritis. Int J Cardiol 1998; 66(S)1: S191-94.
30. Sivakumar S, Sivakumar MR. Intimal medial thickness in Takayasu arteritis and change with treatment. Rheumatology. 2010; 49(suppl 1): i123–i134.
31. Sharma S, Talwar KK. Carotid intima-medial thickness as a marker of disease activity in Takayasu's arteritis : Int J Cardiol. 2006 Apr 14; 108: 385-90.
32. Sivakumar S, Arjundas D, Sivakumar MR. Use of Transcranial Doppler for the detection of severe carotid stenosis. Neurology. 2010; 74(suppl 2): A228.
33. Hoffman, GS, Merkel, PA, Brasington RD, Lenschow DJ & Liang P. Anti-tumor necrosis factor therapy in patients with difficult to treat Takayasu arteritis. Arth Rheu. 50, 2296-2304 (2004).
34. Molloy ES, Langford CA, Clark TM, Gota CE & Hoffman GS. Anti-tumour necrosis factor therapy in patients with refractory Takayasu arteritis: long-term follow-up. Ann. Rheum. Dis. 67, 1567-69 (2008).
35. Joseph G, George PV, Pati PK, Chandy ST. Feasibility of angioplasty and stenting for abdominal aortic lesions adjacent to previously stented visceral artery lesions in patients with Takayasu arteritis. Cardiovasc Intervent Radiol. 2007 Mar-Apr; 30(2): 293-96.
36. Lagneau P, Michel JB, Vuong PN. Surgical Treatment of Takayasu's Disease. Ann Surg 1987; 205: 157-66.
37. Garg SK, Mohan S, Kumar S. Diagnostic value of 3D contrast-enhanced magnetic resonance angiography in Takayasu's arteritis – a comparative study with digital subtraction angiography. Eur Radiol. 2011 Aug; 21(8): 1658-66.
38. Paul Bacon, Rajappa Sivakumar, Debashish Danda, RamnathMisra Vasculitis assessment and Takayasu aorto-arteritis Indian Journal of Rheumatology 2012 September; Vol.7, No. 3.
39. Ramnath Misra, Debashish Danda, Sivakumar Rajappa, et al on behalf of the Indian Rheumatology Vasculitis (IRAVAS) group. Development and initial validation of the Indian Takayasu Clinical Activity Score (ITAS2010. Rheumatology 2013; 52 (Oct): 1795-1801 .doi:10.1093/ rheumatology/ket128.

SECTION 9

Gout and other Crystal Arthropathies

GOUT AND HYPERURICEMIA 1

Binoy Paul

Gout is a clinical syndrome occurs as a result of deposition of monosodium urate monohydrate crystals from hyperuricemic body fluids. The crystals may be deposited in a joint leading to acute inflammatory response or in soft tissues like cartilage causing no inflammation.

History

Gout is one of the oldest joint diseases known to humanity. History of gout claimed the attention of many medical historians unlike other joint disorders[1,2]. The word gout is derived from the Latin word "gutta" meaning drop. It is based on the ancient belief that gouty arthritis occur due to instillation of malevolent humour or toxic fluid into the joint of the victim drop by drop by the evil spirits. Hippocrates made several observations about gout popularly known as aphorisms of gout. (Table 1)[3] in 3rd century BC, Galen described tophi. Crystals in gouty tophi were first demonstrated by Antony Van Leeuwenhock in 1679. It was in 1848, Sir Alfred Garrod demonstrated hyperuricemia as the basic cause of gout. He described a semi quantitative thread test for measurement of urate. In 1961, McCarty and Hollander established the association between gouty arthritis and articular crystal deposition.

Table 1. Hippocratic aphorisms of gout

Eunuchs do not take gout nor become bald,
A women does not take gout unless her menses be stopped
A young man does not take gout unless he indulges in coitus
In gouty affection inflammation subsides in 40 days

Pathogenesis

Uric acid is the end product of purine degradation in human. Purines are derived either from diet or by de novo synthesis. The purine nucleotides when broken down first converted to a monoxide called hypoxanthine which then metabolized to a deoxy purine, xanthine. The final conversion of xanthine leads to production of uric acid which is a trioxypurine[4]. Lower animals possess an enzyme called uricase which converts insoluble uric acid to soluble allantoin and carbon dioxide. During the evolutionary process humans and greater apes lost this enzyme, hence further metabolism of uric acid is not possible in human body. Uric acid has important physiological role, it is a powerful antioxidant removing the damaging reactive oxygen metabolites.

The level of elevated uric acid is the key biochemical precursor of gout, the higher the persons urate level, greater the chance of gouty arthritis. Hyperuricemia can be defined as a serum urate concentration greater than 7 mg/dl in males and 6 mg/dl in females. The lower normal level of uric acid in females is believed to be due to the uricosuric effects of oestrogen.

Most of the uric acid produced in the body is eliminated through urine or intestine. The kidney excretes two third of the uric acid turn over and rest via saliva, gastric juice, pancreatic secretion or bowel (Fig. 1).

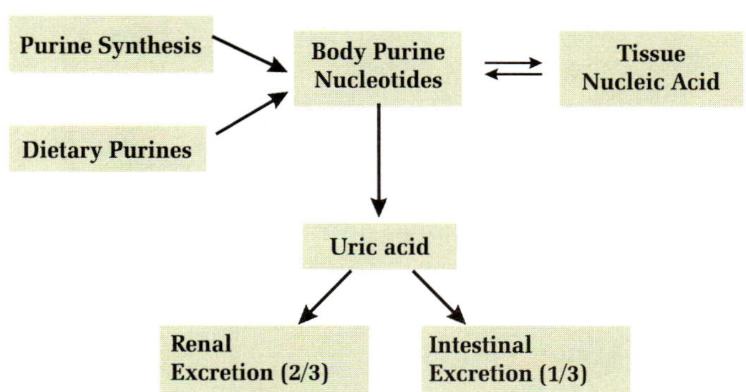

Fig. 1 : Human purine metabolism.

Most investigations support a four component model for renal handling of uric acid[5]. There is 100% filtration at glomeruli, pre-secretory reabsorption of 98-100% at the proximal convoluted tubule by active transport, tubular secretion of around 50% and post secretory reabsorption probably at collecting duct of 40-50% (Fig. 2). The existence of a discrete fourth component (post secretory reabsorption) is controversial[6,7]. Around 5-10% of the uric acid is excreted finally through the urine as the final net absorption is approximately 90-95%.

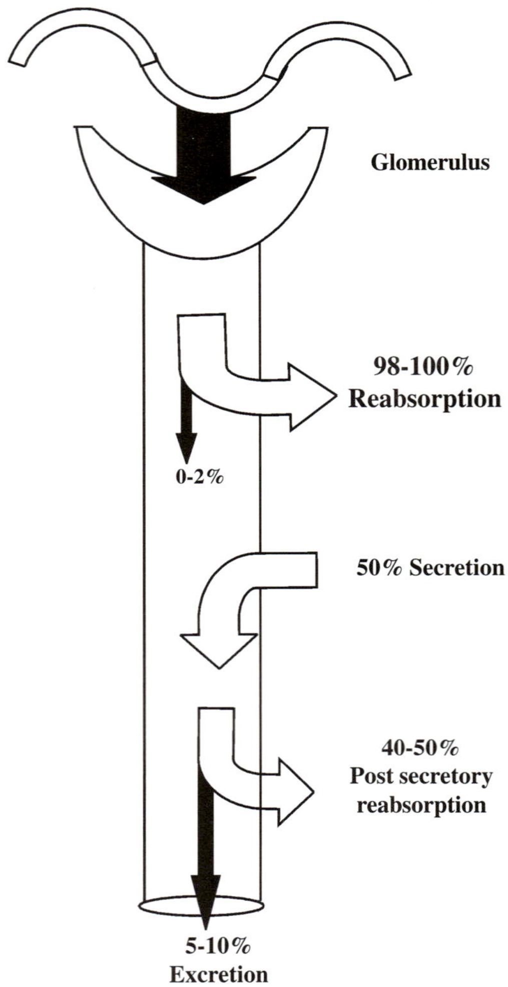

Fig. 2 : Four component model of renal urate transport.

Classification

Based on the aetiology gout is classified into **primary gout** (of unknown cause) and **secondary gout** (due to an underlying illness leading to hyperuricemia).

Primary gout

The exact defect in the pathogenesis of primary gout is still not known, but is likely to be due to a genetic defect in the renal urate handling. Around 90% of these patients are found to have a decreased renal clearance of uric acid and in 10% over production of uric acid is attributed as the cause of hyperuricemia.

Secondary gout

The various causes of hyperuricemia leading to secondary gout are given in Table 2.

Table 2. Causes of hyperuricemia and secondary gout

Increased urate production:

- Inherited enzyme defects
  - Hypoxanthine-guanine phosphoribosyl transferase deficiency (Lesch Nyhan Syndrome)
  - Glucose 6 phosphatase deficiency
  - Phosphoribosyl pyrophosphate synthetase overactivity
- Myeloproliferative disorders
- Psoriasis
- Haemolytic diseases
- Malignancies, Tumour lysis syndrome
- High purine diet
- Alcohol

Decreased renal clearance

- Renal : Chronic renal failure, polycystic kidney disease, lead nephropathy
- Endocrine : Hyperparathyroidism, Hypothyroidism, Diabetes Insipidus
- Metabolic : Lactic acidosis, Ketoacidosis, Starvation, Severe dehydration
- Metabolic syndrome : Obesity, Dyslipidemia, Hypertension.
- Drugs : Diuretics, Low dose aspirin, Pyrazinamide, Cyclosporin, Ethambutol
- Miscellaneous : Sarcoidosis, Toxemia of pregnancy, Down syndrome

Epidemiology

Gout is the most common cause of inflammatory arthritis in men aged more than 50 years in the western world affecting approximately 1-2% of adult male[4]. It is rare in children and premenopausal women. After menopause due to the lack of oestrogen uric acid excretion will be reduced, thereby putting women at equal risk of gout as men. Of the gout patients above the age of 60 years, 50% are women.

The exact incidence of gout in India is not clear. The ILAR COPCORD study from the rural population of western Maharashtra shows a prevalence of 0.1%[8]. The prevalence is likely to be high in urban India. The incidence of gout in younger population in India is also high. A study from South India shows that 15.8% of the affected patients were below the age of 30 years[9].

The incidence and prevalence of gout is on the rise. Studies show that prevalence of gout has doubled over the last two decades[10,11]. The increasing longevity, dietary trends, obesity, metabolic syndrome, hypertension, increased use of low dose aspirin, increased survival after myocardial infarction, end stage renal disease and major organ transplantations are important factors that are responsible for the changing demographics.

Clinical features

The clinical stages of gout can be divided under three headings. Acute gouty arthritis, intercritical gout and chronic tophaceous gout.

Acute gouty arthritis

The classic symptoms of acute gouty arthritis are recurrent attacks of acute, markedly painful monoarticular or oligoarticular inflammation affecting the peripheral joints of lower limb. Studies from South India showed that first metatarso phalangeal joint (MTPJ) is the most common joint involved in the course of the disease (91.3%). However, ankle joint is the commonest site of initial episode (32%) followed by the first MTPJ (32%)[12]. The other joints involved in decreasing frequency are ankle, tarsal joints, knees, fingers, elbow or wrist. Hip, spine and shoulders are usually spared. Most patients know when an acute episode is imminent. Due to the prodromal mast cell degranulation and release of histamine, many patients report local itching and burning before the onset of acute arthritis. Acute episodes often begin in the early morning often between 2 am and 7 am. The episodes build to a peak over several hours with intense pain and increase sensitivity of overlying skin. Fever is common and more likely to occur with polyarticular episodes. Complete resolution of symptoms is seen within a week even without treatment[13]. Bursae which are near to or communicate with joints may also be the sites of inflammation. Of these, olecranon and prepatellar bursitis are the most common ones. Polyarticular presentation of first attack of gout is uncommon. But can occur in postmenopausal females, chronic alcoholics, hypertensive males on diuretics, gout accompanying myeloproliferative disorders and in patients using cyclosporine.

Intercritical (Interval) gout

The interval between two gouty attacks is called intercritical gout. The patient will be totally asymptomatic, but uric acid crystals may be demonstrated by aspiration of synovial fluid from previously inflamed joint. Progressive bony erosions can also develop during this period which can be demonstrated by serial radiographs of the affected joint. The duration of intercritical gout varies. Without diet control, lifestyle modification or drugs, most patients experience a second episode of arthritis within 2 years.

Chronic tophaceous gout

When recurrent acute gout and hyperuricemia go untreated, the condition may be evolved from a picture of sporadic acute mono or oligo arthritis, through recurrent poly arthritis to persistent low grade joint inflammation, joint deformity and deposition of urate crystals to form visible tophi. Tophi occur in around one third of chronic gout patients often 5-10 years after the first episode. The tophaceous deposits develop around the joints of the feet, finger joints, elbows, Achilles tendon and pinna of the ear or within olecranon and prepatellar bursae. The relative mild discomfort of chronic gout may be punctuated by episodes of acute arthritis. Even though acute attacks are more common in the joints of feet, it is in the hands where evidence of chronic gout will be more prominent.

Non-rheumatological involvement in hyperuricemia

Persistently high uric acid levels can lead to urolithiasis, urate or uric acid nephropathy[14]. Urolithiasis precedes the onset of gouty arthritis in many patients with hyperuricemia. The uric acid may act as a nidus on which other crystals like calcium oxalate can precipitate. Uric acid stones can develop in individuals without other manifestations of gout, only 20% of whom are hyperuricemic. Urate nephropathy (urate nephrosis) is a late manifestation of severe hyperuricemia due to deposition of monosodium urate crystals in the medullary interstitium and pyramids leading to chronic renal insufficiency. Uric acid nephropathy is due to precipitation of uric acid in the renal tubules or collecting duct causing obstruction of urine flow and acute renal failure. This usually occurs in aggressive chemotherapy for malignancies or untreated hyperuricemia patients developing severe dehydration or acidosis.

Differential diagnosis

Differential diagnosis of acute gout includes palindromic rheumatism, septic arthritis, pseudogout and haemarthrosis. Even though septic arthritis can be excluded by joint aspiration and bacteriological work up, palindromic rheumatism often causes confusion. Palindromic rheumatism is more common in females. Many of them have a positive family history of rheumatoid arthritis, the upper limb joints are often affected from the early attacks onwards and around one third have positive rheumatoid factor. AntiCCP can also be positive. The joint fluid shows features of acute inflammation with notable absence of uric acid crystals. Response to colchicine is poor and attacks abate with NSAIDs.

Haemarthrosis and pseudo gout can also be readily excluded by synovial fluid examination.

Chronic tophaceous gout can mimic chronic rheumatoid arthritis with rheumatoid nodules, nodal osteoarthritis, psoriatic arthritis or scleroderma. Demonstration of uric acid crystals from tophus and radiological examination of hands helps in the differentiation.

Investigations

Acute or chronic joint symptoms or signs associated with hyperuricemia are not sufficient to establish a diagnosis of gout as both musculoskeletal pain and unrelated hyperuricemia are common in general population. Further more in around one third of patients have normal uric acid level during the acute attack.

Blood tests

During the acute episodes, rise in ESR or CRP and neutrophilic leucocytosis are common, with the ESR frequently exceeding 50mm/1st hr. The serum uric acid level is high only in around two thirds of patients during acute attack hence, normal uric acid level will not exclude the diagnosis of gout. Although hyperuricemia is not a requirement for the diagnosis of gout and its presence in patients with arthritis does not necessarily establish the diagnosis. The risk of gout increases with the degree and duration of hyperuricemia.

Demonstration of urate crystals

Demonstration of needle shaped negatively birefringent crystals in the synovial fluid or from the tophi by polarizing light microscopy is considered as the gold standard for the diagnosis of gout. Ordinary light microscope can be adapted to serve as a polarizing microscope using first order red filter made of ordinary glass slides and transparent cellophane tapes[15]. Intracellular monosodium urate crystals are highly suggestive of acute gout. Extracellular urate identifiable in synovial fluid or from tophi are seen in intercritical gout or chronic tophaceous gout.

Imaging

Radiography in acute initial attacks show only soft tissue swelling of the affected joint. The radiographic hallmarks of long standing gout are those of an asymmetric inflammatory erosive arthritis often accompanied by soft tissue nodules with retention of normal joint space. Sharply punched out oval or round defects situated in the marginal area of joint often with overhanging margins are also classical of gout[16]. Periarticular osteopenia is not seen unlike chronic rheumatoid disease. Tophi may be seen as soft tissue mass of intermediate density, sometimes calcified. In advanced cases, joint destruction can be extensive with subluxation and bony ankylosis of interphalangeal joints and metacarpophalangeal joints of hand and feet. On ultrasonography (USG), monosodium urate crystals may appear within the synovial fluid as a "snow storm appearance" or deposited on superficial articular cartilage as the "double contour sign"[17]. Collection of MSU crystals within tophus can also be identified by USG. Conventional CT scan can visualize tophi and bony erosion in gout and also helpful in scoring the bone erosions[18]. Dual energy CT (DECT) is a new technique, which has a potential to demonstrate MSU crystal deposition[19]. MRI also demonstrate joint effusion, synovitis, tendon disorder, tophus, cartilage disorder or bone oedema in gout[20].

Urinary uric acid excretion

Estimation of 24 hour urinary uric acid excretion is helpful in identifying under excretors or over producers. Around 90% of patients with primary gout are under excretors. For adults ingesting a purine free diet, a total excretion upto 600mg/day and on normal diet up to 800mg/day is within normal range. A value of more than 1000mg/day clearly suggests overproduction (value of 800-1000mg/day is borderline) and a value below 600mg/day suggests under excretion.

There is no requirement to estimate urine uric acid excretion in gout as a routine practice except in children and young females with gout who have normal renal function. These patients may require further investigations of purine metabolism or genetic studies to find out the different enzymatic abnormalities causing gout.

Diagnosis

A definite diagnosis of gout is best established by aspiration and demonstration of uric acid crystals from synovial fluid or tophus. But when crystal examination is negative or not possible two different criteria has been proposed for the presumptive diagnosis.

A triad of acute monoarthritis, hyperuricemia and dramatic response to colchicines is highly diagnostic in the absence of demonstrating the crystals[21,33].

Alternatively the American College of Rheumatology also proposed diagnostic criteria for acute gout (Table 3).

Table 3. ACR Criteria for classification of acute gouty arthritis

The presence of urate crystal in the synovial fluid or tophus proved by chemical means or polarized light microscopy

OR

Presence of 6 of the following 12 Clinical, Laboratory and Radiographic phenomenon

- More than one attack of acute arthritis
- Maximal inflammation developed within 1 day
- Attack of monoarticular arthritis
- Joint redness – observed
- Painful or swollen first metatarsophalangeal joint
- Unilateral attack involving first metatarsophalangeal joint
- Unilateral attack involving tarsal joint
- Suspected tophus
- Hyperuricemia
- Asymptomatic swelling within the joint (Radiograph)
- Subcortical cyst without erosion (radiograph)
- Negative culture of joint fluid for microorganisms during attack of joint inflammation

Treatment

Principle goals of management

1. Treat the acute attack promptly and effectively to alleviate the pain and restore the inflamed joint to useful function quickly.
2. Prevent recurrences of acute gouty arthritis.
3. Prevent or reverse complications due to crystal deposition in joints, kidneys or other sites.
4. Correct the contributory factors of hyperuricemia like obesity, dyslipidemia or hypertension.

Asymptomatic hyperuricemia

There are many individuals with uric acid level higher than that of normal controls (more than 7mg in males and 6mg in females) who are totally asymptomatic and hyperuricemia is detected incidentally. Few of them can have musculoskeletal pain but not directly related to hyperuricemia. But mere presence of hyperuricemia is not an indication for specific hypo uricemic treatment. But a meticulous search for the cause of hyperuricemia should be made. In more than 70% of hyperuricemia patients, an underlying cause can be readily found out by history, physical examination and a few investigations. Metabolic syndrome, renal and thyroid diseases, alcoholism and drugs are the usual causes. Correction of the underlying cause often brings back the uric acid values to normal levels. Currently there is no evidence of benefit to be obtained from treating sustained asymptomatic hyperuricemia unless it is severe (more than 12mg/dl) or in situations where there is acute or massive urate over production. Hence, it seems prudent not to treat asymptomatic hyperuricemia with specific antihyperuricemic agents unless symptoms attributable to hyperuricemia like arthritis, urolithiasis or nephropathy develops. The rare exceptions to this rule include individuals with a known hereditary cause of uric acid over production or at potentially high risk for acute uric acid nephropathy as in tumour lysis.

Management of acute gout

Acute gouty attacks may be successfully terminated by Colchicine, Non steroidal anti-inflammatory drugs (NSAIDs) or Corticosteroids.

Colchicine

Colchicine is the ideal drug in patients where the diagnosis of gout is not confirmed. The latest recommendation is to start colchicine 1mg (2 x 0.5mg) as a loading dose, followed by 0.5mg one hour later, if needed one more (0.5mg) after twelve hours; continued colchicine (upto 0.5mg thrice a day) until acute attack resolves[31,32]. Colchicine acts by inhibiting the action on neutrophils. It retards the adhesiveness and motility of neutrophils and prevents chemotaxis. It also down regulates TNF- receptors on macrophages and inhibits mast cell histamine release.

NSAIDs

They are superior to colchicines in terms of speed of onset of action and are useful if the diagnosis of gout is already confirmed. They are relatively contraindicated in patients with renal insufficiency, hypertension, peptic ulcer or congestive heart failure. The most useful drug is Indomethacin (50mg 6 hourly) in most of the patients. The high dose should be maintained for 2 days or until the pain settles and then reduce to 50mg 8hourly. Once the acute attack subsides completely, the dose may be further reduced to 25mg 8 hourly or 75mg daily (of the sustained release preparations). Other NSAIDs like Naproxen (750mg/day) and Etoricoxib (120mg/day), Diclofenac 50mg 8hrly are equally effective in most of the studies.

Corticosteroids

Corticosteroids are preferred only if NSAIDs and colchicine are contraindicated or not effective. Intraarticular steroids are very useful in treatment of acute gout limited to a single joint or bursae. High dose oral corticosteroids (Prednisolone 30mg daily tapered off in 7-10 days) may be used in polyarticular gout.

Parenteral Methyl prednisolone (100-150mg IM/IV) and ACTH (25-40 IU IM or S/C) are useful in poly articular gout following surgery, renal or hepatic insufficiency or congestive heart failure. The dose can be repeated 12 hourly for 1-3 days in incomplete responders.

Anti-Interleukin I (IL-1) therapy

Blocking IL-1 is a novel concept in the treatment of acute gout because MSU crystals stimulate the inflammasome in gout leading to IL-1b secretion. IL-1 inhibitors prevent IL-1b secretion via this mechanism and also block IL1 secretion by macrophages via a toll like receptor dependent mechanism. The anti IL1 drugs like ***Anakinra, Canakinumab and Rilonacept*** have been studied in patients with acute gout[22,23,24]. Subcutaneous injection of Anakinra 100mg daily was found to be well tolerated and very effective in acute gout flares when used in patients who were unable to take other gout medications because of renal impairment, allergies, gastrointestinal haemorrhage and history of renal stones. Pain decreased in 79% on average after three injections. No side effects were noted during the trial and clinical examination showed complete resolution in 9 out of 10 patients on day 3.

Phase III studies have been completed using both Canakinumab and Rilonacept in management of acute gout flares. Canakinumab 150mg subcutaneously was found to be superior to Triamcinolone 40mg intra muscularly in gouty arthritis patients in flares. Rilonacept 80-120mg subcutaneously weekly was given on initiation of urate lowering therapy to prevent acute gout flares for a period of 16 weeks and found to be highly effective in preventing flares as compared with placebo.

Anti IL1 treatments are highly effective in terminating gout flares in those unresponsive to NSAIDs and colchicines and also reducing the risk of flares in initiation of urate lowering therapy; the effect is long lasting over several weeks. However, they are extremely expensive when compared with traditional agents and therefore their place in the management of gout is still to be determined.

Treatment of chronic gout

Around 60% patients with an attack of acute gout will have a second episode within one year and 78% within two years. Hence identification and correction of hyperuricemia is highly essential in these patients.

High purine diet like red meat, sea food, pulses are better avoided. Alcohol increases uric acid production and decreases uric acid excretion. Beer which has the highest content of guanosine further increases uric acid production. Patients are advised abstinence from all forms of alcohols. Vitamin C and bing cherries in diet are found to reduce the urate levels.

Other causes of hyperuricemia like hypertension, obesity, dyslipidemia, use of low dose aspirin and diuretics should also be addressed. Maintaining a urine flow of at least 1400 ml/day will also facilitate urate excretion.

Table 4. Correctable factors contributory to hyperuricemia

- Obesity
- Hypertriglyceridemia
- High purine diet
- Alcohol consumption
- Uncontrolled hypertension
- Diuretic or low dose aspirin therapy
- Suboptimal urine flow (<1400ml/day)

Pharmacotherapy

Optimal treatment of chronic gout requires long term reduction of serum urate concentration to the lower half of normal uric acid range in order to eliminate acute gouty attacks, reduce tophi and prevent ongoing joint or organ damage. The urate lowering therapy should not be commenced during acute attacks as flares are common when there is a sudden lowering of serum uric acid. Concomitant use of oral colchicine (0.5 mg 12 hourly) or low dose NSAIDs (Naproxen 250mg bd/Indomethacin 25mg bd) for a period 2-3 months will reduce the probability of occurrence of flares.

Table 5. Indications of long term urate lowering drug therapy

- More than 2-3 attacks of gout per year
- Chronic tophaceous or polyarticular gout
- Uric acid over production due to genetic enzyme defects
- Urate urolithiasis or nephropathy
- Adjunct to cytotoxic therapy in malignancies
- Radiological examination revealing established erosions or lytic lesions
- Patients with co-morbidities where attacks are difficult to treat

The drugs used for chronic gout can be divided in three categories[25]

1. **Uricostatic drugs** – e.g.: Allopurinol, Oxipurinol, Febuxostat.

2. **Uricosuric agents** – Probenecid, Sulfinpyrazone, Benzbromerone, Losartan, Amlodipine, Fenofibrate.
3. **Uricolytic drugs** : Uricase, Rasburicase, PEG uricase.

Uricostatic drugs

Allopurinol as a header is a xanthine oxidase inhibitor and is the urate lowering drug of choice in around 95% of cases. It is useful in both urate over producers and under excretors, has convenient once daily dosing, useful in patients with renal calculi, cheep and readily available. It is better to start with a low dose of 100mg daily 4-8 weeks after subsidence of acute gouty arthritis. The dose can be stepped up to 100mg per week to an initial maintenance dose of 300mg day. The uric acid level will fall in 2-4 weeks time to lower half of normal reference range in majority of patients. An occasional patient may need a dose of 600-900mg day. Once the target uric acid level is achieved, the measurement can be spaced out to 3-6 month intervals. Maintain adequate hydration and alkalinize the urine by giving sodium bicarbonate if the uric acid level is very high. There are numerous potential drug interactions particularly with azathioprine, theophylline and oral anticoagulants. Around 2% patients can develop hypersensitivity to allopurinol. Other uncommon side effects are fever, bone marrow suppression, hepatitis and sarcoid like reactions.

Oxipurinol:

The active metabolite of allopurinol has been tried with limited success in patients sensitive to allopurinol. Its use is limited by the lack of availability, poor gastrointestinal absorption and a high prevalence of cross reactivity with allopurinol.

Febuxostat:

This new drug is a nonpurine xanthine oxidase inhibitor with high oral absorption (85%) and high affinity to serum proteins. It has a half life of 4-9 hours. It is metabolized by liver and renal excretion is insignificant. It selectively inhibit both oxidized and reduced forms of xanthine oxidase without inhibiting other enzymes in purine metabolism. The major adverse effects of Febuxostat are hepatic (transaminase elevation), nausea and dizziness. It is contraindicated in patients treated with azathioprine, mercaptopurine or theophylline which are metabolized by xanthine oxidase. The initial dose of febuxostat is 40mg/daily taken with or without food. If the target serum uric acid level (around 5mg/dl) is not achieved dosage may be increased to 80mg and to 120mg/day. Like allopurinol, the drug should be started after subsidence of acute attack of gout and prophylactic therapy with low dose colchicines or an NSAID is required to reduce the incidence of paradoxical gout flares. It can be safely given without dosage reduction in patients with renal impairment and moderate hepatic impairment.

Uricosuric agents

Uricosuric agents are useful to correct hyperuricemia and gout due to defective renal excretion of urates. The drugs are contraindicated if there is a history of urolithiasis or if there is poor urine volume. They are less effective in presence of renal disease with low glomerular filtration rate. Urine should be alkalinized to avoid uric acid crystal deposition Adequate hydration to maintain daily urine volume above 1400 ml/day is also needed. Probenecid is a well tolerated uricosuric agent. The starting dose is 0.5gm/day and gradually increases to 0.5gm two to three times daily. Sulfinpyrazone is three to six times more potent than probenecid and more effective patients with renal dysfunction. The initial dose is 50mg twice daily and increased up to 200mg twice daily.

Benzbromarone another uricosuric agent which is a good alternative to allopurinol in patients with impaired renal function. It is an option for patients where gout has been difficult to control with allopurinol alone in renal transplant patients on cyclosporine and patients with significant renal failure. However Benzbromarone is metabolized by cytochrome P450 and was withdrawn from widespread use because of its hepatotoxicity. Today it is used only on a named patient basis, in much of Europe. It is a highly effective drug in reducing serum urate level and can be used in patients with renal transplant and in combination with xanthine oxidase inhibitors.

Lesinurad is an novel uricosuric agent and has been studied in combination with allopurinol in refractory gout. The combination of Lesinurad with allopurinol is superior to monotherapy with allopurinol alone. The combination is well tolerated.

Benzbromerone, Lesinurad and sulfinpyrazone are not available in India and probenecid is of limited supply. Hence there is increasing interest in the uricosuric effects of commonly used anti-hypertensives like Losartan and Amlodipine and lipid lowering drugs like Fenofibrate. Losartan, an angiotensin converting enzyme inhibitor has moderate uricosuric properties by decreasing the renal tubular reabsorption of uric acid[26]. It is particularly useful as an adjunct therapy in patients with gout and hypertension. It is also useful to control the hyperuricemia caused by thiazides. The usual dose needed is 50mg/day. Amlodipine, calcium channel blockers at the dose of 5-10mg/day is also found to be useful in reducing the urate levels in patients with cyclosporine induced hyperuricemia[27]. Fenofibrate, the lipid lowering agent increases the renal urate clearance 25-30% reduction of uric acid level is possible with a dose of 200mg/day and a good choice in patients with gout and dyslipidemia.

Uricolytic agents

Humans are prone to hyperuricemia due to lack of functional uricase (Urate Oxidase), unlike lower animals. Natural uricase from Aspergillus flavus effectively reduces uric acid levels in human but is highly antigenic. A biosynthesized recombinant A. flavus uricase (Rasburicase) is available in US for tumor lysis syndrome in patients allergic to allopurinol[27]. The drug dramatically reduce serum uric acid when gives intravenously. The use is limited to a single course due to the high chance of potential immunogenicity. The adverse effects include anaphylaxis, methaemoglobinemia and haemolysis in patients with Glucose 6 Phosphate dehydrogenase deficiency.

Polyethylene glycol formulated uricase (PEG-uricase) has the advantages of prolonging the drug half life and decreasing antigenicity[27,28]. The mean termination half-life of PEG-uricase is 2 weeks compares to 19hrs for Rasburicase. The most common adverse events of PEG uricase are inducing an acute flare and infusion reactions like nausea, vomiting, respiratory symptoms, myalgia and rash, but not anaphylaxis[29]. PEG-uricase is another potentially powerful agent for treating refractory gout in those who are unable to tolerate other drugs. It could have a role in debulking tophi in advanced gout before switching to another agent for maintenance treatment[30].

Complicated gout

When gout is associated with various co-morbidities, it makes gout more complicated and difficult to manage. Following are some examples of complicated gout and guidelines for management.

Gout in organ transplant recipients

Hyper uricemia is present in approximately 80% of renal transplant recipients who are on cyclosporine-A and approximately 10% of them develop acute gout. Cyclosporine markedly reduces the renal excretion of uric acid and the chronic use of diuretics further elevate uric acid in these patients. The incidence of gout in lung and heart transplants on cyclosporine is also significantly increased. The gouty arthritis in these patients is more severe than primary gout. Poly articular presentation is common and tophi occurs early in them.

NSAIDS are not safe as they may result in sodium and water retention and have a negative effect on renal function. Toxicity of colchicine especially myopathy is high in patients receiving cyclosporine and hence not a safe option. Many transplant recipients developing acute gout are already on systemic steroids as a part of immuno suppressive therapy as a result, the treatment options for acute attacks are really limited. In acute mono articular gout, intra articular steroids can be tried (after excluding sepsis by synovial fluid study). High dose oral/parenteral steroids or intra muscular ACTH (40 IU) can be tried in poly articular gout. If the response is poor one may be forced to rely more heavily on the use of analgesics (morphine or tramadol) in this setting. Recently uric acid oxidase has been successfully tried in lowering the uric acid levels. Further attacks of gout on transplant recipients can be prevented by using non-cyclosporine immuno modulator drugs and minimizing the use of diuretics. Uricosuric agents like probenecid can be given if the renal function is normal. Both allopurinol and febuxostat have potential severe interaction with azathioprine as it is metabolized by xanthine oxidase which is inhibited by these drugs. The dose of allopurinol should be lowered to one third or one fourth of the original dose. It is better to restrict the dose of allopurinol to 50-100mg/day in these patients. As there are no published data on concomitant use of azathioprine and febuxostat the combination is better avoided.

Gout in patients with chronic renal failure

In CRF with gout NSAIDs are contraindicated and loading dose of colchicine is better avoided. Intra articular steroids are useful in monoarthritis while systemic steroids are helpful in those with poly articular attack.

Long term prophylaxis with low dose colchicine is better avoided especially in those with advanced renal failure due to the chance of colchicine toxicity. Instead febuxostat can be started. Low dose oral corticosteroids can be given in the initial few months in order to prevent acute flares.

Gout in patients with peptic ulcer or recent upper gastrointestinal haemorrhage

It is not uncommon that patients develop upper gastrointestinal haemorrhage secondary to NSAIDs therapy or active peptic ulcer during their gouty attacks, or develop an acute attack shortly after an intra abdominal surgery. If nothing is allowed per mouth, intra articular or systemic steroids can be tried. NSAIDs and appropriate anti ulcer drugs (misoprostol or proton pump inhibitor) or a specific cox-2 inhibitor like Etoricoxib can be used with precaution for short periods along with allopurinol for prevention of recurrent attacks.

Gout in patients allergic to allopurinol

Use of allopurinol in patients who are allergic to sulphonamides compound should be cautioned or avoided if possible. Around 2% of individual taking allopurinol develop hypersensitivity reactions and 20% of them are severe. When allopurinol sensitivity is encountered, one can use febuxostat. The other options are uricosuric agents like probenecid, losartan or fenofibrate. Patients are advised to avoid drugs that might increase serum uric acid level such as thiazides or salicylates and put on a strict diet restriction. Although uricase is another option in severe allopurinol sensitivity it is not available in India.

Gout in patients with chronic liver disease

Gout can occur in patients with chronic liver disease often precipitated by excessive alcohol consumption. Use of NSAIDs for acute attack should be cautioned as there is increased risk of gastrointestinal haemorrhage. Loading dose oral colchicine is also not safe. Stoppage of alcohol intake is crucial. Parenteral steroids are useful during acute attack and allopurinol can be given for prophylaxis.

Gout as a part of metabolic syndrome

Hyperuricemia occuring as a part of metabolic syndrome along with obesity, hypertension, hyperlipidemia, diabetes mellitus and coronary artery disease is not uncommon in day today practice. In this situation, interaction of various drugs the patient is taking for these medical disorders and their side effects should be of concern. Low dose aspirin is well known to cause hyperuricemia. Allopurinol can prolong the half life of anticoagulant warfarin. Febuxostat has no interaction with warfarin and can be safely combined.Colchicine increases the toxicity of statins especially rabdomyolysis or myoglobinuria. In hypertension with hyperuricemia, diuretics are better avoided. Losartan, amlodipine and fenofibrate are better due to their uricosuric effects.

Finally, treatment of gout particularly those with tophaceous deposits usually requires long term prophylactic NSAIDs or colchicine to prevent acute attacks and years of hypouricemic therapy in order to deplete urate deposition in the body. Patients with multiple medical problems usually take many medications and occasionally have short life expectancy. One should weigh between the benefit of a prolong use of prophylactic and hypouricemic agents against the potential drug interactions and the adverse reactions.

Further Reading :

1. Nuki G. Treatment of crystal arthropathy. History and advances. Rheum Dis Clin N Amer 2006; 32: 333-57.
2. Paul BJ. History of Rheumatology. What are the lessons for us? Ind J Rheum 2006; 1: 123-28.
3. Pasero G, Marsen P. Hippocrates and Rheumatology. Clin Exp Rheum 2004; 22: 687-89.
4. Pande I. An update on gout. Indian J Rheum 2006; 1: 60-65.
5. Maesaka JK, Fishbane S. Regulation of renal urate excretion: a critical review. Am J Kidney Dis 1998; 32: 917-20.
6. Mc Lean L. The pathogenesis of gout. In : Hochberg MC, Silman AJ, Smolen JS, Weinblatt ME, Weisman MH (Eds), Rheumatology 3rd Ed. Edinburgh, Mosby-Elsevier Ltd., 2003; 1903-18.
7. Mount B, Kwon CY, Zandi-Najad K. Renal urate transport. Rheum. Dis Clin N Amer. 2006; 32: 313-31.
8. Chopra A, Patil J, Billempelly V, Relwani J, Tandle HS. Prevalence of rheumatic diseases in a rural population in Western India. A WHO-ILAR COPCORD study. J Assoc Phys Ind 2001; 49: 240-46.
9. Danda D, Mathew A, Mathew J, Vasugi Z. Clinical profile of young onset gout.J Ind Rheum Assoc. 2005; 13: 51-53.
10. Mikuls TR, Farrar JT, Bilker WB, at el. Gout epidemiology: results from the UK General Practice Research Database, 1990-1999, Ann Rheum Dis 2005; 64: 267-72.
11. Wallace KL, Riedel AA Joseph-Ridge N, Wortmann R. Increasing prevalence of gout and hyperuricemia over 10 years among older adults in a managed care population. J Rheum 2004; 21: 1582-87.
12. PaulBJ,MujeebRT,SudheeshT. Clinical Study of Gout in North Kerala, Ind. J. of Rheumatology, 2009.4:149-52.
13. Suresh E. Diagnosis and management of gout : a rational approach. Post Graduate Med J 2005; 81: 572-79.
14. Philip J. Crystal induced arthropathies. In Pispathi PK (Ed). Manual of Rheumatology 2nd Ed. Mumbai Indian Rheumatology Assoc 2002; pp 130-50.
15. Owen DS. A cheap and useful compensated polarizing microscope. N Eng J Med 1971; 285: 1152.
16. Johnny U, Monu V, Pope TL Jr : Gout A clinical and radiological review Radio. Clin N Am 2004 ,42:169-84.
17. Thiele RG. Role of ultrasound and other advanced imaging in the diagnosis and management of gout. Curr. Rheumatol. Rep. 2011; 13: 146-53.
18. Dalbeth N, Doyle A, Boyer L. Development of a computed tomography method of scoring bone erosions in patients with gout. Validation and Clinical Implications. Rheumatology (Oxford) 2011; 50: 410-46.
19. Desai A, Peterson JJ, Garner HW et al. Clinical utility of dual energy CT for evaluation of tophaceous gout. Radiographics, 2011; 21: 1365-75.
20. Dalbeth N, Doyle A, McQueen FM. Imaging in gout: insight into the pathological features of disease. Curr. Opin. Rheumatol. 2012; 24: 132-38.
21. Wortman LR. Gout and hyperuricemia. In Firestein GS et al (Ed). Kelleys Text book of Rheumatology 8th ed. Saunders Elsevier 2009; 1481-1506.
22. So A, Desmedt T, Revas S et al. A pilot study of IL1, inhibition by anakinra in acute gout. Arthritis Res Ther. 2007; 9-R 28.
23. Brown JP, So A, Dikranian A et al. Long term efficacy and safety of canakinumab versus triamacinolone acetanide in acute gouty arthritis. In: ACR Conference Proceedings, 2011; Abstract 1019.
24. Schumacher HR, Evans RR, Birbara CA et al. Rilonacept for prevention of gout flares associated with uric acid lowering therapy : response rate across three phase III clinical trials In: ACR Conference proceedings, 2011; Abstract 1024.
25. Paul BJ : Crystal arthritis- Past, present and future. JK Science 2006; 8; 232-34.
26. Sica DA, Schoolworth AC. Uric acid and Losartan. Curr Opin Nephrol Hypertens 2002; 475-82.
27. Schumacher HR Jr, Chen LX. Newer therapeutic approaches : Gout Rheum. Dis Clin N Amer 2006; 32: 235-44.
28. Rider TG, Jordan KM: The modern management of gout. Rheumatology (Oxford) 2010:49(1): 5-14.
29. Sherman MR, Saifir MGP, Perez-Ruiz F. PEG uricase in the management of treatment resistant gout and hyperuricemia. Adv. Drug. Deliver. Rev. 2008; 60: 59-68.
30. Jordan KM. Up-to-date management of gout. Curr. Opin. Rheumatology. 2012; 24: 149-51.
31. Khanna D et al. 2012 American college of Rheumatology guidelines for the management of gout. Part 1 : Systematic non-pharmacologic and pharmacologic therapeutic approaches to hyperuricemia. Arthritis Care Res 2012; 64: 1431-46.
32. Khanna D et al. 2012 American college of Rheumatology guidelines for the management of gout. Part 2 : Therapy and anti-inflammatory prophylaxis of acute gouty arthritis. Arthritis Care Res 2012; 64: 1447-61.
33. Nicola Dalbeth, Jaap Fransen, Tim Jansen et al. New classification criteria for gout: a frame work for progress. Rheumatology 2013; 52: 1748-53.

CPPD AND OTHER CRYSTAL RELATED ARTHRITIS

2

Renu Saigal

Introduction:

Crystal-induced arthropathies are a group of disorders that involve deposition of crystals into joints and soft tissues, resulting in articular and periarticular inflammation and injury (Table-1). Calcium deposition diseases (CDD) are characterized with disordered calcification of cartilage and/or periarticular soft tissue. CDD are usually associated with aging, degenerative joint disease, genetic and metabolic disorders. Most common calcium containing crystals are calcium pyrophosphate dihydrate (CPPD) and basic calcium phosphate (BCP) crystals which include carbonated-substituted hydroxyapatite, tricalcium phosphate, octacalcium phosphate and magnesium whitlockite.

the surface of chondrocytes. Crystal formation is enhanced by locally elevated high ionic product (Calcium ion X Inorganic pyrophosphate) or factors in the cartilage matrix that promote crystal formation. Augmented extracellular PPi occur around chondrocytes, appears to result from overactivity of one or more of a group of nucleoside triphosphate pyrophosphohydrolase (NTPPPH) enzymes. These enzymes are capable of catalyzing the production of PPi by the extracellular hydrolysis of nucleoside triphosphates, especially ATP. In addition plasma cell membrane glycoprotein-1 (PC-1) has the greatest effect on extracellular PPi levels and matrix calcification in mineralizing cells. One function of extracellular PPi (ePPi) appears to provide the substrate for CPPD crystal formation in the immediate environment of chondrocytes and bind to and inhibit the growth of basic calcium phosphate (BCP) crystals. In addition, some of the excess PPi production may take place intracellularly through NTPPH. The enhanced ecto-NTPPPH activity results from vesicles which are selectively rich in NTPPPH activity.

TABLE 1. Classification of crystal related arthropathies

Kind of Crystal	Acute Condition	Chronic Condition
Monosodium urate (MSU)	Acute gouty arthritis	Urate arthropathy, chronic gout
Calcium pyrophosphate (CPP)	Acute Pseudogout	Pyrophosphate arthropathy
Hydroxyapatite (HA)	Acute calcific periarthritis	OA and destructive arthropathy*
Oxalate (in renal failure)	Acute arthritis	Chronic arthritis
Cholesterol (in RA and OA)	Acute arthritis and acute bursitis*	Chronic arthritis and bursitis*

* association only, no causative relation.

CPPD disease :

CPPD disease is heterogeneous group of diseases, which affects mainly fibrocartilage and hyaline cartilage of joints and is characterized by deposition of calcium pyrophosphate (CPP) crystals. CPP associated arthritis was found to be third most common inflammatory arthritis in one study in Italian population. CPP crystals in articular tissues have been found in 10-15% persons aged 65-75 years and in 30-50% of adults over 85 years[1]. Chondrocalcinosis may be seen in 36% between 75-84 years, 44% in those more than 85 years of age.

Few case reports are available from India but in general, awareness about this entity is less appreciated among physicians and orthopedicians[2,3].

Pathophysiology of CDD: (fig. 1)

CPPD crystal formation occurs almost exclusively in the articular and periarticular tissue, usually near

Other factors that may contribute to excess PPi and crystal formation include decreased activity of tissue non-specific alkaline phosphatase [which normally hydrolyses PPi to inorganic phosphate (2Pi)], degenerating cellular debris, abnormal matrix collagen, and the local influence of growth factors (transforming growth factor and insulin-like growth factor). The primary stimulus of ePPi production is transforming growth factor (TGF-α). TGF-α enhances the expression of NTPPPH and ANK (a transmembrane protein) all of which increase the extracellular transport of PPi.

In normal chondrocytes, insulin-like growth factor-1 (IGF-1) inhibits TGF-β from elevating chondrocyte extracellular PPi. Thus, the decreased IGF-1 responsiveness in OA leads to sustained elevation of PPi generation because of active TGF-α.

Other stimuli for ePPi production include ascorbate, retinoic acid and thyroid hormone.

Role of the human homolog of the 'ank' gene (ANKH) in familial CPPD disease has been found in pedigrees for this PPi transport channel with the chondrocalcinosis phenotype[4].

Mutation of the ANK anion transporter gene on the chondrocyte plasma membrane in hereditary form leads to gaining of function and more transport of PPi in extracellular space.

In addition, mutations in or just upstream from the chromosome 5p locus of ANKH have also been identified in individuals with idiopathic or sporadic CPPD deposition disease[5,6].

Hereditary form is also characterized by diminished activity of the protective pyrophosphate hydrolases leading to decreased hydrolysis of PPi to inorganic phosphate, and increased transmembrane transport of intracellular PPi of chondrocyte to its extracellular matrix.

Metabolic disorders (Fig. 1) result in increased ePPi and alters cartilage matrix in such a way that CPPD crystals are formed over it.

CPPD activate NALP3-dependent inflammasome assembly in synovial mononuclear phagocytes and neutrophils, an action that is associated release of the pro-inflammatory cytokine IL-1beta.

BCP crystals stimulate synoviocyte proliferation and produce inflammatory cytokine (IL-1, TNF-α, PGE2 and metalloproteases). BCP crystals also directly induce chondrocytes to produce nitric oxide and undergo apoptosis. Metalloproteases production and chondrocyte apoptosis contribute to cartilage destruction. A vicious cycle is formed as Chondrocyte apoptosis leads to more formation of BCP crystals.

Annexin 5 might be involved in calcium influx induced by BCP crystals.

Etiology and disease associations

Aging

Risk of CPPD doubles every decade between 45 and 85 years of age independently of other risk factors. OA cartilage encourages deposition of calcium crystals CPP as well as BCP. Conversely, CPPD could be a primary factor that causes or amplifies joint damage in OA[7].

Trauma

Risk of CPPD increases after meniscectomy[8].

CPPD disease may be categorized as:

1. Sporadic 2. Familial 3. Secondary (Metabolic and endocrinal)

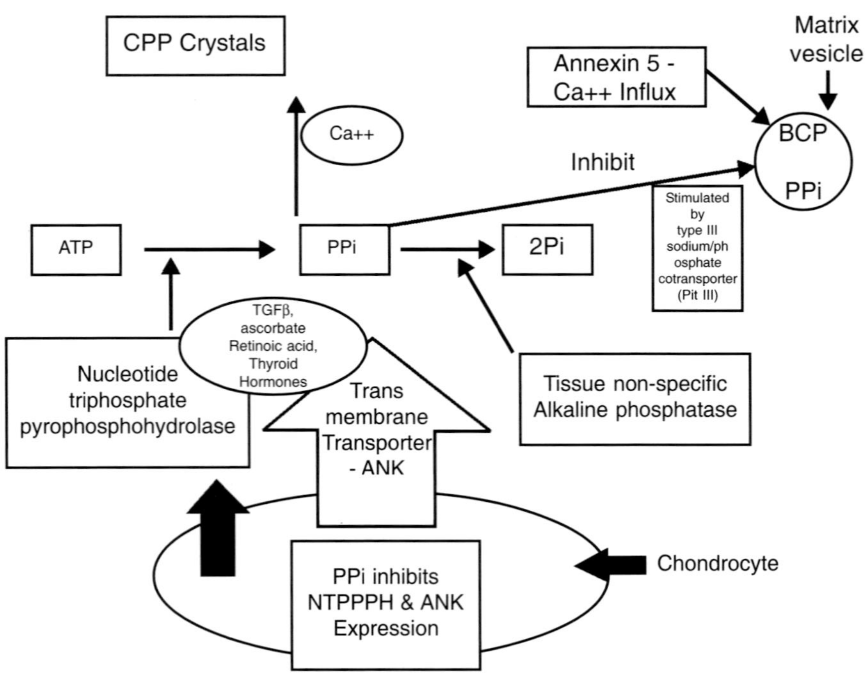

Fig. 1 : The pathogenesis of CPPD disease.

Sporadic/Idiopathic

This is most common form of CPPD disease. Age is the strongest attributable factor.

Familial CPPD

CPPD deposition can occur as a monogenic disorder with varying phenotypes characterized by florid early onset (<45 years) and polyarticular pattern. Two genetic locations are recognized, one (CCAL1) related to mutation in long arm of chromosome 8 (8q) and associated with severe OA and the other (CCLA2) resulting from a mutation on the short arm of chromosome 5 (5p) occurring from families from the UK, France and Argentina. ANKH (ankylosis human) gene which codes for ANK, is the responsible gene.

Two main clinical types are reported, the first characterized by early-onset, florid, polyarticular chondrocalcinosis and the second by late-onset oligoarticular with arthritis resembling sporadic pyrophospate arthropathy.

Secondary CPPD (metabolic, endocrinal) :

1. Hemochromatosis
2. Primary hyperparathyroidism
3. Hypophosphatasia
4. Hypomagnesaemia[9] : Secondary to
   - Diuretic use
   - Short bowel syndrome secondary to radiation enteritis or mesenteric artery thrombosis.
   - Gitelman's syndrome

CPPD has been reported after intraarticular hyaluronate[10] with hypocalcaemia following parathyroidectomy and intravenous pamidronate[11].

Clinical presentations

Earlier CPPD disease was classified on the basis of clinical presentations. Recently a new nomenclature has been suggested by European league against rheumatism (Eular) which avoided the 'pseudo' term. (Table -2).

Table 2. Nomenclature and classification of CPPD disease.

Earlier nomenclature (clinical classifications):

Pseudogout (type A)
Pseudo rheumatoid arthritis (type B)
Pseudo-osteoarthritis (with acute attacks - type C; without inflammation - type D),
Lanthanic or asymptomatic (type E)
Pseudoneuropathic (type F).

EULAR Classification:

1. Asymptomatic CPPD (Lanthanic)
2. OA with CPPD (Pseudo-osteoarthritis)
3. Acute CPP crystal arthritis (Pseudogout)
4. Chronic CPP crystal inflammatory arthritis or Chronic pyrophosphate arthropathy (Pseudo rheumatoid arthritis)

(Previous nomenclature has been given in parentheses)

Asymptomatic CPPD (LANTHANIC)

There are no apparent clinical manifestations, only isolated cartilage calcification (CC) in hyaline and/ or fibro-cartilage or osteoarthritis (OA) with CC may be seen. It is most common form of CPPD disease.

Osteoarthritis (OA) with CPPD (Pseudo OA)

Earlier used term pseudo-OA is CPPD associated arthropathy with joint degeneration, seen in 50% of patients with CPPD disease. It often targets knee joints, with chronic symptoms like pain, stiffness and functional limitation (Type D) and/or acute attacks of pseudogout (Type C) induced by CPP crystals in about half of these patients. OA with CPPD is associated with more inflammatory symptoms and signs, an atypical distribution (e.g. radiocarpal, midcarpal, metacarpophalangeal (2nd & 3rd), glenohumeral, elbow, hip, hindfoot or mid-foot involvement and more patellofemoral compartment involvement[12] and prominent cyst (Fig. 2) and osteophyte formation on radiographs as compared to OA without CPPD and tends to affect elderly women (Box - 1).

Box 1. Clinical and radiologic features of chronic pyrophosphate arthropathy which help in differentiating from primary OA.

- Involvement of wrist, elbows, MCP and shoulder joints
- Affection of isolated parts of joints within a joint e.g. radiocarpal joint in the wrist, patellofemoral joint in the knee joint, talonavicular and talocalcaneal joints in the ankle.
- Characterized by large subchondral cysts. (Fig. 2)
- Association with calcific tendonitis, chondrocalcinosis and periarticular soft tissue calcific deposition
- Associated with rapid joint destruction with intraarticular bony fragments

Differing reports are available regarding CPP disease as a poor prognostic marker, Reuge et al found that more patients with OA and CPPD disease required knee replacement surgery than patients with primary OA without crystal deposits[13] while other studies showed slowing of radiographic degenerative changes and a negative association between baseline

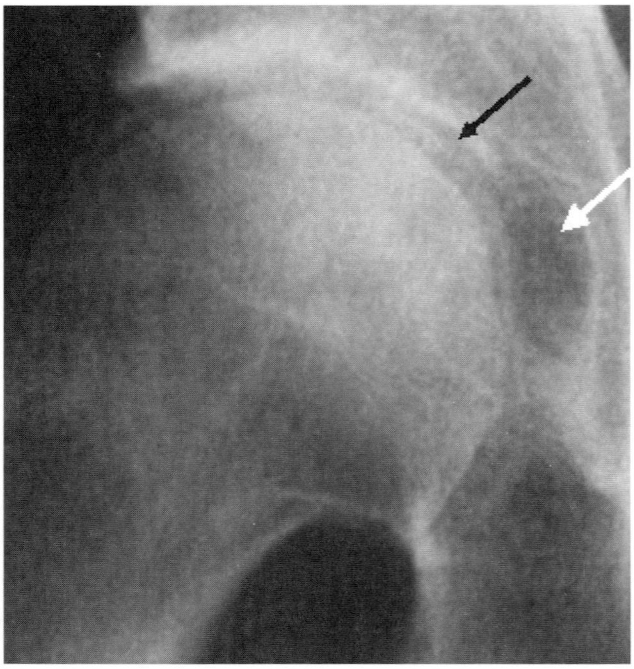

Fig. 2 : CC of Hip Joint (Black arrow) with large subchondral cyst (White arrow).

CC and risk for cartilage loss by MRI in these patients[14,15]. Patients with OA with CPPD tend to undergo knee arthroplasty at an older age than patients with non-CPPD OA[16].

Acute CPP crystal arthritis (Pseudogout)

The rapid development of severe joint pain, swelling and tenderness that reaches its maximum within 6-24 h, especially with overlying erythema, is highly suggestive of acute crystal inflammation though not specific for acute CPP crystal arthritis. It often involves knee (50%), wrist or shoulder joints of a patient over age 65 years while first metatarsal joint is more commonly involved in gout. Acute monoarthritis in CPPD disease has to be distinguished from septic or reactive arthritis. Though polyarticular acute flares are not rare. Fever, leucocytosis with increased polymorphs and erythrocyte sedimentation rate can occur during acute attacks of pseudogout, in such cases pyogenic arthritis, osteomyelitis should be excluded.

These inflammatory symptoms occur because of lysis of polymorphonuclear white cells that have ingested CPP crystals. These attacks are self limited and last approximately 3-4 days[17]. Although cluster attacks may last for several weeks. This form is seen in 25% of patients. Provocative factors are stress response to intercurrent illness,trauma, surgery, arthroscopic lavage for knee osteoarthrtitis (OA) with pre-existing CC[18], bisphosphonates[19] and repeated intra-articular hyaluronan injections[20] and after GM-CSF[21]. The presence of radiographic CC and advanced age increases this likelihood, but definitive diagnosis needs to be crystal proven.

Chronic CPP Crystal inflammatory arthritis or chronic pyrophosphate arthropathy (Pseudo rheumatoid arthritis)

It occurs in 5% or less of patients with symptomatic CPP crystal deposition disease. This type of CPP arthritis is non-erosive, involves multiple joints (oligo or polyarhtritis) with attacks of articular inflammation, which may last up to several months and may be associated with elevation of CRP and ESR. It may resemble rheumatoid arthritis and other chronic inflammatory joint diseases in older adults. Radiographs will show CC, but the diagnosis should be crystal proven. Clinical picture closely resembles rheumatoid arthritis with significant morning stiffness, fatigue, symmetric inflammation of proximal interphalangeal and metacarpophalangeal joints[22], synovial thickening, local oedema and restricted joint motion due to inflammation or flexion contracture.

Pseudoneuropathic Joints

This is a severe destructive arthropathy without an underlying neurologic disorder. It occurs in fewer than 5% patients with CPPD crystal deposits that involves knee joints. Presence of CC can aid in diagnosis of these patients.

Atypical/periarticular CPPD

CPPD crystals can also form at tendon insertions. Carpal tunnel syndrome, periarthritis of shoulder, tendinitis, tenosynovitis, and bursitis may occur[23-25]. Unusual joints like sacroiliac joints, sternoclavicular joints may be involved. Spinal involvement may occur leading to spinal stenosis or bony ankylosis resembling ankylosing spondylitis or diffuse idiopathic skeletal hyperostosis. Tophaceous pseudogout due to deposition of CPP crystals may lead to neural impingement syndromes, space filling lesions or pseudotumour especially in temporomandibular joint, spine and the hand which resembles rheumatoid nodules or tophi have been described[26]. Tumoral hand joint swelling diagnosed by FNAC has been reported from India while six patients with CPPD disease involving the ligamentum flavum of the cervical and thoracic spine has also been reported[27,28].

Various other conditions may be associated with chondrocalcinosis and pseudogout (Box - 2).

Box 2. Conditions associated with chondrocalcinosis and pseudogout.

Strong association
Hyperparathyroidism
Hypophosphatasia
Hypomagnesaemia
Haemochromatosis

Possible association
Gout
Ochronosis
Wilson's Disease
Hypothyroidism
X-linked hypophosphatemic rickets
Familial hypocalciuric hypercalcemia
AcromegalyFamilial hypocalciuric hypercalcemia
Acromegaly
Haemochromatosis

Screening for associated diseases

Patients with a diagnosis of CPPD crystal deposition disease may be associated with hemochromatosis, hyperparathyroidism, hypomagnesemia, hypophosphatasia, hypothyroidism, and familial hypocalciuric hypercalcemia therefore following should be measured: serum calcium, phosphorus, magnesium, alkaline phosphatase, ferritin, iron and total iron binding capacity, serum transferrin, thyroid-stimulating hormone and Free T4.

Diagnosis

Definite diagnosis

The detection of CPPD crystals in tissue or synovial fluid by polarizing light microscopy remains the gold standard for the diagnosis of CPPD disease. CPPD crystals should be seen under polarized light microscope or phase contrast microscope, even one or two crystals which are parallelepipedic or rhomboid, long or short rods or small squares ranging from 2-20μm in length, with absent or weak positive birefringence in SF or biopsied tissue is confirmatory evidence of CPPD disease (Fig. 3). In contrast to monosodium urate (MSU) crystals CPPD crystals are more difficult to detect as these are smaller in size, polymorphic and are more weakly or not birefringent.

In inflamed joints during acute attack phagocytosed intracellular crystals in polymorphonuclear leukocytes are always present. CPP crystals may be detected in SF from asymptomatic knees which have never been inflamed and also in between inflammatory episodes.

Examination of SF should be carried out immediately after joint aspiration as identification of the crystals diminishes as the time between joint aspiration and microscopic examination increases and cellular components will deteriorate rapidly when outside the body. Cell integrity is especially important for detecting smaller intracellular CPPD crystals. Another source of error is positively birefringent material other than CPP crystals may be identified in SF, unless care is taken to clean and remove dust from the microscope slide and cover slip. Corticosteroid and cholesterol crystals may be misidentified as CPPD crystals. EULAR recommendations are refinement of an optimal protocol including training and standards for identification of CPPD crystals in joint fluids[29].

CPPD crystals are positively birefringent (i.e. blue when the long axis of the crystal is parallel to the axis of slow rotation of light of the red plate compensator and yellow when the long axis of the crystal is perpendicular to the axis of the compensator) (Fig. 4).

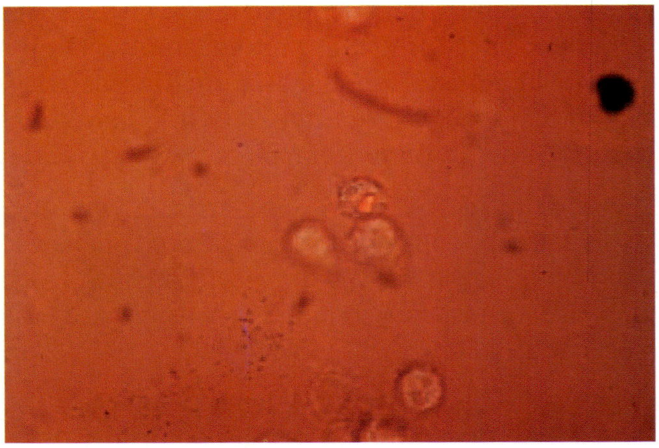

Fig. 3 : Polymorphic, Rectangular, Square, Rod-like (Weak positively birefringent) crystals seen both intracellularly (proportional to inflammation) & extracellularly.
(with permission from www.uptodate.com)

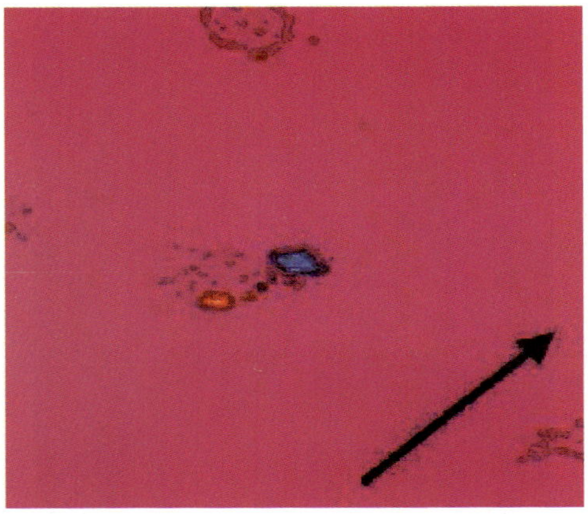

Fig. 4 : CPPD crystals with compensated polarized microscopy. The black arrow indicates the direction of the compensator. Crystals parallel to the compensator are blue, while those perpendicular to the compensator are yellow.
(Reference and available at : www.medscape.com)

During acute attack : SF is inflammatory, with 10,000-50,000 WBCs/μL, more than 90% of which are neutrophils. Glucose levels are usually normal. SF may be turbid and haemorrhagic.

Imaging

Plain film radiography

Radiographic CC supports the diagnosis but its absence does not exclude it. CC may be seen as punctate or linear radiodensities in fibrocartilage and hyaline or articular cartilage but may also be seen in ligaments, bursae, joint capsules and tendons with lesser frequency (Fig. 5-8).

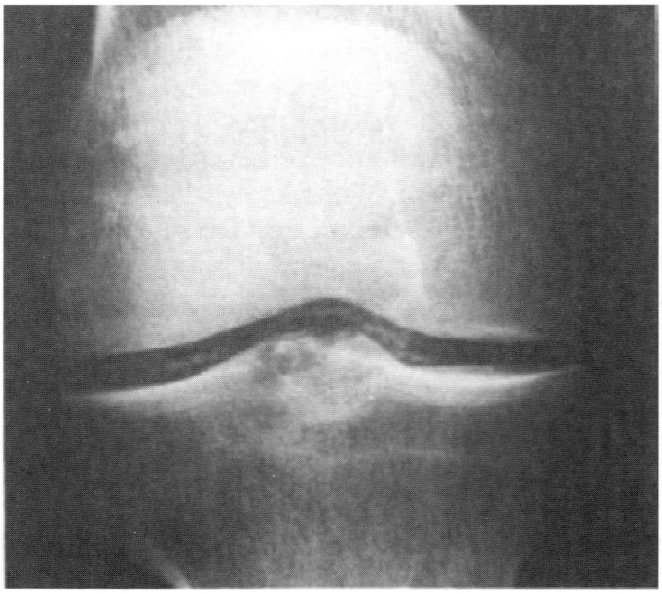

Fig. 5 : Meniscal calcification in knee joint.

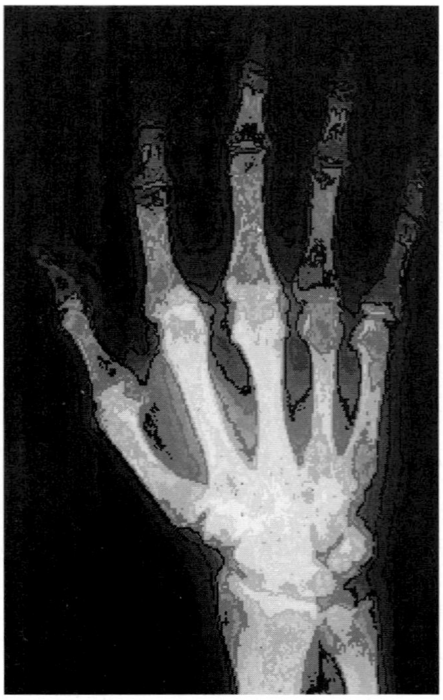

Fig. 6 : CC of triangular ligament of wrist joint.

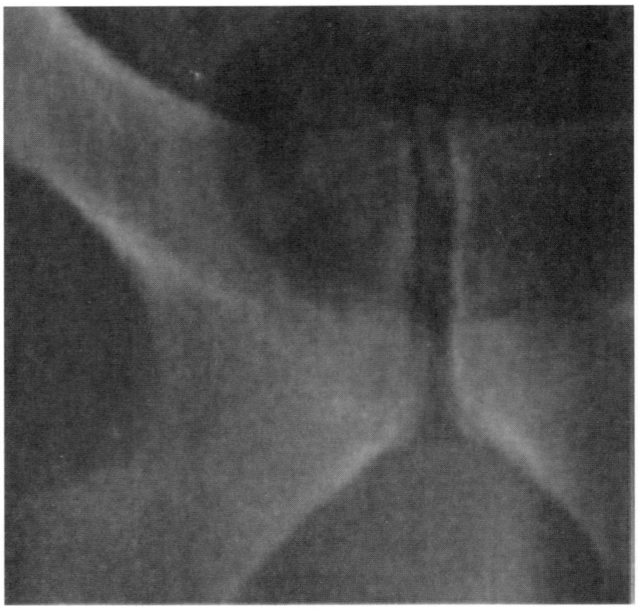

Fig. 7 : CC of symphysis Pubis.

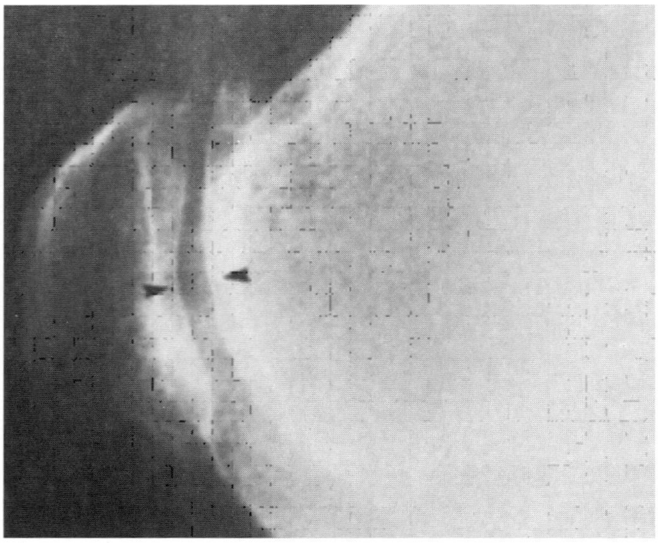

Fig. 8 : CC Patellofemoral joint - Characteristically involved in CPPD Disease.

Cartilages affected are menisci of the knee, the symphysis pubis, the triangular discs of the distal radioulnar joint, and the glenoid and acetabular labra. CPPD crystal deposits in hyaline cartilage frequently appear as a radiopaque line paralleling the surface of the underlying bone.

Other radiological features are squared-off bone ends and hook-like osteophytes in the metacarpo phalangeal (MCP) joints, particularly in the second and third MCP joints, such changes are common in hemochromatosis-associated CPPD.

Severe patellofemoral joint space narrowing, especially with a wrapped-around deformity of the patella on the femur may be seen and this finding is seen in hyperparathyroidism with or without CPPD crystal deposition.

Discordance between SF CPP crystal positivity and CC may occur as:

1. BCP crystals may produce CC.
2. Inability to identify radiological CC when there is significant cartilage loss while CPP crystals may be seen.
3. Under detection of small number of non-refringent or weakly birefringent CPP crystals.

Ultrasonography (US)

US demonstrates a high specificity and good sensitivity for detection of the CPP crystals, using microscopic analysis as the positive standard[30]. US appears to be more sensitive (100%) than x-rays (82%) in identifying CPPD[31].

Various patterns may be seen:

- Thin hyperechoic bands paralleling the articular cartilage surface.
- Small punctate hyperechoic spots, referred to as a "punctate pattern" often with acoustic shadowing, most often found in images of fibrocartilage of the wrist and in tendons.
- Nodular hyperechoic, homogeneous deposits in bursae and articular recesses.

Treatment

Treatment should be according to patient characteristics, risk factors and comorbidities but usually is inadequate as currently no treatment modifies CPP crystal formation or dissolution.

Asymptomatic CPPD : Should not be treated unless it is secondary to hyperparathyroidism or hemochromatosis, treatment of these metabolic diseases prevents further end organ damage.

Treatment of acute CPP arthritis:

Non-pharmacological measures :

1. Ice packs
2. Rest to the joint affected.
3. Joint aspiration to remove crystals
4. Joint immobilization.

Intra- articular CS

Long-acting glucocorticoids (Triamcinolone acetonide 40 mg for large joints with 1 ml of 2% Lignocaine) for mono or oligo-articular CPPD disease are sufficient to suppress inflammation. Joint immobilization for 3-5 days is advised after intra-articular injection for prolonging local anti-inflammatory effect of CS.

Cochicine 0.5 mg 3-4 times a day is also used. Loading dose of 1 mg may be given. In elderly, co-morbidities and deranged renal functions may limit its use. It can also be given for prevention of acute attacks in 0.5-1 mg/day dosage.

Oral/parenteral CS or ACTH:

May be given in a short tapering regime if[46]:

- NSAIDs, colchicine are contraindicated because of renal dysfunction
- Joint is not amenable to intra-articular CS injection
- There is polyarticular involvement

Treatment of chronic CPP crystal inflammatory arthritis:

1. Oral NSAID (with proton pump inhibitors)
2. Colchicine (0.5-1.0 mg daily)
3. Low-dose corticosteroid
4. Methotrexate (5-10mg/week) and hydroxy-chloroquine have been used with success in patients refractory to conventional treatment[32].

Treatment of associated co-morbidities

Oral calcium supplements to suppress PTH in hyperparathyroidism, magnesium replacement (300 mg/day) in patient with hypomagnesaemia[33] and appropriate therapy for hemochromatosis is beneficial.

Newer drugs

Blockade of the NLRP3 (cryopyrin) inflammasome interleukin 1α pathway may offer a new treatment strategy for crystal-associated arthritis[34].

Intra-articular injection of radiocolloid (Yttrium-90): In one small double-blind, placebo-controlled randomized control trial in patients with symptomatic knee OA plus CPPD has been found to be beneficial[35,36].

Probenecid (ANKH anion channel blockade)[37], **polyphosphates[38], phosphocitrate[39] trans-glutaminase 2 :** There is considerable interest, but no relevant data, on the possibility of using these drugs to modulate pyrophosphate levels and influence CPP crystal formation and dissolution, either by local intra-articular or systemic administration.

Basic calcium phosphate (BCP)-hydroxyapatite deposition disease

Calcifications due to BCP contain mixtures of carbonate-substituted hydroxyapatite and octacalcium phosphate. Because these ultramicroscopic crystals are nonacidic calcium phosphates, the term basic calcium phosphate (BCP) is more precise than apatite.

Basic calcium phosphate (BCP) crystal deposition is associated with several clinical manifestations, ranging from asymptomatic status to destructive arthropathy known as 'Milwaukee shoulder syndrome'. These crystals are difficult to identify because of their small size and lack of easily available detection methods.

In patients with osteoarthritis 30-60% may show BCP crystals in SF and may contribute to low grade inflammation and cartilage destruction[40]. Acute calcific periarthritis-tendinitis, bursitis, enthesitis may be seen in these patients along with periarticular soft tissue calcifications.

Shoulder involvement is common, and involvement of the first metatarsophalangeal joint ('pseudopodagra'), primarily affecting young women, has been described. Other commonly involved joints are knee and greater trochanters, elbows, wrists and digits (erosive OA).

Milwaukee shoulder-knee syndrome

Severe destructive often bilateral arthropathy of the shoulder and of the knee associated with BCP crystals has been described primarily in older women, and leads to pain and functional impairment. It is associated with rotator cuff degeneration and glenohumeral joint instability[41,42] (Fig. 9).

Large skin hematoma is readily visible around the shoulder joint at the time of acute attacks. Characteristically, large joint effusions are present, and synovial fluid is frequently blood tinged.

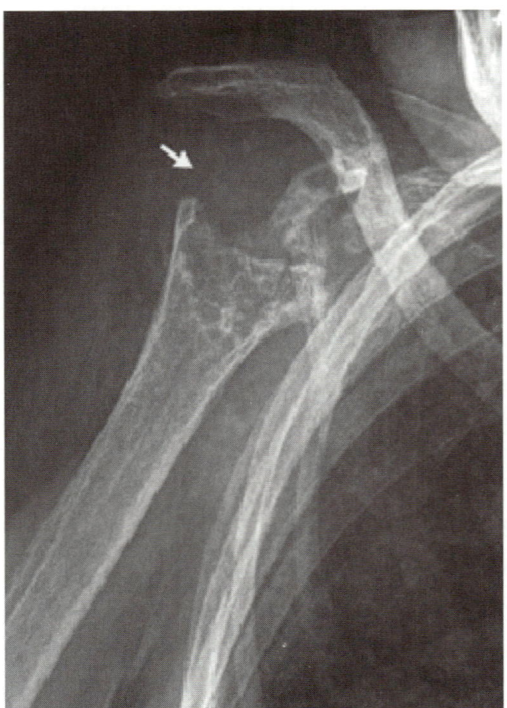

Fig. 9 : Milwaukee Shoulder[41]

Radiologic changes

Glenohumeral degeneration, intra & periarticular calcification, rotator cuff degeneration and rupture, and superior displacement of the humeral head may be seen.

Computerized tomography

CPPD and BCP crystal deposition around the odontoid process may be seen in CT scans and may produce acute cervical pain and stiffness with fever[43] (crowned Dens Syndrome).

Synovial fluid

SF is serosanguineous and contains few cells. Under regular light microscopy, hydroxyapatite crystals may appear as clumps or as intracellular "shiny coins" and are not birefringent under polarized microscopy so can be mistaken for dirt or debris. These are visible under light microscopy with alizarin red staining, wet preparation of patients SF shows extracellular and intracellular (in degenerating cells) BCP crystals as bright red[44]. (Fig. 10) but other calcium-containing particulates may also be stained.

Transmission electron microscope shows the clumps of dark needle-shaped crystals of BCP, dense globular structures of various sizes, and isolated crystals are 7-25 nm in diameter, shiny, non-birefringent , intra or extracellular which stain purplish with Wright's stain .

Radiograph diffraction and biochemical analysis may be needed for definitive identification of these crystals.

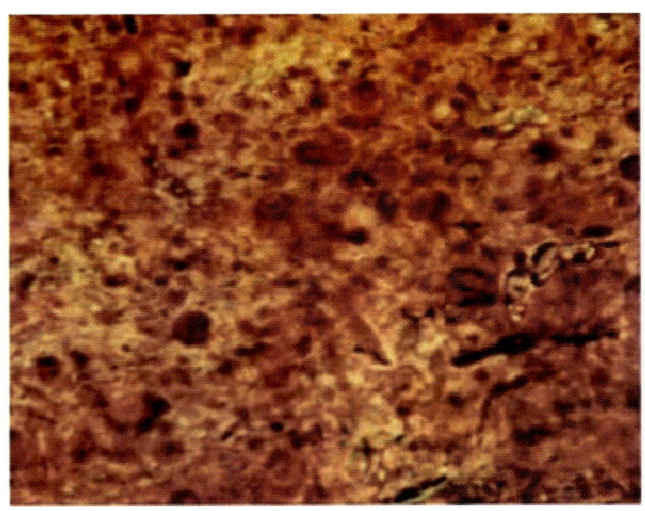

Fig. 10 : Wet preparation of the patient's synovial fluid shown intact and degenerating cells with intracellular and extracellular inclusions.

Treatment :

Non-pharmacological

- Ice packs
- Rest
- Removal of crystals by aspiration

Pharmacological

- NSAIDs
- Colchicine (for 2 weeks)
- Oral or local steroids

Acute periarthritis: Besides above measures, physiotherapy is the treatment of choice. In presence of CRF phosphate levels should be decreased.

Pulsed Ultrasound for calcific tendinitis has been shown to be beneficial[45].

Potent inhibition of HA crystal formation by PPi analogue phosphocitrate and bisphosphonates has been undergoing trials, more data are required before these are recommended for routine use.

Other crystal arthritides

Salient features of other important crystal arthritides are given in Table 3.

TABLE 3. Other Crystal Arthritides

Crystals	Association	Clinical picture	Crystal Characteristics	SF
Oxalate	*Renal Failure *Amyloid induced CRF in RA	Acute or chronic Knees, hands, ankles, feet, tendon sheaths & Bursae Intervertebral disc	Bipyramidal or envelope like shape, 5-30μm, strongly birefringent. EM : Dense & Foamy	Total leukocytes < 2000 cell/ cumm
Depot CS Crystals	Recent I/A Inj (1st 8 hours following Inj) After triamcinolone hexacetonide	destruction Joint injected	Irregular, rods & squares with intense positive or negative birefringence	
Lipid Crystals: *Cholesterol	RA or OA, recent hemarthrosis or trauma	Joints & Olecranon bursa	**Cholesterol Monohydrate :** Broad plates with notched corners, **Anhydrate cholesterol :** rod-shaped, helical birefringent crystals, ranging from 2 to 20 μm	Glossy golden appereance 10-40,000 leukocytes/ cumm
***Lipid liquid crystals**	Type II or IV Hyperlipoprotein-emias, Antecedent Infection	Monoarthritis : Knee joints most commonly involved, Tenosynovitis	Maltese cross like microspherules	
Rice Bodies	Chronic Synovitis e.g. RA		Huge semicircular balloons with birefringent edges.	Macroscopi-cally, they look like edible rice
Protein crystals : IgG Cryoprecipi-table paraprotein			Immunoglobulin crystals - pleomorphic, > 60μm or small squares, hexagon rods, stain with methylene blue	
Charcot Leyden crystals (lysophospho-lipase)	In eosinophil laden fluids		Spindle shaped	
Hematoidin crystals	Hb breakdown product		Golden Rhomboids on light microscopy	

Summary

Crystal induced arthropathies are important cause of inflammatory arthropathies.

CDD viz CPPD and BCP crystal deposition diseases affect elderly population and often confused with OA.

CPPD disease in younger population should alarm physician to search for secondary causes.

Clinical presentations vary from asymptomatic to acute/ chronic inflammatory arthropathies.

BCP crystal arthropathy may lead to destructive arthropathy in elderly.

Radiolography and US demonstrate CC.

Crystal demonstration by polarizing/electron microscope is the gold standard diagnostic investigation.

Treatment of these disorders is symptomatic and usually inadequate.

Acknowledgements :

Dr. Abhishek Agrawal (Associate Professor Medicine Department, SMS Medical College) for all the help in preparing this manuscript.

Further Reading :

1. Schumacher HR, Chen XL: Gout and other crystal associated arthropathies. In Harrison's Principles of Internal Medicine (17th Edition) Wiener C, Fauci AS, Braunwald E et al. (Eds). McGraw-Hill, NY, USA 1-5 (2008).
2. Das SK, Das SK; Mishra K. K.G. Medical College, Lucknow, India. CPPD Crystal deposition disease and asia. Journal of Indian Rheumatism Association. 2000 Jun; 8(2): 68-69.
3. Fonseca A, Crasta J, Swamy M. Calcium Pyrophosphate Dihydrate Crystal Deposition Disease-A Case Report. Journal of Clinical and Diagnostic Research [serial online] 2010 February [cited: 2010 February 1]; 3: 2079-82.
4. Williams CJ, Zhang Y, Timms A, Bonavita G, Caeiro F, Broxholme J et al. Autosomal dominant familial calcium pyrophosphate dihydrate deposition disease is caused by mutation in the transmembrane protein ANKH. Am J Hum Genet 2002; 71: 985.
5. Pendleton A, Johnson MD, Hughes A, Gurley KA, Ho AM, Doherty M, et al. Mutations in ANKH cause chondrocalcinosis. Am J Hum Genet 2002; 71: 933.
6. Hughes AE, McGibbon D, Woodward E, Dixey J, Doherty M. Localisation of a gene for chondrocalcinosis to chromosome 5p. Hum Mol Genet 1995; 4: 1225.
7. Dieppe P, Doherty M. Which comes first: Crystals or osteoarthritis? J Rheumatol 1983; 10: 42-43.
8. Doherty M, Watt I, Dieppe PA. Localised chondrocalcinosis in post-meniscectomy knees. Lancet 1982; 1: 1207-10.
9. Richette P, Ayoub G, Lahalle S, Vicaut E, Badran AM, Joly F et al. Hypomagnesemia associated with chondrocalcinosis: a cross-sectional study. Arthritis Rheum 2007; 57: 1496-501.
10. Luzar MJ, Altawil B. Pseudogout following intraarticular injection of sodium hyaluronate. Arthritis Rheum 1998; 41: 939.
11. Geelhoed GW, Kelly TR. Pseudogout as a clue and complication in primary hyperparathyroidism. Surgery 1989; 106: 1036.
12. Richette P, Bardin T, Doherty M: An update on the epidemiology of calcium pyrophosphate dehydrate crystal deposition disease. Rheumatology (Oxford) 48(7), 711-715 (2009) & EULAR Recommendatins for calcium pyrophosphate deposition Ann Rheum Dis 2011; 70: 563-70.
13. Reuge L, Van Linthoudt D, Gerster JC. Local deposition of calcium pyrophosphate crystals in evolution of knee osteoarthritis. Clin. Rheumatol. 2001; 20(6): 428-31.
14. Doherty M, Dieppe P, Watt I. Pyrophosphate arthropathy. a prospective study. Br j Rheumatol. 1993; 32(3): 189-196.
15. Neogi T, Nevitt M, Niu j, LaValley MP, Hunter DJ, Terkeltaub R et al. Lack of association between chondrocalcinosis and increased risk of cartilage loss in knees with osteoarthritis: results of two prospective longitudinal magnetic resonance imaging studies. Arthrits Rheum. 2003; 54(6): 1822-28.
16. Viriyavejkul P, Wilairatana V, Tanavalee A, Jaovisidha K. Comparison of characteristics of patients with and without calcium pyrophosphate dehydrate crystal deposition disease who underwent total knee replacement surgery for osteoarthritis. Osteoarthritis Cartilage. 2007 Feb;15(2):232-35.
17. Arveux I, Strutynsky C, Camus A, et al. Acute articular chondrocalcinosis episodes in the elderly subject. On a prospective evaluation in short-stay geriatric care and a literature review. Revue Geriatr 1999; 24: 63-68.
18. Pasquetti P, Selvi E, Righeschi K, Fabbroni M, De Stefano R, Frati E et al. Joint lavage and pseudogout. Ann Rheum Dis. 2004; 63(11): 1529-30.
19. Wendling D, Tisserand G, Griffond V, Saccomani C, Toussirot E, Tisserand G. Acute psuedogout after pamidronate infusion. Clin. Rheumatol. 2008; 27(9): 1205-06.
20. Halverson PB, Derfus BA: Calcium crystal induced-induced inflammation. Curr. Opin. Rheumatol 2001; 13(3): 221-24.
21. Sandor V, Hassan R, Kohn E. Exacerbation of pseugout by granulocyte colony stimulating factor. Ann Intern Med. 1996; 125: 781.
22. Saadeh C: Calcium pyrophosphate deposition disease (2006) retrieved from: http://emedicine.medscape.com/article/330936-overview. (as accessed on 25 May 2012).
23. Gerster JC, Lagier R, Boivin G. Achilles tendinitis associated with chondrocalcinosis. J Rheumatol 1980; 7: 82-88.
24. Gerster JC, Lagier R. Upper limb pyrophosphate tenosynovitis outside the carpal tunnel. Ann Rheum Dis 1989; 48: 689-91.

25. Jean CG, René L, Georges B. Olecranon bursitis related to calcium pyrophosphate dihydrate crystal deposition disease. Clinical and pathologic study. Arthritis Rheum 1982; 25: 989-96.
26. Yamakawa K, Iwasaki H, Ohjimi Y, Kikuchi M, Iwashita A, Isayama T et al. Tumoral calcium pyrophosphate dehydrate deposition disease. A clinicopathologic analysis of five cases. Pathol. Res. Pract. 2001; 197(7): 499-506.
27. Kishore B, Khare P, Gupta RJ, Gupta C, Khare V. Tumoral calcium pyrophosphate dihydrate crystal deposition disease: a rare diagnosis by fine-needle aspiration. Diagn Cytopathol. 2010 Jan; 38(1): 47-50.
28. Morgan MP, McCarthy GM. Signaling mechanisms involved in crystal-induced tissue damage. Curr Opin Rheumatol. 2002 May; 14(3): 292-7. Review.
29. Muthukumar N, Karuppaswamy U. Tumoral calcium pyrophosphate dehydrate deposition disease of the ligamentum flavum. Neurosurgery. 2003 Jul; 53(1): 103-8; discussion 108-09.
30. Zhang W, Doherty, Bardin T, Barskova V, Guerne PA, Jansen TL et al. EULAR recommendations for Calcium pyrophosphate deposition. Part I: terminology and diagnosis. Ann Rheum Dis. 2011; 70: 563-70.
31. Filippou G, Frediani B, Gallo A, Menza L, Falsetti P, Baldi F et al. A "new" technique for the diagnosis of chondrocalcinosis of the knee: sensitivity and specificity of high-frequency ultrasonography. Ann Rheum Dis. 2007 Aug; 66(8): 1126-28.
32. Chollet-Janin A, Finckh A, Dudler J, Guerne PA: Methotrexate as an alternative therapy for chronic calcium pyrophosphate deposition disease: an exploratory analysis. Arthritis Rheum. 2007; 56(2), 688-92.
33. Frediani B, Filippou G, Falsetti P, Lorenzini S, Baldi F, Acciai C, et al. Diagnosis of calcium pyrophosphate dehydrate crystal deposition disease: ultrasonographic criteria proposed. Ann Rheum Dis. 2005 Apr; 64(4): 638-40.
34. Runeberg L, Collan Y, Jokinen EJ, Lähdevirta J, Aro A. Hypomagnesemia due to renal disease of unknown etiology. Am J Med. 1975 Dec; 59(6): 873-81.
35. Zhang W, Nuki G, Moskowitz RW, Abramson S, Altman RD, Arden NK et al. OARSI recommendations for the management of hip and knee osteoarthritis: part III: Changes in evidence following systematic cumulative update of research published through January 2009. Osteoarthritis Cartilage. 2010 Apr; 18(4): 476-99.
36. Doherty M, Dieppe PA. Effect of intra-articular yttrium-90 on chronic pyrophosphate arthropathy of the knee. Lancet 1981; 2: 1243-46.
37. Ho AH, Johson MD, Kingsley DM. Role of the mouse ank gene in control of tissue calcification and arthritis. Science 2000; 289: 165-70.
38. Cini R, Chindamo D, Catenaccio M, Lorenzini S, Selvi E, Nerucci F, et al. Dissolution of calcium pyrophosphate crystals by polyphosphates: an in vitro and ex vivo study. Ann Rheum Dis. 2001 Oct; 60(10): 962-67.
39. A Wierzbicki, H.S Cheung, Molecular modeling of inhibition of crystals of calcium pyrophosphate dihydrate by phosphocitrate, Journal of Molecular Structure: THEOCHEM.1998Volume 454, Issues 2-3, 28 November 1998, Pages 287-97.
40. Jaovisidha K, Rosenthal AK. Calcium crystals in osteoarthritis. Curr Opin Rheumatol 2002; 14(3): 298-302.
41. McCarty DL, Halverson PB, Carrera CF, Brewer BJ, Kozin F. "Milwaukee shoulder" - association of microspheroids containing hydroxyapatite crystals, active collagenase, and neutral protease with rotator cuff defects. I. Clinical aspects. Arthritis Rheum 1981; 24: 464-73.
42. Genta MS, Gabay C. Images in clinical medicine. Milwaukee shoulder. N Engl J Med. 2006 Jan 12; 354(2): e2.
43. Scutellari PN, Galeotti R, Leprotti S, Ridolfi M, Franciosi R, Antinolfi G. The crowned dens syndrome: evaluation with CT imaging. Radiol Med 2007; 112: 195-207.
44. Rosenthal AK, Mandel N. Identification of crystals in synovial fluids and joint tissues. Curr Rheumatol Rep 2001; 3: 11-16.
45. Ebenbichler GR, Erdogmus CB, Resch KL, Funovics MA, Kainberger F, Barisani G et al.: Ultrasound therapy for calcific tendinitis of the shoulder. N. Engl. J. Med. 1999; 340(20): 1533-38.
46. Zang W, Doherty M, Bardin Tet al. EULAR recommendations for CPPD. Part 2 : Management. Ann Rheum Dis 2011; 70: 571-75.

SECTION 10

Osteoarthritis

OSTEOARTHRITIS 1

Siddharth Kumar Das

Osteoarthritis (OA) is the commonest joint disease affecting human beings and an important cause of disability. It is a disease of the joint as a whole in which all parts of the joint are involved. There is focal loss of cartilage with accompanying periarticular bone response in the form of subchondral bone sclerosis and attempted new bone formation by way of bony over-growths called osteophytes. This may be preceded by bone marrow edema and meniscal damage, and may be associated with inflammation of the synovium. OA usually presents, clinically, as joint pain and crepitus in the elderly. Radiologically it is characterised by decreased joint space (cartilage space), osteophytes, eburnation and a variety of deformities that develop as the disease progresses. On MRI it is characterized by presence of bone marrow lesions, meniscal damage, focal cartilage damage, and osteophytes. Its progress is variable and in some patients it is relentless, leading to joint failure (end stage joint disease) necessitating joint replacement.

Epidemiology

According to World Health Organization (WHO) OA is the second commonest musculoskeletal problem in the world population (30%) after back-pain (50%). The reported prevalence of OA from a study in rural Pune is 5.78%. Our data shows a prevalence of 4.2% in rural areas and 11.5% in urban areas of Lucknow. OA is an important cause of disability.

The major risk factors associated with the knee joint OA seen in a population study were: age, female sex, obesity, non smoking, occupational knee bending, physical labor and chondrocalcinosis. Symptomatic and radiographic OA increases with age. The age related increase is more apparent in women. The disease follows a more severe course in women with greater symptoms, more extensive involvement, and increased prevalence of knee and hand joint involvement. In contrast, men have an increased prevalence of hip OA in western countries.

Obesity precedes OA and is not its secondary consequence. Obese persons have a higher risk of developing OA that is progressive in its course. Weight reduction is shown to be an effective primary as well as secondary disease prevention strategy (5% reduction in weight decreases the risk for symptomatic OA by 50%). Effect of obesity on OA is not necessarily due to weight induced mechanical damage but also because of the secretion of adipokines from the adipose tissue.

Association between habitual overuse of a joint and development of OA has not been adequately studied. Occupational groups that overuse particular joints have an increased risk of OA in those joints. Occupational knee bending may be associated with OA. Prolonged squatting for day-to-day activities is associated with OA. No convincing data is available to show the increased risk of generalised OA following sports activities. However, progression of OA is said to be associated with faster pace per mile.

Chondrocalcinosis imparts a small risk, independent of age, for development of OA knee. Chondrocalcinosis was thought to be rare in India. Although no epidemiological data is available, our recent experience suggests that it is not common.

Indians are said to have more knee OA as compared to western people, while hip OA is comparatively uncommon in India. This could be related to excessive squatting involved in our day-to-day activities, especially amongst women.

Joint trauma is an important risk factor for OA. Meniscectomy is associated with very high risk for secondary OA. Quadriceps weakness is believed to be an important risk factor. Vitamin D deficiency, possibly acting through production of quadriceps weakness or through undefined mechanisms, has been said to be associated with progression of OA. The role of Vitamin D in OA is controversial, with many papers suggesting some role and others showing no role. Vitamin D supplementation does not seem to be beneficial in individual patients.

Evidences for the genetic basis of OA are increasing. Heberden's nodes have been shown to be clustered in families with a dominant inheritance. Twin studies show 65% concordance for developing OA. Certain familial forms of OA associated with spondyloepiphyseal dysplasias are linked to a point mutation in the gene coding for type II collagen (COL2A1). An endemic variety of secondary OA has been reported from Western Karnataka (Handigodu disease) that in all probability is due to a spondyloepiphyseal dysplasia occurring in a particular community. No specific genetic abnormality has been found so far. Hand joint OA is being linked to HFE gene mutations. However the common HFE variants C282Y and H63D are not associated with primary OA of the hip or knee. Genetic basis of generalised OA is further strengthened by its association with HLA A1, B8 and α1 Antitrypsin M2 phenotype. Genome wide scans suggest possibility of susceptibility to be linked to chromosome 2q in the IL-1 and IL-1Ra gene region. This cluster may harbour susceptibility for knee OA but not for hip OA. Defect in the mouse progressive ankylosis gene (ANK gene) is associated with hydroxyapatite crystal deposition and extensive calcification and OA in the mouse similar to that seen in humans. Mutations to this gene have also been found in humans.

Pathogenesis

Normal hyaline cartilage consists of chondrocytes (1-2%) embedded in extracellular matrix composed of water, type II collagen and proteoglycan. The cartilage remains stable with active degeneration and regeneration occurring in equilibrium. OA results from excessive degeneration compared to regeneration.

A simplified scheme of pathogenesis is given in Fig. 1. The chondrocyte secretes numerous inactive matrix metalloproteinases (MMPs) that include stromelysin, collagenase, gelatinase etc., which can destroy the whole cartilage framework. These are activated by plasmin and activated stromelysin. Chondrocytes also release tissue plasminogen activator (tPA) that converts blood or chondrocyte-derived plasminogen to active plasmin. This cascade is modulated by tissue inhibitor of MMPs (TIMP) and plasminogen activation inhibitor (PAI). Interleukin-1 (IL-1) plays a pivotal role in OA pathogenesis. It increases the synthesis and secretion of latent or inactive MMPs and tPA; suppresses proteoglycan synthesis and hence the cartilage repair; decreases TIMP1 and PAI synthesis; and may be involved in proliferation of osteoblast-like cells leading to osteophyte formation.

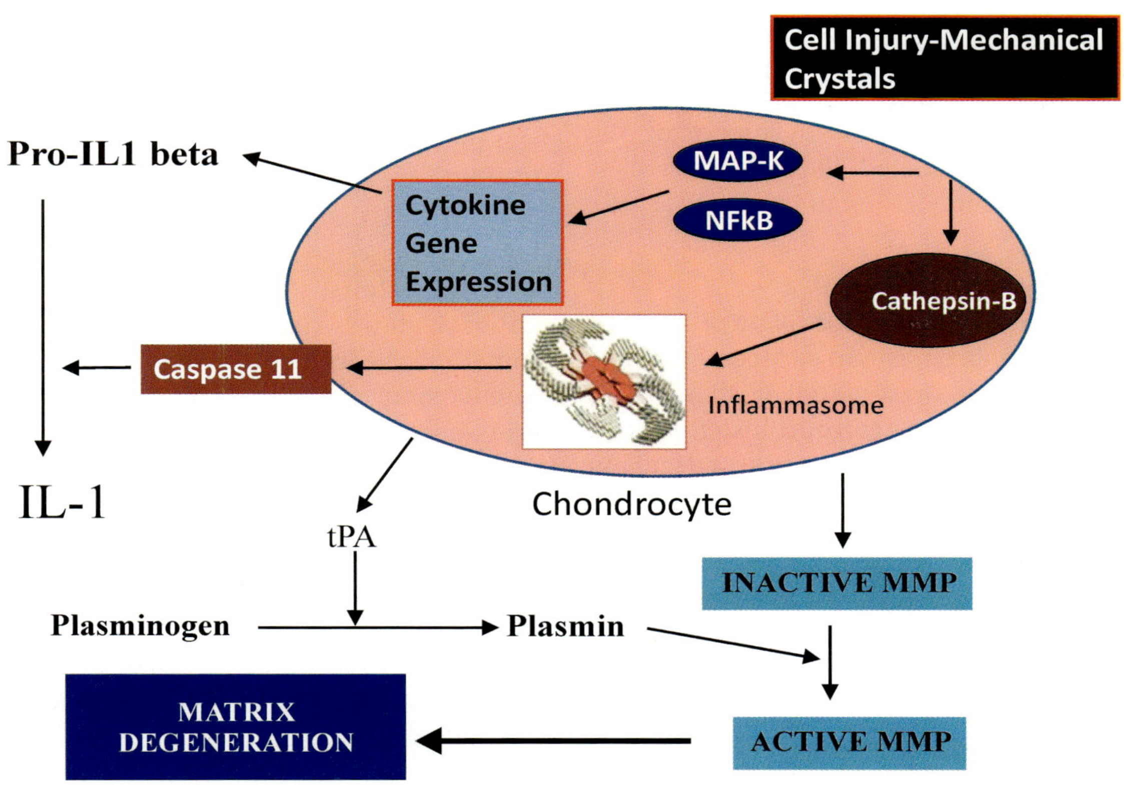

Fig. 1 : A Simplified scheme of Pathogenesis

Previously OA was considered to be an inherently non-inflammatory disorder. In our experience in a tertiary care hospital, about 25-30% patients have clinical inflammation at the time of presentation. There are reports that 80-90% patients with symptomatic knee OA have synovitis seen on MRI. It is known that inflammation worsens OA. Damage to chondrocytes and other cells due to crystals, mechanical stress, etc. starts the process. There is stimulation of the NFkB and MAPK pathways leading to activation of cytokine genes, which results in formation of Pro-IL1b. Cell damage also leads to production of large sized chondrocytes containing Phagolysosomes that contain Cathepsin B and other proteolytic enzymes. These enzymes not only directly lead to matrix degeneration but also stimulate the Inflammasome. Activated Inflammasomes release caspase 1. Caspase 1 is also called Pro-IL1b converting enzyme. This enzyme converts Pro-IL1b to IL1b. IL1b is a powerful mediator of inflammation. IL1b in turn stimulates the chondrocytes leading to cartilage degeneration as indicated above.

Matrix repair is stimulated by insulin-like growth factor (IGF-1), fibroblast growth factor (FGF) and transforming growth factor-β (TGF-β). These factors act by increasing proteoglycan synthesis. In early stages there is matrix and cartilage degeneration on the one hand and active chondrocyte replication with enhanced biosynthesis on the other hand. This leads to a state of homeostasis, known as compensated OA, in which both repair and degeneration are balanced. After a few years, the repairable process is exhausted. This leaves cartilage degradation unopposed leading to progressive OA.

Morphology

Earliest identifiable changes are loss of proteoglycans and decreased metachromasia in cartilage. There is focal loss of cartilage, associated with reactive proliferation of chondrocytes to form clusters of large sized chondrocytes. These chondrocytes are rich in lysosomes. These chondrocytes release Collagen I. Hyaline cartilage contains Collagen II and Collagen I is inferior for the purpose of hyaline cartilage. Progression of these changes leads to breach of surface integrity, fissures, pitting, flaking of cartilage and development of vertical clefts (fibrillations). These fissures deepen and may expose the subchondral bone, which becomes ivory-like due to thickening and vascularisation (eburnation). Associated bone growth in subchondral region leads to sclerosis, while growth in joint margin leads to the formation of osteophytes.

Pain in Osteoarthritis

The cartilage is devoid of nerve endings hence the pain in OA is believed to originate from the capsule, synovium, periosteum, subchondral bone, ligaments, muscles etc. One of the theories is that when periosteum is stretched at the time of osteophyte formation there is pain. Our studies suggest that pain in OA may be in part due to presence of inflammation in the joint. It has also been seen that the symptom of pain waxes and wanes with development and resolution of bone marrow lesions seen on MRI. Bone marrow lesions are considered a hall mark of Osteoarthritis on MRI.

Natural history

The course of OA is highly variable, with radiological progression seen in one-third to two-third of patients. Improvements are rare. Symptoms may progress, improve, or may even be arrested spontaneously and do not correlate well with radiographic progression. Patients with multiple affected joints have more rapid progression of OA. Advanced age and obesity are also associated with a more rapid progression of OA.

Clinical features

Pain in or around the involved joints is the cardinal symptom. There may be a delay of many years in the appearance of pain after the start of cartilage degeneration. Usually it starts insidiously as intermittent localised deep ache, often aggravated by joint use and relieved by rest, in an asymmetric fashion and then becoming symmetric. The pain becomes persistent and in advanced cases night pain may be very disturbing. Knee joint involvement is typically associated with pain on descending or ascending stairs, squatting for toilet, climbing into autorickshaws, rickshaws, buses etc. There may sometimes be locking or buckling of knee joint. Hip involvement causes problems of gait and pain in the groin or medial part of thigh.

Stiffness in joints on first awakening (morning stiffness) in OA usually lasts 5-30 minutes (average 20 min). Stiffness occurring after periods of rest or inactivity (gelling) is frequent and lasts for a few minutes. Functional impairment in OA is highly variable. It depends on associated muscle wasting and weakness, and on radiological severity of the disease.

Signs of OA include bony swelling, crepitus (sensation of bone rubbing against bone on movement), restricted movement, deformity of joint, muscle weakness and muscle wasting. Occasionally, the joints like knee show signs of inflammation and effusion. Commonly involved joints in OA are the knee, the hand joints (distal interphalangeal, first carpo-metacarpal and proximal interphalangeal joints) and spinal apophyseal joints, while hip, acromioclavicular and sternoclavicular involvement are uncommon. The carpal joints, the elbow, ankle, glenohumeral and the intertarsal joints are rarely involved in OA.

In the knee joint OA, crepitus is common and may be observed in over 90% of the patients. A bowleg deformity is seen in over 50% of patients. Signs of local inflammation often with joint effusion, are commonly seen in patients visiting tertiary care centres. This may probably be due to more symptomatic patients being referred to such centres. Such patients are likely to have superimposed microcrystal-induced inflammation.

Subsets of OA

OA for convenience can be categorised into four major subsets, though subsets may overlap with each other.

Nodal generalised OA (GOA) : Nodal generalised OA is a well-recognised subset seen 10 times more often in obese women mainly in the middle-elderly age group. It is characterised by distal and to a lesser extent proximal interphalangeal (IP) joint involvement, Heberden's nodes (bony enlargement of distal IP joint), Bouchard's nodes (bony enlargement at proximal IP joints) and familial clustering.

Erosive OA : It is characterised by involvement of the interphalangeal joints of the hands often with florid inflammation and erosive changes that lead to deformities and ankylosis. A small proportion (15%) may evolve into seropositive rheumatoid arthritis (RA).

Isolated knee Joint OA : It is the most common form of OA. The knee joint may be the only joint involved. It usually starts unilaterally in the beginning but becomes bilateral over a period of time. Obesity, female gender and knee bending are the important risk factors. It may involve predominantly medial femorotibial, lateral femorotibial or patellofemoral compartment. The involvement of patellofemoral component is tested by eliciting pain when patella is compressed against femur on active extension of knee (Shrug test).

Spinal OA : Apophyseal joint involvement is the only true form of OA of spine. Pain in bending backwards is the typical physical sign of such involvement. Intervertebral disk degeneration with osteophyte formation is considered an integral part of OA. It is common in the lumbar (lumbar spondylosis) and the cervical (cervical spondylosis) regions, the sites of maximum spinal motion. It gives rise to local pain and/or radicular pain. Diffuse idiopathic skeletal hyperostosis (DISH) comprises of flowing calcification along the anterior longitudinal ligaments and disk margins. DISH and OPLL (ossification of posterior longitudinal ligament) are considered to be variant forms of OA.

Investigations

The earliest feature is joint space narrowing, which in the knee is best demonstrated in weight-bearing x-rays. Joint space narrowing seen in standardised x-rays is also used as a measure of progression of OA. Osteophytes are the hallmark of the disease. They indicate relatively advanced disease. Their presence is considered important for diagnosis. Subchondral bone sclerosis, subchondral cysts, deformity, loose bodies and calcification (linear due to calcium pyrophosphate dihydrate (CPPD) deposition, spotty due to hydroxyapatite deposition) are the other changes. Radiograph of a knee joint showing osteoarthritic changes are given in Fig. 2.

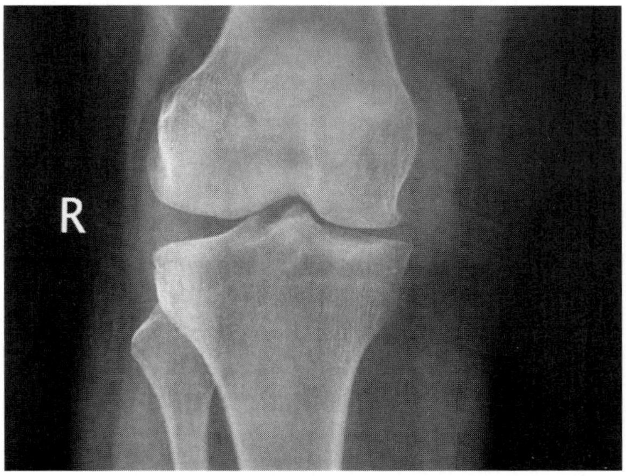

Fig. 2 : Showing osteoarthritis of the right knee with a) space reduction in medial compartment, b) eburnation of medial tibial surface, and c) small osteophytes (Grade 1) on medial aspect of tibial and femoral condyle.

CT scan is useful in assessing the axial joints. Ultrasound is useful in assessing changes in periarticular structures like Baker's cysts. Magnetic resonance imaging (MRI) is rapidly developing into a tool for evaluation of cartilage in Osteoarthritis. Cartilage defects - fibrillations and surface defects, cartilage quality, meniscal and ligamentous damage, bone marrow edema (lesion), and synovitis and synovial effusion can be seen on MRI. Bone marrow edema, cartilage defects, meniscal tears, osteophytes, subchondral cysts, sclerosis, synovitis and synovial effusions seen on MRI correlate with pain in the knee joint. MRI is being more commonly used to evaluate OA. MRI can also be used to non invasively study the progress of osteoarthritis and response to medicines.

Synovial fluid examination usually shows mild leucocytosis (< 2000/mm^3) with mononuclear cell predominance. In some patients with inflammation, counts may be higher. Synovial fluid may show the presence of crystals in 30-70% patients depending on the technique used. The crystals found most commonly are CPPD. However hydroxyapatite (HA) or basic calcium phosphate (BCP) crystals may also be detectable.

Serum osteocalcin levels, a non - collagenous marker for bone formation, and serum or urinary C-telopeptide of type I collagen (a collagenous marker for bone formation) are decreased in OA. Whereas serum cartilage oligomeric protein (COMP), a non-collagenous marker, and urinary C telopeptide of type II collagen, a marker of cartilage turnover, are increased in osteoarthritis. Their role in diagnosis or following up patients to assess response has not been defined so far.

Diagnosis

The diagnosis of OA is essentially clinico-radiological. It is well known that early OA may not show radiological changes, while only 50-60% of subjects with OA by radiograph are clinically symptomatic. For epidemiological and research purposes, American College of Rheumatology has revised classification criteria for OA of different patterns of joint involvements. An early diagnosis of Osteoarthritis may be attempted by MRI using the signs of meniscal damage, cartilage defect, bone marrow lesion, and osteophytes. Presence of any two changes may be used to make an early diagnosis of osteoathritis.

In the presence of inflammation, differentiation of OA (especially the generalised primary OA of nodular variety) from inflammatory joint disorders like RA could be difficult. However, topography (pattern) of the joint involvement, the laboratory investigations and the x-ray findings help in clinching the diagnosis. RA is associated with more signs of inflammation and commonly affects MCP, wrist, carpal and other peripheral joints and also upper cervical region of the spinal cord. Generalised OA involves PIP, DIP, the first carpo-metacarpal joints in the hands and cervical and lumbar regions of the spinal cord. Erosive form of OA is particularly difficult to differentiate from RA. A positive rheumatoid factor and an elevated ESR helps in diagnosing RA. Patients having mixed disease further complicate the picture. OA changes may follow joint damage due to RA. It must be remembered that joints actively involved in RA seldom show osteophytes or minimal osteophytes. Isolated knee joint pain, crepitus, inflammation, and positive rheumatoid factor is still likely to be OA rather than RA.

Management

The accepted management of OA is palliative aimed at minimising pain, optimising function and reducing disability. Improvements in our understanding of the aetiopathogenesis of OA have raised the possibility of rational therapeutic approaches directed at either retarding the disease progression and/or enhancing normal repair, the concept of disease modification. Management of OA requires a comprehensive approach involving non-pharmacological, pharmacological and surgical therapies.

Non-pharmacological therapy

Among non-pharmacological measures, patient education is important. The main emphasis should be on joint protection measures like avoidance of poor posture and limitation of excess joint loading. This may require some alteration in performance of day-to-day activities (e.g. change of toilet seat from the Indian style to Western style; avoiding weight loading movements of the joints e.g. avoidance of squatting, sitting on low chairs, sofas, floor and climbing stairs; use of walking stick, etc.). Weight reduction is desired in obese patients. Studies suggest that significant weight reduction of up to 20% in extremely obese patients may result in both symptom relief and modification of disease process. When the disease is relatively quiescent, low intensity weight bearing activity is safe and beneficial for muscle strength, mobility and a sense of well-being. But when the disease is active with significant pain and inflammation, patients need to use good judgment and not to overstress with exercise. Quadriceps strengthening exercises are important. Supportive devices are often an adjunct to other modalities. Devices like canes, crutches, orthotic shoes and knee braces when used properly, lead to pain relief with improved function. A ¼ inch lateral heel raise or shoe insert is frequently helpful in varus deformity of knees.

Application of superficial heat on periarticular tissue by hot packs and paraffin baths alleviates pain, relieves muscle spasm and decreases joint stiffness. Deep heat by ultrasound and diathermy may damage cartilage and should be avoided. Heat increases inflammation hence ice packs are more helpful for acutely inflamed knee or bursae.

Pharmacological management

Nonsteroidal anti-inflammatory drugs (NSAIDs) are commonly prescribed for treatment of both pain and inflammation in OA. Their use is limited by their gastric and renal toxicity especially in the elderly. Some studies have shown acetaminophen (paracetamol) 1 gm 2-3 times a day (max 3 gm/day) to be equally effective as other NSAIDs in OA with the advantage of fewer gastric side effects. However in meta-analysis Paracetamol is relatively weak in providing relief from pain. NSAIDs like aspirin and indomethacin which have been implicated in cartilage degeneration are best avoided in OA. Newer selective cyclooxygenase-II (COX-2) inhibitors like meloxicam, celecoxib, etoricoxib etc. have been shown to be chondroneutral and more effective with fewer NSAID-related side effects.

Capsaicin cream on topical application gives relief from pain and tenderness by depleting substance P in sensory nerve endings and may be used. Absorption of locally applied NSAIDs may be erratic but may have a strong psychological value.

Chondroitin and glucosamine are nutraceutical agents whose use and effectiveness have been debated. There are conflicting reports in literature. glucosamine sulphate is given in doses of 1500 mg per day and chondroitin sulphate 1200 mg/day. They may provide symptomatic benefit in some patients with knee OA. There are number of trials showing disease modification using glucosamine sulphate and chondroitin sulphate. The sulphate salt of Glucosamine marketed by only one pharmaceutical company only has been shown to be consistently effective in OA and also been shown to provide disease modification. However the Hydrochloride salt has never been shown to benefit patients. Diacerein in doses of 50 mg twice daily with meals is being used in osteoarthritis. Diacerein has a small, consistent benefit in improvement of pain. It is believed to be an inhibitor of IL1. Intra-articular (IA) steroids are employed in patients with knee OA who have effusion. It is one of the most effective therapies for pain in OA. Since repeated steroid injections may lead to cartilage damage, it is recommended that the same joint should not be injected more than 3 times a year. The incidence of joint infection following IA steroid is low. Visco-supplementation with intra-articular hyaluronan injection is a promising therapy, providing symptomatic relief to about 60-70% patients with knee OA; especially those without effusion. Hyaluronan is given as 3-5 (depending on molecular weight) intra-articular injections at weekly intervals. Local reaction may occur in up to 8% patients. Considering that IL1 is an important cytokine for development and progression of OA, Anakinra has been used intra-articularly. Early studies have shown that its use is safe but a single injection with a half life of a few hours is not effective in reducing pain. Anti TNF agents have also been used but more studies are required.

One center from India has reported that patients presenting with significant inflammation with demonstrable CPPD crystals in joint fluid have better symptomatic relief when treated with colchicine. Even in patients not presenting with inflammation, colchicine produces significant symptomatic relief. Colchicine is used in a dose of 0.5 mg twice daily.

Tetracyclines chelate metals and thus inactivate MMPs. Since MMPs are important in the degradative pathway, doxycycline has been used for management of OA. One study has shown disease modification but no effect on symptoms.

Chondrocyte transplants directly or in scaffolds, have been shown to regenerate near-normal hyaline cartilage and may be useful for small lesions of the cartilage after knee injury but are not currently routinely recommended in treatment of OA.

Surgical management

Joint replacement must be considered in any patient presenting with end-stage joint disease with pain not responding to conventional treatment. Total Knee replacement is opted by patients who are older and have weakness, more difficulty in getting up, more problems in climbing stairs and impaired extension of knee. Total hip and knee joint replacements have improved the quality of life and decreased the morbidity of OA. Many studies report excellent long term results in over 90% of patients undergoing joint replacements. Other surgical therapies available for OA are loose body removal, osteotomy, arthrodesis, arthroplasty, synovectomy, debridement etc. with varying grades of efficacies.

Prevention of osteoarthritis

Weight reduction is the only known method of reducing the risk of symptomatic osteoarthritis. Lifestyle changes, like less squatting and less stair climbing, are likely to help. Some of the therapies described for OA possibly reduce the progression of OA and hence are more likely to be beneficial for prevention rather than for treatment. An early diagnosis of OA is now possible using MRI. It may even be possible to identify pre-OA and take preventive measures. However, one has to realise that if prevention of OA is to be attempted the effort has to start in twenties and thirties, as we know that the seeds of OA are sown early.

Further Reading :

1. Doherty M: Clinical feature and subset characterisation. In Doherty M (ed.) OA, London Wolfe, 1994; 61-83.
2. Felson DT: The epidemiology of knee OA: results from the Framingham OA study. Seminars in Arthritis and Rheumatism 1990; 20(3 suppl): 42-50.
3. Watt I: Radiology and Imaging. In Doherty M (ed.) OA, London Wolfe, 1994; 85-114.
4. Cochrane Database Syst Rev 2005 Apr 18; (2): CD002946.
5. Lane NE, Buckwalter JA: Exercise: A cause of OA? Rheumatic Dise Clin Nor Am 1993; 19: 617-32.
6. McCarty DJ: Arthropathies associated with calcium containing crystals. Hosp Pract Ed 1998; 21: 109-15, 118-20.
7. Koopman WJ, Moskowitz RW: Clinical and laboratory findings in OA. In (Ed) Arthritis and Allied Conditions. A Text Book of Rheumatology, 13th Ed. Williams & Wilkins Company, 1997; 105: 1985-2012.
8. Das SK, Farooqi A : Osteoarthritis. Best & Research Clin Rhem 2008; 22: 657-759.
9. Marc Hochberg, Roy Altman, Karine April et al. American College of Rheumatology 2012 Recommendations for the Use of Nonpharmacologic and Pharmacologic Therapies in Osteoarthritis of the Hand, Hip and Knee. Arthritis Care & Research 2012; 64: 465–74.

SECTION 11

Osteoporosis

EPIDEMIOLOGY, PATHOPHYSIOLOGY AND DIAGNOSIS OF OSTEOPOROSIS

1

Rohini Handa

Introduction

Osteoporosis is a common metabolic bone disease characterized by reduced bone mass and micro-architectural deterioration in bone tissue that results in an increased susceptibility to fragility fractures. Common sites for fragility fractures are the spine, hip and the wrist. Osteoporosis is a significant public health problem in many parts of the world including India. The importance of osteoporosis can be gauged by the fact that a woman's life time risk of hip fracture equals the combined risk of breast, uterine and ovarian cancers, and the risk of dying from hip fracture equals mortality from breast cancer. Men too are not spared. They account for nearly 20-30% of all hip fractures that occur, and one third of these men do not survive more than a year. The economic consequences of osteoporosis continue to mount as the life span of populations around the world increases. Health Technology Assessment, a multi-disciplinary field of policy analysis that examines the medical, social, economic, and ethical implications of medical technology in health care is being increasingly applied to osteoporosis[1]. The disease, unfortunately, has attracted little attention and even less action in our country. The effectiveness of treatment strategies currently available mandates that osteoporosis be diagnosed and treated much before the occurrence of complications like fracture.

WHO definition

The World Health Organisation (WHO) defines osteoporosis on the basis of bone mass measurement which is compared to the peak bone mass in a young adult of the same race and sex (T-score)[2]. The difference between the patient's value and the norm is expressed in terms of standard deviations (SD) below the mean. Osteopenia is defined as bone mineral density (BMD) between -1 to 2.5 SD below the young adult reference mean. Osteoporosis is BMD -2.5 or more SD below the young adult reference mean. Established or severe osteoporosis is defined as T-score below -2.5 in presence of one or more fragility fractures. In general, fracture risk approximately doubles for each -1 SD decrease in T-score.

It is to be emphasized that the WHO classification is derived from studies of white postmenopausal women and is not applicable to men, children or premenopausal women. According to International Society of Clinical Densitometry (ISCD) guidelines these definitions may also be used for men older than 50 years[3,4]. The ISCD recommends the use of Z-scores in premenopausal women, men younger than 50 years, and children. Z-scores compare individual BMD measurements to age-matched reference populations. A Z-score lower than -2 is defined as "below the expected range for age". The ICSD official position states that the diagnosis of osteoporosis in children, adolescents and men below 50 should NOT be made on the basis of densitometric criteria alone. Fracture risk assessment is preferred in these groups.

Data from India show that BMD in Indians may be different from the Caucasians and the western reference values may not be applicable to Indians. The widespread vitamin D deficiency in India may also have an impact on BMD. BMD estimation carried out in healthy young individuals from Indian paramilitary forces with normal bone and mineral biochemical values revealed that 35%-50% of men and 14%-32% of women were osteopenic at different sites, while an additional 10% of men had osteoporosis of the lumbar spine despite optimal nutrition, good sunlight exposure and regular physical exercise[5]. This underscores the urgent need to have normative values of BMD from our country.

Epidemiology

Ageing of the world's population has meant an increase in the number of osteoporotic individuals. Current estimates put the number of women affected worldwide at > 200 million. Most of the epidemiologic information comes from the West. In white populations about 50% of women and 20% of men older than 50 years are likely to get a fragility fracture in their remaining lifetime. Studies from Europe reveal that more than 30% of European women aged ≥ 50 years have osteoporosis with lifetime fracture incidence of 14%, 11% and 13% for hip, vertebral and distal forearm respectively[6]. Annually, more than 1.3 million osteoporotic fractures occur in the USA while the number of such fractures in Europe in 2000 was 3.79 million[7].

Hip fractures are commoner in women (~80% of hip fractures) as compared to men. Fracture rates adjusted for age are highest in Scandinavian and North American populations with lower rates in Asian and Latin American populations. Rates seem to be lower in rural areas than in urban areas in any country[8]. Even if age-adjusted incidence rates for hip fractures were to remain stable, the number of hip fractures worldwide are estimated to rise from 1.7 million in 1990 to 6.3 million in 2050[8]. Geographical variation in the prevalence and incidence of vertebral fractures is less than that for hip fractures. The prevalence of vertebral fractures in men and women is almost equal, likely due to occupation associated trauma in men. Unlike hip fractures that result primarily from a fall, most vertebral fractures result from routine activities like bending or lifting light objects. Falls account for only one fourth of vertebral fractures. Vertebral fractures are often clinically silent. However, one vertebral fracture leads to a tenfold increase in risk of subsequent vertebral fractures. The epidemiology of wrist fractures differs from hip and vertebral fractures. Most wrist fractures occur in women, 50% of whom are older than 65 years. The incidence in men is low and does not increase much with age[8].

No population based statistics are available from India. However, estimates by expert groups like the Osteoporosis Society of India place the number of osteoporosis patients in India at 26 million approximately (2003 figures), with the numbers projected to increase to 36 million by 2013[9]. Detection and treatment before the occurrence of first fracture is uncommon in India. Several factors are responsible for this state of neglect: the long held belief that osteoporosis is a disease of the west and does not exist in India, the notion that osteoporosis is an inevitable consequence of ageing which can neither be treated nor prevented, preoccupation of the health planners with infectious diseases, limited availability of diagnostic facilities and paucity of epidemiologic data[10]. The low life expectancy at birth seen a few decades ago meant that many Indians, in the past, did not live long enough to develop osteoporosis. Osteoporosis was thought to be an exotic disease, seldom suspected, rarely diagnosed. With ageing of the Indian population this has changed. According to the 2011 census, the population of India is 1.2 billion of which the aged (60+) population is 7.5%. This translates to more than 90 million elderly in India.

Studies from India have shown that osteoporotic fractures usually occur 10-20 years earlier in Indians compared to Caucasians[11-14]. Some of these studies were done nearly 50 years ago, and none have used DXA to confirm osteoporosis. One hypothesis that has been advanced to explain the early onset of osteoporosis in Indians is that a dietary deficiency of calcium, beginning early in life, leads to a lower peak bone mass and consequently osteoporosis at an earlier age in Indians[15]. Another contributory factor could be subclinical vitamin D deficiency causing malabsorption of calcium without overt osteomalacia. However, this hypothesis needs validation in rigorously designed studies. Another interesting observation has been the high proportion of male patients among hip fracture cases in India[11-14] which is contrary to what is seen in other areas. Lower hospital attendance and health service utilization by women in India, especially the elderly, may be the main reason behind the skewing of data.

Pathophysiology

Bone tissue comprises of an extracellular matrix within which bone mineral comprising chiefly of hydroxyapatite is deposited. The matrix is composed of type 1 collagen, proteoglycans, and non collagenous proteins like osteocalcin, osteopontin etc. Bone cells include osteoclasts, osteoblasts and osteocytes. Osteoclasts are bone resorbing cells derived from hematopoietic cells of monocyte macrophage lineage while osteoblasts are derived from pluripotent stromal stem cells. Osteocytes are terminally differentiated osteoblasts. Osteoporosis is characterized by excessive bone loss due to abnormalities in the bone remodeling cycle, normally a balanced process where resorption by osteoclasts is coupled with bone formation by osteoblasts. Despite the fact that cortical bone comprises 90% of the skeleton, the trabecular (cancellous) bone is more affected by osteoporosis because it has 20 times greater number of remodeling units/volume. The trabeculae decrease in number and size. Both genetic and environmental factors contribute to peak bone mass. Candidate genes linked to BMD include those coding for Vitamin-D receptor, estrogen receptor, collagen Iα1 and LDL receptor-related protein 5 [LRP5]. However, the link between gene polymorphisms and fracture risk is poor.

RANK (receptor activator of NFκB), RANK ligand (RANKL) and OPG are the three key cytokines that regulate osteoclast recruitment and function. Osteoclasts express RANK while osteoblasts express RANKL constitutively on their cell surface. RANKL, a member of the TNF superfamily of ligands and receptors, is essential for the differentiation, activation and survival of osteoclasts. Osteoprotegerin (OPG), a soluble decoy receptor secreted by osteoblasts, blocks the interaction between RANKL and RANK and serves as a physiological regulator of bone turnover. Bone loss results both from estrogen deficiency as well as by estrogen independent, age-related mechanisms like secondary hyperparathyroidism and reduced mechanical loading. Estrogen deficiency results in a remodeling imbalance, increase in bone turnover and increased osteoclastogenesis. There is increased production of pro-inflammatory cytokines like IL-1 and TNF. TNF stimulates M-CSF (macrophage colony stimulating factor) and RANKL production. RANKL interacts with its cognate receptor RANK to increase osteoclast generation[16]. RANKL and IL-1 also increase osteoclast survival by preventing apoptosis.

A lot of attention is currently being paid to the Wnt signaling pathway and LRP5, a member of the LDL receptor family[17-19]. The LRP5 and Wnt/ß-catenin play a central role in bone mass accrual and normal distribution. The LRP5 is a co-receptor of Wnt located on the osteoblast membrane between two other receptors, Frizzled and Kremen. Frizzled and LRP5 bind to Wnt, thereby stabilizing ß-catenin and activating bone formation. The Wnt signaling pathway has antagonists/inhibitors like Dickkopf protein (Dkk) and SOST gene product, sclerostin. Dkk binds to Kremen and LRP5, the latter undergoes internalization and cannot bind Wnt leading to degradation of ß catenin and to inhibition of bone formation.

Better understanding of the disease pathophysiology has led to development of several targeted therapies for osteoporosis. Denosumab, a humanized monoclonal antibody to RANKL, mimics the function of OPG and is approved for treatment of osteoporosis. Similarly, inhibitors of sclerostin are a potential therapeutic target. Other agents in the pipeline are cathepsin K inhibitors. Cathepsin K is a tissue-specific cysteine protease expressed by osteoclasts that degrades bone and cartilage matrix proteins, including type 1 collagen.

Diagnosis

Osteoporosis typically produces no symptoms. Often the first manifestation is a low impact fracture, defined as a fracture resulting from trauma equal to or less than fall from standing height. Fractures commonly occur in distal fore arm, vertebrae and hips, though any bone can be involved. Vertebral crush fractures lead to spinal deformities, kyphosis (Dowager's hump), loss of height and a protuberant abdomen. Isolated involvement of upper dorsal vertebrae should arouse suspicion of causes other than osteoporosis. Patients may complain of early satiety due to abdominal compression and breathlessness. Hip fractures are associated with excess mortality and are the most dreaded consequence of osteoporosis.

Baseline investigations in a patient with osteoporosis should include blood counts, renal and liver function tests, serum calcium, phosphorus, alkaline phosphatase and urinary calcium excretion. In postmenopausal osteoporosis, these hematological and biochemical investigations are within normal limits and serve to rule out secondary causes.

In absence of tools to measure bone quality, bone mass has become the surrogate marker used to define osteoporosis. Table 1 provides a listing of various techniques for assessing bone mass. The current gold standard to measure BMD is DXA (dual energy x-ray absorptiometry). The correct terminology as approved by ICSD is DXA not DEXA. DXA can be used for lumbar spine, proximal femur, forearm and even for assessment of total body mineral composition. The radiation risk is negligible. From a clinical stand point it is pertinent to mention that the presence of one or more fragility fractures is sufficient to classify a patient as severe osteoporosis even if the BMD is normal. Peripheral DXA (pDXA) can measure BMD at peripheral sites like heel, phalanges and forearm. It is recommended that pDXA measurements should be interpreted with device-specific upper and lower thresholds.

The correlation of BMD at different sites is modest and BMD at a specific site is the best predictor of fracture at that particular site. DXA should ideally be performed at 2 sites: hip and spine (anteroposterior). Osteophytes may interfere with BMD measurement in the AP view in the elderly. Lateral spine DXA, though less precise because accurate positioning is difficult, is less affected by spinal degenerative disease. The lumbar spine, due to greater quantity of trabecular bone, is more sensitive to the changes in BMD that occur in response to drug treatment. The contraindications for bone densitometry include pregnancy, recent gastrointestinal contrast studies and radionuclide tests. DXA results may be inaccurate in presence of osteomalacia, osteoarthritis, vascular calcification, previous fracture etc.

There are differences in calibration between DXA machines of different companies and an individual patient's bone density reading can differ by as much as 12% when different DXA machines are used. Therefore, whenever possible, follow-up measurements for a given patient should use the same scanning procedure and the same instrument as for the original measurement. Most treatments lead to an increase in BMD of 1-6% over 3 years. The small magnitude of change in response to treatment necessitates that repeat BMD measurements be spaced fairly far apart, usually 2 years or more. An increase in BMD of at least 3-4% is required as "the least significant difference" that exceeds the error of the measurement. Although BMD is useful in guiding decisions to initiate treatment, subsequent changes in BMD do not fully explain reductions in fracture risk. Over reliance solely on BMD should be avoided. Fracture protection benefit may be realized even before BMD gains are detected. No change in BMD should not be taken to imply that therapy is not working because treatment may have prevented bone loss. DXA is also now utilized to pick up vertebral fractures. Vertebral Fracture Assessment (VFA) refers to densitometric spine imaging performed for the purpose of detecting vertebral fractures.

Quantitative ultrasound (QUS) is an inexpensive modality of measuring BMD. It may provide some idea about bone quality. The parameters measured include speed of sound (SOS) and broadband ultrasound attenuation (BUA) at the calcaneus. Different systems yield different values which are not comparable. The current role of QUS is as a screening procedure. It cannot be used to diagnosis osteoporosis or make treatment decisions. Low QUS is an independent risk factor for fracture in post-menopausal women over 65 years of age.

Measurement of bone quality

Unlike BMD measurements that are well standardized, techniques to measure bone quality in vivo are in a state of development. These include quantitative ultrasonography, magnetic resonance (MR) imaging and MR spectroscopy, multidetector CT, and high-resolution peripheral quantitative (HR-pQ) CT[20].

Who should be subjected to bone mass measurements

Resource constraints do not permit universal screening of all postmenopausal women in most countries. The ISCD recommends screening for osteoporosis in:

- Women aged 65 and older.
- Postmenopausal women under age 65 with risk factors: medical disorders like RA, patients on long term corticosteroids (> 3 months), current smoking, thinness (BMI < 21), parents with history of fragility fracture and previous history of fragility fracture.
- Women during the menopausal transition with clinical risk factors for fracture, such as low body weight, prior fracture, or high-risk medication use.
- Men aged 70 and older.
- Men under age 70 with clinical risk factors for fracture.
- Adults with a fragility fracture.
- Adults with a disease or condition associated with low bone mass or bone loss.
- Adults taking medications associated with low bone mass or bone loss.
- Anyone being considered for pharmacologic therapy.
- Anyone being treated, to monitor treatment effect.
- Anyone not receiving therapy in whom evidence of bone loss would lead to treatment.
- Women discontinuing estrogen should be considered for bone density testing according to the indications listed above.

Osteoporosis in special situations

Osteoporosis in men

The incidence of hip and spine fractures in men increases with age and is about one third of that in women. Men often have vertebral deformities associated with trauma earlier in life. In men, the increase in hip fractures tends to occur later in life, and a higher proportion of men have definable secondary causes. Screening for osteoporosis in older men who do not have fractures may be justified now that preventive therapy is available.

Juvenile osteoporosis

Juvenile osteoporosis a rare, self-limiting disease begins between the ages of 8 and 14 years with back pain and vertebral compression. If bone turnover is high, antiresorptive or anti inflammatory agents may be beneficial. Spontaneous remission usually occurs, and the disorder usually does not lead to permanent deformity.

Idiopathic osteoporosis

Osteoporosis with no obvious secondary cause in premenopausal women or younger men is called idiopathic osteoporosis. The term is not used consistently and patients so defined include individuals with both high and low bone turnover. Some patients have a transient, self-limited condition, whereas others have a progressive and disabling disease.

Osteoporosis in pregnancy

Osteoporosis in pregnancy is rare and may represent bone disease that has been present before pregnancy. Ordinarily, maternal bone loss is modest, presumably because the high levels of estrogen protect the skeleton and high levels of 1, 25(OH) D stimulate calcium absorption. Bone loss can occur, however, particularly in adolescents. Lactation is associated with transient bone loss but is not a risk factor for osteoporosis.

Localized osteoporosis

Immobilization is the most common cause of localized osteoporosis. Regional migratory osteoporosis can occur without immobilization, particularly in the lower extremities. This phenomenon may be associated with local inflammation or autonomic dysfunction with vasomotor changes and hyperesthesia, a syndrome called reflex sympathetic dystrophy.

Secondary osteoporosis

The division of osteoporosis into primary and secondary forms is somewhat arbitrary. For example, patients with diseases that lead to hypogonadism early in life are considered to have secondary osteoporosis, whereas osteoporosis in women with natural menopause and older men with low sex hormone levels is termed primary.

Although most postmenopausal women and older men do not have a definable secondary cause, the few who do can be treated more effectively.

There are many causes of secondary osteoporosis namely Cushing's syndrome, hyperthyroidism, hypogonadism in males, malabsorption secondary to inflammatory disease, corticosteroids therapy, anticonvulsants, aluminum containing antacids, chronic renal failure and malnutrition.

The perimenopausal transition is a good opportunity to initiate discussion about risk factors or osteoporosis and to consider further evaluation. Universal screening of all postmenopausal women is not possible even in developed countries. Those at risk need screening. A careful history and physical examination should be performed to identify risk factors for osteoporosis. A low Z-score increases the suspicion of a secondary disease. Height loss > 2.5 to 3.8 cm (> 1 to 1.5 in.) is an indication for radiography to rule out asymptomatic vertebral fractures, as is the presence of significant kyphosis or back pain, particularly if it begins after menopause. Conventional radiographs are an insensitive way to assess osteoporosis since bone loss becomes apparent only when mass has decreased by about 30-50%. DXA is ideal; measurements should be made at two sites: femoral neck and lumbar spine.

Biochemical markers of bone turnover

These provide an integrated assessment of global disease activity in contrast to DXA which is regional. Biochemical markers for osteoblastic activity include serum bone-specific alkaline phosphatase, serum osteocalcin and the peptide cleavage products of procollagen synthesis – type I amino terminal propeptide (PINP) and carboxy terminal propeptide of type I procollagen (PICP). Markers for the osteoclastic activity include urinary hydroxyproline, urinary collagen cross links (pyridinoline as well as deoxypyridinoline), urinary cross linked C telopeptide of type I collagen, urinary cross linked N telopeptide of type I collagen, urinary hydroxylysine glycosides, serum bone sialoprotein, serum C-terminal pyridinium cross linked telopeptide domain of Type I collagen (S-ICTP) and, serum tartarate resistant acid phosphatase (TRAP). The commonly used resorption marker is N-telopeptide (NTx), which can be measured in urine and in serum. Biochemical markers do not help in diagnosis of osteoporosis or in prediction of fracture risk. These markers are usually employed to monitor response to treatment. Bone resorption markers decrease relatively faster than bone formation markers. In general, a decrease of 30% after 3 months of treatment with antiresorptives is considered a treatment response. Bone formative agents like teriparatide increase markers of bone formation as well as bone resorption due to physiological coupling of bone formation with resorption. Biochemical assessment has the potential advantage of obviating the need for repeated BMD measurements. A demonstrable fall in these markers may indirectly help to increase patient compliance. An emerging view is that DXA may be used for initial diagnosis and urinary markers for assessing response to therapy. Drawbacks include lack of consensus on which marker(s) to choose and as high as 20% variability in repeat measurements.

Fracture risk assessment

Bone densitometry has several limitations apart from availability, cost, and lack of normative reference values. It does not capture bone quality. Therefore, for a clinician fracture risk assessment assumes greater importance than BMD alone. The WHO in 2008 has come out with an algorithm for 10-year absolute fracture risk called FRAX™[21]. This can give a 10-year probability of fracture risk even if BMD values are not available. Risk factors included in FRAX™ (www.shef.ac.uk/FRAX) are age, sex, glucocorticoid use, secondary osteoporosis, rheumatoid arthritis, family history, prior fragility fracture, low BMI, current smoking, alcohol consumption > 3 units/day and femoral neck BMD. Falls were not included as their risk is unlikely to be modified by a pharmaceutical intervention. The FRAX algorithms are suitable for men and women between 40-90 years. Four FRAX™ assessment models have been constructed: the 10 years probability of hip fracture, with and without BMD at the femoral neck, and the 10 years probability of other osteoporotic fractures, with and without BMD at the femoral neck. Other osteoporotic fractures include forearm, clinical spine and humerus. FRAX models are available for several countries including Singapore. The Singapore model has been developed for 3 ethnic groups: Chinese, Malay and Indians. Based on this, a surrogate model for India is available and can be accessed at http://www.shef.ac.uk/FRAX/tool.aspx?country=51. Treatment decisions, however, have to be individualized based on the fracture risk and are not provided by FRAX™.

FRAX tool risk factors	
Current age	Gender
Low BMI	Alcohol 3 or more drinks /day
Previous fracture	Parental history of hip fracture
Femoral neck BMD	Rheumatoid arthritis
Secondary osteoporosis	Current smoking
Steroid > 5mg/day for > m/ever	

Conclusions

Physician awareness is important for detection of osteoporosis before the first fracture. BMD determination by DXA is currently the gold standard for diagnosis. Recent elucidation of pathophysiologic mechanisms has opened vistas for novel therapies.

Key points

- Osteoporosis is emerging as a major public health problem in India because of an increase in the elderly population
- DXA is the current gold standard for determining BMD
- DXA does not capture bone quality
- The WHO fracture risk assessment tool (FRAX™) aids quantification of fracture risk and can be used even if DXA is not available.

Table 1. Various modalities for measurement of bone density.

Technique	Comments
Conventional skeletal radiography	Insensitive method. Detects loss greater than 30-50% of bone mass.
Photon absorptiometry (single and dual)	Requires radionuclide source of photons. Largely replaced by x-ray absorptiometry.
Digital x-ray radiogrammetry (DXR)	Measures BMD at metacarpals using high resolution radiographs. Precision of the technique has improved with the introduction of computerized analysis of a digitized radiograph of the hand using information from the three middle metacarpals. DXR results correlate with axial and appendicular DXA results. Useful in risk assessment in areas where geography and population size restrict access to DXA. Advantages: low cost, operator independent.
Single energy x-ray absorptiometry	Useful only for peripheral sites, like forearm or heel. Cannot be used for axial skeleton.
Dual energy x-ray absorptiometry (DXA)	Current gold standard. Useful both for axial and appendicular skeleton. Can measure total body mineral content. Drawback: It is a two-dimensional measurement which measures density/area (in grams per square cm). Areal BMD is influenced by bone size. Overestimates fracture risk in individuals with small body frame whose areal BMD is lower. In obese individuals superimposed soft tissue results in elevated BMD readings.
Quantitative computed tomography	Drawbacks: cost and high radiation dose (0.06-2.9 mSv) against 1-50 μSv of DXA. May be used in very small or large patients or in older individuals with advanced degenerative changes like DISH (diffuse idiopathic skeletal hyperostosis).
Ultrasonography	Screening procedure used for cancellous bone in heel and phalanges. Advantages: low cost, no radiation, portable.
Magnetic Resonance Imaging	Provides exceptional resolution of internal structure of cancellous bone. Provides no direct information on BMD. At present a research procedure.

Further Reding :

1. Hiligsmann M, Kanis JA, Compston J, et al. Health Technology Assessment in Osteoporosis. Cacif Tissue Int 2013. Advance publication Mar 21.
2. World Health Organisation. Assessment of fracture risk and its application to screening for postmenopausal osteoporosis: report of a WHO Study Group. WHO Technical Report Series 843; 1994: 1-29.
3. Baim S, Binkley N, Bilezikian JP, et al. Official positions of the International Society for Clinical Densitometry and executive summary of the 2007 ISCD Position Development Conference. J Clin Densitom 2008; 11: 75-91.
4. Lewiecki EM, Gordon CM, Baim S, et al. Special report on the 2007 adult and pediatric Position Development Conferences of the International Society for Clinical Densitometry. Osteoporos Int 2008; 19: 1369-78.
5. Tandon N, Marwaha RK, Kalra S, Gupta N, Dudha A, Kochupillai N. Bone mineral parameters in healthy young Indian adults with optimal vitamin D availability. Natl Med J India 2003; 16: 298-302.
6. Dennison E, Cooper C. Epidemiology of osteoporotic fractures. Horm Res 2000; 54 (Suppl 1): 58-63.
7. Kanis JA, Johnell O. Requirements for DXA for the management of osteoporosis in Europe. Osteoporos Int 2005; 3: 229-38. Epub 2004 Dec 24.
8. Sambrook P, Cooper C. Osteoporosis. Lancet 2006; 367: 2010-18.
9. Dey AB, Handa R. Action Plan Osteoporosis. Consensus statement of an expert group. Osteoporosis Society of India, New Delhi, 2003.
10. Handa R. Management of Osteoporosis: The Indian perspective. Clinical Calcium 2004; 14: 100-05.
11. Nordin BEC. International patterns of osteoporosis. Clin Orthop 1966; 45: 17-30.
12. Gupta AK, Samuel KC, Kurian PM, Rallan RC. Preliminary study of the incidence and etiology of femoral neck fractures in Indians. Indian J Med Res 1967; 55: 1341-48.
13. Vaishnava H, Rizvi SNA. Frequency of osteomalacia and osteoporosis in fractures of proximal femur. Lancet 1974; 1: 676-77.
14. Sankaran B. Clinical studies: Incidence of fracture neck of femur and inter-trochanteric fractures in three Delhi hospitals. In: Osteoporosis. Sankaran B. 2000. 9-18. New Delhi.
15. Gupta A. Osteoporosis in India - The nutritional hypothesis. Natl Med J India 1996; 9: 268-74.
16. Boyle WJ, Scott Simonet W, Lacey DL. Osteoclast differentiation and activation. Nature 2003; 423: 337-42.
17. May P, Woldt E, Matz RL, Boucher P. The LDL receptor-related protein (LRP) family: an old family of proteins with new physiological functions. Ann Med. 2007; 39: 219-28.
18. Diarra D, Stolina M, Polzer K et al. Dickkopf-1 is a master regulator of joint remodeling. Nat Med 2007; 13: 156-63.
19. Goldring SR, Goldring MB. Eating bone or adding it: the Wnt pathway decides. Nat Med 2007; 13: 133-34.
20. Link TM. Osteoporosis Imaging: State of the Art and Advanced Imaging. Radiology 2012; 263: 3-17.
21. McCloskey E, Kanis JA. FRAX updates 2012. Curr Opin Rheumatol 2012; 24: 554-60.

PRINCIPLES IN THE PREVENTION AND TREATMENT OF OSTEOPOROSIS

2

V Rajamani

Osteoporosis is currently an important public health issue due to its widespread prevalence and high socio-economic impact. Bone undergoes a continuous remodelling process mediated by osteoblasts and osteoclasts. In postmenopausal osteoporosis a negative bone balance occurs as a result of bone resorption taking place at a faster rate than bone formation. Several guidelines provide recommendations for prevention and treatment of osteoporosis.

Goals of therapy

- Prevention of fractures
- Maximization of peak bone mass
- Prevention of age-related and secondary causes of bone loss
- Preservation of the structural integrity of the skeleton
- Improvement in quality of life

USE OF NON-PHARMACOLOGIC THERAPY

Lifestyle modification

Exercise, reducing caffeine and alcohol intake and quitting smoking are all lifestyle modifications that can optimize peak bone mass, minimize bone loss and ultimately prevent fractures. These measures are recommended to people of all ages.

Physical activity

Exercise is one of the most important approaches to preventing fractures in people at risk of osteoporosis[18]. Patients with osteoporosis need to maintain as vigorous an exercise program as is compatible with their bone fragility status. Exercise improves bone strength, muscular performance, joint flexibility and balance to decrease the risk of falls and fractures. It produces a small increase in bone density (0.5-3%) and decreases the frequency of hip fractures in the elder population. Exercise such as walking, jogging and stair climbing improve bone density in healthy sedentary postmenopausal women. Swimming is not considered a weight-bearing exercise. Exercise programs that include weight bearing activity three or four times per week for 45 minutes or weight lifting two or three times per week for 20-30 minutes are beneficial. Bicycling, water aerobics and chair exercises may help maintain bone density in patients with restricted activity.

Cigarette smoking

Tobacco use is directly harmful to bone and is an independent risk factor for osteoporosis. All smokers should be encouraged to quit. Smoking is associated with decreased sex hormone concentrations, lower body weight, early menopause, increased bone turnover markers, decreased calcium absorption and reduced bone mineral density.

Alcohol and caffeine

Excessive alcohol should be discouraged. Excessive caffeine consumption may be associated with increased calcium excretion, higher rates of bone loss and modest increases in fracture risk[1]. Alcoholics are more susceptible to osteoporosis because of poor nutrition, impaired calcium and vitamin D metabolism and increased risk of falls.

Fall prevention

Personal and home safety precautions should be reviewed with all patients at risk of falls. Medications that are associated with an increased risk of falls such as hypnotics, anxiolytics and antidepressants etc. should be avoided. Devices such as external hip protectors that reduce hip fractures should be comfortable and cosmetically acceptable. Personal and home safety precautions should be reviewed with all patients at risk of falls.

Protein intake

Osteoporosis in many fragile elderly individuals having low protein intake will probably not respond well to pharmacotherapy until their nutritional status is repaired. This is seen in patients with hip fracture, whose outcomes have been shown to improve dramatically with protein supplementation.

Treatment should be initiated in

- patients with hip or vertebral fractures
- patients with Bone Mineral Density (BMD) T score of equal to or less than - 2.5
- postmenopausal women and men aged 50 years and older with osteopenia and an elevated fracture risk as calculated by FRAX

While T-score remains the standard for diagnosing osteoporosis, BMD misses other risk factors which are necessary to properly evaluate the fracture risk of patients. The use of additional factors in fracture risk assessment is also advantageous in cases where BMD cannot be determined and in helping to decide the necessity of BMD assessments. FRAX is a diagnostic tool used to evaluate the 10 years probability of major osteoporotic fracture (hip, spine, humerus or wrist) and the 10 years probability of hip fracture for a patient.

- 10% or less considered as low risk
- 10-20% as medium risk and
- more than 20% or a T score -2.5 or history of a fragility fracture as high risk

The National Osteoporosis Foundation incorporated the 10 years absolute probability of fracture into their guidelines for the treatment and prevention of osteoporosis. With FRAX patients who are at high risk of fracture but would not be candidates for preventative therapy using traditional T-score assessment can be identified.

Management following assessment:

- Low risk - reassure and reassess in 5 years or less depending on the clinical context.
- Intermediate risk - measure BMD and recalculate the fracture risk to determine whether an individual's risk lies above or below the intervention threshold.
- High risk - can be considered for treatment without the need for BMD, although BMD measurement may sometimes be appropriate, particularly in younger postmenopausal women.

FRAX tool risk factors	
Current age	Gender
Low BMI	Alcohol 3 or more drinks/day
Previous fracture	Parental history of hip fracture
Femoral neck BMD	Rheumatoid arthritis
Secondary osteoporosis	Current smoking
Steroid > 5mg/day for > m/ever	

Current treatment options include bisphosphonates, calcitonin, parathyroid hormone, raloxifene, strontium and denosumab. Alendronate, risedronate, zoledronic acid and teriparatide reduce the risk of vertebral, hip and other nonvertebral fractures. Ibandronate, raloxifene do not affect the risk of nonvertebral fractures.

Bisphosphonates

Bisphosphonate drugs act to inhibit bone resorption.(Table 1&2) Bisphosphonates have a half-life >10 years. Nitrogen-containing bisphosphonates bind to hydroxyapatite and block the enzyme farnesyl pyrophosphate synthase in bone tissue and inhibit osteoclast-mediated bone resorption. Alendronate, risedronate, ibandronate and zoledronic acid are approved for the treatment of osteoporosis.

Alendronate: Orally 70 mg once a week is effective for both the treatment and prevention of osteoporosis in postmenopausal women and for men with osteoporosis. It reduces the risk of vertebral fracture by approximately 50 percent, hip and wrist fractures by approximately 30 percent. It is well tolerated and effective for at least 10 years. Stopping this after 5 years may be reasonable, as there appears to be a residual benefit on BMD and fractures for up to 5 years. Stopping bisphosphonate in low-risk women (eg, no history of fracture and T-scores better than -2.5) as long as they are followed carefully by BMD and assessment of risk factors. For women at highest risk for fracture (T-score below -3.5), continue up to 10 years, as BMD and fracture benefits are maintained with no increased risk of adverse events[1,2].

Risedronate: Orally 35 mg once a week. The risk of vertebral and nonvertebral fractures was reduced by 41% and 39%, respectively. When discontinued the beneficial effects on BMD and markers of bone turnover appear to revert partially or completely within one year. The risk of upper GI side effects appears to be low[3].

Ibandronate: Ibandronate sodium is approved for the prevention and treatment of postmenopausal osteoporosis. Ibandronate reduces the incidence of vertebral fractures by about 50% over three years. A once-monthly 150 mg oral formulation is available[4].

IV Bisphosphonates: Are alternatives for patients who cannot tolerate oral bisphosphonates.

Zoledronic acid:[5] (5 mg) administered yearly must be infused over 15 minutes. The beneficial effects of a three year course are maintained in the subsequent three years. In patients at high risk for fractures (eg, existing vertebral fracture or femoral neck BMD T-score <-2.5 after an initial course of therapy), continuing beyond three years must be considered in light of the potential risks of long-term therapy. Not recommended if e GFR <35ml. Zoledronic acid is associated with myalgia, fever, and headache. Serum creatinine should be assessed before and after use due to reports of renal failure after treatment. It is important to adequately hydrate before the infusion. Ibandronate 3 mgs once every three months given as a 30 second intravenous injection.

Side effects

Renal: All bisphosphonates are not recommended if e GFR <35ml.

Gastrointestinal: reflux, pill esophagitis, oesophageal ulcers and rarely oesophageal stricture. Use of a nonsteroidal anti-inflammatory drug increases the risk. Bisphosphonates should not be used in patients with Barrett's oesophagus[6].

Musculoskeletal: Some patients have severe musculoskeletal pain. This is rare and may sometimes not resolve completely with stopping the bisphosphonate.

Ocular: Rarely pain, blurred vision, conjunctivitis, uveitis and scleritis.

Osteonecrosis of the jaw: ONJ has been reported with all four bisphosphonates but rare. The incidence is less than 1 per 1,000 with an oral bisphosphonate and four times higher in individuals with risk factors receiving a parenteral bisphosphonate[7]. Risk factors include dental extractions, dental implants, poorly fitting dentures, steroids, smoking, pre-existing dental disease, cancer and anti-cancer therapy, oral bone manipulating surgery and duration of exposure to bisphosphonate treatment.

Advise: Good oral hygiene and regular dental visits. Some recommend discontinuing oral bisphosphonates for three months prior to performing the dental surgery if patient has been on it for 3 years. Restart when the bone has healed. There is no data to suggest that this would lower ONJ risk. Bisphosphonates stay in bone for years. There will be no adverse consequences on BMD or fracture risk.

Atrial fibrillation: The risk of AF from oral bisphosphonates is small. Care should be taken when giving IV bisphosphonates for patients with underlying heart disease and/or a history of AF[8].

Atypical femur fractures

Bisphosphonates suppress osteoclast activity. Prolonged therapy can lead to over suppression of bone turnover and increased skeletal fragility. While the ultimate strength of the bone improves, its toughness decreases by as much as 20% due to the accumulation of micro damage and lack of effective remodelling within the bone. The decrease in toughness may lead to failure in areas with high tensile forces, such as the sub trochanteric and diaphyseal regions of the femur. Cases of atypical fracture (particularly sub trochanteric) seen in younger women treated for several years. Some data suggest that long-term use >5 years of bisphosphonates increases the risk of atypical fractures, but the absolute risk is low. A large dual-centre review has established the incidence of atypical femoral fractures at 7% of the study population, 81% of whom had been on bisphosphonate treatment for a mean of 4.6 years. Hence awareness of this possibility is necessary so that symptomatic patients (mid-thigh pain) can be appropriately investigated. X Rays show cortical thickening and scalloping of the lateral cortex in the diaphyseal region of the femur prior to fracture. MRI shows bone oedema in this area. Prophylactic fixation of insufficiency fractures recommended in patients with a history of long-term bisphosphonate use, persistent thigh pain and radiographic changes. Although drug holidays have been recommended, they have not been proven to prevent these fractures[9].

Flu-like symptoms: low grade fever, myalgia and arthralgia can occur within 24 to 72 hours of intravenous infusion. Paracetomol relieves the symptoms.

Hypocalcaemia may be transient and asymptomatic. This can be severe in vitamin D deficient individuals. Vitamin D deficiency should be treated prior to bisphosphonate infusions.

Teriparatide (Recombinant PTH 1-34)[10,11]

1-34 Teriparatide produced using recombinant DNA technology retains all of the biologic activity of the intact parathyroid hormone peptide. It is an anabolic therapy that results in significant increases in bone mineral density and a reduction in fracture risk. It is recommended for the treatment of severe osteoporosis in both men and women. Sustained elevation of parathyroid hormone leads to bone resorption. Intermittent administration of recombinant human PTH has been shown to stimulate bone formation more than resorption. PTH 1-34 stimulates preosteoblasts to mature into bone-forming osteoblasts. Bone formation begins within the first month of PTH treatment and peaks six to nine months after initiation. BMD gains occur in the first few months, although antifracture efficacy is evident only after six months of treatment. BMD changes with PTH begin to level off after 18 months. Increases in bone formation occur more quickly than bone resorption, leading to marked increase in lumbar BMD whereas the radius shows either no change or a decrease in areal BMD within the first year of treatment. The actions of PTH on skeletal compartments are site-specific. In addition to improvements in BMD, PTH reduces the risk of spine fractures in postmenopausal women with osteoporosis.

PTH administration decreased vertebral fracture rates by 65 to 75%. Relative percent change in BMD per individual may vary because of significant heterogeneity in skeletal responsiveness. A study reported a 53% relative risk reduction in nonvertebral fragility fractures. The percentage of vertebral fracture reduction that could be attributed to the improvement in spine BMD was only 30 to 40 percent. The non-BMD determinants of bone strength could include changes in bone geometry, microarchitecture, remodelling rate and/or collagen/mineral matrix properties. Teriparatide does not significantly accelerate fracture repair. There is no additional benefit of combining PTH with bisphosphonates. This reduces the anabolic effect of teriparatide in both women and men. PTH is a treatment option in patients who cannot tolerate or who fail bisphosphonate therapy. Teriparatide increases BMD in women previously treated with bisphosphonates, although the improvement may be less than in women not previously exposed to bisphosphonates. After PTH treatment is discontinued, an antiresorptive preferably a bisphosphonate, should be used to preserve or increase gains in BMD acquired with PTH.

Dose: 20 mcg/day administered subcutaneously as daily injections.

Side effects: orthostatic hypotension, nausea, myalgia and arthralgia, headache and hypercalcemia. Serum uric acid increases with teriparatide and may precipitate an attack of gout in some. Patients with uric acid levels greater than 7.5 mg/dl probably should not be treated with PTH until their uric acid is controlled. Teriparatide is not recommended for use beyond 2 years due to lack of safety and efficacy data. Osteosarcoma was observed in rats exposed to teriparatide at much higher doses.

Denosumab

Denosumab is a humanized monoclonal antibody against RANK. Osteoblasts express RANKL (receptor activator of nuclear factor kappa B ligand) a member of the TNF superfamily of ligands and receptors. Osteoclast precursors express RANK. Cell contact induces osteoclast development. RANKL is blocked by osteoprotegerin, produced by osteoblasts. Denosumab reduces bone resorption by inhibiting the formation, function and survival of osteoclasts. It improves bone mineral density of the lumbar spine and hips and reduces the incidence of new vertebral, hip and non-vertebral fractures in postmenopausal women. It reduces the incidence of vertebral fractures by about 68%, hip fractures by about 40% and non-vertebral fractures by about 20% over 3 years. Denosumab is recommended as a treatment option for the primary prevention in postmenopausal women at increased risk of fractures and for the secondary prevention of osteoporotic fragility fractures in postmenopausal women at increased risk of fractures who are unable to comply with, have an intolerance or contraindication to bisphosphonates. The reduction in vertebral fracture noted with denosumab is similar to the reductions reported for teriparatide and intravenous zoledronic acid and greater than that reported for alendronate. It is also indicated for treatment of osteoporosis in men at high risk of fracture, prevention of bone loss in women with breast cancer treated with aromatase inhibitors, men with prostate cancer on gonadotropin releasing hormone agonists. It is effective in postmenopausal women who previously were treated with bisphosphonates. When denosumab treatment is stopped, bone loss can be rapid and alternative agents should be considered to maintain BMD. It is associated with significantly greater reductions in biochemical markers of bone turnover than alendronate.

Adverse effects: musculoskeletal pain, hypercholesterolemia, cystitis, eczema, cellulitis and flatulence. In some trials there were a greater number of infections requiring hospitalization (eg, diverticulitis, pneumonia, atypical pneumonia, appendicitis, cellulitis and labyrinthitis). Pancreatitis has also been reported. Hypocalcaemia must be corrected before starting denosumab. Denosumab has been associated with the development of ONJ (two cases) and atypical femur fractures. The overall rates of adverse events of neoplasm were similar between treatment groups in trials.

Dose: 60mg in a 1 mL solution subcutaneous injection every six months[12].

Selective oestrogen receptor modulators

Selective Oestrogen Receptor Modulators reproduce the beneficial effects of oestrogens on the skeletal system, without the negative effects on the breast and endometrium due to their selective binding to ER. Raloxifene is a selective oestrogen receptor modulator (SERM). It is used for the prevention and treatment of osteoporosis. It increases bone mineral density and reduces the risk of vertebral fractures. It antagonizes oestrogen receptors in some tissues and agonizes oestrogen receptors at other tissues, such as bone tissue. Raloxifene is administered as a once-daily oral formulation. Side effects include hot flashes and increase in thromboembolic events. Raloxifene reduces the risk of vertebral fractures by about 30% in patients with a prior vertebral fracture and by about 55% in patients without a prior vertebral fracture over 3 years. Raloxifene reduces breast cancer risk. Raloxifene is an alternative for women who are unable to tolerate oral or intravenous bisphosphonates[13].

Dose: 60 mg/day taken at any time of the day with calcium and vitamin D3.

Strontium ranelate

Strontium ranelate has anabolic and antiresorptive properties and has been shown to reduce vertebral fracture, and to a lesser extent, non-vertebral fracture, but the mechanism is unclear[17]. Incorporation of strontium into the crystal structure replacing calcium may be part of its mechanism of effect. A substantial proportion of the increase in BMD during strontium therapy is due to the physical effect of strontium accumulation in bone tissue therefore, is not indicative of greater fracture risk reduction. Strontium ranelate and raloxifene are recommended as alternative treatment options for the primary and secondary prevention of osteoporotic fragility fractures in postmenopausal women: who are unable to comply with or have a contraindication to bisphosphonates or who also have a combination of T-score -3.5 or below on average, age and number of independent clinical risk factors for fracture.

Side effects: Diarrhoea, venous thromboembolism and pulmonary embolism, severe skin reactions.

Calcitonin[14]

Calcitonin-salmon an antiresorptive, affects BMD by inhibiting the activity of osteoclasts. Calcitonin has positive effects on spinal BMD, but has not been shown to significantly affect hip BMD. It has a modest effect on BMD and weak antifracture efficacy compared with bisphosphonates and parathyroid hormone. In addition it has been shown to alleviate pain from vertebral osteoporotic compression fractures. The mechanism of calcitonin-produced analgesia remains unclear, but it has been proposed to include beta endorphins and prostaglandins. It is available as nasal spray for the treatment of osteoporosis. It should be administered daily in alternating nostrils. Common side effects are local, and include nasal irritation and occasionally epistaxis. There are recent recommendations that marketing of calcitonin salmon for the treatment of osteoporosis in women greater than 5 years after menopause be stopped citing concerns about a possible cancer risk, mainly basal cell carcinomas of the skin. The risk was low and needed further investigation. The drug could benefit a subset of patients.

Calcium

Lifelong adequate calcium intake is necessary for the acquisition of peak bone mass and subsequent maintenance of bone health. Women older than 50 years should consume at least 1,200 mg/day of elemental calcium. Intakes in excess of 1,200 to 1,500 mg/day have limited potential for benefit and may increase the risk of developing kidney stones or cardiovascular disease. Increasing dietary calcium intake usually decreases urinary oxalate excretion. High calcium intake within limits typically reduces stone risk. Calcium obtained from dietary sources appears to be better than calcium from supplements in regard to lowering stone risk. Calcium-rich foods include milk, (daily intake of three glasses is an effective supplement), cheese, yogurt, and broccoli, green leafy vegetables, kale, dried fruit, nuts and pulses (peas, beans and lentils). Maximal absorption of calcium occurs at 500 to 600 mg/dose. Calcium supplementation should be divided throughout the day, usually twice daily, to achieve maximal absorption. In general, calcium carbonate (taken with food) provides 40% elemental calcium and calcium citrate (empty stomach) 21% of elemental calcium. Common side effects are constipation, bloating and flatulence. Calcium supplements can interfere with absorption of iron, tetracyclines and fluoroquinolones and bisphosphonates. They should not be taken together.

Vitamin D

Vitamin D deficiency is a common problem in India due to several factors including high fibre diet, genetic factors, inadequate sun exposure, pollution hampering the ultraviolet rays to adequately synthesize vitamin D in the skin and cultural and traditional habits. Given the prevalence of vitamin D inadequacy in the population, failure to ensure calcium supplementation in bisphosphonate-treated patients could expose them to risk of potentially serious hypocalcaemia. If deficiency is present, correct with weekly dose of 60,000 IU for 8 weeks. One must be careful not to over treat D deficiency without actually monitoring levels. The current recommendations of 2000 IU of vitamin D per day in the diet should be adhered to avoid deficiency in the Indian population.

Response to therapy

Serial BMD that is stable or improving is evidence for a treatment response. If BMD decreases on treatment, look for compliance to therapy, inadequate gastrointestinal absorption, inadequate intake of calcium and vitamin D and evaluate for secondary causes of bone loss. When the patient is taking the drug correctly and the change in BMD is small, continue the same treatment and repeat BMD one year later. If the lack of response is due to intolerance or poor absorption, change to intravenous zoledronic acid. If there is a decline in the BMD >5 percent change to IV bisphosphonate. In patients with BMD T score < -3.5 and in those who continue to fracture on oral bisphosphonates change to teriparatide. There is currently no consensus on how long to continue bisphosphonate therapy. It is suggested in those with a stable BMD, no previous fractures and low risk for fracture, stopping alendronate or risedronate after five years and zoledronic acid after three years. The drug should be resumed if BMD decreases on at least two consecutive DXA. After five years those at high risk of fracture therapy can be continued.

Osteoporosis in men[16]

In males, testosterone is aromatized to estradiol. Both free and bioavailable plasma levels of testosterone and estradiol decrease with age in males and fracture risk is associated with low estradiol levels. Testosterone may increase muscle mass and prevent fractures related to falls. Younger hypogonadal males should be treated with testosterone to attain peak bone mass and increase BMD. Lifestyle-related risk factors (smoking, alcohol consumption, lack of physical activity and low body weight) should be addressed. The antifracture efficacy of antiresorptive and anabolic treatment for osteoporosis has not been documented in larger studies. Bisphosphonates (alendronate, risedronate, ibandronate and zoledronic acid), nasal calcitonin, denosumab and teriparatide appear to have similar effects in males and females. Trials using PTH in men show that teriparatide can increase spine and hip BMD. There are no trials for fracture outcome in men, but teriparatide is approved for the treatment of this condition in men. The antiresorptive drugs should be used in males with (a) BMD T-score less than -2.5 and one or more risk factors, (b) hip and vertebral fractures, (c) glucocorticoid-induced osteoporosis, (d) patients receiving androgen deprivation therapy. Anabolic treatment should be used in more severe spinal fracture cases, including glucocorticoid-induced osteoporosis. Men should take 1000-1200mg calcium daily ideally from dietary sources, supplemented if necessary. Men with vitamin D levels <30ng/ml should receive supplementation. Men with osteoporosis should engage in weight-bearing exercises 3-4 sessions per week[19].

Steroid induced osteoporosis[15]

A rapid decline in BMD begins within the first 3 months steroid use and peaks at 6 months, followed by a slower, steady loss with continued use. Bone density alone may not be the sole reliable diagnostic approach for some patients receiving steroids, because fractures may occur independently of a decline in bone mass.

Limitations to FRAX

Clinical risk factors contributing to FRAX do not take into account dose response. The risk associated with steroid use is dose related. Those receiving higher doses are likely to have a greater absolute fracture risk than estimated. FRAX uses only the bone density value for the hip. Patients receiving steroids frequently lose bone mass in the spine before the hip, leading to a possible underestimation of fracture risk. FRAX alone cannot replace clinical judgment in risk stratification.

Glucocorticoid induced osteoporosis should be treated promptly and effectively[20].All patients should receive vit D and calcium adequately. Initial treatment would be with bisphosphonates. Teriparatide is considered in the presence of risk factors or low BMD.

Postmenopausal women and men age > 50 years

If starting steroids for > 3 months, or on steroids for at least 3 months.

Low Risk

Steroid <7.5mg/day: no pharmacologic treatment recommended.

Steroid 7.5mg/day: alendronate, risedronate or zoledronic acid.

Medium Risk

Alendronate or Risedronate for any dose of glucocorticoids.

Zoledronic acid for < 7.5 mg/day prednisone.

High Risk

Steroid < 5 mg/day for <1 month: alendronate, risedronate or zoledronic acid.

Steroid any dose of glucocorticoids used for > 1 month: alendronate, risedronate, zoledronic acid or teriparatide.

The guidelines for Premenopausal women and men age <50 years are different.

Counselling

Smoking avoidance, limiting excessive alcohol intake, weight-bearing activities, calcium and vitamin D intake and supplementation, fall risk assessment. Consideration for vertebral fracture assessment or radiographic imaging of the spine and obtaining baseline and follow up BMD measurement.

Newer Drugs

Odanacatib is being evaluated as an oral weekly formulation in both men and women with osteoporosis. Odanacatib blocks cathepsin K, the major enzyme in osteoclasts that is responsible for breakdown of existing bone tissue. It is part of a new generation of designer drugs where the function of osteoclast is disrupted by targeting this enzyme that plays a vital role in bone resorption. That it is able to do so without killing the osteoclast or curtailing its formation is a point of difference between odanacatib and other antiresorptive currently in use, such as the bisphosphonates and denosumab. Odanacatib, taken weekly, is considered more convenient than the bisphosphonate. Odanacatib has shown potential in previous trials to improve BMD at the spine and hip among postmenopausal women. This drug is likely to gain a place in osteoporosis treatment. Results of the trials are awaited.

Table 1. Effect of Antiresorptive Agents on Spine BMD.

Studies		Spine BMD (%) over placebo
Raloxifene-Extended MORE	n=7705 (4-year)	2.6
Alendronate-FIT1	n=2027 (3-year)	6.2
Alendronate-FIT2	n=4432 (4-year)	6.8
Risedronate-VERT-NA	n=2458 (3-year)	6.5
Risedronate-VERT-MN	n=1226 (3-year)	5.9
Ibandronate-BONE	n=2946 (3-year)	4.2
Teriparatide-FPT	n=1637 (21-month)	9
BMD increase cannot totally explain the observed clinical anti-fracture efficacy		

Table 2. Anti-fracture efficacy of antiresorptive agents.

	% risk reduction in vertebral fracture (1-year)	% risk reduction in vertebral fracture (3-year)	% risk reduction in non-vertebral fracture (3-year)
Raloxifene	50	30	NS
Alendronate	-	47	20
Risedronate	63	45	40
Ibandronate	68	52	69
Teriparatide	65 (21-month)	54 (21-month)	
NOT direct comparison study			

Further Reading :

1. Black DM, Thompson DE, Bauer DC, et al. Fracture risk reduction with alendronate in women with osteoporosis: the Fracture Intervention Trial. J Clin Endocrinol Metab. 2000; 85: 4118-24.
2. Black DM, Schwartz AV, Ensrud KE, et al. Effects of continuing or stopping alendronate after 5 years of treatment: the Fracture Intervention Trial Long-term Extension (FLEX): a randomized trial. JAMA. 2006; 296: 2927-38.
3. Harris ST, Watts NB, Genant HK, et al. Effects of risedronate treatment on vertebral and nonvertebral fractures in women with postmenopausal osteoporosis: a randomized controlled trial. Vertebral Efficacy With Risedronate Therapy (VERT) Study Group. JAMA 1999; 282: 1344.
4. Chesnut CH, Ettinger MP, Miller PD, et al. Ibandronate produces significant, similar antifracture efficacy in North American and European women: new clinical findings from BONE. Curr Med Res Opin 2005; 21: 391.
5. Black DM, Delmas PD, Eastell R, et al. Once-yearly zoledronic acid for treatment of postmenopausal osteoporosis. N Engl J Med. 2007; 356: 1809-22 HORIZON.
6. Oral osteoporosis drugs (bisphosphonates): drug safety communication-potential increased risk of esophageal cancer. FDA. July 21, 2011.
7. Khosla S, Burr D, Cauley J, et al. Bisphosphonate-associated osteonecrosis of the jaw: report of a task force of the American Society for Bone and Mineral Research. J Bone Miner Res 2007; 22: 1479.
8. Bunch TJ, Anderson JL, May HT, et al. Relation of bisphosphonate therapies and risk of developing atrial fibrillation. Am J Cardiol 2009; 103: 824.
9. Meier RP, Perneger TV, Stern R, et al. Increasing Occurrence of Atypical Femoral Fractures Associated With Bisphosphonate Use Atypical Femoral Fractures and Bisphosphonate Use. Arch Intern Med 2012; 172: 930.
10. Effect of parathyroid hormone (1-34) on fractures and bone mineral density in postmenopausal women with osteoporosis. N Engl J Med. 2001 May 10; 344(19): 1434-41.
11. Therapy of osteoporosis in men with teriparatide J Osteoporos. 2011; 2011: 463675. doi: 10.4061/2011/463675. Epub 2011 Oct 25.
12. Denosumab for prevention of fractures in postmenopausal women with osteoporosis. N Engl J Med 2009; 361(8): 756-65. Freedom trial Cummings SR, San Martin J, McClung MR, Siris ES, Eastell R, Reid IR, et al.
13. Severity of prevalent vertebral fractures and the risk of subsequent nonvertebral fractures.

    Results from the MORE trial. Delmas PD, Genant HK, Crans GG, Stock JL.
14. Prospective, randomized trial of nasal spray calcitonin in postmenopausal women with established osteoporosis: Chestnut CH, Silverman SL, Andriano K, Genant H, Gimona A, Harris A, et al, The PROOF Study. Am J Med 2000; 109: 267-76.
15. ACR 2010. Recommendations for the Prevention and Treatment of Glucocorticoid-Induced Osteoporosis Arthritis Care & Research Vol. 62, No. 11, November 2010, 1515-26. DOI 10.1002/acr.20295.
16. Kurland ES, Cosman F, McMahon DJ, et al. Parathyroid hormone as a therapy for idiopathic osteoporosis in men: effects on bone mineral density and bone markers. J Clin Endocrinol Metab 2000; 85: 3069.
17. Reginster JY, Seeman E, De Vernejoul MC, et al. Strontium ranelate reduces the risk of nonvertebral fractures in postmenopausal women with osteoporosis: Treatment of Peripheral Osteoporosis (TROPOS) study. J Clin Endocrinol Metab 2005; 90: 2816.
18. Exercise for preventing and treating osteoporosis in postmenopausal women. Cochrane Database Syst Rev 2011; CD000333. Howe TE, Shea B, Dawson LJ, et al.
19. Watts NB, Adler RA, Bilezikian JP, et al. Osteoporosis in Men: An Endocrine Society Clinical Practice Guideline. J Clin Endocrinol Metab 2012; 97: 1802-22.
20. V Krishnamurthy, Aman Sharma, Amita Aggarwal, Umakumar, Sanjiv Amin, U R K Rao et al. Indian rheumatology association guidelines for management of glucocorticoid induced osteoporosis. Indian J of Rheumat 2011; 6: 68-75.

SECTION 12

Rheumatic Diseases of Childhood

CLASSIFICATION AND APPROACH TO JUVENILE IDIOPATHIC ARTHRITIS

1

Amita Aggarwal

Juvenile idiopathic arthritis (JIA) is the commonest cause of chronic arthritis in childhood. Its prevalence varies from 1 to 4 per thousand children among Caucasians. Its incidence is around 9-14 per 100,000 children. As the name suggests it is characterized by onset of arthritis before 16th birthday, which lasts for at least 6 weeks and has no known cause. Since it includes different forms of idiopathic arthritis it has varied manifestations with variable outcome in different subsets.

Classification

Classification criteria for juvenile arthritis have undergone series of changes to facilitate uniformity across the world. In the beginning we had two separate criteria one being followed in North America and Canada called the American College of Rheumatology (ACR) criteria[1] and another being used in Europe the so called EULAR criteria[2].

In ACR criteria (Table 1) juvenile arthritis is called juvenile rheumatoid arthritis (JRA). The duration of 6 weeks helps to exclude transient, post-traumatic and other causes of arthritis. The onset type is based on the disease manifestations and number of joints involved in the first 6 months of disease.

Table 1. ACR criteria

Arthritis in one or more joints Duration of disease ≥ 6 weeks Onset before 16 years of age Exclusion of all known causes
Onset type: defined on basis of type of disease in the first 6 months Systemic onset (Presence of systemic features) Oligoarticular disease (< or = 4 joints involved) Polyarticular disease (> 4 joints involved)
Modified from reference [1]

The EULAR criteria also includes spondylo-arthropathy (juvenile ankylosing spondylitis, psoriatic arthritis, arthritis associated with inflammatory bowel disease) and requires the arthritis to be present for at least 12 weeks before inclusion. They did not define any exclusions and termed juvenile arthritis as juvenile chronic arthritis (JCA). In EULAR classification Rheumatoid factor positive polyarticular disease was called juvenile rheumatoid arthritis (JRA) whereas ACR criteria had used JRA term for all patients of juvenile arthritis. This difference in nomenclature led to some confusion thus the pediatric standing committee of International league of association of Rheumatology Societies (ILAR) held discussion among a group of rheumatologists across the world in Durban in 1997 and proposed a new classification for juvenile arthritis. This classification is called the ILAR classification[3] and attempts to group clinically distinguishable diseases as distinct entities rather than put them all under an umbrella term. They tried to define homogenous, mutually exclusive groups with clear cut inclusion and exclusion criteria. It also takes into account the extra-articular manifestation of disease and removes the discrepancy between JRA/JCA of 6 or 12 weeks. The basic aim was to have categories that would facilitate research and help in predicting outcome. A further revision of ILAR criteria was proposed after 4 years in Edmonton which included deletion of probable systemic onset disease as well as addition of boys > 6 years as one criteria in ERA[4] (Table 2).

Table 2. ILAR classification

Disease	Criteria	Exclusions*
Systemic arthritis	Quotidian fever for at least 2 weeks Evanescent, non fixed erythematous rash Arthritis In the absence of arthritis, 1 and 2 (above) plus any two of: Generalized lymphadenopathy Hepatosplenomegaly Serositis	1, 2, 3, 4
Polyarthritis RF-	Arthritis of > 5 joints during the first 6 months of disease	1, 2, 3, 4, 5
Polyarthritis RF+	Arthritis of > 5 joints during the first 6 months of disease Positive RF on at least two occasions 3 months apart	1, 2, 3, 5
Oligoarthritis Extended oligoarthritis	Arthritis of 1-4 joints during the first 6 months of disease Affecting a total of more than 4 joints after the first 6 months of disease	1, 2, 3, 4, 5
Enthesitis related arthritis	Arthritis and enthesitis or, in the absence of enthesitis, arthritis and at least 2 of: Sacroiliac joint tenderness Inflammatory spinal pain HLA-B27, onset of arthritis in a male over 6 years of age Family history of at least one of the following: i) Anterior uveitis with pain, redness or photophobia, ii) Spondyloarthropathy confirmed by a rheumatologist iii) Inflammatory bowel disease Anterior uveitis with pain, redness, or photophobia	1, 4, 5
Psoriatic arthritis	Arthritis and psoriasis or, in the absence of psoriasis, arthritis and a family history of psoriasis in parents or siblings and either: Dactylitis in the patient or Nail abnormalities (pitting or onycholysis) in the patient	2, 3, 4, 5
Undifferentiated Arthritis	Arthritis that fulfills criteria in no category or in 2 or more of the above categories	

*

1. Psoriasis or a history of psoriasis in the patient or first degree relative.
2. Arthritis in an HLA-B27 positive male beginning after the 6th birthday.
3. Ankylosing spondylitis, enthesitis related arthritis, sacroiliitis with inflammatory bowel disease, Reiter's syndrome, or acute anterior uveitis, or a history of one of these disorders in a first-degree relative.
4. The presence of IgM rheumatoid factor on at least 2 occasions at least 3 months apart.
5. The presence of systemic JIA in the patient.

The three classifications are shown below to highlight the differences (Table 3).

Table 3. Comparison of three classification system

	ACR[6]	EULAR[7]	ILAR[8]
Term	Juvenile rheumatoid arthritis (JRA)	Juvenile chronic arthritis (JCA)	Juvenile idiopathic arthritis (JIA)
Age of onset	< 16 years	< 16 years	< 16 years
Duration of arthritis	6 weeks	12 weeks	6 weeks
Exclusions	Yes	No	Yes
Subtypes	Systemic onset Oligoarticular Polyarticular	Systemic onset Oligoarticular Polyarticular (JRA) Juvenile ankylosing spondylitis Juvenile psoriatic arthritis Arthritis associated with Inflammatory bowel disease	Systemic onset Oligoarticular, ERA Polyarticular RF+ Polyarticular RF - Psoriatic arthritis Enthesitis related arthritis Others
Spondyloarthropathy	Excluded	Included	Included

Presently, ILAR classification is the most widely used classification worldwide.

Approach to diagnosis

As there is no diagnostic test for JIA, diagnosis is primarily clinical and based on symptoms and signs. Investigations help in exclusion of other conditions which can mimic JIA, assessment of degree of inflammation and in subtype classification.

The cardinal manifestation of JIA is arthritis thus, in any child presenting with joint pain, we need to find out if he has arthritis or not. Sometimes pain arising from periarticular tissues, bone or muscles can be misdiagnosed as arthritis. Presence of joint swelling, limitation of movement and joint tenderness suggest arthritis. Presence of pain worse in the morning associated with early morning stiffness, absence of history of trauma, presence of fatigue or fever suggest inflammatory arthritis. The first symptoms of arthritis in a young child may be irritability, decreased physical activity and inability to move the limb. Child may refuse to go to school or play. Further he may insist on being carried in the lap. Rarely the child may not have any pain but the parent's notice a limp.

Distribution and number of joints involved helps to define the subtype of JIA. In oligoarticular disease knee and ankle are the most involved joints whereas in enthesitis related arthritis along with knee and ankle foot and hip joints are also involved in an asymmetrical fashion. In contrast, in polyarticular disease there is symmetrical involvement of both small and large joints of upper and lower limbs. In systemic onset disease any number of joints may be affected. In India ERA is the commonest subtype seen[5,6] thus any boy beyond the age of 6 years presenting with lower limb arthritis must be examined for entheseal site pain. The common entheseal sites are tendoachillies, quadriceps tendon, iliac crest, anterior and posterior superior iliac spine and greater trochanter of femur.

Salient features of different forms of juvenile arthritis are summarized in the Table 4:

Children with JIA may have fatigue, low grade fever, anorexia but in a child with systemic onset JIA (SoJIA) the systemic manifestations are very prominent and include fever exceeding 38.5°C at least once or twice a day with normal or subnormal temperatures in between. During the episode of fever the child complains of myalgias, arthralgias and may have an evanescent, erythematous, macular rash on trunk. As the illness continues the child may develop generalized lymphadenopathy, hepatosplenomegaly, anemia and weight loss. Arthritis can occur concurrently or may follow fever by few months.

Table 4. Different clinical features of JIA

	Oligoarthritis	ERA	RF negative polyarthritis	RF positive polyarthritis	Systemic onset
Relative frequency					
Worldwide	30-60%	3-11%	10-30%	5-10%	10-15%
India	21%	36%	17%	12%	8%
Age (years)	1-4	8-14	1-6	8-16	Any
Sex ratio	F >> M	M >> F	F > M	F >> M	F = M
Number of joints involved	≤ 4	Any	> 4	> 4	Any
Systemic features	Absent	Absent	+	+	+++
Enthesopathy	Absent	Present	Absent	Absent	Absent
Rash	-	–	–	–	++
Uveitis	Chronic anterior uveitis (20%)	Acute anterior (25%)	–	–	–
RF	Rare	Rare	Rare	15-20%	Rare
ANA	60%	Rare	20%	15-20%	Rare
HLA B27	–	70%	–	–	–
Complications	Blindness Leg length discrepancy	Spondylo-arthropathy	– –	Joint destruction	Amyloidosis, ankylosis of joints
Prognosis	Good	Good	Fair	Poor	Fair

The other manifestations of SoJIA can be non-suppurative pharyngitis, and serositis. Rarely, a patient may present for the first time with macrophage activation syndrome, a serious complication of SoJIA. It is characterized by high continuous fever, thrombocytopenia, anemia, liver dysfunction, coagulopathy and rapid downhill course. If not treated early it is usually fatal.

In psoriatic arthritis, the major manifestation in a child is the presence of oligoarthritis with dactylitis. Later the joint manifestations can be oligoarticular or polyarticular. In a very young child there may not be any skin manifestations but the family history may be present. The skin manifestations usually precede the development of arthritis but can also follow it after few years.

In India, ERA is the commonest subtype comprising 40-44% of patients seen at a tertiary care hospital. Oligoarticular disease is seen in less than 10% of children whereas polyarthritis is seen in 30-40% and systemic onset in 15-26%[5,6]. The population based prevalence was different with low prevalence of systemic onset JIA[7] (Table 4).

Differential diagnosis

The diagnosis of JIA should be made only after excluding other causes of arthritis. The differential diagnosis includes infections, malignancy, metabolic diseases and other connective tissue diseases.

In India the most important diagnosis to exclude is acute rheumatic fever. An arthritis of >12 weeks duration, involvement of small and large joints, presence of cervical spine involvement and presence of hepatosplenomegaly or lymphadenopathy favors the diagnosis of JIA whereas, presence of subcutaneous nodules, erythema marginatum, carditis and, migratory arthritis favors the diagnosis of acute rheumatic fever. Elevated ASO titers are not diagnostic of acute rheumatic fever and can be found in 20-25% of children with JIA.

Chronic monoarthritis in a child is often diagnosed as tuberculosis. Presence of enthesopathy, low backache and radiological involvement of other joints suggests a diagnosis of JIA. In a young girl below 5 years of age, monoarthritis of knee joint is more likely to be due to oligoarticular JIA rather than TB. Presence of ANA will further support diagnosis of oligo JIA. In case of doubt, synovial fluid analysis and biopsy should be done before diagnosing tuberculosis.

In children presenting with systemic features like fever, lymphadenopathy, hepato-splenomegaly, and anemia along with arthritis careful exclusion of acute leukemia should be done. In some children arthritis may develop months before appearance of lymphoblasts in the periphery. Presence of bone pains, severe joint pains with minimal joint swelling, sternal tenderness, lymphocytosis, thrombocytopenia, elevated serum lactate dehydrogenase levels should alert a physician to the possibility of leukemia.

Similarly findings of polyarthralgia, lymphocytosis, thrombocytopenia can also be seen in SLE. A child with juvenile dermatomyositis can also present with fever, rash, aches and pains and inability to move about, only a careful examination of muscles would help in clinching the diagnosis.

Investigations

Routine blood investigations suggest chronic inflammation as reflected by anemia, neutrophilic leucocytosis, thrombocytosis, raised ESR and CRP. This acute phase response is most marked in SoJIA and least in oligoarticular disease. Mild rise in transaminase levels may be seen.

Autoantibodies are useful in defining rheumatoid factor positive subtype of JIA. RF can be induced during infection thus it should be positive twice at a gap of 12 weeks to be taken as RF positive. Anti-nuclear antibody (ANA) is positive in oligoarticular (70%) and polyarticular disease (25%). Presence of ANA increases the risk of chronic uveitis and warrants 3 monthly eye checkup. However in India ANA positivity is uncommon in polyarticular disease. Markedly elevated ferritin and low fibrinogen levels are useful in suspecting a diagnosis of macrophage activation syndrome, which can be confirmed on bone marrow examination.

Radiographs are not very helpful in early disease, but they may help in excluding other causes like osteomyelitis, scurvy, rickets, septic arthritis, and leukemia. During later phase of the disease soft tissue swelling, juxta articular osteopenia, ankylosis and erosions may be present.

High resolution ultrasound and MRI of joints are helpful in early detection of synovitis and erosions. Detection of asymptomatic hip joints disease in a child with unilateral hip joint pain helps in excluding infection and suggests a diagnosis of ERA. Further it helps to exclude Perthe's disease, transient synovitis, minor fractures etc.

Synovial fluid analysis and biopsy is only needed in a child with persistent monoarthritis to exclude chronic infection. SoJIA or Still's disease diagnosis will be discussed in the next chapter.

Further Reading :

1. Cassidy JT, Levinson JE, Bass JC, Baum J, Brewer EJ et al. A study of classification criteria for the diagnosis of juvenile rheumatoid arthritis. Arthritis Rheum 1986; 29: 274-81.
2. Prieur AM, Ansell BM, Bardfeld R, Bhettay E, Bojkinov I, Denieskiewics K, Fantini F, Halvelka S, Hoyeraal HM, Jais JP, et al. Is onset type evaluated during the first 3 months of disease satisfactory for defining the sub-groups of juvenile chronic arthritis? A EULAR Cooperative Study (1983-1986). Clin Exp Rheumatol 1990; 8: 321-25.
3. Petty RE, Southwood TR, Baum J, Bhettay E, Glass DN, Manners P, et al. Revision of the proposed classification criteria for juvenile idiopathic arthritis: Durban, 1997. J Rheumatol 1998; 25: 1991-94.
4. Petty RE, Southwood TR, Manners P, et al ILAR Classification of JIA: Second Revision, Edmonton, 2001. J Rheumatol 2004; 31: 390-92.
5. Aggarwal A, Misra R. Juvenile chronic arthritis in India: Is it different from that seen in the Western countries? Rheumatol International 1994; 14: 53-56.
6. Sawhney S. Spectrum of juvenile idiopathic arthritis in a tertiary care centre in northern India. Indian Journal of Rheumatology 2006; 1: 3-6.
7. Kunjir V, Venugopalan A, Chopra A. Profile of Indian patients with juvenile onset chronic inflammatory joint disease using the ILAR classification criteria for JIA: a community-based cohort study. J Rheumatol. 2010; 37: 1756-62.

JUVENILE IDIOPATHIC ARTHRITIS

2

C P Rajendran

Variety, the spice of life is what is encountered in the thought provoking entity of childhood arthritides, to which Juvenile Idiopathic Arthritis (JIA) as it is presently termed belongs. It still remains a diagnosis of exclusion, with a lean towards clinical diagnosis remaining the mainstay, revealing the heterogenous nature of the presentation.

Evolution of the concept of inflammatory polyarthritis in childhood was by Cornil in 1864.

Diamond Berger in 1890 reported acute onset, large joint involvement, remissions, exacerbations and normal growth pattern disturbances with a good prognosis.

It was Sir Robert Still, who by his exemplary clinical bedside skilful observation discovered the disease with an insidious onset before the second dentition in girls.

The acute onset presented with lymphadenopathy, fever, splenomegaly, pleuritis, pericarditis, early contracture, muscle atrophy, early cervical spine involvement, '*perhaps more than one disease with a different aetiology*'.

Features which were not observed were rash and articular pain. Hence the name Still's disease for acute systemic onset Juvenile Idiopathic Arthritis.

The present definition of JIA is arthritis of unknown aetiology that begins before the sixteenth birthday and persists for at least six weeks, other known conditions being excluded.

Epidemiology

It is universal and constitutes 2.7% to 5.2% of all Rheumatoid Arthritis (RA) with onset before fifteen years.

The incidence is 2 to 20 per 100,000 and the prevalence 16 to 150 per 100,000.

Systemic onset juvenile idiopathic arthritis (SoJIA) constitutes 10% to 20% of all Juvenile idiopathic arthritis[3].

In India JIA constituted 41% among patients reported from a single tertiary referral centre[4]. It peaks between one and three years of age in females, whereas in males it is bimodal with the first peak at two years and the second between eight and ten years[1].

Etiopathogenesis

It is an autoimmune disease involving the innate immune system, T cells, immune complexes, cytokines etc., with complex genetic traits, involving both HLA and non HLA genes.

The possible causes include aberrant immunologic regulation, psychogenic stress, trauma, hormonal abnormalities and infection.

The different subtypes each have a different aetiological background.

Both the IL2RA/CD25 gene and VTCN1 gene have been implicated as JIA susceptibility loci.

Sola et al found the synovium to contain messenger ribonucleic acid (mRNA) for vascular endothelial growth factor and orgiopoitein 1, as well as for their receptors.

Oligoarthritis

CD8+ T cell activation with a THO/THI bias in synovial fluid has been observed. Early onset oligoarthritis disease with a polyarticular course is associated with TCR allele vβ 6.1 in HLA QA*0101 positive children. TCR vβ 20 allele is associated with the oligoarticular type. IFNγ is found in the synovial fluid. HLA-A2 and IL-1A2 is associated with early onset, oligoarthritis DR1, DPBI*0201 is a risk factor for extended oligoarthritis.

Systemic onset juvenile idiopathic arthritis (SoJIA)

SoJIA is associated with a mixed Th cell response. There is a global "T cell regulating effect" with decrease in CD4 + T cells and increased and decreased CD8 cells. B cells are increased in SoJIA. Single nucleotide polymorphism for IL-6 and MIF along with increased sE-selectin have been observed[5].

The triggers of Systemic - onset JIA may be infection. When initiated, SJIA is driven by innate proinflammatory cytokines e.g. IL-1 and IL-6, Endogeneous Toll-like receptor ligands, including S100 proteins, probably synergize with cytokines to perpetuate inflammation. These and other findings show that SoJIA is an autoinflammatory condition.

Polyarticular:

TCR vβ8 alleles have been associated with polyarticular types. IL-1, IL-6, IL-8 and TNFα are increased. HLA-DR1 and HLA-DR4 are strong associations. DR4 is also associated with anti CCP antibodies and with more severe disease progression.

Enthesitis related arthropathy (ERA):

It is associated with IFNγ in the synovial fluid, as well as with increased levels of IL-1, IL-6 and TNFα: HLA- B27 associated with JAS.

Other observations include in general increased expression of RANK - the Receptor Activator for NFkB, IL-4 in synovial T cells and IL-12 in synovial fluid. Outcomes for chronic arthritis are defined by TNFα G → A -238 and -308 polymorphisms, as are TAP polymorphisms with a prediction for early onset disease.

Autoantibodies in JIA are many and usually an epiphenomenon except in vasculitis where immune complexes participate in vascular inflammation. They are directed against histones, keratin, perinuclear, RA33, ANCA, cardiolipin, Type-I, II and IV collagen, cartilage like protein, IgA antigliadin antibodies, CCP, filaggrin etc., to name of few.

Hormonal imbalances with decreased 17β estradiol, sulphated DHEA and synovial testosterone with increased serum prolactin levels have been observed. Predisposition to infection e.g. parvovirus 19, EBV - molecular mimicry, Rubella, HIV, Influenza, post MMR and Rubella vaccination, IgA deficiency and hypogamma globulinemia, decreased C2 complement have been postulated, as has been trauma.

Psychological factors comprising neuroendocrine gene polymorphism and autonomic nervous system dysregulation in response to stress have been observed.

Classification

Classifications are many and have evolved over the years through the following names:

1. Juvenile rheumatoid arthritis (JRA) framed by the American College of Rheumatology (ACR) 1973.
2. Juvenile Chronic Arthritis (JCA) coined by the European League against Rheumatism (EULAR) in 1977.
3. Juvenile Idiopathic Arthritis proposed by the pediatric Task force of the International League of Associations for Rheumatology (ILAR), initially formulated in Santiago, then Durban, revised in Edmonton in 2001 and is still undergoing a series of suggestions to make it precise.

The requisites to make a provisional diagnosis of juvenile arthritis are that

1. The age of onset should be under 16 years.
2. Arthritis meaning joint swelling or effusion or two of the following - limited joint movement, tenderness and warmth, not due to primary mechanical disorders or other identifiable causes.
3. Exclusion of other forms of juvenile arthritis.

Definitions

Onset type

Constellation of clinical signs present during the first 6 months of the illness.

Course

Course which the disease takes after the first 6 months of the illness.

Polyarticular type

Arthritis involving five or more joints during the first 6 months of the disease. In order to determine the onset subtype, each joint is counted separately, except the joints of the cervical spine, carpus, tarsus and metatarso phalangeal joints, each of which is counted as one joint.

Oligoarticular / pauciarticular

Arthritis involving four or fewer joints during the first 6 months of the disease. Monoarthritis is included in this onset group.

Positive test for rheumatoid factor (RF)

At least 2 positive results, in an accredited laboratory 3 months apart, during the first 6 months of disease.

Quotidian fever

Fever that rises to > 39°C once a day and returns to ≤ 37°C between fever peaks[2] (Fig. 1).

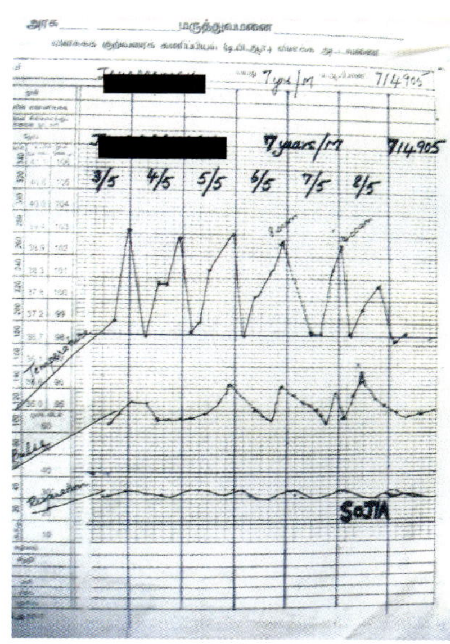

Fig. 1 : Quotidian pattern of fever in a 7 year old boy with systemic onset. JIA showing the late peak in the evening upto 39°C and touching normal, lasting more than 3 days.

Table 1. Classification of childhood arthritis

ACR	EULAR	ILAR
Juvenile rheumatoid arthritis	Juvenile chronic arthritis	Juvenile idiopathic arthritis
Systemic	Systemic	Systemic
Polyarticular	Polyarticular JCA	Polyarticular RF - Negative Polyarticular RF - Positive
Pauciarticular	Pauciarticular	Oligoarticular Persistent Extended
	Juvenile psoriatic arthritis Juvenile ankylosing spondylitis	Psoriatric arthritis Enthesitis related arthritis Other arthritis

Table 2. Characteristics of the ACR, EULAR and ILAR classification of childhood arthritis

Characteristics	ACR	EULAR	ILAR
Basis of classification	Clinical onset and course	Clinical and serologic (RF) onset only	Clinical and serologic (RF) onset and course
Onset types	Three	Six	Seven
Course subtypes	Nine	None	Two
Age at onset of arthritis	≤ 16 yrs	≤ 16 yrs	≤ 16 yrs
Duration of arthritis	≥ 6 wks	≥ 3 months	≥ 6 wks
Includes "JAS"	No	Yes	Yes
Includes "JPsA"	No	Yes	Yes
Includes "IBD"	No	Yes	Yes
Includes reactive arthritis	No	No	No
Exclusion of other disease	Yes	Yes	Yes

The three different classifications of childhood Arthritis are given in the (Table -1).

JAS - Juvenile Ankylosing Spondylitis

JPsA - Juvenile Psoriatic Arthritis

IBD - Inflammatory Bowel Disease

RF - Rheumatoid Factor

An ACR criteria is simple and validated. Psoriatic arthritis and JAS are not included. In EULAR classification, the term JRA is reserved for positive rheumatoid factor (RF). JAS, JPsA and IBD are included. There is no definition of RF positivity and no criteria for JAS and JPsA.

In ACR and EULAR mutually exclusive categories of childhood arthritis have not been well delineated (Table 2).

The ILAR classification has been introduced for better delineation, research purposes, maintaining homogeneity and representing mutually exclusive idiopathic childhood arthritis categories based both on clinical and laboratory features. In the second Edmonton revision 2001, there are seven categories (Table 3) with exclusions being designated a, b, c, d and e.

Exclusions

1. Psoriasis or history of psoriasis in the patient or first degree relative.
2. Arthritis in a HLA-B27 positive male beginning after the sixth birthday.
3. Ankylosing spondylitis, sacroiliitis with inflammatory bowel disease, Reiter's syndrome, actue uveitis or a history of one of these disorders in a first degree relative or a history of one of these disorders in a first degree relative.
4. Presence of IgM rheumatoid factor on two occasions at least three months apart.
5. Presence of systemic JIA in the patient.

These are applied in each category and are subject to change as per the availability of new data.

Table 3. ILAR criteria - Edmonton second revision

Categories	Definitions	Exclusions
Systemic	Arthritis in one or more joints with or preceded by fever of at least two weeks duration, quotidian for at least 3 days and accompanied by one or more of the following: 1) Evanescent (non fixed) erythematous rash 2) Generalised lymph node enlargement 3) Hepatomegaly and/or splenomegaly 4) Serositis	a, b, c, d
Oligoarthritis	1) Arthritis affecting 1 to 4 joints during the first 6 months of disease, persistent, affecting not more than 4 joints throughout the disease course. 2) Extended oligoarthritis: affecting a total of more than 4 joints after the first 6 months of disease.	a, b, c, d, e
Polyarthritis (RF Negative)	Arthritis affecting 5 or more joints during the first 6 months of disease RF-negative	a, b, c, d, e
Polyarthritis (RF Positive)	Arthritis affecting 5 or more joints, during the first 6 months of disease. RF - positive, on at least two occasions three months apart, during the first 6 months of disease.	a, b, c, e
Psoriatic arthritis (PsA)	Arthritis and psoriasis or arthritis and at least 2 of the following 1) Dactylitis 2) Nail pitting or onycholysis 3) Psoriasis in a first degree relative	b, c, d, e
Enthesitis related arthritis ERA	Arthritis and enthesitis or arthritis or enthesitis with at least 2 of the following 1) Presence of or a history of sacroiliac joint tenderness and/or inflammatory 2) Presence of HLA-B27 antigen. 3) Onset of arthritis in a male over 6 years of age 4) Acute (symptomatic) anterior uveitis 5) History of ankylosing spondylitis, ERA, sacro ilitis with IBD, Reiter's syndrome or acute anterior uveitis in a first degree relative.	a, d, e
Undifferentiated arthritis	Arthritis that fulfils criteria in no category or in 2 or more of the above categories	

For future improvement a number of descriptors have been proposed. These are age at onset, further description of arthritis (large/small joints, symmetry, upper or lower limbs, and individual joint involvement) course of disease, ANA positivity, chronic or acute anterior uveitis and HLA allelic associations, among which ANA positivity has received the maximum attention.

The ILAR criteria represent not only a change in terminology but in definition also[2].

Identification of arthritis in a child

Diagnosing arthritis is not difficult when there are overt symptoms like decreased ability to run, limping, intermittent complaints of pain and or swelling often worse in the morning leading on to inability to get up from the bed or exacerbated by intermittent illness or fatigue.

Difficulties arise when symptoms are subtle like incessant crying in infants, refusal to feed, others being fever, increased irritability, "falling off" of the child's growth curve for both height and weight, bruising due to falls, undiagnosed fracture in untreated arthritis and emotional regression. The other causes of childhood joint pain are growing pains, psychogenic pain, fibromyalgia, complex regional pain syndromes, avascular necrosis, chondromalacia patellae, osteochondroses, hypermobility and interited or congenital syndromes.

Systemic onset JIA (SoJIA)

This subset constitutes 10% to 20% of JIA. It occurs throughout childhood and affects both sexes equally. The systemic symptoms appear long before the actual arthritis becomes manifest. They include fever, the classic evanescent rash, reduced appetite, weight loss, severe arthralgia, myalgia, lymphadenopathy, hepatosplenomegaly polyserositis and severe anaemia (Fig. 2). It can progress to oligoarthritis with a good prognosis or a polyarthritis with erosion and a poor prognosis. SoJIA with HLA DR4 positivity is associated with polyarthritis (ANA positivity is rare around 5% to 10% as is RF positivity which is around 2%). Hip involvement usually bilateral is seen in 50% of cases (Fig. 3).

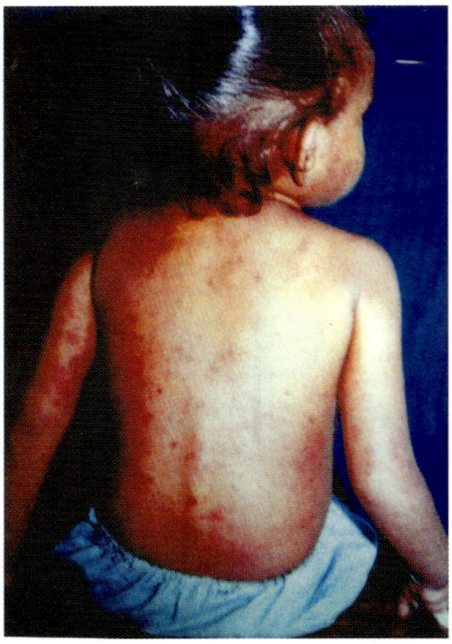

Fig. 2 : Typical salmon pink evanescent macular rash of systemic onset JIA in a 3 year old girl.

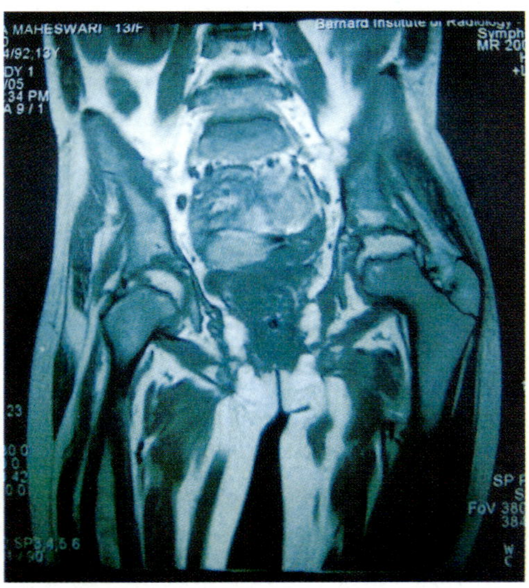

Fig. 3 : MRI of the hip joints in a 13 year old girl with long standing unremitting systemic onset JIA showing severe bilateral hip arthritis and avascular necrosis of both femoral heads. There is superior subluxation of the left femoral head.

Macrophage activation syndrome (MAS) originally thought to be a devastating complication of systemic JIA, is actually a variant of Hemophagocytic lymphohistiocytosis (HLH)[8]. MAS and SoJIA are different ends of the same spectrum. MAS is characterized by acute onset of fever, hepatosplenomegaly, lymphadenopathy, hepatic failure with encephalopathy, purpura, bruising, mucosal bleeding, thrombocytopenia, elevated liver enzymes, mild hypoalbuminemia, elevated triglycerides and ferritin. Prothrombin time (PT) and partial thromboplastin time (PTT) are prolonged with decrease in blood levels of Vit. K dependent clotting factors. Fibrin degradation products are found in the plasma. The ESR is low due to hypofibrinogenemia induced by consumptive coagulopathy. The pattern of fever changing from irregular to continuous and the high ESR becoming normal in spite of persisting systemic features are important indicators of MAS. Active phagocytosis by macrophages and histiocytes are seen in bone marrow examination. Similar subclinical abnormalities have been observed in SoJIA for e.g. increased ferritin levels and coagulation abnormalities suggesting an occult MAS in SoJIA.

In HLA as in MAS it is the innate immune system comprising the monocyte/macrophage/histocyte lineage and Natural Killer (NK) cells which is also prominent in SoJIA. Defective NK cells, abnormal perforin level expression and cytotoxic CD8+ T lymphocyte function, hence T cell proliferation resulting in IFNγ and GM-CSF production, sustained macrophage activation leading to TNFα, IL-1 and IL-6 production have been observed. Increased expansion of dendritic cells results in increased IL-1 activity in SoJIA. The increased IL-6 levels in SoJIA correlate with thrombocytosis, microcytic anaemia, growth retardation and osteopenia[3].

Defective apoptosis has been suggested by the association of MAS and Kikuchi's necrotizing lymphadenitis in patients with SoJIA[3].

Hence the role of IL-1 receptor antagonist (Anakinra/Kineret) and IL-6 antagnostis in SoJIA[16].

Oligoarthritis

This is the most common and least severe, comprising 40% to 60% of the patients. It affects less than four joints, most often the knees followed by the ankles and elbows (Fig. 4).

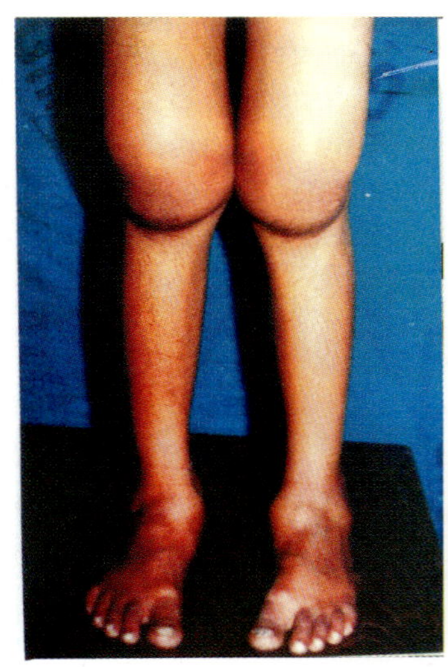

Fig. 4. Oligoarthritis in a 7 year old boy showing bilateral knee and ankle arthritis.

Recently isolated severe and disabling primary hip disease representing a separate oligoarthritis subtype of JIA has been described. Systemic manifestations are absent. In 50% of the children it is knee monoarthritis. Asymptomatic uveitis is seen in 20% of children. Deformities are less, growth is not affected and remissions occur. There are two course subtypes : 1. Persistent. 2. Extended.

Exceptional findings include

a) ESR greater than 40mm/hr or high platelet count (> 6,00,000/mm^3).

b) Hemoglobin less than 10.5g%.

c) Involvement of any joint other than the knee, toes, PIPs, MCPs and elbow. Ankle, wrist and elbow involvement are unusual. Cervical spine involvement, usually painless indicates extended oligoarticular, polyarticular or systemic involvement. Wrist, temporomandibular joint, cervical spine, hip involvement and uveitis (Fig. 5) indicate an aggressive course requiring aggressive therapy[7].

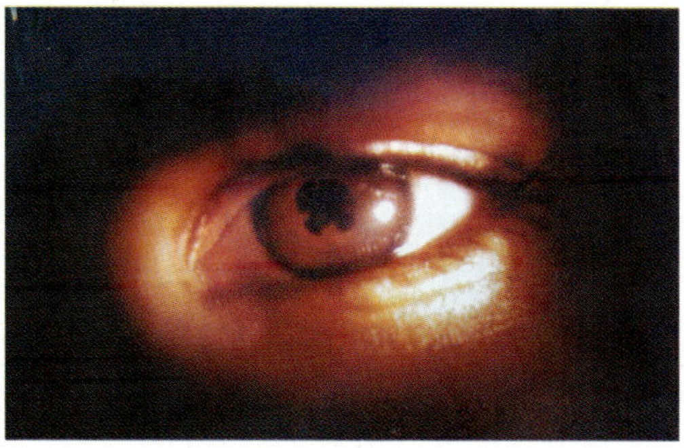

Fig. 5 : Left eye of a 3 year old girl showing an irregular pupil, due to adhesion of the iris to the anterior surface of the lens secondary to chronic anterior uveitis.

Polyarthritis

This constitutes about 40% of all cases of JIA. It is common upto the age of 3 or after 10 years. Both large and small joints especially PIPS, 2nd and 3rd MCPs and DIPs can be involved. 50% have spine or hip involvement. Chronic uveitis is rare. Some have low grade fever, hepatosplenomegaly or lymphadenopathy. Diagnosis in early adolescence especially in RF seropositive girls, indicates a severe disease with extra articular features, nodules and erosions behaving like adult rheumatoid arthritis.

Compared with adults with RF positive rheumatoid arthritis, however children are at less risk for rheumatoid lung involvement and vasculitis.

The seronegative subtype polyarticular variety occurs throughout childhood and is less severe (Fig. 6).

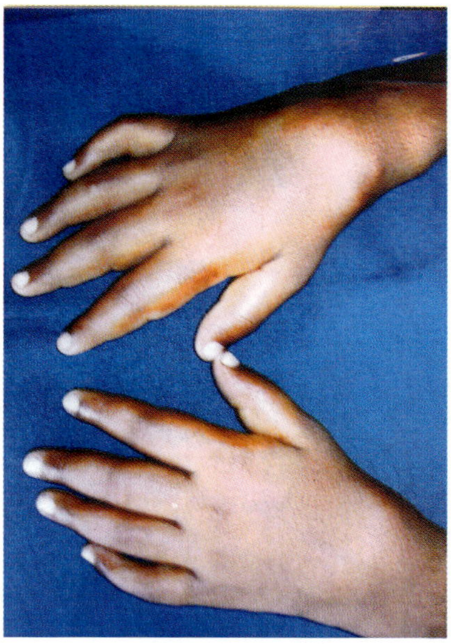

Fig. 6 : Symmetric polyarthritis affecting the MCP, PIP and radiocarpal joints in a 5 year old girl.

Arthritis of small joints in a child with incomplete ossification can be made out by the closer intercarpal distance. Knee disease is severe and erosive. Temporomandibular (TM) joint involvement can cause facial asymmetry, deviation of jaw and micrognathia. Decrease in the interincisor distance to less than 4 cm causes inanition. Headache and compromise of middle ear function can occur in TM joint involvement. ANA positive patients have a high incidence and HLA-B27 positive patients have a low incidence of TM joint involvement (Table 4).

Enthesitis related arthritis (ERA)

ERA is associated with increased frequency of spinal involvement, with more than half having significant disease on reaching adulthood with a poorer physical outcome compared to oligo and polyarticular types (Fig.7).

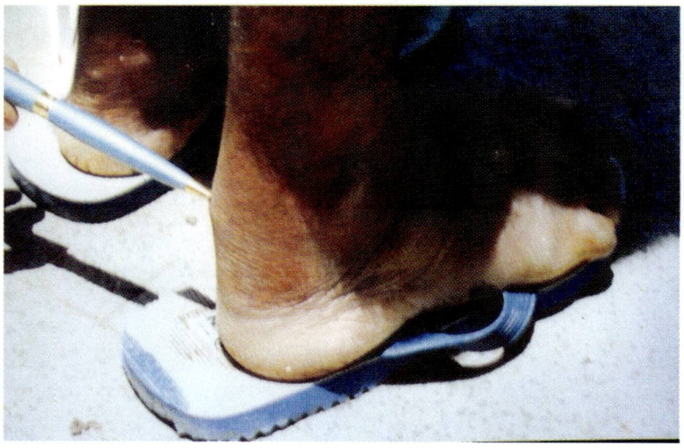

Fig. 7 : Tendoachilles tendinitis in a 14 year old boy with enthesits related arthritis

Table 4. Differential diagnosis

Systemic Onset	Oligoarthritis	Polyarthritis
Rheumatic fever[22] Infective endocarditis	Septic arthritis Reactive arthritis	Systemic Lupus Erythematosus (Fig.13) Dermatomyositis (Fig.14)
Leukemia	Hemophilia	Spondyloarthropathies
Systemic Lupus Erythematosus	Tuberculosis	Scleroderma (Fig.12)
Juvenile dermatomyositis	Villonodular synovitis	Serum sickness
Vasculitis (Fig.10)		Leukemia (Fig.11) Immune deficient states

For Juvenile ankylosing spondylitis (JAS), enthesitis is a predominant manifestation, differentiating it from other forms of JIA[8].

Peripheral joints are involved in 82% and distal joints are affected more than proximal, with midtarsitis, first MTP arthritis being common as is shoulder involvement, with sparing of hands. Lumbosacral disease occurs only later[8].

Aggressive treatment is required for JAS especially in patients with family history of similar disease, presence of HLA DRBI*08, absence of HLA DPB1*02, early hip, ankle arthritis and persistently elevated ESR within the first 6 months[9].

Juvenile psoriatic arthritis (JPsA)

The age of onset is biphasic with one peak at 2 years and one again late in childhood. Younger females exhibit dactylitis and small joint involvement with anti nuclear antibodies. Progression to polyarticular disease is more common in younger children, who take a longer time to achieve remission (Fig. 8, 9).

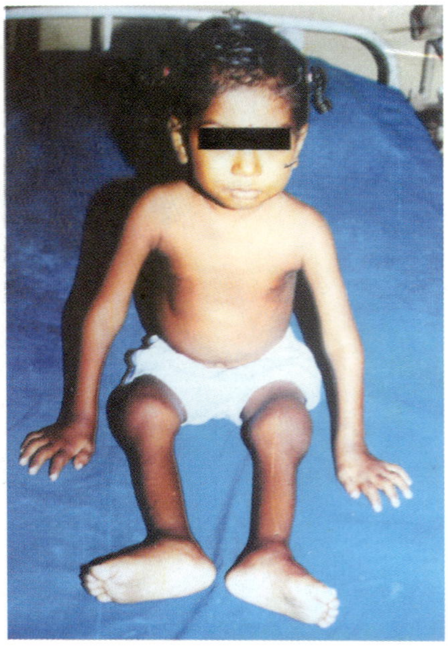

Fig. 8. Polyarthritis in a 3 year old girl - seronegative for rheumatoid factor

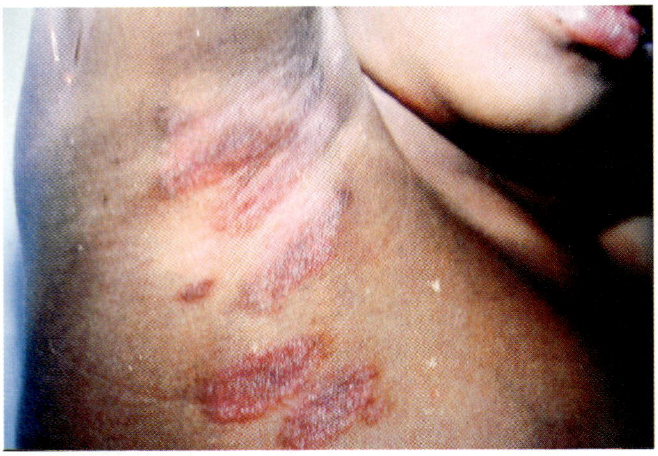

Fig. 9. Well demarcated erythematous scaly lesions of psoriasis in a 3 year old child with psoriatic arthritis.

In contrast it is enthesits, axial joint disease and persistent oligoarthritis in older children[10].

It is mediated by activated CD8+T cells, TNFα, NF-kB, IL-1β, IL-2R and IL-8.

Asymptomatic uveitis is common in 15% to 20%, with ANA being positive demanding an aggressive immunomodulatory therapy with intraocular and intravitreal implants.

Growth retardation – in JIA is due to decreased serum IGF-1 and IGFBP-3. Hence growth hormone treatment is effective in linear growth[13].

Differential diagnosis

Investigations

Laboratory investigations are only an aid to clinical diagnosis

1. Anaemia.
2. Leucocytosis- increased neutrophils - especially in SoJIA.
3. Increased serum ferritin - Systemic activity.
4. Thrombocytosis - In exacerbations and long standing disease.
5. ESR - elevation - Useful but not totally reliable.
6. CRP - elevation - more reliable.

7. Increased serum immunoglobulins - Correlates with disease activity.
8. ASO - One third of SoJIA have elevated ASO titre due to inflammation and polyclonal B cell activation- anamnestic reaction.
9. RF
   - High IgM levels are seen in active disease with severe radiological deformities[12].
   - High IgA levels are seen with active disease in polyarticular onset[12].
   - Positive in older, polyarticular erosive disease with HLA. DW4 (DRB10401) and DW14 (DRB*404) specifities.

   Hidden RF
   - Seen in 68% to 75% of children.
   - IgM 19S antiglobulins detected by acid elution of IgM containing fractions of serum from a gel filtration column.
10. Anti CCP antibodies
    - Present in RF positive patients with polyarthritis[13]
    - Predicts erosive joint disease in polyarthritis[6]
11. ANA - More useful than RF.
    - mostly IgG, less commonly IgM and IgA.
    - HEp-2 cells show homogenous and speckled pattern
12. Complement

    C3 - increased in active disease

    C3d - increased

    - especially in SoJIA
13. Lipids - High triglycerides
    - reduced plasma cholesterol
    - reduced high density lipoproteins
    - reduced platelet activating factor (PAF) acetylhydrolase activity
14. Synovial fluid
    - cells, can be reduced or very high upto one lakh.
    - complement not as low as in adults
    - IgG and IgM RF with hidden RF's present

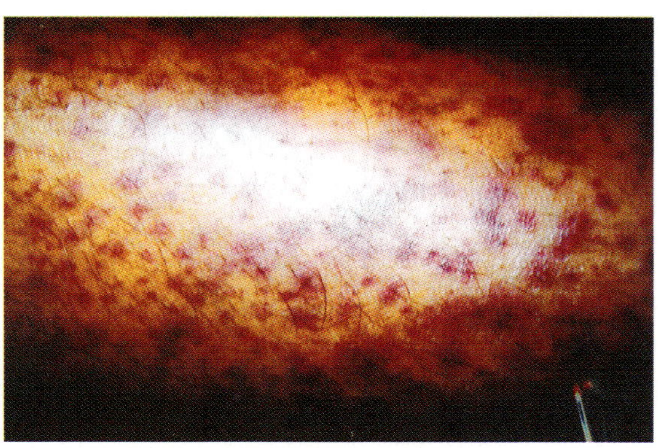

Fig. 10. Purpuric rashes of vasculitis - Henoch Schonlein purpura in a 14 year old boy.

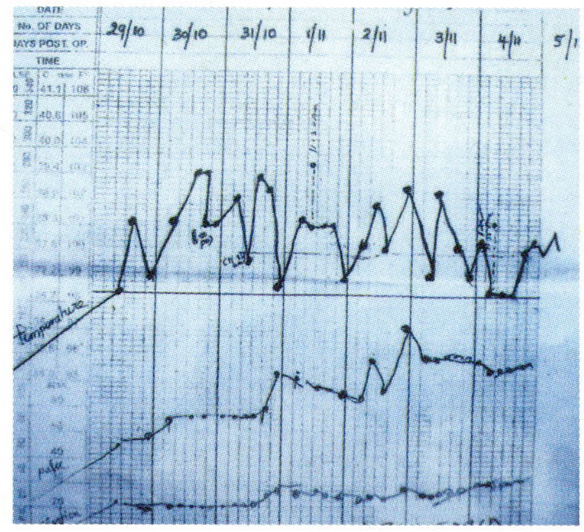

Fig. 11. Fever pattern in a 13 year old boy with Leukemia. The temperature does not touch normal every day.

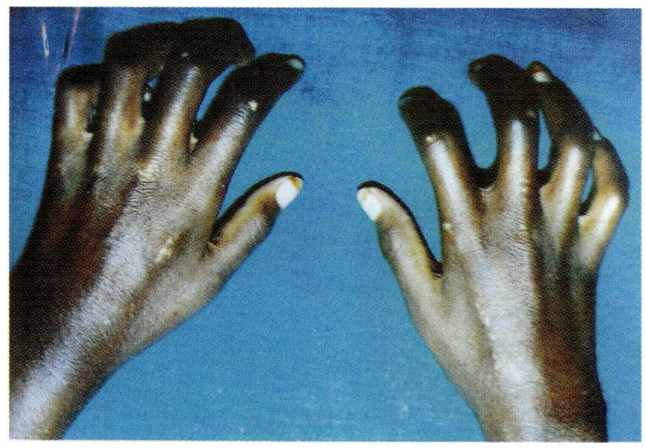

Fig. 12. Sclerodactyly with flexion conractures and pigmentary changes in a 12 year old boy with diffuse cutaneous systemic sclerosis.

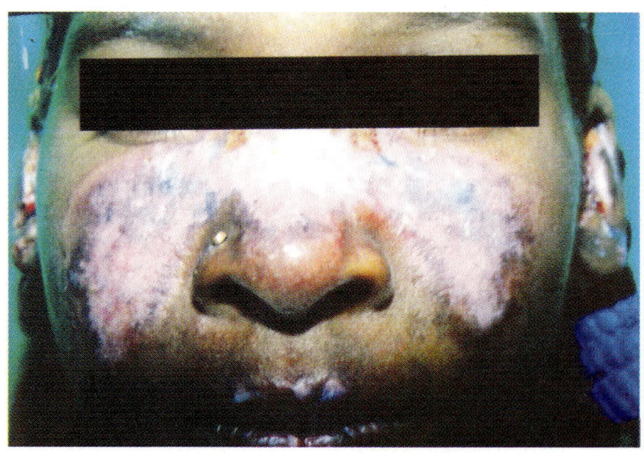

Fig. 13 : Typical butterfly malar rash of Systemic Lupus Erythematosus in a 7 year old girl.

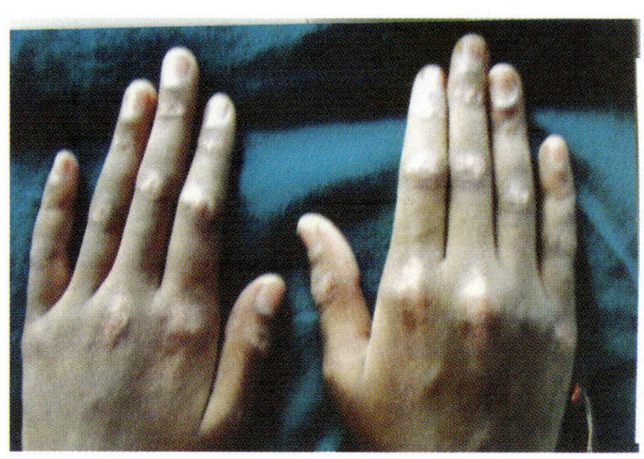

Fig. 14. Gottron's papules over the MCP, PIP and DIP joints in a 10 year old girl with Dermatomyositis.

Imaging

1. Radiography.
2. Ultrasound, - for early erosions, high power Doppler for vascularity.
3. Magnetic Resonance Imaging (MRI) for early synovial changes (intravenous Gadolinium enhanced) (Fig. 15).

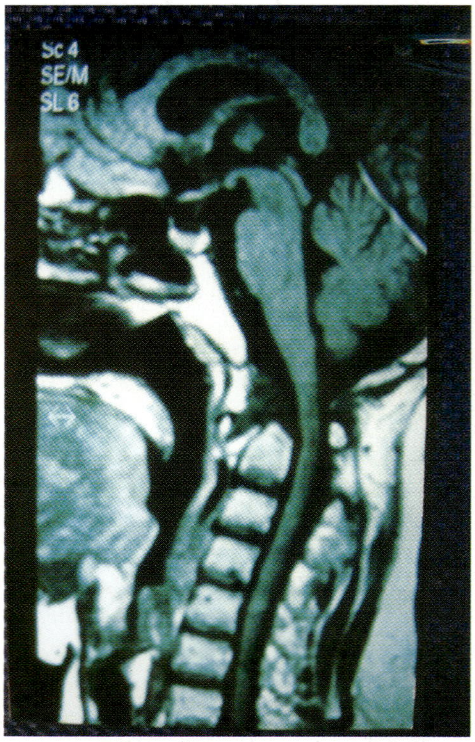

Fig. 15 : T2 weighted MRI image of the cervical spine in long standing RF seropositive polyarthritis showing an inflammatory mass between C1 anterior arch and odontoid on the upper cervical cord.

4. Dexa for osteopenia.

Conventional radiography (recently digital x-rays) reveal soft tissue swelling, widened intraarticular space, periosteal new bone apposition, growth arrest lines, ballooning of femoral condyles and later erosions, subluxation and osteoporosis.

Peculiarities are DIP involvement, ulnar carpal deviation and radial MCP deviation.

The predominant changes in the cervical spine are in the apophyseal joints and upper cervical segments. X-rays should be taken after precautions, in the lateral - neutral, flexion and extension views and open mouth anteroposterior views (Fig. 16).

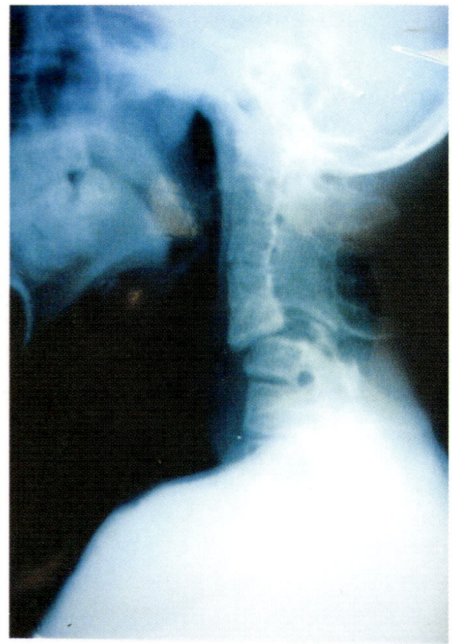

Fig. 16 : X-ray of the cervical spine in a 16 year old boy with systemic onset JIA since 7 years of age showing fusion of the upper cervical vertebrae, subaxial subluxation of C5 over C6, fusion of the apophyseal joints and osteoporosis of the cervical vertebrae.

The normal upper limit of A-0 distance is 4mm. Atlantoaxial subluxation is common. Bony fusion at C2 C3 level with fusion of adjacent spinous processes occurs.

A small C4 may be seen in early onset disease and short body stature. Apophyseal joint ankylosis occurs at multiple levels. Arthritis mutilans deformity is associated with severe changes in the cervical spine[14].

Treatment

Objectives

Immediate:

1. Prevent and relieve discomfort
2. Preserve function
3. Prevent deformities
4. Prevent and control inflammation

Long-term:

- Prevent or minimize side effects of disease and treatment.
- Promote growth and development
- Promote rehabilitation.
- Promote education

A multi disciplinary team approach with the above goals should include a pediatric rheumatologist, clinician, nurse, social worker, physical therapist, occupational therapist, psychologist and also consultations with orthopedic surgeons, dentist or dietician.

Physical and occupational therapy

Aim

1. Relieve pain - Heat
   - Cold
2) Retrieve joint function - Splints
   - Exercises
   - ROM, Passive, Active
   - Cervical collar
3. Regain posture - Prone lying exercises
4. Rest and exercise adequately.
5. Reconstruct gait with modified foot wear.
6. Relaxation training, guided imaging and biofeedback.

Table 5. Uveitis screening

No uveitis

Oligoarticular onset

- Initial visit
- Every 3 months for 2 years
- Every 4 to 6 months for 7 years
- Once a year thereafter

Polyarticular onset

- Initial visit
- Every 4 to 6 months for 5 years
- Once a year thereafter

Systemic onset

- Initial visit
- Once a year thereafter

With Uveitis

With inactive ocular disease

- Once in every 3 months

With active ocular disease

- as indicated by response to therapy.

Orthopedic surgery

Has a limited but important role and includes synovectomy, soft tissue surgery and reconstructive surgery.

Table 6. Medical treatment

Oligo	Poly	Systemic	ERA	PsA
NSAID ↓ IA Steroid ↓ Hydroxy chloroquine ↓ Methotrexate ↓ IM Gold Sulphasalazine ↓ sTNFα P75:FP	NSAID ↓ Methotrexate ↓ Prednisolone (IV methyl for flare +/- oral steroids) or IA steroid ↓ Hydroxy chloroquine ↓ IM Gold sulphasalazine Leflunomide ↓ sTNFα p 75:FP	NSAID ↓ Prednisolone IV steroid pulse for flare +/- oral steroids or IA steroid ↓ Methotrexate ↓ Cyclosporin A Thalidomide IV immunoglobulin ↓ sTNFα p75: FP ↓ IL-I RA ↓ IL-6 antagonist	NSAID ↓ Sulphasalazine ↓ TNF α Blocker + Steroid (Intralesional)	NSAID ↓ methotrexate Cyclosporine A ↓ TNF α blocker + Skin Treatment
	(Poly and Systemic ↘ ↙) Autologous stem cell transplantation			

Vitamin D and calcium prophylaxis in all

Table 7. Drugs

Name	Dose	Preparation	Special comments
NSAIDs			
Naproxen (FDA Approved)	15 mg to 20 mg/kg/day (2 divided doses)	125 mg/5ml 250/500mg Tabs	Cutaneous Pseudo Porphyria
Tolemetin (FDA Approved)	25 to 30 mg/kg/day 3 Divided Doses		
Ibuprofen (FDA Approved)	35 mg/kg/day 3 to 4 divided doses	200 mg/400 mg tablets. 100mg/5ml suspension (45 mg/kg/day)	
Sulindac			Less nephrotoxic
Indomethacin	1 to 3 mg/kg/day Maximum 125 mg/ Day in divided doses		fever and pericarditis of SoJIA.
Analgesics Acetaminophen			
DMARDs			
Methotrexate	0.35 to 0.65 mg/kg/week orally Maximum 1 mg/kg/week. Above 15 mg/week subcutaneous injection (can be used orally)		Second line drug Folic acid - 1 mg/day, supplementation
Cyclosporin A	3 to 5 mg/kg/day two divided doses	Oral liquid / Capsules	MAS
Sulphasalazine	12.5 mg/kg/day upto 50 mg/kg	500 mg tabs	
Hydroxychloroquine	5 mg-6mg/kg/day	200 mg	Children above 7 years
Parenteral Gold	5mg IM Test Dose 0.75 mg to 1 mg/kg/week (maximum 50 mg/week)	Aurothiomalate / Aurothioglucose	Dose to be reduced after 6 months.
Newer Agents			
Biologics			
TNF α p 75:FP **(Etanercept)**	0.4 mg/kg/sc twice a week		25mg per injection + Methotrexate + NSAID
Inadequate clinical response	0.8 to 1.8 mg/kg/sc/wk		Polyarticular Juv. PsA Refractory ERA
Infliximab (Recombinant monoclonal human IgG antibody to TNFα)	3 mg/kg iv infusion 0, 2, 6 and then every 8 weeks +/- MTX		Ongoing trials in polyarticular JlA.
Anakinra **(Recombinant Human IL-IRA)**	1mg/kg to 100mg/day/sc		Refractory SoJIA Competitively inhibits IL-l binding to IL-IR.
IL-1 TRAP (Rilonacept) (Fusion Protein - Human extracellular receptor domain, Fc. Portion of human IgG1)	2 to 4 mg/kg/wk/sc		More effective than anakinra
IL-6 Inhibitor Tocilizumab	8 mg/kg/every other week		SoJIA
Other Agents			
Leflunomide	3 mg to 10 mg/day		Polyarticular JIA

Name	Dose	Preparation	Special comments
STEROIDS			
Prednisolone (Low dose)	0.1 to 0.2 mg/kg		- bridge therapy - moderate to severe polyarthritis
High Dose	0.25 mg to 1 mg/kg/day (maximum 40 mg/day)		Severe disease
IV Pulse Methyl Prednisolone	10 mg to 30 mg/kg		Very severe disease
Intra-articular steroid	Triamcinalone hexacetonide / Triamicinalone acetonide		Monoarticular
Others	IV immunoglobulin, thalidomide, statins cyclophosphamide, growth hormone, nutritional supplements and recombinant erythropoietin.		

Outcome

Primary outcome measured is the ACR 30 - Pedi 30. Pedi 30 includes the following:

- Swollen joints (66), Tender joints (68), Patient's global assessment by VAS (0-100mm), Physician's global assessment, Assessment of physical function by patient, parent or guardian - by CHAQ and ESR.

Assessment

- 30% improvement from baseline in 3 out of the 6 measures.
- 30% worsening in one out of the 6 measures[17].

Secondary outcome measures are ACR 50, ACR 70 and CRP.

Remission criteria

Preliminary clinical remission criteria have been validated for selected categories of JIA using the OMERACT filter and a Delphi questionnaire survey where, ID means Inactive disease, no active arthritis, fever, rash, serositis, splenomegaly, generalized lymphadenopathy secondary to JIA or active uveitis, normal ESR, CRP and physician's global assessment of disease activity.

CRM means clinical remission on medication, 6 continuous months of inactive disease on medication

CR means clinical remission off medication, 12 months of inactive disease off medication.

The select categories are oligo (persistent and extended), polyarticular (RF +ve and RF -ve) and SoJIA[17].

- Remission rates in oligoarthritis range from 36% to 84%.
- Systemic or polyarthritis are less likely to remit
- Polyarthritis on follow up have radiology erosions.
- Early onset is associated with more disabilities.
- ANA positivity is associated with less disability.
- RA with shared epitope is associated with less disability in SoJIA.
- 70% to 90% have a satisfactory outcome.
- 10% to 20% enter into adulthood with moderate to severe functional disabilities.

Perspective

Polyarticular, RF negative was the commonest type observed in India, followed by ERA in Chennai and oligoarticular in Kolkata, as compared with Italy where Oligoarthritis was the commonest. In India ERA was common type in some rheumatology centre (Table 8).

Uveitis and ANA positivity were rare in India.

HLA-A2 allele was associated commonly with SoJIA and ERA in our centre.

Hidden Rheumatoid factors were less in India when compared with Turkey (56%).

9 patients with polyarticular JIA have received Infliximab at 3 mg/kg every 8 weeks, with no adverse effects or tuberculosis[19].

The horizon as it appears now, as far as JIA is concerned is not very far, as disparities and heterogeneities are being solved, with new insights into the classification, genetic associations, newer immunological methods, and the advent of Biologics especially in rediscovering SoJIA. Vision

Table 8 : Perspective global and Indian - ILAR criteria

Salient Features	Chennai C. P. Rajendran (4 years) 2002-2006	Kolkata D. sircar (1 year) 2003-2004	Italy stabile A (10 years)
Total number	495	50	85
SoJIA	105 (21.2%)	11 (22%)	28.2%
Oligoarthritis	75 (15.2%)	14 (28%)	55.3%
Persistent	71		
Extended	4		
Polyarticular	175 (35.3%)	15 (30%)	11.8%
RF Negative	134	10	
RF Positive	41	5	
Psoriatic	8 (1.6%)	9 (18%)	1
ERA	131 (26.46%)	1	1
Undifferentiated	1 (0.20%)	-	2

2007 will see JIA taking giant strides in all aspects and it will no longer be "***Alice in wonderland and the mad Hatter***!"

However, the 2011 American College of Rheumatology recommendations for the treatment of Juvenile idiopathic arthritis have other suggestions to offer[20]. They applied the established Research and Development/University of California at Los Angeles (RND/UCLA) Appropriateness Method to derive recommendations.

These recommendations describe five JIA treatment groups. They are

1) History of arthritis of four or fewer joints.
2) History of arthritis of five or more joints.
3) Active sacroiliac arthritis.
4) Systemic arthritis with active systemic features and without active arthritis.
5) Systemic arthritis with active arthritis and without active systemic features.

Specific features for poor prognosis are described for each category, as well as disease activity levels.

The recommendations focus on the initiation and safety monitoring of therapeutic agents used in the treatment of JIA, specifically NSAIDs, Intra-articular glucocorticoid injections, nonbiologic disease-modifying antirheumatic drugs (DMARDs), biologic DMARDs, and systemic glucocorticoids. Medications used "off label" as part of common and widely accepted practice were considered.

Treatment algorithms have been given for all the five categories as enlisted in the Tables below (9, 10, 11, 12 and 13 - obtained from ACR 2011 Guidelines).

DMARDs

- Weaning methotrexate after a duration of remission of 12 months on treatment was not superior to a duration of 6 months after achieving remission.
- The biomarker S100 MRP A8/A14 turned out to be a valuable prognostic marker with a higher relapse rate in patients with higher blood levels at the time of discontinuation of treatment.
- Conventional therapy with methotrexate, leflunomide or sulfasalazine is often not successful in ameliorating disease, especially in patients with polyarticular and systemic - onset JIA.
- Approximately 50% of the clinical disease activity persists despite prolonged treatment.

Biologics

Etanercept

Efficacy of Etanercept in polyarticular disease refractory to treatment lasted for 8 years. The therapeutic goal 'remission', meaning an 'inactive disease', is reached after 2 years of treatment in approximately half of the Registry patients. A prospective Phase IV study demonstrated safety and efficacy of once weekly application of etanercept with a dose of 0.8mg/kg/week (maximum dose 50 mg/week).

Adalimumab

Adalimumab seems valuable not only for treatment of arthritis but also for chronic recurrent anterior uveitis, which emerged in approximately 20% of JIA patients especially in oligoarticular JIA, seronegative polyarthritis and juvenile psoriatic arthritis.

Table 9

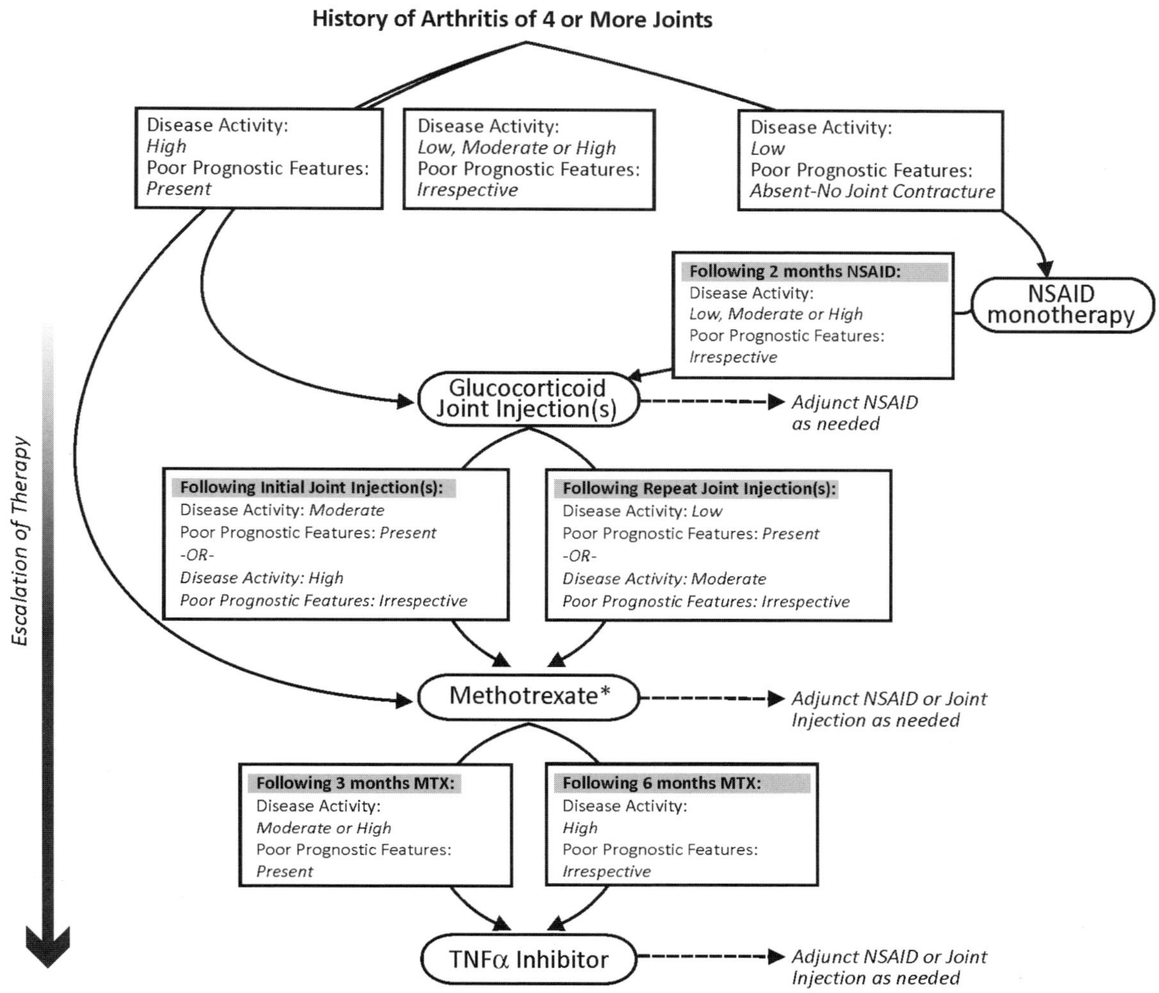

Table 10

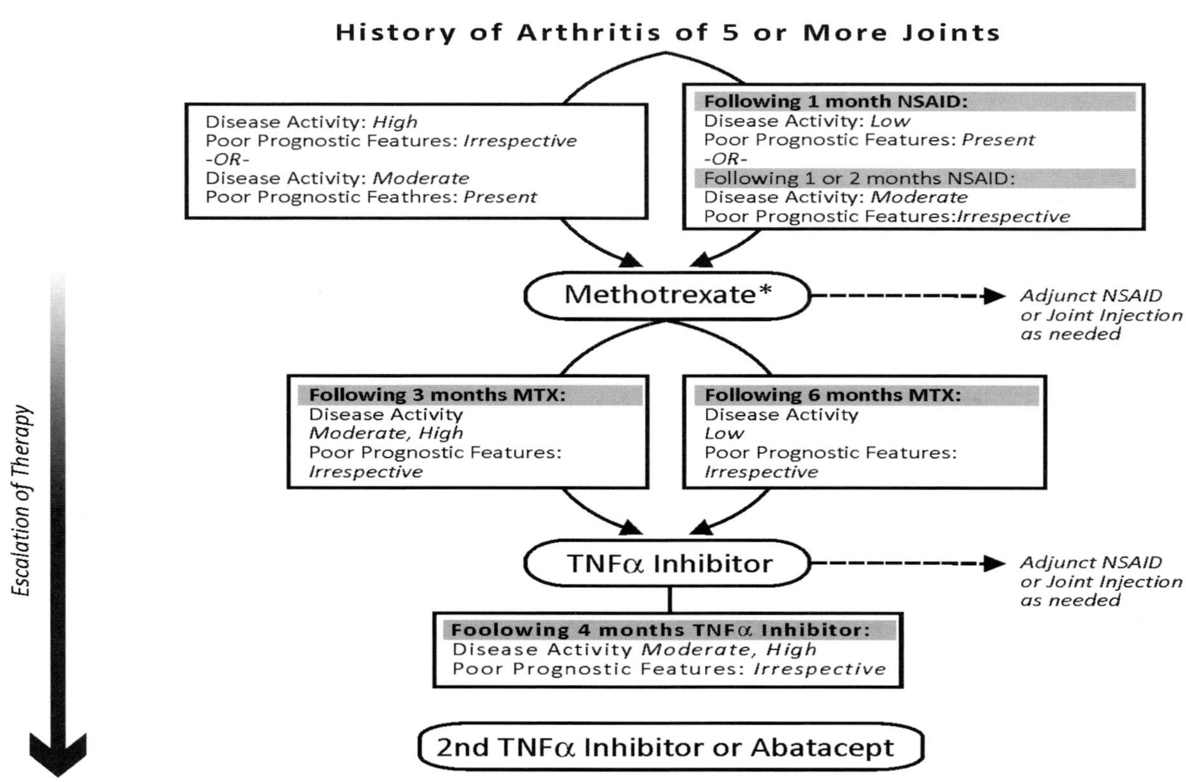

Table 11. Initiation of TNFα inhibitor is recommended for patients who satisfy the following key clinical parameter criteria:

Prior Treatment	Disease Acitivity	Features of Poor Prognosis
Adequate trial of NSAIDs	High	Present
Methotrexate for 3 months	High	Irrespective
Methotrexate for 3 months	Moderate	Present
Methotrexate for 6 months	Moderate	Absent
Sulfasalazine for 3 months	Moderate to high	Irrespective
Sulfasalazine for 6 months	Low	Present

Table 12

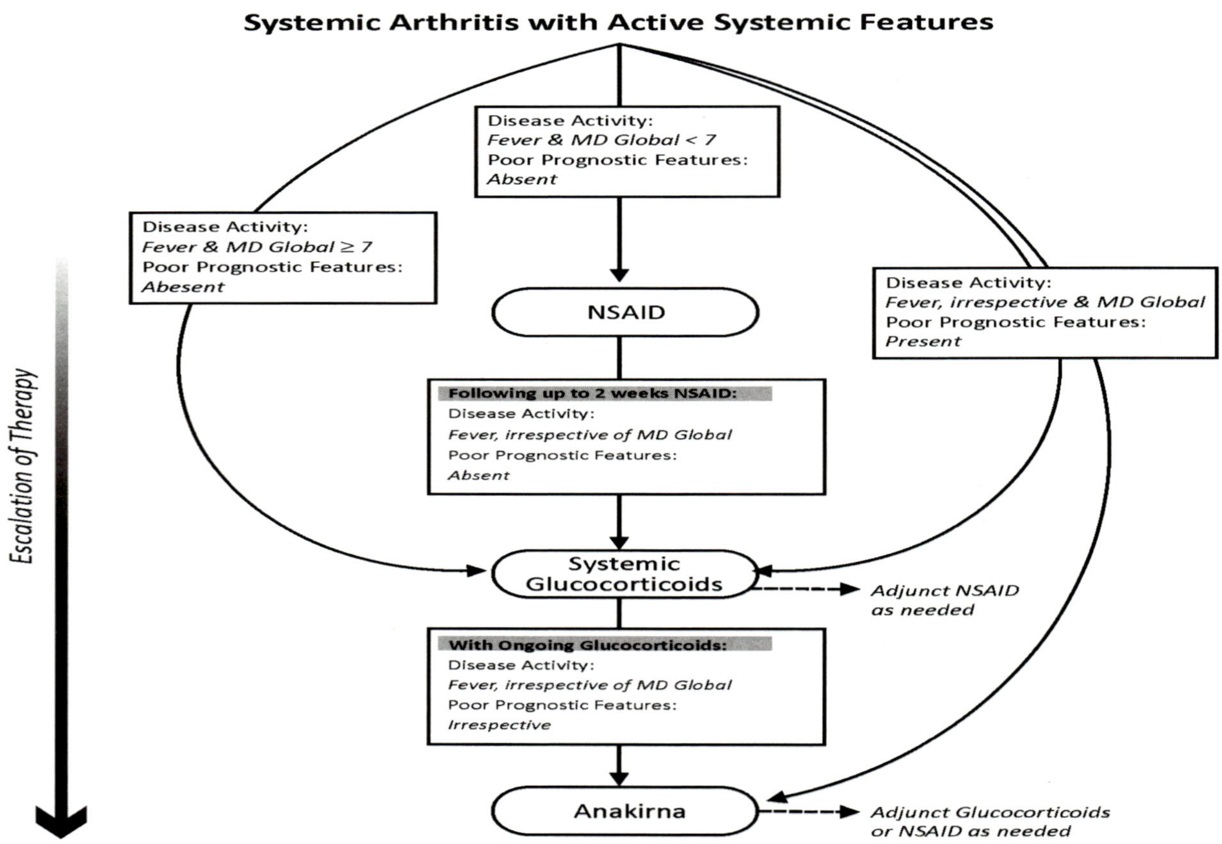

Table 13

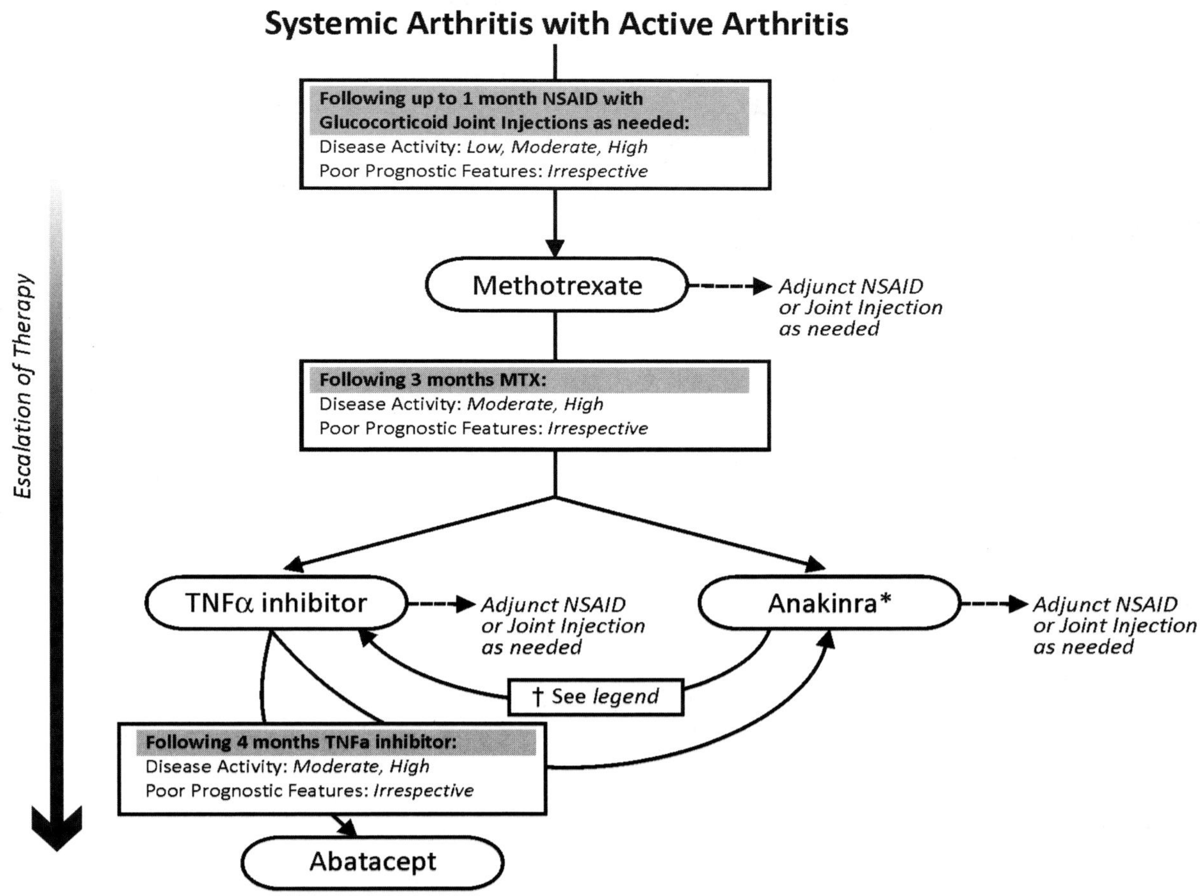

Infliximab

Infliximab may have a special role in JIA patients with uveitis or in patients who are not compliant to subcutaneous therapy.

Golimumab and Certolizumab

Certolizumab pegol, a PEGylated Fab fragment of a humanized anti-TNF antibody, was tested in Phase III trials in Crohn's disease and in rheumatoid arthritis[18]. Certolizumab pegol doses not possess a Fc-region. Therefore, cell-mediated cytotoxicity is not possible. This could possibly result in a decreased risk for infectious diseases compared with full TNF-antibodies. However it may be of value for treatment during pregnancy, since without a Fc-fragment it will not cross the placental border.

Golimumab is a completely humanized monoclonal anti-TNF antibody for subcutaneous application binding both soluble and membrane bound forms of TNF-α. So far it is studied and approved for the treatment of rheumatoid arthritis and adult ankylosing spondylitis while a controlled multicenter trial in JIA (GoKids) is still ongoing.

Anakinra

Anakinra has very good results in the short term, but these may not be sustained. Guidelines suggest that if systemic features are prominent anakinra has the preference, but if arthritis is prominent etanercept may be a good option also.

Tocilizumab

Tocilizumab seems highly effective in treating systemic-onset JIA. Tocilizumab was approved for the treatment of systemic and polyarticular JIA in Japan in 2008.

Rilonacept

Rilonacept is so far not approved for the treatment of JIA.

Canakinumab

Canakinumab a human IL-1 antibody with prolonged plasma half-life, binds selectively to IL-1 without interfering with IL-1RA. It is administered as a subcutaneous injection once monthly. It is effective in the treatment of systemic JIA.

Abatacept

Abatacept has been approved for treatment of JIA patients of 6 years or older with polyarticular arthritis in patients refractory or intolerant to TNF inhibitors.

Rituximab

Rituximab has been tried in systemic and polyarticular JIA. A decrease in disease activity was observed within 6-8 weeks, with a decline in systemic and articular manifestations as well as laboratory parameters of inflammation.

To summarize as per the ACR 2011 guidelines

- First evidence has been presented for a 'window of opportunity' giving preference to an early institution of an effective drug therapy.
- Methotrexate followed by TNF inhibitors are recommended for polyarticular JIA.
- For methotrexate it has been shown that in patients reaching remission on drug for at least 6 months duration, discontinuation is an option.
- Discontinuation of methotrexate treatment in patients with stable clinical inhibitors of IL-1 or IL-6 are considered to be superior while in those patients with prominent arthritis TNF inhibitors are still a valuable option.
- Abatacept is an approved option for JIA patients refractory to treatment with TNF inhibition, while B-cell depletion with rituximab as an off-label alternative is restricted to exceptional patients with resistant and intolerable disease activity.

Key point

1. Juvenile Idiopathic Arthritis in India is different from the West.
2. MRI Imaging is essential as early asymptomatic cervical spine involvement can occur.
3. Systemic onset Juvenile Idiopathic Arthritis is a separate entity with a different pathogenesis requiring a different treatment.

Further Reading :

1. Cassidy James T, Petty RE, Southwood TR: Chronic Arthritis in Childhood, Polyarthritis, Oligoarthritis, Systemic Arthritis, Psoriatic Arthritis. Cassidy James T, Petty RE. Text book of Pediatric Rheumatology Fifth Edition, Elsevier Saunders, Philadelphia, 2005: 2006-311, 318-31.
2. Petty RE, Southwood TR, Manners P, Baum J, Glass DN, Goldenberg J et al. International league of Associations for Rheumatology; Classification of Juvenile Idiopathic Arthritis: Second Revision, Edmonton, 2001. The Journal of Rheumatology 2004; 31:2. 390-92.
3. Ramanan RV, Grom AA. Does SoJIA belong under JIA. Rheumatology 2005; 44(11): 1350-53.
4. Sawhney S, Magatha-es CS. Pediatric Rheumatology - A globalperespective. Best Practice and Research Clinical Rheumatology 2006; Vol.20, No.2: 201-321.
5. Chen C-Y, Tlao-CH, Cu LS, Yang MH, Kuo ML, Huang J.L. Comparison of Soluble adhesion molecules in Juvenile idiopathic arthritis between the active and remission stages. Annals of the Rheumatic diseases 2002; 61: 167-70.
6. Kasapcopur O, Altun S, Aslan M, Karaarslan S, Kamburoglu - Goksel A, Sarba S et al. Diagnostic accuracy of anti-cyclic citrullinated peptide antibodies in juvenile idiopathic arthritis. Annals of the Rheumatic Diseases 2004; 63: 1687-89.
7. Matar - Al, MJ, Petty RE, Tucker LB, Malleson PN, Schroeder ML, Cabral DA. The early pattern of joint involvement predicts disease progression in children with oligoarticular (panciarticular) juvenile rheumatoid arthritis. Arthritis & Rheumatism 2002 Oct; 46(10): 2708-15.
8. Cassidy James T, Petty RE, Juvenile Ankylosing Spondylitis. Cassidy James T, Petty RE. Textbook of Pediatric Rheumatology Fifth Edition, Elsevier Saunders, Philadelphia, 2005: 312-17.
9. Flato Berit, Maria Anna, Vold Hoffman, Reiff Andreas, Forre Oystein, Lien Gunnhild et al., Long Term Outcome and Prognostic factors in Enthesitis - Related Arthritis. Arthritis and Rheumatism. Vol. 54, No.11, 2006 Nov. 3573-82.
10. Stoll ML, Zurakowski D, Nigrovic LE, Nicholi DP, Sundel PR, Nigrovic PA. Patients with JPsA comprise two distinct populations. Arthritis and Rheumatism Vol.54, No.11, 2006 Nov: 3564-72.
11. quarta Laura, Corrado Addolonata, Melillo Nadia, Cantatore Paolo Francesco. JIA - an update on clinical and therapeutic approaches. Ann Ital Med. Int 2005; 20: 211-17.
12. Ferreira RA, Silva CH, Silva DA, Sopelete MC, Kiss MH, Minoo JR et al. Is measurement of IgM and IgA rheumatoid factors (RF) in juvenile rheumatoid arthritis clinically useful? Rheumatol. Int. 2006 29 Sep (Epub).
13. Low JM, Chauhan AK. Determination of anti CCP antibodies in sera of patints with JIA. Journal of Rheumatology 2004 Sep; 31(9): 1829-33.
14. Laiho K, Savolainen A, Kantiainen H, Kekhi P, Kauppi M. Cervical spine in JCA; Cervical spine in Mutilant JCA. Joint Spine 2001 Oct 68(5): 425-29.
15. Lovell DJ, Reiff A, Jones OY, Schneider R, Nocton J, Stein LD. Long term safety and efficacy of etanercept in children with polyarticular course JRA. Arthritis Rheumatism 2006 June, 54(6): 1987-94.
16. Iriyogen, Patricia, Olson, Judyann, Hom, Christine et al. Treatment of Systemic Onset Juvenile Idiopathic Arthritis with Anakinra. Pediatric Rheumatology Online Journal. sOjia.htm 2007 May. 1-9.
17. Wallace CA, Ravelli A, Huang B, Gianni EH. Preliminary validation of clinical remission criteria using OMERACT filter for select categories of JIA. Journal of Rheumatology 2006, April 33(4): 789-95.
18. Kumar Ashok. Experience with anti-tumor necrosis factor - therapy in India. Aplar Journal of Rheumatology, Vol.9, Number 2, 2006 August: 136-41(6).
19. Sircar D, Ghosh B, Ghosh Alakendu, Haldar S. Juvenile Idiopathic Arthritis. Indian Pediatrics, 2006, May, Vol. 43: 429-33.
20. Timothy Beukelman, Nivedita M. Patkar, Kenneth G. Saag et al. 2011American College of Rheumatology Recommendations for the treatment of JIA, Arthritis Care and Research, April 2011: Vol. 63. No. 4, 465-82.
21. Sawhney S. Juvenile idiopathic arthritis: Classification, clinical features, and management. Indian Journal of Rheumatology. 2012; 7(1): 11-21.
22. Vijay Vishwanathan. Acute Rheumatic fever. Indian Journal of Rheumatology. 2012; 7(1): 36-43.

PRINCIPLES AND MANAGEMENT OF JIA

3

Sujata Sawhney

Introduction

Juvenile idiopathic arthritis (JIA) is the commonest rheumatologic condition that afflicts children. This condition has many subcategories of arthritis included under its umbrella and the presentation of these patients is thus varied[1].

The clinical spectrum of patients is protean and may be

- A preschool female toddler who has a painless swollen knee joint noted by a vigilant parent: oligoarticular JIA-OJIA.
- A ten year old male child with high spiking fevers, an evanescent rash that peaks at the spike of the fever and wrist arthritis: Systemic onset JIA-SoJIA.
- A three year old child with symmetric arthritis of the wrists, small joints of the hands and the knees: Rheumatoid factor negative poly JIA-PJIA, RF negative.
- A 15 year old girl with symmetric small joint arthritis of the hands, wrists and elbows: Seropositive JIA: RF positive JIA.
- A 9 year old boy with pain and swelling of an ankle joint, the first metatarsophalangeal joint and buttock pain: Enthesitis related arthritis-ERA.
- A twelve year old girl with a swollen wrist and dactylitis of two digits with a history of psoriasis: Psoriatic arthritis.

Clearly children with such varied clinical presentations should be and indeed are treated differently.

It is important to understand that the JIA criteria are to *classify* children with idiopathic arthritis are not *diagnostic* criteria. The diagnosis still remains one of exclusion, such that each subcategory of the disease has different differential diagnoses to be considered depending upon the age, sex, joint distribution, extra articular features, analysis of the complete blood count etc[2].

Management issues in JIA

This chapter will be discussed under the following headings:

- Principles of management of JIA
- Objective assessment
- Medications available
- Management of various types of JIA
- Specific pediatric issues
  - Growth
  - Deformities
  - Osteoporosis
  - Uveitis
  - When should methotrexate be stopped?
  - Schooling
  - Psychological issues, compliance
  - Prognosis
- Transition of the adolescent with JIA to adult care
- Conclusions

Principles of management of JIA

The aims of good management of JIA are conveniently divided into short and long term goals.

In the short term it is important to

- Try and achieve *complete* disease control. It is now realized that the more rapidly inflammation can be controlled, the less likely it is that there will be permanent sequel. An adequate response is no longer acceptable. The possibility of absolute control of inflammation is a goal that should be pursued. Medically, though the major focus is on arthritis, other extra articular complications need consideration e.g. uveitis, serositis, growth retardation, and osteopenia[3].
- Preserve vision in the child with ANA positive oligoarthritis and uveitis. This subgroup of children is found in India and standard screening for uveitis is essential and must be built into the care plan of these children[4].
- Prevent pain and discomfort.

In the long term it is important to

- Maintain the child in complete disease control
- Prevent joint destruction and deformities
- Promote growth and development
- Ensure psychological well being
- Carefully follow radiological disease progression to follow erosive disease changes
- Limit side effects of drugs
- Rehabilitate the child with deformities[5,6]

Team approach

These aims can be fulfilled if the child with inflammatory joint disease has access to a multidisciplinary team with inputs from a pediatric rheumatologist, who will liaise with the local general pediatrician or GP, ophthalmologist, nurse specialist, physiotherapist, occupational therapist, orthopedic surgeon, podiatrist, clinical psychologist and social worker. Each member of the team has defined roles, and depending upon the clinical situation a member adopts the key role. The therapists are the key personnel to restore function and strength of affected joints and musculature. Together they plan a treatment program that incorporates a range of exercises, stretches for the joints, and activities of daily living. The occupational therapist has two important roles: to provide custom made splints to maintain joint position especially for the wrist and the knee joint, and assist children whose disability requires modification of the environment. Children and adolescents can improve aerobic endurance through participation in weight bearing physical conditioning programs without any disease exacerbation or increased pain. They can also achieve decreased joint signs and symptoms through increased physical activity. Lastly the therapist is usually the key person to educate the parents, and school personnel to ensure integration of therapy goals into the child's daily routine[7].

The challenge is to formulate a rational management plan for each child. Even though it is well known that about one third of patients achieve remission over a long term, the prognostic factors are not well known at disease onset. Hence early referral to a tertiary level centre and careful disease monitoring with attention to three areas: disease activity, damage and radiographic progression must be built into day to day management of each child with JIA[8]. The challenge in our country is to identify the team members under one roof and ensure that the child has a practical and comprehensive care pathway.

Objective assessment

In order to achieve complete disease control it is important that pediatric rheumatologists have easy to use clinical tools that help to accurately and objectively

- Assess disease activity in children with JIA.
- Define inactive disease and remission.

Tools to assess disease activity

This would be equivalent to the DAS 28 scoring for rheumatoid arthritis patients and BASDAI for the patients with ankylosing spondylitis. Currently there are no uniform scoring criteria for JIA, as the disease is very heterogeneous.

Outcome measures are used for children with JIA, more for clinical trials than day to day clinical practice. These are critical objective parameters designed to judge whether or not the patient has improved. An international consensus conference proposed "The definition of improvement" to assess disease response. This definition employs a core set of six response variables: global assessment of the severity of disease by the physician, global assessment of overall well-being by the patient or parent, number of "active" joints (joints with swelling not due to deformity or joints with limitation of motion and with pain, tenderness or both), number of joints with limitation of motion, functional ability by a validated test, and erythrocyte sedimentation rate. To meet the definition of improvement at a scheduled visit the patient should have a 30 percent improvement from base line in at least three of the six response variables, and worsening of 30 percent or more in no more than one of the six response variables[9,10]. These improvements are similar to the ACR 30/50/70 for adult Rheumatoid Arthritis.

Juvenile arthritis disease activity (JADAS) score:

Up until a few years ago no numeric tool (the equivalent of DAS 28) was available for routine use in pediatrics across all disease subcategories. Recently a composite disease activity score for JIA, the JADAS has been recently developed and now defined for disease activity as well. The JADAS includes 4 measures: physician global assessment of disease activity, parent/patient global assessment of well-being, active joint count, and erythrocyte sedimentation rate. A score of 0-57 is given for a 27 joint count. It has revealed strong responsiveness to clinical change which is not true for the Childhood health assessment questionnaire (CHAQ). This may be used as a standard tool in the clinic over the next few years. Recently, cut off values have been proposed to match inactive disease. A cut off of 1 equates inactive disease[11,12].

Remission/inactive disease

Two types of clinical remission are described.

Inactive disease: Is the term used for children with JIA who have no active arthritis, systemic features or uveitis, normal ESR and a physician score that suggests no active disease.

Clinical remission on medication: Is the presence of inactive disease for six months on medication.

Clinical remission: Is the term reserved for children with JIA who have inactive disease for 12 months after discontinuing all medications[13].

These are important concepts as children often flare during the course of their disease and over the long term it is possible to see the time they have spent with inactive disease. In fact, in newer trials in pediatric rheumatology such as the Trial of early aggressive therapy in polyarticular juvenile idiopathic arthritis (TREAT) trial, inactive disease has been used as primary end point underscoring the importance of no disease versus improvement as defined by ACR PEDI 30/50/70 described above[14].

Medications available

There are basically five modalities of treatment that can be used to treat children with JIA. They are

- NSAIDs
- Steroids
- DMARDs: Disease modifying anti rheumatic drugs
- BRMs: Biologic response modifiers
- ASCT: Autologous stem cell transplantation

In the eighties and nineties children with JIA had a poor long term outcome because of late recognition and few pharmacologic choices. The three therapeutic advances in the last decade that have made a substantial difference to the quality of life and outcome of children with JIA are

- Intra articular injections of triamcinalone hexacetonide
- Weekly methotrexate
- Use of twice weekly Etanercept and the availability of several other BRMS[33]

NSAIDs

The choice of NSAIDs is empiric; Clinical trials of NSAIDs by the PRCSG (Pediatric Rheumatology Collaborative Study Group) in the U.S.A. concluded that 65% of the children who were going to respond would do so in the first 4 weeks. If a child does not respond it is logical to use an NSIAD from a chemical class different to the one used earlier. 25-33% of patients with oligoarthritis show a significant response to NSAIDs alone. Aspirin is not used commonly any longer for the following reasons - the need for frequent dosing, greater frequency of liver enzyme abnormality, and the association of Reye's syndrome[15,16]. Only ibuprofen and nimesulide are available in the syrup form in India. In the vast majority of children who tolerate non selective cox inhibitors well, there is no added advantage of using a cox2 inhibitor. With the availability of biologics and routine use of methotrexate, the number of patients' currently using non-steroidal anti-inflammatory drugs has reduced over the years. On a review of charts in a US centre in 2003, 53% used an anti-inflammatory dose compared to 35% in 2008[17].

Steroids

These are very potent drugs and can be used in JIA in many ways: Oral, Intravenous and also intraarticular. The use of systemic steroids is limited primarily because of their well known side effect profile with long term use. Oral low dose steroids are used sometimes to help patients with severe poly JIA. Both IV and oral steroids are used to treat disease flares in SoJIA[8]. It is important to use the lowest possible dose for a short interval and not let the high disease activity be masked by long term steroids[18,19].

Triamcinalone hexacetonide is the preferred intraarticular steroid preparation due to the long-term and predominantly local effects, with a reported median duration of improvement for a period of 74 weeks. In 60% of patients the response persists for at least 6 months, in 45% there is no inflammation for at least one year. Early and continued use of intraarticular steroid has been recently shown to be associated with less leg length discrepancy in young children with oligoarticular JIA[20]. This agent is not available in India, where we use Triamcinalone Acetonide.

Side effects of intraarticular steroids are very few. It is a very safe procedure that may need appropriate sedation in a child under 7 or 8 years of age. Problems after a joint injection are:

- Periarticular subcutaneous atrophy.
- Asymptomatic calcification.

There is no report of it being associated with any joint or cartilage damage[21,22].

Disease modifying anti rheumatic drugs (DMARDs)

Methotrexate

Introduction: Has been established as a treatment for JIA in 1992 after a controlled trial showed efficacy at 10mg/m2[23]. This drug is the first choice across most centers in the world.

Dose and route: The route can be oral or injectable, and the safety profile is very good. Most children tolerate this medicine very well and in younger children doses at 1mg/kg/week are commonly used. The maximum dose is usually restricted to 25-30 mg/week. Patients respond within three months, very occasionally is the response delayed beyond this. There is recent data to support the use of higher dose of methotrexate as published recently in the TREAT trial where a weekly dose of upto 40mg was given to children with poly JIA[14]. The poorest response with methotrexate is in patients with systemic onset JIA (SoJIA). Folic acid at a dose of 1mg/day is often given to children receiving methotrexate and appears to lessen the toxicity and increase the tolerability of the drug.

Side effects, precautions and monitoring: Precautions to be exercised while the patient is on methotrexate are as follows: no live vaccination, avoidance of alcohol, and no pregnancy while on the drug. Regular monitoring of hemoglobin, white cell count, platelets and liver function tests every 10-12 weeks is mandatory. Many patients develop nausea and emesis a few years after taking methotrexate; this is because of accumulation of polyglutamated methotrexate intracellularly[24].

Leflunomide

Introduction: Is another DMARD that has been recently been shown to be as effective as methotrexate. It can be used alone or in combination with other DMARDs.

Dose: between 5-20mg per day depending upon the size of the child.

Precautions: The female patients who start this drug should be strictly counseled against pregnancy as this is highly teratogenic. Some children may develop diarrhea while on it. The liver function and complete blood count needs to be checked every 10-12 weeks[25].

Sulfasalazine

Introduction: This medication is effective in children with enthesitis related arthritis and interestingly contraindicated in the active phase of systemic onset JIA, where it has been reported to cause macrophage activation syndrome.

It is a combination of sulphapyridine and 5 amino salicylic acid, and thus has a cross reactivity with other sulpha containing compounds.

Dose and route: It is orally administered in two divided doses per day. The starting dose is 10mg/kg /per day, built up to 50mg/kg/day.

Side effects and precautions: This drug can cause side effects in 11-31% of patients: nausea, rash, elevated liver enzymes and cytopenias. As with methotrexate the complete blood counts and liver enzymes need monitoring every 8 weeks while the patient is on sulphasalazine[26].

Hydroxychloroquine

Introduction: An anti malarial, it has been used in children with JIA since 1951. Is a mild but safe DMARD. There is only one placebo controlled trial that has shown the drug to be effective in reducing pain on movement[27].

Dose and Route: In children the dose is 6mg/kg/day, given as a single daily dose.

Side effects and monitoring: Ophthalmic screening every six months to one year is recommended in children who take this drug. This is to screen for retinal toxicity. It can cause a "grey tone change" to the skin color in Asian patients. It may rarely cause nausea, dyspepsia and abdominal discomfort in some children[7].

Other DMARDs: Only a brief mention will be made of these agents

Cyclophosphamide

It is a powerful alkylating agent and a place only for the treatment of severe SoJIA where it has been used in combination with IV methyl prednisone boluses[28].

Cyclosporine

This drug inhibits the early phase of T cell activation and production of several cytokines. It has been combined with methotrexate, shown to reduce fever in SoJIA, be effective for uveitis and specifically for treatment of Macrophage Activation Syndrome. It is used only occasionally in clinical practice. The use is limited by the side effect profile which includes increased facial hair, gingivitis, hypertension and nephrotoxicity[29].

Thalidomide

Has both anti TNF and anti angiogenic properties. It has a place in the treatment of refractory SoJIA. Used in a dose of up to 5mg/Kg/day it has been shown to be effective in > 50% of patients with refractory SoJIA. Two important side effects are somnolence and peripheral neuropathy; the latter is irreversible and needs to be monitored[30].

Biologics

Introduction

There are many cellular and molecular mechanisms that participate in inflammation in patients with juvenile arthritis. Specific biologic agents have been developed that can target one or more steps involved in the immune response[31].

Interference with cytokines

T-cell and monocyte derived cytokines are responsible for clinical features seen in JIA such as arthritis, enthesitis, and for systemic features like fever. Cytokines are thus important targets for therapeutic manipulation. The cytokines that can be manipulated in clinical practice are those that have an important role to play in both the initiation and perpetuation of the rheumatoid process. The three main cytokines that can be manipulated are:

Tumor necrosis factor

IL-1
IL-6

Agents that block TNF alpha are excellent tools to reduce disease activity, improve function and even retard or reverse structural damage. Three anti-TNF agents available are etanercept, infliximab, and adalimumab. The use, dose and side effects of each are detailed in Table 1. Double blind, randomized controlled studies have shown all three agents to be an effective therapy in patients with juvenile idiopathic arthritis (JIA). Currently these drugs are given to patients who do not respond to an adequate trial with methotrexate[32-35].

Generic Issues with Anti-TNF agents

Safety issues are a concern because of the ubiquitous role of TNF.

- To date the only consistent adverse event seen with etanercept has been injection site reactions. There should be caution, however, with using etanercept in patients with a serious infection, or recurrent infections or patients with untreated or latent tuberculosis.
- There has been a recent controversy about the role of TNF blockers and malignancies. The FDA reported that there is evidence that treatment with TNF blockers in children may increase the risk of malignancy. However, the cases were confounded by the potential risk of malignancy associated with underlying illnesses and the use of concomitant immunosuppressants; therefore, a clear causal relationship could not be established. The FDA had 48 cases of malignancy in children treated with TNF blockers, half were lymphomas[36]. The German Biologics registry identified five of 1290 patients to have a malignancy over an eight year observation period. All patients had been exposed to multiple cytotoxics in addition to TNF blockade[37]. Thus it is important that patients on TNF inhibitors be followed up carefully and screened appropriately over the long term. Some rare neurological and hematological events have been noted.
- Active TB may develop soon after initiation of treatment with infliximab. Thus before prescribing the drug physicians should screen patients for latent tuberculosis infection or disease. Screening is usually with X-Ray Chest, PPD and CT chest if needed. The risk of TB is more with infliximab than with etanercept as it binds to both the soluble and membrane bound TNF[38,39].

Table 1 : Anti TNF agents for JIA[34]

Name	Etanercept	Infliximab	Adalimumab
Chemical content	Fully humanized	Mouse chimeric	Fully humanized
Chemical structure	P-75 soluble TNF receptor protein fused to human Fc region of human IgG_1	Chimeric IgG_1 Anti TNF alpha antibodies – mouse antibodies and constant region of human antibody.	Recombinant human IgG_1 monoclonal antibody
Binds to	Soluble TNF alpha	Soluble TNF alpha and membrane bound TNF alpha	Soluble TNF alpha and membrane bound TNF alpha
Half life	Four days	Weeks	2 weeks
Route	Subcutaneous	Intra venous (IV)	Subcutaneous
Dose	0.4mg/kg/dose/twice weekly or 0.8mg/kg once weekly	3-6 mg/kg/dose	24 mg/m^2
Frequency	Bi weekly or weekly	At 0, 2, 6 weeks and then every eight weeks	Every 2 weeks
Methotrexate use	Optional	Recommended (to prevent anti infliximab antibodies)	Optional
Risk of Tuberculosis	+	++	++

- Infliximab a mouse chimeric antibody, is known to give infusion reactions and must be given in a hospital setting with adequate monitoring facilities. Though in the western world their use is open ended, in India these agents are prohibitively expensive and are used to reduce disease burden and are very seldom continued for years on end. The rough cost of etanercept therapy for a 30 kg child is 4.16 lakh rupees for one year, for infliximab is 3.24 lakh rupees per year + incidental hospital charges for nine admissions in the first year. Adalimumab is not yet available in India.

Anakinra

This is interleukin-1 receptor antagonist (IL-1Ra). It is a naturally occurring acute phase anti inflammatory protein part of the IL-1 super gene family. It is an important physiologic regulator of IL-1 induced inflammatory activity. Anakinra is a human recombinant form of IL-1 Ra that is produced by recombinant technology in E. coli. It is given daily by subcutaneous injection. In pediatrics it is used in a dose of 0.1 mg/kg/day and it has been shown to be effective in etanercept resistant SoJIA patients[40]. This is currently unavailable in India.

Anti IL-6 receptor antibody

There is evidence that IL-6 is important in the clinical features of SoJIA. Anti IL-6 receptor antibody (MRA) is a genetically engineered humanized monoclonal antibody. There is evidence for the use of tocilizumab, an IL 6 blocker for both SJIA and PJIA. This drug has recently received FDA approval in children with SoJIA in USA. It is used in a dose of 12mg/kg in children with a weight $<$ 30 kg and 8 mg/kg in children with weight $>$ 30 kg. The systemic features respond within a few weeks, arthritis abates more slowly[41,42].

Tolerance induction

Trials of oral type II collagen have been tried in adults and children, as there is documented auto immunity to type II collagen in both adult and juvenile arthritis. Studies in humans have shown variable results, but with advances in other areas this has not been further studied.

Inhibition of MHC/antigen/T cell receptor (TCR) interaction

This is also called the tri molecular complex. If the initiating antigen was known an immunization program could be developed to prevent disease. Another possible route of attack is to block the MHC sites by anti MHC antibodies. This is currently under study.

Inhibition of cellular function and cell to cell interaction

T-cells are critical in initiating the rheumatoid process in both children and adults. Another promising approach has been to block co-stimulatory interactions between T and B cells, for example by inhibiting B7 pathway with CTLA-4Ig. Initial results in adult patients look promising[31]. A selective T-cell co-stimulation modulator Abatacept has been shown to be very effective in children with polyarticular JIA both in the short term and long term, upto 21 months[43,44].

B Cell blockade: The Anti CD 20 molecule Rituximab is very effective in adult Rheumatoid Arthritis. There is no class I evidence for its use in JIA, however there are recent reports of the effectiveness of Rituximab in both PJIA and SoJIA which need to be replicated in other centres[45,46].

Stem cell transplantation in JIA

Autologous haemopoietic stem-cell transplantation (AHSCT) has been described as a possible treatment for severe autoimmune disease refractory to conventional treatment. The first four children with severe forms of juvenile idiopathic arthritis to receive AHSCT were reported in 1999, Collaborative European trails with strict entry criteria and pre transplant conditioning are ongoing. Patient selection is critical prior to undertaking this modality of treatment where the mortality has now come down from 14% to 5%[47].

Management of various sub types of JIA [3,7,8,31,48]

OJIA

Mild disease can be treated with NSAIDs or Intraarticular steroids. With functional limitations both modalities are started together. If only NSAIDs are used alone, marked improvement should occur in 4 weeks and resolution of the involved joints occurs in 3 to 4 months. Once the patient is in clinical remission on drugs (6 months) the NSAIDs may be discontinued.

If intraarticular steroids are used resolution occurs in a week. Adequate, age appropriate sedation administered by a skilled team is a prerequisite.

Disease modifying anti rheumatic drugs (DMARD's) are generally not recommended for this group, although occasionally methotrexate may be useful for difficult to control uveitis or when damage to a critical joint threatens function (wrist/hip). If the disease extends, methotrexate should be considered for the treatment. OJIA is not always a benign disease. Persistent elevation of ESR, long standing anemia, or difficult to control arthritis and or uveitis are all indications for aggressive treatment/DMARD use[49].

Polyarticular JIA

The patient with mild disease can be treated with NSAIDs and Hydroxychloroquine. The vast majority of children will need methotrexate. Low dose steroids which are tapered over a few months are very useful in many patients and the benefits outweigh the risks. TNF antagonists are reserved for the methotrexate resistant patients.

SoJIA

The patients who have this disease vary significantly from each other. There are some children with raging fevers and barely discernible arthritis, others with severe arthritis and then a group that have a life threatening condition called macrophage activation syndrome where the child with SoJIA develops high grade continuous fever, a marked drop in the platelets and ESR and features of coagulopathy. A child with very mild disease may improve with NSAIDs alone over a period of a few weeks. A vast majority of the children will require steroids that can be given orally at a dose of 1-2 mg/kg/day slowly tapered off over a period of months or even IV methylprednisolone at a dose of 30 mg/kg/day, maximum 500mg given daily for 2-3 days. Methotrexate has a good effect on the arthritis but its role in reducing the systemic features of this condition is not well understood. Thalidomide may be used in the resistant patient where it controls both the arthritis and the systemic features. The resistant patients are a challenge to manage and strategies that are beneficial include use of pulsed methyl prednisolone, thalidomide and biologics. The biologic agents used in this subcategory are Anakinra and tocilizumab; TNF blockers have the least efficacy in this subcategory.

Cyclosporine has a dramatic effect in children with macrophage activation syndrome and may also be used to control systemic features in the difficult to control patients.

Enthesitis related arthritis

The components of this disease that need attention are

- Acute anterior uveitis
- Peripheral joint disease
- Enthesitis
- Axial joint disease

If the patient has <4 peripheral joints without critical joint involvement such as hip disease they may be treated with NSAIDs and intraarticular steroids. With hip joint disease most children would get methotrexate as for polyarticular disease. Sulfasalazine is also effective in these patients. With axial disease the child would need regular NSAIDs for weeks together. If there is no improvement in three months TNF blockers should be used. Acute uveitis may be treated with topical steroids, and in recurrent uveitis the child will benefit with long term methotrexate or infliximab[50].

Psoriatic arthritis (PsA)

Asymmetric involvement of small joints especially the DIP joint and dactylitis is characteristic of psoriatic arthritis. Significant nail pitting often precedes arthritis. The skin and joint disease may not always follow the same course. In addition to local skin treatment NSAIDs/intra-articular steroids are used for localized disease involving a few joints, and methotrexate is used for aggressive disease involving multiple joints[51].

American College of Rheumatology guidelines for the management of JIA

The ACR has chosen to describe the pathways of care for five groups of children: Oligo disease, poly disease, SoJIA with systemic features, SoJIA with articular disease and finally, the child with axial disease. Each category of disease has been divided into low, moderate and high disease activity and poor prognostic markers have been listed for each category. The keys points that have not been covered in the ACR recommendations are:

- The indications for systemic glucocorticoids for the treatment of synovitis were not considered, owing to a lack of published evidence.
- The treatment of uveitis, enthesitis, and macrophage activation syndrome was not considered.

Role of folic acid, weaning of methotrexate/biologic response modifiers (BRM)/abnormal liver function tests have not been discussed. The guidelines are heavily weighted in favour of use of biologics within 3-6 months of ongoing disease in most categories which is not practical for the majority of Indian children with JIA. Each category is stratified per the prognostic features with details available on the application of these recommendations for the practicing clinician[52,63].

Specific pediatric issues

Growth

Significant chronic disease in childhood often impacts the growth in children and thus all children with JIA should be placed on a growth chart and periodic assessment of their height velocity should be calculated. Not just the disease but the drugs used to control severe inflammation such as steroids can permanently stunt the height of children. The best strategy to maximize growth is aggressive disease control, nutritional support and judicious yet minimal use of steroids.

Deformities

Unlike adults, children with arthritis have a growing skeleton and are vulnerable to developing deformities and flexion contractures early in the disease process. Thus a child with a single knee involvement may develop a knee flexion contracture, overgrowth of the involved limb and ultimately scoliosis if left untreated. Early appropriate multi disciplinary team management is the answer to these issues.

Osteoporosis

There are many factors that adversely affect bone mass in children with JIA. Active arthritis, medications used in arthritis, especially steroids also have a known osteopenic effect. Decreased physical fitness and participation in organized sport, in addition to poor vitamin D and calcium intake contribute to the low bone mineral density (BMD). Interventional studies are lacking, and until sound evidence is available the following strategies are generally employed to optimize bone mass in children with JIA: aggressive control of disease activity, avoidance of corticosteroid use, and optimizing physical activity and calcium intake.

Bisphosphonates - alendronate has been recently shown to be effective in treating secondary osteoporosis in juvenile idiopathic arthritis. This drug crosses the placenta and is highly teratogenic. It is thus mandatory for patients to avoid pregnancy during the duration of treatment and for up to six months after discontinuation. Finally, growth hormone also improves the growth retardation and osteoporosis, particularly where the disease is stable but not in remission[53,54].

Uveitis

Standard treatment of uveitis is the use of topical methyl prednisolone and mydriatics to prevent synechiae. Methotrexate and other DMARDS have been tried; benefit is seen with methotrexate at higher doses, when used subcutaneously. Uveitis in JIA is silent and painless except in children with enthesitis related arthritis that have acute painful anterior uveitis. Thus, it is mandatory that patients with JIA be electively screened for uveitis at regular intervals, as untreated uveitis is known to cause visual loss and morbidity[20,55]. Recently it has been shown that infliximab is effective in treatment of uveitis.

When should methotrexate be stopped?

There is paucity of published data to support evidence-based decisions in this area. Most authors suggest that discontinuation of methotrexate when treatment induced remission has persisted for less than 1 year frequently results in return of arthritis within 6 months of drug discontinuation. NSAIDS and other medications are discontinued prior to attempting withdrawal of methotrexate[24]. Newer biomarkers will aid in better judgment re the risk of relapse for the patients. Recent data suggest that MRP8/14 serum concentrations correlate closely with disease activity and therefore might be an additional measurement for monitoring anti-inflammatory treatment of individual patients with SJIA. MRP8/14 serum concentrations are the first predictive biomarkers indicating subclinical disease activity and stratifying patients at risk of relapse during times of clinically inactive disease[56].

Schooling

Attending school full time and participating in all school activities is an achievable goal in all children. An occupational therapist is a useful member of the team to liaise with the school authorities regarding seating, pen grips, physical education etc. It is important that the disease be in remission so that the child has the opportunity to attend regular full time school which will serve two purposes: education and peer interaction which in turn help to make the child a balanced adult.

Psychological issues, disease education and compliance:

This is as important as the medical care. Important issues while looking after this area are:

- The child should be strongly encouraged to attend school regularly and children should determine their own level of activity.
- Participation in recreational activities and camps with other children with arthritis is very beneficial. As children grow up vocational evaluation is very important for the adolescent with arthritis.
- Disease education which is culturally appropriate must be built into the care plan of the child with arthritis. Verbal information should be followed on with written information about both the disease and drugs given to the child. The family must be part of the decision making tree for the child and should be appropriately counseled about the disease activity and be given realistic expectations for the future.
- Psychological assessment for the child with chronic disease is a routine process in the western

world, but is not common place in India. Children and families should be assessed for depression and inability to cope with the disease. If there are any concerns they should be addressed by a trained counselor/psychologist.

- Compliance with hospital visits, drug intake and following exercise regime are issues that must be addressed periodically

Prognosis

The prognostic criteria are not well defined for juvenile arthritis. Some features do predict ongoing disease activity at follow up. They are: young age at onset, positive rheumatoid factor, long duration of elevated ESR and polyarthritis[57]. Schneider et al studied children with systemic onset juvenile idiopathic arthritis (SoJIA) and suggested that children with SoJIA who at six months of disease had persistent systemic features and a platelet count of more than six lakhs were likely to have severe destructive arthritis and a poor functional outcome[58].

Data from India suggests that after a median follow-up of 10 years up to 60% of patients have active disease. Certainly a major contributor here is delayed referral and delayed institution of disease modifying agents[59].

Transition of the adolescent with JIA to adult care

JIA is a chronic disease, with a one third of patients carrying the disease into their adult years. Delayed adolescence both physically and emotionally is being recognized more widely and adolescent centered services to aid transition to adulthood have a major role to play in the long term care of the patient with JIA. Transition to adult care is a process that begins in the adolescent age. Attention to vocational skills, independent living skills, and self-advocacy warrant careful care and planning. Compliance with the treatment regime including medication, blood monitoring, exercises, splint usage and regular visits to health professionals are demanding on the child and family. Multi-disciplinary teams under one roof are ideally suited to deliver this care. Patient and parent education and incorporation of the needs of the child and family members in planning care are essential to ensure that the patient adheres to the plan[60,61].

Conclusion

Development of new therapies has dramatically increased our ability to treat children with JIA, and hopefully induce remission in many children. There is though, lack of evidence based treatment for some JIA subtypes. Effect of early aggressive therapy on the disease, early use of BRMs and potential of combination induction therapy needs to be studied. Multi center control trials will provide this much needed evidence in the near future[62].

It is important for pediatric rheumatologists in turn to objectively assess patients, look at the core set criteria in these patients at set points in time, and guide therapy with the aim to achieve remission as per standard definitions recently validated for use for JIA patients.

Strategies likely to help children with JIA in our country are:

- Formal inclusion of pediatric rheumatology core curriculum both at the graduate and postgraduate levels in medical schools in India.
- Increased awareness amongst general pediatricians and orthopedic surgeons about the prevalence of JIA in India and the urgency of early referral of these patients to specialist centers.
- Establishment of several tertiary level centers with good skill and expertise to direct care for these children.

Summary table

1. JIA is a group of heterogeneous diseases; it remains a diagnosis of exclusion.
2. Each subcategory of the disease should be managed per the age of the child, number of joints involved and the criticality of involved joints.
3. The disease should be assessed objectively and regularly assessed with the aim of achieving clinical remission in shortest possible time.
4. Early diagnosis and appropriate management by a multidisciplinary team gives the best possible outcome.

Further Reading :

1. Petty RE, Southwood TR, Baum J, Bhettay E, Glass DN, et. al. Revision of the proposed classification criteria for juvenile idiopathic arthritis: Durban, 1997 [see comments]. J Rheumatol. 1998; 25(10): 1991-94.
2. Sawhney S. Juvenile idiopathic arthritis: Classification, clinical features, and management. Indian Journal of Rheumatology. 2012; 7(1): 11-21.
3. Cassidy JT, Petty RE. Chronic Arthritis in Childhood. In text book of pediatric rheumatology. 2005, 5th edn. pg. 206-260. Elsevier Saunders.
4. Sawhney S. Spectrum of juvenile idiopathic arthritis in a tertiary care centre in Northern India. Indian Journal of Rheumatol. 2006; 14(1): 3-6.
5. Wallace CA. On beyond methotrexate treatment of severe juvenile rheumatoid arthritis. [Review] Clin Exp Rheumatol. 1999; 17(4): 499-504.
6. Rossi F, Di Dia F, Galipo O, Pistorio A, Valle M, Magni-Manzoni S, et al Use of the Sharp and Larsen scoring methods in the assessment of radiographic progression in juvenile idiopathic arthritis. Arthritis Rheum. 2006 15; 55(5): 717-23.
7. Wallace CA. Current management of juvenile idiopathic arthritis. Best Pract Res Clin Rheumatol. 2006 20(2): 279-300.
8. Ravelli A, Martini A. Juvenile idiopathic arthritis. Lancet. 2007 3; 369(9563): 767-78.
9. Giannini EH, Ruperto N, Ravelli A, Lovell DJ, Felson DT, Martini A. Preliminary definition of improvement in juvenile arthritis. Arthritis Rheum 1997; 40: 1202-09.
10. Giannini EH, Lovell DJ, Felson DT, Goldsmith CH. Preliminary core set of outcome variables for use in JRA clinical trials. Arthritis Rheum 1994; 37: Suppl: S428.
11. Consolaro A, Ruperto N, Bazso A, Pistorio A, Magni-Manzoni S, Filocamo G, et al, for the Paediatric Rheumatology International Trials Organisation. Development and validation of a composite disease activity score for juvenile idiopathic arthritis. Arthritis Rheum 2009; 61: 658-66.
12. Consolaro A, Bracciolini G, Ruperto N, Pistorio A, Magni-Manzoni S, Malattia C, et al. Remission, minimal disease activity and acceptable symptom state in juvenile idiopathic arthritis. Arthritis & Rheumatism. 2012: doi: 10.1002/art. 34373.
13. Wallace CA, Ruperto N, Giannini E; Childhood Arthritis and Rheumatology Research Alliance; Pediatric Rheumatology International Trials Organization; Pediatric Rheumatology Collaborative Study Group. Preliminary criteria for clinical remission for select categories of juvenile idiopathic arthritis. J Rheumatol. 2004; 31(11): 2290-94.
14. Wallace CA, Giannini EH, Spalding SJ, Hashkes PJ, O'Neil KM, Zeft AS, et al. Trial of early aggressive therapy in polyarticular juvenile idiopathic arthritis. Arthritis & Rheumatism. 2012; 64(6): 2012-21.
15. Kvien TK. Hoyeraal HM, Sandstad B. Naproxen and acetylastliclic acid in the treatment of pauciarticular and polyarticular juvenile rheumatoid arthritis: assessment of tolerance and efficacy in a single center 24-weeks double-blind parallel study. Scand J Rheumatol. 1984; 13: 342-50.
16. Giannini EH, Cawkwell GD. Drug treatment in children with juvenile rheuamatoid arthritis. Pediatr Clin North Am. 1995; 42: 1099-1125.
17. Kochar R, Walsh K, Jain A, Spalding S, Hashkes P. Decreased use of non-steroidal anti-inflammatory drugs for the treatment of juvenile idiopathic arthritis in the era of modern aggressive treatment. Rheumatol Int. 2011; 10. 1007/s00296-011-2084-87.
18. Kahn P. Juvenile idiopathic arthritis - an update on pharmacotherapy. Bulletin of the NYU hospital for joint diseases. 2011; 69(3): 264-76.
19. Beresford MW, Baildam EM. New advances in the management of juvenile idiopathic arthritis-1: non-biological therapy. Arch Dis Child Educ Pract Ed 2009; 94: 144-50.
20. Sherry DD. What's new in the diagnosis and treatment of juvenile rheumatoid arthritis. [Review] J Pediatr Orthop. 2000; 20(4): 419-20.
21. Padeh S, Passwell JH. Intraarticular corticosteroid injections in the management of children with chronic arthritis. Arthritis Rheum. 1998; 41: 1210-14.
22. Hagelberg S, Magnusson B, Jenner G, Andersson U. Do frequent corticosteroid injections in the knee cause cartilage damage in juvenile chronic arthritis? Long-term follow-up with MRI. J Rheumatol. 2000; 27(suppl 58): 95.
23. Giannini EH, Brewer EJ, Kuzmina N, Shaikov A, Maximov A, Vorontsov I, et al.Methotrexate in resistant juvenile rheumatoid arthritis. Results of the U.S.A.-U.S.S.R. double-blind, placebo-controlled trial. The Pediatric Rheumatology Collaborative Study Group and the Cooperative Children's Study Group. N Engl J Med. 1992 16; 326(16): 1043-49.
24. Wallace CA. The use of methotrexate in childhood rheumatic diseases. Arthritis Rheum. 1998; 3: 381-91.
25. Silverman E, Mouy R, Spiegel L Jung LK, Saurenmann PK, et al. Leflunomide or methotrexate for Juvenile Rheumatoid Arthritis. N Engl J Med. 2005; 352(16): 1655-66.
26. Burgos-Vargas R, Vazquez-Mellado J, Pacheco-Tena C, Hernandez-Garduno A, Goycochea-Robles MV. A 26 week randomised, double blind, placebo controlled exploratory study of sulfasalazine in juvenile onset spondylo-arthropathies. Ann Rheum Dis. 2002; 61(10): 941-42.
27. Brewer EJ, Giannini EH, Kuzmina N, Alekseev L. Penicillamine and hydroxychloroquine in the treatment of severe juvenile rheumatoid arthritis. Results of the U.S.A.-U.S.S.R. double-blind placebo-controlled trial. N Engl J Med. 1986 15; 314(20): 1269-76.
28. Shaikov AV, Maximov AA, Speransky AI, Lovell DJ, Giannini EH, Solovyev SK. Repetitive use of pulse therapy with methylprednisolone and cyclophosphamide in addition to oral methotrexate in children with systemic juvenile rheumatoid arthritis--preliminary results of a longterm study. J Rheumatol. 1992; 19(4): 612-16.
29. Mouy R, Stephan JL, Pillet P, Haddad E, Hubert P, Prieur AM. Efficacy of cyclosporine A in the treatment of macrophage activation syndrome in juvenile arthritis: report of five cases. J Pediatr. 1996; 129(5): 750-54.
30. Lehman TJ, Schechter SJ, Sundel RP, Oliveira SK, Huttenlocher A, Onel KB. Thalidomide for severe systemic onset juvenile rheumatoid arthritis: A multicenter study. J Pediatr. 2004; 145(6): 856-57.
31. Laxer RM. Pharmacology and Drug Therapy. In Cassidy JT, Petty RE, Text book of Rheumatology. 2005 5th edn. Pg. 76-141, Elsevier Saunders.

32. Schmeling H, Mathony K, John V, Keysser G, Burdach S, Horneff G. A combination of etanercept and methotrexate for the treatment of refractory juvenile idiopathic arthritis: a pilot study. Ann Rheum Dis. 2001; 60(4): 410-12.

33. Lovell DJ, Giannini EH, Reiff A, Cawkwell GD, Silvermann ED, Nocton JJ, et al. Etanercept in Children with Polyarticular Juvenile Rheumatoid Arthritis. N Engl J Med 2000; 342: 763-9.

34. Lovell DJ, Ruperto N, Goodman S et al: Preliminary data from the study of adalimumab in children with juvenile idiopathic arthritis (JIA). Arthritis Rheum. 2004; 40: S436-37.

35. Gerloni V, Pontikaki I, Gattinara M, et al: Efficacy of repeated intravenous infusions of an anti-tumor necrosis factor alpha monoclonal antibody, infliximab, in the persistently active, refractory juvenile idiopathic arthritis. Results of an open-label prospective study. Arthritis Rheum. 2005; 52: 548-53.

36. Diak P, Siegel J, La Grenade L, Choi L, Lemery S, McMahon A. Tumor necrosis factor ? blockers and malignancy in children: Forty-eight cases reported to the food and drug administration. Arthritis & Rheumatism. 2010; 62(8): 2517-24.

37. Horneff G, Foeldvari I, Minden K, Moebius D, Hospach T. Report on malignancies in the German juvenile idiopathic arthritis registry. Rheumatol. 2011; 50(1): 230-36.

38. Keane J, Gershon S, Wise RP, Mirabile-Levens E, Kasznica J, Schwieterman WD, Siegel JN, Braun MM. Tuberculosis associated with infliximab, a tumor necrosis factor alpha-neutralizing agent. N Engl J Med. 2001 11; 345(15): 1098-104.

39. Botsios C. Safety of tumour necrosis factor and interleukin-1 blocking agents in rheumatic diseases. Autoimmunity Reviews. 2005; 4(3): 162-70.

40. Irigoyen PI, Olson J, Hom C, Ilowite NT: Treatment of systemic onset juvenile rheumatoid arthritis with Anakinra. Arthritis Rheum. 2004; 40: S437-38.

41. Yokota S, Imagawa T, Mori M, et al. Efficacy and safety of tocilizumab in patients with systemic-onset juvenile idiopathic arthritis: a randomised, double-blind, placebo-controlled, withdrawal phase III trial. Lancet 2008; 371: 998-1006.

42. Imagawa T, Yokota S, Mori M, Miyamae T, Takei S, Imanaka H, et al. Safety and efficacy of tocilizumab, an anti-IL-6-receptor monoclonal antibody, in patients with polyarticular-course juvenile idiopathic arthritis. Modern Rheumatol. 2012; 22(1): 109-15.

43. Ruperto N, Lovell DJ, Quartier P, Paz E, Rubio-Perez N, Silva CA, et al. Abatacept in children with juvenile idiopathic arthritis: a randomised, double-blind, placebo-controlled withdrawal trial. Lancet 2008; 372: 383-91.

44. Ruperto N, Lovell DJ, Quartier P, et al. Long-term safety and efficacy of abatacept in children with juvenile idiopathic arthritis. Arthritis Rheum 2010; 62: 1792-802.

45. Kasher-Meron M, Uziel Y, Amital H. Successful treatment with B-cell depleting therapy for refractory systemic onset juvenile idiopathic arthritis: a case report. Rheumatol. 2009. 48: 445-66.

46. Alexeeva E, Valieva S, Bzarova T, Semikina E, Isaeva K, Lisitsyn A, et al. Efficacy and safety of repeat courses of rituximab treatment in patients with severe refractory juvenile idiopathic arthritis. Clinical Rheumatology. 2011; 30(9): 1163-72.

47. Wulffraat NM, Vastert B, Tyndall A. Treatment of refractory autoimmune diseases with autologous stem cell transplantation: focus on juvenile idiopathic arthritis. Bone Marrow Transplant. 2005; 35 Suppl 1: S27-29.

48. Sawhney S. Recent advances in management of JIA. In Advances in Pediatrics. Editors Dutta A, Sachdeva A. 944-953. 1st ed 2006. Jaypee publishers.

49. Lehman TJA. Oligoarticular JIA: Is it a benign disease? Pediatric Rheumatology Online Journal. 2005; 3(3): 144-46.

50. Petty RE, Malleson P. Spondyloarthropathies of childhood. Pediatr Clin North Am. 1986; 33(5): 1079-96.

51. Woo P, Wedderburn LR. Juvenile chronic arthritis. The Lancet . 1998; 351(9107) 969-73.

52. Beukelman T, Patkar NM, Saag KG et al. American college of rheumatology recommendations for the treatment of juvenile idiopathic arthritis: initiation and safety monitoring of therapeutic agents for the treatment of arthritis and systemic features. Arthritis Care Res. 2011; 63: 465-82.

53. Rabinovich CE. Bone mineral status in juvenile rheumatoid arthritis. [Review] J Rheumatol. 2000; 27 Suppl 58: 34-37.

54. Bianchi ML, Cimaz R, Bardare M, Zulian F, Lepore L, et al. Efficacy and safety issues of alendronate for the treatment of osteoporosis in diffuse connective diseases in children. Arthritis Rheum. 2000; 43(9): 1960-66.

55. Weiss AH, Wallace CA, Sherry DD. Methotrexate for resistant chronic uveitis in children with juvenile rheumatoid arthritis. [see comments]. J Pediatr. 1998; 133(2): 266-68.

56. Holzinger D, Frosch M, Kastrup A, Prince FHM, Otten MH, Van Suijlekom-Smit LWA, et al. The Toll-like receptor 4 agonist MRP8/14 protein complex is a sensitive indicator for disease activity and predicts relapses in systemic-onset juvenile idiopathic arthritis. Annals of the Rheumatic Diseases. 2012; 71(6): 974-80.

57. Ravelli A, Martini A. Early predictors of outcome in juvenile idiopathic arthritis. Clin Exp Rheumatol. 2003; 21(5 Suppl 31): S89-93.

58. Spiegel LR, Schneider R, Lang BA, Birdi N, Silverman ED, Laxer RM, Stephens D, Feldman BM. Early predictors of poor functional outcome in systemic-onset juvenile rheumatoid arthritis: a multicenter cohort study. Arthritis Rheum. 2000; 43(11): 2402-09.

59. Aggarwal A, Agarwal V, Danda D, Misra R. Outcome in juvenile rheumatoid arthritis in India. Indian Pediatr 2004; 41(2): 180-84.

60. Athreya BH. A general approach to management of children with rheumatic diseases. In : Cassidy JT, Petty RE. Textbook of Pediatric Rheumatology. W.B.Saunders. Philadelphia. 2001: 190-211.

61. Zak M, Pedersen FK. Juvenile chronic arthritis into adulthood: a long term follow up study. Rheumatology. 2000; 39: 198-204.

62. Hashkes PJ Laxer RM. Medical treatment of juvenile arthritis. JAMA, 2005; 294(13): 1671-1884.

63. Ringold S., Weiss PF Beukelman T, et al .2013 Update of the 2011 American College of Rheumatology Recommendations for the Treatment of Juvenile Idiopathic Arthritis: Recommendations for the Medical Therapy of Children with Systemic Juvenile Idiopathic Arthritis and Tuberculosis Screening among Children Receiving Biologic Medications. Arthritis & Rheumatism 2013; 65(Oct): 2499-2512. doi: 10.1002/art.38092.

KAWASAKI DISEASE 4

Surjit Singh, Nutan Kamath

Kawasaki disease (KD) is an acute, self limited, systemic inflammatory, medium vessel childhood vasculitic illness, due to a pathologic immune response to any of a variety of environmental or infectious triggers, to which certain individuals are predisposed, probably by virtue of their genetic constitution. This disease has a potential to cause severe complications, and significant morbidity and mortality. Expeditious treatment can largely prevent these complications, underscoring the importance of early and accurate diagnosis. The diagnosis based on clinical criteria (Table 1) and correct identification of KD can be as exacting a challenge today, as it has been for almost 40 years.

This vasculitis bears the eponym *Kawasaki disease* due to the painstaking description of a novel febrile illness in 50 children first described by Tomisaku Kawasaki from Japan as acute mucocutaneous lymph node syndrome in 1967. Although KD has probably existed for more than 100 years, only with publication of diagnostic criteria has the condition been widely recognized.

Table 1. Criteria for diagnosis of kawasaki disease

Fever more than 5 days **+** **At least 4 of the following 5 clinical signs not explained by another disease process**
1. Bilateral conjunctival injection.
2. Changes in the oropharyngeal mucous membranes (including one or more of: injected or fissured lips, strawberry tongue, injected pharynx).
3. Changes in the peripheral extremities, including erythema and/or edema of the hands and feet (acute phase) or periungual desquamation (convalescent phase).
4. Polymorphous rash, primarily truncal; nonvesicular.
5. Cervical lymphadenopathy with at least one node > 1.5 cm.

Epidemiology

KD is an illness of early childhood. The overall incidence of the disease is ~10-20/100,000 children/ year in the United States and much higher in Asia. The annual incidence in children of Japanese ancestry is near 250 per 100,000 children < 5 years old. The condition is probably under-diagnosed and under-reported in India. A recent hospital based study from Chandigarh has shown that the incidence of KD is at least 4.54 cases per 100,000 children below 15 years of age. We have further shown that KD is now the commonest vasculitic disorder in children.

In the United States and Japan, the disease is more common during winter and early spring months; boys outnumber girls by ~1.5 to 1.7:1. In Chandigarh, KD is seen to occur more commonly during the months of May and October, with a distinct nadir in February. Around 85% of affected patients in the USA and Japan are under the age of 5 years, whereas India has an apparently higher proportion of older children. This may probably be because of the fact that cases of KD in infants and younger children may be getting missed at present.

Rates of recurrence and familial occurrence of KD are best documented in literature from Japan. (~3% and ~1% respectively). The risk of occurrence in twins is ~13%. Within one year of onset of the first case in the family, the rate in a sibling is 2.1%; ~50% of the second cases develop within 10 days of the first case. Higher rates of KD in the siblings of index cases and twins suggest a possible role for genetic predisposition that interacts with exposure to the etiologic agent in the environment.

Reported associations of KD with antecedent respiratory illness, exposure to carpet cleaning fluids, preexisting eczema, using a humidifier and living near a standing body of water have not been consistently confirmed.

Coronary artery aneurysms (CAA) or ectasia develop in ~15% to 25% of untreated children. The case fatality rate from KD in Japan is 0.08%. Virtually all deaths in patients with KD result from its cardiac sequelae. The peak mortality occurs 15 to 45 days after the onset of fever. However, sudden death from myocardial infarction may sometimes occur many years later in individuals who may have had CAA and stenosis due to KD in early childhood. Many cases of fatal and nonfatal myocardial infarction in young adults have been attributed to 'missed' KD in childhood.

Etiology and pathogenesis

Etiology remains unknown, although clinical and epidemiological features strongly suggest an infectious transmissible childhood disease. A self limited, generally non-recurring illness characterized by fever, rash, enanthem, conjunctival injection, and cervical lymphadenopathy fits well with an infectious trigger or etiology. The laboratory features also suggest infection. However bacterial and viral cultures and serological methods, as well as animal inoculation, have failed to identify an infectious cause.

Efforts to associate KD with exposure to drugs or environmental pollutants as toxins, pesticides, chemicals and heavy metals have failed, although clinical similarities between KD and acrodynia (mercury hypersensitivity) are notable.

The pathogenesis of coronary arteritis is still not clear, but endothelial cell activation, CD68+ monocyte/ macrophages, CD8+ (cytotoxic) lymphocytes, and oligolclonal IgA plasma cells appear to be involved. Enzymes including matrix metalloproteinases may be involved in the development of aneurysmal dilatation. Vascular endothelial growth factor (VEGF), monocyte chemotactic and activating factor (MCAF or MCP-1), tumor necrosis factor and various interleukins appear to play a role in the vasculitic process.

Hence a reasonable working hypothesis is that KD is a stereotyped, pathological immune response to one or a variety of environmental or infectious triggers, to which certain individuals are predisposed by virtue of their genetic constitution. Predilection for childhood onset may reflect the presence of developmental antigens that are targets for inflammatory response only early in life, subtle maturational defects in immune responsiveness or the timing of exposure to environmental triggers.

A novel hypothesis has been preferred recently. Dr. Jane Burns (San Diego, USA) and her team of researchers have analysed of the epidemiology of KD in 3 geographically distinct areas, viz. Japan, Hawaii, and San Diego. The authors have noted a peculiar pattern wherein KD cases have been linked to "large-scale wind currents originating in central Asia and traversing the north Pacific". These wind currents are said to carry a "viable, UV-resistant bacteria present in the upper troposphere", over long distances through the wintertime development of a "duct" that sweeps from Asia to the western North Pacific and an associated trans-Pacific transport across the North Pacific from Japan to Hawaii and southern California. These results suggest that it may be possible to predict the occurrence of KD in a given geographical area, even without knowledge of the putative etiologic agent responsible for KD, based on changes in tropospheric conditions.

Genetics of KD

Onouchi and colleagues had previously performed genome-wide linkage studies (GWLS) of affected sib pairs of KD and mapped several candidate loci for susceptibility genes. From these loci, they identified SNPs in ITPKC at 19q13.2 and CASP3 at 4q35, which confer risk for KD in Japanese and Caucasian children. To identify additional KD risk loci, the workers have recently performed a genome-wide association study (GWAS). Association of genome-wide significance level was achieved at 4 loci, including those of ITPKC and CASP3. Similar GWAS have also been reported from Singapore and Taiwan. Several new novel loci (e.g. COPB2) have now been identified.

Clinical features

Disease course

The disease is divided into three phases:

1. Acute febrile period of approximately 10-14 days.
2. Subacute phase of approximately 2-4 weeks, ending with a return to normal of the platelet count and erythrocyte sedimentation rate.
3. Convalescent or recovery period lasting months or years, during which time vessels affected by the disease undergo healing, remodelling, and scarring.

There are no precise markers of the transition from one phase to another, and current therapy has altered this three phase pattern of disease considerably.

Diagnosis

Diagnosis of KD is based on a typical temporal sequence of clinical features, none of them individually having any diagnostic significance. These clinical features evolve over a period of days, and the entire spectrum may not be seen at any one particular point of time. This difficulty is further compounded by the fact that there is no diagnostic laboratory test. However it is essential to make a diagnosis in the acute phase as appropriate treatment in this phase results in a significant reduction in coronary complications. Criteria for diagnosis (Table 1) have virtually remained unchanged in the 40 years since the disorder was initially described. *It cannot be overemphasized that the dignosis of KD is entirely clinical.* The criteria are only meant to be used in helping the clinician arrive at a diagnosis. Further at least 10% of children with characteristic CAA of KD never actually fulfill criteria for KD. Strict application of the criteria helps in distinguishing KD from its mimics.

Younger children with KD are typically quite irritable and reluctant to be examined; older children may be less so. Fever typically is high spiking and remittent, with peak temperatures generally > 39°C.

Conjunctival injection (not conjunctivitis), which spares the zone immediately around the iris (limbus), is more prominent on the bulbar than the tarsal conjunctiva. The presence of a conjunctival discharge and especially purulence, corneal ulceration and conjunctival edema suggest an alternate diagnosis. Mild acute iridocyclitis or anterior uveitis may be noted with a slit lamp. Pharyngeal injection is typically nonfocal, without discrete lesions or tonsilar exudates. Lymphadenopathy (LAP) is typically unilateral, often firm, nonfluctuant and nontender, confined to anterior cervical region, especially nodes overlying the sternocleidomastoid muscles. Diffuse LAP with axillary and inguinal nodes or splenomegaly is not typical of KD.

Rash is located on the trunk and may spread to the face, extremities and perineum. It can be polymorphic - scarlatiniform, macular, papular, multiforme and purpuric. *However, it is almost never vesicular*. It usually fades without residue. Pastia's lines (i.e. faint petechial lines in skin folds of the anti cubital fossae, inguinal and axillary regions) suggest scarlet fever rather than KD. Perineal desquamation (especially when perianal in localization) is characteristically seen in the first few days of KD. The other type of desquamation occurs typically 10 days or more after start of fever, and should be sought at the tips of the fingers and toes just below the distal edge of the nails (*periungual desquamation*). As skin peeling is a late manifestation especially in USA and Japan, it is useful more for retrospective confirmation of the diagnosis rather than for making therapeutic decisions. In Indian children, however, it is believed that peeling of skin is seen to coincide with the acute phase of fever and often within 10 days.

Cardiovascular manifestations can be prominent in the acute phase. Although coronary aneurysms are the most significant cardiovascular complication of KD, other cardiac complications can also occur. These include myocarditis (some degree of which is believed to occur in all patients of KD), valvulitis (usually mitral), pericardial effusion and development of aneurysms in the systemic arteries. Myocarditis may manifest as tachycardia out of proportion to the degree of fever. ECG abnormalities such as prolonged PR interval, ST-T segment changes and decreased voltage of R waves may also be suggestive of myocarditis. Pericarditis with small pericardial effusions occur in approximately 25% of acute KD patients. Valvular disease, predominantly mitral regurgitation, occurs in approximately 1%. Certain clinical factors are predictive of an increased risk for coronary disease - these include prolonged fever, recurrence of fever following an afebrile period of more than 48 hours, arrhythmias other than first- degree heart block, male gender, age less than 1 year and cardiomegaly.

KD is a systemic vasculitis and may involve several other organ systems. Clinical features often seen in the condition include arthralgia/arthritis, myositis, transient unilateral peripheral facial nerve palsy, high frequency sensorineural hearing loss (25 to 35 dB), aseptic meningitis, peripheral aneurysms and hemophagocytic syndrome. Arthritis in the first week of illness tends to involve small interphalangeal as well as large weight-bearing joints; after tenth day of illness favours large weight bearing joints, especially knees and ankles. KD is also known to present with pulmonary infiltrates, pleural effusion, diarrhea, vomiting, abdominal pain, hydrops of gall bladder, paralytic ileus, acute surgical abdomen, hepatomegaly, testicular swelling, redness and crusting at the sight of BCG inoculation. It is important to be aware of these varied clinical features lest the condition be confused with an entirely unrelated disease. In fact it is the presence of these additional clinical features which makes the diagnosis of this condition so challenging.

Transverse furrows across fingernails (Beau's lines) may be seen during the convalescent phase of KD. These are present for several days and tend to grow out with the nail.

Laboratory findings

There are no laboratory tests that are pathognomonic of KD. *This fact cannot be overemphasized*. However, certain findings may be of help in arriving at a diagnosis. A neutrophilic leukocytosis predominance is common in the acute phase. Elevations in sedimentation rate and in other acute-phase reactants are almost universally present in the first week of illness and may persist for 4 to 6 weeks. The platelet count is generally normal in the first week of illness. Thrombocytosis, often in excess of 5-10 x 10^5/cu.mm, is only seen after the second week of illness is over. This is an acute phase reactant and is rather characteristic of KD, but is usually not there to help the clinician in the crucial period of the first 7-10 days of fever.

Echocardiography is used for detection of coronary aneurysms and in experienced hands has a sensitivity and specificity of more than 95%. Coronary angiography is almost never required for delineation of the aneurysms.

Incomplete/atypical kawasaki disease

Like any other medical condition KD can sometimes present with atypical features. It may be extremely difficult to make a correct diagnosis of this condition under these circumstances. Children presenting with fever and fewer than 4 of the other features have been labelled as having "atypical" or "incomplete" KD. The problem is that these children are also at risk for coronary aneurysms. Atypical KD is believed to be more common in young infants and clinical recognition of such cases is quite difficult.

The diagnosis of IKD can be a contentious issue. It is for this reason that echocardiography has been recommended for the work-up of prolonged fever (say, more than 7 days) in young children, especially when there is laboratory evidence of systemic inflammation, and no other explanation for the febrile illness. Laboratory findings of atypical/ incomplete KD appear to be similar to those of classic cases.

Differential diagnosis

Differential diagnosis of KD includes viral infections - measles, Epstein Barr virus and adeno virus which share the signs of mucocutaneous inflammation but with less evidence of systemic inflammation and lack the extremity changes seen in KD. Toxin-mediated illness, especially β hemolytic streptococcal infection and toxic shock syndrome lack the ocular and articular involvement typical of KD. Drug reactions like Stevens-Johnson syndrome or serum sickness can mimic KD, but with subtle differences in the ocular and mucosal manifestations.

Approach to diagnosis

Making a diagnosis of KD is difficult due to absence of certainty and concern for atypical cases. Observation of borderline cases for a few days is often the best course of action as the so called 'incomplete' case may evolve with time into a 'complete' case. Spontaneous improvement and resolution of fever suggests a viral illness while its persistence or development of additional signs point to KD. Children may present with only fever and a unilateral enlarged lymph node. The rash and mucosal changes that follow may be mistaken for a drug reaction - often to an antimicrobial administered for a presumed bacterial lymphadenitis. Sterile pyuria is well known in KD. It can be mistaken for a partially treated urinary tract infection. The young infant may present with fever, rash and cerebrospinal fluid pleocytosis and may be misdiagnosed as viral/bacterial meningitis. Occasionally a child may present with an acute abdomen, with consideration for surgical intervention. Ophthalmologic consultation may help in picking up uveitis which is commonly seen in KD. Echocardiography is used to look for early coronary involvement or depressed myocardial function, though only 10% of aneurysms develop before the tenth day of illness.

KD and the heart

A significant proportion (upto 20-25%) of untreated KD patients may develop coronary abnormalities, including diffuse dilatation (coronary diameter > 4mm) and aneurysm formation. This process commonly occurs from 10 days to 4 weeks after the onset of symptoms. It is important to understand that an echocardiographic study done earlier in the course of the illness may miss these findings. Development of aneurysms more than 6 weeks after the onset of illness is also uncommon. With appropriate treatment, these aneurysms tend to regress over the next few months. Patients who do not have complete resolution of aneurysms may show either a decrease in the size of aneurysm or go on to develop coronary stenosis. This coronary stenosis may be complicated by premature atherosclerosis and lead to significant coronary obstruction and myocardial ischemia later in life. According to the Japanese workers myocardial infarction that occurs under these circumstances may be indistinguishable from that seen in association with primary atherosclerosis. It is also said that larger aneurysms (i.e. more than 8 mm), and especially the ones in the more proximal regions of the coronaries, are less likely to regress than the others. In such cases thrombosis and rupture can also occur. In patients with persistent aneurysms, coronary artery stenosis can develop as long as 10-21 years after the acute episode of KD.

KD is said to be the commonest cause of myocardial infarction in children. Myocardial infarction in children presents with atypical symptoms as compared to adults. The main presenting complaints are uneasiness, vomiting, shock and abdominal pain. It is important to note that chest pain may not be a significant feature in young children. In almost one-third of children, the infarcts may be asymptomatic. However the ECG and cardiac enzyme changes are like those in adults.

Treatment

Intravenous immunoglobulin (IVIG) is the drug of choice and is very effective when given in the first 10 days of illness. It reduces the chances of development of coronary abnormalities from 20-25% to less than 2%. IVIG results in a rapid defervescence of fever and normalization of the acute phase reactants. IVIG also improves myocardial function in KD patients having myocarditis. It is given as a single dose intravenous infusion of 2gm/kg administered over several hours. While the mechanism of action of IVIG in patients with KD is largely unknown, it is believed that it results in downregulation of the cytokine cascade. Some patients may require additional doses of IVIG. Infliximab, a monoclonal antibody against TNF-α, has also been used in such circumstances.

Aspirin is administered in patients with KD for its anti-inflammatory and antithrombotic effects. During the acute phase of illness, aspirin is administered at 40-60 mg/kg/d given every 6 hours. After the fever has settled, the dose of aspirin is reduced to 3 to 5 mg/kg/day (anti-platelet dose). Children with KD who develop coronary artery aneurysms are kept on long term aspirin. It may be noted that KD is the only vasculitic disorder in which use of glucocorticoids has been relatively contraindicated, although there is a role for carefully considering this therapeutic option under special circumstances.

A repeat echocardiogram is obtained at 2-3 weeks and again at 6-8 weeks following the onset of illness. Aspirin can be discontinued after the sedimentation rate and platelet counts have normalized (this usually takes 6-8 weeks) and the echocardiograms are reported to be normal. If echocardiography done at 6-8 weeks shows evidence of coronary dilatation and/or aneurysms, low dose aspirin needs to be continued indefinitely and follow-up investigations (coronary angiography/dobutamine stress echocardiography) are then mandatory.

Further management of KD patients with aneurysms is dependent on the severity of coronary disease. Patients with a single small aneurysm should receive long-term aspirin and avoid physical sports. Patients with giant/multiple aneurysms are usually put on long-term warfarin along with aspirin.

Mortality

KD was associated with a mortality rate of 1-2% in the pre-IVIG era. This has dropped to 0.08% with improved recognition and appropriate therapy of the disease in the acute phase. Deaths are most common 2 to 12 weeks after the onset of the illness and are usually secondary to the coronary aneurysms and complications thereof.

Summary

Kawasaki disease is a medium vessel vasculitis and presents as an acute febrile illness of young children. KD is, by no means, a rare disease. It has been reported from all parts of the world and is being increasingly recognized in India. KD is believed to be the commonest vasculitic disorder of children and is now recognized as the most common cause of acquired heart disease in children in the United States and Japan. Clinical and epidemiological features of KD support an infectious cause, but the etiology remains elusive. KD causes significant coronary artery disease that may lead to myocardial infarction and sudden death. For the pediatrician, it is important to diagnose KD as early as possible because a significant proportion of untreated patients develop coronary artery abnormalities. These coronary abnormalities can have life-long sequelae. Intravenous immunoglobulin is the treatment of choice and is highly effective in preventing the coronary complications associated with the disease.

Further Reading :

1. Petty RE, Cassidy JT. Kawasaki disease. In: Cassidy JT. 1. Petty RE, Cassidy JT. Kawasaki disease. In: Cassidy JT. Petty RE (eds). Textbook of Pediatric Rheumatology 5th ed. Philadelphia; WB Saunders Company, 2005: 521-38.
2. Kawasaki T. Acute febrile mucocutaneous syndrome with lymphoid involvement with specific desquamation of the fingers and toes in children. Arerugi 1967; 16: 178-222.
3. Singh S. Kawasaki Disease: A clinical dilemma. Indian Pediatrics 1999; 36: 871-75.
4. Newburger JW, Takahasi M. Gerber MA, Gewitz MH, Tani LY, Burns JC et al, Diagnosis, Treatment and Long-Term Management of Kawasaki Disease : A Statement for Health Professionals from the Committee on Rheumatic Fever, Endocardidits and Kawasaki Disease, Council on Cardiovascular Disease in the Young. American Heart Association. Pediatrics 2004; 114(6): 1708-33.
5. Singh S, Bansal A, Gupta A, Kumar RM, Mittal BR. Kawasaki disease - a decade of experience from Northern India. Int Heart J 2005; 46: 679-89.
6. Muta H, Ishii M, Sakaue T. et al. Older Age is a Risk Factor for the Development of Cardiovascular Sequelae in Kawasaki Disease. Pediatrics 2004; 144: 751-54.
7. Mitra A, Singh S, Devidayal, Khullar M. Serum lipids in North Indian children treated for Kawasaki disease. Int Heart J 2005; 46: 811-17.
8. Oechslin EC, Arbenz U, Mayer K. Giant and fusiform aneurysms of coronary arteries following early and adequate treatment of suspected Kawasaki disease. Heart 2004; 90: 1437.
9. Kato H, Ichinose E, Kawasaki T: Myocardial infarction in Kawasaki disease. Clinical analyses in 195 cases. J Pediatr 1986; 108: 923-27.
10. Narayanan SN, Krishna Veni, Sabiranathan K. Kawasaki disease. Indian Pediatr 1997; 34: 139-43.
11. Kato H, Ichinose E, Yoshioka F, Takechi T, Matsunaga S, Suzuki K, et al. Fate of coronary aneurysm in Kawasaki disease: Serial coronary angiography and long-term follow-up study. Am J Cardiol 1982; 49: 1758-66.
12. Mitra S, Singh S, Grover A, Kumar L. A child with prolonged pyrexia and peripheral desquamation: Is it Kawasaki Disease? Indian Pediatr 2000; 37: 786-89.
13. Nakano H, Ueda K, Saito A, Nojima K. Repeated quantitative angiograms in coronary arterial aneurysms in Kawasaki disease. Am J Cardiol 1985; 56: 846-51.
14. Singh S, Kansra S. Kawasaki disease. Natl Med J India 2005; 18: 20-24.
15. Kushner HI, Macnee R, Burns JC. Impressions of Kawasaki syndrome in India. Indian Pediatr. 2006; 43(11): 939-42.
16. Kobayashi T, Saji T, Otani T, Takeuchi K, Nakamura T, Arakawa H et al. Efficacy of immunoglobulin plus prednisolone for prevention of coronary artery abnormalities in severe Kawasaki disease (RAISE study): a randomised, open-label, blinded-endpoints trial. Lancet 2012; 379: 1613-20.
17. Rodo X, Ballester J, Cayan D, Melish ME, Nakamura Y, Uehara R et al. Association of Kawasaki disease with tropospheric wind patterns. Nature SCIENTIFIC REPORTS |1 : 152| DOI: 10.1038/srep00152.
18. Printz BF, Sleeper LA, Newburger JW, Minich LL, Bradley T, Meryl S. Cohen MS et al. Noncoronary Cardiac Abnormalities Are Associated With Coronary Artery Dilation and With Laboratory Inflammatory Markers in Acute Kawasaki Disease. J Am Coll Cardiol 2011 doi: 10.1016/j.jacc.2010.08.619.
19. Singh S, Aulakh R, Bhalla AK, Suri D, Manojkumar R, Narula N, Burns JC. Is Kawasaki disease incidence rising in Chandigarh, North India? Arch Dis Child 2011; 96: 137-40.
20. Aggarwal P, Singh S, Suri D, Narula N, Manojkumar R. Symptomatic myocarditis in Kawasaki disease. Indian Journal of Pediatrics 2012; 79(6): 813-14.
21. Singh S, Aulakh R, Bhalla AK, Suri D, Manojkumar R, Narula N, Burns JC. Is Kawasaki disease incidence rising in Chandigarh, North India? Archives of Disease in Childhood 2011; 96(2): 137-40.
22. Ghelani SJ, Singh S, Manojkumar R. QT interval dispersion in North Indian children with Kawasaki disease without overt coronary artery abnormalities. Rheumatology International 2011; 31: 301-05.
23. Chaudhuri K, Ahluwalia TS, Singh S, Binepal G, Khullar M. Polymorphism in the promoter of the CCL5 gene (CCL5G-403A) in a cohort of North Indian children with Kawasaki disease. A preliminary study. Clinical and Experimantal Rheumatology 2011; 29 (1 Suppl. 64): S 126-30.
24. Kashyap R, Mittal BR, Bhattacharya A, Manojkumar R, Singh S. Exercise myocardial perfusion imaging to evaluate inducible ischaemia in children with Kawasaki disease. Nuclear Medicine Communications 2011; 32(2): 137-41.
25. Singh S, Aulakh R. Kawasaki disease and Henoch Schonlein purpura: changing trends at a tertiary care hospital in north India (1993-2008). Rheumatology International 2010; 30(6): 771-74.
26. Nishad P, Singh S, Sidhu M, Malhi P. Cognitive and behaviour assessment following Kawasaki disease - a study from North India. Rheumatology International 2010; 30(6): 851-54.
27. Ghelani S, Singh S, Manojkumar R. Endothelial dysfunction in a cohort of North Indian children with Kawasaki disease without overt coronary artery involvement. Journal of Cardiology 2009; 53: 226-31.
28. Ling XB, Lau K, Kanegaye JT, Pan Z, Peng S, Ji J et al. A diagnostic algorithm combining clinical and molecular data distinguishes Kawasaki disease from other febrile illnesses. BMC Medicine 2011, 9: 130.
29. Singh S, Kawasaki T. Kawasaki disease - an Indian perspective. Indian Pediatrics 2009; 46(7): 563-71.
30. Son MB, Gauvreau K, Burns JC, Corinaldesi E, Tremoulet AH, Watson VE. Infliximab for intravenous immunoglobulin resistance in Kawasaki disease: A retrospective study. J Pediatr 2011; 158: 644-49.
31. Singh S, Gupta D, Suri D, Kumar RM, Ahluwalia J, Das R, Varma N. Thrombocytopenia as a presenting feature of Kawasaki disease: a case series from North India. Rheumatology International 2009; 30(2): 245-48.
32. Singh S, Gupta MK, Bansal A, Kumar RM, Mittal BR. A comparison of the clinical profile of Kawasaki disease in children from Northern India above and below 5 years of age. Clinical and Experimental Rheumatology 2007, 25: 654-57.
33. Jimenez-Fernandez S, Trofimenko V, Fick, Jr, DB, Phan HAL, Linz PE, Nayak K et al. Prevalence of Kawasaki disease in young adults with suspected myocardial ischemia. Circulation. 2012; 125:2447-53.

SECTION 13

Infection and Arthritis

SEPTIC ARTHRITIS INCLUDING TUBERCULOSIS 1

Kaushik Bhojani, Shubhada Kalke

Introduction

Septic arthritis is infection in a joint. Infection of the bone is called osteomyelitis. It can either set off an immunological reaction causing arthritis or it can directly affect the bone and joint.

Septic arthritis caused by pyogenic bacteria is a medical emergency. Thus infection is an important cause of morbidity and increased mortality in rheumatology.

Definition

Septic arthritis is defined as pathologic inoculation by live organisms of the joint by direct or hematogenous route rather than immunological response to pathogens as in reactive arthritis[1].

Typically a single joint is involved however sometimes there can be simultaneous affection of a few jonts. Polyarthritic involvement is rare and is usually an immunologically mediated response to a systemic infection such as subacute bacterial endocarditis, Hepatitis B etc.

Epidemiology

Etiology

Reported incidence of septic arthritis is 10-29 cases per 100,000 populations. Mortality rate, despite advances in therapy is approximately 10% in large series and up to one-third of cases experience residual functional impairment[2].

The infection can be bacterial, viral and less frequently fungal or parasitic. Bacterial infections usually have an acute presentation. A little more indolent presentation would favour the possibility of a mycobacterial or a fungal etiology.

Non Gonococcal Septic Arthritis (Bacterial)

- The most common organism is Staph. aureus in both monoarticular and polyarticular infections, streptococci and H. influenzae are more frequently associated with polyarticular infections[3].
- Staphylococci account for 60% cases of septic arthritis (80% of these are due to Staph aureus) and streptococci are seen in 20%.
- The incidence of Streptococcus pneumoniae infections and H influenzae is 2% each[1].
- Gram negative organisms are seen in the elderly. Though E coli is the most common, Pseudomonas aeruginosa, Proteus mirabilis, Serratia marcesens and Klebsiella pneumoniae are also reported.
- H influenzae requires effective reticuloendothelial system for clearance and hence, is commonly seen in immunocompromised states like alcoholism, asplenia, hypogammaglobulinaemia and Systemic Lupus Erythematosus (SLE).
- Pneumococcal infections are also known to be more frequent in alcoholics and those with defective humoral immunity.
- Salmonella arthritis is a rare complication of salmonellosis. Patients with SLE or sickle cell anaemia are at increased risk because the chronic salmonella carrier state is more common in these conditions[4].
- Intravenous (IV) drug abusers are predisposed to a variety of microorganisms: Staph. and Streptococcal infections from their own flora and gram negative organisms from the drugs and the injection apparatus.
- Anaerobes (5%) are seen with trauma and penetrating wounds.
- Arthritis due to Listeria monocytogenes, Streptobacillus moniliformis, and Pasteurella multocida is rare.

Pathophysiology

Bacteria can reach the joints by three routes

1) Haematogenous:
   - This is the commonest route. The synovium is highly vascular and lacks a limiting membrane. This permits access of bacteria to the synovial fluid
   - Presence of inflammation (e.g. RA) or foreign material such as a prosthetic joint increases likelihood of bacterial colonization.
2) Direct entry into the joint space can occur by penetrating injury or iatrogenic conditions like intra-articular injection, arthroscopy and arthrotomy.
3) Direct spread from an adjacent focus of infection- e.g. osteomyelitis, or psoas or gluteal abscess.

Once infection is established, bacteria multiply in the synovium and the synovial fluid. They are phagocytosed by polymorphonuclear leucocytes and an inflammatory reaction develops. Polymorphs release cytokines and proteinases leading to a rapid degradation of cartilage and the formation of abscesses. Significant loss of cartilage develops within 2-3 weeks. This rapid destruction necessitates early diagnosis and treatment. Chronic inflammation may develop in untreated or partially treated infections and leading to further damage and loss to cartilage and subchondral bone[1].

Clinical features

- As mentioned earlier the majority of such patients would present with monoarthritis. Knee is the commonest joint involved followed by hip. (50% and 20% respectively)[1].
- Other joints involved are shoulder, wrist, elbow and ankle albeit in a lesser frequency.
- Polyarticular involvement may be seen in cases of Rheumatoid Arthritis (RA) and immunosuppressed patients due to any cause.
- Patient presents with acute onset of pain, swelling, erythema, warmth and tenderness of the affected joint with a decreased range of motion.
- Fever may be present and may be high grade at times but may also be absent in patients on immunosuppressive therapy as is usual in our specialty.
- History of preceding infection involving skin, respiratory and urinary tract may be present.
- IV drug abusers are usually young and without pre-existing joint disease. Presentation in them is often more insidious with a predilection for the axial joints, and periarticular soft-tissue abscess or osteomyelitis is frequently seen[5].
- Sacroiliac and sternoclavicular joints are very uncommon except in IV drug users.
- Involvement of small joints of the fingers is rare and usually due to a direct inoculation

Meningococcal arthritis

Arthritis complicates about 2-10 per cent of cases of acute meningococcaemia and may present in any of the following manner.

- Polyarthralgias, polyarthritis or tenosynovitis.
- Purulent joint effusions that are usually mono or oligoarticular during the acute phase of their meningococcal infection. Cultures of synovial fluid can be positive or negative. In some cases, these effusions may be slow to resolve.
- A small number of patients may develop articular manifestations only after 1-2 weeks, when the patient has responded to antibiotics.
- Cultures of synovial fluid are negative but immune complexes are detectable in blood and synovial fluid. This eventually resolves without further antibiotic therapy.
- In primary meningococcal arthritis, patients present with purulent arthritis without signs of acute meningococcaemia. The knee and ankle are most commonly involved. Cultures of synovial fluid are positive in 90 per cent of cases[1].

Gonococcal arthritis

- Gonnococcal arthritis usually affects young, healthy sexually active females more commonly than males.
- Patients with disseminated gonococcal infection present with an acute illness with fever, rigors, pustular rash, tenosynovitis and migratory polyarthralgias or polyarthritis with non-purulent effusions of 3-4 days duration.
- Some patients present with a purulent arthritis (commonly knee monoarthritis) that resembles non-gonococcal arthritis[1].
- To culture N. gonorrhoeae from the joint, synovial fluid should be immediately spread on a pre-warmed, chocolate-agar plate and incubated in an atmosphere enriched with carbon dioxide, or sent to the laboratory on transport media[1].

Spiorochetal arthritis

- **Syphilitic arthritis**

Secondary syphilis may be associated with arthritis particularly affecting the knees and ankles. Sometimes other joints such as the shoulders, wrists and Sacroiliac joints may also be involved.

Usually follows an indolent course and usually responds to penicillin. Sometimes a neuropathic joint may occur due to sensory loss in cases of Tabes dorsalis[6].

- **Lyme disease**

Arthritis is a late manifestation of Lyme disease. It presents as an acute, often recurrent, monoarthritis commonly affecting the knee joint. It should be suspected if the history of previous similar episodes, history of tick exposure and rash, and residence in an endemic area are present. Though synovial fluid leucocyte count may be markedly elevated, bacterial cultures are negative and the diagnosis is confirmed by serologic testing. Lyme disease is not a common cause of monoarthritis in India.

Septic arthritis in children

Bacteria causing septic arthritis in children vary with age.

- In hospital acquired neonatal septic arthritis, staphlyococcus and gram negative bacilli are common agents, while in community acquired infection, streptococci are commonly implicated.
- Between two months to two years of age H influenzae is responsible for 50% of septic arthritis. After Hib vaccination, this is no more case in USA. Presentation is similar to adults.

Hip is the most commonly involved joint in children and coexistent osteomyelitis is more often present in children and is associated with poorer outcome. A history of non-penetrating trauma may be present prior to septic arthritis[7,8].

Diagnosis

- Joint infection should be suspected in monoarticular joint affection until proven otherwise especially in the presence of risk factors (Table 1).
- Severe pain and tenderness should arouse suspicion.
- Aspiration of the affected joint followed by fluid analysis and culture is necessary to confirm the diagnosis.

Laboratory findings

1) ESR and CRP levels are markedly elevated and can be used as a marker for response to treatment.
2) Peripheral white cell and platelet counts may be elevated.
3) Blood culture is positive in one third of patients.
4) Synovial fluid examination :
   - Necessary whenever infection is suspected and should precede antibiotic treatment. Typically fluid is cloudy and may sometimes be frankly purulent with leucocyte count of over 50,000/mm with greater than 90 per cent polymorphonuclear leucocytes. In partially treated patients, the count may not be so high[3].
   - Proteins and LDH are elevated[9].
   - Synovial fluid glucose may be decreased in about 50% of cases, but by itself does not distinguish infection from inflammation.
   - Gram's stain is positive in 50-65% of cases of non-gonococcal arthritis and culture is usually positive[8].
   - Fluid should be examined for crystals as gout and pseudogout can mimic septic arthritis.

Imaging

- Plain radiograph should be done to rule out osteomyelitis. Periarticular soft-tissue swelling or joint effusion (separation of fat pads) is seen in early stages. Subsequently periarticular osteoporosis, joint space narrowing and erosions develop. Plain radiographs may also help in differential diagnosis by demonstrating a fracture, chondrocalcinosis, gouty erosions, avascular necrosis and pre-existing joint disease. Infection should be suspected when radiographs reveal damage in joints within few weeks[10].
- Ultrasonography and CT are useful in aspirating deep seated joints[11].
- MRI may detect early osteomyelitis in non responding infection especially of axial joint[12]. Both MRI and CT are best suited to monitoring progress[13].
- Gallium and indium scans are of most value when infection of a prosthetic joints is suspected, where they may distinguish an infected from an uninfected, painful loose prosthesis.
- Indium scan though more laborious, is more sensitive and specific and results are available in 24 hours[14].
- PET scan is one of the more recent imaging tools and is fast gaining popularity. In fact it is considered to be a better modality than even MRI in certain situations such as in degenerative

spinal disease or in post operative cases with metal implants[15].

Differentiating infection from inflammation may sometimes be difficult even with all the aids.

Differential diagnosis

Crystal induced arthritis

- Previous history of gout or pseudogout attacks, presence of tophi, first MTP joint affection, presence of risk factors such as diuretic therapy, alcohol abuse, or renal insufficiency, metabolic syndrome etc are helpful.
- Identification of crystals in the synovial fluid by polarizing microscopy is necessary for confirming the diagnosis.
- Occasionally infection may also complicate gout.

Haemarthrosis

- Usually a history of trauma is elicitable
- Risk factors like anticoagulants or sickle cell disease are useful
- Frank haemorrhagic joint aspirate is diagnostic.

Rheumatoid arthritis

- Though commonly presents as symmetric polyarthritis with a female predilection
- It can rarely start as monoarthritis and thus mimic septic arthritis.
- Confirmation is usually by synovial fluid aspiration with a negative and histopathology of synovial membrane (Fig. 1).
- Serological tests such as rheumatoid factor and cyclic citrullinated peptide may provide supportive evidence.

Psoriatic arthritis:

- Usually a history of psoriasis is present.
- Skin rash
- Nail changes such as onycholysis and nail Pitting may be present
- Arthritis presentation is highly variable ranging from polyarticular symmetric arthritis to asymmetric oligoarthritis along with sacroilitis resembling spondyloarthropathy
- DIP inflammation if present is characteristic

Reactive arthritis:

- This illness is usually known to occur in young males and they present with predominantly lower limb asymmetric oligoarthritis[6].
- Typically patients are very sick looking at presentation. Some of them may possess the HLA-B27 gene.
- History of respiratory, urinary or gut infection few weeks prior is important.
- Associated manifestations include eye or genitourinary inflammation and oral ulcers.
- Triad of urethritis and conjunctivitis along with the arthritis was known as Reiter's syndrome earlier.

Adult Still's Disease

- The triad of rash, arthritis and high-spiking quotidian fever is characteristic.
- Rash is typically evanescent.
- There may be lymphadenopathy and hepato-splenomegaly and even serositis in some cases.
- Peripheral leucocyte count is high and ESR, CRP, LDH levels are elevated.
- Ferritin levels are very high.
- The diagnosis is by exclusion of other causes.

Neuropathic joint

Rapidly destructive joint changes may raise suspicion of infection. The presence of peripheral neuropathy with sensory deficit and marked fragmentation of periarticular bone with little joint effusion are typical features.

Table 1. Risk factors for septic arthritis

- Older age.
- Pre-existing joint disease - Rheumatoid arthritis, osteoarthritis.
- Prosthetic joint.
- Low socio-economic status.
- Intravenous drug abuse.
- Alcoholism.
- Diabetes.
- Previous intra-articular steroid injection.
- Cutaneous ulcers.
- Other immunocompromised states - Human immunodeficiency virus infection, chronic liver disease.
- Immunosuppressants.

Management

- Early diagnosis by aspiration of the joint followed by appropriate antibiotic treatment as per gram stain and culture report and adequate drainage are essential for good outcome.
- The choice of antibiotic is guided by gram staining or the expected/likely organism initially and subsequently based on culture sensitivity reports. The guideline for antibiotic therapy is given in Table 2.

Table 2. Guidelines for antibiotic use in septic arthritis

Organism isolated	Antibiotic suggested	Dose
Staph aureus	Cloxacillin Flucloxacin Ceftriaxone	0.5-1 gm 6 hrly 1-2 g 6 hrly 1-2 g 12 hrly
Methicillin resistant staph aureus	Vancomycin	1 gm every 12 h
Staph epidermidis	Vancomycin	1 gm every 12 h
Streptococci	Penicillin G Ceftriaxone	2.5 million U every 4 h 1-2 g every 12 h
Enterococci	Penicillin G and Gentamycin	2.5 million U every 4h and 1 mg/kg every 8 h
N. gonorrhoae	Ceftriaxone	1-2 g every 12 h
N. meningitidis	Penicillin G	2.5-5 million U every 6 h
H. influenzae ampicillin resistance suspected	Ampicillin Cefotaxime Ceftriaxone	2 g every 6 h 1 g every 8 h 1-2 g every 12 h
Enterobacteriaceae	Cefotaxime	2 g every 8h
Pseudomonas	Ceftazidime + Amikacin or Ciprofloxacin	2 g every 8 h
Borrelia burgdoferi (Lyme disease)	Doxycycline Amoxicillin Ceftriaxone	100 mg PO twice a day 500 mg PO QID 2g /d

- There is no consensus on duration of therapy[16].
- Intravenous antibiotics should be continued until signs of joint are significantly decreased and synovial cultures are sterile.
- Oral antibiotics can then be used to complete the course of treatment, which is usually about 2 weeks in uncomplicated cases but should be extended to 4 weeks or more where
  - i) There was a significant delay in starting treatment
  - ii) Slow response
  - iii) Infection with virulent organisms such as Staph. aureus or Gram-negative bacilli,
  - iv) Immunosuppressed patients and
  - v) Concurrent osteomyelitis[1,17].
- Infected joints can be drained by closed needle aspiration, tidal irrigation, arthroscopy or athrotomy.
  - i) Closed-needle aspiration avoids the expense of more invasive procedures and is a satisfactory approach as long as the joint can be completely drained and there is clear evidence of improvement[18,19].
  - ii) Arthroscopy has the advantage as it can combine good visualization of the joint space and articular cartilage and facilitate effective drainage and irrigation. It also avoids the post-operative joint stiffness and slow functional improvement sometimes associated with arthrotomy[20].
  - iii) Arthrotomy is required for deeper joints

Regardless of the mode of drainage, early mobilization of the joint is important in regaining maximal function. Passive range of motion exercises should be started as soon as possible.

Non bacterial infectious arthritis

Viral arthritis

- Viruses like parvovirus, rubella, hepatitis B & C mumps, chickenpox, chikungunya and HIV.
- Viruses can induce arthritis by direct toxic effect or immune mediated mechanisms.
- Management comprises of treatment of the cause and symptomatic treatment of arthritis with anti-inflammatory drugs.

Fungal arthritis

It is rare and mostly associated with immuno-suppression.

- Actinomycosis, aspergillus, candida, coccidioidomycosis, blastomycosis and sporotrichosis have been reported.

- In RA, candida has been found in prosthetic joint.
- Antifungal agents like Amphoterecin B, Ketoconazole or Fluconazole along with surgical debridement as indicated constitute therapy.

Parasitic arthritis

Even in endemic areas, direct parasitic involvement of the musculoskeletal system is relatively rare.

- The reported protozoa are Blastocystis hominis, Cryptosporidium, Entamoeba histolytica, Giardia lamblia, Isospora belli. Toxoplasma gondii and Ancylostoma. Ascaris dracunculus, Filariae, Taenia, Toxocariasis, Strongyloides, Schistosoma, Hydatid are the helminths.
- Treatment of the cause can initially aggravate musculoskeletal symptoms before recovery.

Special situations

Septic Arthritis complicating rheumatoid arthritis

Patients with RA are at increased risk for septic arthritis. About 10% of cases of pyogenic arthritis are associated with RA[21]. There are many reasons why this happens such as i) Chronically inflamed joints. ii) Corticosteroid usage. iii) Vasculitic ulcers. iv) use of Biological agents such as TNF Inhibitors. IL1 and 6 Inhibitors etc. v) Diabetes mellitus. Use of TNF inhibitors has also led to an increased risk of Tuberculosis in our Population and hence by extension an increase in the risk of mycobacterial septic arthritis[6]. Identifying joint infections in the patient with RA can be difficult.

- Fever and leucocytosis may be absent[22]. The erythrocyte sedimentation rate is of limited value in distinguishing infection from disease flare.
- Both infection and disease flare can be mono or polyarticular.
- Infection should be suspected in RA if there is sudden worsening of one or more joints, presence of distant site of infection and when pain and tenderness are more than expected.
- Joint infections in RA patients tend to have a poorer outcome than in other groups with pyogenic arthritis.
- The mortality rate in reported series is about 20 per cent and more than double in patients with polyarticular involvement[22].

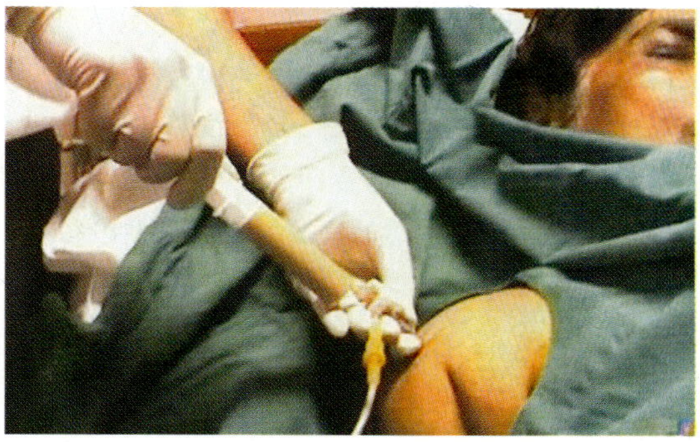

Fig. 1 : Showing aspiration of purulent fluid from shoulder joint in patient with rheumatoid arthritis.

Prosthetic joint infection

Though the incidence of infection after hip or knee arthroplasty has reduced, to less than 1 per cent of cases with treatment of pre-operative infections, improved surgical technique, and prophylactic antibiotics, it remains a major cause of failed joint replacement surgery.

- The incidence of infection for prosthetic joint revisions is about 5-10 times that of primary procedures[23].
- Infection should be suspected in cases of new or increasing prosthetic joint pain. Cultures should be obtained at all revision operations.
- Prosthetic joint infections present with joint pain, fever, and erythema.
- Swelling of the joint may or may not occur. Leucocytosis is not seen.
- An elevated ESR (Erthrocyte Sedimentation rate) or CRP (C reactive protein) is almost always present.
- About 25% of prosthetic joint infections are seen in the first three months after surgery usually as a result of bacteria introduced at the time of surgery or wounds complicated by haematoma, stitch abscess, dehiscence or infection[24].
- The prosthetic joint is at risk of infection at all times, due to haematogenous seeding from a distant site infection.
- The onset of pain is usually insidious, and local and systemic signs of joint infection are absent. This may result in delay in diagnosis.
- Differentiation from loosening is important. Bone scan is useful. Joint aspiration also may help. Staph Epidermidis is the most common organism in infected prosthesis.

Management of infected prosthetic joints

- Removal of the prosthesis, cement and all necrotic bone and soft tissue.

Exceptions include :

i) Very early post-operative infections in which the joint can be effectively drained and irrigated or the polypropylene component of knee prosthesis can be changed.

ii) Patient unfit for surgery due to poor medical status and those infected with an organism of low virulence. In this case, chronic suppression of infection by antibiotics has at times been successful in salvaging the prosthesis and preventing further infectious complications.

- After 6 weeks of IV antibiotics a prosthesis can be re-implanted and with a better functional outcome. Arthrodesis may be required in few cases[1].

Tuberculous arthritis

Osteoarticular tuberculosis accounts in 1-3% of tuberculosis patients. In 50% of them, this is associated with concurrent pulmonary tuberculosis. Infection can reach synovium. i) by haematogenous route. ii) contiguous spread and reactivation of old focus of infection. There is granulomatous reaction and formation of pannus that erodes the cartilage, the overlying capsule and tendons. Rarely, a sinus may form.

Clinical features

- Onset is typically insidious.
- It is usually monoarticular though polyarticular involvement may occur in high risk groups.
- Pre existing arthritis or old trauma and immuno compromised states like alcoholism, prolonged use of corticosteroids, and immunodeficiency diseases are important predisposing factors.
- Weight bearing joints are more frequently affected, and microtrauma to cartilage is thought to predispose to the infection. The joints commonly involved are the hip, knee, ankle, sacroiliac, wrist, and shoulder in order of frequency.
- Localized pain is the commonest presentation, usually associated with swelling and functional impairment.
- As infection progresses, pararticular muscle develops significant wasting and joint destruction occurs at a pace slower than pyogenic arthritis.
- Eventually flexion contractures, joint deformity and sometimes discharging sinus(es) can develop. The end-result may be fibrous or bony ankylosis.
- Systemic symptoms such as fever, night sweats, weight loss and malaise may or may not be present.
- Other manifestation of osteoarticular tuberculosis are Pott's spine, dactylitis and Poncet's disease.
- Spine is involved in 50% of cases:

  i) dorsal and lumbar regions are the most common sites,

  ii) Sacroiliac joint may be involved. (Fig. 2 and Fig. 3).

  iii) The infection may begin and remain confined to a vertebral body, may lead to its collapse and kyphotic deformity or gibbus may develop.

  iv) It may begin with end plate of the vertebral body with early involvement of the adjacent disc and the next vertebral end plate.

  v) Anterior lesions develop under the anterior spinal ligament to involve several vertebrae. Pedicles and laminae are affected more commonly; whereas the transverse or spinous processes are affected infrequently.

  vi) The Paravertebral abscess is common. It may remain localized or track along tissue planes. For example, in the lumbar spine, the abscess may track along the psoas muscle and present as a swelling in groin or thigh and from thoracic vertebrae it may follow the course of a rib and present anteriorly on the chest wall. The local symptoms are pain, muscle spasm, and limitation of movement.

  vii) Pain is characteristically worse during sleep.

  viii) Occasionally kyphosis or a cold abscess may be the first manifestation.

- Small joints, metaphysis of long and short bones, tendon sheaths of hands and feet, bursae, muscles and deep fasciae are rarely affected. Tuberculous dactylitis is the variant of tuberculous osteomyelitis affecting the short tubular bone of the hands and feet.
- Radiographs typically show a central, lytic, cystic, and expansile lesion (spina ventosa).
- Poncet's disease is a polyarthritis associated with visceral tuberculosis without bacteriological involvement of the joint. It may be considered as reactive arthritis to tuberculoprotein[25,26].

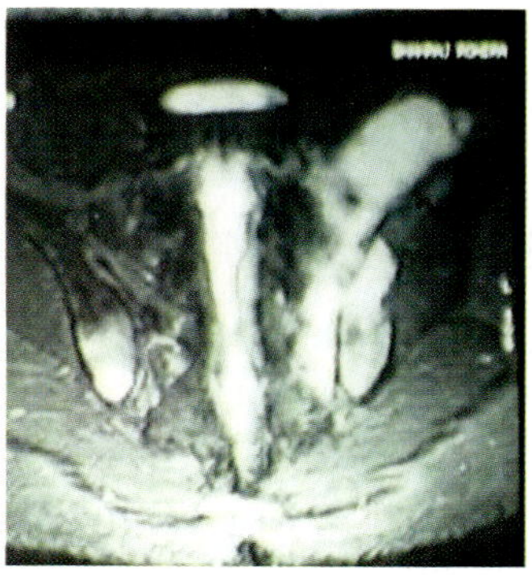

Fig. 2 : MRI showing sacroiliac involvement in tuberculous arthritis.

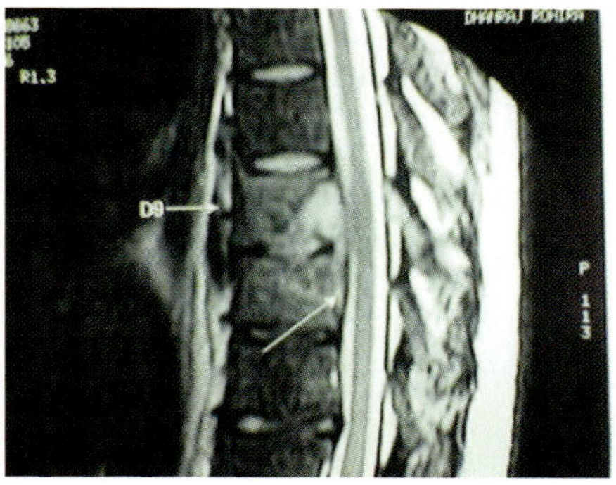

Fig. 3 : MRI of the same patient showing tuberculous lesion at D9. CT guided aspiration confirmed Koch's and culture showed sensitive bacteria.

Laboratory findings

- Essential to confirm the diagnosis.
- Definitive diagnosis needs demonstration of AFB in synovial tissue and/or culture of the organism.
- Synovial fluid yields positive results in up to 80 percent of patients with tuberculous arthritis.
- Synovial biopsy showing caseating granulomas is highly suggestive.
- Adenosine deaminase (ADA) levels have limited value as they are elevated in inflammatory arthritis[27].
- Evidence of active pulmonary or extra-pulmonary tuberculosis and strongly positive Mantoux test are supportive evidence.
- ESR and CRP may be elevated.
- PCR and serological tests do not have diagnostic specificity.

Imaging

Radiological changes are non specific.

- Chest x-ray may show pulmonary tuberculosis.
- Joint x-ray may show osteopenia, joint space narrowing, cyst formation, erosion, destruction of joint and ankylosis depending on progression.
- CT scan is useful to assess bony lesions and for guided biopsies.
- MRI is useful in evaluating soft tissue lesions, cold abscess, intramedullary and extradural lesions.

Differential diagnosis: includes RA, septic arthritis, brucellosis and malignant disease. Presence of erythema nodosum, parotid gland swelling and red eye (uveitis and chorioretinitis) may lead to confusion with other connective tissue diseases.

Treatment: same as for tuberculosis elsewhere.

- Isoniazid (5mg/kg), Rifampicin (10mg/kg), Ethambutol (15mg/kg) and Pyrazinamide (25mg/kg) daily for two months followed by Isoniazid and Rifampicin for 4-6 months in non resistant cases as recommended by WHO.
- Some believe that treatment for 9-12 months is necessary.

Key points

- Septic arthritis presents with acute onset and tuberculous arthritis with insidious onset.
- Staph. aureus is the most common organism.
- Pain, erythema, swelling, tenderness and limitation of the movement of a single large joint, commonly knee is a typical presentation.
- Synovial fluid analysis before starting antibiotic is crucial.
- Early diagnosis, adequate drainage and appropriate, prompt antibiotic therapy is essential for good outcome.

Further Reading :

1. Miller ML. Pyogenic arthritis in adults. Oxford textbook of rheumatology (3rd edition) Oxford University Press 2004; 6.2.1: 591-98.
2. Kaandorp CJE., Dinant HJ., van de Laar M.A.F.J., et al. (b) Incidence and sources of native and prosthetic joint infection: a community based prospective survey. Ann Rheum Dis. 1997; 56: 470-75.
3. Dubost J. et al. Polyarticular septic arthritis. Med 1993; 72: 296-310.
4. Medina F, Fraga A and Lavalle C. Salmonella septic arthritis in systemic lupus erythematosus. The importance of the chronic carrier state. J Rheum 1989; 16: 203-08.
5. Chandrasekar PH. and Narula AP. Bone and joint infections in intravenous drug abusers.Rev Infect Dis 1986; 8: 904-11.
6. Lawrence C. Madoff. Infectious Arthritis. Harrison's Rheumatology (Second Edition) 243-53.
7. Goldenberg DN. Bacterial arthritis. Textbook of rheumatology (5th edn) WB Saunders company 1997; 90: 1435-49.
8. Douglas S., Swanson and Mary Anne Jackson. Pyogenic arthritis in children Oxford textbook of rheumatology (3rd edition) Oxford University Press 2004; 6.2.7: 637-44, 2: 599-603.
9. Riordan T, Doyle D, Tabaqchali S. Synovial fluid lactic acid measurement in the diagnosis and management of septic arthritis. J Clin Patho 1982; 35: 390-94.
10. V. Krishnamurthy. Infection and arthritis. Manual of rheumatology. Indian Rheumatology Assoc 1999; 7: 131-40.
11. Jacob D, Cyteval C, Moinard M. Interventional sonography. J Radiol. 2005; 86 (12.2): 1911-23.
12. Tehranzadeh J, Wang F, Mesgarzadeh M. Magnetic resonance imaging of osteomyelitis. Crit Rev Diagn Imag 1992; 33: 495-534.
13. Wilson DJ. Soft tissue and joint infection. Eur Radiol 2004; 14: Suppl 3: E64-71.
14. Molina-Murphy IL, Palmer EL, Scott JA, et al. Polyclonal, nonspecific 111 In-IgG scintigraphy in the evaluation of complicated osteomyelitis and septic arthritis. Qt J N Med 1999; 43: 29-37.
15. Martin Gotthardt, Chantal P. Bleeker-Rovers, Otto C. Boerman, Wim J.G. Oyen. Imaging of Inflammation by PET, conventional Scintigraphy, and Other Imaging Techiques J Nucl Med 2010; 51(12): 1937-49.
16. Mathews CJ, Kingsley G, Field M et al. Management of septic arthritis: a systematic review. Ann Rheum Dis 2007; 66: 440-45.
17. Joshi VR. Joint infection. J Gen Med 2004; 16: 31-32.
18. Cavalier R, Herman MJ, Pizzutillo PD, et al. Ultrasound-guided aspiration of the hip in children: a new technique. Clin Orthop Relat Res 2003; 415: 244-47.
19. Ike RW. Tidal irrigation in septic arthritis of the knee: A potential alternative to surgical drainage. J Rheum 1993; 20: 2104-11.
20. Thiery JA. Arthroscopic drainage in septic arthritis of the knee: a multicenter study. Arth 1989; 5: 65-69.
21. Cooper C, Cawley, M.I.D. Bacterial arthritis in an English health district: a 10 year review. Ann Rheum Dis 1986; 45: 458-63.
22. Gardner GC, Weisman, MH. Pyoarthrosis in patients with rheumatoid arthritis: a report of 13 cases and a review of the literature from the past 40 years. Am J Med 1990; 88: 503-11.
23. Wymenga AB, van Horne JRA, Muytjens HL, et al. Perioperative factors associated with septic arthritis after arthroplasty. Acta Ortho Scand 1992; 63: 665-71.
24. Inman RD, Gallegos KV, Brause BD, et al. Clinical and microbial features of prosthetic joint infection. Am J Med 1984; 77: 47-53.
25. Amin SN. Mycobacterial diseases. Oxford textbook of rheumatology (3rd edition) Oxford Univ Press 2004; 6.2.7: 637-44.
26. C Balakrishanan. Tuberculosis and leprosy as rheumatic diseases. Bom Hosp J1997; 39: 296-301.
27. Pettersson T, Klockars M, Weber TH, et al. Adenosine deaminase activity in joint effusions. Scand J Rheum 1988; 17: 365-69.

BRUCELLA ARTHRITIS

2

Yojana Gokhale

Introduction

Osteoarticular brucellosis is the most frequent, localized complication of brucellosis, a world wide zoonosis. Due to severe rheumatism associated with this febrile illness or due to osteoarticular complications of the disease, these patients can present to a rheumatologist. Rheumatologists should be well versed with clinical and radiological features of brucellosis as well as its lab diagnosis.

Epidemiology

Brucellosis is a common zoonotic disease of world wide distribution. Only 17 countries in the world are declared brucella free by the World Health Organization. It is an important economic problem and a serious health hazard in Gulf countries, Mediterranean region, Indian subcontinents, Mexico and South-Central America. The disease is transmitted to man by consumption of contaminated milk or milk products or contact with infected cattle. In some parts of India 19-39% buffalos are seropositive for brucellosis[1,2]. With non-availability of pasteurized milk in many parts of the country and farming being the main occupation in rural India both these modes of transmission are possible. Many residents of urban slums go back to villages during the monsoon and have contact with domesticated animals. Infection can also be acquired during foreign travel (Gulf countries) or consumption of soft Mexican cheese. The organisms enter human body through skin, conjunctiva, gut or respiratory mucosa. Brucellosis is reported from all parts of India, Mumbai[3], Rajasthan[4], Hyderabad[5], Kerala[6].

Microbiology

Brucellae are gram negative aerobic, cocco-bacilli. They are fastidious organisms, have long incubation period (4-6 weeks), grow well on Castaneda biphasic (solid and liquid) medium. There are four subtypes which cause human infection. B. melitensis (sheep, goat, camel), B. abortus (cow, buffalo, camel, horses), B. suis (pigs), B. cannis (dogs). B. melitensis is most virulent. Brucellae are killed by boiling or pasteurization of milk but survive in biological material for long periods at low temperature. Brucellosis is a sexually transmitted disease in cattle, so a single infected bison can infect the whole herd. The infected animal should be culled to protect the herd. This is not done for obvious reasons and the infected animal keeps spreading the infection to man through infected milk and also to the rest of the herd. In cattle it causes recurrent abortion.

Immunity and pathogenesis

Exposure to brucellosis generates both humoral and cell mediated immune response. Antibodies promote clearance of extracellular brucellae. Polymorpho-nuclear cells engulf the organism. These intra cellular organisms can establish persistent intracellular infection. Initial replication of brucellae takes place within cell of the lymph nodes draining the point of entry. Subsequent haematogenous spread may result in chronic localizing infection at almost any site, although the reticuloendothelial system, musculoskeletal and the genito urinary system are most frequently involved. Local tissue response may include abscess or granuloma formation with or without necrosis or caseation. IgM antibodies appear by one week of acquiring the infection and persist for 9-12 months. The IgG antibodies appear by 4 weeks and persist beyond 18-24 months. Blocking antibodies appear in over 50% patients by 3-6 months of acquiring the infection. They are responsible for false negative standard 'Tube Agglutination Test' by 3-6 months time[7].

Clinical features of brucellosis

Brucellosis is a febrile illness which can be mistaken for any other febrile illness that is endemic locally e.g. malaria, typhoid or tuberculosis. This is because manifestations of brucellosis are non specific. The fever is accompanied by sweats which may be drenching. Malaise, myalgia, arthralgia, backache are common associated symptoms. The patients look well even at height of fever. Later fever appears only on exertion and subsides on rest. Most patients get night sweats. During acute stage there may be enlargement of the lymph nodes, liver or spleen. The patient is prone to relapses and the disorder is followed by a protracted convalescence with some form of severe rheumatism.The incubation period is 1-8 weeks. History of exposure to domesticated animals or consumption of unpasturized milk or milk products is available in 70-80% cases.

Bone and joint brucellosis

Osteoarticular localization was recognized as early as 1861 by Martson. The frequency of Bone and joint involvement in brucellosis ranges from 2-85%[8] as reported by different authors. It is mostly related to brucella species, being most frequent in B. melitensis followed by B. suis and B. abortus respectively. They may occur at any age. Arthralgia and arthritis due to brucellosis is more common in childhood than in older age groups, whereas brucella spondylitis is more common in adults than in children.

Brucella spondylitis

It occurred in 50% patients of brucellosis in pre-antibiotic era. Most common site of affection is the lumbar spine (L4), but dorsal as well as cervical spine may be involved. Multiple vertebral sites may also be involved. The lesion generally starts at the superior end plate of the fourth lumbar vertebra producing an erosion which may heal by sclerosis producing what is known as 'Pedro Pons' sign (Fig. 1) on lateral radiograph. The lesion may spread to involve the entire vertebra or even the adjacent vertebra. There can be collapse of vertebra (Fig. 2) radiologically simulating tuberculous spondylitis. Paravertebral, epidural as well as psoas abscess can develop.

Clinical features[8,3]

Patients with brucella spondylitis present with low back pain, gradual or sudden onset, radiating to the leg, resembling prolapsed intervertebral disc. The pain is aggravated by walking and relieved by rest. Patient may become bedridden due to severe pain. The affected vertebral spine is usually tender. There is paraspinal muscle spasm. The straight leg raising test may be positive. Deformities and neurological deficit are rare as compared to that in tuberculous spine, but do occur due to delay in diagnosis in our country[3]. History of febrile illness in 3-6 months preceding the onset of back pain is available in most patients[3].

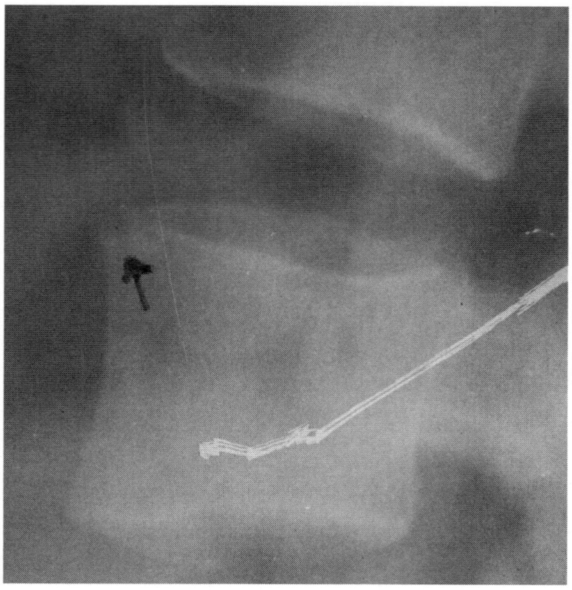

Fig. 1 : Pedropons' sign, erosion at anterior superior angle of L4.

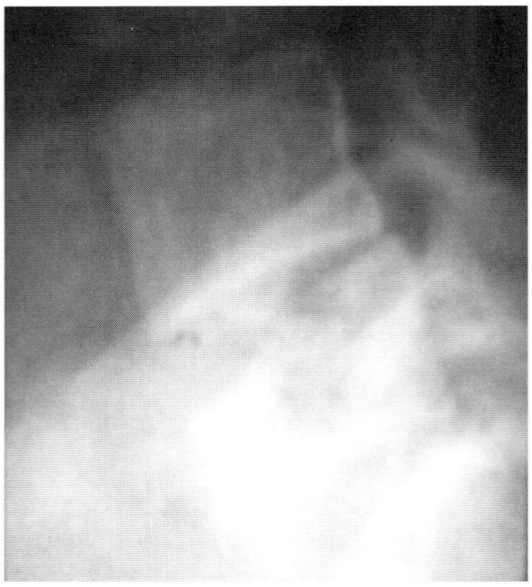

Fig. 2 : Brucella spondylitis, X-ray showing L5 collapse.

Diagnosis

There are no characteristic features of brucella spondylitis as such. A history of animal contact, consumption of raw milk or milk products, visit to gulf countries and febrile illness in a patient with back pain should arouse suspicion of brucella spondylitis. The plain radiograph may be normal in early stage. Scintigraphy using Tm99 is particularly useful in such cases. It shows increased tracer uptake at affected vertebral sites and/or the sacroiliac joint (Fig. 3). Bacteriological studies or blood culture are gold standard but the yield is poor in spondylitis, especially in afebrile patients. Hence one must rely on serological tests. IgM, IgG antibodies to brucella can be detected in the patients' serum by ELISA. It is highly sensitive and specific. The 'tube

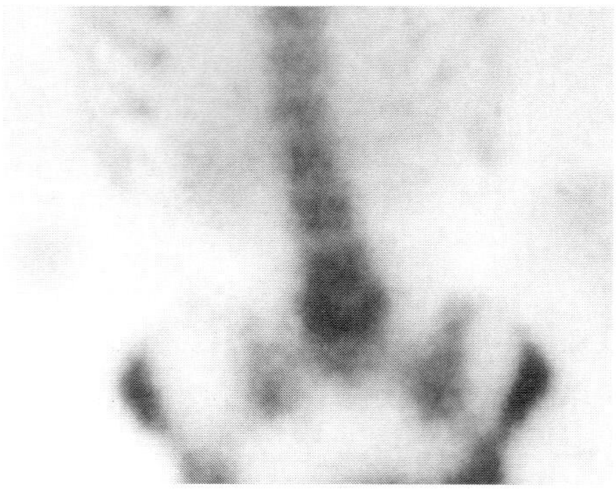

Fig. 3 : Bone scan : increased uptake in L5 and left sacroiliac joint.

agglutination test' (TAT) may be false negative due to blocking antibodies after 3-6 months of acquiring the infection. Exact titers above which TAT should be considered positive in India is not known. Titers more than 1:80 or above may be considered positive. PCR based assays for brucella are reported to be very specific and sensitive. Blood culture by classical Castanida method takes 4-6 weeks to grow brucella. With automated culture systems like BACTEC culture on blood[9], synovial fluid[10], CSF results are available with in 3-7 days. Other laboratory investigations tests like the white cell count may be normal and the erythrocyte sedimentation rate may be normal or raised.

Radiological investigations

Very early in the disease, upto 3 months, the plain radiograph may be normal. Earliest lesion is an area of bone destruction at the discovertebral junction. This may have a rim of sclerosis. An anterior osteophyte resembling that due to degenerative spondylitis of the lumbar spine is a common finding due to old healed lesion. In advanced cases there is vertebral body collapse and decreased intervertebral disc space (Fig. 4). Paravertebral soft tissue may be seen.

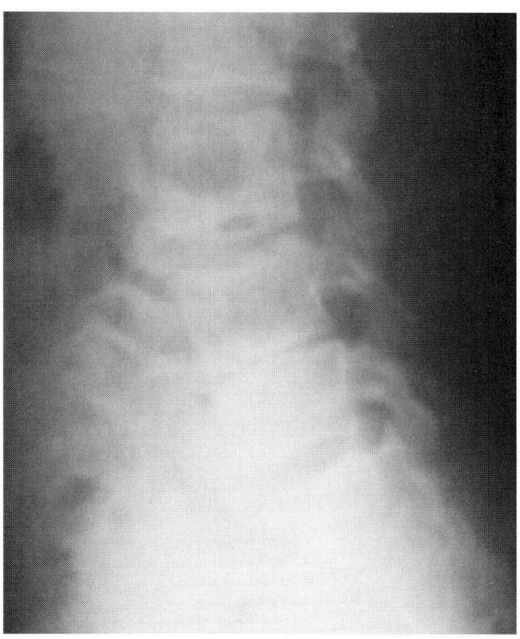

Fig. 4 : Brucella spondylitis, x-ray, L4-5 discitis.

An MRI scan in early lesion depicts areas of low signal intensity at the superior end plate in T1 images, they become hyper intense on T2 images. Such lesions may be seen at multiple vertebral levels (Fig. 5). Para spinal or epidural or psoas abscesses may be seen.

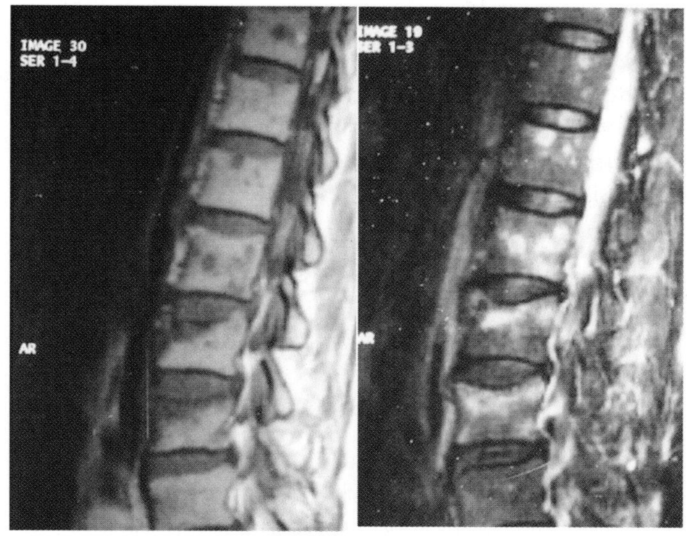

Fig. 5 : MRI scan (a) T1 image showing hypo intense marrow signals, (b) T2 image showing hyper intense marrow signals. The patient was extensively investigated for myeloma, lympho reticular malignancy and metastasis in a cancer hospital. Her serological test for brucellosis was strongly positive and there was excellent clinical response to anti brucella therapy, along with improvement in MRI.

Histopathological features

Tissue obtained by image guided percutaneous biopsy or at the time of surgery may reveal non specific cellular infiltration with or without granuloma formation. Occasionally caseous necrosis may be present. Cultures of aspirate may rarely grow brucellae.

Brucella sacroiliitis[8,11,3]

Involvement of sacroiliac joints comprises a large proportion of patients of brucella arthritis. It is reported in 13-52% patients of brucellosis. Acute sacroiliitis due to brucellosis occurs in the early febrile stage of infection. It presents as unilateral pain over the sacroiliac joint, fever with chills and sweats. The sacroiliac maneuvers are very painful, as also the hip movements and the straight leg raising test. It lasts for few weeks to resolve spontaneously. All acute unilateral sacroiliitis should be investigated for brucellosis. In few cases opposite sacroiliac joint may subsequently get affected. Sacroiliitis may be associated with lumbar spondylitis with low back pain simulating ankylosing spondylitis. Brucella organisms may be cultured from sacroiliac joint aspirate.

Differential diagnosis of brucella spondylitis and sacroiliitis

1. **Spinal tuberculosis:** Clinically as well as radiologically spinal tuberculosis can mimic brucella arthritis brucella spondylytis.

Brucellosis has a predilection for the lumbar spine where as tuberculosis affects the dorsal spine more often. Spinal deformities are common in tuberculosis. Positive brucella serology is useful in making the diagnosis of brucella spodylitis.

2. **Prolapsed intervertebral disc:** Patients of brucella spondylitis experience low back pain radiating to leg and SLR can be positive. The radiograph may be normal in early brucella spondylitis. An abnormal bone scan and positive brucella serology can aid in the diagnosis.
3. **Degenerative lumbar spondylosis:** Anterior osteophytes in the elderly patients' radiogram may simulate healed old brucella spondylitis. Such changes on lateral radiograph are common in elderly patients in endemic areas.
4. **Ankylosing spondylitis:** Sacroiliitis and osteophytes due to healed superior end plate lesion may mimic ankylosing spondylitis.
5. **Malignancy or secondary deposits:** Back pain, abnormal tracer uptake in bones and the spine on bone scan and abnormal marrow signals on the MRI may be mistaken for malignancies like myeloma or secondaries.
6. **Staphylococcal osteomyelitis of the spine**.
7. **Actinomycosis**.

Brucella arthritis[8,11]

Arthralgia is reported in over 80% patients of brucellosis. Bone scans in patients of brucellosis with musculoskeletal symptoms show abnormal uptake at multiple sites. Arthritis may be monoarthritis usually affecting the large joints or migratory polyarthritis. The synovial fluid may grow Brucella. The radiographs may be normal in early cases. Bone scans would show increased tracer uptake. Septic and destructive arthritis of one joint may occur due to brucellosis. If there is delay in diagnosis destructive changes and radiological abnormalities, such as joint space narrowing occur. Brucella arthritis is easily diagnosed in countries where brucellosis is endemic. It is more common in children than adults. There are not many reports on brucella arthritis of joints other than sacroiliac joint from India.

Tenosynovitis, bursitis and osteomyelitis due to brucellosis are reported.

Treatment of osteoarticular brucellosis[8]

Principles

1. Always use combination therapy (2 or 3 drugs).
2. Duration of therapy should be 6-12 weeks or more.
3. Paraspinal abscesses do respond to medical line of treatment. Surgery should be reserved for patients with neurological deficit or relapse due to abscess[8].

Anti brucella drugs

Doxycycline 100 mg bid, Rifampicin 900 mg daily, Ciprofloxacin 500gm bid, Co-trimoxazole 10mg/ kg of TMP (i.e. 3 tablets bid), Inj. Streptomycin 1gm daily for 3 weeks. Ceftrioxone is also effective against brucellae.

Relapse

It is known to occur in 20% cases. Its diagnosis is clinical. One can document rising titers of antibodies. Treatment is like fresh attack.

Prevention

Using pasteurized milk is the best method of prevention. For occupations where brucellosis is an occupational hazard use of gloves, masks, gowns should be promoted. Animal disease can be prevented by culling infected cattle and immunization of the new born and the uninfected animals.

- Diagnosis of brucella spondylitis and sacroiliitis should be considered in patients with chronic backache with history of animal contact or consumption of raw milk or milk products or travel to Gulf countries and in all patients with acute unilateral sacroiliitis.
- IgM, IgG antibodies towards brucella by ELISA are highly sensitive and specific.
- Combination anti-brucella drugs (2 or 3) should be used for 6-12 weeks for treatment.

Further Reading :

1. M.N. Brahmabhatt, R.N. Varasada, C.D. Bhong and J.B. Nayak, Seroprevelance of Brucella Spp. In buffalos in the central Gujrat region of India, Buffalo Bulletin 2009; 28: 73-75.
2. Chauhan H.C., B.S. Chandel and N.M. Shah. Seroprevalence of brucellosis in buffaloes of Gujarat. Indian Vet. J. 2000, 77: 1105-06.
3. Gokhale Y A, Ambardekar A G, Bhasin A, et al, Brucella spondylitis and sacroiliitis in the general population in Mumbai, J Assoc Phy Ind 2003; 51: 659-66.
4. Kochar D K , Agarwal N, Jain N, et al, Clinical profile of Neurobrucellosis - A report of 12 cases from Bikaner (North, - West India), J Assoc Phy Ind 2000; 48: 376-80.
5. Prasad S, R Rao, et al, Brucellosis : A cause of backache, J Ind Rheum Assoc 1994; 2: 27-31.
6. Deepak S, S G Bronson, Sibi, et al, Brucella isolated from bone marrow, J Assoc Phy Ind 2003; 51: 717-18.
7. Ariza J , Pelliacel T, Pallares R et al. Specific antibody profile in human brucellosis, Clinical Infectious diseases, 1992; 14: 131-40.
8. Madkour MM. Bone and joint brucellosis. In Madkour MM. Brucellosis 1st ed, Butterworth London. 1981; 90-104.
9. P Maleknejad, FB Hashemi, B Fatollahzadeh, S Jafari, H Peeri Dogaheh, Direct Urease Test and Acridine Orange Staining on Bactec Blood Culture for Rapid Presumptive Diagnosis of Brucellosis, Iranian J Publ Health, 2005; 34: 52-55.
10. Pablo Y, Nechama P, Joseph, Use of BACTEC 9240 Blood Culture System for Detection of Brucella melitensis in Synovial Fluid, J Clinical Microbiology, 2001; 39: 738-39.
11. Eliseo Pascual, Brucella arthritis. In Oxford Text book of Rheumatology 3rd ed. Oxford University Press, London, UK. 2004; 937-44.

ARTHRITIS IN LEPROSY

3

Vikas Agarwal, Anupam Wakhlu

Introduction

Leprosy is a chronic granulomatous infectious disease caused by *Mycobacterium leprae* with skin and nerves being predominantly affected. As of 1985 leprosy was endemic in 122 countries with a global prevalence of 12 per 10,000. After the world health assembly accepted the goal of elimination of leprosy as a public health problem (prevalence rate of less than one case per 10,000 persons) by 2000 AD[1], widespread use of multi drug treatment (MDT) has reduced the disease burden dramatically. Leprosy has been eliminated from 108 out of the 122 endemic countries. In 2002, 70% of the world's registered leprosy patients were in India. The world health organization's April 2005 figures show South-East Asia (SEA) Region as the only region where the prevalence is above the elimination goal. India alone represents 80% of prevalence and 88% of new detected cases in the SEA Region. As of June 2006, the prevalence in India is 0.84 per 10,000, but nine states/union territories of the country still have not achieved the elimination target of the World Health Assembly[2]. India is considered to be the epicenter of the problem, where 260 000 of the 408 000 people diagnosed across the world in 2004 reside[3] states of Uttar Pradesh and Bihar has the highest number of cases (Fig. 1).

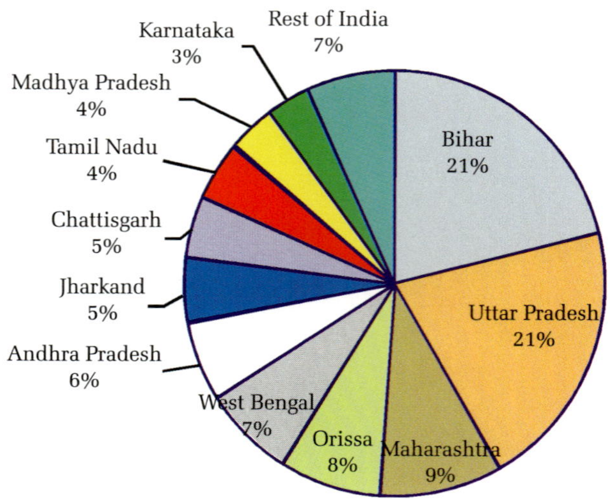

Fig. 1 : Registered Prevalence Per State, India. April 2003.

Epidemiology

The frequency of arthritis in leprosy is difficult to estimate because the prevalence varies from 1-78%[4-6]. The prevalence of arthritis was only 1% (27 patients) in a study on 2500 Indian patients with leprosy[7]. Cossermelli observed arthritis as the first manifestation of leprosy in 12 out of 44 patients with arthritis[8] while Gibson reported arthralgia/arthritis as the first manifestation of leprosy in 39% (12/31) of patients in lepra reaction[9]. However, in another series of 18 unselected patients, 14 had arthritis[10] whereas Atkin et al reported that one-third of 66 consecutive leprosy outpatients had an inflammatory arthritis[5]. A recent review has suggested that rheumatic symptoms occur in as many as 75% of leprosy patients and are the third most common reason for hospital admission[11]. In a recent report from Nagpur, authors have reported rheumatic manifestation in 61.4% patients (54.3% having arthritis and 17.4% having soft tissue rheumatism)[12].

Clinical features

The classical clinical presentation of leprosy is in form of hypo or anesthetic and anhidrotic macules, patches or plaques or widespread papulo-nodular lesions, with or without paresthesias or loss of nerve function in the distribution of mononeuropathy or mononeuritis multiplex/polyneuropathy and blindness.

Rheumatological manifestations of leprosy are less common/less commonly recognized than the classical dermatological and neurological presentation. Joint involvement has been recognized as early as 600 BC in Chinese literature[13]. For the sake of pathophysiologic clarity, arthritis or rheumatological manifestations in leprosy can be divided into three broad categories. (a) Arthritis as a consequence of direct infiltration of synovium (b) Reactional states with arthritis or rheumatological manifestations (c) Manifestations arising as a consequence of complications such as peripheral neuropathy eg. (Charcot's or neuropathic joints). Though neuropathic joints and septic arthritis are well known types of articular involvement in leprosy, inflammatory arthritis is much more common during the reactional states. The pattern of arthritis may vary from monoarticular involvement to symmetric polyarticular disease mimicking rheumatoid arthritis (RA) and or oligoarticular involvement with/without sacroiliitis or enthesitis mimicking seronegative spondylo-arthropathy.

Charcot's arthritis involving weight bearing joints of the lower limbs is usually seen in around 10% of the cases of leprosy[14]. Characteristically, these patients have had long-drawn disease course with peripheral neuropathy and the joints are grossly damaged and deformed with little pain.

Till recently, inflammatory arthritis was not recognized as an important feature of leprosy[15]. However since the first report (1964) by Lele *et al*, inflammatory arthritis has been recognized as a characteristic presentation of leprosy, especially during reactional states. They described 13 cases of leprosy that developed acute painful symmetric polyarthritis involving hand joints and mimicked RA[7]. In 6 cases, synovial biopsy showed granulomatous reaction without evidence of M. leprae infiltration. Modi *et al* reported 5 years later from South India, 27 cases of leprosy (16 lepromatous, 6 borderline, 3 tuberculoid and 2 pure neuritic) presenting with similar polyarthritis mimicking RA. Of these 27 cases, 21 had lepra reaction. Synovial biopsy (23 cases) was reported as non-specific in 10, healed scarred lesions in 9; however, in 4 cases, M. leprae could be demonstrated. In 6 cases of arthritis with no reactional state, therapy with dapsone aggravated the arthritis and led to involvement of the newer joints[16]. Similar cases were reported from different parts of the globe (Table 1). Gibson *et al* reported the presence of arthritis in 20 out of 31 (65%) patients with lepra reaction[9]. Arthritis in type 1 (12/20) and type 2 reactions (8/20) could not be differentiated. Lele and Modi observed involvement of small joints of hands in 22 (81%) cases and most were in reaction. The duration of arthritis also was only 1 month in most patients[16].

Type II reaction, erythema nodosum leprosum (ENL), is an immune complex mediated reaction seen predominantly in patients with borderline lepromatous or polar lepromatous leprosy. The cytokine profile is typically T_{H2} type (increased IL-6, IL-10). It usually occurs after first 6 months of the anti-leprosy therapy, although leprosy presenting with ENL as the first manifestation is well documented. Arthritis in Type I, reversal reaction, had similar presentation like ENL. However, the underlying pathogenic process is exaggerated CD4+ cell mediated immune response to M. leprae antigens. The cytokine profile is typically T_{H1} type (increased TNF-α, IFN-γ, IL-2, IL-4). It commonly occurs in the first six months of antileprosy therapy. However, unique to Indian subcontinent is the spontaneous onset of lepra reactions that has been reported recently[17]. This further adds to the diagnostic dilemma. Reactional state associated arthritis usually begins abruptly and is associated with constitutional symptoms, fever and weight loss, and other features of reaction, new/worsening of cutaneous lesions or paresthesias. It characteristically involves small joints of hands and wrist and may mimic RA. It may last for few weeks to months. Patients may have repeated episodes of such type of arthritis (Box 1). Occasionally patients with reactional states may present with insidious onset chronic symmetric polyarthritis mimicking RA.

Table 1. Reaction associated acute arthritis

Author, year	No. of patients	Type of arthritis	Joints involved	Synovial histology
Lele et al 1965[7]	13	Acute symmetric polyarthritis	Wrists, PIP, knee	6, granuloma + No AFB
Modi et al 1969[16]	27	Acute symmetric polyarthritis	Wrists, PIP, ankle	AFB + (4)
Chandrasekaran et al 1993[15]	7	Acute symmetric polyarthritis	MCP, PIP, ankle, knee	ND
Gibson et al 1994[9]	20	Acute symmetric polyarthritis	Wrist, MCP, PIP, ankle, MTP	AFB – (2)

Box 1. Characteristics of arthritis in reactional states

- May precede, accompany or follow the reaction
- Acute in onset
- Severe pain, morning stiffness and disability
- Parallels the duration and course of reaction
- Responds to antileprosy and anti-inflammatory therapy especially glucocorticoids
- Mostly resolve without sequelae
- Synovial biopsies do not show M. leprae
- Radiographs do not show erosions

Another presentation reported during reactional states is swollen hand and feet syndrome. In 1980, Albert and colleagues from California reported rheumatic manifestations on 15 patients with leprosy requiring hospitalization. Authors described 3 cases with acute onset painful edema over the dorsum of the hands with marked restriction of the movements and nodules along the extensor tendons. Biopsy of the nodules showed granulomatous reaction with positivity for M. leprae. All these 3 patients responded to antileprosy and glucocorticoid therapy[6]. At our centre, we have seen ten such cases in last 10 years (Fig. 2).

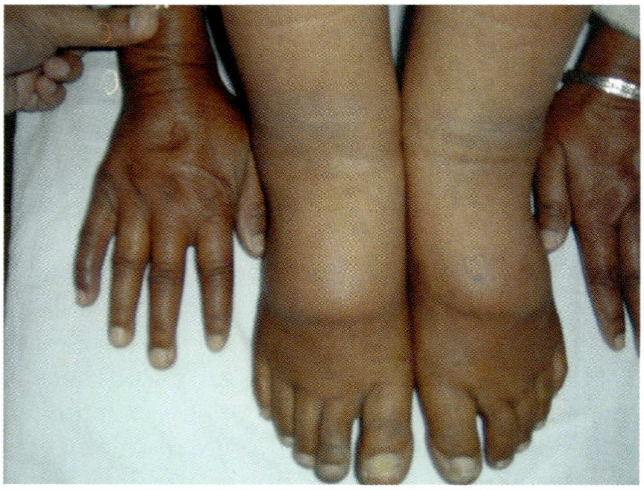

Fig. 2 : Swollen hand and feet syndrome.

In 1987[18] and 1989[5], Atkins et al reported insidious onset chronic symmetric polyarthritis mimicking RA but not associated with lepra reaction. Reaction independent chronic RA-like arthritis was described first among patients from Papua New-Guinea[18] where 31 of the 55 patients had chronic inflammatory polyarthritis. Later, in another study from Egyptian leprosy colony[5] 20 of 33 patients were reported to have reaction independent chronic symmetric polyarthritis. The largest series of reaction independent arthritis, 39 out of 44 patients, was reported by Cossermalli *et al*[8]. Arthritis in majority of their patients was chronic (mean duration 11 years) and RA-like in distribution (Box 2). Leprosy was diagnosed more than 10 years ago in most of their patients and was currently inactive in 19 patients. There was no difference in the prevalence of arthritis among patients with and without active leprosy. Chronic RA-like presentations are the most likely to be misdiagnosed. Atkins *et al* has reported boutonnière and swan neck deformities in their patients[5]. Though these patients had considerable relief with antileprosy therapy, their arthritis never resolved completely. Few of the patients had erosive arthritis just like RA (Table 2).

Box 2.Characteristics of chronic arthritis in leprosy

- Insidious onset
- Seen in all forms of leprosy
- Usually polyarticular and symmetric
- May be asymmetric and oligoarticular
- Spontaneous exacerbations and remissions
- Responds to antileprosy therapy but rarely resolves completely
- Radiographs may show erosions
- Synovial biopsy may show M. leprae/granuloma

Another presentation reported by Cossermalli *et al* was sacroiliitis (seen in 64% of their patients) and tenosynovitis[8]. Other rheumatological manifestations like cryoglobulinemic vasculitis, dermatomyositis, enthesitis and isolated tenosynovitis[19] (Fig. 3) have also been reported.

Recently, we have reported five patients with pure neuritic leprosy presenting with arthritis, tenosynovitis with/without paresthesias and thickened nerves[20]. Nerve biopsy in all the cases was confirmatory. Pure neuritic (PN) form of leprosy (no skin manifestations) is common in India (4.6%) and Nepal (8.7%)[21]. Three of our patients with RA-like

Table 2. Reaction independent chronic arthritis

Author, year	No. of patients	Type of arthritis	Joints involved	Radiology
Atkins et al 1987[17]	31	Symmetrical polyarthritis	Wrists, MCP, PIP, Knee, MTP	ND
Atkins et al 1989[5]	20	Symmetrical polyarthritis	Wrists, MCP, PIP, Knee, MTP	Erosive disease (n=12)
Cossermelli et al 1998[8]	39	Symmetrical polyarthritis	Wrists, MCP, PIP, Knee, MTP	Erosive disease

MCP; metacarpophalangeal, PIP; proximal interphalangeal, MTP; metatarsophalangeal, ND; Not done

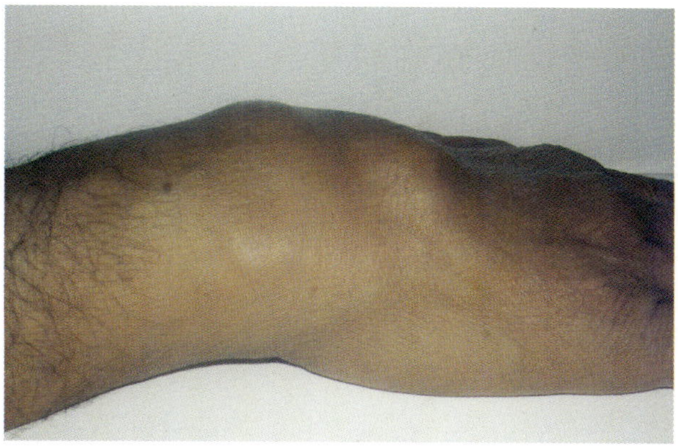

Fig 3: Isolated tenosynovitis of extensor tendons of the hand.

presentation were on disease-modifying drugs. Patients with PN leprosy lack skin manifestations and the only clue may be from the presence of tenosynovitis and neuropathy.

Some case scenarios observed by the authors exemplify the capability of leprosy to mimic presentations of common rheumatological disorders.

1) **Reactive arthritis like:** A young male presented with an isolated acute onset inflammatory knee arthritis 2 weeks after a diarrheal illness. Inflammatory parameters were raised. Everything pointed to a reactive arthritis till synovial fluid reveals AFB. Wade Fite stain for lepra bacilli was positive in synovial fluid. Subsequent clinical examination reveals a thickened lateral popliteal nerve. The diagnosis was confirmed by a sural nerve biopsy. Patient recovered completely with symptomatic and antileprosy therapy.

2) **Chronic monoarthritis like:** A 25 years female presented with progressively increasing pain in the left elbow for the last 5-6 months. There were no systemic complaints. Examination revealed tenderness in the medial aspect of the elbow joint with swelling and marked tenderness in the posteromedial lower arm region. Ultrasound examination revealed an anechoic linear collection extending from the cubital tunnel to the lower part of the arm as described above. A tubercular collection was suspected in the elbow and arm region. Attempts to aspirate this collection led to radiating paresthesias in the ulnar nerve distribution. A very thick ulnar nerve swelling was then suspected. Repeat USG examination revealed an ulnar nerve with cross sectional area of 22 mm^2 (upper normal of 7.5 mm^2) and diameter of 49 mm (Fig. 4). Further examination revealed 25% sensory loss in the nerve distribution and a thickened lateral popliteal nerve. Sural nerve biopsy confirmed leprosy. Patient responded dramatically to initiation of steroids with antileprosy therapy.

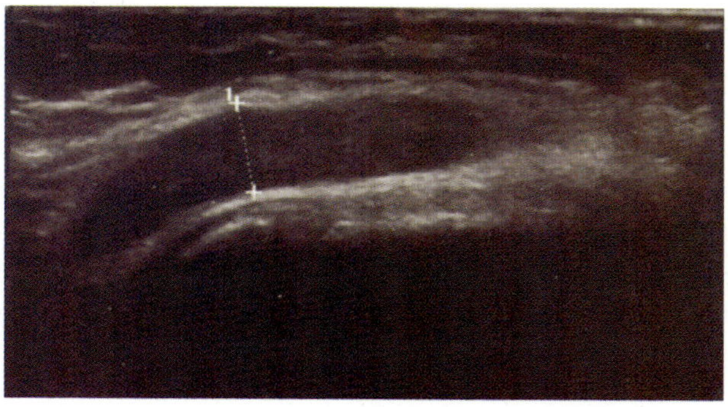

Fig. 4 : Markedly thickened Ulnar nerve mimicking localized collection.

3) **Sarcoid like:** A 35 year male presented with a 3 month history of ankle arthritis and recurrent typical attacks of erythema nodosum (Fig. 5). He denied any sensory loss. Radiograph of chest did not reveal any evidence of lymphadenopathy and there was no evidence of pulmonary/extra-pulmonary Tuberculosis at that time. Biopsy of the subcutaneous nodules revealed evidence of both septal and nodular panniculitis, which led to a suspicion of leprosy. Bilateral lateral popliteal nerves were thickened without definite sensory loss. There were no skin lesions. Sural nerve biopsy confirmed leprosy and the patient completely recovered with therapy.

5a

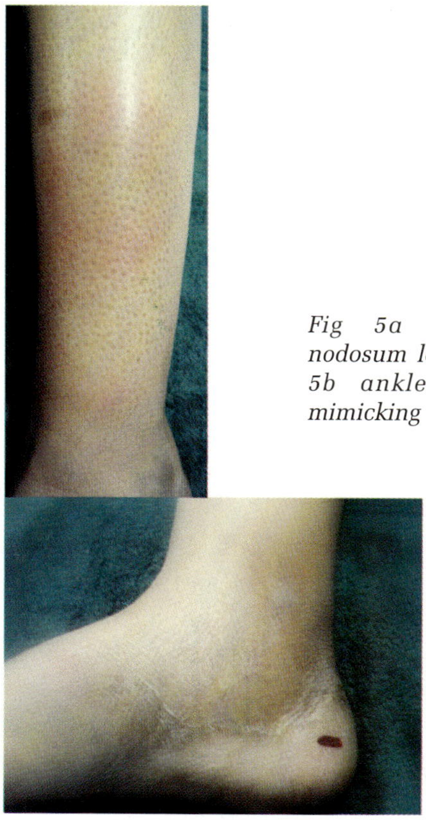

5b

Fig 5a Erythema nodosum lesions with 5b ankle arthritis mimicking Sarcoidosis

4) **Relapsing polychondritis like:** A forty year old woman presented with complaints of recurrent episodes of pain in the ear, throat and nose intermittently for the last 3 years. These episodes used to respond well to NSAID's and steroids given by local practitioners. Ear involvement was limited to bilateral erythematous painful involvement of the pinna (Fig. 6). Pain in the throat region was limited to tenderness of the thyroid cartilage and tracheal rings. There was also some change in the quality of voice. Nasal involvement was characterized by tenderness of the cartilaginous nasal septum. A tentative diagnosis of relapsing polychondritis was entertained. However, closer examination of the ear revealed involvement of the cartilage of the ear as well as the pinna, which is typically spared in RP. A thickened great auricular nerve prompted a search for leprosy. Slit skin smear was positive for lepra bacilli. The patient responded well to multibacillary treatment along with NSAID's and steroids.

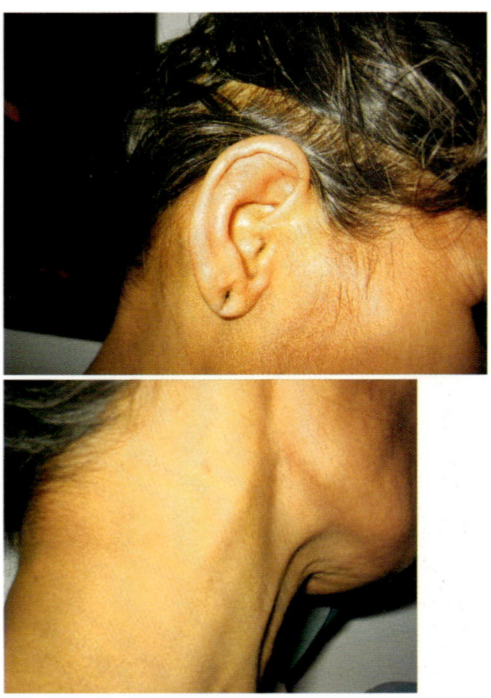

Fig. 6 : Auricular chondritis with thickened right greater auricular nerve mimicking Relapsing polychondritis

5) **Psoriatic arthritis like:** A twenty-five year old male presented with a seronegative asymmetrical small and large joint polyarthritis of 9 month duration. There was no history of back pain and no skin lesions of psoriasis. Nails appeared dystrophic and brittle (Fig. 7). On enquiry, the patient revealed that he had actually chewed the nails off but had not felt much pain. Note that the nails are involved in the ring and little finger, which is in the ulnar nerve distribution. Ulnar nerve was found to be thickened and there was significant sensory loss in that distribution. Abdominal exam revealed a hypohidrotic, hypoanesthetic, hairless patch consistent with leprosy. Skin biopsy confirmed the same. The patient responded well to therapy.

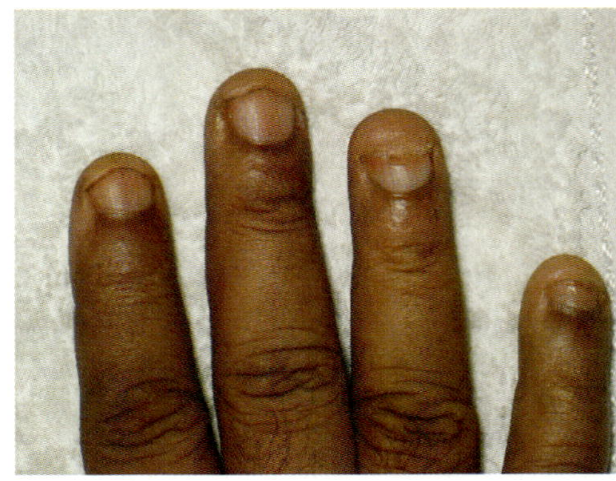

Fig. 7 : Dystrophic nails.

6) **Vasculitis like:** A fifty year female presented with ulcerative, superficially gangrenous and vasculitis like lesions (Fig. 8) of the lower limbs for the last one month. There was associated pyrexia, asymmetrical oligoarthritis and anorexia. She also had loss of peripheral sensations in the lower limbs. Skin biopsy revealed ulcerated dermis and epidermis with presence of foamy macrophages and evidence of vasculitis. Wade fite stain was positive for lepra bacilli. The patient was diagnosed as a case of Lucio phenomenon and was treated with MDT and steroids. Patient showed gradual improvement.

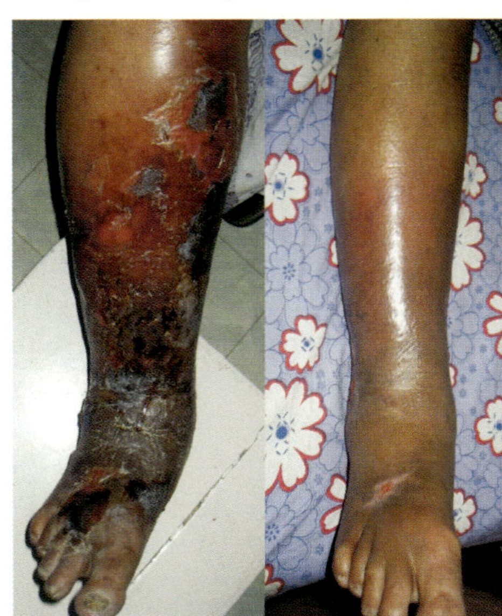

Fig. 8 : Lucio phenomenon mimicking necrotizing vasculitis (before and after MDT).

Table 3 Characteristics of different forms of arthritis in leprosy

	Onset	symmetric	Oligo/ polyarthritis	Response to		Joint erosions	*M. leprae* in synovium
				Antimicrobial	Glucocorticoid		
Lepra reaction	Acute	Yes	Yes	Poor	Good	No	No
Chronic arthritis	Insidious	Yes	Yes	Good	Not much	Yes	Yes
Swollen hand	Acute	Yes	Yes	Poor	Good	No	No
Pure neuritic*	Acute to insidious	Yes	Yes	Yes	Yes	No	No

* M. leprae can only be demonstrated in the nerve biopsy. May present with tenosynovitis alone.

Diagnosis

Clinically, attributing arthritis in a patient of leprosy with characteristic neuro-cutaneous lesions is easy; however sometimes arthritis may be the presenting complaint (Table 3). In such cases careful attention to skin involvement; hypo/anaesthetic macules or patches or erythematous macules, nodules, with history of paresthesias, thickened and tender peripheral nerves and or sensory and motor neuropathy may help in suggesting the diagnosis of leprosy. Histopathologically, synovium in chronic polyarthritis patients may demonstrate granuloma, epithelioid cells and M. leprae. Pernambuco *et al*[22] described 25 cases of arthritis from whom 32 synovial fluid samples, 20 synovial and three bone biopsies were obtained to provide a spectrum of change from non-inflammatory to turbid effusions and a synovium which was usually inflammatory and contained mycobacteria in nine samples taken from six patients.

Radiological features of leprosy related arthritis has been studied in different series. Juxta articular erosions were seen in 12 (43%) out of 28 available radiographs[5]. Modi and Lele reported erosions in only 3 out of 27 patients with acute arthritis and leprosy[16]. Periosteitis, bone resorption and deformed joints have been reported[13]. Cossermaeli *et al* reported Grade 1 to 3 sacroiliitis in 35 out of 55 radiographs[8].

Recently, a new serological test for detection of antibodies to the M. leprae specific PGL-I has been developed. These antibodies are present in 90% of patients with untreated lepromatous disease, but only 40-50% of patients with paucibacillary disease, and 1-5% of healthy controls[23]. Skin smears, taken to detect intradermal acid fast bacilli, have high specificity, but low sensitivity, because about 70% of all leprosy patients may be smear negative[24].

Treatment

The first-line drugs against leprosy are rifampicin, clofazimine, and dapsone. According to World Health Organization (WHO), all patients should receive a multidrug combination with monthly supervision. For paucibacillary disease Rifampicin 600 mg once monthly and dapsone 100 mg daily for 6 months and for multibacillary leprosy, rifampicin 600 mg and clofazimine 300 mg once a month along with clofazimine 50 mg and dapsone 100 mg daily for 24 months and for paucibacillary single lesion, rifampicin 600 mg, ofloxacin 400 mg, and minocycline 100 mg single dose is enough[25].

The treatment of reactions associated arthritis is aimed at controlling acute inflammation, easing pain, and reversing nerve damage. Corticosteroids along with multidrug therapy (MDT) regimen form the backbone of management of reactions. Prednisolone administration should be started at the dose of 1mg/kg/day and gradually tapered by 5 mg every 2-4 weeks depending upon the improvement in the signs and symptoms. Patients with borderline-tuberculoid reactions commonly need corticosteroids for 3-4 months, whereas those with borderline-lepromatous reactions may need treatment for 6 months. Recovery is less in patients with pre-existing impairment of nerve function or with chronic or recurrent reactions. Sometimes type 2 reaction ENL may not respond to MDT and prednisolone; in such cases, clofazimine can be used at 300 mg daily for several months or thalidomide (TNF-α blocker) (400 mg daily) is better than steroids and is the drug of choice for young men with severe ENL. Pentoxifylline, which inhibits TNF-α production, has been used to treat ENL but was inferior to both thalidomide and steroids. Monoclonal antibodies targeting the TNF-α molecule seem logical choice for therapy of severe refractory ENL but needs more critical evaluation before it can be recommended routinely[23].

Conclusion

Leprosy, by itself, as a consequence of its reactional states or as a consequence of its complications may mimic completely a number of rheumatological disorders, in addition to the classical forms of articular involvement already described. It requires a high index of suspicion coupled with good clinical examination to clinch the diagnosis.

Arthritis, tenosynovitis, may be the presenting manifestations and may mimic RA or spondyloarthropathy. Increased awareness about articular manifestations and careful attention to cutaneous lesions and thickened peripheral nerves may help in clinching the diagnosis of leprosy.

Further Reading :

1. World Health Assembly. Leprosy resolution WHA 44.9, Forty-fourth World Health Assembly, May 13, 1991.
2. The World Health Organization - India website information on leprosy (Accessed June 21, 2006, at http://www.whoindia.org/EN/Section3/Section122_1213.htm).
3. Making progress towards leprosy elimination. Editorial. Lancet 2006; 367: 276.
4. Bonvoisin B, Martin JM, Bouvier M, Bocquet B, Boulliat J, Duivon JP. Articular manifestations in leprosy. Sem Hop 1983; 59: 302-05.
5. Atkin SL, el-Ghobarey A, Kamel M, Owen JP, Dick WC. Clinical and laboratory studies of arthritis in leprosy. Br Med J. 1989; 298: 1423-25.
6. Albert DA, Weisman MH, Kaplan R. The rheumatic manifestations of leprosy (Hansen's disease). Medicine 1980; 59: 442-48.
7. Lele RD. Sainani GS, Sharma KD et al. Leprosy presenting as rheumatoid arthritis. J Assoc Physicians India 1965; 13: 275-77.
8. Cossermelli-Messina W, Festa Neto C, Cossermelli W. Articular inflammatory manifestations in patients with different forms of leprosy. J Rheumatol 1998; 25: 111-19.
9. Gibson T, Ahsan Q & Hussein K. Arthritis of leprosy. British Journal of Rheumatology 1994; 33: 963-966.
10. Alcocer J, Herrera R, Lavelle C, Guidino J, Fraga A. Inflammatory arthropathy in leprosy. Arthritis and Rheumatism 1979; 22: 587.
11. Pernambuco J, Cossermelli-Messina W. Rheumatic manifestations of leprosy: clinical aspects. J Rheumatol 1993; 20: 897-898.
12. Vengadakrishnan K, Saraswat PK, Mathur PC. A study of rhematological manifestations of leprosy. Ind J Dermatol Venereol Leprol 2004; 70: 76-78.
13. Messner RP. Arthritis due to mycobacteria and fungi. In Arthritis and Allied conditions. (eds) McCarthy and Febiger DL. Philadelphia, PA. 1979. pp1380.
14. Horibe S, Tada K, Nagano J. Neuroarthropathy of the foot in leprosy. J Bone Joint Surg Br 1988; 70: 481-85.
15. Chandrasekaran A. Infection and arthritis, our experience, a retrospective and prospective study. J Ind Rheumat Assoc 1993; 1: 7-20.
16. Modi TH, Lele RD. Acute joint manifestations in leprosy. J Assoc Physicians India 1969; 17: 247-54.
17. Prasad S, Misra R, Aggarwal A, Lawrence A, Haroon N et al. Leprosy revealed in Rheumatology clinic: a case series. Int J Rheum Dis 2013, DOI: 10.1111/j. 1756-185X.2012.01819.x
18. Atkin SL, Welbury RR, Stanfield E, Beavis D, Iwais B, Dick WC. Clinical and laboratory studies of inflammatory polyarthritis in patients with leprosy in Papua New Guinea. Ann Rheum Dis 1987; 46: 688-90.
19. Agarwal V, Wakhlu A, Aggarwal A, Pal L, Misra RN. Tenosynovitis as the presenting manifestation of Leprosy. J Ind Rheumat Assoc 2002; 10: 69-70
20. Haroon N, Agarwal V, Aggarwal A, Kumari N, Krishnani N, Misra R. Arthritis as presenting manifestation of pure neuritic leprosy-a rheumatologist's dilemma. Rheumatology (Oxford). 2007; 46: 653-56.
21. Mahajan PM, Jogaikar DG, Mehta JM. A study of pure neuritic leprosy: clinical experience. Indian J Lepr 1996; 68: 137-41.
22. Pernambuco J, Opromolla D, Tolentino M, Fleury R. Arthritis in lepromatous Hansen's disease. International Journal of Leprosy 1979; 47: 353-54.
23. Britton WJ, Lockwood DNJ. Leprosy. Lancet 2004; 363: 1209-19.
24. Lockwood DN. Leprosy elimination: a virtual phenomenon or a reality? BMJ 2002; 324: 1516-18.
25. WHO. Expert Committee on Leprosy, 7th Report, 1998: 1-43.

RHEUMATIC MANIFESTATIONS OF HIV INFECTION 4

Ajay Wanchu, Neha Garg

Musculoskeletal disorders are common with Human Immunodeficiency Virus (HIV) infection and the prevalence increases in the late stages of disease. These disorders can result in significant morbidity and may impair the patient's quality of life. Various musculoskeletal syndromes that can affect HIV infected patients include reactive arthritis, infective arthritis, diffuse infiltrative lymphocytic syndrome (DILS) and myositis. These manifestations are categorized in Table 1. The approximate incidence of various rheumatic diseases in HIV infected patients is mentioned in Table 2.

Table 1: Classification of rheumatic disorders in HIV infected patients.

Manifestations occurring anytime in the course of HIV disease
- HIV associated arthralgias and arthritis including crystal induced arthritis
- HIV associated seronegative arthritides, including reactive, psoriatic undifferentiated spondyloarthritis
- Diffuse infiltrative lymphocytic syndrome (DILS)
- Vasculitis
- Myopathy
- Bone disorders

Manifestations of advanced HIV disease (result of CD4 helper T-cell dysfunction)
- Septic arthritis
- Osteomyelitis
- Pyomyositis

Manifestations associated with HIV treatment
- Osteoporosis
- Avascular necrosis (AVN)

Manifestations associated with immune reconstitution syndrome (IRIS)
- Rheumatoid arthritis (RA)
- Systemic lupus erythematosus (SLE)
- Sarcoidosis

Table 2: Relative frequency HIV Disease manifestations[52]

Disease Manifestation	Percentage
Arthralgia	5.5
Septic arthritis	1
Osteomyelitis	0.9
Avascular necrosis	0.7
SLE	0.3
Psoriasis	0.2
Rheumatoid arthritis	0.1
Polymyositis	0.1
Scleroderma	0.1
Psoriatic arthritis, reactive arthritis,	0
Ankylosing spondylitis, Sjogren's syndrome	0

Several important facts need to be taken into consideration when approaching an HIV infected patient with a musculoskeletal syndrome. All musculoskeletal syndromes that occur in non-HIV infected patients can occur in HIV infected patients[1]. HIV infection can alter clinical presentation and course of a pre-existing rheumatic disorder. Thus reactive arthritis, inflammatory arthritis or musculoskeletal infections may be more severe in their presentation and course in HIV infected persons. In other instances, musculoskeletal infections may be more insidious and subtle in onset. The chances of an opportunistic infection as a cause for a musculoskeletal complaint depend on the stage of the HIV disease. During early stages (CD4 count above 300), opportunistic infections are unlikely, although resistance to common bacterial pathogens may be reduced. For the most part, diagnostic tests and treatment regimens for musculoskeletal syndromes are similar to non HIV infected patients, except that the HIV related medications may result in side effects and interactions that affect differential diagnosis and limit therapeutic options.

Manifestations occuring anytime in the course of HIV disease

Several rheumatic manifestations occur throughout the course of HIV disease. These include seronegative spondyloarthritides, DILS, vasculitis and myopathy. The development of some specific rheumatic disorders and paucity of others in HIV infected patients is consistent with the subgroup of these disorders as being immune-mediated.

HIV-associated arthralgia: The frequency of otherwise unexplained arthralgia has been reported to be over 40%[2]. Usual sites of involvement are knees, shoulders and elbows. Most individuals presenting with arthralgia alone do not develop inflammatory joint disease and they are usually seronegative for rheumatoid factor and ANA. Course is usually self limiting within 6 weeks of onset of joint symptoms. The most appropriate treatment, along with non-narcotic analgesics such as paracetamol, is reassurance.

HIV-associated arthritis: Arthritis has been reported to occur in more than 10% patients in some reports[3]. It was initially described as oligoarthritis, predominantly affecting lower extremities that tend to be self-limiting, lasting less than 6 weeks. It can be distinguished from HIV associated reactive arthritis in that there is no mucocutaneous involvement, enthesopathy and there is no association with HLA-B27 or any other known genetic factor. The main distinguishing features between HIV associated arthritis and reactive arthritis are presented in Table 3. Synovial fluid cultures are typically sterile. There may be a direct role of HIV as shown by the isolation of HIV from a synovial fluid sample and electron microscopy showing particles resembling retrovirus. X-rays of the affected joints are usually normal. Treatment of this disorder includes nonsteroidal anti-inflammatory drugs (NSAIDs) and, in more severe cases, low-dose glucocorticoids. Rarely these patients may require hydroxychloroquine (HCQ) or sulfasalazine (SSZ).

Spondyloarthritides: Prevalence of seronegative spondyloarthritides (including reactive arthritis, psoriatic arthritis and undifferentiated arthritis) has increased in last few years and is reported to be 0.4 to 10%. Clinical features are similar to non HIV patients but with less uveitis and axial skeleton involvement. They also have frequent extra-articular manifestations like conjunctivitis, urethritis, circinate balanitis and keratoderma blennorrhagicum. Treatment usually includes *highly active antiretroviral therapy* HAART, sulfasalazine (SSZ) and TNF-α inhibitor therapy as described below.

Reactive arthritis (ReA): Reactive arthritis (formerly known as Reiter's syndrome), the first rheumatic syndrome reported in patients with HIV infection, can be a severe illness, but whether it occurs more frequently in HIV infected patients than in the general population is controversial. Two small retrospective studies of HIV infected patients reported a prevalence of ReA syndrome between 5% and 10%, which is 100 to 200 times higher than the expected prevalence in the general population[4,5]. Another study of more than 1,000 homosexual men found the frequency of reactive arthritis to be only 0.3 to 0.5%, and there was no difference in the frequency between HIV positive and HIV negative subjects[6]. Between 63% and 75% of HIV infected patients with reactive arthritis have HLA-B27 cell-surface antigen, a percentage that does not differ significantly from 70% to 80% usually reported for white persons with reactive arthritis without HIV infection.

The most common presentation is that of a seronegative peripheral arthritis mainly involving the lower extremities, usually accompanied by enthesitis (sausage toes or fingers, Achilles tendinitis, and plantar fasciitis). Multi-digit dactylitis, especially in the upper extremities, is common and may be relatively painless. Frank synovitis is less common but may occur at the ankle, subtalar, metatarsophalangeal, and interphalangeal joints of the feet. Knee involvement is common, often asymmetric, and without radiologic changes. Hip involvement is uncommon. Synovitis of the wrist, elbow and shoulder is uncommon but may lead to contractures and joint fusion. Enthesopathy may occur at the medial and lateral epicondyles, rotator cuff, or flexor tendons of the digits. The axial skeleton is usually not involved, and although radiographs may reveal sacroiliitis, clinical sacroiliitis is uncommon. Mucocutaneous features especially keratoderma blennorrhagicum and circinate balanitis are common. Psoriasiform skin rashes are also common and can be extensive. It can be difficult to distinguish HIV associated reactive arthritis from psoriatic arthritis[7]. Urethritis occurs as often as it does in HIV negative reactive arthritis. Constitutional features of Reiter's syndrome, including weight loss, malaise, lymphadenopathy, and diarrhoea, are difficult to distinguish from features of adrenal HIV disease. Clinicians should recommend HIV testing for patients with Reiter's syndrome whose behaviour puts them at an increased risk for HIV infection. It is important to differentiate this entity from HIV associated arthritis. The major differentiating features are shown in Table 3.

Table 3 : Contrasting features of HIV associated arthritis and reactive arthritis[53]

Feature	HIV arthritis	HIV associated reactive arthritis
Joint involvement	Variable	Asymmetric oligoarthritis
Enthesopathy	Absent	Present
Mucocutaneous involvement	Absent	Present
Synovial fluid white blood cells	500-2,000/μL	2,000-10,000/μL
Organisms in synovium	HIV	Chlamydia
HLA - B27	Absent	Present
Course	Self-limited	Chronic/relapsing

The treatment is similar to that of HIV negative patients with reactive arthritis. NSAIDs are the mainstay of treatment. Indomethacin is often recommended, not only for its efficacy, but also for its unique inhibition of HIV replication observed *in vitro*. Phenylbutazone is rarely used now because of its high side effect profile, but can be useful in refractory cases. Patients frequently have an inadequate response to NSAIDs alone, and these drugs are frequently used in combination with second line drugs. Sulfasalazine has been shown to be effective in some studies at doses of 1.5 to 2g/day. Methotrexate was initially thought to be contraindicated because of its immunosuppressive effect and in early reports it was associated with the development of *Pneumocystis carinii* pneumonia and other opportunistic infections. Subsequently, with careful monitoring of HIV viral loads and CD4 counts, and the patient's clinical status, more recent studies have shown a role for methotrexate in the treatment of reactive arthritis and psoriatic arthritis occurring in the setting of HIV infection. Hydroxychloroquine has also been reported to be useful not only in treating HIV-associated reactive arthritis but also in reducing HIV replication as evidenced in various *in vitro* and *in vivo* studies[8]. TNF α blockers have been commonly used in patients with spondyloarthritis without HIV. However, fears of worsening immunosuppression have limited their use in HIV infection. There is one case report of improvement in refractory reactive arthritis associated with HIV with infliximab treatment. However, this agent should be used with extreme caution and only in patients with otherwise refractory HIV associated reactive arthritis and spondyloarthropathy. It would be prudent to wait for more studies before TNF blockers can be routinely recommended in patients with HIV associated reactive arthritis.

Psoriasis and psoriatic arthritis: Psoriatic arthritis with or without psoriasis occurs in HIV infected persons with prevalence probably being the same as that in non HIV infected persons (1 to 2%)[9]. The severity of the HIV associated psoriasis and psoriatic arthritis tends to be worrisome.

The foot and ankle are the commonest and most severe sites of inflammation. Intense enthesopathy and dactylitis, especially in the feet, usually accompany the arthritis. The enthesopathy can be a major cause of disability. Frank synovitis and synovial effusions are less frequent, but can occur at the ankle and subtalar, metatarsophalangeal, and interphalangeal joints of the feet. Sacroiliac and spine involvement may also occur. The radiologic appearance of these joints may mimic classic psoriatic arthritis, with "pencil-in-cup" deformities and osteolysis, even in the absence of frank psoriasis.

Nail involvement is a common presenting symptom of arthritis, especially in the distal interphalangeal joints of the hands and feet. Many patients with psoriatic skin manifestations or onycholysis have only these findings and do not meet the criteria for the diagnosis of psoriatic arthritis. Nail involvement occurs in most patients who present with inflammatory articular symptoms. There is a high clinical correlation between skin and joint involvement, and joints may develop erosive changes and crippling deformities. Patients with HIV infection and psoriatic arthritis fall into one of the two patterns of disease: either the articular disease is sustained and aggressive, progressing to joint erosions, or it is characterized by mild and intermittent joint involvement.

The treatment is similar to that for non-HIV infected patients. NSAIDs, such as indomethacin (75 to 150 mg per day), can be used initially for the joint symptoms, but results have not been encouraging. There are a few data on second-line agents such as gold, methotrexate, and azathioprine. There are reports that phenylbutazone (100mg 3 times per day) is an effective drug for this arthritis, and neutropenia has not been a problem even in patients receiving zidovudine concurrently. Patients in whom psoriatic skin or joint disease does not respond to NSAIDs may be treated with phenylbutazone (100 mg 3 times a day) or sulfasalazine (1 to 2 g per day). Etretinate may also be helpful. The use of psoralen and pulsed UV-A phototherapy (PUVA) has helped the skin and joints of some HIV infected individuals with psoriatic arthritis. Antiretroviral treatment has also been found to be helpful.

Undifferentiated spondyloarthropathy: This is a frequently encountered problem where there are features of reactive arthritis or psoriatic arthritis in patients who do not otherwise develop full-blown disease, such as enthesopathy (plantar fasciitis, achilles tendinitis). The treatment is symptomatic (NSAIDs, intralesional corticosteroid injections, when indicated), and sulfasalazine in patients with more extensive disease.

Gout: There is higher prevalence of hyperuricemia to about 42% in HIV patients. Annual incidence of gout in this population is 0.5%. Uncontrolled HIV replication and rapid cell turnover could be related to hyperuricemia[10]. Other possible mechanisms could be mitochondrial toxicity from some anti-retrovirals like stavudine and didanosine.

Diffuse infiltrative lymphocytic syndrome (DILS): This is a syndrome superficially resembling Sjögren's syndrome has been found in patients infected with HIV and is characterized by massive parotid enlargement and xerostomia. It can appear at any stage of HIV disease (at any CD4 level and the prevalence varies between 3 to 7.8% in different series[11].

The primary pathogenic association is reflected as a distinct host immune response in people with the HLA-DR5 phenotype[12]. The association with HLA-DRB1 alleles expressing the *ILEDE* amino acid sequence in the third diversity region, usually *HLA-DRB1*1102, DRB1*1301, and DRB1*1302* has been associated with delayed progression to AIDS among patients with DILS. This is due to delay in the evolution of the HIV-1 virus from the less aggressive M-tropic strain to the more rapidly replicating T-tropic strain by a more effective CD8 lymphocyte response. Pathologically, there is focal sialadenitis, similar to that seen in Sjogren's syndrome, although there is less destruction of the salivary glands. CD8+ lymphocytes constitute most of the inflammatory infiltrate, unlike that seen in primary (non-HIV-associated) Sjogren's syndrome. Also, the SSA and SSB antibodies are usually negative, another feature which can help in distinguishing between the two diseases.

The classic defining presentation of DILS is a painless parotid enlargement that is often massive. The mean duration between the discovery of HIV seropositivity and onset of symptoms in one study was 3.4 years. Submandibular and lacrimal gland enlargement often occurs together and this enlargement is accompanied by sicca symptoms in more than 60% patients. In addition to glandular manifestations, there are several extraglandular features in these patients (Table 4).

Significantly, the reported frequency of lymphocytic interstitial pneumonitis (LIP) was as high as 25 to 50 percent in patients with DILS. Since the introduction of protease inhibitors, this complication occurs much less frequently. Extra-glandular manifestations like renal tubular acidosis, lymphoma and polymyositis, in patients with DILS, are usually associated with a poor prognosis. The major features distinguishing this disorder from Sjogren's syndrome and another mimic in the form of Hepatitis C are shown in Table 5.

Table 4 : Extra-glandular manifestations among patients with DILS

Neurological
- 7th cranial nerve palsy
- Aseptic lymphocytic meningitis
- Peripheral neuropathy

Pulmonary
- Lymphocytic interstitial pneumonitis (LIP)

Gastrointestinal
- Lymphocytic hepatitis

Renal
- Renal tubular acidosis
- Interstitial nephritis

Hematological
- Lymphoma

Musculoskeletal
- Peripheral arthritis
- Polymyositis

Table 5. Differences between DILS and Sjogren's syndrome[11]

Variables	Sjogren's syndrome	DILS	Hepatitis C
Sicca	Present	Present	Present
Glandular Manifestations	Moderate parotid enlargement	Moderate to severe parotid enlargement	Mild to moderate parotid enlargement
Extraglandular Manifestations	Mainly pulmonary, gastrointestinal, renal and neurologic involvement	Mainly musculoskeletal, pulmonary,gastrointestinal and neurologic involvement	Mainly gastrointestinal and musculoskeletal involvement
Infiltrating lymphocytic Phenotype	CD4	CD8	CD4
Autoantibodies	High frequency of rheumatoid factor, ANA, anti-Ro/SSA & anti-La/SSB	Low frequency of rheumatoid factor, ANA and anti-Ro mostly absent	High frequency of rheumatoid factor and ANA, anti-Ro/SSA present, though less frequently than primary
HLA association	B8, DR2, DR3 and DR4	B45, -B49, -B50 DR11 (DRS) and DRw6 (DRBI*1301 *1302)	SS DR11 (DRS)

To establish the diagnosis the suggested criteria are: the patient (1) must be HIV seropositive (2) must have bilateral salivary gland enlargement or xerostomia persisting for more than 6 months and (3) must have salivary or lacrimal gland lymphocytic infiltration as confirmed by histology or Ga scintigraphy, in the absence of granulomatous or neoplastic enlargement[12]. CT scanning has also been used to determine the extent of glandular swelling and evaluate parotid cysts and possible salivary glandular malignancy. It has been suggested that all patients with painless parotid gland enlargement and sicca symptoms should be tested for HIV.

Management of these patients depends upon the extent of the manifestations. If glandular swelling does not worry patients and the sicca symptoms are mild or even absent, then, reassurance and observation are enough. Pilocarpine, in doses of 5-10mg thrice daily, may be needed for sicca symptoms. In addition, regular dental care is required. One report suggests that moderate doses of corticosteroids (up to 30 to 40mg of prednisolone per day) are useful in treating both the glandular swelling and sicca symptoms of DILS without adversely affecting the frequency of opportunistic infections, raising the viral loads or depressing the CD4 counts[13]. The effect is usually transient. LIP may need higher doses of corticosteroids (up to 60mg/day of prednisolone), sometimes for extended periods. Antiretroviral therapy has been shown to be beneficial in glandular swelling and sicca symptoms.

Vasculitis: Every type of vasculitis, including small, medium and large vessel vasculitis, has been described in patients with HIV infection[14]. Patients presenting with vasculitis should be tested routinely for HIV, hepatitis B and hepatitis C infection. In one study 23% patients with symptomatic HIV disease had vasculitis[15]. These include hypersensitivity vasculitis, polyarteritis nodosa, Henoch-Schonlein purpura, Kawasaki disease both in children and adults, giant cell arteritis, granulomatosis with polyangiitis (formerly known as Wegener's Granulomatosis), cryoglobulinemic vasculitis and primary angiitis of the central nervous system (PACNS) and infectious vasculitides[16].

Polyarteritis nodosa (PAN): PAN seems to be the most common form in HIV infected individuals. Recently HIV associated PAN has been described to differ significantly from classical (hepatitis B associated) PAN and that associated with hepatitis C virus infection[17] (Table 6).

Cerebral vasculitis or primary angiitis of central nervous system (PACNS): Among HIV patients with stroke, PACNS are a cause for cerebrovascular disease about 13% of the times. It can occur at anytime during the course of HIV disease, immediately after HIV infection, or when CD4 count falls below 200 copies/μl.

Cryoglobulinemic vasculitis: Prevalence of cryoglobulinemic vasculitis (mostly Type II and III mixed cryoglobulins[18]) in HIV positive patients ranges from 1 to 27% but it is more frequent with HIV and hepatitis co-infection. It is mostly asymptomatic with HIV infection alone and HIV infection does not seem to play a significant role in cryoglobulin production[19]. In patients with HIV and hepatitis C co-infection, cryoglobulinemic vasculitis presents with palpable purpura, mononeuritis multiplex, arthralgias and nephritic syndrome. It responds well to hepatitis C treatment with ribavirin and interferon. Most recently, rituximab has been successfully used in the treatment of cryoglobulinemic vasculitis with or without hepatitis C[20,21]. It remains to be seen whether rituximab can be used in cryoglobulinemic vasculitis associated with HIV with or without HCV co-infection.

Table 6 : Differences between different forms of polyarteritis nodosa[17]

	Classical PAN	Non-classical PAN	Non-classical PAN
Association	HBV	HIV	HCV
Symptoms	Waxing, waning	Acute	Acute
Organs involved	Renal, GI and CNS	Skin, joints	Skin and others like HBV
Response to therapy	Fair (steroids and cytotoxic)	Excellent (mostly steroids only; HAART)	Good (Rituximab); relapses common; Peg IFN + Ribavarin
Severity	Severe	Mild	Moderate
Relapses	Common	Uncommon	Very common
Prognosis	Poor (35% mortality)	Excellent	Fair (10% mortality)

Infectious vasculitis: Vasculitis resulting from Cytomegalovirus (CMV) infection can result in prominent gastrointestinal and cutaneous manifestations and its inclusion bodies can be demonstrated by microscopy. In such cases, antiviral agents like ganciclovir or foscarnet should be started and any concomitant immunosuppressive therapy reduced. Any drugs that can cause hypersensitivity vasculitis must be withdrawn. Life-threatening complications affecting the lungs, kidneys or the nervous system should be treated with steroids and immunosuppressives. Antiretroviral drugs may help and additional therapy in the form of prophylaxis for herpes zoster and Pneumocystis pneumonia must be instituted.

Myopathy: Skeletal muscle involvement may occur at all stages of HIV infection and may be classified as follows: (1) HIV associated myopathies and related conditions; (2) muscle complications of antiretroviral therapy; (3) opportunistic infections and tumor infiltrations of skeletal muscle; and (4) rhabdomyolysis[22,23]. In one series of skeletal muscle biopsies performed in adult patients with HIV infection at autopsy there was atrophy in most and inflammatory infiltrates, necrosis and infection in few[24].

Polymyositis: HIV associated polymyositis typically occurs early in HIV infection. The usual presentation is that of a subacute, progressive proximal muscle weakness occurring in the patient with elevated creatine kinase. Auto antibodies like anti-Jo-1 and anti-Mi-2 are absent. Electromyography shows myopathic motor unit potentials with early recruitment and full interference patterns, fibrillation potentials, positive sharp waves, and complex repetitive discharges suggesting an irritative process. Microscopic examination of muscle biopsies reveals interstitial inflammatory infiltrates of variable intensity associated with degenerating and regenerating myofibrils, like those seen in polymyositis without HIV1. Concomitant vasculitis is rare. As seen in patients who are HIV seronegative, the predominant cell populations are CD8+ T cells and macrophages invading or surrounding healthy muscle fibres in the endomysial distribution. The etiopathogenesis of the disorder is not clear. It is controversial whether the virus directly leads to inflammatory myopathy. The treatment of HIV associated polymyositis is similar to that for other inflammatory myopathies. Both creatine kinase and muscle weakness usually respond to glucocorticoids[25]. Refractory cases might need immunosuppressive agents such as methotrexate or azathioprine. These should, however, be used with caution with careful monitoring of the patient's clinical status, CD4+ counts and HIV viral load. Concomitant DILS may be present in up to half the patients. Dermatomyositis, inclusion body myositis and nemaline rod myopathy[26] are also seen very rarely in association with HIV.

Rhabdomyolysis: HIV can also be complicated by rhabdomyolysis. This can be caused by HIV infection itself, as a consequence of interaction between protease inhibitors and statins (cytochrome p450 mediated interaction leading to high levels of statins), or as an adverse other drugs like integrase inhibitors (raltegravir) or from hypersensitivity to abacavir.

HIV wasting syndrome: It should be distinguished from other causes of generalised weight loss in HIV patients from infections, lactic acidosis etc. Muscle biopsy may show diffuse atrophy or type II atrophy, neurogenic atrophy without much inflammation.

Other forms of muscle disease in HIV-positive individuals have included myalgias, myasthenic syndrome, chronic fatigue and fibromyalgia. HIV associated myasthenia is not always associated with antibodies to acetylcholine receptors and could be a transient phenomenon. Treatment is based on antiretroviral therapy, combined with, if necessary, pyridostgimine, prednisone or intravenous immunoglobulin[27].

Myopathy related HAART: With the introduction of antiretroviral drugs in the treatment of HIV infection, especially zidovudine (AZT), there have been reports of a diffuse myopathy[28]. It is characterized by insidious onset of myalgias, muscle tenderness, and proximal muscle weakness after a mean duration of 11 months therapy with the drug. It tends to be dose related, being associated with mitochondrial dysfunction. Both clinically, by EMG and by muscle biopsy, it is difficult to distinguish it from HIV-associated polymyositis, although inflammatory infiltrates tend to be much less severe and sometimes absent in patients receiving zidovudine. Ragged red fibres may be seen suggesting mitochondrial dysfunction. A rare life threatening syndrome with severe hepatic steatosis, lactic acidosis, mitochondrial myopathy and pancreatitis resembling Reye's syndrome in children, has been described with stavudine in combination another nucleoside analogue and a protease inhibitor. Consequently, in any HIV infected individual who presents with an elevated CK, especially when symptoms of myalgia or muscle weakness are present, the drug should be discontinued for 4 weeks and the patient re-evaluated before EMG or muscle biopsies are undertaken.

Infectious myositis (pyomyositis): Pyomyositis can be an important cause of muscle problems in patients with HIV infection[29]. Organisms implicated have included *Staphylococcus aureus*, *Salmonella enteritidis*, Microsporum, and Toxoplasma. Their management is on lines similar to HIV uninfected individuals. Extra nodal non-Hodgkin's lymphoma has been described in HIV patients and presents with rapidly growing muscle mass with fever. Histopathological examination shows a lymphomatous proliferation destroying the muscle tissue. Magnetic resonance imaging can be useful in distinguishing lymphomatous involvement of muscle from other causes of limb swelling in HIV patients. Kaposi sarcoma can also involve the muscle.

Manifestations of advanced HIV disease

During later stages of infection patients are predisposed to a several infectious disorders. These might be either localized or systemic in nature and may primarily or secondarily involve the musculoskeletal system. The profile of organisms infecting the joints can also vary with high risk behaviour patterns or underlying conditions associated with HIV infection. Thus, intravenous drug users are more likely to be infected with *Staphylococcus aureus*, *Pseudomonas aeruginosa*, and *Candida albicans*.

As described in the previous section, musculoskeletal infections include septic arthritis, osteomyelitis and pyomyositis. Data is conflicting regarding the incidence of septic arthritis among the HIV infected. There is no good data to show that bacterial infections of bones or joints due to the usual infective agents, such as *S. aureus* are more frequent in patients with HIV infection[30]. A large retrospective case series[31] of 4023 HIV-infected patients studied between 1985 and 1996 found that *S. aureus* was the commonest infective agent isolated. The incidence was not related to CD4 counts suggesting that parenteral drug use rather than HIV infection per se predisposed to the joint infections. Even so, at CD4 counts dropping below 100/μl, atypical mycobacteria can be isolated from some joints. Most common among these are M. hemophilum and M. kansasii. Studies from Africa have shown that septic arthritis occurred much less frequently than HIV-associated aseptic arthritis[32]. On the other hand, there are studies to show an increased incidence of septic arthritis among HIV-infected haemophiliacs compared with those who are uninfected[33]. Management of these conditions is similar to that of HIV uninfected individuals. Some form of joint drainage may be required.

Osteomyelitis typically occurs at very low CD 4 counts. The most common organism remains Staphylococcus aureus. It can often be polymicrobial. As well, after instituting HAART, pre-existing quiescent mycobacteria like M. avium intracellulare can cause symptomatic osteomyelitis.

Other infections of relevance include those by atypical mycobacteria and fungi. Among the former the common species implicated are *Mycobacterium avium intracellulare* complex (MAI), *M. kansasii*, *M. haemophilum*, *M. terrae*, and *M. fortuitum*. *M. haemophilum* is implicated in more than half of the cases, and *M. kansasii* accounts for an additional 25%. Other lesions seen in about half the patients are cutaneous nodules, ulcers, and draining sinus tracts[33]. These infections occur late in the course of HIV, usually when the CD4+ T lymphocyte count is less than 100/ul. Clarithromycin has been reported as effective in most atypical mycobacterial bone and joint infections along with standard antituberculous therapy.

At CD4 counts less than 100/μL there is a high risk of development of fungal infections, which can have oligoarticular or polyarticular arthritis and may also result in tendon sheath effusion. These infections may be especially difficult to eradicate and may need chronic suppressive antifungal therapy.

Rheumatic manifestations associated with anti-retroviral drugs

Different anti retroviral drugs are associated with a host of side effects which include some rheumatic manifestations as well[34]. These are listed in Table 7. These include HIV associated bone disorders including avascular necrosis (AVN), osteoporosis and osteomalacia.

Osteonecrosis: Risk of osteonecrosis is approximately 100 fold higher in HIV positive individuals compared to general population. Prevalence of hip osteonecrosis is 4.4% and 11% of them needed total hip arthroplasty in a study[35]. Suggested additional risk factors for development of osteonecrosis in HIV positive patients are thought to be dyslipidemia from protease inhibitors, chronic inflammation, corticosteroids use for immune reconstitution, and presence of anti-cardiolipin antibodies.

Table 7 : Rheumatologic manifestations of anti-retroviral drugs[34]

Medication	Class	Manifestations
Zidovudine, Stavudine	NRTI	Myopathy, Myalgias, Rhabdomyolysis
Tenofovir	NRTI	Osteopenia
Lamivudine, Emtricitabine	NRTI	Arthralgias, Myalgias, Rhabdomyolysis
Efavirenz	NNRTI	Arthralgias, Myopathy, Myalgias
Nevirapine	NNRTI	Arthralgias
Nelfinavir, Lopinavir, Atazanavir	PI	Osteopenia, osteoporosis, AVN
Saquinavir, Ritonavir	PI	Myalgias, Arthralgias, Arthritis, Osteopenia, Osteoporosis
Indinavir	PI	Adhesive Capsulitis shoulder, TM joint pain, Osteopenia, Osteoporosis, AVN
Enfuritide	Fi	Myalgias

Osteopenia and osteoporosis: Osteoporosis is more prevalent in HIV patients than general population. Underlying mechanisms may involve HIV- mediated immune activation, cytokine release and perhaps protease inhibitor use as well. It remains unclear whether HIV associated osteopenia and osteoporosis increase risk of fractures and whether bisphosphonate therapy can prevent it[36].

Osteomalacia: Osteomalacia in HIV patients can occur as a consequence of tenofovir toxicity[37]. Tenofovir inhibits replication of mDNA in proximal tubules and may induce a Fanconi like syndrome causing phosphate loss and osteomalacia. It should be suspected when there is low serum phosphate with high alkaline phosphate in patients with bone complaints. Treatment includes discontinuation of tenofovir and administration of phosphate.

Manifestations of immune reconstitution inflammatory syndrome (IRIS)

Immune reconstitution is a well recognized condition which is manifested by clinical deterioration after initial institution of HAART and is thought to be related to sudden activation or 'reconstitution' of immune system leading to an exaggerated immune response against pathogens or self antigens. Incidence varies from 3 to 25%[38]. The response of various rheumatic disorders to the occurrence of HIV infection has been the subject of several reports. The manifestations vary widely and depend on the pathogen inciting the response and the organ system involved. Some distinct clinical entities are discussed here.

Sarcoidosis: Most patients with symptomatic sarcoidosis have CD4 counts more than 200/μl. It usually manifests with marked increase in CD4 counts following HAART therapy[39]. It should be distinguished from various opportunistic fungal and mycobacterial infections due to similar presentations including pulmonary symptoms and lymphadenopathy. Management includes corticosteroids, immunosuppressive medications. Rarely, in severe cases HAART may need to be temporarily discontinued until control of symptoms.

Rheumatoid arthritis: Initial studies indicated that rheumatoid arthritis underwent a remission with advancing HIV infection[40]. Subsequently, there has been a suggestion that active rheumatoid arthritis can coexist with HIV infection[41]. In HIV, with decreasing CD4 counts, there is a decrease in Th1 response with concomitant decrease in interferon-gamma signature and increase in IL-4 and IL-10. With instituting anti-retroviral treatment these responses are reversed leading to worsening or, even new onset rheumatoid arthritis[42]. Radiographic progression of RA disease has been reported despite advanced HIV disease and lack of clinical activity[43]. Also, increased titres of rheumatoid factor and/or anti-CCP antibodies are found in advanced HIV infection before immune reconstitution and are unreliable markers for diagnosis. After immune reconstitution the specificity of anti-CCP approaches that of general population[44].

Systemic lupus erythematosus: SLE has been reported to coexist with HIV infection[45,46] Initially it was suggested that HIV infection could reduce lupus activity[40]. Later studies, however, have demonstrated certain subsets of patients have a more rapid decline in the CD4 numbers with rapid progression to clinical AIDS[47]. Like RA, SLE can become active with institution of anti-retroviral therapy[48]. Conversely, SLE can also have some consequences on HIV disease. For example, sera from SLE patients have been found to have antibodies that react to HIV in western blot assay leading to a false positive assay[49].

Therefore it is recommended to confirm positive HIV antibody tests by antigen or nucleic acid detection in the setting of connective tissue disease. In addition all patients with SLE should be tested for HIV before starting immunosuppressive medications.

Anti-rheumatic treatment in HIV patients

Treatment of rheumatic diseases in HIV-positive patients is similar regardless of HIV positivity, except for a need for greater caution with immunosuppressive medications and greater vigilance for opportunistic infections in patients with advanced immunosuppression (CD4 $< 200/\mu l$). As a general rule, restrict treatment with immunosuppressive medications to those with CD4 $> 200/\mu l$ and viral load $< 60,000$ copies/μl at the start of immunosuppressive therapy. NSAID's can be used as in HIV sero-negative patients. Indomethacin and hydroxychloroquine (HCQ) may even have some antiviral activity. HCQ was as effective as AZT in a dose of 800 mg/d in controlling viral replication. Methotrexate (MTX) may be used with careful CD4 and viral load monitoring. Sulfasalazine (SSZ) can be used in seronegative spondyloarthritides and does not promote HIV replication. TNF-α blockade may sometimes be needed especially in psoriatic arthritis (PsA) treatment where these are shown to have a good response. It is very important to beware of intercurrent infections. TNF-α inhibitors may be safely used in patients with Hepatitis B and C co-infection.

Serologic abnormalities in patients with HIV

Involvement of the humoral (B-cell) arm of the immune system is reflected by the presence of auto antibodies in the circulation due to spontaneous B cell proliferation and in some instances development of B cell lymphomas[50]. A partial list of B cell defects in HIV infection related to rheumatologic manifestations is shown in Table 8. The most common laboratory abnormality is polyclonal hyperglobulinemia, seen in up to 45% HIV infected individuals. Low titres of rheumatoid factor and antinuclear antibodies have been described in up to 17% of HIV infected patients. IgG anticardiolipin antibodies have been seen in up to 95% patients with AIDS. Anti-2 glycoprotein-1 antibody is seen in up to 47% patients where as lupus anticoagulant is generally not seen. Despite the presence of these antibodies, the full picture of anti phospholipid syndrome is not more common in HIV patients than general population[51]. Antinuclear neutrophilic cytoplasmic antibodies (ANCAs), both c-ANCA and p-ANCA, have also been detected in the serum of HIV positive individuals, in frequencies of up to 43% by ELISA and 18 percent by indirect immunofluorescence. In most instances these serological abnormalities are rarely of clinical significance other than false positive assays.

Table 8 : Consequences of B cell defects related to rheumatic defects.

Consequences of B cell defects
Polyclonal hypergammaglobulinemia
Circulating immune complexes
Anti neutrophil cytoplasmic antibodies
Autoantibodies
• Antinuclear antibodies
• IgG and IgA rheumatoid factors
• Anti phospholipids
• Anti platelet
• Anti RBC
• Anti lymphocyte surface antigen
• Anti neuronal cells
• Anti neurotransmitters
• Anti myelin associated proteins.

Summary

An entire spectrum of rheumatic diseases occurs in HIV-positive individuals, with most of them being typical of what is seen in non HIV populations. Others may be particular to the setting of HIV (DILS, HIV associated arthritis). Treatment of the HIV infection is part of the management of many of these disorders. Antiretroviral agents improve the clinical symptoms of HIV-associated rheumatic diseases. Some drugs used in the treatment of the rheumatic diseases (indomethacin, hydroxychloroquine) have antiretroviral properties too. With the availability of highly active antiretroviral agents the effects that these agents have on the presentation, course and outcome of the HIV associated rheumatic diseases remain to be determined.

Further Reading :

1. Wanchu A. Rheumatic manifestations of HIV infection. APLAR Journal of Rheumatology 2004; 7: 189-91.
2. Berman A, et al. Rheumatic manifestations of human immunodeficiency virus infection. Am J Med 1988; 85: 59-64.
3. Louthrenoo W. Rheumatic manifestations of human immunodeficiency virus infection. Curr Opin Rheumatol. 2008; 20: 92-99.
4. Fischer H, et al. Reiter's syndrome and the acquired immune deficiency syndrome. Arthritis Rheum 1985; 4: S22.
5. Winchester R. AIDS and the rheumatic diseases. Bull Rheum Dis 1990; 39: 1-10.
6. Hochberg MC, et al. HIV infection is not associated with Reiter's syndrome: Data from the Johns Hopkins Multicenter AIDS Cohort Study. AIDS 1990; 4: 1149-51.
7. Reveille JD, Williams FM. Rheumatologic complications of HIV infection. Best Practice & Research Clinical Rheumatology 2006; 20; 1159-79.
8. Sperber K, et al. Inhibition of human immunodeficiency virus type I replication by hydroxychloroquine in T cells and monocytes. AIDS Res Human Retroviruses 1993; 9: 91.
9. Espinoza LR, et al. Psoriatic arthritis and acquired immunodeficiency syndrome. Arthritis Rheum 1988; 31: 1034-40.
10. Manfredi R, Mastroianni A, Coronado OV, Chiodo F. Hyperuricemia and progression of HIV disease. J Acquir Immune Defic Syndr Hum Retrovirol 1996; 12: 318-19.
11. Basu D, Williams FM, Ahn CW, Reveille JD. Changing spectrum of the diffuse infiltrative lymphocytosis syndrome. Arthritis Rheum. 2006; 55: 466-72.
12. Itescu S, et al. Diffuse infiltrative lymphocytosis syndrome: A disorder occurring in human immunodeficiency virus-1 infection that may present as a sicca syndrome. Rheum Dis Clin North Am 1992; 18: 683-97.
13. Williams FM, et al. Short term moderate dose prednisone in the diffuse infiltrative lymphocytosis syndrome (DILS): A prospective study. Arthritis Rheum 1997; 40: S92.
14. Kaye B. Rheumatologic manifestations of HIV infection. Rev Allergy Immunol. 1996; 14: 385-416.
15. Gherardi R, et al. The spectrum of vasculitis in human immunodeficiency virus-infected patients. A clinicopathologic evaluation. Arthritis Rheum 1993; 36: 1164-74.
16. Pagnoux C, Cohen P, Guillevin L. Vasculitides secondary to infections. Clin Exp Rheumatol. 2006; 24 (2 Suppl 41): S71-81.
17. Patel N, Patel N, Khan T, Espinoza R. HIV infection and clinical spectrum of associated vasculitides. Curr Rheumatol Rep. 2011; 6: 506-12.
18. Dimitrakopoulos AN, Kordossis T, Hatzakis A, Moutsopoulos HM. Mixed cryoglobulinemia in HIV-1 infection: the role of HIV-1. Ann Intern Med 1999; 130: 226-30.
19. Scotto G,Cibelli DC, Saracino A, Prato R et al. Cryoglobulinemia in subjects with HCV infection alone, HIV infection and HCV/HIV coinfection. J Infect 2006; 52: 294-99.
20. Michael C, Sneller, Zonghui Hu, and Carol A. Langford. A Randomized Controlled Trial of Rituximab Following Failure of Antiviral Therapy for Hepatitis C Virus-Associated Cryoglobulinemic Vasculitis. Arthritis and Rheum. 2012; 64: 835-42.
21. De Vita S, Quartuccio L, Isola M, Mazzaro C, Scaini P et al. A Randomized Controlled Trial of Rituximab for the Treatment of Severe Cryoglobulinemic Vasculitis. Arthritis and rheum. 2012; 64: 843-85.
22. Authier FJ, Chariot P, Gherardi RK. Skeletal muscle involvement in human immunodeficiency virus (HIV)-infected patients in the era of highly active antiretroviral therapy (HAART). 2005 Sep; 32(3): 247-60.
23. Zell SC, Nielsen S. Clinical correlates to muscle biopsy findings in HIV patients experiencing fatigue: a case series. J Int Assoc Physicians AIDS Care (Chic Ill). 2002; 1: 90-94.
24. Wrzolek MA, et al. Skeletal muscle pathology in AIDS: An autopsy study. Muscle Nerve 1990; 13: 508-15.
25. Johnson RW, Williams FM, Kazi S, Dimachkie MM, Reveille JD. Human immunodeficiency virus-associated polymyositis: a longitudinal study of outcome. Arthritis Rheum 2003; 49: 172-78.
26. Dalakas MC, et al. Progressive nemaline (rod) myopathy associated with HIV infection. N Engl J Med 1987; 317: 1602-03.
27. Strong J, Zochodne DW. Seronegative myasthenia gravis and human immunodeficiency virus infection: response to intravenous gamma globulin and prednisone. Can J Neurol Sci 1998; 25: 254-56.
28. Dalakas MC, et al. Mitochondrial myopathy caused by long-term zidovudine therapy. N Engl J Med 1990; 322: 1098-105.
29. Al-Tawfiq JA, Sarosi GA, Cushing HE. Pyomyositis in the acquired immunodeficiency syndrome. South Med J. 2000; 93: 330-34.
30. Hughes RA, et al. Septic bone, joint and muscle lesions associated with human immunodeficiency virus infection. Br J Rheumatol 1992; 31: 38-41.
31. Vassilopoulos D, Chalasani P, Jurado RL, Workowski K, Agudelo CA. Musculoskeletal infections in patients with human immunodeficiency virus infection. Medicine 1997; 76: 284-94.
32. Blanche P, et al. Acute arthritis and human immunodeficiency virus infection in Rwanda. J Rheumatol 1993; 20: 2123-7.
33. Hirsch R, et al. Human immunodeficiency virus-associated atypical mycobacterial skeletal infarctions. Semin Arthritis Rheum 1996; 25: 347-56.
34. Ghanem KG. Current Diagnosis & Treatment: Rheumatology 2007; 396-404.
35. Miller KD, Masur H, Jones EC, Joe GO et al. High prevalence of osteonecrosis of the femoral head in HIV-infected adults. Ann Intern Med 2002; 137: 17-25.
36. Glesby MJ. Bone disorders in human immunodeficiency virus infection. Clin Infect Dis 2003; 37(Suppl 2): S91-5.
37. Parsonage MJ, Wilkins EG, Snowden N, Issa BG, Savage MW. The development of hypophosphataemic osteomalacia with myopathy in two patients with HIV infection receiving tenofovir therapy. HIV Med 2005; 6: 341-46.
38. French MA. HIV/AIDS: immune reconstitution inflammatory syndrome: a reappraisal. Clin Inf Dis 2009; 48: 101-7.
39. Foulon G, Wislez M, Naccache JM, Blanc FX et al. Sarcoidosis in HIV-infected patients in the era of highly active antiretroviral therapy. Clin Infect Dis 2004; 38: 418-25.

40. Bijlsma JS, et al. Does AIDS "cure" rheumatoid arthritis. Ann Rheum Dis. 1988; 47: 350-52.
41. Orustein MH, et al. A re-examination of the relationship between active rheumatoid arthritis and the acquired immunodeficiency syndrome. Arthritis Rheum 1995; 38: 1701-15.
42. Azeroual A, Harmouche H, Benjilali L, Mezalek ZT, Adnaoui M, Aouni M, Maaouni A. Rheumatoid arthritis associated to HIV infection. Eur J Int Med 2008; 19: e34-35.
43. Muller-Ladner U, Kriegsmann J, Gay RE, Koopman WJ, Gay S, Chatham WW. Progressive joint destruction in a human immunodeficiency virus-infected patient with rheumatoid arthritis. 1995; 38: 1328-32.
44. Du Toit R, Whitelaw D, Taljaard JJ, du Pleiss L, Esser M. Lack of specificity of anti cyclic citrullinated peptide antibodies in advanced HIV infection. J Rheumatol. 2011 Jun; 38(6): 1055-60. Epub 2011 Feb 15.
45. Bambery P, Deodhar SD, Malhotra HS, Sehgal S. Blood transfusion related HBV and HIV infection in a patient with SLE. Lupus 1993; 2: 203-05.
46. Wanchu A, A Sud, S Singh, P Bambery. Human Immunodeficiency Virus Infection in a patient with Systemic Lupus Erythematosus. J Assoc Phy Ind 2003; 51: 1102-04.
47. Hentges F, et al. Prolonged clinically asymptomatic evolution after HIV-1 infection is marked by the absence of complement C4 null alleles at the MHC. Clin Exp Immunol 1992; 88: 237-39.
48. Hazarika I, Chakravarty BP, Dutta S, Mahanta N. Emergence of manifestations of HIV infection in a case of systemic lupus erythematosus following treatment with IV cyclophosphamide. Clin Rheumatol 2006; 25: 98-100.
49. Jindal R, Solomon M, Burrows L. False positive tests for HIV in a woman with lupus and renal failure. NEJM. 1993; 328(17): 1281-82.
50. De Milito A. B lymphocyte dysfunctions in HIV infection. Curr HIV Res. 2004; 2: 11-21.
51. Loizou S, Singh S, Wypkema E, Asherson RA. Anticardiolipin, anti-beta(2)-glycoprotein I and antiprothrombin antibodies in black South African patients with infectious disease. Ann Rheum Dis 2003; 62: 1106-11.
52. Yao Q, Frank M, Glynn M, Altman RD. Rheumatic manifestations in HIV-1 infected in-patients and literature review. Clin Exp Rheumatol 2008; 26; 799-06.
53. Mody GM, Parke FA, Reveille JD. Articular manifestations of human immunodeficiency virus infection. Best Pract Res Clin Rheumatol 2003; 17: 265-87.

SECTION 14
Miscellaneous

SARCOIDOSIS 1

Paul Antony, R Gulati, V S Negi

Sarcoidosis is a multisystem granulomatous disorder of unknown etiology which predominantly involves the lungs in more than 90% cases but can also involve other organs in the body[1,2,3]. It is characterized by the formation of non-caseating granuloma in the tissues. The disease is often acute or subacute and self limiting, but a chronic waxing and waning course over years is also seen.

History

Sarcoidosis was first described by Sir Jonathan Hutchinson as 'relapsing iritis of inherited gout' in a patient with raised purple skin lesions. Later, Caesar Boeck coined the term 'sarkoid' based on the histologic appearance of the skin lesions which he thought resembled sarcoma[1]. He is credited with recording the first case of sarcoidosis in 1875. In 1909 Heerfordt described 'uveoparotid fever', with uveitis, enlarged salivary glands, and facial nerve palsy. Mikulicz in 1937 described salivary and lacrimal gland involvement in sarcoidosis. Sven Lofgren in 1942 recorded a syndrome of erythema nodosum, febrile arthropathy and bilateral hilar adenopathy, now well known as 'Lofgren Syndrome'.

Epidemiology

Sarcoidosis is a relatively common disease affecting both sexes, all the age groups, races and geographic locations[4]. Organ involvement depends on the race, age, and sex of the patient. Females are slightly more susceptible and are more likely to have eye lesions, neurologic involvement, and erythema nodosum[5]. Men are more likely to have hypercalcemia. Chronic skin lesions, eye, liver, bone marrow, and lymph node involvement are more common in Blacks. There is a remarkable difference in the prevalence of sarcoidosis in various ethnic and racial groups. The disease is most common in US African-Americans (10:1 to 17:1) while whites are more often affected in Europe[6]. Most patients present in the 20 to 40 age group. Sarcoidosis favors non-smokers unlike many other lung diseases[7].

Sarcoidosis in India

Because of widespread tuberculosis, the exact prevalence and incidence of sarcoidosis in India is not known. Until 1981, only 75 cases of systemic sarcoidosis had been reported in India. Since then it has been increasingly recognized and diagnosed. Pattern of presentation, tissue involvement, and results of laboratory and radiologic investigations are reported to be similar to those in other parts of the world[8].

Etiology

The wide variability in incidence and prevalence of the disease may be due to a number of factors. These can be broadly classified as:

Genetic factors: The first reported association was between HLA class I antigen (HLA-B8) and acute sarcoidosis[9]. Subsequently, HLA class II antigens, encoded by HLA-DRB1 and DQB1 alleles have been consistently associated with sarcoidosis[10,11]. Pocket 9 of HLA-DQ and pocket 4 of HLA-DR are the most important regions involved in this association[12]. Genes encoding tumor necrosis factor α (TNF-α), interferon-γ and chemokine receptors are logical candidates, but their associations with sarcoidosis has not yet been confirmed[11,12]. To date, two GWAS studies for loci associated with sarcoidosis have shown strongest linkage with chromosomes 3p and 6p in Caucasians and 5p and 5q in American Blacks[13,14].

Environmental factors: As sarcoidosis most commonly involves the lungs, eyes and skin, the search for environmental causes has focused on exposure to airborne antigens. Some of the earliest studies of sarcoidosis reported association with exposure to irritants found in rural settings, such as emission from wood-burning stoves and tree pollen[8]. Later, association between sarcoidosis and exposure to inorganic particles, insecticides and moldy environments was reported[9,10,11]. Izbicki et al documented increased incidence of sarcoidosis among New York City Fire Department rescue workers involved in the 2001 World Trade Center disaster[15].

Remarkable similarity of sarcoid granulomata to those caused by granuloma producing organisms, especially mycobacteria, point towards a putative environmental antigen. However, extensive studies using sensitive techniques such as PCR or T-cell receptor (TCR) repertoire analysis have consistently failed to provide any proof of such an etiology. Recently, much interest has been generated by the isolation of Propionibacterium granulosum DNA in a significant proportion of patients from Europe and Japan[4,7].

Pathology and pathogenesis

Morphology of granuloma

Sarcoid granuloma is typically composed of a central follicle made of tightly packed epithelioid cells and T lymphocytes surrounded by an area of fibroblasts, and a peripheral loose area rich in lymphocytes. They are typically non-necrotic and tightly packed. CD4+ T lymphocytes predominate within the follicle, and CD8+ T and B lymphocytes in the peripheral area. Early fibrotic changes may be present at the periphery. The Langhans giant cells are seen with intracellular Schaumann bodies (concretions of protein and calcium) and Asteroid bodies (stellate inclusions)[2,3].

Pathogenesis

The granuloma formation begins with an antigenic stimulus, followed by MHC Class II mediated T-cell and macrophage activation. There is induction of Th1 response with elaboration of IL2, INFγ and TNFα and subsequent formation of the granulomatous lesions. The periphery of the granuloma is an area of transit where freshly recruited monocytes are continually entering. They then penetrate the central follicle, where they differentiate into epithelioid cells. After their stay within the follicle, epithelioid cells leave the granuloma via the peripheral zone and are assimilated into the surrounding structures. Sarcoid granulomata may persist for several months or years, but usually resolve with fibrous scarring[2,3,4].

The similarity between sarcoidosis and tuberculosis is unmistakable. A number of other persistent antigens like beryllium, silicon, etc also elicit a similar tissue response. This led to the concept of sarcoidosis as an atypical response to mycobacterial and/or other antigens (Table 1). However, highly sensitive techniques have failed to demonstrate the presence of mycobacterial DNA in sarcoid tissues[2,3,5]. Clear evidence of contribution of indigenous P. acnes to the formation of granulomatous lesions in mice lung has been reported recently. Further studies are needed to confirm its role in triggering granulomatous response.

Table 1. Antigens implicated in the pathogenesis of Sarcoidosis

Infectious	Chemical
Bacteria	**Inorganic**
Mycobacterium tuberculosis	Aluminium
Propionibacterium acne	Zirconium
Borrelia burgdorferi	Talc
Mycoplasma	Pollen
	Insecticides
Viruses	**Organic**
Herpes	Inhalants
EBV	Pine tree pollen
CMV	
Coxsackie B virus	Clay
	Jobs involving radiation
	Cotton ginning
	Raising birds

The immune paradox

Sarcoidosis is a disorder mediated by excessive local T helper cell activity. However, the hallmark of sarcoidosis is the intense local inflammation accompanied by a state of anergy. Causes include relative lymphopenia and suppressor CD8+ over activity which happens late[15]. This phenomenon is proposed to be due to the dysfunction of T regulatory (Treg) CD4+/CD25 (bright)/FoxP3+ cells which are unable to inhibit the secretion of TNFα in the involved organ. On the other hand, the expanded population of Treg cells in the periphery leads to a state of anergy. Thus, sarcoidosis patients in the presence of intense local inflammation are paradoxically predisposed to opportunistic infections and cancer. It is proposed that the increased cancer risk would result from chronic inflammation and subsequent tissue damage in affected organs. Treg/Th1 imbalance could also be contributory[6,9]. This paradoxical immune response brings up the question of whether there is an overall exaggerated response or rather a deficiency in regulatory cells resulting in an unregulated Th1 activity. The immunological abnormalities seen in different organs are described in Table 2.

Table 2. Immunological abnormalities in sarcoidosis

Lung and other active granuloma sites

- Overactive CD4 T cells (Th1 type) releasing IFNγ, IL12, IL18 and IL2.
- Excessive expression of Th1 chemokine and cytokine receptors (CXCR3, CCR5, IL-12R and IL-18R) in conjunction with decreased expression of the Th2 chemokine receptors (CXCR4 and CCR4) provides further evidence of the Th1 activity
- Oligoclonal expansion of αβ TCR pool

Peripheral sites

- Circulating PBMC have no Th1 predominance and occasionally show Th2 phenotype
- Decreased expression of natural killer cell inhibitory receptors on CD8+T cells, which impair the ability to control cell-mediated response
- Increased production of macrophage derived fibrogenic cytokines favouring fibrosis.

Clinical features of sarcoidosis

The clinical manifestations of sarcoidosis are protean, ranging from an abnormal chest radiograph in an asymptomatic individual to severe multi-organ involvement Table 3. Chronic dry cough is the most frequent respiratory symptom. Chronic dyspnoea is rare and usually associated with advanced pulmonary fibrosis. Peripheral lymph nodes, eyes and skin are the major extrathoracic sites affected. Constitutional symptoms can be disabling. Erythema nodosum or acute uveitis sometimes heralds the onset of sarcoidosis[2,3,7].

A significant proportion (20 to 40%) present with acute or subacute sarcoidosis of few weeks duration with constitutional symptoms. Some present with cough, dyspnea, retrosternal discomfort and/or polyarthritis.

The insidious form (40-70%) develops over months and usually presents with respiratory complaints. Approximately 10% also have symptoms referable to other organ systems and frequently develop chronic course with permanent damage.

Table 3. Organ involvement in sarcoidosis

Sr. No.	Organ involved	Presenting (%)	Cumulative (%)
1	Constitutional	25	33-70
2	Lymphadenopathy	10-20	40
3	Pulmonary	25-50	95
4	Cutaneous	3	15-30
5	Joint disease	1-14	4-38
6	Parotid glands	< 6	< 7
7	Uveitis	5	10-20
8	Hepatosplenomegaly	4	5-20*
9	Heart	< 1	5-10*
10	Neurologic	< 1	5-10
11	Muscle	< 1	1-5*
12	Bone	< 1	1-13
13	Gastrointestinal	< 1	< 1
14	Kidneys	Rare	Rare

*At autopsy, 20%-67% have heart involvement; on biopsy, 50%-80% have muscle and liver involvement.

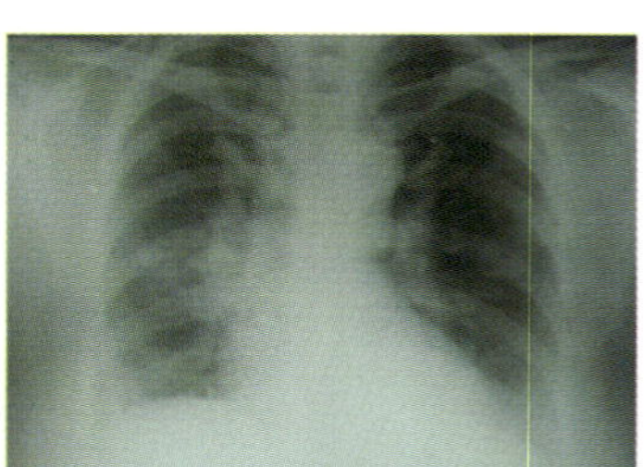

Fig 1a. Hilar adenopathy in Sarcoidosis: plain x ray

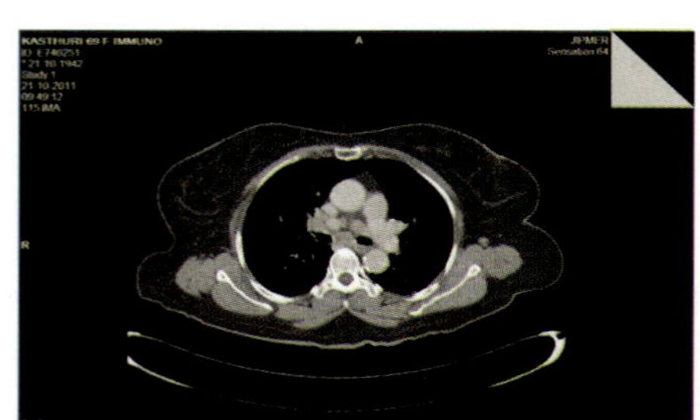

Fig 1b. Hilar adenopathy in Sarcoidosis: HRCT

Intrathoracic manifestations

Pulmonary involvement is the most common visceral manifestation, occurring in 95% of patients[7]. The clinical spectrum ranges from asymptomatic hilar adenopathy (50-60%) to interstitial lung disease with alveolitis (Fig. 1a, Fig. 1b)[8]. Pleural effusions are rare (< 5%). Endobronchial involvement is found on biopsy in 50% and may lead to airway stenosis (10%). Symptoms of lung disease include dry cough (30%), dyspnea (28%), and chest pain (15%). Wheezing can occur in patients with endobronchial involvement. Hemoptysis is rare and occurs primarily in patients with fibrosis and cavities filled with aspergillomas. Lung crackles are heard in only 20%, and clubbing is rare[16].

Ocular

Ocular involvement can occur in upto 25% cases with 5% having it as the initial presentation. Acute bilateral anterior granulomatous uveitis is a common manifestation. A baseline slit lamp examination is warranted as one third of the patients may have asymptomatic eye involvement. Other ophthalmologic manifestations include interstitial keratitis, posterior uveitis, pars planitis ("snowballing" on slit lamp examination), scleral plaques, lacrimal gland enlargement (15 -28%), optic neuropathy, and corneal/conjunctival nodules[1,3,7,19].

Cutaneous involvement

Skin involvement occurs in up to 30% of cases. Skin lesions can be classified as sarcoid specific and non-specific. Specific lesions demonstrate granulomata on biopsy. The most common specific skin manifestation is hyperpigmented maculopapular lesion (2-5 mm) commonly located on the face, nape of the neck, and upper back. Other specific lesions include skin plaques and annular lesions, which are associated with chronic disease course and poor prognosis (Fig. 2). Scars and tattoos can become infiltrated with sarcoidosis[16]. Some patients have subcutaneous or deep dermal involvement. Lupus pernio is the most characteristic skin lesion. It is associated with a chronic progressive, upper and lower airway disease and bony involvement. Non specific lesions like erythema nodosum are also inflammatory but with no granuloma.

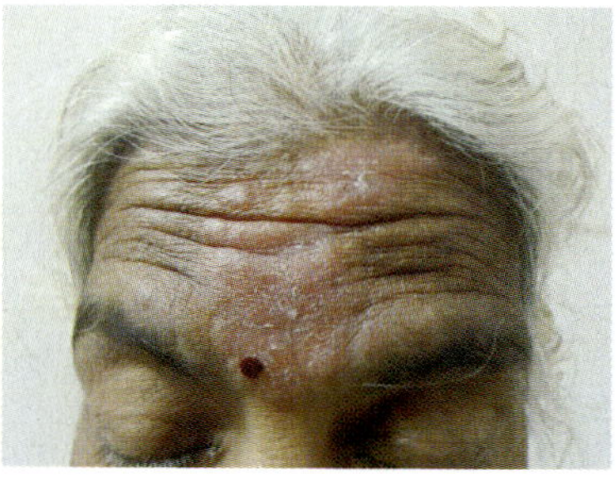

Fig. 2 Cutaneous involvement in sarcoidosis

Cardiac involvement

Manifestations include conduction disturbance, papillary muscle dysfunction, infiltrative cardiomyopathy with congestive heart failure, cor pulmonale due to severe restrictive lung disease and pericarditis. Baseline ECG is mandatory in all cases as 50% cardiac sarcoid will have ECG changes[17,21]. Cardiac rhythm (holter) monitoring and Echo-cardiography may be required in suspected cardiac sarcoid. Myocardial scintigraphy with thallium-201 can show areas of decreased uptake at rest, corresponding to granulomatous/fibrous infiltration. These areas decrease in size with stress imaging, differentiating it from ischemic coronary artery disease. Magnetic resonance imaging (MRI) with gadolinium enhancement, gallium-67 scans, and fasting 18F-fluorodeoxyglucose-labeled positron emission tomography (FDG-PET) can show increased uptake, which may predict response to immunosuppressive therapy. Biopsy is usually non contributory[20].

Nervous system involvement

Symptomatic neurological involvement occurs in only 5% of sarcoidosis with autopsy involvement in upto 60%. It occurs early in the disease course and may be the sole manifestation. Sarcoidosis has a predilection for the base of the brain. Unilateral VII cranial nerve palsy is the most common manifestation[18],[19]. Other features of neurosarcoidosis include leptomeningeal involvement (25%), other cranial neuropathies, hypothalamic and pituitary lesions, localized mass lesion, myelopathy, seizure, psychiatric and cognitive dysfunction. Peripheral neuropathy mainly chronic sensorimotor variety occurs in 8.5-24% of patients with neurosarcoidosis.

Other manifestations

Bone marrow is involved in 15-40% patients. Splenomegaly is seen in 5-10%. Asymptomatic liver involvement may be seen in 60-90% and cholestatic in 20-30%. Kidney involvement is rare and leads to nephrocalcinosis if chronic.

'Lofgrens syndrome' is a symptom complex of erythema nodosum, arthritis and hilar lymphadenopathy. Co-occurence of fever, parotid enlargement, anterior uveitis and facial palsy is named as Heerfordt-Waldenstrom syndrome[22].

Musculoskeletal involvement

Bony involvement is seen in 3-13% cases and involves hands and feet. A high prevalence of radiographically evident osseous disease is associated with chronic progressive form, pulmonary and skin involvement. Lacy reticular appearance of the bones is characteristic. Osteopenia and osteoporosis occur in up to two third patients which may be further aggravated by treatment with steroids[21,22].

Acute sarcoid arthritis is a symmetrical polyarthritis, beginning in the feet and ankles and spreading to knees, proximal interphalangeal joints, metacarpophalangeal joints, wrists, and elbows. Symmetric ankle arthritis is characteristically seen in sarcoidosis and has a sensitivity of 95% and specificity of 92% for the diagnosis. It is usually self limiting with good prognosis. In 6%, it remains active or recurs requiring DMARD therapy. Dactylitis with soft tissue swelling can be seen.

Bone lesions include variable-sized cysts and punched out lesions or lattice like changes, often at the ends of affected bones. The lesions are cortical with preserved periosteum. Symptomatic myopathy is seen in only 5% cases, marked by proximal muscle weakness and tenderness. Granulomatous inflammation and neurogenic atrophy of muscles has also been reported[7,10,12,13].

Childhood sarcoidosis

Sarcoidosis is relatively rare in the pediatric population. The course and prognosis of the disease in children is different compared to adults, and may correlate with the mode of onset and the extent of the disease[23]. The true incidence and prevalence is unknown because of the rarity of the disease. There is no sex predominance in childhood sarcoidosis unlike adults[24].

There are two distinct forms of childhood sarcoidosis. Older children present with a multisystem disease similar to adults with frequent lymphadenopathy and pulmonary involvement along with systemic signs and symptoms, such as fever, malaise and weight loss[24]. In contrast, early-onset disease (< 5 years) is characterized by the triad of rash, uveitis, and arthritis[25]. It may sometimes be asymptomatic and remain undiagnosed.

Radiological abnormalities are seen in ~90% childhood cases, bilateral hilar lymph node enlargement being the most common finding. Ocular involvement occurs in nearly half the patients. Choroidal and conjunctival granulomata are the typical and most common manifestation[26]. Hence, slit lamp examination is mandatory in the evaluation of childhood sarcoidosis[27]. Multiple types of skin lesions and rash are present in more than 80%. Multiple joint swellings in 15-58% patients, especially in younger children may be confused with juvenile idiopathic arthritis. Parotid gland is frequently found to be enlarged. Renal involvement is uncommon and cardiac disease is rare[28]. Serum angiotensin converting enzyme (ACE) levels, though non-specific are useful in following disease activity and effect of therapy. Reference values for serum ACE are age-dependent and healthy children have ACE levels 40-50% higher than in adults. The diagnosis is considered after ruling out tuberculosis, fungal and other infectious diseases, lymphomas and other neoplastic diseases and JIA in sarcoid arthritis.

Prednisolone, in doses of 1-2 mg/kg is the mainstay of treatment used to decrease inflammation[29]. Methotrexate may be used if there is no response to steroids[30]. Hydroxychloroquine may be used to treat skin problems and nonsteroidal anti-inflammatory agents for arthritis. Topical steroids may be used for rashes. Experience with newer therapies like TNF-alpha inhibitors e.g. infliximab is very limited in children. Overall prognosis is considered to be slightly better than in adults.

Sarcoidosis in elderly

Clinical presentation differs in the elderly. An alteration of general health (asthenia, anorexia, weight loss) is the most common manifestation. The intra-thoracic form is most common. Improvement/ stabilization is seen in upto 80% of the elderly patients.

Sarcoidosis in pregnancy

Sarcoidosis does not affect pregnancy adversely, but the disease may worsen after childbirth. The incidence of spontaneous abortion, miscarriage, and congenital fetal abnormalities is not increased.

Diagnosis

Unfortunately there is no single diagnostic test for sarcoidosis. Demonstration of granulomata with exclusion of other diseases, forms the basis of diagnosis. A diagnostic algorithm is provided below (Fig. 3). Initial evaluation of patients is summarized in Table 4.

Table 4. Initial evaluation of patients with sarcoidosis

Initial assessment

- History and physical examination (environmental or occupational exposure and family history)
- Biopsy of affected organ, with special stains and culture of specimen
- Posteroanterior and lateral chest radiographs
- Pulmonary function tests - spirometry with bronchodilator, total lung capacity and diffusion capacity
- Electrocardiography
- Complete ophthalmologic evaluation
- Complete blood count with platelet count and measurement of serum calcium, creatinine, alkaline phosphatase, alanine aminotransferase and aspartate aminotransferase levels
- Measurement of serum level of angiotensin-converting enzyme
- 24 hour urine calcium

Tests for assessment of involved organs:

- Heart - Holter monitoring, echocardiography, cardiac PET, MRI, and EPS
- Lung - right-heart catheterization for pulmonary hypertension
- Central nervous system - MRI, cerebrospinal fluid analysis

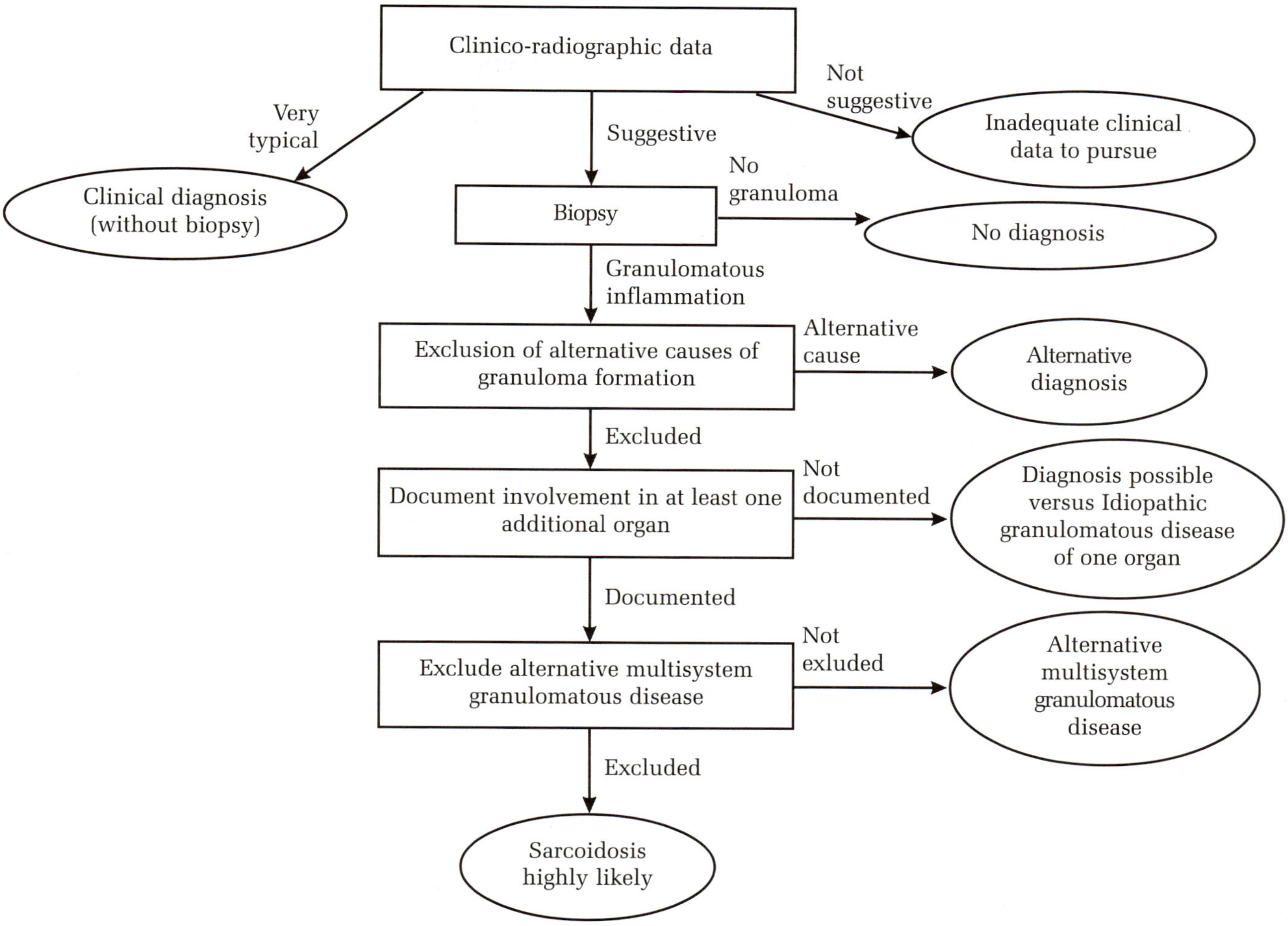

Fig. 3 : Diagnostic algorithm for sarcoidosis

The usual pulmonary involvement is that of interstitial lung disease, with reduced DLCO and restrictive pattern on pulmonary fuction tests. X ray abnormalities are commonly visualized. High resolution CT is the investigation of choice to delineate the extent and pattern of lung involvement. Stages of pulmonary sarcoidosis based on CT findings are shown in Table 5. Bronchoalveolar lavage reveals increased proportion of predominantly Th1 lymphocytes with oligoclonal expansion. Gallium 67 scan shows diffuse increase in uptake. Transbronchial biopsy with demonstration of sarcoid granulomata is essential for diagnosis.

Table 5. Stages of sarcoidosis

Stage	Chest radiograph characteristics	Frequency (%)	Spontaneous remission (%)
0	Normal	5-10	-
I	Bilateral hilar lymphadenopathy (BHL)	50	55-90
II	BHL and parenchymal infiltrates	25	40-70
III	Parenchymal infiltrates without BHL	15	10-20
IV	Signs of fibrosis	5-10	0-5

There is anergy to common test antigens such as candida or PPD. Kveim-Sitzbach test shows a positive reaction with formation of sarcoid granulomata at site of inoculation. However, due to non availability, this test is rarely performed.

Scalene lymphnodes and liver are not ideal sites of biopsy due to common occurrence of nonspecific granulomata at these sites[13,18].

Though investigations are important, sarcoidosis is essentially a diagnosis of exclusion. The tissues obtained should be submitted for microbiologic examination and cultures for tuberculosis and fungus. HIV is a close mimic and should be ruled out before making a diagnosis of sarcoidosis. The differential diagnosis of sarcoidosis is given in Table 6.

In Stage I disease initially follow-up must be every six months and then annually if the disease is stable. In Stage II, III and IV - initially every three to six months followed by annual monitoring indefinitely. Serious extra-pulmonary involvement warrants more frequent monitoring. The patient should be monitored for at least three years after cessation of therapy for potential relapse.

Disease activity measures

There are no definite markers for disease activity in sarcoidosis. Serum ACE level is popular but loosely correlates with total body granuloma burden[31]. However, it is not sensitive or specific enough to be used for screening or diagnosis. ACE gene has a common insertion-deletion polymorphism that influences serum levels[32]. Role of ACE as a tool for assessing disease activity is also limited. During treatment with corticosteroids, ACE levels fluctuate in an unpredictable fashion and frequently do not correlate with the severity of organ involvement[33,34]. Other markers, such as lysozyme, chitotriosidase, neopterin, cytokines and chemokines (sIL2R etc), have not been sufficiently evaluated to recommend widespread adoption at present but may outperform ACE to assess disease activity[35,36].

Table 6. Differential diagnosis of sarcoidosis based on organ system involvement

Lung	Lymph Node	Liver
Tuberculosis	Tuberculosis	Tuberculosis
Atypical mycobacteriosis	Atypical mycobacteriosis	Brucellosis
Cryptococcosis	Brucellosis	Schistosomiasis
Aspergillosis	Toxoplasmosis	Primary biliary cirrhosis
Histoplasmosis	Kikuchi's disease	Crohn's disease
Coccidioidomycosis	Cat-scratch disease	Hodgkin's disease
Blastomycosis	Sarcoid reaction in regional lymph nodes to carcinoma	Non-Hodgkin's lymphoma
Pneumocystis carinii	Hodgkin's disease	GLUS syndrome
Mycoplasma	Non-Hodgkin's lymphoma	
Hypersensitivity pneumonia	GLUS syndrome	**Bone marrow**
Pneumoconiosis (beryllium titanium, aluminum)		Tuberculosis
Drug reactions	**Skin**	Histoplasmosis
Foreign body aspiration	Tuberculosis	Infectious mononucleosis
Chronic interstitial pneumonia including lymphocytic interstitial pneumonia	Atypical mycobacteriosis	Cytomegalovirus
Necrotizing sarcoid granuloma	Fungal infection	Hodgkin's disease
Wegener's granulomatosis	Rheumatoid nodules	Non-Hodgkin's lymphoma
	Foreign body reaction beryllium, zirconium, (tattooing, paraffin)	GLUS syndrome
		Miscellaneous
		Crohn's disease
		Giant cell myocarditis
		GLUS syndrome
		TASS syndrome

Follow up

Follow up is required for all patients with abnormal X-rays, during and after completion of treatment as 70% can have a relapse. Clinical status, especially lung functions should be monitored. The follow up includes assessment of physiological function based on initial organ involvement and further testing in the event of new symptoms or physical findings. Evaluation is also required for monitoring side effects of therapy for example, bone densitometry for corticosteroid use and annual ophthalmologic examination for hydroxychloroquine use.

Therapy of sarcoidosis

Indications for treatment:

Sarcoidosis in many instances is a self limiting illness. Treatment is required for central nervous system, eye and heart involvement (ATS/ERS/WASOG guidelines)[37]. Refractory hypercalcemia and lupus pernio warrant immunosuppressive therapy[38]. Debilitating constitutional symptoms are another indication. Isolated minor pulmonary manifestations may not need therapy. Treatment is indicated in Stage III or IV disease, symptomatic disease or progression of lesions[1,3,14,15]. With other forms, the patients can be followed up in an expectant manner.

Treatment options

Therapy is needed to inhibit antigen presentation, suppress granuloma formation, enhance antigen clearance and inhibit fibrosis (Fig. 4, Table 7).

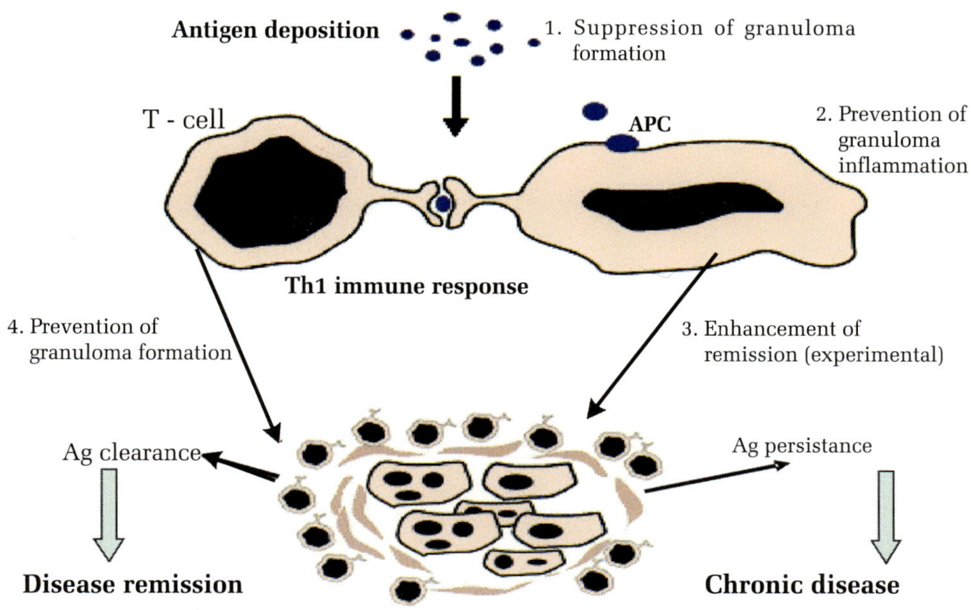

Fig. 4 : Pathogenesis of sarcoidosis and steps of intervention

Table 7. Drugs used in the treatment of sarcoidosis and their mode of action

Aim of treatment	Drugs	Mode of action
Suppress initiation of granuloma formation[1]	Antibiotics: minocycline, dapsone, clofazimine	Inhibit putative antigens like propionibacterium, mycobacterium; ? anti-inflammatory effect
	Antimalarials: hydroxychloroquine	Lysosomotropic; prevent antigen presentation
Prevent progressive granulomatous inflammation[2,4]	Steroids	Inhibit proinflammatory cytokine expression; decrease NFκB; increases TGF β & apoptosis of activated T cells
	Azathioprine	RNA/DNA inhibitor, inhibits T and B cell proliferation; greater suppression of cell mediated immunity
	Methotrexate	Inhibits lymphocyte proliferation and trafficking of lymphocytes to sites of inflammation
	Cyclosporine	Inhibits signal 1 from TCR complex; decreases IL2 production
	Mycophenolate mofetil	Inhibition of inosine monophosphate dehydrogenase
	Infliximab/Adalimumab	Inhibit TNF α
	Pentoxifylline	Decreases TNF α and IL-12
	Thalidomide	Complex effects on IL 12 & other cytokines
Enhance remission of granuloma[3]	Experimental therapies: Anti-amyloid therapy Drugs acting against trimolecular complex Anti TLR2 therapy	Increased antigen clearance

Table 8. Initial therapy of sarcoidosis according to organ involvement and clinical status[32,37]

Organ system involved	Clinical features	Treatment
Lungs	Dyspnea plus FEV1, FVC < 70% cough, wheezing	Prednisone, 20-40 mg/day Inhaled corticosteroid
Eyes (even if asymptomatic)	Anterior uveitis Posterior uveitis Optic neuritis	Topical corticosteroid Prednisone, 20-40 mg/day Prednisone, 20-40 mg/day
Skin	Lupus pernio Plaques, nodules Erythema nodosum	Prednisone, 20-40 mg/day Hydroxychloroquine, 400 mg/day Thalidomide, 100-150 mg/day Methotrexate, 10-15 mg/wk Prednisone, 20-40 mg/day Hydroxychloroquine, 400 mg/day NSAIDs
CNS	Cranial-nerve palsies Intracerebral involvement	Prednisone, 20-40 mg/day Prednisone, 40 mg per day Azathioprine, 150 mg/day Hydroxychloroquine, 400 mg/day
Heart	Complete heart block Ventricular fibrillation, tachycardia Decreased LVEF (< 35%)	Pacemaker AICD Prednisone, 30-40 mg/day AICD
Liver	Cholestatic hepatitis with constitutional symptoms	Prednisone, 20-40 mg/day Ursodiol, 15 mg/kg of body weight per day
Joints & muscles	Arthralgias Granulomatous arthritis Myositis, myopathy	NSAID Prednisone, 20-40 mg/day Prednisone, 20-40 mg/day
Hypercalciuria and hypercalcemia (even if asymptomatic)	Kidney stones, fatigue	Prednisone, 20-40 mg/day Hydroxychloroquine, 400 mg/day

The treatment and outcome of sarcoidosis can be grouped in three broad and partially overlapping groups:

Acute disease resolves within 2 to 5 years from time of diagnosis. Residual fibrosis may persist. Treatment strategy for acute sarcoidosis is given in Fig. 5.

Chronic disease persists beyond 5 years after diagnosis. Treatment of chronic sarcoidosis is given in Fig. 6.

Refractory disease – chronic disease that worsens despite adequate systemic therapy.

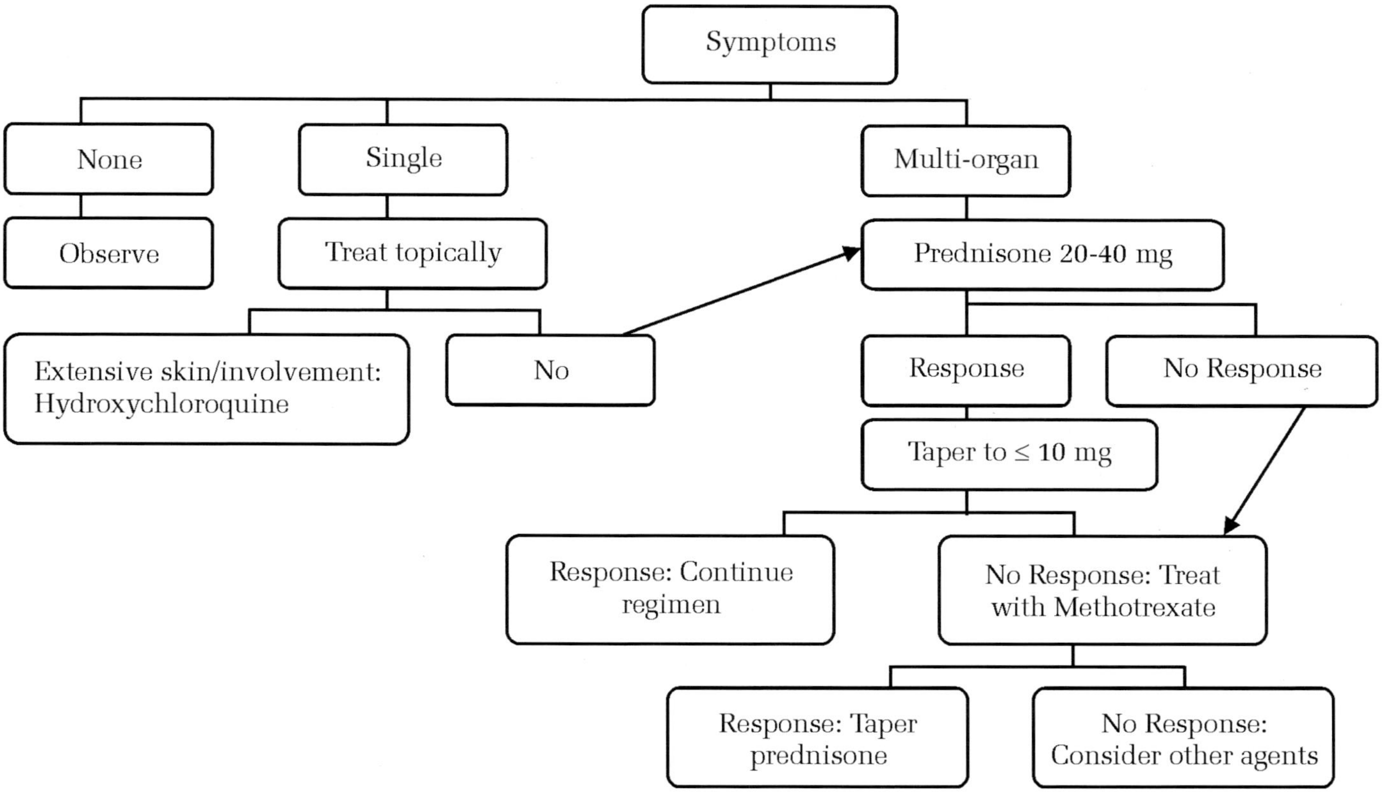

Fig. 5 : Treatment strategy for acute sarcoidosis[39]

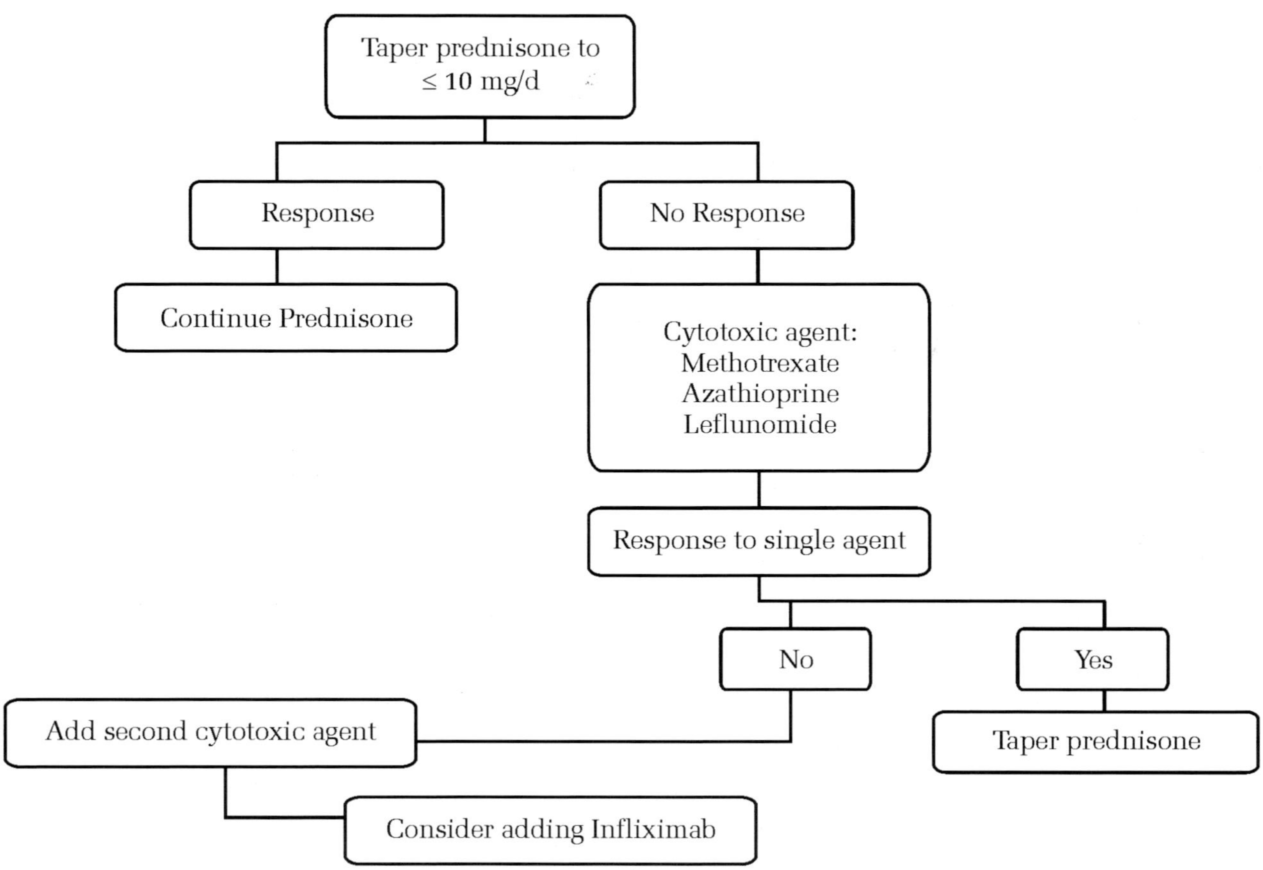

Fig. 6 : Treatment of chronic sarcoidosis[34,39]

Prognosis

The overall prognosis for people with sarcoidosis is good. The organ involvement occurs early with less than 25% developing new symptoms after 2 years of disease[40]. Two thirds of cases resolve spontaneously, one third have chronic disease of which about 5% results in fatality. Death is most commonly a result of progressive pulmonary fibrosis leading to respiratory failure followed by advanced myocardial or neurologic involvement. Although no biological markers exist to accurately predict outcome, certain traits and features may influence prognosis, (Table 9):

Table 9. Prognostic factors in sarcoidosis

Adverse prognostic indicators	Good prognostic factors
Age of onset > 40 years	Erythema nodosum
Black race	Acute presentation like Lofgren's syndrome
Requiring early treatment	
	HLA B8, A1 B8 and DR3
Extrapulmonary involvement	
Cardiac	G6PD deficiency
Neurologic (except isolated cranial nerve palsy)	
Lupus pernio	
Splenomegaly	
Hypercalcemia	
Osseous disease like bone cysts	
Pulmonary involvement	
Stage 3-4 CXR	
Pulmonary hypertension	
Significant lung function impairment	
Moderate to severe dyspnea on presentation	
BAL neutrophilia at presentation	

Further Reading :

1. Boeck C. Multiple benign sarcoid of the skin. J Cutan Genitourin Dis 1899; 17: 543-50.
2. Newman LS, Rose CS, Maier LA. Sarcoidosis. NEJM 1997; 336: 1224-34.
3. Statement on sarcoidosis. Am J Respir Crit Care Med 1999; 160(2): 736-55.
4. Rybicki BA, Major M, Popovich J Jr, Maliarik MJ, Iannuzzi MC. Racial differences in sarcoidosis incidence: a 5-year study in a health maintenance organization. Am J Epidemiol 1997; 145: 234-41.
5. Pietinalho A, Hiraga Y, Hosoda Y, Löfroos AB, Yamaguchi M, Selroos O. The frequency of sarcoidosis in Finland and Hokkaido, Japan: a comparative epidemiological study. Sarcoidosis 1995; 12: 61-67.
6. Baughman RP, Teirstein AS, Judson MA, et al. Clinical characteristics of patients in a case control study of sarcoidosis. Am J Respir Crit Care Med 2001; 164: 1885-89.
7. Milman N, Selroos O. Pulmonary sarcoidosis in the Nordic countries 1950- 1982: epidemiology and clinical picture. Sarcoidosis 1990; 7: 50-57.
8. Gupta S, Dutta S, Kar S, et al. Sarcoidosis in different regions of India: an analytical study. Sarcoidosis 1992; 9 (Suppl 1): 271-72.
9. Brewerton DA, Cockburn C, James DC, James DG, Neville E. HLA antigens in sarcoidosis. Clin Exp Immunol 1977; 27: 227-29.
10. Rossman MD, Thompson B, Frederick M, et al. HLA-DRB1*1101: a significant risk factor for sarcoidosis in blacks and whites. Am J Hum Genet 2003; 73: 720-35.
11. Iannuzzi MC, Maliarik MJ, Poisson LM, Rybicki BA. Sarcoidosis susceptibility and resistance HLA-DQB1 alleles in African Americans. Am J Respir Crit Care Med 2003; 167: 1225-31.
12. Rybicki BA, Maliarik MJ, Poisson LM, Iannuzzi MC. Sarcoidosis and granuloma genes: a family-based study in African- Americans. Eur Respir J 2004; 24: 251-57.
13. Schürmann M, Reichel P, Müller-Myhsok B, Schlaak M, Müller-Quernheim J, Schwinger E. Results from a genome-wide search for predisposing genes in sarcoidosis. Am J Respir Crit Care Med 2001; 164: 840-46.

14. Iannuzzi MC, Iyengar SK, Gray-McGuire C, et al. Genome-wide search for sarcoidosis susceptibility genes in African Americans. Genes Immun 2005; 6: 509-18.

15. Izbicki G, Chavko R, Banauch GI, Weiden MD, Berger KI, Aldrich TK, Hall C, Kelly KJ, Prezant DJ. World Trade Center "sarcoid-like" granulomatous pulmonary disease in New York City Fire Department rescue workers. Chest. 2007 May; 131(5): 1414-23.

16. Scadding JG. Prognosis of intrathoracic sarcoidosis in England: a review of 136 cases after five years' observation. BMJ 1961; 2: 1165-72.

17. Drakea WP, Newman LS. Mycobacterial antigens may be important in sarcoidosis pathogenesis. Cur Opin Pulm Med 2006; 12: 359-63.

18. Kellinghaus C, Schilling M, Lüdemann P. Neurosarcoidosis: clinical experience and diagnostic pitfalls. Eur Neurol 2004; 51: 84-88.

19. Bonfioli AA, Orefice F. Sarcoidosis. Semin Ophthalmol 2005; 20: 177-82.

20. Uemura A, Morimoto S, Hiramitsu S, Kato Y, Ito T, Hishida H. Histologic diagnostic rate of cardiac sarcoidosis: evaluation of endomyocardial biopsies. Am Heart J 1999; 138: 299-302.

21. Grimm W, Alter P, Maisch B. Arrhythmia risk stratification with regard to prophylactic implantable defibrillator therapy in patients with dilated cardiomyopathy: results of MACAS, DEFINITE, and SCDHeFT. Herz 2004; 29: 348-52.

22. H Visser, K Vos, E Zanelli et al. Sarcoid arthritis: clinical characteristics, diagnostic aspects, and risk factors. Ann Rheum Dis 2002; 61: 499-504.

23. Hunninghake GW, Costable U, Ando M et al: ATS/ERS/WASOG statement of sarcoidosis. American Thoracic Society/European Respiratory Society/World Association of Sarcoidosis and other Granulomatous Disorders. Sarcoidosis Vasc Diffuse Lung Dis 1999; 16: 149-73.

24. Hoffman AL, Milman N, Byg KE: Childhood sarcoidosis in Denmark 1979-1994: incidence, clinical features and laboratory results at presentation in 48 children. Acta Paediatr 2004; 93: 30-36.

25. Hafner R, Vogel P: Sarcoidosis of early onset. A challenge for the pediatric rheumatologist. Clin Exp Rheumatol 1993; 11: 685-91.

26. Hoover DL, Khan JA, Giangiacomo J: Pediatric ocular sarcoidosis. Surv Ophthalmol 1986; 30: 215-28.

27. Fauroux B, Clement A: Paediatric sarcoidosis. Paediatr Resir Rev 2005; 6: 128-33.

28. Pattishall EN, Strope GL, Spinola SM, Denny FW: Childhood sarcoidosis. J Pediatr 1986, 108: 169-77. Beneteau-Burnat B, Baudin B, Morgant G, Baumann FC, Giboudeau J: Serum angiotensin-converting enzyme activity in normal children and in those with sarcoidosis. Clin Chem 1990; 36: 344-46.

29. Pattishall EN, Kendig EL Jr: Sarcoidosis in children. Pediatr Pulmonol 1996; 22: 195-203.

30. Gedalia A, Molina JF, Ellis GS Jr, Galen W, Moore C, Espinoza LR. Low-dose methotrexate therapy for childhood sarcoidosis. J Pediatr 1997; 130: 25-29.

31. Shorr AF, Torrington KG, Hnatiuk OW. Endobronchial involvement and airway hyperreactivity in patients with sarcoidosis. Chest 2001; 120: 881-86.

32. Fisher KA, Serlin DM, Wilson KC, Walter RE, Berman JS, Farber HW. Sarcoidosis-associated pulmonary hypertension: outcome with long-term epoprostenol treatment. Chest 2006; 130: 1481-88.

33. Yanardag H, Pamuk ON, Pamuk GE. Lupus pernio in sarcoidosis: clinical features and treatment outcomes of 14 patients. J Clin Rheumatol 2003; 9: 72-76.

34. Scott GC, Berman JM, Higgins JL Jr. CT patterns of nodular hepatic and splenic sarcoidosis: a review of the literature. J Comput Assist Tomogr 1997; 21: 369-72.

35. Handa T, Nagai S, Miki S, et al. Incidence of pulmonary hypertension and its clinical relevance in patients with sarcoidosis. Chest 2006; 129: 1246-52.

36. Judson MA. Hepatic, splenic, and gastrointestinal involvement with sarcoidosis. Semin Respir Crit Care Med 2002; 23: 529- 41.

37. Hunninghake GW, Costabel U, Ando M, et al. ATS/ERS/WASOG statement on sarcoidosis. American Thoracic Society/European Respiratory Society/WorldAssociation of Sarcoidosis and Other Granulomatous Disorders. Sarcoidosis Vasc Diffuse Lung Dis 1999; 16(2): 149-73.

38. Iannuzzi MC, Rybicki BA, Teirstein AS. Sarcoidosis NEJM 2007; 357: 215-65.

39. Robert P. Baughman, Ulrich Costabel, Ronald M. duBois Treatment of Sarcoidosis. Clin Chest Med 2008; (29): 533-48.

40. Judson MA, Baughman RP, Thompson BW, Teirstein AS, Terrin ML, Rossman MD, Yeager H Jr, McLennan G, Bresnitz EA, DePalo L, Hunninghake G, Iannuzzi MC, Johns CJ, Moller DR, Newman LS, Rabin DL, Rose C, Rybicki BA, Weinberger SE, Knatterud GL, Cherniak R; ACCESS Research Group. Two year prognosis of sarcoidosis: the ACCESS experience. Sarcoidosis Vasc Diffuse Lung Dis. 2003 Oct; 20(3): 204-11.

RHEUMATIC DISEASES IN PREGNANCY AND LACTATION

2

Dhanita Khanna, Arun Shrivastava

Rheumatological disorders are common in clinical practice. Several of these occur in young or middle aged females. It is therefore not uncommon to see women harboring rheumatic conditions presenting with pregnancy or seen for various pregnancy related issues. Such patients may not initially present to a rheumatologist but may be seen by an obstetrician, an internist or a family physician. At times, a rheumatological disease may actually become apparent for the first time during a pregnancy.

Pregnancy in a patient with an autoimmune rheumatic disorder is usually a high risk event and several factors need to be considered for proper management of these cases.

Broadly, issues pertaining to autoimmune rheumatic disorders and pregnancy may be divided into :

1. Fertility
2. Rheumatic diseases affecting pregnancy (outcome)
3. Pregnancy affecting rheumatic condition
4. Drugs in rheumatic diseases and their effect on pregnancy and lactation

We shall primarily focus the following review on prototype rheumatic disease, systemic lupus erythematous (SLE). However, we will also see the impact of other common rheumatic disorders like rheumatoid arthritis (RA), as well as rare diseases like mixed connective tissue disease (MCTD) in such a setting.

Pregnancy is immunologically an interesting natural event, wherein the immune system of the woman tolerates the fetal presence even though the latter contains half of the paternal proteins, which should normally have been immunogenic in the lady. Several hypotheses have been put forth to explain this "phenomenon" of fetal (read *paternal*) tolerance, one important being conversion of T helper (TH) lymphocyte cytokines, T_H1 to T_H2, which influences the pregnancy favorably[1]. Needless to mention, SLE, a T_H2 mediated disease is likely to exacerbate during pregnancy, while RA, a predominantly T_H1 disease, usually ameliorates.

Various autoimmune processes are known to influence lupus activity; perhaps the role of humoral immunity is most widely recognized and possibly the most important. In this relation, different autoantibodies (Aab) in SLE, like anti-ds DNA, anti ribosomal P, anti Sm, anti Ro etc are all associated with lupus activity, albeit in different organ systems. Although, the role of these abnormal proteins as direct causal or an epiphenomenon is debatable (but not a matter of discussion in this review), the importance of these cannot be underestimated even if they may involve a major help from the cellular (T cell mediated) arm of immunity. The synthesis of these Aab is under the influence of various cytokines, which comprise the T_H2 arm of the cytokine interactive pathway of disease causation.

Antiphospholipid syndrome[2](APS) is another interesting multisystem Aab associated syndrome, characterized primarily by bland vasculopathy (cf SLE; a small/ medium vessel vasculitis). APS may be associated with other connective tissue disorders including SLE (secondary APS) or may be a primary disease[4] (Primary APS; PAPS). Bad obstetric history including recurrent fetal losses is one of the major manifestations of this syndrome.

We shall see a brief overview of some of these autoantibodies with respect to pregnancy loss in the following :

Autoantibodies in rheumatic diseases associated with pregnancy loss

1. Phospholipid (PL)/ PL-protein binding Aab (Classical aPL)
2. Non-classical PL binding Aab
3. Other Aab

PL/ PL-protein binding Aab

Antiphospholipid antibodies (aPL) are a group of Aabs which bind to negatively charged PL, either directly or through a PL-protein complex. Anticardiolipin antibodies (aCL IgG/M/A) and lupus anticoagulant (LA) are the most commonly tested aPL in suspected cases with primary or secondary APS. Other Aab like anti-phosphatidylserine ab (aPS) and anti-phosphatidylethanolamine ab (aPE) may also be seen in APS but are available for testing less commonly[10]. Aabs directed against PL binding proteins like anti beta2 glycoprotein1 (GP1) ab, anti prothrombin ab and anti annexinV ab may also be tested. Cardiolipin is an antigen derived from bovine heart and is not present in humans; still aCL is the most commonly tested aPL in practice which may reflect presence of another implicating aPL in the patient through cross reactivity between cardiolipin and another phospholipid in human blood.

aPL may cause pregnancy loss, pre-term delivery, small for gestational age baby, as well as still births 2. aPL may be responsible for poor pregnancy outcome through various postulated mechanisms:

a. Promotion of hypercoagulable state in placental bed (possibly responsible for mid-late trimester pregnancy loss);
b. Interfering with trophoblast growth and invasion (possibly responsible for early trimester fetal loss);
c. Hampering with trophoblast function (possibly responsible for early trimester fetal loss).

aPL may be either directed against individual PL/ protein/ or a complex of the two. Most of these proteins are natural anticoagulant plasma proteins also present in placental bed, while most of these negatively charged phospholipids are normally present on the inner leaflet of cell membrane sequestered inside the cell and therefore not available for immunogenic reactions. Perturbations of cell membranes, for example during placental villous formation, may be responsible for exteriorization of these anionic PLs, which may invite immune reaction. The normal anticoagulant plasma proteins are responsible to bind to these PLs and maintain anticoagulant state in placental bed[6]. Aabs directed against these proteins or against PL-protein complexes obviously reverse this and promote pro-coagulant state, responsible partly for poor fetal outcome.

Various aPL have been described in patients with APS and SLE (Table 1).

Table 1 : aPL in pregnancy loss

Autoantibody	PL	Protein
aCL (3)	Cardiolipin	Beta 2 GP 1
aPS (4)	Phosphatidyl serine	Prothrombin Annexin V Beta 2 GP 1
aPE (5)	Phosphatidyl ethanolamine	Kininogen (HMW/LMW)
Anti beta 2 GP1 (3)	—	GP1
Anti annexin V (6)	—	Annexin V
LA	?	Beta 2 GP 1 Prothrombin

aCL: anti cardiolipin ab, aPS: anti phosphatidylserine ab, aPE: anti phosphatidylethanolamine ab, GP 1: Glycoprotein 1.

A detailed discussion on mechanism of action of these antibodies and APS can be seen in Section 8 Chapter 3 (Antiphospholipid antibody syndrome).

Non classical aPL

Anti thromboplastin Aab and anti mitochondrial M5 type Aab have both been described in patients with SLE and fetal loss. The M5 Aab are directed against a mitochondrial membrane PL while the former is directed against thromboplastin complex. The exact pathogenesis of pregnancy loss related to these Aabs is unclear.

Other Aabs

A number of non-PL Aab have been described in patients with pregnancy loss in SLE and APS in the past few years. They may however, be markers of reproductive failure rather than real causatives. More studies are needed to ascribe them a pathogenic role. Some of these Aab are anti laminin 1 ab, anti thyroglobulin ab, anti corpus luteum ab, anti prolactin ab, anti lymphocytotoxic ab, etc. Other Aab like anti Ro, La, U1 RNP also contribute to poor pregnancy outcome in various rheumatic diseases and are described subsequently under individual disease heads.

Autoimmunity and infertility

Despite the fact that a number of cases where immune related causes for primary infertility are presumed are managed with imunosuppression with steroids, immunoglobulins (IVIg), etc, there is no consensus in whether anti sperm ab, anti leucocyte ab, aPL or other immune reactions cause primary or secondary infertility.

Similarly, most of the rheumatological disorders cause pregnancy complications rather than infertility. The possible exception is the use of drugs like cyclophosphamide in rheumatological conditions, which may cause male as well as female infertility. However, menstrual abnormalities may be more common in females with active rheumatological disease which may be a contributing factor for the establishment of successful pregnancy.

Infertile patients with rheumatological disorders, if offered assisted reproduction facility under expert care, may result in establishment of pregnancy. Many of such cases may however have complications during pregnancy especially due to multiple gestations as a result of assisted reproduction. In these cases, for example SLE, disease flares are likely.

Rheumatological disorders and pregnancy

Systemic lupus erythematosus (SLE)

SLE is a prototype rheumatological disease where pregnancy related issues can be best discussed[7]. Data accumulated over past several decades has helped us to understand this problem better.

Fertility

Active lupus activity may be associated with menstrual abnormalities. Fertility however, is mostly normal except in long standing or aggressive disease. Anti Ro and anti U1 RNP antibodies may be markers of subclinical ovarian problems. Specific Aab like anti corpus luteum ab may also be responsible.

Table 2 : SLE symptom mimics in pregnancy

Manifestation	Pregnancy event	SLE event
Skin	Chloasma	Malar rash
Joints	Non-specific pains	Arthralgias
Hematological	Dilutional/ deficiency anemia Hemolytic anemia (HELLP/PE) Thrombocytopenia (HELLP/PE) High ESR (normal/ Anemia)	Anemia Hemolytic anemia (SLE/APS) Thrombocytopenia(SLE/APS) High ESR (activity)
Pedal edema	Dependent/ pressure/ deficiency	Renal lupus/ APS
Weight gain	Normal	Anasarca
Vomiting	Morning sickness	Renal lupus
Hypertension	Volume expansion/ PE/ E	Renal lupus
Seizures	E	CNS/ APS
Proteinuria	Normal pregnancy/ PE	Renal lupus
Azotemia	PE/E	Renal lupus
Poor fetal outcome	Multiple causes	APS/ SLE activity

E-Eclampsia, PE-Pre-eclampsia, HELLP syndrome- Hemolysis, Elevated Liver enzymes, Low Platelet counts

Effect of pregnancy on SLE

Besides the various Aab in SLE described above, hormonal changes in pregnancy include increase levels of estrogens, decreased androgens and increased prolactin. These hormonal changes are also known to contribute to SLE flares. SLE flares may involve minor or major organs and may be at times difficult to differentiate from a pregnancy associated event (Table 2).

Most of the data now suggests that SLE is likely to flare with the onset of pregnancy. SLE flares may be seen in about one thirds to half of patients, of which nearly half occur during immediate post partum period, while rest are divided equally amongst various trimesters. The chances of flare are maximum with pregnancy onset within 3-6 months of active lupus, active nephritis at conception and azotemia with serum creatinine of more than 2.0 mg%. The chances of pregnancy induced hypertension and pre-eclampsia are also more in such patients. Most of the flares described are minor organ[7], possibly due to better management options and early intervention for major organ disease available now, but SLE onset during pregnancy (seen in about 20% of pregnancies) may be associated with severe renal disease with anuric renal failure. These patients may even require termination of pregnancy and dialysis support. Lupus nephritis during pregnancy may manifest with proteinuria, hypertension (HTN) and azotemia which may be confused with pre-eclampsia. Furthermore, pregnant females may normally have mid-trimester proteinuria in 20-25% making further confusion from SLE related renal event. HELLP syndrome seen uncommonly in otherwise pregnant women may also mimic a lupus flare. Although it may be difficult to differentiate lupus nephritis from preeclampsia, high anti ds DNA antibodies and low complement proteins C3 and C4 may be markers of lupus nephritis while hyperuricemia may suggest pre-eclampsia. Some studies have also described role of alternative complement pathway activation markers like Ba, and a low CH50/Ba has been found to be associated with lupus flare[8]. SLE nephritis during pregnancy has been classified as proliferative nephritis (40-70%) followed by nephritic syndrome (10-40%), pre-eclampsia (5-35%), acute renal failure (0-33%) and chronic renal failure (0-19%). Secondary APS associated with SLE may present with renal vein thrombosis. Post partum lupus flares are also common and seen in about 40-50%. With better knowledge of lupus management, the maternal mortality now has been brought down to < 1% from about 20% in 1950s.

Effect of SLE on pregnancy

SLE can have multiple effects on pregnancy outcome ranging from (intrauterine growth retardation) IUGR and pre term birth, to fetal loss and still birth. These are predominantly influenced by active lupus at conception, lupus nephritis and Aabs especially aPL. The pregnancy outcome that is, the newborn may also be susceptible to various lupus related complications labeled (neonatal lupus erythematosus, NLE). Active lupus is also supposed to influence fertility. Spontaneous multiple abortions in normal pregnancy are only about 1-12% which is increased in lupus pregnancy (14-35%). Neonatal deaths are however, not increased in lupus pregnancies. With active nephritis, 40% abortions and 12% intrauterine fetal deaths have been noted compared from 28% and 2% respectively in those without nephritis[9]. Quiescent SLE leads to an 88% success rate compared to 64% with active lupus nephritis. Nephrotic syndrome and creatinine of >1.5mg% may be associated with only about 50% successful pregnancy outcome. aPLs are one of the major contributors of poor pregnancy outcome. These antibodies can lead to multiple spontaneous fetal losses especially in the second and the third trimesters. In various studies, mean fetal loss in SLE with aPL is about 37-48% compared from SLE without aPL (9-18%). Recurrent spontaneous abortions (2 or more) in SLE with aCL are significantly higher compared from aCL negative lupus patients[10]. The maximum chance of poor fetal outcome is seen in patients with both LA and aCL positivity and those with high titres of aCL IgG. Some studies have shown high values of alpha fetoprotein (AFP) in maternal blood in second trimester in APS patients with SLE to be associated with more fetal deaths[3]. Pregnancy is contra-indicated in active glomerulonephritis (especially diffuse proliferative glomerulonephritis), nephrotic syndrome, moderate/ severe hypertension and high serum creatinine levels.

Problems in neonates

Mothers harboring IgG Aab against SS A (Ro), SS B (La) and U1 RNP antigens have been shown to cause neonatal lupus syndrome (NLE). This is manifested by minor manifestations like skin rash, fever, arthralgias and by major problems like congenital heart block (CHB) and other cardiac problems like cardiomyopathy, pericardial effusion and myocarditis. NLE may be seen most commonly in mothers with SLE and primary Sjogren's syndrome. In SLE patients, anti Ro is seen in 30-40% while anti La is seen in 10-20%.

Prevalence of NLE in live births of SLE pregnancies is about 1%. Overall risk of a child of lupus patient having CHB is 1:60, which may increase to 1:20 in presence of Anti Ro ab[11]. In view of various studies describing NLE showed that cutaneous lupus is seen in about 34% of NLE while CHB may be seen in more than 50% of NLE (which is usually incomplete heart block). CHB usually occurs abruptly few weeks before term however may occur any time beyond 20 weeks. Close fetal monitoring in mothers positive for implicating Aab after 18-20 weeks especially with fetal echo, non-stress test and doppler ultrasonography may detect fetal complications early. Abnormal end diastolic umbilical artery pressure in 2nd trimester on doppler ultrasonography and notched uterine artery at second trimester may predict fetal CHB. CHB may require confirmation of diagnosis in-utero by fetal echo and management by pacemaker (even intrauterine) along with steroids. Cutaneous lesions usually manifests as photosensitive sub acute cutaneous lupus like rash (mostly annular rash). The rash lasts for less than 6 months, coinciding with the disappearance of maternal Aab from the infant's blood. Co-occurrence of cutaneous LE and CHB is uncommon.

Investigating a lupus pregnancy

Baseline

- Blood count (CBC), Urine complete with microscopy, renal function (RFT), blood sugars
- VDRL, aPTT, aPLs (aCL, LAC)
- Anti ds DNA, anti Ro, anti La, cryoglobulin, complements, proteins C3, C4
- 24 hr urine protein, 24 hr urine creatinine, if nephritis present

Monthly

- CBC, Urine, RFT, anti ds DNA, complements, protiens C3, C4

Fetal monitoring

- Fetal echo and non-stress test (after 20 wk, where indicated)
  Intrauterine Echo, stress test (where indicated)

Treatment guidelines for a lupus pregnancy

1. Avoid Cytotoxic drugs in1st trimester.
2. Low dose aspirin (<150 mg daily) till 36 week for patients with HTN/ risk for PE/ suspected APS.
3. Stress steroid dose for patients on prolonged steroids.
4. Emergency Caesarian Section for severe preeclampsia, fetal distress, abnormal NST.
5. Post-partum follow up for monitoring lupus flare, local infection (e.g. episiotomy site), urinary tract infection, pneumonia.
6. Use of steroids: 7.5-10 mg prednisolone daily for prophylaxis against occurrence of lupus flares,

to 50-60 mg for active disease. For fetal distress associated with presence of anti Ro/La antibody, steroids like dexamethasone to be considered (not metabolized by placental enzymes).

7. Use of NSAIDs: May be used in early pregnancy. Avoid full dose NSAIDs especially near term. They may cause increased labor due to decreased uterine contractility. They inhibit platelet aggregation which may lead to postpartum bleeds. They may also lead to in utero closure of ductus arteriosus.
8. Antimalarials: Use for minor lupus.
9. Disease flare in mother: Treat as another non-pregnant SLE as per need.
10. Disease flare in fetus: In utero fetal cardiac problems management by fluorinated glucocorticoids. Neonatal cardiac complications may require permanent pacemaker. Cutaneous manifestations usually require topical medications and usually wane within few months.

Following up a lupus pregnancy

- Monthly till 20 week
- 2-3 weekly after 20 week especially with aPL, anti Ro
- BP monitoring monthly initially, weekly after 34 weeks.

Treatment guidelines in APS pregnancy:

Low dose aspirin with/ without heparin may be required (oral anticoagulants are teratogenic) (Table 3). Use supplemental calcium along with heparin and presence of other sites of APS may also guide therapy in such pregnant patients including use of steroids.

Future therapies for APS: Several new therapeutic approaches for APS are emerging like combination antiaggregant therapy (low dose aspirin plus clopidogrel or dipyridamole), oral antifactor Xa (rivaroxaban, apixaban), direct thrombin inhibitors (dabigatran), B-cell depletion (rituximab), statins (fluvastatin, rosuvastatin) and hydroxychloroquine. Of these, the new drugs that pregnant women can use are dipyridamole and hydroxychloroquine. Combination treatment with aspirin plus dipyridamole have shown higher efficacy than has aspirin alone in patients with stroke. Such combination can also be considered in patients in whom warfarin is not effective or safe post pregnancy. Hydroxychloroquine directly inhibits the binding of antiphospholipid antibody-2-glycoprotein-1 complexes to phospholipid surfaces. Hydroxychloroquine should be considered for an adjuvant antithrombotic therapy in patients with systemic lupus erythematosus with antiphospholipid antibodies and patients with primary antiphospholipid syndrome.

Other issues

Various traditional indices like SLEDAI, SLAM, LAI etc used for monitoring SLE activity are useful for non-pregnant lupus. Several modifications in original indices have been incorporated after removing items mimicking pregnancy event, for use in pregnancy. These scales await validation. Until such time, clinical experience may be more useful to differentiate the two.

Table 3. General management guidelines for APS

aPL	Gestational state	None	ASA	Heparin	Combination
Moderate/ high titre aCL/ LA + aCL	Primi	√	√*		
	Multi Last pregnancy live	√	√*		
	Multi Last pregnancy loss (< 15 wk)		√		
	Multi Last pregnancy loss (> 15 wk)				√
Low titre aCL	Any	√			

* Optional

Rheumatoid arthritis (RA):

Most studies suggest, RA ameliorates during pregnancy. In fact, women who never had children are more likely to have risk of RA development than those who had children[12]. However, disease flares are most commonly seen within 6 weeks post partum. Breast feeding may be associated with increased risk of RA onset after delivery or post partum flares[13].

Disease remissions as well as post partum flares cannot be predicted by disease duration, functional class or titres of rheumatoid factor. Pregnancy losses are not significantly increased in RA and no special monitoring is recommended.

Systemic sclerosis (SSc)

Early reports of association of parity with development of SSc have not been substantiated. Fertility is usually not affected. Disease may exacerbate during pregnancy only in about 10-20%, especially in minor manifestations like Raynaud's phenomenon, arthritis and skin thickening. Patients may present with more symptoms of heartburn and breathlessness. Poor maternal outcome may be seen in patients with major organ manifestation like cardiopulmonary involvement. Pre eclampsia and renal crisis may be seen in patients with diffuse disease and hypertension. Steroids may precipitate renal crisis and are best avoided. Fetal outcomes are usually good, although preterm births are seen in diffuse disease. Fetal losses may be more common in late diffuse disease.

Undifferentiated connective tissue diseases (UCTD)

Large studies are not available and data is confined to case reports or small series. Fertility is normal or near normal. Disease flares may be seen in about one fourth of the patients during pregnancy or post partum. Evolution into specific disease like SLE may also be likely during or even after pregnancy. Most of the flares are of minor organ[14]. Most pregnancies end up in live births although minor complications during pregnancy may be seen like hypertension, premature birth etc. Neonatal complications are rare.

Myositis (DM/PM)

Pregnancies during disease are rare since most patients present either after 45 year of age or in childhood. Childhood onset disease even in long term remission may have disease flares during pregnancy[15]. Childhood onset disease and inactive disease has better outcome in terms of live birth compared to adult onset disease and active disease. Onset of disease during pregnancy has worse prognosis. Neonatal complications are unknown.

Mixed connective tissue disease (MCTD)

In this uncommon disease, available literature suggests that most of the times there is no significant influence of pregnancy on disease or vice-versa

Vasculitides

Takayasu's arteritis (TA) occurs in young ladies and so pregnancies are not uncommon. Disease does not seem to relapse but complications like hypertension, eclampsia, cardiac and renal failure may be seen. While fetal loss is not significantly increased, chances of IUGR and premature delivery are high especially with severe disease activity[17].

Other vasculitides are very rare and pregnancy issues are described only in anecdotal case reports or small series. Any significant conclusion may not be derived from them. Possibly, the behavior of the disease and pregnancy may be like SLE. Case of onset of polyarteritis nodosa during pregnancy has been described to have grave maternal prognosis.

Seronegative spondyloarthropathies (SSA)

SSAs include psoriatic arthritis, reactive arthritis and IBD associated arthritis, besides the prototype Ankylosing spondylitis (AS). Although, AS is predominantly male dominated disease, cases of clinical deterioration may be seen in pregnant females suffering from the disease. Spinal disease is usually unchanged or may be exacerbated if spinal disease was active during conception, while peripheral articular or extra articular manifestations may be suppressed during gestation but flare post partum. Psoriatic and reactive arthritis may be seen in females as much as in males. In such cases, pregnancy is usually not affected adversely neither disease flares are noticed.

Inflammatory bowel disease in pregnancy

Fertility is normal in ulcerative colitis (UC) but slightly lower in Crohn's disease (CD); latter may be possibly due to local inflammatory changes in ovaries or fallopian tubes, local fistulae or poor general health. Active CD at conception has more chances of persistence or deterioration of IBD. Exacerbations are frequent immediate post-partum. No poor pregnancy outcome is seen if IBD remains in remission. Active disease at conception has higher risk of spontaneous abortions which may be 40-60% for severe disease

Other complications include higher risk with caesarian section, vaginal fistula formation for vaginal delivery and inadequate healing of episiotomy. Steroids, aminosalicylates and sulfasalazine (SSZ) are safe in usual doses. Folate supplementation is recommended for patients on SSZ. SSZ causes transient reversible azoospermia and may contribute to male infertility. Azathioprine and Cyclosporine are safe except that they may cause low birth weight. Methotrexate is absolutely contra-indicated. Regarding anti TNF agents, experience has yet to accumulate.

Drugs in pregnancy

Cyclophosphamide (Cyc)

This is an alkylating agent widely used in rheumatology practice especially for treatment of aggressive conditions like lupus nephritis. It causes significant gonadal toxicity. In men, it may cause permanent azoospermia, which may be prevented to some extent by using testosterone which may reduce germinal cell activity and hence reduced sensitivity to cyclophosphamide. In females, cyclophosphamide causes follicular death by damaging rapidly dividing granulosa cells, which provide estrogen and progesterone to developing follicles. This stimulates pituitary gonadotropin production, which further enhances recruitment of follicles into pool of maturing follicles and hence further susceptibility to cyclophosphamide.

In SLE patients, the risk of sustained amenorrhea is dependent on age of initiation of cyclophosphamide therapy and the cumulative dose received. Safe factors include short monthly courses (< 7 pulses) and age < 25 years. Maximum risk occurs at > 15 pulses or > 32 years[19]. Route of administration is not important but oral toxicity is more, possibly because of more cumulative dose. In younger pre-menopausal females, potential predictors of sustained amenorrhea may be SLE disease of > 5 years, presence of antibodies against Ro and U1 RNP[20]. Ovarian failure may be prevented or managed through the use of gonadotropin releasing hormone agonist, cryopreservation of oocytes, and embryo cryopreservation.

Cyclophosphamide is absolutely contraindicated in pregnancy due to its teratogenic potential.

Steroids

Doses less than 7.5mg daily are considered safe in most cases. In fact, this dose is employed for maintaining disease remission in many cases of SLE. Higher doses may be required for major disease flares, and are associated with complications especially osteoporosis in pregnancy setting. Patients on prolonged steroids during pregnancy should be given stress doses at delivery. Since placental enzyme metabolize most of the steroids to a significant extent, chances of their entry into fetal circulation are rare. However, fluorinated steroids like dexamethasone and betamethasone are not metabolized by placenta and are therefore employed to treat fetal CHB. Risk of IUGR, prematurity, diabetes and hypertension, besides cushingoid facies and cataract are often seen with steroids. Congenital fetal abnormalities are rare.

Non steroidal anti-inflammatory agents (NSAIDs)

They may inhibit uterine contractility and prolong labor. Fetal anomalies like cleft lips and palates, skeletal and vascular malformations are not described in humans. Higher doses of aspirin can cause in-utero closure of ductal arteriosus and fetal acidosis and should be avoided. Lower doses however act as platelet anti-aggregants and are used in PDA, pre-eclampsia and APS.

Antimalarials

Retinal degeneration, Fallot's tetralogy, etc are not seen in humans in doses used in rheumatological disorders. hydroxychloroquine is now widely used in rheumatology practice during pregnancy.

Azathioprine (Aza)

No evidence of association with fetal anomalies is described in humans. However, chances of prematurity and neonatal cytopenias may be higher.

Biological therapy[21]

Data is sparse on the use of biologics in pregnant and lactating women as they are seldom included in drug trials. There are no controlled studies on use of biologics in pregnancy.

All biologics contain Fc portion of IgG which is actively transferred through the placenta by the FcR on the trophoblast from the beginning of the 2nd trimester and continues till term.

Amongst the TNF inhibitors, there is a difference in the transplacental passage with maximum for monoclonal antibodies (infliximab, adalimumab), less for fusion proteins (etanercept) and least for the Fc part of IgG (certolizumab). Hence TNF inhibitors should be discontinued as soon as pregnancy is recognized or at least prior to 30 weeks of gestation. However there is minimal excretion in the breast milk hence there is less impact in lactating mothers.

Data for abatacept, tolizumab and anakinra is not sufficient hence must be discontinued prior to conception.

For Rituximab no adverse effect in children exposed preconception or in early pregnancy has been seen in some newer studies. However in the 2nd and 3rd trimester it causes B cell depletion in foetus with unknown long term side effects. Hence discontinuation at least 6 months prior to conception is advocated considering its half life of 36 hrs.

With use of biologics in pregnancy risk of impaired immune response or increased susceptibility to infections in exposed children cannot be ruled out. Controlled studies are needed for all biologics to prove their safety. At present continuous use of biological therapy throughout pregnancy cannot be recommended.

Other drugs in pregnancy

Methotrexate (MTx) and leflunomide (lef) are absolutely contraindicated in view of teratogenic potential. Pregnancy should be planned after adequate washout period only. Drug elimination procedures are recommended for lef before contemplating pregnancy. However, a few recent studies have shown that there is no substantial increased risk of adverse pregnancy outcome in women who undergo cholestyramine elimination procedure in early pregnancy compared with controls[22]. However till more conclusive data is available, women receiving leflunomide should be counseled against pregnancy. Sulfasalazine (SSZ) is relatively safe drug and may be used in cases of RA and SSA requiring second line therapy during pregnancy.

For other details on drugs in pregnancy (see Table 4).

Drugs during lactation

1. Antimalarials : Safe
2. D Penicillamine : Data insufficient
3. SSZ : Safe
4. Aza : Cautious use; May cause cytopenias
5. Methotrexate : Avoid
6. Cyclosporine : Possibly safe; data insufficient
7. Cyc : Unsafe
8. NSAIDs : Preferably avoid
9. Leflunomide: avoid
10. Biological therapy: avoid

Table 4 : Summary of rheumatological diseases in pregnancy

Patient with	Fertility	Pregnancy course / outcome	Disease activity	Safe drugs	Unsafe drugs
SLE/APS	↓ Active disease	↑ fetal loss (aPL) ↑ pre-term, IUGR (aPL or active ds) ↑ PIH, PE if active ds ↑ serum uric acid NLE (anti Ro/ La)	↑ if GN, SLE onset in pregnancy, active ds in preceding 6 months. Most flares minor ↑ anti ds DNA, C3/C4,↓ CH50/Ba	Steroids HCQ Aza Heparin Low dose ASA	Cyc MTx Oral anti coagul NSAID#
RA	Normal	Normal	↓ ↑ post partum	Steroids HCQ SSZ	MTx Anti TNF Leflunomide NSAID#
SSc	Normal	↑ pre term/ fetal loss (diffuse ds) ↑ PE (Diffuse ds, renal ds, HTN)	↑ Minor flares		D Pen* ACE-I Steroids
UCTD*	Normal	Mostly normal	↑ or evolution into specific ds or normal	Steroids HCQ	Cyc
DM/ PM*	Normal	↑ fetal complications (active ds, adult onset ds, ds onset in pregnancy)	↑ flares even in childhood onset ds in remission	Steroids IVIg HCQ	Cyc MTx
MCTD*	Normal	Normal	Normal	As above	As above
TA	Normal	↑ PE, HTN ↑ IUGR, pre mature delivery (active ds)	Normal	Steroids Aza	MTx Cyc
Other vasculitis	*	*	*	As above	As above
IBD arthritis	↑ in active CD	↑ fetal loss (active ds)	↑ (active ds) ↑ post partum	ASA Steroids SSZ Aza CsA	Cyc MTx Anti TNF*
Other SSAs	Normal (Several are male dominant)	Normal*	↓ PsA Spinal flares in AS*	Steroids SSZ	NSAID# MTx

* Data insufficient # May be safe in 1st half of pregnancy

Further Reading :

1. Wegmann TG, Lin H, Guilbert L, et al. Bidirectional cytokine interactions in the maternal-fetal relationship: is successful pregnancy a TH2 phenomenon. Immunol Today 1993; 14: 353-56.
2. Levine JS, Branch DW, Rauch J. The antiphospholipid syndrome. N Engl J Med 2002; 346: 752-63.
3. Lee RM, Emlen W, Scott JR, et al. Anti beta 2 GP1 antibodies in women with recurrent spontaneous abortion, unexplained fetal death and antiphospholipid syndrome. Am J Obstet Gynecol 1999; 181: 642-48.
4. McNeil HP, Chesterman CN, Krilis SA. Immunological and clinical importance of anti phospholipid antibodies. Adv Immunol 1991; 49: 193-280.
5. Karmochkine M, Berard M, Piette JC, et al. Anti phosphatidyl ethanolamine antibodies in SLE. Lupus 1993; 2: 157-60.
6. Rand JH, Wu XX. Antibody mediated disruption of the annexin V antithrombotic shield: a new mechanism for thrombosis in the aPL syndrome. Thromb haemost 1999; 82: 649-55.
7. Derksen RH, Bruinse HW, de Groot PG, et al. Pregnancy in SLE: a prospective study. Lupus 1994; 3: 149-55.
8. Julkunen H, Joihidainen T, Kaaja R, et al. Fetal outcome in lupus pregnancy: a retrospective case study of 242 pregnancies in 112 patients. Lupus 1993; 2: 125-31.
9. Gimovsky ML, Montoro M, Paul RH. Pregnancy outcome in women with systemic lupus erythematosus. Obstet Gynecol. 1984; 63: 686-92.
10. Kutteh WH, Lyda EC, Abraham SM, et al. Association of anticardiolipin antibodies and pregnancy loss in women with systemic lupus erythematosus. Fertil Steril 1993; 60: 449-55.
11. Ramsey-Goldman R, Hom D, Deng J-S, et al. Anti SS-A antibodies and fetal outcome in maternal SLE. Arth Rheum 1986; 29: 1269-73.
12. Nelson JL, Ostensen M. Pregnancy and rheumatoid arthritis. Rheum Dis Clin North Am 1997; 23: 195-212.
13. Barrett JH, Brennan P, Fiddler M, et al. Breast-feeding and postpartum relapse in women with rheumatoid and inflammatory arthritis. Arth Rheum 2000; 43: 1010-15.
14. Mosca M, Neri R, Bombardieri S. Undifferentiated connective tissue diseases (UCTD): a review literature and a proposal for preliminary classification criteria. Clin Exp Rheum 1999; 17: 615-20.
15. Gutierrez G, Dagnino R, Mintz G. Polymyositis/ Dermatomyositis and pregnancy. Arth Rheum 1984; 27: 291-94.
16. Lundberg I, Hedfors E. Pregnancy outcome in patients with high titre anti RNP antibodies. A retrospective study of 40 pregnancies. J Rheum 1991; 18: 359-62.
17. Sharma BK, Jain S, Vasishta K. Outcome of pregnancy in Takayasu's arteritis. Int J Cardiol 2000; 75 (Suppl 1): S159-S62.
18. Moser M, Okun N, Mayes D, et al. Crohn's disease, pregnancy and birth weight. Am J Gastroenterol 2000; 95: 1021-26.
19. Boumpas DT, Austin HA 3rd, Vaughan EM, et al. Risk of sustained amenorrhea in patients with SLE receiving intermittent pulse cyclophosphamide therapy. Ann Intern Med 1993; 119: 366-69.
20. Ioannidis JP, Katsifis GE, Tzioufas AG, et al. Predictors of sustained amenorrhea from pulsed intravenous cyclophosphamide in premenopausal women with SLE. J Rheum 2002; 29: 2129-35.
21. Ostensen M and Forger F. Treatment with biologics of pregnant patients with rheumatic diseases. Curr Opin Rheumatol 2011; 23: 293-98
22. Chanbers CD et al. Birth outcomes in women who have taken leflunomide during pregnancy. Arthritis Rheum 2010; 62: 1494-503

FIBROMYALGIA

3

Able Lawrence, Rajkiran Dudam

Introduction

Descriptions of patients who would be classified as Fibromyalgia date from the 19th century and were known by different names like "neurasthenia", "fibrositis" and was widely considered to be psychogenic in origin[1]. Recognition of sleep abnormalities in patients with fibromyalgia in the 1970s led to greater recognition for the disease[2] and to the formulation of the 1990 American College of Rheumatology Criteria[3]. Prevalence of Fibromyalgia in the general population is approximately 2%. It is more commoner in women. 3.4% of women have fibromyalgia as compared to only 0.5% of men.

Pathogenesis

Fibromyalgia is a disorder of abnormal pain processing[4-6]. It is characterized by reduced pain thresholds (hyperalgesia) and pain with normally innocuous stimuli (allodynia). The widespread pain perceived by patients is of central origin and is a result of generalized decrease in pain threshold. This wide spread decrease in pain threshold is analogous to allodynia in neuropathic pain[29] and occurs in response to modalities tested namely pressure, heat/ cold, chemical and electrical stimulation. Normal physiological sensations are processed as pain in patients with fibromyalgia. Patients generally recall stressful events prior to the onset of fibromyalgia. The neurological mechanisms of decreased pain threshold involve windup and central sensitization[5]. Windup is the temporal summation of pain on repetitive stimulation of peripheral nociceptive afferents and has been shown to occur at lower frequency of stimulation in fibromyalgia. Multiple mechanisms are involved in the generation of central pain. Activation of the NMDA (N-methyl D Aspartate) receptors mediate the windup and antagonist of NMDA, ketamine has been shown to decrease/ameliorate fibromyalgia pain at sub anesthetic dosages.

Autonomic abnormalities have been well described in patients with fibromyalgia and are consistent with a generalized increase in sympathetic adrenergic tone and a blunted stress response[7]. Sub diaphragmatic vagotomy in rats leads to a state of generalized decrease in pain threshold similar to fibromyalgia[8]. This state is reversed by denervation or removal of adrenal medulla. Chronic adrenergic stimulation increases the sensitivity of peripheral nociceptive afferents[8]. The major pathway for the catabolism of epinephrine in the periphery is mediated by the enzyme COMT (Catechol O-methyl transferase). Polymorphisms in COMT alter the enzymatic efficiency[7]. Normal healthy women having the low efficiency allele has been shown to have lower pain threshold. There is evidence for the association of COMT gene polymorphisms with susceptibility to fibromyalgia[7].

In animal models of neuropathic pain, increased expression of the $\alpha 2\delta$ subunit voltage gated calcium channel and it's blockade by gabapentin has been demonstrated to reduce allodynia[9]. Pregabalin which belongs to the same class as gabapentin has been shown to reduce pain, probably rebalancing neurotransmitters involved in central pain amplification- reducing substance p and glutamate levels and restoring norepinephrine levels in fibromyalgia[10,30]. Fibromyalgia may be a heterogenous disease with multiple underlying mechanisms involved in different patients. This has its implications. No drug can be expected to be effective in all patients.

Clinical features (Table 3)

Fibromyalgia is a multidimensional syndrome characterized by widespread musculoskeletal pain associated with considerable distress, generalized fatigue, and disturbance of sleep, anxiety and depression[1]. A majority of patients also have chronic fatigue syndrome, temporo-mandibular pain syndrome, headache and irritable bowel syndrome[1]. Pain usually starts axially in the neck, shoulder or back and then spread to involve the rest of the body[3,11,12]. Sometimes, the disease starts after traumatic or stressful events in the patient's life. The pain described as deep, widespread, gnawing and variable. Pain is usually rated as more severe than the pain of Rheumatoid arthritis by patients[13]. Most patients have sleep disturbances with difficulty in initiation of sleep or early awakening. The wide spread pain is also associated with stiffness which can be worse in the morning. Most patients have poor tolerance of physical activity and exercise. Depression and anxiety are common in patients presenting with fibromyalgia[1]. While the depression is more common in patients of fibromyalgia who attend clinics, the life time risk of major depressive illness in fibromyalgia patients from the community is similar to those of normal patients[14]. On the other hand anxiety disorders especially post-traumatic stress disorder and obsessive compulsive disorder were more common in fibromyalgia patients in the community as compared to controls[14]. 'Affective spectrum disorders' are the group of disorders which coexist with one another. These include FM, Irritable bowel syndrome (IBS), chronic fatigue syndrome (CFS), depression and anxiety disorders and migraine. These are all associated with psychological distress (Table 4).

Diagnosis (Table 1,2 & Fig. 1)

The diagnosis of fibromyalgia is made in patients who present with persistent (more than 3 months), widespread and generalized pain who are other wise normal on physical examination by the demonstration of tender points in a majority of 18 sites as defined by the ACR criteria[3,28]. Pain must be present axially and peripherally, upper and lower part of the body as well as on the left and right of the body. If the patient has localized pain restricted to a part of the body, she is diagnosed to have a regional pain syndrome and not fibromyalgia. While the ACR criteria, included control points, pain in fibromyalgia is due to a generalized decrease in pain threshold and patients with very severe fibromyalgia pain may show tenderness at the control points as well. Patient may have tender points other than those included in the ACR criteria list. Presence of any other diagnosis like SLE or Rheumatoid arthritis is not exclusion for fibromyalgia and patients with chronic painful conditions have a higher risk of fibromyalgia as compared to normal individuals. While a haemogram, ESR, urinalysis, chemistry including calcium, vitamin D and alkaline phosphatase and thyroid function may be done, immunological markers like anti nuclear antibodies and rheumatoid factor should not be tested in patients being evaluated for fibromyalgia without a specific indication[13].

Benign joint hypermobility syndrome is a heritable condition where hypermobility of joints (as tested by Beighton score) is associated with pain in multiple joints[15]. Since joint hypermobility is common in the Indian population[15,16], it is fairly common to see patients with fibromyalgia and hypermobility. Benign hypermobility should be considered in the differential diagnosis of fibromyalgia and characteristically these patients lack the tender points. Others include vitamin D deficiency, primary generalized osteoarthritis, polymyalgia rheumatica, rheumatoid arthritis, systemic lupus erythematosus, Sjogren syndrome, inflammatory muscle diseases, myopathies, hypothyroidism, cushings syndrome, infections (HCV, HIV), drugs (statins) and malignancies.

Management of fibromyalgia (Table 5)

Most patients with fibromyalgia are considerably distressed by their symptoms, which is compounded by the apparent difficulty of their family members and health care providers to comprehend their symptoms. Gaining the trust of the patient and reassuring the patient that you can understand the patient's suffering and distress goes a long way in managing these patients who would have been to several doctors in search of relief. Once the trust of the patient is achieved, patients need to be educated tactfully regarding the benign nature of the disease. Patients must be encouraged to resume their normal life and participate in physical activities like exercise including aerobic exercises, strength training, yoga, etc.

Drug therapy

There are no medications which are highly effective or curative for the management of fibromyalgia. The objective of drug therapy is to relieve the symptom domains of pain, sleep, mood, stiffness and overall quality of life. While NSAIDs are widely prescribed and self medicated by the patients, there is no evidence for their benefit in randomized placebo controlled trials. Randomized controlled trials have established the efficacy of Tramadol in relieving the pain of fibromyalgia and in improving the quality of life[17,18]. Tricyclic antidepressants have formed the mainstay of the pharmacotherapy. While efficacy of tricyclic antidepressants has been established in meta-analyses of randomized controlled trials[19,20], overall efficacy is modest. Amitriptyline is initiated at dosages of 10mg to 25 and increased up to 50mg as tolerated. Tricyclic antidepressants are not effective in improving mood in patients with depression at doses used in fibromyalgia. The selective norepinephrine reuptake inhibitors (SNRI) must be used in patients with depression since they are better tolerated and have faster onset of antidepressant effects as compared to tricyclics. SNRIs Duloxetine[21,22] and Milnacipran[23] have been shown to improve symptoms of fibromyalgia in patients with and without depression.

Selective serotonin reuptake inhibitors fluoxetine[24], paroxetine[25] and citalopam[26] have been studied in relatively small randomized controlled trials. While these trials showed trend towards improvement in the treatment arm, larger placebo controlled trials are needed to establish their efficacy in patients without depression. Caution must be exercised while prescribing the fluoxetine along with tramadol due to the potential for drug interaction.

The muscle relaxant drug cyclobenzaprine has been studied for its benefit in fibromyalgia. Meta-analysis of randomized controlled trials[27] showed superior efficacy in relieving pain with an odds ratio of 3 (1.6 to 5.6). The efficacy is modest with a NNT (number needed to treat) of 4.8 for one patient to show improvement. Pregabalin is a ligand for the α2δ voltage gated calcium channels in the spinal cord which are up regulated in the setting of neuropathic pain[9]. It is already used for the treatment of pain in diabetic neuropathy and post herpetic neuralgia and became the first drug approved by FDA for the treatment of fibromyalgia. Pregabalin has been shown to improve fibromyalgia[10,31] when used in the dose of 300 to 450mg/d (in two divided doses). Pregabalin is initiated in the dosage of 75mg twice daily and escalated as required. The common side effects are drowsiness and giddiness. Patients must be explicitly cautioned against driving or operating machinery. Some patients develop generalized pitting edema and weight gain[10].

Most patients continue to have symptoms of fibromyalgia even several years after diagnosis[13]. In the absence of highly effective therapy, a combination of counseling, exercise and lifestyle changes with judicious trial of available medications should be used to relieve the distressing symptoms of the patient and help patient cope with the illness.

Table 1

1990 ACR criteria for diagnosis of fibromyalgia 2 major criteria
1. Presence of widespread pain for at least 3 months, above and below diaphragm, right and left side of the body and trunk and extremities.
2. Presence of tenderness at a minimum of 11 amongst possible 18 designated locations when the tenderness is elicited by digital pressure of 4kg applied with thumb.

Table 2

Sites for elicitation of tender points by 1990 ACR criteria
1. Insertion of the suboccipital muscle.
2. Mid upper trapezium muscle.
3. Origin of the supraspinatus muscle.
4. Under the lower sternomastoid muscle.
5. Near the second costochondral junction.
6. 2 cm distal to the lateral epicondyle.
7. At the prominence of the greater trochanter.
8. Upper outer quadrant of the buttock.
9. At the medial fat pad of the knee.

Sites for control points
1. Mid forehead
2. Junction of proximal two-third and distal third of forearm.
3. Thumbnail

Table 3

Cardinal features of fibromyalgia
• Generalized body ache
• Psychological distress
• Sleep disturbance
• Fatigue
• Normal on physical evaluation

Table 4

Conditions associated with fibromyalgia
• Chronic Fatigue Syndrome
• Depression
• Irritable Bowel Syndrome
• Migraine, Headache
• Irritable bladder, interstitial cystitis.
• Temporomandibular joint pain

Table 5

Management of fibromyalgia
Non pharmacological
• Patient Education
• Exercise and Physiotherapy
• Cognitive Behavioral therapy
Pharmacological
• Tricyclic antidepressants e.g. amitriptyline 10-50 mg.
• Selective serotonin reuptake inhibitors; citalopam, paroxetine
• Selective norepinephrine reuptake inhibitors; duloxetine, milnacipran.
• α2δ ligands, pregabalin, gabapentin.
• Muscle relaxants, cyclobenzaprine
• Analgesic; tramadol
• Serotonin receptor 5HT3 antagonists, tropisetron

Tender point locations for fibromyalgia; occiput; low cervical; trapezius; supraspinatus; second rib; lateral epicondyle; gluteal; greater trochanter; knee.

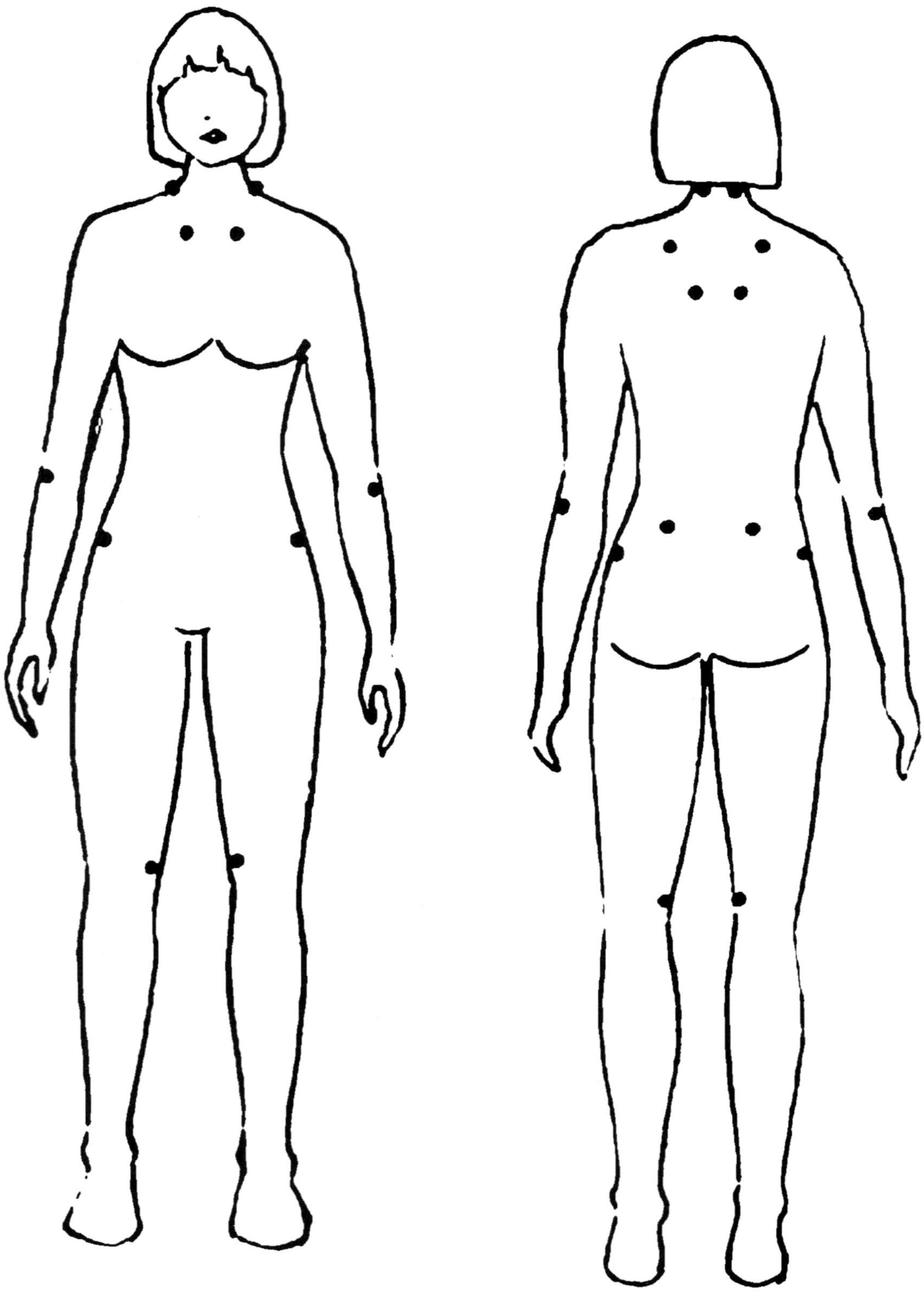

Fig. 1

Further Reading :

1. Mease P, Fibromyalgia syndrome: Review of clinical presentation, pathogenesis, outcome measures and treatment. J Rheum 2005; 32 Suppl 75: 6-21.
2. Smythe HA, Moldofsky H. Two contributions to understanding the fibrositis syndrome. Bull Rheum Dis 1978; 26: 928-31.
3. Wolfe F, Smythe HA, Yunus MB, et al. The American College of Rheumatology 1990 criteria for the classification of fibromyalgia. Report of the multicenter criteria committee. Arths Rheum 1990; 33: 160-72.
4. Rowbotham MC. Is Fibromyalgia a neuropathic pain syndrome? J Rheum 2005; 32 Suppl 75: 38-40.
5. Price DD, Staud R. Neurobiology of Fibromyalgia syndrome. J Rheum 1995; 32 Suppl 75: 22-28.
6. Levine JD, Reichling DB. Fibromyalgia: The nerve of that disease. J Rheum 2005; 32 Suppl 75: 29-37.
7. Manuel Martinez-Lavin. Stress, the stress response system and Fibromyalgia. Arth Res Ther 2007; 9(2) (doi: 10.1196/ar2146) (author pre-print).
8. Khasar SG, Miao FJP, Janig F, et al. Vagotomy induced enhancement of mechanical hyperalgesia in the rat is sympathoadrenal mediated. J Neurosc 1998; 18: 3043-49.
9. Luo ZD, Chaplan SR, Higuera ES, et al. Upregulation of dorsal root ganglion a2d calcium channel subunit and it's correlation with allodynia in spinal nerve injured rats. J Neurosc 2001; 21: 1865-75.
10. Crofford LJ, Rowbotham MC, Mease PJ, et al. Pregabalin for the treatment of fibromyalgia syndrome: Results of a randomized, double blind, placebo controlled trial. Arth Rheum 2005; 52: 1264-73.
11. Wolfe F, Ross K, Anderson J. Prevalence and characteristics of fibromyalgia in the general population. Arth Rheum 1995; 38: 19-28.
12. Yunus M, Masi AT, Calabro JJ, et al. Primary fibromyalgia (fibrositis): clinical study of 50 patients with matched normal controls. Semin Arth Rheum 1981; 11: 151-71.
13. Goldenberg DL. Fibromyalgia and related syndromes. In: Hochberg MC, Silman AJ, Smolen JS, Weinblatt ME and Weisman MH editors. Rheumatology 3rd Ed Mosby 2003; 702.
14. Raphael KG, Janal MN, Nayak S, et al. Psychiatric comorbidities in a community sample of women with fibromyalgia. Pain 2006; 124: 117-25.
15. Lawrence A. Benign joint Hypermobility syndrome. J Indi Rheum 2005; 13: 150-55.
16. Kumar A, Wadhwa S, Acharya P, et al. Benign joint hypermobility syndrome: a hospital-based study from northern India. Indian J Rheumatol 2006; 1: 8-12.
17. Bennet RM, Kamin M, Karim R, et al. Tramadol and acetaminophen combination tablets in the treatment of fibromyalgia pain: a double blind, randomized, placebo controlled study. Am J Med 2003; 114: 537-45.
18. Bennet RM, Schein J, Kosinski MR, et al. Impact of fibromyalgia pain on health related quality of life before and after treatment with tramadol/acetaminophen. Arth Rheum 2005; 53: 519-27.
19. O'Malley PG, Balden E, Tomkins G, et al. Treatment of fibromyalgia with antidepressants. J Gen Intern Med 2000; 15: 659-66.
20. Arnold LM, Keck PE, Welge JA. Antidepressant treatment of Fibromyalgia: A meta-analysis and review. Psychosomatics 2000; 41: 104-13.
21. Arnold LM, Lu Y, Crofford LJ, Wohlreich M, et al. A double blind, multicenter trial comparing duloxetine with placebo in the treatment of fibromyalgia patients with or without major depressive disorder. Arth Rheum 2004; 50: 2974-84.
22. Arnold LM, Rosen B, Pritchett YL, et al. A randomized, double-blind, placebo controlled trial of duloxetine in the treatment of women with fibromyalgia with or without major depressive disorder. Pain 2005; 119: 5-15.
23. Vitton O, Gendreau M, Gendreau J, et al. A double-blind, placebo controlled trial of milnacipran in the treatment of fibromyalgia. Hum Psychopharmacol Clin Exp 2004; 19: S27-35.
24. Arnold LM, Hess EV, Hudson JI, et al. A randomized, placebo controlled, double blind, flexible dose study of fluoxetine in the treatment of women with Fibromyalgia. Am J Med 2002; 15: 191-97.
25. Patkar AA, Masand PS, Krulewicz S, et al. A randomized controlled trial of controlled release paroxetine in Fibromyalgia. Am J Med 2007; 120: 448-54.
26. Anderberg UM, Marteinsdottir I, von Knorring L. Citalopam in patients with Fibromyalgia: a randomized, double-blind, placebo-controlled study. Eur J Pain 2000; 4: 27-35.
27. Tofferi JK, Jackson JL, O'Malley PG. Treatment of fibromyalgia with cyclobenzaprine: a meta-analysis. Arth Rheum 2004; 51: 9-13.
28. Wolfe F, Clauw DJ, Fitzcharles MA, et al. The American College of Rheumatology preliminary diagnostic criteria for fibromyalgia and measurement of symptom severity. Arthritis Care Res. 2010; 62: 600-10.
29. Costigan M, Scholz J, Woolf CJ. Neuropathic pain: a maladaptive response of the nervous system to damage. Annu Rev Neurosci. 2009; 32: 1-32.
30. Costigan M, Scholz J, Samad T, Woolf CJ. Pain. In: Siegel GJ, Albers RW, Brady ST, Price DL, eds. Basic Neurochemistry: Molecular, Cellular and Medical Aspects. 7th ed. Burlington, MA: Elsevier Academic Press; 2006: 927-38.
31. Owen RT. Pregabalin: its efficacy, safety and tolerability profile in fibromyalgia syndrome. Drugs Today. 2007; 43: 857-63.

SOFT TISSUE RHEUMATISM

4

C Lakshmikanth, S Pal

Introduction

"Better armed with technique than a judgement ..." Walshes.

Initially, rheumatology developed as an organ system based subspeciality of medicine devoted to diseases of articular origin. Subsequently, it was realized that this concept was insufficient as the structure like bursae, tendons, sheaths, tendon insertions, are equally affected by the same condition. The term non-articular or soft tissue rheumatism is now accepted for these conditions. Diagnosis of these conditions is often imprecise and management remains largely empirical.

Non-articular or soft tissue rheumatism involves structures outside the synovial lining e.g. bursae, muscles, tendon and joint capsule etc. In non-articular disorders, pain is often localized along the side of joint or along anatomic structures, or in relation to the muscles and soft tissue. Clinical and laboratory features of systemic disorder are lacking. Some of the common subtypes of non articular rheumatism are listed in Table 1 and Table 2 depicts some important causes of these disorders[1].

Dixon had described soft tissue rheumatism as a vast frontier land, ill defined, its features purely categorized and far from internationally agreed[2].

Table 1. Types of soft tissue rheumatism

Fibromyalgia	Referred pain syndrome
Bursitis	Reflex dystrophy
Tenosynovitis	Nerve entrapment syndrome
Myositis	Muscle-tendon junction syndrome
Vasculitis	Enthesopathy i.e. bone-tendon junction syndrome
Phlebitis	Panniculitis

Table 2. Causes of soft tissue rheumatism

Trauma
Acute
Chronic
Due to underlying arthritis
Periarticular calcific deposits
Anxiety
Depression

Etiopathogenesis

These conditions may be associated with a systemic illness. Very often they occur in absence of any systemic illness, when chronic repetitive low-grade trauma, excessive unaccustomed use at work may be responsible. This might be resulting in partial interruption of blood supply, which in turn leads to poor healing, degeneration and chronicity. Sprains and strains of traumatic origin are more common in the younger population. Painful shoulder has the highest frequency in localized form while it ranks fifth among regional rheumatic diseases. Advancing age and female sex are risk factors for shoulder problems.

Economic effects

Non-articular rheumatism represents the most commonly encountered rheumatic cause of sickness and absence from work, resulting in heavy loss in revenue. Behind these huge losses are great pain, misery and disruption of family life.

Classification

There is no single classification, which is accepted all over the world. A practical and useful classification of the conditions included under the term soft tissue rheumatism is as shown in Table 3.

Table 4. lists the major structures and their corresponding lesions involved in localized soft tissue lesions. The anatomical structure involved can be defined by a meticulous examination. The localized soft tissue lesions which are quite commonly seen are given in Table 5.

A few of these conditions included under the term soft tissue rheumatism merit a brief description.

Table 3. Classification of soft tissue rheumatism

Localized	
Epicondylitis	Enthesitis
Tendinitis	Tenosynovitis
Bursitis	Adhesive capsulitis
Ganglion	Calcific periarthritis
Rotator cuff syndrome	Fasciitis
Dupuytren's contracture	Regional myofascial pain
Generalized	
Fibromyalgia	Myofascial pain syndrome
Chronic fatigue syndrome	Psychogenic rheumatism

Table 4. Structure lesion relationship

Structure	Lesions
Tendon and tendon Sheaths	Tendinitis, Peritendinitis, Degeneration, rupture, Tenosynovitis
Tenoperiosteal junction	Enthesopathies, apophysitis
Bursae	Bursitis-acute and chronic
Fasciae	Fasciitis, Dupuytren's contracture
Ligament	Sprain and tear

Table 5. Common sites of lesions

Shoulder	Rotator cuff disorders, capsulitis of joint.
Elbow	Lateral and medial epicondylitis, olecranon bursitis.
Wrist and hand	Carpal tunnel syndrome, Dupuytren's contracture, de Quervain's tenosynovitis.
Trunk	Costochondritis, trapezius muscle spasm, non-specific neck and low back pain.
Hip	Ischial and trochanteric bursitis, meralgia paraesthetica.
Knee	Bursitis and ligament strain.
Ankle and foot	Achilles tendinitis, peritendinitis, plantar fasciitis, bursitis around calcaneus.

Bursitis

Bursitis is localized pain and swelling of a bursa caused by inflammation. It can be due to condition like RA, gout, infection or repetitive trauma[3]. Shoulder pain could be caused by subacromial bursitis, which is often caused by trauma due to frequent shoulder movements. Pain and swelling over the extensor aspect of the elbow can be due to olecranon bursitis. The prepatellar and infrapatellar bursa at the knees can be inflamed causing a painful swelling around knees. These are often caused by local injury as occurs in constant kneeling. The prepatellar bursitis is also called the housemaid's knee.

Tenosynovitis

Inflammation of the synovial tendon sheath, leading to pain and swelling is called tenosynovitis. It may be seen as part of a systemic disorder like RA, or may be due to infections like tuberculosis, leprosy. A very common problem seen in clinical practice is the "trigger finger". This is caused by a nodule composed of fibrous tissue forming in the tendon sheath of the flexor tendons of the fingers or the thumb. The nodule can be palpated in the palm just proximal to the MCP joint on the volar aspect. The nodule interferes with the normal tendon gliding and gives rise to triggering or the locking effect. This effect may be intermittent. It is often seen in OA but could also be part of RA, psoriatic arthritis or apatite crystal deposition disease. Other common forms of tenosynovitis are biceps tenosynovitis, supraspinatus tenosynovitis and de Quervain's tenosynovitis. Sonography of tendons is found useful in diagnosis[4].

Enthesitis

Enthesitis is inflammation at the site of insertion of a tendon into the bone. This is very characteristic of seronegative spondyloarthropathy e.g. Reiter's syndrome, but can also be caused by gout, sports injuries etc. The common forms of enthesitis are :

1. **In the upper extremities**
   a) Lateral epicondylitis - Tennis elbow
   b) Medial epicondylitis - Golfer's elbow
2. **In the lower extremities**
   a) Gluteus medius enthesitis
   b) Sartorius and semi-tendinous enthesitis (SSE)
   c) Tendoachilles enthesitis — This is observed especially in ankylosing spondylitis and reactive arthritis.
   d) Plantar Fasciitis — This is also commonly observed in SSA.

Management

The mainstay of management of the above conditions is relief of pain using heat/cold, counterirritant, local analgesics and NSAIDs supplemented by local intralesional injections of corticosteroids. If infection is suspected to be the underlying cause of the bursitis, percutaneous needle aspiration of the bursa must be performed and the fluid suitably examined to rule out a possibility of infection. If infection is detected, appropriate antibiotic therapy should be administered. Rest and protection of the affected area by immobilization, to prevent further injury to the part, is advocated. Compression with a firm bandage also helps to reduce the swelling.

Painful shoulder

More than 90% of lesions resulting in a painful shoulder are due to extra capsular soft tissue lesions. Shoulder joint is a multi-axial joint that allows greater mobility and is supported by various ligaments and muscles. Majority of painful, non-traumatic conditions about the shoulder joints are caused by tendinitis of rotator cuff. Usually a good history, clinical examination, radiology and MRI helps in the diagnosis[5].

Periarticular condition affecting the shoulder can be grouped into those with and without capsulitis. In capsulitis, there is a generalized restriction of movement, and if the capsule is not involved, passive joint movement are not effected while active movements may be limited by pain or weakness. The management consists of rest, NSAIDs, intra-lesional steroids and physiotherapy.

A condition called "frozen shoulder" with global restriction of active and passive movements in all planes; may be associated with diabetes, thyroid disease and pulmonary disorder. These patients may develop a secondary sympathetic dystrophy, shoulder hand syndrome, which consists of a frozen shoulder with a painful swollen, cold and dystrophic looking hand. This condition may progress until the patient is left with a non-functional hand Sudeck's atrophy.

Repetitive strain syndrome

It is a chronic pain syndrome affecting neck and arm region occurring in activities requiring control of posture, often of a repetitive nature, having psychological factors that contribute to the syndrome. The clinical features include chronic pain in neck, chest wall, arm, hand, inability in work performance previously done without impairment, with variable upper limb swelling with poor grip strength and taut proximal muscles. It is associated with poor sleep pattern, along with mood changes and mild sympathetic dystrophy.

Hypermobile joints

This syndrome consists of a common disorder in which otherwise healthy individuals present with articular symptoms for which no explanation is forthcoming other than joint hypermobility (Tables 6 and 7). These patients are prone to muscle and ligament tears and traumatic and degenerative changes of joints with recurrent dislocations. The clinical effects depend on degree of mobility. The Beighton's diagnostic criteria for the benign joint hypermobility syndrome has a specificity of 93% and sensitivity of 93%[6,7]. Management of benign joint hypermobility syndrome is reassurance by the rheumatologist.

Table 6. Nine-point Beighton hypermobility score

	R	L
The ability to		
1. Passively dorsiflex the 5th metacarpophalangeal joint to 90°	1	1
2. Oppose the thumb to the volar aspect of the ipsilateral forearm	1	1
3. Hyperextend the elbow to 10°	1	1
4. Hyperextend the knee to 10°	1	1
5. Place hands flat on the floor without bending the knees.	1	
Maximum Total	**9**	

One point may be gained for each side for maneuvers 1-4, so that the hypermobility score will have a maximum of 9 points if all are positive.

Table 7

Revised diagnostic criteria for the benign joint hypermobility syndrome shows the presence of two major criteria, or one major and two minor criteria, or four minor criteria. Two minor criteria will suffice where there is an unequivocally affected first-degree relative. BJHS is excluded by presence of Marfans or Ehlers-Danlos syndrome.

Major criteria

1. Beighton score of 4/9 or greater (either currently or historically).
2. Arthralgia for longer than 3 months in 4 or more joints.

Minor criteria

1. Beighton score of 1, 2, or 3/9 (0, 1, 2 or 3 if aged 50+).
2. Arthralgia (> 3 months) in 1-3 joints, or back pain (> 3 months), spondylosis, spondylolisthesis.
3. Dislocation/subluxation in more than one joints, or in one joint on more spondylosis, spondylolisthesis.
4. Soft tissue Rheumatism > 3 lesions (e.g. epicondylitis, tenosynovitis, bursitis).
5. Marfanoid habitus tall, slim, span/height ratio > 1.03 upper : Lower segment ratio < 0.89 arachnodactyly (+ Steinberg/wrist signs).
6. Abnormal skin hyperextensibility, thin skin papyraceous scarring.
7. Eye signs : drooping eyelids or myopia of epimongoloid slant.
8. Varicose veins or hernia or uterine/rectal prolapse.

Fibromyalgia

Introduction

Fibromyalgia (FM), an enigmatic polysymptomatic long lasting disease, is not uncommon but an underdiagnosed disorder, affecting mostly women, characterized by aches all over the body (pain amplification) tenderness at particular sites (tender points) and also other symptoms like fatigue, sleep disturbances [non-restorative sleep-sleep abnormalities may contribute to increased pain perception], morning stiffness, paraesthesia, headache and irritable bowel syndrome.

Etiopathogenesis

The cause of FM is not known. The factors like viral infection, an immunological reaction, trauma and physical or emotional stress [negative emotions-anger or sadness] are considered (Fig.1). Higher levels of day time sleepiness have been shown to be correlated with an increased number of tender points in patients with FM - according to prospective study women in age group of 45 yrs with associated past sleep problems had developed FM than the patients with no sleep problems and age group of 20-44 with or without sleep disturbances.

Familial aggregation of FM has been reported implicating genetic susceptibility[8]. Changes in the function of immune system, serum levels of serotonin, peripheral pain neurotransmitters and substance P are also considered as etiological factors for FM. Anti-smooth muscle antibodies and anti striated muscle antibodies have been reported[9]. The pathophysiology of FM remains unclear, but a lot of inroads have been made particularly with regard to the neuro-endocrine-immunological axis.

Diagnosis

FM is neither a diagnosis of exclusion nor an exclusive diagnosis. The 1990 ACR guidelines for the diagnosis of FM specify in addition to a history of widespread pain of three months, eleven out of eighteen specific tender point sites (thumb palpation)[10].

Other modalities for diagnosis and for differentiating from other mimicking syndromes.

1. Immune neuroendocrinal thermal assay.
2. Single photon emission computed tomography.
3. CT Scan.
4. Sleep wave cycle study [alpha wave intrusion correlating with overnight pain which are typical of awake states but are not specific findings in FM]. - higher levels of day time sleepiness have been shown to be correlated with an increased number of tender points in patients with FM.
5. Falling oxygen saturations.
6. Disruption of growth hormone-Insulin growth factor-1 (muscle pain and poor sleep).
7. Increased concentration of homocysteine in CSF, (decreased CSF B12).
8. Sleep disturbances preceed mood disorders like depression, decreased serotonin levels seen in patients with fibromyalgia.

Whenever the diagnosis of fibromyalgia is made, conditions listed below should be ruled out as they could have generalized aches and pains as their presenting feature or could occur concurrently with fibromyalgia.

1. **Inflammatory rheumatic disorder**
   Rheumatoid arthritis
   Polymyalgia rheumatica
   Systemic lupus erythematosus
   Polymyositis
   Sjogren's syndrome
2. **Neurological disorder**
   Negative emotions - anger or sadness, maniac, depression
   Myopathy
   Myasthenia gravis
3. **Endocrine disorders**
   Hypothyroidism
   Osteomalacia
4. **Infections**
   HIV
   Lyme disease
   Post viral myalgia/arthralgia
5. **Paraneoplastic syndrome**

Management (Table 8)

Both fluoxetine and amitriptyline are effective for FM and they work better in combination than either medication alone[11]. Even though tricyclic antidepressants play a major role in the management of FM, lifestyle alteration and education (e.g. typical recondition and exercise) as well as behavior modification are also important for a successful treatment. The antidepressant venlafaxine may be helpful in some of the FM patients[12].

Management of non restorative sleep in FM -

Non pharmacological therapies like removing bad sleep habits waking up and going to sleep at same time everyday increase day time physical activity cognitive behavioural interventionare helpful.

Pharmacological therapy like pregabalin milnacaprin SSRI"S - duloxetine resulted in reduced FM related pain, mood disorders, sleeping difficulties. FM

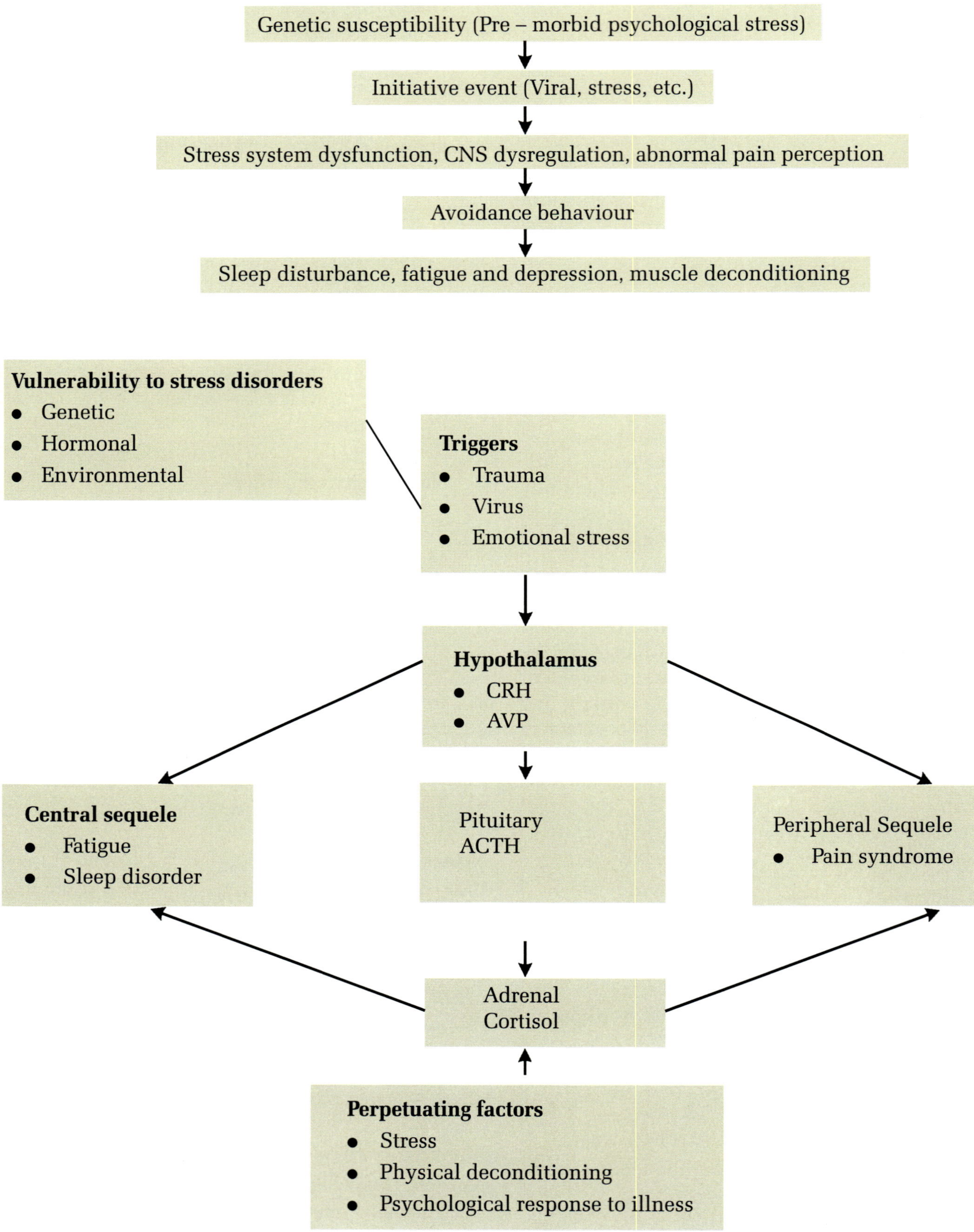

Fig. 1. Fibromyalgia

patients however do not typically have global improvement in response to therapies such as benzodiazepines. Dopamine agonists-rapinirole and pramipraxole used in treatment of restless leg syndrome may improve non-restorative sleep; but SSRIs and TCAs worsen restless leg syndrome.

Negative emotions are managed by forgiveness, education, anger management and grief counselling cognitive behavioural therapy and exercise training for high risk patients with FM. 60-70% of women receiving therapy showed improvement in pain, fatigue and functional disability.

Women with FM and low insulin like growth factor (IGF) levels experience an improvement in overall symptomatology and number of tender points after nine months of daily growth hormone therapy suggesting a secondary growth hormone deficiency (Table 8)[13].

Prognosis

A comprehensive team approach to management of patients with FM is necessary to get a good result.

Conclusion

Though FM is not common in India as in developed countries, it is more prevalent in the high society and economically advanced sections of society than the poorer sections. It is necessary to create an awareness of this disease among general practitioners, physicians, Orthopaedic surgeons and psychiatrists.

[Conclusions can be made about the effects of negative emotions on pain but not vice versa .women with FM showed increased in general level of clinical pain with emotions of anger and sadness whereas women without FM showed reduced threshold for feeling of pain as well as level of pain tolerance with anger or sadness].

Myofascial pain syndrome (MPS)

This condition is characterized by the presence of trigger points and a reference zone of perceived pain which is aggravated during palpation and is characteristically ill-defined and dull. Trigger points are localized areas of deep muscular tenderness located in a taut band in the muscle. The patient therefore presents with pain in a particular region of the body. These trigger points have been described in almost all patients. They can be demonstrated at any age but are common from thirty to fifty years. MPS is a chronic muscle pain disorder with focal areas of tenderness (Trigger points), it is a common cause of chronic orofacial pain seen in patients with temporomandibular (TMJ) disorders[14]. The etiopathogenesis of this condition is unknown. The treatment of this condition consists of identification of the precipitating factor and correcting it. Stretching the specific muscle by appropriate exercise can help. In addition, local injection with lignocaine 0.5 to 1% into the trigger point can provide relief. This condition has much better prognosis than fibromyalgia and most of the patients improve except for a few who develop chronic pain.

Summary

Soft tissue rheumatism refers to pain caused by inflammatory lesions of burse, tendon sheaths, musculotendinous junctions and joint capsules.

Generalized soft tissue lesions present with aching pain in joints and muscles as in fibromyalgia, the hypermobility syndrome, and repetitive strain syndromes related to occupation.

A careful history, and examination for areas of local tenderness, often suggests the diagnosis. Investigations may be required only to exclude other pathology.

Table 8. Treatment modalities for FM.

Intervention	Pain	Pain Threshold	Fatigue	Sleep	Mood	Well being
Amitriptyline/Cyclovenzaprine/Fluoxetine	↓		↓	↑		
Fluoxetin +Amitriptyline	↓			↑		↑
Ketanserin	↓	↑	↓	↑	↑	
Exercise	↓	↑	↑		↑	↑
Acupuncture	↓	↑		↑		↑
Biofeedback	↓	↑		↑	↑	↑
Hypnotherapy	↓		↓	↑	↑	↑
Cognitive behavioral therapy	↓			↑	↑	

Localized soft tissue lesions may cause pain around larger joints like the shoulder, the elbow, the hip and the knee, but may also involve the wrist, the hand, and heel.

Pain in soft tissue rheumatism may also arise from entrapment neuropathies as in the carpal tunnel syndrome, meralgia paraesthetica, and the tarsal tunnel syndrome.

Patients with soft tissue rheumatism may get significant relief from a balanced rest and physiotherapy programme, and from NSAIDs. Occasionally, local steroid injections may be required.

Familial aggregation of FM has been reported implicating genetic susceptibility.

Though FM is not common in India as in developed countries, it is more prevalent in the high society and economically advanced sections of society than the poorer sections.

In the recent contest Juvenile STR is on raised. The drug pregabalin has been tried in this condition.

Some of the local soft tissue rheumatism like mouse mobile syndrome in IT professionals, ghost syndrome in cricketers, Nomophobia (the fear of being without mobile phones).

Drugs like nefazodone and venlafaxine are also being tried.

Further Reading :

1. Pispati PK, Soft tissue rheumatism in: API Text Book of Medicine. EDS. Sainani GS et al 6th edition, J Assoc Phy Ind 1999: 16: 1075-76.
2. Dixon A. Soft tissue rheumatism. Clin Rheum Dis, Saunders, Philadelphia.
3. Cea- pereiro JC, garcia-Meijide J, Mera varela A, et al. A comparison between septic bursitis caused by staphylococcus aureus and those caused by other organisms. Clin Rheum 2001; 20: 10-14.
4. Rasnoissen OS. Sonography of tendons. Scand J Med Sci Sports 2001; 10: 360.
5. Neviaser RJ, Neviaser TJ. The frozen shoulder: diagnosis and management Clin Ortho 1987; 233: 59-61.
6. Child AH. Joint hyper mobility syndrome: inherited disorders of Collagen synthesis. Rheum 1986; 13: 239-42.
7. Beighton POH, Graham R, Bird HA. Hyper mobility of joints, 3rd ed. London, Berlin, New york: springer-verlag 1999.
8. Buskela D, New M L. Haza N I, et al. Familial aggregation in fibromyalgia syndrome. Arth Rheum 1996; 26: 605-11.
9. Jacobsen S, Hoyer-Madsen M, Danneskiold B, et al. Screening of autoantibodies in patients with fibromyalgia syndrome and matched control group APMIS 1990; 97: 665-68.
10. Wolfe FF, Smythe HA, Yunus MD, et al. The American College of Rheumatology 1990. Criteria for the classification of fibromyalgia. Report of the multicentre criteria committee. Arth Rheum 1990; 33: 160-72.
11. Goldenberg DL, Mayaski M. Randomised double blind cross over trail of fluoxetine and amitriptyline in the treatment of fibromyalgia. Arth Rheum 1996; 39: 1852-59.
12. Duight MM. An open clinical trial of venlafaxine treatment of FM. Psychosomatics 1988; 39: 14-17.
13. Bennett RM, Randomized double blind placebo controlled studies of growth hormone in the treatment of FM. AmJ Med 1998; 104: 227-33.
14. Saxen MA. Myofascial pain syndrome: characteristics, diagnosis and treatment J Indian Dent Assoc 1998; 77: 9-12.

5 HYPERMOBILITY JOINT SYNDROME

Bidyut Kumar Das

Introduction

Hypermobililty of the joints has been recognized since the time of Hippocrates and is commonly appreciated as a contributing factor to excellence, for certain professions like gymnastics, acrobatics, and ballet dancing. The abnormal range of movement across the joints in these individuals separate them into a different category. The joint mobility is determined by the surrounding ligamentous laxity. Majority of these persons are asymptomatic. When joint hypermobility is associated with musculoskeletal pain, without evidence of systemic rheumatologic disorder, it is described as joint hypermobility syndrome (HMS)[1] It is a common problem and believed to be the forme fruste of the severe forms of heritable disorders of the connective tissue (HDCT) like Marfan's syndrome, Ehlers-Danlos syndrome and osteogenesis imperfecta[2]. HMS is clinically similar in profile to the hypermoblity variety of Ehlers-Danlos syndrome (previously known as EDS-Type III) and many authorities believe that they are similar diseases[3]. HMS is more common and has a benign natural history, therefore, it is currently referred to as benign joint hypermobility syndrome (BJHS). The associated musculoskeletal manifestations are mild and varied, and primarily related to joints and surrounding soft tissue. Extra-articular manifestations involving other systems, especially neurophysiologic disturbances, have been reported adding another dimension to the evolving concept and understanding of the disease process[4]. Unfortunately, BJHS still remains a less understood entity[5].

Epidemiology

The prevalence of joint hypermobility is highly variable in the general population and depends on the geographical area surveyed and the criteria chosen for diagnosis. Age, sex and ethnic background are important factors contributing to the hypermobility status of the population analysed[6]. There are wide ethnic variations among North American and European Caucasian adults the prevalence is between 5-15%, in Middle East Asia and Western Africa the prevalence is as high as 25-35%[7]. Hypermobility is more common among Asians than black Africans who in turn are more mobile than Caucasians[6]. There are no population based studies in India. However, in a small survey of a group of students in Kerala, South India, hypermobility was diagnosed in 55% of the students examined[8].

Joint laxity is maximal at birth .Children have greater mobility than adults[9]. Many studies have confirmed that mean joint scores are higher among children and adolescents than adults but it diminishes steadily throughout childhood and more slowly during adult life[10]. Women demonstrate more mobility of joint than men at all ages. Pauciarticular hypermobility is more prevalent than the generalized variety[11]. Even a single hypermobile joint can present with any or all of the consequences of hypermobility[12]. Majority of subjects with hypermobility disorder are asymptomatic[13], and it is difficult to predict which individual will eventually suffer from BJHS. Although the syndrome is a common finding in clinical practice, its exact incidence and prevalence in different age, sex and ethnic groups is not known[7]. There is generally a lack of understanding of this disease entity even in developed countries[5] leading to mismanagement of these patients.

Etiopathogenesis

BJHS is believed to be a heritable disorder of the connective tissue. The genetic basis of hypermobility has been demonstrated by pedigree studies indicating its dominant mode of inheritance[6]. Hypermobility syndrome may have a generalised defect in connective tissue as evidenced by its multi-systemic affection. There is overlap of clinical features between BJHS and HDCT (heritable disorders of connective tissue) which indicates that it could be a member of the HDCT, family barring the severity of affection which is mild in BJHS[2]. Many authorities believe that BJHS and the hypermobile form of Ehlers-Danlos Syndrome (EDS) are similar disorders.

The precise genetic defect underlying BJHS is unknown. It is presumed that mutation in the collagen gene, as in EDS, affect the cross linking structure of collagen resulting in laxity of the tissue[14]. It is also suggested that hypermobility could involve multiple genes for the phenotype and degree of hypermobility[15]. Current studies have focused on Tenascin (TNX), a putative gene, whose mutation has been linked to hypermobility in experimental animal model and also to the hypermobile phenotype of EDS[16,17]. Family members, heterozygous for the gene, have clinical features similar to BJHS[18]. The protein expressed by the gene is very large and associated with collagen. It is ubiquitously expressed in tissues[19] including nerves and vasculature. It functions in cell matrix interactions but it is not precisely known how they contribute to the expression of hypermobility.

Clinical profile

The clinical features are variable and present a picture of acute and recurrent joint and soft tissue lesions. Problems often commence in childhood and continue into adult life. The role of hypermobility in the pathogenesis of most lesions associated with BJHS is not well established and there are no longitudinal studies establishing a causal association between hypermobility and musculoskeletal lesions[7].

Broadly, clinical features may be divided into a) articular/periarticular features b) extra articular features. Table I provides details of the various lesions associated with BJHS. Most symptoms are mild and self limiting but some of them are recurrent leading to chronicity.

Table 1. Clinical profile of hypermobility syndrome

JOINT
Arthralgia
Subluxation
Dislocation
Chronic low grade synovitis
Osteoarthritis
SOFT TISSUE
Enthesitis
Tenosynovitis
Bursitis
Rotator cuff syndrome
Back pain
SKELETON
Marfanoid habitus
Osteopenia
Stress fractures
SKIN
Hyperextensible
Striae/ scarring
NEUROPHYSIOLOGIC DEFECTS
Impairment of joint proprioception
Reduced analgesic effect of lidocaine
Dysautonomia
ABDOMINAL WALL AND VISCERA
Abdominal hernia
Uterine prolapse
Cystocoele
Rectocoele
VASCULATURE
Varicose veins
Bruisability

Several studies point towards an increased frequency of arthralgia in patients with hypermobility syndrome[7]. Lax joints are unstable and prone to sublux or dislocate and also vulnerable to trauma. Overuse of these joints may contribute to recurring injuries which may lead to low grade chronic synovitis and degenerative joint problem like osteoarthritis (OA). Many studies suggest an increased prevalence of OA among subjects with hypermobile joints, but the relationship has not been clearly established[7]. Soft tissues are also less resilient to the effects of overuse. There is increased frequency of soft tissue rheumatism (STR) like sprains, enthesitis, tenosynovitis, bursitis and rotator cuff tendinitis[20].

Many patients with BJHS complain of low back pain. It is mostly postural and higher in those professions involving prolonged static postures like sitting or standing[21]. Low backache may also be associated with spondylosis and spondylolisthesis observed in association with BJHS.

Extra-articular manifestations of the syndrome involve the cutaneous, genitourinary, peripheral and autonomic nervous system. Skin hyper-extensibility with resulting scars, urogenital prolapse, varicose veins and hernia may sometimes be seen in BJHS[6]. It is believed that musculotendinous weakness, as a result of defect in the connective tissue of the abdominal wall and pelvic floor, predisposes to increased risk of hernia[22] and genitourinary prolapse[23]. Similarly, weakness in the collagen tissue of the vasculature leads to increased risk of varicose veins.

Compounding the frequent painful episodes in patients with BJHS are neuro-physiological abnormalities like diminished joint proprioception[4,24] which may be an additional contributing factor to enhanced pain perception. Besides, patients complain of reduced anaesthetic effect of lidocaine[25] and spells of dysautonomia[26] like syncopal attacks palpitation and orthostatic hypotension. The mechanisms involved for these neurological abnormalities are not known.

Many patients of BJHS also suffer from chronic pain syndrome like fibromyalgia and bouts of depression resulting in diagnostic and management problems. It is therefore imperative that those diagnosed with fibromyalgic disorder[27] and/ or depression[28] should be meticulously evaluated for the presence of hypermobility syndrome.

Diagnosis

There is no biochemical or genetic marker at present to confirm the diagnosis. It is entirely clinical and based on certain validated criteria. There are two aspects to the diagnosis of BJHS: the presence of hypermobility of the joint, and specific signs and symptoms.

The first quantitative system for analyzing joint hypermobility was suggested in 1964 which was subsequently modified in 1973 and widely accepted as Beighton score[10] (Table 2) .It uses an arbitrary score of four or more as a cut off for hypermobility. It is quick and easy to perform. The scoring system rests on the ability of the individual to perform certain passive manoeuvres in nine determined sites resulting in a nine point scoring system. There is no single diagnostic test for hypermobility. Fig. 1, (A-E) demonstrates the manoeuvres to be performed at the designated joints. Beighton scoring system for hypermobility is a validated and reproducible clinical assessment tool[29]. But there are shortcomings in this system. It is not known whether the cut off levels (Beighton score of 4/9) are universally valid across all age, sex and ethnicity. In case of pauciarticular involvement the hypermobility of the affected joints could pass unnoticed if they are not evaluated.

Table 2. Modified Beighton score

Assessment site	Right	Left
Hyperextension of 5th MCP joint (Fig A)	1	1
Thumb touching the forearm (Fig B)	1	1
Hyperextension of elbow >10° (Fig C)	1	1
Hyperextension of knee joint >10° (Fig D)	1	1
Hand touching the floor with full palms with knees extended (Fig E)	1	
Maximum possible score	9	
Hypermobility present if total score ≥ 4		

The diagnosis of BJHS is based on clinical assessment of signs and symptoms included under the revised Brighton criteria 1998 comprising of major and minor points (Table 3)[30]. The inclusion of the minor criteria is being questioned in the absence of validated causal association with hypermobility. The reproducibility of this diagnostic criteria is at present unknown[29]. However, in the hands of experienced examiners, both Beighton test and Brighton criteria have been found to be reproducible[29].

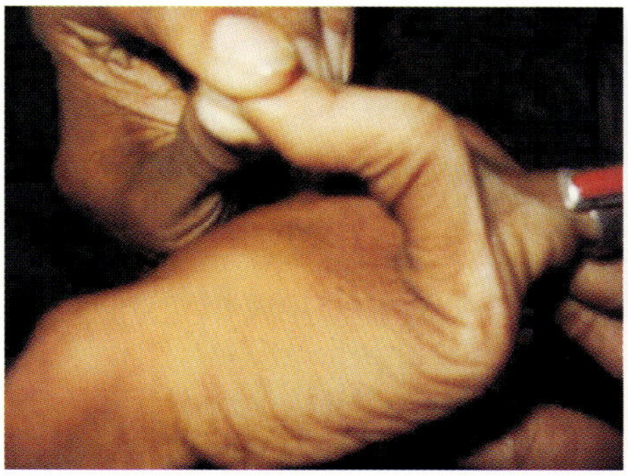

Fig 1A. Passive dorsiflexion of the little finger beyond 90⁰

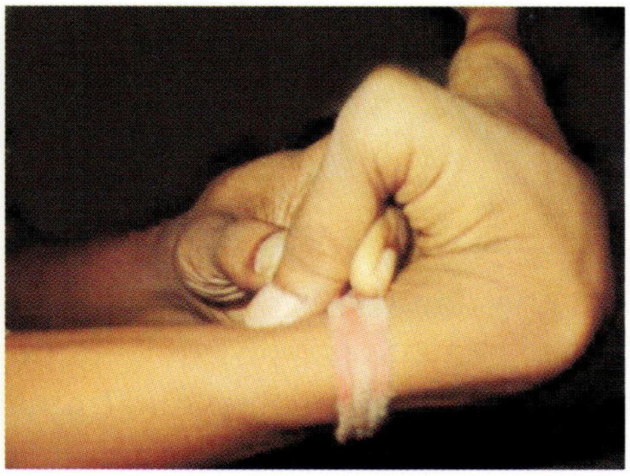

Fig 1B. Passive apposition of the thumb to the flexor aspect of the forearm.

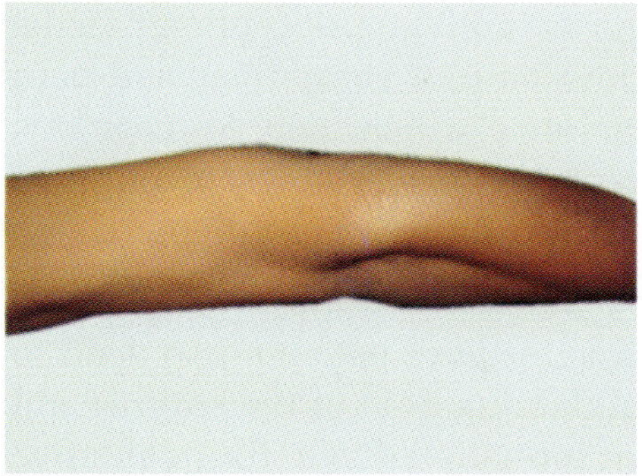

Fig. 1C. Hyperextension of the elbow beyond 10⁰

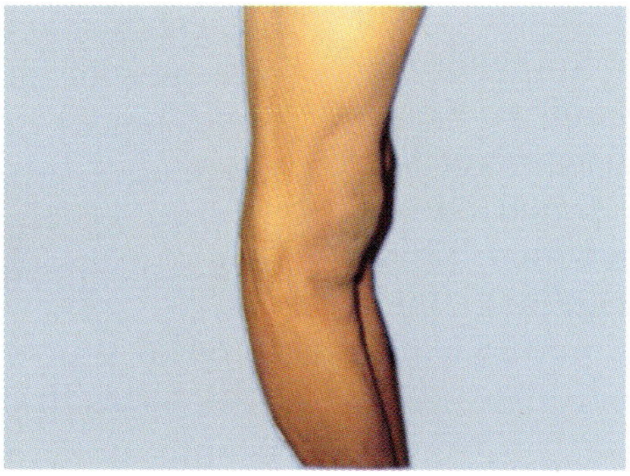

Fig. 1D. Hyperextension of the knee beyond 10⁰

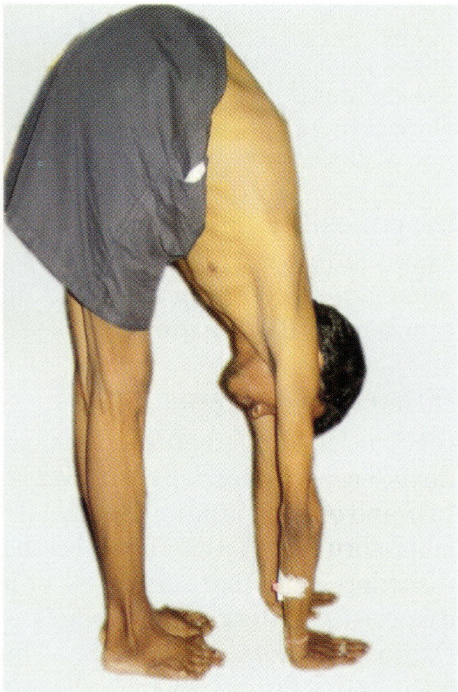

Fig E : Forward flexion of the trunk with the knees straight, with palms resting on the floor.

Table 3. Revised diagnostic criteria (Brighton 1998)

Major criteria

- Beighton score more than or equal to 4
- Arthralgia > 3 months in > 4 joints

Minor criteria

- Beighton score of 1 to 3
- Arthralgia > 3 months in 1 to 3 joints or back pain > 3 months,
- Spondylosis, spondylolysis or spondylolisthesis
- Soft tissue rheumatism > 3 lesions (epicondylitis, tenosynovitis,bursitsis)
- Marfanoid habitus (span: height ratio > 1.03 or upper segment lower segment ratio < 0.89)
- Abnormal skin : striae, hyperextensibility, papyraceous scars,thin skin
- Eye signs : drooping eyelids, myopia or antimongoloid slant
- Varicose veins, hernia, uterine / rectal prolapse

BHS if 2 major OR 1 major + 1 minor OR 4 minor
Criteria 1 major and 1 minor as well as 2 major and 2 minor mutually exclusive
2 minor sufficient if first degree relative present.
Exclusions : Ehlers Danlos (other than type III or hypermobility type) and Marfan's syndrome.

Table 4. Cardinal features of HDCT

Site	Marfan's syndrome	Ehlers-Danlos syndrome	Osteogenesis imperfecta
Joints	Hypermobility	Hypermobility	Hypermobility
Skeleton	Marfanoid habitus	Osteoporosis	Osteoporosis
Skin	Hyperextensibility	Hyperextensibility	Hyperextensibility
Eyes	Ectopia Lentis		Blue sclera
Vasculature	Aortic Dilatation	Intracranial aneurysm	–
Mitral valve	Prolapse	Prolapse	Prolapse

hypermobility be included as part of the clinical examination of patients with episodic and recurring musculoskeletal pain to avoid missing significant numbers of BJHS patients.

Differential diagnosis

BJHS has to be distinguished from more severe genetic disorders of connective tissue like Marfan's syndrome, EDS and osteogenesis imperfecta[2]. There are no reliable laboratory tests that can establish the diagnosis of these disorders. But there are well enumerated clinical diagnostic criteria for Marfan's syndrome, Ehlers Danlos syndrome and BJHS. The cardinal clinical features that distinguish these disorders from each other has been outlined in Table 4.

Certain acquired diseases may present with polyarticular laxity of joints, namely, acromegaly, hyperparathyroidism, rheumatic fever (Jaccoud's arthropathy) and chronic alcoholism[6]. They can be clearly identified from history and clinical examination.

Management

BJHS is a difficult condition to treat. There are no randomized controlled trials comparing the efficacy of existing modalities of treatment[7]. Most recommendations are based on theoretical assumptions.

Management begins with a correct diagnosis based on detailed clinical evaluation and exclusion of more serious connective tissue disorders. Assurance to the patient regarding the benign nature of the problem and long term counselling helps in avoiding unnecessary investigations and inappropriate medications. In view of the fact that there is an inherent laxity of ligaments which makes them prone to acute injury, patients are advised to avoid activities that may precipitate pain.

Physiotherapy forms the mainstay of treatment[6], Exercise programme tailored to the individual needs and directed at the affected tissue helps in symptomatic improvement. But it needs to be continuous and under supervision of a trained physiotherapist to avoid aggravation of injury. Sustained physiotherapy stabilizes lax joints and consequently reduces pain[31]. Isometric exercises improves muscle tone and coordination, while a mode of therapy called closed kinetic chain exercise has been shown to improve joint proprioception[32]. In the presence of chronic pain, a comprehensive pain management programme which includes cognitive behavioural therapy has been shown to improve quality of life. These pains respond poorly to analgesics and non steroidal anti inflammatory drugs.

Surgical procedures in these patients may lead to difficulties in maintaining hold of the sutures and in subsequent scar formation due to the qualitative defect in the skin[6].

Further Reading :

1. Kirk JA, Ansell BM, Bywaters EG. The hypermobility syndrome: musculoskeletal complaints associated with generalized joint hypermobility. Ann Rheum Dis 1967; 26: 419-25.
2. Grahame R. Joint hypermobility and genetic collagen disorders: are they related? Arch Dis Child 1999; 80: 188-91.
3. Beighton P, Paepe AD, Steinman B, et al. Ehlers-Danlos Syndromes; Revised Nosology,Villefranche,1997. AmJ of Med Gen 1998; 77: 31-37.
4. Mallik AK, Ferrel WR, McDonald AG, et al.Impaired proprioceptive acuity at the proximal interphalangeal joint in patients with the hypermobility syndrome. Br J Rheum 1994; 33: 631-37.
5. Grahame R and Bird H. British consultant rheumatologist's perceptions about the hypermobility syndrome: a national survey. Rheum 2001; 40: 559-62.
6. Grahame R. Hypermobility Syndrome. EDS Hochberg MC, Silman AJ, Smolen JS, Weinblatt ME and Weisman MH. Rheumatology. 3rd edition, Mosby, Philadelphia, USA 2003; 2171-75.
7. Remvig L, Jensen DV and Ward RC. Epidemiology of general joint hypermobility and basis for the proposed criteria for benign joint hypermobility syndrome: review of the literature. J of Rheum 2007; 34: 804-09.
8. Lawrence A. Benign Hypermobility Syndrome. J Ind Rheum Assoc 2005; 13: 150-55.
9. Biro F, Gewanter HL, Baum J. The hypermobility syndrome. Paediatr 1983; 72: 701-06.
10. Beighton P, Solomon L, Soskolne CL. Articular mobility in an African population. Ann Rheum Dis 1973; 32: 413-18.
11. Larsson L-G, Baum J, Mudholkar GS. Hypermobility: Features and differential incidence between the sexes. Arth Rheum 1987; 30: 1426-30.
12. Larsson L-G, Baum J, Mudholkar GS, et al. Hypermobility: prevalence and features in a Swedish population. Br J Rheum 1993; 32: 116-19.
13. Grahame R. 'The Hypermobility syndrome'. Ann Rheum Dis 1990; 49: 199-200.
14. Jau-Ren Mao and Bristo J. The Ehlers Danlos syndrome: On beyond collagen. J Clin Invest 2001; 107: 1063-68.
15. Murray KJ and Woo P. Benign joint hypermobility in childhood. Rheum(Oxford) 2001; 40: 489-91.
16. Zweers MC, Kucharekova M and Schalkwijk J. Tenascin-X: a candidate gene for benign joint hypermobility syndrome and hypermobility type Ehlers Danlos syndrome? Ann Rheum Dis 2005; 64: 504-05.
17. Schalkwijk J, Zweers MC, Steijlen PM, et al. A recessive form of the Ehlers- Danlos syndrome caused by Tenascin X deficiency. N Engl J Med 2001; 345: 1167-75.
18. Zweers MC, Bristow J, Steijlen PM, et al. Haploinsufficiency of TNXB is associated with hypermobility type of Ehlers-Danlos syndrome. Am J Hum Genet 2003; 73: 214-17.
19. Matsumoto K, Saga Y, Ikemura T, et al. The distribution of Tenascin X is distinct and often reciprocal to that of Tenascin C. J Cell Bio 1994; 125: 483-93.
20. Hudson N, Fitzcharles MA, Cohen M, et al. The association of soft tissue rheumatism and hypermobility. Br J Rheum 1998; 37: 382-86.
21. Larsson L-G, Baum J, Mudholkar GS, et al.Benefits and liabilities of hypermobility in the back pain disorders of industrial workers. J Intern Med 1995; 238: 461-67.
22. Skoumal M, Haberhauer G, Mayr H. Concomitant diseases in primary joint hypermobility syndrome. Med Klin (Munich) 2004; 99: 585-90.
23. Norton PA, Baker JE, Sharp HC, et al. Genito urinary prolapse and joint hyper mobility in women. Obstet Gynaecol 1995; 85: 225-28.
24. Hall MG, Ferrell WR, Sturrock RD, et al. The effect of the hypermobility syndrome on knee joint proprioception. Br J Rheum 1995; 34: 121-25.
25. Arendt-Nielsen L, Kaalund P, Bjerring P, et al. Insufficient effect of local analgesics in Ehlers-Danlos Type III patients (connective tissue disorder). Acta Anaesthesiol Scand 1990; 34: 358-61.
26. Gazit Y, Nahir AM, Grahame R, et al. Dysautonomia in the joint hypermobility syndrome. Am J Med 2003; 115: 33-40.
27. Acasuso-Diaz M, Collantes-Estevez E. Joint hypermobility in patients with fibromyalgia syndrome. Arth Care Res 1998; 11: 39-42.
28. Grahame R. Pain, distress and joint hyperlaxity. Joint Bone Spine 2000; 67: 157-63.
29. Remvig L, Jensen DV and Ward RC. Are diagnostic criteria for general joint hypermobility and benign hypermobility syndrome based on reproducible and valid tests? A review of the literature. The J of Rheumatol 2007; 34: 798-803.
30. Grahame R, Bird HA, Child A. The revised (Brighton 1998) criteria for the diagnosis of benign joint hypermobility syndrome (BJHS). J Rheum 2000; 27: 1777-79.
31. Barton LM, Bird HA. Improving pain by the stabilization of hyperlax joints. J Orthop Rheum 1996; 9: 46-51.
32. Ferrel WR, Tennant N, Sturrock RD, et al. Amelioration of symptoms by enhancement of proprioception in patients with joint hypermobility syndrome. Arth Rheum 2004; 50: 3323-28.

METABOLIC BONE DISEASE 6

C M Adya, Rajkiran Dudam, C Shivashankar

Introduction

The skeleton is one of the largest organ systems in the body. It consists of a large mineralized matrix and a small but highly active cellular fraction. The skeleton serves the dual functions of maintaining the structure of the body and providing a storehouse of minerals and protein. Imbalance between the structural and the storage functions of the skeleton can be important in the pathogenesis of metabolic bone disease.

Mineralization

Bone mineral is made up of small, imperfect hydroxyapatite crystals, which contain carbonate, magnesium, sodium and potassium. Mineralization requires calcium, phosphate and alkaline phosphatase. This process is impaired in vitamin D deficiency and hypophosphatasia.

Osteoblasts *are responsible for bone formation while osteoclasts lead to bone resorption.* Adult bones also contain less pluripotent bone cell precursors or committed osteoprogenitors (or preosteoblastes) e.g. osteoblasts in early development and during repair produce woven bone whereas more mature osteoblasts produce lamellar bone.

An orderly sequence of bone resorption and formation is termed *remodeling*. In young adults, this cycle is tightly coupled and the amount of new bone formed by osteoblasts is equal to the amount that is resorbed by osteoclasts. Postmenopausal and age related bone loss begins when resorption increases and formation no longer keeps pace. The activation of new remodeling sites may also increase with age. Serum alkaline levels reflect osteoblastic activity.

Remodeling can be activated by both systemic and local factors and probably serves several physiologic functions. Changes in mechanical force can activate remodeling to improve the skeletal strength. Remodeling may serve to repair microdamage, particularly in cortical bone, and this may explain the fact that remodeling is sustained in the aging skeleton. However, loss of osteocytes with age may impair this response. Systemic hormones influence bone remodeling.

Calcium regulating hormones

- Parathyroid Hormone (PTH)

PTH acts on bone to stimulate resorption but does not act on osteoclasts in the absence of cells of the osteoblastic lineage; moreover, PTH receptors are abundant on osteoblasts but not on osteoclasts.

- Vitamin D

The hormonal form of vitamin D-1, 25 (OH) is necessary for intestinal calcium and phosphorus absorption and mineralization.

- Calcitonin

Calcitonin inhibits bone resorption by acting directly on the osteoclast but appears to play a small role in the regulation of bone turnover in adults.

Growth hormone

Deficiency and excess of growth hormone have marked effects on skeletal growth. Growth hormone increases both circulating and local levels of IGF-I, which mediates the skeletal effects of growth hormone

Glucocorticoids

Glucocorticoids exert biphasic effects on bone formation and resorption. In vivo, glucocorticoids may increase bone resorption indirectly by diminishing calcium absorption and producing secondary hyperparathyroidism. Low levels of glucocorticoids increase osteoclastic activity in organ culture, whereas high levels inhibit it, perhaps by decreasing the production of cytokines and prostaglandins. However, glucocorticoids act in the long term in vitro and in vivo to inhibit bone formation, to decrease osteoblast replication and differentiation, to decrease IGF-I synthesis, and to increase osteoblast apoptosis.

Thyroid hormones

In children, hyperthyroidism is associated with increased skeletal growth and hypothyroidism results in decreased growth. Thyroid hormones are crucial for cartilage growth and differentiation and enhance the response to growth hormone.

Insulin

Normal skeletal growth depends on an adequate amount of insulin. Excess insulin production by the foetuses of mothers with uncontrolled diabetes results in excessive growth of the skeleton and other tissues, and undertreated diabetes mellitus impairs skeletal growth and mineralization.

Gonadal hormones

Both estrogens and androgens are critical for skeletal development and maintenance. Bone cells contain estrogen and androgen receptors, but it is difficult to demonstrate direct effects of gonadal steroids on bone formation or resorption in cell and organ culture. Gonadal hormones are crucial for the pubertal growth spurt, and estrogen is necessary for epiphyseal closure. Deficiency of estrogen or androgen increases bone resorption in vivo, possibly by increasing local synthesis or sensitivity to cytokines, such as IL-1, TNF, or IL-6, and to prostaglandins. Androgens can increase bone formation in vivo. The effect of estrogens on bone formation is less clear.

Metabolic bone diseases

Metabolic bone diseases may be classified as under

- Parathyroid related
- Renal Osteodystrophy
- Osteoporosis
- Bone disease due to nutritional, gastrointestinal and hepatic disorders
- Hereditary metabolic and dysplastic skeletal disorders
- Paget's disease
- Vitamin D deficiency,Osteomalacia and Rickets

Parathyroid related

Primary hyperparathyroidism is a disorder characterized by excessive production of parathyroid hormone and hypercalcemia. The hypercalcemia is caused by the dual actions of parathyroid hormone, which stimulate calcium release from bone during the course of osteoclast-mediated bone resorption and conserves calcium filtered through the renal glomerulus.

The original studies of Albright and Reifenstein' characterized primary hyperparathyroidism as disease of "bones and stones". In some patients, primary hyperparathyroidism was complicated by nepрholithiasis, nephrocalcinosis and renal insufficiency. In other patients, a distinctive bone disease predominated. Bone pain, fracture and skeletal deformity were not uncommon. Subperiosteal resorption of the digits, skull and long bones, diffuse osteopenia and brown tumors described the bone disease of primary hyperparathyroidism as "osteitis fibrosa cystica". The most specific radiologic sign of the disease is subperiosteal resorption occurring in typical locations such as the radial aspect of the second and third phalanges, the distal clavicles and the classic "salt and pepper" appearance of the skull seen in severe disease. This sign is still considered to be pathognomonic for primary hyperparathyroidism. Another classic radiologic aspect of severe parathyroid bone disease is cyst formation, completing the terminology "osteitis fibrosa cystica". These cystic defects may become filled with fibrous tissue. If the highly vascularized fibrous tissue contains zones of hemorrhage, blood pigment will accumulate. These changes impart a reddish-brown hue to the tissue, accounting for the term "brown tumor". Since this lesion is reactive rather than neoplastic, the brown tumor is not a tumor at all.

The loss of the lamina dura around the roots of the teeth, an inconstant finding in primary hyperparathyroidism, is yet another radiologic manifestation of marked cortical resorption in this disease. In primary hyperparathyroidism, the dynamic celluar process of bone remodeling is altered by a high turnover state. Primary hyperparathyroidism is believed to have its main deleterious effect on cortical rather than cancellous bone. Increased bone turnover is the hallmark of skeletal involvement in primary primary hyperparathyroidism. The increased bone turnover in primary hyperparathyroidism results in a catabolic effect on the skeleton. It has become clear that excess parathyroid hormone has a highly selective action on different components of the skeleton with deficit of cortical bone and increase in the amount of cancellous bone.

Renal osteodystrophy

"Renal osteodystrophy" a term introduced over 45 years ago, is still used to describe any bone disease occurring in a patient with renal disease. In actuality a very wide spectrum of bone disease can occur to renal failure patients, ranging from states of markedly impaired bone formation and mineralization (such as low-turnover bone disease, osteomalacia) to states of markedly increased bone turnover (hyperparathyroidism) and aluminum toxicity. Unique types of metabolic bone disease may also occur in patients with renal disease but with a normal glomerular filtration rate (GFR), as exemplified by the post renal transplant state, renal tubular acidosis and the nephritic syndrome. Patients with renal disease are also subject to the same factors that can affect bone metabolism in normal individuals: menopausal status, immobilization, thyroid abnormalities, ethanol intake and so forth. The situation becomes even more complex when one realizes that the blood tests and radiographic studies usually used to evaluate metabolic bone disease are not always accurate reflections of bone disease in renal patients. Therefore, it is not surprising that renal osteodystrophy is one of the leading indications for bone biopsy, which is the most accurate way to diagnose bone disease in the renal patient. A transiliac bone biopsy performed after double tetracycline labeling with bone histomorphometric examination and histochemical aluminum staining must be regarded as the "gold standard" for assessing uremic bone disease. In renal osteodystrophy, serum calcium is low to low normal, raised inorganic phosphate, alkaline phosphate and urea/creatinine and normal levels of 25OHD, 1,25 (OH) 2D and PTH.

Roentgenographic studies are frequently performed in renal failure patients and a plethora of roentgenographic finding is noted. Abnormal bony texture, subperiosteal resorption and patchy osteosclerosis of the spine giving it a "rugger jersey" appearance are common features. Avascular necrosis is also seen, but many of these patients have received high doses of corticosteroids with renal transplants. Dietary phosphate restriction, calcium carbonate administration, reduction and preferably elimination of aluminum exposure coupled with judicious use of 1, 25 (OH) 2D can prevent or slow the development of osteodystrophy. Renal replacement therapy also corrects the metabolic abnormalities.

Osteoporosis

Osteoporosis is by far the most common metabolic bone disease and is becoming an increasing public health problem in most countries including India. It is discussed at length in Section 11.

Vitamin D deficiency - Osteomalacia and rickets

Vitamin D deficiaency leads to mineralization defects, in growing skeleton it causes Rickets and in mature skeleton Osteomalacia. Vitamin D is crucial in maintaining serum calcium and phosphorus concerntrations. Vitamin D is obtained on sunlight exposure, 7 dehydrocholesterol is converted to previtamin D3 which is unstable and quickly converted to Vitamin D3 which is ejected out of epidermal cell into extracellular space and later enters circulation. Vitamin D is converted to 25OHD in liver and followed by 1,25(OH)2D in kidneys, the active form which enters the circulation and travels to target tissues enhancing the calcium and phosphorus absorption.

Vitamin D deficiency causes decrease in intestinal calcium absorption that leads to transient decrease in the ionized calcium. This is recognized by the calcium sensor in the parathyroid glands, resulting in increased expression and production of parathyroid hormone (PTH). PTH conserves calcium by increasing tubular reabsorption of calcium throughout the proximal tubule and distal convoluted tubule. PTH stimulates the kidneys to produce 1,25(OH)2D, which in turn increases intestinal calcium absorption. When these actions are inadequate to maintain the blood ionized calcium levels, the PTH and 1,25(OH)2D will interact with their receptors on osteoblasts to increase the expression of RANKL, which in turn mobilizes osteoclast precursors to become mature osteoclasts to remove calcium and phosphorus from the skeleton. Thus unless there are no longer any calcium deposits in the skeleton that can be mobilized, the blood level of calcium is usually normal or low normal in vitamin D deficiency. Only when the skeleton is completely depleted of calcium hypocalcemia be observed. PTH also leads to increased excretion of phosphate thus leading to imbalance in calcium phosphorus product and affecting the mineralization of osteoid. Thus patients with vitamin d deficiency have low levels of serum vitamin D, high PTH, normal or elevated serum alkaline phosphatase,normal or low serum calcium and phosphorus levels.

In countries like India, Bangladesh, Africa where children are adequately exposed to sunlight but have inadequate diet calcium, where daily intake is less than 200 iu or diet rich in phytates and oxalates which interfere with calcium absorption can also lead to Rickets with normal vitamin D levels and low calcium levels.

X-linked hypophosphatemic rickets and autosomal dominant phosphatemic rickets are caused by either an increase in the production of fibroblast growth factor 23(FGF23) and other phosphatonins or a decrease in their catabolism resulting in severe phosphate wasting into the urine.

Oncogenic osteomalacia is due to a benign or malignant tumor production of FGF23/ phosphatonins. The result is increased phosphate wasting into the urine causing severe hypophosphatemia.

Fanconi syndrome is a disorder of the renal proximal tubules resulting in decreased reabsorption of phosphorus, glucose and amino acids. These findings are accompanied by metabolic acidosis secondary to proximal tubular bicarbonate wasting (type II renal tubular acidosis). These patients have hypophosphatemia, hyperphosphaturia, glycosuria and generalized amino aciduria, hypobicarbonatemia and excessive bicarbonate excretion in the urine. The serum calcium is usually normal with an elevated alkaline phosphatase and a normal PTH and 25(OH)D but an inappropriately low or low-normal serum 1,25(OH)2D (Children with Fanconi syndrome have classic x-ray evidence of rickets, whereas adults have radiographic changes consistent with osteomalacia). Patients who have hypophosphatemia in a fasting blood specimen should be evaluated for Fanconi syndrome.

Fluoride has been effective in preventing dental caries in part because the fluoride is incorporated into the dentin, reducing dental caries activity. Fluoride stimulatesosteoblastic activity in low concentrations. And at highconcentrations, fluoride not only stimulates osteoblastic activity but gets incorporated into the skeleton as a calcium fluoride hydroxyapatite crystal. Due to high atomic weight than OH that it replaces, the skeleton appears on x-ray to be hyperdense and there is increase in BMD. It is well documented that ingestion of food and water with a high fluoride content between 3 and 16 ppm is associated with increased bone mineral density that leads to stiffness, rigidity and limitation of movement at the spine and can also lead to skeletal deformities and fractures. Patients who are exposed to high dietary fluoride and are on a low-calcium diet can develop severe skeletal deformities. In children they have classic features of rickets at their epiphyseal plates and in adults osteomalacia with Looser zones. Histologically, the bone biopsy shows widened osteoid seams, poorly mineralized new bone formation, and areas of hypermineralization.

Antiepileptic medication, HIV medication, bisphosphonates (etiondronate) and glucocorticoids can cause osteomalacia.

X rays reveals Looser zones which are radiolucent lines that are often penetrating through the cortex perpendicular to the shaft and are most often seen in the medial cortices of the femurs and in the pelvis and ribs. In children, rickets/osteomalacia there is widening of the physis with fraying, cupping and splaying of the metaphases. The diaphyses of the long bones often appear to be less dense (i.e., osteopenic and show thinning of the cortices with periosteal new bone formation).

Children with rickets often have skeletal abnormalities including a rachitic rosary, bowed or knocked knees, frontal bossing, soft fontanels, delayed tooth eruption, dental caries and severe muscle weakness.

In adults, osteomalacia patients complain aching bone pains or discomfort. They often note that the bone discomfort is worse when sitting or lying in bed. They also have mild proximal muscle weakness and myalgia. tenderness can be elicited with some force on the sternum, radius, ulna or anterior tibia often resulting in wincing bone discomfort. The closest mimic is fibromyalgia.

Vitamin D deficiency is defined as a serum 25(OH)D less than 20 ng/mL. However, to maximize intestinal calcium absorption and to minimize circulating PTH levels, it is desirable to maintain a 25(OH)D greater than 30 ng/mL. To achieve this, at least 1000 to 2000 IU of vitamin D daily or its equivalent is necessary. For every 100 IU of vitamin D2 or vitamin D3 ingested, the blood level of 25(OH)D increases by 1 ng/mL Most individuals have a circulating blood level of 25(OH)D between 15 and 25 ng/mL.

To treat Vitamin D deficiency 60,000 iu (available as granules, soft gel capsules, chewable tablets) weekly for 8 to 12 weeks will improve the serum levels of above 30 ng/ml. later it can be supplemented once in 2 weeks or monthly once or 2000 iu daily. Recommendation for calcium intake is 800 milligrams per day for children younger than the age of 12 years, 1300 milligrams per day for teenagers, 1000 milligrams for adults aged 19 to 50 years and 1200 milligrams per day for adults older than the age of 50.

Bone disease due to nutritional, gastrointestinal and hepatic disorders

Patients with gastro intestinal and hepatic diseases represent a large group at high prevalence of nutritional deficiencies. This is due, in part, to high prevalence of nutritional deficiencies in these patients, but also is a consequence of the underlying digestive disease and of medications used in its treatment. In recent years in patients of liver transplantation increased prevalence of metabolic bone disease is seen.

Hereditary metabolic and dysplastic skeletal disorders

There is considerable diversity of hereditary metabolic and dysplastic skeletal disorders. Some of these conditions are mere radiographic curiosities; others are significant clinical problems. Some feature distinctive derangements in mineral homeostasis; others are diagnosed by X-ray from characteristic disturbances of skeletal growth, modeling, and remodeling. Hypophosphatasia, bone sclerosing dysplasias (osteopterosis, progressive diaphyseal dysplasias, and pyknodystosis) and McCune - Albright syndrome are some of the diseases seen in this group.

Paget's disease

Paget's disease is uncommon in India. It may affect as many as 3% of adults older than 40 years of age; it is often asymptomatic and usually progresses slowly. The primary abnormality in Paget's disease is the localized, uncontrolled formation of large, highly active osteoclasts. The initial lesion is an increase in bone resorption. The response to this resorption, particularly in bones that are subject to mechanical force, is an intense but chaotic increase in osteoblastic activity. Pagetic bone often distributes in a heterogeneous pattern throughout the skeleton. The pelvis, femur, spine, skull and tibia are most commonly involved whereas hands and feet are rarely affected. The most common symptom is bone pain at the site of pagetic involvement. Bowing of the legs is common and pathologic fractures can occur. Vertebral involvement can cause kyphosis and compression of the spinal cord. Neural changes can also result from vascular steal because of the high blood flow to the lesion. The most common consequence of Paget's disease of the skull is hearing loss, which can be both conductive and neurosensory. Extensive involvement of the base of the skull can produce basilar impression and, rarely, brain stem compression. The hypervascular bone results in heart failure. Bisphosphonates are the cornerstone of successful long term therapy.

Conclusion

Metabolic bone diseases and its consequences cause considerable morbidity and mortality among both men and women. Due to better understanding of the pathophysiology of these diseases, newer and specific therapeutic options have become available. This has resulted in aggressive approach to these diseases rather than accepting these problems as inevitable consequences of aging.

Further Reading :

1. Bringhurst FR, Demay MB, Krave SM, at el, Bone and Mineral Metabolism in health and Disease. In: Kasper DL, Fauci AS, Braunwald E et al(editors). Harrison's Principles of Internal medicine, 16th edition. Mcgraw Hill Newyork. 2005; 2238-49.
2. Williams Text book of Endocrinology. Larsen, Kronenberg, Melmed, Polonsky.
3. Endocrinology and Metabolism Clin of N Am Vol 32, No 1 Mar 2003.
4. Growth hormone and IGF research. 15(2005) S 42- S47. Craig B Langman.
5. Holick MF. Osteomalacia and Rickets. Hochberg text book of Rheumatology, 5th Ed, 2011; 1997-2005.

CARDIOVASCULAR RISK IN RHEUMATIC DISEASES

7

Firdaus Fatima, U R K Rao, C Satyavati

Introduction

Patients with Rheumatic diseases (RD) are at an increased risk of death compared with age and gender matched controls without inflammatory joint disease[1,2,3]. Premature Cardiovascular Disease (CVD) is the leading cause of morbidity and mortality. In the past decade there is increased awareness of CV involvement in RD. Cardiovascular disease can be in the form of coronary heart disease (CHD), cerebrovascular disease and peripheral vascular disease. Cardiac involvement in RD has a varied presentation from being asymptomatic or mild to severe or life threatening. There can be pericardial, myocardial, valvular and conducting system abnormalities. The pathogenesis of CVD is multi-factorial and is thought to result due to an interaction between inflammation, metabolic factors, disease related factors and the treatments used[4]. This chapter will focus on cardiovascular risk in RD with a focus on traditional cardiovascular (CV) risk factors in rheumatoid arthritis (RA) and systemic lupus erythematosus (SLE), and other RD.

Rheumatic diseases are associated with premature atherosclerosis which in turn results in ischemic heart disease in young individuals. Inflammation is thought to have a major role in all stages of development of atherogenesis[5]. Chronic inflammation is evidenced by raised inflammatory markers namely ESR and C-creative protein (CRP) which are known to predict CVD. CRP is involved in all steps of atherogenesis including initiation, progression and destabilization of an atheromatous lesion. The factors causing atherosclerosis in autoimmune rheumatic diseases are age, gender, hypertension, diabetes mellitus (DM), metabolic syndrome, hyperlipidaemia, active disease, use of glucocorticosteroid (GCS) of more than 7.5 mg/day and high cumulative dose and biomarkers such as ESR, CRP, pro inflammatory high density lipoprotein (HDL), oxidized low density lipoprotein (LDL) and hyper homocystinemia. Systemic inflammation and traditional CV risk factors together play a role in the development of atherosclerosis.

The risk factors for the development of CVD are called traditional or classic CV risk factors. They can be further differentiated into modifiable and non modifiable traditional CV risk factors as mentioned in Table 1. Mortality from CVD, among the general population is predicted by traditional CV risk factors. Many studies have observed increased prevalence of these risk factors in RD[6,7,26,27] and others have found the occurrence similar to the general population[4,8]. A recent meta-analysis has confirmed the role of traditional CV risk factors in increased CVD observed in RA[9].

Table 1 : Traditional CV risk factors

Non modifiable	Modifiable
Age	Hypertension
Male gender	Diabetes mellitus
Family history of premature CVD	Increasing body mass index
	Dyslipidaemia
	Smoking

Traditional CVD risk factors remain the main focus of attention in the prevention and treatment of CVD. In general population these risk factors are strong predictors of CVD outcome including mortality and morbidity. All the CV risk prediction scores utilize traditional CVD risk factors for the estimation of 10 year probability of CHD.

Cardiovascular risk assessments

Traditional CV risk factors form the basis of CV risk estimation by various methods. These methods are used to predict the percentage risk of a CV event in the next ten years. Many countries have adopted different scoring systems. Worldwide, the most commonly used CV risk estimation method is the Framingham CV risk assessment[10]. Other CV risk assessment methods are the Joint British CV risk estimation scores, the systemic coronary risk evaluation model (SCORE), Reynolds risk score and the recent QRISK prediction scores. Apart from the QRISK prediction score all the other CV event prediction algorithms have been developed using data derived from the Framingham Heart Study.

For ethnic minorities especially individuals from Indian subcontinent it is estimated that the CV risk is higher than predicted and a multiplication factor of 1.4-1.5 may be used.

Metabolic syndrome

Metabolic syndrome (MS) is a cluster of cardiovascular risk factors. The clustering of metabolic abnormalities occur in the same individual and increase CV risk more than the sum of the individual components[11]. The MS is also called insulin resistance syndrome, syndrome X or Reavens syndrome. This syndrome was first described in 1988 by Reaven who found that multiple risk factors like dyslipidaemia, hypertension, and hyperglycaemia cluster together in CVD and called it syndrome X[12]. However, abdominal obesity was not included. There are different definitions of metabolic syndrome. The most commonly used definitions are provided by 1) world health organization (WHO), 2) the national cholesterol education program adult treatment panel III (NCEP ATPIII), and 3) the international diabetes federation (IDF). It was demonstrated that the odds ratio using different definitions of metabolic syndrome NCEP ATPIII (OR 2.00), WHO (OR 1.73) and IDF (OR 1.69) was almost similar in predicting incident CHD which was independent of ethnic origin, age, gender, family history of CVD, smoking, T2DM (type 2), CHD, and non high density lipoprotein (HDL) cholesterol[13]. The NCEP ATP III is simpler and the most widely used in clinical practice.

Three out of the following five criteria are to be met for an individual to be classified as metabolic syndrome NCEP ATPIII[14](Table 2).

Table 2 : Criteria for metabolic syndrome

Risk factor	Cut off point
Waist circumference	> 102cm (>40 inches) in males >88cm (>35inches) in females
Triglycerides	>150mg / dl
Low HDL	Male < 40mg / dl, Female < 50mg / dl
Blood pressure	> 130/85 mm of Hg or an antihypertensive
Fasting blood sugar	> 110 mg/dl or diabetes

The presence of metabolic syndrome increases an individual's risk of type 2 diabetes, and also the risk for CVD. Individuals with MS are at an increased risk for CHD[15].

Rheumatoid arthritis

Rheumatoid arthritis (RA) is associated with disability, deformity and reduced life expectancy. Increased mortality associated with RA has been recognized for fifty years[1]. Mortality in patients with RA is increased when compared to the general population[16,17]. The lifespan of RA patients is decreased approximately by 3-18 years[18]. One of the major causes of death in RA is due to CVD [3] which is responsible for about fifty percent of premature deaths[18,19]. This increased CV mortality is multifactorial and is attributed to traditional CV risk factors, novel risk factors, systemic inflammation due to RA and medicines used in treatment of RA[20]. Despite advances in treatment, mortality in RA has not changed over the past 50 years and no gender difference is observed[21].

Rheumatoid factor (RF) has an impact on mortality. Increased deaths were observed in RF positive patients compared to RF negative patients and the general population. The mortality trend in RF negative patients may be similar to the general population[22]. Anti cyclic citrullinated peptide (CCP) was also found to be strongly associated with mortality[23]. RA patients have different presentation of ischemic heart disease (IHD) when compared to general population. They may have unrecognized silent IHD and no typical chest pain before sudden death[24].

There is increased prevalence of CV risk factors in RA[26,27]. Cardiac involvement in RA can be in the form of pericarditis, endocardial rheumatoid nodules and non specific mitral and aortic valve involvement. These are mostly asymptomatic and can be a finding on echo. However, these abnormalities are not responsible for increase in mortality. The increased mortality is due to premature atherosclerosis causing IHD and heart failure. RA patients have more severe CHD and are more likely to have multiple vessel involvement compared to general population [16].

Non modifiable CV risk factors

Increasing age, male gender and family history of CVD are non modifiable traditional CV risk factors. The risk of CHD increases with increasing age[14]. Male gender is an independent risk factor for CVD. Family history of premature CHD is defined as CHD in first degree relative male less than 55 years and female less than 65 years[14]. Family history of CVD was more frequently reported in RA patients when compared to controls from the general population.

Modifiable traditional CV risk factors

Hypertension (HTN) is one of the modifiable traditional risk factor and a major contributor in the development of CVD. A systolic reading above 140 mm of Hg and a diastolic reading above 90 mm of Hg is considered as hypertension in accordance with guidelines from the British Hypertension society and National cholesterol education programme - Adult treatment panel (NCEP ATP)[14,28]. However, it has been documented that as little as a 10 mm of Hg rise in diastolic BP doubles the risk of development of heart attack, stroke, renal failure and heart failure. In individuals aged less than 50 years, diastolic BP is the most important predictor of an adverse outcome[29]. According to the Framingham Heart study the CV risk was weakly associated with diastolic BP but a much stronger CV event association was observed with systolic BP above 120 mm of Hg[30]. It is also shown that obesity and advancing age are strongly associated with HTN.

Hypertension was reported in 33% of participants in the QUEST-RA (Questionnaires in standard monitoring of patients with RA) study which was conducted in RA patients from 15 countries across Europe and America[25]. Many studies reflect that hypertension may be present in patients with RA as a co morbid condition or evolve in the course of the disease[31,32], for which the use of non steroidal anti inflammatory drugs (NSAIDs) and glucocorticosteroids may be responsible. There is a possibility of hypertension being unrecognized or undertreated in RA.

Diabetes is another modifiable risk factor for CVD. Diabetes is also an important constituent of the metabolic syndrome. Not many studies have explored the prevalence of diabetes in RA. Most of the published literature reports diabetes as part of the CV risk assessments. The QUEST RA study reported an 8% prevalence of diabetes in Caucasian RA patients[25]. However, prevalence of diabetes was reported to be increased in RA when compared to general population[9]. Chronic use of glucocorticoids may interfere with glucose metabolism and insulin sensitivity as well as being associated with adverse body fat accumulation[33]. When compared to low grade inflammation, high grade inflammation in RA was found to increase insulin resistance and reduce beta cell function resulting in impaired fasting blood glucose and diabetes[34].

Smoking is an important constituent of CV risk assessment and is a modifiable CV risk factor. Increased mortality is associated with smoking in RA and the general population[35]. There is an increased prevalence of smoking in RA[9].

Dyslipidaemia is an important risk factor for CVD in the general population[14]. Active RA is associated with dyslipidaemia. The characteristic change being low total cholesterol, LDL cholesterol and disproportionate lowering of HDL cholesterol which results in an unfavourable atherogenic index (ratio of total cholesterol to HDL cholesterol). Lipids have an inverse relationship with disease activity[36].
The increased prevalence of CVD in RA has led to an association with metabolic syndrome. Inflammation is an important feature in obesity, type 2 DM and in patients with chronic inflammatory diseases like RA. High prevalence of MS (44%) was observed in middle aged RA case control study and there was a correlation observed between increasing level of RA disease activity and MS[39].

Studies have correlated insulin resistance and atherosclerosis in RA patients. Hypertension, insulin resistance and triglycerides were associated with carotid intima media thickness (IMT) and presence of plaque[34]. High prevalence of metabolic syndrome was observed in RA patients with long standing disease.

Increased waist circumference is an important constituent of MS. RA patients demonstrated 4 times increased likelihood of having high waist circumference even after adjusting for BMI. Rheumatoid cachexia is an abnormal body composition with increase fat mass and reduced muscle mass. These changes can occur with a normal or increased BMI. Active disease is frequently (30%) associated with rheumatoid cachexia[40]. Loss of muscle mass is attributed to increased circulation of pro inflammatory cytokines especially tumour necrosis factor alpha (TNF) resulting in increased disease activity and reduced physical activity which further reduce the muscle mass and increase the fat mass[37,38,41]. There is a high prevalence of traditional CV risk factors and MS in patients with rheumatic diseases[42].

Systemic lupus erythematosus

Systemic Lupus Erythematosus (SLE) patients exhibit a bimodal pattern of mortality which is due to active disease and infection resulting in early death within the first year and late death after five years of diagnosis is caused by cardiovascular diseases. The cardiac involvement in SLE is a significant cause of morbidity and mortality with a prevalence of more than 50%. All cardiac structures can be involved - pericardium, myocardium, endocardium, coronary arteries and conduction tissue. Pericardial disease is the commonest manifestation and myocardial involvement is less common. The incidence of coronary artery disease (CAD) is high in women with SLE, which is highest between 35- 44 years of age. Compared to age and gender matched controls SLE patients have shown to have increased risk of CVD, which occurs a decade earlier. CVD occurs due to a combination of traditional, non-traditional and disease related factors. Disease related factors associated with CVD are older age at diagnosis, high damage score, long duration of disease and treatment with glucocorticosteroids. Inflammation and antiphospholipid antibodies are important non-traditional factors. Over the years with adequate treatment the survival rates and organ damage in SLE has improved, but CVD is responsible for 30% organ damage in patients with a disease of fifteen years. The 10 year estimated relative risk of CVD is five to eight times higher in patients with SLE when compared to age and gender matched individuals from the general population[43]. Though SLE patients have higher prevalence of CV risk factors the increased risk of CVD is not explained by them. Certain other factors like chronic inflammation, renal impairment, premature menopause, treatment with corticosteroids etc play a role. Relationship between traditional CV risk factors and carotid plaque development and progression has also been observed[44].

Traditional CV risk factors such as hypertension, diabetes and dyslipidemia are more prevalent in SLE. Prevalence of smoking in SLE patients and general population is almost similar but it is the only risk factor predictive of a CV event. Hypertension is 2.5 times more common in patients with lupus when compared to general population and is a predictor of CVD and mortality. Diabetes is also more commonly associated with SLE but is not found to be a predictor of future CV events[45].

Dyslipidaemia in the form of raised triglycerides and VLDL and low HDL and LDL levels has been observed[46]. Hypercholesterolemia is not commonly observed in SLE patients but when present there is an increase in the risk of MI by 18 times[47]. SLE patients tend to acquire traditional CV risk factors despite the disease being more common in younger individuals mostly women. Even though the Framingham risk score underestimates the 10 year probability of developing CVD, traditional CV risk factors should not be ignored and SLE treatment should also focus on control of blood sugar, blood pressure, lipid abnormalities and smoking cessation.

Other rheumatic diseases

There is limited data regarding CVD in other rheumatic diseases. Patients with Psoriatic arthritis (PsA) exhibit increased prevalence of hypertension, diabetes and obesity when compared to general population. PsA patients have an increased CVD risk factors, congestive cardiac failure, myocardial infarction and peripheral vascular disease[48]. Chronic inflammation also plays a role in the development of CVD. Endothelial dysfunction and increased carotid intima media thickness are associated with raised ESR and CRP. Mortality studies have reported conflicting results, some studies have shown increased mortality[49] while others found it to be the same as in general population[50].

Hypertension and diabetes are more frequently associated with PsA and are also associated with increased IMT. The incidence of new onset diabetes is more in patients with PsA compared with matched controls and it is associated with increased disease severity[51]. Smoking is found to be increased in PsA but its association with CV mortality and morbidity is understudied. Dyslipidaemia in the form of low HDL cholesterol and high total cholesterol and HDL cholesterol ratio is observed. The increased prevalence of CVD in PsA is comparable to RA[52].

Ankylosing Spondylitis (AS) is associated with excess CV mortality. Cardiac manifestations can occur in the form of Aortic disease, mitral regurgitation, conduction abnormalities, myocardial and pericardial involvement. Risk of myocardial infarction is high in patients with AS[53]. Hypertension is found to be more prevalent in AS when compared to general population. There are conflicting reports regarding lipids. Not much data is there on diabetes and smoking. Risk factors for CVD especially diabetes, hypertension and dyslipidemia are common in patients with AS, RA and Ps A when compared to controls[7].

In primary Vasculitis, systemic inflammation and vessel ischaemia result in vessel stenosis and occlusion. Compared to controls, patients with systemic vasculitis have an increased carotid artery IMT but the traditional CV risk factors are similar[54]. Hypertension is found to be present in more than fifty percent of individuals with Takayasu disease[55]. Use of corticosteroids in high dose, for long time, inflammation and remodelling of blood vessels and dyslipidemia are to be considered in the development of CVD in patients with vasculitis[56].

There is increased occurrence of arterial events in Inflammatory myositis which is attributed to dyslipidemia and high blood pressure. There is scarcity of data addressing CVD in Sjogren's syndrome and systemic sclerosis.

CVD in Indians with rheumatic diseases

India is the second most populated country in the world with a population 1.21 billion (as per 2011 census). Indian population amounts to 17.31% of the world population. The prevalence of CVD is high in Indians. When compared to Caucasians it occurs a decade early is more severe and follows a malignant course. WHO has predicted CV mortality in Indians to be highest in the world by 2020.

Traditional CV risk factors are more frequent and more than one risk factor is present[57]. Diabetes is seen in south asian Indians about a decade earlier when compared to other Asians and Caucasians[58]. India has the highest number of diabetics in the world[59]. An estimated 32 million people suffer from diabetes and the number is projected to increase to 69.8 million by the year 2025. Smoking rate is reported to be lower among South Asian Indians compared to European populations[60]. Similar reports are there from South Asian Indians from United States of America. Data regarding smoking in Indians is limited. Smoking is almost nonexistent in South Asian Indian women but they have CVD comparable to men[61].

When compared to Caucasians, Indians tend to have lower HDL cholesterol and higher triglycerides[62]. South Asian Indians are known to have a high prevalence of dyslipidaemia in the usual lipid pattern with elevated total cholesterol, high triglycerides and low HDL cholesterol[63]. There is a strong family history of premature CHD[64].

Indians with RA have a very high prevalence of CV risk factors[65,66]. The traditional CV risk factors are high despite the population being predominantly female, younger and smoking history was negligible[65]. Compared to controls RA patients had a fourfold increased odd of having elevated risk of CHD event in ten years[67]. RA patients are also at an increased risk of sub clinical atherosclerosis compared to controls[68,69]. Sub clinical atherosclerosis measured by ultrasound demonstrated increased carotid IMT in Indian patients with RA[70,71]. However, there are no studies addressing CV mortality in the context of RA.

The survival of SLE patients is reduced[72]. Patients with SLE have increased prevalence of CVD. Compared to controls they have increased IMT, carotid plaques and peripheral vascular disease which is affected by dyslipidaemia[73].

Two traditional CV risk factors ie BMI and LDL cholesterol were found to be increased in patients with SpA compared to RA[74].There is scarcity of Indian data on CVD in other rheumatic diseases.

Preventive measures

Patients and family physicians should be made aware of the risk of increased CVD in rheumatic diseases. Physicians, rheumatologists and cardiologists should be aware of the gravity of the problem and be particularly proactive in identifying and treating CV risk factors. Increased vigilance and screening for risk factors is of particular importance. Regular monitoring of blood pressure, frequent blood sugar estimation and annual lipid profile should be advocated. The EULAR have given evidence based recommendations for management of CVD in patients with RA and other rheumatic diseases[75]. Most of the elevated risk factors like diabetes, hypertension, dyslipidaemia and high waist:hip ratio, all clustering to increase the prevalence of metabolic syndrome can potentially be improved by lifestyle modification such as increasing mobility and exercise. Aggressive treatment of rheumatic diseases to reduce the inflammatory burden should be advocated. Assessment of MS and 10 year of CHD risk in patients with rheumatic diseases should be done at regular intervals. As there are no guidelines for Indian patients with RD the EULAR recommendations can be followed[75].

Further Reading :

1. Cobb S, Anderson F & Bauer W. Length of life and cause of death in rheumatoid arthritis. N. Engl. J. Med. 1953; 249: 553-56.
2. Urowitz MB. et al. The bimodal mortality pattern of systemic lupus erythematosus. Am. J. Med. 1976; 60: 221-25.
3. Wolfe F. et al. The mortality of rheumatoid arthritis. Arthritis Rheum. 1994; 37: 481-94.
4. Del Rincon ID, Williams K, Stern MP, Freeman GL, Escalante A. High incidence of cardiovascular events in a rheumatoid arthritis cohort not explained by traditional cardiac risk factors. Arthritis Rheum, 2001; 44: 2737-45.
5. Frostegard J: Rheumatic diseases: insights into inflammation and atherosclerosis. Arterioscler Thromb Vasc Biol 2010, 30: 892-93.
6. MJ Roman and JE Salmon, "Cardiovascular manifestations of rheumatologic diseases", Circulation. 2007; 116(20): 2346-55.
7. Han C, Robinson DWJ, Hackett MV, Paramore LC, Fraeman KH, Bala MV. Cardiovascular disease and risk factors in patients with rheumatoid arthritis, psoriatic arthritis and ankylosing spondylitis. J Rheumatol. 2006; 33: 2167-72.
8. Solomon DH, Curhan GC, Rimm EB, Cannuscio CC, Karlson EW. Cardiovascular risk factors in women with and without rheumatoid arthritis. Arthritis Rheu., 2004; 50: 3444-49.
9. Boyer JF, Gourraud PA, Cantagrel A, Davignon JL, Constantin A. Traditional cardiovascular risk factors in rheumatoid arthritis: a meta-analysis. Joint Bone Spine. 2011; 78: 179-83.
10. Wilson P, D'Agostino RB, Ralph B, Levy D, Belanger AM, Silbershatz H. Prediction of Coronary Heart Disease Using Risk Factor Categories. Circulation. 1998; 97: 1837-47.
11. Reilly MP and Rader DJ. The metabolic syndrome: more than the sum of its parts? Circulation 2003; 108: 1546-51.
12. Reaven GM. Banting lecture. Role of insulin resistance in human disease. Diabetes. 1998; 37: 1595-607.
13. Lorenzo C, Williams K, Hunt KJ, Haffner SM. The National Cholesterol Education Program - Adult Treatment Panel III, International Diabetes Federation and World Health Organization definitions of the metabolic syndrome as predictors of incident cardiovascular disease and diabetes. Diabetes Care. 2007; 30: 8-13.
14. Executive Summary of the Third Report of the National Cholesterol Education Program (NCEP) Expert Panel on Detection, Evaluation and Treatment of High Blood Cholesterol in Adults (Adult Treatment Panel III). JAMA. 2001; 285: 2486-97.
15. Lakka HM, Laaksonen DE, Lakka TA, Niskanen LK, Kumpusalo E, Tuomilehto J, Salonen JT. The metabolic syndrome and total and cardiovascular disease mortality in middle-aged men. JAMA. 2002; 288: 2709-16.
16. Cobb S, Anderson F, Bauer W. Length of life and cause of death in rheumatoid arthritis. N Engl J Med. 1953; 249: 553-56.
17. Gabriel SE, Crowson CS, Kremers HM, Doran MF, Turesson C, O'Fallon WM, Mattesson EL. Survival in rheumatoid arthritis: a population-based analysis of trends over 40 years. Arthritis Rheum. 2003; 48: 54-58.
18. Van Doornum S, Mccoll G and Wicks IP. Accelerated atherosclerosis: an extraarticular feature of rheumatoid arthritis? Arthritis Rheum. 2002; 46: 862-73.
19. Naz SM and Symmons DP. Mortality in established rheumatoid arthritis. Best Pract Res Clin Rheumatol. 2007; 21: 871-83.
20. Avina-Zubieta, J.A.E.A. Risk of cardiovascular mortality in patients with rheumatoid arthritis: a meta-analysis of observational studies. Arthritis Rheum. 2008; 59: 1690-97.
21. Meune C, Touze E, Trinquart L and Allanore Y. Trends in cardiovascular mortality in patients with rheumatoid arthritis over 50 years: a systematic review and meta-analysis of cohort studies. Rheumatology (Oxford), 2009; 48: 1309-13.
22. Gonzalez A, Icen M, Kremers HM, Crowson CS, Davis JM, 3 RD, Therneau TM, Roger VL & Gabriel SE. Mortality trends in rheumatoid arthritis: the role of rheumatoid factor. J Rheumatol, 2008; 35: 1009-14.
23. Farragher TM, Goodson NJ, Naseem H, Silman AJ, Thomson W, Symmons D & Barton A. Association of the HLA-DRB1 gene with premature death, particularly from cardiovascular disease, in patients with rheumatoid arthritis and inflammatory polyarthritis. Arthritis Rheum, 2008; 58: 359-69.
24. Maradit-Kremers H, Nicola PJ, Crowson CS, Ballman KV & Gabriel SE. Cardiovascular death in rheumatoid arthritis: a population-based study. Arthritis Rheum, 2005; 52, 722-32.
25. Naranjo A, Sokka T, Descalzo MA, Calco-Alen J, Horslev-Petersen K, Luukkainen RK, Combe B, Burmester GR, Delvin J, Ferraccioli G, Morelli A, Hoekstra M, Majdan M, Sadkiewicz S, Belmonte M, Holmqvist AC, Choy E, Tunc R, Dimic A, Bergman M, Toloza S & Pincus T. Cardiovascular disease in patients with rheumatoid arthritis: results from the QUEST-RA study. Arthritis Res Ther, 2008; 10: R30.
26. Erb N, Pace AV, Douglas KM, BanksANKS, M. J. & KITAS, G. D. Risk assessment for coronary heart disease in rheumatoid arthritis and osteoarthritis. Scand J Rheumatol, 2004; 33: 293-99.
27. Brady SR, De Courten B, Reid CM, Cicuttini FM, De Courten MP & Liew D. The Role Of Traditional Cardiovascular Risk Factors Among Patients With Rheumatoid Arthritis. J Rheumatol, 2009; 36: 34-40.
28. Bryan Williams NRP, Morris J Brown, Mark Davis, Gordon T Mcinnes, John F Potter, Peter S Sever, Simon Mcg Thom, the BHS Guidelines Working Party, For The British Hypertension Society. Guidelines From The BHS. British Medical Journal 2004; 328: 634-40.
29. Mcentegart A, Capell HA, Creran D, Rumley A, Woodward M and Lowe GDO. Cardiovascular risk factors, including thrombotic variables, in a population with rheumatoid arthritis. Rheumatology (Oxford), 2001; 40: 640-44.
30. Kannel WB. Elevated systolic blood pressure as a cardiovascular risk factor. The American Journal of Cardiology, 2000; 85: 251-55.
31. Goodson N. Coronary artery disease and rheumatoid arthritis. Curr Opin Rheumatol, 2002; 14: 115-20.
32. Panoulas VF, Metsios GS, Pace AV, John H, Treharne GJ, Banks MJ & Kitas GD. Hypertension in rheumatoid arthritis. Rheumatology (Oxford), 2008; 47: 1286-98.

33. Dessein PH, Joffe BI, Stanwix AE, Christian BF & Veller M. Glucocorticoids and insulin sensitivity in rheumatoid arthritis. J Rheumatol, 2004; 31: 867-74.

34. Dessein PH, Tobias M & Veller MG. Metabolic syndrome and subclinical atherosclerosis in rheumatoid arthritis. J Rheumatol, 2006; 33: 2425-32.

35. Goodson NJ, Farragher TM & Symmons DP. Rheumatoid factor, smoking and disease severity: associations with mortality in rheumatoid arthritis. J Rheumatol, 2008; 35: 945-49.

36. Kitas GD, Situnayake RD. Dyslipidaemia and rheumatoid arthritis Annals of the Rheumatic Diseases 1997; 56: 341-42.

37. Kitas GD & Gabriel SE. Cardiovascular disease in rheumatoid arthritis: state of the art and future perspectives. Ann Rheum Dis, 2001; 70: 8-14.

38. Stavropoulos-Kalinoglou A MG, Koutedakis Y, et al. Redefining overweight and obesity in rheumatoid arthritis patients. Ann Rheum Dis, 2007; 66: 1316-21.

39. Karvounaris SA, Sidiropoulos PI, Papadakis JA, Spanakis EK, Bertsias GK, Kritikos HD, Ganotakis ES & Boumpas DT. Metabolic syndrome is common among middle-to-older aged Mediterranean patients with rheumatoid arthritis and correlates with disease activity: a retrospective, cross-sectional, controlled, study. Ann Rheum Dis, 2007; 66:28-33.

40. Giles JT, Ling SM, Ferrucci L, Bartlett SJ, Andersen RE, Towns M, Muller D, Fontaine KR & Bathon JM. Abnormal body composition phenotypes in older rheumatoid arthritis patients: association with disease characteristics and pharmacotherapies. Arthritis Rheum, 2008; 59: 807-15.

41. Summers GD, Metsios GS, Stavropoulos-Kalinoglou A & Kitas GD. Rheumatoid cachexia and cardiovascular disease. Nat Rev Rheumatol, 2010; 6: 445-51.

42. Sidiropoulos PI, Karvounaris SA & Boumpas DT. Metabolic syndrome in rheumatic diseases: epidemiology, pathophysiology and clinical implications. Arthritis Res Ther, 2008; 10: 207.

43. Urowitz MB, Ibañez D, Gladman DD. Atherosclerotic vascular events in a single large lupus cohort: prevalence and risk factors. J Rheumatol 2007; 34: 70-75.

44. Thompson T, Sutton-Tyrrell K, Wildman RP, Kao A, Fitzgerald SG, et al. Progression of carotid intima-media thickness and plaque in women with systemic lupus erythematosus. Arthritis Rheum 2008; 58: 835-42.

45. Bruce IN, Urowitz MB, Gladman DD, Ibanez D, Steiner G. Risk factors for coronary heart disease in women with systemic lupus erythematosus: the Toronto Risk Factor Study. Arthritis Rheum 2003; 48: 3159-67.

46. Kashef S, Ghaedian MM, Rajaee A, Ghaedian A. Dyslipidaemia during the active course of systemic lupus erythematosus in association with anti-double-stranded DNA (anti-dsDNA) antibodies. Rheumatol Int 2007; 27: 235-41.

47. Fischer LM, Schlienger RG, Matter C, Jick H & Meier CR. Effect of rheumatoid arthritis or systemic lupus erythematosus on the risk of first-time acute myocardial infarction. Am. J. Cardiol. 2004; 93: 198-200.

48. C Han, DW Robinson Jr, MV Hackett, LC Paramore, KH Fraeman and MV Bala, "Cardiovascular disease and risk factors in patients with rheumatoid arthritis, psoriatic arthritis, and ankylosing spondylitis," Journal of Rheumatology, 2006; vol.33, no. 11, 2167-72.

49. Y Ali, BDM Tom, CT Schentag, VT Farewell and D Gladman, "Improved survival in psoriatic arthritis with calendar time", Arthritis and Rheumatism, 2007; vol. 56, no. 8, 2708-14.

50. CE Buckley, CR Cavill, GJ Taylor et al. "Mortality in psoriatic arthritis-a single-center study from the UK", Journal of Rheumatology, 2010; vol. 37, no. 10, 2141-44.

51. Khalid U, Hansen PR, Gislason GH, Lindhardsen J, Kristensen SL < Winther SA, Skoy L, Torp-Pedersen and Ahlehoff O. Psoriasis and new onset diabetes: A Danish nationwide cohort study. Diabetes care August 1, 2013; 36(8): 2402-07.

52. Cardiovascular comorbidities in patients with psoriatic arthritis: a systematic review. Anna Jamnitski, Deborah Symmons, Mike J L Peters, Naveed Sattar, Iain MciInnes, Michael T Nurmohamed Ann Rheum Dis 2013; 72: 211-16.

53. Sylvian Mathieu, Laure Gossec, Maxime dougados and Martin Soubrier. Cardiovascular Pro?le in Ankylosing Spondylitis: A Systematic Review and Meta-Analysis. Arthritis Care & Research April 2011; vol. 63, no. 4, 557-63.

54. De Leeuw K, Sanders JS, Stegeman C, Smit A, Kallenberg CG, Bijl M. Accelerated atherosclerosis in patients with Wegener's granulomatosis. Ann Rheum Dis 2005; 64: 753-59.

55. Kerr GS, Hallahan CW, Giordano J, Leavitt RY, Fauci AS, Rottem M, et al. Takayasu's arteritis. Ann Intern Med 1994; 120: 919-29.

56. Lucia Kazelian. Autoimmune disease promoting coronary artery disease in women. Rev Argent Cardiol 2013; 81: 330-34.

57. Enas EA. Coronary Artery Disease epidemic in Indians: A cause for alarm and call for action. J Ind Med Assoc. 2000; 98: 694-702.

58. Mather HM, CN, Fuller JH. Mortality and morbidity from diabetes in South Asians and Europeans: 11 year follow-up of the Southall Diabetes Survey, London, UK. Diabetic Med 1998; 15: 53-59.

59. Gupta R, MA. Type 2 diabetes in India: regional disparities. Br J Diab Vas Dis 2007; 7: 12-16.

60. Agyemang C, SK, Tromp N, Bhopal R, Zaninotto P, Unwin N, Nazroo J, Kunst AE. A cross-national comparative study of smoking prevalence and cessation between English and Dutch South Asian and African origin populations: the role of national context. Nicotine Tob Res, 2010; 12: 557-66.

61. Enas EA, SA, Juturu V, Gupta R. Coronary artery disease in women. Indian Heart J 2001; 53: 282-92.

62. Mckeigue PM, SB, Marmot MG. Relation of central obesity and insulin resistance with high diabetes prevalence and cardiovascular risk in South Asians. Lancet 1991; 337: 382-86.

63. Anand SS, YS, Vuksan V, Devansen S, Teo KK, Montaue PA, Kelemen E, Guo T, Lonn E, Gerstein H, Hegele RA, Mcqueen M. Differences in cardiovascular disease between ethnic groups are not explained by conventional risk factors or atherosclerosis. The Study of Health Assessment and Risk in Ethnic Groups (SHARE). Lancet, 2000; 356: 279-84.

64. Enas EA, Yusuf S & Mehta JL. Prevalence of coronary artery disease in Asian Indians. Am J Cardiol, 1992; 70: 945-49.

65. Firdaus Fatima, URK Rao, RJ Moots, NJ Goodson Prevalence of cardiovascular Risk Factors in Patients with Rheumatoid Arthritis: Results from a South Asian population. Arthritis Rheum 2007; 58 S: 1014.

66. Agarwal D, Malaviya AN. A study of the conventional cardiovascular disease (CVD) risk factors in rheumatoid arthritis (RA) among Indians. Ind J Rheumatol. 2013; 8: 19-23.

67. Fatima Firdaus, Uppuluri Rao R K, Moots Robert J, Goodson Nicola. Raised Traditional Cardiovascular Risk Factors in Indians with Rheumatoid Arthritis. Arthritis Rheum 2009; (S10): 948.

68. Mahajan V, Handa R, Kumar U, Sharma S, Gulati G, Pandey RM, Lakshmy R. Assessment of Atherosclerosis by Carotid Intimomedial Thickness in Patients with Rheumatoid Arthritis. J Assoc Physic India 2008; 56: 587-90.

69. Handa R. Rheumatoid Arthritis - At the Cross Roads of Inflammation and Atherosclerosis! J Assoc Physic India 2013; 61: 529.

70. Grover S, Sinha RP, Singh U, Tewari S, Aggarwal A, Misra R. Subclinical atherosclerosis in rheumatoid arthritis in India. J Rheumatol 2006; 33: 244-47.

71. Adhikari MC, Guin A, Chakraborty S, Sinhamahapatra P, Ghosh A. Subclinical atherosclerosis and endothelial dysfunction in patients with early rheumatoid arthritis as evidenced by measurement of carotid intima-media thickness and flow-mediated vasodilatation: an observational study. Semin Arthritis Rheum. 2012; 41: 669-75.

72. Kumar A, Malaviya AN, Singh RR, Singh YN, Adya CM, Kakkar R. Survival in patients with systemic lupus erythematosus in India. Rheumatol Int 1992; 12: 107-09.

73. Bhatt SP, Handa R, Gulati GS, et al. Peripheral vascular disease in systemic lupus erythematosus. Lupus 2007; 16: 720-23.

74. Traditional cardiovascular (CVD) risk factors in patients with ankylosing spondylitis-spondyloarthritis (AS-SpA) as compared to rheumatoid arthritis (RA). D Agarwal, R Rawat, S Garg, S Kapoor, AN Malaviya Indian Journal of Rheumatology. 2008; 3 (3): S38 Nov.

75. Peters MJ et al. EULAR evidence-based recommendations for cardiovascular risk management in patients with rheumatoid arthritis and other forms of inflammatory arthritis. Ann. Rheum. Dis. 2010; 69: 325-31.

PARENCHYMAL PULMONARY DISEASE AND PULMONARY ARTERY HYPERTENSION IN CONNECTIVE TISSUE DISEASES

8

Uma Kumar

Pulmonary manifestations of connective tissue diseases (CTD) are very diverse because not only lung parenchyma (interstitial lung disease) but also, airways (bronchiolitis and bronchiectasis), pulmonary vasculature (pulmonary hypertension) and respiratory muscles may get affected as shown in Table 1. Pulmonary involvement usually occurs in well-established CTD but occasionally it is the first manifestation of the disease[1,2].

Homma et al[3] observed that 19% of the patients with interstitial lung disease (ILD)developed CTDs over a period of one to eleven years.

Among various types of pulmonary involvement in CTD, ILD remains most common and serious form of pulmonary involvement.

is idiopathic ILD of equivalent severity except in patients with rheumatoid arthritis having usual interstitial pneumonia[5,6,7]. This could be due to early detection and relative slow progression of parenchymal disease in CTD-ILD. Also, patients with CTD are on immunosuppressive therapy which could affect survival.

Therefore, it is argued that presence of mixed histological pattern must raise the possibility of forme fruste of CTD even if patient has no evidence of CTD at that point of time[8].

It has been seen that up to 25% of ILD occurs in the context of undifferentiated CTD, which makes the distinction with idiopathic interstitial pneumonias even more difficult[9]. A particular CTD is more often associated with certain type of ILD as shown in Table 2.

Table 1: Pattern of respiratory involvement in CTD

	Airways	Pleura	Vascular	Interstitial lung disease
Systemic sclerosis	-	-	++++	++++
Rheumatoid arthritis	++	++	+	+++
Polymyositis/Dermatomyositis	-	-	++	++++
Sjogren's Syndrome	+++	+	+	+++
Systemic lupus erythematosus	+	++++	+	+

CTD-ILD

ILD is a group of diffuse parenchymal disorders with variable clinical, radiographic, and histopathological features. Almost all patients with CTD are at risk of ILD, some are more likely to be associated with it than others but the prevalence of connective tissue disease associated interstitial lung disease (CTD-ILD) is not known. In an Indian study, the most common pulmonary involvement observed on chest radiography and High Resolution Computed Tomography (HRCT) of thorax was ILD (38.5%) while pulmonary arterial hypertension (PAH) was observed in 14.9% patients with CTD[4].

Majority of available data is on idiopathic interstitial pneumonias which are being applied to CTD-ILD. But it is realized that CTD-ILD is different from idiopathic variety as it usually shows combination of histopathological patterns and response to immunosuppressive therapy is favourable. CTD-ILD is associated with more favourable prognosis than

Many patients with CTD have subclinical ILD (abnormalities on high resolution CT in asymptomatic individuals)[10]. Also to complicate the picture further, clinical presentation of CTD- ILD is non-specific; therefore detailed clinical examination is very important in diagnosing pulmonary parenchymal involvement. Patients with CTD require thorough assessment to exclude other causes of parenchymal lung disease e.g. pulmonary infection, toxic effect of drug, environment exposure etc. History of dyspnoea, cough, chest pain, hemoptysis, progression of symptoms, occupational history, family history, past history, drug history should all be noted. Most of the drugs commonly used in the treatment of CTD have one or the other pulmonary toxicity known or drugs may mask or unmask respiratory diseases. ILD can occur from infections that develop as a consequence of immune suppression.

Detailed cardiovascular and respiratory system examination must be done. One may find clubbing, increased respiratory rate, crackles, rhonchi/wheeze, decreased breath sounds or pleural rub, loud P2, right ventricular heave or evidence of right heart failure.

Bronchoscopy - Fiberoptic bronchoscopy provides direct visualization of proximal airways which helps in diagnosing parenchymal lung disorders including granulomatosis with polyangiitis, sarcoidosis, and polychondritis etc. Bronchoalveolar lavage (BAL) helps in diagnosing small airways involvement/ infection/haemorrhage and suggesting disease activity in interstitial lung disease. But there is no role of repeated BAL to show on going disease activity/or response to treatment.

Table 2. Frequency and histopathological pattern of ILD in CTD

Rheumatic disease	Frequency of ILD (%)	Characteristic histopathologic pattern	Radiological characteristics
Systemic sclerosis	45 (clinically significant)	NSIP	Increased reticular markings, ground glass opacities,bibasilar prominence
		UIP	Peripheral and bibasilar reticulonodular opacities with honeycombing
Rheumatoid arthritis	20 to 30	UIP, NSIP	As above
		OP	Patchy airspace consolidation, ground glass opacities
		DAD	Diffuse ground glass opacities
Polymyositis/ dermatomyositis	20 to 50	NSIP, UIP COP	As above
Sjogren's syndrome	Up to 25	NSIP	As above
		LIP	Thin walled cysts, ground glass opacities, centrilobular nodules
		OP	As above
		UIP	As above
		DAD	As above
Systemic lupus erythematosus	2 to 8	NSIP, DAD, LIP, OP	As above
		AIP	Ground glass opacities
Mixed connective tissue disease	20 to 60	NSIP	As above

Abbreviations: NSIP: non-specific interstitial pneumonia, UIP: usual interstitial pneumonia, OP: organizing pneumonia, DAD: diffuse alveolar damage, LIP: lymphocytic interstitial pneumonia, AIP: Acute interstitial pneumonia

Pulmonary parenchymal involvement may be asymptomatic and even pulmonary function tests are normal in early stages; therefore a high index of suspicion is required to detect it early.

Investigations

Exercise testing - As mentioned earlier routine lung functions may be normal in early stages. Cardiopulmonary exercise testing (CPET) is a simple screening technique which provides objective evidence of pulmonary involvement. It can differentiate between pulmonary and cardiac origin of dyspnoea depending on oxygen uptake, carbon dioxide production and anaerobic threshold but the facility is not readily available.

Pulmonary function test (PFT) - Along with spirometry, lung volume, diffusing lung capacity and blood gases must be performed. It is important to note that these tests can be normal in early stages of pulmonary disease.

BAL fluid cytokine/chemokines as markers reflecting disease severity and predicting outcome in ILD is still a research area.

Thoracoscopic and open lung biopsy are generally not required. It is preferred in patients where diagnosis is in doubt because of atypical presentation or malignancy is suspected. Video-assisted thoracoscopy (VAT) is better as it allows visualization of lung, causes less pain, shorter recovery and sample taken from more than one lobe. In critically ill patients open lung biopsy is preferred. But decision to perform lung biopsy must be weighed against risk versus benefit, whether biopsy will result in any change in treatment.

Echocardiography - It is a reliable non-invasive method for detecting PAH and it may be particularly useful for the early detection and monitoring. But still, right heart catheterization remains the gold standard for diagnosing PAH.

Chest x-ray - It may be normal in early ILD but other abnormalities may be recognized (pleural effusion/ thickening, mediastinal lymphadenopathy, prominent pulmonary artery conus etc.)

Air space consolidation in patients with CTD may be due to pneumonia, pulmonary haemorrhage or lupus pneumonitis. Pulmonary nodules are seen in granulomatosis with polyangiitis and rheumatoid arthritis.

CT chest - High resolution CT chest is highly sensitive and specific for diagnosing diffuse pulmonary parenchymal disease and correlate well with pathologic findings, therefore lung biopsy is not advocated in all patients with ILD. Various patterns of ILD described in CTD are shown in Table 2.

Role of Positron emission tomography (PET) and serological markers is still not clearly defined.

Treatment

There are no guidelines regarding initiating immunosuppressive therapy for CTD-ILD but available data suggest a benefit to therapy in some patients. Before taking therapeutic decision to initiate therapy, it is important to know (1) underlying CTD (2) extent and rate of progression of ILD (3) likelihood of response based on radiological patterns (4) age of patient (5) ability of the patient to comply with treatment (6) therapies patient is already on. It is important to identifying appropriate group of patients who will benefit with treatment, so as to avoid unnecessary drug toxicity.

Following should be considered in all patients with pulmonary manifestations:

1. Vaccinations.
2. Pulmonary physical rehabilitation programme.
3. Assessment of patient with long term oxygen therapy
4. Appropriate patients should be considered for lung transplantation.

Now we will discuss parenchymal involvement in commonly occurring CTDs:

Systemic sclerosis (SSc)

Pulmonary disease is the leading cause of mortality in SSc. Prevalence of ILD varies from 40-80% depending on the method of diagnosis. Median survival for patients with SSc-ILD ranges from 5-8 years[11].

ILD is thought to be initiated by microvascular injury which leads to endothelial and epithelial cell damage resulting in release of cytokines, chemokines and growth factors; coagulation cascade is also activated. This finally results in activation of fibroblasts. Ethnicity, antibody subset and type of SSc influences occurrence of ILD in SSc. The presence of anti-topoisomerase antibody (anti Scl-70) is strongly associated with development of ILD while anti-centromere antibody seems to be protective. DAD and OP are rare in SSc.

Various immunosuppressants have been used to treat SSc-ILD, with variable results.

Scleroderma Lung Study showed that oral cyclophosphamide (2 mg/kg/day) for one year slowed progression of ILD but benefit was not maintained at two years[12].

A Cochrane metaanalysis did not demonstrate any statistically significant improvement in pulmonary function but European League Against Rheumatism recommends use of cyclophosphamide in SSc-ILD[13,4].

Another drug, mycophenolate mofetil has shown promising results with stabilization and improvement of pulmonary functions. A randomized double-blind trial comparing mycophenolate mofetil and oral cyclophosphamide in patients with SSc-ILD is on-going[15,16].

Recently several reports are published on the use of Rituximab in SSc-ILD with promising results[17,18].

Current recommendation for treatment include low dose steroids (10mg/day), and monthly cyclophosphamide pulse for 6 months followed by either azathioprine or mycophenolate mofetil with low dose corticosteroids. High dose of steroids must be avoided in SSc because of risk of scleroderma renal crisis.

Studies have shown beneficial effect of haematopoietic stem cell transplantation but further studies are required to establish its definite therapeutic role[19]. Pulmonary parenchymal involvement in individual CTD is discussed below:

Systemic lupus erythematosus (SLE)

Pulmonary involvement is frequent in SLE, and can affect the pleura, pulmonary vasculature, and parenchyma. Pulmonary parenchymal abnormalities mostly occur due to bacterial infections although opportunistic infections occur with increased frequency. Pulmonary haemorrhage is a serious life threatening complication. Pulmonary fibrosis is less common in SLE than in rheumatoid arthritis or systemic sclerosis. Other manifestations of SLE are loss of lung volume related to diaphragmatic dysfunction, pulmonary oedema. Acute lupus pneumonitis is an uncommon manifestation of SLE and it remains the diagnosis of exclusion. Interstitial lung disease and pulmonary hypertension may be seen in approximately 30% SLE patients. An unusual syndrome referred to as shrinking lung syndrome has been described in SLE patients with progressive dyspnoea.

Rheumatoid arthritis(RA)

Lung disease is the second leading cause of death in RA. Pulmonary complications of RA include pleural effusion, nodular lung disease, diffuse interstitial fibrosis, pulmonary vasculitis, alveolar hemorrhage, obstructive pulmonary disease, and infections. Pulmonary nodules can appear before, with, or after the onset of RA either singly or in clusters. Diffuse interstitial fibrosis occurs mostly in patients who have subcutaneous nodules and is usually slowly progressive. Pulmonary vasculitis, one of the least common pulmonary manifestations of RA is seen in patients with severe RA. Patients with RA and lung lesions had a high incidence of bronchitis and bronchiectasis. Pulmonary manifestations of RA that are more likely to become symptomatic include ILD and pulmonary artery hypertension. Pulmonary involvement in patients with RA may also result from the drugs used in therapy.

Polymyositis/dermatomyositis

Pulmonary involvement is seen in approximately 40% patients with PM/DM and it contributes significantly to morbidity and mortality. The most common manifestation is aspiration pneumonia secondary to pharyngeal muscle weakness. Other parenchymal abnormalities include ILD, bronchiolitis obliterans organizing pneumonia and diffuse alveolar damage. The presence of ILD markedly influences the disease course in inflammatory myositis. The strongest predictive factor for occurrence of ILD is the presence of autoantibodies to aminoacyltRNAsynthetase, most commonly anti- Jo-1[20, 21,22].

Sjögren's syndrome (SS)

Involvement of exocrine glands in the upper respiratory tract frequently leads to dryness of the nasal passages and bronchi.The most common manifestation associated with Sjögren syndrome is pulmonary fibrosis. There is an increased prevalence of lymphocytic interstitial pneumonitis involving the lower lobes.

Mixed connective tissue disease (MCTD)

Pulmonary involvement is a common complication of MCTD but it is not usually clinically evident early in the course of the disease. ILD and PAH are rather serious pulmonary complications of MCTD. The prognosis of ILD may be better in MCTD than in SSc. The elevated levels of immune complexes and increased complement consumption indicated that immune complex mediated alveolo-capillary membrane damage and tissue injury might play a role in the pathogenesis of ILD in MCTD.

Granulomatosis with polyangiitis

The pulmonary manifestations of granulomatosis with polyangiitis are diverse, ranging from asymptomatic lung nodules to fulminant alveolar hemorrhage. Nodules are usually multiple and bilateral, and often result in cavitations.Localized or diffuse areas of air-space consolidation may be present. Involvement of the tracheal or bronchial walls usually consists of mucosal or submucosal granulomatous thickening. The infiltrates, which may wax and wane, are often misdiagnosed initially as pneumonia.

Churg-Strauss syndrome

The lung is the most common organ involved with more than 90% patients having histories of asthma. Quite often transient peripherally distributed areas of consolidation are observed. Nodules may occur, but cavitation is rare. Other, less common manifestations include interlobular septal thickening, and bronchial wall thickening.

Ankylosing spondylitis

50-85% patients have pulmonary abnormalities in ankylosing spondylitis which include apical fibrosis, ILD, emphysema, bronchiectasis and pleural thickening. Quite often apical bullae with cavitation is mistaken for tuberculosis.

Pulmonary artery hypertension

It is a hemodynamic abnormality characterized by elevated resting mean pulmonary artery pressure ≥ 25 mm Hg with a normal pulmonary capillary wedge pressure of ≤ 15 mm Hg and pulmonary vascular resistance > 3 Wood units on right heart catheterization[23].

It is a serious progressive life threatening complication of many CTDs, more frequently associated are SSc, SLE and MCTD, also to a lesser extent in RA, dermatomyositis and primary Sjogren syndrome. The updated clinical classification of pulmonary hypertension given at the 4th World Symposium in 2008 (Dana Point, California) classifies connective tissue disease associated PAH (CTD-PAH) in the first category[24]. The detailed classification is given in Table 3. Severity of PAH is assessed by NYHA/WHO functional classification of pulmonary hypertension[25].

have lower cardiac output and tend to have worse survival. 1- , 2- and 3- year survival rates were reported as 45, 35 and 28 % respectively[31]. In a US registry of PAH patients wherein one-year survival was 86% among patients with CTD-PAH compared to 93% in patients with idiopathic PAH ($P < 0.0001$)[32].

Table 3 : Dana Point 2008 classification of pulmonary hypertension

1.	**Pulmonary arterial hypertension (PAH)**
1.1.	Idiopathic PAH
1.2.	Heritable 1.2.1. BMPR2 (bone morphogenetic protein receptor type 2) 1.2.2. ALK1 (activin receptor-like kinase type 1), endoglin (with or without hereditary hemorrhagic telangiectasia) 1.2.3. Unknown
1.3.	Drug- and toxin-induced
1.4.	Associated with 1.4.1. Connective tissue diseases 1.4.2. HIV infection 1.4.3. Portal hypertension 1.4.4. Congenital heart diseases 1.4.5. Schistosomiasis 1.4.6. Chronic hemolytic anemia
1.5.	Persistent pulmonary hypertension of the newborn
1. 6.	Pulmonary veno-occlusive disease (PVOD) and/or pulmonary capillary hemangiomatosis (PCH)
2.	**Pulmonary hypertension owing to left heart disease**
2.1.	Systolic dysfunction 2.2. Diastolic dysfunction 2.3. Valvular disease
3.	**Pulmonary hypertension owing to lung diseases and/or hypoxia**
3.1.	Chronic obstructive pulmonary disease 3.2. Interstitial lung disease 3.3. Other pulmonary diseases with mixed restrictive and obstructive pattern 3.4. Sleep-disordered breathing 3.5. Alveolar hypoventilation disorders 3.6. Chronic exposure to high altitude 3.7. Developmental abnormalities
4.	**Chronic thromboembolic pulmonary hypertension (CTEPH)**
5.	**Pulmonary hypertension with unclear multifactorial mechanisms**
5.1.	Hematologic disorders: myeloproliferative disorders, splenectomy 5.2. Systemic disorders: sarcoidosis, pulmonary Langerhans cell histiocytosis: lymphangio leiomyomatosis, neurofibromatosis, vasculitis 5.3. Metabolic disorders: glycogen storage disease, Gaucher disease, thyroid disorders 5.4. Others: tumoral obstruction, fibrosing mediastinitis, chronic renal failure on dialysis

Epidemiology

In a population based study in Japan, 2.6 % of over 3500 patients with CTD had PAH[26]. In the NIH registry, among 236 cases of unexplained PAH, about 8% was associated with CTD[27]. Prevalence of PAH in SSc patients has been widely reported to vary from 5-35%. A study from India has shown a prevalence of 32%[28]. PAH has been described in 0.5-14% patients with SLE[29]. The prevalence of PAH has been reported as 5.02 % and 0.56% in patients with MCTD and polymyositis/dermatomyositis respectively in a study from Japan[30].

There is considerable variability in the epidemiological data depending on variability in the patient population, definition and the method used for assessing PAH. Females are more commonly affected in CTD-PAH. However, compared to idiopathic PAH, patients with CTD -PAH are older,

Pathogenesis

It results from imbalance between vasodilator/ antiproliferative and vasoconstrictor/proliferative mechanisms in pulmonary vasculature[33]. Persistent pulmonary vasoconstriction and lumen obliteration of small- and medium-sized arteries and arterioles in association with the formation of plexiform lesions and in situ thrombosis is implicated in the pathophysiology of PAH. PAH due to chronic pulmonary thromboembolism - on the basis of lung scintigraphy is divided into two groups. The first is chronic macro-thromboembolism that obstructs arteries larger than the segmental pulmonary artery and can be distinguished from idiopathic PAH. The second is chronic micro-thromboembolism which cannot be distinguished from idiopathic PAH, and is often seen in antiphospholipid antibody syndrome. PAH due to peripheral pulmonary artery vasculitis is seen in SLE and Takayasu's arteritis[34].

Advancements in the understanding of the molecular mechanisms involved in PAH suggest that endothelial dysfunction plays a key role in the genesis and progression of PAH[35]. In addition, it is widely accepted that immune and/or inflammatory mechanisms could play a significant role in PAH genesis or progression in patients with CTD. Thus, immunosuppressive therapy should be better evaluated in that setting.

Clinical features

History

The earliest symptom of PAH is typically dyspnea on exertion. Occurrence of lower extremity edema and angina generally represents more advanced disease, likely due to right ventricular failure and reduced coronary blood flow to a hypertrophied right ventricle respectively.

Physical examination

Jugular venous pressure is raised, with a large a wave. A low volume carotid pulse is appreciable. There may be a left parasternal heave, an ejection click and flow murmur in the same area, a closely split second heart sound with a loud pulmonic component, and a fourth heart sound of right ventricular origin. Late in the course, signs of right ventricular failure may be present.

Investigations

Electrocardiogram - presence of P pulmonale and findings of right ventricular hypertrophy and strain pattern.

Radiological investigations: Chest X-Ray shows evidence of enlarged pulmonary artery and obliteration of retrosternal space on lateral view chest X- ray. CT pulmonary angiogram is required to rule out pulmonary thromboembolism.

Echocardiography - It remains the most important noninvasive investigation in diagnosis of PAH. A moderate-high correlation (0.57-0.93) has been reported between transthoracic echo estimation of right ventricular systolic pressure (RVSP) and systolic pulmonary artery pressure (PAP) measured via right heart catheterization.

DLCO - The low and decreasing DLCO (in patients without ILD) portends a poor outcome in all types of PAH.

Exercise testing - 6-minute walk test is often used as a primary outcome in assessment of new PAH therapies and tracking of disease progression. Decreased oxygen saturation and walk distances predict increased mortality.

Cardiac catheterization - It remains the gold standard for the detection of PAH.

Brain natriuretic peptide (BNP) and N-terminal-pro-BNP have prognostic importance in patients with PAH. Lower levels at baseline and improvement during follow-up are associated with better outcomes.

Management

Currently available therapies increase exercise tolerance, improve hemodynamic parameters and provide survival benefit[36,37].

Once the patient is diagnosed with PAH and initiated on specific therapy, he should be assessed to monitor response to therapy at 3-6 months intervals depending on the severity of disease. Although a general assessment of the patient will provide important information whether the patient has improved symptomatically, objective evaluation is very important. 6 minutes' walk test (6MWT), NYHA/WHO functional class, PAP and cardiac output are used for objectively assessing efficacy of the treatment. Assessment by serum biomarkers and right heart function may also be helpful. Patients with NYHA/ WHO functional class II and III can be started on oral treatment and parenteral prostanoids are considered suitable in patients with WHO class IV or those with poor prognostic markers in class III. If response to treatment is suboptimal, switching or adding another drug is suggested. There is very little evidence to refer to in deciding 'which drug is to be used when'. Several RCTs of combination therapy are ongoing.

The guidelines suggested by the European Respiratory Society and the European Society of Cardiology provide an evidence based approach to the management of PAH in CTD (as shown in the Fig. 1)[38].

Right heart failure, if present, is treated with oral anticoagulant medications, diuretics, digoxin, along with inhaled oxygen and other supportive therapy. In cases where medical therapy fails or cannot be given lung transplantation must also be considered.

Lifestyle modification

Avoidance of high altitudes (> 1500-2000 meters) is advisable to avoid hypoxia-induced pulmonary vasoconstriction. These patients may require supplemental oxygen when travelling by plane. Heavy physical activity should be avoided and it should be limited to symptom free level. Since infections may be fatal, steps to avoid them by means of pneumococcal and influenza vaccinations are also

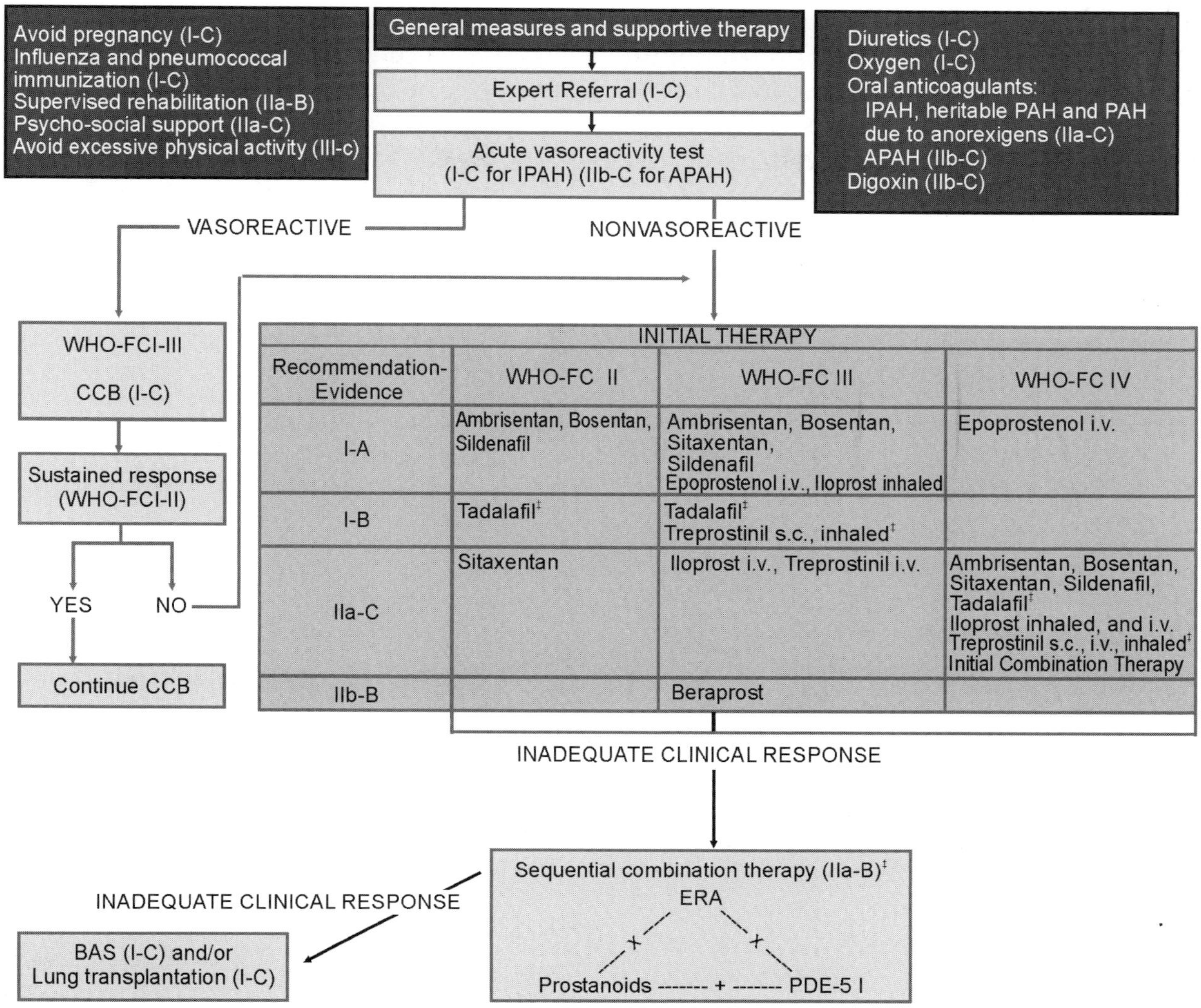

Fig. 1- Evidence-based treatment algorithm for pulmonary arterial hypertension patients (for group 1 patients only)

† To maintain arterial blood O_2 pressure ≥ 8 kPa (60 mmHg). ‡ Under regulatory review in the European Union. §IIa-C for WHO-FC II. APAH, associated pulmonary arterial hypertension; BAS, balloon atrial septostomy; CCB, calcium channel blocker; ERA, endothelin receptor antagonist; IPAH, idiopathic pulmonary arterial hypertension; PDE5 I; phosphodiesterase type-5 inhibitor; WHO-FC, World Health Organization functional class[38]. (Adapted from reference Nakanishi N. 2009 ESC/ERS Pulmonary Hypertension Guidelines and Connective Tissue Disease. Allergology International. 2011; 60: 419-424)

recommended. Infections and anaemia should be promptly treated. Pregnancy is also not recommended in women with PAH because of the high rate of associated mortality.

Medical treatment

Calcium channel blockers-No evidence exists for this treatment in CTD- PAH, and because of the potential negative inotropic effect, treatment should not be started without a positive acute vasoreactive test.

PDE-5 inhibitors - Two oral phophodiesterase-5 inhibitors (sildenafil and tadalafil) improved exercise, functional capacity and hemodynamics in large well conducted RCTs[39,40]. Side effects experienced were, headache, flushing, dyspepsia and epistaxis.

Endothelin (ET) receptor antagonists - Two oral ET receptor antagonists (bosentan and ambrisentan) are currently approved. Bosentan (which blocks both ETA and ETB receptors) increases exercise capacity and improve symptoms and pulmonary hemodynamics in patients with PAH[39]. Side effect profile includes headaches, hypotension and hepatic dysfunction. Ambrisentan (selective ETA receptor antagonist) has been shown to improve symptoms, exercise capacity, pulmonary hemodynamics and time to clinical worsening in patients with PAH[41]. Both are available in India.

Prostacyclins - Prostacyclins are the treatment of choice in patients with severe PAH but their nonavailability, high cost and adverse effects preclude their use. As of now, 4 prostacyclin analogs are licensed for use in PAH : epoprostenol, treprostinil, iloprost (in the USA and some European countries) and beraprost (in Japan and Korea) but these are not available in India.

Immunosupressive therapy - A recent trial has shown benefit with intensive immunosuppressive therapy using combination of cyclophosphamide and glucocorticoids on top of conventional vasodilator therapy, which markedly improved pulmonary hemodynamics and long term prognosis of patients with CTD-PAH other than SSc-associated PAH. Future studies involving larger numbers of patients are needed in this respect[42].

Lung transplantation (usually double lung), which has overall five year survival of around 50%, is indicated in severe PAH or inoperable CTEPH if medical management fails. Atrial septostomy decompresses the pressure loaded heart and may be useful in patients with severe PAH who have intractable syncope but not severe hypoxaemia, or as a bridge to transplantation.

Despite advances in understanding the pathogenesis and progress in the management of PAH, it still remains a life threatening condition. Recent data have shown efficacy for drugs targeting existing pathways, such as macitentan (endothelin receptor antagonist), riociguat (nitric oxide pathway via soluble guanylatecyclase stimulation), and Selexipag, (a prostacyclin receptor agonist). Despite promising effects on pulmonary haemodynamics the tyrosine kinase inhibitor (imatinib) has recently been withdrawn because of safety concerns. Studies with stem cell therapy are ongoing with promising results.

Key points

1. CTD-ILD is different from idiopathic variety in terms of mixed histological pattern and favourable response to immunosuppressive therapy.
2. Presence of mixed histological pattern must raise possibility of forme fruste of CTD even if patient has no evidence of CTD at that point of time.
3. Screening for PAH and early treatment is important to improve quality of life and prognosis.
4. Specific therapies are available for treating CTD-PAH.

Further Reading :

1. Colby TV. Pulmonary pathology in patients with systemic autoimmune diseases Clin Chest Med 1998; 19(4): 587-612.
2. Fischer A, Solomon JJ, du Bois RM, et al. Lung disease with anti-CCP antibodies but not rheumatoid arthritis or connective tissue disease. Respir Med 2012; 106: 1040-47.
3. Homma Y, Ohtsuka Y, Janimura K. Can interstitial pneumonia as the sole presentation of collagen vascular disease be differentiated from idiopathic interstitial pneumonia? Respiration 1995; 62: 248-51.
4. Gaude G S, MahishaleV, Srivastva A. Pulmonary Manifestations in Connective Tissue Disorders: Hospital-based Study at a Tertiary Care Hospital. Indian J Chest Dis Allied Sci 2009; 51: 145-51.
5. Park JH, Kim DS, Park IN, et al. Prognosis of fibrotic interstitial pneumonia: idiopathic versus collagen vascular disease-related subtypes. Am J RespirCrit Care Med 2007; 175: 705-11.
6. Karakostas KX, Constantopoulos SH, Moutsopoulos HH. Idiopathic pulmonary fibrosis and pulmonary fi brosis in diffuse systemic sclerosis: two fibroses with different prognoses. Respiration 1997; 64: 81-85.
7. Wells AU, Cullinan P, Hansell DM, et al. Fibrosing alveolitis associated with systemic sclerosis has a better prognosis than lone cryptogenic fibrosing alveolitis. Am J RespirCrit Care Med 1994; 149: 1583-90.
8. Cottin V. Interstitial lung disease: are we missing formes frustes of connective tissue disease? EurRespir J 2006; 28: 893-96.
9. Mittoo S, Gelber AC, Christopher-Stine L, Horton MR,Lechtzin N, Danoff SK. ascertainment of collagen vascular disease in patients presenting with interstitial lung disease. Respir Med 2009; 103: 1152-58.
10. Doyle TJ, Hunninghake GM, Rosas IO. Subclinical interstitial lung disease: why you should care. Am J RespirCrit Care Med 2012; 185: 1147-53.
11. Altman RD, Medsger TA Jr, Bloch DA , Michel BA. Predictors of survival in systemic sclerosis (scleroderma). Arthritis Rheum. 1991; 34(4): 403-13.
12. TashkinDP, Elashoff R, Clements PJ. Scleroderma Lung Study Research Group . Cyclophosphamide versus placebo in scleroderma lung disease . N Engl J Med. 2006; 354(25): 2655-66.
13. Nannini C, West CP, Erwin PJ, Matteson EL. Effects of cyclophosphamide on pulmonary function in patients with scleroderma and interstitial lung disease: a systematic review and meta-analysis of randomized controlled trials and observational prospective cohort studies. Arthritis Res Ther. 2008; 10(5): R124.
14. Kowal-Bielecka O, Landewé R, Avouac J, Chwiesko S, Miniati I, Czirjak L et al. EULAR recommendations for the treatment of systemic sclerosis: a report from the EULAR Scleroderma Trials and Research group (EUSTAR). Ann Rheum Dis. 2009; 68(5): 620-28.
15. Gerbino AJ, Goss CH, Molitor JA. Effect of mycophenolate mofetil on pulmonary function in scleroderma-associated interstitial lung disease. Chest. 2008; 133(2): 455-60.

16. Zamora AC, Wolters PJ, Collard HR, Connolly MK, Elicker BM, Webb WR et al. Use of mycophenolate mofetil to treat scleroderma-associated interstitial lung disease. Respir Med. 2008; 102(1): 150-55.
17. McGonagle D, Tan AL, Madden J, Rawstron AC, Rehman A, Emery P et al. Successful treatment of resistant scleroderma-associated interstitial lung disease with rituximab. Rheumatology (Oxford). 2008; 47(4): 552-53.
18. YooWH. Successful treatment of steroid and cyclophosphamide-resistant diffuse scleroderma-associated interstitial lung disease with rituximab. Rheumatol Int 2012; 32(3): 795-98.
19. Burt RK, Shah SJ, Dill K, Grant T, Gheorghiade M, Schroeder J et al. Autologous non-myeloablative haemopoietic stem-cell transplantation compared with pulse cyclophosphamide once per month for systemic sclerosis (ASSIST): an open-label, randomised phase 2 trial. Lancet. 2011; 378(9790): 498-506.
20. Castelino FV, Varga J. - Interstitial lung disease in connective tissue diseases: evolving concepts of pathogenesis and management. Arthritis Research & Therapy 2010; 12: 213. doi:10.1186/ar3097.
21. Chen IJ, Jan Wu YJ, Lin CW, et al. - Interstitial lung disease in polymyositis and dermatomyositis. Clin Rheumatol 2009; 28: 639-46.
22. Fujisawa T, Suda T, Nakamura Y, et al. - Differences in clinical features and prognosis of interstitial lung diseases between polymyositis and dermatomyositis. J Rheumatol 2005; 32: 58.
23. McGoon M, Gutterman D, Steen V, Barst R, McCrory DC, Fortin TA, et al. Screening, early detection, and diagnosis of pulmonary arterial hypertension. Chest. 2004; 126: 14S-34S.
24. Simonneau G, Robbins I, Beghetti M, Channick RN, Delcroix M, Denton CP, et al. Updated clinical classification of pulmonary hypertension. J Am Coll Cardiol 2009; 54: S43-54.
25. British Cardiac Society Guidelines and Medical Practice Committee. Recommendations on the management of pulmonary hypertension in medical practice. Heart 2001; 86(Suppl 1): 1-13.
26. Yoshida S, Katayama M. Pulmonary hypertension in patients with connective tissue diseases. Nippon Rinsho 2001; 59: 1164-67.
27. Rich S, Dantzker DR, Ayres SM, Edward HB, Brundage BH, Detre KM, et al. Primary pulmonary hypertension. A national prospective study. Ann Intern Med 1987; 107: 216-223.
28. Kumar U, Ramteke R, Yadav R, Ramam M, Handa R, Kumar A. Prevalence and predictors of pulmonary artery hypertension in systemic sclerosis. J Assoc Physic India 2008; 56: 413-17.
29. Haas C. Pulmonary hypertension associated with systemic lupus erythematosus. Bull Acad Natl Med 2004; 188: 985-97.
30. Shunji Yoshida. Pulmonary Arterial Hypertension in Connective Tissue Diseases. Allergology International. 2011; 60: 405-09.
31. Galiè N, Manes A, Farahani KV, Pelino F, Palazzini M, Negro L et al. Pulmonary arterial hypertension associated to connective tissue diseases. Lupus. 2005; 14(9): 713-17.
32. Chung L, Liu J, Parsons L, Hassoun PM, McGoon M, Badesch DB et al. Characterization of connective tissue disease-associated pulmonary arterial hypertension from REVEAL: identifying systemic sclerosis as a unique phenotype. Chest 2010, 138: 1383-94.
33. Farber HW, Loscalzo J. Pulmonary arterial hypertension. N Engl J Med 2004; 351: 1655-65.
34. Shunji Yoshida. Pulmonary Arterial Hypertension in Connective Tissue Diseases. Allergology International. 2011; 60: 405-09.
35. Budhiraja R, Tuder R M, Hassoun P M. Endothelial dysfunction in pulmonary hypertension. Circulation 2004; 109: 159-65.
36. Galie N, Manes A, Negro L, Palazzini M, Bacchi-Reggiani ML, Branzi A: A meta-analysis of randomized controlled trials in pulmonary arterial hypertension. Eur Heart J 2009, 30: 394-403.
37. Williams MH, Das C, Handler CE, Akram MR, Davar J, Denton CP, Smith CJ, Black CM, Coghlan JG: Systemic sclerosis associated pulmonary hypertension: improved survival in the current era. Heart 2006, 92: 926-32.
38. Nakanishi N. 2009 ESC/ERS Pulmonary Hypertension Guidelines and Connective Tissue Disease. Allergology International. 2011; 60: 419-24.
39. Galie N, Ghofrani HA, Torbicki A, Barst RJ, Rubin LJ, Badesch D, et al. Sildenafil citrate therapy for pulmonary arterial hypertension. N Engl J Med 2005; 353: 2148-57.
40. Galie N, Brundage BH, Ghofrani HA, Oudiz RJ, Simonneau G, Safdar Z, et al. Tadalafil therapy for pulmonary arterial hypertension. Circulation 2009; 119: 2894-903.
41. Channick RN, Simonneau G, Sitbon O, Robbins IM, Frost A, Tapson VF. Effects of the dual endothelin-receptor antagonist bosentan in patients with pulmonary hypertension: a randomised placebo-controlled study. Lancet. 2001; 358(9288): 1119-23.
42. Fukomoto Y, Shimokawa H. Recent Progress in the Management of Pulmonary Hypertension.Circ J 2011; 75: 1801-10.

EMERGENCIES IN RHEUMATOLOGY PRACTICE

9

Firdaus Fatima, V Shantaram

Introduction

Emergencies in rheumatological diseases may be encountered by rheumatologists, physicians or specialists in different fields of medicine. Multisystem problems involving heart, brain, lungs, kidneys, gastrointestinal, vascular and haemopoietic systems, eyes, skin and joints may present in various combinations and in a catastrophic manner causing considerable morbidity and mortality if not adequately managed. Besides, the disease itself may land the patient into emergencies. Early recognition and management of events may be life saving or at least help in preventing vital organ damage[1].

Rheumatological emergencies may be classified into following two categories.

1. Those occurring as complications of the disease process.
2. Iatrogenic: Drug induced disorders.

Table 1 depicts a few rheumatological emergencies organ wise and Table 2. enlists emergencies caused by certain rheumatic disorders[2].

Table 1: Rheumatological emergencies organ-wise

Central nervous system
- Encephalopathy
- Psychosis
- Cranial neuritis
- Seizures
- Intracranial bleed
- Cortical venous thrombosis

Eye
- Optic neuritis
- Iridocyclitis
- Scleritis
- Central retinal artery/vein occlusion
- Corneal melting
- Scleromalacia perforans

Cardiovascular
- Pericardial effusion
- Pericarditis, myocarditis, endocarditis
- Conduction abnormalities
- Myocardial infarction
- Impending gangrene with digital ischaemia
- Neonatal lupus

Lungs
- Pleural effusion
- Crico arytenoid arthritis
- Pneumonitis (lupus, infective or hypersensitive)
- Pulmonary hemorrhage, embolism
- Adult respiratory distress syndrome (ARDS)
- Vanishing lung syndrome
- Spontaneous pneumothorax in systemic sclerosis

Gastrointestinal
- Gastritis
- Gastro intestinal bleeding
- Pancreatitis

Kidney
- Acute glomerulonephritis
- Acute interstitial nephritis
- Diffuse arteritis

Haematological
- Thrombocytopenia
- Aplastic anaemia
- Leucopenia
- Haemolysis

Skin
- Exfoliative dermatitis
- Steven Johnson syndrome
- Toxic epidermal necrolysis

Bone and joints
- Septic arthritis
- Osteoporotic fractures
- Acute reactive arthritis
- Acute polyarthritis
- Acute gout
- Acute synovial rupture
- Atlanto axial dislocation
- Acute low back ache
- Avascular necrosis

Obstetric
- Pregnancy loss
- Flare of disease- lupus
- Foetal cardiac conduction abnormalities

Table 2: Rheumatological emergencies - disease-wise

Rheumatoid arthritis (RA)
- Atlanto-axial dislocation
- Scleromalacia perforans
- Vasculitis
- Infections
- Acute exacerbation of synovitis

Seronegative sponyloarthropathy (SpA)
- Iridocyclitis
- Cauda equina syndrome

Systemic lupus erythematosus (SLE)
- Seizures, psychosis, encephalopathy
- Pericarditis, myocarditis, endocarditis
- Pneumonitis, ARDS
- Acute glomerulonephritis
- Vasculitis
- Hypertensive crisis
- Acute pancreatitis
- Polyserositis
- Infections

Antiphospholipid antibody syndrome(APAS)
- Stroke
- Acute myocardial infarction
- Retinal vessel thrombosis
- Pulmonary embolism and infarction
- Thrombocytopenia
- Placental ischaemia and foetal loss
- Catastrophic APL syndrome

Vasculitis
- Cerebral vasculitis
- Optic neuritis
- Uveitis
- Mesenteric vasculitis
- Acute nephritis
- Hypertensive crisis

Systemic sclerosis and mixed connective tissue disease (MCTD)
- Digital vasculitis and ischaemia
- Scleroderma renal crisis

Inflammatory myositis (poly/dermatomyositis)
- Respiratory failure

Crystal-induced arthropathies
- Acute gout
- Acute interstitial nephritis

Arthritis related to infections
- Septic arthritis
- Reactive arthritis

Osteoporosis
- Fracture

Miscellaneous disorders
- Haemophilic arthropathy
- Rupture of Baker's cyst

Vasculitis

Necrotising vasculitis involving both small and medium sized arteries producing target organ ischaemia and eventual vital organ failure consequent to inflammation, thrombosis and occlusion of the blood vessel represents common basis for most emergencies in rheumatology. Vasculitis and its consequences may be primary or sole manifestation of a disease or it may be secondary to rheumatic diseases such as RA, SLE and non rheumatic diseases like sarcoidosis, malignancy and severe bacterial infections[3]. Medium vessel vasculitis like polyarteritis nodosa (PAN), anti neutrophil cytoplasmic antibody (ANCA) associated vasculitis can present with digital gangrene. Treatment is to be started after ruling out infection. Aggressive immuno-suppression is required using intravenous (IV) methyl prednisolone, IV immunoglobulin followed by oral steroids, cyclophosphamide or mycophenolate mofetil is required.

Lupus flare

Flare of existing lupus can be due to stress, exposure to sunlight, infection, withdrawal of steroid and pregnancy[4,17]. The precipitating factor of a flare should be identified and treated accordingly. A patient with flare of lupus can present with features of increased skin rash, arthritis, renal, serositis, vasculitic, or neuropsychiatric manifestations. Severe hematologic manifestations in the form of cytopenias are also seen. In a febrile lupus patient infection should be ruled out. Features differentiating infection and a flare are (i) Fever with chills is usually suggestive of infection (ii) Leukocytosis and raised C-reactive protein suggests infection (iii) high C3 C4 complement levels are seen in infections. However, it is to be remembered that sometimes infection and a flare can coexist making the differentiation difficult. Best treatment of lupus flares is by withdrawal of the precipitating factor and increasing the corticosteroid dose. Life threatening emergencies in SLE are listed in Table 3.

Table 3: Life threatening manifestations of SLE - Responses to glucocorticoids (GCS)

a. **Manifestations usually responsive to high doses of GCS**
   - Severe dermatitis of subacute cutaneous lupus
   - Acute polyarthritis
   - Polyserositis
   - Myocarditis
   - Lupus pneumonitis
   - Glomerulonephritis
   - Haemolytic anemia
   - Thrombocytopenia
   - Diffuse CNS syndrome
   - Lupus crisis (high fever and prostration)

b. **Manifestations not often responsive to GCS**
   - Thrombosis
   - Scarred end stage renal disease
   - Membranous glomerulonephritis
   - Psychosis not related to SLE

Scleroderma renal crisis

Scleroderma renal crisis constitutes one of the medical emergencies in rheumatology[5]. 70-80% of the renal crisis occurs in patients with diffuse cutaneous systemic sclerosis in the first 4-5 years of the disease. It is more often seen in winters especially in patients on steroids. Most of the patients have abnormal blood pressure but normotensive crisis may also be seen (10% of cases). Renal involvement in scleroderma renal crisis is characterized by acute onset of renal failure, abrupt onset of moderate to marked hypertension and urine sediment that is usually normal or reveals only mild proteinuria with few cells or casts. In addition to acute hypertension there is rising serum creatinine which may not become normal. Scleroderma renal crisis is thought to be due to severe Raynaud's of renal arteries. Risk factors for development of scleroderma renal crisis are diffuse skin disease, unexplained anemia, new cardiac event (congestive heart failure, pericardial effusion), high dose steroid, anti RNA polymerase III antibody. The features of this syndrome are accelerated hypertension, dyspnoea, headache, seizures, rapidly progressive renal failure, thrombocytopenia, microangiopathic hemolytic anemia, congestive heart failure, pericardial effusions, fundoscopic changes and urinary abnormalities. Scleroderma renal crisis can be fatal if not diagnosed and managed promptly. ACE inhibitors are useful in hypertensive and normotensive crisis. Careful management of hypertension, tapering and withdrawal of steroids to avoid addisonian crisis is needed. Dialysis and renal transplant may also be needed in patients with irreversible kidney damage.

Catastrophic antiphospholipid syndrome

Catastrophic antiphospholipid syndrome is a devastating emergency. It usually presents as multiorgan failure[6,7]. Antiphospholipid syndrome can be primary or secondary to connective tissue diseases, malignancy and infections. Pulmonary, gastrointestinal, cardiac, renal and neurological involvement is common. If three or more organs are involved in antiphospholipid antibody syndrome it is called catastrophic antiphospholipid syndrome. Precipitating factors are surgery, infections, oral contraceptive use and withdrawal of anticoagulants. Antiphospholipid antibody, lupus anticoagulant or anti beta 2 glycoprotein 1 antibody can be positive. Patients with antiphospholipid antibodies may present as emergencies owing to vascular occlusion in both arterial and venous system. Unexplained sudden limb ischaemia, acute pulmonary thromboembolism, acute myocardial infarction, acute cerebrovascular accident, acute bowel infarction or intestinal ischaemia, acute Budd Chiari syndrome, epilepsy, severe haemolytic anaemia and thrombocytopenia, abortion and other obstetric complications may be the presenting feature of primary antiphospholipid antibody syndrome. Stroke at a young age is one of the commonest manifestations of antiphospholipid syndrome[8,9]. SLE, RA and other rheumatological entities may have antiphospholipid antibodies with thrombotic complications superadded on the background features which characterize their specific diagnosis. Catastrophic vascular occlusion leading to multiorgan failure is usually sudden, diagnostically confusing and immediately life threatening. Despite adequate treatment mortality is high.

A clinical diagnosis in a patient with rheumatological emergency can be made by thorough history and clinical examination. After clinical diagnosis full laboratory workup (Table 4) should be done in a comprehensive manner to confirm the diagnosis and objectively evaluate the various organ dysfunctions.

Table 4 : Laboratory work up

Clinical pathology and microbiology	
Haemoglobin	Complete urine examination
Leucocytes, platelets, eosinophils	(Proteinuria, haematuria, granular casts)
ESR, plasma viscosity	24 hrs urinary proteins
Cryoglobulin screening	Urine, Blood, Synovial fluid cultures
Biochemistry	
Blood sugar	
Urea, creatinine, electrolytes, uric acid	
Liver function tests	
Coagulation profile, lupus anticoagulant	
Immunology	
Antinuclear factor (ANF)	
Rheumatoid factor (RF)	
IgG (Polyclonal)	
Rising ANCA titers or conversion from ANCA negative to ANCA positive	
Serum IL-2 receptor assay	
Anti-GBM (anti - glomerular basement membrane) antibody	
Antiphospholipid antibodies- IgG, IgM, IgA	
Coombs' test	
Serology	
Hepatitis B and C, HIV antibody	
Imageology and Radiology	
Chest X-ray, abdominal ultrasound,	
CT scan/MRI	
ECHO cardiography	
Nuclear medicine	
Lung perfusion scan,	
Renal perfusion scan	
Isotope venogram	

General principles of management of rheumatological emergencies

The primary aim of treatment is to save life and prevent irreversible organ damage by instituting suitable and aggressive therapy. It is useful to have a team approach involving anesthetist, nephrologists, cardiologist and other relevant specialists. Basic emergency therapy common to all organ involvement is the one directed to suppress the immunologic damage[10]. Before starting aggressive treatment with high dose of glucocorticosteroid infections must be excluded. Co morbid conditions like hypertension, diabetes mellitus, obesity, osteoporosis and psychiatric disorders should also be considered.

Aggressive immunosuppressive therapy- pulse therapy is useful. The regimen consists of : IV Methylprednisolone 0.5 g/day for 3 days with IV Cyclophosphamide 0.5 g-1 g, single dose. Repeat pulse therapy may be given every 3-4 weeks till remission is achieved. It is safe and effective. Alternate regimens available are: Oral prednisolone 1 mg/kg/day + IV cyclophosphamide once in month OR Oral prednisolone 1 mg /kg/day+ Azathioprine 2 mg/kg/day, Plasmapheresis[11] is useful with short term benefit.

Other appropriate therapy

Anticoagulation

Patient coming with complications consequent to arterial and venous occlusions as in primary or secondary antiphospholipid syndromes require effective anticoagulant therapy with IV heparin followed by oral anticoagulation. Acute pulmonary vascular disease with pulmonary hypertension owing to vasculitis and thrombosis requires full dose of anticoagulation.

Anticoagulation therapy consists of

a) IV drip of Heparin 1000 IU/hr for first 24 to 48 hours (dose adjusted according to APTT): OR Low molecular weight heparin which is free of side effects like platelet reduction, osteoporosis and does not require monitoring by coagulation tests.
b) Followed by warfarin oral anticoagulant therapy to be continued on long term basis and dose adjusted with prothrombin time level (INR 2.3).

New therapies for aggressive SLE include IV gamma globulins, mycophenolate mofetil and immunoablation with autologous stem cell transplantation. Trials with B-lymphocyte depletion by rituximab have shown promising results[12].

Antihypertensive drug therapy

Hypertensive crisis may be the presenting feature of some collagen vascular diseases due to involvement of kidney. Most common among them is lupus nephritis which warrants not only adequate control of blood pressure but immunosuppressive therapy also. PSS/PAN are other important causes. Angiotensin converting enzyme (ACE) inhibitors are particularly valuable in SLE, PSS and other nephritic disorders. Calcium channel blocker, post-synaptic alpha blockers and diuretics are also helpful. Intravenous sodium nitroprusside or nitroglycerine may be given in accelerated/malignant hypertension to prevent target organ damage of brain, heart and kidneys. In aortoarteritis presenting with severe reno vascular hypertension, balloon angioplasty and surgery are useful.

Acute cardiac emergencies

Pericarditis is common in collagen vascular diseases like SLE but massive pericardial effusion presenting with cardiac tamponade is rare. Once diagnosed by clinical and echocardiographic means, pericardiocentesis along with steroid therapy should be extended. However, if on analysis of pericardial fluid infective process like tuberculosis is suggested anti-Koch's therapy should be given. Acute myocarditis with congestive heart failure and conduction and rhythm problems require intensive therapy. The cornerstone in management represents steroid immunosuppressive therapy along with cyclophosphamide and IV immonoglobulins if required. Acute myocardial infarction may be the presenting feature of antiphospholipid antibody syndrome, PAN and SLE. Besides routine therapy, anticoagulants are important to salvage these patients. Anticoagulation is also required in cardiomyopathy to prevent thrombus formation and thromboembolism.

Endocarditis may be encountered in some patients with SLE (Libman Sacks) and anticardiolipin antibody syndrome with sterile blood culture. Peripheral embolism from left atrial clot may occur resulting in ischaemia in other organs. Anticoagulation is needed in these patients. If infective basis is suspected full course antibiotic therapy should be extended guided by blood culture and sensitivity study.

Neurological emergencies

Commonest presentations are due to vasculitis that may manifest as encephalopathy, focal or multifocal CNS disturbances, hemorrhagic or ischaemic stroke and myelitis. Cranial neuropathies single or multiple may occur. Characteristic rheumatologic disorders associated with such presentation are Behcet's disease, SLE and PAN while they are distinctly rare in RA and SSA. The value of corticosteroid and other immunosuppressive agents is established as the basic pathogenetic mechanism in CNS damage is immune mediated. Pulse therapy may be needed in serious cases. However, in certain subset of patients with antiphospholipid antibody syndrome cortical micro infarcts due to thrombotic occlusions occur and anticoagulant therapy is indicated. Stenting of the occluded vessel may also be needed. Sometimes vasculitic and thrombotic process may co-exist and in acute stages it is safer to treat both. In the management of CNS lupus before starting steroids it is important to rule out infection by doing a CSF examination. The treatment comprises anti-coagulation with high dose immunosupression with pulse IV methylprednisolone or IV immunoglobulin. If seizure is the presenting feature, antiepileptic drugs have to be employed. Acute psychiatric illness may be the presenting feature of lupus cerebral vasculitis where anti psychotic drugs are used besides steroid and immunosuppression. However if patients with SLE on steroid therapy develop abnormal behavior, steroid induced psychosis has to be considered. Steroid psychosis usually develops within a week of starting high dose steroids.

Ocular emergencies

Ischaemic optic neuropathy due to vasculitis may be observed in temporal arteritis, PAN and SLE while retinal vasculitis is observed in Behcet's syndrome. Immediate steroid therapy using high dose prednisone (1 mg/kg/day) may save vision in the vast majority of subjects. Painful red eye due to conjunctivitis or iridocyclitis may be encountered in seronegative spondyloarthropathies (SSA) and Behcet's syndrome. Granulomatous scleritis is observed in rheumatoid arthritis. Iridocyclitis is a major problem in children with pauciarticular arthritis and positive antinuclear antibodies. It can lead to blindness from secondary glaucoma, cataract formation and band keratopathy. These patients should be handled by taking help from ophthalmologists. Use of local drugs like mydriatics and corticosteroid eye drops beside systemic steroids are recommended. Uveitis usually responds to local steroids and systemic steroids may not be needed. Scleromalacia perforans is an extremely rare complication of RA; it requires urgent medical and surgical intervention.

Emergencies involving peripheral joints and the spine

Septic arthritis

Septic arthritis is a medical emergency that is associated with a high mortality and joint destruction or deformity if left untreated. It is associated with high spiking fever, severe swelling and redness of the involved joint which is usually a large joint. The presentation is usually monoarticular involving large joints; polyarticular involvement is also seen e.g. gonococcal arthritis and associated with active bacteremia. Septic arthritis is particularly observed in elderly patients, diabetics, chronically debilitated or immunocompromised subjects or RA. Patients with prosthetic joints are especially susceptible to infection. Usual bacterial pathogen is Staphylococcus or Streptococcus but a wide variety of microbes may be involved like Gram-negative bacteria and anaerobes, mycobacteria or even fungi.

Synovial fluid gram stain and cultures will be helpful in confirming the diagnosis and identifying the organism. Management involves immediate implementation of appropriate antibiotics given parenterally for the first 2 weeks, followed by oral antibiotics for at least a further 4 to 6 weeks. Usually prolonged antibiotic regime is required in RA patients where there is high chance of sequestration of organism due to altered anatomy and bone erosions. Joint drainage done by closed needle aspiration is needed as long as effusion is detectable. Tissue debris of clots may be removed by sterile saline irrigation. Septic joint is one of the most preventable and successfully treatable rheumatologic emergencies.

Acute arthritis

Acute arthritis may start de novo or as an exacerbation of chronic arthritis like rheumatoid arthritis. Anatomically and functionally deranged joints are susceptible to septic and mechanical complications. Acute arthritis can be mono, oligo or poly arthritis.

Acute mono arthritis can be due to septic arthritis, gout or trauma. Acute oligo or polyarthritis can be due to viral arthritis, reactive arthritis, Reiter's syndrome, rheumatic fever, HIV, disseminated gonococcal infection etc. Infection has to be ruled out. These conditions have to be differentiated from rheumatoid arthritis by blood tests and X ray[13]. Treatment consists of non steroidal anti-inflammatory drugs (NSAIDS), low dose corticosteroids and disease modifying drugs when needed. In acute exacerbation of chronic arthritis such as rheumatoid arthritis adequate management with NSAIDS, low dose steroids, increased DMARDS and addition of a biologic response modifier is needed.

Acute gout

Acute gout is one the commonest rheumatological emergencies. The patient, usually a male, presents with acute pain in one of the lower limb joints, commonly the first metatarsophalangeal joint. The presentation is usually with monoarthritis (single joint) or oligoarthritis (2-4 joints). Acute gout is common in older individuals, mostly men. The acute stage is managed with NSAIDs, colchicine and corticosteroids. Allopurinol should not be started in the acute phase. When the acute phase settles allopurinol is started and is to be continued lifelong.

Atlantoaxial and sub-axial subluxation

Atlantoaxial subluxation (AAS) in RA results from damage to bones, ligaments and cartilages caused by invasive rheumatoid synovial tissue. Proliferating synovial tissue with its associated inflammation can weaken or damage the transverse ligament and erode or fracture the odontoid process which can lead to subluxation. It can be precipitated by whiplash injury or trauma to the neck. Furthermore, destruction of the lateral atlantoaxial facet, atlantooccipital joints and their associated ligaments contribute to the neurologic complications of rheumatoid cervical spine.

In all varieties of cervical spine involvement spinal cord compression may be potentially life-threatening. In addition to manifestations from direct compression of nervous system structures, signs and symptoms of vascular insufficiency to neural tissues can be evident e.g. vertebrobasilar insufficiency. MRI may play a prominent role in identifying the patients with impending neural compression who will require prompt therapeutic intervention.

Atlantoaxial instability can be conservatively managed by using Philadelphia collar. Fusion with or without decompression is the usual procedure adopted. Post operative mortality is less than 10%. Care should be taken during surgery to avoid forceful neck movements which can result in dislocation, fracture and death. Pain reduction with surgery is almost certain while neurologic improvement occurs in most individuals.

Acute synovial rupture

This is a fairly common complication of synovitis that may be precipitated by muscle exertion. It is classically seen in RA patients and some patients with osteoarthritis having Baker's cyst in the popliteal fossa as a complication of knee joint disease, and patients with early knee effusion are particularly at risk. Rarely wrist, shoulder and elbow joints may also be involved. Synovial fluid from ruptured Baker's cyst usually trickles down into calf, but rupture of popliteal cyst must be carefully distinguished from deep vein thrombosis as the erroneous use of anticoagulants in the former will lead to hemorrhage into the joints and calf leading to secondary muscle necrosis with subsequent contracture and rarely Sudeck's atrophy.

Diagnosis is mainly clinical and may be confirmed by an arthrogram or ultrasonogram. Management includes rest to the joints, elevation of legs and aspiration of joints along with intraarticular corticosteroid injection to prevent fluid reaccumulation. Early mobilization and muscle strengthening exercises are needed when acute symptoms subside.

Avascular necrosis

Avascular necrosis may present with acute excruciating pain in the joint. It is commonly seen in patients on long term steroids and SLE. It can also be seen in antiphospholipid antibody syndrome. Treatment is adequate analgesia, bone grafting or joint replacement.

Low backache

Low back ache (LBA) can commonly present as a medical emergency[14]. Individuals especially at risk of LBA include manual labourers, truck drivers, material handlers, medical care providers eg. who shift trolleys and patients, persons caring for infants (repeated lifting from cribs, play pens etc.), weight lifters, dancers, divers, and gymnasts. The notable thing about acute LBA is its favourable natural history where 90% cases recover within 6 weeks.

The first task of the clinician should be to rule out medical emergencies, viz., cauda equina syndrome, epidural abscess, rupture of aortic aneurysm or dissection of aorta. It is important to know if the pain is inflammatory in nature. The clinician should look for saddle anaesthesia, bladder/bowel involvement, asymmetric loss of reflexes, pulse inequality, hypotension, or circulatory instability. Adequate use of NSAIDs is required for pain relief. In acute flare of spondyloarthropathy low dose corticosteroid, increase in dose of DMARDs, addition of another DMARD or biologic response modifiers like etanercept and infliximab is needed. A neurological compromise or features of cord compression warrants urgent CT myelogram or MRI and neurosurgical intervention. Cardiovascular catastrophes warrant angiography and surgery.

Osteoporotic vertebral collapse can present with acute backache with or without neurological deficit. Pain management is by NSAIDs; salmon calcitonin nasal spray is also used. In some patients vertebroplasty may be needed.

Macrophage activation syndrome

Macrophage activation syndrome is a newly recognised life threatening complication of rheumatic diseases. It is more frequently seen in adult onset Stills disease and systemic onset juvenile idiopathic arthritis, systemic lupus erythematosus and other connective tissue disorders[15]. The condition is triggered by viral infection especially Ebstien Barr, Cytomegalo virus and drugs such as parenteral gold, sulphasalazine, methotrexate and TNF blockers like etanercept and anti IL 1 receptor therapy (anakinra). Macrophage activation syndrome is characterized by non remitting high fever, pancytopenia, hepatosplenomegaly, lymphadenopathy, increased liver enzymes, high triglycerides, high lactic dehydrogenase, low sodium levels and abnormal coagulation profile. Treatment is mainly with parental high dose corticosteroids. Use of IV immunoglobulins, cyclophosphamide, plasma exchange and etoposide have shown conflicting results.

Emergencies due to iatrogenic complications

Non-steroidal anti-inflammatory drugs (NSAIDs) are the most widely used amongst all pharmacologic agents in rheumatological practice. Their major drawbacks include gastrointestinal side effects due to gastropathy and the recently recognized enteropathy. The risk is cumulative and dose-related. Serious gastrointestinal side effects due to NSAIDs can be prevented by concomitant use of proton pump inhibitors, oral PGE2 analogue misoprostol or H2-receptor antagonists.

There are several other complications associated with the use of NSAIDs that may pose as serious emergencies. These include exfoliative dermatitis or vesiculobullous eruptions (especially with azapropazone, fenbufen), blood dyscrasias (especially with phenylbutazone) and allergic asthmatic attack (mainly caused by aspirin due to prostaglandin inhibition).

Renal failure may occasionally be precipitated with the use of NSAIDs especially in elderly patients and those with impaired renal function. Reduced glomerular filtration is the result of altered prostaglandin dependent renal blood flow which may lead to renal papillary necrosis. Acute adrenal insufficiency may be precipitated by corticosteroid withdrawal in patients who have received prednisolone in doses as low as 3 mg/day for a prolonged duration or within 7 to 28 days with doses of 20 to 30 mg/day. Recovery of the hypothalamic-pituitary-adrenal axis may take as long as 1 year. One must be particularly cautious about this condition during the stress of intercurrent illness or surgery and adequate corticosteroid coverage is necessary. Drug induced emergencies are enumerated in Table 5.

Disease-modifying anti-rheumatoid drugs (DMARDs) and immunomodulators are mainstay of the treatment in rheumatological practice. Exfoliative dermatitis and Steven-Johnson syndrome are the dreaded complications with the use of gold, penicillamine, sulphasalazine and other DMARDs. Gold salts and penicillamine are not used now a days . The major drawback in the long-term use of DMARDs is the wide variety of side effects associated with their use if not properly monitored. Long term use of these agents can cause complications like sepsis, neutropenia, anaemia, thrombocytopenia etc. Haematological toxicity is the most serious complication of DMARDs and immuno-suppressive agents; hence regular blood monitoring is required. Profound leucopoenia may lead to fatal infections while thrombocytopenia may lead to fatal bleeding. Treatment of such conditions includes appropriate supportive measures, blood component therapy and bone marrow stimulation with growth factors. Methylprednisolone in bolus doses may be life saving in such cases.

A rare dangerous side effect of methotrexate is interstitial pneumonitis or "methotrexate lung".

Table 5: Drugs - Emergencies

Nonsteroidal anti-inflammatory drugs
- Acute gastritis
- Perforated / bleeding peptic ulcer
- Acute enteritis
- Analgesic nephropathy
- Acute interstitial nephritis
- Hypersensitivity reactions
- Thrombocytopenia
- Pancytopenia
- Steven Johnson syndrome
- Erythema multiforme
- Toxic epidermal necrolysis
- Acute interstitial nephritis

Glucocorticoids
- Psychosis
- Acute Addisonian crisis due to withdrawal

Disease modifying antirheumatic drugs
- Hypersensitivity reactions
- Thrombocytopenia
- Aplastic anaemia
- Exfoliative dermatitis
- Steven Johnson syndrome

Biologic response modifiers
- Infusion reactions
- Infections

Rheumatological emergencies are not uncommon in clinical practice. The emergencies as described can be life threatening and in some instances, referrals to other specialists may be needed[16]. Therapy should be instituted quickly and aggressively.

Further Reading :

1. Kumar A, Marwaha V, Grover R. Emergencies in rheumatology. J Ind Med Assoc. 2003; 101: 520-26.
2. Rao U R K, Shantaram V. Rheumatolological emergencies Supplement to JAPI . 2006; 54: 58-61.
3. Gonzalez-Gay MA, Garcia-Porrua C. Epidemiology of the vasculitides. Rheum Dis Clin N Am. 2001; 27: 729-49.
4. Clowse M E B. Lupus Activity in Pregnancy. Rheum Dis Clin Nor Am. 2007; 33: 237-52.
5. Steen VD. Scleroderma renal crisis. Rheum Dis Clin Nor Am. 2003; 29: 315-33.
6. Ricard Cervera, Asherson RA, Josep Font. Catastrophic Antiphospholipid Syndrome. Rheum Dis Clin Nor Am. 2006; 32: 575-90.
7. Asherson R A et al. Catastrophic antiphospholipid syndrome: international consensus statement on classification criteria and treatment guidelines. Lupus. 2003; 12(7): 530-34.
8. Singh K, Gaina M, Shome DK et al. The association of anti phospholipid antibodies with ischemic stroke and myocardial infarction in young and their correlation: A preliminary study. JAPI. 2001; 49: 527-29.
9. Nagaraja D, Christopher R, Manjari T. Anticardiolipin antibodies in ischemic stroke in the young: Indian experience. J Neurol Sci. 1997; 159: 137-42.
10. Bansal VK, Beto JA. Treatment of lupus nephritis: A metaanalysis of clinical trials. Am J Kidney Dis.1997; 29: 193-99.
11. Lewis EJ, Hunsicken LG, Lan SP, Rohde RD, Lachin JM. A controlled trial of plasmapheresis therapy in severe lupus nephritis. N Engl J Med. 1992; 326: 1373-79.
12. Perez M V, CastroB H, Saharopulos O P et al. Clinical and immunological effects of Rituximab in patients with lupus nephritis refractory to conventional therapy: a pilot study. Arthritis Res Ther. 2006; 8: R83.
13. John D. Reveille et al. Spondyloarthritis: update on pathogenesis and management. The Am J of Medicine. 2005; 118(6) : 592-603.
14. Handa R, Aggarwal P, Wali J P. Rheumatological emergencies in clinical practice. J Ind Acad Clin Med. 2000; 5: 135-41.
15. Ravelli A. Macrophage activation syndrome. Curr Opin Rheumatol. 2002; 14 (5): 548-52.
16. Smith EC, Berry H, Scott DL. The clinical need for an acute rheumatology referral service Br J Rheumatol. 1996; 35: 389-91.
17. Joshi V R, Bhogani K. Emergencies in systemic lupus erythrmatosus. Post Grad Mad 2004; 18: 343-47.

RHEUMATOLOGICAL MANIFESTATIONS OF SYSTEMIC DISEASES

10

Milind Nadkar, Vikram Londhey

Introduction

The differential diagnosis of most of the rheumatological symptoms consist of some or the other systemic non-rheumatological diseases. Hence, it is important not to overlook a systemic cause of a rheumatological symptom. These symptoms may either be as a presenting feature of the systemic disease or present during the course of the disease e.g. haemarthrosis may be the initial presenting feature of haemophilia, whereas adhesive capsulitis may be seen in a known diabetic patient. A rheumatologist or a primary care physician must be conversant with the various systemic diseases which can affect the musculoskeletal system. Table 1 enumerates the common systemic diseases having rheumatological manifestations; however one must be aware that this list cannot be complete as many uncommon diseases may also have rheumatological manifestations. Also, one must be aware that many drugs produce musculoskeletal symptoms as their side effects.

Systemic diseases can have protean rheumatological manifestations. Table 2 lists various rheumatological manifestations with common systemic diseases. Some of the systemic diseases with rheumatological manifestations are discussed here. However, infections including HIV, connective tissue diseases, and drug related side effects are not discussed here

Table 1. Systemic diseases having rheumatological manifestations

Endocrine disorders	**Renal diseases**	**Neurological diseases**	**Others**
Diabetes mellitus	Chronic kidney disease	Neuropathies	Haemochromatosis
Hypothyroidism	Renal transplant	Myopathies	Amyloidosis
Hyperthyroidism	Renal tubular acidosis	Focal neurological deficits	Sarcoidosis
Hypoparathyroidism	**Malignancies**		Porphyria
Hyperparathyroidism	Primary malignancies	Hemiplegia/paraplegia	
Acromegaly	Tumour lysis syndrome	Parkinsonism	
Cushing's syndrome	Paraneoplastic syndromes		
Addison's disease			
Haematological diseases	**Inborn errors of metabolism**		
Haemophilia	Homocysteinuria		
Sickle cell disease	Alkaptonuria		
Thalassaemia	Mucopolysaccharidosis		
Leukaemias	Fabry's disease		
Lymphomas			
Myelomas			

* Infections including HIV, connective tissue diseases, heritable collagen diseases and drug related side effects are not discussed here.

Table 2. Rheumatological manifestations in systemic diseases

Clinical feature	Systemic diseases
Monoarthritis	Haemophilia, Paget's disease
Polyarthritis	Sickle cell disease, hypothyroidism, acromegaly, amyloidosis, sarcoidosis, haemochromatosis, ochronosis
Adhesive capsulitis	Diabetes mellitus, thyroid disorders
Pseudogout	Diabetes mellitus
Hypertrophic osteoarthropathy	Cyanotic congenital heart diseases, cirrhosis of liver, inflammatory bowel disease, thyroid acropachy
Neuropathic joints	Neuropathies, tabes dorsalis, diabetes mellitus
Muscle weakness	Hypothyroidism, hyperthyroidism, hypoparathyroidism, hyperparathyroidism, acromegaly, cushing's syndrome

Erythema nodosum	Ulcerative colitis, sarcoidosis, regional enteritis, malignancies
Pyoderma gangrenosum	Inflammatory bowel disease, leukaemia, myelofibrosis
Reflex sympathetic dystrophy syndrome	Hemiplegia, syringomyelia, peripheral neuropathy diabetes mellitus
Relapsing polychondritis	Thyroiditis, ulcerative colitis, psoriasis
Carpal tunnel syndrome	Hypothyroidism, acromegaly, diabetes mellitus
Meralgia paraesthetica	Diabetes mellitus
Osteomalacia	Vitamin D deficiency, Crohn's disease, malabsorption syndrome, chronic kidney disease, hepatobiliary diseases, pancreatic insufficiency, fanconi syndrome
Osteoporosis	Cushing's syndrome, hyperparathyroidism, hypogonadism, hypopituitarism, thyrotoxicosis, multiple myeloma, leukaemias, lymphoma, insulin-dependent diabetes mellitus, bowel diseases, mastocytosis
Osteonecrosis	Haemoglobinopathies, haemophilia, coagulopathies, sickle cell disease, cushing's syndrome, chronic liver disease, sepsis, alcoholism, pancreatitis, post-organ transplant, chronic dialysis

Endocrine disorders

Most of the endocrine disorders have some or the other musculoskeletal manifestations[1,2].

Diabetes mellitus

Diabetes mellitus has a plethora of musculoskeletal manifestations. These are enumerated in Table 3[3,4].

Table 3. Rheumatological manifestations in diabetes mellitus

- Carpal tunnel syndrome/tarsal tunnel syndrome
- Adhesive capsulitis (frozen shoulder) involving shoulder
- Trigger finger/thumb
- Charcot's arthropathy (neuropathic joints)
- Stiff hand syndrome 'Prayer sign'
- Diffuse idiopathic skeletal hyperostosis (DISH)
- Dupuytren's contracture
- Reflex sympathetic dystrophy syndrome (RSDS)
- Amyotrophy
- Osteomyelitis/ septic arthritis/ septic bursitis
- Predisposition to gout and calcium pyrophosphate deposition disease (CPPD)

Carpal tunnel syndrome (CTS) may be seen in 25% of diabetics[5]. Nocturnal paraesthesia is the predominant feature which may be confused with diabetic peripheral neuropathy. Timely surgical intervention is required to prevent thenar muscle atrophy.

Adhesive capsulitis or frozen shoulder is a common feature seen in diabetics. It may be associated with calcific tendinitis, painful tendinitis and bursitis and reflex sympathetic dystrophy syndrome (RSDS).

Stiff hand syndrome also called limited joint mobility syndrome or cheiroarthropathy is seen in type 1 and type 2 diabetes. Excessive glycosylation of collagen in the skin and periarticular structures and decreased collagen degeneration and removal are the pathogenetic mechanism which results in thick and inelastic tissue. Permanently contracted fingers at metacarpophalangeal (MCP) and proximal interphalangeal (PIP) joints with thick shiny skin are typical.

Trigger finger or thumb is a painful condition due to flexor tenosynovitis frequently seen in diabetes having pathogenetic mechanism similar to stiff hand syndrome. Dupuytren's contractures can occur alone early in the course of diabetes or may be seen with stiff hand syndrome.

Diffuse idiopathic skeletal hyperostosis (DISH) also called Forestier disease is commonly seen in type 2 obese diabetic patients. It manifests by proliferation of new bone formation at joint margins commonly seen in the axial skeleton. The postulated pathogenetic mechanism is related to hyperlipidaemia and other metabolic factors than diabetes itself.

Charcot joints or neuropathic arthropathy is commonly seen in diabetes with sensory neuropathy, causing damage and derangement of any weight-bearing joint like knee, ankles and small joints of the feet. The onset is usually sudden following a trivial trauma. The radiological findings are out of proportion to the degree of pain.

Osteomyelitis of the foot is a common problem in long standing diabetes which does not draw attention of the patient till it is advanced. This is due to peripheral sensory neuropathy, foot injuries, pressure ulcers and susceptibility to infection which occur due to poor foot care. Septic arthritis and septic bursitis caused by *Staphylococcus aureus* and Gram-negative organisms is also a common problem in diabetics.

Both gout and pseudogout are seen more frequently in diabetics as compared to general population.

Thyroid disorders

Osteopenia, Osteoporosis, pretibial myxoedema, thyroid acropachy, adhesive capsulitis, periarthritis and proximal muscle weakness are the common musculoskeletal disorders seen in hyperthyroidism[5-8].

Hashimoto's thyroiditis and Grave's disease have a strong relationship with other rheumatic diseases.

Arthritis seen in hypothyroidism may mimic rheumatoid arthritis, gout or pseudogout on clinical presentation. This arthropathy typically resolves with correction of hypothyroid state. Flexor tenosynovitis and thickening of transverse carpal ligament may result in carpal tunnel syndrome which is commonly seen in hypothyroidism.

Myopathy of hypothyroidism resembles polymyalgia rheumatica or polymyositis depending on whether stiffness or weakness is the predominant feature. Hoffman syndrome seen in patients with myxoedema is characterised by muscular weakness and stiffness with increased muscle mass.

Parathyroid disorders

Hyperparathyroidism
Rheumatic manifestations of hyperparathyroidism are listed in Table 4[1,2].

Table 4. Rheumatic manifestations of hyperparathyroidism

- Pseudogout/ chondrocalcinosis associated with CPPD
- Polyarthritis mimicking rheumatoid arthritis
- Shoulder joint arthropathy
- Tendon rupture and lax ligament
- Hyperuricaemia
- Subperiosteal bone resorption and Brown tumour
- Osteoporosis
- Proximal muscle weakness

Shoulder joint arthropathy occurs in hyperparathyroidism. It gives rise to intra-articular and periarticular erosion on the head of humerus. Some patients may be asymptomatic in spite of radiographic changes. The common articular involvement in hyperparathyroidism may resemble rheumatoid arthritis (RA), pseudogout or gout.

In primary hyperparathyroidism, Brown tumour which consists of osteoclasts-osteoclastomas and chondrocalcinosis is seen.

Hypoparathyroidism

Spondylitis is a feature of hypoparathyroidism. Sacroiliac joint is typically spared and ankylosis is secondary to extensive calcification of paraspinal ligaments. Hypocalcemia caused by hypoparathyroidism can be associated with myopathy, rhabdomyolysis and neuromyotonia[9]. Subcutaneous calcifications can occur in patients with hypoparathyroidism particularly around the hips and shoulders. It can remain asymptomatic or present as painful calcific periarthritis.

Acromegaly

Musculoskeletal manifestations of acromegaly are listed in Table 5[10]. The pathogenetic mechanism of these manifestations is the result of chronic overstimulation of osteoblasts, osteoclasts, fibroblasts, chondrocytes and muscle cells by growth hormone. Absence of morning stiffness is characteristic of acromegalic arthropathy.

Table 5. Musculoskeletal manifestations of acromegaly

- Bony proliferation
- Synovial thickening
- Bursal hyperplasia
- Backache
- Premature osteoarthritis
- Cartilage hyperplasia
- Hypermobility
- Carpal tunnel syndrome
- Proximal muscle weakness
- Raynaud's phenomenon

Cushing's syndrome

Osteonecrosis, osteoporosis and steroid myopathy are common occurrences in Cushing's syndrome.

Adrenal insufficiency

Myalgias, arthralgias and back pain can be seen in chronic adrenal insufficiency.

Haematological disorders

Haemophilia

The articular disease in haemophiliacs includes acute haemarthrosis, subacute or chronic arthritis and end-stage haemophilic arthropathy, septic arthritis and muscle and soft-tissue haemorrhage[11,12,22]. Nearly all patients experience acute bleeding in the joints, a complication of haemophilia requiring therapeutic intervention, and when recurrent leads to chronic deforming arthritis.

Acute haemarthrosis

Acute haemarthrosis first occurs in childhood when child begins to walk. As the child grows older the frequency diminishes, but may persist till adulthood. Stiffness, warmth and severe pain in the affected joint are seen due to distension of joint capsule. If haemostasis is achieved early after the onset of haemarthrosis, joint function may return to normal within 24 hours. If haemostasis is not achieved early or if haemorrhage is more advanced, joint function may take as long as 10-14 days for recovery.

Subacute/ chronic arthritis

Recurrent haemarthrosis may lead to joint abnormalities which persist in intervals between bleeding episodes. The affected joint has chronic swelling, minimal or no pain and is slightly warm. Chronic synovitis may persist with or without joint effusion. There is restriction of movement and flexion deformity in the joint involved.

End-stage haemophilic arthropathy

This has features of osteoarthritis (OA) and RA. The joint is enlarged, 'knobby' due to osteophytes, has restriction of movement, fibrous ankylosis, subluxated, malaligned and lax.

Septic arthritis

This is usually monoarticular commonly involving the knee joint but polyarticular disease can also occur. Articular pain that does not improve with replacement of factor VIII/IX or increasing warmth in the joint should raise the possibility of septic arthritis. Recurrent joint aspiration, arthropathy, IV drug abuse and HIV disease are the risk factors for septic arthritis in haemophiliacs. Staphylococci, pneumococci and *Salmonella* are the most common causative organisms in haemophiliacs.

Muscle and soft tissue haemorrhage

Bleeding in muscles and soft tissue is common in haemophiliacs. Iliopsoas, gastrocnemius and volar compartment of forearm are the common sites. These may manifest as acute groin pain, equinus deformity from heel cord contracture and flexion deformities of wrist and fingers, respectively. Simple muscle cysts, compartment syndrome, subperiosteal or interosseous haemorrhage leading to pseudotumour are the complications commonly seen.

Sickle cell disease

The musculoskeletal manifestations of sickle cell disease are listed in Table 6.

Table 6. Musculoskeletal manifestations of sickle cell disease

Hand foot syndrome	Dramatic diffuse symmetrical tender warm swelling of hands, feet or both. More common in children.
Hyperuricaemia and gout	Polyarticular attacks with fever to be differentiated from sickle cell crisis. More common in adults.
Osteomyelitis	*Salmonella*, Gram-negative bacilli and *Staphylococcus aureus* are the common organisms.
Septic arthritis	*Staphylococcus aureus, Escherichia coli, Enterobacter and Salmonella.*
Joint effusion	Knees and ankles are commonly involved.
Erosions and chronic inflammation	Chronic synovitis with cartilage destruction or erosions at wrists, metacarpal heads and calcanei.
Bone infarction	Localised deformities and failure of bone growth is seen.
Aseptic necrosis (avascular necrosis)	Head of femur and humerus and tibial condyles are common sites.
Bony trabecular changes	Expansion of marrow producing wider intertrabecular spaces with coarser trabeculations mostly evident in axial skeleton. Coarsened vertebral body trabeculae, deep central cup-like indentation, osteophytes, vertebral fusion
Others	Bony ankylosis of hip/spine, protrusio acetabuli, acute localised muscle swelling with necrosis.

Thalassaemia

Diffuse ankle and foot tenderness with ankle effusions. Leg length discrepancies, limb deformities and metaphyseal abnormalities are seen. Arthropathy due to iron overload may occur. Hyperuricaemia, gout and septic arthritis can occur in thalassaemia.

Neurological disorders

The various neurological disorders which present with musculoskeletal disorders include the following.

Neuropathies may be seen as entrapment neuropathies e.g. carpal tunnel syndrome, tarsal tunnel syndrome or vasculitic neuropathy like foot drop or wrist drop as seen in RA vasculitis. Hereditary sensorimotor neuropathy (Charcot Marie Tooth) may present with pes cavus.

Painful stiff shoulder, posture deformities like torticollis, scoliosis and hand foot dystonias are common in Parkinson's disease. Neuropathic joints and swollen fingers may be seen in syringomyelia. Shoulder pain may be presenting feature of neurologic amyotrophy.

Focal neurologic deficits may be followed by reflex sympathetic dystrophy syndrome (RSDS) of the affected limb.

Renal diseases

Chronic renal failure gives rise to various musculoskeletal manifestations. Hypocalcaemia, due to chronic kidney disease leads to secondary hyperparathyroidism which has a complex effect on bones leading to various skeletal disorders collectively known as 'renal osteodystrophy'. Osteitis fibrosa cystica due to high level of PTH and Osteomalacia due to vitamin D deficiency or aluminium retention is commonly seen. Occasionally, secondary hyperparathyroidism may be severe enough to cause extraskeletal calcifications, fractures and bone pain.

Chronic kidney disease also causes hyperuricaemia but rarely these patients develop gout.

Renal tubular acidosis can manifest with rickets in children or osteomalacia in adults due to hypophosphataemia and failure to synthesise 1, 25(OH)2 D as a result of systemic acidosis.

Transplant arthropathy

A self-limiting, pseudo-inflammatory arthropathy commonly involving knees, ankles, hips or wrists presenting as joint pain, tenderness, effusion, periarticular oedema and erythema occurs in post-renal transplant patients on high levels of cyclosporine[13,14].

Malignancies

There are a variety of ways in musculoskeletal manifestations associated with malignancy. The pathogenesis of this complex relationship is incompletely understood. However recognition of these manifestations is necessary for the efficient diagnosis[15,16].

Osseous metastasis, synovial sarcoma, leukaemia and paraproteinaemic arthropathies are commonly seen with primary malignancies. 95% of primary malignancies in bone are metastatic carcinomas which spread by haematogenous route prostate, breast, kidney and thyroid.

A primary or metastatic neoplasm in juxta-articular site or synovium may mimic mono or oligoarthritis. Villonodular synovitis a benign neoplasm commonly affecting knee may cause juxta-articular lytic lesion mimicking a primary bone neoplasm. Metastasis in small joints of hands may mimic RA or gout. Synovial sarcomas are located in periarticular tissues commonly in knee, foot and ankle and thigh.

Musculoskeletal symptoms may precede leukaemia and lymphoma in the form of bone pain or arthralgia which is usually asymmetrical[17]. The articular symptoms are due to periarticular infiltration rather than synovial involvement. Multiple myeloma may present as RA-like polyarthritis or like degenerative joint disease.

Table 7. Rheumatological manifestations of paraneoplastic syndromes

Connective Tissue Disorders	Malignancy
Carcinomatous polyarthritis	Multiple types of solid tumours including breast
Lupus-like syndrome	Lymphoproliferative disorders
Necrotising vasculitis	Lymphoreticular disorders
Cryoglobulinaemia	Plasma cell dyscrasias, monoclonal gammopathies, lymphoproliferative tumours.
Immune complex diseases	Hodgkin's disease
Reflex sympathetic dystrophy syndrome (RSDS)	Pancoast tumour, ovarian and endometrial cancer
Polymyalgia rheumatica	Solid tumours, lymphoproliferative disorders
Panniculitis	Pancreatic cancer
Osteogenic osteomalacia	Solid tumours and tumours of mesenchymal origin
Erythromelalgia	Myeloproliferative disorders
Hypertrophic osteoarthropathy	Lung cancer, osteogenic sarcoma, Hodgkin's disease, nasopharyngeal cancer
Carcinomatous neuromyopathy	Small cell carcinoma of lung, stomach and ovarian cancer
Dermatomyositis	Ovarian and gastric cancer

Tumour lysis syndrome causing hyperuricaemia is a well known cause of secondary gout.

Paraneoplastic syndromes

Musculoskeletal involvement can be seen as a part of paraneoplastic process. The rheumatological manifestations of paraneoplastic syndromes are highlighted in Table 7[18].

Sarcoidosis

The joints, bones and the muscles can be affected in sarcoidosis in the following way[19,20].

Joints

Arthralgias, arthritis and periarthritis are the predominant features. Arthralgias with synovitis, periarthritis or both are a prominent feature of acute form of sarcoidosis. Joint swelling and arthralgias are usually accompanied by erythema nodosum, but may occur before hilar lymphadenopathy and may occur in the absence of erythema nodosum.

In acute sarcoidosis, 65% patients may have articular features. Acute sarcoid arthritis is typically symmetrical, migratory or additive polyarthritis involving knees, ankles, wrists, elbows and PIP joints. Periarthritic lesions with overlying warmth and erythema are prominent around knees and ankles. This inflammation subsides within weeks but may persist as long as 4 months with little residual joint damage.

Episodic articular/periarticular inflammation may be seen in chronic sarcoidosis. Chronic sarcoid arthropathy is relatively uncommon. It commonly manifests as monoarticular or oligoarticular arthropathy affecting knees, ankles, shoulders, wrists and hips without destruction of joints. Rarely it may present as polyarticular destructive variant of sarcoid involving hands and wrists, with characteristic osseous lesions. This is seen in patients with multisystem disease. Joint fluid aspiration typically reveals mononuclear cell infiltration in the synovial fluid.

In children, it manifests as painless synovial proliferation involving one or more large joint. Uveitis is more severe and may be the only presenting feature in the absence of joint or pulmonary disease.

Bone

Bone involvement is seen in 5% patients with sarcoidosis. Osseous lesions are frequently asymptomatic. Lytic lesions are more common appearing as cortical defects or cysts in the phalanges and metacarpal or metatarsal bones. Digital osseous lesions may be associated with diffuse dactylitis and destruction of subchondral bone resulting in joint destruction. Pathological fractures may be seen due to osseous involvement of long bones. Overlying nodular cutaneous lesions of lupus pernio may be associated with osseous lesions. Back pain or neurologic deficits may be seen due to nerve root or spinal cord compression.

Muscle

About 50-80% patients have skeletal muscle involvement. Focal pain and tenderness, muscle cramps, palpable nodules or pseudohypertrophy involving proximal muscles of upper or lower limbs are the clinical features which may mimic polymyositis or the patient may be asymptomatic. Chronic myopathy may have palpable hard, indurated muscles with calcification of muscles surrounding soft tissue.

The muscle involvement may be of two types; nodular and myopathic. The nodular type often involves extremities and causes solitary or multiple nodules. The myopathic type involves muscles symmetrically and diffusely without causing an intra-muscular mass.

Metabolic diseases

Amyloidosis

Amyloid deposition may present with carpal tunnel syndrome, nail dystrophy, alopecia and amyloid arthropathy; whereas Aβ2 M amyloid leads to carpal tunnel syndrome, persistent joint effusion, spondyloarthropathy and cystic bone lesions.

Haemochromatosis

The spectrum of joint disease in haemochromatosis ranges from arthralgia to symmetric polyarthritis, although unilateral disease, especially of the weight-bearing joints has been described[21]. Hands, wrists, knees, hips, ankles, feet and shoulders are the joints commonly involved. Progressive degenerative arthropathy with joint space narrowing, subchondral sclerosis and cyst formation is typical.

Chondrocalcinosis of hands and wrists is common. However, widespread chondrocalcinosis in elbows, shoulders, hips, knees, ankles, feet, pubic symphysis, intervertebral discs and periarticular soft tissues is also described. Chondrocalcinosis in haemochromatosis is associated with CPPD or hydroxyapatite crystals. There is squaring of metacarpal heads and index and middle finger MCP joints are commonly affected causing the so called 'victory' or 'peace' sign.

The arthropathy of haemochromatosis often results in pain and stiffness prior to enlargement of the joint. Limitation of movement, increased pain with use, progressive deformity and bone spur formation are seen in the final stages. Synovial thickening, warmth and erythema are characteristically absent.

The arthropathy of haemochromatosis needs to be differentiated from RA, OA and idiopathic CPPD.

Porphyria

Cutaneous sclerosis mimicking scleroderma may be seen in patients with porphyria cutanea tarda.

Inborn errors of metabolism

Homocystinuria may mimic Marfan's syndrome in its skeletal features. There is arachnodactyly, scoliosis, dislocation of lens and juvenile osteoporosis.

Ochronosis, pigmentation of articular cartilage, early osteoarthritis and calcification of intervertebral discs is seen in alkaptonuria.

Hypermobility seen in Marfan's syndrome, Ehlers-Danlos syndrome and Larsen's syndrome in 12-19% individuals results in joint laxity predisposing periarticular structures to injury leading to pain around one or several joints.

Mucopolysaccharidosis results in hypomobility of the joints thereby causing joint contractures mainly in fingers and other joints. It may present with claw hand. Degenerative changes and flexion contractures of fingers, severe burning in fingers and toes are the common findings in patients suffering from Fabry's disease. Secondary osteonecrosis of weight-bearing joints like hip may be seen.

Gaucher's disease may present with hip and knee pain. Polyarthralgias, pathological fractures of long bones and vertebral compression are also seen. Severe degenerative hip disease as a result of vascular necrosis and collapse of femoral head is not uncommon.

Key points

- Diverse systemic illnesses can give rise to musculoskeletal symptoms and a primary physician should be aware of them so that he is not misled in the diagnosis.
- Almost all the derangements of the endocrine organs produce a plethora of musculoskeletal manifestations.
- Diabetes mellitus and hypothyroidism must be kept as an important differential diagnosis for most of the rheumatological conditions
- Sickle cell disease and haemophilia can affect bones and joints in multiple ways
- Chronic kidney disease can produce musculoskeletal manifestations mostly through secondary hyperparathyroidism.
- Malignancies and paraneoplastic syndrome produce a variety of features.
- Metabolic conditions though rare, are also an important consideration.
- Sarcoidosis and systemic infections including HIV also affect musculoskeletal system in major way.

Further Reading :

1. Jacobs-Kosmin D, DeHoratius RJ. Musculoskeletal manifestations of endocrine disorders. Curr Opin Rheumatol 2005; 17: 64-69.
2. Lockshin MD. Endocrine origins of rheumatic disease. Diagnostic clues to interrelated syndromes. Postgrad Med 2002; 111: 87-88, 91-92.
3. Parada-Turska J, Majdan M. The musculoskeletal system in diabetic patients. Postepy Hig Med Dosw (Online) 2005; 59: 236-44.
4. Crispin JC, Alcocer-Varela J. Rheumatologic manifestations of diabetes mellitus. Am J Med 2003; 114: 753-57.
5. de Rijk MC, Vermeij FH, Suntjens M, van Doorn PA. Does a carpal tunnel syndrome predict an underlying disease? J Neurol Neurosurg Psychiatry 2007; 78: 635-37.
6. Soy M, Guldiken S, Arikan E, Altun BU, Tugrul A. Frequency of rheumatic diseases in patients with autoimmune thyroid disease. Rheumatol Int 2007; 27: 575-77.
7. Cakir M, Samanci N, Balci N, Balci MK. Musculoskeletal manifestations in patients with thyroid disease. Clin Endocrinol (Oxford) 2003; 59: 162-67.
8. Duyff RF, Van den Bosch J, Laman DM, van Loon BJ, Linssen WH. Neuromuscular findings in thyroid dysfunction: a prospective clinical and electrodiagnostic study. J Neurol Neurosurg Psychiatry 2000; 68: 750-55.
9. Hong Yan Wen, H.Ralph Schumacher Jr, Li Yun Zhang. Parathyroid Disease. Rheum Dis Clin N Am. 2010; 36: 647-64.
10. Podgorski M, Robinson B, Weissberger A, Stiel J, Wang S, Brooks PM. Articular manifestations of acromegaly. Aust N Z J Med 1988; 18: 28-35.
11. Roosendaal G, Lafeber FP. Pathogenesis of haemophilic arthropathy. Haemophilia 2006; 12 Suppl 3: 117-21.
12. Hoots WK. Pathogenesis of haemophilic arthropathy. Semin Haematol 2006; 43(1 Suppl 1): S18-22.
13. Abbott KC, Kimmel PL, Dharnidharka V, Oglesby RJ, Agodoa LY, Caillard S. New-onset gout after kidney transplantation: incidence, risk factors and implications. Transplantation 2005; 80: 1383-91.
14. Stamp L, Ha L, Searle M, O'Donnell J, Frampton C, Chapman P. Gout in renal transplant recipients. Nephrology (Carlton) 2006; 11: 367-71.
15. Gonçalves M, Terreri MT, Barbosa CM, Len CA, Lee L, Hilário MO. Diagnosis of malignancies in children with musculoskeletal complaints. Sao Paulo Med J 2005; 123: 21-23.
16. Abu-Shakra M, Buskila D, Ehrenfeld M, Conrad K, Shoenfeld Y. Cancer and autoimmunity: autoimmune and rheumatic features in patients with malignancies. Ann Rheum Dis 2001; 60: 433-41.
17. Kiltz U, Brandt J, Zochling J, Braun J. Rheumatic manifestations of lymphoproliferative disorders. Clin Exp Rheumatol 2007; 25: 35-39.
18. Fam AG. Paraneoplastic rheumatic syndromes. Baillieres Best Pract Res Clin Rheumatol 2000; 14: 515-33.
19. Sato T, Tsuru T, Hagiwara K, Miyashita K, Matsuno H, Goto A, Oritsu M, Hamanaka T, Akiyama O. Sarcoidosis with acute recurrent polyarthritis and hypercalcaemia. Intern Med 2006; 45: 363-68.
20. Zisman DA, Shorr AF, Lynch JP. Sarcoidosis involving the musculoskeletal system. Semin Respir Crit Care Med 2002; 23: 555-70.
21. Van Linthoudt D, So AK. Miscellaneous metabolic diseases. Curr Opin Rheumatol 1996; 8: 262-5.
22. Deepti Suri, Sapna Oberoi. Arthropathy in Heamatological disorders in children. Indian Journal of Rheumatology. 2012; 7(1): 44-51.

SURGICAL ASPECTS IN RHEUMATOLOGY 11

Raghava Dutt Mulukutla, Aditya Kapoor

Introduction

The last few decades are testament to the rapid progress that has been made in the treatment of arthritides. The advent of newer anti rheumatoid therapies such as monoclonal antibodies, cytokines and the enormous strides made in the field of molecular biology have greatly contributed to understanding and controlling these disorders. Despite such progress a number of patients with pain and deformities require surgery. Surgery plays a great role in improving the quality of life of these patients, and varies from simple procedures such as correction of toe deformities to extensive reconstructions in spine surgery. It is of course imperative before undertaking any surgical procedure that the surgeon and the rheumatologist carefully assess the patient and discuss the treatment strategies. Controlling the disease as much as possible with medication and improving the general well being before undertaking major surgical procedures goes a long way in ensuring a good surgical result for the patient.

Rheumatoid arthritis (RA)

Knee

Patients with RA are usually much younger than their counterparts who have osteoarthritis. The bone is of a poorer quality due to the combination of the disease and steroid usage. It has been suggested that the prostaglandins released by the rheumatoid synovium further cause subchondral resorption[1]. Rheumatoid knees are usually valgus knees with severe flexion deformities. In addition poor soft tissue presents a challenge in balancing the knees. There is also an increased incidence of infection following total knee replacement which has been postulated to be as high as 3 times than that associated with osteoarthritis[2,3]. The common surgical procedures in the knee are 1. Synovectomy and 2. Total Knee Replacement.

Synovectomy is done when the articular cartilage is not completely destroyed and the hypertrophied synovium is not responding to medication, and helps to reduce the pain and stiffness. Synovectomy can be done by an open procedure or can be done by the minimal access technique via the Arthroscope. Arthroscopic synovectomy and debridement of the joint has the advantage of early rehabilitation after surgery.

Total knee replacement is indicated when there is gross destruction of the joint with persistent pain, stiffness and deformity. Patients with rheumatoid arthritis require total knee arthroplasty much earlier. Modern prosthesis however have increased longevity and thus surgery can safely be performed early. In cases with gross instability and poor bone quality stemmed implants or hinged prosthesis are usually preferred. In addition patients with gross articular deficiency do better with metallic augments rather than those with bone grafts. Patients with severe flexion deformities require extensive capsular release. Great caution must be exercised in patients with severe flexion deformities as a large distal femoral cut produces mid flexion instability. Severe valgus deformities require release of the lateral structures using a pie crust method and plication of lax medial structures. Failure to do so results in a recurrence of the valgus deformity. There remains considerable debate as to whether cruciate retaining prosthesis is better than posterior cruciate substituting[4,5]. It seems logical that given attenuation of the posterior cruciate ligament in these patients a cruciate substituting replacement is the implant of choice (Fig. 1).

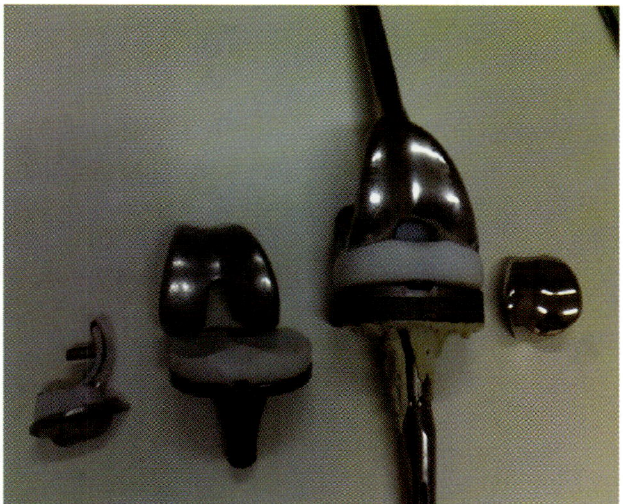

Fig. 1 : Different knee implants

Noteworthy is the fact that patients with rheumatoid arthritis have lesser expectations following knee replacement and often do not persist with physiotherapy (Fig. 2).

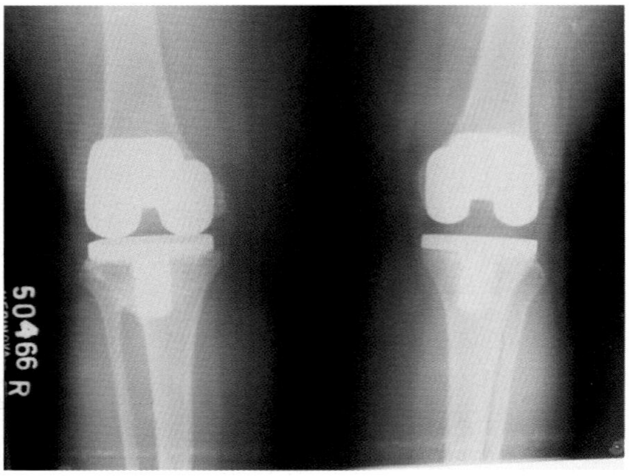

Fig. 2 : Rheumatoid with bilateral total knee replacements

Finally surgery at any age improves the quality of life by decreasing the pain and improving the range of movement in the joints. However, as in any major surgical procedure, the overall assessment of the patient, the risks and rewards must be carefully assessed in these patients[6].

Hip

Hip replacement is indicated to relieve pain and also to increase the range of movement. These surgeries are also offered to patients with other collagen disorders such as Ankylosing spondylitis, Lupus, Psoriatic arthritis and in patients with Still's disease[7]. Most of the Rheumatoid patients are on steroids which cause further weakening of the already osteoporotic bone leading to higher risk of fractures and infection in these patients. There could be partial absorption of head and protrusio is common[8]. Patients with rheumatoid arthritis get excellent pain relief and in those with involvement of both hip and knee, the hip should be operated first. Traditionally cemented total hip replacements were considered the gold standard with bone grafts, anti-protrusio cages and tantalum meshes. However as these patients are usually young a higher rate of loosening has been observed and hybrid total hip arthroplasty with an uncemented cup and cemented stem has been advocated. Recent evidence does not show much difference between cemented and uncemented prosthesis in these patients provided there is sufficient bone stock[9]. In those patients with poor soft tissue and musculature and those who are prone to dislocation it is worth considering constrained liners.

Rheumatoid hand

Rheumatoid deformities lead to impairment of grip, grasp, and pinch, often leaving the patient unable to carry out activities of daily living. The surgical options range from synovectomy, tenosynovectomy, tendon realignment, reconstructive surgery or arthroplasty to arthrodesis. The deformities are usually bilateral, with wrist and metacarpo phalangeal joints being mostly affected.

Tenosynovitis

Symptoms of tenosynovitis often occur before those of intra-articular disease. The affected sites are the dorsal and volar aspects of the wrist and the volar aspect of the digits. Synovitis of the tendons can cause pain, dysfunction, and eventual rupture of tendons. Rest, steroid injections and medical management often induce remission. As dorsal tenosynovitis becomes more apparent, the likelihood of tendon rupture increases. Tenosynovectomy is usually recommended if symptoms do not improve after 4-6 months of medical therapy. After a tenosynovectomy, tendon rupture rarely occurs and complications are infrequent, however postoperative adhesions may occur.

Flexor tenosynovitis in the fingers

Synovial proliferation leads to rheumatoid nodules on tendons which results in trigger finger.

Four types of trigger finger occur in RA.

Type 1 is similar to nonrheumatoid stenosing tenosynovitis, in which the tendons catch at the first annular pulley during flexion secondary to small, localized hyperproliferation of the synovium.

Type 2 nodules form in the distal palm and cause the finger to lock in flexion.

Type 3 nodules on the flexor digitorum profundus (FDP) tendon near the second annular pulley (over the proximal phalanx) lock the finger in extension.

Type 4 trigger finger results from generalized tenosynovitis within the fibro-osseous canal. Active motion is more restricted than passive motion, and contracture and stiffness result.

Flexor tenosynovectomy and nodule excision are recommended for all types of tenosynovitis and trigger finger.

Tendon ruptures

The 2 main causes of tendon rupture are chronic attrition of the tendon over a bony prominence and synovial invasion of the tendon.

Extensor tendon ruptures

Commonly affected extensor tendons are the extensor pollicis longus, extensor digiti minimi, and extensor digitorum communis. Significant deformity with or without functional loss is an indication for repair by end-to-end, tendon graft, or tendon transfer methods to avoid flexion contractures[10,11]. Tendon transfer with the extensor indicis or extensor carpi radialis longus is the preferred method[11].

Flexor tendon ruptures

The flexor pollicis longus is the most common flexor tendon to be affected. The treatment of most FDP ruptures is best with a tendon graft or arthrodesis at the DIP joint. FDS tendon rupture alone does not require repair. In the wrist and palm, the ruptured tendon can be sutured to an adjacent tendon, but within the flexor sheath this is not an option. Tendon grafts have shown poor results[10]. Staged reconstruction has had mixed results. A further option in older patients with severe disease may be fusion of the PIP and DIP joints in a functional position.

Metacarpophalangeal (MCP) joint arthritis

Non operative treatment is advised initially for the MCP joints. Synovectomy is indicated in patients refractory to 6-9 months of medical treatment and with persistent MP joint synovitis with minimal joint deformity. Additionally, extensor tendon relocation with synovectomy is indicated for those with early volar subluxation and ulnar drift.

A variety of Arthroplasty techniques (Swanson)[12,13], arthrodesis[14] and tendon transfers are some of the procedures that are routinely carried out giving good cosmetic and functional results.

Arthroplasties of the metacarpophalangeal joint include reconstruction of the soft tissues only or involve complete joint replacement. Soft-tissue arthroplasty usually incorporates synovectomy with reconstruction of the radial collateral ligament or tendon realignment. Intrinsic transfer of the extensor tendons from the ulnar side to the radial side of the adjacent finger to increase stability is another surgical option. Intrinsic release to alleviate intrinsic tightness eliminates the dorsal digital expansion tightness that is one of the subluxing forces on the metacarpophalangeal joint. Joint replacement is indicated when pain is not controlled or function is impaired.

Intrinsic plus deformity:

Due to tightness of intrinsic muscles wherein proximal interphalangeal cannot be flexed while the metacarpophalangeal joint is fully extended and often associated with ulnar deviation of fingers. This deformity is treated with intrinsic release.

Swan neck deformity

Hyper extension of the Proximal Interphalangeal (PIP) joint and flexion of the Distal Interphalangeal (DIP) joint, with flexion at times of the MCP joint.

Nalebuff classified swan-neck deformities into the following 4 types, on the basis of PIP joint mobility[15]. This classification also aids in treatment of these deformitites

- Type I - PIP joints are flexible in all positions.
- Type II - PIP joint flexion is limited in certain positions.
- Type III - PIP joint flexion is limited in all positions.
- Type IV - PIP joints are stiff and have a poor radiographic appearance.

Type I deformities are treated with a silver ring splint, dermodesis or DIP joint fusion and PIP joint flexor tenodesis.

Type II deformities require intrinsic release or for those with severe involvement of the MCP joints, silicone implant arthroplasty is performed, combined with the rebalancing of the intrinsics and the long extensor tendons[16,17].

Type III deformities require 1 or more procedures, including PIP joint manipulation, lateral band mobilization and flexor tenosynovectomy or tenolysis.

Type IV deformities need a salvage-type procedure-namely, arthrodesis or arthroplasty. Fusion is useful for the index and middle fingers, because these digits need lateral stability when opposed to the thumb during pinch. Arthroplasty is recommended for the ring and small fingers, where mobility aids grasp.

Button hole (Boutonniere) deformity

Flexion deformity of the PIP joint takes place with secondary extension of DIP joint and MCP joint. With mild boutonniere deformity a slight extensor lag (10-15°) is present at the PIP joint with slight hyperextension at the DIP joint and no involvement of the MP joint. Treatment consists of extensor tenotomy. In moderate boutonniere deformity, flexion deformity at the PIP joint increases to 30-40° and the MCP joint begins to hyperextend. Reconstruction of the extensor mechanism by shortening the central slip and moving the lateral bands dorsally is indicated to restore PIP extension and hand function. In patients with severe deformity arthrodesis or arthroplasty of the metacarpophalangeal joints is required.

Thumb deformities

Nalebuff described 5 types of deformities of the thumb

Type I thumb disorder is a boutonniere deformity.

Type II thumb deformity involves hyperextension at the MP joint, the IP joint and the CMC joint is often subluxed.

The type III thumb deformity is a swan-neck deformity.

The type IV deformity is analogous to skier thumb or gamekeeper's thumb.

The type V deformity is identical to the type 3 deformity but does not involve adduction of the first metacarpal.

The type VI deformity involves isolated IP joint and/ or MCP joint destruction with subluxation,

The IP joint should be fused at 0-20° of flexion. MCP joint arthroplasty is recommended if the CMC joint or IP joint is to be fused. CMC joint arthroplasties are a good option for significant deformities of the thumb if the disease initiated at the CMC joint.

Wrist

This involves synovitis with ulnar subluxation and carpal collapse. Treatment includes synovectomy, excision of the ulna by Darrachs or Milch procedure or the Sauve Kapandje procedure. Severe disease is treated by arthroplasty or arthrodesis.

Elbow

Synovectomy and Excision of Radial head are traditional procedures performed in these patients. Total elbow replacement is considered only after medical treatment has not been successful and when synovectomy is not considered as an option. The aim of surgery is to restore function through relief of pain and restoration of motion and stability[18]. Newer designs are being introduced to decrease the incidence of loosening, subluxation etc. to less than 5% overall[19].

Shoulder

In patients with Rheumatoid Arthritis, there is usually wear of Glenoid with subsequent medialisation of the shoulder joint. There is also in addition attenuation of the rotator cuff. Shoulder replacement is a good option in those patients who are not responding to medication and the joint has become painful with restricted movement[20]. In cuff deficient arthropathy Reverse Shoulder prosthesis is employed. Long term results of shoulder resurfacing are satisfactory. If both the elbow and shoulder are affected, simultaneous replacements can be offered to patients.

Rheumatoid foot

Vanio[21] found that out of 955 adult patients with rheumatoid arthritis, 89% had symptomatic arthritis of feet. Most of these patients require simple surgery especially in the forefoot as over three fourths of these patients have forefoot involvement. The deformities in these disorders can be categorized in to 1. forefoot 2. mid foot and 3. hind foot

Forefoot

Hallux Valgus, subluxation or dislocation of the MTP joints, clawing of toes, plantar callosities, and hammer toes are some of the conditions that one comes across in these patients[22]. Most of these problems can be addressed by prescribing appropriate foot wear with wide toe box and use of inserts and pads at appropriate areas. However, if these simple methods do not control the pain and discomfort, the surgeon has no choice but to offer surgery. Forefoot arthroplasty is an excellent procedure for deformity of the forefoot with pleasing cosmetic results. Different techniques of forefoot Arthroplasty are described with satisfactory results ranging from 80 to 90%. Correction of flexion deformities of proximal interphalangeal joints, a simple procedure gives good symptomatic relief and helps in better ambulation of these patients.

Midfoot

Involvement of the mid foot leading to deformity is rather infrequent. Involvement of 1st metatarsal and medial cuneiform can be dealt with by arthrodesis[23].

Hindfoot

Valgus of the heel secondary to involvement of the subtalar and talonavicular joints with involvement of the interosseous and bifurcate ligaments, talonavicular ligament and capsule force the heel into Valgus. Associated weakness or rupture of the Posterior Tibial tendon is also seen in these patients leading to loss of plantar arch and pronation of foot. Tendon transfers are not indicated in these deformities and selective arthrodesis such as subtalar arthrodesis or triple arthrodesis as indicated by the involvement of joints gives a good stable hind foot with relief from pain.

Ankle arthroplasty is useful for patients with tibio talar arthritis. The results of ankle arthroplasty have improved considerably with the newer prosthesis like the the Scandinavian total ankle replacement system. However long term results of ankle arthroplasty are not comparable to hip and knee replacements and data collection on these will clarify the long term prognosis and subsequent revision rate.

Rheumatoid spine

Instability of cervical spine is the most dangerous complication of RA. According to Lipson[24] the incidence of instability is 43-86%. He noted that the incidence of neck pain is 40-88% and incidence of neurological symptoms and signs is 7-34%.The worst of the problems in these patients is Cervical Myelopathy as mortality is very high in this group of patients.

Three types of instability take place in the cervical spine

1. Basilar impression or Atlantoaxial impaction.
2. Atlantoaxial instability.
3. Subaxial subluxation.

Basilar impression is settling of skull on C1 and C1 on the C2 vertebra. This is primarily due to erosive arthritis and bone loss. Basilar impression can also lead to vertebral arterial thrombosis. Atlantoaxial subluxation (AAS) can be anterior, posterior, or lateral and the incidence is 11-46%. Subluxation greater than 10-12 mm is clinically significant and an indication of complete disruption of ligaments in this area. Subaxial subluxation (SAS) results from synovitis of the facet joints and also due to erosion of end plates. Pain and neurological deficit is an indication for surgery in these patients. Menzes et al[25] suggest fusion for instabilities that are reducible and decompression anteriorly or posteriorly combined with fusion when indicated for these complex problems.

Ankylosing spondylitis

Osteotomy of spine was originally described by Smith Petersen in the year 1945 to correct flexion deformity of spine. When the deformity is severe the patient's vision is limited and has difficulty in looking forwards and his gaze is mainly limited to a field near his feet and around that area. It is quite difficult to walk; respiration almost completely diaphragmatic, gastrointestinal problems are common. Surgery not only improves these functions but also improves the appearance of the patient and thereby improving the quality of life.

Osteotomy is done both at lumbar level and at cervical level and various techniques are described. With the availability of rod and screw constructs and various newer methods of fixation of cervical spine[26], there is no need for bed rest and the patient can be mobilised soon after surgery. The osteotomy is fraught with danger of neurological deficit and deaths have been reported due to rupture of aorta especially in the correction of Thoraco lumbar kyphosis. However the complication rates have come down significantly with newer techniques.

Arthritis of hips is quite common in patients with Ankylosing Spondylitis and joint replacement surgery helps these patients by relieving their pain and improving their mobility[27] (Fig. 3 &4). Protrusio acetabuli commonly occurs in these patients. Post operatively in patients with ankylosing spondylitis re-ankylosis may occur and presents a considerable problem.

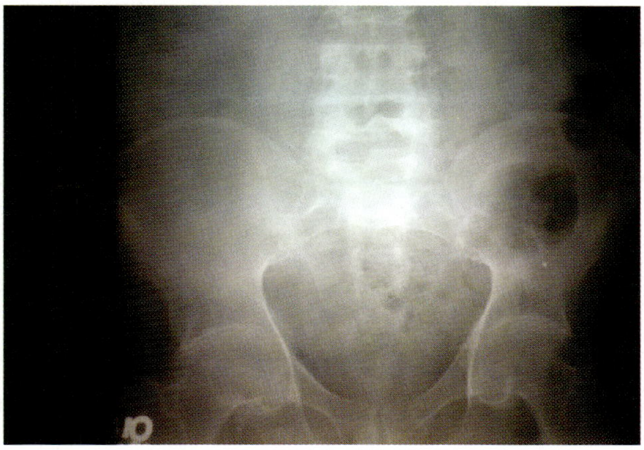

Fig. 3 : Ankylosing spondylitis affecting both hips

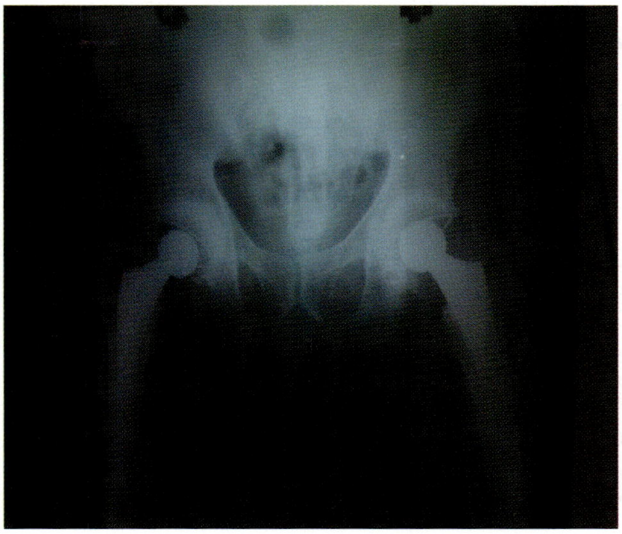

Fig. 4 : Bilateral total hip replacement

Surgery in osteoarthritis

Surgery in osteoarthritis has become standardized over the last three decades. Significant advances are made in joint designs which are lasting longer, with more and more younger patients being subjected to these procedures.

Hip

Primary osteoarthritis of the hip is rare in India and secondary arthritis is much more common. A variety of hip replacements including the traditional cemented implants as well as uncemented implants are used today (Fig. 5).

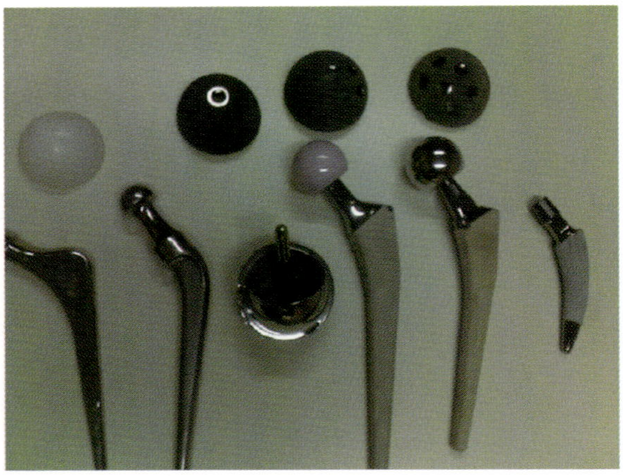

Fig. 5 : Different hip prosthesis

The consensus is that for younger patients uncemented replacements are suitable. According to data from the National Joint Registry(NJR) from England and Wales and from the Scandinavian Arthroplasty Register cemented and uncemented hips have a equally good survival. Metal on metal joint resurfacing (Fig. 6) a procedure where minimal bone resection is done has had mixed results and data from the NJR reveals a high failure rate. Additionally metal on metal large diameter bearing surfaces (Fig. 7) have had problems with metal ions and pseudotumour formation leading to their decreased use. Ceramic on Ceramic bearing surfaces (Fig. 8) have shown excellent long term results. Fragility and stripe wear of the bearing surfaces still remains a problem.

Newer materials like Ultra High Molecular Weight Poly Ethylene have shown encouraging results. A recent innovation has been the Mid Hip Prosthesis and a mobile bearing hip which of course require long term data analysis.

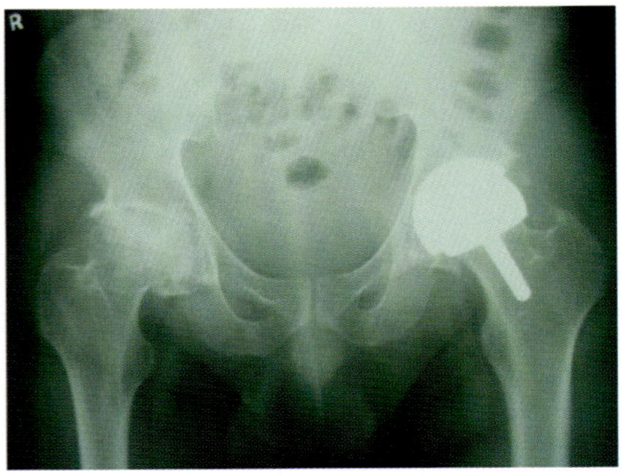

Fig. 6 : Metal on metal resurfacing (currently out of favour)

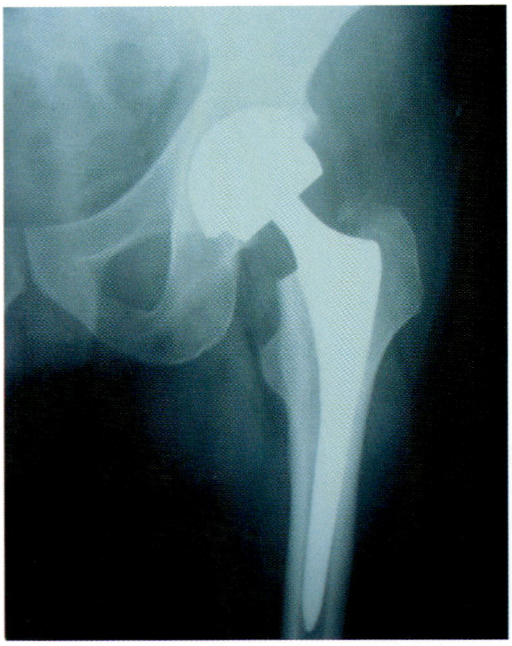

Fig. 7 : Metal on metal large diameter bearing

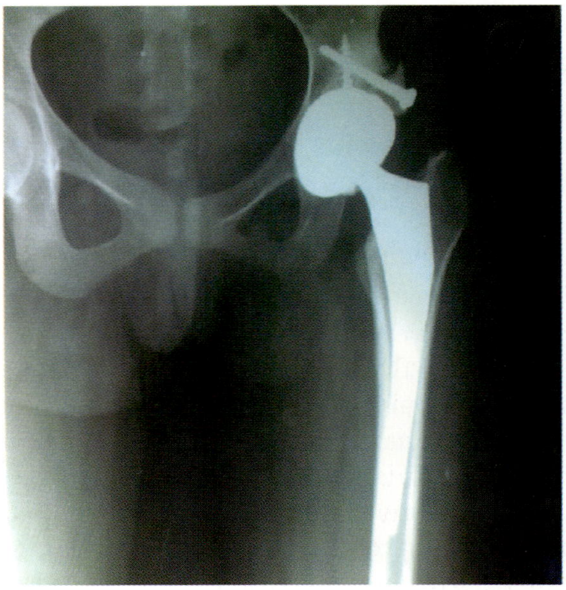

Fig. 8 : Ceramic on ceramic uncemented THR for secondary osteoarthritis

Knee

Various surgical options are available for these patients. For those with a uni compartmental arthritis, with varus deformity, the procedure of High Tibial Osteotomy gives good pain relief. In those with locking of the knee due to loose bodies, arthroscopic removal gives excellent results. Arthroscopy as a procedure is also used to debride and wash out the joint and trim meniscal tears. Joint replacement surgery has come of age, with replacement of patellar surface alone in those with patello femoral arthritis and unicompartmental replacement[28] when only the medial compartment is involved. Total Knee replacement has remained the gold standard for those with tricompartmental arthritis, not responding to conservative measures. Newer designs which suit the female knee and mobile bearing knees which improve the range of flexion are in the armamentarium of the surgeon today. Total knee replacement gives excellent pain relief following surgery with good long term results.

Data from the National Joint Registry reveals that fixed bearing prosthesis do better than mobile bearing and cruciate retaining prosthesis have increased survival rates as compared to cruciate substituting. Also cemented tibia have been found better than uncemented. A high failure rate has been ascribed to unicompartmental prosthesis. This may be due to disease progression than actual prosthesis failure itself. Computer navigation as well as custom prosthesis aim to increase the long term survivorship of implants, however definitive long term results are awaited.

Osteoporosis

Though any long bone may be fractured, surgery in osteoporosis is mostly done for fractures around the wrist, shoulder, hip and the spine. In the last 15 years newer techniques of fracture fixation are developed which include reduction of the fracture under anaesthesia and fixation of these fractures with percutaneous screws rods and wires. This way the morbidity and mortality associated with long open procedures are avoided in these patients who are elderly and suffer from other co-morbid conditions[29]. Newer plate and screw systems which prevent loosening of the implants giving better fixation of osteoporotic fractures are now being routinely used by orthopaedic surgeons.

Percutaneous vertebroplasty and kyphoplasty

Percutaneous Vertebroplasty is the method of injecting bone cement into a Vertebra under local anaesthesia. Though used for a variety of conditions such as haematological malignancies, secondaries of spine etc. it is most commonly used in osteoporotic vertebral compression fractures. This method has gained popularity in the last 15 years though was first done by Dearamond[30] of France in 1987. A needle of appropriate diameter is introduced under local anaesthesia under image intensifier control via the pedicle or by the side of the vertebral body. Once the needle is ideally placed in the pathological vertebra liquid bone cement is injected. The patient can be mobilised within a few hours and the procedure can be done as a day case. In Kyphoplasty, a balloon is first introduced in to the fractured vertebra and inflated. This results in expanding the compressed vertebra and help regain its shape. Cement when injected after expanding the vertebra, retains the shape of the vertebra. Pain relief is significant and sustained at about 70-80% at one year[31]. Mechanism of action : The solidified cement provides immediate strength to the vertebra thereby relieving the pain[32]. The heat generated while the cement is being set is supposed to damage the nerve endings, another explanation for the pain relief (Figs. 9, 10, 11).

A variety of surgical techniques such as cement augmentation of pedicle screws, expanding screws and hydroxy apatite coated screws (Fig. 12, 13) are used in patients whose spines are osteoporotic. Multi segment fixations, use of cross connectors (Fig. 14 & 15) prophylactic vertebroplasties and various other techniques are used by spine surgeon today in patients with rheumatoid arthritis and other metabolic disorders.

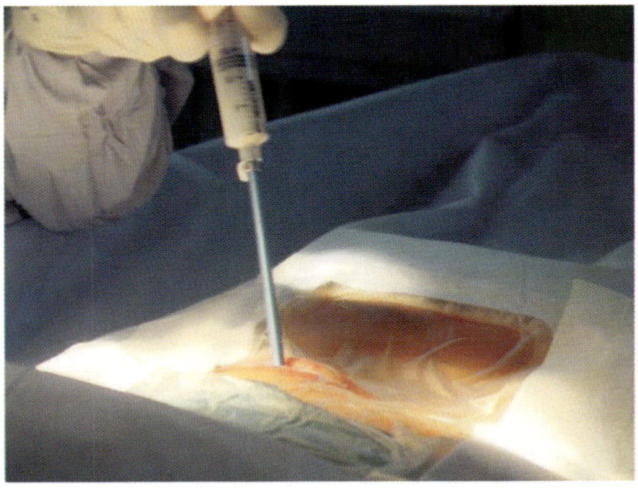

Fig. 9 : Cement being injected for vertebroplasty

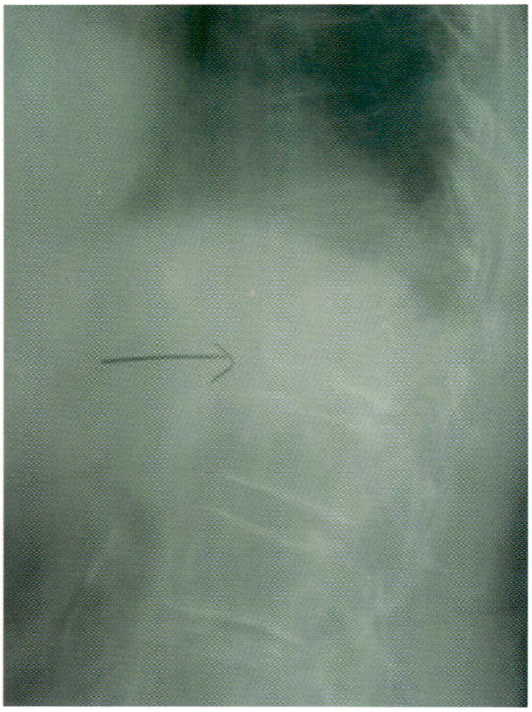

Fig. 10 : Osteoporotic vertebral fracture

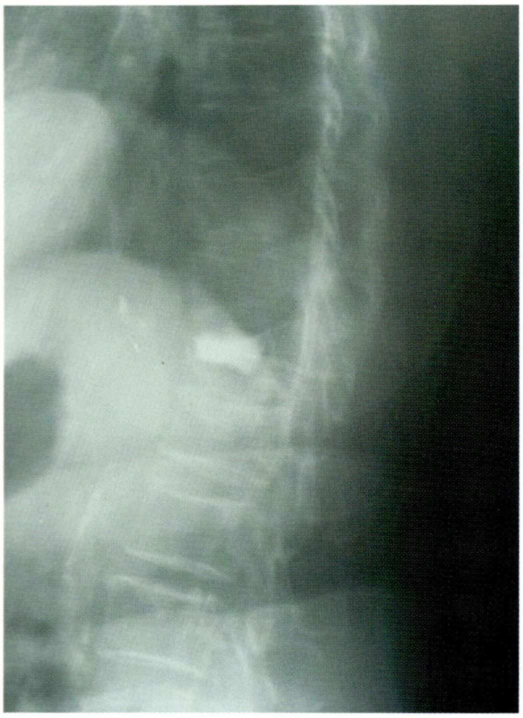

Fig. 11 : Following vertebroplasty

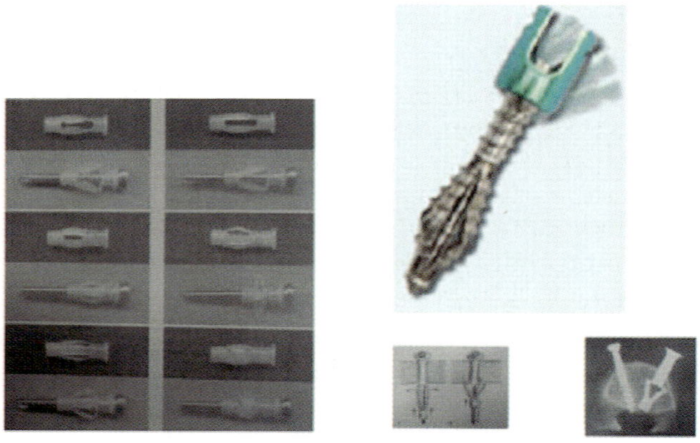

Fig. 12 : Expanding pedicle screws

Fig. 13 : Hydroxyapatite coated screws

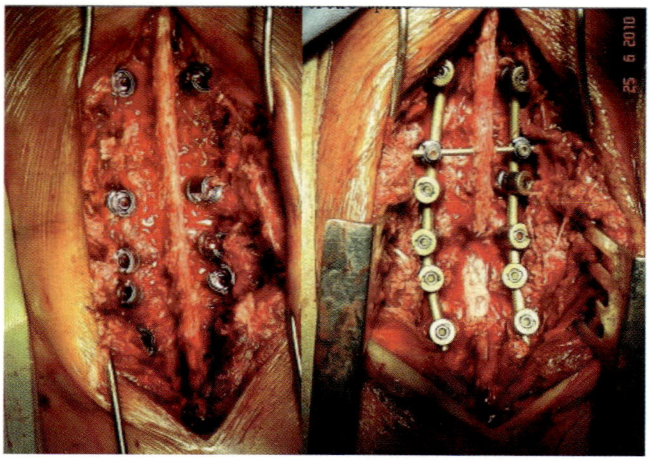

Fig. 14 : Multisegment fixation and use of cross connectors in osteoporotic spine

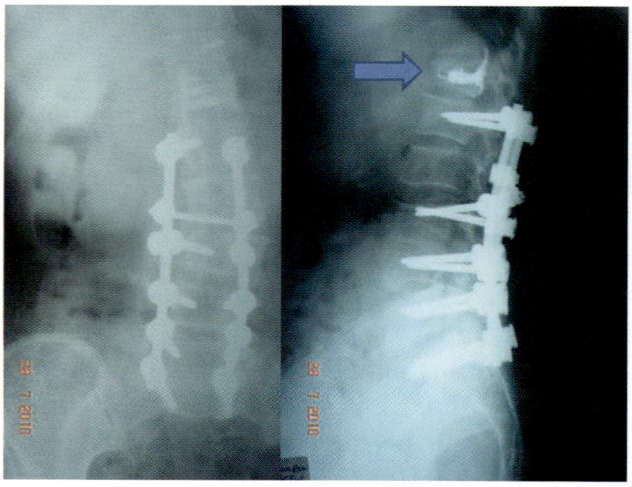

Fig. 15 : Prophylactic vertebroplasty

Conclusion

A vast variety of surgical options now exist in the armamentarium of the surgeon when conservative measures fail. As newer techniques, biomaterials and further understanding of the disease and reparative process at the cellular level evolves the outcome following surgery is likely to improve. However the rush to embrace newer methods by the surgeon must be balanced by evidence based outcome.

Further Reading :

1. Chmell MJ, Scott RD. Total knee arthroplasty in patients with rheumatoid arthritis. An overview. Clin Orthop Relat Res. 1999; (366): 54-60.
2. Poss R, Thornhill TS, Ewald FC, Thomas WH, Batte NJ, Sledge CB. Factors influencing the incidence and outcome of infection following total joint arthroplasty. Clin Orthop Relat Res. 1984; (182): 117-126.
3. Wilson MG, Kelley K, Thornhill TS. Infection as a complication of total knee-replacement arthroplasty. Risk factors and treatment in sixty-seven cases. J Bone Joint Surg Am. 1990; 72: 878-883.
4. Laskin RS, O'Flynn HM. The Insall Award. Total knee replacement with posterior cruciate ligament retention in rheumatoid arthritis. Problems and complications. Clin Orthop Relat Res. 1997; (345): 24-28.
5. Scott RD, Sarokhan AJ, Dalziel R. Total hip and total knee arthroplasty in juvenile rheumatoid arthritis. Clin Orthop Relat Res. 1984; (182): 90-98.
6. Adam RF, Noble J: Primary total knee arthroplasty in the elderly, J Arthroplasty 1994; 9: 495.
7. Arden GP, Ansell BM, Hunter MJ: Total Hip replacement in Juvenile chronic polyarthritis and ankylosing spondylitis, Clin Orthop 1972; 84-130.
8. Ebert FR, Hussain S, Krackow KA: Total hip Arthroplasty for protrusio acetabuli; a 3-9 year follow up of the Heywood Technique, Orthopaedics 1992; 15: 17.
9. Rob E. Zwartelé, Suzanne Witjes, H. Cornelis Doets, Theo Stijnen, and Rudolf G. Pöll Cementless total hip arthroplasty in rheumatoid arthritis: a systematic review of the Literature. Knee Surg Relat Res. 2012 March; 24(1): 1-6.
10. Nalebuff EA, Feldon PG, Millender LH. Rheumatoid Arthritis in the Hand and Wrist. In: Green DP, Green's Operative Hand Surgery. 2nd ed. New York, NY: Churchill Livingstone; 1988.
11. Lee DM, Weinblatt ME. Rheumatoid arthritis. Lancet. Sep 15 2001; 358(9285): 903-11.
12. Swanson AB: Flexible implant Arthroplasty for arthritic finger joints: rationale, technique, and results of treatment, J bone and Joint Surg. 1972; 54-A: 435.
13. Wilson YG, Stykes PJ, Niranjan NS: Long term follow up of Swanson's silastic arthroplasty of metacarpophalangeal joints in RA, J Hand surg 1993; 18-B: 81.
14. Stern PJ, Gates NT et al: Tension and arthrodesis of small joints in the hand J Hand Surg 1993; 18-A: 194.
15. Nalebuff EA. The rheumatoid swan-neck deformity. Hand Clin. May 1989; 5(2): 203-14.
16. Bickel KD. The dorsal approach to silicone implant arthroplasty of the proximal interphalangeal joint. J Hand Surg Am. Jul-Aug 2007; 32(6): 909-13.
17. Sirotakova M, Figus A, Jarrett P, Mishra A, Elliot D. Correction of swan neck deformity. In rheumatoid arthritis using a new lateral extensor band technique. J Hand Surg Eur Vol. Dec 2008; 33(6): 712-6.
18. Goldberg VM, Figgie HEIII, Inglis AE et al : Current concepts review : Total elbow Arthroplasty, J Bone Joint surg 1988; 70-A: 778.
19. Morrey BF: Elbow reconstructive surgery. In Chapman MW, Editor : Operative Orthopaedics, Philadelphia, 1988, JB Lippincot.
20. Neer CSII: The Rheumatoid shoulder. In Cruess RL, Mitchell NS, editors : Surgery of Rheumatoid arthritis, Philadelphia 1971, JB Lippincott.
21. Vanio K: The rheumatoid foot: a clinical study with pathologic and roentgenological comments, Ann Chir Gynaecol 1956; 45: 1.
22. Richardson EG, Campbell's Operative Orthopaedics Edin 9 Volume 2 pages 1998; 1785-1804.
23. Thompson FM, Mann RA: Arthtides, In Mann RA, Coughlin MJ, editors: Surgery of the foot and ankle, ed 6, St Louis, 1993, Mosby.
24. Lipson SJ: Rheumatoid arthritis of Cervical Spine, Clin Orthop 1984; 182: 143.
25. Menzes AH,Van Gilder JC, Graf CJ, et al: Craniocervical abnormalities: a comprehensive surgical approach, J Neurosurj 1980; 53: 444.
26. Simmons EH: The surgical correction of flexion deformity of the Cervical Spine in Ankylosing spondylitis, Clin Orthop 1972; 86: 132.
27. Kilgus DN, Amstuz HC, Wolgin MA, et al: Joint replacement surgery for ankylosed Hips. J Bone and Joint Surg 1990; 72-A: 45.
28. Barns CL, Scott RD: Unicompartmental knee Arthroplasty, AAOS Instr Course Lect 1993; 42: 309.
29. Silvermann SL. The Clinical consequences of vertebral compression fracture. Bone 1992; 13: S27-31.
30. Galibert P, Dearmond H, Rosat P, et al. Preliminary note on the treatment of Vertebral angioma by Percutaneous acrylic Vertebroplasty. Neurochirugie 1987; 33: 166-68.
31. MartinJB, Jean B, Sugiu K, et al. Vertebroplasty: Clinical experience and follow up results. Bone 1999; 25: 11S-5S.
32. Dean JR, Ison KT, Gishen P. The strengthening effect of Percutaneous Vertebroplasty. Clin Radiol 2000; 55: 471-76.

RHEUMATOLOGY AND INDIAN SYSTEMS OF MEDICINE

12

Wagh Shrikant, Datta kumar

Introduction

Indian systems of Medicine include Ayurveda, Yoga, Unani, Siddha and Homeopathy (controlled by Department of AYUSH, Govt. of India) and their training is imparted through institutions and universities approved by government. A large survey of 45000 patients from 35 selected districts across India showed that about one third patients preferred indigenous medical systems (18.7% Ayurveda, 1% Siddha/Unani) and Homeopathy (12.7%) for common ailments[1]. A survey by Govt. of Delhi indicates that 13.8% of all patients (n=212) attending Ayurveda dispensaries are those of arthritis (next to GI disorders 15%). Indian studies in patients of rheumatoid arthritis (RA) indicate usage of complementary and alternative medicine (CAM) in 40-72% cases. *Ayurveda* (28%), *Homeopathy* (20%), *Yoga* (17%) and *Pranayama* (12%) are most preferred alternative therapies in rheumatoid arthritis. Qualitative and quantitative research in CAM therapies is growing continuously though it is difficult or impossible to conduct standard randomized controlled trials in certain forms of therapies. Integration of scientifically proven safe and effective interventions with conventional management may be possible as more evidence becomes available. The American College of Rheumatology position statement recommends that healthcare providers should be informed about common CAM modalities, based upon appropriate scientific evaluation, and should be able to discuss them knowledgeably with patients.

Soul-mind-body medicine forms basis of most complementary therapies. Traditional medicine views human beings as a part of universe. Changes in universe affect human beings directly or indirectly. Disease is an imbalance between body and the milieu and therapy is directed to reestablish this balance. Emphasis on disease prevention rather than treatment, maximization of inherent healing capacity, individualized diagnosis and management are some of the peculiarities of these therapies.

The patient factor

Patients use CAM therapies for variety of reasons[2]. Users of alternative medicine can be disease-specific users (intending to treat a disease) or general health users ('good for health'). Sources of advice for use of CAM include friends and relatives, healthcare professionals, printed material including books and periodicals, television and Internet. Pharmaco-epidemiological studies have indicated various factors that influence patient decision to use CAM (Table 1).

Table 1. Factors influencing decision of CAM usage

Ethnicity e.g. non Hispanic whites in RA	Female sex
Higher income	Age under 55 years
Higher literacy	Earlier age of diagnosis
Perceived 'no side effects' of treatment	Longer duration of disease
'No cure' in Allopathy	Poorer functional status
Lower cost of treatment	Higher level of pain
	Sleep disturbances

Patients can use multiple therapies at the same time. Many CAM users use therapies as self-medication and OTC purchase rather than approaching a registered therapist. Some patients continue drug usage without a follow up visit with the practitioner. Most patients do not reveal usage of CAM to the treating physician unless specifically enquired into. Patient satisfaction with CAM, though considerable, does not necessarily mean improvement in disease condition.

Adverse events and related issues

A) Heavy metal contamination: Lead (*Naga*), mercury (*Parada*) and arsenic (*Hartala, Manshila, Somal*) are the metals of concern found in indigenous drugs. Mercury is a component of innumerable drugs whereas lead is used in antidiabetic, aphrodisiac, nerve-tonics and other preparations. Lead is an important constituent of *Mahayograj guggulu* and *Vatavidhwansa*, commonly used by Ayurveda practitioners for arthritis. It may be mentioned that arsenic trioxide is anti-arthritic as it induces apoptosis of fibroblast-like synoviocytes in experimental rheumatoid arthritis[3].

   Ayurveda texts describe various purification processes for raw metals to make them more biocompatible. Metal containing drugs must be used in small doses under expert supervision for a limited period of 2-3 months. Inappropriate detoxification and prolonged administration by way of self-medication seem to be determining factors. Lead is found, as in case of vegetables, in pure herbal (nonmetal) compounds too. Environmental contamination (pesticides, fuel, paints, rubber, plastic, etc.) is also an important factor in evaluation of safety of herbomineral compounds.

B) Potential drug interactions: Ginger and garlic have antiplatelet and other anticoagulant effects. Increase in risk of NSAIDs or steroid related bleeding needs further evaluation.

C) Adverse effects: These are summarized in Table 2. (Figs. 1 and 2)

Table 2. Known adverse effects of herbs used in Ayurveda

Guggulu Commiphora mukul	Hyperacidity, alopecia, impotence, skin rashes
Erand (Castor) Ricinus communis	Irritant purgative
Bhallataka Semicarpus anacardium	Hyperacidity, skin eruptions
Haritaki Terminalia chebula	Diarrhoea, abortion
Vatsanabh Aconitum ferox	Aconite related effects
Karaskar Strychnos nux vomica	Strychnine related effects
Dhattura Dhattura stramonium	Scopolamine related effects

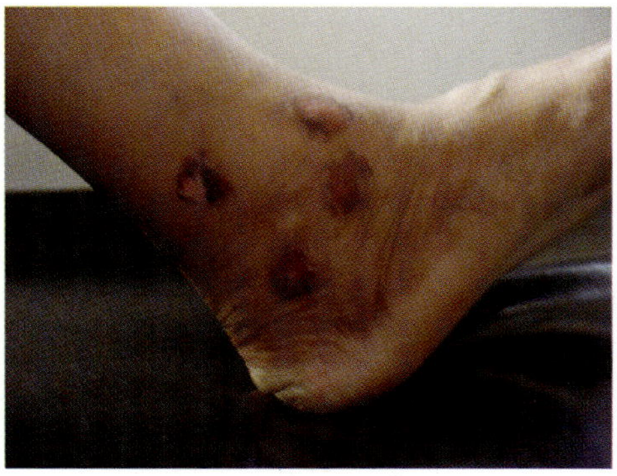

Fig. 1

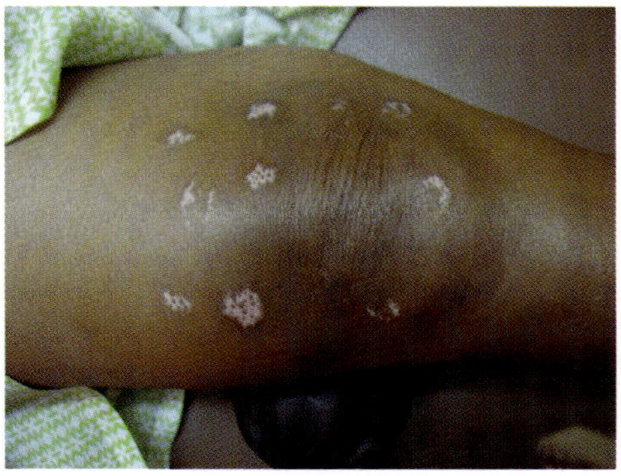

Fig. 2

D) Adulteration: Many reports indicate presence of steroids in Ayurveda products. *Charak* calls malpracticing physicians as *Chadmachar* (dishonest, deceptive, fraudulent, feigned physicians-quacks).

E) Cost of therapy: Cost of CAM therapy varies widely and higher prices do not necessarily mean more effective treatment. This leads to wastage of valuable resources in some cases.

F) Delay in diagnosis: Delay in seeking medical help while pursuing alternative therapies can be detrimental and may at times be life-threatening. This is especially true in early inflammatory rheumatic diseases.

The following write-up deals with relevant fundamentals and therapies along with important evidence in each of AYUSH systems. More details are available in literature devoted to these sciences. Other therapies such as music therapy, aroma therapy, hydrotherapy, massage, magnet therapy, etc. could not be included due to technical reasons. Many clinical and experimental studies in Indian medical systems are published in CAM journals, regional language publications, non-indexed Indian scientific journals, post-graduate and doctoral thesis and other literature. Though substantial, this work is difficult to locate in net-searches.

Ayurveda

Ayurveda (Sanskrit: *Ayu*-life, *Veda*-to know) is a knowledge system to understand life[4]. Its purpose is to protect health, prevent diseases and restore life. The fundamentals of Ayurveda (Table 3) are based on ancient Indian philosophical systems such as *Sankhya*, *Vaisheshik* and Yoga. *Charak Samhita* (600 BC) and Sushrut Samhita (500 BC) are the oldest Ayurveda texts available today.

Table 3. Fundamental concepts of Ayurveda

General	*Lokasammita-purusha, Panchmahabhoota*
Physiology	*Dosha-dhatu-mala, Agni, Srotas*
Pathology	*Aam, Viruddhadosha, Viikrutadushya, Srotovaigunya, Srotorodha*
Pharmacology	*Dravya-guna-karma , Rasa-veerya-vipak-prabhav*
Therapeutics	*Swasthavrutta, Rasayana, Vaajikarna, Sanshodhana, Sanshamana*

Fundamental principles

Every matter in the universe is made up five perceivable fundamental principles (*Pancha-mahabhoota*) viz. *Pruthvi*-Earth, *Aap*-water, *Teja*-fire,

Vayu-air, and *Aakash*-space. Human body, too, consists of these fundamental elements. The living beings have, in addition, *chetana tatva* made up of *manas* (mind) and *aatman* (spirit). Physiology in Ayurveda is explained in terms of *dosha*, *dhatu* and *mala*. *Vata*, *pitta* and *kapha* are the three *doshas* (vital biophysiological forces) which govern every living body. *Vata-dosha*, the most important one, is manifested by all bodily movements and functions. *Pitta-dosha* is manifested by metabolism and biotransformation while *kapha-dosha* is manifested by cohesiveness of body constituents. Health is balance of these physiological functions whereas disease implies the reverse. 7 basic tissues (*dhatu*) support and sustain human body viz. *rasa*-plasma, *rakta*-blood/red cells, *mamsa*-muscles, *meda*-adipose tissue, *asthi*-bone, *majja*-marrow and *shukra*-sperm/ ovum. These are derived from diet and are sequentially formed one after another. Perfect health of all tissues is also called as strength and is responsible for *vyadhi-kshamatva* (immunity). *Mala* are excretables separated from body tissues.

The maintenance of human body depends on the nature of ingested food. *Agni* (biological fire) with its various components is responsible for digestion, metabolism and assimilation of food. The entire process of bio-transformation and transportation takes place in innumerable body channels (*srotas*) - gross or subtle. Disease process is initiated and sustained by derangement of *agni* and loss of competence of *srotas*.

Aam (undigested matter) is a toxic substance generated in the human body due to several factors including improperly digested food, inadequately biotransformed macro molecules in circulation and inefficiently cleared waste products. Inadvertently ingested or injected non food substances also have potential to produce *aam* in human system. *Aam*, which can be correlated with antigenic substances or antigen-antibody complexes, is considered to be central pathogenic factor in many diseases.

Pain is a feature of dominant *vata dosha*. *Vata* governs all bodily functions. It tends to increase by various factors including diet, lifestyle, climate and age. It can get aggravated or vitiated secondary to other diseases and mental or physical trauma. *Vata* aggravating factors include bitter, pungent and astringent food items, stale and stored food substances, hectic lifestyle and frequent traveling, cool and dry climate, and advancing age.

Several classical diagnostic methods described in Ayurveda are essentially clinical. Emphasis is on assessing involvement of *dosha*, *dhatu*, *mala* and *mahabhoot*. The balance between *rugna-bala* (resistance power and immunological status) and *vyadhi-bala* (virulence of causative factors and severity of pathogenesis) determines the prognostic outcome. Human constitution (*prakruti*) is also considered before determination of individualized therapeutic modalities. A physician needs to be well-trained, knowledgeable, wise and intelligent to use clinical methods of diagnosis and monitoring therapy.

Management of diseases aims at harmonious balance between *doshas*, *dhatus* and *malas*. Principles of management include:

1. *Nidaana-parivarjan* - avoiding causative and precipitating factors
2. *Sanshamana* - restoration the physiological homeostasis through drugs, dietary (*Ahara*) and behavioral (*Vihara*) modifications
3. *Sanshodhana* - expelling out increased and vitiated *dosha* or toxins
4. *Rasayana* - reparative and rejuvenative therapy

Rheumatology in Ayurveda

360 bones and 210 joints in human body are described in Sushruta Samhita. Joints are of two types - *chala* (diarthrosis) and *sthir* (fixed); which are further classified into pivot joints, gomphosis, sutures, etc. Synovial membrane (*sleshmadhara kala*), which secretes synovial fluid (*sleshaka kapha*) is also described. Clinical features of arthritis viz. pain, swelling and stiffness are described as *sandhigata vata*. Several such clinical syndromes mentioned in literature can be classified under the category of rheumatic diseases. Table 4 enlists these diseases comparable with conventional rheumatic conditions. *Sandhigata vata* and *aamvata* are prototypes of noninflammatory and inflammatory arthropathies respectively. *Vatarakta* is another clinical condition described in most of the classical texts. Some clinical features of *vatarakta* are akin to those of acute gouty arthritis whereas others are comparable with flares in systemic connective tissue diseases.

Table 4. Modern rheumatologic disorders in Ayurveda

Sandhigatavata	Degenerative arthritides
Aamvata	Inflammatory arthritides
Vatarakta	Inflammatory connective tissue diseases, gout
Kroshtrukasheersha	Infective arthritis, Charcot joint
Grudhrasi	Sciatica
Avabahuk	Frozen shoulder
Vishvachi	Brachial neuralgia
Vatakantaka	Heel pain
Manyastambha	Torticolis, cervical spondylosis
Katishool	Low back pain

Aamvata

Aam is generated in an individual with suboptimal metabolic capacity (including genetic predispositions) when exposed to incompatible food and contradictory physical activities. It gets further vitiated by *tridosha* (dooshit aam - equivalent of immune complexes) which, under the influence of *vata*, tends to get lodged particularly at the sites of *sleshma sthaana* (connective tissues and joints) to produce *aamvata* (immunoinflammatory rheumatic diseases). In advanced cases (*atyartha vidagdha*), arteries and blood vessels are also influenced (*dhamni pratipadhyate*).

Six treatment modalities are advised for *aamvata*. These are *langhana* (fasting), *swedana* (fomentation), *tikta-deepan-katu dravya* (bitter, hot, agni-promoting drugs), *virechana* (laxation), *sneha-paan* (oleation) and *basti* (medicated enemas). The former three modalities are primarily targeted against *aam* whereas the later three are targeted against *vata* component of the disease. These treatment modalities are recommended in a sequential order for early disease. In chronic and advanced cases, these modalities can be implemented in combinations and not necessarily in sequential order. *Langhana*, *virechana* and *basti* are contraindicated in certain circumstances. Several formulations are indicated for *aamvata* in classical texts. Table 5 gives list of selected formulations commonly used in *aamvata*.

Table 5. Ayurveda formulations commonly used in *Aamvata*

Formulation	Nature	Dose
Yogaraj Guggulu	Herbal; tablets/ pills	500mg-1gm t.i.d.
Arogyavardhini	Herbo-mineral; tablets/pills	250-500mg t.i.d.
Eranda shunthi	Herbal; oil Powder	5-7.5ml o.d. 2-3gm o.d.
Sinhananda Guggulu	Herbal; tablets/ pills	250-500mg t.i.d.
Vishatinduka vati	Herbal; tablets/ pills	125-250mg t.i.d.
Rasnasaptak kwatha	Herbal; Decoction	10-20ml b.i.d.
Ajmodadi churna	Herbal; powder	2-4gm b.i.d.
Khandashun-thyavaleha	Herbal; Electuary	4-5gm o.d.
Bhallatak ksheerpak	Herbal, milk processed	1.5-6gm (incremental)

Sandhigata vata

Sandhigata vata is a prototype of non-inflammatory, degenerative joint diseases. Aggravating factors including aging lead to derangement of *vata dosha* which influences the *asthi dhatwagni* (bone and cartilage metabolism). This adversely affects the metabolic homeostasis of cartilage leading to degradation (*hanti sandheen*) of joints. Continuation of *vata* aggravating factors accelerates the process of 'joint failure'.

Management of *sandhigata vata* includes avoidance of vata aggravating factors (*nidaana parivarjan*) and external as well as internal treatment. External treatment is essentially local oil massage, medicated steam or other form of heat, and application of medicaments. Massage is avoided in inflammatory effusions and medicines recommended for *aamvata* are used. *Basti* of decoctions and oils is also frequently recommended. Oral medications have *anti-vata* and *asthi-rasayana* (bone and cartilage promoting and protective) properties. Table 6 enlists commonly prescribed formulations in *sandhigata vata*.

Table 6. Ayurveda formulations commonly used for Sandhigata vata

Formulation	Nature	Dose
Gokshuradi Guggulu	Herbal, tablets/ pills	250mg- 1gm t.i.d.
Nirgundi tail	Herbal, local application	As per area
Dashamoola kwath	Herbal, decoction	10-20ml b.i.d.
Ashwagandha choorna	Herbal, powder	2.5-5gm b.i.d.
Chandraprabha vati	Herbo-mineral, tablets/pills	250mg b.i.d./t.i.d.
Dashanga lepa	Herbal, local application	As per area
Trayodashanga Guggulu	Herbal, tablets/ pills	250mg-1gm t.i.d.
Bhallatakasav	Herbal, Decoction	10-20ml b.i.d.
Abha Guggulu	Herbal, tablets/ pills	250-750mg t.i.d.
Mahavata-vidhvansa	Herbo-mineral, tablet/pills	125-250mg b.i.d./ t.i.d
Mahayograj Guggulu	Herbo-mineral; pills/tablets	125-250mg b.i.d./ t.i.d

Inflammatory rheumatic conditions

Innumerable studies have been carried out on various plants and formulations. Though a review[5] of 7 RCTs failed to show efficacy of Ayurveda therapies, many plants have shown anti-inflammatory and other effects necessitating further evaluation. Three Indian studies[6] including those under NIMITLI (New Millennium Indian Technology Initiative) used combination of herbs. These herbal compounds have shown modest efficacy in RA with improvement in joint pain, swelling and other ACR variables. The plants used included *Ashwagandha* (Withania somnifera), *Guggulu* (Boswellia serrata), *Shunthi* (Zingiber officinale), *Haridra* (Curcuma longa), *Methika* (Trigonella foenum-graecum), *Atasi* (Linnum usitatissimum), *Gokshura* (Tribulus terrestris) and *Mareecha* (Piper nigrum). One polyherb drug resulted in 44% ACR 20 response as against 51% with hydroxychloroquin. A study of RA patients showed Ayurveda drug to be as effective as methotrexate[7]. *Guggulu*, *Shunthi*, *Haridra*, *Ashwagandha*, and *Bhallataka* (Semicarpus anacardium) are some of the plants that have shown anti-inflammatory effect due to downregulation of proinflammatory cytokines[8]. Many plants have immunomodulator effects too[9]. *Nirgundi* (Vitex negundo), *Pippali* (Piper longum), *Rasna* (Inula helenium), etc are anti-inflammatory and require further evaluation.

Fasting (*Langhana*) followed by *sansarjan-krama* (gradual reintroduction of food) is an important and early intervention recommended for inflammatory rheumatic diseases especially RA. Modification of diet by eliminating certain food items is also a common practice. A systematic review of fasting followed by vegetarian diet in RA showed statistically and clinically significant beneficial long-term effect. Possible use of this modality needs to be studied further with randomized long term studies[10].

Swedana (fomentation of various types) is an important form of arthritis therapy. Infrared sauna reduces pain, stiffness and fatigue in patients with RA and ankylosing spondylitis. 40-50% reduction in levels of inflammatory cytokines was observed after mild whole body hyperthermia in patients with ankylosing spondylitis[11].

Rectal administration of drugs (*basti* - one of the *panchakama*) is particularly useful in *Vata*-disorders. Two main types, *nirooha* (evacuation) and *anuwasana* (retention of oils/ghee prepared with plants), are used in inflammatory and non-inflammatory arthritis respectively. They may be used in various combinations using different herbs according to *dosha*-dominance. Rectal administration leads to faster drug action as it avoids pH variations of gastrointestinal tract and bypasses enterohepatic circulation. Significant improvement ($p < 0.001$) observed in pain, swelling and stiffness with use of basti in 12 patients of RA[12].

Noninflammatory rheumatic conditions

A polyherb preparation[6] (used in RA *vide supra*) showed significant efficacy (pain and function) and safety at 32 weeks in knee osteoarthritis (OA). Use of *Shunthi* and *Guduchi* (Tinospora cordifolia) did not show significant beneficial effects in knee OA. Pain relief was equivalent to celecoxib with addition of *Guggulu* to this combination (NMITLI study). Promising effect of pain relief with *Guggulu* is due to inhibition of 5-lipooxygenase, an enzyme in leukotriene pathway[13]. Castor (*Eranda* - Ricinus communis) oil is as effective as diclofenac for pain relief in knee OA[14]. Plants like Amalaki (Phyllanthus emblica)[15] and *Haridra* (Curcumin)[16] and a *guggulu*-formulation (*Triphala Guggulu*) are chondroprotective and can facilitate chondrogenesis[17].

Leech therapy (a type of *panchkarma*) lead to significant pain relief, reduction in joint stiffness and improved function in patients with advanced knee OA[18].

Metabolic bone diseases

Asthi-kshaya (loss/waning of bones) is a prototype of osteoporosis and other metabolic bone diseases. A proper diet is essential for healthy bones as all *dhatus* including *Asthi* are manufactured from food by respective *agni* (parathormone, vitamin D, etc). Senility leads to *asthi-kshaya* due to *vata-dominance*. Other factors such as female hormones and advantages of exercise are also indicated in literature[19]. Loss of hair, nail abnormalities, dryness, bone pains, joint laxity and fatigue are described as clinical features. The therapy is directed to osteoid with oral or rectal (*basti*) oleation and use of bitter drugs such as *Guduchi*, *Nimba* (Azadirachta indica), *Ativisha* (Aconotum heterophylum) and marine products like *Praval* (Coral), *Shankh* (Conch shell), *Shukti* (Pearl oyster), etc. Many of these plants contain significant quantities of phytoestrogens where as marine products are cheap sources of natural calcium.

Other Indian systems

Unani medicine

Unani (also called as *hikmat*) Medicine is based on theory of four elements (air, water, fire and earth) in Universe and four humors (blood, phlegm, yellow bile and black bile) in the body. Temperament (*mizaj*) of one individual differs from another due to qualitative differences of humors. Power of self preservation (*quwwat-e-mudabbira*) maintains humoral balance. Diseases are caused by imbalance of humors. Therapy is directed to regain *quwwat-e-mudabbira*. Regimental therapy (*ilaj bit tadbeer* including *hajamat*-cupping, massage, cauterization, etc.), diet and drugs (90% herbal) are used in treatment of medical conditions.

India is the world leader of Unani Medicine as *hikmat* evolved mainly in India and adjoining countries. Many Ayurvedic drugs, therefore, find place in Unani medicine. *Guggulu*, *Ashwagandha*, ginger and turmeric are included in Unani pharmacopia. National formulatory of Unani Medicines includes 7 drugs for use in joint disorders including RA, gout and sciatica (*yograj gugal*, garlic, nux vomica, colchicum, etc). Clinical trial of UNIM-301 (coded) along with drug-fomentation and oil has shown promising results in a large number (n= 7529) of cases of arthritis (*waja ul mafasil*) including RA though further validation is necessary. A small study (n=20) showed encouraging results of cupping in RA[20].

Siddha medicine

Siddha medicine[21] is another ancient system of Dravidian origin prevalent in Tamil speaking parts of India. *Varma* branch of Siddha medicine is based on 100 vital body points and deals with traumatology. The principles of this system, both fundamental and applied, appear to be similar to those in Ayurveda.

The classification of majority of diseases is based on clinical symptoms or disequilibrium of humors (*doshas*). Use of metals and minerals is predominant in Siddha therapeutics though herbal and animal products are also used. Siddha system classifies metals and minerals according to their properties. Elaborate processing of these minerals with herbs, use of *muppu* (a natural salt) and use of proper adjunct with the drug are some of the peculiarities.

Despite growing global concern about metal content of drugs, Siddha system claims to offer remedies for arthritis (*keel-vayu*) and spinal disorders. Various herbs are described for use in arthritis[22]. Clinical trials have been performed using drugs like *Gauri Chintamani*, *Lingachendooram*, *Vettumaran gutika*, a polyherb compound and other formulations in RA (*valiazhal keelvayu*).

Homeopathy

Homeopathy[23] has significant presence in India. Arthritis stands 9th amongst 10 most common conditions treated by homeopaths.

Homeopathy believes that disease is intimately associated with life of an individual and is a result of internal concealed causes. Illness occurs only when resistance within the body is disturbed. The entire individual person rather than organ systems needs to be treated for effective cure. Effects of drugs on various healthy individuals (Proving) are recorded in Materia-Medica (Repertory). A patient is accurately assessed on the basis of all symptoms including psychological and matched with similar picture in Materia-Medica. Drugs are used on the principle of 'like cures like' though finding a right remedy can be highly challenging in a given patient. The goal is to provoke vital force within the body with minimum possible dose to elevate body systems and eliminate disease. Therapy thus differs amongst patients of same conventional diagnosis in classical homeopathy. Avoidance of 'antidotes' such as coffee, mint, camphor, acupuncture, etc. is also important. Concurrent use of modern drugs is usually discouraged on the basis of antidoting.

Remedies prescribed for a particular ailment without detailed assessment of individual patient is known as 'nonclassical' homeopathy. Combination drugs are available for ailments such as arthritis pain. *Apis*, *Belladona*, *Bryonia*, *Colchicum*, *Kali-Carb*, *Ledam Pal*, *Naitr Mur*, *Pulsatilla*, and *Rhux-tox* are some of the drugs used for arthritis.

Although homeopathy claims success in treatment of autoimmune disorders, it is not generally advised in long standing arthritis with damaged cartilage. Clinical studies are difficult due to individualization of treatment in classical homeopathy and challenges in selection of drugs. A double-blind prospective study of classical homeopathy treatment in RA showed significant improvement than placebo controls over 3 month period. Significant improvement in pain and sleep was observed in patients with fibromyalgia. 4 good quality studies in RA showed mixed results[24] whereas one study in OA did not show positive results. It has been suggested that clinical benefits in RA may be attributable to consultation process and not to the homeopathic drug[25].

Nature cure therapies

Naturopathy is drug free, noninvasive system of primary health care based on theory of self healing capacity of the body and principles of prevention of diseases by way of natural methods and healthy living. While treating the whole person alongwith physical, psychological and social factors, the physician allocates more time towards patient education. A good doctor-patient relationship, thus, is of immense value in this form of therapy. Naturopaths are expected to have basic scientific knowledge of prevention and treatment of diseases. Based on common philosophy and principles of naturopathy, an individual practitioner can use a single form of intervention or a combination of various modalities such as diet therapy, mind-body medicine (meditation, hypnosis, etc), hydrotherapy, physical therapy, herbal medicine as well as Ayurveda, homeopathy, Chinese medicine, etc.

Manual and manipulation techniques such as massage therapy, osteopathic medicine, and chiropractic are widely practiced in India for various rheumatologic conditions. Trained manual medical practitioners can be of immense help in certain rheumatologic conditions. Manual techniques appear to be beneficial for back and neck pain (but not in ankylosing spondylitis) when compared with no treatment or placebo. Studies have shown mixed results in several conditions such as carpal tunnel syndrome, osteoarthritic knee pain, fibromyalgia, ankylosing spondylitis and diffuse idiopathic skeletal hyperostosis[26].

Yoga

Yoga (Sanskrit: union of body, mind and soul) is one of the six great philosophies of India. Ayurveda has included Yoga as a part of ideal life style. Yoga is classified as mind-body medicine (enhance capacity of mind to affect bodily functions and symptoms) and has comprehensive potential for development of physical, mental and social health. It needs to be practiced in 8 stages (*Ashtang* - eight fold - Yoga) one after another for all round development of personality. These 8 stages are *Yama* (self-restraint, moral attitudes towards society), *Niyama* (self regimen, good individual observances), *Asana* (postures), *Pranayama* (control of breathing and thoughts), *Pratyahar* (conscious withdrawal from senses), *Dharana* (self absorption - concentration), *Dhyana* (meditation) and *Samadhi* (trans).

Various methods of Yoga having common ultimate goal of *Samadhi*, are broadly classified into 4 streams viz. *Karmayoga* (work, service), *Bhaktiyoga* (devotion, worship), *Dyananayoga* (philosophy, wisdom) and *Rajayoga* (mind control, meditation). *Hathayoga* (*Ha:* sun, *Tha:* moon; balance of two qualities) is a part of *Rajayoga* and controls mind through *Asana*, *Pranayama*, *Mudra* (voluntary action) and *Bandh* (voluntary action to generate positive pressure). Yoga also undertakes cleansing techniques (similar to *panchakarma* in Ayurveda) without use of any drugs.

A typical Yoga session (usually one hour) consists of warm up exercises (*Surya Namaskar*) followed by 8-10 *asanas* with intermittent rest, *pranayama*, meditation and *Aum* recitation one after other. Proper breathing techniques are emphasized throughout the session. Yoga needs to be practiced everyday as a way of life and a good teacher is required at least for initial training.

Most *asanas* have a dynamic (assumption of posture and coming back) and static (maintenance of posture for a length of time) component. *Asanas* are not mere stretching exercises. It is an effortless assumption of a stable and comfortable posture as well as mind without any unpleasant sensation of pain. Most *Asanas* are corrective in nature and are useful in reconditioning of the musculoskeletal system by slow and smooth movements. Some *Asanas* are for relaxation (e.g. *Shavasana*) or meditation (e.g. *Padmasana*).

Yoga is advised for various musculoskeletal disorders. International Association of Yoga Therapists (www.iayt.org) advocates Yoga for arthritis and allied conditions. Yoga and Tai-chi (Chinese counterpart of Yoga combining deep breathing and relaxation with slow and gentle movements while maintaining good posture) sessions form a part of training programs for lupus patients.

Mechanical actions of Yoga, such as changes in cell shape due to mechanical and fluid pressure, can have physiologic effects at cellular level and might alter joint function[27]. Yoga increases muscle strength and flexibility. It helps alleviating anxiety and depression. It down-regulates hypothalamic-pituitary-adrenal axis, suppresses sympathetic nervous system and increases vagal tone. IL-6 and C-reactive protein levels are lower in regular Yoga practitioners[28]. Reduction in stress related inflammatory responses is a beneficial effect of Yoga in inflammatory rheumatic conditions. Clinical studies including few good quality ones, indicate utility of Yoga in RA, OA (knee, hand), low back pain, and carpal tunnel syndrome. Yoga has positive effects on pain and pain-related disability[29]. Two studies in fibromyalgia showed significant reduction in pain and anxiety.

Incorrectly performed *asanas* can cause problems in several ways though no adverse events are reported. Some of the *asanas* such as *Padmasana* and *Vajrasana* (extreme knee flexion), can be difficult to perform and may actually be deleterious in cases of knee osteoarthritis.

Acupuncture

Chinese medicine is based on theories of *yin* (feminine, dark, cold, passive, negative factors) and *yang* (opposite factors- similar to pairs of qualities in Ayurveda) and presence of *chi* (*Chetna* in Ayurveda, Vital force in homeopathy) as physiological basis of living body. *Chi* flows in human body by various pathways called meridians (*Srotas* in Ayurveda, *Nadi* in Yoga) and symptoms of disease signal problems in the flow of *chi*. Circulation of *chi* within the body is influenced by five elements (similar to Ayurveda except wood and metal in place of *Vayu* and *Akash*) and health is harmoniously balanced cyclic interaction of these elements. Diseases result from deficient, excessive or stagnated *chi*. Most of the musculoskeletal disorders are classified as *Bi* (Vata dosha in Ayurveda) disorders. The treatment is person-specific rather than disease-oriented and is directed towards normalization of *chi*.

Chinese medicine has many varieties and traditions different from each other. Traditional Chinese Medicine uses medicinal herbs supplemented by acupuncture. Acupuncture[30] involves insertion of thin, sterile, stainless steel needles into points (on meridians) on the skin up to epidermis. Electrical stimulation or herbal extracts may be applied to these needles. Acupuncture, too, can be of various types such as classical, Japanese, moxibustion (burning of moxa herb on acupoint), laser beam therapy, acupressure-shiatsu (massage techniques, no needles), auricular-wrist-scalp acupuncture, etc.

Acupuncture is quiet popular in India and 60% of patients attending acupuncture clinics are claimed to be those with musculoskeletal disorders. Acupuncture trials are difficult to evaluate due to questionable correctness of procedure, problems in blinding and low overall quality. Published clinical studies indicate efficacy of acupuncture in low back pain, chronic neck pain, chronic knee pain, tennis elbow, shoulder pain, OA, fibromyalgia and in reducing drug induced nausea. Statistically significant and clinically relevant improvement is seen in noninflammatory rheumatic conditions such as OA, low back pain and lateral elbow pain[31]. Pain control following acupuncture was found to be better than controls (sham or no acupuncture) in cases with OA, shoulder pain and neck-back pain[32].

Acknowledgement

The authors wish to thank various experts in their respective fields for providing help in preparing manuscript of earlier version.

Dr. A. B. Vaidya , Dr. Aswinikumar Raut (Ayurveda), Dr. Babu Joseph (Naturopathy), Dr. K. C. Shah, Dr. Kedar Longani(Homeopathy), Hakim Abdul Malik Asmi (Unani), Dr. P. B. Lohiya (Acupuncture).

Further Reading :

1. Singh P, Yadav RJ, Pandey A. Utilization of indigenous systems of medicine and homeopathy in India. Indian Jour Med Res. 2005; 122: 137-42.
2. C. Ramos-Ramus, A. Raut. Complementary and alternative practices in Rheumatology. Best Pract Res Clin Rheumatology 2008; 22(4): 741-57.
3. Mei Y, Zheng Y, Wang H, Gao J, Liu D, Zhao Y, Zhang Z. Arsenic Trioxide induces apoptosis of fibroblast-like synoviocytes and represents anti-arthritic effect in experimental model of rheumatoid arthritis. Jour Rheum 2011; 38: 36-43.
4. Singh R.H. The holistic principles of Ayurvedic Medicine First Ed. Chaukhamba Sanskrit Pratishthan, Delhi 2003.
5. Park J, Ernst E. Ayurvedic medicine for rheumatoid arthritis: a systematic review. Semin Arthritis Rheum 2005; 34(5): 705-13.
6. Chopra A, Saluja M, Tillu G. Ayurveda - modern medicine interface: A critical appraisal of studies of Ayurvedic medicines to treat osteoarthritis and rheumatoid arthritis Jour Ayurveda Integr Med 2010; 1 (3) 190-98.
7. Furst D., Venkatramn M M, McGann M, Manohar P R, Booth-LaForce C, Sarin R et al. Double-blind randomized controlled pilot study comparing classic Ayurvedic medicine, methotrexate, and their combination in Rheumatoid Arthritis. Jour Clin Rheum 2011; 17(4): 185-92.
8. Aggarwal B B, Prasad S, Reuter S, Kannappan R, Yadev V. R, Park B, et al. Identification of Novel Anti-inflammatory Agents from Ayurvedic Medicine for Prevention of Chronic Diseases: "Reverse Pharmacology" and "Bedside to Bench" Approach. Curr Drug Targets. 2011; 12(11): 1595-1653.
9. Singh V K, Sharma P K, Dudhe R, Kumar N. Immuno-modulatory effects of some traditional medicinal plants. J. Chem. Pharm. Res. 2011; 3(1): 675-84.
10. Muller H, Wilhelmi de Toledo F, and Resch K. L. Fasting followed by vegetarian diet in patients with rheumatoid arthritis: a systematic review. Scand J Rheumatol 2001; 30: 1-10.
11. Tarner I H, Müller-Ladner U, Uhlemann C, Lange U. The effect of mild whole-body hyperthermia on systemic levels of TNF-alpha, IL-1beta, and IL-6 in patients with ankylosing spondylitis Clinical Rheumat 2009; 28 (4): 397-402.
12. Baria R, Joshi N, Pandya D. Clinical efficacy of Panchamuladi Kaala Basti (enema) in the management of Amavata (Rheumatoid Arthritis). AYU 2011; 32: 90-94.
13. Williams F M K, Spector T D. A new 5-lipoxygenase inhibitor seems to be safe and effective for the treatment of osteoarthritis Nature Rev Rheumat 2009; 5: 132-33.
14. Medhi B, Kishore K, Singh U, Seth S D. Comparative clinical trial of castor oil and diclofenac sodium in patients with osteoarthritis. Phytotherapy Research 2009; 23: 1469-1473.
15. Sumantran V N, Kulkarni A, Chandwaskar R, Harsulkar A, Patwardhan B, Chopra A, Wagh U V. Evidence-Based Compl Alt Med 2008; 5: 329-35.
16. Buhrmann C, Mobasheri A, Matis U, Mehdi S. Curcumin mediated suppression of nuclear factor-?B promotes chondrogenic differentiation of mesenchymal stem cells in a high-density co-culture microenvironment. Arthritis Res Therapy 2010; 12: R127.
17. Sumantran V N, Kulkarni A A, Harsulkar A, Wele A, Koppikar S J, Chandwaskar R et al. Hyaluronidase and collagenase inhibitory activities of the herbal formulation Triphala guggulu Jour Biosciences 2007; 32: 755-61.
18. Andereya S, Stanzel S, Maus U, Mueller-Rath R, Mumme T, Siebert C H et al. Assessment of leech therapy for knee osteoarthritis. A randomized study Acta Orthopaedica 2008; 79: 235-43.
19. Wagh S. Osteoporosis and Asthi-kshaya - Pathophysiology and principles of management. Deerghayu International. 2008; 24: 9-15.
20. Anjum N, Jamil S, Hannan A, Akhtar J, Ahmad B. Clinical efficacy of Hijamat (cupping) in Waja-ul-Mafasil (Arthritis). Ind J Traditional Knowledge. 2005; 4: 412-15.
21. Subbarayappa B V, Siddha Medicine: an overview. Lancet 1997; 350: 1841-44.
22. Wilson E, Rajamanickam G V, Vyas N, Agarwal A, Dubey G P. Herbs used in Siddha Medicine for Arthritis - A Review. Ind Jour Traditional Knowledge 2007; 6: 678-86.
23. Bell I R. Homeopathic Medicine and Rheumatologic Disorders. Horwitz R, Muller D (Ed) Integrative Rheumatology, 1st Ed, Oxford University Press, New York. 2011; 192-207.
24. Jonas WB, Kaptchuk TJ, Linde K. A critical overview of homeopathy. Ann Int Med, 2003; 138: 393-99.
25. Brien S, Lachance L, Prescott P, McDermott C, Lewith G. Homeopathy has clinical benefits in rheumatoid arthritis patients that are attributable to the consultation process but not the homeopathic remedy: a randomized controlled clinical trial. Rheumatolgy Oxford, 2011; 50: 1070-82.
26. Rowland-Seymour A, Jernberg J B. Manual Medicine in Rheumatologic Disorders Horwitz R, Muller D (Ed) Integrative Rheumatology, 1st Ed, Oxford University Press, New York. 2011; 93-107.
27. Garfinkel M, Yoga as a Complementary Therapy. Geriatric Aging. 2006; 9: 190-94.
28. Kiecolt-Glaser J K, Christian L, Praston H, Houts C R, Malarkey W B, emery C F, Glaser R. Stress, Inflammation and Yoga Practice. Psychosom Med. 2010; 72: 113-21.
29. Uhlig T. Tai Chi and yoga as complementary therapies in rheumatologic conditions. Best Practice and Res. Clin. Rheumatology. 2012; 26: 387-98.
30. Pyne D, Shenker N G, Demystifying Acupuncture. Rheumatology Oxford 2008; 47: 1132-76.
31. Ernst E, Lee M S. Acupuncture for rheumatic conditions: an overview of systematic reviews. Rheumatology Oxford 2010; 49: 1957-61.
32. Vickers A J, Cronin A M, Maschino A C, Lewith G, MacPherson H, Nadine E. et al Acupuncture for Chronic Pain - Individual Patient Data Meta-analysis ONLINE FIRST Arch Intern Med. 2012; (): 1-10. doi:10.1001/archinternmed.2012.3654 (Abstract).

APPENDIX

Classification/Diagnostic Criteria

Compiled by : Dr. Shrikant Wagh

1 Rheumatoid Arthritis ACR/EULAR 2010, ARA 1987
2 Spondyloarthritis (ESSG, ASAS) - Axial 2009, Peripheral 2011
3 Psoriatic Arthritis (CASPAR) 2006
4 Gout ACR
5 Osteoarthritis - Hands and Knees ACR
6 Systemic Lupus Erythematosus SLICC 2012
7 Anti-phospholipid Antibody Syndrome 2006
8 Sjogren's Syndrome ACR 2012
9 Systemic Sclerosis ACR/EULAR 2013
10 Polymyositis-Dermatomyositis
11 Mixed Connective Tissue Disease
12 Fibromyalgia ACR 2010
13 Joint Hypermobility

1. Rheumatoid Arthritis

The 2010 American College of Rheumatology/European League Against Rheumatism classification criteria for Rheumatoid Arthritis (for newly presenting patients)

Score

Target population: Patients who

1) have at least 1 joint with definite clinical synovitis (swelling)[1]
2) with the synovitis not better explained by another disease[2]

Classification criteria for RA (score-based algorithm: add score of categories A-D; a score of ≥ 6/10 is needed for classification of a patient as having definite RA)[3]

A. Joint involvement[4]	
1 large joint[5]	0
2-10 large joints	1
1-3 small joints (with or without involvement of large joints)[6]	2
4-10 small joints (with or without involvement of large joints)	3
>10 joints (at least 1 small joint)[7]	5
B. Serology (at least 1 test result is needed for classification)[8]	
Negative RF and negative ACPA	0
Low-positive RF or low-positive ACPA	2
High-positive RF or high-positive ACPA	3
C. Acute-phase reactants (at least 1 test result is needed for classification)[9]	
Normal CRP and normal ESR	0
Abnormal CRP or abnormal ESR	1

D. Duration of symptoms[10]	
< 6 weeks	0
≥ 6 weeks	1

1. The criteria are aimed at classification of newly presenting patients. In addition, patients with erosive disease typical of rheumatoid arthritis (RA) with a history compatible with prior fulfillment of the 2010 criteria should be classified as having RA.

   Patients with long-standing disease, including those whose disease is inactive (with or without treatment) who, based on retrospectively available data, have previously fulfilled the 2010 criteria should be classified as having RA.
2. Differential diagnoses differ in patients with different presentations, but may include conditions such as systemic lupus erythematosus, psoriatic arthritis and gout. If it is unclear about the relevant differential diagnoses to consider, an expert rheumatologist should be consulted.
3. Although patients with a score of less than 6/10 are not classifiable as having RA, their status can be reassessed and the criteria might be fulfilled cumulatively over time.
4. Joint involvement refers to any swollen or tender joint on examination, which may be confirmed by imaging evidence of synovitis. Distal interphalangeal joints, first carpometacarpal joints and first metatarsophalangeal joints are excluded from assessment. Categories of joint distribution are classified according to the location and number of involved joints, with placement into the highest category possible based on the pattern of joint involvement.
5. 'Large joints' refers to shoulders, elbows, hips, knees and ankles.
6. 'Small joints' refers to the metacarpophalangeal joints, proximal interphalangeal joints, second to fifth metatarsophalangeal joints, thumb interphalangeal joints and wrists.
7. In this category, at least one of the involved joints must be a small joint; the other joints can include any combination of large and additional small joints, as well as other joints not specifically listed elsewhere (eg, temporomandibular, acromioclavicular, sternoclavicular, etc.).
8. Negative refers to international unit (IU) values that are less than or equal to the upper limit of normal (ULN) for the laboratory and assay; low-positive refers to IU values that are higher than the ULN but three of less times the ULN for the laboratory and assay; high-positive refers to IU values that are more than three times the ULN for the laboratory and assay. When rheumatoid factor (RF) information is only available as positive or negative, a positive result should be scored as low-positive for RF.
9. Normal/abnormal is determined by local laboratory standards.
10. Duration of symptoms refers to patient self-report of the duration of signs or symptoms of synovitis (eg, pain, swelling, tenderness) of joints that are clinically involved at the time of assessment, regardless of treatment status.

(ACPA: anti-citrullinated protein antibody. CRP: C-reactive protein. ESR: erythrocyte sedimentation rate.)

Daniel Aletaha et al, 2010 Rheumatoid arthritis classification criteria: an American College of Rheumatology/European League Against Rheumatism collaborative initiative. Ann Rheum Dis 2010; 69: 1580-88.

1987 revised criteria for the classification of rheumatoid arthritis (ARA)*

	Criterion	Definition
1.	Morning stiffness	Morning stiffness in and around the joints, lasting at least 1 hour before maximal improvement
2.	Arthritis of 3 or more joint areas	At least 3 joint areas simultaneously have had soft tissue swelling or fluid (not bony overgrowth alone) observed by a physician. The 14 possible areas are right or left PIP, MCP, wrist, elbow, knee, ankle, and MTP joints
3.	Arthritis of hand joints	At least 1 area swollen (as defined above) in a wrist, MCP or PIP Joint
4.	Symmetric arthritis	Simultaneous involvement of the same joint areas (as defined in 2) on both sides of the body (bilateral involvement of PIPs, MCPs or MTPs is acceptable without absolute symmetry)
5.	Rheumatoid nodules	Subcutaneous nodules, over bony prominences or extensor surfaces or in juxtaarticular regions, observed by a physician
6.	Serum rheumatoid factor	Demonstration of abnormal amounts of serum rheumatoid factor by any method for which the result has been positive in < 5% of normal control subjects
7.	Radiographic changes	Radiographic changes typical of rheumatoid arthritis on posteroanterior hand and wrist radiographs, which must include erosions or unequivocal bony decalcification localized in or most marked adjacent to the involved joints

* For classification purposes, a patient shall be said to have rheumatoid arthritis if he/she has satisfied at least 4 of these 7 criteria. Criteria 1 through 4 must have been present for at least 6 weeks. Patients with 2 clinical diagnoses are not excluded. Designation as classic, definite or probable rheumatoid arthritis is not to be made.

Arnett FC et al. The American Rheumatism Association 1987 revised criteria for the classification of rheumatoid arthritis. Arthritis & Rheumatism 1988; 31: 315-24.

2. Spondyloarthritis

European Spondyloarthropathy Study Group (ESSG) criteria

Inflammatory spinal pain*

OR

Synovitis: asymmetrical or predominantly in the lower limbs

AND

One or more of the following:

1. Positive Family history
2. Psoriasis
3. Inflammatory Bowel Disease
4. Urethritis, cervicitis, or acute diarrhea within one month before arthritis
5. Buttock pain alternating between right and left gluteal areas
6. Enthesopathy
7. Sacroiliitis

Specifications

Variable	Definition
Inflammatory spinal pain*	History or present symptoms of spinal pain in back, dorsal, or cervical region, with at least four of the following: (a) onset before age 45, (b) insidious onset, (c) improved by exercise, (d) associated with morning stiffness, (e) at least 3 months duration
Synovitis	Past or present asymmetric arthritis or arthritis predominantly in the lower limbs
Family history	
Psoriasis	Past or present psoriasis diagnosed by a doctor
Inflammatory bowel disease	Past or present Crohn's disease or ulcerative colitis diagnosed by a doctor and confirmed by radiographic examination or endoscopy
Alternating buttock pain	Past or present pain alternating between the right and left gluteal regions
Enthesopathy	Past or present spontaneous pain or tenderness at examination at the site of the insertion of the Achilles tendon or plantar fascia
Acute diarrhea	Episode of diarrhoea occurring within 1 month before arthritis
Urethritis/cervicitis	Non-gonococcal urethritis or cervicitis occurring within 1 month before arthritis
Sacroiliitis	Bilateral grade 2-4 or unilateral grade 3-4, according to the following radiographic grading system: 0=normal, 1=possible, 2=minimal, 3=moderate and 4=ankylosis

ASAS criteria for classification of axial spondyloarthritis

(To be applied in patients with chronic back pain and age at onset of back pain less than 45 years)

Sacroiliitis on imaging plus one or more SpA feature[1]

OR

HLA-B27 plus 2 or more other SpA features[2]

[1]SpA Features

1. Inflammatory back pain (IBP)
2. Arthritis
3. Enthesitis (heel)
4. Uveitis

[2]Sacroiliitis on Imaging

1. Active (acute) inflammation on MRI highly suggestive of sacroiliitis associated with SpA
2. Definite radiographic sacroiliitis according to modified NY criteria

5. Dactylitis
6. Psoriasis
7. Crohn's/colitis
8. Good response to NSAIDs
9. Family history for SpA
10. HLA-B27
11. Elevated CRP

Specifications

Clinical criterion	Definition
IBP	IBP according to experts: four out of five of the following parameters present: (1) age at onset, 40 years, (2) insidious onset, (3) improvement with exercise, (4) no improvement with rest, (5) pain at night (with improvement upon getting up)
Arthritis	Past or present active synovitis diagnosed by a doctor
Family history	Presence in first-degree or second-degree relatives of any of the following: (a) ankylosing spondylitis, (b) psoriasis, (c) uveitis, (d) reactive arthritis, (e) inflammatory bowel disease
Psoriasis	Past or present psoriasis diagnosed by a doctor
Inflammatory bowel disease	Past or present Crohn's disease or ulcerative colitis diagnosed by a doctor
Dactylitis	Past or present dactylitis diagnosed by a doctor
Enthesitis	Heel enthesitis: past or present spontaneous pain or tenderness at examination at the site of the insertion of the Achilles tendon or plantar fascia at the calcaneus
Uveitis anterior	Past or present uveitis anterior, confirmed by an ophthalmologist
Good response to NSAIDs	At 24-48 h after a full dose of NSAID the back pain is not present anymore or much better
HLA-B27	Positive testing according to standard laboratory Techniques
Elevated CRP	CRP above upper normal limit in the presence of back pain, after exclusion of other causes for elevated CRP concentration
Sacroiliitis by x rays	Bilateral grade 2-4 or unilateral grade 3-4, according to the modified New York criteria
Sacroiliitis by MRI	Active inflammatory lesions of sacroiliac joints with definite bone marrow oedema/osteitis suggestive of sacroiliitis associated with spondyloarthritis

J Sieper et al The Assessment of SpondyloArthritis International Society (ASAS) handbook: a guide to assess spondyloarthritis Ann Rheum Dis 2009; 68(Suppl II): ii1-ii44.

Classification criteria for peripheral spondyloarthritis (ASAS) 2010

Arthritis or Enthesitis or Dactylitis

plus 1 or more of:
1. Psoriasis
2. Inflammatory bowel disease
3. Preceding infection
4. HLA-B27
5. Uveitis
6. Sacroiliitis on imaging (radiographs or MRI)

OR

plus 2 or more of the remaining:
1. Arthritis
2. Enthesitis
3. Dactylitis
4. IBP in the past
5. Positive family history for SpA

The criteria are applicable to patients with peripheral arthritis (usually predominantly of the lower limbs and/or asymmetric arthritis), and/or enthesitis, and/or dactylitis.

M Rudwaleit et al The Assessment of SpondyloArthritis International Society classification criteria for peripheral spondyloarthritis and for spondyloarthritis in general Ann Rheum Dis 2011; 70: 25-31.

3. Psoriatic Arthritis

Classification Criteria for Psoriatic Arthritis (CASPAR) - 2006

To meet the CASPAR criteria, a patient must have inflammatory articular disease (joint, spine, or entheseal) with 3 or more points from the following 5 categories:

1.	Evidence of Psoriasis (one of the three)	Current psoriasis	psoriatic skin or scalp disease present today as judged by a rheumatologist or dermatologist
		Personal history of psoriasis	A history of psoriasis that may be obtained from a patient, family physician, dermatologist, rheumatologist or other qualified health care Provider
		Family history of psoriasis	History of psoriasis in a first- or second-degree relative according to patient report
2.	Psoriatic nail dystrophy		Includes onycholysis, pitting, and hyperkeratosis observed on current physical examination.
3.	Negative test for rheumatoid factor		By any method except latex but preferably by enzyme-linked immunosorbent assay or nephelometry, according to the local laboratory reference range
4.	Dactylitis (one of the two)	Current	Swelling of an entire digit
		History	Recorded by a rheumatologist
5.	Radiography	Evidence of juxtaarticular new bone formation, appearing as ill-defined ossification near joint margins (but excluding osteophyte formation) on plain radiographs of the hand or foot.	

William Taylor et al. Classification Criteria for Psoriatic Arthritis Development of New Criteria From a Large International Study Arthritis & Rheumatism 2006; 54(8): 2665-73.

4. Gout

American College of Rheumatology preliminary criteria for the classification of the acute arthritis of primary gout

Gout may be diagnosed if one of the following criteria is present:

1) Monosodium urate crystals in synovial fluid, or
2) Tophi confirmed with crystal examination, or
3) At least six of the following findings:
   1. Asymmetric swelling within a joint on a radiograph
   2. First metatarsophalangeal joint is tender or swollen (i.e., podagra)
   3. Hyperuricemia
   4. Maximal inflammation developed within one day
   5. Monoarthritis attack
   6. More than one acute arthritis attack
   7. Redness observed over joints
   8. Subcortical cysts without erosions on a radiograph
   9. Suspected tophi
   10. Synovial fluid culture negative for organisms during an acute attack
   11. Unilateral first metatarsophalangeal joint attack
   12. Unilateral tarsal joint attack

Wallace SL et al. Preliminary criteria for the classification of the acute arthritis of primary gout. Arthritis Rheum 1977; 20: 896.

5. Osteoarthritis

Radiologic and clinical criteria for hand and knee osteoarthritis (ACR)

Hand Osteoarthritis - Clinical Criteria

Hand pain, aching, or stiffness and 3 or 4 of the following features:

1. Hard tissue enlargement of 2 or more of 10 selected joints
2. Hard tissue enlargement of 2 or more DIP joints
3. Fewer than 3 swollen MCP joints
4. Deformity of at least 1 of 10 selected joints

* Ten selected joints include bilateral second and third distal interphalangeal joints, second and third proximal interphalangeal joints, and first carpometacarpal joint.

Altman R et al. The American College of Rheumatology criteria for the classification and reporting of osteoarthritis of the hand. Arthritis Rheum1990; 33: 1601-10.

Knee Osteoarthritis - Criteria

Clinical and laboratory	Clinical and radiographic	Clinical
Knee Pain Plus at least 5 out of 9: 1. Age > 50 years 2. Stiffness < 30 minutes 3. Crepitus 4. Bony Tenderness 5. Bony enlargement 6. No palpable warmth 7. ESR < 40 mm/hour 8. RF < 1:40 9. SF OA*	Knee Pain Plus at least 1 out of 3: 1. Age > 50 years 2. Stiffness < 30 minutes 3. Crepitus Plus Osteophytes	Knee Pain Plus at least 3 out of 6: 1. Age > 50 years 2. Stiffness < 30 minutes 3. Crepitus 4. Bony Tenderness 5. Bony enlargement 6. No palpable warmth

*SF OA = synovial fluid signs of OA (clear, viscous, or white blood cell count < 2,000/mm^3)

R. Altman et al. The American College of Rheumatology criteria for the classification and reporting of osteoarthritis of the knee. Arthritis Rheum 1986; 29: 1039-49.

6. Systemic Lupus Erythematosus

Clinical and immunologic criteria for Systemic Lupus Erythematosus

The SLICC classification system 2012*

Classify a patient as having SLE if he or she satisfies 4 of the clinical and immunologic criteria used in the SLICC classification criteria, including at least one clinical criterion and one immunologic criterion,

OR

if he or she has biopsy-proven nephritis compatible with SLE in the presence of ANAs or anti-dsDNA antibodies.

Clinical criteria

1	Acute cutaneous lupus, including:	Lupus malar rash (do not count if malar discoid) Bullous lupus Toxic epidermal necrolysis variant of SLE Maculopapular lupus rash Photosensitive lupus rash OR subacute cutaneous lupus (nonindurated psoriaform and/or annular polycyclic lesions that resolve without scarring, although occasionally with postinflammatory dyspigmentation or telangiectasias)
2	Chronic cutaneous lupus, including:	Classic discoid rash • Localized (above the neck) • Generalized (above and below the neck) Hypertrophic (verrucus) lupus Lupus panniculitis (profundus) Mucosal lupus

		Lupus erythematosus tumidus Chillblains lupus Discoid lupus/lichen planus overlap
3	Oral ulcers	Palate (Buccal, Tongue) OR nasal ulcers in the absence of other causes, such as vasculitis, Behcet's disease, infection (herpes virus), inflammatory bowel disease, reactive arthritis, and acidic foods
4	Nonscarring alopecia	diffuse thinning or hair fragility with visible broken hairs in the absence of other causes such as alopecia areata, drugs, iron deficiency, and androgenic alopecia
5	Synovitis involving 2 or more joints	characterized by swelling or effusion OR tenderness in 2 or more joints and at least 30 minutes of morning stiffness
6	Serositis	Typical pleurisy for more than 1 day OR pleural effusions OR pleural rub Typical pericardial pain (pain with recumbency improved by sitting forward) for more than 1 day OR pericardial effusion OR pericardial rub OR pericarditis by electrocardiography in the absence of other causes, such as infection, uremia, and Dressler's pericarditis
7	Renal	Urine protein-to-creatinine ratio (or 24-hour urine protein) representing 500 mg protein/24 hours OR red blood cell casts
8	Neurologic	Seizures Psychosis Mononeuritis multiplex in the absence of other known causes such as primary vasculitis Myelitis Peripheral or cranial neuropathy in the absence of other known causes such as primary vasculitis, infection, and diabetes mellitus Acute confusional state in the absence of other causes, including toxic/metabolic, uremia, drugs
9	Hemolytic anemia	
10	Leukopenia OR Lymphopenia	Leukopenia ($< 4{,}000/mm^3$ at least once) in the absence of other known causes such as Felty's syndrome, drugs, and portal hypertension OR Lymphopenia ($<1{,}000/mm^3$ at least once) in the absence of other known causes such as corticosteroids, drugs, and infection
11	Thrombocytopenia ($<100{,}000/mm^3$)	at least once in the absence of other known causes such as drugs, portal hypertension, and thrombotic thrombocytopenic purpura

Immunologic criteria

1 ANA level above laboratory reference range

2 Anti-dsDNA antibody level above laboratory reference range (or > 2-fold the reference range if tested by ELISA)

3 Anti-Sm: presence of antibody to Sm nuclear antigen

4 Antiphospholipid antibody positivity as determined by any of the following:
Positive test result for lupus anticoagulant
False-positive test result for rapid plasma reagin
Medium- or high-titer anticardiolipin antibody level (IgA, IgG, or IgM)
Positive test result for anti-2-glycoprotein I (IgA, IgG, or IgM)

5 Low complement : Low C3, Low C4, Low CH50

6 Direct Coombs' test in the absence of hemolytic anemia

Criteria are cumulative and need not be present concurrently.
*SLICC = Systemic Lupus International Collaborating Clinics

Michelle Petri et al. Derivation and Validation of the Systemic Lupus International Collaborating Clinics Classification Criteria for Systemic Lupus Erythematosus. Arthritis & Rheumatism 2012; 64(8): 2677-86.

7. Antiphospholipid Antibody Syndrome

Revised classification criteria for the antiphospholipid syndrome
International consensus 2006

Antiphospholipid antibody syndrome (APS) is present if at least one of the clinical criteria and one of the laboratory criteria that follow are met. Classification of APS should be avoided if less than 12 weeks or more than 5 years separate the positive aPL test and the clinical manifestation.

Clinical Criteria

1	Vascular thrombosis[1]	One or more clinical episodes[2] of arterial, venous, or small vessel thrombosis, in any tissue or organ. Thrombosis must be confirmed by objective validated criteria (i.e. unequivocal findings of appropriate imaging studies or histopathology). For histopathologic confirmation, thrombosis should be present without significant evidence of inflammation in the vessel wall.
2	Pregnancy morbidity	(a) One or more unexplained deaths of a morphologically normal fetus at or beyond the 10th week of gestation, with normal fetal morphology documented by ultrasound or by direct examination of the fetus, OR (b) One or more premature births of a morphologically normal neonate before the 34th week of gestation because of: (i) eclampsia or severe preeclampsia defined according to standard definitions, or (ii) recognized features of placental insufficiency[3], OR (c) Three or more unexplained consecutive spontaneous abortions before the 10th week of gestation, with maternal anatomic or hormonal abnormalities and paternal and maternal chromosomal causes excluded. In studies of populations of patients who have more than one type of pregnancy morbidity, investigators are strongly encouraged to stratify groups of subjects according to a, b, or c above.

Laboratory criteria**

1 Lupus anticoagulant (LA) present in plasma, on two or more occasions at least 12 weeks apart, detected according to the guidelines of the International Society on Thrombosis and Haemostasis (Scientific Subcommittee on LAs/ phospholipid-dependent antibodies).

2 Anticardiolipin (aCL) antibody of IgG and/or IgM isotype in serum or plasma, present in medium or high titer (i.e. > 40 GPL or MPL, or > the 99th percentile), on two or more occasions, at least 12 weeks apart, measured by a standardized ELISA.

3 Anti-beta 2 glycoprotein-I antibody of IgG and/or IgM isotype in serum or plasma (in titer >the 99th percentile), present on two or more occasions, at least 12 weeks apart, measured by a standardized ELISA, according to recommended procedures.

1. Coexisting inherited or acquired factors for thrombosis are not reasons for excluding patients from APS trials. However, two subgroups of APS patients should be recognized, according to: (a) the presence, and (b) the absence of additional risk factors for thrombosis.

   Indicative (but not exhaustive) such cases include: age (> 55 in men, and > 65 in women) and the presence of any of the established risk factors for cardiovascular disease (hypertension, diabetes mellitus, elevated LDL or low HDL cholesterol, cigarette smoking, family history of premature cardiovascular disease, body mass index equal to or more than 30 kg m2, microalbuminuria, estimated GFR < 60 ml min-1, inherited thrombophilias, oral contraceptives, nephrotic syndrome, malignancy, immobilization, and surgery. Thus, patients who fulfill criteria should be stratified according to contributing causes of thrombosis.

2. A thrombotic episode in the past could be considered as a clinical criterion, provided that thrombosis is proved by appropriate diagnostic means and that no alternative diagnosis or cause of thrombosis is found. Superficial venous thrombosis is not included in the clinical criteria.

3. Generally accepted features of placental insufficiency include: (i) abnormal or non-reassuring fetal surveillance test(s), e.g. a non-reactive non-stress test, suggestive of fetal hypoxemia, (ii) abnormal Doppler flow velocimetry waveform analysis suggestive of fetal hypoxemia, e.g. absent end-diastolic flow in the umbilical artery, (iii) oligohydramnios, e.g. an amniotic fluid index of 5 cm or less, or (iv) a postnatal birth weight less than the 10th percentile for the gestational age.

**Investigators are strongly advised to classify APS patients in studies into one of the following categories: I, more than one laboratory criteria present (any combination); IIa, LA present alone; IIb, aCL antibody present alone; IIc, anti-beta 2 glycoprotein-I antibody present alone.

Miyakis S et al. International consensus statement on an update of the classification criteria for definite antiphospholipid syndrome (APS). J Thromb Haemost 2006; 4: 295-306.

8. Sjogren's Syndrome

Classification criteria for Sjogren's syndrome (SjS) - 2012

The classification of SjS, which applies to individuals with signs/symptoms that may be suggestive of SjS, will be met in patients who have at least 2 of the following 3 objective features:

1. Positive serum anti-SSA/Ro and/or anti-SSB/La or positive rheumatoid factor and ANA titer 1:320.
2. Labial salivary gland biopsy exhibiting focal lymphocytic sialadenitis with a focus score 1 focus/ $4\ mm^2$ (1).
3. Keratoconjunctivitis sicca with ocular staining score 3 (assuming that individual is not currently using daily eye drops for glaucoma and has not had corneal surgery or cosmetic eyelid surgery in the last 5 years) (2).

Prior diagnosis of any of the following conditions would exclude participation in SjS studies or therapeutic trials because of overlapping clinical features or interference with criteria tests:

a. History of head and neck radiation treatment
b. Hepatitis C infection
c. Acquired immunodeficiency syndrome
d. Sarcoidosis
e. Amyloidosis
f. Graft versus host disease
g. IgG4-related disease

(1) Using histopathologic definitions and focus score assessment methods as previously described.

(2) Using ocular staining score as previously described.

S. C. Shiboski et al, American College of Rheumatology Classification Criteria for Sjogren's Syndrome: A Data-Driven, Expert Consensus Approach in the Sjogren's International Collaborative Clinical Alliance Cohort. Arthritis Care & Research 2012; 64(4): 475-87.

9. Systemic Sclerosis

Criteria for the classification of Systemic Sclerosis (SSc) - ACR/EULAR 2013*

Item	Sub-item	Weight
Bilateral skin thickening of fingers extending proximal to MCP joints		9
Skin thickening of the fingers[1]	Puffy fingers	2
	Whole Finger, distal to MCP	4
Finger tip lesions[1]	Digital Tip Ulcers	2
	Pitting Scars	3
Telangiectasia		2
Abnormal nailfold capillaries		2
Pulmonary arterial hypertension and/or interstitial lung disease[2]		2
Raynaud's phenomenon		3
Scleroderma related antibodies[3]		3

* These criteria are applicable to any patient considered for inclusion in an SSc study. The criteria are not applicable to patients with skin thickening sparing the fingers or to patients who have a scleroderma-like disorder that better explains their manifestations (e.g., nephrogenic sclerosing fibrosis, generalized morphea, eosinophilic fasciitis, scleredema diabeticorum, scleromyxedema, erythromyalgia, porphyria, lichen sclerosis, graft-versus-host disease, diabetic cheiroarthropathy).

[1]Count the higher score only. [2]Maximum score is 2. [3]Maximum score is 3.

The total score is determined by adding the maximum weight (score) in each category. Patients with a total score of 9 or more are classified as having definite SSc.

Definitions of items/subitems

Skin thickening	Skin thickening or hardening not due to scarring after injury, trauma, etc.
Puffy fingers	Swollen digits-a diffuse, usually nonpitting increase in soft tissue mass of the digits extending beyond the normal confines of the joint capsule. Normal digits are tapered distally with the tissues following the contours of the digital bone and joint structures. Swelling of the digits obliterates these contours. Not due to other causes such as inflammatory dactylitis.
Fingertip ulcers or pitting scars	Ulcers or scars distal to or at the proximal interphalangeal joint not thought to be due to trauma. Digital pitting scars are depressed areas at digital tips as a result of ischemia, rather than trauma or exogenous causes.
Telangiectasia	Telangiectasiae are visible macular dilated superficial blood vessels, which collapse upon pressure and fill slowly when pressure is released. Telangiectasiae in a scleroderma-like pattern are round and well demarcated and found on hands, lips, inside of the mouth, and/or are large mat-like telangiectasiae. Distinguishable from rapidly filling spider angiomas with central arteriole and from dilated superficial vessels.
Abnormal nailfold capillary pattern consistent with SSc	Enlarged capillaries and/or capillary loss with or without pericapillary hemorrhages at the nailfold. May also be seen on the cuticle.
Pulmonary arterial hypertension	Pulmonary arterial hypertension diagnosed by right-sided heart catheterization according to standard definitions.
Interstitial lung disease	Pulmonary fibrosis seen on high-resolution computed tomography or chest radiography, most pronounced in the basilar portions of the lungs, or occurrence of "Velcro" crackles on auscultation, not due to another cause such as congestive heart failure.
Raynaud's phenomenon	Self-reported or reported by a physician, with at least a 2-phase color change in finger(s) and often toe(s) consisting of pallor, cyanosis, and/or reactive hyperemia in response to cold exposure or emotion; usually one phase is pallor.
SSc-related autoantibodies	Anticentromere antibody or centromere pattern seen on antinuclear antibody testing, anti-topoisomerase I antibody (also known as anti-Scl-70 antibody), or anti-RNA polymerase III antibody. Positive according to local laboratory standards.

Frank van den Hoogen, Dinesh Khanna, et al 2013 Classification Criteria for Systemic Sclerosis

An American College of Rheumatology/European League Against Rheumatism Collaborative Initiative.

Arthritis & Rheumatism 2013; xx:xx (to be published in November 2013)

10. Polymyositis and Dermatomyositis

Diagnostic Criteria for Polymyositis and Dermatomyositis (Bohan & Peter)

1. Symmetric proximal muscle weakness with or without dysphagia and respiratory muscle weakness developing over weeks or months.
2. Elevation of serum skeletal muscle enzymes (creatine kinase and aldolase), transaminases and lactate dehydrogenase.
3. The electromyographic triad of
   a. short, small, polyphasic motor unit potentials;
   b. fibrillations, positive sharp waves, and insertional irritability; and
   c. bizarre, high frequency repetitive discharges.
4. Muscle biopsy specimen abnormalities of degeneration, regeneration, necrosis, phagocytosis and interstitial mononuclear infiltrate.
5. Typical skin rash of dermatomyositis, including a heliotrope rash, Gottron's sign and Gottron's papules.

Interpretation

Diagnosis	Polymyositis	Dermatomyositis
Definite	4 Criteria	3-4 Criteria plus rash
Probable	3 Criteria	2 Criteria plus rash
Possible	2 Criteria	1 Criterion plus rash

Bohan A, Peter JB. Polymyositis and dermatomyositis. N Engl J Med 1975; 292: 344-47, 403-07.

11. Mixed Connective Tissue Disease

Criteria for Diagnosis of Mixed Connective Tissue Disease (MCTD) (Alarcon-Segovia)

Clinical Criteria

1 Swollen hands
2 Acrosclerosis with or without proximal Systemic Sclerosis (SSc)
3 Raynaud's phenomenon
4 Myositis: based on laboratory and/or histological findings
5 Synovitis

Serologic Criterion

1 Anti-RNP-antibodies with a titer of > 1:1600 at the hem-agglutinin assay

Diagnosis of MCTD is based on serologic criterion plus 3 out of 5 criteria.

Criteria 4 and 5 are also required to distinguish MCTD from SSc.

Alarcon-Segovia D, Mixed connective tissue disease and overlap syndromes. Clin Dermatol 1994; 12(2): 309-16.

12. Fibromyalgia

Fibromyalgia Diagnostic Criteria (ACR 2010)

Criteria

A patient satisfies diagnostic criteria for fibromyalgia if the following 3 conditions are met:

1) Widespread pain index (WPI) > or = 7 and symptom severity (SS) scale score > or = 5 or WPI 3-6 and SS scale score > or = 9.
2) Symptoms have been present at a similar level for at least 3 months.
3) The patient does not have a disorder that would otherwise explain the pain.

Ascertainment

1. **WPI:** note the number areas in which the patient has had pain over the last week. In how many areas has the patient had pain? Score will be between 0 and 19.

   Shoulder girdle (left and right), Upper arm (left and right), Lower arm (left and right), Hip (buttock, trochanter) - left and right, Upper leg (left and right), Lower leg (left and right), Jaw (left and right), Chest, Abdomen, Upper back, Lower back, Neck

2. **SS scale score:** 1. Fatigue 2. Waking unrefreshed 3. Cognitive symptoms.
   For the each of the 3 symptoms above, indicate the level of severity over the past week using the following scale:
   0 = no problem
   1 = slight or mild problems, generally mild or intermittent
   2 = moderate, considerable problems, often present and/or at a moderate level
   3 = severe: pervasive, continuous, life-disturbing problems
   Considering somatic symptoms in general, indicate whether the patient has:*
   0 = no symptoms
   1 = few symptoms
   2 = a moderate number of symptoms
   3 = a great deal of symptoms

The SS scale score is the sum of the severity of the 3 symptoms (fatigue, waking unrefreshed, cognitive symptoms) plus the extent (severity) of somatic symptoms in general. The final score is between 0 and 12.

* Somatic symptoms that might be considered: muscle pain, irritable bowel syndrome, fatigue/tiredness, thinking or remembering problem, muscle weakness, headache, pain/cramps in the abdomen, numbness/tingling, dizziness, insomnia, depression, constipation, pain in the upper abdomen, nausea, nervousness, chest pain, blurred vision, fever, diarrhea, dry mouth, itching, wheezing, Raynaud's phenomenon, hives/welts, ringing in ears, vomiting, heartburn, oral ulcers, loss of/change in taste, seizures, dry eyes, shortness of breath, loss of appetite, rash, sun sensitivity, hearing difficulties, easy bruising, hair loss, frequent urination, painful urination, and bladder spasms.

Frederick Wolfe et al. The American College of Rheumatology Preliminary Diagnostic Criteria for Fibromyalgia and Measurement of Symptom Severity. Arthritis Care & Research 2010; 62(5): 600-10.

13. Joint Hypermobility

Beighton score for joint hypermobility syndrome

One point is gained for each side of the body for the first four manoeuvres listed below, such that the hypermobility score is a maximum of 9 if all are positive.

1. Passive dorsiflexion of the fifth metacarpophalangeal joint to 90° or more
2. Opposition of the thumb to the volar aspect of the ipsilateral forearm
3. Hyperextension of the elbow to 10° or more
4. Hyperextension of the knee to 10° or more
5. Placing of hands flat on the floor without bending the knees (1 point only)

1998 Brighton criteria for classification of joint hypermobility syndrome

Joint hypermobility syndrome is diagnosed in the presence of:

1. 2 major criteria;
2. 1 major criterion plus 2 minor criteria; or
3. 4 minor criteria.

Two minor criteria will suffice where there is an unequivocally affected first degree relative.

The syndrome is excluded by the presence of Marfan's or Ehlers-Danlos syndromes

Major criteria

1. Beighton score of 4 (either currently or previously)
2. Arthralgia for longer than three months in four or more joints

Minor criteria

1. Beighton score of 1, 2, or 3 (0, 1, 2, or 3 if aged > 50 years)
2. Arthralgia in 1-3 joints or back pain or spondylosis, spondylolysis and/or spondylolisthesis
3. Dislocation in more than 1 joint or in one joint on more than one occasion
4. Three or more soft tissue lesions (e.g. epicondylitis, tenosynovitis, bursitis)
5. Marfanoid habitus (tall, slim, ratio of span to height greater than 1.03 and/or ratio of upper segment to lower segment less than 0.89, arachnodactyly)
6. Abnormal skin: striae, hyperextensibility, thin skin, papyraceous scarring
7. Eye signs: drooping eyelids, myopia, or antimongoloid slant
8. Varicose veins, hernia, or uterine or rectal prolapsed
9. Mitral valve prolapsed

Beighton P et al. Articular mobility in an African population. Ann Rheum Dis 1973; 32: 413-18.

Grahame R, Bird HA, Child A. The revised (Brighton 1998) criteria for the diagnosis of benign joint hypermobility syndrome (BJHS). J Rheumatol 2000; 27: 1777-79.

INDEX

A

B

C

D

E

F

G

H

I

J

P

R

U

V

W

X

Y

Z